FRESHNEY'S CULTURE OF ANIMAL CELLS

FRESHNEY'S CULTURE OF ANIMAL CELLS

A MANUAL OF BASIC TECHNIQUE AND SPECIALIZED APPLICATIONS
Eighth Edition

Amanda Capes-Davis
Children's Medical Research Institute

Westmead, Australia

Address correspondence to: PO Box 4671, North Rocks, NSW 2151, Australia

R. Ian Freshney
Institute of Cancer Sciences

University of Glasgow

Reviewing Editors

Robert J. Geraghty
Cancer Research UK Cambridge Institute

University of Cambridge

Raymond W. Nims
RMC Pharmaceutical Solutions, Inc

WILEY Blackwell

This eighth edition first published 2021
© 2021 John Wiley & Sons, Ltd

Edition History
1e; (Blackwell Publishing Ltd, 1986) 2e; (Blackwell Publishing Limited, 1996) 3e; (Blackwell Publishing Ltd, 2007) 4e: (2000) 5e: (2005) 6e: (2010) 7e: (Wiley Blackwell, 2016)

The right of Amanda Capes-Davis and R. Ian Freshney to be identified as the authors of this work has been asserted in accordance with law.

Registered Offices
John Wiley & Sons, Inc., 111 River Street, Hoboken, NJ 07030, USA
John Wiley & Sons Ltd, The Atrium, Southern Gate, Chichester, West Sussex, PO19 8SQ, UK

Editorial Office
9600 Garsington Road, Oxford, OX4 2DQ, UK

For details of our global editorial offices, customer services, and more information about Wiley products visit us at www.wiley.com.

Wiley also publishes its books in a variety of electronic formats and by print-on-demand. Some content that appears in standard print versions of this book may not be available in other formats.

Library of Congress Cataloging-in-Publication Data

Names: Capes-Davis, Amanda, author. | Capes-Davis, Amanda, editor. |
    Freshney, R. Ian. Culture of animal cells.
Title: Freshney's culture of animal cells : a manual of basic technique and
    specialized applications / Amanda Capes-Davis, Children's Medical
    Research Institute, The University of Sydney, R. Ian Freshney, Institute
    of Cancer Sciences, University of Glasgow ; reviewing editors, Robert J.
    Geraghty, Cancer Research UK Cambridge Institute, University of
    Cambridge, Raymond W. Nims, RMC Pharmaceutical Solutions, Inc.
Other titles: Culture of animal cells
Description: Eighth edition. | Hoboken, NJ : Wiley-Blackwell, 2021. |
    Revised edition of: Culture of animal cells : a manual of basic
    technique and specialized applications by R. Ian Freshney. Seventh
    edition. [2016]. | Includes bibliographical references and index.
Identifiers: LCCN 2020029671 (print) | LCCN 2020029672 (ebook) | ISBN
    9781119513018 (hardback) | ISBN 9781119513032 (adobe pdf) | ISBN
    9781119513049 (epub)
Subjects: LCSH: Tissue culture – Laboratory manuals. | Cell
    culture – Laboratory manuals.
Classification: LCC QH585.2 .F74 2021 (print) | LCC QH585.2 (ebook) |
DDC
    571.5/38 – dc23
LC record available at https://lccn.loc.gov/2020029671
LC ebook record available at https://lccn.loc.gov/2020029672

Cover Design: Wiley
Cover Images: courtesy of Dr. Christophe Leterrier

Set in 10/12pt BemboStd by SPi Global, Chennai, India
Printed and bound by CPI Group (UK) Ltd, Croydon, CR0 4YY

C9781119513018_140125

The manufacturer's authorized representative according to the EU
General Product Safety Regulation is Wiley-VCH GmbH, Boschstr.
12, 69469 Weinheim, Germany, e-mail: Product_Safety@wiley.com.

In memory of R. Ian Freshney

1938–2019

whose legacy lives on
in his family, friends, colleagues,
and every laboratory where a copy of this book is found

Contents

Foreword

This edition of *Freshney's Culture of Animal Cells* marks the passing of a giant in the world of cell and tissue culture. Dr R. Ian Freshney passed away in January 2019 at the age of 80 after a long and distinguished career as a scientist and educator.

Ian Freshney learned his tissue culture in the laboratory of Dr John Paul, himself an author of a classic textbook. Working from the 1970s onwards, Ian played an important role in the development of our modern culture techniques. He also developed a passion for cell and tissue culture, leading to many colleagues coming to him to discuss their cell culture challenges. This passion became the defining characteristic of his career and would lead to numerous journal articles, books, and training courses.

Ian was an excellent teacher – articulate, charismatic, and hugely knowledgeable. His expertise was noticed by his peers and led to an unexpected opportunity. In 1980, Paul Chapple and his colleagues at the W. Alton Jones Cell Science Center were looking for a good textbook. They asked Ian if he would be interested to work on a new book to act as a practical handbook for cell and tissue culture. The result was the first edition of *Culture of Animal Cells*. More than 40 years later, the book remains in print and has grown in significance with each new edition. Most laboratories I have visited have "Freshney" somewhere on their bookshelves.

Ian asked me to work with him on the eighth edition of the book in 2016. At that point, we had collaborated together for almost a decade on a register of misidentified cell lines, now curated by the International Cell Line Authentication Committee (ICLAC). Ian knew that I loved his book, which I had used myself when learning cell culture as a student. He also knew that I shared his passion for tissue culture in all its diversity. He was incredibly gracious in allowing me to make changes and I hope the result does us both proud – and our colleagues, whose contributions have extended the reach of this book far beyond our own experiences.

When I tell my colleagues that I am working on a cell culture textbook, they usually ask: do people still use a textbook for cell culture? It is a valid question. A quick search with my favorite search engine brings up an abundance of cell culture tips, tricks, and videos online. Better still, many laboratories offer in-house training that pairs young students with experienced practitioners. But cell culture is more than a technique. Cell culture is a field in its own right that underpins some of the most exciting advances in modern medicine. A field deserves a textbook that helps us to understand cell culture and come to grips with its practical challenges.

Practical training and online videos will teach you how to do cell culture, but a textbook will teach you why. You need to understand the reasons why a technique is done in a particular way if you wish to adapt it for a new application. The "why" is usually built up over many years of publications in different laboratories. A textbook author will look at those publications and spin them into a factual story that (hopefully) makes sense to someone who is new to that particular technique. An author should also do his or her best to understand the practical questions that readers may ask when they encounter a technique for the first time.

Each edition of this book brings new techniques and applications for cell culture. The eighth edition has seen some remarkable advances, including iPSCs, CRISPR, and CAR T-cell therapy – all explored in later chapters. With many new advances and confusing terminology, it can be hard for newcomers to find the information that they need. For the eighth edition, we have divided the book into Parts to help with navigation. If you are new to cell culture and need to learn aseptic technique, you can go straight to Part IV, which is focused on handling cultures. Part II, which is focused on regulatory requirements, is also important reading so that you understand the core concepts that underpin safety and ethics. Parts I and III, which focus on the biology of cultured cells and how to select media and substrates, may not be required initially but will become important at some stage in the future.

Although the "how" of cell culture has changed a great deal over time, the "why" has been remarkably consistent. For example, Chapter 20 is dedicated to cloning and selection of cell populations. Cloning was first performed successfully by Katherine Sanford, Wilton Earle, and their colleagues in 1948. Their technique is not commonly used these days, but it was developed to provide a restricted environment that allowed the accumulation of growth factors that would nourish their clonal populations. The same principles are used in today's microfluidic platforms.

As a newcomer, I was told that I would either love or hate cell culture. For some of us, it is an ordeal to be endured for a particular outcome. For others, handling becomes a meditative process and we discover the engaging quirks of our chosen cultures. Whichever group you belong to, I hope that this book eases your way and keeps you company as you come to grips with new techniques and applications. It continues to offer a treasure trove of cell culture wisdom from Ian and I, and from our colleagues to explore whenever you need it.

Amanda Capes-Davis

Acknowledgments

It is no easy task to understand the advances that have occurred with each new edition of this book; indeed, it is only possible through the input of many colleagues. We are grateful to our reviewing editors: Bob Geraghty and Ray Nims for the current edition, and Carl Gregory and Stefan Przyborski for the previous edition. We also thank all our colleagues who reviewed individual chapters, contributed protocols or minireviews, or responded to questions or requests for data. For this edition, we are grateful to Amos Bairoch, Ngoc Chau, Jeremy Crook, Paul de Sousa, Sue Eccles, Adam Engler, Norbert Fusenig, Ruowen Ge, Daniela Grimm, Lily Huschtscha, Tiina Jokela, Fumio Kasai, Arihiro Kohara, Mark LaBarge, Madeline Lancaster, Christophe Leterrier, Saroj Mathupala, Hyo Eun Moon, Elsa Moy, Yukio Nakamura, Roger Reddel, Thibault Robin, Toshiro Sato, Christine Smythe, Hans-Jürgen Stark, Rachel Steeg, Greg Sykes, Jerilyn Timlin, Ornella Tolhurst, Masahiro Tomita, Cord Uphoff, Robert Utama, Jan van der Valk, Maria Vinci, Jatin Vyas, and Kaye Wycherley. We also thank the suppliers of tissue culture equipment and reagents who contributed images and responded to questions.

Some of the data presented here come from colleagues who worked with Ian over the years including Sheila Brown, Ian Cunningham, Lynn Evans, Margaret Frame, Elaine Hart, Mohammad Zareen Khan, Alison Mackie, Carol McCormick, John McLean, Alistair McNab, Alex MacPhee, Diana Morgan, Alison Murray, Irene Osprey, and Natasha Yevdokimova. Other data come from Amanda's "cell culture home" at Children's Medical Research Institute (CMRI), led by Roger Reddel. Amanda is grateful for Roger's encouragement and for the cell culture wisdom and expertise of Elsa Moy, Samath Pen, and George Theodosopoulos at CellBank Australia, and of Melinda Power, Kirsten Steiner, Josh Studdert, and Mila Tittel at CMRI's Operations Unit. Members of the International Cell Line Authentication Committee (ICLAC) are warmly acknowledged for their vast store of knowledge of cell line misidentification.

We are indebted to our colleagues at ECACC for data and much helpful discussion, including Isobel Atkin, Ed Burnett, Jim Cooper, David Lewis, Chris Morris, and Peter Thraves. We are also indebted to our colleagues at ATCC for their many contributions, and particularly to the late Rob Hay, who was a constant source of help and advice. Rob and Ian had been friends since doing their doctorates together in Glasgow on opposite sides of the same bench. Other colleagues who have helped and advised over the years include John Maxwell Anderson, Robert Auerbach, Bob Brown, Mike Butler, Kenneth Calman, Paul Chapple, Peter del Vecchio, Don Dougall, Sergey Federoff, Mike Gabridge, Roland Grafström, David I. Graham, Richard Ham, Stan Kaye, Nicol Keith, Dan Lundin, John Masters, Peter McHardy, Wally McKeehan, Rona McKie, Anne-Marie McNicol, Stephen Merry, John Paul, Jane Plumb, Yvonne Reid, John Ryan, Jim Smith, David G. T. Thomas, Peter Vaughan, Charity Waymouth, and Paul Workman.

We have been fortunate to receive excellent advice and support from the editorial staff of John Wiley & Sons. We would also like to acknowledge with sincere gratitude all those who have taken the trouble to write to the authors or to John Wiley & Sons with constructive criticism on previous editions. It is pleasant and satisfying to hear from those who have found this book beneficial, but even more important to hear from those who have found deficiencies, which we can then attempt to rectify. For those who feel more comfortable giving feedback in the digital world, Amanda can be contacted via Twitter @Cell_Detective.

Finally, we would like to thank our families for all their support and practical help. Ian's daughter Gillian and son Norman were involved in the preparation of the first edition of the book, many years ago. Amanda's daughter Renee acted as a sounding board for the latest edition regarding the complexities that face the newcomer to tissue culture, and Amanda's mother Nerida Capes was a constant source of encouragement. Above all, we would like to thank our spouses and closest colleagues: Mary Freshney and Darryn Capes-Davis. Both have contributed material to the book as experts – Mary as a scientist and Darryn as an engineer – and have offered hours of help and discussion.

Amanda Capes-Davis and R. Ian Freshney

Abbreviations

Abbreviations are provided in full when first used in each chapter. Abbreviations for organizations, societies, and committees are not listed here; most appear in a single chapter only.

2D	two-dimensional
3D	three-dimensional
AAV	adenovirus-associated virus
ACE	angiotensin converting enzyme
*Ac*NPV	*Autographa californica* multiple nucleopolyhedrovirus
AFLP	amplified fragment length polymorphism
AFSC	amniotic fluid stem cell
AIDS	acquired immunodeficiency syndrome
ALARA	as low as reasonably achievable
ALT	alternative lengthening of telomeres
ASC	adipose-derived stem cell
ATP	adenosine triphosphate
BAE	bovine aortic endothelium
BAEE	Nα-benzoyl-L-arginine ethyl ester
BCT	B-cell targeting (hybridoma technique)
BEGM	bronchial epithelial growth medium
BEV	baculovirus expression vector
BLV	bovine leukemia virus
BME	Basal Medium (Eagle's)
BMP	bone morphogenetic protein
BMS	building management system
bp	base pairs
BPE	bovine pituitary extract
BPyV	bovine polyomavirus
BrdU	5′-bromo-2′-deoxyuridine
BSA	bovine serum albumin
BSC	biological safety cabinet
BSL	biosafety level
BVDV	bovine viral diarrhea virus
CAD	computer-aided design
CAFC	cobblestone area-forming cell
calcein-AM	calcein-acetoxymethyl
CAM	(1) chorioallantoic membrane; (2) cell adhesion molecule (usually with additional prefix, e.g. NCAM [neural])
cAMP	cyclic adenosine monophosphate
CAR	chimeric antigen receptor (used in CAR T-cells)
Cas	CRISPR-associated protein
CBMP	cell-based medicinal product
CCD	charge-coupled device
CCLE	Cancer Cell Line Encyclopedia
CD	clusters of differentiation
CDK	cyclin-dependent kinase

cDNA	complementary DNA
CEA	carcinoembryonic antigen
CFC	colony forming cell
CFR	Code of Federal Regulations (United States)
CFU	colony forming unit
CG	chorionic gonadotropin (prefix may be used to indicate species, e.g. hCG [human])
CGH	comparative genomic hybridization (sometimes with additional prefix, e.g. aCGH [array])
CHO	Chinese hamster ovary
CITES	Convention on International Trade in Endangered Species of Wild Fauna and Flora
CK	creatine kinase
CL	containment level
CLASTR	cell line authentication using STR
CMC	carboxymethylcellulose
CMF	calcium- and magnesium-free saline
CMOS	complementary metal oxide semiconductor
CMV	cytomegalovirus
CNS	central nervous system
CNTF	ciliary neurotropic factor
CNV	copy number variation
CO1	cytochrome c oxidase I
COG	cost of goods
COSHH	Control of Substances Hazardous to Health (United Kingdom)
COSMIC	Catalogue of Somatic Mutations in Cancer
CRISPR	clustered regularly interspaced short palindromic repeat
crRNA	CRISPR RNA
CS	calf serum
CSC	cancer stem cell
CSF	colony-stimulating factor
DAF	diacetyl fluorescein
DAG	diacylglycerol
DAPI	4′,6-diamidino-2-phenylindole dihydrochloride
dATP	deoxyadenosine triphosphate
DBSS	dissection balanced salt solution
dCTP	deoxycytidine triphosphate
DEAE	diethylaminoethyl
dGTP	deoxyguanosine triphosphate
DIC	differential interference contrast
DMEM	Dulbecco's modified Eagle's medium
DMSO	dimethyl sulfoxide
DNA	deoxyribonucleic acid
DNase	deoxyribonuclease
dNTP	deoxynucleotide triphosphate
DO	dissolved oxygen
DPBS	Dulbecco's phosphate-buffered saline
DPBS-A	Dulbecco's phosphate-buffered saline solution A (without Ca^{2+} and Mg^{2+})
DSB	double-strand break
dTTP	deoxythymidine triphosphate
EBSS	Earle's balanced salt solution
EBV	Epstein-Barr virus
ECAP	eChick Atlas Project
ECGS	endothelial cell growth supplement
ECIS	electric cell-substrate impedance sensing
ECM	extracellular matrix
EDC	1-ethyl-3-(3-dimethylaminopropyl) carbodiimide hydrochloride
EDTA	ethylene diamine tetra-acetic acid
EGF	epidermal growth factor
EGTA	ethylene glycol tetra-acetic acid
EHS	Engelbreth-Holm-Swarm (mouse sarcoma)
ELISA	enzyme-linked immunosorbent assay
ELN	electronic laboratory notebook
EMT	epithelial–mesenchymal transition
EPC	endothelial progenitor cell
EpiSC	epiblast stem cell
EPL	early primitive ectoderm-like (stem cell)
EPO	erythropoietin
ESC	embryonic stem cell (prefix may be used to indicate species, e.g. hESC [human]; mESC [mouse])
FAB-SC	stem cell dependent on FGF-2, activin, and BIO (an inhibitor of glycogen synthase kinase-3β)
FACS	fluorescence-activated cell sorting
FAP	fibroblast-activation protein
FBS	fetal bovine serum
FeLV	feline leukemia/sarcoma virus

FGF	fibroblast growth factor
FISH	fluorescence *in situ* hybridization (prefix may be used, e.g. m-FISH [multiplex])
FITC	fluorescein isothiocyanate
FSH	follicle-stimulating hormone
FSP	fibroblast surface protein
g	radial acceleration relative to gravity
G6PD	glucose-6-phosphate dehydrogenase
GABA	γ-aminobutyric acid
Gal-C	galactocerebroside
GALV	gibbon ape leukemia virus
GCCP	Good Cell Culture Practice
G-CSF	granulocyte colony-stimulating factor
GDPR	General Data Protection Regulation (European Union)
GDSC	Genomics of Drug Sensitivity in Cancer (cell line panel)
GFAP	glial fibrillary acidic protein
GFOGER	glycine-phenylalanine-pyrrolysine-glycine-glutamic acid-arginine (amino acid motif)
GFP	green fluorescent protein
GHS	Globally Harmonized System of Classification and Labeling of Chemicals
GLP	Good Laboratory Practice
GM-CSF	granulocyte-macrophage colony-stimulating factor
GMEM	Glasgow Minimum Essential Medium
GMF	glia maturation factor
GMO	genetically modified organism
GMP	Good Manufacturing Practice
GPDH	glycerol phosphate dehydrogenase
GR	normalized growth rate inhibition
GRGDS	glycine-arginine-glycine-aspartic acid-serine (amino acid motif)
gRNA	guide RNA
GRP	gastrin-releasing peptide
h	human (used as prefix before various cell types, hormones, genes, etc.)
HAT	hypoxanthine, aminopterin, and thymidine (selective medium)
HBSS	Hanks's balanced salt solution
HBV	hepatitis B virus
HDAC	histone deacetylase
HDR	homology-directed repair
HEPA	high-efficiency particulate air (filter)

HEPES	4-(2-hydroxyethyl)-1-piperazine-ethanesulfonic acid
HES	hydroxyethyl starch
HGCC	human glioblastoma cell culture (cell line panel)
HGF	hepatocyte growth factor
HHV	human herpesvirus
HIF	hypoxia inducible factor
HIPAA	Health Insurance Portability and Accountability Act (United States)
HIV	human immunodeficiency virus
HMBA	hexamethylene-*bis*-acetamide
HMEC	human mammary epithelial cell
hPSCreg	Human Pluripotent Stem Cell Registry
HPV	human papillomavirus
HSA	human serum albumin
HSC	hematopoietic stem cell
HSPG	heparan sulfate proteoglycan
HTLV	human T-cell lymphotropic virus
HUVEC	human umbilical vein endothelial cell
IFN	interferon
IGF	insulin-like growth factor
IgG	immunoglobulin G
IgSF	immunoglobulin superfamily
IL	interleukin
IMDM	Iscove's modified Dulbecco's medium
iN	induced neuronal (cell)
IPA	isopropanol
iPSC	induced pluripotent stem cell
IQ	installation qualification
IR	infrared
IT	information technology
ITR	inverted terminal repeat
ITS	insulin, transferrin, and selenium (medium supplement)
IVF	*in vitro* fertilization
kb	kilobase
KGF	keratinocyte growth factor
KSHV	Kaposi's sarcoma herpesvirus
LCL	lymphoblastoid cell line
LDH	lactate dehydrogenase
LIF	leukemia inhibitory factor
LIMS	laboratory information management systems

LL	leukemia-lymphoma
LTC-IC	long-term culture initiating cell
m	mouse (used as prefix before various cell types, hormones, genes, etc.)
M199	Medium 199
mAb	monoclonal antibody
MAC	*Mycobacterum avium-intracellulare* complex
MACS	(1) magnet-activated cell sorting; (2) multipotent adult stem cell
MAPC	multipotent adult progenitor cell
M-CSF	macrophage colony-stimulating factor
MEF	mouse embryonic fibroblast
MEM	Minimum Essential Medium (Eagle's)
MET	mesenchymal–epithelial transition
MIAMI	marrow-isolated adult multilineage inducible (cell)
MIATA	minimal information about T-cell assays
miRNA	microRNA
MLV	murine leukemia virus
MMLV	Moloney murine leukemia virus
MMR	mismatch repair
mPES	modified polyethersulfone
MPS	massively parallel sequencing
mRNA	messenger RNA
MSC	mesenchymal stromal cell
MSH	melanocyte-stimulating hormone
MSI	microsatellite instability
MST	maximal serial transfer (of explants)
MTT	3-(4,5-dimethylthiazol-2-yl)-2,5-diphenyltetrazolium bromide
NCI-60	National Cancer Institute-60 (cell line panel)
NEAA	non-essential amino acids
NGF	nerve growth factor
NGS	next-generation sequencing
NHBE	normal human bronchial epithelial (cell)
NHEJ	non-homologous end joining
NHPP	National Hormone and Peptide Program
NHS	*N*-hydroxysulfosuccinimide
NSCLC	non-small cell lung carcinoma
OEC	olfactory ensheathing cell
OQ	operational qualification
PAM	protospacer-adjacent motif
PAT	process analytical technologies

PBS	phosphate-buffered saline
PC	polycarbonate
PCNA	proliferating cell nuclear antigen
PCR	polymerase chain reaction
PCV	packed cell volume
PD-ECGF	platelet-derived endothelial cell growth factor
PDGF	platelet-derived growth factor
PDL	population doubling level
PDMS	polydimethylsiloxane
PDT	population doubling time
PDX	patient-derived xenograft
PE	polyethylene
PEG	polyethylene glycol
PEI	polyethyleneimine
PEM	polyelectrolyte multilayer
PES	polyethersulfone
PET	polyethylene terephthalate
PGA	polyglycolic acid
γPGA	poly-γ-glutamic acid
PHA	phytohemagglutinin
pHEMA	poly(2-hydroxyethyl methacrylate)
PKC	protein kinase C
PLA	polylactic acid
PLD	phospholipase D
PMA	phorbol 12-myristate 13-acetate (also known as TPA)
PMMA	polymethyl methacrylate
PPE	personal protective equipment
PPLO	pleuropneumonia-like organism (mycoplasma)
PPTP	Pediatric Preclinical Testing Program (cell line panel)
PQ	performance qualification
PSC	pluripotent stem cell (prefix may be used to indicate species, e.g. hPSC [human])
PTFE	polytetrafluoroethylene
PVA	polyvinyl alcohol
PVC	polyvinyl chloride
PVDF	polyvinylidene difluoride
PVP	polyvinylpyrrolidone
PW	purified water
QA	quality assurance
Q-banding	Quinacrine banding

QbD	quality by design
QC	quality control
Q-PCR	quantitative polymerase chain reaction
RBC	red blood cell
RCCS	Rotary Cell Culture System
RCL	replication competent lentivirus
REACH	Registration, Evaluation, Authorisation and Restriction of Chemicals (European Union)
RGD	arginyl-glycyl-aspartic acid (amino acid motif)
RNA	ribonucleic acid
RNA-seq	RNA sequencing
RNP	ribonucleoprotein
RO	reverse osmosis
ROCK	Rho-associated coiled-coil containing kinase (also referred to as "Rho kinase")
rpm	revolutions per minute
RRID	Research Resource Identifier
rRNA	ribosomal RNA
RT-PCR	reverse transcriptase polymerase chain reaction
RVD	repeat variable diresidue
SARS	severe acute respiratory syndrome
SCF	stem cell factor
SCLC	small cell lung carcinoma
SCNT	somatic cell nuclear transfer
SDS	safety data sheet
SKY	spectral karyotyping
S-MEM	spinner modification of Minimum Essential Medium
SMP	standard microbiological practice
SMRV	squirrel monkey retrovirus
SNP	single nucleotide polymorphism
SOP	standard operating procedure
SP	special practice
SPW	semi-purified water
SSEA	stage-specific embryonic antigen
SSFV	spleen focus-forming virus
ssmAb	stereospecific monoclonal antibody
SST	stereospecific targeting (hybridoma technique)
STAP	stimulus-triggered acquisition of pluripotency (now retracted)
STR	short tandem repeat
SV40	simian virus 40
SV40 TAg	SV40 large T antigen
SWATH-MS	sequential-window acquisition of all theoretical fragments mass spectrometry
TAE	Tris base, acetic acid, EDTA (buffer)
TALE	transcription activator-like effector
TALEN	transcription activator-like effector nuclease
TC	thermal conductivity
TDLU	terminal ductal lobular unit
TE	Tris base, EDTA (buffer)
TEER	transepithelial electrical resistance
TERT	telomerase reverse transcriptase (prefix may be used to indicate species, e.g. hTERT [human])
T_g	glass transition temperature
TGF	transforming growth factor
TLM	telomere lengthening mechanism
TNF	tumor necrosis factor
TOC	total organic carbon
TPA	12-O-tetradecanoylphorbol-13-acetate (also known as PMA)
tPA	tissue plasminogen activator
TR	telomerase RNA (prefix may be used to indicate species, e.g. hTR [human])
tracrRNA	*trans*-activating crRNA
ts	temperature-sensitive
TSH	thyroid-stimulating hormone
U	units
ULA	ultra-low attachment (microwell plate)
uPA	urokinase plasminogen activator
UPW	ultrapure water
USD	US dollar
UV	ultraviolet
VEGF	vascular endothelial growth factor
VSEL	very small embryonic-like (stem cell)
vWF	von Willebrand factor
XMRV	xenotropic murine leukemia virus-related virus
XTT	2,3-bis-(2-methoxy-4-nitro-5-sulfophenyl)-2H-tetrazolium-5-carboxanilide
ZFN	zinc-finger nuclease

Book Navigation

ON THE COVER

Front Cover. A glial cell in culture, visualized using three-dimensional stochastic optical reconstruction microscopy (3D-STORM). The cell was labeled for actin and color-coded for depth (blue to red, −400 to +400 nm). The visible part of the cell is around 35 μm × 35 μm.

Back Cover. COS-7 cells in culture, visualized using structured illumination confocal microscopy. Cells were labeled for DNA (DAPI, blue), clathrin-coated pits (clathrin light chain, yellow), microtubules (α-tubulin, magenta), and actin (cyan). Each cell is around 75 μm × 60 μm.

Both images were contributed by Christophe Leterrier, Aix-Marseille Université, CNRS, INP UMR7051, Neuro-Cyto, Marseille, France.

MAIN TEXT

The book is divided into Parts, which contain chapters on related topics. For example, Part IV focuses on handling cultures and includes aseptic technique, primary culture, subculture (passaging), and cryopreservation. Each Part has a list of key learning outcomes that can be used for training. Chapters are divided into numbered sections; they also contain protocols and minireviews. Protocols in the main text of the book are indicated by the prefix "P"; for example, searching for "Protocol P12.1" brings up a protocol for aseptic technique when handling flasks in a biological safety cabinet (BSC). Minireviews are indicated by the prefix "M"; for example, searching for "Minireview M9.1" brings up a minireview on hypoxic cell culture. Each chapter has its own lists of Suppliers and References. Supplier websites were current at the time of writing but are prone to change over time. Supplier information can be used as key words in online searches if websites are no longer current.

COMPANION WEBSITE

This book is accompanied by a companion website:
www.wiley.com/go/freshney/cellculture8.

The website contains the Supplementary Material, which because of space constraints can only appear in electronic form. The Supplementary Material is grouped by chapter and indicated by the prefix "S;" for example, "Supp. S24.1" brings up a protocol for the culture of epidermal keratinocytes. The Supplementary Material is listed at the end of each chapter, between the list of Suppliers and the References.

PART I

Understanding Cell Culture

After reading the following chapters in this part of the book, you will be able to:

(1) **Introduction:**

  (a) Find the meaning of terms used throughout the book, starting with "tissue culture" and "cell culture."

  (b) Describe the historical development of tissue culture techniques and list some examples of important technical innovations.

  (c) List some of the advantages of tissue culture and its applications.

  (d) List some of the limitations of tissue culture and how these may be addressed.

(2) **Biology of Cultured Cells:**

  (a) Describe some of the structures and molecules that are required for cell adhesion and motility to occur.

  (b) List the phases of the cell cycle and describe their roles in cell division.

  (c) List the germ layers that develop during vertebrate embryogenesis and provide examples of cell types that arise from each layer.

  (d) Explain the meaning of stem cell terminology, including "pluripotent stem cell" (PSC), and the relationship between potency and differentiation.

  (e) Recognize various forms of cell death, particularly apoptosis.

(3) **Origin and Evolution of Cultured Cells:**

  (a) Explain why cultures are inherently heterogeneous, looking at both their origins and their evolution with handling.

  (b) Discuss the changes in genotype that may be found in culture, including chromosomal aberrations and loss of heterozygosity (LOH).

  (c) Discuss the changes in phenotype that may occur and the key variables in growth conditions that may influence the culture phenotype.

  (d) Explain the meaning of the terms "senescence," "immortalization," and "transformation."

  (e) List the characteristics of transformation, particularly the changes in growth that occur with transformation and how those may be assessed.

Freshney's Culture of Animal Cells: A Manual of Basic Technique and Specialized Applications, Eighth Edition. Amanda Capes-Davis and R. Ian Freshney.
© 2021 John Wiley & Sons Ltd. Published 2021 by John Wiley & Sons Ltd.
Companion website: www.wiley.com/go/freshney/cellculture8

CHAPTER 1

Introduction

1.1 TERMINOLOGY

1.1.1 Tissue Culture and Cell Culture

Tissue culture, in its various guises, has been in use for more than a century (Harrison 1907; Burrows 1910). Tissue culture methodology was initially devised to study the behavior of animal cells, free of systemic variations that might arise *in vivo* during normal homeostasis and under the stress of an experiment. As the term implies, fragments of tissue were originally used to provide a source of cells; this form of culture is commonly known as "explant" culture. Culture of cells within and from such tissue fragments came to dominate the field for more than 50 years (Fischer 1925; Parker 1961). Because culture conditions were initially suboptimal, growth was restricted to the radial migration of cells from the explant, with occasional mitoses in the outgrowth.

Throughout the first three decades of tissue culture, efforts were made to induce an "immortal" line of cells that would continue to replicate outside the tissue fragment. Success was initially reported by Alexis Carrel, who maintained a culture for more than 30 years that was originally derived from embryonic chick heart (Carrel 1912). Carrel's achievement could not be reproduced by other laboratories; it was subsequently shown that this cell type does not survive in culture for more than a year (Hayflick 1965). In retrospect, Carrel's "immortal" cells were probably an artifact due to cross-contamination (see Chapter 17) (Witkowski 1980). The first continuous cell lines were established by Wilton Earle and his colleagues at the National Cancer Institute (NCI). Earle's "L" strain was initiated in 1941 and cloned to give rise to the L-929 cell line (see Figure 1.1) (Earle et al.

1943; Sanford et al. 1948). After a further decade of effort, George Gey and colleagues succeeded in establishing the HeLa cell line from a young woman with cervical carcinoma (Gey et al. 1952; Skloot 2010). L-929 and HeLa were used to develop many of the culture conditions and techniques that are still in use today (Ham 1974).

"Tissue culture" is used throughout this book as a generic term, referring to the culture of cells that were initially derived from a tissue sample. The word "culture" implies that cells can undergo replication outside the organism. In some senses, "tissue culture" is an historical term. Most of the explosive expansion in the field in the second half of the twentieth century was made possible by the use of dispersed cell cultures and, in particular, by the availability of L-929, HeLa, and other cell lines (see Figure 1.2). It is not surprising that the term "cell culture" has become increasingly popular within the scientific literature. However, dispersed cell cultures make up only one type of culture. "Tissue culture" has taken on a deeper meaning in the last few years, thanks to the development of three-dimensional (3D) cultures. These 3D models are truly tissue culture in all senses of the term; for example, organoid cultures develop complex structures that reflect the behavior of tissues within the original organism (see Figure 1.3).

1.1.2 Sources of Terminology

Although we have tried to use consistent terminology throughout this book to reduce confusion, the meanings of some terms may be controversial. The Tissue Culture Association (now the Society for In Vitro Biology) developed

Freshney's Culture of Animal Cells: A Manual of Basic Technique and Specialized Applications, Eighth Edition. Amanda Capes-Davis and R. Ian Freshney.
© 2021 John Wiley & Sons Ltd. Published 2021 by John Wiley & Sons Ltd.
Companion website: www.wiley.com/go/freshney/cellculture8

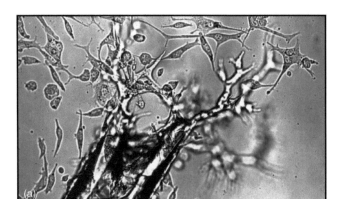

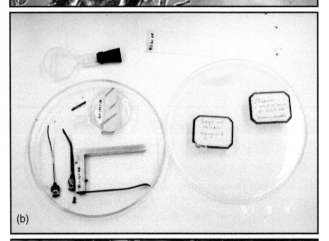

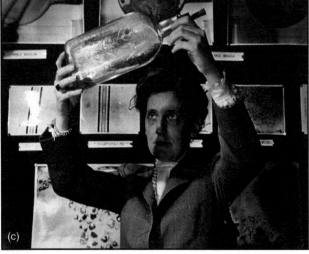

Fig. 1.1. Establishment of the L-929 cell line. (a) Cloning of the L strain to give L-929 (also known as NCTC clone 929), as photographed by Wilton Earle. Single cells were isolated and cultured in sealed glass capillary tubes. If proliferation was observed, the tube was broken and the cells allowed to spill out. (b) Equipment used to perform single cell isolation. The Carrel flask, shown top left, was later modified in Earle's laboratory for large-scale tissue culture. (c) Virginia Evans, who worked with Earle to develop the first defined substrates for tissue culture, examining growth within a modified tissue culture flask ("T-flask"). *Source*: (a) National Cancer Institute, image AV-4300-4382; (b) Office of NIH History and Stetten Museum, item 91.0001.161; (c) US National Library of Medicine, image 101393939. Public domain.

consensus terminology in an effort to reduce confusion and improve communication (Schaeffer 1990). This book primarily follows the consensus terminology, as do several guidelines on Good Cell Culture Practice (GCCP) (Coecke et al. 2005; Geraghty et al. 2014). Terms that are particularly controversial or confusing are discussed at the beginning of the relevant chapter; for example, "cell line" and "cell strain" are discussed at the beginning of Chapter 14. In some cases, a term is not included in the consensus terminology and the meaning may vary across the field. These terms are clarified and source references provided in a relevant chapter; for example, "organoid culture" is discussed in Chapter 27. Tissue culture-related terms are listed at the end of this book (see Appendix A). Other terms (e.g. from molecular biology) may be found elsewhere (Cammack et al. 2008).

1.2 HISTORICAL DEVELOPMENT

1.2.1 Substrates and Media

L-cells were initially grown in clotted plasma that was bathed in a mixture of horse serum, chick embryonic extract, and Earle's balanced salt solution (EBSS) (Earle et al. 1943). These conditions were almost completely undefined (apart from the EBSS), leading to great difficulty in generating reproducible results. The first defined substrate was not introduced until 40 years after the first experiments, when Virginia Evans and Wilton Earle used perforated cellophane to culture L-cells (see Table 1.1) (Evans et al. 1947). Cells grew exceptionally well on the cellophane, which provided a 3D matrix, since the cells grew through the perforations in the sheet. The cellophane matrix could also be extensively folded and used for scale-up in 3D culture (Sanford and Evans 1982). Once the culture adapted to growing on the cellophane, the cells would usually grow on the glass floor of the flask, leading to increasing use of glass (and later plastic) for cell culture (Evans and Sanford 1978). The further development of culture vessels and substrates is described in Chapter 8.

The increasing number of cells grown *in vitro* (which translates to "in the glass") meant that a standardized medium formulation was necessary to provide consistent nutritional requirements. The first medium formulation, Medium 199, was developed for the culture of chick embryonic fibroblasts (Morgan et al. 1950). Although it was effective for some cell types and applications, Medium 199 did not support the extensive proliferation that was seen with "natural" media containing serum and chick embryo extract (Morgan et al. 1955). A variety of other formulations were developed to address this concern, particularly for use with L-929 and HeLa (Ham 1974). The most popular were Eagle's Basal Medium (BME) and Minimum Essential Medium (MEM), which supported growth of a wide variety of cell lines (Eagle 1955, 1959). The development and use of defined media are described in Chapter 9.

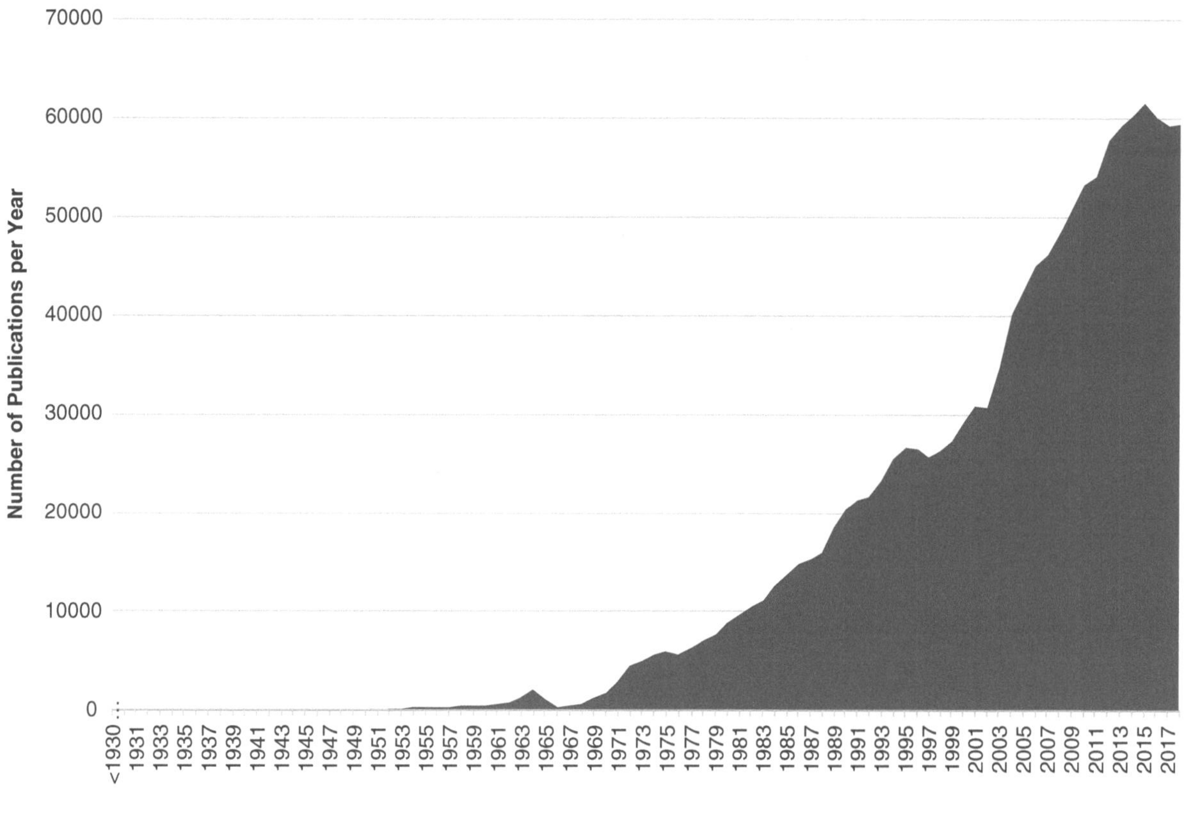

Fig. 1.2. Growth of cell and tissue culture. A search was performed in the PubMed database using the text string ("tissue culture" OR "cell culture" OR "cultured cells" OR "cell line"). A total of 1 395 112 items were found and are displayed by year of publication. Results shown here date from before 1930 to 2018; the years 2019–2020 (n = 32 279) were incomplete and were thus omitted. More than half of these results (n = 809 041) could be found using a single term "cell line."

Although many of the classic media formulations were intended for use without embryo extract or serum, it has been difficult to remove serum entirely, in what Honor Fell once described as a "tiresome fact" of tissue culture (Fell 1972). Extensive progress has since been made in developing serum-free culture media, thanks to the efforts of Richard Ham, Gordon Sato, and colleagues in the 1970s and 1980s (Ham and McKeehan 1978; Barnes and Sato 1980). More recently, we are indebted to James Thomson and other modern pioneers who have developed defined media for stem cell culture (Ludwig et al. 2006; Chen et al. 2011). The development and use of serum-free media are described in Chapter 10.

1.2.2 Primary Cultures and Cell Lines

The development of suitable substrates and media meant that cells could be grown at a larger scale and for longer time periods, requiring transfer from one culture vessel to another. This process is known as subculture or passage. Before the first subculture is performed, the sample is known as a primary culture. Cells in a primary culture may have migrated from an explant or may be released from it through disaggregation (see Figure 1.4). Disaggregation of tissue fragments and plating out of the dispersed cells was initially performed in 1916 using trypsin, although this enzyme was not adopted for subculture until some years later (Rous and Jones 1916; Dulbecco 1952). Once a primary culture has been passaged, it becomes a cell line, at least according to the consensus terminology (see Section 1.1.2). More information on the changes that occur throughout the lifespan of a culture can be found in Chapter 3. Procedures for generating primary cultures and passaging cell lines are described in Chapters 13 and 14.

Subculture (particularly the first subculture to form a cell line) implies (i) an increase in the total number of cells over several generations (population doublings) and (ii) the ultimate predominance of cells or cell lineages with a high proliferative capacity, resulting in (iii) increased uniformity in the cell population. These are all advantages, but there are also disadvantages that must be considered (see Table 1.2). Subculture results in an increased risk of contamination by microorganisms or by other cultures. Prevention of contamination relies on good aseptic technique, which is discussed in Chapter 12.

With continued subculture, cells may display changes in their genetic makeup (genotype) or their observable physical characteristics (phenotype). Cell lines must be preserved at early passage if these problems are to be avoided. Procedures for cryopreservation and cell banking are described in Chapter 15; validation and characterization of cell lines are discussed in Chapters 16–19.

Many cell lines cease to grow after repeated subculture. This was originally thought to be due to poor culture conditions, until Leonard Hayflick and Paul Moorhead

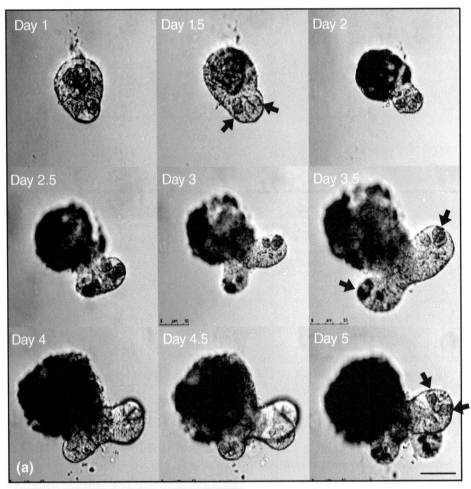

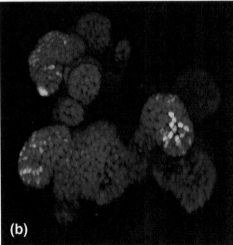

(a-b) Intestinal Organoids. (a) Time course of growth for a single intestinal organoid, showing the development of crypt-like domains. The arrows indicate granule-containing Paneth cells; the scale bar represents 50 μm. Differential image contrast (DIC) image. (b) Intestinal organoids after 3 weeks in culture. Green indicates Lgr5-GFP stem cells, which are localized at the tips of the crypt-like domains; cells have been counterstained with ToPro-3 (red). Reconstructed 3D confocal image. Source: [Sato et al. 2009], DOI 10.1038/nature07935, reproduced with permission of Springer Nature.

Fig. 1.3. Examples of organoid culture.

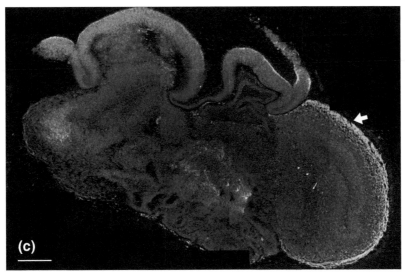

(c) Cerebral Organoid. Cerebral organoids can be generated using a multistep procedure (see Fig. 27.12). Sectioning and immunohistochemistry reveal complex morphology, with heterogeneous regions containing neural progenitors (Sox2, red) and neurons (Tuj1, green; see arrow). DNA is indicated by Hoechst staining (blue). The scale bar represents 200 μm. Source: [Lancaster et al. 2013], DOI 10.1038/nature12517, reproduced with permission of Springer Nature.

Fig. 1.3. (Continued)

proved that normal fibroblasts consistently ceased to divide after a certain number of generations (now known as the Hayflick limit) (Hayflick and Moorhead 1961). A subset of cell lines continue to divide even after the Hayflick limit has been reached and are known as continuous or immortalized cell lines. Hilary Koprowski and colleagues discovered that simian virus 40 (SV40) could be used to induce "transformation" of the culture, which was associated with an immortal lifespan (Koprowski et al. 1962; Girardi et al. 1965). Study of oncogenic viruses led to a better understanding of the genes that were required for immortalization and how to deliver them to the cell with greater specificity. Gene delivery, editing, and immortalization are discussed in Chapter 22.

Some cell types are more difficult to grow than others, leading to the need for optimized culture conditions. The feeder layer, which was developed by Theodore Puck and Philip Marcus for cell cloning, was a particularly important innovation in this area (Puck and Marcus 1955; Marcus et al. 2006). Increasing familiarity with the feeder layer and its interaction with other cell types led to the first cellular therapies, with Howard Green and colleagues using feeder layers to culture human keratinocytes for the treatment of burns (Green et al. 1979; Green 2008). In retrospect, Green's work represented the first use of stem cell therapy, since keratinocyte expansion depends on the maintenance of epidermal stem cells (Hynds et al. 2018). Feeder layers were then used for the culture of embryonic stem cells (ESCs) and for

induction of pluripotency in adult cells (Evans and Kaufman 1981; Martin 1981; Thomson et al. 1998; Takahashi and Yamanaka 2006; Takahashi et al. 2007). Although stem cells can now be grown feeder-free, feeder layers continue to play an important role in the culture of specific cell types. Procedures for the culture of stem cells (including induction of pluripotency), specific cell types, and tumor cells are described in Chapters 23–25. Induction of differentiation is a challenge for many cell types and is discussed in Chapter 26.

1.2.3 Organ, Organotypic, and Organoid Culture

Although cell lines are a mainstay of most tissue culture laboratories, concerns have been expressed for many years regarding their suitability to model *in vivo* behavior. Perhaps the earliest concern was expressed by David Thomson, who observed that dispersed cells displayed uncontrolled growth compared to cells that are maintained *in situ* (Thomson 1914). This observation was the first step in what came to be known as "organ culture" (see Figure 1.4). Rather than trying to establish a cell line, the proponents of organ culture aimed to study cells in their original environment with minimal changes to the tissue architecture. Organ culture techniques were pioneered by Thomas Strangeways, Honor Fell, and colleagues at the Strangeways Research Laboratory (Strangeways and Fell 1925; Fell and Robison 1929; Fell 1972). Organ culture was technically challenging but was believed to be the best way to study physiological behavior,

TABLE 1.1. Technical innovations in tissue culture development.

Year of publication	Technical innovation	Book section where innovation discussed	References
1907	Primary explant of frog embryo nerve fiber in a hanging drop, leading to outgrowth *in vitro*	1.1.1, 27.4.1	Harrison (1907)
1910	Use of plasma clot to culture tissue fragments; culture of chick embryonic tissue fragments (explants)	1.1.1	Burrows (1910)
1916	Trypsinization of explants	1.2.1	Rous and Jones (1916)
1923	Development of Carrel flasks	8.5.3	Carrel (1923)
1925	Differentiation of embryonic tissue in organ culture	1.2.3, 27.8	Strangeways and Fell (1925)
1928	Time-lapse recording of live cell behavior	18.4.3	Canti (1928)
1929	Use of watch-glass method for organ culture	1.2.3	Fell and Robison (1929)
1933	Development of the roller tube	28.3.1	Gey (1933)
1936	Culture of poliovirus in primary embryonic nervous tissue	1.3	Sabin and Olitsky (1936)
1943	Establishment of the first continuous cell line (L)	1.1.1	Earle et al. (1943)
1947	Culture of L-cells on the first defined substrate	1.2.1	Evans et al. (1947)
1948	Capillary cloning of L-cells to give L-929	1.1.1	Sanford et al. (1948)
	Use of antibiotics in culture	9.6.2	Keilova (1948)
1949	Use of glycerol as a cryoprotective agent	15.1.1	Polge et al. (1949)
	Use of the Coulter principle for automated cell counting	19.1.2	Coulter (1949)
1950	Development of Medium 199 for chick embryo fibroblasts and organ culture	1.2.1, 9.4	Morgan et al. (1950)
1951	Culture of cells in three dimensions using cellulose sponge, folded cellophane, or glass rings	1.2.1, 27.5	Earle et al. (1951); Leighton (1951)
	Use of a hemocytometer for counting cultured cells	19.1.1	Sanford et al. (1951)
1952	Establishment of the first human cell line (HeLa)	1.1.1, 1.3	Gey et al. (1952)
	Use of trypsin for subculture	1.2.2	Dulbecco (1952)
	Organotypic culture performed by inducing dissociated cells to reaggregate in 3D culture	1.2.3	Moscona and Moscona (1952)
1953–1955	Mass production of HeLa cells for polio vaccination program	1.3	Syverton and Scherer (1953)
1953	Demonstration of contact inhibition by time-lapse cinematography	3.6.2	Abercrombie and Heaysman (1953)
	Culture of cells on either side of a membrane filter (leading to the development of filter well inserts)	27.5.2	Grobstein (1953)
1954	Culture of cells in suspension; development of shaker culture	8.6.2, 14.7	Earle et al. (1954); Owens et al. (1954)
	Development of the modified Carrel flask (the "T-flask")	8.5.3	Earle and Highhouse (1954)

(continued)

TABLE 1.1. (*continued*)

Year of publication	Technical innovation	Book section where innovation discussed	References
1955	Production of irradiated feeder layers for use in cloning	8.4, 20.6.2	Puck and Marcus (1955)
	Development of BME for culture of L-929 and HeLa cells	9.4	Eagle (1955)
1956	Culture of cells on rat tail collagen	8.3.3	Ehrmann and Gey (1956)
	Demonstration of mycoplasma contamination	16.4	Robinson et al. (1956)
1957	Use of collagenase for disaggregation of mammary glands	13.5.3	Lasfargues (1957)
	Development of spinner culture	28.2.1	McLimans et al. (1957)
	Development of "Friend cells" as a model for erythroid differentiation	24.5.3, 26.3.1	Friend (1957)
1958	Use of stirred-tank systems for scale-up of mammalian culture	28.2.2	Ziegler et al. (1958)
1959	Use of dimethyl sulfoxide (DMSO) as a cryoprotective agent	15.1.1	Lovelock and Bishop (1959)
	Demonstration of interspecies contamination	17.3.4	Rothfels et al. (1959)
1961	Definition of finite lifespan of normal human fibroblasts	1.2.2, 3.2.1	Hayflick and Moorhead (1961)
1962	Transformation of cells obtained from tissue explants using SV40	22.3.1	Koprowski et al. (1962)
	Development of Grace's medium for insect culture	24.6.2	Grace (1962)
1963	Establishment of the WI-38 cell line for viral vaccine production	1.3	Hayflick (1963)
	Development of protocols to establish 3T3 mouse embryonic fibroblast (MEF) cell lines	3.2.1	Todaro and Green (1963)
1964	Cloning of cells in suspension using soft agar	20.3.1	Macpherson and Montagnier (1964)
	Formation of embryoid bodies; cloning of embryonal carcinoma cells to assess pluripotency	23.2	Kleinsmith and Pierce (1964)
1965	Development of serum-free media and defined growth factors for cloning and for normal and specialized cell types	10.2, 10.3	Ham (1965); Barnes and Sato (1980)
1966	Colony formation by hematopoietic cells	20.3	Bradley and Metcalf (1966); Ichikawa et al. (1966)
1967	Demonstration of intraspecies cross-contamination with HeLa	17.3.1	Gartler (1967)
	Isolation of lymphoblastoid cell lines (LCLs)	22.3.1	Moore et al. (1967)
	Culture of cells on microcarriers	27.5.4, 28.3.3	van Wezel (1967)
1968	Use of a "biohazard hood" to prevent infection during microbiological procedures	6.3.4	Coriell and McGarrity (1968)
1969	Transfer of human tumor cells to "nude" mice	25.4.5	Rygaard and Povlsen (1969)

(*continued*

TABLE 1.1. (*continued*)

Year of publication	Technical innovation	Book section where innovation discussed	References
1970	High density culture using perfusion	28.2.3	Kruse et al. (1970)
1971	Culture of mouse epidermal keratinocytes	24.2.1	Fusenig (1971)
	Production of feeder layers using mitomycin C	20.6.2	Macpherson and Bryden (1971)
	Culture of tumor cells as spheroids in suspension	25.4.4, 27.4.1	Sutherland et al. (1971)
1972	Primary culture of hepatocytes disaggregated in collagenase	24.2.6	Leffert and Paul (1972)
	Perfused culture in hollow fibers	27.5.3	Knazek et al. (1972)
1973	Transfection using calcium phosphate for gene delivery	22.1.1	Graham and van der Eb (1973)
1974	Use of Giemsa banding for detection of misidentified cell lines	17.3.4	Nelson-Rees et al. (1974)
1975	Culture of human epidermal keratinocytes using feeder layers of 3T3 mouse fibroblasts	1.2.2, 24.2.1	Rheinwald and Green (1975)
	Culture of hepatocytes on floating collagen membranes	8.3.3	Michalopoulos and Pitot (1975)
	Production of monoclonal antibodies by hybridomas (mouse myelocyte–splenocyte hybrids)	24.5.4	Köhler and Milstein (1975)
1977	Culture of normal human fibroblasts under hypoxic conditions	5.3.1	Packer and Fuehr (1977)
	Characterization of matrix from Engelbreth-Holm-Swarm (EHS) mouse sarcoma (resulting in Matrigel® and other products)	8.3.2	Orkin et al. (1977)
	Detection of mycoplasma by DNA fluorescence	16.4.1	Chen (1977)
	Suspension cloning using Methocel	20.3.2	Dao et al. (1977)
1979	Growth of human epidermal keratinocytes for use as skin grafts	1.2.2, 24.2.1	Green et al. (1979)
1981	Culture of mouse embryonic stem cells (mESCs)	23.2.1	Evans and Kaufman (1981); Martin (1981)
1982	Use of electroporation for gene delivery	22.1.3	Neumann et al. (1982)
1983	Immortalization of human diploid fibroblasts by SV40	22.3.1	Huschtscha and Holliday (1983)
	Development of baculovirus expression vectors (BEVs)	22.1.4	Smith et al. (1983)
1984	Induction of differentiation in pluripotent embryonal carcinoma cells	23.2	Andrews et al. (1984)
	Development of stereolithography (later used for 3D bioprinting)	28.7.3	Hull (1984)
1986	Development of confocal microscopy	18.4.4	Petran et al. (1986)
1987	Use of lipofection for gene delivery	22.1.2	Felgner et al. (1987)
1990	Use of retroviral vectors for gene delivery	22.1.4	Miller et al. (1990)
1991	Culture of human adult mesenchymal stromal cells (MSCs)	23.5, 23.6	Caplan (1991)

(*continued*

TABLE 1.1. (*continued*)

Year of publication	Technical innovation	Book section where innovation discussed	References
1992	Culture of primary breast epithelial cells within an EHS matrix, resulting in 3D structures	27.5.1	Petersen et al. (1992)
1993	Generation of chimeric antigen receptor (CAR) T-cells	24.5.5	Eshhar et al. (1993)
1996	Development of zinc finger nucleases (ZFNs)	22.2.1	Kim et al. (1996)
1998	Culture of human embryonic stem cells (hESCs)	23.2.2	Thomson et al. (1998)
1999	Use of short tandem repeat (STR) profiling for authentication of cell lines	17.3.2	Tanabe et al. (1999); Masters et al. (2001)
	Large-scale culture of suspension cells in disposable bioreactor using wave-like agitation	28.2.2	Singh (1999)
2006	Induction of pluripotency in mouse embryonic and adult fibroblasts using defined factors	23.3	Takahashi and Yamanaka (2006)
	Establishment of hESC cultures using defined conditions	23.4.2	Ludwig et al. (2006)
2007	Induction of pluripotency in human adult fibroblasts	23.3	Takahashi et al. (2007)
	Use of DNA barcoding for species identification in cell lines	17.3.3	Cooper et al. (2007)
2008	Development of cortical structures in spheroid microplates	27.6	Eiraku et al. (2008)
2009	Establishment of intestinal organoids from adult stem cells in the intestinal crypt	27.6	Sato et al. (2009)
2010	Use of CAR T-cells for treatment of lymphoma	24.5.5	Kochenderfer et al. (2010)
2011	Development of chemically defined conditions for culture of human induced pluripotent stem cells (iPSCs)	23.4.2	Chen et al. (2011)
2012	Use of clustered regularly interspaced short palindromic repeat (CRISPR) and CRISPR-associated (Cas) proteins for gene editing	22.2.3	Jinek et al. (2012)
	Development of conditional reprogramming using Rho kinase (ROCK) inhibitor and 3T3 feeder layers	22.3.4	Liu et al. (2012)
2013	Establishment of cerebral organoids that model human brain development	27.6	Lancaster et al. (2013)

particularly in complex structures such as the nervous system and the developing embryo.

In 1952, Aron Moscona performed a classic experiment that would eventually lead to today's 3D culture models (Moscona and Moscona 1952). Using techniques that he learned at the Strangeways Research Laboratories, Moscona used trypsin to dissociate cells from several chick embryonic organs, and then placed the cells in close association in the hollow of a ground slide – a technique known as the "watch-glass" method (Fell 1972). The dissociated cells reaggregated to form structures that reflected their tissues of origin. This type of culture, where cells from multiple lineages are brought together to recapitulate the original tissue, is referred to here as organotypic culture (see Figure 1.4). Organotypic culture has provided new prospects for the study of cell interaction among discrete, defined populations of homogeneous and potentially genetically and phenotypically defined cells. It has also provided exciting opportunities to develop "tissue equivalent" models for toxicity testing, such as keratinocyte/fibroblast co-culture systems, and to build constructs for tissue engineering (Stark et al. 1999; Vunjak–Novakovic and Freshney 2006).

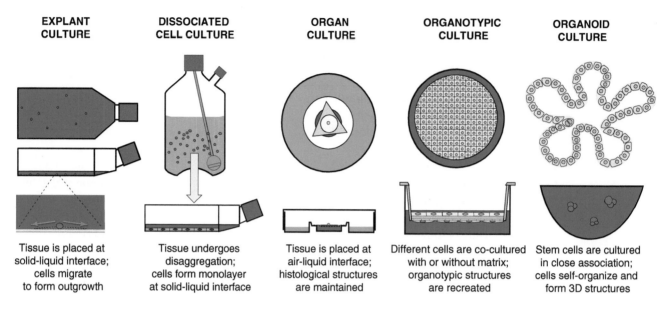

EXPLANT CULTURE	DISSOCIATED CELL CULTURE	ORGAN CULTURE	ORGANOTYPIC CULTURE	ORGANOID CULTURE
Tissue is placed at solid-liquid interface; cells migrate to form outgrowth	Tissue undergoes disaggregation; cells form monolayer at solid-liquid interface	Tissue is placed at air-liquid interface; histological structures are maintained	Different cells are co-cultured with or without matrix; organotypic structures are recreated	Stem cells are cultured in close association; cells self-organize and form 3D structures

Fig. 1.4. Types of tissue culture. Additional terms may be used to describe 3D culture; see Figure 27.1.

TABLE 1.2. Subculture.

Advantages	Disadvantages
Propagation	Trauma of enzymatic or mechanical dissociation
More cells	Overgrowth of unspecialized or stromal cells
Possibility of cloning	Selection of cells adapted to culture
Expanded stocks	Loss of differentiated properties
Increased homogeneity	Genetic instability
Characterization, authentication	Change in relative abundance of clonal populations
Cryopreservation	Increased risk of microbial contamination
Distribution to multiple laboratories	Increased risk of cross-contamination

In 2008, Yoshiki Sasai and colleagues discovered that ESCs could spontaneously self-organize to form polarized cortical tissues if they were grown in close association in a spheroid microplate – an environment that is strikingly similar to the "watch-glass" method (Eiraku et al. 2008). Such "organoid cultures" can be grown from pluripotent stem cells (PSCs) or adult stem cells. The potential of the latter approach has been shown by Hans Clevers and colleagues, in a series of elegant studies on LGR5-positive stem cells in the intestinal crypt and in other epithelial tissues (see Figure 1.3a, b) (Clevers 2016). This type of culture has tremendous possibilities for personalized therapy. For example, it is now possible to develop a biobank of organoids from patients with colorectal carcinoma, and use these cultures for genomic characterization and personalized drug screening (van de Wetering et al. 2015).

1.3 APPLICATIONS

Initially, the development of cell culture owed much to the needs of two major branches of medical research: the cultivation of viruses and the understanding of neoplasia. The standardization of conditions and cell lines for the cultivation of viruses undoubtedly provided much impetus to the development of modern tissue culture technology, particularly the production of large numbers of cells that were suitable for biochemical and molecular analysis. This and other technical improvements, which were made possible by the commercial supply of reliable media and sera and by the greater control of contamination with antibiotics and laminar flow equipment, made tissue culture more accessible and resulted in a broad range of applications (see Figure 1.5). The field has moved from being exploratory research, conducted by a few individuals, to becoming a major tool in cell and molecular biology,

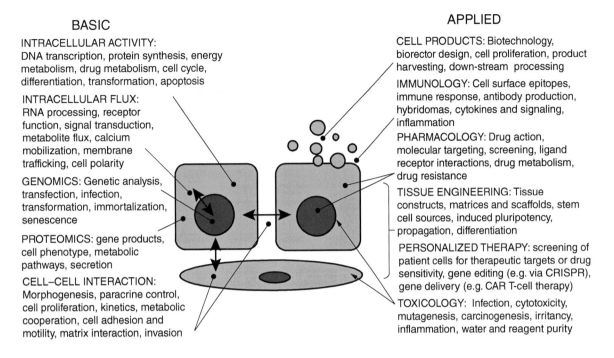

BASIC

INTRACELLULAR ACTIVITY:
DNA transcription, protein synthesis, energy metabolism, drug metabolism, cell cycle, differentiation, transformation, apoptosis

INTRACELLULAR FLUX:
RNA processing, receptor function, signal transduction, metabolite flux, calcium mobilization, membrane trafficking, cell polarity

GENOMICS: Genetic analysis, transfection, infection, transformation, immortalization, senescence

PROTEOMICS: gene products, cell phenotype, metabolic pathways, secretion

CELL–CELL INTERACTION:
Morphogenesis, paracrine control, cell proliferation, kinetics, metabolic cooperation, cell adhesion and motility, matrix interaction, invasion

APPLIED

CELL PRODUCTS: Biotechnology, biorector design, cell proliferation, product harvesting, down-stream processing

IMMUNOLOGY: Cell surface epitopes, immune response, antibody production, hybridomas, cytokines and signaling, inflammation

PHARMACOLOGY: Drug action, molecular targeting, screening, ligand receptor interactions, drug metabolism, drug resistance

TISSUE ENGINEERING: Tissue constructs, matrices and scaffolds, stem cell sources, induced pluripotency, propagation, differentiation

PERSONALIZED THERAPY: screening of patient cells for therapeutic targets or drug sensitivity, gene editing (e.g. via CRISPR), gene delivery (e.g. CAR T-cell therapy)

TOXICOLOGY: Infection, cytotoxicity, mutagenesis, carcinogenesis, irritancy, inflammation, water and reagent purity

Fig. 1.5. Tissue culture applications. Applications are broadly divided into "basic" and "applied," but in many cases, tissue culture is applied in both areas.

virology, bioengineering, and industrial pharmaceutics on a scale that would have astonished the early workers.

Vaccine development is perhaps the oldest tissue culture application and can be used to illustrate its effect on society. Tissue culture was initially used to grow poliovirus in the 1930s (Sabin and Olitsky 1936). In 1952, the newly established HeLa cell line was found to act as a good host for poliovirus and was mass produced for evaluation of the Salk polio vaccine (see Figure 1.6) (Brown and Henderson 1983). Primary cultures were initially used to prepare the vaccine itself, but it was quickly recognized that this resulted in unacceptable safety concerns (Koprowski et al. 1962). The WI-38 cell line was established in an effort to provide a safer alternative and selected, after extensive testing, for the production of polio vaccine and, subsequently, for rubella and other vaccines (Hayflick et al. 1962; Plotkin et al. 1969). Recently, it was estimated that 450 000 deaths have been averted in the United States (and roughly 10.3 million globally) due to vaccines developed using WI-38 cells (Olshansky and Hayflick 2017). More than 50 human viral vaccines have now been manufactured using WI-38 and other cell lines (Gallo-Ramirez et al. 2015). The Vero cell line, which was derived from a female monkey (*Chlorocebus sabaeus*), is highly susceptible to many viruses and can now be used for rapid vaccine development against emerging pathogens. Vero cells have been used to cultivate H5N1 influenza virus, severe acute respiratory system (SARS) coronaviruses, and other global threats to public health (Barrett et al. 2017).

Other applications that have arisen from tissue culture and are now widespread include (i) *in vitro* fertilization (IVF), which developed from early advances in embryo culture (Edwards 1996); (ii) manufacture of tissue grafts and other cellular therapies, starting with Howard Green's skin transplants and extending to today's 3D constructs (Hynds et al. 2018); (iii) generation of hormones, growth factors, antibodies, and other biological products (Sato et al. 2010); (iv) *in vitro* diagnostic techniques such as amniocentesis; and (v) *in vitro* toxicity testing for pharmaceuticals, medical devices, and numerous other products (Shukla et al. 2010). Although the benefits of these applications are clear, tissue culture has had a complex impact on society in a broader sense. Concerns regarding lack of consent for the establishment of cell lines in the early years of tissue culture have led to clear ethical requirements regarding the use of human tissue for research (Beskow 2016). Ethical issues will continue to arise in relation to gene editing, organoid culture, and other emerging applications (Munsie and Gyngell 2018). The basic principles that underlie safety and ethics in tissue culture are discussed in Chapter 6.

1.4 ADVANTAGES OF TISSUE CULTURE

1.4.1 Environmental Control

Two major advantages of tissue culture (see Table 1.3) are the ability to control the physiochemical environment (including

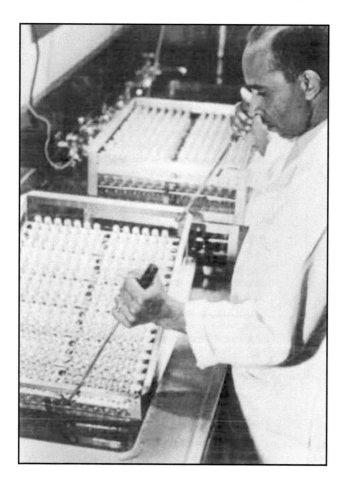

Fig. 1.6. Mass production of HeLa cells for polio vaccine testing. James C. Harris of the Tuskegee Institute, preparing culture tubes for use in field trials of the polio vaccine. The stainless-steel racks shown were designed by William F. Scherer to minimize handling of individual tubes. By July 1955, the Tuskegee team had shipped approximately 600 000 cultures to laboratories across the United States. *Source*: Brown and Henderson (1983), reproduced with permission of Oxford University Press.

pH, temperature, osmotic pressure, and O_2 and CO_2 tension) and the physiological conditions. The former can be controlled very precisely, whereas the latter must be kept relatively constant but cannot always be defined where cell lines still require supplementation of the medium with serum or other poorly defined constituents. These supplements are prone to batch variation and contain undefined elements such as hormones and other stimulants and inhibitors. The identification of some of the essential components of serum, together with a better understanding of factors regulating cell proliferation, has made the replacement of serum with defined constituents feasible, as explained in Chapters 9 and 10.

The role of the extracellular matrix (ECM) is important but subject to similar limitations as the use of serum – that is, the matrix is often necessary, but not always precisely defined. Prospects for defined ECM have improved, however, as its

individual constituents have been identified and manufactured. It is now possible to develop "designer matrices" that provide an optimized microenvironment for specific cell types. Although matrix design requires specialized expertise, most laboratories now have access to a range of ECM coatings and mimetic treatments, as described in Chapter 8.

1.4.2 Homogeneity and Characterization

Tissue samples are invariably heterogeneous. Replicates, even from one tissue, vary in their constituent cell types. By contrast, cultured cell lines assume a homogeneous (or at least uniform) constitution after one or two passages. The cells are randomly mixed at each transfer and the selective pressure of the culture conditions tends to produce a homogeneous culture of the most vigorous cell type. Hence, subdivided cultures will act as biological replicates, allowing further expansion, cryopreservation, characterization, or use in experiments. Homogeneity can be increased by performing cell cloning or by selecting or separating cells within the culture based on specific characteristics. These techniques allow the characteristics of the culture to be better understood and preserved for as long as suitable frozen stocks are retained in cryostorage. Cryopreservation and cell line characterization are discussed in Chapters 15 and 18. Techniques for cell cloning, selection, and separation are discussed in Chapters 20 and 21.

1.4.3 Economy, Scale, and Automation

Exposure of a culture to a drug or other reagent can occur at a lower, and more precisely defined, concentration and with direct access to the cell. Consequently, less reagent is required than for injection *in vivo*, where > 90% may be lost by distribution to tissues other than those under study and by metabolism and excretion. Cell lines can be scaled up to increase the number of cells that are produced (e.g. for manufacture of biological products at industrial scale) or the number of replicates (e.g. for high-throughput screening of therapeutic products). Large-scale procedures can also be automated for increased speed, reproducibility, or convenience. Scale-up of cell culture processes, including high-throughput screening and automation, is described in Chapter 28.

1.4.4 Replacement of *In Vivo* Models

As the various types of tissue culture have become more extensively characterized, they have gained acceptance as models for toxicology testing alongside *in vivo* animal models. Cell lines can be used to detect environmental pollutants and to test various commercial products for hazardous properties, including pharmaceuticals and personal care products such as shampoos and cosmetics. For example, the rainbow trout cell line RTgill-W1 is sensitive to many environmental toxins, making it a suitable model for toxicity testing of environmental samples (Lee et al. 2009). Panels of tumor cell

TABLE 1.3. Advantages of tissue culture.

Category	Advantages
Physicochemical environment	Control of pH, temperature, osmolality, and dissolved gases
Physiological conditions	Control of hormone, growth factor, and nutrient concentrations
Microenvironment	Regulation of matrix, cell–cell interaction, and gaseous diffusion
Cell line homogeneity	Replicates can be prepared; cell cloning or other techniques can be used to increase homogeneity
Preservation	Cells can undergo cryopreservation and be stored in cryofreezers
Characterization	Genomic analysis, immunostaining, and other forms of characterization are easily performed
Validation	Purity and authenticity can be specifically tested to demonstrate lack of contamination or misidentification
Certification	Safety and efficacy testing can be performed to confirm that a cell line is fit for purpose, e.g. for use in vaccine production
Replicates and variability	Quantitation is easy and statistical analysis is usually straightforward; multiple replicates are easily generated
Reagent saving	Reduced volumes, direct access to cells, lower costs
Control of $C \times T$	Ability to define dose, concentration, and time (duration of exposure)
Automation and mechanization	Scale-up and automation can be performed using bioreactors, high-throughput liquid handling systems, or robotic platforms
Scale	Culture volumes from a few microliters to 20 000 l
Time saving	Assay time can be reduced at least by an order of magnitude
Reduction of animal use	Toxicity testing of pharmaceutics, cosmetics, etc. can be performed using validated culture systems

lines have been assembled for high-throughput screening of anticancer drugs, resulting in the discovery of new therapeutic substances. For example, ovarian cell lines played an important role in the development of cisplatin and related drugs for ovarian carcinoma (Kelland et al. 1992).

The development of 3D culture models, with a more accurate replication of the *in vivo* cell phenotype, has increased the accuracy of *in vitro* modeling and the relevance of targeted metabolic pathways. This field has been particularly driven by expressions of concern by many community members and advocacy groups over the use of animals for scientific purposes. The need for replacement, reduction, and refinement (the 3Rs) was proposed by William Russell and Rex Birch more than 50 years ago (Russell and Birch 1959). Although Russell and Birch foresaw that tissue culture could replace animal models, cell lines and the other models available in their era could not be used to study complex physiological processes such as inflammation. Over time, the continued push for alternative testing has led to the development of organotypic "tissue equivalent" models that have undergone validation for use in toxicity testing (Sheasgreen et al. 2009). These systems can be used to replace animals in some assays, such as the Draize test for acute skin and eye irritation. Alternative systems have become more widely used and promoted following the adoption of regulations to limit animal-based testing in the European Union and elsewhere. The role of tissue culture in toxicity testing is discussed in Chapter 29.

1.5 LIMITATIONS OF TISSUE CULTURE

1.5.1 Quality and Expertise

Cell lines are the commonest cultures to be found in today's research laboratories. Although the advantages of cell lines are obvious, their development has brought with it a number of unforeseen problems (see Table 1.4). The observation that differentiated properties were lost in culture alerted early workers to the problems of dedifferentiation and selection and the fact that cell lines may be genetically, as well as phenotypically, unstable. Unfortunately, the early workers could not predict that cell lines would (i) provide an ideal substrate for the growth of mycoplasma, and (ii) extend the risks of cross-contamination and misidentification that were already familiar to microbiologists. Both problems continue to be frequently overlooked by today's workers, despite access to modern detection methods. This book and a number of expert guidelines on GCCP reinforce the need for validation to minimize the risks of microbial contamination, misidentification, and cross-contamination (Coecke et al. 2005; Geraghty et al. 2014; OECD 2018). More information on these topics and

TABLE 1.4. Limitations of tissue culture.

Category	Disadvantaged or limitations
Quality	Microbial contamination or cross-contamination may occur, e.g. due to contact with aerosols from other cultures
	Chemical contamination may arise, e.g. due to use of shared glassware from experimental work
Genotype	Continuous cell lines display genomic instability
	Passaging leads to clonal evolution and changes in genotype
Phenotype	Selective overgrowth may occur with unwanted cell types
	Passaging leads to clonal evolution and changes in phenotype
	Conditions favor uncontrolled proliferation over differentiation
Expertise	Aseptic technique must be used to minimize contamination risks
	Training is necessary to teach GCCP
	Validation testing is necessary to detect some forms of contamination
Laboratory	A separate sterile handling area is required
	Biological containment may require laboratory certification
	Waste may require decontamination as part of safe disposal
Equipment	A biological safety cabinet (BSC) is typically used, based on risk assessment and requirements for laminar airflow
	Basic equipment is required for tissue culture
Reagents	Reagents, substrates, etc. must be sterile and of suitable purity
	Procedures for preparation and sterilization must be adhered to
Quantity	Scale-up or high-throughput applications require capital investment
Cost	Serum, growth factors, and other reagents can be expensive
	Disposable plasticware is expensive and results in increased waste
Application	Geometry and microenvironment changes cell function
	Difficult to grow some species and cell types
	Difficult to model ingestion, absorption, distribution, metabolism, and excretion in pharmacokinetic studies

problems with reproducibility when using cell culture models can be found in Chapters 7, 16, and 17.

Tissue culture requires training and expertise for such problems to be avoided. Handling must be performed under strict aseptic conditions, because animal cells grow much less rapidly than many of the common contaminants, such as bacteria, molds, and yeasts. Contaminants may include pathogens, so containment is also necessary for safe handling of living cultures. Unlike microorganisms, cells from multicellular animals do not normally exist in isolation and, consequently, are not able to sustain an independent existence without the provision of a complex environment simulating blood plasma or interstitial fluid. These conditions imply a level of skill and understanding on the part of the operator in order to appreciate the requirements of the system and to diagnose problems as they arise. Hence, tissue culture should not be undertaken casually to run one or two experiments, but requires proper training, strict control of procedures, and a controlled environment. The design and layout of a tissue culture laboratory is discussed in Chapter 4, and the necessary equipment and safety requirements are described in Chapters 5 and 6. Training and problem solving are explored in Chapters 30 and 31.

1.5.2 Quantity and Cost

A major limitation of cell culture is the expenditure of effort and materials that goes into the production of relatively little tissue. A realistic maximum per batch for most small laboratories (with two or three people doing tissue culture) might be 1–10 g (wet weight) of cells. With a little more effort and the facilities of a larger laboratory, 10–100 g is possible; > 100 g implies industrial pilot-plant scale, a level that is beyond the reach of most laboratories but is not impossible if special facilities are provided. If industrial manufacture is performed, it is possible to generate kilogram quantities of cells.

The cost of producing cells in culture, excluding capital and labor costs, is about 10 times that of animal tissue. When the world's first tissue culture grown hamburger was eaten in London in 2013, the creator, Mark Post of Maastricht, reckoned it cost about €250 000! If large amounts of tissue (> 10 g) are required, the reasons for providing them by culture must be very compelling. For smaller amounts of tissue (~ 10 g), the costs are more readily absorbed into routine expenditure, but it is always worth considering whether assays or preparative procedures can be scaled down or automated. Automation is likely to come with a significant up-front cost but is worthwhile if costs are balanced by long-term savings due to reduced manual labor. Microscale and nanoscale assays can often be quicker (because of reduced manipulation times, volumes, etc.) and may be more readily automated. Such assays typically involve microfluidic systems, which are discussed in Chapter 21. Scale-up and automation are explored in Chapter 28.

1.5.3 Limited Species and Cell Types

Ross Harrison chose the frog as his source of tissue for initial experiments in tissue culture (Harrison 1907), presumably because it was a cold-blooded animal and, consequently, incubation was not required. Because tissue regeneration is more common in lower vertebrates, Harrison may also have felt that growth was more likely when compared to mammalian tissue. Although his technique initiated a new wave of interest in the cultivation of tissue *in vitro*, few later workers were to follow his example in the selection of species. The accessibility of different tissues, many of which grew well in culture, originally made the embryonated hen's egg a favorite choice. However, the well-established genetic background of the mouse, the success of experimental animal husbandry (particularly with genetically pure strains of rodents), and the development of transgenic mouse technology led to the selection of this animal as a favorite species.

Once the first human cell line had been established, human tissue became a favorite source of cells, helped later by the classic studies of Leonard Hayflick on the finite lifespan of cells in culture (Hayflick and Moorhead 1961) and the preference of virologists and molecular geneticists to work with human material. However, it has been difficult to establish cell lines from some human tissues. Over time, serum-free selective media have been optimized for epidermal keratinocytes, bronchial epithelium, vascular endothelium, and many other cell types. Some of these selective media are available commercially, although the cost remains high relative to the cost of regular media. Techniques have also been developed to extend the lifespan of certain cell types, such as the use of feeder layers for keratinocyte cultures and for conditional reprogramming of some epithelial cells (Rheinwald and Green 1975; Liu et al. 2012).

Analysis of data from the Cellosaurus knowledge resource shows that although cell lines have been established from more than 660 species, most are derived from a small subset of species (Bairoch 2018). As of January 2020, the most common species were (i) human (*Homo sapiens*, 87 495 cell lines); (ii) mouse (*Mus musculus*, 20 817 cell lines); (iii) rat (*Rattus norvegicus*, 2131 cell lines); (iv) Chinese hamster (*Cricetulus griseus*, 1736 cell lines); and (v) domestic dog (*Canis lupus familiaris*, 680 cell lines). More than 370 species are represented by only one or two cell lines (all data: personal communication, Amos Bairoch). This distribution is perhaps not surprising; each species may require an investment in time and effort before culture conditions are optimized for its growth. However, the effort is worthwhile when future

applications are considered. For example, culture conditions have been optimized for the establishment of cell lines from the black flying fox (*Pteropus alecto*) (Crameri et al. 2009). Bats are important reservoirs of infection for viruses that may cross the species barrier, such as SARS coronaviruses (Ge et al. 2013). Recently, snake cells were successfully grown in organoid culture, resulting in the development of 3D glandular structures that produce venom *in vitro* (Post et al. 2020). These snake venom gland organoids will allow research to be performed on species where venom has been difficult to obtain *in vivo*.

1.5.4 Limited Understanding of the Cell and its Microenvironment

When the first major advances in cell line propagation were achieved in the 1950s, many workers observed loss of the phenotypic characteristics typical of the tissue from which the cells had been isolated. This effect was blamed on dedifferentiation, a process that was assumed to be the reversal of differentiation. Gordon Sato demonstrated that this initial finding was largely due to selective overgrowth by fibroblasts, and would go on to develop enrichment techniques and serum-free selective media to avoid this problem (Sato et al. 2010). However, it remains true that a culture's differentiated properties are often lost under normal conditions of serial propagation. It is not clear how this happens. Either the differentiated cells dedifferentiate when they start to proliferate or, more likely, the culture becomes dominated by undifferentiated precursor cells with greater proliferative capacity. Continuous cell lines may also be affected by genotypic instability, resulting from their unstable aneuploid chromosomal constitution. All tissue culture practitioners should understand that their cultures undergo evolution with continued handling, resulting in unforeseen consequences that may threaten the reproducibility of experimental work. These consequences can be almost entirely avoided if the worker freezes the culture early in its lifespan using cell banking procedures. The changes that are likely to occur throughout the lifespan of a culture are described in Chapter 3. Cryopreservation and cell banking are explained in Chapter 15. In some cases, it is possible to induce differentiation within a culture even when dedifferentiation is believed to have occurred; this topic is discussed in Chapter 26.

Most of the differences in cell behavior between cultured cells and their counterparts *in vivo* come from the dissociation of cells from their 3D geometry *in situ* and their propagation on a two-dimensional (2D) substrate. Specific cell interactions that are characteristic of the histology of the tissue are lost and the matrix that surrounds the cell is replaced by a foreign substrate that fails to mimic its chemical composition or its physical properties. As Mina Bissell once observed, "half the secret of the cell is outside the cell" (Bissell 2016). Many of the technical innovations described in this book come from tissue culture pioneers who have increased our understanding of how cells behave *in vivo* and what they require for a more physiological environment. The discoveries made by these scientists – from Ross Harrison onwards – are the bedrock on which today's exciting discoveries are built.

REFERENCES

Abercrombie, M. and Heaysman, J.E. (1953). Observations on the social behaviour of cells in tissue culture. I. Speed of movement of chick heart fibroblasts in relation to their mutual contacts. *Exp. Cell Res.* 5 (1): 111–131. https://doi.org/10.1016/0014-4827(53)90098-6.

Andrews, P.W., Damjanov, I., Simon, D. et al. (1984). Pluripotent embryonal carcinoma clones derived from the human teratocarcinoma cell line Tera-2. Differentiation in vivo and in vitro. *Lab. Invest.* 50 (2): 147–162.

Bairoch, A. (2018). The Cellosaurus, a cell-line knowledge resource. *J. Biomol. Tech.* 29 (2): 25–38. https://doi.org/10.7171/jbt.18-2902-002.

Barnes, D. and Sato, G. (1980). Methods for growth of cultured cells in serum-free medium. *Anal. Biochem.* 102 (2): 255–270.

Barrett, P.N., Terpening, S.J., Snow, D. et al. (2017). Vero cell technology for rapid development of inactivated whole virus vaccines for emerging viral diseases. *Expert Rev. Vaccines* 16 (9): 883–894. https://doi.org/10.1080/14760584.2017.1357471.

Beskow, L.M. (2016). Lessons from HeLa cells: the ethics and policy of biospecimens. *Annu. Rev. Genomics Hum. Genet.* 17: 395–417. https://doi.org/10.1146/annurev-genom-083115-022536.

Bissell, M.J. (2016). Thinking in three dimensions: discovering reciprocal signaling between the extracellular matrix and nucleus and the wisdom of microenvironment and tissue architecture. *Mol. Biol. Cell* 27 (21): 3205–3209. https://doi.org/10.1091/mbc.E16-06-0440.

Bradley, T.R. and Metcalf, D. (1966). The growth of mouse bone marrow cells in vitro. *Aust. J. Exp. Biol. Med. Sci.* 44 (3): 287–299. https://doi.org/10.1038/icb.1966.28.

Brown, R.W. and Henderson, J.H. (1983). The mass production and distribution of HeLa cells at Tuskegee Institute, 1953–1955. *J. Hist. Med. Allied Sci.* 38 (4): 415–431.

Burrows, M.T. (1910). The cultivation of tissues of the chick-embryo outside the body. *JAMA* 55 (24): 2057–2058. https://doi.org/10.1001/jama.1910.04330240035009.

Cammack, R., Atwood, T., Campbell, P. et al. (2008). *Oxford Dictionary of Biochemistry and Molecular Biology*, 2e. Oxford: Oxford University Press.

Canti, R.G. (1928). Cinematograph demonstration of living tissue cells growing in vitro. *Arch. Exp. Zellforsch.* 6: 86–97.

Caplan, A.I. (1991). Mesenchymal stem cells. *J. Orthop. Res.* 9 (5): 641–650. https://doi.org/10.1002/jor.1100090504.

Carrel, A. (1912). On the permanent life of tissues outside of the organism. *J. Exp. Med.* 15 (5): 516–528.

Carrel, A. (1923). A method for the physiological study of tissues in vitro. *J. Exp. Med.* 38 (4): 407–418.

Chen, G., Gulbranson, D.R., Hou, Z. et al. (2011). Chemically defined conditions for human iPSC derivation and culture. *Nat. Methods* 8 (5): 424–429. https://doi.org/10.1038/nmeth.1593.

Chen, T.R. (1977). In situ detection of mycoplasma contamination in cell cultures by fluorescent Hoechst 33258 stain. *Exp. Cell. Res.* 104 (2): 255–262. https://doi.org/10.1016/0014-4827(77)90089-1.

Clevers, H. (2016). Modeling development and disease with organoids. *Cell* 165 (7): 1586–1597. https://doi.org/10.1016/j.cell.2016.05.082.

Coecke, S., Balls, M., Bowe, G. et al. (2005). Guidance on good cell culture practice. a report of the second ECVAM task force on good cell culture practice. *Altern. Lab. Anim.* 33 (3): 261–287.

Cooper, J.K., Sykes, G., King, S. et al. (2007). Species identification in cell culture: a two-pronged molecular approach. *In Vitro Cell. Dev. Biol. Anim.* 43 (10): 344–351. https://doi.org/10.1007/s11626-007-9060-2.

Coriell, L.L. and McGarrity, G.J. (1968). Biohazard hood to prevent infection during microbiological procedures. *Appl. Microbiol.* 16 (12): 1895–1900.

Coulter, W. H. (1949). Patent US2656508A: means for counting particles suspended in a fluid. https://patents.google.com/patent/us2656508a/en.

Crameri, G., Todd, S., Grimley, S. et al. (2009). Establishment, immortalisation and characterisation of pteropid bat cell lines. *PLoS One* 4 (12): e8266. https://doi.org/10.1371/journal.pone.0008266.

Dao, C., Metcalf, D., Zittoun, R. et al. (1977). Normal human bone marrow cultures in vitro: cellular composition and maturation of the granulocytic colonies. *Br. J. Haematol.* 37 (1): 127–136.

Dulbecco, R. (1952). Production of plaques in monolayer tissue cultures by single particles of an animal virus. *Proc. Natl Acad. Sci. U.S.A.* 38 (8): 747–752. https://doi.org/10.1073/pnas.38.8.747.

Eagle, H. (1955). Nutrition needs of mammalian cells in tissue culture. *Science* 122 (3168): 501–514. https://doi.org/10.1126/science.122.3168.501.

Eagle, H. (1959). Amino acid metabolism in mammalian cell cultures. *Science* 130 (3373): 432–437.

Earle, W.R. and Highhouse, F. (1954). Culture flasks for use with plane surface substrate tissue cultures. *J. Natl. Cancer Inst.* 14 (4): 841–851. https://doi.org/10.1093/jnci/14.4.841.

Earle, W.R., Schilling, E.L., Stark, T.H. et al. (1943). Production of malignancy in vitro IV. The mouse fibroblast cultures and changes seen in the living cells. *J. Natl. Cancer Inst.* 4 (2): 165–212.

Earle, W.R., Schilling, E.L., and Shannon, J.E. Jr. (1951). Growth of animal tissue cells on three-dimensional substrates. *J. Natl. Cancer Inst.* 12 (1): 179–193.

Earle, W.R., Schilling, E.L., Bryant, J.C. et al. (1954). The growth of pure strain L cells in fluid-suspension cultures. *J. Natl. Cancer Inst.* 14 (5): 1159–1171.

Edwards, R.G. (1996). The history of assisted human conception with especial reference to endocrinology. *Exp. Clin. Endocrinol. Diabetes* 104 (3): 183–204. https://doi.org/10.1055/s-0029-1211443.

Ehrmann, R.L. and Gey, G.O. (1956). The growth of cells on a transparent gel of reconstituted rat-tail collagen. *J. Natl. Cancer Inst.* 16 (6): 1375–1403.

Eiraku, M., Watanabe, K., Matsuo-Takasaki, M. et al. (2008). Self-organized formation of polarized cortical tissues from ESCs and its active manipulation by extrinsic signals. *Cell Stem Cell* 3 (5): 519–532. https://doi.org/10.1016/j.stem.2008.09.002.

Eshhar, Z., Waks, T., Gross, G. et al. (1993). Specific activation and targeting of cytotoxic lymphocytes through chimeric single chains consisting of antibody-binding domains and the gamma or zeta subunits of the immunoglobulin and T-cell receptors. *Proc. Natl Acad. Sci. U.S.A.* 90 (2): 720–724. https://doi.org/10.1073/pnas.90.2.720.

Evans, M.J. and Kaufman, M.H. (1981). Establishment in culture of pluripotential cells from mouse embryos. *Nature* 292 (5819): 154–156.

Evans, V.J. and Sanford, K.K. (1978). Development of defined media for studies on malignant transformation in culture. In: *Nutritional Requirements of Cultured Cells* (ed. H. Katsuta), 149–194. Tokyo and Baltimore: Japan Scientific Societies Press and University Park Press.

Evans, V.J., Earle, W.R., Schilling, E.L. et al. (1947). The use of perforated cellophane for the growth of cells in tissue culture. *J. Natl Cancer Inst.* 8 (3): 103–119.

Felgner, P.L., Gadek, T.R., Holm, M. et al. (1987). Lipofection: a highly efficient, lipid-mediated DNA-transfection procedure. *Proc. Natl Acad. Sci. U.S.A.* 84 (21): 7413–7417.

Fell, H.B. (1972). Tissue culture and its contribution to biology and medicine. *J. Exp. Biol.* 57 (1): 1–13.

Fell, H.B. and Robison, R. (1929). The growth, development and phosphatase activity of embryonic avian femora and limb-buds cultivated in vitro. *Biochem. J.* 23 (4): 767–784. https://doi.org/10.1042/bj0230767.

Fischer, A. (1925). *Tissue Culture: Studies in Experimental Morphology and General Physiology of Tissue Cells in Vitro*. London: William Heinemann.

Friend, C. (1957). Cell-free transmission in adult Swiss mice of a disease having the character of a leukemia. *J. Exp. Med.* 105 (4): 307–318. https://doi.org/10.1084/jem.105.4.307.

Fusenig, N.E. (1971). Isolation and cultivation of epidermal cells from embryonic mouse skin. *Naturwissenschaften* 58 (8): 421. https://doi.org/10.1007/bf00591536.

Gallo-Ramirez, L.E., Nikolay, A., Genzel, Y. et al. (2015). Bioreactor concepts for cell culture-based viral vaccine production. *Expert Rev. Vaccines* 14 (9): 1181–1195. https://doi.org/10.1586/14760584.2015.1067144.

Gartler, S.M. (1967). Genetic markers as tracers in cell culture. *Natl Cancer Inst. Monogr.* 26: 167–195.

Ge, X.Y., Li, J.L., Yang, X.L. et al. (2013). Isolation and characterization of a bat SARS-like coronavirus that uses the ACE2 receptor. *Nature* 503 (7477): 535–538. https://doi.org/10.1038/nature12711.

Geraghty, R.J., Capes-Davis, A., Davis, J.M. et al. (2014). Guidelines for the use of cell lines in biomedical research. *Br. J. Cancer* 111 (6): 1021–1046. https://doi.org/10.1038/bjc.2014.166.

Gey, G.O. (1933). An improved technic for massive tissue culture. *Cancer Res.* 17 (3): 752–756. https://doi.org/10.1158/ajc.1933.752.

Gey, G.O., Coffman, W.D., and Kubicek, M.T. (1952). Tissue culture studies of the proliferative capacity of cervical carcinoma and normal epithelium. *Cancer Res.* 12: 264–265.

Girardi, A.J., Jensen, F.C., and Koprowski, H. (1965). SV40-induced transformation of human diploid cells: crisis and recovery. *J. Cell. Comp. Physiol.* 65: 69–83.

Grace, T.D. (1962). Establishment of four strains of cells from insect tissues grown in vitro. *Nature* 195: 788–789.

Graham, F.L. and van der Eb, A.J. (1973). A new technique for the assay of infectivity of human adenovirus 5 DNA. *Virology* 52 (2): 456–467. https://doi.org/10.1016/0042-6822(73)90341-3.

Green, H. (2008). The birth of therapy with cultured cells. *Bioessays* 30 (9): 897–903. https://doi.org/10.1002/bies.20797.

Green, H., Kehinde, O., and Thomas, J. (1979). Growth of cultured human epidermal cells into multiple epithelia suitable for grafting. *Proc. Natl Acad. Sci. U.S.A.* 76 (11): 5665–5668. https://doi.org/10.1073/pnas.76.11.5665.

Grobstein, C. (1953). Morphogenetic interaction between embryonic mouse tissues separated by a membrane filter. *Nature* 172 (4384): 869–870.

Ham, R.G. (1965). Clonal growth of mammalian cells in a chemically defined, synthetic medium. *Proc. Natl Acad. Sci. U.S.A.* 53: 288–293.

Ham, R.G. (1974). Nutritional requirements of primary cultures. a neglected problem of modern biology. *In Vitro* 10: 119–129.

Ham, R.G. and McKeehan, W.L. (1978). Development of improved media and culture conditions for clonal growth of normal diploid cells. *In Vitro* 14 (1): 11–22.

Harrison, R.G. (1907). Observations on the living developing nerve fiber. *Proc. Soc. Exp. Biol.* 4 (1): 140–143.

Hayflick, L. (1963). A comparison of primary monkey kidney, heteroploid cell lines, and human diploid cell strains for human virus vaccine preparation. *Am. Rev. Respir. Dis.* 88 (Suppl): 387–393. https://doi.org/10.1164/arrd.1963.88.3P2.387.

Hayflick, L. (1965). The limited in vitro lifetime of human diploid cell strains. *Exp. Cell. Res.* 37: 614–636.

Hayflick, L. and Moorhead, P.S. (1961). The serial cultivation of human diploid cell strains. *Exp. Cell. Res.* 25: 585–621.

Hayflick, L., Plotkin, S.A., Norton, T.W. et al. (1962). Preparation of poliovirus vaccines in a human fetal diploid cell strain. *Am. J. Hyg.* 75: 240–258. https://doi.org/10.1093/oxfordjournals.aje.a120247.

Hull, C. W. (1984). Patent US4575330A: apparatus for production of three-dimensional objects by stereolithography. https://patents.google.com/patent/us4575330a/en.

Huschtscha, L.I. and Holliday, R. (1983). Limited and unlimited growth of SV40-transformed cells from human diploid MRC-5 fibroblasts. *J. Cell Sci.* 63: 77–99.

Hynds, R.E., Bonfanti, P., and Janes, S.M. (2018). Regenerating human epithelia with cultured stem cells: feeder cells, organoids and beyond. *EMBO Mol. Med.* 10 (2): 139–150. https://doi.org/10.15252/emmm.201708213.

Ichikawa, Y., Pluznik, D.H., and Sachs, L. (1966). In vitro control of the development of macrophage and granulocyte colonies. *Proc. Natl Acad. Sci. U.S.A.* 56 (2): 488–495. https://doi.org/10.1073/pnas.56.2.488.

Jinek, M., Chylinski, K., Fonfara, I. et al. (2012). A programmable dual-RNA-guided DNA endonuclease in adaptive bacterial immunity. *Science* 337 (6096): 816–821. https://doi.org/10.1126/science.1225829.

Keilova, H. (1948). The effect of streptomycin on tissue cultures. *Experientia* 4: 483.

Kelland, L.R., Jones, M., Abel, G. et al. (1992). Human ovarian-carcinoma cell lines and companion xenografts: a disease-oriented approach to new platinum anticancer drug discovery. *Cancer Chemother. Pharmacol.* 30 (1): 43–50. https://doi.org/10.1007/bf00686484.

Kim, Y.G., Cha, J., and Chandrasegaran, S. (1996). Hybrid restriction enzymes: zinc finger fusions to Fok I cleavage domain. *Proc. Natl Acad. Sci. U.S.A.* 93 (3): 1156–1160.

Kleinsmith, L.J. and Pierce, G.B. Jr. (1964). Multipotentiality of single embryonal carcinoma cells. *Cancer Res.* 24: 1544–1551.

Knazek, R.A., Gullino, P.M., Kohler, P.O. et al. (1972). Cell culture on artificial capillaries: an approach to tissue growth in vitro. *Science* 178 (4056): 65–66.

Kochenderfer, J.N., Wilson, W.H., Janik, J.E. et al. (2010). Eradication of B-lineage cells and regression of lymphoma in a patient treated with autologous T cells genetically engineered to recognize CD19. *Blood* 116 (20): 4099–4102. https://doi.org/10.1182/blood-2010-04-281931.

Köhler, G. and Milstein, C. (1975). Continuous cultures of fused cells secreting antibody of predefined specificity. *Nature* 256 (5517): 495–497. https://doi.org/10.1038/256495a0.

Koprowski, H., Ponten, J.A., Jensen, F.C. et al. (1962). Transformation of cultures of human tissue infected with simian virus SV40. *J. Cell Comp. Physiol.* 59 (3): 281–292. https://doi.org/10.1002/jcp.1030590308.

Kruse, P.F. Jr., Keen, L.N., and Whittle, W.L. (1970). Some distinctive characteristics of high density perfusion cultures of diverse cell types. *In Vitro* 6 (1): 75–88. https://doi.org/10.1007/bf02616136.

Lancaster, M.A., Renner, M., Martin, C.A. et al. (2013). Cerebral organoids model human brain development and microcephaly. *Nature* 501 (7467): 373–379. https://doi.org/10.1038/nature12517.

Lasfargues, E.Y. (1957). Cultivation and behavior in vitro of the normal mammary epithelium of the adult mouse. *Anat. Rec.* 127 (1): 117–129.

Lee, L.E., Dayeh, V.R., Schirmer, K. et al. (2009). Applications and potential uses of fish gill cell lines: examples with RTgill-W1. *In Vitro Cell. Dev. Biol. Anim.* 45 (3–4): 127–134. https://doi.org/10.1007/s11626-008-9173-2.

Leffert, H.L. and Paul, D. (1972). Studies on primary cultures of differentiated fetal liver cells. *J. Cell Biol.* 52 (3): 559–568. https://doi.org/10.1083/jcb.52.3.559.

Leighton, J. (1951). A sponge matrix method for tissue culture; formation of organized aggregates of cells in vitro. *J. Natl Cancer Inst.* 12 (3): 545–561.

Liu, X., Ory, V., Chapman, S. et al. (2012). ROCK inhibitor and feeder cells induce the conditional reprogramming of epithelial cells. *Am. J. Pathol.* 180 (2): 599–607. https://doi.org/10.1016/j.ajpath.2011.10.036.

Lovelock, J.E. and Bishop, M.W. (1959). Prevention of freezing damage to living cells by dimethyl sulphoxide. *Nature* 183 (4672): 1394–1395.

Ludwig, T.E., Levenstein, M.E., Jones, J.M. et al. (2006). Derivation of human embryonic stem cells in defined conditions. *Nat. Biotechnol.* 24 (2): 185–187. https://doi.org/10.1038/nbt1177.

Macpherson, I. and Bryden, A. (1971). Mitomycin C treated cells as feeders. *Exp. Cell. Res.* 69 (1): 240–241.

Macpherson, I. and Montagnier, L. (1964). Agar suspension culture for the selective assay of cells transformed by polyoma virus. *Virology* 23: 291–294.

Marcus, P.I., Sato, G.H., Ham, R.G. et al. (2006). A tribute to Dr. Theodore T. Puck (September 24, 1916–November 6, 2005). *In Vitro Cell. Dev. Biol. Anim.* 42 (8–9): 235–241. https://doi.org/10.1290/0606039A.1.

Martin, G.R. (1981). Isolation of a pluripotent cell line from early mouse embryos cultured in medium conditioned by teratocarcinoma stem cells. *Proc. Natl Acad. Sci. U.S.A.* 78 (12): 7634–7638.

Masters, J.R., Thomson, J.A., Daly-Burns, B. et al. (2001). Short tandem repeat profiling provides an international reference standard for human cell lines. *Proc. Natl Acad. Sci. U.S.A.* 98 (14): 8012–8017. https://doi.org/10.1073/pnas.121616198.

McLimans, W.F., Davis, E.V., Glover, F.L. et al. (1957). The submerged culture of mammalian cells; the spinner culture. *J. Immunol.* 79 (5): 428–433.

Michalopoulos, G. and Pitot, H.C. (1975). Primary culture of parenchymal liver cells on collagen membranes. Morphological and biochemical observations. *Exp. Cell. Res.* 94 (1): 70–78. https://doi.org/10.1016/0014-4827(75)90532-7.

Miller, D.G., Adam, M.A., and Miller, A.D. (1990). Gene transfer by retrovirus vectors occurs only in cells that are actively replicating at the time of infection. *Mol. Cell. Biol.* 10 (8): 4239–4242. https://doi.org/10.1128/mcb.10.8.4239.

Moore, G.E., Gerner, R.E., and Franklin, H.A. (1967). Culture of normal human leukocytes. *JAMA* 199 (8): 519–524.

Morgan, J.F., Morton, H.J., and Parker, R.C. (1950). Nutrition of animal cells in tissue culture; initial studies on a synthetic medium. *Proc. Soc. Exp. Biol. Med.* 73 (1): 1–8.

Morgan, J.F., Campbell, M.E., and Morton, H.J. (1955). The nutrition of animal tissues cultivated in vitro. I. A survey of natural materials as supplements to synthetic medium 199. *J. Natl Cancer Inst.* 16 (2): 557–567.

Moscona, A. and Moscona, H. (1952). The dissociation and aggregation of cells from organ rudiments of the early chick embryo. *J. Anat.* 86 (3): 287–301.

Munsie, M. and Gyngell, C. (2018). Ethical issues in genetic modification and why application matters. *Curr. Opin. Genet. Dev.* 52: 7–12. https://doi.org/10.1016/j.gde.2018.05.002.

Nelson-Rees, W.A., Flandermeyer, R.R., and Hawthorne, P.K. (1974). Banded marker chromosomes as indicators of intraspecies cellular contamination. *Science* 184 (4141): 1093–1096. https://doi.org/10.1126/science.184.4141.1093.

Neumann, E., Schaefer-Ridder, M., Wang, Y. et al. (1982). Gene transfer into mouse lyoma cells by electroporation in high electric fields. *EMBO J.* 1 (7): 841–845.

OECD (2018). *Guidance Document on Good In Vitro Method Practices (GIVIMP)*. Paris: OECD Publishing.

Olshansky, S.J. and Hayflick, L. (2017). The role of the WI-38 cell strain in saving lives and reducing morbidity. *AIMS Public Health* 4 (2): 127–138. https://doi.org/10.3934/publichealth.2017.2.127.

Orkin, R.W., Gehron, P., McGoodwin, E.B. et al. (1977). A murine tumor producing a matrix of basement membrane. *J. Exp. Med.* 145 (1): 204–220. https://doi.org/10.1084/jem.145.1.204.

Owens, O.v.H., Gey, M.K., and Gey, G.O. (1954). Growth of cells in agitated fluid medium. *Ann. N.Y. Acad. Sci.* 58 (7): 1039–1055. https://doi.org/10.1111/j.1749-6632.1954.tb45891.x.

Packer, L. and Fuehr, K. (1977). Low oxygen concentration extends the lifespan of cultured human diploid cells. *Nature* 267 (5610): 423–425.

Parker, R.C. (1961). *Methods of Tissue Culture*, 3e. London: Pitman Medical.

Petersen, O.W., Ronnov-Jessen, L., Howlett, A.R. et al. (1992). Interaction with basement membrane serves to rapidly distinguish growth and differentiation pattern of normal and malignant human breast epithelial cells. *Proc. Natl Acad. Sci. U.S.A.* 89 (19): 9064–9068. https://doi.org/10.1073/pnas.89.19.9064.

Petran, M., Hadravsky, M., Benes, J. et al. (1986). In vivo microscopy using the tandem scanning microscope. *Ann. N.Y.*

Acad. Sci. 483: 440–447. https://doi.org/10.1111/j.1749-6632.1986.tb34554.x.

Plotkin, S.A., Farquhar, J.D., Katz, M. et al. (1969). Attenuation of RA 27-3 rubella virus in WI-38 human diploid cells. *Am. J. Dis. Child.* 118 (2): 178–185. https://doi.org/10.1001/archpedi.1969.02100040180004.

Polge, C., Smith, A.U., and Parkes, A.S. (1949). Revival of spermatozoa after vitrification and dehydration at low temperatures. *Nature* 164 (4172): 666.

Post, Y., Puschhof, J., Beumer, J. et al. (2020). Snake venom gland organoids. *Cell* 180 (2): 233–247. https://doi.org/10.1016/j.cell.2019.11.038.

Puck, T.T. and Marcus, P.I. (1955). A rapid method for viable cell titration and clone production with HeLa cells in tissue culture: the use of X-irradiated cells to supply conditioning factors. *Proc. Natl Acad. Sci. U.S.A.* 41 (7): 432–437.

Rheinwald, J.G. and Green, H. (1975). Serial cultivation of strains of human epidermal keratinocytes: the formation of keratinizing colonies from single cells. *Cell* 6 (3): 331–343.

Robinson, L.B., Wichelhausen, R.H., and Roizman, B. (1956). Contamination of human cell cultures by pleuropneumonialike organisms. *Science* 124 (3232): 1147–1148. https://doi.org/10.1126/science.124.3232.1147.

Rothfels, K.H., Axelrad, A.A., Siminovitch, L. et al. (1959). The origin of altered cell line from mouse, monkey, and man as indicated by chromosome and transplantation studies. *Proc. Can. Cancer Conf.* 3: 189–214.

Rous, P. and Jones, F.S. (1916). A method for obtaining suspensions of living cells from the fixed tissues, and for the plating out of individual cells. *J. Exp. Med.* 23 (4): 549–555.

Russell, W.M.S. and Birch, R.L. (1959). *The Principles of Humane Experimental Technique*. London: Methuen.

Rygaard, J. and Povlsen, C.O. (1969). Heterotransplantation of a human malignant tumour to "nude" mice. *Acta Pathol. Microbiol. Scand.* 77 (4): 758–760.

Sabin, A.B. and Olitsky, P.K. (1936). Cultivation of poliomyelitis virus in vitro in human embryonic nervous tissue. *Proc. Soc. Exp. Biol. Med.* 34 (3): 357–359. https://doi.org/10.3181/00379727-34-8619c.

Sanford, K.K. and Evans, V.J. (1982). A quest for the mechanism of "spontaneous" malignant transformation in culture with associated advances in culture technology. *J. Natl Cancer Inst.* 68 (6): 895–913.

Sanford, K.K., Earle, W.R., and Likely, G.D. (1948). The growth in vitro of single isolated tissue cells. *J. Natl Cancer Inst.* 9 (3): 229–246.

Sanford, K.K., Earle, W.R., Evans, V.J. et al. (1951). The measurement of proliferation in tissue cultures by enumeration of cell nuclei. *J. Natl Cancer Inst.* 11 (4): 773–795.

Sato, G.H., Sato, J.D., Okamoto, T. et al. (2010). Tissue culture: the unlimited potential. *In Vitro Cell. Dev. Biol. Anim.* 46 (7): 590–594. https://doi.org/10.1007/s11626-010-9315-1.

Sato, T., Vries, R.G., Snippert, H.J. et al. (2009). Single Lgr 5 stem cells build crypt-villus structures in vitro without a mesenchymal niche. *Nature* 459 (7244): 262–265. https://doi.org/10.1038/nature07935.

Schaeffer, W.I. (1990). Terminology associated with cell, tissue, and organ culture, molecular biology, and molecular genetics. Tissue Culture Association Terminology Committee. *In Vitro Cell. Dev. Biol.* 26 (1): 97–101.

Sheasgreen, J., Klausner, M., Kandarova, H. et al. (2009). The mat tek story – how the three Rs principles led to 3-D tissue success! *Altern. Lab. Anim.* 37 (6): 611–622. https://doi.org/10.1177/026119290903700606.

Shukla, S.J., Huang, R., Austin, C.P. et al. (2010). The future of toxicity testing: a focus on in vitro methods using a quantitative high-throughput screening platform. *Drug Discovery Today* 15 (23–24): 997–1007. https://doi.org/10.1016/j.drudis.2010.07.007.

Singh, V. (1999). Disposable bioreactor for cell culture using wave-induced agitation. *Cytotechnology* 30 (1–3): 149–158. https://doi.org/10.1023/A:1008025016272.

Skloot, R. (2010). *The Immortal Life of Henrietta Lacks*. New York: Crown Publishers.

Smith, G.E., Summers, M.D., and Fraser, M.J. (1983). Production of human beta interferon in insect cells infected with a baculovirus expression vector. *Mol. Cell. Biol.* 3 (12): 2156–2165.

Stark, H.J., Baur, M., Breitkreutz, D. et al. (1999). Organotypic keratinocyte cocultures in defined medium with regular epidermal morphogenesis and differentiation. *J. Invest. Dermatol.* 112 (5): 681–691. https://doi.org/10.1046/j.1523-1747.1999.00573.x.

Strangeways, T.S.P. and Fell, H.B. (1925). Experimental studies on the differentiation of embryonic tissues growing in vivo and in vitro. I. The development of the undifferentated limb-bud (a) when subcutaneously grafted into the post-embryonic chick and (b) when cultivated in vitro. *Proc. R. Soc. Lond. B.* 99: 340–366. https://doi.org/10.1098/rspb.1926.0017.

Sutherland, R.M., McCredie, J.A., and Inch, W.R. (1971). Growth of multicell spheroids in tissue culture as a model of nodular carcinomas. *J. Natl Cancer Inst.* 46 (1): 113–120.

Syverton, J.T. and Scherer, W.F. (1953). Applications of strains of mammalian cells to the study of animal viruses. *Cold Spring Harbor Symp. Quant. Biol.* 18: 285–289. https://doi.org/10.1101/sqb.1953.018.01.041.

Takahashi, K. and Yamanaka, S. (2006). Induction of pluripotent stem cells from mouse embryonic and adult fibroblast cultures by defined factors. *Cell* 126 (4): 663–676. https://doi.org/10.1016/j.cell.2006.07.024.

Takahashi, K., Tanabe, K., Ohnuki, M. et al. (2007). Induction of pluripotent stem cells from adult human fibroblasts by defined factors. *Cell* 131 (5): 861–872. https://doi.org/10.1016/j.cell.2007.11.019.

Tanabe, H., Takada, Y., Minegishi, D. et al. (1999). Cell line individualization by STR multiplex system in the cell bank found cross-contamination between ECV304 and EJ-1/T24. *Tissue Cult. Res. Commun.* 18 (4): 329–338. https://doi.org/10.11418/jtca1981.18.4_329.

Thomson, D. (1914). Controlled growth en masse (somatic growth) of embryonic chick tissue in vitro. *Proc. R. Soc. Med.* 7: 71–75.

Thomson, J.A., Itskovitz-Eldor, J., Shapiro, S.S. et al. (1998). Embryonic stem cell lines derived from human blastocysts. *Science* 282 (5391): 1145–1147.

Todaro, G.J. and Green, H. (1963). Quantitative studies of the growth of mouse embryo cells in culture and their development into established lines. *J. Cell Biol.* 17: 299–313.

van de Wetering, M., Francies, H.E., Francis, J.M. et al. (2015). Prospective derivation of a living organoid biobank of colorectal cancer patients. *Cell* 161 (4): 933–945. https://doi.org/10.1016/j.cell.2015.03.053.

van Wezel, A.L. (1967). Growth of cell-strains and primary cells on micro-carriers in homogeneous culture. *Nature* 216 (5110): 64–65. https://doi.org/10.1038/216064a0.

Vunjak-Novakovic, G. and Freshney, R.I. (2006). *Culture of Cells for Tissue Engineering*. Hoboken, NJ: Wiley-Liss.

Witkowski, J.A. (1980). Dr. Carrel's immortal cells. *Med. Hist.* 24 (2): 129–142. https://doi.org/10.1017/s0025727300040126.

Ziegler, D.W., Davis, E.V., Thomas, W.J. et al. (1958). The propagation of mammalian cells in a 20-liter stainless steel fermentor. *Appl. Microbiol.* 6 (5): 305–310.

CHAPTER 2

Biology of Cultured Cells

The biology of the cell is a vast topic, whether that cell is located in the living organism (*in vivo*) or "in the glass" (*in vitro*) – a description used for cells grown outside the body since the earliest days of tissue culture (Fischer 1925). However, an understanding of cell biology is essential for tissue culture to be truly successful and for experimental work to be reliable and reproducible. This chapter aims to briefly discuss some of the essential concepts that are required for later chapters in this book, with particular reference to cell adhesion, cell division, cell fate, and cell death. Fortunately, other textbooks are available that explore these topics at far greater depth (Alberts et al. 2015; Gilbert and Barresi 2016). If you cannot find a good textbook on cell and developmental biology in your local library or bookstore, these benchmark publications are accessible through the National Center for Biotechnology Information (NCBI) website (NCBI 2019).

2.1 THE CULTURE ENVIRONMENT

The validity of the cultured cell as a model of physiological function *in vivo* has frequently been criticized (Hynds et al. 2018). Often, the cell does not express the correct phenotype because the tissue microenvironment has changed (see Section 3.4). Cell–cell and cell–matrix interactions are reduced *in vitro* because the cells typically lack the three-dimensional (3D) architecture and microenvironment found *in vivo*, and many hormonal and nutritional stimuli are absent (see Sections 26.3.4, 26.3.5). This creates an environment that favors the spreading, migration, and proliferation of unspecialized progenitor cells, rather than the expression of differentiated functions. The culture environment influences cellular behavior via (i) the degree of contact with other cells; (ii) the nature of the substrate on or in which the cells grow – solid, as on plastic or other rigid matrix; semisolid, as in a gel such as collagen or agar; or liquid, as in a suspension culture; (iii) the substances that the cells secrete to add to the substrate, resulting in the formation of the extracellular matrix (ECM); (iv) the physicochemical and physiological constitution of the culture medium; (v) the constitution of the gas phase; and (vi) the incubation temperature. The provision of the appropriate microenvironment, including substrate adhesion, growth factor concentration, and cell interaction, is fundamental to the expression of specialized functions and is discussed in later chapters (see Sections 26.3.3, 27.3).

2.2 CELL ADHESION

Most cells from solid tissues grow as monolayers that are adherent to the culture substrate. Whenever such cells are transferred to a new vessel, they will need to attach and spread out on the substrate before they start to proliferate. Originally, it was found that cells preferred to attach and spread on glass that had a slight net negative charge (see Section 8.1) (Rappaport et al. 1960). In retrospect, this finding is rather puzzling; the mammalian cell membrane contains a higher proportion of anionic molecules in its inner leaflet, resulting in a net negative charge and (one would expect) an electrostatic repulsion near a negatively charged surface (Weiss and Zeigel 1971). If the early studies are indeed correct, it

Freshney's Culture of Animal Cells: A Manual of Basic Technique and Specialized Applications, Eighth Edition. Amanda Capes-Davis and R. Ian Freshney.
© 2021 John Wiley & Sons Ltd. Published 2021 by John Wiley & Sons Ltd.
Companion website: www.wiley.com/go/freshney/cellculture8

is possible that the negatively charged glass surface attracts positively charged ions from the medium, resulting in an "electrical double layer" that is more suitable for cell adhesion (Guo et al. 2017). This hypothesis may be supported by the fact that polylysine, which has a positive charge, is commonly used to improve cell adhesion as a coating or (less commonly) as a supplement to the culture medium (see Section 8.3.5) (Yamane 1978). It is unlikely that charge alone is sufficient to mimic the complex interactions between a cell and the surrounding ECM, but charge alterations probably allow cell attachment and spreading, and under these conditions the cells may be capable of producing their own matrix.

Cells adhere to each other and to the ECM through various adhesion molecules (see Sections 2.2.2, 2.2.4) (Alberts et al. 2015). Thus, it seems likely that cells will attach and secrete ECM constituents before further spreading can occur. These matrix constituents will adhere to the charged substrate, and the cells will then bind via specific receptors. Different cell types will spread out and grow in different ways, resulting in a characteristic appearance (morphology) under the microscope (see Section 14.4.1). Typically, adherent cells are described as "epithelial-like" or "fibroblast-like" (see Section 18.1.2) (Reid 2017). Epithelial-like cells have a rounded shape and tend to grow in patches with a "cobblestone" appearance.

These cells appear to require cell–cell adhesion for optimum survival and growth, which is consistent with their behavior *in vivo* (see Figure 2.1). Fibroblast-like cells have an elongated shape that is bipolar or stellate in appearance. Usually, these cells migrate individually at low densities and their main requirement is for substrate attachment and cell spreading.

Some cells can proliferate without the need for adhesion, resulting in a life in suspension within the culture medium. Suspension culture was initially used for hematopoietic cells, resulting in the establishment of more than 600 leukemia-lymphoma cell lines (Drexler and MacLeod 2010). Growth in suspension is a useful way to increase the number of cells under cultivation and is now used for the scale-up of many cell types, using specialized vessels (see Section 8.6.2). Other cells may associate with each other in preference to the culture substrate, resulting in the formation of multicellular aggregates or "spheroids." Spheroid culture is a form of 3D culture, which is discussed later in this book (see Section 27.4.1).

Changes in adhesion and growth may arise as cultures are handled *in vitro* and can have ominous implications. Cells may grow on top of each other in a disorderly manner, due to failure to regulate their movement and proliferation in close proximity (loss of contact inhibition). They may also

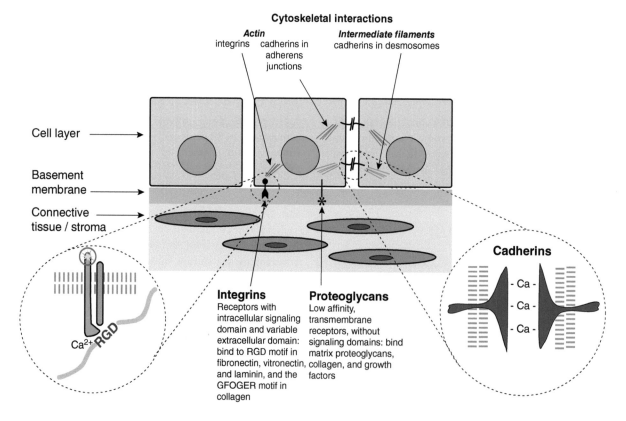

Fig. 2.1. Cell adhesion. Diagrammatic representation of a layer of epithelial cells above connective tissue (stroma) containing fibrocytes, separated from it by the basement membrane. Cadherins are depicted between like cells, while integrins and proteoglycans are shown between the epithelial cells and the ECM.

develop an increased ability to grow independent of the culture substrate (anchorage independent growth). These changes are often associated with *in vivo* tumor formation and other signs of malignant behavior, a phenomenon known as transformation (see Section 3.6). Adhesion and cell–cell interactions are clearly important biological properties in culture that may have broader implications for cellular behavior within the body.

2.2.1 Intercellular Junctions

Although some cell adhesion molecules are diffusely arranged in the plasma membrane, others are organized into intercellular junctions. These intercellular junctions can be broadly divided into (Alberts et al. 2015):

(1) ***Occluding junctions*, e.g. *tight junctions*.** These act to seal the space between cells, such as secretory cells in an acinus or duct or endothelial cells in a blood vessel. Any molecules traveling from the apical to basal surface (and vice versa) must therefore pass through the cell in a regulated fashion.

(2) ***Anchoring junctions*, e.g. *desmosomes and adherens junctions*.** These hold epithelial cells together or attach cells to the ECM. Desmosomes may be distributed throughout the area of plasma membranes in contact (see Figure 2.2a), while adherens junctions are often associated with tight junctions at the apical end of lateral cell contacts (see Figure 2.2b).

(3) ***Communicating junctions*, e.g. *gap junctions*.** These allow ions, nutrients, and small signaling molecules such as cyclic adenosine monophosphate (cAMP) to pass between cells for communication and other functions (Herve and Derangeon 2013).

As epithelial cells differentiate in confluent cultures, they can form an increasing number of desmosomes and (if some morphological organization occurs) complete junctional complexes that contain desmosomes, adherens junctions, and tight junctions (see Figure 2.2b). Enzymatic disaggregation of the tissue, or of an attached monolayer culture, will digest some of the ECM and may even degrade some of the extracellular domains of transmembrane proteins, allowing cells to become dissociated from each other. Chelating agents such as ethylene diamine tetra-acetic acid (EDTA) are often added because binding to ECM constituents through cadherins and integrins depends on the divalent cations Ca^{2+} and Mg^{2+}. Thus, chelating agents may enhance disaggregation. Epithelial cells are generally more resistant to disaggregation, as they tend to have tighter junctional complexes holding them together. Endothelial cells may also express tight junctions in culture, especially if left at confluence for prolonged periods on a preformed matrix, and can be difficult to dissociate. Mesenchymal cells, which are more dependent on integrin–matrix interactions for intercellular bonding, are more easily dissociated.

2.2.2 Cell Adhesion Molecules

Cadherins. Cadherins are Ca^{2+}-dependent transmembrane glycoproteins that are primarily involved in interactions between homologous cells, e.g. in epithelial monolayers (see Figure 2.1). These proteins are homophilic, i.e. homologous molecules in opposing cells interact with each other. Classical cadherins act via adherens junctions, where they connect to the actin cytoskeleton and have a signaling, as well as structural, role (Hartsock and Nelson 2008; Saito et al. 2012). E-cadherin is typically expressed in epithelial cells and is important for culture of human pluripotent stem cells (hPSCs) and for spheroid formation

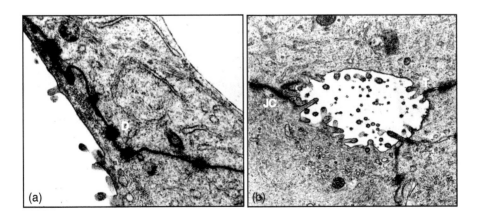

Fig. 2.2. Intercellular junctions. Electron micrograph of CA-KD cells, an early-passage culture from an adenocarcinoma metastasis to the brain (primary site unknown). Cells were grown on petriPERM dishes (see Section 8.6.1). (a) Desmosomes (D) between two cells in contact, magnification × 28 000; (b) canaliculus showing tight junctions (T) and junctional complex (JC), magnification × 18 500. *Source*: Images courtesy of Carolyn MacDonald.

in many cell types (Lambshead et al. 2013; Cui et al. 2017). N-cadherin is typically expressed in fibroblasts and may be expressed inappropriately in some cancer cells, in association with epithelial–mesenchymal transition (EMT) (see Section 26.2.5) (Rygaard et al. 1992; Wheelock et al. 2008). Other, non-classical cadherins act via the desmosomes, such as desmoglein and desmocollin. Desmosomal cadherins connect to the intermediate filament cytoskeleton and, like the classical cadherins, form cell–cell anchoring junctions which can promote sorting among cells to allow assembly into tissues (Saito et al. 2012).

Integrins. Integrins are transmembrane receptors that are primarily involved in interactions with the ECM (see Figure 2.1). Each integrin is a heterodimer with one α- and one β-subunit; the extracellular domains of these subunits are highly polymorphic, resulting in considerable diversity. These proteins are heterophilic, i.e. they can bind to various ECM molecules such as fibronectin, laminin, and entactin. For about a third of the integrin family, binding occurs via the arginyl-glycyl-aspartic acid (RGD) motif (Ruoslahti 1996; Barczyk et al. 2010). This motif is often used to improve substrate adhesion, although others are also important, e.g. the glycine-phenylalanine-pyrrolysine-glycine-glutamic acid-arginine (GFOGER) motif is used for collagen binding (see Section 8.3.4) (Barczyk et al. 2010). Within the cell, integrins provide a mechanical link between the ECM and the cytoskeleton. They are important for cell spreading and interact with intracellular activators such as talin and vinculin to form focal adhesions and to regulate the strength of cell attachment (Frame and Norman 2008; Fuhrmann and Engler 2015). Integrins also have a role as signaling receptors, where they help to regulate cell cycle progression and apoptosis (Schwartz and Assoian 2001; Stupack and Cheresh 2002). Integrin-mediated degradation of the ECM occurs during tumor invasion; some integrins have been used as markers of tumor progression, e.g. integrin $\alpha2\beta1$ with melanoma (Dowling et al. 2008).

Immunoglobulin superfamily (IgSF). Members of the immunoglobulin superfamily (IgSF) include intercellular adhesion molecule (ICAM), liver cell adhesion molecule (LCAM), and neural cell adhesion molecule (NCAM). IgSF members all carry immunoglobulin-like domains, which comprise two β sheets held in a "sandwich" structure (Hynes 1999). These proteins are Ca^{2+}-independent and may be homophilic (e.g. interaction between NCAM in neural synapses) or heterophilic (e.g. interaction between ICAM and integrins in immunological synapses). IgSF proteins are often inappropriately expressed during cancer development; NCAM is expressed in almost all small cell lung carcinoma samples, along with N-cadherin (Rygaard et al. 1992).

Other adhesion molecules. Two other transmembrane protein families, claudins and occludins, are important for the formation of the tight junction (Hartsock and Nelson 2008).

Like the cadherins, claudins and occludins are homophilic transmembrane proteins which bind tightly to each other across the gap between adjacent cells. They interact with intracellular scaffold proteins which, in turn, interact with desmosomes or adherens junctions. Claudin and occludin proteins help to control the tight junction's ion selectivity and permeability and are thus important for its barrier function (Hartsock and Nelson 2008). The tight junction is also important for cell polarity (Shin et al. 2006).

Transmembrane proteoglycans can interact with ECM constituents such as collagen or other proteoglycans. Some transmembrane and soluble proteoglycans may act as low-affinity growth factor receptors and may stabilize, activate, and/or translocate the growth factor to the high-affinity receptor, participating in its dimerization (Schlessinger et al. 1995; Yevdokimova and Freshney 1997; Forsten-Williams et al. 2008).

2.2.3 Cytoskeleton

The cytoskeleton has been described as a cellular shapeshifter that can alter its physical form and shape to accommodate the immediate needs of the cell (Leterrier et al. 2017). It is a complex and beautiful structure, as shown by the image on the front cover of this edition (courtesy of C. Leterrier), where a cultured glial cell has been labeled for actin. The cytoskeleton has three main components (Leterrier et al. 2017):

(1) **Actin filaments (F-actin).** These are contractile structural filaments, approximately 8 nm in diameter, which are composed of actin monomers (G-actin). Actin filaments regulate cell shape and motility, are attached to the plasma membrane and nucleus, and are responsible for transferring signals from integrins (bound to the ECM) to the nucleus in a manner capable of altering gene transcription. The density and location of actin filaments is regulated by polarized polymerization and depolymerization. They depolymerize when the cell rounds up, to divide and repolymerize as the cell spreads after division. Actin filaments may assemble with other molecules such as myosin II to form larger contractile structures, including stress fibers and muscle myofibrils (Tojkander et al. 2012).

(2) **Intermediate filaments.** These are rigid or semi-rigid filaments, approximately 10 nm in diameter, which vary in their composition, e.g. cytokeratin in epithelial cells, vimentin in fibroblasts, and glial fibrillary acidic protein (GFAP) in astrocytes (see Section 18.3.1). Intermediate filaments act as a rigid skeleton against the contractility of actin filaments to regulate cell shape.

(3) **Microtubules.** At 24 nm, these are the largest of the cytoskeletal elements and mediate cell and organelle motility as well as providing structural support (de Forges et al. 2012). Microtubules act in conjunction with actin and intermediate filaments to extend pseudopodia

during cell migration. Like actin, their length and distribution are regulated by polarized polymerization and depolymerization, resulting in transport of organelles throughout the cell cytoplasm. Chromatids migrate along microtubules arranged in a spindle during mitosis, and mitochondria are distributed within the cell attached to microtubules. Inhibitors of tubulin polymerization (e.g. colchicine or vinblastine) block cells at metaphase in cell division by inhibiting spindle formation and will also inhibit cell migration and invasion.

Cell adhesion molecules are attached to elements of the cytoskeleton. The attachment of integrins to actin filaments via linker proteins is associated with reciprocal signaling between the cell surface and the nucleus. Cadherins can link to the actin cytoskeleton in adherens junctions, mediating changes in cell shape and morphogenesis (Yap et al. 2015). Desmosomes, which also employ cadherins, link to the intermediate filaments – in this case, cytokeratins – via an intracellular plaque, which has a structural as well as a signaling role. All of these molecules and interactions are tightly regulated and their dysregulation results in various human diseases (Lee and Dominguez 2010).

2.2.4 Extracellular Matrix (ECM)

Intercellular spaces in tissues are filled with ECM, which provides a scaffold and a microenvironment that is dynamic and unique to that particular tissue (Hynes 2009; Frantz et al. 2010). ECM is essentially made up of fibrous proteins, proteoglycans, and water. Fibrous proteins include various collagens, fibronectins, laminins, and elastins (Halper and Kjaer 2014). Proteoglycans include perlecan, aggrecan, decorin, and lumican; these molecules are hydrophilic, leading to the formation of a hydrated gel (Frantz et al. 2010). Despite these common features, ECM composition varies considerably depending on the cell type and tissue under study, e.g. elastin content varies from a few percent in skin to more than 70% in some ligament structures (Halper and Kjaer 2014). The complexity of the ECM plays a significant role in the regulation of the phenotype of the cells attached to it. A dynamic equilibrium exists in which the cells attached to the ECM control its composition and, in turn, the composition of the ECM regulates the cell phenotype (Kleinman et al. 2003; Chen et al. 2007). Hence a proliferating, migratory fibroblast will require a different ECM from a differentiating epithelial cell or neuron.

Cultured cells will generate their own ECM and can produce highly specialized structures, e.g. co-culture of keratinocytes and dermal cells can lead to formation of a basement membrane (see Figure 27.15e, f) (Smola et al. 1998). However, some cultures cannot generate the microenvironment that they require for proliferation or expression of differentiated functions. In those cases, exogenous ECM can be added to the substrate as a coating (see Section 8.3.2). One of the most common sources of exogenous ECM is a preformed matrix generated by the Engelbreth-Holm-Swarm (EHS) mouse sarcoma, resulting in Matrigel® and other ECM products (see Table 8.1). Matrigel is often used to encourage differentiation and morphogenesis in culture and generates a lattice-like network with epithelial or endothelial cells (see Figure 26.6c). Individual ECM constituents may also be added, e.g. collagen (see Section 8.3.3). These animal-derived products are not well defined and there is a risk that undetected pathogens may be present. An increasing number of alternatives are available, such as recombinant proteins or ECM mimetic treatments (see Section 8.3.4; see also Table 8.1).

Although the individual ECM constituents are important, the physical properties of the matrix must also be considered (Engler et al. 2006; Kshitiz et al. 2012). Recent advances in tissue engineering have resulted in the development of new substrates that allow precise "tuning" of their mechanical properties. For example, the elasticity of the substrate can be altered to be more consistent with the elasticities of various tissues (see Figure 2.3a, b). Mesenchymal stromal cells (MSCs) that are grown on gels with different elasticities develop distinctly different phenotypes (see Figure 2.3c). For example, neurogenic markers are expressed at 0.1–1 kPa, myogenic markers at 8–17 kPa, and osteogenic markers at 25–40 kPa (Engler et al. 2006).

2.2.5 Cell Motility

Movement of cells on their substrate is essential for tissue culture, e.g. allowing cells to migrate away from a tissue explant (see Section 13.4). Scientists have been recording the movement of cells in culture for more than 90 years (Canti 1927; Stramer and Dunn 2015). Movement is most apparent with macrophages and granulocytes; among adherent cells *in vitro*, the most motile are fibroblasts at a low cell density (when cells are not in contact) and the least motile are dense epithelial monolayers. Fibroblasts migrate as individual cells with a recognizable polarity of movement. Epithelial cells, unless transformed, tend not to display random migration as polarized single cells. When seeded at a low density, they will migrate until they make contact with another cell and then stop. The cessation of movement when cells make contact with each other is known as contact inhibition and may be lost in culture as part of transformation (see Section 3.6.1). Eventually, epithelial cells will accumulate in patches and the whole patch may show signs of coordinated movement (Casanova 2002).

Animal cells usually display a "crawling" motion that uses substrate adhesion and rearrangement of the cytoskeleton to achieve directional movement (Ananthakrishnan and Ehrlicher 2007). The edge of the cell that is closest to the direction of movement undergoes restructuring of its actin network to extend a flat leading edge, known as a lamellum (Ponti et al. 2004). The cell then adheres to the substrate at the leading edge and releases its adhesion points at the trailing edge. Contractile forces are generated by the adhesion system

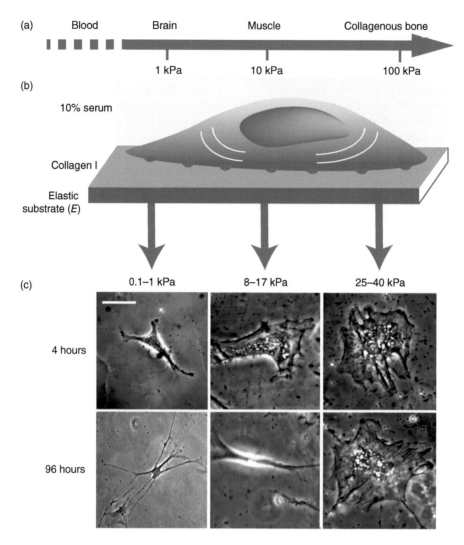

Fig. 2.3. Tissue elasticity and differentiation. (a) Solid tissues exhibit a range of stiffness, as measured by the elastic modulus, *E*. (b) The elasticity of a substrate on which cells are grown can be varied through crosslinking, covalent attachment to collagen I, and control of thickness. (c) MSCs develop different phenotypes when grown on substrates of different elasticity. Cells develop increasingly branched, spindle, or polygonal shapes when grown on matrices respectively in the range typical of $\sim E_{\mathrm{brain}}$ (0.1–1 kPa), $\sim E_{\mathrm{muscle}}$ (8–17 kPa), or stiff crosslinked–collagen matrices (25–40 kPa). Scale bar represents 20 μm. *Source*: Adapted from Engler, A. J., Sen, S., Sweeney, H. L., et al. (2006). Matrix elasticity directs stem cell lineage specification. Cell 126 (4):677-89. doi: 10.1016/j.cell.2006.06.044, reproduced with permission of Elsevier.

and the cytoskeleton to propel the cell forward and control its shape (Ananthakrishnan and Ehrlicher 2007; Burnette et al. 2014). Cross talk occurs between the actin network and the microtubules, resulting in the rapid reorganization of the cytoskeleton and the development of cellular asymmetry (Rodriguez et al. 2003). Both systems are regulated by the Rho family of small GTPases; Rho kinase (ROCK) inhibitors can be used for conditional reprogramming or as "survival factors" for stem cell culture, suggesting a major role for Rho and the cytoskeleton in cell survival and immortalization (see Sections 22.3.4, 23.4.1).

2.3 CELL DIVISION

2.3.1 Cell Cycle

Most cells in the body do not actively undergo cell division (Williams and Stoeber 2012). However, a different environment exists *in vitro*, where conditions are usually designed to favor proliferation (see Section 9.1). Mammalian cells in continuous culture typically divide every one to three days (MacLeod et al. 2011). The cell division cycle is made up of four sequential phases (see Figure 2.4) (Alberts et al. 2015):

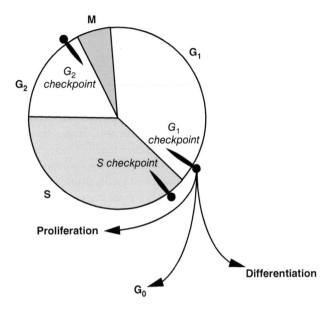

Fig. 2.4. Cell cycle. The cell cycle is divided into four phases: G_1, S, G_2, and M. Cell cycle checkpoints are shown at G_1, S, and G_2 (the mitotic spindle checkpoint during M is not shown). The cell cycle can be arrested in G_1 by the action of pRB and other cell cycle inhibitors. Depending on the nature of the signals received, the cell may withdraw temporarily from the cycle and enter G_0. From G_0, the cell may re-enter the cycle or progress toward differentiation. The cell cycle can also be arrested at other checkpoints in S and G_2 by cell cycle inhibitors such as p53.

(1) **M phase (Mitosis).** In this phase, the cell divides into two daughter cells. The chromatin condenses into chromosomes, and the two individual chromatids, which make up the chromosome, segregate to each daughter cell. The segregation of sister chromatids is under the control of the mitotic spindle. A checkpoint exists at the spindle to halt cell division if problems are detected.

(2) **G_1 phase (Gap 1).** A major checkpoint exists during G_1 (shortly before the S phase), which allows the cell to check whether the conditions are right for cell division. The cell can either commit to another cycle of cell division or withdraw from the cell cycle at this point. Withdrawal may be reversible (G_0) or irreversible.

(3) **S phase (Synthesis).** In this phase, the DNA replicates. Replication can be detected by adding DNA precursors such as [³H]thymidine or bromo-deoxyuridine (BrdU). A checkpoint exists during S phase to pick up problems with replication e.g. stalling of the replication forks (Giono and Manfredi 2006).

(4) **G_2 phase (Gap 2).** A major checkpoint exists during G_2 (before progression to the M phase), which allows assessment of DNA integrity and the completeness of DNA replication. If problems are detected, cell cycle arrest occurs, allowing DNA repair or cell death if repair is impossible (see Section 2.5).

2.3.2 Control of the Cell Cycle

Entry into the cell cycle is regulated by signals from the environment. Low cell density leaves cells with free edges and renders them capable of spreading, which permits their entry into the cycle in the presence of mitogenic growth factors (see Sections 9.5.2, 10.4.1). High cell density inhibits the proliferation of normal cells, though not transformed cells (see Section 3.6.2). Inhibition of proliferation is initiated by cell contact and is accentuated by crowding and the resultant changes in the shape of the cell.

Progression through the cell cycle is driven by the cyclins and cyclin-dependent kinases (CDKs) (Williams and Stoeber 2012). Cyclins are typically upregulated by signal transduction cascades, which in turn are activated by phosphorylation of the intracellular domain of the receptor when it is bound to a mitogenic growth factor (see Figure 2.5a). Cyclins work in partnership with CDKs to regulate different phases of the cell cycle. For example, cyclin D1, D2, or D3 work with CDK4 or CDK6 to promote entry into the cell cycle (Malumbres 2014). Persistent mitogen stimulation leads to accumulation of cyclin D-dependent kinases within the nucleus, where they collaborate with cyclin E and CDK2 to phosphorylate pRb (Sherr and McCormick 2002).

Members of the RB family allow the cell cycle to continue when they are phosphorylated but will act to block entry into S phase when they are in their active, hypophosphorylated forms (see Figure 2.5b). These proteins bind to various E2F transcription factors to repress the expression of genes that are necessary for DNA replication (Sherr and McCormick 2002). Members of the INK4 family such as p16[INK4a] also participate in this pathway. The INK4a proteins inhibit the activity of cyclin D-dependent kinases, preventing phosphorylation of pRB family proteins. Both pRb and p16[INK4a] have been extensively studied in their role as tumor suppressors. Loss of function of pRb or p16[INK4a], or overexpression of cyclin D1 or CDK4, occurs in many different tumor types (Sherr and McCormick 2002).

DNA damage leads to activation of multiple checkpoints, largely through the actions of p53, which plays a central role in safeguarding the integrity of the genome (see Figure 2.5c) (Jin and Levine 2001). p53 is a transcription factor that can respond to signals from many different pathways, e.g. hypoxia or oncogene activation. It is normally present at low levels in resting cells but becomes stabilized and activated when it is freed from its binding partner, MDM2. This may occur through post-translational modification of p53 or through the action of p14[ARF] (in humans; p19[ARF] in mice), which inhibits MDM2 (Harris and Levine 2005). Once activated, p53 induces expression of the CDK inhibitor p21[Waf1/Cip1], which induces cell cycle arrest at G_1. p53 may also induce cell cycle arrest at the G_2/M checkpoint or the S phase checkpoint (Giono and Manfredi 2006). Depending on the stimulus, p53 activation can lead to cell cycle arrest, DNA repair, apoptosis, or inhibition of angiogenesis and metastasis (Jin and Levine 2001). Most human tumors carry a mutation

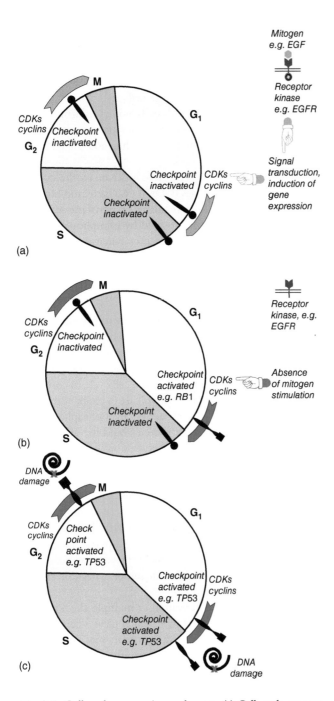

Fig. 2.5. Cell cycle progression and arrest. (a) Cell cycle progression. A mitogenic growth factor such as epidermal growth factor (EGF) interacts with its receptor to initiate signal transduction, leading to upregulation of cyclins, CDKs, and other genes. pRb and p53 remain inactive, allowing the cell cycle to proceed through checkpoints at G_1, S, and G_2. (b) Cell cycle arrest due to loss of mitogen stimulation. pRb becomes activated and initiates arrest at the G_1 checkpoint. The cell does not proceed into S phase and withdraws from the cell cycle. (c) Cell cycle arrest due to DNA damage. p53 is activated and initiates arrest at multiple checkpoints. The cell withdraws from the cell cycle for DNA repair or to induce cell death.

in either the *p53* tumor suppressor gene (> 50%) or other genes involved in the p53 pathway (Sherr and McCormick 2002).

Much of the evidence for these steps in the control of cell proliferation has emerged from studies of oncogene and tumor suppressor genes in cultures of tumor cells, aiming to develop new therapeutic approaches for cancer (McDonald and El-Deiry 2000). The immediate benefit, however, has been a better understanding of the factors required to regulate cell proliferation in culture. Other benefits have included the identification of genes that enhance cell proliferation, some of which can be used to immortalize finite cell lines (see Section 22.3; see also Minireview M3.1).

2.4 CELL FATE

Cell division is often described as "symmetric" or "asymmetric" (Morrison and Kimble 2006). These terms refer to the fate of the two daughter cells following cell division. Daughter cells can retain the ability to proliferate or make a commitment to differentiate. Cell division may thus result in (i) symmetric division into two proliferating cells; (ii) symmetric division into two committed cells; or (iii) asymmetric division into a proliferating cell and a committed cell (Bavle 2014). These different cell fates are used during embryonic development and to renew cell populations in the adult organism, e.g. in the hematopoietic system. Problems with cell fate play a major role in diseases such as cancer, where the tumor type is closely linked to the origin of the cell. Malignant tumors are often less differentiated than benign lesions, suggesting a reversal of the differentiation signals that are put in place during development (Lytle et al. 2018).

2.4.1 Embryonic Lineages

During early vertebrate development, the inner cell mass of the embryo develops three germ layers: (i) an outer layer, known as the ectoderm; (ii) an inner layer, known as the endoderm; and (iii) a middle layer, known as the mesoderm (Gilbert and Barresi 2016). As the embryo continues to develop, the various cell types that will be found within the body develop from the three germ layers (Young and Black 2004). The ectoderm gives rise to the outer surface epithelia, i.e. the epidermis, buccal epithelium, and outer cervical epithelium. During neurulation, the ectoderm also gives rise to the neuroectoderm, which becomes the neural system (central and peripheral neurons and glia), some neuroendocrine cells, and melanocytes. The endoderm gives rise to the epithelium of the gut and associated organs such as lung, liver, and pancreas. The mesoderm gives rise to the embryonic mesenchyme, which becomes the connective tissue (supporting tissues such as bone, cartilage, and muscle), vascular tissue (endothelium, smooth muscle, and pericytes), and the hematopoietic system. While most epithelial cells derive from the ectoderm or endoderm, some are mesodermal in origin (e.g. the kidney tubules and the mesothelium lining the body cavity).

During organogenesis in the embryo, the tissues derived from the primitive germ layers become associated in a process of mutual induction of differentiation. For example, the lung epithelium arises from cells in the endoderm (in a bud in the primitive gut), which are induced to become tracheal, bronchial, and alveolar cells with secretory, lining, and respiratory functions. This occurs under the inductive influence of the associated mesenchyme, which is itself induced by the endodermally derived cells to become fibrous and elastic connective tissue and smooth muscle. Each individual organ is thus comprised of different tissues and each tissue is made up of different cell types, which may arise from different germ layers. Each differentiated cell within the adult organism can trace its lineage back to a specific embryonic stem cell (ESC), allowing scientists to develop a "fate map" that sets out its developmental history (Woodworth et al. 2017).

2.4.2 Stem Cells and Potency

A stem cell can be defined as a cell that can perpetuate itself through self-renewal and generate differentiated cells (Lytle et al. 2018). Stem cells can use asymmetric cell division to produce daughter cells with different fates, depending on intrinsic factors (e.g. positioning of polarity factors within the cell) or extrinsic factors (e.g. signaling molecules that act as external cues) (Morrison and Kimble 2006; Williams and Fuchs 2013). Stem cells may also use symmetric cell division to expand their number. Such cells may shift from primarily symmetric division during embryonic development (expanding the stem cell pool) to primarily asymmetric division in mid to late gestation (expanding the number of differentiated cells) (Morrison and Kimble 2006).

As cells become committed to differentiation, they display more restricted potency, which can be defined as the range of commitment options that are available (Smith 2006). Various terms are used to describe the potency of a stem cell, which may cause confusion if they are used interchangeably. For this book, the following terms are used (Smith 2006; NIH 2019):

(1) ***Totipotent Stem Cell.*** Cells can give rise to the entire organism, including the three germ layers, and supporting extra-embryonic tissues such as the placenta. The zygote (fertilized egg) is the classic example of a totipotent stem cell.

(2) ***Pluripotent Stem Cell (PSC).*** These cells can form all tissues of an organism (derived from the three germ layers) but are not sufficient to support its full development, e.g. due to lack of extra-embryonic tissues. The ESC is a classic example *in vivo*, but some other cell types are also considered to be pluripotent, including induced pluripotent stem cells (iPSCs), which have been manipulated *in*

vitro to increase their level of potency (see Sections 2.4.4, 23.1).

(3) ***Multipotent Stem Cell.*** These cells can give rise to multiple cell types or lineages but cannot produce lineages from all three germ layers. The hematopoietic stem cell (HSC) is a classic example that gives rise to lymphoid and myeloid lineages. Some lineages may give rise to sublineages, e.g. the myeloid lineage gives rise to erythrocytes (red blood cells), neutrophils, monocytes/macrophages, basophils, eosinophils, and megakaryocytes (platelets). Stem cells in these sublineages may also be described as oligopotent (able to form two or more lineages within a tissue) or unipotent (able to form only a single lineage).

(4) ***Progenitor Cell.*** This term is less well defined than the preceding ones and is used here to refer to multipotent cells that are committed to a specific lineage but still retain the ability to divide. Historically, the concept of the progenitor cell is linked to its ability to form colonies *in vitro* (see Sections 20.2, 20.3) (Yoder 2004).

(5) ***Precursor Cell.*** Like the progenitor cell, this term is not well defined and is used here to refer to any ancestral cell in the same lineage as the cell type of interest.

2.4.3 Differentiation

Stem cells were originally examined by studying teratoid tumors (which contain cells from all three germ layers) and the rescue of blood cell production within the adult organism (Solter 2006; Clevers and Watt 2018). Over time, it became clear from these models that stem cells give rise to a tree-like hierarchy of increasingly differentiated cell types (see Figure 2.6a). Each "branch" of that "tree" can be regarded as a lineage that becomes committed to a particular cell type. The position of a cell within that lineage determines its levels of potency and differentiation. Stem cells give rise to progenitor cells, which then give rise to fully committed cells that exit the cell cycle. This classic hierarchy does not necessarily apply to all tissue types; different tissues may have different needs for maintenance and repair. For example, cells in the adult epidermis retain the capacity for self-renewal (allowing epidermal cells to be expanded *in vitro* to treat burns) but a single epidermal stem cell has been difficult to identify, suggesting that stem cell functions are distributed over a larger population (Clevers and Watt 2018).

Differentiation can be defined as the process by which cells, tissues, or organs acquire specialized characteristics. Cells that are described as "differentiated" express a phenotype that is characteristic of the functionally mature cell *in vivo*. However, there is a tendency for cultures to lose their differentiated properties, particularly if they are handled over an extended time period (see Sections 3.4.1, 26.2.3). Historically, the inability of cultures to express the characteristic *in vivo* phenotype has been linked to dedifferentiation (see Figure 2.6b), but the exact cause is often unclear. Various mechanisms may be responsible, including (i) loss of

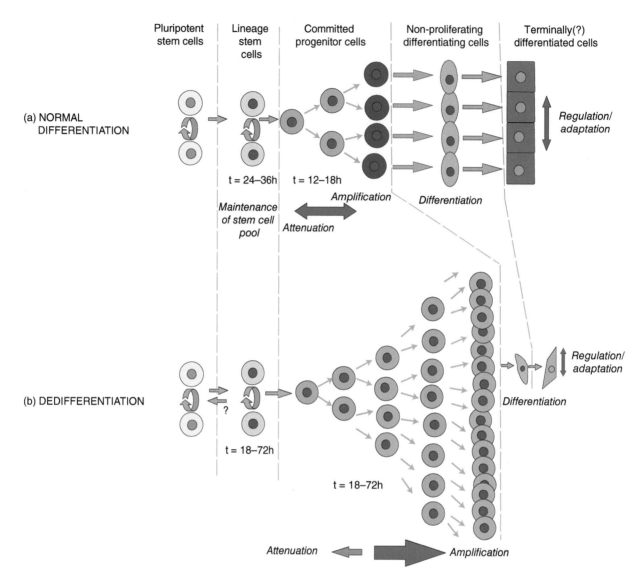

Fig. 2.6. Differentiation from stem cells. (a) *In vivo*, a small stem cell pool gives rise to a proliferating progenitor compartment that produces the differentiated cell pool. (b) *In vitro*, differentiation is limited by the need to proliferate, and the population becomes predominantly progenitor cells, although stem cells may also be present. Pluripotent stem cells (far left) have also been cultured from some tissues, but their relationship to the tissue stem cells is as yet unclear. Culture conditions select mainly for the proliferating progenitor cell compartment of the tissue or induce cells that are partially differentiated to revert to a progenitor status.

expression of differentiated properties due to absence of the appropriate inducers (e.g. hormones or interactions with other cells or the ECM), which is potentially reversible; (ii) selection of the incorrect lineage; (iii) overgrowth of terminally differentiated cells, which have reduced proliferative capacity, by undifferentiated cells of the correct lineage; or (iv) reversion of the differentiated cell to a more primitive precursor or stem cell. Practical steps to induce differentiation are discussed in a later chapter (see Section 26.4).

2.4.4 Control of Potency and Differentiation

Considering the various mechanisms that may be responsible for dedifferentiation, it seems reasonable to assume that cultures are made up of cells at various stages of differentiation – stem cells, progenitor cells, and mature differentiated cells. Cultures are likely to contain cells at all of these stages in equilibrium with one another. The equilibrium will be determined by the origin of the culture (e.g. whether it comes from a malignant tumor), its age, and the user's selection of growth and handling parameters, e.g. whether cells are grown at low or high density (see Figure 2.7).

Cultures can be manipulated or "reprogrammed" to increase their level of potency by introducing a set of defined factors (Smith 2006). Use of defined factors to generate iPSCs is described later in this book (see Section 23.3). Cultures can also be manipulated to increase their level of differentiation (see Chapter 26). The conditions required for induction of differentiation – a high cell density, enhanced cell–cell and cell–matrix interaction, and the presence of various differentiation factors (see Section 26.3) – may often be antagonistic to cell proliferation and vice versa. If differentiation is required, it may be necessary to define two distinct sets of conditions – one to optimize proliferation and one to optimize differentiation. Surprisingly, in view of the concept that differentiation is dysfunctional in malignant cells, tumor cell lines have provided some of the best models for induction of differentiation (see Section 26.2.1; see also Table 26.1).

The culture environment has a key role to play in the control of potency and differentiation. It has been recognized for many years that specific cellular functions are retained longer when the 3D structure of the tissue is retained. This was originally achieved using organ culture, where a tissue fragment or slice was cultured in its entirety (see Section 27.8). Unfortunately, organ cultures cannot be propagated, must be prepared *de novo* for each experiment, and are difficult to characterize compared to monolayer cultures. The solution to these difficulties has been to develop new techniques and technologies for 3D culture. Cells may be grown on (or in) scaffolds that mimic the unique properties of their microenvironment *in vivo* (see Sections 8.2.2, 8.6.5). This has led to the development of hydrogels that are rationally designed for specific tissue types, for use in tissue engineering (see Section 27.5). Cells may also be encouraged to associate with each other using scaffold-free systems that encourage multicellular aggregates to form (see Section 27.4).

2.4.5 Lineage Commitment

Progression from a stem cell to a particular pathway of differentiation traditionally implied an increase in commitment, which was regarded as the point where a cell or its progeny could no longer transfer to a separate lineage. As the stem cell became more committed to a particular lineage, its descendants expressed a more limited and specialized repertoire of properties. This commitment to differentiation was thought to be irreversible past a certain point, which was referred to as terminal differentiation. The term implied that a cell had progressed down a particular lineage to a point at which the mature phenotype was fully expressed and beyond which the cell could not progress; it also implied that the cell was unable to re-enter the cell cycle.

This classic model of lineage commitment was developed from study of the hematopoietic system, where the HSC gives rise to progenitor cells that are increasingly committed to lymphoid and myeloid lineages. Once commitment has occurred, the myeloid progenitor cell cannot change lineage to adopt lymphoid characteristics; it also cannot revert to a stem cell that is capable of giving rise to both lineages.

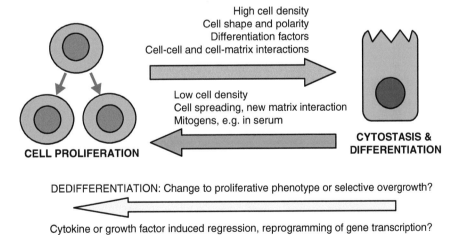

Fig. 2.7. Differentiation and proliferation. Cells in culture can be considered to be in a state of equilibrium between cell proliferation and differentiation. Normal culture conditions (low cell density, mitogenic growth factors) will favor cell proliferation, while high cell density and addition of differentiation factors will induce differentiation. The position of the equilibrium will depend on culture conditions. Dedifferentiation of the cells in the culture may be due to the effect of growth factors or cytokines inducing a more proliferative phenotype or overgrowth of a precursor cell type. With chemical induction or genetic manipulation, reprogramming of gene expression may shift the phenotype from a differentiated cell to a proliferating precursor or stem cell.

However, although the classic model has been useful, it is unlikely to apply to all tissues. Stem cell function appears to be distributed over a broad population of cells in some tissues, allowing differentiated cells to convert to a more potent cell type when required (Clevers and Watt 2018). If irreversible commitment exists outside the hematopoietic system, it must occur much later than previously thought; some precursor cells can revert to stem cells with multilineage potential, and even fully mature cells can be made to revert to stem cell status with appropriate genetic or epigenetic manipulation

(see Section 23.3), showing that "it ain't necessarily so" when it comes to terminal differentiation.

The classic model of lineage commitment is also unlikely to apply to tumors, which demonstrate abnormal differentiation (see Section 26.2.1). For example, the K-562 cell line was isolated from a myeloid leukemia but was subsequently shown to be capable of erythroid differentiation (see Table 26.1) (Andersson et al. 1979). Rather than being a committed myeloid progenitor converting to an erythroid lineage, the tumor probably arose in the stem cell type that gives rise to

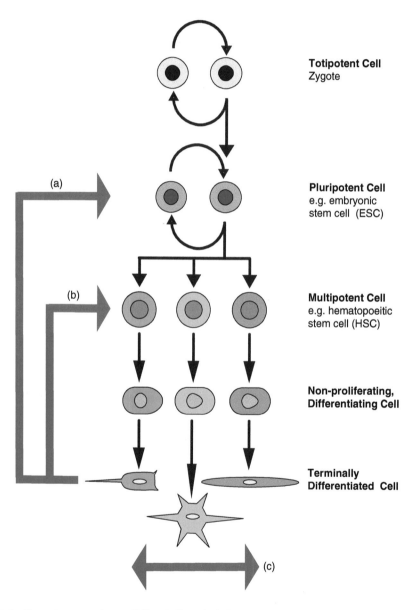

Fig. 2.8. Commitment and reversibility. Cells with different levels of potency and differentiation are shown here in diagrammatic form. At the top of the hierarchy is the totipotent cell (the zygote). Its descendants become progressively less potent and more committed to specific lineages. However, cells may revert to a more potent precursor cell through (a) induction of pluripotency; (b) reversion to a multipotent "progenitor" cell; or (c) transdifferentiation to a different cell type. Each mechanism may be feasible but will vary in its degree of difficulty.

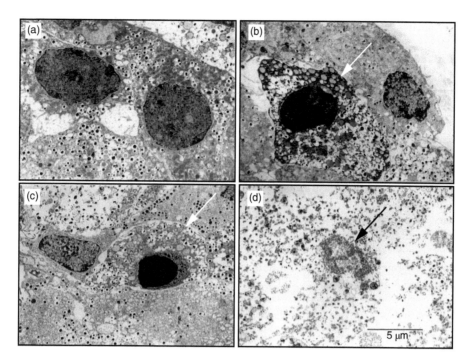

Fig. 2.9. Pancreatic islet cell death. Rat pancreatic islet cells in primary culture were exposed to polyinosinic-polycytidylic acid (poly IC; 50 μg/ml), interferon-γ (IFN-γ; 150 U/ml), or both for 48 hours to induce cell death. Cell morphology was examined by electron microscopy: (a) with no treatment; (b) with exposure to poly IC; and (c, d) with exposure to poly IC and IFN-γ. White arrows in (b, c) point to apoptotic cells, while the black arrow in (d) points to a cell that has undergone necrosis, resulting in loss of plasma membrane and nuclear membrane integrity. Scale bar as shown. *Source*: Scarim, A. L., Arnush, M., Blair, L. A., et al. (2001). Mechanisms of beta-cell death in response to double-stranded (ds) RNA and interferon-gamma: dsRNA-dependent protein kinase apoptosis and nitric oxide-dependent necrosis. Am J Pathol 159, reproduced with permission of Elsevier.

both erythroid and myeloid lineages. For some reason, as yet unknown, continued culture favored erythroid differentiation rather than the myeloid features seen in the original tumor.

2.4.6 Lineage Plasticity

Many claims have been made regarding cells transferring from one lineage to another, dating back to the 1930s (Moen 1935). The earliest claims to be substantiated related to the eye, where it was discovered that clonal cell lines from chick retina were able to form lens-like structures *in vitro* (Eguchi and Okada 1973). Lens regeneration can occur in amphibians using a similar mechanism *in vivo*, recruiting cells from the cornea (in frogs) or the pigment epithelial cells of the dorsal iris (in newts) (Henry and Tsonis 2010). These cells demonstrate lineage plasticity, allowing them to switch to a more potent state in response to physiological demands or insults (Smith 2006). The simplest explanation for lineage plasticity is that a mature, differentiated cell can revert to a precursor cell with greater potency (see Figure 2.8). However, direct conversion of one cell type to another can also occur without reversion to a pluripotent state (see Section 26.2.4). Fibroblasts can be converted to myoblasts using the

transcription factor *MyoD*, and various cell types can be converted to induced neuronal (iN) cells using a combination of three factors (Tanabe et al. 2015). The restriction of potency to develop various lineages has been likened to a set of marbles rolling down a hill into one of several valleys, which represent the various cell fates (Waddington 1957; Takahashi and Yamanaka 2016). Lineage conversion may arise by pushing one of the marbles back up the hill to a state of increased potency, before it is directed into another valley. Direct conversion from one cell type to another may arise by pushing the marble from one valley to another; this may be more difficult, but is achievable if the correct factors or conditions are supplied (Wang and Unternaehrer 2019).

2.5 CELL DEATH

Cell death causes obvious changes in morphology that can be seen during routine observation (Crowley et al. 2016). In some cases, cell death is accidental due to environmental insults, e.g. temperature extremes or toxic chemicals. However, the cell can also induce a regulated form of death using various molecular mechanisms. The terms used to describe cell death have been developed over many years and are often based on cell morphology. Updated terminology, based on molecular mechanisms, has been published by the Nomenclature Committee on Cell Death (Galluzzi et al. 2018). Commonly used terms in this book include:

(1) *Apoptosis.* Cells typically develop chromatin condensation and fragmentation, cytoplasmic shrinkage, and plasma membrane blebbing (see Figures 2.9, 29.2c, d) (Scarim et al. 2001). Some cells fragment into small apoptotic bodies, while others undergo secondary necrosis. Apoptosis can be divided into intrinsic and extrinsic forms. Intrinsic apoptosis induces mitochondrial changes, while extrinsic apoptosis uses "death receptors" in the plasma membrane to transmit external signals. Both forms of apoptosis share a common "execution pathway" involving caspase 3. Further subtypes exist; for example, anoikis is a form of intrinsic apoptosis that is triggered by loss of attachment to the ECM (Galluzzi et al. 2018).

(2) *Necrosis.* The plasma membrane of the cell becomes swollen and extends large bubble-like projections that eventually rupture, resulting in cell lysis (Crowley et al. 2016; Silke and Johnstone 2016). Necrosis is traditionally linked to accidental cell death, but some forms of regulated cell death also give rise to necrotic morphology (Galluzzi et al. 2018).

(3) *Autophagy.* Cells undergoing autophagy-dependent cell death display large, vacuole-like structures. However, it may be unclear if autophagy is the cause of cell death or is simply seen in association with it (Silke and Johnstone 2016). Autophagy is a process that is used by the normal cell to recycle cellular components. It is also seen during cell stress (e.g. during nutrient deprivation *in vitro*), when it may promote cell survival rather than cell death (Gregory et al. 2016).

Although the appearance of the cell is helpful to assess cell death, it may also cause confusion. A cell in the process of mitosis has a rounded morphology that can be difficult to distinguish from apoptosis or necrosis (Crowley et al. 2016). Morphological observation is not sufficient to truly understand what is happening within a culture; other assays must be performed to look for loss of viability, determine which form of regulated cell death has occurred, and examine specific pathways (see Section 29.2.3). All cell-based assays have limitations as well as benefits and it is important to keep these in mind when selecting methods or interpreting data.

REFERENCES

Alberts, B., Johnson, A., Lewis, J. et al. (2015). *Molecular Biology of the Cell*, 6e. New York: Garland Science.

Ananthakrishnan, R. and Ehrlicher, A. (2007). The forces behind cell movement. *Int. J. Biol. Sci.* 3 (5): 303–317.

Andersson, L.C., Jokinen, M., and Gahmberg, C.G. (1979). Induction of erythroid differentiation in the human leukaemia cell line K562. *Nature* 278 (5702): 364–365. https://doi.org/10.1038/278364a0.

Barczyk, M., Carracedo, S., and Gullberg, D. (2010). Integrins. *Cell Tissue Res.* 339 (1): 269–280. https://doi.org/10.1007/s00441-009-0834-6.

Bavle, R.M. (2014). Mitosis at a glance. *J. Oral Maxillofac. Pathol.* 18 (Suppl 1): S2–S5. https://doi.org/10.4103/0973-029X.141175.

Burnette, D.T., Shao, L., Ott, C. et al. (2014). A contractile and counterbalancing adhesion system controls the 3D shape of crawling cells. *J. Cell Biol.* 205 (1): 83–96. https://doi.org/10.1083/jcb.201311104.

Canti, R. G. (1927). Cells in tissue culture (normal and abnormal). https://wellcomecollection.org/works/yygn7b8s (accessed 18 September 2019).

Casanova, J.E. (2002). Epithelial cell cytoskeleton and intracellular trafficking V. Confluence of membrane trafficking and motility in epithelial cell models. *Am. J. Physiol. Gastrointest. Liver Physiol.* 283 (5): G1015–G1019. https://doi.org/10.1152/ajpgi.00255.2002.

Chen, S.S., Fitzgerald, W., Zimmerberg, J. et al. (2007). Cell–cell and cell–extracellular matrix interactions regulate embryonic stem cell differentiation. *Stem Cells* 25 (3): 553–561. https://doi.org/10.1634/stemcells.2006-0419.

Clevers, H. and Watt, F.M. (2018). Defining adult stem cells by function, not by phenotype. *Annu. Rev. Biochem.* 87: 1015–1027. https://doi.org/10.1146/annurev-biochem-062917-012341.

Crowley, L.C., Marfell, B.J., Scott, A.P. et al. (2016). Dead cert: measuring cell death. *Cold Spring Harb. Protoc.* 2016 (12) https://doi.org/10.1101/pdb.top070318.

Cui, X., Hartanto, Y., and Zhang, H. (2017). Advances in multicellular spheroids formation. *J. R. Soc. Interface* 14 (127) https://doi.org/10.1098/rsif.2016.0877.

Dowling, P., Walsh, N., and Clynes, M. (2008). Membrane and membrane-associated proteins involved in the aggressive phenotype displayed by highly invasive cancer cells. *Proteomics* 8 (19): 4054–4065. https://doi.org/10.1002/pmic.200800098.

Drexler, H.G. and MacLeod, R.A. (2010). History of leukemia-lymphoma cell lines. *Hum. Cell* 23 (3): 75–82. https://doi.org/10.1111/j.1749-0774.2010.00087.x.

Eguchi, G. and Okada, T.S. (1973). Differentiation of lens tissue from the progeny of chick retinal pigment cells cultured in vitro: a demonstration of a switch of cell types in clonal cell culture. *Proc. Natl Acad. Sci. U.S.A.* 70 (5): 1495–1499.

Engler, A.J., Sen, S., Sweeney, H.L. et al. (2006). Matrix elasticity directs stem cell lineage specification. *Cell* 126 (4): 677–689. https://doi.org/10.1016/j.cell.2006.06.044.

Fischer, A. (1925). *Tissue Culture: Studies in Experimental Morphology and General Physiology of Tissue Cells In Vitro*. London: William Heinemann.

de Forges, H., Bouissou, A., and Perez, F. (2012). Interplay between microtubule dynamics and intracellular organization. *Int. J. Biochem. Cell Biol.* 44 (2): 266–274. https://doi.org/10.1016/j.biocel.2011.11.009.

Forsten-Williams, K., Chu, C.L., Fannon, M. et al. (2008). Control of growth factor networks by heparan sulfate proteoglycans. *Ann. Biomed. Eng.* 36 (12): 2134–2148. https://doi.org/10.1007/s10439-008-9575-z.

Frame, M. and Norman, J. (2008). A tal(in) of cell spreading. *Nat. Cell Biol.* 10 (9): 1017–1019. https://doi.org/10.1038/ncb0908-1017.

Frantz, C., Stewart, K.M., and Weaver, V.M. (2010). The extracellular matrix at a glance. *J. Cell Sci.* 123 (Pt 24): 4195–4200. https://doi.org/10.1242/jcs.023820.

Fuhrmann, A. and Engler, A.J. (2015). The cytoskeleton regulates cell attachment strength. *Biophys. J.* 109 (1): 57–65. https://doi.org/10.1016/j.bpj.2015.06.003.

Galluzzi, L., Vitale, I., Aaronson, S.A. et al. (2018). Molecular mechanisms of cell death: recommendations of the nomenclature committee on cell death 2018. *Cell Death Differ.* 25 (3): 486–541. https://doi.org/10.1038/s41418-017-0012-4.

Gilbert, S.F. and Barresi, M.F. (2016). *Developmental Biology*, 11e. Sunderland, MA: Sinauer Associates.

Giono, L.E. and Manfredi, J.J. (2006). The p53 tumor suppressor participates in multiple cell cycle checkpoints. *J. Cell. Physiol.* 209 (1): 13–20. https://doi.org/10.1002/jcp.20689.

Gregory, S., Swamy, S., Hewitt, Z. et al. (2016). Autophagic response to cell culture stress in pluripotent stem cells. *Biochem. Biophys. Res. Commun.* 473 (3): 758–763. https://doi.org/10.1016/j.bbrc.2015.09.080.

Guo, S., Zhu, X., and Loh, X.J. (2017). Controlling cell adhesion using layer-by-layer approaches for biomedical applications. *Mater. Sci. Eng. C Mater. Biol. Appl.* 70 (Pt 2): 1163–1175. https://doi.org/10.1016/j.msec.2016.03.074.

Halper, J. and Kjaer, M. (2014). Basic components of connective tissues and extracellular matrix: elastin, fibrillin, fibulins, fibrinogen, fibronectin, laminin, tenascins and thrombospondins. *Adv. Exp. Med. Biol.* 802: 31–47. https://doi.org/10.1007/978-94-007-7893-1_3.

Harris, S.L. and Levine, A.J. (2005). The p53 pathway: positive and negative feedback loops. *Oncogene* 24 (17): 2899–2908. https://doi.org/10.1038/sj.onc.1208615.

Hartsock, A. and Nelson, W.J. (2008). Adherens and tight junctions: structure, function and connections to the actin cytoskeleton. *Biochim. Biophys. Acta* 1778 (3): 660–669. https://doi.org/10.1016/j.bbamem.2007.07.012.

Henry, J.J. and Tsonis, P.A. (2010). Molecular and cellular aspects of amphibian lens regeneration. *Prog. Retin. Eye Res.* 29 (6): 543–555. https://doi.org/10.1016/j.preteyeres.2010.07.002.

Herve, J.C. and Derangeon, M. (2013). Gap-junction-mediated cell-to-cell communication. *Cell Tissue Res.* 352 (1): 21–31. https://doi.org/10.1007/s00441-012-1485-6.

Hynds, R.E., Vladimirou, E., and Janes, S.M. (2018). The secret lives of cancer cell lines. *Dis. Model. Mech.* 11 (11) https://doi.org/10.1242/dmm.037366.

Hynes, R.O. (1999). Cell adhesion: old and new questions. *Trends Cell Biol.* 9 (12): M33–M37.

Hynes, R.O. (2009). The extracellular matrix: not just pretty fibrils. *Science* 326 (5957): 1216–1219. https://doi.org/10.1126/science.1176009.

Jin, S. and Levine, A.J. (2001). The p53 functional circuit. *J. Cell Sci.* 114 (Pt 23): 4139–4140.

Kleinman, H.K., Philp, D., and Hoffman, M.P. (2003). Role of the extracellular matrix in morphogenesis. *Curr. Opin. Biotechnol.* 14 (5): 526–532.

Kshitiz, Park, J., Kim, P. et al. (2012). Control of stem cell fate and function by engineering physical microenvironments. *Integr. Biol. (Camb.)* 4 (9): 1008–1018.

Lambshead, J.W., Meagher, L., O'Brien, C. et al. (2013). Defining synthetic surfaces for human pluripotent stem cell culture. *Cell Regen. (Lond.)* 2 (1): 7. https://doi.org/10.1186/2045-9769-2-7.

Lee, S.H. and Dominguez, R. (2010). Regulation of actin cytoskeleton dynamics in cells. *Mol. Cells* 29 (4): 311–325. https://doi.org/10.1007/s10059-010-0053-8.

Leterrier, C., Dubey, P., and Roy, S. (2017). The nano-architecture of the axonal cytoskeleton. *Nat. Rev. Neurosci.* 18 (12): 713–726. https://doi.org/10.1038/nrn.2017.129.

Lytle, N.K., Barber, A.G., and Reya, T. (2018). Stem cell fate in cancer growth, progression and therapy resistance. *Nat. Rev. Cancer* 18 (11): 669–680. https://doi.org/10.1038/s41568-018-0056-x.

MacLeod, R.A., Kaufmann, M., and Drexler, H.G. (2011). Cytogenetic analysis of cancer cell lines. *Methods Mol. Biol.* 731: 57–78. https://doi.org/10.1007/978-1-61779-080-5_6.

Malumbres, M. (2014). Cyclin-dependent kinases. *Genome Biol.* 15 (6): 122.

McDonald, E.R. 3rd and El-Deiry, W.S. (2000). Cell cycle control as a basis for cancer drug development (review). *Int. J. Oncol.* 16 (5): 871–886.

Moen, J.K. (1935). The development of pure cultures of fibroblasts from single mononuclear cells. *J. Exp. Med.* 61 (2): 247–260.

Morrison, S.J. and Kimble, J. (2006). Asymmetric and symmetric stem-cell divisions in development and cancer. *Nature* 441 (7097): 1068–1074. https://doi.org/10.1038/nature04956.

NCBI (2019). Bookshelf. https://www.ncbi.nlm.nih.gov/books (accessed 12 April 2019).

NIH (2019). Stem cell information: glossary. https://stemcells.nih.gov/glossary.htm (accessed 30 April 2019).

Ponti, A., Machacek, M., Gupton, S.L. et al. (2004). Two distinct actin networks drive the protrusion of migrating cells. *Science* 305 (5691): 1782–1786. https://doi.org/10.1126/science.1100533.

Rappaport, C., Poole, J.P., and Rappaport, H.P. (1960). Studies on properties of surfaces required for growth of mammalian cells in synthetic medium. I. The HeLa cell. *Exp. Cell Res.* 20: 465–479.

Reid, Y.A. (2017). Best practices for naming, receiving, and managing cells in culture. *In Vitro Cell. Dev. Biol. Anim.* 53 (9): 761–774. https://doi.org/10.1007/s11626-017-0199-1.

Rodriguez, O.C., Schaefer, A.W., Mandato, C.A. et al. (2003). Conserved microtubule–actin interactions in cell movement and morphogenesis. *Nat. Cell Biol.* 5 (7): 599–609. https://doi.org/10.1038/ncb0703-599.

Ruoslahti, E. (1996). RGD and other recognition sequences for integrins. *Annu. Rev. Cell Dev. Biol.* 12: 697–715. https://doi.org/10.1146/annurev.cellbio.12.1.697.

Rygaard, K., Moller, C., Bock, E. et al. (1992). Expression of cadherin and NCAM in human small cell lung cancer cell lines and xenografts. *Br. J. Cancer* 65 (4): 573–577.

Saito, M., Tucker, D.K., Kohlhorst, D. et al. (2012). Classical and desmosomal cadherins at a glance. *J. Cell Sci.* 125 (Pt 11): 2547–2552. https://doi.org/10.1242/jcs.066654.

Scarim, A.L., Arnush, M., Blair, L.A. et al. (2001). Mechanisms of beta-cell death in response to double-stranded (ds) RNA and interferon-gamma: dsRNA-dependent protein kinase apoptosis and nitric oxide-dependent necrosis. *Am. J. Pathol.* 159 (1): 273–283. https://doi.org/10.1016/s0002-9440(10)61693-8.

Schlessinger, J., Lax, I., and Lemmon, M. (1995). Regulation of growth factor activation by proteoglycans: what is the role of the low affinity receptors? *Cell* 83 (3): 357–360.

Schwartz, M.A. and Assoian, R.K. (2001). Integrins and cell proliferation: regulation of cyclin-dependent kinases via cytoplasmic signaling pathways. *J. Cell Sci.* 114 (Pt 14): 2553–2560.

Sherr, C.J. and McCormick, F. (2002). The RB and p53 pathways in cancer. *Cancer Cell* 2 (2): 103–112.

Shin, K., Fogg, V.C., and Margolis, B. (2006). Tight junctions and cell polarity. *Annu. Rev. Cell Dev. Biol.* 22: 207–235. https://doi.org/10.1146/annurev.cellbio.22.010305.104219.

Silke, J. and Johnstone, R.W. (2016). In the midst of life – cell death: what is it, what is it good for, and how to study it. *Cold Spring Harb. Protoc.* 2016 (12) https://doi.org/10.1101/pdb.top070508.

Smith, A. (2006). A glossary for stem-cell biology. *Nature* 441 (7097): 1060–1060. https://doi.org/10.1038/nature04954.

Smola, H., Stark, H.J., Thiekotter, G. et al. (1998). Dynamics of basement membrane formation by keratinocyte–fibroblast interactions in organotypic skin culture. *Exp. Cell Res.* 239 (2): 399–410. https://doi.org/10.1006/excr.1997.3910.

Solter, D. (2006). From teratocarcinomas to embryonic stem cells and beyond: a history of embryonic stem cell research. *Nat. Rev. Genet.* 7 (4): 319–327. https://doi.org/10.1038/nrg1827.

Stramer, B.M. and Dunn, G.A. (2015). Cells on film – the past and future of cinemicroscopy. *J. Cell Sci.* 128 (1): 9–13. https://doi.org/10.1242/jcs.165019.

Stupack, D.G. and Cheresh, D.A. (2002). Get a ligand, get a life: integrins, signaling and cell survival. *J. Cell Sci.* 115 (Pt 19): 3729–3738.

Takahashi, K. and Yamanaka, S. (2016). A decade of transcription factor-mediated reprogramming to pluripotency. *Nat. Rev. Mol. Cell Biol.* 17 (3): 183–193. https://doi.org/10.1038/nrm.2016.8.

Tanabe, K., Haag, D., and Wernig, M. (2015). Direct somatic lineage conversion. *Philos. Trans. R. Soc. Lond. Ser. B Biol. Sci.* 370 (1680): 20140368. https://doi.org/10.1098/rstb.2014.0368.

Tojkander, S., Gateva, G., and Lappalainen, P. (2012). Actin stress fibers – assembly, dynamics and biological roles. *J. Cell Sci.* 125 (Pt 8): 1855–1864. https://doi.org/10.1242/jcs.098087.

Waddington, C.H. (1957). *The Strategy of the Genes.* London: George Allen & Unwin.

Wang, H. and Unternaehrer, J.J. (2019). Epithelial–mesenchymal transition and cancer stem cells: at the crossroads of differentiation and dedifferentiation. *Dev. Dyn.* 248 (1): 10–20. https://doi.org/10.1002/dvdy.24678.

Weiss, L. and Zeigel, R. (1971). Cell surface negativity and the binding of positively charged particles. *J. Cell. Physiol.* 77 (2): 179–186. https://doi.org/10.1002/jcp.1040770208.

Wheelock, M.J., Shintani, Y., Maeda, M. et al. (2008). Cadherin switching. *J. Cell Sci.* 121 (Pt 6): 727–735. https://doi.org/10.1242/jcs.000455.

Williams, G.H. and Stoeber, K. (2012). The cell cycle and cancer. *J. Pathol.* 226 (2): 352–364. https://doi.org/10.1002/path.3022.

Williams, S.E. and Fuchs, E. (2013). Oriented divisions, fate decisions. *Curr. Opin. Cell Biol.* 25 (6): 749–758. https://doi.org/10.1016/j.ceb.2013.08.003.

Woodworth, M.B., Girskis, K.M., and Walsh, C.A. (2017). Building a lineage from single cells: genetic techniques for cell lineage tracking. *Nat. Rev. Genet.* 18 (4): 230–244. https://doi.org/10.1038/nrg.2016.159.

Yamane, I. (1978). Development and application of a serum-free culture medium for primary culture. In: *Nutritional Requirements of Cultured Cells* (ed. H. Katsuta), 1–19. Tokyo and Baltimore: Japan Scientific Societies Press and University Park Press.

Yap, A.S., Gomez, G.A., and Parton, R.G. (2015). Adherens junctions revisualized: organizing cadherins as nanoassemblies. *Dev. Cell* 35 (1): 12–20. https://doi.org/10.1016/j.devcel.2015.09.012.

Yevdokimova, N. and Freshney, R.I. (1997). Activation of paracrine growth factors by heparan sulphate induced by glucocorticoid in A549 lung carcinoma cells. *Br. J. Cancer* 76 (3): 281–289.

Yoder, M.C. (2004). Blood cell progenitors: insights into the properties of stem cells. *Anat. Rec. A Discov. Mol. Cell Evol. Biol.* 276 (1): 66–74. https://doi.org/10.1002/ar.a.10133.

Young, H.E. and Black, A.C. Jr. (2004). Adult stem cells. *Anat. Rec. A Discov. Mol. Cell Evol. Biol.* 276 (1): 75–102. https://doi.org/10.1002/ar.a.10134.

CHAPTER 3

Origin and Evolution of Cultured Cells

It is often assumed that cultures remain the same, regardless of their history and handling. Classic cell lines such as HeLa and MCF-7 are used by many laboratories for extended periods of time and shared with colleagues. Does this affect their validity as research models? A number of studies have demonstrated that widely used cell lines harbor significantly different populations, which evolve over time in a way that affects the behavior of the culture. Analysis of the HeLa cell line and its derivatives (sourced from 13 different laboratories) has demonstrated large-scale copy number variation (CNV) between different samples, including major chromosomal variations (Liu et al. 2019). Another study, looking at the MCF-7 cell line, found genetic variability in samples from the same batch of vials, sourced from a cell repository (Kleensang et al. 2016). Genetic variation in MCF-7 is associated with the presence of multiple subclones, which vary in their abundance between different samples of the same cell line (Ben-David et al. 2018). These variations can have important functional consequences. Genetic variability in the MCF-7 cell line was associated with changes in behavior in metabolic studies and differences in drug responsiveness when cells were tested against 321 anti-cancer compounds (Kleensang et al. 2016; Ben-David et al. 2018).

Today's genomic studies shine a new light on an old problem. Genetic variation *in vitro* has been studied for more than 60 years, resulting in longstanding concerns over the handling of cells in culture (Hsu and Moorhead 1957; Moorhead 1974). This chapter looks at the key events in the life of a culture and associated changes in biology; the practical consequences of these changes will be explored in later chapters.

3.1 ORIGIN OF CULTURED CELLS

3.1.1 Sample Origin

Even the smallest biological sample contains a complex assortment of cells. For example, single-cell analysis of cultures from adult human brain samples (removed during neurosurgery) identified at least five cell types within the culture: astrocytes, neurons, microglia, endothelial cells, and astrocytes (Spaethling et al. 2017). Samples of human skin contain epithelial cells (keratinocytes and Merkel cells) and many other cell types including fibroblasts, melanocytes, and immune cells (Rognoni and Watt 2018). Each cell type may display different properties in culture, depending on its location of origin. For example, skin fibroblasts can display different morphologies and behavior in culture, depending on their anatomical site or their location within the dermis (Castor et al. 1962; Lynch and Watt 2018). Such variations can have consequences for today's specialized applications. Fibroblasts from the papillary dermis in the skin support the formation of a stratified epithelium in three-dimensional (3D) culture more effectively than fibroblasts from the reticular dermis that lies beneath (Lynch and Watt 2018).

Mesenchymal cells, derived from fibroblasts within the stroma or its vascular elements, will usually overgrow the other cell types if serum is present and selective culture conditions are not used (see Section 20.7). This has given rise to some very useful models (see Table 14.2), including finite cultures (e.g. WI-38, MRC-5, IMR-90) and continuous cell lines (e.g. BHK-21, Chinese hamster ovary [CHO], L-929). However, fibroblast overgrowth represents one of the major challenges of tissue culture since its inception — namely,

Freshney's Culture of Animal Cells: A Manual of Basic Technique and Specialized Applications, Eighth Edition. Amanda Capes-Davis and R. Ian Freshney.

how to prevent the loss of the more fragile, slower-growing specialized cells such as hepatic parenchyma or epidermal keratinocytes. Inadequate culture conditions are largely to blame for this problem, and considerable progress has been made to develop selective media and substrates for the maintenance of many specialized cell lines (see Section 10.1; see also Chapter 24). This has succeeded to the extent that many specialized cell types are available commercially (see Table 24.1).

Different biological samples contain cells with different abilities to replicate and to express a differentiated phenotype. It may be useful to think of a cell culture as being an equilibrium between stem cells, progenitor cells, and mature differentiated cells (see Sections 2.4.3, 2.4.4). The proportion of cells at different stages will be determined by the source of the culture and by other factors, such as the culture conditions. Cell lines derived from the embryo may contain a higher proportion of stem cells and progenitor cells and be capable of greater self-renewal than cultures from adults. Cultures from tissues undergoing continuous renewal *in vivo* may still contain stem cells, which will have a prolonged lifespan under the appropriate conditions (e.g. epidermis, intestinal epithelium, and hematopoietic cells). Cultures from tissues that renew solely under stress may contain only committed progenitor cells with a limited lifespan (e.g. muscle). Thus, the behavior of the cultured cell is defined not only by its lineage *in vivo* but also by its position in that lineage (stem cell, progenitor cell, or mature differentiated cell).

3.1.2 Disease Origin

A malignant neoplasm (cancer) is prone to genomic instability and clonal evolution, which adds greater complexity when working with tumor samples (Ben-David et al. 2019). Genomic instability is an important characteristic of cancer development, leading to a mutant genotype that enables sustained proliferation, resistance to cell death, and other hallmarks of cancer (Hanahan and Weinberg 2011). Malignant tumors contain subclonal populations with different genetic variations, resulting in clonal evolution as the different populations compete for space and resources within the tumor (Nowell 1976; Merlo et al. 2006). Clonal evolution may result in greater diversity, where cells display increased levels of genetic, epigenetic, phenotypic, and functional heterogeneity (Maley et al. 2017). It may also result in lesser diversity if a subclone becomes increasingly dominant within the tumor because its genotype confers a selective advantage (Kasai et al. 2016). Genomic instability and clonal evolution affect all cultures, particularly cell lines, which may inherit these characteristics from their disease of origin or acquire similar behavior in culture through transformation (see Sections 3.2.2, 3.6).

3.2 EVOLUTION OF CELL LINES

3.2.1 Phases of Cell Cultivation

Primary culture. Cell culture is a selective process that favors some cells at the expense of others (see Table 3.1). Primary culture is the first phase of cell cultivation and is usually performed by allowing cells to migrate from a tissue fragment (primary explantation) or by performing tissue disaggregation using an enzymatic or mechanical method of dispersal (see Figures 1.4, 13.1). In primary explantation (see Section 13.4), selection occurs by virtue of the cells' capacity to migrate from the explant. Macrophages often migrate from the tissue fragment first, followed by fibroblasts and then epithelial cells; some specialized cells remain immobile within the explant and thus do not appear within the culture (Paul 1975). Tissue disaggregation (see Sections 13.5, 13.6) results in a different form of selection. Only those cells that survive the dispersal method and adhere to the substrate (or that survive in suspension) will persist in the primary culture.

If the primary culture is maintained for more than a few hours, further selection will occur. Cells that are capable of proliferation will increase, some cell types will survive but not increase, and yet others will be unable to survive under the particular conditions of the culture. Hence, the relative proportion of each cell type will change and will continue to do so until (in the case of monolayer cultures) all the available

TABLE 3.1. Selection in cell line development.

Stage	Factors influencing selection	
	Primary explantation	Tissue disaggregation
Isolation	Mechanical damage	Mechanical or enzymatic damage
Primary culture	Adhesion of explant; differential outgrowth (migration) from the explant; cell proliferation	Cell adhesion and spreading; cell proliferation
First subculture	Trypsin sensitivity; nutrient, hormone, and substrate limitations; proliferative ability	
Propagation as a cell line	Relative growth rates of different cells; selective overgrowth of one lineage	
	Nutrient, hormone, and substrate limitations	
	Effect of cell density on predominance of normal or transformed phenotype	
Senescence; transformation	Normal cells die out; transformed cells overgrow	

culture substrate is occupied. Both cell population changes and adaptive modifications occur continuously throughout the culture period, making it difficult to select a time when the culture may be regarded as homogeneous or stable.

Finite cell lines. Subculture (transfer of cells or tissue from one culture vessel to another) marks the point at which a primary culture becomes a cell line, at least according to consensus terminology (see Section 14.1) (Schaeffer 1990). With each successive subculture, the component of the population with the ability to proliferate most rapidly will gradually predominate, and non-proliferating or slowly proliferating cells will be diluted out. This is most apparent after the first passage, in which differences in proliferative capacity are compounded with varying abilities to withstand the trauma of trypsinization and transfer (see Section 13.2). By the third passage, the culture usually becomes more stable and the most resilient, rapidly proliferating cell types will become predominant.

Many cultures have a limited replicative lifespan and are described here as finite cell lines, e.g. normal human fibroblasts (see Figure 3.1a) (Schaeffer 1990). The lifespan of a cell line can be estimated by counting the cumulative number of population doublings that occur within the culture (see Figure 3.1e; see also Section 14.2.2). Finite cell lines proliferate until they reach a reproducible population doubling level (PDL), which usually occurs at 20–100 PDL for normal human cells. Cells cease to divide at that point, although they may remain viable for up to 18 months thereafter. This phenomenon is known as senescence and the number of population doublings at which it occurs is known as the Hayflick limit (Hayflick and Moorhead 1961). The Hayflick limit remains consistent when a cell line is grown under the same culture conditions, but will vary between cultures depending on their species, cell lineage, and culture conditions.

Continuous cell lines. A subset of cultures (notably from rodents and human tumors) can spontaneously become established as continuous cell lines, which continue to proliferate and are essentially immortal. This is particularly likely with mouse cells, and to a lesser degree with rat and Syrian hamster cells (Newbold 2002). Much of the early work to understand this phenomenon was performed using mouse embryonic fibroblasts (MEFs). Protocols were developed to establish MEF cell lines, resulting in the various 3T3 cell lines, which were established by plating 3×10^5 cells per dish and performing subculture every three days (Todaro and Green 1963). Immortalization may also be induced in normal diploid rodent cells as a rare event following exposure to chemical carcinogens (Newbold 2002).

Human tumor cells may spontaneously give rise to continuous cell lines, but the rate of establishment will vary with the tumor type. In one early study of 200 tumors, continuous cell lines were established from only 6% of samples (Giard et al. 1973). Success rates were highest with squamous cell

(epidermoid) carcinoma and lowest with breast carcinoma. In contrast, most normal human cells do not spontaneously continue to proliferate and will undergo senescence, even after exposure to carcinogens (Newbold 2002). These cells may be genetically modified to extend their lifespan, e.g. using simian virus 40 (SV40) or its large T antigen (TAg) (see Section 22.3). After genetic modification, cells usually continue to proliferate for a limited period of time before entering crisis, which is characterized by decreased proliferation and increased cell death (see Figure 3.1b, c). A minority of cells (1 in 10^7 human fibroblasts) emerge from crisis to become immortal (see Figure 3.1d, e) (Huschtscha and Holliday 1983).

Culture conditions and handling are important when establishing a continuous cell line. This is perhaps best seen with human epidermal keratinocytes, where improvements in culture conditions have led to conditional immortalization or "reprogramming" (see Section 22.3.4). Improved culture conditions may be successful for this cell type because of its self-renewal capacity *in vivo*; epidermal keratinocytes are known to undergo spontaneous immortalization (Fusenig and Boukamp 1998), and it is possible that the culture conditions have simply allowed the stem cells within the culture to survive and proliferate. However, there is no doubt that optimization of culture conditions can have a major impact on cell line establishment (Ince et al. 2015). Validation testing is also important when establishing a continuous cell line. Decreased proliferation during senescence and crisis may allow a contaminating cell to rapidly overgrow the culture, resulting in a misidentified cell line (see Chapter 17). Always put aside a portion of the original tissue sample as a reference for later comparison.

3.2.2 Clonal Evolution

Continuous cell lines may appear homogenous, especially when compared to the original sample from which they were established, but this is not the case when their behavior is examined more closely. Forty years of cell-based research has emphatically shown that cell lines contain multiple clonal populations that vary in their behavior (Westermark 1978; Ben-David et al. 2018). For example, the rat hepatoma cell line H4-II-E-C3 consists of at least four subclones that differ in their constitutive and glucocorticoid-induced levels of tyrosine aminotransferase (see Figure 3.2). Looking at human cell lines, the lymphoma cell line U-2932 contains at least two subpopulations with different cell surface markers and unique chromosomal rearrangements; similar cell surface markers were present in the original tumor tissue (Quentmeier et al. 2013). Further analysis of B-cell lymphoma cell lines indicated that 12% of cultures carried detectable clonal populations, which could be sorted by flow cytometry using different cell surface markers; authentication testing was performed to exclude cross-contamination (see Chapter 17) (Quentmeier et al. 2016).

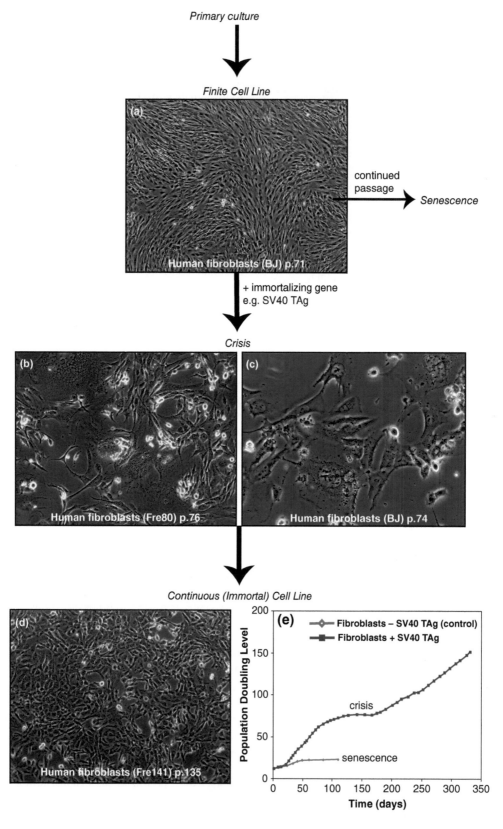

Fig. 3.1. Cultivation of human fibroblasts. Images of human fibroblasts (a) at early passage as a finite cell line; (b, c) after introduction of SV40 TAg, resulting in extension of lifespan and subsequent crisis; (d) at later passage as a continuous cell line. (e) Graph showing cumulative population doublings with and without SV40 TAg. Images and data show representative behavior at each phase of cultivation from multiple cell lines. Cell lines include BJ (a, c) from foreskin and Fre80 (b), Fre141 (d), and Fre98 (e) from breast. Images were captured using a phase contrast microscope with ×4 (a, d), ×10 (b), and ×20 (c) objectives. *Source*: Courtesy of Lily Huschtscha, Cancer Research Unit, CMRI.

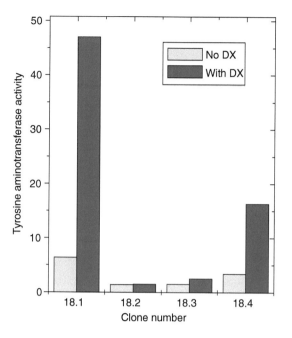

Fig. 3.2. Clonal variation. Variation in tyrosine aminotransferase activity among clonal populations of a rat minimal-deviation hepatoma cell strain, H-4-II-E-C3. Cells were cloned; clone 18 was isolated, grown up, and recloned; and four second-generation clones were assayed for tyrosine aminotransferase activity, with and without pretreatment of the culture with dexamethasone (DX). Light gray bars, basal level; dark blue bars, induced level. *Source*: data courtesy of J. Somerville.

Clonal populations evolve with ongoing handling, most likely through clonal selection. Cytogenetic and sequence analysis of a cell line from endometrial adenocarcinoma, Ishikawa 3-H-12, identified multiple cell populations with different chromosomal rearrangements (Kasai et al. 2016). Interestingly, some variants became less abundant with passaging; cells with 45 chromosomes increased from 60 to 86% during extended culture, due to a decrease in cells with 46 chromosomes. This suggests a model of clonal variation where certain populations expand with passaging to dominate the culture (see Figure 3.3). "Bottlenecks" may arise that select for certain populations, e.g. overdilution of cell lines after passaging will favor populations that have a growth advantage (Drexler and MacLeod 2010). It is tempting to speculate that genetic heterogeneity has been downplayed as an important issue because laboratories have tended to work with late passage cultures where a smaller number of clonal populations have become dominant. The push to use authenticated cell lines at earlier passage may have contributed to the current awareness of clonal heterogeneity.

3.3 CHANGES IN GENOTYPE

The genotype of a cell line refers primarily to its genetic composition (see Section 18.2). The term was coined for the study of heredity, which assumes that the genome remains stable over time. This assumption is largely true for normal human cells, which are usually diploid and genetically stable when placed into culture (Hayflick 1965). However, when cultures develop the ability to proliferate indefinitely, they often develop abnormal karyotypes, along with an array of other genetic changes (see Section 3.6) (MacLeod et al. 2011). Continuous cell lines are likely to display highly abnormal karyotypes and marked genetic instability (Wenger et al. 2004; Saito et al. 2011; Kasai et al. 2016). This is not surprising, considering that many of these cell lines come from malignant tumors and are thus prone to genomic instability from inception; others acquire such changes as part of transformation (see Sections 3.1.2, 3.6).

The extent of *in vitro* genetic abnormality will vary with the species and tissue type. Most age-related tumors in humans are carcinomas and display highly abnormal karyotypes in culture; murine tumors are more likely to be mesenchymal and highly abnormal karyotypes are less common (Rangarajan and Weinberg 2003). Differences between mice and humans are likely to relate to telomere length, telomerase activity (which is present in most mouse cells but absent in most adult human somatic cells), and differing uses of p53 and pRB pathways (Wright and Shay 2000; Forsyth et al. 2002).

3.3.1 Chromosomal Aberrations

Although virtually every cell in the animal has the normal diploid set of chromosomes, this is more variable in culture. Continuous cell lines are usually heteroploid, i.e. there is a wide range in chromosome number among individual cells in the population (see Figure 3.4). Most tumor cultures show signs of aneuploidy, i.e. deviations from the normal complement of chromosomes (Biedler 1975; Roschke et al. 2003). Chromosomal aberrations are also common due to structural rearrangements (see Figure 3.5). These are particularly obvious in tumor cells but also occur in cell lines that are derived from normal tissue. For example, the Vero cell line from normal monkey kidney contains at least one chromosomal fusion (between chromosomes 7 and 24) and seven translocations (Osada et al. 2014). Although the mouse karyotype is made up exclusively of small telocentric chromosomes, many continuous murine cell lines carry metacentric chromosomes due to Robertsonian fusion of the telomeres (see Figure 3.5a).

Specific chromosomal rearrangements may be found in certain tumors. The first such rearrangement to be

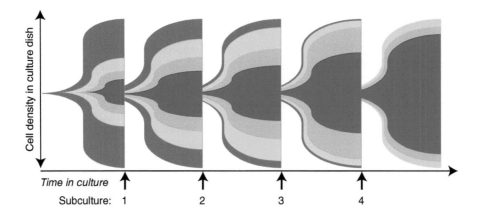

Fig. 3.3. Model for *in vitro* clonal evolution. Cell lines are made up of heterogeneous clones with different genomic variations, resulting in different proliferation rates. Repeated passaging leads to clones with higher proliferation rates becoming dominant, resulting in a change in composition. Here, clones are indicated by different colors. The outer clone is predominant at early passage but gradually declines in abundance with repeated passage and finally disappears from the cell line, allowing replacement by the inner clone. *Source*: Kasai, F., Hirayama, N., Ozawa, M., Iemura, M., & Kohara, A. (2016). Changes of heterogeneous cell populations in the Ishikawa cell line during long-term culture: Proposal for an in vitro clonal evolution model of tumor cells. Genomics, 107(6), 259–266. © 2016 Elsevier.

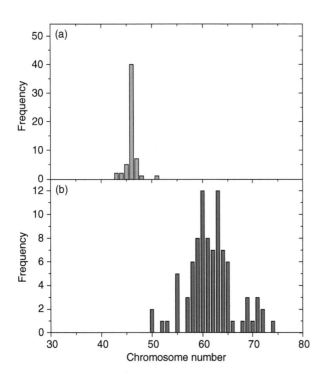

Fig. 3.4. Chromosome numbers of finite and continuous cell lines. (a) A normal human glial cell line; (b) a continuous cell line from human metastatic melanoma.

1999; MacLeod et al. 2008). Many hematological malignancies and childhood sarcomas have associations with specific chromosomal rearrangements, but the list is broadening to other tumor types (Mitelman et al. 2007). A database can be used to search for associations between chromosomal rearrangements and specific tumors, which is valuable for cell line characterization and confirmation of neoplasia (Mitelman et al. 2019).

Certain chromosomal aberrations are more common in culture and may be cell-line specific, which is important for cell line characterization (see Section 18.2.2) (Tsuji et al. 2010). Some are unique to individual cell lines and their derivatives, e.g. HeLa carries unique "marker" chromosomes that can be used for authentication (see Section 17.3.1). Other changes are shared by many different cultures; approximately a third of male cell lines lose part or all of the Y chromosome *in vitro*, as assessed by the presence of Y-specific amelogenin (Yu et al. 2015). Loss of the Y chromosome occurs in many different cancer types *in vivo* and may lead to increased tumorigenicity (Vijayakumar et al. 2005). Chromosomal changes are not confined to cancer cells; trisomy of chromosome 12 is common in human pluripotent stem cells (hPSCs), where it is associated with increased proliferation and tumorigenicity (Ben-David et al. 2014). Genetic instability in stem cell cultures is an important topic in its own right and is discussed in a later chapter (see Sections 23.4.1, 23.7).

3.3.2 Genomic Variation

An increasing amount of sequence data have been assembled for widely used cell lines, leading to release of the full genome for CHO-K1, HeLa, Vero, and other cell lines (Xu et al.

documented was the Philadelphia chromosome in chronic myeloid leukemia, which is now known to be a reciprocal translocation between chromosomes 9 and 22 (Nowell and Hungerford 1960). This translocation results in fusion of *BCR* and *ABL* genes and is retained *in vitro* (Drexler et al.

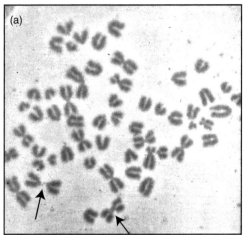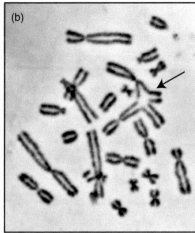

Fig. 3.5. Chromosome aberrations. (a) P2 cells (a clone of L-929 mouse fibroblasts) showing multiple telomeric fusions, with two marked by arrows. (b) Y5 Chinese hamster cells showing recombination event between two dissimilar chromosomes. This cell would be unlikely to survive. *Source*: R. Ian Freshney.

2011; Adey et al. 2013; Landry et al. 2013; Lewis et al. 2013; Osada et al. 2014). Overall, genomic data show that cell lines continue to display genetic changes that are consistent with their sample and disease of origin (see Section 3.1) (Barretina et al. 2012; Domcke et al. 2013). For example, the metastatic melanoma COLO-829 cell line displays a mutational signature that is typical of ultraviolet light exposure and is absent from a lymphoblastoid cell line (LCL) established from the same donor (Pleasance et al. 2010). However, ongoing culture results in many additional genomic changes, including:

(1) *Loss of heterozygosity (LOH).* LOH is perhaps the most common genetic change *in vitro* and was first observed using isoenzyme analysis (see Section 17.3.1) (Dracopoli and Fogh 1983). LOH can be extensive; 23% of HeLa Kyoto is described as homozygous and 55.3% of the HuH-7 cell line is believed to be affected by LOH (Landry et al. 2013; Kasai et al. 2018). These changes may have major functional consequences. A 9-Mb deletion in chromosome 12 of the Vero cell line is associated with loss of type I interferon, which probably explains this cell line's high susceptibility to viruses and other microbial insults in culture (Osada et al. 2014).

(2) *Copy number variation (CNV).* Many cell lines exhibit CNV, ranging from the domain level to large chromosomal segments and even whole chromosomes (Liu et al. 2019). In addition to reduced copy number (LOH), duplication of genomic sequence is common, e.g. HeLa and its derivatives have a copy number of three at many loci, consistent with its triploid karyotype (Adey et al. 2013; Landry et al. 2013). CNV progressively accumulates as HeLa cells are passaged (Liu et al. 2019).

(3) *Accumulation of mutations.* Many cultures have a higher spontaneous mutation rate *in vitro*, perhaps associated with their high rate of cell proliferation and defective DNA surveillance genes. Genomic analysis has shown that mutations are widespread; 301 752 new single nucleotide polymorphisms (SNPs) were identified when the CHO cell line was modified to give the antibody-producing C0101 cell line, representing 9% of its total SNPs (Lewis et al. 2013). Loss of p53 function was the most common genetic event during immortalization of thyroid carcinoma cell lines, and is likely to occur in other cell types (Landa et al. 2019).

(4) *Gene fusions.* RNA sequencing of 675 human cancer cell lines identified more than 2000 previously undocumented gene fusions (Klijn et al. 2015). It was not clear from this study (which did not have access to original tumor samples) whether the fusions occurred in the primary tumor or during culture, but a large number of fusion events affected protein kinases and were likely to be functionally relevant.

(5) *Allelic drift.* Some cell lines display high levels of genetic instability, resulting in allelic drift when they are tested for authenticity, e.g. using short tandem repeat (STR) profiling. This is particularly likely in cell lines with microsatellite instability (MSI) (see Section 17.4.2). Changes in STR loci have been found in different clonal populations, suggesting that their relative abundance is determined by clonal evolution with passaging (see Section 3.2.2).

Genomic analysis is a powerful tool that can reveal important information about widely used cell lines. For example, genomic sequence has been used to map the integration of human papillomavirus-18 (HPV-18) on chromosome 8 in the HeLa cell line, which is relevant to its origin from cervical carcinoma (Adey et al. 2013; Landry et al. 2013). However, publication of genomic data can have ethical concerns if it affects the privacy of the person from which the cell line was established. Always assess any risks to donor privacy before releasing genomic information (see Supp. S6.4).

3.4 CHANGES IN PHENOTYPE

The phenotype of a cell line refers to its physical appearance and other observable characteristics, including finer measures of assessment, e.g. of specialized functions (Fisch 2017). A primary culture generally shows a close resemblance to the cells in the parent tissue and will probably continue to express specialized functions, particularly when the culture becomes confluent, while also retaining diversity. Retention of these characteristics during subculture requires the use of selective conditions: (i) to retain the correct cell lineage; (ii) to favor proliferation within this lineage; and (iii) to allow for subsequent application of inducing conditions which will favor the expression of the differentiated phenotype. Thus, the culture conditions will have a major impact on the culture's subsequent phenotype. Clonal evolution will also have an important role (see Section 3.2.2), resulting in phenotypic drift as the culture is passaged.

Phenotype has traditionally been studied using tissue- or disease-specific markers, e.g. antibody panels that are used for immunocytochemistry (Quentmeier et al. 2001). These are often not sufficient to identify a particular cell type with a high degree of confidence. For example, the "melanoma-specific" antibody HMB-45 (which detects the GP100 protein and is used for melanoma diagnosis) stains a small number of other tumor types and does not stain 25% of melanoma cell lines (Gown et al. 1986; Bonetti et al. 1989; Quentmeier et al. 2001). Such markers are tumor-specific in the clinic because they are combined with the patient's clinical history, tumor histology, and other factors that together enable an accurate diagnosis. However, markers are becoming more informative as "omic" studies generate large datasets for exploration; a recent study used sequential-window acquisition of all theoretical fragments mass spectrometry (SWATH-MS) to quantify more than 5000 proteins across 14 HeLa samples (Liu et al. 2019).

3.4.1 Phenotypic Variation

Most studies agree that cell lines typically retain a degree of similarity to their parent tissues, provided controlled culture conditions are used (Wang et al. 2006; Gazdar et al. 2010; Mouradov et al. 2014). However, this is not always the case; some cell lines express a "variant" phenotype that does not correspond to what is expected for that tissue type. A good selection process is essential when choosing cell lines as

models (Mohseny et al. 2011). Changes in phenotype that must be considered include:

(1) **Heterogeneity across clonal populations.** Continuous cell lines carry multiple clonal populations and have heterogeneous genomes (see Sections 3.2.2, 3.3.2). This translates to variations in protein number and abundance between different samples of the same cell line (Liu et al. 2019). As a result, the different samples of the same cell line may display heterogeneous phenotypes, e.g. variable drug response profiles.

(2) **Loss of differentiated characteristics.** Dedifferentiation is commonly reported in culture, especially for highly passaged cell lines. For example, a recent study of thyroid carcinoma cell lines showed that all 60 cell lines were dedifferentiated, regardless of the tumor of origin (Landa et al. 2019). However, this is a somewhat complex topic; loss of differentiation may represent permanent loss of function or a temporary adaptation to the culture conditions (see Section 26.2.3).

(3) **Gain of differentiated characteristics.** A subset of cell lines gain a phenotype that is different to their reported tissue type or disease diagnosis. For example, analysis of 50 small cell lung carcinoma cell lines showed that 15 were "variant" cell lines with unexpected morphology or biochemical markers (Gazdar et al. 1985). Five of these cell lines came from variant tumors, while three cultures acquired their phenotypes *in vitro*. This may represent clonal evolution of variant populations in the parental tumors, but it is possible that cell lines acquired variant phenotypes through other mechanisms. Epithelial–mesenchymal transition (EMT) is commonly seen in cancer cell lines and is discussed in a later chapter (see Section 26.2.5).

(4) **Common expression signatures.** Concerns have been expressed that cancer cell lines bear more resemblance to each other than they do to their tissues of origin (Gillet et al. 2013). This can be seen with the NCI-60 cell line panel; analysis of multidrug resistance genes showed that the ovarian cell lines in the panel displayed strikingly different expression patterns compared to tissue samples (Gillet et al. 2011). A subset of genes was upregulated across the entire cell line panel, regardless of tissue type; this subset was associated with response to environmental stress. Genes may also be upregulated in association with EMT (see Section 26.2.5) (Klijn et al. 2015).

(5) **Common phases of cell cultivation.** Finite cultures undergo a characteristic change in phenotype when senescence or crisis occurs, regardless of the cell type (see Figure 3.1b,c). The transition to senescence is irreversible once the cell exits the cell cycle. Similarly, immortalization and transformation can have dramatic effects on the phenotype of a cell line that are likely to be irreversible (see Sections 3.5, 3.6).

3.4.2 Phenotype and Culture Conditions

It is essential to use clearly defined and standardized culture conditions when studying cell lines and their phenotypes. For example, the lung cell line A549 displays characteristics typical of alveolar type II cells at early passage but has more variable characteristics at later passage (Lieber et al. 1976; Cooper et al. 2016). A549 cells can be induced to differentiate by using Ham's F12 medium instead of Dulbecco's modified Eagle's medium (DMEM) and by maintaining cultures at high density for 25 days (see Figure 3.6) (Cooper et al. 2016). These simple changes are sufficient to induce key markers of differentiation and to increase the similarity of gene expression when compared to primary alveolar type II cells. A549 cells may also be induced to differentiate when grown in filter wells and co-cultured with lung fibroblasts (Speirs et al. 1991).

Key variables to consider when defining the culture conditions include (i) the substrate on which the cells are grown, e.g. the use of coatings to mimic the extracellular matrix (ECM) (see Section 8.3); (ii) the culture vessel used (see Section 8.5); (iii) the physicochemical environment, including the pH, atmospheric conditions, and temperature (see Section 9.2); (iv) the medium formulation (see Sections 9.7, 10.6.1); (v) the presence of hormones or other supplements to induce differentiation (see Section 26.3); and (vi) the geometry of the culture (see Section 27.3). The vital missing factor in two-dimensional (2D) geometry is cell–cell interaction and the reciprocal signaling capacity that it entails, both directly and in the creation of the correct matrix for a specific microenvironmental niche (see Sections 2.1, 2.2). 3D culture allows cells to associate in dense groups and sets up the correct geometry to create diffusion gradients, where a factor produced by the colony will reach its highest concentration in the center and will be lost from the edge by diffusion. Conversely, an exogenous factor will have the

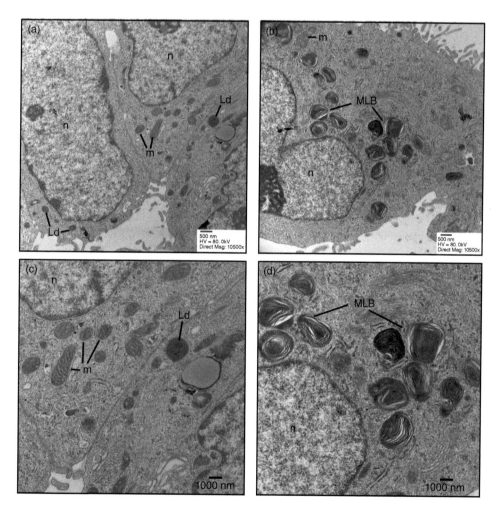

Fig. 3.6. Induction of differentiation in A549 cells. Electron micrographs of A549 cells, grown in Ham's F12 medium for 11 days (a, c) or 21 days (b, d). At day 11, lipid droplets can be seen (Ld); at day 21, multilamellar bodies (MLB) have developed that resemble those seen in primary alveolar epithelial type II cells. Nuclei (n) and mitochondria (m) are also visible. *Source*: Cooper et al. (2016), https://doi.org/10.1371/journal.pone.0164438, licensed under CC BY 4.0.

highest concentration at the edge and lowest in the center. Gradients, long shown to be essential for morphogenesis, can be established simply by altering the geometry of the culture (Warmflash et al. 2014). All of these variables, if understood, may be used to create the correct environment for induction of differentiation (see Section 26.3).

The capacity to express differentiated markers under the influence of inducing conditions may mean that the cells being cultured are mature and only require induction to continue synthesizing specialized proteins. Alternatively, it is possible that the culture contains a high proportion of precursor cells that are capable of proliferation but retain a commitment to that lineage, which becomes evident when the correct inducing conditions are applied (see Sections 2.4.3, 2.4.4). If cell culture represents an equilibrium between stem cells, progenitor cells, and mature differentiated cells, its equilibrium would shift according to the environmental conditions. Routine serial passage at relatively low cell densities would promote cell proliferation and constrain differentiation, whereas high cell densities, low serum, and the appropriate hormones or cytokines would promote differentiation and inhibit cell proliferation (see Figure 2.7).

3.5 SENESCENCE AND IMMORTALIZATION

Minireview M3.1. **Minireview: Senescence and Immortalization** Contributed by Roger R. Reddel, Cancer Research Unit, Children's Medical Research Institute, Faculty of Medicine and Health, The University of Sydney, Westmead, NSW 2145, Australia.

Human somatic cells divide only a finite number of times before entering a state referred to as senescence, which is characterized by permanent exit from the cell division cycle and morphological changes (see Figure 3.7) (Campisi 2013). In contrast, long-term survival of a species depends on its germ line being functionally immortal.

Cellular senescence occurs during normal embryonic development, and during aging. It also occurs in response to various environmental and therapeutic agents such as cytotoxic cancer chemotherapy. The motivation for most studies of senescence to date has been to gain insights into aging or into cancer treatment (Nardella et al. 2011) because (i) almost all human cancers contain cell populations which have become immortalized, so understanding how this occurs may identify new targets for development of therapies (Reddel 2000); and (ii) a substantial proportion of biomedical research depends on the availability of *in vitro* cell culture models, and it is convenient, and sometimes essential, to use cultures that do not eventually become senescent.

Mechanisms of Senescence and Immortalization

The proliferative capacity of normal somatic cells is limited by shortening of their chromosome ends (i.e. the telomeres) which occurs with each cell division cycle (see Figure 3.8). When their telomeres have shortened sufficiently, cells undergo replicative senescence (Allsopp et al. 1992). Other stimuli, including specific chemical compounds and excessive signaling from oncogenes, can also cause senescence.

Cells can bypass senescence if the p53 and pRB/p16^{INK4a} tumor suppressor pathways are inactivated, and the cells will continue dividing until their telomeres become so short that they lose their capacity to prevent end-to-end fusion of chromosomes, resulting in genomic instability and cell death (i.e. culture crisis). A small minority of the cells undergo spontaneous genetic changes that activate a telomere lengthening mechanism (TLM) and immortalization. The two known TLMs are telomerase activation and alternative lengthening of telomeres (ALT). Telomerase is

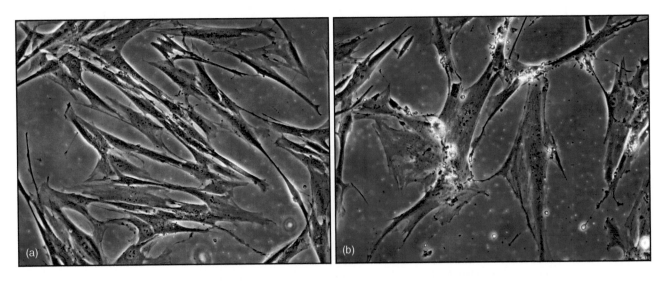

Fig. 3.7. Morphological changes in senescence. Young (a) and senescent (b) fibroblasts. Phase contrast, 20× objective. *Source*: Courtesy of Wei-Qin Jiang and Roger Reddel, Cancer Research Unit, CMRI.

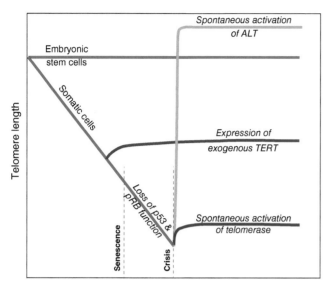

Fig. 3.8. Telomere length. The chart indicates telomere length as a function of proliferation (shown as population doublings) in embryonic stem cells (immortal), normal somatic cells (mortal) which become senescent, somatic cells with loss of p53 and pRB tumor suppressor function, which eventually enter culture crisis, and cells immortalized by spontaneous activation of telomerase or ALT, or by activation of telomerase by expression of exogenous TERT. *Source*: courtesy of Roger Reddel, Cancer Research Unit, CMRI.

a reverse transcriptase enzyme complex which synthesizes telomere sequences *de novo*, and contains several subunits including a reverse transcriptase protein (hTERT) and an RNA (hTR) with a template region for reverse transcription of the 5′-TTAGGG-3′ telomere repeat sequence (Blackburn 2005). ALT is a recombination-mediated DNA replication process in which telomeres are lengthened by using other telomeric DNA as a copy template for synthesis of new telomeric DNA (Cesare and Reddel 2010).

Methods for Obtaining Senescent Cultures

In principle, the simplest procedure for obtaining senescent cultures is to passage cells of a normal finite cell line repeatedly until they become senescent. In practice, however, this may be difficult because of the time required and because the cells may not all reach their proliferative limit simultaneously. It may therefore be more useful to induce senescence rapidly and synchronously by overexpressing an oncogene such as mutant HRAS (Serrano et al. 1997).

Methods for Obtaining Immortalized Cell Cultures

(1) *Cancer cell lines.* It is often possible to establish an immortalized culture from cancer specimens, although it is much easier to do this for some types of cancer (e.g. small cell

carcinoma of the lung) than others (e.g. prostate cancer). Presumably, available cell culture conditions suit some cancer cell types much better than others. Many cancer cell lines can be obtained from international cell repositories (see Table 15.3).

(2) *Embryonic stem cells.* Human embryonic stem cells (hESC) are pluripotent cells derived from the inner cell mass of five- to seven-day blastocysts, which have the ability to form a continuous cell line (Hasegawa et al. 2010). In some jurisdictions a license is required to create an hESC line (see Section 6.4.2). Established hESC lines may be obtained from repositories (see Table 15.3).

(3) *Induced pluripotent stem cells.* Induced pluripotent stem cells (iPSC) can be obtained from adult human cells by manipulating expression of various genes (e.g. *POU5F1*, *GLIS1*, *SOX2*, and *KLF4*) or addition of chemicals (e.g. ALK5 and MEK inhibitors) to reprogram them to an ESC-like state (Hirschi et al. 2014). iPSC lines can differentiate into many normal cell types but will proliferate indefinitely if not allowed to differentiate.

(4) *Viral oncogenes.* Classic techniques for immortalizing cells use viral oncogenes. These methods are still useful for obtaining cell lines from specific postnatal cell types and for studying the genetic events involved in the immortalization process. Infection of blood samples with Epstein-Barr virus (EBV) results in the generation of LCLs, which are used as an essentially limitless source of DNA from individuals for genetic studies. Telomerase is activated by an unknown mechanism 45–160 population doublings after EBV infection (Counter et al. 1994). In addition, immortalization can be facilitated by oncogenes of the SV40 early region and the E6 and E7 genes of oncogenic HPVs (especially HPV-16). These oncoproteins cause bypass of senescence by abrogating functions of p53 and pRB, allowing the cells to continue proliferating until culture crisis. Rare cells emerging from crisis have spontaneously activated a TLM and are immortalized. SV40-transformed fibroblasts activate either ALT or telomerase, whereas SV40-transformed epithelial cells and HPV-transformed cells are more likely to activate telomerase.

(5) *TERT transduction.* Telomerase activity can be induced in human cells by expression constructs encoding human telomerase reverse transcriptase (hTERT), the catalytic subunit of telomerase (Bodnar et al. 1998). Cells immortalized in this way retain more normal characteristics than cells immortalized following expression of viral oncoproteins (Jiang et al. 1999; Toouli et al. 2002).

(6) *Conditional reprogramming.* Co-culture of human normal or tumor epithelial cells with irradiated mouse J2 fibroblast feeder cells in the presence of an Rho kinase (ROCK) inhibitor (Y-27632) allows the cells to be propagated indefinitely *in vitro*, while retaining the ability to undergo differentiation when the ROCK inhibitor is withdrawn (see Section 22.3.4) (Liu et al. 2017). Unlike

ESC and iPSC, conditionally reprogrammed cells do not express high levels of Sox2, Oct4, Nanog, or Klf4 proteins (Suprynowicz et al. 2012). The conditionally reprogrammed cells have telomerase activity, and if derived from normal cells retain normal karyotypes (Chapman et al. 2010). The growth of fibroblasts and other stromal cells tends to be suppressed under these conditions, although some modifications to the protocol may ameliorate this (Liu et al. 2017).

Conclusions

Further investigation of the molecular biology of senescence and immortalization is expected to aid our understanding of aging and cancer. Advances in the technology of iPSC production and induction of differentiation, or the use of conditionally reprogrammed cells, will make it possible to obtain sufficient quantities of many essentially normal cell types, and this is likely to displace the use of cancer cell lines and virally immortalized cell lines as surrogates for *in vitro* studies of normal cells.

3.5.1 Intrinsic Control of Senescence

Studies of human fibroblasts have demonstrated that cells must overcome senescence and crisis if they are to become immortal (see Figure 3.1). Three independent mechanisms must be altered in normal human fibroblasts: the p53 pathway, pRB/p16^{INK4a} pathway, and a telomere shortening pathway (Wright and Shay 2000). SV40 TAg is often used to induce immortalization and is known to bind p53 and pRB (see Section 22.3.1). However, the low incidence of cells emerging from crisis indicates that additional genetic changes are required (Huschtscha and Holliday 1983). The finite lifespan of cells in culture is regulated by a group of 10 or more dominantly acting senescence genes, the products of which negatively regulate cell cycle progression (Goldstein et al. 1989; Sasaki et al. 1994). It is likely that one or more of these genes negatively regulates the expression of telomerase. Telomerase is expressed in germ cells and has moderate activity in stem cells, but is repressed in most normal human adult somatic cells (Forsyth et al. 2002; Campisi 2013). Deletions and/or mutations within the hTERT promoter or in other senescence-related genes, or overexpression or mutation of one or more oncogenes that override the action of the senescence genes may allow cells to escape from the negative control of the cell cycle and re-express telomerase.

3.5.2 Extrinsic Control of Senescence

Before the physiological basis of senescence and immortalization was understood, it was widely assumed that cells ceased to proliferate because of inadequate culture conditions (Shay and Wright 2000). We now know that senescence is triggered by a biological event; shortening of the telomeres at the chromosomal ends essentially acts as a "counting mechanism" to limit the cell's replicative capacity (Hayflick 1998). However, there are many disparities between *in vitro* senescence and the replication of cells in the body (Rubin 2002). It is now clear that cultures can undergo *in vitro* senescence in response to a number of extrinsic factors, depending on the culture conditions used (Wright and Shay 2002). Causes of *in vitro* senescence include (i) telomere shortening (see Minireview M3.1); (ii) genomic damage; (iii) epigenomic damage; (iv) excessive mitogenic stimulation; and (v) activation of tumor suppressor pathways such as p53/p21^{CIP1} and pRB/p16^{INK4a} (Campisi and d'Adda di Fagagna 2007; Campisi 2013). Many of these events occur as a result of culture stress. Thus, growing cells under conditions that reduce stress-associated signaling can extend their lifespan *in vitro*. This can be done by using hypoxic culture conditions (see Minireview M9.1) (Wright and Shay 2002). Growing human diploid fibroblasts in reduced oxygen (1–12%) extends their lifespan beyond the Hayflick limit, and reintroduction of atmospheric (21%) oxygen rapidly induces cellular senescence (Packer and Fuehr 1977; Saito et al. 1995).

3.6 TRANSFORMATION

The meaning of the term "transformation" has changed over time. In microbiology, where the term was first used in this context, it refers to the introduction and stable genomic integration of foreign DNA into the cell, resulting in genetic modification (Schaeffer 1990). Other terms are now used for this process in animal cells, such as "gene delivery", "transfection,", or "transduction" (see Section 22.1). Transformation in animal cells has been broadly defined as a heritable change that results in acquisition of an altered phenotype (Schaeffer 1990). Different forms of transformation have been described based on their characteristics, e.g. "neoplastic transformation" refers to the ability to form malignant tumors in animals (Schaeffer 1990). However, these distinctions are not always preserved and such terms are often vague and poorly defined. For this book, "transformation" refers to an *in vitro* phenotype with specific characteristics (see Section 3.6.1) that is associated with *in vivo* tumor formation and malignant behavior. Transformation may be induced by an oncogenic virus, various oncogenes, or exposure to ionizing radiation or chemical carcinogens. It may also occur spontaneously.

3.6.1 Characteristics of Transformation

In vitro transformation results in a phenotype that is common to many cell types, regardless of their origins. Transformed cells have four predominant characteristics: (i) genetic instability (see Section 3.3); (ii) immortalization (see Section 3.5); (iii) aberrant growth control (see Section 3.6.2); and (iv) tumorigenicity (see Section 3.6.3). The term transformation implies that all four characteristics are present. Additional changes may be observed that contribute to the transformed phenotype, such as disruption in cell polarity (see Table 3.2).

TABLE 3.2. Properties of transformed cells.

Property	Assay	Resources for more information
Genetic instability		
Heteroploidy	Cytogenetic analysis; DNA content by flow cytometry	Protocol P17.2; Figure 3.4; MacLeod et al. (2011)
Aneuploidy	Cytogenetic analysis	Protocol P17.2; Figure 3.5; MacLeod et al. (2011)
Chromosomal structural rearrangements	G-banding; fluorescence *in situ* hybridization (FISH)	MacLeod et al. (2011)
High spontaneous mutation rate	Genomic analysis	Lewis et al. (2013); Landa et al. (2019)
Overexpressed or mutated oncogenes	Immunostaining; sequencing; expression analysis, e.g. microarrays	Supp. S18.3
Deleted or mutated suppressor genes	Immunostaining; sequencing; expression analysis, e.g. microarrays	Supp. S18.3
Immortalization		
Continued proliferation	Growth beyond 100 PDL	Minireview M3.1; Figure 3.1
Aberrant growth control		
Loss of contact inhibition	Microscopic observation for altered behavior at high density	Figure 3.9a, b
Growth on confluent monolayers of homologous cells	Microscopic observation for transformation foci	Figure 3.9c
Reduced density limitation of growth	Growth curve for saturation density; measurement of growth fraction at saturation density	Section 19.3.3; Figures 3.9d, 19.7; Khan et al. (1991)
Reduced population doubling time (PDT)	Growth curve for PDT	Section 19.3.3; Figure 19.7
Increased cloning efficiency	Dilution cloning in limiting serum	Protocols P19.4, P20.1
Low serum requirement	Dilution cloning in limiting serum	Protocols P19.4, P20.1
Production of autocrine growth factors	Immunostaining; dilution cloning in limiting serum with blocking antibody or peptide inhibitor	Protocols P19.4, P20.1; Supp. S18.3
Anchorage independence	Cloning in soft agar or Methocel	Protocols P20.3, P20.4
Structural changes		
Disruption in cell polarity	Immunostaining; polarized transport in filter wells	Protocol P27.2; Supp. S18.3
Modified actin cytoskeleton	Immunostaining	Supp. S18.3; Figure 3.10g–i
Loss of cell surface-associated fibronectin	Immunostaining	Supp. S18.3
Altered expression of cell adhesion molecules (CAMs), cadherins, integrins	Immunostaining; expression profiling	Supp. S18.3
Modified extracellular matrix (ECM)	Immunostaining; expression profiling	Supp. S18.3
Tumorigenicity and malignancy		
Tumor development	Xenograft in suitable mouse model	Giovanella et al. (1972); Sharkey and Fogh (1984)
Invasiveness	CAM assay; 3D invasion assay, e.g. filter well invasion assay	Section 27.5.2; Figure 3.12; Brunton et al. (1997); Deryugina and Quigley (2008)
Angiogenesis	CAM assay; 3D angiogenesis assay, e.g. tube formation assay	Supp. S27.2; Figure 3.12e
Enhanced protease secretion, e.g. plasminogen activator	Plasminogen assay; expression analysis	Frame et al. (1984); Kaphle et al. (2019)
Expression of angiogenic factors, e.g. VEGF	ELISA; expression analysis, e.g. quantitative RT-PCR	Buchler et al. (2004); Beckermann et al. (2008)

The mouse NIH 3T3 cell line is often used as a model for *in vitro* transformation (Jainchill et al. 1969; Rubin 2017). If NIH 3T3 cultures are maintained at a low cell density and are not allowed to remain at confluence for any length of time, they display contact inhibition, which can be defined as the inhibition of movement and proliferation when cells are in contact with one another (Ribatti 2017). When confluence is reached, cells that remain sensitive to contact inhibition and density limitation will stop dividing (see Section 19.3.1, "Plateau Phase"). However, if cells remain at high density for extended periods, transformed foci may spontaneously appear that exhibit aberrant growth control (see Figures 3.9c, 18.1v–x). These foci begin to pile up and will ultimately overgrow the culture. The fact that these cells are not apparent at low densities or when confluence is first reached suggests that they arise *de novo*; studies of NIH 3T3 cells have demonstrated progressive loss of contact inhibition, resulting in areas of hyperplasia that were more permissive for transformation (Rubin 2017).

Although these studies are relevant to tumor development (Rubin 2017), it could be argued that they are based on a culture artifact. NIH 3T3 was selected specifically for its high levels of contact inhibition; this cell type is often found to be tumorigenic and loses its value as a feeder layer over time (Jainchill et al. 1969; Edington et al. 2004). More recent studies have explored *in vitro* transformation using cells from older mice of a different strain (Padilla-Nash et al. 2012, 2013). As with NIH 3T3 cells, transformation was a multistep process (Padilla-Nash et al. 2012). Three distinct stages of transformation occurred, in which cells were (i) pre-immortal, with retention of contact inhibition; (ii) immortal, with increased proliferation; and (iii) transformed, with loss of contact inhibition and anchorage independence (see Figure 3.10). The length of time at each stage varied with the tissue type. These cultures displayed progressive genetic instability and changes in gene expression that were remarkably similar to those found in human tumorigenesis, demonstrating that *in vitro* transformation can be used as a model for cancer development (Padilla-Nash et al. 2013).

Immortalization *per se* does not imply the development of aberrant growth control and tumor development. A number of immortal cell lines retain contact inhibition, are anchorage dependent, and are not tumorigenic. However, some aspects of growth control are likely to be abnormal and genomic instability is likely to increase. Immortalized cell lines often lose the ability to differentiate, but there are reports of telomerase-induced immortalization of keratinocytes and skeletal muscle satellite cells without abrogation of p53 activity; these models retain the ability to differentiate (Dickson et al. 2000; Wootton et al. 2003). A comparison of immortalized human mammary cells with SV40 TAg or hTERT (see Minireview M3.1) suggested that hTERT immortalization gave a more stable genotype and better differentiation (Toouli et al. 2002).

Genetic instability appears to both promote and depend on *in vitro* transformation (Fearon and Vogelstein 1990; Pinto and Clevers 2005). The ability of a cell line to grow continuously probably reflects its capacity for genetic variation, allowing subsequent selection. Genetic variation often involves the deletion or mutation of the *TP53* gene, which would normally arrest cell cycle progression if DNA were to become mutated, and activation of a telomere maintenance mechanism (see Minireview M3.1). Many continuous cell lines develop a subtetraploid chromosome number and marked genomic variation, as discussed earlier in this chapter (see Section 3.3).

3.6.2 Aberrant Growth Control

Changes in growth with transformation. Aberrant growth control *in vitro* is closely linked to tumor development *in vivo* (Abercrombie 1979; Ribatti 2017). This was originally discovered during study of contact-insensitive 3T3 cultures, which were likely to produce tumors in living mice, while contact-sensitive cultures typically did not (Aaronson and Todaro 1968). Various changes in growth have been documented and used as the basis for cell transformation assays (see Table 3.2), including:

(7) **Loss of contact inhibition.** This is particularly obvious in MEFs and some other rodent cultures (Creton et al. 2012). Loss of contact inhibition is considered an essential feature of *in vitro* transformation and its link to tumor development has been thoroughly explored using the 3T3 cell lines (Rubin 2017). Loss of contact inhibition is detected by morphological observation, looking for a disoriented monolayer of cells, or by the development of transformed foci within the regular pattern of normal surrounding cells (see Figure 3.9).

(8) **Increased saturation density.** Human glioma cells show disorganized growth and reduced density limitation of proliferation, resulting in a higher saturation density at the plateau phase of the growth curve compared to normal glial cells (see Figure 3.11; see also Figure 19.10) (Westermark 1973; Freshney et al. 1980). Similar results have been seen for other cell types (Eagle et al. 1970; Khan et al. 1991).

(9) **Increased cloning efficiency.** Cloning efficiency refers to the percentage of cells that form clonal populations at low density (see Section 20.1). Cloning efficiency is usually poor for primary cultures and finite cell lines but increases following *in vitro* transformation (Eagle et al. 1970).

(10) **Reduced serum dependence.** Transformed cells have a lower serum dependence when compared to their normal counterparts (Eagle et al. 1970). Cells typically grow well at lower serum concentrations (e.g. 0.1–0.5%) and are likely to achieve higher population densities in higher serum concentrations (e.g. 20%).

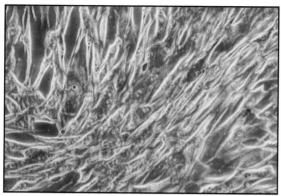

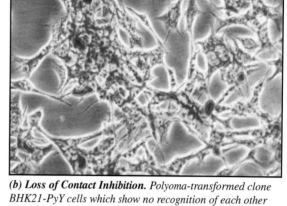

(a) Contact Inhibition. Late log-phase cultures of BHK21-C13 Cells tend to assume a parallel orientation and will not overgrow each other. Phase contrast; 10x objective. Source: R. Ian Freshney.

(b) Loss of Contact Inhibition. Polyoma-transformed clone BHK21-PyY cells which show no recognition of each other and grow randomly over each other. Phase contrast; 10x objective. Source: R. Ian Freshney.

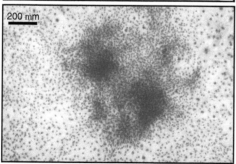

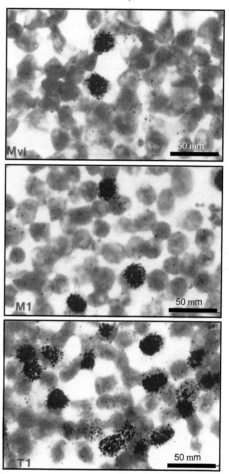

(c) Focus Formation in 3T3 Cells. A monolayer of 3T3-Swiss albino cells, left at confluence for 3 weeks; foci of transformed cells can be seen escaping contact inhibition. Top, whole flask; bottom, 4x objective. Scale bars as indicated. Giemsa stained. Source: R. Ian Freshney.

(d) [^{3}H]thymidine Incorporation in Mv1Lu mink lung cells. Mink lung cells, Mv1Lu, were labeled with [3H]thymidine for 1 h, fixed, coated with autoradiographic emulsion, and stained with Giemsa. Mv1 (top) is the control cell line, Mv1Lu; M1 (middle) is Mv1Lu transfected with the myc oncogene; and T1 (bottom) is Mv1Lu transfected with mutant ras. T1 cells were tumorigenic and had a statistically significant (p<0.001) increase in labeling index compared to Mv1Lu. Source: courtesy of M. Z. Khan.

Fig. 3.9. Changes in morphology with transformation.

Pre-immortal   Immortal   Transformed

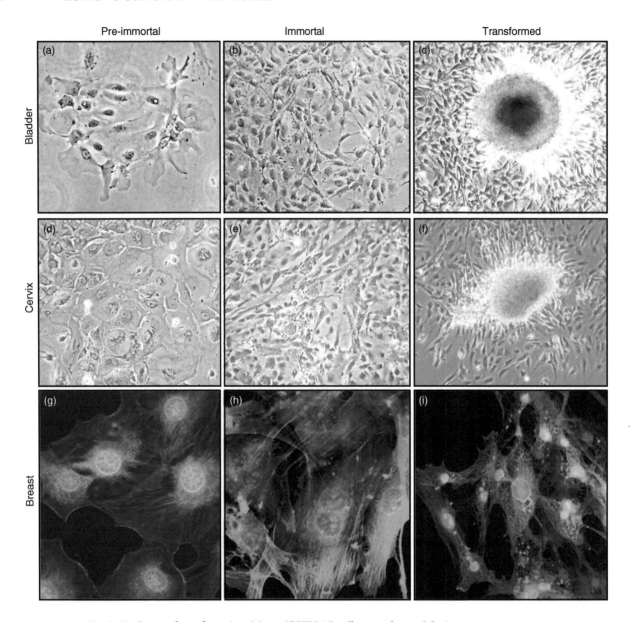

Fig. 3.10. Stages of transformation. Mouse (C57BL/6) cells were observed during spontaneous transformation. Cultures came from bladder (a–c), cervical (d–f), and mammary (g–i) epithelium. Sequential stages were observed and designated as pre-immortal (a, d, g), immortal (b, e, h), and transformed (c, f, i). In addition to phase contrast images (a–f), mammary epithelial cells were stained for actin and nuclei were counterstained with 4′,6-diamidino-2-phenylindole dihydrochloride (DAPI) (g–i). Actin fibers were more disorganized at the immortal stage, compared to pre-immortal cells, and became concentrated into dense figures at the transformed stage. *Source*: Padilla-Nash, H. M., Hathcock, K., McNeil, N. E., et al. (2012). Spontaneous transformation of murine epithelial cells requires the early acquisition of specific chromosomal aneuploidies and genomic imbalances. Genes Chromosomes Cancer 51 (4):353–74. doi: 10.1002/gcc.21921, reproduced with permission of John Wiley and Sons.

(11) ***Anchorage-independent growth.*** Transformed cells develop a greater ability to grow independently from the culture substrate, either in stirred suspension culture or suspended in semisolid media (Macpherson and Montagnier 1964). Cloning in soft agar or Methocel continues to be used as an assay for transformation and can be used to assess clonogenicity of tumor stem cells, provided the cell type clones readily in suspension (see Section 20.3). Anchorage independence is also associated with metastatic potential (Mori et al. 2009).

Mechanisms of aberrant growth control. Why does aberrant growth control occur? Transformed cultures are

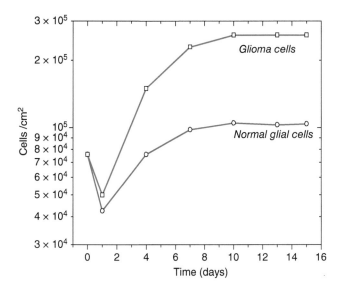

Fig. 3.11. Density limitation of cell proliferation. The difference in plateaus (saturation densities) attained by cultures from normal brain (circles, lower line) and a glioma (squares, top line). Cells were seeded onto 13-mm coverslips, and 48 hours later the coverslips were transferred to 90-mm Petri dishes containing 20 ml growth medium, to minimize exhaustion of the medium.

assumed to have acquired some degree of autonomous growth control by overexpression of normal or mutant oncogenes or by deletion or inactivation of normal or mutant suppressor genes. Contact inhibition is known to require several tumor suppressor genes, including *NF2* and *STK11* (Hanahan and Weinberg 2011). *NF2* encodes the Merlin protein, which acts to couple cell-surface adhesion molecules such as E-cadherin to the epidermal growth factor (EGF) receptor and other receptor tyrosine kinases (Curto and McClatchey 2008). *STK11* encodes an epithelial polarity protein that organizes epithelial structure and has a role in the maintenance of tissue integrity (Hanahan and Weinberg 2011).

Transformed cells may lack specific cell adhesion molecules such as liver cell adhesion molecule (LCAM), which can be transfected back into the cell to regenerate the non-transformed phenotype (Mege et al. 1988). Cells may overexpress other molecules such as neural cell adhesion molecule (NCAM), where the extracellular domain is subject to alternative splicing, e.g. in small cell lung carcinoma (Patel et al. 1989; Rygaard et al. 1992). The expression of various integrins and their degree of phosphorylation can also change, resulting in altered regulation of gene transcription, cytoskeletal interactions, and substrate adhesion; these changes result in an altered relationship between cell spreading and cell proliferation (see Section 2.2.1).

Transformed cells have a lower serum dependence due, in part, to their secretion of growth factors and other mitogens. These factors can be collectively described here as autocrine growth factors. Implicit in this definition is that (i) the cell

produces the factor; (ii) the cell has receptors for the factor; and (iii) the cell responds to the factor by entering mitosis. Some autocrine growth factors may have an apparent transforming activity on normal cells (Richmond et al. 1985). For example, transforming growth factor-α (TGF-α) may cause non-transformed cells to adopt a transformed phenotype and grow in suspension; unlike true transformation, the effect is reversible when the factor is removed (Todaro and De Larco 1978). Co-expression of TGF-α with SV40 TAg or MYC has a synergistic effect on tumor development, suggesting that the various factors act together to provide a more permissive environment for transformation (Sandgren et al. 1993).

Anchorage-independent growth is likely to occur through deregulation of anoikis, which is a particular form of apoptosis that occurs due to disruption of interactions between the normal cell and the ECM (Frisch and Francis 1994; Taddei et al. 2012). The ECM normally acts to suppress anoikis through integrin-mediated signal transduction. Anoikis is an important mechanism that prevents normal epithelial cells from spreading to other locations; mesenchymal cells are usually more resistant to this mechanism (Taddei et al. 2012). Transformed cells develop a resistance to anoikis through a variety of mechanisms, including changes in integrin expression and secretion of collagen, which essentially provides a supportive ECM at the site of metastasis (Burnier et al. 2011). Thus, assays for anchorage-independent growth may be useful for the study of metastatic behavior (Mori et al. 2009).

3.6.3 Tumorigenicity and Malignancy

Transformation is a multistep process that often culminates in the production of cells that demonstrate malignant behavior, both *in vitro* and when returned to a living organism. Malignant behavior is traditionally studied by injecting cell lines into the thymus-deficient "nude" mouse (or a similar rodent model) to generate a xenograft (Giovanella et al. 1972). Early studies demonstrated that human cancer cells and continuous cell lines produced tumors in these mice, with a histological appearance that was usually consistent with the original tumor type, whereas normal cells did not (Sharkey and Fogh 1984).

Although mouse xenograft studies are extremely useful, their relationship with human cancer is a complex one. Many tumor biopsies and cell lines fail to produce tumors as xenografts; those that do take frequently fail to metastasize, although they may be invasive locally (Sharkey and Fogh 1984). Patient-derived xenografts undergo clonal evolution in the mouse that is different to human tumor evolution and display genomic instability that is comparable to that of cell lines (Ben-David et al. 2017). The mouse model may have other unintended consequences. Spontaneous cell–cell fusion can occur between human tumor cells and mouse stromal cells in xenograft models, where it contributes to tumor progression (Jacobsen et al. 2006).

The relationship between cancer and *in vitro* transformation is also complex. Tumor cell lines tend to have a higher

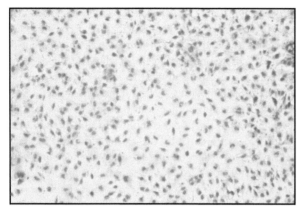

(a) Contact inhibited confluent monolayer of normal human fetal instestinal cell line FHs 74 Int. Giemsa stained; 10x objective. Source: R. Ian Freshney.

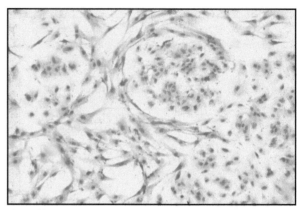

(b) ***Infiltration of Normal Monolayer.*** *Human glioma cell line WLY infiltrating FHs 74 Int (shown in [a]). Giemsa stained; 10x objective. Source: R. Ian Freshney.*

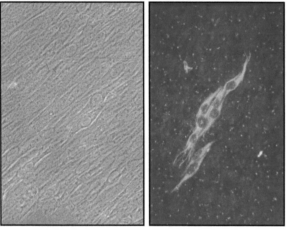

(c) ***Non-Small Cell Carcinoma Cell Line Infiltrating Normal Fibroblasts.*** *Left panel: phase contrast. Right panel: indirect immunofluorescence for cytokeratin revealing the tumor cells migrating in parallel with the fibroblasts. 40x objective. Source: R. Ian Freshney.*

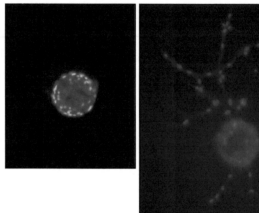

(d) ***In vitro Angiogenesis Assay (see Supp. S27.2).*** *Left panel: control showing endothelial cells forming a monolayer on the surface of Cytodex™ 3 microcarrier beads. Right panel: tube formation after endothelial cells are treated with 10 ng/mL FGF. DAPI stained. Source: Courtesy of V. Subramanian.*

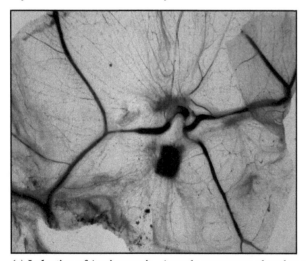

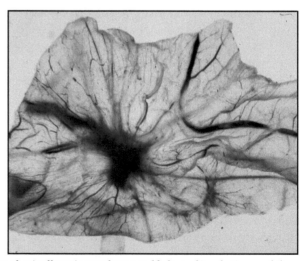

(e) ***Induction of Angiogenesis.*** *A crude extract was placed on the chorioallantoic membrane at 10 days of incubation, and the membrane was removed 2 weeks later. Left panel: normal glial cell extract. Right panel: crude extract of Walker 256 carcinoma cells. Source: Courtesy of Margaret Frame.*

Fig. 3.12. In vitro assays for invasiveness and angiogenesis.

take rate *in vivo* compared to surgical specimens (Sharkey and Fogh 1984). Such cell lines (which are presumably already transformed) can undergo further transformation *in vitro* with an increased growth rate, reduced anchorage dependence, more pronounced aneuploidy, and immortalization. A series of steps is required for malignant transformation, but these steps need not be coordinated or interdependent and they may not necessarily be individually tumorigenic. Cells will undergo selective pressure based on their environment, which means that *in vitro* events may not necessarily follow the same sequence as tumor progression in the body.

Various *in vitro* assays have been developed to explore specific properties associated with malignant neoplasia. Some of these assays provide indicators that are more readily quantified than *in vivo* assays. Neoplastic properties that can be used for *in vitro* assays include (see also Table 3.2):

(1) **Invasiveness.** Tumor cells develop the ability to penetrate the basement membrane by forming invadopodia, which also enable cell migration and degradation of the ECM (Eddy et al. 2017). Invasiveness can be assessed using the chick embryo chorioallantoic membrane (CAM) assay (see Figure 3.12e). CAM assays are also useful to study metastasis and angiogenesis; the chick embryo does not develop a fully functioning lymphoid system until late in embryogenesis, making it naturally immunodeficient at the time the assay is performed (Deryugina and Quigley 2008). However, the push to reduce animal usage and improve reproducibility has led to an increasing use of 3D *in vitro* models, such as filter well invasion assays (see Figure 3.13) (Brunton et al. 1997).

(2) **Expression of proteolytic enzymes.** Transformed cells often demonstrate an increase in proteolytic enzymes. The plasminogen activation system appears to be active in virtually all cancer types, mediated by tissue plasminogen activator (tPA) or urokinase plasminogen activator (uPA), and can be measured *in vitro* (Dano et al. 2005). For example, plasminogen activator levels are higher in glioma cultures compared to normal brain cells and are associated with increased migration and invasion of collagen hydrogels (Frame et al. 1984; Kaphle et al. 2019). Because proteolytic activity may be associated with the cell surface of many normal cells and is absent on some tumor cells, an equivalent normal cell must be used as a control.

(3) **Angiogenesis.** Tumor cells are able to stimulate the formation of new blood vessels, resulting in "sprouting" of new vessels from relatively early in cancer development (Hanahan and Weinberg 2011). Acquisition of the angiogenic phenotype is associated with genetic instability and the subsequent activation of mutant oncogenes and loss of tumor suppressor genes (Rak et al. 2002). Blood vessel formation can be assessed using the CAM assay, where a fluorescent-tagged lectin can be injected to visualize the vascular network (Deryugina and Quigley 2008). A variety of 3D techniques are also used to study angiogenic sprouting, including migration of endothelial cells from microcarrier beads into a suitable matrix (see Figure 3.12d) and formation of microvessels in a micromolded hydrogel network (Rodenhizer et al. 2018).

(4) **Expression of angiogenic factors.** Tumor cells release a number of mitogens that control or contribute to angiogenesis, particularly vascular endothelial growth factor (VEGF) and members of the fibroblast growth factor (FGF) family (see Table 10.1) (Hanahan and Weinberg 2011). Some of these factors can be assessed *in vitro*, e.g. VEGF can be measured in cell lysates and culture supernatant (Buchler et al. 2004; Beckermann et al. 2008).

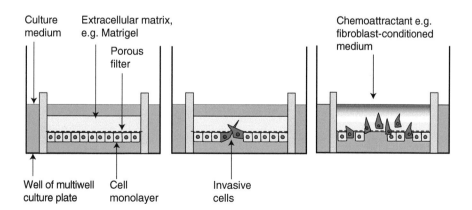

Fig. 3.13. Filter well invasion. Cells plated on the underside of the filter migrate through the filter into growth factor-depleted Matrigel® in the well of the filter insert, encouraged by the addition of a chemoattractant, such as fibroblast-conditioned medium, to the upper side of the Matrigel. This diagram is based on a published method (Brunton et al. 1997). *Source*: based on Brunton, V. G., Ozanne, B. W., Paraskeva, C., & Frame, M. C. (1997). A role for epidermal growth factor receptor, c-Src and focal adhesion kinase in an in vitro model for the progression of colon cancer. Oncogene, 14(3), 283–293.

3.7 CONCLUSIONS: ORIGIN AND EVOLUTION

At the beginning of this chapter, the question was asked: if cell lines are used for extended periods of time, does this affect their validity as research models? The answer to this question is clearly "yes.". Tumor cells display genomic variation, which persists despite the selective pressures that occur in the various phases of cell cultivation. Cellular phenotypes evolve over time due to the cell's ability to adapt to its environment. Culture artifacts occur such as *in vitro* transformation, resulting in aberrant growth control and malignant behavior. Scientists may overlook (or even welcome) these changes because transformed cells grow more rapidly and are therefore easier to work with.

Despite these caveats, cell lines continue to be important for scientific discovery. For example, leukemia and lymphoma cell lines were instrumental in developing imatinib (Gleevec), which targets the Philadelphia chromosome (see Section 3.3.1) (MacLeod et al. 2008). Even culture artifacts can be useful; studies of *in vitro* transformation have played an important part in understanding cancer dynamics (Rubin 2017). Rather than simply concluding that cell lines are unreliable, it would be better to make culture handling more reliable and reproducible and culture conditions more physiological.

REFERENCES

Aaronson, S.A. and Todaro, G.J. (1968). Basis for the acquisition of malignant potential by mouse cells cultivated in vitro. *Science* 162 (3857): 1024–1026.

Abercrombie, M. (1979). Contact inhibition and malignancy. *Nature* 281 (5729): 259–262.

Adey, A., Burton, J.N., Kitzman, J.O. et al. (2013). The haplotype-resolved genome and epigenome of the aneuploid HeLa cancer cell line. *Nature* 500 (7461): 207–211. https://doi.org/10.1038/nature12064.

Allsopp, R.C., Vaziri, H., Patterson, C. et al. (1992). Telomere length predicts replicative capacity of human fibroblasts. *Proc. Natl Acad. Sci. U.S.A.* 89 (21): 10114–10118.

Barretina, J., Caponigro, G., Stransky, N. et al. (2012). The cancer cell line encyclopedia enables predictive modelling of anticancer drug sensitivity. *Nature* 483 (7391): 603–607. https://doi.org/10.1038/nature11003.

Beckermann, B.M., Kallifatidis, G., Groth, A. et al. (2008). VEGF expression by mesenchymal stem cells contributes to angiogenesis in pancreatic carcinoma. *Br. J. Cancer* 99 (4): 622–631. https://doi.org/10.1038/sj.bjc.6604508.

Ben-David, U., Arad, G., Weissbein, U. et al. (2014). Aneuploidy induces profound changes in gene expression, proliferation and tumorigenicity of human pluripotent stem cells. *Nat. Commun.* 5: 4825. https://doi.org/10.1038/ncomms5825.

Ben-David, U., Ha, G., Tseng, Y.Y. et al. (2017). Patient-derived xenografts undergo mouse-specific tumor evolution. *Nat. Genet.* 49 (11): 1567–1575. https://doi.org/10.1038/ng.3967.

Ben-David, U., Siranosian, B., Ha, G. et al. (2018). Genetic and transcriptional evolution alters cancer cell line drug response. *Nature* 560: 325–330. https://doi.org/10.1038/s41586-018-0409-3.

Ben-David, U., Beroukhim, R., and Golub, T.R. (2019). Genomic evolution of cancer models: perils and opportunities. *Nat. Rev. Cancer* 19 (2): 97–109. https://doi.org/10.1038/s41568-018-0095-3.

Biedler, J.L. (1975). Chromosome abnormalities in human tumor cells in culture. In: *Human Tumor Cells in Vitro* (ed. J. Fogh), 359–394. Boston, MA: Springer.

Blackburn, E.H. (2005). Telomeres and telomerase: their mechanisms of action and the effects of altering their functions. *FEBS Lett.* 579 (4): 859–862. https://doi.org/10.1016/j.febslet.2004.11.036.

Bodnar, A.G., Ouellette, M., Frolkis, M. et al. (1998). Extension of life-span by introduction of telomerase into normal human cells. *Science* 279 (5349): 349–352.

Bonetti, F., Colombari, R., Manfrin, E. et al. (1989). Breast carcinoma with positive results for melanoma marker (HMB-45). HMB-45 immunoreactivity in normal and neoplastic breast. *Am. J. Clin. Pathol.* 92 (4): 491–495.

Brunton, V.G., Ozanne, B.W., Paraskeva, C. et al. (1997). A role for epidermal growth factor receptor, c-Src and focal adhesion kinase in an in vitro model for the progression of colon cancer. *Oncogene* 14 (3): 283–293. https://doi.org/10.1038/sj.onc.1200827.

Buchler, P., Reber, H.A., Buchler, M.W. et al. (2004). Antiangiogenic activity of genistein in pancreatic carcinoma cells is mediated by the inhibition of hypoxia-inducible factor-1 and the down-regulation of VEGF gene expression. *Cancer* 100 (1): 201–210. https://doi.org/10.1002/cncr.11873.

Burnier, J.V., Wang, N., Michel, R.P. et al. (2011). Type IV collagen-initiated signals provide survival and growth cues required for liver metastasis. *Oncogene* 30 (35): 3766–3783. https://doi.org/10.1038/onc.2011.89.

Campisi, J. (2013). Aging, cellular senescence, and cancer. *Annu. Rev. Physiol.* 75: 685–705. https://doi.org/10.1146/annurev-physiol-030212-183653.

Campisi, J. and d'Adda di Fagagna, F. (2007). Cellular senescence: when bad things happen to good cells. *Nat. Rev. Mol. Cell. Biol.* 8 (9): 729–740. https://doi.org/10.1038/nrm2233.

Castor, C.W., Prince, R.K., and Dorstewitz, E.L. (1962). Characteristics of human "fibroblasts" cultivated in vitro from different anatomical sites. *Lab. Invest.* 11: 703–713.

Cesare, A.J. and Reddel, R.R. (2010). Alternative lengthening of telomeres: models, mechanisms and implications. *Nat. Rev. Genet.* 11 (5): 319–330. https://doi.org/10.1038/nrg2763.

Chapman, S., Liu, X., Meyers, C. et al. (2010). Human keratinocytes are efficiently immortalized by a Rho kinase inhibitor. *J. Clin. Invest.* 120 (7): 2619–2626. https://doi.org/10.1172/JCI42297.

Cooper, J.R., Abdullatif, M.B., Burnett, E.C. et al. (2016). Long term culture of the A549 cancer cell line promotes multilamellar body formation and differentiation towards an alveolar type II pneumocyte phenotype. *PLoS One* 11 (10): e0164438. https://doi.org/10.1371/journal.pone.0164438.

Counter, C.M., Botelho, F.M., Wang, P. et al. (1994). Stabilization of short telomeres and telomerase activity accompany immortalization of Epstein-Barr virus-transformed human B lymphocytes. *J. Virol.* 68 (5): 3410–3414.

Creton, S., Aardema, M.J., Carmichael, P.L. et al. (2012). Cell transformation assays for prediction of carcinogenic potential: state of the science and future research needs. *Mutagenesis* 27 (1): 93–101. https://doi.org/10.1093/mutage/ger053.

Curto, M. and McClatchey, A.I. (2008). Nf2/Merlin: a coordinator of receptor signalling and intercellular contact. *Br. J. Cancer.* 98 (2): 256–262. https://doi.org/10.1038/sj.bjc.6604002.

Dano, K., Behrendt, N., Hoyer-Hansen, G. et al. (2005). Plasminogen activation and cancer. *Thromb. Haemost.* 93 (4): 676–681. https://doi.org/10.1160/TH05-01-0054.

Deryugina, E.I. and Quigley, J.P. (2008). Chick embryo chorioallantoic membrane model systems to study and visualize human tumor cell metastasis. *Histochem. Cell Biol.* 130 (6): 1119–1130. https://doi.org/10.1007/s00418-008-0536-2.

Dickson, M.A., Hahn, W.C., Ino, Y. et al. (2000). Human keratinocytes that express hTERT and also bypass a p 16 (INK4a)-enforced mechanism that limits life span become immortal yet retain normal growth and differentiation characteristics. *Mol. Cell. Biol.* 20 (4): 1436–1447.

Domcke, S., Sinha, R., Levine, D.A. et al. (2013). Evaluating cell lines as tumour models by comparison of genomic profiles. *Nat. Commun.* 4: 2126. https://doi.org/10.1038/ncomms3126.

Dracopoli, N.C. and Fogh, J. (1983). Loss of heterozygosity in cultured human tumor cell lines. *J. Natl Cancer Inst.* 70 (1): 83–87.

Drexler, H.G. and MacLeod, R.A. (2010). History of leukemia-lymphoma cell lines. *Hum. Cell* 23 (3): 75–82. https://doi.org/10.1111/j.1749-0774.2010.00087.x.

Drexler, H.G., MacLeod, R.A., and Uphoff, C.C. (1999). Leukemia cell lines: in vitro models for the study of Philadelphia chromosome-positive leukemia. *Leuk. Res.* 23 (3): 207–215.

Eagle, H., Foley, G.E., Koprowski, H. et al. (1970). Growth characteristics of virus-transformed cells. Maximum population density, inhibition by normal cells, serum requirement, growth in soft agar, and xenogeneic transplantability. *J. Exp. Med.* 131 (4): 863–879.

Eddy, R.J., Weidmann, M.D., Sharma, V.P. et al. (2017). Tumor cell invadopodia: invasive protrusions that orchestrate metastasis. *Trends Cell Biol.* 27 (8): 595–607. https://doi.org/10.1016/j.tcb.2017.03.003.

Edington, K.G., Berry, I.J., O'Prey, M. et al. (2004). Multistage head and neck squamous cell carcinoma. In: *Culture of Human Tumor Cells* (eds. R. Pfragner and R.I. Freshney), 261–288. Hoboken, NJ: Wiley-Liss.

Fearon, E.R. and Vogelstein, B. (1990). A genetic model for colorectal tumorigenesis. *Cell* 61 (5): 759–767.

Fisch, G.S. (2017). Whither the genotype–phenotype relationship? An historical and methodological appraisal. *Am. J. Med. Genet. C Semin. Med. Genet.* 175 (3): 343–353. https://doi.org/10.1002/ajmg.c.31571.

Forsyth, N.R., Wright, W.E., and Shay, J.W. (2002). Telomerase and differentiation in multicellular organisms: turn it off, turn it on, and turn it off again. *Differentiation* 69 (4–5): 188–197. https://doi.org/10.1046/j.1432-0436.2002.690412.x.

Frame, M.C., Freshney, R.I., Vaughan, P.F. et al. (1984). Interrelationship between differentiation and malignancy-associated properties in glioma. *Br. J. Cancer* 49 (3): 269–280.

Freshney, R.I., Sherry, A., Hassanzadah, M. et al. (1980). Control of cell proliferation in human glioma by glucocorticoids. *Br. J. Cancer* 41 (6): 857–866.

Frisch, S.M. and Francis, H. (1994). Disruption of epithelial cell–matrix interactions induces apoptosis. *J. Cell. Biol.* 124 (4): 619–626.

Fusenig, N.E. and Boukamp, P. (1998). Multiple stages and genetic alterations in immortalization, malignant transformation, and tumor progression of human skin keratinocytes. *Mol. Carcinog.* 23 (3): 144–158.

Gazdar, A.F., Carney, D.N., Nau, M.M. et al. (1985). Characterization of variant subclasses of cell lines derived from small cell lung cancer having distinctive biochemical, morphological, and growth properties. *Cancer Res.* 45 (6): 2924–2930.

Gazdar, A.F., Gao, B., and Minna, J.D. (2010). Lung cancer cell lines: useless artifacts or invaluable tools for medical science? *Lung Cancer* 68 (3): 309–318. https://doi.org/10.1016/j.lungcan.2009.12.005.

Giard, D.J., Aaronson, S.A., Todaro, G.J. et al. (1973). In vitro cultivation of human tumors: establishment of cell lines derived from a series of solid tumors. *J. Natl Cancer Inst.* 51 (5): 1417–1423.

Gillet, J.P., Calcagno, A.M., Varma, S. et al. (2011). Redefining the relevance of established cancer cell lines to the study of mechanisms of clinical anti-cancer drug resistance. *Proc. Natl Acad. Sci. U.S.A.* 108 (46): 18708–18713. https://doi.org/10.1073/pnas.1111840108.

Gillet, J.P., Varma, S., and Gottesman, M.M. (2013). The clinical relevance of cancer cell lines. *J. Natl Cancer Inst.* 105 (7): 452–458. https://doi.org/10.1093/jnci/djt007.

Giovanella, B.C., Yim, S.O., Stehlin, J.S. et al. (1972). Development of invasive tumors in the "nude" mouse after injection of cultured human melanoma cells. *J. Natl Cancer Inst.* 48 (5): 1531–1533.

Goldstein, S., Murano, S., Benes, H. et al. (1989). Studies on the molecular-genetic basis of replicative senescence in werner syndrome and normal fibroblasts. *Exp. Gerontol.* 24 (5–6): 461–468.

Gown, A.M., Vogel, A.M., Hoak, D. et al. (1986). Monoclonal antibodies specific for melanocytic tumors distinguish subpopulations of melanocytes. *Am. J. Pathol.* 123 (2): 195–203.

Hanahan, D. and Weinberg, R.A. (2011). Hallmarks of cancer: the next generation. *Cell* 144 (5): 646–674. https://doi.org/10.1016/j.cell.2011.02.013.

Hasegawa, K., Pomeroy, J.E., and Pera, M.F. (2010). Current technology for the derivation of pluripotent stem cell lines from human embryos. *Cell Stem Cell* 6 (6): 521–531. https://doi.org/10.1016/j.stem.2010.05.010.

Hayflick, L. (1965). The limited in vitro lifetime of human diploid cell strains. *Exp. Cell. Res.* 37: 614–636.

Hayflick, L. (1998). A brief history of the mortality and immortality of cultured cells. *Keio J. Med.* 47 (3): 174–182.

Hayflick, L. and Moorhead, P.S. (1961). The serial cultivation of human diploid cell strains. *Exp. Cell. Res.* 25: 585–621.

Hirschi, K.K., Li, S., and Roy, K. (2014). Induced pluripotent stem cells for regenerative medicine. *Annu. Rev. Biomed. Eng.* 16: 277–294. https://doi.org/10.1146/annurev-bioeng-071813-105108.

Hsu, T.C. and Moorhead, P.S. (1957). Mammalian chromosomes in vitro VII. Heteroploidy in human cell strains. *J. Natl Cancer Inst.* 18 (3): 463–471.

Huschtscha, L.I. and Holliday, R. (1983). Limited and unlimited growth of SV40-transformed cells from human diploid MRC-5 fibroblasts. *J. Cell Sci.* 63: 77–99.

Ince, T.A., Sousa, A.D., Jones, M.A. et al. (2015). Characterization of twenty-five ovarian tumour cell lines that phenocopy primary tumours. *Nat. Commun.* 6: 7419. https://doi.org/10.1038/ncomms8419.

Jacobsen, B.M., Harrell, J.C., Jedlicka, P. et al. (2006). Spontaneous fusion with, and transformation of mouse stroma by, malignant human breast cancer epithelium. *Cancer Res.* 66 (16): 8274–8279. https://doi.org/10.1158/0008-5472.CAN-06-1456.

Jainchill, J.L., Aaronson, S.A., and Todaro, G.J. (1969). Murine sarcoma and leukemia viruses: assay using clonal lines of contact-inhibited mouse cells. *J. Virol.* 4 (5): 549–553.

Jiang, X.R., Jimenez, G., Chang, E. et al. (1999). Telomerase expression in human somatic cells does not induce changes associated with a transformed phenotype. *Nat. Genet.* 21 (1): 111–114. https://doi.org/10.1038/5056.

Kaphle, P., Li, Y., and Yao, L. (2019). The mechanical and pharmacological regulation of glioblastoma cell migration in 3D matrices. *J. Cell. Physiol.* 234 (4): 3948–3960. https://doi.org/10.1002/jcp.27209.

Kasai, F., Hirayama, N., Ozawa, M. et al. (2016). Changes of heterogeneous cell populations in the Ishikawa cell line during long-term culture: proposal for an in vitro clonal evolution model of tumor cells. *Genomics* 107 (6): 259–266. https://doi.org/10.1016/j.ygeno.2016.04.003.

Kasai, F., Hirayama, N., Ozawa, M. et al. (2018). HuH-7 reference genome profile: complex karyotype composed of massive loss of heterozygosity. *Hum. Cell* 31 (3): 261–267. https://doi.org/10.1007/s13577-018-0212-3.

Khan, M.Z., Spandidos, D.A., Kerr, D.J. et al. (1991). Oncogene transfection of mink lung cells: effect on growth characteristics in vitro and in vivo. *Anticancer Res.* 11 (3): 1343–1348.

Kleensang, A., Vantangoli, M.M., Odwin-DaCosta, S. et al. (2016). Genetic variability in a frozen batch of MCF-7 cells invisible in routine authentication affecting cell function. *Sci. Rep.* 6: 28994. https://doi.org/10.1038/srep28994.

Klijn, C., Durinck, S., Stawiski, E.W. et al. (2015). A comprehensive transcriptional portrait of human cancer cell lines. *Nat. Biotechnol.* 33 (3): 306–312. https://doi.org/10.1038/nbt.3080.

Landa, I., Pozdeyev, N., Korch, C. et al. (2019). Comprehensive genetic characterization of human thyroid cancer cell lines: a validated panel for preclinical studies. *Clin. Cancer Res.* 25: 3141–3151. https://doi.org/10.1158/1078-0432.CCR-18-2953.

Landry, J.J., Pyl, P.T., Rausch, T. et al. (2013). The genomic and transcriptomic landscape of a HeLa cell line. *G3 (Bethesda)* 3 (8): 1213–1224. https://doi.org/10.1534/g3.113.005777.

Lewis, N.E., Liu, X., Li, Y. et al. (2013). Genomic landscapes of Chinese hamster ovary cell lines as revealed by the *Cricetulus griseus* draft genome. *Nat. Biotechnol.* 31 (8): 759–765. https://doi.org/10.1038/nbt.2624.

Lieber, M., Smith, B., Szakal, A. et al. (1976). A continuous tumor-cell line from a human lung carcinoma with properties of type II alveolar epithelial cells. *Int. J. Cancer* 17 (1): 62–70.

Liu, X., Krawczyk, E., Suprynowicz, F.A. et al. (2017). Conditional reprogramming and long-term expansion of normal and tumor cells from human biospecimens. *Nat. Protoc.* 12 (2): 439–451. https://doi.org/10.1038/nprot.2016.174.

Liu, Y., Mi, Y., Mueller, T. et al. (2019). Multi-omic measurements of heterogeneity in HeLa cells across laboratories. *Nat. Biotechnol.* 37: 314–322. https://doi.org/10.1038/s41587-019-0037-y.

Lynch, M.D. and Watt, F.M. (2018). Fibroblast heterogeneity: implications for human disease. *J. Clin. Invest.* 128 (1): 26–35. https://doi.org/10.1172/JCI93555.

MacLeod, R.A., Nagel, S., Scherr, M. et al. (2008). Human leukemia and lymphoma cell lines as models and resources. *Curr. Med. Chem.* 15 (4): 339–359.

MacLeod, R.A., Kaufmann, M., and Drexler, H.G. (2011). Cytogenetic analysis of cancer cell lines. *Methods Mol. Biol.* 731: 57–78. https://doi.org/10.1007/978-1-61779-080-5_6.

Macpherson, I. and Montagnier, L. (1964). Agar suspension culture for the selective assay of cells transformed by polyoma virus. *Virology* 23: 291–294.

Maley, C.C., Aktipis, A., Graham, T.A. et al. (2017). Classifying the evolutionary and ecological features of neoplasms. *Nat. Rev. Cancer* 17 (10): 605–619. https://doi.org/10.1038/nrc.2017.69.

Mege, R.M., Matsuzaki, F., Gallin, W.J. et al. (1988). Construction of epithelioid sheets by transfection of mouse sarcoma cells with cDNAs for chicken cell adhesion molecules. *Proc. Natl Acad. Sci. U.S.A.* 85 (19): 7274–7278.

Merlo, L.M., Pepper, J.W., Reid, B.J. et al. (2006). Cancer as an evolutionary and ecological process. *Nat. Rev. Cancer* 6 (12): 924–935. https://doi.org/10.1038/nrc2013.

Mitelman, F., Johansson, B., and Mertens, F. (2007). The impact of translocations and gene fusions on cancer causation. *Nat. Rev. Cancer* 7 (4): 233–245. https://doi.org/10.1038/nrc2091.

Mitelman, F., Johansson, B., and Mertens, F. (2019). Mitelman database of chromosome aberrations and gene fusions in cancer. https://cgap.nci.nih.gov/Chromosomes/Mitelman (accessed 22 February 2019).

Mohseny, A.B., Machado, I., Cai, Y. et al. (2011). Functional characterization of osteosarcoma cell lines provides representative models to study the human disease. *Lab. Invest.* 91 (8): 1195–1205. https://doi.org/10.1038/labinvest.2011.72.

Moorhead, P.S. (1974). How long can one safely work with cell lines? *In Vitro* 10: 143–148.

Mori, S., Chang, J.T., Andrechek, E.R. et al. (2009). Anchorage-independent cell growth signature identifies tumors with metastatic potential. *Oncogene* 28 (31): 2796–2805. https://doi.org/10.1038/onc.2009.139.

Mouradov, D., Sloggett, C., Jorissen, R.N. et al. (2014). Colorectal cancer cell lines are representative models of the main molecular subtypes of primary cancer. *Cancer Res.* 74 (12): 3238–3247. https://doi.org/10.1158/0008-5472.CAN-14-0013.

Nardella, C., Clohessy, J.G., Alimonti, A. et al. (2011). Pro-senescence therapy for cancer treatment. *Nat. Rev. Cancer* 11 (7): 503–511. https://doi.org/10.1038/nrc3057.

Newbold, R.F. (2002). The significance of telomerase activation and cellular immortalization in human cancer. *Mutagenesis* 17 (6): 539–550.

Nowell, P.C. (1976). The clonal evolution of tumor cell populations. *Science* 194 (4260): 23–28.

Nowell, P.C. and Hungerford, D.A. (1960). Chromosome studies on normal and leukemic human leukocytes. *J. Natl Cancer Inst.* 25: 85–109.

Osada, N., Kohara, A., Yamaji, T. et al. (2014). The genome landscape of the African green monkey kidney-derived Vero cell line. *DNA Res.* 21 (6): 673–683. https://doi.org/10.1093/dnares/dsu029.

Packer, L. and Fuehr, K. (1977). Low oxygen concentration extends the lifespan of cultured human diploid cells. *Nature* 267 (5610): 423–425.

Padilla-Nash, H.M., Hathcock, K., McNeil, N.E. et al. (2012). Spontaneous transformation of murine epithelial cells requires the early acquisition of specific chromosomal aneuploidies and genomic imbalances. *Genes Chromosomes Cancer* 51 (4): 353–374. https://doi.org/10.1002/gcc.21921.

Padilla-Nash, H.M., McNeil, N.E., Yi, M. et al. (2013). Aneuploidy, oncogene amplification and epithelial to mesenchymal transition define spontaneous transformation of murine epithelial cells. *Carcinogenesis* 34 (8): 1929–1939. https://doi.org/10.1093/carcin/bgt138.

Patel, K., Moore, S.E., Dickson, G. et al. (1989). Neural cell adhesion molecule (NCAM) is the antigen recognized by monoclonal antibodies of similar specificity in small-cell lung carcinoma and neuroblastoma. *Int. J. Cancer* 44 (4): 573–578.

Paul, J. (1975). *Cell and Tissue Culture*, 5e. Edinburgh: Churchill Livingstone.

Pinto, D. and Clevers, H. (2005). Wnt, stem cells and cancer in the intestine. *Biol. Cell* 97 (3): 185–196. https://doi.org/10.1042/BC20040094.

Pleasance, E.D., Cheetham, R.K., Stephens, P.J. et al. (2010). A comprehensive catalogue of somatic mutations from a human cancer genome. *Nature* 463 (7278): 191–196. https://doi.org/10.1038/nature08658.

Quentmeier, H., Osborn, M., Reinhardt, J. et al. (2001). Immunocytochemical analysis of cell lines derived from solid tumors. *J. Histochem. Cytochem.* 49 (11): 1369–1378. https://doi.org/10.1177/002215540104901105.

Quentmeier, H., Amini, R.M., Berglund, M. et al. (2013). U-2932: two clones in one cell line, a tool for the study of clonal evolution. *Leukemia* 27 (5): 1155–1164. https://doi.org/10.1038/leu.2012.358.

Quentmeier, H., Pommerenke, C., Ammerpohl, O. et al. (2016). Subclones in B-lymphoma cell lines: isogenic models for the study of gene regulation. *Oncotarget* 7 (39): 63456–63465. https://doi.org/10.18632/oncotarget.11524.

Rak, J., Yu, J.L., Kerbel, R.S. et al. (2002). What do oncogenic mutations have to do with angiogenesis/vascular dependence of tumors? *Cancer Res.* 62 (7): 1931–1934.

Rangarajan, A. and Weinberg, R.A. (2003). Opinion: comparative biology of mouse versus human cells: modelling human cancer in mice. *Nat. Rev. Cancer* 3 (12): 952–959. https://doi.org/10.1038/nrc1235.

Reddel, R.R. (2000). The role of senescence and immortalization in carcinogenesis. *Carcinogenesis* 21 (3): 477–484.

Ribatti, D. (2017). A revisited concept: contact inhibition of growth. From cell biology to malignancy. *Exp. Cell. Res.* 359 (1): 17–19. https://doi.org/10.1016/j.yexcr.2017.06.012.

Richmond, A., Lawson, D.H., Nixon, D.W. et al. (1985). Characterization of autostimulatory and transforming growth factors from human melanoma cells. *Cancer Res.* 45 (12 Pt 1): 6390–6394.

Rodenhizer, D., Dean, T., D'Arcangelo, E. et al. (2018). The current landscape of 3D in vitro tumor models: what cancer hallmarks are accessible for drug discovery? *Adv. Healthc. Mater.* 7 (8): e1701174. https://doi.org/10.1002/adhm.201701174.

Rognoni, E. and Watt, F.M. (2018). Skin cell heterogeneity in development, wound healing, and cancer. *Trends Cell Biol.* 28 (9): 709–722. https://doi.org/10.1016/j.tcb.2018.05.002.

Roschke, A.V., Tonon, G., Gehlhaus, K.S. et al. (2003). Karyotypic complexity of the NCI-60 drug-screening panel. *Cancer Res.* 63 (24): 8634–8647.

Rubin, H. (2002). The disparity between human cell senescence in vitro and lifelong replication in vivo. *Nat. Biotechnol.* 20 (7): 675–681. https://doi.org/10.1038/nbt0702-675.

Rubin, H. (2017). Dynamics of cell transformation in culture and its significance for tumor development in animals. *Proc. Natl Acad. Sci. U.S.A.* 114 (46): 12237–12242. https://doi.org/10.1073/pnas.1715236114.

Rygaard, K., Moller, C., Bock, E. et al. (1992). Expression of cadherin and NCAM in human small cell lung cancer cell lines and xenografts. *Br. J. Cancer* 65 (4): 573–577.

Saito, H., Hammond, A.T., and Moses, R.E. (1995). The effect of low oxygen tension on the in vitro-replicative life span of human diploid fibroblast cells and their transformed derivatives. *Exp. Cell. Res.* 217 (2): 272–279. https://doi.org/10.1006/excr.1995.1087.

Saito, S., Morita, K., Kohara, A. et al. (2011). Use of BAC array CGH for evaluation of chromosomal stability of clinically used human mesenchymal stem cells and of cancer cell lines. *Hum. Cell* 24 (1): 2–8. https://doi.org/10.1007/s13577-010-0006-8.

Sandgren, E.P., Luetteke, N.C., Qiu, T.H. et al. (1993). Transforming growth factor alpha dramatically enhances oncogene-induced carcinogenesis in transgenic mouse pancreas and liver. *Mol. Cell. Biol.* 13 (1): 320–330.

Sasaki, M., Honda, T., Yamada, H. et al. (1994). Evidence for multiple pathways to cellular senescence. *Cancer Res.* 54 (23): 6090–6093.

Schaeffer, W.I. (1990). Terminology associated with cell, tissue, and organ culture, molecular biology, and molecular genetics. Tissue culture association terminology committee. *In Vitro Cell Dev. Biol.* 26 (1): 97–101.

Serrano, M., Lin, A.W., McCurrach, M.E. et al. (1997). Oncogenic ras provokes premature cell senescence associated with accumulation of p 53 and p16INK4a. *Cell* 88 (5): 593–602.

Sharkey, F.E. and Fogh, J. (1984). Considerations in the use of nude mice for cancer research. *Cancer Metastasis Rev.* 3 (4): 341–360.

Shay, J.W. and Wright, W.E. (2000). Hayflick, his limit, and cellular ageing. *Nat. Rev. Mol. Cell Biol.* 1 (1): 72–76. https://doi.org/10.1038/35036093.

Spaethling, J.M., Na, Y.J., Lee, J. et al. (2017). Primary cell culture of live neurosurgically resected aged adult human brain cells and single cell transcriptomics. *Cell Rep.* 18 (3): 791–803. https://doi.org/10.1016/j.celrep.2016.12.066.

Speirs, V., Ray, K.P., and Freshney, R.I. (1991). Paracrine control of differentiation in the alveolar carcinoma, A549, by human foetal lung fibroblasts. *Br. J. Cancer* 64 (4): 693–699.

Suprynowicz, F.A., Upadhyay, G., Krawczyk, E. et al. (2012). Conditionally reprogrammed cells represent a stem-like state of adult epithelial cells. *Proc. Natl Acad. Sci. U.S.A.* 109 (49): 20035–20040. https://doi.org/10.1073/pnas.1213241109.

Taddei, M.L., Giannoni, E., Fiaschi, T. et al. (2012). Anoikis: an emerging hallmark in health and diseases. *J. Pathol.* 226 (2): 380–393. https://doi.org/10.1002/path.3000.

Todaro, G.J. and De Larco, J.E. (1978). Growth factors produced by sarcoma virus-transformed cells. *Cancer Res.* 38 (11 Pt 2): 4147–4154.

Todaro, G.J. and Green, H. (1963). Quantitative studies of the growth of mouse embryo cells in culture and their development into established lines. *J. Cell Biol.* 17: 299–313.

Toouli, C.D., Huschtscha, L.I., Neumann, A.A. et al. (2002). Comparison of human mammary epithelial cells immortalized by simian virus 40 T-antigen or by the telomerase catalytic subunit. *Oncogene* 21 (1): 128–139. https://doi.org/10.1038/sj.onc.1205014.

Tsuji, K., Kawauchi, S., Saito, S. et al. (2010). Breast cancer cell lines carry cell line-specific genomic alterations that are distinct from aberrations in breast cancer tissues: comparison of the CGH profiles between cancer cell lines and primary cancer tissues. *BMC Cancer* 10: 15. https://doi.org/10.1186/1471-2407-10-15.

Vijayakumar, S., Garcia, D., Hensel, C.H. et al. (2005). The human Y chromosome suppresses the tumorigenicity of PC-3, a human prostate cancer cell line, in athymic nude mice. *Genes Chromosomes Cancer* 44 (4): 365–372. https://doi.org/10.1002/gcc.20250.

Wang, H., Huang, S., Shou, J. et al. (2006). Comparative analysis and integrative classification of NCI60 cell lines and primary tumors using gene expression profiling data. *BMC Genomics* 7: 166. https://doi.org/10.1186/1471-2164-7-166.

Warmflash, A., Sorre, B., Etoc, F. et al. (2014). A method to recapitulate early embryonic spatial patterning in human embryonic stem cells. *Nat. Methods* 11 (8): 847–854. https://doi.org/10.1038/nmeth.3016.

Wenger, S.L., Senft, J.R., Sargent, L.M. et al. (2004). Comparison of established cell lines at different passages by karyotype and comparative genomic hybridization. *Biosci. Rep.* 24 (6): 631–639. https://doi.org/10.1007/s10540-005-2797-5.

Westermark, B. (1973). The deficient density-dependent growth control of human malignant glioma cells and virus-transformed glia-like cells in culture. *Int. J. Cancer* 12 (2): 438–451.

Westermark, B. (1978). Growth control in miniclones of human glial cells. *Exp. Cell. Res.* 111 (2): 295–299.

Wootton, M., Steeghs, K., Watt, D. et al. (2003). Telomerase alone extends the replicative life span of human skeletal muscle cells without compromising genomic stability. *Hum. Gene Ther.* 14 (15): 1473–1487. https://doi.org/10.1089/104303403769211682.

Wright, W.E. and Shay, J.W. (2000). Telomere dynamics in cancer progression and prevention: fundamental differences in human and mouse telomere biology. *Nat. Med.* 6 (8): 849–851. https://doi.org/10.1038/78592.

Wright, W.E. and Shay, J.W. (2002). Historical claims and current interpretations of replicative aging. *Nat. Biotechnol.* 20 (7): 682–688. https://doi.org/10.1038/nbt0702-682.

Xu, X., Nagarajan, H., Lewis, N.E. et al. (2011). The genomic sequence of the Chinese hamster ovary (CHO)-K1 cell line. *Nat. Biotechnol.* 29 (8): 735–741. https://doi.org/10.1038/nbt.1932.

Yu, M., Selvaraj, S.K., Liang-Chu, M.M. et al. (2015). A resource for cell line authentication, annotation and quality control. *Nature* 520 (7547): 307–311. https://doi.org/10.1038/nature14397.

PART II

Laboratory and Regulatory Requirements

After reading the following chapters in this part of the book, you will be able to:

(4) *Laboratory Design and Layout:*

(a) Describe the general design requirements that apply when designing a new or modifying an existing laboratory space for tissue culture.

(b) Explain the purpose of laboratory containment and provide examples of primary and secondary containment measures.

(c) List the essential areas that must be included in a tissue culture laboratory.

(d) Discuss key considerations that apply when positioning a biological safety cabinet (BSC) in a sterile handling area or a cryofreezer in a cryostorage area.

(e) Explain the purpose of contingency planning and list the key priorities that should be considered when such plans are developed.

(5) *Equipment and Materials:*

(a) Explain the various priorities that apply when acquiring tissue culture equipment.

(b) List the essential equipment items that are required for tissue culture.

(c) List the consumables that will be required at the BSC for use during tissue culture.

(d) Discuss the parameters that should be considered when selecting a new BSC and the differences between the various classes of BSC.

(e) Discuss the advantages and disadvantages of the CO_2 incubator and comment on alternative equipment that may be used if a CO_2 incubator is not available.

(6) *Safety and Bioethics:*

(a) Describe the various levels of safety regulations that apply to tissue culture laboratories and how to find specific requirements that apply to your location.

(b) List examples of commonly occurring hazards in tissue culture laboratories.

(c) Perform a risk assessment and discuss how to control the resulting risks.

(d) Discuss the source of biohazard risk in living cultures and how that risk is classified in different countries.

(e) Describe how to safely dispose of waste from a tissue culture laboratory.

(f) Summarize ethical principles that apply when obtaining tissue samples for use in the laboratory and provide examples of requirements that must be met.

(g) Discuss how to obtain informed donor consent for use of human tissue in research.

(7) *Reproducibility and Good Cell Culture Practice:*

(a) Discuss the causes of irreproducible research and list specific actions that can be taken to improve reproducibility in each area.

(b) Summarize the key principles that make up Good Cell Culture Practice (GCCP).

(c) Explain the meaning of "cell line provenance" and describe the cell line information that should be included when reporting for publication.

(d) Discuss the validation testing that should be performed for all cell lines.

(e) Explain the purposes of quality assurance (QA) and quality control (QC).

(f) Describe how replicate samples may be collected.

Freshney's Culture of Animal Cells: A Manual of Basic Technique and Specialized Applications, Eighth Edition. Amanda Capes-Davis and R. Ian Freshney.
© 2021 John Wiley & Sons Ltd. Published 2021 by John Wiley & Sons Ltd.
Companion website: www.wiley.com/go/freshney/cellculture8

CHAPTER 4

Laboratory Design and Layout

The design of tissue culture laboratories is governed by two central considerations: asepsis and containment. Asepsis is required to prevent contamination and relies on control of airflow to minimize exposure to dust and liquid droplets. Containment is required to prevent exposure of workers, bystanders, or the external environment to any hazardous biological agents (pathogens) that may be present. The development of laminar flow cabinets, followed by biological safety cabinets (BSCs), greatly simplified approaches to these problems and allowed the utilization of general laboratory areas for tissue culture (see Sections 4.2.1, 5.1.1, 12.2.2). However, the location of a tissue culture area must still be carefully chosen to ensure that asepsis and containment can both be achieved and maintained.

Scientists spend many hours working in a tissue culture laboratory. It is important to take time to plan the design and layout, asking users what they require from the space and how to optimize the layout. Contingency planning for laboratory incidents and broader disasters should also be considered as part of design, resulting in better integration of engineering control devices that help to ensure worker safety and sample integrity (see Sections 4.3, 6.1; see also Supp. S6.1).

4.1 DESIGN REQUIREMENTS

4.1.1 General Design Considerations

The design and layout of a tissue culture laboratory depends on the type and scale of operations, the number of users, and the space available. The layout of a small, self-contained laboratory for two to three workers (see Figure 4.1) will differ from a medium-sized laboratory for five to six workers (see Figures 4.2, 4.3) or a large laboratory shared by 20–30 workers (see Figure 4.4). The layout will also vary depending on the number of rooms available. If multiple rooms are available, some may be reserved for specific tasks. For example, additional rooms may be used to provide a dedicated microscope room, a quarantine area, or a warmroom for incubation (see Figure 4.3; see also Section 4.2.2). If additional rooms are available for storage, preparation, sterilization, and washup (see Figure 4.4), these may be shared by more than one tissue culture laboratory.

Some general principles apply regardless of size. It is essential to have a dedicated tissue culture laboratory with an adjacent preparation area, or a number of smaller ones with a common preparation area, rather than to perform tissue culture alongside regular laboratory work with only a BSC for protection. A separate facility gives better contamination protection and allows tissue culture stocks to be kept separate from regular laboratory reagents and glassware. Physical separation becomes essential for containment if cells have a higher level of risk compared to other laboratory samples (see Sections 4.1.3, 6.1.1). Always perform a risk assessment when planning a new tissue culture laboratory, ensuring that safety is included as an integral part of facility design (see Section 6.1.2) (OECD 2018).

The distribution of equipment in the tissue culture laboratory should be arranged such that items of equipment that create drafts and dust are located as far as possible away from the sterile handling area. Similarly, equipment that may harbor fungal contamination should be arranged well away from the sterile handling area, preferably in another room.

Freshney's Culture of Animal Cells: A Manual of Basic Technique and Specialized Applications, Eighth Edition. Amanda Capes-Davis and R. Ian Freshney.
© 2021 John Wiley & Sons Ltd. Published 2021 by John Wiley & Sons Ltd.
Companion website: www.wiley.com/go/freshney/cellculture8

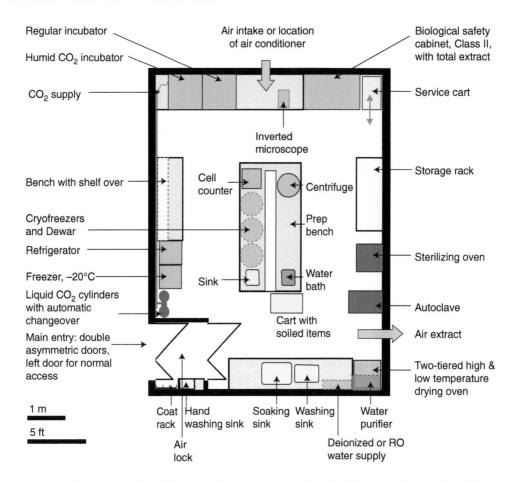

Fig. 4.1. Small tissue culture laboratory. Suggested layout for simple, self-contained tissue culture laboratory for use by two or three persons. Dark-shaded areas represent movable equipment, lighter shaded areas fixed or movable furniture.

Hence refrigerators and freezers, which generate drafts from compressor operation and may harbor fungal contamination in areas of water condensation, should be distant from the BSC. Microscopes and cell counters may be located close to the BSC, which will also be the most convenient location. A centrifuge can be located near the BSC, provided the chamber is sealed when running to minimize aerosols, but should not share a bench with a microscope because of potential vibration. If all associated activities must share the same room (see Figure 4.1), washup, bottling, and sterilizing should be at the opposite end of the room from the BSC, and items brought from outside, such as cylinders and liquid nitrogen Dewars, should be located near the door. The door itself should be on the opposite end from the BSC to minimize air currents and through traffic.

If possible, the tissue culture laboratory should be separated from the preparation, washup, and sterilization areas (see Figures 4.3, 4.4). This modular approach will also help to manage construction costs, as it is possible to tackle each area separately (Wesselschmidt and Schwartz 2011). If you have a large tissue culture laboratory with a separate washup and sterilization facility, it will be convenient to have this on the same floor as, and adjacent to, the clean room, with no steps to negotiate, so that carts or trolleys may be used. Across a corridor is probably ideal (see Figure 4.4; see also Sections 4.2.4, 4.2.5). Try to imagine the flow of traffic (people, reagents, carts, etc.) and arrange for minimum conflict, easy access to stores, good access for replenishing stocks without interfering with sterile work, and easy removal of waste and soiled items.

Decisions on laboratory design must be made with reference to your budget, the available space, and local regulatory requirements (e.g. safety regulations; see Section 6.1.2). If a conversion of existing facilities is contemplated, there will be significant structural limitations. Choose the location carefully to avoid space constraints and awkward projections into the room that will limit flexibility and air flow. For major renovations or new construction, seek out an architect who can work with you to make the best use of the space and comply with regulatory requirements.

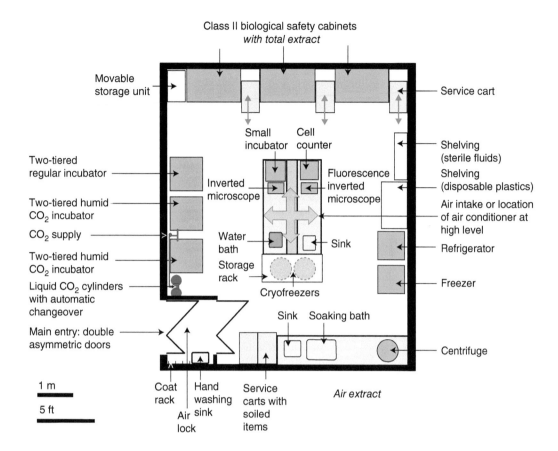

Fig. 4.2. Medium-sized tissue culture laboratory. Suitable for five or six persons, with washing-up and preparation facility located elsewhere. Dark-shaded areas represent movable equipment, light-shaded areas movable or fixed furniture.

4.1.2 User Requirements

Essential areas and equipment. Regardless of size, certain areas and equipment are common to all tissue culture laboratories. The following areas and equipment should be considered:

(1) Hand washing sink(s) at entry to laboratory.
(2) Sterile handling area containing one or more BSCs for culture handling procedures (see Section 4.2.1).
(3) Equipment benches or area adjacent to the sterile handling area with centrifuge, cell counter, and inverted microscope. Preferably, the microscope should not share a bench with the centrifuge.
(4) Carts for servicing sterile handling area.
(5) Vacuum pump or similar system for withdrawing and discarding media.
(6) Waste containers for the various laboratory waste streams (see Section 6.3.6), including solid and liquid biological waste, which require containment and decontamination as part of safe disposal.
(7) Incubators suitable for culture of sealed flasks, vented flasks, dishes, and plates. Large-scale incubation (e.g.

bioreactors, roller bottles) may also require a large incubator or warmroom (see Section 4.2.2).
(8) Provision of CO_2 and other gas mixtures for incubation of cultures.
(9) Quarantine area suitable for initial culture when samples are received from outside the laboratory, until observation and testing confirm that cultures are free from contamination (see Section 4.2.3).
(10) Preparation area for preparation and bottling of liquids, packaging of plastics, etc. (see Section 4.2.4).
(11) Water purification equipment to generate purified water, including ultrapure water (UPW), which is necessary for medium preparation (see Section 11.5.1).
(12) Washup area for cleaning glassware and apparatus (see Section 4.2.5).
(13) Sterilization equipment using steam, dry heat, and filtration methods.
(14) Cold storage area at temperatures of $4\,°C$ (working strength and $10×$ concentrated media), $-20\,°C$ (serum, glutamine, antibiotics), $-80\,°C$ (DNA, RNA, protein), and $-196\,°C$ (cryofreezer storage using liquid nitrogen to maintain frozen stocks). Space for backup equipment should also be considered (see Section 4.2.6).

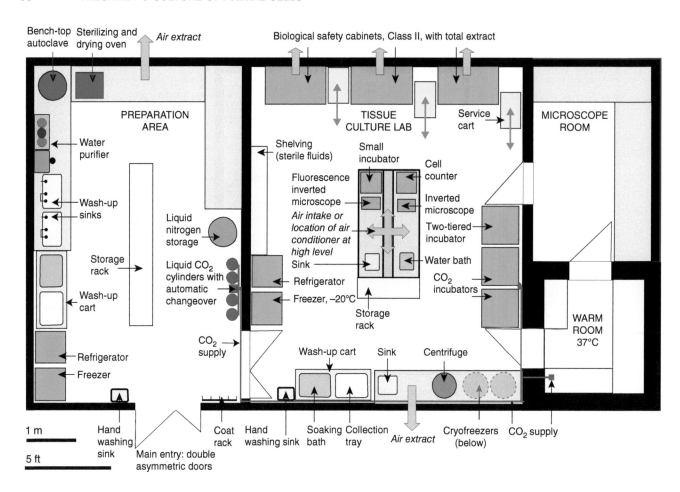

Bench-top autoclave · Sterilizing and drying oven · *Air extract* · Biological safety cabinets, Class II, with total extract

PREPARATION AREA

Water purifier
Wash-up sinks
Storage rack
Wash-up cart
Refrigerator
Freezer

Liquid nitrogen storage
Liquid CO_2 cylinders with automatic changeover
CO_2 supply

TISSUE CULTURE LAB

Shelving (sterile fluids)
Small incubator
Cell counter
Fluorescence inverted microscope
Inverted microscope
Air intake or location of air conditioner at high level
Two-tiered incubator
Sink
Refrigerator
Water bath
CO_2 incubators
Freezer, –20°C
Storage rack

Service cart

MICROSCOPE ROOM

WARM ROOM 37°C

1 m
5 ft

Hand washing sink
Main entry: double asymmetric doors
Coat rack
Hand washing sink
Wash-up cart
Soaking bath
Collection tray
Air extract
Sink
Centrifuge
Cryofreezers (below)
CO_2 supply

Fig. 4.3. Tissue culture laboratory with adjacent preparation room. Medium-sized tissue culture lab (see Figure 4.2), but with attached preparation area, microscope room, and 37 °C warmroom. Scale 1 : 100.

(15) Room temperature storage area for consumables and some reagents (e.g. plastics, glassware, gloves, water, salt solutions).

It can be a challenge to develop a layout that suits the available space and provides access to essential areas and equipment. The layout required for laboratory areas is dealt with below (see Section 4.2) and the necessary equipment in the next chapter (see Chapter 5). Access to more specialized areas and equipment will vary with the user's needs. For example, high power microscopy with fluorescence may be needed for characterization (see Section 18.4). Typically, these need not be located in a sterile area and may be positioned elsewhere in shared equipment areas or core facilities.

User parameters. Although there are clearly many common elements, the design must also take the user's specific needs into consideration. User parameters that may determine the scale, design, and layout of these facilities include:

(1) **Intended use.** What type of work do you intend to perform? What equipment must be located inside the room?

Does your work have any specialized requirements that must be considered? For example, a cell repository may require a tissue culture suite with a shared anteroom connected to a series of small cubicles, each with a single BSC, to allow individual handling of cultures and minimize the risk of cross-contamination.

(2) **Biohazard risk.** What biological samples will be brought into the room, and what risk will those materials carry? If workers intend to use samples for genetic modification or growth of biological agents, will those applications modify the risk level? These questions will help to determine the level of containment required once the laboratory is operational. Physical containment can have major implications for laboratory design and engineering (see Section 4.1.3).

(3) **Number of users.** How many people will work in the facility, how long will they work each week, and what kinds of culture will they perform? These considerations determine how many BSCs will be required (based on whether people can share hoods or whether they will require a hood each for most of the day) and whether

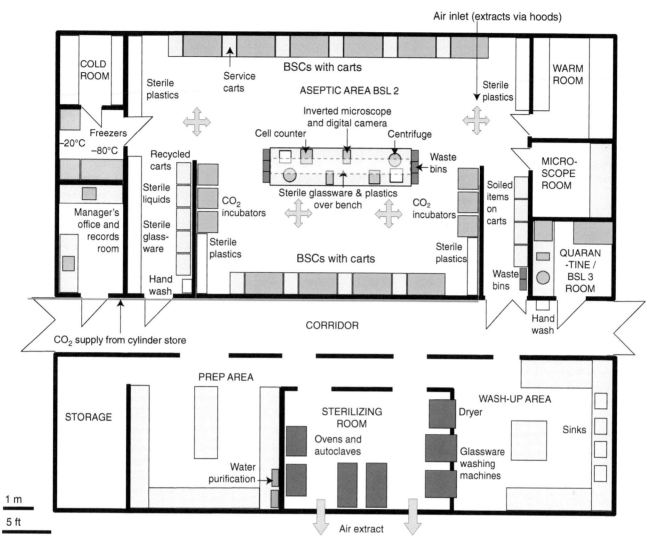

Fig. 4.4. Large tissue culture laboratory. Suitable for 20–30 persons. Adjacent washing-up, sterilization, and preparation area. Dark-shaded areas represent equipment, light-shaded areas fixed and movable furniture. Scale 1 : 200.

a large area will be needed to handle bioreactors, animal tissue dissections, or large numbers of cultures. As a rough guide, 12 BSCs in a communal facility can accommodate 50 people with intermittent requirements; extended or continuous use will reduce the capacity proportionally.

(4) *Space.* How much space is available and how should it be assigned to the various areas? The largest space should be given to the tissue culture laboratory, which must accommodate BSCs, microscopes, centrifuges, and incubators, with sufficient space for stocks of reagents and plasticware. The second largest space should be used for service areas including washup, preparation, and sterilization. The third largest space is for storage, and the fourth is for incubation if a warmroom is to be included

(see Section 4.2.2). A reasonable estimate is 4 : 2 : 1 : 1, in the order just presented.

(5) *Ergonomics.* How will people utilize the various areas, and can the layout minimize awkward positions and repetitive actions? Ergonomic issues are important when choosing equipment (see Section 6.1.4) but should also be considered as part of laboratory design. For example, storage shelves above head height should be positioned to allow step ladder access and should not be intended for heavy items.

(6) *Location of service areas.* Facilities for washing up and for sterilization should be located (i) close to the aseptic area that they service; and (ii) on an outside wall to allow for the possibility of heat extraction from ovens and steam vents from autoclaves. Give your washup, sterilization,

and preparation staff a reasonable visual outlook if possible; they usually perform fairly repetitive duties, and a pleasant view can help to make their tasks a little more enjoyable.

(7) **Storage.** What is the scale of the work contemplated and how much storage space will this require for disposable plastics and other reagents? Separate and distinct areas must be provided for sterile and non-sterile items. What proportion of the work will be cell line work, with its requirement for storage in liquid nitrogen?

(8) **Access.** Make sure that doorways are wide enough and high enough and that ceilings have sufficient clearance to allow the installation of equipment such as BSCs (which may need additional space for ductwork), incubators, and autoclaves. Provide space for access to equipment and installed services during maintenance. Will people require access to the animal facility for animal tissue? If so, ensure that tissue culture is reasonably accessible to, but not contiguous with, the animal facility and space is provided for a double change of lab coats.

These questions will enable you to decide what size of facility you require and what type of accommodation – one or two small rooms (see Figures 4.1, 4.2), or a suite of rooms incorporating washup, sterilization, one or more aseptic areas, an incubation room, a dark room for fluorescence microscopy, a refrigeration room, and storage (see Figure 4.3). If a larger-scale facility is required, it may be preferable to separate the tissue culture facility proper from ancillary support areas (see Figure 4.4). If the washup becomes a shared facility, care must be taken to separate tissue culture glassware from chemical glassware and to ensure a clear distinction is made between sterile and non-sterile items.

While a small facility may conveniently locate liquid nitrogen cryofreezers, Dewars, and CO_2 cylinders in or near the laboratory, larger facilities will benefit from having these located in purpose-built storage (see Section 4.2.6). Positioning of liquid nitrogen is determined by a number of factors. The location should allow easy access to delivery vehicles and ongoing supply during a natural disaster or other incident (see Section 4.3). The physical distance between liquid nitrogen supply tanks and cryofreezer storage is also important. The greater the distance between supply tanks and cryofreezers, the more evaporation will occur, resulting in increased costs for liquid nitrogen delivery. Larger facilities will benefit from installing a piped gas system to supply multiple incubators with CO_2 using a single supply tank. When installing any gas supply, regardless of size, a risk assessment must be performed to determine whether engineering controls are needed to manage safety risks (see Sections 4.1.4, 6.1.1).

4.1.3 Regulatory Requirements

Construction. It will be necessary to contact the appropriate planning and building control authority if new construction, major reconstruction, or major renovation is required. This is usually administered by the local government authority (city, region, or state) and the approach will generally be made by the architect or institutional planning and development department. All institutions are required to comply with relevant local and national regulations; design requirements will also vary with the institution or application. For example, the US National Institutes of Health (NIH) Office of Research Facilities has published a Design Requirements Manual that applies to all NIH-owned, leased, operated, and funded buildings and facilities (NIH 2016). It is the responsibility of the person in charge of the facility to liaise with the architect, design team, or institutional department to determine the relevant compliance requirements, ensure that compliance is met, and understand how this may impinge on work practices within the laboratory.

Safety and containment. All tissue culture laboratories must comply with local regulations to ensure that laboratory work is performed with due regard to safety and biological samples are contained for the protection of the worker, any additional bystanders, and the external environment. Biohazard risk must be assessed as part of the design process, resulting in a risk rating that determines the level of physical containment and biosafety that will apply (see Tables 6.3, 6.4). Although containment is often seen as a regulatory requirement, containment measures are evidence-based and rely on data and lessons learned from past laboratory incidents (Kimman et al. 2008).

Physical containment can be divided into primary and secondary measures (Phillips and Runkle 1967). Primary containment measures include laboratory practices and equipment (e.g. BSCs, cryofreezer, sealed centrifuge rotor) that act as the first line of defense against exposure. Secondary containment measures consist of building features that act as a further line of defense if primary containment is breached. Building features that contribute to containment include walls and floors, sealed windows, lockable doors, air handling and ventilation systems, and access to handwashing facilities and equipment for waste decontamination.

As the estimated risk rating increases, secondary containment becomes more important (Kimman et al. 2008). High containment laboratories apply multiple containment measures; for example, laboratories at Biosafety Level 4 (BSL-4) (see Table 6.4) may be designed as "suit laboratories" where workers wear full-body positive-pressure suits (CDC and NIH 2009). Containment at this level is complex and its effectiveness can be difficult to assess. Experience with severe acute respiratory syndrome (SARS) suggests that high containment measures can be effective but are highly dependent on safe handling and storage procedures (Li et al. 2005; Lim et al. 2006). Effective containment at the source is the most critical part of biosafety at all levels, through strict adherence to good microbiological practice combined with appropriate use of containment equipment (van Soolingen et al. 2014). While laboratory design is important for secondary containment, it falls to the laboratory manager and senior scientists to train workers in containment of biological agents at the source.

The tissue culture laboratory may need to be separated from other laboratory areas. Containment requires that the contents of the tissue culture laboratory do not escape to adjacent areas, while asepsis requires that the contents of other laboratories do not enter the tissue culture room. If the material being handled is potentially biohazardous but the work needs to remain sterile, then some sort of buffer area will be required, e.g. an airlock, anteroom, or corridor. As noted for containment, asepsis is highly dependent on behavior and training – for example, teaching workers to wash their hands when they enter the tissue culture laboratory.

There are many different hazards in a tissue culture laboratory apart from biohazards (see Figure 6.1). As part of designing a tissue culture laboratory, it is important to perform a risk assessment to consider what hazards may apply and how the associated risks should be managed (see Section 6.1.1; see also Supp. S6.1) (OECD 2018). Risk management may include installing engineering control measures during construction, which is always preferable to retrofitting an existing facility, if possible.

4.1.4 Engineering Requirements

Air handling and ventilation. Air handling is a challenging requirement for tissue culture laboratories because of the combined requirements of asepsis and containment. Ideally, a tissue culture laboratory should be supplied by air that has passed through a high-efficiency particulate air (HEPA) filter and is at positive pressure relative to surrounding work areas, avoiding influx of contaminated air from outside. If containment is required from surrounding areas, the tissue culture laboratory should be at negative pressure relative to those areas. To satisfy both requirements, a positive-pressure buffer zone (e.g. using a corridor as shown in Figure 4.4) or a series of pressure differentials across the tissue culture suite may be used. Complex pressure differentials require specialized expertise. It is important to discuss requirements for asepsis and containment with architects and consulting

engineers, who will put complex requirements into practice. Ductwork requirements for BSCs (see Section 4.2.1) must also be discussed to ensure that they are incorporated into the overall airflow design. Air balances will need to be fine-tuned during construction and throughout the life of the facility. HEPA filters must be changed at regular intervals; clogging of filters over time makes it more difficult to maintain pressure differentials.

Control of temperature and humidity is challenging for many tissue culture laboratories. BSCs have large motors that may generate 200–800 W per device (Webber 2008); incubators and other equipment will add to the overall heat load. Heat load can be managed by regulating room temperature and airflow, and by varying the proportion of fresh versus recirculated air and the number of air changes per hour. Humidity can be managed by careful positioning of the sterile handling area (e.g. away from sources of condensation) and by regulating room conditions and airflow. In some laboratories, airflow rates are reduced outside working hours to improve energy efficiency, with manual override switches for workers to turn on air conditioning if it is required. However, such arrangements are usually not possible in tissue culture laboratories, where optimal environmental conditions must be maintained continuously.

Gas supply and monitoring. Tissue culture laboratories require a supply of CO_2 and may require other gases, depending on the culture conditions used (see Sections 9.2.3 and 9.2.4). Some laboratories may choose to gas individual flasks, requiring an additional supply of compressed air and access to an electronic gas blender or mixer. However, most laboratories will install incubators that regulate gas levels in the incubator chamber. Gas may be supplied using cylinders or a system that connects multiple incubators to a central supply (see Figure 4.5). A central supply that is piped to multiple incubators is typically a more reliable source compared to individual gas cylinders. Although expensive to install, piped supplies cover their costs over time if sufficient users will benefit from their installation. As a rough guide, two to three people will only require a few cylinders, 10–15 will probably benefit from a piped supply from a bank of cylinders, and for more than 15 it will pay to have a storage tank as a central supply of CO_2. Centralized supply systems should have at least two storage vessels, with one acting as a backup tank connected to an automated changeover system that will switch between tanks when needed (see Figure 4.5a-b).

Rooms where gas cylinders or piped gas systems are used require a risk assessment to ensure that any safety concerns are incorporated into room design (see Section 6.1.1). CO_2 is an asphyxiating gas due to displacement of oxygen in ambient

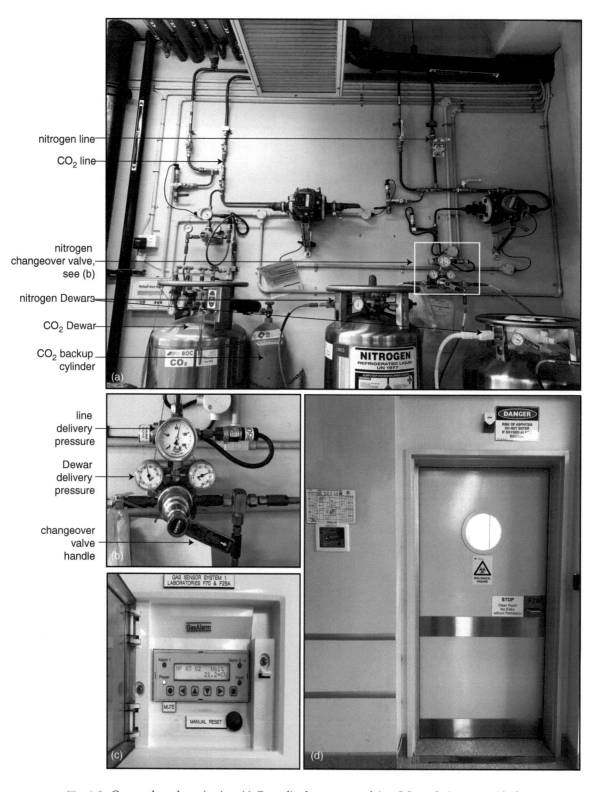

Fig. 4.5. Gas supply and monitoring. (a) Centralized systems supplying CO_2 and nitrogen, with the latter supplied as a gas arising from evaporation of liquid nitrogen; (b) close-up of the automated changeover system for nitrogen supply (white box in [a]), with switchover governed by pressure differentials between the two vessels; (c) gas monitoring panel displaying O_2; (d) tissue culture laboratory with restricted access, showing a gas monitoring panel in place outside the door with visible alarm. *Source*: Amanda Capes-Davis, courtesy of CMRI.

air; CO_2 is also toxic in its own right (Langford 2005). The level of risk will depend on the amount of gas within the system, the volume of air within the room, and the rate of air exchange. Piped supply systems contain a large volume of gas, resulting in a higher level of risk if a leak occurs. In some cases, it may be advisable to install gas sensors to monitor oxygen and CO_2 levels. Sensors should be connected to a monitoring system that can trigger both audible and visible alarms when oxygen levels decrease or CO_2 levels increase above normal levels (see Figure 4.5). Piped gas systems may have solenoid cut-off valves installed to isolate gas supply once an alarm is active.

Rooms where liquid nitrogen is stored or handled require careful risk assessment to ensure that asphyxia risks can be managed. Asphyxiation is a safety concern for liquid nitrogen because of its expansion ratio; 1 l of liquid nitrogen releases 683 l of nitrogen gas, resulting in a rapid decrease in ambient oxygen. Anyone who enters a low oxygen environment is at risk of asphyxiation, with possible loss of consciousness after several breaths. The level of risk will depend on a number of factors, including the amount of liquid nitrogen within the room, the volume of air, and the airflow rate. The risk may also increase in high containment laboratories, where negative pressure gradients may accelerate gas buildup (Finkel 2007). Typically, oxygen monitoring and continuous ventilation should be installed in any room where liquid nitrogen is held. Gas sensors should be carefully positioned and regularly calibrated. Sensors should be connected to a monitoring system that will trigger audible and visible alarms when oxygen levels decrease to suboptimal levels (see Figure 4.5c–d). Alarms should trigger changes in ventilation to assist with dispersal of nitrogen gas and to allow oxygen levels to rapidly return to normal. Users of these rooms should be trained to leave immediately when an alarm becomes active and stay out until the alarm is resolved.

Power, water, and other services. A tissue culture laboratory requires a number of services to be installed. Some may need engineering control measures to manage safety concerns. Services include:

(1) *Power supply.* Power is always underestimated, in terms of both the number of outlets and the number of amps required per outlet. Assess carefully the equipment that will be required, assume that both the number of appliances and their power consumption will treble within the life of the building in its present form, and try to provide sufficient power for their needs. Essential equipment (e.g. refrigerators and freezers) should be connected to essential power circuits, which are backed up by an auxiliary generator. Typically, the power supply will automatically switch over to the backup generator to supply essential equipment if the main power fails for a predetermined time period. The time period used may vary between institutions (from a few seconds to a few minutes), depending on the frequency and duration of minor power glitches. Power arrangements may include engineering controls to manage safety concerns. For example, some equipment items have an increased risk of electric shock or fire. To manage these risks, "power kill" or "emergency isolation" switches may be installed for workers to use in the event of an incident.

(2) *Water supply.* Tissue culture laboratories should include separate hand washing sinks and laboratory sinks that are supplied with hot and cold laboratory water (see Section 4.2.5). Adequate floor drainage should be provided in the preparation and washup areas, with a slight fall in floor level from the tissue culture lab to the washup. Access to purified water is also needed (see Sections 5.4.1, 11.5.1). In some facilities, purified water is piped to laboratory sinks. However, pipework can build up dirt and algae, requiring routine testing and decontamination.

(3) *Vacuum supply.* A vacuum line can be useful for evacuating culture flasks and reduces the number of repetitive actions when handling multiple flasks. It is important for vacuum systems to be correctly installed. A collection vessel must be present with an additional trap flask to prevent fluid, vapor, or other contaminants from entering the vacuum line and pump; a hydrophobic filter should be positioned to protect the pump itself (see Figure 5.6). The vacuum pump must also be protected against the line being left open inadvertently. This can be accomplished via a pressure-activated foot switch that closes when no longer pressed. Given the difficulty in cleaning out vacuum lines and the repair or replacement cost if the central pump becomes contaminated, it is better to provide individual pumps at each workstation (see Figures 5.1, 5.6), or one pump between two workstations.

(4) *Computer infrastructure.* The modern tissue culture laboratory needs computers to be installed in association with digital cameras, cell counters, and other equipment (see Section 5.2.3). Although images and other data can be stored locally on computer hard drives, this represents a risk to data integrity; many laboratories prefer to connect all computers to a shared server so that data can be stored in a central location. Equipment monitoring and remote data backup are also needed as part of contingency planning (see Section 4.3.1). Ideally, the infrastructure for all these systems will be installed during construction.

Access and security. Most containment laboratories have restricted access arrangements in place. For lower containment laboratories, restricted access to the building may be sufficient, with a closed door to separate the tissue culture laboratory from other areas. Higher containment laboratories usually have more stringent access requirements (see Tables 6.4 and 6.5); for example, access may be limited to authorized workers using a key card or similar system. Some laboratories may also handle pathogens or toxins that represent a biosecurity risk. Such agents should undergo risk assessment so that any additional biosecurity requirements can be incorporated into facility design (see Supp. S6.3).

Although restricted access arrangements may be in place, it is important to plan ahead for an emergency situation (see Section 4.3.1). Doors should have release buttons installed so that users can override access restrictions from the inside of the facility in case of a safety incident. Access arrangements should allow emergency services personnel to enter the area during a fire or other emergency situation.

Fittings and furnishings. All tissue culture rooms should be designed for easy cleaning, with a rounded cove between wall and ceiling and wall and floor; coved finishes eliminate crevices, which are difficult to clean (HSE 2001). Floors should be covered in a material that is smooth, easily cleanable, and resistant to disinfectants (e.g. vinyl). A slight fall in the floor level should be allowed toward a floor drain, which should be located outside the door of the room (i.e. well away from the sterile handling area). This arrangement allows liberal use of water when the floor is washed, but, more importantly, it protects equipment from damaging floods if sinks or equipment overflow. Furniture should fit tightly to the floor, with a cove at the base of each unit, or be suspended from the bench with a space left underneath that is sufficient to allow cleaning.

Lighting systems should provide sufficient illumination for the task with minimal shadowing and glare (Kosniewski and Fiander 2013). Incandescent (tungsten) lighting is gradually being withdrawn in favor of more energy-efficient sources of lighting such as the fluorescent tube and light emitting diode (LED) fixture. However, fluorescent light is a major cause of degradation of tissue culture medium (see Section 8.8). Installation of fluorescent lighting may be necessary for greater energy efficiency, but lighting should be considered when designing a warmroom or coldroom so that exposure of medium to fluorescent light or daylight is minimal.

4.2 LAYOUT OF LABORATORY AREAS

Division of tissue culture laboratories into different areas is beneficial for the user and an important part of design for microbiological safety (Phillips and Runkle 1967). Six areas must be provided as a minimum requirement to enable sterile handling, incubation, preparation, washup, sterilization, and storage (see Table 4.1). Additional areas are desirable if a larger space is available. If only a single room can be used, create a "sterility gradient;" the clean area for sterile handling should be located at one end of the room, farthest from the door, and washup and sterilization facilities should be placed at the other

TABLE 4.1. Requirements for tissue culture areas.

Minimum requirements	Desirable features	Useful additions
Laboratory space that is separate from animal house and microbiological labs	Positively pressured, filtered air (air-conditioning)	Piped CO_2 gas supply with gas monitoring and alarms
Sterile handling area that is clean and quiet, and with no through traffic	Service cart (portable)	Compressed air and other gas supplies
Space for BSCs	Equipment bench or area adjacent to culture area	Vacuum line
Space for incubator(s)	Separate quarantine area or room	Separate warmroom with temperature recorder
Access to preparation area	Separate preparation room	Separate containment room (could double as quarantine room)
Access to washup area with sterilization facilities	Separate washup room with sterilization facilities	
Sinks for hand washing	Separate room for cold storage including cryofreezers	Separate microscope room (could convert to darkroom)
Access to backup generator power	Separate gas cylinder store	Liquid nitrogen storage tank(s) and separate storeroom for cryofreezers
Access to purified water	Computers installed alongside equipment with network connectivity	
Space for cryofreezer(s) with oxygen monitoring and alarms		Storeroom for bulk plastics and other room-temperature consumables
Access to storage areas that can hold:	Equipment monitoring and alarms	
Liquids: ambient, 4 °C, −20 °C Chemicals: ambient, 4 °C, −20 °C; can share with liquids, but keep chemicals in sealed container over desiccant Glassware, plastics (shelving) Small items (drawers) Specialized equipment (slow turnover), cupboard(s) Refrigerators and freezers with space for cold storage CO_2 cylinders		

end, with preparation, storage, and incubation in between. The preparation area should be adjacent to the washup and sterilization areas, and storage and incubators should be readily accessible to the sterile handling area (see Figures 4.1, 4.4). Several tissue culture laboratories may prefer to share preparation, washup, and sterilization areas. Sharing of services allows sterile handling and incubation to proceed in a smaller space while still providing adequate support areas.

4.2.1 Sterile Handling Area

Sterile work should be located in a quiet part of the tissue culture laboratory that is restricted to tissue culture (i.e. not shared with chemical work or with other organisms such as bacteria, yeast, or protozoa). There should be no through traffic or other disturbance that is likely to cause dust or drafts. The work area, in its simplest form, consists of a plastic laminate-topped bench, preferably plain white or neutral gray, to facilitate the observation of cultures and dissection, and to allow an accurate reading of pH when phenol red is used as an indicator. Nothing should be stored on the handling bench and any shelving above should be used only in conjunction with sterile work (e.g. for holding pipettes and instruments). The bench should be either freestanding (away from the wall) or sealed to the wall with a plastic sealing strip or mastic sealant.

Biological safety cabinets (BSCs) and laminar flow hoods. BSCs are employed for sterile work in most tissue culture laboratories because of the need for asepsis and containment. Historically, asepsis was the primary concern for tissue culture laboratories (Caputo 1988); tissue culture was performed on a bench in a separate room or cubicle (see Figure 12.2a, b). The introduction of laminar flow hoods with sterile air blown onto the work surface afforded greater control of sterility at a lower cost than providing a separate sterile room (see Section 12.2). Although these horizontal laminar flow hoods (also known as "clean work stations") may be useful to prepare medium or other non-toxic culture reagents, they do not provide containment. Filtered air is blown across the work surface directly at the operator, exposing him or her to aerosols that are inevitably generated during handling of cultured cells (see Figure 4.6a). The BSC overcomes this problem, providing both asepsis and containment (Kruse et al. 1991). Cabinet design was modified over time to filter intake and exhaust air and modify airflow so that it is directed vertically within the cabinet (see Figure 4.6b).

BSC classes and types are described in the next chapter (see Section 5.1.1). Most tissue culture laboratories perform sterile work using a Class II BSC; future reference to a BSC will imply a Class II BSC unless specified otherwise. Although some procedures may not require a Class II BSC, it is better to assume that such a cabinet may be required and incorporate into the design for the sterile handling area. More information can be found later in this book on using a BSC for

containment and maintaining good aseptic technique within a BSC (see Sections 6.3.4, 12.4.1).

Some BSCs are hard-ducted to vent exhaust air outside the laboratory (see Table 5.2). Installation of ductwork to the BSC has some advantages (e.g. decontamination with formaldehyde is facilitated if it is required) but it can be detrimental to room airflow. If the BSC is hard-ducted, it should be run continuously to avoid variation in airflow or pressure balances when the cabinet is turned on and off. Running a BSC continuously is beneficial to maintain asepsis but it will increase power consumption and heat load inside the room. Think about the work that you intend to perform, assess the associated risks (see Section 6.1.1), and refer to Standards and other safety regulations to determine whether hard ducting is required (British Standards Institute 2005). Always look for the option that provides the best risk management or flexibility for future work, provided it is feasible for the space and your budget.

BSCs should be installed as part of the construction contract; upgrading to a BSC after construction can be expensive and disruptive because cabinets have significant space and ventilation requirements. Each BSC should be carefully positioned to minimize cross drafts, which can impair performance (Rake 1978). There should be no through traffic and no air currents from adjacent equipment. Cabinets should have a lateral separation of at least 500 mm (2 ft), to allow access for maintenance and to minimize interference in airflow between hoods. Additional separation may be required if the air exhaust is located on one side of the cabinet; there may be a choice between top and side exhaust when purchasing the unit. If BSCs are facing each other, there should be a minimum of 3000 mm (10 ft) between the fronts of each cabinet. Detailed requirements for BSC installation and use have been developed (British Standards Institute 2005); these may vary based on your location. Always check local safety regulations to ensure that the positioning of the BSC is compliant with regulatory requirements (OECD 2018).

Service cart. A small service cart, trolley, or folding flap with a minimum size of 500 mm (18 in.) should be provided beside each BSC to hold materials that may be required but are not in immediate use. Service carts should have wheels so that they can be positioned for easy access and removed for cleaning and restocking. In some laboratories, carts are used to remove used and soiled materials from the work area and exchanged for a clean freshly stocked cart each time the operator changes.

Equipment bench or area. It is often convenient to position a bench for a microscope, cell counter, and other instruments near the sterile handling area. The equipment bench may be used to divide the area or separate it from the other end of the lab (see Figures 4.1, 4.4). The bench may also be used for other accessory equipment, such as a small centrifuge. The equipment bench may double as a

(a)

(b)

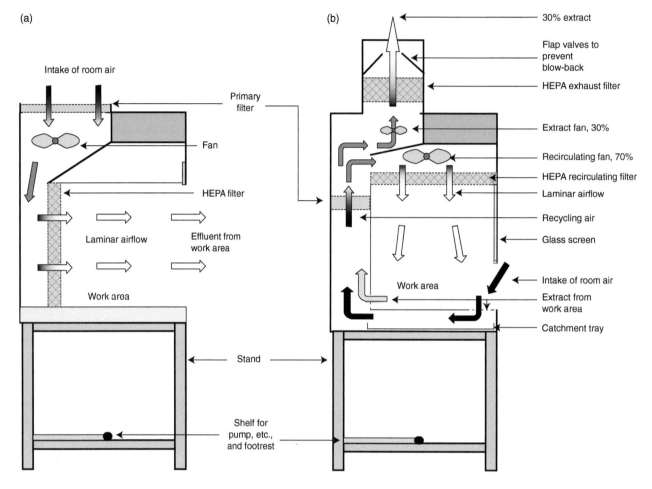

Intake of room air

Primary filter

Fan

HEPA filter

Laminar airflow

Effluent from work area

Work area

Stand

Shelf for pump, etc., and footrest

30% extract

Flap valves to prevent blow-back

HEPA exhaust filter

Extract fan, 30%

Recirculating fan, 70%

HEPA recirculating filter

Laminar airflow

Recycling air

Glass screen

Intake of room air

Extract from work area

Catchment tray

Work area

Fig. 4.6. Airflow in laminar flow hoods and BSCs. (a) Laminar flow hood with horizontal airflow, resulting in asepsis but not containment; (b) BSC with vertical laminar airflow and HEPA filters, resulting in both asepsis and containment (see also Figure 6.7). Large arrows denote direction of airflow.

storage area for sterile glassware, plastics, pipettes, screw caps, syringes, etc., in drawer units below and open shelves above.

If a digital camera or electronic cell counter is used, it may be necessary to locate a computer alongside the microscope and connect the computer to the organization's data storage system. A separate microscope area or room may be desirable if there is sufficient space, allowing darkroom photography of fluorescent markers without disturbing other occupants of the room (see Figures 4.3, 4.4).

4.2.2 Incubation Area

The requirement for cleanliness is not as stringent for incubation areas as it is for sterile handling. However, it is important to maintain clean air, a low disturbance level, and minimal traffic to avoid dust, spores, and the drafts that carry them. Incubation is usually performed using separate incubators, but some laboratories prefer a thermostatically controlled warm-room with sufficient space to hold a number of culture vessels. Decisions about incubation will be determined by the size and

type of culture vessels required, the temperature, gas phase, humidity, and proximity to the work space.

Incubators. Incubators take up a relatively small amount of space if only one or two are required. Many brands supply stacking kits that allow two to three incubators to be stacked in tiers during installation (see Figure 5.14a). The same effect can be produced by positioning incubators on wheeled trolleys that are custom-built at the correct height and width. This approach can be used to squeeze a number of incubators into a relatively small space. However, a large number of incubators results in rapid accumulation of heat (see Section 4.1.4), is expensive, and requires more time to maintain and clean. It is important to consider how many incubators you will need. As a rough guide, $0.2\,\mathrm{m}^3$ (200 l, 7 ft^3) of incubation space with $0.5\,\mathrm{m}^2$ (6 ft^2) shelf space is needed per person.

Standalone incubators have many advantages. They allow laboratories to maintain different cell lines or experiments using different culture conditions. Most incubators that are used for cell culture provide a humid environment by using a water pan or by injecting humid air into the incubator chamber. Incubators used for cell culture typically regulate gas levels by injecting gas from a cylinder or piped supply system (see Section 5.3.1; see also Figure 4.5). Although regulation of gas levels may not be needed for all experiments, it is required for many cell lines and should be considered during design of the incubation area.

Warmrooms. Standalone incubators have some disadvantages, including contamination of the water pan and a tendency to lose heat and recover slowly when the incubator door is opened. Disadvantages can be minimized through good incubator design, but laboratories may wish to consider alternatives for some applications. A warmroom allows large numbers of sealed culture vessels to be incubated more efficiently. This option is particularly useful for large numbers of sealed flasks, but cannot be utilized for open plates and dishes, which require a humid CO_2 incubator. A warmroom is also useful if scale-up and automation results in oddly shaped vessels or configurations that will not fit into the incubator chamber (see Chapter 28). Many scale-up and automation devices are supplied with their own insulation and temperature controls, but this is not always the case.

Warmrooms require careful thermostatic control with dual heating circuits and safety thermometers. They may also require cooling, or at least ventilation, particularly if heat-generating equipment is to be used (e.g. stirrers and rollers). More information on designing a warmroom is provided as Supplementary Material online (see Supp. S4.1).

4.2.3 Quarantine Area

If sufficient space is available, a separate room may be designated for quarantine or a different level of containment. Newly arrived cell lines or biopsies can be handled in the quarantine area until they are shown to be free of contamination, particularly mycoplasma and viral pathogens (see Sections 16.2, 16.4, 16.5). Ideally, the quarantine room should be separate from the rest of the tissue culture area with its own BSC, incubators, freezer, refrigerator, centrifuge, supplies, and waste disposal. If this is not possible, quarantine may be achieved by setting aside an incubator specifically for quarantine cultures and by handling such cultures at the end of the working day.

It may be possible to use the quarantine area for cultures with increased risk levels (e.g. cell lines that are known to carry pathogens). It is important to perform a risk assessment for any live culture and apply the containment measures prescribed for that level of risk (see Section 6.3.3). Higher containment levels may have significant impact on the design of the quarantine area; the quarantine room may need to be at negative pressure compared to the main handling area, and a higher class BSC may be required with a separate extract and pathogen trap (see Section 6.3.4; see also Figure 6.7).

4.2.4 Preparation Area

The need for a preparation area and the size of such an area will be determined by the laboratory's approach to sourcing cell culture media. Extensive preparation of media in small laboratories can be avoided if there is a reliable source of commercial culture media. Although a large enterprise (approximately 50 people doing tissue culture) may still find it more economical to prepare its own media, smaller laboratories may prefer to purchase ready-made media. These laboratories would need only to prepare reagents, such as water, salt solutions, and ethylene diamine tetra-acetic acid (EDTA), bottle these reagents, and package screw caps and other small items for sterilization. In that case, although the preparation area should still be clean and quiet, sterile handling is not necessary, as all the items will be sterilized.

If reliable commercial media are difficult to obtain, the preparation area should be large enough to accommodate a coarse and a fine balance, a pH meter, and, if possible, an osmometer (see Figure 9.8). Bench space will be required for dissolving and stirring solutions and for bottling and packaging various materials, and additional ambient temperature and refrigerated shelf space will be needed. If possible, a laminar flow hood or BSC should be provided in the sterile area for filter sterilization and bottling (see Section 4.2.1). Note that horizontal laminar flow hoods do not protect the operator and are not suitable for antibiotics or hazardous chemicals, due to the potential for inhalation of powder or aerosols. Incubator space will also be required for quality control of sterility (i.e. incubation of media samples for sterility testing).

Heat-stable solutions and equipment can be autoclaved or dry-heat sterilized using ovens or other equipment located at the non-sterile end of the preparation area. Both streams then converge on the storage areas (see Figure 4.3). Water purification equipment may also be located in the preparation area, allowing easy access to various grades of water purity (see Section 11.5.1).

4.2.5 Washup Area

Sinks. All tissue culture laboratories must have access to a sink for hand washing. Some tissue culture areas will have a separate sink for rinsing laboratory glassware even if a separate washup and sterilization area is positioned elsewhere. However, sinks can be a source of microbial contamination and some users prefer to position them outside the tissue culture laboratory. If you are designing a lab from scratch, it is important to carefully position the sink and decide on the size that you want. Stainless steel or polypropylene are best – the former if you plan to use radioisotopes and the latter for hypochlorite disinfectants.

Sinks should be deep enough to allow soaking, washing, and rinsing of all your glassware (except pipettes and the largest bottles) without having to stoop too far to reach into them. A sink that is 400 mm (15 in.) wide × 600 mm (24 in.) long × 300 mm (12 in.) deep × 900 mm (3 ft) high is about right. It is better to be too high than too low – a short person can always stand on a raised step to reach a high sink, but a tall person will always have to bend down if the sink is too low. A raised edge around the top of the sink will contain spillage and prevent the operator from getting wet when bending over the sink. The raised edge should go around behind the taps (faucets) at the back. Each laboratory sink should receive a supply of hot and cold laboratory water and purified water (e.g. reverse osmosis [RO] water); sinks in washup areas may have a combined hot-and-cold mixer and a hose connection for a rinsing device.

Washup. The need for an extensive washup area has been minimized by the move to disposable plasticware, but there will still be a need for some glassware washing as well as handling of used plastics, which may need to be autoclaved. In larger facilities, washup and sterilization are often performed in the same area, allowing materials to be sterilized after washup and before their return to the sterile handling area. Washup and sterilization facilities are best situated outside the tissue culture laboratory, as the humidity and heat that they produce may be difficult to dissipate without increasing the airflow above desirable limits.

The washup area should have space for soaking glassware and accommodate an automatic washing machine. Soaking baths or sinks should be deep enough so that all used glassware (except the largest bottles and cylinders) can be totally immersed in detergent during soaking, but not so deep that the weight of the glass is sufficient to break smaller items at the bottom. There should be plenty of bench space for handling baskets of glassware, sorting glassware by size or function, and packaging and sealing packs for sterilization. Trolleys or carts are often useful for collecting dirty glassware and redistributing clean stocks (but remember to allocate parking space for them!).

Sterilization. Sterilization methods that must be considered include steam (autoclave), dry heat (oven), and filtration (see Chapter 11). Autoclaves have additional requirements for power and heat dispersal that should be included in the design process; ideally, autoclaves and ovens should be located in a separate room if possible (see Figures 4.3, 4.4). If the sterilization facilities must be located in the tissue culture laboratory, place them nearest the air extract and farthest from the sterile handling area, with an efficient extraction fan.

4.2.6 Storage Area

Storage can be divided into two different types: cold and room-temperature storage. Cold storage has implications for the user's safety and for maintenance of sample integrity, requiring engineering controls to be incorporated during the design process (see Sections 4.1.4, 4.3.2). Room-temperature storage does not have these concerns but should still be carefully planned so that the user can quickly and easily find everything they need to perform their work. For both types of storage, sterile and non-sterile items should be kept separate and all items should be clearly labeled.

Cold storage. Cold storage is required for the following items:

(1) Sterile liquids at 4 °C (media), −20 °C (reagents such as serum, trypsin, glutamine, etc.), or −80 °C (heat-labile reagents where defrost cycles may be detrimental).
(2) Biological materials at −20 °C or −80 °C (DNA, RNA, protein, etc.).
(3) Cryofreezers containing cryopreserved cells and tissue samples maintained using liquid nitrogen in liquid or vapor phase (see Section 15.2.3).
(4) Liquid nitrogen to replenish cryofreezers. Liquid nitrogen can be stored:
   a) in Dewars on the floor (small volume, e.g. 25 l);
   b) in a large storage vessel on a trolley or wheeled castors (medium volume, e.g. 180 l);
   c) in storage tanks with provision to fill Dewars locally and/or provide liquid nitrogen using a piped supply (large volume, e.g. 500 l). Tanks may be positioned in a room with adequate ventilation or outdoors in secure, weatherproof housing (see Figure 4.7).

If option (c) above is selected, central liquid nitrogen storage tanks should ideally be positioned near the cryofreezers that they supply. The storage tanks can then be connected to cryofreezers using a minimal length of pipework, reducing unnecessary loss of liquid nitrogen through evaporation. Pipework should be insulated for the same reason. If a piped supply is used, cryofreezers must have the capacity to regulate liquid nitrogen supply. In a conventional cryofreezer, where liquid nitrogen is located in the cryovial storage chamber, the regulator will measure upper and lower levels to start and stop the fill process. Dual temperature probes should be included near the top and at the center of the storage chamber to monitor the storage temperature (see Section 15.1.3). It is advisable to add an independent temperature probe to the storage chamber, connected to an independent equipment alarm system (see Section 4.3.2).

❖ *Safety Note. There is a risk of asphyxiation due to reduced oxygen in any room where liquid nitrogen is stored or handled (see Section 4.1.4).*

Refrigerators and freezers may be positioned near cryofreezers and liquid nitrogen tanks, near room-temperature storage, or in the corridor adjacent to the tissue culture laboratory. The latter is particularly useful for warm climates, allowing their heat load to be dispersed away from laboratory areas, provided corridors are not obstructed (see Figure 4.8a). Where space is limited to a single room, refrigerators and freezers should be located toward the non-sterile end of the lab, as the doors and compressor fans create dust and drafts and may harbor fungal spores. These devices also require maintenance and periodic defrosting, which is best separated from the sterile handling area.

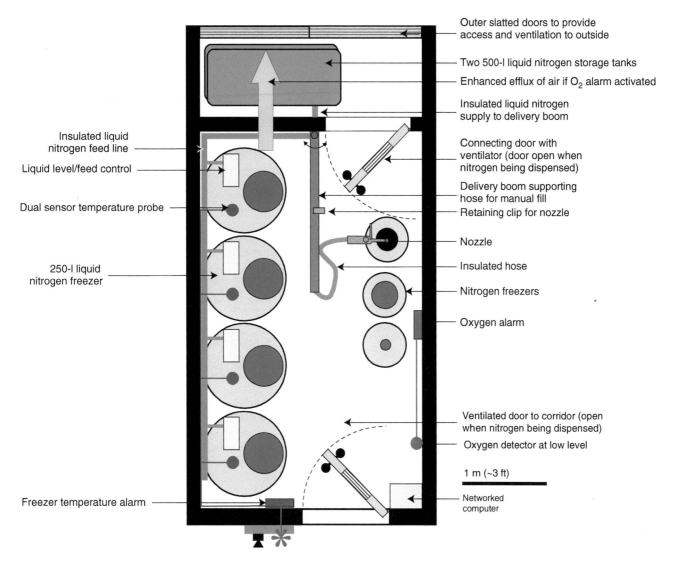

Outer slatted doors to provide access and ventilation to outside

Two 500-l liquid nitrogen storage tanks

Enhanced efflux of air if O_2 alarm activated

Insulated liquid nitrogen supply to delivery boom

Connecting door with ventilator (door open when nitrogen being dispensed)

Delivery boom supporting hose for manual fill

Retaining clip for nozzle

Nozzle

Insulated hose

Nitrogen freezers

Oxygen alarm

Ventilated door to corridor (open when nitrogen being dispensed)

Oxygen detector at low level

1 m (~3 ft)

Networked computer

Insulated liquid nitrogen feed line

Liquid level/feed control

Dual sensor temperature probe

250-l liquid nitrogen freezer

Freezer temperature alarm

Fig. 4.7. Liquid nitrogen store and cryostore. The liquid nitrogen store is best located on an outer wall with ventilation to the outside and easy access for deliveries. If the cryostore is adjacent, cryofreezers may be filled directly from an overhead supply line and flexible hose. Doors are left open for ventilation during filling; a wall-mounted oxygen alarm with a low-mounted detector sounds if oxygen falls below a safe level.

Some laboratories include a walk-in coldroom. A coldroom may be required if cold preparation or isolation procedures are used and will typically give more storage per cubic meter. However, the utilization of that space is important – how easy is it to clean and defrost, and how well can space be allocated to individual users? Several independent refrigerators will occupy more space than the equivalent volume of coldroom but are likely to be easier to manage and maintain in the event of failure. If a coldroom is to be installed, the walls should be smooth and easily cleaned, and the racking should be on castors to facilitate moving for cleaning. Coldrooms should be cleaned out regularly to eliminate old stock, and the walls and shelves should be washed with an antiseptic detergent to minimize fungal contamination. A walk-in freezer room is also possible but usually not recommended – it is difficult and unpleasant to clear out and the contents become difficult to relocate when maintenance is required.

As a rough guide, you will need 200 l (~ 8 ft³) of 4 °C storage and 100 l (~ 4 ft³) of −20 °C storage per person. The volume per person increases with fewer people. Thus, one person may need a 250-l (10-ft³) refrigerator and a 150-l (6-ft³) freezer. Of course, these figures refer to storage space only; allowance must be made for access and working space in a walk-in room. Some facilities will have specialized requirements to consider. For example, a biobank may install large

Fig. 4.8. Storage areas. (a) Positioning of refrigerators and freezers in corridors to reduce heat load in laboratory areas. Note that recessed alcoves are used to preserve the corridor as a thoroughfare. It is important to check that equipment locations comply with local safety regulations; for example, equipment storage may not be possible in corridors used for emergency evacuation. (b) Room temperature storage area, using a compactus with adjustable shelves to maximize storage. *Source*: Amanda Capes-Davis, courtesy of CMRI.

−80 °C freezers that allow automation of storage processes. These large-volume units must provide their own redundant systems in case of compressor failure or loss of electrical supply (Baird et al. 2013).

Room-temperature storage. Room-temperature storage is required for the following items:

(1) Sterile liquids at room temperature (salt solutions, water, etc.).
(2) Sterile and non-sterile glassware, including media bottles and pipettes.
(3) Sterile disposable plastics (e.g. culture flasks, multiwell plates, Petri dishes, centrifuge tubes, vials, and syringes).
(4) Screw caps, stoppers, etc., both sterile and non-sterile.
(5) Apparatus such as filters, both sterile and non-sterile.
(6) Protective personal equipment (PPE), including gowns or lab coats and gloves.
(7) Plastic bags and other disposable items.
(8) Cylinders for carbon dioxide, nitrogen, or other gases if a central supply system is not in use (see Section 4.1.4).

❖ *Safety Note. Gas cylinders should be tethered to the wall or bench in a rack (see Figure 6.5).*

Storage areas 1–6 should be within easy reach of the sterile handling area and allow ready access for both withdrawal and replenishment of stocks, keeping older stocks to the front. Double-sided units are useful because they may be restocked from one side and used from the other. A storage "compactus" (mobile shelving unit) is also useful to maximize storage space (see Figure 4.8b). Remember to allocate sufficient space for storage, as doing so will allow you to make bulk purchases, thereby saving money, and, at the same time, reduce the risk of running out of valuable stocks at times when they cannot be replaced.

❖ *Safety Note. It is essential to have a lip on the edge of both sides of a shelf if the shelf is at a high level and glassware and reagents are stored on it. This prevents items being accidentally dislodged during use and when stocks are replenished.*

4.3 DISASTER AND CONTINGENCY PLANNING

Laboratory design and layout become critically important when a disaster occurs. For example, if a freezer malfunctions and causes a fire, its impact will be determined by the presence of emergency equipment (e.g. fire alarms), access to fire exits (allowing workers to evacuate safely and emergency service personnel to enter), and the number of freezers in the laboratory (allowing backup of samples across multiple freezers). Laboratory design should include planning for disasters, which can be defined as events that prevent an organization from carrying on normal business operations (Mische and Wilkerson 2016).

Many research organizations perform disaster planning as a regulatory requirement. For example, colleges and universities in the United States may be required to perform disaster planning based on the National Incident Management System (NIMS) (FEMA 2018). Incident management systems are developed in response to many different threats and hazards, whether natural (e.g. fire, flood, or earthquake) or manmade (e.g. building-wide power loss, large chemical spill). Such events may result in large-scale emergencies that threaten worker safety, sample and data integrity, and ongoing laboratory operation.

Small-scale incidents must also be considered. For example, thawing of a freezer containing irreplaceable research material may not threaten worker safety, but it can have a major impact on sample integrity and ongoing projects. Such events are relatively common. Analysis of equipment data at a research institute during 2016–2017 showed that, on average, seven to eight freezers and refrigerators failed outside working hours during a 12-month period. All incidents were detected using an equipment alarm system and sample integrity was preserved by transferring samples to backup equipment (data courtesy of Amanda Capes-Davis and Darryn Capes-Davis, CMRI).

4.3.1 Contingency Plans and Priorities

The term contingency planning is commonly used to cover both large-scale disasters and smaller-scale laboratory incidents. Contingency planning involves anticipating a specific hazard based on specific events and known risks. The contingency planning process can be broken down into three simple questions (IFRC 2012). What is going to happen? What are we going to do about it? What can we do ahead of time to get prepared?

Determining exactly what is going to happen is impossible. Natural disasters are low probability events and may not adhere to contingency plans, no matter how much time and effort is invested in their development (Bjugn and Hansen 2013). However, it is useful to perform a risk assessment based on the likelihood and severity of various incident scenarios, using a similar process to the safety risk assessment (see Section 6.1.1). The risk assessment process helps to identify priorities for risk mitigation, which can be defined as sustained actions that are taken to reduce or eliminate long-term risk to life and property from the hazards that are identified (Mische and Wilkerson 2016).

The following priorities are particularly important for tissue culture laboratories, with each area requiring specific mitigation strategies:

(1) **Safety.** The safety of workers, visitors, and bystanders is the highest priority in any crisis or emergency situation (National Research Council 2011). Most countries have regulatory requirements for emergency evacuation procedures and equipment. Always comply with these requirements, including evacuation drills – such drills are critically important in emergency situations, where there is very little time for thought. Be aware of the locations of the nearest fire exits and safety equipment, including emergency alarm points, safety showers, eyewash stations, and first aid kits.

(2) **Animal welfare.** The safety and wellbeing of research animals is the second highest priority during any crisis or emergency situation (National Research Council 2011). If cells are obtained from research animals or used in xenograft models, this becomes an important consideration for tissue culture personnel.

(3) **Sample integrity.** Research samples and reagents must be maintained during a crisis and throughout the post-crisis recovery period. Cultures in incubators are difficult to maintain for extended amounts of time, particularly if the crisis results in limited access to the laboratory. The highest priority for sample integrity is to preserve samples and reagents in cold storage. Cryofreezers must be carefully positioned during layout of the laboratory so that there is good access for liquid nitrogen deliveries but minimal risk of flooding. If warnings are issued before a natural disaster, workers should top up liquid nitrogen in cryofreezers and carry out emergency checks (e.g. of essential supplies and generator fuel). Otherwise, sample integrity relies on engineering controls including backup generators and equipment alarm systems (see Sections 4.1.4, 4.3.2). If these measures fail, it is important to minimize loss of irreplaceable samples. Offsite storage is offered by some cell repositories as a service or may be arranged as an exchange between two laboratories in different locations. Deposit of novel cell lines at a cell repository is another important way to maintain sample integrity; this minimizes the risk of contamination and ensures that unique cell lines are preserved (see Section 15.6).

(4) **Data integrity.** Integrity of electronic systems is increasingly important during crisis management. Research data are obviously needed for ongoing work, but many laboratories now rely on electronic access to phone systems, contact information for laboratory personnel, and hazardous substance manifests for emergency service personnel. The highest priority for data integrity is to preserve access to emergency communication and information during the crisis period. Contact information for key personnel should be printed and posted

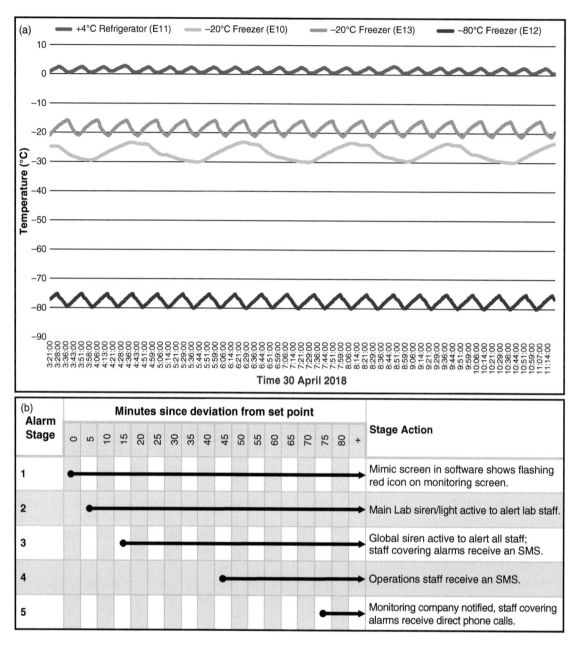

Fig. 4.9. Equipment monitoring and alarm systems. (a) Temperature monitoring data over an 8-h period, showing typical temperature fluctuations for a refrigerator, two −20 °C freezers, and a −80 °C freezer; (b) preprogrammed sequence of events following deviation from a temperature set point, resulting in local, global, and remote alarms. *Source*: Courtesy of Darryn Capes-Davis, CMRI.

throughout the laboratory so that it is easily accessible in a crisis situation. Research data should be maintained until access is required post crisis. Always store electronic information in a secure central location, with offsite storage and backup to minimize loss of research data.

(5) **Research continuity.** After the initial incident has been addressed, an extended period of recovery may follow post crisis. For example, looking at the Christchurch, New Zealand earthquakes of 2010–2011, power was not restored to the city for five days after the main event and computer and telephone connections were not restored for five months (Morrin and Robinson 2013). It is impossible to plan for all scenarios because each crisis will vary in its impact. However, it is important to think about which samples and procedures should be restored as high priority. Separate laboratory areas may mean that some operations can be preserved following a laboratory incident.

4.3.2 Equipment Monitoring and Alarms

Cold storage facilities are critical infrastructure for any tissue culture laboratory and represent an important focus for contingency planning (Mische and Wilkerson 2016). Failure of freezers or refrigerators may affect a single device (e.g. due to compressor failure) or multiple devices (e.g. due to loss of electrical supply). In some cases, one cause can lead to the other – for example, a faulty device can overload an electrical circuit, resulting in failure of multiple devices due to power loss. It is essential to comprehend the value of the samples that are stored in these devices and develop monitoring and alarm systems that preserve their viability and integrity.

Many pieces of equipment and high containment areas have local alarms that become active following breakdown in temperature or ventilation control. Always pay attention to these alarms, which convey essential information for safety and sample integrity. However, it is important to be aware that equipment alarms are not always reliable. This is illustrated by a freezer failure that occurred in a brain biobank during 2012 (Baird et al. 2013). The summary readout on the front panel indicated "System is OK," while the equipment warmed to room temperature, resulting in loss of unique tissue samples. To manage this risk, a separate temperature probe should be added to monitor temperature in each device independently.

An equipment alarm system is highly desirable to monitor independent temperature probes and generate automated alarms when variables exceed their set points. Such systems also generate data that provide useful information on the conditions within cold storage equipment (see Figure 4.9a). Equipment alarm systems are typically customized for specific organizations, as it is difficult to provide a "one size fits all" approach. It may be possible to expand an organization's Building Management System (BMS); the BMS is typically used to monitor environmental conditions within the laboratory (e.g. room temperature, pressure differentials), which is a close fit to the requirements of an equipment alarm system (Keenan 2007). Temperature and alarm information should be accessible remotely, allowing on-call staff to respond to alarms while away from the building. A preprogrammed sequence may be used to control alarms, starting with a local siren and light before progressing to building-wide and remote alarms (see Figure 4.9b).

Many laboratories do not have the resources and expertise required to develop a customized equipment alarm system. However, even simple measures can be effective. On-call staff can perform temperature monitoring checks and respond to alarms outside working hours. The correct procedures to carry out and the person or persons to contact should be posted beside the equipment or alarm panel and brought to the notice of all staff (e.g. during induction of new workers; see Section 30.1.2). Whatever the method used, it is essential that everyone becomes involved. The need for involvement should be made clear as it is to everyone's benefit. All members of the organization should be trained to respond to alarms, either to take responsibility themselves or to "pass the baton" to a more appropriate person. It is in everyone's interest to see that the correct person is informed and correct procedures are carried out.

Supp. S4.1 Designing a Warmroom.

REFERENCES

Baird, P.M., Benes, F.M., Chan, C.H. et al. (2013). How is your biobank handling disaster recovery efforts? *Biopreserv. Biobank.* 11 (4): 194–201. https://doi.org/10.1089/bio.2013.1142.

Bjugn, R. and Hansen, J. (2013). Learning by erring: fire! *Biopreserv. Biobank.* 11 (4): 202–205. https://doi.org/10.1089/bio.2013.0020.

British Standards Institute (2005). BS 5726:2005. Microbiological safety cabinets. London: British Standards Institute.

Caputo, J.L. (1988). Biosafety procedures in cell culture. *J. Tissue Cult. Methods* 11 (4): 223–227.

CDC and NIH (2009). Biosafety in microbiological and biomedical laboratories. https://www.cdc.gov/labs/bmbl.html (accessed 11 March 2019).

FEMA (2018). National Incident Management System. Department of Homeland Security. https://www.fema.gov/national-incident-management-system (accessed 25 April 2018).

Finkel, E. (2007). Research safety. Inquest flags little-known danger of high-containment labs. *Science* 316 (5825): 677. https://doi.org/10.1126/science.316.5825.677.

HSE (2001). The management, design and operation of microbiological containment laboratories. www.hse.gov.uk/pubns/books/microbio-cont.htm (accessed 28 March 2018).

IFRC (2012). Contingency planning guide. http://www.ifrc.org/pagefiles/40825/1220900-cpg%202012-en-lr.pdf (accessed 25 April 2018).

Keenan, D. (2007). Development of an equipment monitoring system. Lab Manager. http://www.labmanager.com/laboratory-technology/2007/10/development-of-an-equipment-monitoring-system#.wuba3ohub-h (accessed 25 April 2018).

Kimman, T.G., Smit, E., and Klein, M.R. (2008). Evidence-based biosafety: a review of the principles and effectiveness of microbiological containment measures. *Clin. Microbiol. Rev.* 21 (3): 403–425. https://doi.org/10.1128/cmr.00014-08.

Kosniewski, J. and Fiander, B. (2013). The lighting balancing act for laboratories. Forensic Magazine. https://www.forensicmag.com/article/2013/08/lighting-balancing-act-laboratories (accessed 25 April 2018).

Kruse, R.H., Puckett, W.H., and Richardson, J.H. (1991). Biological safety cabinetry. *Clin. Microbiol. Rev.* 4 (2): 207–241.

Langford, N.J. (2005). Carbon dioxide poisoning. *Toxicol. Rev.* 24 (4): 229–235.

Li, L., Gu, J., Shi, X. et al. (2005). Biosafety level 3 laboratory for autopsies of patients with severe acute respiratory syndrome: principles, practices, and prospects. *Clin. Infect. Dis.* 41 (6): 815–821. https://doi.org/10.1086/432720.

Lim, W., Ng, K.C., and Tsang, D.N. (2006). Laboratory containment of SARS virus. *Ann. Acad. Med. Singap.* 35 (5): 354–360.

Mische, S. and Wilkerson, A. (2016). Disaster and contingency planning for scientific shared resource cores. *J. Biomol. Tech.* 27 (1): 4–17. https://doi.org/10.7171/jbt.16-2701-003.

Morrin, H.R. and Robinson, B.A. (2013). Sustaining a biobank through a series of earthquake swarms: lessons learned from our New Zealand experience. *Biopreserv. Biobank.* 11 (4): 211–215. https://doi.org/10.1089/bio.2013.0033.

National Research Council (2011). Prudent practices in the laboratory: handling and management of chemical hazards, updated version. https://www.ncbi.nlm.nih.gov/books/nbk55878 (accessed 25 April 2018).

NIH (2016). Design requirements manual. https://www.orf.od .nih.gov/policiesandguidelines/biomedicalandanimalresearch facilitiesdesignpoliciesandguidelines/pages/designrequirements manual2016.aspx (accessed 2 April 2017).

OECD (2018). *Guidance Document on Good in Vitro Method Practices (GIVIMP)*. Paris: OECD Publishing.

Phillips, G.B. and Runkle, R.S. (1967). Laboratory design for microbiological safety. *Appl. Microbiol.* 15 (2): 378–389.

Rake, B.W. (1978). Influence of crossdrafts on the performance of a biological safety cabinet. *Appl. Environ. Microbiol.* 36 (2): 278–283.

van Soolingen, D., Wisselink, H.J., Lumb, R. et al. (2014). Practical biosafety in the tuberculosis laboratory: containment at the source is what truly counts. *Int. J. Tuberc. Lung Dis.* 18 (8): 885–889. https://doi.org/10.5588/ijtld.13.0629.

Webber, B. A. (2008). University of Michigan field study of Class II Biological Safety Cabinet energy consumption costs. www .thermofisher.com.au/uploads/file/scientific/applications/ equipment-furniture/university-of-michigan-study-of-class- ii-biological-safety-cabinet-energy-consumption-costs.pdf (accessed 27 April 2018).

Wesselschmidt, R.L. and Schwartz, P.H. (2011). The stem cell laboratory: design, equipment, and oversight. *Methods Mol. Biol.* 767: 3–13. https://doi.org/10.1007/978-1-61779-201-4_1.

CHAPTER 5

Equipment and Materials

Unless unlimited funds are available, it will be necessary to prioritize the needs of a tissue culture laboratory when acquiring new equipment (see Table 5.1). Equipment may be prioritized as (i) essential, i.e. items that must be used for tissue culture to be done safely and reliably; (ii) beneficial, i.e. items that allow tissue culture to be done in a better, easier, faster, or more efficient manner; and (iii) specialized, i.e. items that are required for some procedures or to enable more sophisticated analyses but are not essential for routine handling. Ergonomic considerations are particularly important for tissue culture laboratories, where procedures are often performed over prolonged periods of time. User comfort will increase efficiency and reduce short-term fatigue, and may reduce musculoskeletal injury over longer periods of time (see Section 6.1.4).

Equipment items are presented here under activity-based subject headings, following the laboratory layout described in the previous chapter (see Section 4.2). This chapter focuses primarily on "routine" equipment used for everyday procedures; specialized equipment is discussed in later chapters. Supplier information may vary considerably from country to country. Use the supplier list at the end of each chapter as a starting point, discuss supplier choices with colleagues, and refer to online resources (Lab Manager 2017; Biocompare 2018).

5.1 STERILE HANDLING AREA EQUIPMENT

5.1.1 Biological Safety Cabinet (BSC)

A biological safety cabinet (BSC) is used to provide asepsis for cultures during handling and containment from any pathogens that may be present (see Section 4.2.1) (WHO 2004; CDC and NIH 2009). Although it is possible to carry out aseptic procedures without laminar flow (see Protocol P12.3) or to do so using a laminar flow hood (also known as a "clean workstation"), neither approach allows for containment of cultures from the worker or the external environment. The BSC provides an aseptic environment but also acts as primary containment for protection of the worker, with the laboratory acting as secondary containment if the BSC fails or a spill occurs (see Section 4.1.3). Although the need for containment is determined by risk assessment for each culture, many laboratories choose to apply consistent procedures to all cultures for ease of operation and to comply with broader safety regulations (see Section 6.1.2).

BSC classes and types. BSCs are divided into three classes (see Table 5.2; see also Figure 6.7). Class II BSCs are designed to provide both asepsis and containment and are used by most laboratories for tissue culture. The other classes of BSC are not routinely used for tissue culture. Class I BSCs do not provide asepsis; airflow in a Class I cabinet is similar to a chemical fume hood, providing protection of the worker and the environment but not the material that is handled within the cabinet. Class III BSCs are used to handle highly infectious agents and provide increased containment for the worker and the environment if this is required following risk assessment (see Sections 6.1.1, 6.3.4). Future reference to a BSC in this chapter will imply a Class II BSC unless noted otherwise.

Class II BSCs are subdivided into four types – A1, A2, B1, and B2 (see Table 5.2). A1 and A2 cabinets recirculate 70%

Freshney's Culture of Animal Cells: A Manual of Basic Technique and Specialized Applications, Eighth Edition. Amanda Capes-Davis and R. Ian Freshney.
© 2021 John Wiley & Sons Ltd. Published 2021 by John Wiley & Sons Ltd.
Companion website: www.wiley.com/go/freshney/cellculture8

TABLE 5.1. Tissue culture equipment priorities.

Item[a]	Purpose
Essential	
Biological safety cabinet (BSC)	Maintain aseptic environment and containment of potential biohazards
Pipette controller(s)	Dispense medium and reagents (1–100 ml)
Pipettor(s)	Dispense small volumes (1 μl–1 ml) accurately and reproducibly
Bench-top centrifuge	Centrifuge cells during trypsinization, thawing, etc.
Inverted microscope	Enable routine observation and assessment of cultures
Camera for inverted microscope(s)	Photograph cultures; may allow video recording
Hemocytometer slide and coverslips	Enable manual counting of cells
CO_2 incubator	Incubate cells under controlled conditions (temperature, humidity, CO_2); use dry incubator if CO_2 incubator is not available
CO_2 supply	Provide CO_2 for incubator; either CO_2 cylinder or a piped system with back-up vessel and changeover device (see Figure 4.5)
Equipment alarm system or portable temperature monitor, gas monitor	Monitor temperature or other equipment data; may generate alarms (see Section 4.3.2)
Water bath	Prewarm medium and serum before handling procedures
Water purification system	Purify water for use in tissue culture or experiments
Balance	Weigh chemicals for reagent and medium preparation
pH meter	Measure pH in prepared media and reagents
Glassware washing machine	Wash reusable glassware and plasticware
Glassware drying oven	Dry reusable glassware and plasticware
Autoclave	Sterilize materials (e.g. bottles, pipette tips), stable liquids, and solutions
4 °C refrigerator	Store media and reagents
−20 °C freezer	Store unstable media, serum, and reagents
−70 °C (or below) freezer	Store biological specimens (DNA, RNA, protein)
Cryofreezer	Preserve cell lines using seed stock approach (see Section 15.5)
Liquid nitrogen storage Dewar	Provide local liquid nitrogen supply (not required if piped supply is used)
Controlled cooling device	Control freezing rate during cryopreservation (see Section 15.2.2)
Beneficial	
Portable service cart	Store frequently used items near the BSC
Portable glassware cart	Transfer soiled glassware from culture area to washup
Automated dispenser	Perform repetitive dispensing of liquids, e.g. for cell counter
Syringe dispenser	Perform repetitive dispensing of small volumes
Peristaltic pump or vacuum pump	Perform repetitive handling of large volumes, filter sterilization
Upright fluorescence microscope	Enable high power observation and assays, e.g. chromosome analysis, mycoplasma detection
Dissecting microscope	Enable embryo culture, and colony counting and picking
Monitor and computer	Enable digital photography, video recording, record keeping
Cell counter	Enable automated cell counting, analysis of size, and other attributes
Dry incubator	Incubate cells at a controlled temperature; enable sterility testing
Hypoxic (tri-gas) incubator	Incubate cells under hypoxic conditions
Hot plate magnetic stirrer	Dissolve chemicals for reagent and medium preparation
Conductivity meter	Enable quality testing of reagents
Osmometer	Enable quality testing of reagents
Sterilizing oven	Sterilize glassware, metals, and heat-resistant plastics; can double as drying oven
Filter sterilization apparatus	Sterilize heat labile solutions (disposables required)
Rate-controlled freezer	Control freezing rate using a preprogrammed sequence

TABLE 5.1. *(continued)*

Item[a]	Purpose
Specialized	
Automated plate dispenser	Perform repetitive liquid handling using microtitration plates
Automated plate reader	Analyze chromogenic endpoints in microtitration assays (ideally also able to read fluorescence and/or luminescence)
Bioreactor	Perform culture under controlled conditions for scale-up, 3D culture, induction of differentiation, etc. (see Section 28.1)
Liquid handling robot	Perform repetitive liquid handling for high-throughput assays
Large-capacity centrifuge (6 × 1 l)	Harvest large-scale suspension cultures
Centrifugal elutriator	Perform cell separation (see Section 21.2.2)
Live cell imaging equipment	Enable cell growth curves and live cell assays
Colony counter	Count colonies in cloning efficiency and survival assays
Flow cytometer	Analyze and sorts cells using various criteria (see Section 21.4)
Micromanipulators	Allow microinjection
Roller rack	Enable scale-up using roller bottle culture (see Section 28.3.1)
Magnetic stirrer	Enable stirrer or spinner culture (see Sections 8.6.2, 28.2.1)

[a]This table assumes that disposable plasticware (tubes, pipettes, flasks, etc.) are used.

TABLE 5.2. BSC classes.

Class	Inflow velocity		Exhausted airflow	Recirculated airflow	Plenum pressure[a]	Exhaust system
	(ft/min)	(m/s)	(%)	(%)		
I	75	0.36	100	0	−ve	Hard duct
II A1	75–100	0.38–0.51	30	70	N/A	Room or thimble connection
II A2	100	0.51	30	70	−ve	Room or thimble connection
II B1	100	0.51	30	70	−ve	Hard duct
II B2	100	0.51	100	0	−ve	Hard duct
III	N/A	N/A	100	0	−ve	Hard duct

Source: Based on R. Ian Freshney and Amanda Capes-Davis (WHO 2004; CDC and NIH 2009).
[a]Pressure in plenum surrounding all contaminated plenums.
N/A, Not applicable; −ve, negative pressure.

of the air and exhaust 30% into the room or via a thimble connection into the regular room ventilation duct. The A2 cabinet provides a zone of negative pressure surrounding the air plenum in the cabinet, making it a safer option if leakage occurs. B1 and B2 cabinets were originally developed to allow safe handling of minute quantities of hazardous chemicals; these cabinets require a dedicated extract duct to discharge exhaust air away from the laboratory (CDC and NIH 2009). B1 cabinets recirculate 30% and exhaust 70%, while B2 cabinets exhaust 100% with no air recirculation (WHO 2004).

BSCs must be certified to ensure that high-efficiency particulate air (HEPA) filters are intact and airflow meets specifications for cabinet design. BSCs are certified when they are installed or moved, and during normal use (at least once a year). Testing is done by a trained and competent person – usually a service engineer – to ensure that equipment complies with the relevant standard. Decontamination must

be performed before the service engineer works with the cabinet; the choice of method will depend on risk assessment (see Section 6.3.5).

• *Safety Note. It is important to familiarize yourself with local and national safety regulations before installing equipment, as legal requirements and recommendations will vary (see Section 6.1.2).*

User parameters. The following requirements should also be considered when selecting a new BSC:

(1) **Number.** Usually, one BSC is sufficient for two to three people.
(2) **Size.** A working surface of 1200 mm (4 ft) wide × 600 mm (2 ft) deep is usually adequate, unless large equipment will be used in the BSC. Check the external dimensions of

the BSC and compare to the floor space available. Make sure that the BSC can fit through the doorway to the laboratory, and check that there is sufficient headroom for venting to the room or for ducting to the exterior. Some BSCs vent from the side instead of the top. This can be an advantage where the ceiling height is low but may make venting to the outside via a separate hard duct more difficult.

(3) *Position.* Will the airflow from other cabinets, the room ventilation, or independent air-conditioning units interfere with the integrity of the work space of the BSC? That is, will contaminated air spill in or aerosols leak out because of turbulence? To meet this condition you will need 3000 mm (10 ft) of face-to-face separation between two BSCs, 1500 mm (5 ft) between the front aperture and a bench or doorway, and 300 mm (1 ft) between the side of the BSC and a wall or column (British Standards Institute 2005).

(4) *Servicing.* When in place, can the BSC be serviced easily? Space will be needed around the equipment for the service engineer during maintenance and certification. (Discuss with the service engineer, not the salesperson!)

(5) *User comfort and ergonomics.* Some cabinets have awkward ducting below the work surface (which leaves no room for your knees), lights or other accessories above that strike your head, or screens that obscure your vision. The person who will use the BSC most should try to simulate normal use by sitting at it before purchasing. Consider the following questions:

Can you get your knees under the BSC while sitting comfortably and close enough to work, with your hands at least halfway into the BSC and without leaning forward?

Can you rest your forearms comfortably when you are not using your hands?

Can stools or chairs be adjusted to improve user comfort?

Is there a footrest in the correct place and can it be adjusted if necessary?

Are you able to see what you are doing without placing strain on your neck?

Is the lighting convenient and adequate? Do you notice any excessive glare or heat coming from the light housing?

(6) *Noise and vibration level.* Noisy BSCs are more fatiguing. Think about overall noise and vibration levels for the room and try to minimize where possible.

(7) *Venting.* BSCs may vent from the top or to the side. Generally, top venting will make it easier to connect to ductwork, while side venting may be required if the ceiling height is limiting.

(8) *Interior.* There should be access for cleaning inside the working area and below the work surface in the event of spillage.

The work surface should be easily removable for cleaning.

A divided work surface is easier to remove but can trap spillage in the crack between the sections.

A perforated work surface is more likely to allow spillage to go through; a solid work surface vented at the front and back is preferable.

The edges of the work surface, when lifted, should be smooth. If the edges are sharp they can cause injury during cleaning.

(9) *Front screen.* The front screen or sash should be able to be raised, lowered, or removed completely to facilitate cleaning and handling of bulky culture apparatus. Remember, however, that a BSC will not give you, the operator, or the culture the required protection if you remove the front screen.

(10) *Ultraviolet (UV) light.* A UV light is included in some BSCs, although their effectiveness has been much debated (see Section 12.2.2). Make sure the coverage is adequate, although it will never reach gaps between and around the edge of the trays of the work surface. UV lights will need regular cleaning and replacement. Be aware that some plastics are UV-sensitive and check with the supplier that the front and side screens, if plastic, will not degrade in UV light.

The choice of a specific brand and model will depend on the laboratory space (see Section 4.2.1) and the user's requirements. BSCs are available as freestanding or bench-top units. Individual freestanding cabinets with lockable castors are preferable, as they separate operators and can be moved around. The ergonomics of the cabinet and chair can have a serious impact on the comfort of the operator. Chairs should be adjustable so that the seat height and back angle can be modified to suit the height of the hood (see Figure 12.3). The user should be able to work comfortably with the chair drawn up close to the front edge of the hood and with adequate foot support.

5.1.2 BSC Services and Consumables

The user will need access to services when using the BSC. These include power for equipment, gas (if required for individual flasks), peristaltic or vacuum pumps for filter sterilization or aspiration of spent medium, and other sterile liquid handling equipment. The layout for services and accessory equipment should be carefully planned (see Figure 5.1). Allow plenty of legroom underneath, with space for pumps, aspirators, and so forth. If services such as power or vacuum are installed through the BSC itself, check that they are accessible for maintenance and cleaning.

A moveable cart can be positioned next to the BSC or moved around as required. These carts conveniently fill the space between adjacent BSCs and are easily removed for cleaning or maintenance. The primary purpose of such a

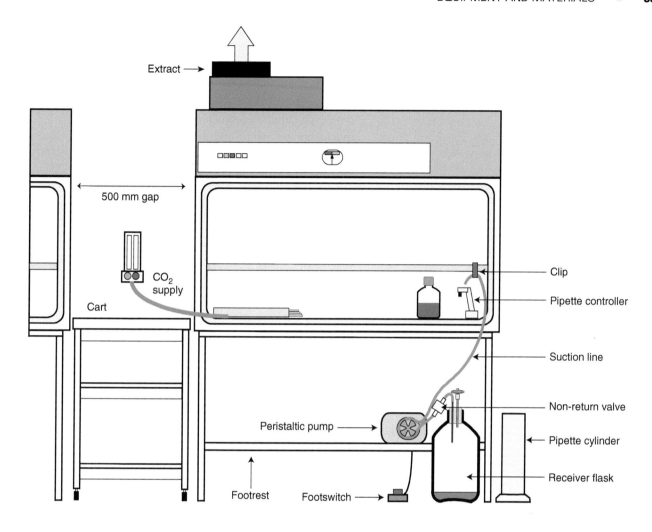

Fig. 5.1. Layout of BSC services. A peristaltic pump, connected to a receiver vessel via a non-return valve, is shown on the right side below the BSC, with a footswitch to activate the pump. The suction line from the pump leads to the work area and delivers spent medium to the receiver flask. The receiver flask is vented via a filter to prevent spread of aerosol into the room. A delivery tube from a gas mixer or cylinder allows gassing of flasks, e.g. for hypoxic conditions.

cart is to provide a convenient surface for frequently used items (see Table 5.3). Stocks of culture vessels, cryovials, and other items can be stored on shelves adjacent to the BSC (see Figure 12.3; see also Sections 8.5, 15.2.1). Shelf space is also required for stable liquids, such as water. Carts may be used to carry materials to and from the BSC and for staff to restock materials. Mobile carts are also useful for clearing soiled glassware and used items from the sterile handling area to the washup area, although this is likely to require a separate, larger cart that is parked at a convenient location (see Figures 4.3 and 4.4).

Tissue culture laboratories now typically use disposable plastic tubes and pipettes instead of glass (see Figure 5.2). Disposable plastics can account for approximately 60% of the tissue culture budget – even more than serum. "Shopping around" will often result in a cheaper price, but do not

be tempted to change products too frequently, and always test a new supplier's product for many different cultures or applications before committing to it (see Section 11.8.2). Plan ahead to estimate how much you will need; you may be able to buy in bulk or combine the purchasing power of several laboratories to obtain discounts on plasticware that you know will be reliable and of good quality.

Disposable plastic flasks and dishes are also widely used in preference to glass and are discussed in a later chapter (see Section 8.5). Choice is determined by (i) the yield (number of cells) required; (ii) whether the culture is grown in monolayer or suspension; and (iii) the sampling regime (see Section 7.6), i.e. are the samples to be collected simultaneously or at intervals over a period of time? Petri dishes are less expensive than flasks, though more prone to contamination and spillage. Depending on the pattern of work and the sterility of the

TABLE 5.3. Consumables at the BSC.

Item	Size	Purpose
Pipettes	1, 2, 5, 10, 25 ml	Dispense and withdraw medium, reagents, and cells
Pipette tips	20, 200, 1000 µl	Dispense and withdraw small volumes
Culture flasks	25, 75, 175 cm^2	Perform primary culture and propagation
Petri dishes	3.5, 5 or 6, 9 or 10 cm	Perform primary culture and propagation, cloning
Multi-well plates	4-, 6-, 12-, 24-, 96-well	Sample replicates
Sterile containers:		Store sterile liquids
Bottles, glass or plastic	100, 500 ml	Prepare and store media
Sample pots	50 ml	Store tissue samples
Universal containers	30 ml	Store various samples
Centrifuge tubes	15, 50, 250 ml	Centrifuge cells, store various samples
Bijou	5 ml	Store small samples
Microcentrifuge tubes	1, 2, 5 ml	Store small samples
Cryovials	1, 2, 5 ml	Store frozen samples, cells
Syringes	1, 2, 5, 10, 25 ml	Withdraw and dispense viscous liquids and small volumes from vials
Syringe needles	21–23 g	Withdraw from septum vials
	19 g	Dispense liquids and cells
Filters:		Sterilize heat-labile liquids
Syringe tip	13, 25, 47 mm	Sterilize small volumes
Bottle-top or flask	47 mm	Sterilize large volumes
Surgical gloves	Small, medium, large	Protect operator from biohazards and hazardous chemicals
Lint-free swabs	50 × 50 mm	Swab down work surface
Paper towels	Various	Mop up spillage
Disinfectants:		Disinfect surfaces, biohazardous waste
Alcohol (ethanol or isopropanol), 70%		Swab down work surface (see Section 12.3.1)
Sodium hypochlorite or other suitable disinfectant		Disinfect waste (see Section 6.3.6)

environment, they are worth considering, at least for use in experiments if not for routine propagation of cell lines. Petri dishes are particularly useful for clonogenic assays, in which colonies must be stained and counted or isolated at the end of an experiment (see Sections 19.4.1, 20.2).

5.1.3 Sterile Liquid Handling Equipment

Simple pipetting is one of the most frequent tasks required in the routine handling of cultures and is an important part of sterile handling (see Section 12.3). Pipetting of liquids requires the user to select the pipette material (e.g. glass or plastic), containers, and controllers. Smaller volumes require pipettors with disposable pipette tips or a needle and syringe. Larger volumes or repetitive pipetting require more ergonomic liquid dispensers. Peristaltic or vacuum pumps are also useful to minimize repetitive dispensing of larger liquid volumes or aspiration of spent medium.

Pipettes and pipette controllers. Serological pipettes are used to dispense specific volumes of liquid in the 1–100 ml range (see Figure 5.2b). Liquid is drawn into the pipette using negative pressure; in the past this was done by mouth pipetting, but this risky practice can now be eliminated by using a pipette controller. Pipettes should be of the blow-out variety, wide tipped for fast delivery, and graduated to the tip, with the maximum point of the scale at the top rather than the tip.

Pipettes can be made from glass (reusable) or plastic (disposable). Most tissue culture laboratories now use plastic pipettes, which are usually made from polystyrene. Glass pipettes are relatively cheap if reused, but they have an increased risk of breakage, contamination, and biohazard risk if washing and sterilization are not performed correctly. Glass pipettes must be collected into disinfectant in pipette cylinders (also known as pipette hods) after use. Clean pipettes must be plugged with cotton or similar material and placed into cans for autoclaving. By contrast, plastic pipettes come prepacked and presterilized and do not have

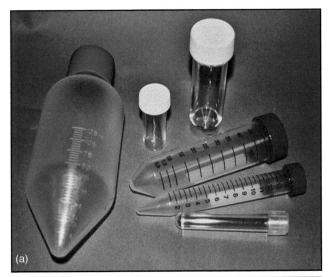

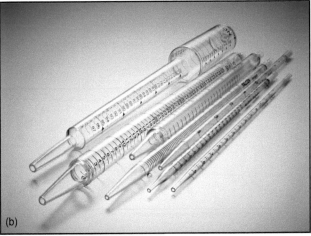

Fig. 5.2. Tubes and pipettes. (a) Centrifuge and sample tubes in 5- to 250-ml sizes. Clockwise from left: 250-ml centrifuge tube (Corning), 5-ml bijou bottle (Sterilin), 30-ml universal container (Sterilin), 50-ml centrifuge tube (BD Biosciences), 15-ml centrifuge tube (BD Biosciences), 5-ml sample tube (BD Biosciences). (b) Serological pipettes in 1- to 100-ml sizes (Falcon). *Source*: (a) R. Ian Freshney; (b) courtesy of Corning Life Sciences with permission.

the safety problems associated with handling chipped or broken glass pipettes. Nor do they have to be washed, which is relatively difficult to do, or plugged, which is tedious. On the downside, they are expensive and slower to use if singly packed. Bulk packaging can result in a high wastage rate unless packs are shared, which is not recommended (see Section 12.3.6). Plastic pipettes also add a significant burden to disposal, particularly if they must be disinfected first.

A variety of pipette controllers can be used with serological pipettes. A rubber bulb (see Figure 12.6a) is cheap and simple to use, but the rubber can quickly become worn or contaminated. A motorized pipette controller offers better speed, accuracy, and reproducibility (see Figure 5.3a). The major determinants in choosing a pipette controller are the

weight and feel of the instrument during continuous use; it is best to try one before purchasing it. Motorized controllers may have a separate or built-in pump and may be mains-operated or rechargeable; many suppliers make spare parts available. Pipette controllers usually have a filter at the pipette insert to minimize the transfer of contaminants. Some filters are disposable, while others are reusable after resterilization (see Figure 12.7 for the proper method of inserting a pipette into a pipetting device).

Pipette cans were previously used for glass pipettes but are worth considering even when using plastic pipettes, particularly if the pipettes are not individually wrapped. Cans keep the pipettes tidy in the BSC and can be closed when not in use. Square-sectioned cans, $75 \times 75 \times 300–400$ mm, are preferable to round, as they stack more easily and will not roll about the work surface. Pipette cans that were previously used for autoclaving are usually aluminum or nickel-plated steel. Cans may have silicone rubber-lined top and bottom ends to avoid chipping glass pipettes during handling.

Pipette cylinders were originally used for glass pipettes but are still useful for the disposal of plastic pipettes. Pipette cylinders are usually made from polypropylene and should be freestanding and distributed around the laboratory, one per workstation. Pipette cylinders are designed to hold disinfectant, allowing decontamination before washing (glass) or disposal (plastic). If pipette cylinders are not available, laboratories will need to substitute another impermeable container or dispose of pipettes directly into biohazard waste containers. If the addition of disinfectant is not feasible, some other form of decontamination must be performed before disposal (see Section 6.3.6).

If precise liquid measurement is not required, a Pasteur pipette or aspirating pipette can be used (e.g. when removing spent culture medium). Glass Pasteur pipettes should be regarded as disposable and discarded not into pipette cylinders but into sharps disposal containers. Plastic Pasteur pipettes are also available.

Pipettors and pipette tips. Pipettors originated from Eppendorf micropipettes used for dispensing 10–200 µl. As the working range now extends up to 5 ml or more, the term "micropipette" is not always appropriate, and the instrument is referred to here as a pipettor. They are available as manual or electronic, single or multichannel (see Figure 5.3b). Most of these devices rely on piston cylinder systems to draw up a precise volume of liquid into a disposable pipette tip. Pipettors should be calibrated regularly to ensure that they remain accurate.

It is assumed that the inside of a pipettor does not displace enough air to compromise sterility, but this may not always be the case. If you are performing serial subculture of a stock cell line (as opposed to a short-term experiment with cells that will not be propagated beyond the experiment), there is a risk that contamination may occur through the pipettor. In that case, you must use a regular plugged

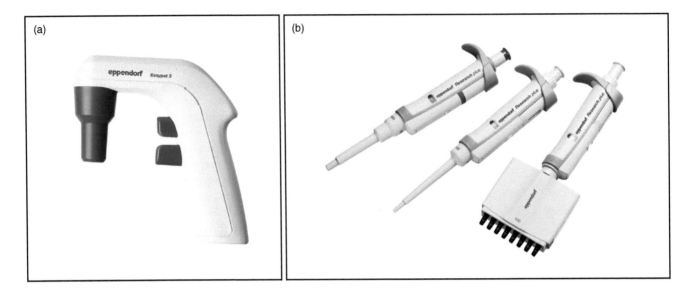

Fig. 5.3. Pipette controller and pipettors. (a) Easypet® three motorized pipette controller; (b) Research® plus pipettors with single and multiple channel options. *Source*: Courtesy of Eppendorf with permission.

pipette with a sterile length that is sufficient to reach into the vessel that you are sampling. If you are using a small enough container to preclude contact from the non-sterile stem, it is permissible to use a pipettor, provided the tip has a filter that minimizes microbial or cross-contamination. Otherwise, you run the risk of contamination from the non-sterile stem or from aerosol or fluid drawn up into the stem. Routine subculture, which should be rapid and secure from microbial and cross-contamination but need not be very accurate, is best performed with serological pipettes. Experimental work, which must be accurate but should not involve further propagation of the cells used, may benefit from using pipettors.

Disposable pipette tips can be bought loose to package and sterilize in the laboratory, or they can be bought prepackaged in racks ready for use. Prepackaged tips are available sterile or "PCR Clean;" the latter does not necessarily mean that the tip is sterile, meaning that sterilization is still required. Loose tips are cheaper, but more labor intensive. Prepacked tips are much more convenient, but considerably more expensive. Some racks can be refilled and resterilized, which presents a reasonable compromise.

Syringes and syringe dispensers. Syringes and needles should not be used extensively in normal handling because of concerns regarding safety, sterility, and shear stress in the needle when cells are handled. However, syringes are used for filtration in conjunction with syringe filter adapters, and, with needles, may be required for extraction of reagents (e.g. drugs or antibiotics) from sealed vials.

Syringes are also used as part of small-volume repetitive dispensers. Liquid is dispensed through incremental movement of a piston in the syringe or by a repeated syringe action with a two-way valve connected to a reservoir (the Cornwall

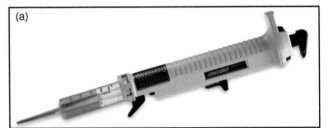

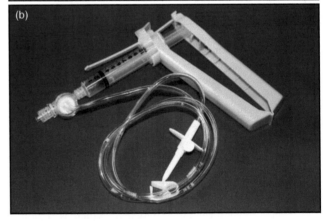

Fig. 5.4. Syringe dispensers. (a) Stepping dispenser operated by incremental movement of syringe piston, activated by thumb button (Repette); (b) repeating syringe dispenser with two-way valve connected to inlet tubing via an inline filter (Cornwall Syringe). *Source*: (a) courtesy of VWR-Jencons; (b) courtesy of Research Laboratory Supply with permission.

Syringe; see Figure 5.4). Sticking of the valves can occur but is minimized by avoiding the drying cycle after autoclaving and flushing the syringe out with serum-free medium or a salt solution before and after use. Syringe dispensers offer an

Fig. 5.5. Automatic dispensers. (a) The Perimatic Premier, suitable for repetitive dispensing and dilution in the 1- to 1000-ml range. If the device is used for sterile operations, only the delivery tube needs to be autoclaved. (b) Zippette bottle-top dispenser, suitable for 1- to 30-ml range; autoclavable. *Source:* Courtesy of VWR-Jencons with permission.

ergonomic way to minimize repetitive actions during experimental work, reducing the risk of musculoskeletal injury.

Peristaltic pump and vacuum pump systems. Repeated delivery of large liquid volumes can be performed using a peristaltic pump. A peristaltic pump dispenser (see Figure 5.5a) is useful if large medium volumes or a large number of replicate volumes are to be dispensed. A peristaltic pump has the advantage that it may be activated via a footswitch, leaving the hands free. Accuracy and reproducibility can be maintained at high levels over a range from 10 to 100 ml. A peristaltic pump dispenser is only worthwhile if a large number of flasks will be handled. Alternatively, liquid may be dispensed using a bottle-top dispenser (see Figure 5.5b). Care must be taken setting up such devices to avoid contaminating the tubing at the reservoir and delivery ends. Only the delivery tube is autoclaved; additional delivery tubes may be sterilized and held in stock, allowing a quick changeover in the event of accidental contamination or changes in culture or reagent.

Suction from a peristaltic pump may be used to remove spent medium or other reagents from a culture flask, and the effluent can be collected directly into disinfectant in a vented receiver (see Figures 5.6, 5.7; see also Section 6.3.6). This approach minimizes the risk of discharging aerosols into the atmosphere. The inlet line should extend further below the stopper than the outlet by at least 5 cm (2 in.), so that waste does not splash back into the vent. Placing a self-closing valve in the suction line allows the pipette tubing or the waste receiver to be replaced without fear of spillage (see Figure 5.6c). Vacuum lines can be very difficult to clean out if they become contaminated and should be disposable, allowing tubing to be replaced on a regular basis.

A vacuum pump may be used downstream from the receiver instead of a peristaltic pump, with the addition of a hydrophobic filter and a trap between the receiver and the pump to minimize the risk of aerosol and waste entering the pump (e.g. Millex Filter Unit, Merck Millipore, #SLFA05000; see Figure 5.6d). Do not draw air through a vacuum pump from a receiver containing hypochlorite solution; the free chlorine will corrode the pump and could be toxic. A non-corrosive disinfectant may be used within the receiver flask or added when the receiver is full for at least 30 minutes before disposal (see Section 6.3.6).

Automated liquid handling. Automated or semi-automated liquid handling systems are increasingly used for cell culture scale-up, replicate sampling, and high-throughput screening. The introduction of microwell plates (see Figure 8.4) has brought with it many automated dispensers, transfer devices, pipetting stations, and robotic systems (see Figures 5.8, 5.9). The range of equipment suitable for microtitration plates includes plate mixers, centrifuge carriers, and plate readers but is so extensive that it cannot be covered here.

Automation results in greater speed, accuracy, and reproducibility but requires an investment in infrastructure that must be justified by a high level of use, preferably as a shared resource. Whether a simple manual system or a complex automated one is chosen, the choice is governed mainly by five criteria:

(1) Ease of use and ergonomic efficiency.
(2) Cost relative to time saved and increased efficiency.
(3) Accuracy and reproducibility in serial or parallel delivery.
(4) Ease of sterilization and effect on accuracy and reproducibility.

Fig. 5.6. Aspiration of medium. (a) Pipette connected via tube to a peristaltic pump being used to remove medium from a flask. (b) Peristaltic pump on the suction line from the BSC in (a) leading to a waste receiver via a peristaltic pump. (c) Self-closing valve for use on suction line to prevent escape of waste when replacing suction line (ThermoFisher Scientific, #6177-0250). (d) Vacuum pump used in place of peristaltic pump as used in (b). Hydrophobic filter and trap used to protect pump from carry-over of fluid. (e) Withdrawal of fluid from multiwell plates with Vacusafe™ vacuum pump. Right adapter can be used with regular pipettes and flasks. Hypochlorite and other corrosive disinfectants cannot be used while operating the pump and must be added afterwards and left for long enough to allow disinfection. *Source*: (a, b) R. Ian Freshney; (c) courtesy of Carl Gregory; (d) Amanda Capes-Davis, courtesy of CMRI; (e) courtesy of Integra Biosciences with permission.

(5) Mechanical, electrical, chemical, biological, and radiological safety.

• *Safety Note. Most automated pipetting devices tend to expel fluid at a higher rate than during normal manual operation and consequently have a greater propensity to generate aerosols. This must be kept in mind when using hazardous substances.*

5.1.4 Centrifuge

Cell suspensions require centrifugation to increase the concentration of cells or to wash off a reagent. Cells sediment satisfactorily at $80–100\,g$ (see Appendix B for comments on conversion from rpm to g). Higher g may cause damage and promote agglutination of the pellet. A small bench-top centrifuge (see Figure 5.10a) installed near the BSC is sufficient

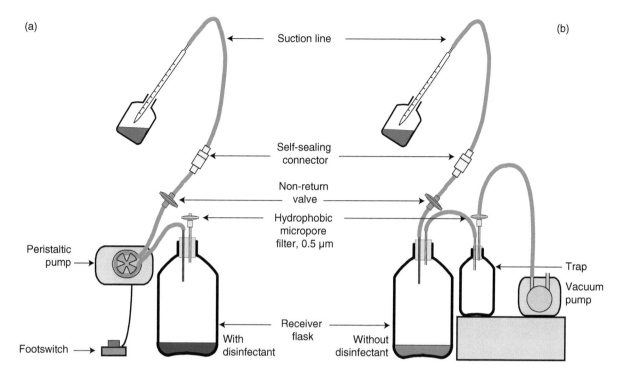

Fig. 5.7. Aspiration pump setup. (a) Aspiration with peristaltic pump. Note footswitch so that pump does not run continuously. (b) Aspiration with vacuum pump. Note trap to reduce risk of contamination of pump. Both systems have a hydrophobic filter to prevent release of aerosol and a non-return valve on the suction line to prevent reflux. A self-sealing connector can also be incorporated into the suction line to allow replacement of the pipette end of the tubing without escape of waste.

Fig. 5.8. Plate handling systems. (a) epMotion® (Eppendorf) semi-automated electronic pipette for parallel 96-well plate processing; (b) Transtar™ (Corning) for seeding, transferring medium, replica plating, and similar manipulations with microtitration plates; (c) MicroFill™ (BioTek Instruments) dispenser for 24- to 384-well plates. *Source*: (a) courtesy of Eppendorf; (b) courtesy of Corning Life Sciences; (c) courtesy of BioTek Instruments with permission.

for most purposes, particularly if it allows multiple rotors to be used. A large-capacity refrigerated centrifuge (e.g. 4×1 or $6 \times 1 \, l$) will be required if large-scale suspension cultures are contemplated (see Figure 5.10b; see also Section 28.2).

Refrigeration is not necessary for most applications and will create a larger footprint, although it can be useful to prevent cell samples overheating if a prolonged spin is required or to perform experimental work.

(a,b) Janus® Liquid Handling Workstation (PerkinElmer). *This workstation is configured for automated handling of samples in 96- and 384-well formats. A number of other formats can be used if required, including test tubes and vials. (a) Samples are handled within an enclosed space to reduce external contamination; (b) the Twister® robot (Caliper LifeSciences) allows up to 240 plates to be processed in a single run.*

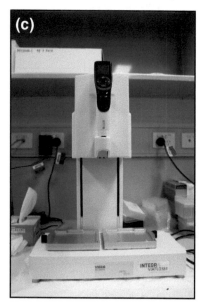

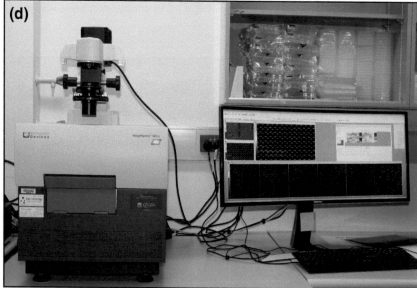

(c) Microplate Pipetting Station. *The VIAFLOW384 (Integra BioSciences) can be used to load tips and perform pipetting across multiple plates. This particular device can be used for 24-, 96-, or 384-plates. See Fig. 5.8 for other plate handling systems.*

(d) Automated Cell Imaging System. *The ImageXpress® Micro XL (Molecular Devices) can be used to capture images from slides, multiwell plates, Transwell inserts, and other formats at 1x to 100x magnification. The system can be used for various cell-based assays, including cytotoxicity and migration assays.*

Source for all images: Amanda Capes-Davis, courtesy of Ngoc Chau and Drug Screening Facility, CMRI.

Fig. 5.9. High-throughput liquid handling and cell-based assays.

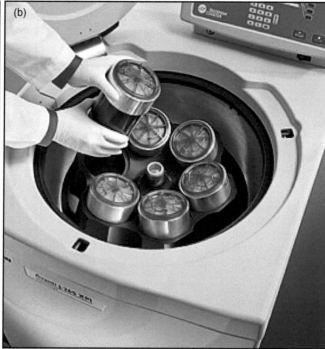

Fig. 5.10. Centrifuges. (a) Small, non-refrigerated, bench-top centrifuge (Eppendorf 5804); a microplate rotor is shown here, although other rotors are available. (b) Large-capacity centrifuge, 6 × 1 l, refrigerated (Beckman Avanti J-26S). *Source*: (a) copyright © 2014 courtesy of Eppendorf AG, Germany; (b) courtesy of Beckman Coulter with permission.

5.2 IMAGING AND ANALYSIS EQUIPMENT

The array of microscopes, counters, and analyzers that are available to study cultured cells can be bewildering. When deciding on equipment it is important to consider what will be used routinely every day and what will be used occasionally or for specialized applications. This chapter looks mostly at the "routine" end of the spectrum. Specialized microscopes and instruments are typically installed away from the tissue culture laboratory in shared core facilities, operated by scientists with specialized expertise; these include microscopes that are used for high-resolution imaging (see Section 18.4.4). Whatever microscope you may require, look for the most ergonomic options (e.g. adjustable for user comfort) and the best-quality optics and camera that you can afford.

5.2.1 Microscopes

Inverted microscope. An inverted microscope is essential for routine observation. It cannot be overemphasized that it is vital to look at cultures regularly to detect morphological changes and microbial contamination (see Sections 14.4.1, 16.3, 18.4.1). Long working-distance phase-contrast optics (condenser and objectives) are required to compensate for the thickness of plastic flasks. The increasing use of fluorescent tags for viewing live cells means that fluorescence optics

should be considered as well. The microscope should have a phototube for a digital camera, linked to a monitor. All these features can be met without requiring a large and expensive research microscope. Microscopes and imaging systems are now available for routine observation that are compact and easy to use without sacrificing image quality (see Figure 5.11).

Make certain that the stage is large enough to accommodate large roller bottles between it and the condenser, if required (see Section 28.3.1), and decide on useful accessories to purchase. For example, an object marker can be inserted in the nosepiece in place of an objective and used to mark the underside of a dish, allowing the user to pick a colony or follow the development of a particularly interesting area (see Section 20.4.1). A microinjection port may be required for micromanipulation (e.g. nuclear transplantation), but this is usually performed in a specialized facility.

Upright microscope. An upright fluorescence microscope may be required for chromosome analysis, mycoplasma detection, or other applications. Select a high-grade research microscope with regular bright-field optics up to 100× objective magnification; phase contrast up to at least 40×, and preferably 100×, objective magnification; and fluorescence optics with epi-illumination and 40× and 100× objectives for immunofluorescence or mycoplasma staining. A 50× water-immersion objective is particularly useful for observing

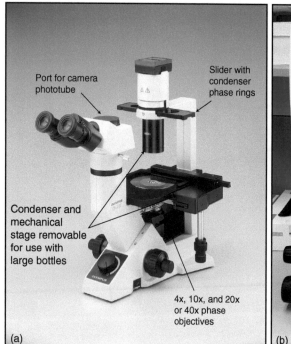
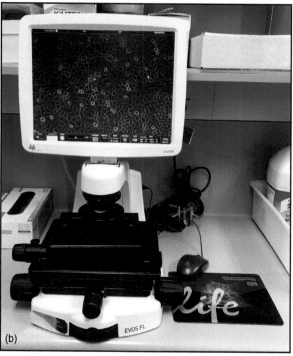

Fig. 5.11. Inverted microscopes. (a) Olympus CKX41 inverted microscope fitted with phase-contrast optics and trinocular head with port for attaching a digital camera; (b) EVOS® FL Imaging System (ThermoFisher Scientific) in use for routine observation. A flask has been positioned on the stage inside a light shield box to minimize background for fluorescent imaging. *Source*: (a) courtesy of Olympus, UK, Ltd with permission; (b) Amanda Capes-Davis, courtesy of CMRI.

routine mycoplasma wet preparations with Hoechst stain (see Protocol P16.4).

Dissecting microscope. Dissection of small pieces of tissue or embryo culture (e.g. for analysis of whole mouse embryos or organs or tissue from small invertebrates) will require a dissecting microscope (Figure 5.12). If available, a dissecting microscope is useful for counting monolayer colonies and is essential for counting and picking small colonies in agar.

Live-cell imaging equipment. Live-cell imaging is increasingly popular and is discussed in a later chapter (see Section 18.4.3). Many laboratories use specialized microscopy facilities for live cell analysis, but some equipment may be available locally for observation or growth assays. Examples include the CytoSMART 2 System (Lonza) and the Incucyte (Sartorius; see Figure 18.4). These systems include a mini-microscope that is installed within the incubator chamber.

5.2.2 Cameras

A still digital camera is sufficient for recording observations, but a charge-coupled device (CCD) camera or complementary metal oxide semiconductor (CMOS) camera will allow

real-time viewing and digital recording (see Sections 18.4.2, 18.4.3). Digital cameras and associated tools for image analysis are important to quantify cell behavior, such as cell signaling and migration, and to localize intracellular components using immunofluorescence. Digital cameras and monitors are also a valuable aid for the discussion of cultures and for training new staff and students (see Supp. S30.8).

For routine digital imaging, choose a high-resolution camera. High sensitivity is helpful for fluorescent images but can lead to overillumination when used for light microscopy; consider what applications you will use most. A 15-megapixel camera gives resolution that should be more than adequate for publication. Color or monochromatic cameras are both available, but color is more expensive and is usually not required for routine use. Consider how the camera will interface with computers – for example, whether USB connections will be required. If a computer is not available, purchase a camera with a separate memory card and download images on a regular basis.

5.2.3 Computer and Monitor

A computer and monitor are useful adjuncts to the digital camera, allowing the user to view images on a large screen in real-time. Computers are usually considered essential for

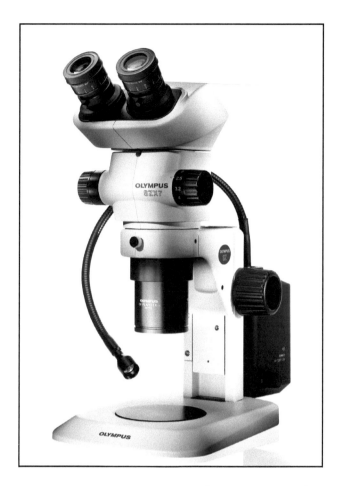

Fig. 5.12. Dissecting microscope. Note the flat base on this Olympus SZX7 microscope, which makes dissection more comfortable. *Source*: Courtesy of Olympus with permission.

shared core facilities, allowing multiple users to capture digital images and transfer data to a central server or to storage in the cloud. Space and budget for computer infrastructure is often more limited in a tissue culture laboratory and there may be additional concerns regarding asepsis. It is important to consider where the computer will be positioned and how to minimize contamination. Keyboards are a potential contamination risk and should be chosen with an eye to easy cleaning. A smaller portable device or tablet may be preferred, but asepsis continues to be a concern; any portable device should be decontaminated when brought into the room or removed from it.

A computer can be useful for record-keeping. Users may enter records for primary culture, cell line maintenance, and experiments using a networked computer or portable device (see Sections 13.8, 14.9). However, many users still prefer to complete records manually in a notebook and transfer to the computer at a later date. Whichever method is used, cell line data are best maintained using an electronic database that can also serve as an inventory control for cryofreezers and a record of shipments into and out of the laboratory (see

Section 15.7; see also Supp. S15.1). In larger laboratories this can simplify purchase and stocking of plastics, reagents, and media.

5.2.4 Cell Counting and Analysis Equipment

Hemocytometer slide. Cultured cells may be counted by a variety of different direct and indirect methods (see Section 19.1). The simplest method is to count cells manually using a hemocytometer slide on an inverted microscope (see Figure 19.1). The hemocytometer slide is the most popular choice for counting cells, used by 71% of 400 respondents in a 2009 survey (Ongena et al. 2010). It is also the cheapest option and allows cell viability to be determined by dye exclusion (see Section 19.1.1; see also Protocols P19.1, P19.2). If used routinely, one slide should be issued per person, each with multiple coverslips.

Automated cell counter. Manual counting may not be suitable if you wish to perform a large number of counts for experimental work or to assess cell size or other characteristics. An automated cell counter (see Section 19.1.2) is a great advantage when more than two or three cell lines are to be counted with replicate samples and is essential for precise quantitative growth kinetics. Most mid-range or top-of-the-range cell counters can measure additional variables such as variability, size, or apoptosis (the latter depending on expression of visible markers). Automated cell counters generally retain their readings and allow data to be saved to a computer or shared network.

Colony counter. Monolayer colonies are easily counted by eye or on a dissecting microscope with a felt-tip pen to mark off the colonies, but if many plates are to be counted, an automated counter will help make the process faster and more accurate. The simplest device uses an electrode-tipped marker pen, which records a count when you touch down on a colony. They often have a magnifying lens to help visualize the colonies. From there, a large increase in sophistication and cost takes you to an automated electronic colony counter, which counts colonies using image analysis software. These counters are very rapid, can discriminate between colonies of different diameters, and can even cope with contiguous colonies (see Section 19.4.2).

5.3 INCUBATION EQUIPMENT

Tissue culture requires precise control of temperature, humidity, and gas levels during periods of incubation. Most laboratories use incubators that regulate temperature and CO_2 tension ("CO_2 incubators"); a humid environment is usually achieved using a water reservoir or a water pan within the chamber. More complex gas regulation can be achieved using a hypoxic incubator that regulates CO_2, nitrogen, and oxygen. However, tissue culture can be performed without these devices if necessary, using a dry incubator or a warmroom (see Supp. S4.1). Scale-up can be achieved using incubator accessories, such as roller racks and magnetic stirrers (see Section 5.3.2). Specialized conditions for scale-up of tissue culture are considered in a later chapter (see Chapter 28).

The size of incubator required will depend on usage – both the number of users and the number of cultures. Five people using only microtitration plates could have 1000 plates (~ 100 000 individual cultures) or 10 experiments each in a modest-sized incubator, while one person doing cell cloning in Petri dishes could fill the incubator with a few experiments. Two smaller incubators are preferable to one large incubator, giving better temperature control, quicker recovery after opening, and easier cleaning and maintenance arrangements for multiple users. In addition, one can be used for frequent access and the other for limited access. A small incubator close to the BSC is useful for short-term incubations (e.g. trypsinization) and a separate small incubator should be set aside for quarantine if a separate quarantine room is not available (see Section 4.2.3).

Incubator choice will depend on the user's requirements and an assessment of how well conditions are maintained within the incubator during a normal working day. If purchasing a new incubator, it helps to "try before you buy" and ask for feedback from other users. Look for data on temperature and gas regulation, recovery times when the incubator door is opened and closed, and freedom from contamination (e.g. ask whether fittings can be removed to allow all parts of the chamber to be wiped down). Maintenance of incubator temperature relies on the presence of either a water jacket or an "air jacket" that uses airflow and temperature on/off regulation to maintain a similar degree of temperature uniformity (Panasonic 2015). Water jackets can maintain a constant temperature throughout multiple door openings and loss of power, but the water makes the incubator heavier and some users report problems with microbial contamination within the jacket. A water-jacketed incubator must also be filled when it is installed and then emptied and refilled when the incubator is moved. An air jacket results in a lighter incubator and a faster recovery when the door is opened and closed, but does not provide insulation in a disaster situation if power to the incubator is lost.

5.3.1 Incubators

Dry incubator. If a dry incubator is to be used for tissue culture, check that it is large enough (~ 50–200 l [1.5–6 ft³] per person) and that it can control temperature to within ±0.2 °C. A safety thermostat should cut off if the incubator overheats or, preferably, should regulate the incubator if the first thermostat fails. The incubator should be resistant to corrosion (e.g. stainless steel, although anodized aluminum is acceptable for a dry incubator) and should be easily cleaned. Dry incubators can have long lifespans and will be useful

Fig. 5.13. Culture chambers. Inexpensive alternatives to CO_2 incubator. Upper shelf, custom-made clear plastic box; lower shelf, anerobic jar (Corning – BD-Biosciences). *Source*: R. Ian Freshney, courtesy of Reeve Irvine Institute.

for many different applications (e.g. for sterility testing of reagents; see Supp. S11.6).

Humidity and gas conditions must be regulated separately if using a dry incubator for tissue culture. Cultures can be incubated in sealed flasks, but some vessels (e.g. Petri dishes or multiwell plates) require a controlled atmosphere with high humidity and elevated CO_2 tension (see Section 9.2.3). The cheapest way of controlling the gas phase is to place the cultures in a separate box, anerobic jar, or culture chamber (see Figure 5.13), gas the container with the correct gas mixture, and then seal it. If the container is not completely filled with dishes, include an open dish of water to increase the humidity inside the chamber. Chambers may be purchased (e.g. C-Chamber, BioSpherix) or custom-made. Vacuum bags (e.g. for food storage) can also be used to achieve and maintain hypoxia for at least four days (Bakmiwewa et al. 2015).

CO_2 incubator. CO_2 incubators (see Figure 5.14) are more expensive, but their ease of use and superior control of CO_2 tension, temperature, and humidity justify the expenditure. CO_2 tension is controlled using a CO_2-monitoring device, which draws air from the incubator into a sample chamber, determines the concentration of CO_2 using a sensor, and injects CO_2 into the incubator to make up any deficiency. Air is circulated around the incubator by natural convection or by using a fan to keep both the CO_2 level and the

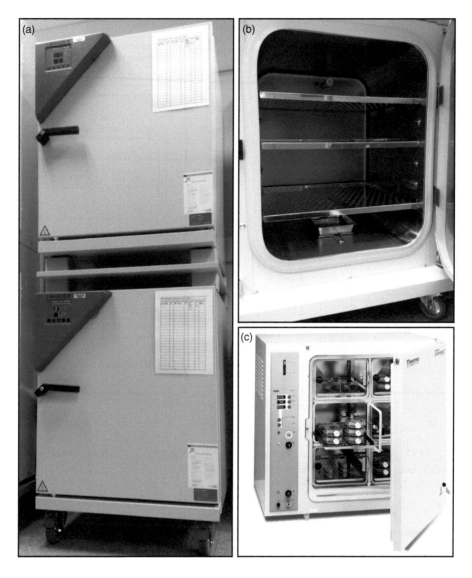

Fig. 5.14. CO_2 incubators. (a) Fan-less CO_2 incubators (Binder) stacked to reduce footprint. (b) Interior of CO_2 incubator (Binder) showing shelving and water tray in place and smooth, easily cleaned interior with removable racking which does not penetrate the stainless-steel lining. (c) Heraeus Cytoperm® 2 CO_2 incubator, with individual compartments to minimize atmospheric disturbance when opening. *Source*: (a, b) Amanda Capes-Davis, courtesy of CMRI; (c) courtesy of ThermoFisher Scientific with permission.

temperature uniform. It is claimed that fan-circulated incubators recover faster after opening, although natural convection incubators can still have a quick recovery time.

A humidified atmosphere is usually achieved by placing a water pan within the chamber (see Figures 5.14b, 5.15). A warm, humidified atmosphere is necessary but increases the risk of microbial contamination, which can be difficult to eliminate (see Section 16.1.3). Heated wall incubators tend to reduce fungal contamination because the walls are more likely to remain dry, even at high relative humidity. Copper or copper alloy surfaces are also reported to reduce contamination rates through release of bactericidal copper ions. Fan circulation may increase the spread of contamination, leading to the use of micropore filtration and laminar airflow in some incubators to inhibit the circulation of microorganisms. However, air circulation is only one route for contamination to arise; condensation or spills are also problematic. Contamination can affect any incubator and is highly dependent on culture practice and whether the incubator is cleaned on a regular basis. Some incubators have high-temperature sterilization cycles, which make cleaning easier and more effective

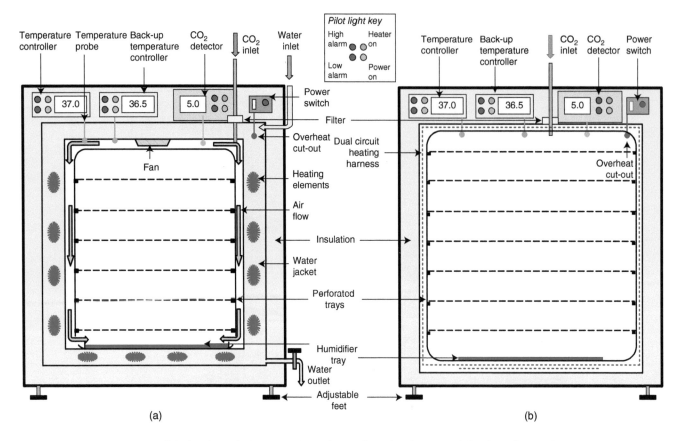

Fig. 5.15. CO$_2$ incubator design. Front view of control panel and section of chamber of two stylized humid CO$_2$ incubators: (a) water jacketed with circulating fan; (b) dry-walled ("air jacketed") with no circulating fan (not representative of any particular manufacturer or model).

and thus represent a particularly worthwhile investment in managing microbial contamination.

Hypoxic incubator. Certain cell types prefer reduced oxygen concentrations relative to ambient air (see Section 9.2.4; see also Minireview M9.1). Although reduced oxygen (hypoxia) is detrimental for the whole organism, it is considered physiological in many tissues, where oxygen concentration may range between 1% and 12% (Geraghty et al. 2014). A low oxygen tension may be required to extend cell lifespan, demonstrate the activity of some anticancer drugs, and for the culture of human embryonic stem cells (hESCs) (Packer and Fuehr 1977; Bertout et al. 2008; Lengner et al. 2010). If variable oxygen may be required, you will need to use a tri-gas incubator or find an alternative way to generate hypoxic conditions (Bakmiwewa et al. 2015). Tri-gas incubators use air, nitrogen, and CO$_2$ to control gas levels. Oxygen levels are decreased (typically to a range between 0.5% and 2%) by increasing the proportion of nitrogen within the culture chamber (Geraghty et al. 2014). Nitrogen gas must be supplied from a separate cylinder or piped gas system; it is important to consider the space and infrastructure required as part of incubator purchase.

Monitoring and maintenance. Monitoring of incubator temperature is beneficial for any concerns to be detected early, hopefully before cultures are affected. It is important to perform independent measurement because readouts can malfunction during equipment failures (see Section 4.3). The best approach is to install an independent equipment alarm system (see Section 4.3.2). If this is not possible, a portable thermometer with ranges from below −70 °C to about +200 °C can be used to check incubators, cold storage equipment, and sterilizing ovens. Some portable thermometers act as data loggers and temperature and humidity data can be downloaded for later analysis (e.g. Omega, #OM-70 series).

Monitoring of CO$_2$ is also important to detect incorrect readings that require calibration. CO$_2$ incubators typically use thermal conductivity (TC) or infrared (IR) CO$_2$ sensors. Calibration requirements vary with the sensor and the incubator model; TC sensors require a humid environment for correct performance, i.e. the water pan should contain water. Always read the manual or check with your supplier to understand monitoring and calibration requirements for each incubator.

Frequent cleaning of incubators – particularly humidified ones – is essential. All shelves and racks should be readily removable without leaving inaccessible crevices, holes, or corners (see Figure 5.14b) and should be autoclavable. Some incubators offer high-temperature cycles for decontamination, which make cleaning easier and more effective. If fungal contamination is present, two cycles of heat decontamination should be performed – the first to kill vegetative organisms and the second to target any spores that survived the first cycle (Geraghty et al. 2014). Other incubators offer UV treatment or HEPA filters to reduce microbial contamination within the water pan or in circulating air. All of these treatments help to reduce contamination, but there is no substitute for regular cleaning (see Sections 12.5.1, 16.1.3).

5.3.2 Incubator Accessories

Shelves. Incubator shelving is usually perforated to facilitate the circulation of air. However, the perforations can lead to irregularities in cell distribution in monolayer cultures, with variations in cell density following the pattern of the perforations on the shelves. The variations may be due to convection currents generated over points of contact relative to holes in the shelf, or they may be related to areas that cool down more quickly when the door is opened. Although no problem may arise in routine maintenance, it may help to place multiwell plates or other vessels on a ceramic tile or metal tray during experiments in which uniform density is important. Further problem solving may be necessary if problems continue; for example, the incubator may have been installed on an uneven surface and its "feet" may require adjustment (see Section 31.8.3).

Roller racks. Roller racks are used to scale up monolayer culture using roller bottles (see Section 28.3.2). The choice of apparatus is determined by the scale (i.e. the size and number of bottles). The scale may be calculated from the number of cells required, the maximum attainable cell density, and the surface area of the bottles (see Table 8.2). A large number of small bottles gives the highest surface area but tends to be more labor intensive in handling, so a usual compromise is to select bottles of approximately 125 mm (5 in.) in diameter and various lengths from 150 to 500 mm (6 to 20 in.). The length of the bottle will determine the maximum yield but is limited by the size of the rack; the height of the rack will determine the number of tiers (i.e. rows) of bottles. Although it is cheaper to buy a larger rack than several small ones, the latter alternative is preferred because it (i) allows a gradual increase in capacity (having confirmed that the system works); (ii) may be easier to locate in a warmroom or incubator; and (iii) allows operations to continue even if one rack requires maintenance. Roller racks may be installed in a warmroom or an incubator that has been specially designed to accommodate roller racks and other roll-in equipment (see Figures 5.16, 28.7).

Magnetic stirrer. A rapid stirring action for dissolving chemicals is available with any stirrer and may benefit from a built-in hotplate. However, different conditions apply to

Fig. 5.16. Roller rack incubator. High-capacity roller bottle rack with dedicated incubator housing. *Source*: Courtesy of Wheaton Scientific with permission.

enzymatic tissue disaggregation or suspension culture (see Sections 13.5, 14.7). When performing cell culture using a magnetic stirrer (i) the motor should not heat the culture (use the rotating-field type of drive or a belt drive from an external motor); (ii) the speed must be controlled down to 50 rpm; (iii) the torque at low rpm should still be capable of stirring up to 10 l of fluid; (iv) the device should be able to maintain several cultures simultaneously; (v) each stirrer position should be individually controlled; and (vi) a readout of rpm should appear for each position. It is preferable to have a dedicated magnetic stirrer for culture work (see Figure 5.17). Any device that is placed into a humid incubator should be resistant to corrosion, be easily cleaned, and generate minimal heat within the chamber.

5.3.3 Water Baths

A water bath is useful to prewarm medium and serum before using for tissue culture. Bottles and tubes can be prewarmed using an incubator, but it takes more time for bottles to

Fig. 5.17. Magnetic stirrer. Four-position BioStir device for suspension culture or tissue disaggregation. *Source*: Courtesy of Wheaton Scientific with permission.

come to the correct temperature in an incubator compared to a water bath. However, the water bath can be a source of microbial contamination. Position the water bath away from the sterile handling area and wipe down prewarmed bottles or tubes before they enter the BSC. Water baths should be cleaned regularly to minimize microbial growth.

5.4 PREPARATION AND WASHUP EQUIPMENT

5.4.1 Water Purification Systems

Water used in tissue culture must be of a very high purity. Purified water is typically divided into three grades or types (see Section 11.5.1). Ultrapure water (UPW, also known as Type 1 water) is used to prepare tissue culture media and other reagents. Water purification systems use a three- or four-stage process to generate UPW (see Figure 5.18). The most important principle is that each stage should be qualitatively different. Reverse osmosis (RO) or distillation may be performed as a baseline, followed by carbon filtration, deionization, and micropore filtration (see Figure 11.6). RO purification is cheaper if you pay the fuel bills; if you do not, distillation has the advantage that water can be heat sterilized.

Pure (Type 2) water is used for some laboratory applications or to feed UPW purification systems. Many deionized systems provide this level of water quality. The deionizer should have a conductivity meter monitoring the effluent, to indicate when the cartridge must be changed. Other cartridges should be dated and replaced according to the manufacturer's instructions. Primary grade (Type 3) water may be used for rinsing glassware and other less-sensitive tasks. RO purification is often used to generate this level of water quality.

More information on water quality and purification methods can be found in a later chapter (see Section 11.5).

5.4.2 Preparation Equipment

Balances. Although most laboratories obtain media that are already prepared, it may be necessary to prepare some reagents in house. Doing so will require a balance (an electronic one with automatic tare) capable of weighing items from around 10 mg up to 100 g or even 1 kg, depending on the scale of the operation. If you are a service provider, it is often preferable to prepare large quantities, sometimes 10× concentrated, so the amounts to be weighed can be quite high. It may prove better to buy two balances, coarse and fine, as the outlay may be similar and the convenience and accuracy are increased.

Hot plate magnetic stirrer. In addition to the stirrers used for suspension cultures and trypsinization, it may be desirable to have a magnetic stirrer with a hot plate to accelerate the dissolution of some reagents. Placing a solution on a stirrer in a warmroom may suffice, but leaving solutions stirring at 37 °C for extended periods can lead to microbial growth, so stable solutions are best stirred at a higher temperature for a shorter time.

pH meter. A pH meter is required for the preparation of media and reagents. Although a phenol red indicator is used to monitor pH in most media and many other reagents, a pH meter is still important, particularly for experimental work (e.g. in preparing cultures for fluorescence assays or in estrogen binding assays where phenol red can interfere). It is also important when preparing stock solutions and for regular quality control checks during preparation of media and reagents.

Conductivity meter. When solutions are prepared in the laboratory, it is essential to perform quality control measures to guard against errors (see Section 11.8). A simple check of ionic concentration can be made with a conductivity meter against a known standard, such as normal saline (0.15 M).

Osmometer. One of the most important physical properties of a culture medium, and one that is often difficult to predict, is the osmolality (see Section 9.2.6). Although the conductivity is controlled by the concentration of ionized molecules, non-ionized particles can also contribute to the osmolality. An osmometer (see Figure 9.8) is therefore a useful accessory to check solutions as they are made up, to adjust new formulations, to compensate for the addition of reagents to the medium, and to act as a second line of quality control. Osmometers usually work by depressing the freezing point of a medium or elevating its vapor pressure. Choose one with a low sample volume (< 1 ml), because you may want to measure a valuable or scarce reagent on occasion, and the accuracy (+10 mosmol/kg) may be less important than the value or scarcity of the reagent.

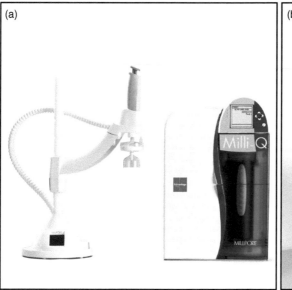
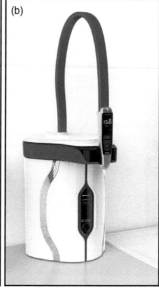

Fig. 5.18. Water purification systems. (a) Millipore Milli-Q; (b) Elga Purelab Chorus 1. Both systems produce UPW for the preparation of media and tissue culture reagents; they can also provide Type 2 or Type 3 water for rinsing or supplying an autoclave. *Source*: (a) Courtesy of Merck Millipore; (b) Courtesy of Elga-Veolia with permission.

5.4.3 Washup Equipment

Probably the best way to produce clean glassware is to have a reliable person doing the washing up, but from the worker's viewpoint, washing up is repetitive and time will be better spent on other tasks. Washing up has been reduced by the move to disposable plasticware, but bottles, beakers, and other items continue to be used and must be washed. A glassware washing machine and drying oven are necessary for most laboratories and may be shared between multiple laboratories in a large building (see Figure 5.19a).

Glassware washing machine. A variety of laboratory glassware washers are available that are quite satisfactory. Look for the following principles of operation:

(1) The machine should be elevated to waist height to reduce bending when loading and unloading the washer. A plinth may be installed for some units (see Figure 5.19a).
(2) A choice of racks should be provided with individual spigots over which you can place bottles, flasks, etc. Open vessels such as beakers will wash satisfactorily in a whirling-arm spray, but narrow-necked vessels need a mobile injector unit that provides individual jets of water. Bottles should rest on a mat or be held in a cradle that protects the neck of the bottle from chipping (see Figure 5.19b).

(3) The pump that forces the water through the jets should have a high delivery pressure, requiring around 1.5–4 kW (2–5 hp), depending on the size of the machine.
(4) Water for washing should be heated to a minimum of 80 °C.
(5) Facility for a deionized water rinse should be available at the end of the cycle. The water should be heated to between 50 and 60 °C; otherwise the glassware may crack after the hot wash and rinse. The rinse should be delivered as a continuous flush, discarded, and not recycled.
(6) Preferably, rinse water from the end of the previous wash cycle should be discarded and not retained for the pre-rinse of the next wash. Discarding the rinse water reduces the risk of carryover when glassware is used for tissue culture or for working with hazardous substances.
(7) The machine should be lined with stainless steel and plumbed with stainless-steel or nylon piping.

Glassware drying oven. A glassware drying oven should be of large capacity, fan driven, and able to reach 100 °C. Reusable plastics such as polypropylene and polystyrene will require a lower temperature of around 50–60 °C (polycarbonate will usually tolerate higher drying temperatures). Temperature settings should be determined by the drying rate; 50–60 °C may be adequate for both glassware and plasticware and will reduce the risk of burns when workers are unloading the oven. A sterilizing oven or dry incubator may be used for glassware drying if the required temperature can be maintained (see Sections 5.3.1 and 5.4.4).

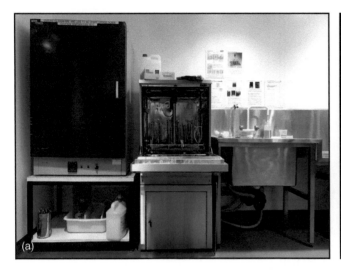

Fig. 5.19. Laboratory glassware washer. (a) Washup area showing drying oven (left), laboratory glassware washer (middle), and sink (right). The glassware washer (Miele) is open to show the interior; it is installed on a plinth to raise it to waist height, improving user comfort. (b) Interior of glassware washer showing mobile injector unit. *Source*: Amanda Capes-Davis, courtesy of CMRI.

Fig. 5.20. Autoclaves. (a) Small, top-loading autoclave (Prestige Medical). (b) Medium-sized (300 l; 10 ft³) laboratory autoclave with square chamber for maximum load. The recorder on the top console is connected to a temperature probe in the bottle in the center of the load (see arrow). (c) Large institute autoclave built into washup room. The recorder is accessible from the back of the autoclave in a separate room. *Source*: (a, b) R. Ian Freshney, courtesy of CRUK Beatson Institute; (c) Amanda Capes-Davis, courtesy of CMRI.

5.4.4 Sterilization Equipment

Autoclave. The autoclave is essentially a steam sterilizer, which means that the simplest and cheapest version will be a domestic pressure cooker that generates 100 kPa (1 atm, 15 lb/in.²) above ambient pressure. However, laboratory autoclaves have additional safety and quality control features.

A small bench-top autoclave (see Figure 5.20a) will give automatic programming and safety locking. A medium-sized freestanding model will have a programmable timer, a choice of pre- and poststerilization evacuation, and a temperature logger (see Figure 5.20b). A large model is likely to be shared by an entire institute and may be built into the washup room (see Figure 5.20c). Medium and large options provide more flexibility and the opportunity to comply with good laboratory practice (GLP) (see Section 7.2.2). Depending on local biosafety regulations, you may be required to autoclave biological waste, so you will need to consider whether this requires an additional autoclave. When installing an autoclave, leave sufficient space around the unit for maintenance and ventilation, provide adequate air extraction to remove heat and steam, and ensure that a suitable drain is available for condensate.

Most small autoclaves come with their own steam generator (calorifier), but larger machines may have a self-contained steam generator, a separate steam generator, or the facility to use a steam line. If high-pressure steam is available on-line, that will be the cheapest and simplest method of heating and pressurizing the autoclave; if not, it is best to purchase a sterilizer complete with its own self-contained steam generator. Such a sterilizer will be cheaper to install and easier to move. With the largest machines, you may not have a choice, as they are frequently offered only with a separate generator. In that case, you will need to allow space for the generator at the planning stage.

Autoclave cycles are usually described as "wet" (for water, balanced salt solutions, etc.) or "dry" (for empty bottles, instruments, etc.). Wet cycles are performed without evacuation of the chamber before or after sterilization. Dry cycles require the chamber to be evacuated before sterilization or the air to be replaced by downward displacement, allowing better access for hot steam. The chamber should also be evacuated after sterilization to remove steam and promote subsequent drying; otherwise items will emerge wet, leaving a trace of contamination from the condensate on drying. Always use deionized or RO water to supply the autoclave, particularly when a "postvac" cycle is not available.

Autoclaves must undergo regular validation to ensure that sterilization is effective. Validation testing may include monitoring of cycle conditions and testing of sterility indicators that are included in autoclave runs (see Section 11.2.7). Apply autoclave tape to goods in each run as a quick check and avoid overfilling bottles, which can affect sterilization.

• *Safety Note. Autoclaves can cause severe burns and are pressurized during operation, resulting in the risk of explosion. Autoclaves should only be operated by trained and competent personnel.*

Sterilizing oven. Although most sterilizing can be done in an autoclave, it is preferable to sterilize glassware by dry heat, avoiding the possibility of chemical contamination from steam condensate or corrosion. Such sterilization will require a high-temperature (160–180 °C) fan-powered oven (e.g. Binder FD or Panasonic MOV series) to ensure even heating throughout the load (see Section 11.3.2). Do not get an oven that is too large for the amount or size of glassware that you use. It is better to use two small ovens rather than one

big one; heating is easier, more uniform, quicker, and more economical when only a little glassware is being used. You are also better protected during breakdowns.

Filter sterilization apparatus. A number of options are available for sterile filtration, depending on the scale of your operation (see Section 11.7). Filtration may be by positive pressure and will require a pump upstream from a pressure vessel, mostly for large volumes of 10 l or more; smaller volumes can be handled with a smaller reservoir and a downstream peristaltic pump. However, most laboratories now use disposable filters ranging in size from 25-mm syringe adapters through 47-mm in line, bottle-top adapters, or filter flasks (see Table 11.6). It is also wise to keep a small selection of larger sizes on hand.

5.5 COLD STORAGE EQUIPMENT

5.5.1 Refrigerators and Freezers

Most tissue culture reagents can be stored at 4 or −20 °C. Some drugs, reagents, or products from cultures (e.g. RNA) require a temperature of −70 to −90 °C, at which point most, if not all, of the water is frozen and most chemical and radiolytic reactions are severely limited. A −70 to −90 °C freezer is also useful for freezing cells within a foam box or commercial freezing container (see Section 15.2.2). Most tissue culture laboratories require a 4 °C refrigerator, −20 °C freezer, and −70 to −90 °C freezer. Access to a coldroom may also be beneficial (see Section 4.2.6).

Refrigerators and freezers are available at domestic, commercial (e.g. catering), and laboratory or hospital grade. Usually, a domestic refrigerator or freezer is quite efficient and cheaper than special laboratory equipment. Domestic refrigerators are available without a freezer compartment ("larder refrigerators"), giving more space and eliminating the need for defrosting. However, if you require 400 l (12 ft³) or more storage, a large hospital, blood bank, or catering refrigerator may be better. The chest type of freezer is more efficient at maintaining a low temperature with minimum power consumption, but vertical cabinets are much less extravagant in floor space and easier to access. If you choose a vertical cabinet type, make sure that it has individual compartments (e.g. six to eight compartments in a 400-l [15-ft³] freezer) with separate close-fitting doors, and expect to pay at least 20% more than for a chest type. Doors should be opaque to shield medium and reagents from light (see Section 9.8).

Autodefrost ("frost-free") refrigerators and freezers have automatic defrost cycles that result in regular temperature oscillations (see Figure 4.9). Although autodefrost freezers may be bad for some reagents (e.g. enzymes or antibiotics), they are used for many tissue culture stocks, whose bulk precludes major temperature fluctuations and whose nature is less sensitive to severe cryogenic damage. Conceivably, serum could deteriorate during oscillations in the temperature of an autodefrost freezer, but in the authors' experience it does not do so in practice (see Section 9.8).

Low-temperature freezers generate a great deal of heat, which must be dissipated for them to work efficiently (or at all). Such freezers should be located in a well-ventilated or air-conditioned area such that the ambient temperature does not rise above 23 °C (see Section 4.2.6). Periodic maintenance will reduce the risk of equipment malfunction and overheating. Freezers build up dust on the outside and ice on the inside, resulting in increased risk of contamination and failure to maintain temperature. Cleaning the freezer's filter and condenser coils, and defrosting the freezer or removing ice from the door, will reduce problems and extend the lifespan of your equipment.

Refrigerators and freezers should be clearly labeled, e.g. to ensure that they are not used to store food. Temperature monitoring should be performed using an independent probe, using an equipment alarm system (see Section 4.3.2) or a separate temperature monitor if more extensive monitoring is not available. Backup refrigerators and freezers should be available in case of equipment failure, with instructions for how to proceed outside working hours.

5.5.2 Cryofreezers

Detailed information on cryostorage containers and equipment can be found in a later chapter (see Section 15.2). In brief, most laboratories will choose storage in liquid nitrogen, with the choice of cryofreezer depending on the size and the type of system required. For a small laboratory, a 35-l cryofreezer with a narrow neck and storage in canes and canisters or in drawers in a rack system should hold about 500–1000 cryovials (see Figure 15.6a–e). Larger freezers may hold > 10 000 cryovials and include models with walls perfused with liquid nitrogen (isothermal tanks), cutting down on nitrogen consumption and providing safe storage without any liquid nitrogen in the storage chamber itself (see Figure 15.6f–h). If selecting an isothermal tank, it is important to establish that no particulate material, water, or water vapor can enter the perfusion system; blockages can be difficult, or even impossible, to clear. It is better to purchase several small freezers than one large freezer; distributing stock among more than one freezer will reduce the risk if one freezer fails.

An appropriate storage vessel should also be purchased to enable a backup supply of liquid nitrogen to be held. The size of the vessel depends on (i) the size of the freezer; (ii) the frequency and reliability of delivery of liquid nitrogen; and (iii) the rate of evaporation of the liquid nitrogen. A 35-l narrow-necked freezer using 5–10 l/week will only require a 25-l Dewar as long as a regular supply is available. Larger freezers are best supplied from a dedicated storage tank or piped system, e.g. a 160-l storage vessel linked to a 320-l freezer with automatic filling and alarm, or a 500-l tank for a larger freezer or for several smaller freezers (see Section 4.2.6; see also Figure 4.7). In general, one or two cryofreezers may be located in the tissue culture laboratory along with a storage

Dewar for top-up, but if the number and/or size of the freezers increases, it will probably be more efficient to store these in a dedicated storage area adjacent to the liquid nitrogen supply tank.

Cryofreezers are used for long-term storage and contain unique and irreplaceable samples. All cryofreezers should be regularly checked for problems (e.g. to remove ice buildup on the lid) and temperature should be monitored using an independent temperature probe (see Section 4.3.2). The need for reliable alarm systems and a regular checking procedure increases once the freezers are removed from everyday observation and scrutiny. "Out of sight, out of mind" is a well-worn saying but still true.

5.5.3 Rate-Controlled Freezer

Cells may be frozen simply by placing them in an insulated box in a freezer at −70 to −90 °C. Insulation is required to control the cooling rate and absorb the latent heat of fusion that is generated during ice crystal formation (see Section 15.1.2). Many laboratories purchase a commercial freezing container or develop a custom-made solution, such as a polystyrene foam packing container or insulating foam tube (see Section 15.2.2). However, a programmable rate-controlled freezer is useful for large batches of cryovials, different cooling rates, or complex programmed cooling curves (see Figure 15.4).

Suppliers.

Supplier	URL
BD Biosciences	http://www.bdbiosciences.com
Beckman Coulter Life Sciences	http://www.beckman.com
Binder	http://www.binder-world.com/us
BioSpherix	http://www.biospherix.com
Cole-Parmer	http://www.coleparmer.com
Corning	http://www.corning.com/worldwide/en/ products/life-sciences/products/cell-culture .html
Elga-Veolia	http://www.elgalabwater.com
Eppendorf	http://www.eppendorf.com/oc-en
Integra Biosciences	http://www.integra-biosciences.com/ global/en
Leica	http://www.leica-microsystems.com
Lonza	http://www.lonza.com
Merck Millipore	http://www.merckmillipore.com
Miele	www.miele.com./en/com/index-pro.htm
Nikon	http://www.nikoninstruments.com/ applications/life-sciences
Olympus	http://www.olympus-lifescience.com/en

Supplier	URL
Omega Engineering	www.omega.co.uk
Panasonic Healthcare (PHCbi)	http://www.phchd.com/eu/biomedical/pharmaceutical
Sartorius	http://www.sartorius.com/en
ThermoFisher Scientific	http://www.thermofisher.com/us/en/home/life-science/cell-culture.html
VWR (Jencons)	https://uk.vwr.com/store/
Wheaton (DWK Life Sciences)	http://wheaton.com
Zeiss	http://www.zeiss.com/microscopy/int/home.html

REFERENCES

Bakmiwewa, S.M., Heng, B., Guillemin, G.J. et al. (2015). An effective, low-cost method for achieving and maintaining hypoxia during cell culture studies. *Biotechniques* 59 (4): 223–224, 226, 228–9. doi: https://doi.org/10.2144/000114341.

Bertout, J.A., Patel, S.A., and Simon, M.C. (2008). The impact of O₂ availability on human cancer. *Nat. Rev. Cancer* 8 (12): 967–975. https://doi.org/10.1038/nrc2540.

Biocompare (2018). The buyer's guide for life scientists. https://www.biocompare.com (accessed 7 May 2018).

British Standards Institute (2005). BS 5726:2005. Microbiological safety cabinets. London: British Standards Institute.

CDC and NIH (2009). biosafety in microbiological and biomedical laboratories. https://www.cdc.gov/labs/bmbl.html (accessed 11 March 2019).

Geraghty, R.J., Capes-Davis, A., Davis, J.M. et al. (2014). Guidelines for the use of cell lines in biomedical research. *Br. J. Cancer* 111 (6): 1021–1046. https://doi.org/10.1038/bjc.2014.166.

Lab Manager (2017). 2018 Product resource guide. Lab Manager 12 (7). http://go.labmanager.com/2018-product-resource-guide (accessed 7 May 2018).

Lengner, C.J., Gimelbrant, A.A., Erwin, J.A. et al. (2010). Derivation of pre-X inactivation human embryonic stem cells under physiological oxygen concentrations. *Cell* 141 (5): 872–883. https://doi.org/10.1016/j.cell.2010.04.010.

Ongena, K., Das, C., Smith, J.L. et al. (2010). Determining cell number during cell culture using the Scepter cell counter. *J. Vis. Exp.* 45 https://doi.org/10.3791/2204.

Packer, L. and Fuehr, K. (1977). Low oxygen concentration extends the lifespan of cultured human diploid cells. *Nature* 267 (5610): 423–425.

Panasonic (2015). Water jacketed vs. air jacketed CO₂ incubators: competitive analysis. https://www.labrepco.com/data/file-downloads/panasonic_water_jacketed_vs_air_jacketed_co2_incubators_1443794033.pdf (accessed 11 April 2018).

WHO (2004). Laboratory biosafety manual. http://www.who.int/csr/resources/publications/biosafety/who_cds_csr_lyo_2004_11/en (accessed 9 May 2018).

CHAPTER 6

Safety and Bioethics

Everyone who works in a tissue culture laboratory must understand how to stay safe in that environment and how to keep others safe from any hazards that may be present. Personnel must also comply with any relevant ethical obligations. The tissue culture laboratory is subject to the same safety and regulatory issues as any workplace but has certain issues specific to the nature of the work. Tissue culture is also performed at varying scales, from small basic research groups to large industrial enterprises. While the principles of safety management remain fundamentally the same, the degree of regulation changes with the work environment. The focus in this chapter is on laboratories that perform research and teaching in academic and independent research institutes. Implications for work in an industrial environment will also be considered, but the regulatory requirements for clinical trials fall outside the scope of this book. The advice given in this chapter is general and should not be construed as complying with any national or international legal requirement. References to formal regulations are provided as a starting point for further investigation, mostly within the United States and United Kingdom. The obligation to generate formal guidelines lies with individual institutions.

6.1 LABORATORY SAFETY

The tissue culture laboratory, like any workplace, can be dangerous. Most people who perform tissue culture are trained to do so safely. However, the laboratory may include students, support staff, or other workers who do not have extensive experience with the laboratory environment and its hazards.

Accidents can affect anyone, even the experienced tissue culture operator, where familiarity can lead to a casual approach when dealing with hazards (Van Noorden 2013).

Tissue culture laboratories have many different hazards (see Table 6.1). Hazards can be identified by inspecting the laboratory, consulting on safety issues, and collecting existing information on workplace hazards. The more esoteric and poorly understood hazards tend to receive the most attention (e.g. genetic manipulation; see Section 6.3.7). However, looking at safety incident data, serious incidents in the tissue culture laboratory are thankfully rare; most minor accidents or injuries appear to be caused by common hazards that can be readily addressed (see Figure 6.1). It is important to approach all safety concerns using a consistent and objective approach, based on the level of risk that may apply in each case.

6.1.1 Risk Assessment

A risk can be broadly defined as the possibility that harm (illness, injury, death, or damage) might occur when exposed to a hazard. It is important to identify potential hazards but, at the same time, to be rational and proportionate in assessing the associated risks. If a risk is not seen as realistic, precautions will tend to be disregarded and the whole approach to safety will be placed in disrepute. Risk assessment is an important principle that is incorporated into most modern safety legislation.

Instructions for performing a risk assessment will vary with your location, but broadly speaking, you need to look at the various hazards associated with the activity, the severity of harm that might arise from each hazard, and the likelihood

TABLE 6.1. Common hazards in tissue culture laboratories.

Category	Item	Risk	Precautions
Sharps hazard	Broken glass	Injury, infection	Dispose of carefully in designated bin
	Broken or damaged pipettes	Injury, infection	Use plastic; discard damaged items; dispose of glass and plastic separately
	Sharp instruments, e.g. scalpel blades	Injury, infection	Handle carefully; discard in sharps bin
	Syringe needles	Injury, infection	Minimize or eliminate use; discard into sharps bin
Falls hazard	Water on floor	Slipping	Check tubing and replace regularly; mop up spills promptly
	Cables, tubing on floor	Leakage, snagging, tripping	Check connections; clip in place; keep away from passage floors
Ergonomic hazard	Prolonged sitting, e.g. microscope use	Musculoskeletal injury	Buy ergonomic equipment; adjust before use; stop and stretch regularly
	Handling heavy items, e.g. gas cylinders	Musculoskeletal injury	Secure in place; use correct lifting technique
Burns	Hot equipment, e.g. ovens, autoclaves	Burns	Post warning notices; provide PPE
	Cold equipment, e.g. metal racks in liquid nitrogen	Frostbite	Provide PPE
Fire	Bunsen burners; flaming, particularly in association with alcohol	Fire, burns, toxic fumes	Keep Bunsen burners out of hoods and do not place under cupboards or shelves; do not return flaming instruments to alcohol
Electrical hazard	Equipment faults	Fire, burns, electric shock	Regular testing and tagging; install "power kill" switches
Biological hazard	Adventitious pathogens	Infection	Screen cultures for likely pathogens; use universal precautions; minimize aerosols; decontaminate waste
	Genetic manipulation	Infection, DNA transfer	Use appropriate containment; follow GMO guidelines
	Propagation of pathogens, e.g. viruses	Infection	Use appropriate containment; minimize aerosols
	Large spills of biological material	Infection	Prepare a "spill kit" with suitable PPE, containment materials, and disinfectant that can be used to clean up a major spill; develop emergency procedures for handling major spills
	Failure of containment, e.g. from faulty BSC	Infection	Perform regular testing and certification of BSCs
Chemical hazard	Hazardous chemicals	Acute toxicity, long-term health problems	Read safety information, e.g. SDS; wear PPE; use the appropriate safety cabinet; dispose of waste correctly
Radiation hazard	Addition of radioisotopes to cultures	Radiation exposure through skin or mucous membranes, ingestion, inhalation	Monitor personnel for exposure; provide screens; use ALARA principle
	Irradiation of cultures	Radiation exposure through X-ray or gamma-ray devices	Monitor personnel for exposure; provide screens; use ALARA principle
Gas hazard	Asphyxia due to oxygen depletion, e.g. in liquid nitrogen storage areas	Confusion, loss of consciousness, death	Monitor gas levels (oxygen); evacuate room when alarm sounds
	Gas toxicity, e.g. CO_2 intoxication	Confusion, loss of consciousness, death	Monitor gas levels (CO_2); evacuate room when alarm sounds
Pressure hazard	Explosion, e.g. vial explosion due to pressure buildup	Injury, infection	Provide PPE including face visor; evacuate area if explosion imminent

ALARA, As low as reasonably achievable; BSC, biological safety cabinet; GMO, genetically modified organism; PPE, personal protective equipment; SDS, safety data sheet.

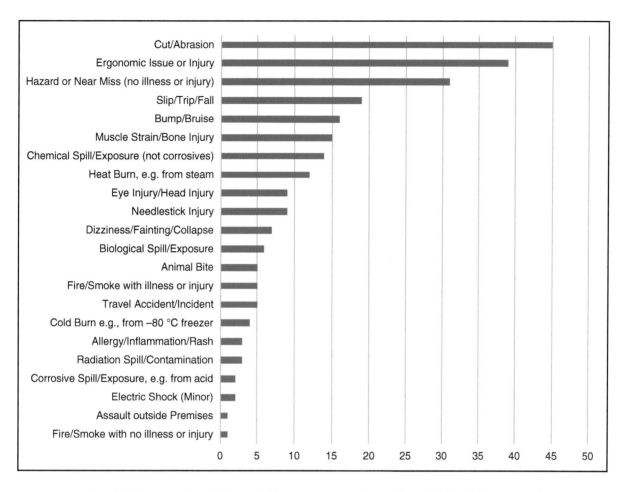

Fig. 6.1. Laboratory safety incidents. Incidents were reported over 10 years (2007–2017) at a research institute. All workers were encouraged to report safety incidents, including near misses, for consultation and further action. Reported incidents are shared for educational purposes, to illustrate the hazards that are commonly encountered in research laboratories. *Source*: Darryn Capes-Davis, courtesy of CMRI.

that harm will occur (see Figure 6.2). Some organizations use a risk matrix or a similar tool to help estimate the level of risk (Winder and Makin 2006; Capes-Davis and Capes-Davis 2018). Always remember that such tools have weaknesses; a risk matrix can oversimplify risk assessment and it can be difficult to compare different types of risks. Refer to your local safety regulations for guidance on risk assessments and consult with your local safety officer, who will be able to advise and assist with performing risk assessments.

Although the nature of the hazard itself may remain constant, the level of risk will vary depending on other modifying factors (see Table 6.2). Modifying factors should be considered when you perform a risk assessment and when you choose control measures to manage the risks. It should be remembered that if, having performed a risk assessment, the resulting level of risk is anything other than low, then procedures and processes will need to be changed in order to reduce the risk. If this is not possible, appropriate control measures will be required. Standard Operating Procedures (SOPs) (see

Section 7.5.1) and Personal Protective Equipment (PPE) (see Section 12.2.4) are examples of risk control measures that are commonly used in tissue culture laboratories. Safety regulations typically set out a "hierarchy" of risk control measures that should be used; more information on this topic is available as Supplementary material online (see Supp. S6.1).

6.1.2 Safety Regulations

The general principles of hazard identification, risk assessment, and control measures apply broadly to all tissue culture laboratories. However, compliance requirements will vary from location to location and overlapping requirements may need to be addressed. Safety regulations exist at a number of different levels (see Figure 6.3). International bodies such as the European Union (EU) develop overarching directives and regulations that set out requirements for member countries and states. National governments enact laws that set out mandatory requirements and penalties; in some areas, governments appoint statutory authorities to

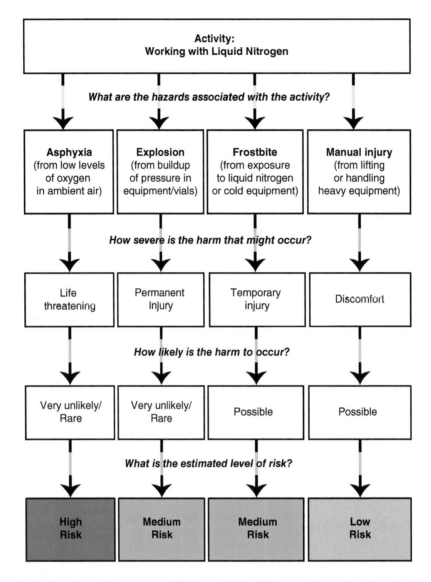

Fig. 6.2. Performing a risk assessment. Working with liquid nitrogen has various hazards, including asphyxia, explosion, frostbite, and manual injury. For each hazard, the level of risk has been estimated using a risk matrix, which considers the severity and likelihood of harm (Winder and Makin 2006; Capes-Davis and Capes-Davis 2018). This example is provided as an illustration only. Always perform your own risk assessment and consider the various hazards and modifying factors that may apply in your location (see Tables 6.1, 6.2).

regulate compliance and develop processes for licensing and accreditation. Similar mechanisms may exist at state level. Expert stakeholder groups (e.g. scientists with expertise in biosafety) develop written standards, codes of practice, manuals, and other guidance documents that set out best practice. Funding bodies, journals, and institutions (e.g. universities) develop guidelines, rules, and policies that apply to their work. Institutional committees may also review and approve specific projects – for example, if using recombinant DNA technology (see Section 6.3.7).

Where do you start if you are looking for safety regulations that apply to your particular project and laboratory? Many

organizations have an office or department that specializes in safety and compliance and hopefully participates in training workers. Senior staff and members of local committees are repositories of local knowledge and often make themselves available to answer questions. Safety regulations at state, national, and international level are usually published online; these may be accompanied by helpful guidelines with explanations and examples. Universities and other organizations often compile these resources on their websites and make additional policies and guidelines available to spell out what you need to do. Finally, best practice documents such as written standards and manuals can be tremendous resources

TABLE 6.2. Modifying factors in risk assessment.

Category	Items affecting risk
Operator	
Experience	Level; relevance; background
Training	Previous training; new requirements
PPE	Adequacy; properly worn (e.g. whether lab coat is buttoned up); regular laundering; repair or disposal when damaged
Facility	
Laboratory space	Sufficient space for equipment and personnel
Laboratory design	Availability of environmental monitoring (e.g. oxygen levels) and safety equipment (e.g. safety shower, eyewash station)
Equipment	
Overall condition	Age; adherence to new safety regulations; maintenance; repair
Suitability for task	Access; sample capacity; level and type of containment
Mechanical stability	Loading; anchorage; balance
Electrical safety	Connections; leakage to ground (earth); proximity of water
Containment	Aerosol generation; toxic fume generation; leakage or spillage from work area or exhaust ductwork; site of effluent and downwind risk
Decontamination	Method; duration; effectiveness; quality control testing
Physical risks	
Excessive heat	Heat load from equipment; dissipation; effect on operator performance
Intense cold	Frostbite; reduced dexterity from numbing
Noise and vibration	Background noise levels; equipment noise; equipment vibration
Light	Background lighting; task lighting; ultraviolet light exposure
Electric shock	Equipment installation; routine testing and tagging; repair
Fire	Faulty electrical wiring; incursion of water near wiring
	Emergency equipment and procedures (e.g. fire exit, evacuation drills)
	Storage of solvents and flammable chemicals (e.g. do not store ether in refrigerators); use of high oxygen gas mixtures
Hazardous chemicals (including gases and volatile liquids)	
Health hazards	Acute effects (e.g. for corrosive substances, asphyxiating gases)
	Chronic effects (e.g. cytotoxic, carcinogenic, mutagenic, and teratogenic substances; allergic reactions)
Reaction with water	Heat generation; superheated liquids
Reaction with solvents	Generation of heat, splashing, explosive mixtures
Volatility	Intoxication; asphyxiation
Generation of powders and aerosols	Inhalation; dissemination
Storage and transport conditions	Lack of security; flooding; instability; container breakage or leakage
Biohazard	
Pathogenicity	Species; host specificity; mode of transmission; viability; availability of vaccine or treatment
Genetic manipulation	Host specificity; vector infectivity; attenuation; conditional activation
Containment	Primary containment (e.g. BSC, handling procedures); secondary containment (e.g. facility design)

TABLE 6.2. (*continued*)

Category	Items affecting risk
Radiation	
Emission	Type; energy; half-life; mode of exposure (e.g. through ingestion); penetration (e.g. through shielding); interaction (e.g. ionization)
Localization on ingestion	DNA precursors, such as [^{3}H]thymidine
Disposal	Solid, liquid, gaseous form; storage to reduce energy; legal limits
Special circumstances	
General health status	Tiredness; discomfort; ability to concentrate
Pregnancy	Risk to fetus, teratogenicity; changes to immune system
Illness	Immunodeficiency
Therapy	Immunosuppression
Cuts and abrasions	Increased risk of absorption
Allergy	Latex gloves; powders (e.g. detergents); aerosols
Elements of procedures	
Scale	Amount used; suitability of containment for that amount; size of equipment and facilities (e.g. problems with access); number of personnel involved
Complexity	Number of steps or stages; number of options; interacting systems and procedures
Duration	Fatigue; prolonged sitting; process time; incubation time; storage time
Number of persons involved	Level of supervision; overcrowding; access to equipment and facilities
Location	Containment; security and access

because they bring together expert working groups to build a consensus on compliance requirements. Standards are available for laboratory safety at national and international level – for example, an International Organization for Standardization (ISO) Standard is available for safety requirements in medical laboratories – ISO 15190:2003(en) (International Organization for Standardization 2003). Standards can be expensive to purchase in bulk, but you may have access to a subscription through a local university or library.

Examples of safety regulations are provided throughout this chapter, primarily from the United States and United Kingdom. Guidance on health and safety is available in the United States from the Occupational Safety and Health Administration (OSHA), and in the United Kingdom from the Health and Safety Executive (HSE). In Europe, safety regulations are typically handled at the national level; each country takes responsibility for compliance with EU Directives (Bielecka and Mohammadi 2014). Information and advice can be obtained from the European Agency for Safety and Health at Work (EU-OSHA, www.osha.europa.eu). Outside Europe, safety guidelines are handled at national level (and in some cases at state level). Compliance with legislation is mandatory. By contrast, best practice documents such as written standards and manuals are typically regarded as advisory in nature. They, and the recommendations in this book, are not intended as regulatory documents and should not be quoted as if they are.

6.1.3 Training

Safety is everyone's responsibility, but it falls on the experienced operator to train newcomers in safe tissue culture procedures and maintain the tissue culture laboratory as a safe environment for all who work there. New personnel must receive an orientation or induction that explains the precautions, rules, and responsibilities that apply in the tissue culture laboratory (see Section 30.1.2). Hands-on training is essential, but written policies and SOPs should also be provided with information on safe practices, regulatory requirements (including biosafety, biosecurity, and bioethics), and validation testing (see Sections 6.3, 6.4, 7.4, 7.5.1). Any changes to regulations or local policies and procedures should be communicated to all personnel as part of ongoing training. One of the most important roles of the institution in providing a safe working environment is to provide the correct training in appropriate laboratory procedures and to ensure that new and existing members of staff are and remain familiar with safety regulations. Consultation mechanisms should be available so that safety concerns can be promptly reported and suitably addressed by workers and management together.

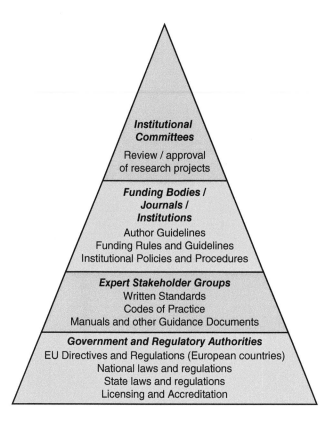

Fig. 6.3. Levels of safety regulation. Sources of safety regulations are broadly classified here into four groups at different levels, along with the types of safety regulation that each group produces. The groups at the base of the triangle produce safety regulations that underpin the work of the groups that are listed toward the apex of the triangle. Feedback will occur between the different levels – for example, guidance documents from expert groups may be cited when new legislation is developed.

Tissue culture laboratories can vary considerably and it should never be assumed that an experienced operator will know how to stay safe in a new environment. Any new member of the unit should be regarded as an apprentice, regardless of their seniority, until they demonstrate that they are technically competent and can work independently within the regulatory structure that applies to that particular laboratory. Training programs and exercises are discussed in a later chapter that will help new tissue culture operators to develop practical skills for core activities (see Section 30.2).

6.1.4 Ergonomics

Ergonomics can be described as "fitting a job to a person" (OSHA 2018a). It aims to prevent soft tissue injuries and musculoskeletal disorders that can be caused by sudden or sustained exposure to repetitive motion, awkward posture, force, noise, vibration, light, or heat. Tissue culture may require operators to hold awkward postures or perform repetitive actions for long periods of time (e.g. using a microscope or pipetting). Many personnel engage in similar activities away from tissue culture, such as prolonged computer work or manual tasks. It is not surprising that soft tissue injuries and musculoskeletal disorders are common in research organizations (see Figure 6.1).

There are some excellent resources available online with information on musculoskeletal disorders and ergonomic programs (EU-OSHA 2018; HSE 2018b; NIOSH 2018). For tissue culture, the following broad principles may be useful:

(1) **Purchase ergonomic equipment.** Biological safety cabinets (BSCs), chairs, computer workstations, and microscopes offer an increasing number of ergonomic features. All ergonomic features should be easily adjustable. Tissue culture equipment is typically shared by several users, making it necessary to adjust settings to suit your needs.

(2) **Set up your space before you start work.** Does your chair enable you to sit comfortably with support for your feet and back? Is your computer screen at eye level? Can you reach everything that you need without stretching or twisting?

(3) **Be aware of your posture.** Is your body in a "neutral" posture while at rest? Neutral posture avoids bending, overextending, or twisting – the body is "ears over shoulders over hips" facing forwards with straight wrists, relaxed fingers, and feet on the floor. A neutral posture helps to reduce muscle tension and improve user comfort.

(4) **Pause and stretch.** Aim for 1- to 2-minute breaks every 20–30 minutes, and a 5-minute break every hour. People should be encouraged to organize their work into suitable time periods, so that they move about at regular intervals and vary the activities that they are undertaking.

(5) **Minimize repetitive activities and manual tasks.** If you cannot eliminate repetitive activities, intersperse with other tasks and use correct handling techniques (EU-OSHA 2018). Repetitive pipetting can be minimized using a multichannel pipette, syringe dispenser, or automated liquid handling equipment (see Section 5.1.3).

(6) **Seek help early.** Treat pain or other symptoms as warning signs. Organizations should encourage workers to report ergonomic issues early so that the work activity can be assessed and changes made. Simple changes (e.g. intermittent stretch exercise programs) can help to reduce symptoms of muscle strain (Marangoni 2010). Always seek medical advice if you are concerned or symptoms persist.

6.2 HAZARDS IN TISSUE CULTURE LABORATORIES

What hazards are specific to the tissue culture laboratory? Biohazard is perhaps the most obvious concern for anyone who is working with tissue or living cells (see Section 6.3.1), but other hazards must be considered. Common hazards are listed earlier in this chapter (see Table 6.1), but it is worth discussing specific tissue culture hazards in more detail to understand the control measures that should be applied.

6.2.1 Needlestick and Sharps Injuries

Accidental handling of sharp items (e.g. broken pipettes, scalpel blades, and syringe needles) is a common and potentially serious hazard for any tissue culture operator. Accidental inoculation from a sharp item can result in transmission of pathogens, causing death or serious illness (see Section 6.3.1). Sharp items awaiting waste disposal are particularly dangerous because all workers involved in waste disposal are unknowingly placed at risk of exposure. With increased national and international initiatives to reduce the amount of single-use plastics disposed of worldwide, it is likely that there will be an increase in the use of glass pipettes in the tissue culture laboratory, with a resulting increase in the risk of sharps type injuries.

Pipettes are most likely to break when they are inserted into a pipette controller or when they are placed into a waste disposal container awaiting disposal. Take care when you are fitting a pipette controller onto a pipette. Check that the neck is sound, hold the pipette as near the end as possible, and apply gentle pressure with the pipette pointing away from your knuckles (see Figure 12.7). Although this is primarily a risk arising from the use of glass pipettes, even plastic pipettes can be damaged and break on insertion, so always check the top of each pipette before use.

The risk of pipette breakage increases when too many pipettes are forced into too small a container while awaiting disposal (see Figure 6.4). Although the risk has been reduced by the move from glass to plastic, plastic pipettes can still break if sufficient force is applied. In addition, if too many pipettes are placed in a container of disinfectant prior to disposal, they might not be exposed to the disinfectant and could therefore be potentially infectious or contain viable genetically modified cells. Always dispose of plastic pipettes into a suitable hard-walled container, made of impermeable plastic that is resistant to laceration and breakage. Glass Pasteur pipettes break very easily, and the shards are extremely hazardous. Discard glass Pasteur pipettes immediately after use into a dedicated sharps disposal container (Gwyther 1990). Disposable plastic pipettes and glass Pasteur pipettes should be discarded into separate containers as they may require different methods for decontamination and disposal.

Avoid using syringes and needles unless they are needed for loading ampoules (use a blunt cannula) or withdrawing fluid from a capped vial. If syringes and needles are used, unsheathe the needle slowly and carefully with the needle pointing away from you. Never re-sheathe a needle or any other sharp item; the closer your hand is to the sharp item, the more likely an injury will be. Always discard directly into a dedicated sharps

Fig. 6.4. Overfilled pipette cylinder. Pipettes protruding from a pipette cylinder as a result of attempted insertion of pipettes after the cylinder is full; those protruding from the cylinder will not soak properly or be disinfected and are prone to breakage when other pipettes are added. *Source*: R. Ian Freshney.

disposal container and do not use these receptacles for general waste.

6.2.2 Hazardous Substances

The most commonly used hazardous substances in the tissue culture laboratory are detergents and disinfectants (see Section 11.1). Detergents – particularly those used in automatic washing machines – are usually caustic, and even when they are not, they can cause irritation to the skin, eyes, and lungs. Liquid concentrates are more easily handled and are often less toxic compared to powdered versions. Use liquid-based detergents in a dispensing device whenever possible and wear gloves and eye protection. Chemical disinfectants such as hypochlorite solution ("bleach") can be highly toxic and should be used cautiously, either in tablet form or as a liquid dispensed from a dispenser. Liquid or tablet formulations may be more expensive, but the cost is worthwhile if safety incidents can be avoided. Substitute with less toxic disinfectants if possible.

Hazardous chemicals are used in some procedures (e.g. cytotoxicity assays). Chemicals that require special attention include (i) dimethyl sulfoxide (DMSO), which is a powerful solvent and skin penetrant and can transport many substances through the skin (Horita and Weber 1964); (ii) cytotoxic substances (able to inhibit or prevent the function of cells *in vitro* or *in vivo*); (iii) carcinogenic substances (able to cause cancer); (iv) mutagenic substances (able to damage DNA); and (v) teratogenic substances (able to cause developmental malformations). Carcinogens, mutagens, and other toxic chemicals are sometimes dissolved in DMSO, increasing the risk of uptake via the skin. DMSO is also able to penetrate protective gloves, including rubber latex and silicone. Nitrile gloves provide a better barrier but may degrade rapidly following exposure. Always look up information on chemical resistance when selecting gloves and perform a risk assessment to select the best control measures that apply to your procedure, including appropriate PPE and containment using a BSC or other suitable equipment (see Section 6.3.4).

A Safety Data Sheet (SDS) should be available for any hazardous chemical that is purchased from a reputable supplier, setting out the hazards associated with that substance and the risk control measures that are effective. In most countries this is now required by law. The United Nations has developed a Globally Harmonized System of Classification and Labeling of Chemicals (GHS) to ensure that countries are consistent in their classification of chemicals, labels, and SDS information (UN 2011). GHS has now been adopted in many countries as part of safety regulations. In the United States, handling of hazardous substances is regulated by the Occupational Safety and Health Administration (OSHA 2018b). In the United Kingdom, guidance on Control of Substances Hazardous to Health (COSHH) is available from the Health and Safety Executive (www.hse.gov.uk/coshh). However, it should be noted that the SDS is not a risk assessment; the information provided should be used to conduct your own assessment.

Laboratories that use radioisotopes and other sources of ionizing radiation must comply with additional safety regulations. Many research laboratories have substituted radioactive materials for less hazardous substances and thus no longer use ionizing radiation. For those who continue to require it, more information on ionizing radiation is available as Supplementary Material online (see Supp. S6.2).

6.2.3 Asphyxia and Explosion

Tissue culture requires a controlled atmosphere and so it is inevitable that a supply of gas will be needed in or near the tissue culture laboratory. Most gases used in tissue culture (CO_2, nitrogen, oxygen) are not harmful in small amounts but can be dangerous if handled incorrectly. If a major leak occurs, there is a risk of asphyxia from CO_2 and inert gases such as nitrogen. Asphyxia has resulted in fatalities in research laboratories, particularly when working with liquid nitrogen or dry ice where large amounts of gas can be released (Gill et al. 2002; Finkel 2007; Kim and Lee 2008). CO_2 is toxic in its own right and symptoms of toxicity may occur before asphyxia develops (Langford 2005).

Always perform a risk assessment before you install or work with any hazardous gas. This includes dry ice and liquid nitrogen – remember that 1 l of liquid nitrogen generates nearly 700 l of nitrogen gas! Risks are best addressed by installing engineering controls during laboratory construction (see Section 4.1.4). Monitoring of oxygen and CO_2 levels should always be performed in rooms where nitrogen and CO_2 are stored in bulk, or where there is a piped supply to the room. Personnel should be trained to evacuate the room immediately if an alarm is activated and stay out until the alarm is resolved. If a person collapses due to hazardous gas levels, anyone who tries to rescue them will also be rapidly affected, resulting in a chain reaction of casualties. Call for emergency services or other personnel who can enter the room safely using an independent air supply.

Gases are usually supplied under pressure. Any pressurized system has a risk of explosion if the pressure builds up due to malfunction or if the equipment is handled inappropriately. Pressurized systems must only be connected or modified by trained and competent personnel. Gas cylinders should be stored upright and properly secured during transport, storage, and use; this will also reduce the risk of crush injuries due to the cylinder falling (see Figure 6.5). A falling cylinder can shear off the regulator, resulting in an extremely hazardous explosive release of pressure. Always use the correct regulator and never use a damaged cylinder. Regulator manufacturers will give their products a set service life, which is typically five years. However, this interval could be less for applications with corrosive gases; regulators will normally carry an inspection or replacement date stamp. To minimize the risk of pressure related accidents, it is important to change gas regulators at the intervals specified by the manufacturer. Keep gas supplies separate from flammable chemicals and ignition sources (e.g. Bunsen burners).

Frozen cryovials or glass ampoules can also explode. When submerged in liquid nitrogen, a pressure differential exists between the outside and the inside of the vial. If it is not perfectly sealed, liquid nitrogen may be drawn in, causing the vial to explode when it is thawed. The problem is particularly severe with glass ampoules but can also occur with plastic cryovials. It can be avoided by storing vials in vapor phase of liquid nitrogen (see Section 15.2.3). Thawing of vials that have been stored submerged in liquid nitrogen should always be performed in a container with a lid, such as a plastic bucket (see Protocol P15.2; see also Figure 15.8). Suitable PPE, including a face shield or safety goggles, must be worn whenever frozen vials are handled.

Fig. 6.5. Gas cylinder clamp. This system fits different sizes of cylinder and can be moved from one position to another if necessary; similar devices are available from most laboratory suppliers. Gas cylinders should not be positioned without a securing clamp and strap. (a) Wrong: cylinders without securing clamp and strap. (b) Right: cylinder clamped onto edge of bench and securing gas cylinder with fabric strap; also shows cuff in place to protect the cylinder valve. *Source*: R. Ian Freshney.

6.2.4 Burns and Frostbite

Because liquid nitrogen has a temperature of −196 °C, direct contact with the liquid (e.g. splashes) or with anything that has been submerged in it rapidly results in cold burns (frostbite). It is often underestimated just how quickly cold burns will occur and how rapidly skin will adhere to a metal surface at this temperature. Gloves are essential and must be thick enough to act as insulation, but flexible enough to allow the manipulation of cryovials (e.g. Cryo-Gloves® or Waterproof Cryo-Grip® gloves, Tempshield).

Heat burns can arise from (i) heat-generating equipment such as autoclaves, ovens, and hot plates; (ii) handling of items that have just been removed from them; (iii) steam from hot liquids or autoclaves; and (iv) naked flames such as a Bunsen burner. It is vital to leave sufficient time for autoclaves and ovens to cool before items are removed, both for the items being sterilized and for the safety of the operator. This cooling time also reduces exposure to superheated liquids, which can occur when liquids are heated in an autoclave or microwave. A superheated liquid is one that is heated to above its boiling point without actually boiling; any disturbance can induce sudden discharge of steam and splashing. Cooling times can be enforced by using a timer on automatic equipment, but for manual equipment, training is needed to ensure that all personnel are aware of the risk and strictly adhere to proper practice. Insulated gloves, safety goggles, and other PPE should always be provided where hot items are being handled.

6.2.5 Fire

The risk of fire is increased during tissue culture procedures if personnel use alcohol for swabbing or sterilization, together with a Bunsen burner for flaming. Keep the two separate; always ensure that alcohol for sterilizing instruments is kept in minimum volumes in a narrow-necked bottle or flask that is not easily upset (see Figure 6.6). Alcohol for swabbing should be kept in a plastic wash bottle or spray bottle and should not be used in the presence of an open flame. When instruments are sterilized in alcohol and the alcohol is subsequently burnt off, care must be taken not to return the instruments to the alcohol while they are still alight. If you are using this technique, keep a damp cloth nearby to smother the flames if the alcohol ignites.

6.2.6 Equipment Hazards

Equipment hazards include fire, electric shock, or burns when working with electrical equipment and musculoskeletal injuries when working with heavy equipment. Other

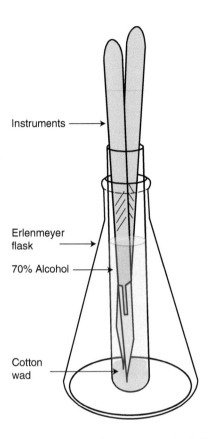

Fig. 6.6. Alcohol sterilization of instruments. The flask has a wide base to prevent tipping, and the center tube reduces the amount of alcohol required, so that spillage, if it occurs, is minimized. *Source*: R. Ian Freshney, from an original idea by M. G. Freshney.

hazards will depend on the specific equipment item and its use. For example, centrifuges, homogenizers, and sonicators may generate toxic fumes or aerosols that require the use of sealed rotors or operation in a safety cabinet. If your group does not have a laboratory management team, a specific person should be appointed to manage all equipment maintenance, electrical safety, and mechanical reliability. This person should ensure that equipment adheres to safety regulations. A specific person should also be put in charge of each item of equipment. Ideally, this should be someone who uses the equipment regularly and is familiar with its operation and requirements. Each "curator" acts as the point of contact for their item of equipment and should oversee its maintenance and operation and train others in its safe use. Personnel should also keep equipment records as an important part of quality assurance (see Section 7.5.4).

Fire and electrical shock are particularly important hazards for research laboratories; equipment may be operated in environments that increase the likelihood of an electrical fault. Many countries require that electrical equipment is regularly tested for faults and tagged to indicate that no faults have been detected. In the United States, the electrical safety of work equipment is dealt with by OHSA (Benson and Reczek

2017). In the United Kingdom, information and guidance on work equipment and machinery is available from HSE (HSE 2018c).

6.3 BIOSAFETY

The terms "biohazard," "biosafety," and "pathogen" are commonly used in tissue culture laboratories. For this book, biohazard is defined as the risk that harm may occur due to exposure to biological material. Biosafety is defined as the containment principles, technologies, and practices that are implemented to prevent unintentional exposure to pathogens and toxins or their accidental release (WHO 2004). Pathogen is defined as a cell culture, microorganism, or other biological material that is hazardous to human health. A pathogen may cause harm through infection, toxicity, or other mechanisms. "Biosecurity" is a related term that can be defined as the institutional and personal security measures that are designed to prevent the loss, theft, misuse, diversion, or intentional release of pathogens (WHO 2004). More information on biosecurity is available as Supplementary Material online (see Supp. S6.3).

Guidance on biosafety is available from various sources. The World Health Organization (WHO) has published its *Laboratory Biosafety Manual* since 1983 (WHO 2004). In the United States, the Centers for Disease Control and Prevention (CDC) and the National Institutes of Health (NIH) have published their *Biosafety in Microbiological and Biomedical Laboratories* since 1984 (CDC and NIH 2009). These documents provide excellent consensus recommendations. However, different countries may vary in their terminology and specific requirements – for example, the term "biological agent" is used in the United Kingdom where it is defined in legislation (HSE 2005). Always refer to local safety regulations for relevant terms and requirements for risk assessment (see Section 6.1.2).

6.3.1 Source of Biohazard Risk

Known pathogens. Many pathogens are grown or maintained using *in vitro* model systems – for example, poliovirus was first produced successfully in tissue culture using human embryonic neural tissue and was later cultivated using HeLa cells (see Figure 1.6). Biological safety issues are clearest when known pathogens are used, because the regulations covering such pathogens are well established. In the United States, the CDC has developed summary statements and containment requirements for many known pathogens (CDC and NIH 2009). In the United Kingdom, the HSE Advisory Committee on Dangerous Pathogens (ACDP) provides an Approved List of biological agents that is used to assign hazard groups, perform risk assessment, and apply control measures (HSE 2013).

Tissue and body fluids. Pathogens may be present as adventitious agents in any form of biological material, including body fluids (e.g. blood) and tissue samples obtained at biopsy, surgery, or autopsy. This is perhaps best documented for human and primate material. For example, a screen for pathogenic viruses in 844 human cell lines identified 17 cases where viruses were believed to come from the donor in association with their disease, and seven cases where viruses were likely to come from the donor but were unrelated to disease (Shioda et al. 2018). Pathogens have also been identified in animal products such as serum, particularly where animals are sourced from parts of the world with a high level of endemic infectious diseases (Grizzle and Polt 1988; Wells et al. 1989; Tedder et al. 1995). Pathogens may be transmitted from rodent colonies, particularly in xenograft facilities, where they may be transmitted from cell lines to immunocompromised mice (Biggar et al. 1977; Lloyd and Jones 1986; Dykewicz et al. 1992).

Risk assessment should consider the most likely pathogens that may be present – for example, human samples may carry hepatitis B, human immunodeficiency virus (HIV), or tuberculosis. However, because of increased international travel and movement of people, unexpected or unlikely pathogens can also be present in a sample. Pathogens in animal populations will vary depending on the species and source (e.g. laboratory or wild populations). For each pathogen, think about its mode of transmission and survival outside the body to determine who is at risk and how harm could arise. The risks associated with tissue and body fluids can be controlled using universal precautions, testing of serum and other animal products, and health monitoring of animal populations. Ideally, all human biospecimens should be tested for adventitious agents before handling and the authority to test should be agreed by the donor on the consent form (see Section 6.4.3). However, samples often need to be processed and cultured quickly, which means that you must proceed before testing is performed or before test results are available. Best practice is to always handle biological material as if a pathogen is present, using universal precautions (Buesching et al. 1989).

Cell lines. Primary cultures and finite cell lines usually carry a risk that is similar to the tissue and body fluids from which they were obtained. Primary cells from blood and lymphoid cells of human or simian origin are considered to carry an increased risk because blood-borne pathogens may be present (HSE 2005). Continuous cell lines that have been well characterized are usually considered to have a lower risk and may be used for many years without incident. However, such cell lines can carry undocumented viruses (Uphoff et al. 2010; Shioda et al. 2018). For example, screening of 577 human cell lines has shown that 3.3% carry retrovirus (Uphoff et al. 2015); this is particularly common in laboratories with access to xenograft facilities (see Section 16.5). Viruses can also be transmitted between cell lines through contamination (Stang et al. 2009). It is not clear how many viruses carry a risk to human health, but all continuous cell lines should be treated with caution in the knowledge that undocumented pathogens may be present. Always perform a risk assessment; if some cells should be handled using a BSC, it is recommended that all cell lines should be handled at a similar level of containment.

Many human continuous cell lines are derived from tumors or have undergone transformation in culture. The long-term consequences of inoculation of these cells are uncertain. Human-to-human cancer transmission has been documented following deliberate inoculation, needlestick injury, and organ transplantation (Southam et al. 1957; Gartner et al. 1996; Desai et al. 2012). Although the risk may be low, needles and other sharps should be avoided in tissue culture where possible, particularly if workers are pregnant or have compromised immune systems. Cell lines should not be generated from people working in the same laboratory; accidental self inoculation would generate no immune response, which could pose a risk particularly if the cell line had been transformed. Laboratory personnel should be aware of cancer transmission alongside other forms of biohazard risk (Lazebnik and Parris 2015).

6.3.2 Biohazard Risk Groups

Most safety regulations include a classification system that groups pathogens or genetically modified organisms (GMOs) at similar risk levels, which is useful to standardize the various risk control measures that may be required. Biohazard risk groups were developed for laboratory work with pathogens and may vary from country to country (Tian and Zheng 2014). The most widely quoted risk classification was developed by WHO; variations on these four risk groups have been incorporated into national legislation in many countries (WHO 2004; OECD 2018). The WHO classification consists of:

(1) *Risk Group 1 (no or low individual and community risk).* A microorganism that is unlikely to cause human or animal disease.

(2) *Risk Group 2 (moderate individual risk, low community risk).* A pathogen that can cause human or animal disease but is unlikely to be a serious hazard to laboratory workers, the community, livestock, or the environment. Laboratory exposures may cause serious infection, but effective treatment and preventive measures are available and the risk of spread of infection is limited.

(3) *Risk Group 3 (high individual risk, low community risk).* A pathogen that usually causes serious human or animal disease but does not ordinarily spread from one infected individual to another. Effective treatment and preventive measures are available.

(4) *Risk Group 4 (high individual and community risk).* A pathogen that usually causes serious human or animal disease and that can be readily transmitted from one individual to another, directly or indirectly. Effective treatment and preventive measures are not usually available.

6.3.3 Biological Containment Levels

The four biohazard risk groups correlate with four biosafety or containment levels. The selection of the appropriate level of biosafety or containment depends on risk assessment. Although risk groups correlate with biosafety or containment levels, they may not agree completely. Always refer to safety regulations for guidance on how pathogens or GMOs are classified and think about any factors that may modify the risks associated with your work (CDC and NIH 2009; HSE 2013; NIH 2016).

Biological containment levels are determined by different variables that include laboratory design, personnel behavior, equipment use, and safety procedures. In the United States, guidance on biosafety levels is available from NIH and CDC (CDC and NIH 2009) (see Table 6.3). In the United Kingdom, guidance on containment levels is available from HSE (HSE 2001, 2006) (see Table 6.4). New biosafety guidance and regulations may be released at any time. Always refer to local safety information, contact your local biological safety committee, and consult the appropriate regulatory authority to understand the compliance requirements that apply to your work.

What level of containment should apply to the tissue culture laboratory? Clearly the decision will depend on risk assessment, but broadly speaking, tissue culture hazards can be considered to be (HSE 2005):

(1) *Low requiring Level 1 containment.* Level 1 containment may be suitable for continuous cell lines that have been tested for pathogens, resulting in a lower risk of adventitious agents. Work can be performed on the open bench, depending on good microbiological technique. This type of work is normally conducted in a specially defined area, which may simply be defined as the "tissue culture laboratory," but which will have Level 1 conditions applied to it.

(2) *Medium requiring Level 2 containment.* Level 2 containment is suitable for finite or continuous cell lines that have not been tested for pathogens, resulting in a higher risk of adventitious agents. Work is typically performed in a BSC, although it may be possible to perform some work on the open bench and use the BSC for procedures that generate aerosols.

(3) *High requiring containment appropriate to the agent and potential risk.* Cell lines that are known to carry pathogens will need containment that is appropriate to their biohazard risk (see Section 6.3.1). For some pathogens, you may be able to use a BSC that offers a higher level of containment in a Level 2 facility. However, for other pathogens, a high

containment laboratory (Level 3–4) will be necessary. Look for guidance on specific pathogens as part of risk assessment (HSE 2013; NIH 2016).

6.3.4 Containment Equipment

The BSC is used in most laboratories for primary containment and serves to protect the operator from aerosols that are generated by tissue culture procedures (WHO 2004; CDC and NIH 2009). Horizontal laminar flow work stations do not provide biohazard protection for the operator or the environment, and fume hoods do not provide sterile protection for the material being handled. BSCs can protect the operator, the environment, and the material being handled; the degree and type of protection will depend on the class of BSC that is used. To briefly summarize BSC classes (see also Section 5.1.1):

(1) *BSC Class I.* Provides biohazard protection for the operator and the environment but does not provide sterile protection for the material being handled. These cabinets are not suitable for use in tissue culture.

(2) *BSC Class II.* Provides biohazard protection for the operator and the environment, and sterile protection for the material being handled. Class II BSCs are divided into A1, A2, B1, and B2 types (see Table 5.2). The Class II A2 cabinet is the most common type used in tissue culture (see Figure 6.7a). Class II type B cabinets were developed to allow containment of minute quantities of hazardous chemicals (CDC and NIH 2009). The effectiveness of chemical containment will depend on the type of chemical, the amount used, and the exhaust arrangements. Additional features may be required for containment of volatile chemicals and pathogens (see Figure 6.7b).

(3) *BSC Class III.* Provides protection for the operator and the environment, and sterile protection for the material being handled. These cabinets are typically used in high containment laboratories. Hand access is via glove ports and materials access is via double doored-hatches (see Figure 6.7c).

All BSCs must be certified as part of a strict maintenance program, with testing by a service engineer at regular intervals (see Section 5.1.1). Always remember that the effectiveness of a BSC depends on the integrity of its airflow (see Section 4.2.1). Airflow can be compromised by cross drafts outside the cabinet, rapid movement when your arms enter or leave the cabinet, and items that are present inside the cabinet. Correct technique when using a BSC will be discussed in a later chapter (see Section 12.4.1).

Other equipment or materials may be used for physical containment. For example, biological materials must be suitably packaged for shipping so that people and the environment are not exposed to biohazard. Strict safety regulations

TABLE 6.3. United States biosafety levels.

	BSL 1 (SMP)	BSL 2 (SP)	BSL 3 (SP)	BSL 4 (SP)
Access	Door for access control, pest control program. Sign with universal biohazard symbol at entrance to laboratory when infectious agents present. Access limited when work in progress but not separated from regular traffic. Most work conducted on open bench tops.	Door should be self-closing with locks. Advise all persons of potential hazards and requirements to meet for entry/exit. Provide medical surveillance for laboratory personnel, offer available immunizations for agents handled in laboratory, consider storage of serum samples.	As BSL 2 + access through two self-closing doors.	As BSL 3 + access through secure locked doors with logbook. Only persons required for scientific or support purposes are authorized to enter. Entry and exit through change and shower rooms, and through airlock for suit laboratory. Door interlocks for autoclaves and fumigation chambers.
Airflow and ventilation	Should apply to most laboratories. Not specified. Windows that open should have screens.	Position BSCs so airflow fluctuations do not interfere with operations. Windows must have screens.	As BSL 2 + ducted air ventilation system with sustained directional airflow and visual monitoring. Air exhaust must not recirculate. Windows must be sealed.	Dedicated and alarmed non-recirculating ventilation system to maintain negative pressure. Exhaust air through HEPA filters. Supply and exhaust fans interlocked. Windows must be break-resistant and sealed.
Cleaning	Easy to clean; spaces between cabinets, equipment, etc.; impervious bench surfaces. Chairs should be covered with non-porous material.	As BSL 1 + routine decontamination of work surfaces and equipment.	As BSL 2 + sealed smooth finish for walls and ceilings. Door spaces and ventilation openings able to be sealed for decontamination.	As BSL 3 + separate zone or building constructed as sealed internal shell with decontamination and backflow prevention for drains, HEPA filters for vents.
PPE and personal hygiene	No eating, drinking, etc. No mouth pipetting. Wear PPE – lab coat, gloves, and protective eyewear. Remove before leaving the laboratory. Sink for hand washing. Wash hands before leaving the laboratory.	As BSL 1 + eyewash facility and provision for laundering of protective clothing.	As BSL 2 + protective clothing with solid front, decontaminate before laundering. Eye and face protection for splashes outside the BSC. Sink for hand washing should be hands-free or automatic.	As BSL 3 + use of laboratory clothing and personal body shower before leaving. Suit laboratory personnel will wear positive-pressure air suit.
BSCs	Not required. Procedures should minimize splashes and aerosols.	Use a BSC for all procedures that may generate aerosols. Class II may recirculate.	Use a BSC for all procedures with manipulation of infectious materials. BSCs should be certified at least annually. Class II may recirculate. Class III must be hard connected.	Use a Class II BSC with positive-pressure air suit (suit laboratory) or a Class III BSC (cabinet laboratory).

TABLE 6.3. *(continued)*

	BSL 1 (SMP)	BSL 2 (SP)	BSL 3 (SP)	BSL 4 (SP)
Equipment	Not specified.	Decontaminate routinely, after spills or splashes, and before repair, maintenance, or removal. Protect vacuum lines with disinfectant traps.	As BSL 2 + equipment producing aerosols should be contained using primary barrier device with HEPA filtration of exhaust. Protect vacuum lines with HEPA filters.	As BSL 3 + equipment and materials enter by double-door autoclave, dunk tank, fumigation chamber, or airlock. All equipment and materials must be decontaminated before removal.
Sharps	Develop and implement policies for safe handling of sharps. Use sharps containers, decontaminate before disposal.	As BSL 1 + restrict to unavoidable use.	As BSL 2.	As BSL 2.
Decontamination	Use appropriate disinfectant. Decontaminate work surfaces after work is complete and after spills or splashes.	As BSL 1 + provide validated decontamination method for all laboratory waste.	As BSL 2.	As BSL 2 + Class III cabinet must be decontaminated using validated gaseous or vapor method.
Disposal	Decontaminate before disposal. Dispose of used gloves with laboratory waste.	As BSL 1 + validated decontamination method.	As BSL 2 + method preferably within facility.	As BSL 3 + decontaminate all material using double-door autoclave or equivalent method. Decontaminate liquid effluent with physical and biological validation.
Storage and transfer	No food storage in laboratory areas or equipment.	As BSL 1 + leak-proof container.	As BSL 2.	Non-breakable, sealed primary and secondary containers. Transfer via dunk tank or equivalent method.
Biosafety manual and training	Training required for laboratory personnel, including exposure precautions and evaluation procedures.	Biosafety manual required, must be adopted as policy, and be accessible. Training required, with proficiency in SMP and SP.	As BSL 2.	As BSL 2 + protocols for emergency situations and high proficiency in SMP and SP.
Accidents and spills	Not specified.	Trained staff should clean up spills. Evaluate incidents immediately and respond as per biosafety manual. Report incidents to supervisor and keep written records.	As BSL 2.	As BSL 2 + have facility available for quarantine and care of illness. Use system to report absences, incidents, and medical surveillance. Report incidents to supervisor and management.
Validation of facility	Not specified.	Validated decontamination method.	As BSL 2.	As BSL 2 + safety of effluent.

Source: Based on R. Ian Freshney and Amanda Capes-Davis, guidance documents (CDC and NIH 2009). Requirements are highly summarized; always refer to guidance documents for detailed information.

BSL, Biosafety level; HEPA, high-efficiency particulate air; SMP, standard microbiological practices, common to all laboratories; SP, special practices, necessary to address the risk of agents requiring increasing levels of containment.

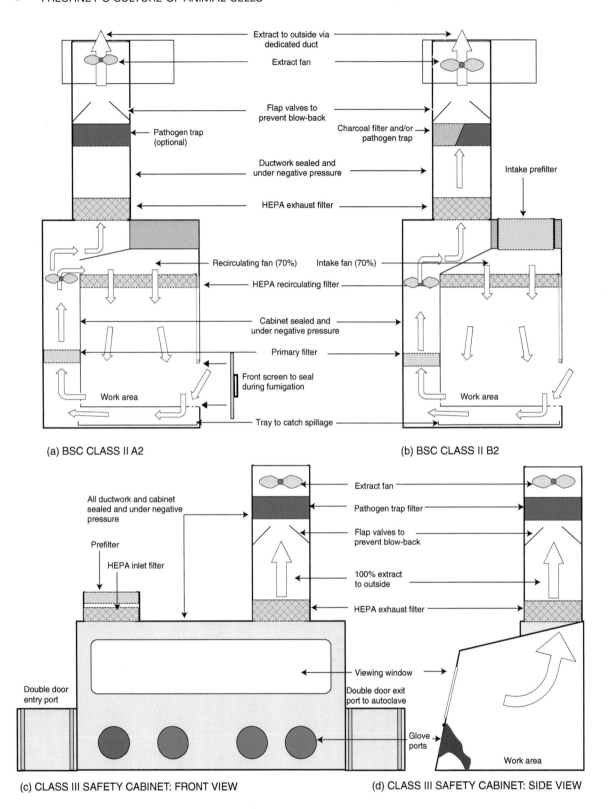

Fig. 6.7. Biological safety cabinet (BSC) design. (a) Class II A2 BSC, recirculating 70% of the air and exhausting 30% of the air via a filter and ducted out of the room through an optional pathogen trap. Air is taken in at the front of the cabinet to make up the recirculating volume and prevent overspill from the work area. (b) Class II B2 BSC with charcoal filters on extract and no recirculating air. (c) Class III non-recirculating, sealed cabinet with glove pockets; works at negative pressure and with air lock for entry of equipment and direct access to autoclave, either connected or adjacent. (d) Side view of Class III cabinet.

TABLE 6.4. United Kingdom containment levels.

	CL 1	CL 2	CL 3	CL 4
Access	Door closed when work in progress.	Access for authorized people only.	As CL 2 + separate lab with observation window. Door locked when lab unoccupied.	As CL3 + pest control. Entry via airlock reached after change facility and shower.
Space	Not specified.	Minimum of 11 m³/person.	As CL 2.	As CL 2.
Airflow	Not specified.	Negative pressure not required unless mechanically ventilated. Air filtration not required.	Negative pressure required. Extract air must be HEPA filtered, supply and extract interlocked.	Dedicated ventilation system maintaining not less than −75 Pa between laboratory and ambient air. Input and extract air must be HEPA filtered.
Cleaning	Easy to clean.	As CL 1 + bench surfaces impervious to water and resistant to acids, alkalis, solvents, and disinfectants.	As CL 2 + floor surface impervious to water and resistant to acids, alkalis, solvents, and disinfectants.	As CL 3 + separate zone or building sealable to permit disinfection (fumigation).
PPE and personal hygiene	Protective clothing worn; check and clean regularly, repair and replace when defective. No eating, drinking, etc. No mouth pipetting.	As CL 1 + wear gloves if required by risk assessment. Hand wash basin near exit, hands free or automatic.	As CL 2 + side or back fastening on protective clothing, autoclave before laundering or disposal. Gloves must be worn.	Complete change of clothing, autoclave after use. Shower before leaving. Suit laboratory personnel will use positive-pressure air suits.
BSCs	Not required but aerosol generation minimized.	BSC or isolator required for procedures generating aerosols. Class I or Class II. Must exhaust through HEPA filter, preferably direct to outside air.	As CL 2 + recirculate exhaust only if two HEPA filters used.	Class III BSC for hazard group 4 agents. Entry point usually has dunk tank; exit point usually has autoclave.
Equipment	Not specified.	Autoclave within the building.	Should contain own equipment and provide autoclave within laboratory suite.	Must contain own equipment and provide double-ended autoclave, preferably from Class III cabinet.
Decontamination	Disinfectant available. Decontaminate work surfaces at regular intervals.	Specified disinfection procedures in place.	As CL 2 + lab sealable to permit disinfection.	As CL 3 + lab sealable to permit fumigation.

TABLE 6.4. (*continued*)

	CL 1	CL 2	CL 3	CL 4
Disposal	Into disinfectant	Waste treated to ensure non-infectious, steam sterilization preferred. Safe collection and disposal.	As CL 2.	All liquid effluent and solid waste inactivated by steam sterilization or equivalent method.
Storage	Not specified.	Safe storage of biological agents.	As CL 2.	As CL 2 + register select agents or toxins.
Training	Not specified.	Written training records if required by risk assessment.	Written training records and list of exposed personnel required.	High level of expertise. Procedures for spills and emergencies required.
Accidents	Report accidents.	As CL 1 + evacuate, allow aerosols to settle, and clean up spill as per spills procedure.	As CL 2 + develop contingency plans, report accidental release.	As CL 3 + lone work not permitted, report incidents to appropriate authorities.
Validation of facility	Facility must meet acceptable standards.	As CL 1 + BSCs, autoclaves, and alarm systems tested against appropriate standards.	As CL 2 + laboratory sealable for disinfection.	Facility must meet regulatory requirements.

Source: Based on R. Ian Freshney and Amanda Capes-Davis, guidance documents (HSE 2001, 2006). Requirements are highly summarized; always refer to guidance documents for detailed information.
CL, Containment level

have been developed for the transport of biohazardous materials by air, road, or other modes of transportation. These are discussed in the Supplementary Material online (see Supp. S15.1).

6.3.5 Decontamination and Fumigation

Decontamination is a physical or chemical process that renders an area, device, item, or material safe to handle (CDC and NIH 2009). Tissue culture procedures generate small droplets that remain on nearby surfaces, resulting in possible exposure to pathogens and microbial contaminants. The BSC should be allowed to run for at least four minutes to allow the cabinet to "purge," removing any suspended particulates. Surfaces should be wiped down carefully using a suitable disinfectant, which will vary depending on the material being handled. Most tissue culture laboratories use 70% alcohol (ethanol or isopropanol) for surface disinfection (see Section 12.3.1).

Surface decontamination is sufficient for many tissue culture procedures. However, decontamination of the BSC or the room may be necessary in some locations (e.g. Level 3–4 laboratories) or at some times (e.g. before BSC service or repair, or following incidents of microbial contamination).

Fumigation is performed using formaldehyde gas, vaporized hydrogen peroxide (VHP), or chlorine dioxide gas (CDC and NIH 2009; Geraghty et al. 2014). However, formaldehyde is an extremely toxic compound. As well as being a sensitizer that can cause allergic reactions, it is also now recognized as a class 1 carcinogen by the WHO International Agency for Research on Cancer and as a class 1B carcinogen and class 2 mutagen by the EU (EU 2008). These classifications will potentially eliminate the use of formaldehyde as a general disinfectant agent in the European laboratory sector, as it will only be approved for sale as a general biocide for limited specialist applications. It is therefore recommended that alternatives such as VHP be used for routine sterilization of BSCs and tissue culture laboratories.

• *Safety Note. Chemicals used in fumigation can be highly toxic. Fumigation should only be performed by trained and competent personnel.*

6.3.6 Waste Disposal and Disinfectants

Waste must be managed appropriately to ensure that people and the environment are not exposed to hazards. Waste from research activities usually consists of small volumes of material that carry different hazards (Rau et al. 2000). Waste management relies on a common set of principles including prevention, reduction, recycling, and safe disposal. Waste disposal is subject to safety regulations (Bielecka and Mohammadi 2014). In the United States, disposal of potentially infectious materials is covered by the OSHA Bloodborne Pathogen Standard (OSHA 2001). In the United Kingdom, guidance on laboratory waste is included as part of a broader approach to healthcare waste (UK Department of Health 2013). Waste disposal requirements will vary depending on your location and the nature of the hazard; always check your local safety regulations.

Waste should be segregated into "waste streams" that require different methods of disposal. In the tissue culture laboratory, waste streams include general waste for disposal (e.g. plastic wrappings), general waste for recycling (e.g. paper and cardboard), solid biological waste (e.g. used plasticware and gloves), and liquid biological waste (e.g. spent tissue culture medium). Waste from cancer research laboratories, in particular, may contain cytotoxic drugs or chemicals and may require a separate waste stream, depending on your local safety regulations (HSE 2019).

Biological waste requires decontamination as part of safe disposal. Solid biological waste is usually autoclaved or incinerated. Liquid biological waste is usually treated using a suitable disinfectant (WHO 2004) and then discarded once decontamination is complete. The choice of disinfectant will depend on risk assessment — for example, whether spores or prions are likely to be present. Spores are produced by fungi and some bacteria (e.g. *Bacillus* species) and are resistant to some disinfectants, particularly at lower concentrations. Prions are difficult to eradicate and may survive autoclaving and prolonged treatment with chemical disinfectants; incineration is the safest method of disposal (WHO 2004; CDC and NIH 2009).

Solutions that release chlorine (hypochlorite solutions or "bleach") are commonly used for decontamination of liquid biological waste. Always look up the manufacturers' instructions and local safety regulations to select a suitable dose and treatment time for decontamination. Hypochlorite solutions are often used at 0.1% (1 g/l) of available chlorine, but this will vary with the degree of soiling; blood spills require 0.5% (5 g/l) of available chlorine (WHO 2004). Hypochlorite solutions are effective and easily washed off reusable items, but they are highly alkaline and can be corrosive to metal (even stainless steel, particularly at welded seams). Soaking baths and cylinders should be made of polypropylene. In some locations, hypochlorite solutions should be inactivated before disposal and this can be done using sodium thiosulfate (Hegde et al. 2012).

- *Safety Note. Hypochlorite solutions can cause burns to skin or eyes, lung damage, and bleaching of clothing. PPE should be worn including eye protection, gloves, and a lab coat or apron.*

Where corrosion needs to be avoided, substitution with less corrosive disinfectants such as Virkon™ (LANXESS – formerly DuPont Deutschland GMbH) should be considered. Virkon is a peroxygen-based disinfectant that has a wide spectrum of activity against most groups of bacteria, fungi and viruses (Hernandez et al. 2000).

6.3.7 Genetically Modified Organisms (GMOs)

The use of recombinant techniques for gene delivery or editing (see Sections 22.1, 22.2) results in a GMO that may carry a different level of risk to the parental material. The potential for GMOs to acquire pathogenic properties has been a longstanding concern, resulting in the development of a consensus approach to genetic modification at the Asilomar Conference (Berg et al. 1975). Containment was put forward as a central component in experimental design, with risk assessment determining the level of containment to be applied. This approach continues to be used because it is effective in managing any associated risks, despite concerns that the risks are theoretical and can result in complicated safety regulations (Petrella 2015).

Safety requirements for GMOs are typically drawn from the stronger body of evidence that has been developed for unmodified pathogenic organisms (Kimman et al. 2008). However, biosafety and genetic modification may be addressed by different safety regulations so there may be differences between the requirements for pathogens and GMOs. Typically, a regulatory body will oversee contained use of GMOs. If you use or generate GMOs, your laboratory may need certification and you may need to notify the regulatory body or seek approval from a local biosafety committee, depending on the level of risk associated with your work. In the United States, guidelines for recombinant DNA research are available from the NIH Office of Science Policy (NIH 2016). In the United Kingdom, guidance on the contained use of GMOs is available from HSE (HSE 2018a). GMO regulations in the United Kingdom were updated in 2014 and include requirements from the EU for contained use of genetically modified microorganisms.

6.4 BIOETHICS

Working with human and other animal tissue presents a number of ethical problems involving acquisition, subsequent handling, and the ultimate use of the material (Hansson 2009). Ethical problems may arise from use of cell lines many years after their establishment. Because all cell lines carry genomic material, there is a risk that publication of genomic data may affect the privacy of donors or their families (Hudson and Collins 2013). Ethical requirements for tissue or cell lines will vary from country to country. Always refer to local ethical regulations and ask for advice from your organization, local ethics committee, or regulatory body.

6.4.1 Ethical Use of Animal Tissue

Cell lines have been established from more than 660 different species, as determined by analysis of cell lines in the Cellosaurus knowledge resource (see Section 1.5.3) (Bairoch 2018). For all species, ethical regulations must ensure that tissue is obtained without causing undue pain or suffering. Ethical guidelines also highlight the need for replacement, reduction, and refinement (the 3Rs) if animals are to be used for scientific purposes (Festing and Wilkinson 2007). Tissue culture can support these goals by developing alternative methods that should dramatically reduce the use of animals in future (Kerecman Myers et al. 2017).

Most countries have legislation or other regulations that govern the care and use of animals for scientific purposes. Typically, research projects that use animals must be assessed and approved before any work commences. The proposed use must be considered justified, and suitable provisions must be set out for their care and wellbeing. This is equally true whether animals are undergoing routine husbandry, experimental conditions in an animal facility, or clinical conditions in a veterinary hospital. A local committee – an Animal Welfare and Ethical Review Body (AWERB) in the United Kingdom, and an Institutional Animal Care and Use Committee (IACUC) in the United States – will usually review projects and oversee animal welfare, while a national regulatory body takes responsibility for licensing and approval processes (Festing and Wilkinson 2007). Many organizations, funding agencies, and journals set out policies on use of animals that will apply to research, funding, and publication (Wiley 2014).

Do ethical regulations apply to all species? A distinction is often made between higher vertebrates and other species, based on the perception that higher vertebrates have sufficient brain capacity and neural organization to feel pain and distress from partway through their embryonic development. However, it is difficult to assess pain and discomfort in some species (Harvey-Clark 2011). Ethical regulations may apply to lower vertebrates and some invertebrates – for example, in Australia the term "animal" includes any live non-human vertebrate (including fish and reptiles) and all cephalopods (NHMRC 2013). Always refer to your local ethical regulations and discuss with the local officer or department who is responsible for safety and compliance.

6.4.2 Ethical Use of Human Tissue

Any research that involves human subjects is subject to ethical requirements, as stated in the Declaration of Helsinki (Carlson et al. 2004). Projects that use human tissue must undergo ethical review in accordance with that country's ethical regulations. A local committee – a Research Ethics Committee

(REC) in the United Kingdom, and an Institutional Review Board (IRB) in the United States – will decide whether the work is reasonable and justified by the possible outcome. The REC or IRB must be contacted before any work with human tissue is initiated. This is best done at the planning stage; most funding bodies will require evidence of ethical approval before awarding funding.

In the United Kingdom, RECs are overseen by the Health Research Authority (HRA) (HRA 2018). Collection of human tissue for use in a "scheduled purpose" (including research) must comply with the Human Tissue Act (2004), which is regulated by the Human Tissue Authority (HTA) (Geraghty et al. 2014). Donor consent (see Section 6.4.3) is the fundamental principle underpinning the legislation, with the use of postmortem human tissue samples being particularly stringently regulated. The HTA have published a series of Codes of Practice, providing guidance to professionals carrying out activities which lie within the remit of the HTA (HTA 2018). Those individuals in the United Kingdom using human tissue samples for a "scheduled purpose" should familiarize themselves with the Codes of Practice relevant to their proposed work. The Human Tissue Act (2004) defines human tissue ("relevant material") as material that consists of, or includes, human cells (Geraghty et al. 2014). Tissue and primary cells are covered by this definition, but cell lines are not. Material that is created outside the body is specifically excluded; once a cell line has been generated, it is assumed that this is constituted by the progeny of the original cells and that the original cells no longer exist. More information is available from the relevant regulatory bodies; a comprehensive summary is included in guidelines for the use of cell lines in biomedical research (Geraghty et al. 2014).

In the United States, IRBs are registered with the Office for Human Research Protections (OHRP, www.hhs.gov/ohrp). Collection of human tissue is regulated under the Federal Policy for the Protection of Human Subjects in Research, usually referred to as the "Common Rule." The Common Rule was recently revised, with changes taking effect in January 2018 (OHRP 2017). The new definition of "human subject" includes a living individual about whom an investigator "obtains, uses, studies, analyzes, or generates identifiable private information or identifiable biospecimens." Under this definition, a cell line might be subject to the Common Rule if the identity of a living individual can be readily ascertained by the investigator or associated with the biospecimen. However, exemptions may also apply – for example, if that information is publicly available.

Ethical restrictions are particularly stringent for projects that use human embryos, human embryonic stem cells (hESCs), or cell lines derived from human embryos. In the United Kingdom, the Human Fertilization and Embryology Authority (HFEA) acts as an independent regulator of *in vitro* fertilization (IVF) and human embryo research, including the use of cell lines derived from human embryos (Geraghty et al. 2014). In the United States, the OHRP sets out guidance regarding research that uses hESCs and germ cells from fetal tissue (OHRP 2002). Regulatory requirements may be complex – for example, the Human Pluripotent Stem Cell Registry (hPSCreg) collects information on cell lines from 25 different countries (Seltmann et al. 2016).

Clearly, ethical requirements for tissue culture will vary depending on your location, donor, and application. Different ethical regulations may apply if you wish to obtain tissue from children, minority populations, vulnerable individuals (e.g. prisoners), or deceased individuals. Always refer to local ethical regulations and seek out advisory groups to help you communicate your research, understand its implications, and form respectful partnerships (Haring et al. 2018). Requirements may come from unexpected directions – for example, privacy laws have implications for sharing research data from cell lines. The topic of donor privacy is discussed in more detail in the Supplementary Material that is available online (see Supp. S6.4).

6.4.3 Donor Consent

All donors must now give informed consent for participation in research projects, in accordance with the Declaration of Helsinki (Carlson et al. 2004). Typically, a consent form will be prepared as part of an ethics application and reviewed by an ethics committee. Additional input may come from your organization, funding body, clinical collaborators, patient support groups, or consumer representatives.

There are various models of consent that may be used. The following list of consent models is quoted from (Beskow 2016):

(1) **Blanket consent.** Individuals are asked to consent to all future research with no limitations or conditions.
(2) **Broad consent.** Individuals are asked to consent to the collection and storage of biospecimens for future unspecified research, which will occur under conditions defined at the time of consent (e.g. oversight or the right to withdraw).
(3) **Categorical consent.** Individuals are asked to consent to the collection and storage of biospecimens for future research use and are offered a checklist of options to stipulate by whom and in what ways they can be used.
(4) **Dynamic consent.** Individuals are provided with an interactive, digital system that allows them to tailor, modify, and update consent choices as their circumstances change and in response to specific studies.
(5) **Study-specific consent.** Individuals are contacted and asked for consent for each research use.

The choice of consent model will depend on local ethical regulations and any additional requirements from your organization, ethics committee, or funding body. Blanket consent and broad consent are probably the most common models for laboratories that collect tissue for later use. An example of a blanket consent form is given here (see Table 6.5). Broad consent is preferred by many ethical practitioners because it allows some limits to be set by the donor; examples are available from regulatory bodies and other sources (Beskow et al. 2010; Lowenthal et al. 2012; NIA 2018). Categorical, dynamic, and study-specific consent models would make it difficult if not impossible for laboratories to distribute cell lines as part of later access arrangements.

There are a number of points to consider when requesting consent for tissue that may lead to cell line generation (Geraghty et al. 2014), regardless of the consent model used. The following questions may help to explore these points further:

(1) **Who can give consent?** Permission may be required from the next of kin (relative or person in a qualifying relationship) if the donor is too unwell or is otherwise unable to give informed consent. Always refer to your local ethics committee for guidance if you wish to obtain consent from children or vulnerable individuals who may not be able to give informed consent. In the United Kingdom, a code of practice is available that gives detailed guidance on how to obtain appropriate, informed, and valid consent (HTA 2017).
(2) **What will the tissue be used for?** A short summary of your project should be prepared to explain what you are doing, why, and what the possible outcome will be, particularly if it is seen to be of medical benefit. Think about how to convey your information in a way that is easily understood by the donor and their family. For example, you might provide an information sheet for the donor to keep that includes contact information for any further questions.
(3) **What other information should be included in the consent form?** Some elements of a consent form are mandatory – for example, to explain any risks that may arise from the procedure (NIA 2018). Refer to your local ethics committee for guidance.
(4) **Will the donor's personal health information be kept confidential?** Privacy and confidentiality should be explicitly covered in the consent form. Who will see the donor's health information? What measures will be used to keep their name and other personal identifiers confidential? Use of genomic data can also be clarified as part of privacy and confidentiality requirements (Knoppers et al. 2011).
(5) **Will additional testing be performed?** The donor should consent to any screening of the tissue for adventitious pathogens and decide whether he or she wishes to be made aware of the outcome of the tests.
(6) **What will happen to any incidental findings?** The donor should be informed whether later incidental findings will be returned to them. If incidental findings are to be returned to the donor, it should be clear how that

TABLE 6.5. Blanket consent form.

This form requests your permission to take a sample of your blood or one or more small pieces of tissue to be used for medical research. This sample, or cell lines or other products derived from it, may be used by a number of different research organizations, or it may be stored for an extended period awaiting use. It is also possible that it may eventually be used by a commercial company to develop future drugs. We would like you to be aware of this and of the fact that, by signing this form, you give up any claim that you own the tissue or its components, regardless of the use that may be made of it. You should also be aware of, and agree to, the possible testing of the tissue for infectious agents, such as AIDS or hepatitis.

I am willing to have tissue removed for use in medical research and development. I have read and understand, to the best of my ability, the background material that I have been given. (If the donor is too unwell to sign, a close relative should sign on his or her behalf.)

Name of donor . . . . *Name of relative*

. . . .

Signature . . . .

Date . . . .

This material will be coded, and absolute confidence will be maintained. Your name will not be given to anyone other than the person taking the sample.

Do you wish to receive any information from this material that relates to your health? **Yes / No**

Signature . . . . *Date* . . . .

Would you like, or prefer, that this information be given to your doctor? **Yes / No**

If yes, name of doctor . . . .

Address of doctor . . . .

information will be conveyed (e.g. if the patient's doctor should be contacted) (Johns et al. 2014).

(7) ***What will happen to any unused tissue or cells?*** Some biobanks provide additional options to address how unused tissue or cells will be handled. For some communities, this is an important part of informed consent (Morrin et al. 2005).

(8) ***Will genetic modification be performed?*** Authority may be needed for subsequent genetic modification or to use cell lines for sensitive applications (Lowenthal et al. 2012). This may require further discussion and consent from the donor.

(9) ***Who owns the tissue and any cells or products that may be derived from it?*** Ownership of tissue or cell lines and patent rights from any commercial collaboration must be clarified as part of the consent process. If commercial use is planned, the donor should be told what financial benefit might be gained; a waiver to commercial rights is usually requested (Geraghty et al. 2014). Any conflict of interest by the scientist or attending physician should be clearly stated (Curran 1991).

(10) ***Are there any conditions associated with the consent?*** For example, a donor may consent to their tissue sample being used for research purposes but may impose certain conditions such as not wanting the sample to be used in animal models. Always keep a record of the wording on the consent for later reference, to ensure that any further use is within the terms of the donor's consent.

Suppliers

Supplier	URL
LANXESS Deutschland GmbH	http://virkon.com/products-applications/disinfectants
Tempshield	http://tempshield.com

Supp. S6.1 Hierarchy of Risk Controls.

Supp. S6.2 Ionizing Radiation.

Supp. S6.3 Biosecurity.

Supp. S6.4 Donor Privacy.

REFERENCES

Bairoch, A. (2018). The Cellosaurus, a cell-line knowledge resource. *J. Biomol. Tech.* 29 (2): 25–38. https://doi.org/10.7171/jbt.18-2902-002.

Benson, L. M. and Reczek, K. (2017). A guide to United States electrical and electronic equipment compliance requirements. https://dx.doi.org/10.6028/nist.ir.8118r1 (accessed 5 June 2018).

Berg, P., Baltimore, D., Brenner, S. et al. (1975). Summary statement of the Asilomar conference on recombinant DNA molecules. *Proc. Natl Acad. Sci. U.S.A.* 72 (6): 1981–1984.

Beskow, L.M. (2016). Lessons from HeLa cells: the ethics and policy of biospecimens. *Annu. Rev. Genomics Hum. Genet.* 17: 395–417. https://doi.org/10.1146/annurev-genom-083115-022536.

Beskow, L.M., Friedman, J.Y., Hardy, N.C. et al. (2010). Developing a simplified consent form for biobanking. *PLoS One* 5 (10): e13302. https://doi.org/10.1371/journal.pone.0013302.

Bielecka, A. and Mohammadi, A.A. (2014). State-of-the-art in biosafety and biosecurity in European countries. *Arch. Immunol. Ther. Exp. (Warsz)* 62 (3): 169–178. https://doi.org/10.1007/s00005-014-0290-1.

Biggar, R.J., Schmidt, T.J., and Woodall, J.P. (1977). Lymphocytic choriomeningitis in laboratory personnel exposed to hamsters inadvertently infected with LCM virus. *J. Am. Vet. Med. Assoc.* 171 (9): 829–832.

Buesching, W.J., Neff, J.C., and Sharma, H.M. (1989). Infectious hazards in the clinical laboratory: a program to protect laboratory personnel. *Clin. Lab. Med.* 9 (2): 351–361.

Capes-Davis, A. and Capes-Davis, D. L. (2018). Assessing risks at work: figshare. https://doi.org/10.6084/m9.figshare.5848590.v1

Carlson, R.V., Boyd, K.M., and Webb, D.J. (2004). The revision of the declaration of Helsinki: past, present and future. *Br. J. Clin. Pharmacol.* 57 (6): 695–713. https://doi.org/10.1111/j.1365-2125.2004.02103.x.

CDC and NIH (2009). Biosafety in microbiological and biomedical laboratories. https://www.cdc.gov/labs/bmbl.html (accessed 11 March 2019).

Curran, W.J. (1991). Scientific and commercial development of human cell lines. Issues of property, ethics, and conflict of interest. *N. Engl. J. Med.* 324 (14): 998–1000. https://doi.org/10.1056/NEJM199104043241419.

Desai, R., Collett, D., Watson, C.J. et al. (2012). Cancer transmission from organ donors – unavoidable but low risk. *Transplantation* 94 (12): 1200–1207. https://doi.org/10.1097/TP.0b013e318272df41.

Dykewicz, C.A., Dato, V.M., Fisher-Hoch, S.P. et al. (1992). Lymphocytic choriomeningitis outbreak associated with nude mice in a research institute. *JAMA* 267 (10): 1349–1353.

EU (2008). Regulation (EC) No. 1272/2008 – classification, labelling and packaging of substances and mixtures (CLP). https://osha.europa.eu/en/legislation/directives/regulation-ec-no-1272-2008-classification-labelling-and-packaging-of-substances-and-mixtures (accessed 6 January 2019).

EU-OSHA (2018). Musculoskeletal disorders. https://osha.europa.eu/en/themes/musculoskeletal-disorders (accessed 6 June 2018).

Festing, S. and Wilkinson, R. (2007). The ethics of animal research. Talking point on the use of animals in scientific research. *EMBO Rep.* 8 (6): 526–530. https://doi.org/10.1038/sj.embor.7400993.

Finkel, E. (2007). Research safety. Inquest flags little-known danger of high-containment labs. *Science* 316 (5825): 677. https://doi.org/10.1126/science.316.5825.677.

Gartner, H.V., Seidl, C., Luckenbach, C. et al. (1996). Genetic analysis of a sarcoma accidentally transplanted from a patient to a surgeon. *N. Engl. J. Med.* 335 (20): 1494–1496. https://doi.org/10.1056/NEJM199611143352004.

Geraghty, R.J., Capes-Davis, A., Davis, J.M. et al. (2014). Guidelines for the use of cell lines in biomedical research. *Br. J. Cancer* 111 (6): 1021–1046. https://doi.org/10.1038/bjc.2014.166.

Gill, J.R., Ely, S.F., and Hua, Z. (2002). Environmental gas displacement: three accidental deaths in the workplace. *Am. J. Forensic Med. Pathol.* 23 (1): 26–30.

Grizzle, W.E. and Polt, S.S. (1988). Guidelines to avoid personnel contamination by infective agents in research laboratories that use human tissues. *J. Tissue Cult. Methods* 11 (4): 191–199. https://doi.org/10.1007/BF01407313.

Gwyther, J. (1990). Sharps disposal containers and their use. *J. Hosp. Infect.* 15 (3): 287–294.

Hansson, M.G. (2009). Ethics and biobanks. *Br. J. Cancer* 100 (1): 8–12. https://doi.org/10.1038/sj.bjc.6604795.

Haring, R.C., Henry, W.A., Hudson, M. et al. (2018). Views on clinical trial recruitment, biospecimen collection, and cancer research: population science from landscapes of the Haudenosaunee (people of the longhouse). *J. Cancer Educ.* 33 (1): 44–51. https://doi.org/10.1007/s13187-016-1067-5.

Harvey-Clark, C. (2011). IACUC challenges in invertebrate research. *ILAR J.* 52 (2): 213–220.

Hegde, J., Bashetty, K., and Krishnakumar, U.G. (2012). Quantity of sodium thiosulfate required to neutralize various concentrations of sodium hypochlorite. *Asian J. Pharm. Health Sci.* 2 (3): 390–393.

Hernandez, A., Martro, E., Matas, L. et al. (2000). Assessment of in-vitro efficacy of 1% Virkon against bacteria, fungi, viruses and spores by means of AFNOR guidelines. *J. Hosp. Infect.* 46 (3): 203–209. https://doi.org/10.1053/jhin.2000.0818.

Horita, A. and Weber, L.J. (1964). Skin penetrating property of drugs dissolved in dimethylsulfoxide (DMSO) and other vehicles. *Life Sci.* 3 (1962): 1389–1395.

HRA (2018). Use of human tissue in research. https://www.hra.nhs.uk/planning-and-improving-research/policies-standards-legislation/use-tissue-research (accessed 15 June 2018).

HSE (2001). The management, design and operation of microbiological containment laboratories. www.hse.gov.uk/pubns/books/microbio-cont.htm (accessed 28 March 2018).

HSE (2005). Biological agents: managing the risks in laboratories and healthcare premises. www.hse.gov.uk/biosafety/biologagents.pdf (accessed 30 May 2018).

HSE (2006). The principles, design and containment of Containment Level 4 facilities. www.hse.gov.uk/pubns/web09.pdf (accessed 22 June 2018).

HSE (2013). The approved list of biological agents. www.hse.gov.uk/pubns/misc208.pdf (accessed 28 March 2018).

HSE (2018a). Genetically modified organisms (contained use). www.hse.gov.uk/biosafety/gmo/index.htm (accessed 8 June 2018).

HSE (2018b). Human factors and ergonomics. www.hse.gov.uk/humanfactors (accessed 6 June 2018).

HSE (2018c). Work equipment and machinery. www.hse.gov.uk/work-equipment-machinery/index.htm (accessed 5 June 2018).

HSE (2019). Safe handling of cytotoxic drugs in the workplace. www.hse.gov.uk/healthservices/safe-use-cytotoxic-drugs.htm (accessed 6 January 2019).

HTA (2017). Code of Practice A: guiding principles and the fundamental principle of consent. www.hta.gov.uk/sites/default/files/files/hta%20code%20a.pdf (accessed 6 January 2019).

HTA (2018). HTA legislation. www.hta.gov.uk/guidance-professionals/hta-legislation (accessed 15 June 2018).

Hudson, K.L. and Collins, F.S. (2013). Biospecimen policy: family matters. *Nature* 500 (7461): 141–142. https://doi.org/10.1038/500141a.

International Organization for Standardization (2003). ISO 15190:2003(en). Medical laboratories – requirements for safety. Geneva: International Organization for Standardization.

Johns, A.L., Miller, D.K., Simpson, S.H. et al. (2014). Returning individual research results for genome sequences of pancreatic cancer. *Genome Med.* 6 (5): 42. https://doi.org/10.1186/gm558.

Kerecman Myers, D., Goldberg, A.M., Poth, A. et al. (2017). From in vivo to in vitro: the medical device testing paradigm shift. *ALTEX* 34 (4): 479–500. https://doi.org/10.14573/altex.1608081.

Kim, D.H. and Lee, H.J. (2008). Evaporated liquid nitrogen-induced asphyxia: a case report. *J. Korean Med. Sci.* 23 (1): 163–165. https://doi.org/10.3346/jkms.2008.23.1.163.

Kimman, T.G., Smit, E., and Klein, M.R. (2008). Evidence-based biosafety: a review of the principles and effectiveness of microbiological containment measures. *Clin. Microbiol. Rev.* 21 (3): 403–425. https://doi.org/10.1128/cmr.00014-08.

Knoppers, B.M., Isasi, R., Benvenisty, N. et al. (2011). Publishing SNP genotypes of human embryonic stem cell lines: policy statement of the International Stem Cell Forum Ethics Working Party. *Stem Cell Rev.* 7 (3): 482–484. https://doi.org/10.1007/s12015-010-9226-2.

Langford, N.J. (2005). Carbon dioxide poisoning. *Toxicol. Rev.* 24 (4): 229–235.

Lazebnik, Y. and Parris, G.E. (2015). Comment on: 'guidelines for the use of cell lines in biomedical research': human-to-human cancer transmission as a laboratory safety concern. *Br. J. Cancer* 112 (12): 1976–1977. https://doi.org/10.1038/bjc.2014.656.

Lloyd, G. and Jones, N. (1986). Infection of laboratory workers with hantavirus acquired from immunocytomas propagated in laboratory rats. *J. Infect.* 12 (2): 117–125.

Lowenthal, J., Lipnick, S., Rao, M. et al. (2012). Specimen collection for induced pluripotent stem cell research: harmonizing the approach to informed consent. *Stem Cells Transl. Med.* 1 (5): 409–421. https://doi.org/10.5966/sctm.2012-0029.

Marangoni, A.H. (2010). Effects of intermittent stretching exercises at work on musculoskeletal pain associated with the use of a personal computer and the influence of media on outcomes. *Work* 36 (1): 27–37. https://doi.org/10.3233/WOR-2010-1004.

Morrin, H., Gunningham, S., Currie, M. et al. (2005). The Christchurch Tissue Bank to support cancer research. *N. Z. Med. J.* 118 (1225): U1735.

NHMRC (2013). Australian code for the care and use of animals for scientific purposes. www.nhmrc.gov.au/guidelines-publications/ea28 (accessed 14 June 2018).

NIA (2018). Informed consent. https://www.nia.nih.gov/research/dgcg/clinical-research-study-investigators-toolbox/informed-consent (accessed 17 June 2018).

NIH (2016). NIH guidelines for research involving recombinant or synthetic nucleic acid molecules. https://osp.od.nih.gov/biotechnology/biosafety-and-recombinant-dna-activities (accessed 8 June 2018).

NIOSH (2018). Ergonomics and musculoskeletal disorders. https://www.cdc.gov/niosh/topics/ergonomics (accessed 6 June 2018).

OECD (2018). *Guidance Document on Good in Vitro Method Practices (GIVIMP)*. Paris: OECD Publishing.

OHRP (2002). Human embryonic stem cells, germ cells, and cell-derived test articles. https://www.hhs.gov/ohrp/regulations-and-policy/guidance/guidance-on-research-involving-stem-cells/index.html (accessed 15 June 2018).

OHRP (2017). Revised Common Rule. https://www.hhs.gov/ohrp/regulations-and-policy/regulations/finalized-revisions-common-rule/index.html (accessed 15 June 2018).

OSHA (2001). 29 CFR 1910.1030 Bloodborne pathogens. In Code of Federal Regulations. Washington, DC: United States Department of Labor.

OSHA (2018a). Ergonomics. https://www.osha.gov/sltc/ergonomics/index.html (accessed 5 June 2018).

OSHA (2018b). Healthcare wide hazards: hazardous chemicals. https://www.osha.gov/sltc/etools/hospital/hazards/hazchem/haz.html (accessed 12 June 2018).

Petrella, B.L. (2015). Biosafety oversight and compliance: what do you mean, I have to fill out another form?! *Curr. Protoc. Microbiol.* 39: 1A.5.1–1A.5.16. https://doi.org/10.1002/9780471729259.mc01a05s39.

Rau, E.H., Alaimo, R.J., Ashbrook, P.C. et al. (2000). Minimization and management of wastes from biomedical research. *Environ. Health Perspect.* 108 ((Suppl 6): 953–977.

Seltmann, S., Lekschas, F., Muller, R. et al. (2016). hPSCreg – the human pluripotent stem cell registry. *Nucleic Acids Res.* 44 (D1): D757–D763. https://doi.org/10.1093/nar/gkv963.

Shioda, S., Kasai, F., Watanabe, K. et al. (2018). Screening for 15 pathogenic viruses in human cell lines registered at the JCRB cell bank: characterization of in vitro human cells by viral infection. *R. Soc. Open Sci.* 5 (5): 172472. https://doi.org/10.1098/rsos.172472.

Southam, C.M., Moore, A.E., and Rhoads, C.P. (1957). Homotransplantation of human cell lines. *Science* 125 (3239): 158–160.

Stang, A., Petrasch-Parwez, E., Brandt, S. et al. (2009). Unintended spread of a biosafety level 2 recombinant retrovirus. *Retrovirology* 6: 86. https://doi.org/10.1186/1742-4690-6-86.

Tedder, R.S., Zuckerman, M.A., Goldstone, A.H. et al. (1995). Hepatitis B transmission from contaminated cryopreservation tank. *Lancet* 346 (8968): 137–140.

Tian, D. and Zheng, T. (2014). Comparison and analysis of biological agent category lists based on biosafety and biodefense. *PLoS One* 9 (6): e101163. https://doi.org/10.1371/journal.pone.0101163.

UK Department of Health (2013). Safe management of healthcare waste. https://www.gov.uk/government/publications/guidance-on-the-safe-management-of-healthcare-waste (accessed 11 June 2018).

UN (2011). Globally Harmonized System of Classification and Labelling of Chemicals (GHS). http://www.unece.org/fileadmin/dam/trans/danger/publi/ghs/ghs_rev04/english/st-sg-ac10-30-rev4e.pdf (accessed 4 June 2018).

Uphoff, C.C., Denkmann, S.A., Steube, K.G. et al. (2010). Detection of EBV, HBV, HCV, HIV-1, HTLV-I and -II, and SMRV in human and other primate cell lines. *J. Biomed. Biotechnol.* 2010: 904767. https://doi.org/10.1155/2010/904767.

Uphoff, C.C., Lange, S., Denkmann, S.A. et al. (2015). Prevalence and characterization of murine leukemia virus contamination in human cell lines. *PLoS One* 10 (4): e0125622. https://doi.org/10.1371/journal.pone.0125622.

Van Noorden, R. (2013). Safety survey reveals lab risks. *Nature* 493 (7430): 9–10. https://doi.org/10.1038/493009a.

Wells, D.L., Lipper, S.L., Hilliard, J.K. et al. (1989). Herpesvirus simiae contamination of primary rhesus monkey kidney cell cultures. CDC recommendations to minimize risks to laboratory personnel. *Diagn. Microbiol. Infect. Dis.* 12 (4): 333–335.

WHO (2004). Laboratory biosafety manual. http://www.who.int/csr/resources/publications/biosafety/who_cds_csr_lyo_2004_11/en (accessed 9 May 2018).

Wiley (2014). Best practice guidelines on publishing ethics: a publisher's perspective. https://authorservices.wiley.com/asset/ethics_guidelines_7.06.17.pdf (accessed 14 June 2018).

Winder, C. and Makin, A. M. (2006). New approaches to OHS risk assessments. CCH Hazard Alert, 20 October 2006, 1–4.

CHAPTER 7

Reproducibility and Good Cell Culture Practice

Tissue culture has evolved into a broad technology that is used for basic research, industrial production, and clinical translation. Many of the fundamental processes are shared among these groups, as are the common problems that accompany the culture of cells and cell lines. Lack of reproducibility has been highlighted as one such problem and in recent years has become a "hot topic" across the broader scientific enterprise. This chapter aims to provide an overview of reproducibility and good practice requirements that may apply to tissue culture and its applications. Common procedures need some level of standardization so that results are reproducible in different locations and at different times. Standardized requirements have been developed for Good Cell Culture Practice (GCCP), Good Laboratory Practice (GLP), and Good Manufacturing Practice (GMP). Good experimental design – for example, when selecting replicate samples – is also important. Materials must be adequately documented and reported, and common problems must be detected through validation testing. Practical consequences of these requirements are discussed in more detail in later chapters of this book.

7.1 REPRODUCIBILITY

7.1.1 Terminology: Reproducible Research

Reproducibility can be defined as the ability to duplicate the results of a prior study using the same materials as were used by the original investigator (National Science Foundation 2015). However, the term "reproducibility" has acquired a broader meaning in life sciences research. In 2011–2012, investigators at Bayer HealthCare and Amgen described their efforts to reproduce preclinical research in-house before investing in candidate drug targets (Prinz et al. 2011; Begley and Ellis 2012). One of these groups was able to generate results that were completely in line with the original publication in only 20–25% of studies (Prinz et al. 2011); the other group confirmed initial findings in only 11% of studies (Begley and Ellis 2012). Although these low reproducibility levels acted as a catalyst for discussion, similar concerns have been expressed in other areas of preclinical research. In a survey of *Nature* authors performed between 2016 and 2017, 86% of respondents agreed there was a "crisis of reproducibility" in science today (*Nature* Editors 2018).

In this chapter, we focus on reproducibility as part of a wider effect to improve the quality and validity of life sciences research. For that effort, "reproducible research" may include (Goodman et al. 2016):

(1) **Methods reproducibility.** The provision of sufficient detail about materials, procedures, and data so that the same procedures could be repeated by others.
(2) **Results reproducibility.** The ability of an independent study to reproduce the same results, if using the same materials and procedures. Results reproducibility is referred to in some studies as "replication."
(3) **Inferential reproducibility.** The ability to draw qualitatively similar conclusions from either an independent study or a reanalysis of the data from the original study.

Freshney's Culture of Animal Cells: A Manual of Basic Technique and Specialized Applications, Eighth Edition. Amanda Capes-Davis and R. Ian Freshney.
© 2021 John Wiley & Sons Ltd. Published 2021 by John Wiley & Sons Ltd.
Companion website: www.wiley.com/go/freshney/cellculture8

7.1.2 Causes of Irreproducible Research

There may be good reasons to explain why a laboratory cannot reproduce an experiment. Techniques and research models are becoming more time-consuming and sophisticated, resulting in the need for greater expertise and increased sensitivity to small variations in experimental conditions (Bissell 2013). However, many of the reported problems with reproducibility raise concerns regarding "sloppy science," where lack of reproducibility is caused by insufficient scientific rigor (Harris 2017). An early study in this field concluded that the majority of published research findings are false (Ioannidis 2005). Subsequent studies on avoidable waste in the production and reporting of research evidence suggested that about 85% of research investment is wasted (Chalmers and Glasziou 2009; Macleod et al. 2014). These studies included cases where the major conclusions from a publication could not be replicated by the original investigators when they were blinded to test samples versus control samples, suggesting that there is more at stake than a failure to reproduce experimental conditions (Begley and Ioannidis 2015).

Concerns about reproducibility and scientific rigor extend to many publications that use continuous cell lines. Incorrect use of cell lines is acknowledged to be an important cause of irreproducible research (Freedman et al. 2015; Neimark 2015; Drucker 2016). For example, cell lines continue to be widely used even when they are known to be misidentified and thus fail to correspond to the original, authentic material (see Figure 7.1). Misidentified cell lines have appeared in an estimated 32 755 journal articles, with the number of publications steadily increasing over time (Horbach and Halffman 2017). Use of misidentified cell lines may lead to irreproducible research, but it leads to broader concerns regarding lack of scientific rigor. Misidentified cell lines are easily avoided, provided authentication testing is performed to detect the problem, and standardized methods for authentication are widely available (Almeida et al. 2016). Authentication testing to detect misidentified cell lines is discussed briefly in this chapter and in more detail later in this book (see Section 7.4.2; see also Chapter 17).

What are the primary causes of irreproducible research? Problems with reproducibility in cell-based research have a number of causes, which can be broadly categorized as:

(1) *Scientific misconduct.* Analysis of 2047 journal retractions showed that 67.4% were attributable to misconduct (Fang et al. 2012). Retractions are not necessarily linked to problems with reproducibility; causes include image duplication, sequence duplication, and plagiarism (Byrne and Labbe 2017). However, misconduct is an important cause of irreproducible research that requires specific investigation. For example, the phenomenon of stimulus-triggered acquisition of pluripotency (STAP) was investigated following inability to reproduce initial findings (De Los Angeles et al. 2015). STAP cells were shown to be contaminated, with a high probability that contamination arose through intent or negligence (Katsura et al. 2014).

(2) *Scientific error.* Analysis of the same dataset of 2047 retractions showed that 21.3% were attributable to error (Fang et al. 2012; Casadevall et al. 2014). Errors included problems with DNA sequencing and cloning, problems with controls, and errors made during analysis. Errors in sample size, data analysis, and data interpretation have also been highlighted by a number of researchers (Lazic et al. 2018).

(3) *Inadequate reporting.* Problems with reporting do not typically result in retraction, but inadequate reporting is nevertheless an important cause of irreproducible research. For example, analysis of 238 publications showed that 54% of research materials were not uniquely identifiable; only 43% of cell lines could be uniquely identified (Vasilevsky et al. 2013). Authors provided cell line names but rarely reported source information, passage number, or contamination status. Cell line names in themselves are not sufficient to uniquely identify a cell line (see Section 7.3.2).

(4) *Material quality.* Contamination and misidentification are major quality concerns for tissue culture and particularly for continuous cell lines (see Chapters 16 and 17). Quality concerns may arise through misconduct, error, or inadequate reporting – for example, failure to report correct cell line information may lead to error when the cell line is used by other laboratories. Cultures must be tested for microbial contamination and authenticity to exclude such quality concerns (see Section 7.4). If testing is not performed before publication and reported by the authors, it is impossible to know whether quality concerns are present in published manuscripts.

(5) *Material handling.* Continued passaging of cell lines can result in marked differences between different samples of the same cell line (Hughes et al. 2007). Differences may arise for a number of reasons including development of novel mutations, selection of clonal populations, or changes in differentiation with prolonged handling or altered growth conditions. Variations in routine handling procedures (e.g. counting cells) may also result in problems with reproducibility; these procedures are often taken for granted, but may have a substantial impact, e.g. when cell lines are used for screening of anticancer drugs (Niepel et al. 2019).

(6) *Material heterogeneity.* Tissue and cell samples are inherently heterogeneous, unless efforts are made to isolate single cells or clonal populations (Hartung et al. 2002). Many samples used for tissue culture are derived from malignant tumors, which evolve over time and can display marked genetic instability (Maley et al. 2017). Culture handling exacerbates these differences, resulting in evolution of clonal populations (Ben-David et al. 2018). As

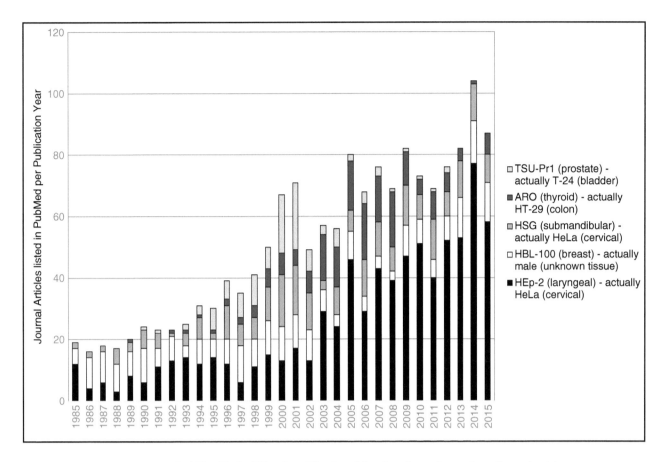

Fig. 7.1. Use of misidentified cell lines in publications. The chart shows the number of journal articles that use five misidentified cell lines (TSU-Pr1, ARO, HSG, HBL-100, and HEp-2) over 30 years (1985–2015). These misidentified cell lines all come from a different tissue type to the one originally reported. The correct cell line identity and tissue type are indicated at the right of the chart. A PubMed search was performed for each cell line, looking for its name in association with the incorrect tissue type (e.g. "HEp-2" AND "laryngeal"). *Source*: Adapted from Fusenig, N. E., Capes-Davis, A., Bianchini, F., Sundell, S., & Lichter, P. (2017). The need for a worldwide consensus for cell line authentication: Experience implementing a mandatory requirement at the International Journal of Cancer. PLOS Biology, 15(4), e2001438.

a result, variations may arise within and between samples that are sufficient to cause irreproducible research.

7.1.3 Solutions to Irreproducible Research

Looking at the list of causes above, there are no easy solutions to irreproducible research. Some practical actions to improve reproducibility in cell-based research are listed in Table 7.1. To be effective, reproducibility initiatives must involve multiple stakeholders and must be practical and achievable for laboratories without a major investment in resources. Who are the stakeholders in reproducibility initiatives? For effective solutions, the answer must be "everyone." Individual scientists, organizations, journals, funding bodies, and regulators must all participate in the effort to increase value and decrease waste in biomedical research (Moher et al. 2016).

Efforts to improve material quality were initially made by individual scientists. One of the trailblazers in this area, Roland Nardone, prepared a white paper in 2007 that set out specific initiatives to address widespread cross-contamination of cell lines (Nardone 2007). Nardone recommended that funding bodies and key journals should require authentication testing of cell lines as a mandatory step for funding or publication. Professional societies should endorse these policies and provide training for their members, while laboratory and academic department heads should ensure that staff are aware of the problem and the testing required to address it.

Many journals have now acted to improve material quality, reproducibility, and scientific rigor. The *International Journal of Cancer* was the first journal to require cell line authentication for publication, commencing in 2010 (Fusenig et al. 2017). Other journals have followed their example to require

TABLE 7.1. Improving reproducibility in cell-based research.

Cause of irreproducible research	Actions to improve reproducibility and rigor
Scientific misconduct	Train scientists in research integrity
	Provide clear and consistent research integrity policies
	Develop detection tools for reuse of content (Byrne and Labbe 2017)
	Develop transparent processes to investigate misconduct
	Reduce pressure to publish
Scientific error	Train scientists in good experimental design and analysis
	Provide platforms for post-publication peer review (Knoepfler 2015)
	Develop sustainable models of open access publication
	Develop transparent processes to report error openly and rapidly
	Reduce stigma from scientific error
Inadequate reporting	Train scientists in reporting requirements for publication
	Provide unique identifiers for materials (Bandrowski et al. 2016)
	Provide platforms for open access to protocols (Teytelman et al. 2016)
	Develop good practice requirements for reporting
	Develop reporting checklists for key information (e.g. ICLAC 2014)
Material quality	Train scientists in GCCP and the need for material validation
	Develop written Standards and guidelines for validation testing
	Develop material Standards for use in validation testing as references
	Include validation testing in reporting checklists
	Require validation testing as a condition of publication and funding
Material handling	Train scientists in GCCP and the need for good handling practice
	Develop good practice requirements for handling
	Store banks of early passage material (see Section 15.5)
	Report medium and handling conditions in protocols and publications
Material heterogeneity	Train scientists in GCCP and the need to understand *in vitro* systems
	Improve characterization of cell lines
	Improve sample diversity in experimental work (Karp 2018)
	Store banks of early passage material (see Section 15.5)
	Report source and passage number in protocols and publications

authentication (see Table 7.2) or to increase reproducibility through related initiatives. For example, many journals published by the Nature Publishing Group have implemented a reporting policy for new submissions (*Nature* Editors 2015). Authors must address a checklist of key factors that contribute to irreproducible research, including authentication of cell lines. Author feedback was sought on the checklist in 2017 (*Nature* Editors 2018); 49% of respondents thought that the checklist had improved the quality of published work, and 78% continued to use the checklist to some degree following publication. Minimum reporting standards for life scientists are currently in draft, which will hopefully provide a consensus approach across multiple journals (Chambers et al. 2019).

Some funding bodies have also acted to improve material quality, reproducibility, and scientific rigor. The Prostate Cancer Foundation was the first funding body to require

cell line authentication, commencing in 2013 (Almeida et al. 2016). In the United States, the National Institutes of Health (NIH) has worked with key stakeholders to develop principles and guidelines for reporting of preclinical research (NIH 2017). The NIH guidelines recommend that the source of the cell line and its authentication and mycoplasma contamination status should be reported. Authentication of key resources, including cell lines, is now addressed specifically in NIH grant applications as part of a broader effort to improve reproducibility through rigor and transparency (NIH 2015). In the United Kingdom, Cancer Research UK requires that specific cell line authentication protocols should be incorporated into experimental work (Cancer Research UK 2017).

Very few organizations or regulatory bodies have developed reproducibility initiatives. There is a clear need for

TABLE 7.2. Journals requiring authentication testing.

Publisher/journal	Requirement for authentication		
	Mandatory testing	Reporting of testing status[a]	Awareness/ encouragement
American Association for Cancer Research publications, e.g. *Cancer Research*		X	
Bentham Science publications, e.g. *Current Biotechnology*	X		
Biochemical Pharmacology		X	
BioMed Central journals, e.g. *BMC Cancer*			X
BioTechniques			X
Cancer		X	
Carcinogenesis		X	
Cell Biochemistry and Biophysics	X		
Cell Biology International	X		
Clinical Orthopaedics and Related Research	X		
Endocrine Society journals, e.g. *Endocrinology*	X		
FASEB Journal	X		
International Journal of Cancer	X		
Investigative Ophthalmology & Visual Science			X
In Vitro Cellular & Developmental Biology – Animal		X	
Journal of Hepatology	X		
Journal of Medicinal Chemistry			X
Journal of Molecular Biology			X
Journal of the National Cancer Institute		X	
Molecular Vision	X		
Nature Publishing Group journals, e.g. *Nature*		X	
Neuro-Oncology		X	
Oncotarget		X	
PLoS ONE			X
Society for Endocrinology journals, e.g. *Journal of Molecular Endocrinology*	X		

Source: Adapted from Fusenig, N. E., Capes-Davis, A., Bianchini, F., Sundell, S., & Lichter, P. (2017). The need for a worldwide consensus for cell line authentication: Experience implementing a mandatory requirement at the International Journal of Cancer. PLOS Biology, 15(4), e2001438.
[a] "Reporting of testing status" means that authors are required to report whether testing has been done. However, authors can choose not to test their cell lines; this increases transparency but does not prove that cell lines are authentic.
This list is based on publisher and journal policies that are available online; policies may be updated at any time. Always refer to specific publication guidelines to determine requirements for authors.

good institutional practice; institutions must provide oversight, policies, and training in research integrity and good research practice (Begley et al. 2015; ICLAC 2019). In the absence of organizational involvement, the responsibility comes back to individual scientists and laboratories to develop their own reproducibility initiatives. Some scientists and laboratories have risen to this challenge, developing simple and practical solutions that should be more widely adopted. For example, the discovery of a misidentified cell line in one laboratory resulted in development of an "in-out" policy where cell lines are tested on arrival at the laboratory and before cell lines or data are released elsewhere (Kniss and Summerfield 2014).

7.2 GOOD PRACTICE REQUIREMENTS

Different laboratories can have very different rules and procedures. Who determines what is "good practice" for tissue culture laboratories, and how should laboratories standardize their procedures to improve reproducibility and rigor? The answers to these questions are determined by consensus and by considering the level of rigor and reproducibility that is required for your specific application. GCCP is a core requirement for all laboratories (including those that practice "basic" research) and acts as a foundation for other, more stringent levels of good practice (e.g. as set out by the Organization for Economic Co-operation and Development [OECD 2018a]).

7.2.1 Good Cell Culture Practice (GCCP)

The concept of GCCP initially came from efforts to address quality concerns associated with the use of cell lines and to standardize *in vitro* procedures (Freshney 2002; Hartung et al. 2002). In the United Kingdom, a committee of cell culture scientists developed a set of guidelines for the use of cell lines in biomedical research. The guidelines, which were first published in 1999 and revised in 2014, highlighted cell culture problems and provided recommendations on how those problems could be identified, avoided, and if possible eliminated (Geraghty et al. 2014). In Europe, a separate GCCP task force was convened by the European Centre for the Validation of Alternative Methods (ECVAM). The task force was established in 1999 and issued a report in 2005 on six key principles of GCCP and their application (Coecke et al. 2005). Although these two groups worked separately, their recommendations complement each other and there is broad consensus across all guidance documents.

The six key principles of GCCP result in a flexible approach that can be applied to any application. GCCP principles can be summarized as follows (Coecke et al. 2005):

(1) **Understanding.** It is important to establish and maintain a sufficient understanding of your *in vitro* system and of relevant factors that could affect it. What changes occur when cells are cultured from a tissue sample and are used to establish a continuous cell line? How do the environment and handling of the culture affect its behavior? What testing is needed to ensure that the culture is a good model for your application? Answering these questions will help to determine the materials that should be used, how those materials should be handled, and the testing that should be performed.

(2) **Quality assurance (QA).** It is also important to assure the quality of all materials and methods to maintain the validity and reproducibility of your *in vitro* system. Validation testing and QA are discussed in more detail in Sections 7.4 and 7.5.

(3) **Documentation.** Sufficient information must be recorded to track the materials and methods used, to permit the repetition of the work, and to enable the target audience to understand and evaluate the work. This documentation is often referred to as the provenance of a cell line. Cell line provenance and reporting for publication are discussed in more detail in Section 7.3.

(4) **Safety.** Adequate measures should be established and maintained to protect individuals and the environment from any potential hazards. Laboratory safety is discussed in a previous chapter (see Section 6.1).

(5) **Compliance.** Compliance with relevant laws and regulations is essential, including ethical guidelines. Bioethics is discussed in a previous chapter (see Section 6.4).

(6) **Training.** Relevant and adequate education and training must be provided for all tissue culture personnel to promote high work quality and safety. Training is discussed later in this book (see Chapter 30).

GCCP applies to all tissue culture laboratories. However, the way in which the principles are applied will vary with the application. Recent advances in cell culture have resulted in "GCCP 2.0," with a new focus on stem cells, toxicology testing, three-dimensional (3D) culture, and microphysiological systems (e.g. organ-on-chip approaches) (Eskes et al. 2017; Pamies et al. 2017, 2018). Experts in specialized cells and applications are also working to develop other consensus guidelines. For example, the International Stem Cell Banking Initiative (ISCBI) has developed consensus guidelines for banking and supply of human embryonic stem cell lines for research purposes (see Section 23.7) (International Stem Cell Banking Initiative 2009).

7.2.2 Good Laboratory Practice (GLP)

GLP was originally developed for non-clinical safety testing of various items, including pharmaceutical products, cosmetic products, and food additives (OECD 1998). Non-clinical safety testing is an important part of drug development that precedes clinical studies in human subjects. Safety testing must meet high standards of quality, reproducibility, and rigor. It is also important to standardize procedures so that testing data can be used for assessment in multiple countries.

GLP refers to a quality system concerned with the organizational process and the conditions under which non-clinical health and environmental safety studies are planned, performed, monitored, recorded, archived, and reported (OECD 1998). GLP regulations were first developed by the United States Food and Drug Administration (FDA) in 1976, following concerns regarding the quality of testing data. The OECD used these GLP regulations to develop a set of principles, which were revised in 1998 to give the Revised OECD Principles of GLP. Participating countries need to establish national compliance monitoring programs to ensure that the GLP principles are met. A system for mutual acceptance of data applies so that test results are accepted for assessment purposes in all participating countries.

Broad guidance on the Revised OECD Principles of GLP and how to implement the principles in a testing laboratory is available from the World Health Organization (WHO 2009) and OECD (2018b). The Revised OECD Principles of GLP focus on 10 key areas (OECD 1998):

(1) The test facility organization and personnel.
(2) The QA program.
(3) The facilities that are available where the studies are performed.
(4) The apparatus, materials, and reagents that are used.
(5) The test systems that are used.
(6) The test and reference items that are employed.

(7) The Standard Operating Procedures (SOPs) that are written for the study.
(8) The performance of the study.
(9) The reporting of study results.
(10) The storage and retention of records and procedures.

In vitro cell-based models are becoming more widely used for safety and efficacy testing (see Chapter 29). For example, 3D organotypic systems have been developed as tissue equivalents for assessment of eye and skin irritation (see Sections 27.7.1, 29.5.1). These and other new advances (e.g. organ-on-chip technologies; see Section 29.5.2) provide exciting opportunities to model whole organs more effectively (Alepee et al. 2014; Eskes et al. 2017). However, the increasing complexity of these systems may lead to problems with standardization and reproducibility. There is a need for additional guidance when applying GLP principles to *in vitro* studies. The OECD has now released a guidance document that describes how to apply Good *In Vitro* Method Practices (GIVMP) for various applications; the document is an important resource when developing new test methods, e.g. OECD test guidelines (OECD 2018a).

GLP typically does not apply to research laboratories. Although GLP principles would improve the quality and reproducibility of experimental work, it is difficult to implement these principles in academic innovative research (Eskes et al. 2017), where procedures are constantly evolving to meet the needs of different research projects. Research laboratories should implement GCCP to ensure that research is suitably rigorous and reproducible.

7.2.3 Good Manufacturing Practice (GMP)

GMP was originally developed for the manufacture of pharmaceutical products. GMP is applied after preclinical safety testing (requiring GLP) to the process of manufacturing a drug for use in patients (WHO 2016). Quality and safety requirements have gradually become standardized over the last century, resulting in introduction of the term "GMP" in 1962 (Arayne et al. 2008). Current GMP (frequently referred to as "cGMP") includes all quality measures that must be applied to ensure that medicinal products are consistently produced and controlled to ensure their safety, efficacy, and quality. GMP also has legal components that cover responsibilities for distribution, contract manufacturing and testing, and responses to product defects and complaints (WHO 2016).

GMP is essential for the manufacture of biological products. The term "biological product" can be broadly defined as an item that contains or is derived from cells or tissues and is produced for use in diagnosis, prevention, or treatment of human disease. The need for GMP when manufacturing biological products was highlighted as early as 1901, when serum used for diphtheria antitoxin was contaminated with tetanus bacilli. This tragic event led to the deaths of 13 children and the introduction of the United States Biologics Control

Act (FDA 2002). A number of biological products are now subject to regulation, including vaccines, blood and plasma products, recombinant therapeutic proteins, and cell-based therapies. The number of cell-based therapies (referred to in Europe as cell-based medicinal products [CBMPs]) has steadily increased since the year 2000. A review of the ClinicalTrials.gov database in 2015 identified 1342 active cell-based therapy clinical trials, with the most common cell types being hematopoietic cells, mesenchymal stromal cells (MSCs), and lymphocytes (Heathman et al. 2015).

Biological products have unique manufacturing challenges (Rayment and Williams 2010). Terminal sterilization is usually not possible as part of the manufacturing process; sterility testing must therefore be performed to detect contamination. Biological products must also be tested for other safety concerns, including the presence of bacterial endotoxins, which act as pyrogenic substances to induce fever or septic shock (Fennrich et al. 2016). Variations in biological material result in variations in the potency of the final product, which may be difficult to assess if a good test method is not available. Finally, it may be difficult to deliver a high-quality, consistent product when working with cells or tissue. For example, the need to scale up cultures for cell-based therapy may lead to loss of the desired phenotype or the appearance of undesirable behavior. Constraints will apply to scale-up and harvesting that depend on the manufacturing technology and the cells used (Heathman et al. 2015). These challenges must all be addressed through a complex and rigorous development process (see Figure 7.2).

In the United States, the FDA's Center for Biologics Evaluation and Research (CBER) is responsible for regulation of biological products to ensure that they are safe and effective, and maintain a high standard of quality. GMP requirements are set out as part of the United States Pharmacopeia and the Code of Federal Regulations Title 21 (21 CFR) (PAS 83: 2012 [British Standards Institution 2012; FDA 2019]). Title 21 includes standards for sterility testing (21 CFR 610.12), endotoxin testing (21 CFR 610.13), and mycoplasma testing (21 CFR 610.30). In Europe, the European Medicines Agency (EMA) co-ordinates inspections to verify compliance with GMP standards and works to harmonize GMP activities in the European Union. The European Directorate for the Quality of Medicines and Healthcare (EDQM) publishes requirements for biological products as part of the European Pharmacopeia (EDQM 2020). Requirements for biological testing are set out in a series of monographs, including sterility (monograph 2.6.1), mycoplasma (monograph 2.6.7), and bacterial endotoxins (monograph 2.6.14).

How do you reconcile different GMP requirements if you are developing a biological product that will be used in more than one country? Harmonization of GMP requirements is a high priority for many organizations, including the WHO, which drafted its first international GMP requirements for biological products in 1991 (WHO 2016). However, despite

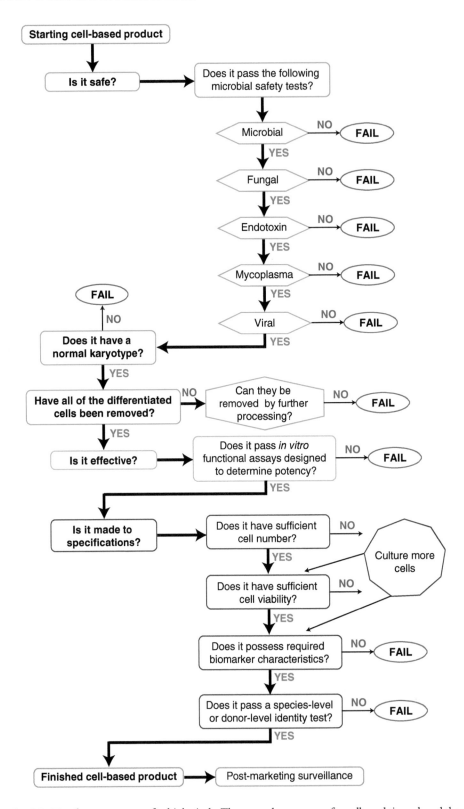

Fig. 7.2. Development process for biologicals. The general test process for cell- and tissue-based therapies is divided into the three main stages of development, including safety, efficacy, and purity of manufacture. In each stage, multiple requirements must be met to allow the product to continue through the next stage of the process. *Source*: Rayment, E.A. and Williams, D.J. (2010), Concise Review: Mind the Gap: Challenges in Characterizing and Quantifying Cell- and Tissue-Based Therapies for Clinical Translation. STEM CELLS, 28: 996–1004. © 2010 John Wiley & Sons.

best efforts, complex regulatory frameworks have developed that can result in different regional GMP requirements. For example, in the United States, cultured cells grown on collagen or synthetic membranes may be regulated as "devices" rather than "biological products," resulting in different compliance requirements (FDA 2018). Guidance documents are available to provide increased clarity regarding international GMP requirements (PAS 83: 2012 [British Standards Institution 2012]). However, GMP compliance is a specialized field in its own right; it is important to bring in quality and regulatory expertise as early as possible if you intend to develop a biological product.

7.3 CELL LINE PROVENANCE

The term "provenance" can be literally defined as the source or origin of a particular item (Freshney 2002). The term is used extensively when referring to antiques and paintings, where this information is essential to prove that items are authentic and of good quality. The term is also applied to finite and continuous cell lines. "Cell line provenance" refers to the records that provide information on a cell line's origins, handling, validation, and behavior. A cell line with a detailed and complete provenance gains value like a piece of antique furniture or a painting.

Cell line provenance is important because it addresses many of the causes of irreproducible research (see Section 7.1.3). Keeping records on a cell line's origins, handling, validation, and behavior allows adequate information on that cell line to be reported in publications. This allows the reader to replicate the experiment and independently assess whether the work reaches valid conclusions. Scientists who understand the importance of cell line provenance can also use it to maximize reproducibility and minimize errors in their own work. For example, laboratories can detect cases of misidentification at an early stage if they include authentication testing as part of provenance information (Gazdar et al. 2010). Cell line provenance is an important part of a QA system under GLP or GMP, although the term used may vary – for example, a human pluripotent stem cell (hPSC) bank may have a Cell Line Master File or a Cell Line History File that performs a comparable function (Stacey 2017).

Collecting records for cell line provenance may seem daunting at first glance. However, it is surprisingly easy if records are kept routinely as part of everyday handling. Many scientists keep a notebook where they document routine maintenance, cryostorage, and experimental observations. If routine procedures are recorded as SOPs, it may only be necessary to refer to the SOP and record the materials used on that particular day. A table can be used to keep records of primary culture or maintenance of a cell line, with a focus on events such as feeding cultures, performing subculture or cryopreservation, or thawing cryovials (see Tables 14.6, 14.7, 15.4, 15.5). All of these records can be kept in electronic

form, which allows better long-term storage. Electronic laboratory notebooks (ELNs) and laboratory information management systems (LIMS) are now available that allow such records to be stored and backed up. Examples include CloudLIMS (www.cloudlims.com), LabArchives (www.labarchives.com), and LabVantage (www.labvantage.com).

7.3.1 Provenance Information

What information should be included as part of the provenance of a cell line? Ideally, provenance information should describe how the cell line was isolated and what has happened to it since isolation. For existing cell lines, some of this information will be available before you start work (either from publications or from a colleague by word of mouth) and will play a part in your decision to use it in the first place. Initial information should be independently verified and added to, once work commences.

Provenance information can be broadly divided into the following areas:

(1) **Cell line origin.** When and where was culture commenced? What information was reported about the donor and the tissue from which the cell line was later established? What method was used to establish a continuous cell line? The Cellosaurus knowledge resource now makes this information easier to find for published cell lines (Bairoch 2018). Searching under the cell line's name or a unique identifier (see Section 7.3.2) will bring up a summary of cell line information and lists of key publications, cell line repositories, and databases with more information.

(2) **Cell line source.** Where did the cell line that is used in your laboratory come from? Because continuous cell lines may be passed from laboratory to laboratory, the source may not be the laboratory where culture was commenced. It is important to obtain the cell line from a reliable source, ideally the originator or a cell repository (see Section 15.6).

(3) **Cell line validation.** What testing is required to ensure that the cell line is a valid model for your application? A reliable source should perform validation testing before they publish or distribute the cell line, which is then used as a baseline for comparison (see Section 7.4; see also Chapters 16, 17). If validation testing has not been performed by your source, you should do so immediately upon receipt of the culture, quarantining the culture until the results of validation are available.

(4) **Cell line characterization.** What are the key characteristics of the cell line that you should examine as part of routine maintenance or experimental procedures? Cell line characterization is addressed in detail in a later chapter (see Chapter 18). Key characteristics include morphology, growth, and expression of tissue-specific markers. Because these characteristics can change with

further handling, it is important to establish an early baseline for comparison (e.g. saving images of the culture at different density; see Section 14.4.1).

(5) ***Cell line handling.*** What culture environment should be used to maintain the cell line and preserve its key characteristics? What procedures should be used for feeding, subculture, and cryopreservation? You can usually obtain this information from a reliable source (e.g. the catalogue and product information provided by a cell repository). Changes in handling can affect the behavior of the cell line, so it is important to keep handling as consistent as possible.

7.3.2 Reporting for Publication

Not all provenance information is suitable for publication. For example, if your laboratory establishes a cell line, you may have access to private health information from the donor that should be kept confidential (see Supp. S6.4). However, it is important to publish sufficient information for readers to identify the cell line and understand its origin, source, and validation testing. A survey of 238 publications found that only 43% provided sufficient information to uniquely identify the cell lines used (Vasilevsky et al. 2013). Unfortunately, cell line names are not sufficient to uniquely identify individual cell lines if they are difficult to use in searches (Sarntivijai et al. 2008). The importance of this finding is highlighted by data from the Cellosaurus knowledge resource (Bairoch 2018). As of January 2020, 423 cell line names were shared by two or more different cell lines (including names with different casing), affecting a total of 909 cell lines (personal communication, Amos Bairoch). Other important cell line information is often omitted from publications, such as source and passage number.

The proliferation of cell lines with identical names has resulted in efforts to develop guidelines for naming new cell lines (see Section 14.2.1). Although these guidelines are important, they cannot address many of the existing conflicts in cell line nomenclature. The best solution for existing cell lines is to provide an additional identifier that is unique and easily searchable. The Resource Identification Initiative has developed Research Resource Identifiers (RRIDs) as unique identifiers for research materials (Bandrowski et al. 2016). RRIDs are generated for cell lines using the Cellosaurus knowledge resource, which collaborates with the Resource Identification Initiative to provide this information (Bairoch 2018).

What information should be reported when cell lines are used in publications? The International Cell Line Authentication Committee (ICLAC) has developed a checklist for reporting cell line information in manuscripts and grant applications (see Table 7.3). Checklists typically focus on key information that should be included in all publications. You will also need to include information on your unique experimental conditions. Guidance is available for reporting experimental protocols in publications (Giraldo et al. 2018), but manuscript word limits may affect the amount of information that you can provide. Method repositories are a good solution, allowing you to release protocols online and cite the full protocol within your manuscript (Teytelman et al. 2016).

Minimal information models may apply to some fields of research and clinical translation. For example, Minimal Information About T-Cell Assays (MIATA) has been developed to promote standardization of T-cell assays (Britten et al. 2011). If such a standard applies to your field, it is important to report your study in a way that complies with the consensus requirement. Journals may cite these requirements in their author guidelines for manuscript submission. A minimum set of reporting standards for all life science research is currently in development (Chambers et al. 2019).

7.4 VALIDATION TESTING

The term "validation" usually refers to the documented act of ensuring that any procedure, process, equipment, material, activity, or system actually gives the expected results with adequate reproducibility (Andrews et al. 2015). For cell lines, the central aim of validation testing is to answer the question: is this cell line worth using? As noted previously (see Section 7.1.3), contaminated or misidentified cell lines are important causes of irreproducible research. Testing for these quality concerns is a central element of GCCP for all tissue culture laboratories, regardless of application. Testing is also essential for GLP and GMP, although the term used may vary – for example, "quality control" may be used instead of "validation." For this book, the term "validation testing" refers to testing of cell lines, while "quality control" refers to testing of cell lines, reagents, vessels, and equipment as part of QA (see Section 7.5). The extent of quality control testing will vary with the application, but "validation testing" of cell lines should be performed at all levels of good practice.

It has been argued that testing of cell lines is not required if validation will not affect the outcome that is being measured. For example, the originator of the Chang liver cell line did not collect provenance information or see the need for validation testing because the cell line was originally used for viral culture (Hall 2017). Chang liver cells are now used in preclinical research as a model for normal liver cells. Unfortunately, testing has shown that Chang liver is misidentified and is actually HeLa. Validation testing enables any laboratory that uses Chang liver to detect this problem and make an informed decision as to whether the cell line is a valid model. This example, and many others, show that validation testing is important for all cell lines because of their widespread use as research models, to detect quality concerns prior to use.

What does validation testing involve? As a minimum, testing for microbial contamination and authenticity should be performed, regardless of application. Additional testing will be needed for some cell types and applications. For example,

TABLE 7.3. Cell line checklist for publication.

Checklist requirement	Example: M14 melanoma cell line[a]
What is the name of the cell line?	M14
What is the Research Resource Identifier (RRID) for the cell line?	RRID:CVCL_1395
Is the cell line known to be misidentified?	No, the cell line is not known to be misidentified (*Source*: ICLAC Register of Misidentified Cell Lines).
Has authentication testing been performed? List the method used and the result. For human cell lines, make the short tandem repeat (STR) profile available with the publication.	Yes, STR profiling was performed on a passage 15 sample. The STR profile corresponded to serum DNA from the donor.
Has mycoplasma testing been performed? List the method used and the result.	Yes, mycoplasma testing was performed using the e-Myco plus Mycoplasma PCR Detection kit. Mycoplasma was not detected.
Is the source of the cell line provided? List the catalogue number if obtained from a cell repository.	Yes, the sample was obtained from the originator as stated in the Materials and Methods.
Is sufficient information provided to replicate experiments that use the cell line? List the passage or population doubling level used for experiments. List the culture conditions including medium, supplements, substrate, and coatings.	Yes, see Materials and Methods and Supplementary Methods.

Source: Adapted from Checklist requirements, International Cell Line Authentication Committee (ICLAC 2014).
[a]This example comes from (Korch et al. 2018) and is provided for educational purposes. Requirements will vary depending on the publication; always refer to the publication's author guidelines for more information.

stem cell lines require additional testing to assess pluripotency, genome integrity, and malignant potential; these requirements are discussed in a later chapter (see Section 23.7). Once validation has been performed, cell lines can be stored in liquid nitrogen until they are required (see Sections 15.4, 15.5). Validation testing of stocks in liquid nitrogen reduces the need for ongoing testing; end users can periodically return to validated stocks to minimize quality and handling concerns. End users may store their own stocks for the duration of a project, but as these stocks are no longer fully validated, they should not be passed on to other users.

7.4.1 Testing for Microbial Contamination

However detailed your records or meticulous your experimental technique, the resulting work is devoid of value (or at least heavily compromised) if the cell line is contaminated with one or more microorganisms. Contamination may be obvious on visual inspection (see Figure 16.2). Where contamination is overt, it is less of a problem, because it can be detected and the culture discarded. However, contamination may not be obvious by visual inspection. Cryptic contamination may occur because (i) cells are routinely maintained in antibiotics; (ii) microorganisms are slow-growing or difficult to detect by eye; (iii) routine testing for such organisms is not performed; or (iv) routine testing is not available, e.g. for some viruses or prions.

Mycoplasma contamination is common in tissue culture laboratories and is difficult to detect by eye, resulting in the need for specific testing. Mycoplasma testing must be performed as part of validation. If mycoplasma is detected, cultures should be discarded unless they are unique and irreplaceable, in which case, treatment may be considered. These topics are explored in a later chapter (see Chapter 16). Other forms of cryptic contamination may require additional testing; for example, slow-growing or unusual microorganisms may require sterility testing using microbiological media (see Section 11.8.1).

The best way to deal with microbial contamination is to avoid it in the first place. Contamination can be avoided by (i) using good aseptic technique (see Section 12.4); (ii) obtaining cell lines from a reliable source, e.g. a cell repository (see Table 15.3); (iii) culturing cells in the absence of antibiotics (see Section 9.6.2); (iv) testing regularly for mycoplasma (see Protocols P16.3, 16.4); and (v) testing periodically for other cryptic microorganisms, based on risk assessment (e.g. viruses; see Section 16.5).

7.4.2 Testing for Authenticity

Misidentification is one of the most severe problems facing tissue culture laboratories. Misidentified cell lines are difficult to detect by eye. Although an experienced cell culture operator may suspect that a problem has occurred, it is impossible to prove or to determine the correct identity of the cell line without specific testing. Failure to test for authenticity can result in multiple publications being affected. Once published, the problem multiplies as publications are cited and other laboratories decide to use the same cell lines for their own work. Authentication testing is one of the most vital aspects of validation and no laboratory, however small, can afford to ignore it.

Authentication testing relies on analysis of genotypic data. Ideally, the cell line's genotype is compared to a reference genotype that has been prepared from donor material (e.g. tissue or blood) that was stored by the original laboratory for this purpose. If donor material is not available, early passages of the same cell line may be used for comparison, but there is a risk that these early passages may themselves be misidentified. This risk is addressed by comparing the cell line's genotype to a database of other cell lines that has been assembled from multiple sources. Comparison of samples from multiple laboratories results in detection of unexpected matches and has been a mainstay of authentication testing for many years (ATCC SDO Workgroup ASN-0002 2010).

Comparison of samples between laboratories means that a consensus method must be used. Short tandem repeat (STR) profiling is the consensus method for authentication of human cell lines and has also been utilized for some non-human cell lines (ATCC SDO Workgroup ASN-0002 2010; Almeida et al. 2016). Single nucleotide polymorphism (SNP) analysis can also be used for authentication. Consensus has not yet been reached on a set of SNP markers for comparison, so this method is typically used for in-house testing with comparison to internal reference samples. Many non-human species do not have test methods available that can compare samples to individual level. Where this is the case, cell lines are tested to species level or strain level. These topics and suitable test methods are explored in a later chapter (see Chapter 17).

Can a cell line's phenotype (e.g. expression of tissue-specific markers) be used for authentication testing? Phenotypic data can provide useful supporting evidence but can also be misleading. For example, the Chang liver cell line expresses liver-specific markers and was previously believed to be a good model for normal liver cells (see Section 7.4). STR profiling has proven unequivocally that Chang liver cells are actually HeLa, which is derived from cervical carcinoma (Hall 2017). Phenotype may be unreliable in cell lines, particularly when cells are grown in a culture environment that encourages rapid growth and dedifferentiation (see Section 26.2.2).

7.5 QUALITY ASSURANCE (QA)

QA focuses on operational standards for all aspects of activities, to ensure that a procedure or a product consistently meets quality requirements (Grizzle et al. 2008). The term "quality management system" may also be used (Grizzle et al. 2015). Quality control testing (including validation testing; see Section 7.4) is used to assess specific quality outcomes and is one component of a QA program. The need for QA will depend on your application. QA is mandatory for laboratories that perform GMP or GLP cell culture. These laboratories typically use quality management systems that comply with the Revised OECD Principles of GLP or with other relevant Standards (ISO 9001: 2015) (ISO 2015a). QA is not mandatory for research laboratories that perform GCCP. However, many research laboratories apply QA principles (i.e. spirit of GMP or GLP) to improve reproducibility and assist with troubleshooting. For example, laboratories that prepare their own medium will perform sterility testing to detect microbial contamination (see Section 11.8.1) and document the results for future reference. Documentation of procedures, testing, materials, and equipment maintenance can form the nucleus of a simple QA program that is achievable by any tissue culture laboratory, regardless of size. The following points apply to research laboratories that are interested to develop a simplified QA program to assist with reproducibility and troubleshooting.

7.5.1 Standard Operating Procedures (SOPs)

SOPs are used to document the protocols and procedures that are used in the laboratory (Grizzle et al. 2015). SOPs should be written in sufficient detail so that anyone who performs that procedure can do so safely and consistently. When used in a QA program, a SOP is typically written by an expert in the procedure and is reviewed by a committee or working group before being approved for release. Once the SOP is released, personnel should follow the procedure exactly as written. If changes are made, the SOP must be revised and reapproved, and a new version released; the change must be communicated to workers and the previous version removed from circulation.

This approach to SOPs can be difficult to apply in research laboratories, where individuals or groups may need to make changes as research progresses or to develop variations for different applications. However, SOPs are useful for tissue culture laboratories (and other shared facilities) where workers need to take a consistent approach to some procedures and equipment. SOPs are also important for laboratory safety and training (see Sections 6.1.3, 30.1.3). For example, anyone who operates an autoclave should have access to a written procedure that describes how to do so safely and how to check that items have been sterilized.

Laboratory personnel should work together to develop a set of SOPs that are essential for safety, training, and consistency. ELNs (see Section 7.3) and method repositories are now available, allowing laboratory SOPs to be developed in collaboration and maintained in electronic form (Teytelman et al. 2016). Together, these SOPs act as a laboratory manual describing how to perform common cell culture procedures, conduct validation testing, and operate equipment. Individuals can use the laboratory SOPs as the basis for their own protocols. Deviation from the SOP should be discouraged unless there is scientific justification; if the change is justified, the deviation should be accurately recorded for future reference. Casual transfer of protocols by word of mouth, and the accidental deviations that result, should be avoided.

7.5.2 Media and Reagents

Most purchased media and reagents are manufactured by companies that comply with quality requirements (e.g. ISO 9001: 2015 certification) (ISO 2015a). This means that media and reagents should be manufactured with due regard for safety and consistency. Quality control testing by the manufacturer is included as part of the product information that is released to the customer. Reagents of animal origin, such as serum and trypsin, require additional screening for viruses and other microbial contaminants (see Section 9.7.1). Refer to the product information to see a summary of the testing that has been performed.

Quality control testing of purchased media and reagents may include cell line "performance" testing. This testing typically focuses on viability, growth, or plating of a small number of cell lines and may not address your specific requirements (e.g. growth of a particular cell line). It is important to perform your own testing; ask for a sample to try before you buy, particularly if you intend to buy in bulk. Testing usually involves analysis of a cell line's growth curve and cloning efficiency (see Section 11.8.2).

Despite all best efforts, media and reagents can experience quality problems. Batch numbers of medium, serum, trypsin, and other reagents should be recorded as part of routine maintenance or experimental procedures. Batch numbers can be traced to the supplier, who can then trace reagent constituents back to their manufacturers. Commercial products often carry barcodes, so a barcode reader may be useful to assist with documentation of reagents in the tissue culture laboratory.

7.5.3 Culture Vessels

As for reagents and media, culture vessels are usually manufactured by companies that comply with quality requirements (e.g. ISO 9001: 2015 certification) (ISO 2015a) and perform quality control testing. Quality control for cell culture plasticware may include testing for growth, cloning efficiency, sterility, pyrogenic agents, and leachable chemicals. Always read the product information to understand what testing has been performed before you purchase. Record any changes in culture vessels as part of routine maintenance or experimental procedures.

Tissue culture vessels can vary between manufacturers – for example, tissue culture plasticware is treated to improve cell adhesion using a number of different methods (see Sections 8.2.1, 8.3.7). As with media and reagents, it is important to test new culture vessels before you buy, and particularly before you buy in bulk. To test a new substrate, grow your cells on it as a regular monolayer in the appropriate medium; limiting concentrations of serum (or serum-free medium) should be used, as high serum concentrations may mask imperfections in the plastic. Analysis of growth rate and cloning efficiency is also important, with comparison to baseline data, e.g. from testing media and reagents (see Section 7.5.2).

Quality concerns include leaching of chemicals from disposable plasticware. Leachable chemicals have been shown to affect cell growth and to disrupt some experimental procedures (Olivieri et al. 2012; Wood et al. 2013). In some cases, the source of the chemical was traced back to plastic bags that were used for culture media; these bags were employed as single-use systems for manufacture of biological products (Wood et al. 2013; Hammond et al. 2014). Addressing this problem for GMP applications requires the entire supply chain to develop suitable testing and quality standards.

7.5.4 Equipment

New equipment should be purchased from companies that comply with quality requirements and assist with installation and training. Once installed, laboratory personnel become responsible for ongoing safety and quality requirements. Always retain equipment manuals for future reference and keep records of equipment maintenance, repair, calibration, and testing. Each item of equipment should have a person nominated as a "curator" to arrange ongoing maintenance and keep equipment records.

Some items of equipment require regular certification to ensure that they can be operated safely and effectively. Biosafety cabinets (BSCs) must be certified to ensure that filters are intact and airflow meets specifications for cabinet design (see Section 5.1.1). If biological waste is decontaminated by autoclaving, the autoclave must be tested to ensure that waste is safe for disposal (see Section 6.3.6). Equipment monitoring and safety testing are discussed elsewhere (see Sections 4.3.2, 6.2.6). Equipment used for GMP-compliant cell culture will require installation qualification (IQ), operational qualification (OQ), and performance qualification (PQ).

7.5.5 Facilities

The tissue culture laboratory may have specific requirements for room temperature, humidity, air quality, and differential pressure. Any specific requirements should be clearly communicated, and monitoring should be performed to ensure that requirements are met. If conditions deviate from quality requirements, the event should be recorded and laboratory personnel should be alerted; repeated non-conformance will require corrective action. Monitoring of environmental conditions is done automatically in some facilities, where a Building Management System (BMS) is used to control and monitor mechanical and electrical infrastructure. The BMS may also be modified to monitor equipment conditions and generate equipment alarms (see Section 4.3.2).

Facilities that operate at GLP or GMP will have more specialized quality requirements. For example, manufacture of cell-based therapies may be performed in a cleanroom that complies with international standards for air cleanliness (ISO 14644-1: 2015) (ISO 2015b). Cleanrooms require specialized expertise from design stage through to validation and ongoing operation. Relevant standards and recommended practices are available from the Institute of Environmental Sciences and Technology (IEST 2018). Always refer to local regulations to determine compliance requirements – for example, GMP requirements for buildings and facilities are included in Code of Federal Regulations Title 21 (Subpart C) (FDA 2019).

7.6 REPLICATE SAMPLING

Replicate sampling does not usually apply to routine handling procedures, where the aim is to minimize variation and retain key characteristics. However, it becomes vitally important for cell-based assays and high-throughput studies. Lack of statistical power in such studies is an important cause of irreproducible research (Begley and Ioannidis 2015). Research findings are commonly based on single studies that provide p-values of less than 0.05; selective or distorted reporting of these studies results in additional bias (Ioannidis 2005). Scientists must make informed decisions regarding the types and numbers of samples that should be used to improve reproducibility and scientific rigor.

7.6.1 Experimental Design

Good experimental design requires careful selection of replicate samples. As an illustration, consider how you would test the hypothesis that breast cancer cell lines grow faster than lung cancer cell lines (Lazic et al. 2018). You could test multiple samples obtained from one breast cancer cell line and one lung cancer cell line (sometimes referred to as "technical replicates"). This approach is not sufficiently robust; the two cell lines may differ in their rates of growth because of many different variables. You could also test single samples obtained from multiple breast cancer cell lines and multiple lung cancer cell lines (sometimes referred to as "biological replicates"). This approach is not sufficiently reproducible; if only one sample is tested from each cell line, the results will vary when the experiment is repeated. The best approach is to design an experiment where you test multiple samples from multiple breast and lung cell lines.

Good experimental design is beyond the scope of this book, which is focused primarily on practical handling procedures. There are some excellent resources available elsewhere for laboratories to refer to. For example, lessons from measurement science can be used to improve confidence in research results and reduce uncertainty (Plant et al. 2014, 2018). Similarly, lessons from meta-analysis can be used to reduce publication bias and improve sample selection (Stanley et al. 2017; Lazic et al. 2018). Lessons from animal studies can be used to improve sample diversity, which can help to make studies more representative and improve external validity (Karp 2018).

7.6.2 Samples and Data

What culture vessels should be used for replicate samples? Many types of culture vessel are available for monolayer culture replicates (see Section 8.5). The choice of vessel is determined by (i) the number of cells required in each sample; (ii) the frequency or type of sampling; (iii) the relative ease of handling; and (iv) the cost of each vessel type. Replicate sampling is most readily performed in multiwell plates if the incubation time is not a variable (see Section 8.5.2). Adhesive film (e.g. AeraSeal™ film or SealMate™ sealing system, Excel Scientific) may be used to seal the plate, reducing evaporation and contamination, and giving a more even performance across the plate (see Figure 8.4). It also allows individual wells or rows to be sampled without opening the rest of the plate.

If samples are collected over a period of time (e.g. daily for five days), the repeated removal of a plate for daily processing may impair growth in the remaining wells. This can be avoided by using single plates for each time point, but multiple plates add up to a significant cost. If budget is an issue, individual tubes can be used. Leighton tubes (plastic, ThermoFisher Scientific #156758; glass, Bellco Glass #1908–1916 125) provide a flat growth surface for culture of multiple samples at relatively low cost, e.g. when dealing with amniocentesis samples (see Supp. S23.4). If the optical quality of the tubes is not critical, flat-bottomed glass specimen tubes or even glass scintillation vials may be suitable. If glass vials or tubes are used, they must be washed as tissue culture glassware (see Section 11.3); they cannot be used for tissue culture after use with scintillant. Sealing large numbers of vials or tubes can become tedious, so some laboratories use vinyl tape rather than screw caps to seal individual tubes. Such tape can be color coded to identify different treatments.

Analysis of data from cultured cells is not necessarily different to any other biological system. However, it is important to be aware that large amounts of data can be produced from cell culture experiments, particularly when dealing with microwell plates in high-throughput assays. A number of companies now market computer programs that display and analyze data from microwell plate assays; these programs include titration curves, enzyme kinetics, and binding assays. Handling tissue culture-derived data will depend on the data to be captured, the scale of the experiment, and the number of parameters. Scale-up of the experiment may lead to very different requirements. For example, cell counting is the accepted method to construct a growth curve for a single cell line under two or three sets of conditions. However, cell counting does not lend itself to multiple growth curves to measure the response of several cell lines to various growth factor combinations. This challenge may be addressed by using a live-cell analysis system or an indirect method of measuring cell number (see Section 19.3.2). Always remember to include management of these datasets as part of your experimental design.

Suppliers

Supplier	URL
Bellco Glass	http://www.bellcoglass.com
Excel Scientific	http://www.excelscientific.com/index.html
ThermoFisher Scientific	http://www.thermofisher.com/us/en/home/life-science/cell-culture.html

REFERENCES

Alepee, N., Bahinski, A., Daneshian, M. et al. (2014). State-of-the-art of 3D cultures (organs-on-a-chip) in safety testing and pathophysiology. *ALTEX* 31 (4): 441–477. https://doi.org/10.14573/altex1406111.

Almeida, J.L., Cole, K.D., and Plant, A.L. (2016). Standards for cell line authentication and beyond. *PLoS Biol.* 14 (6): e1002476. https://doi.org/10.1371/journal.pbio.1002476.

Andrews, P.W., Baker, D., Benvinisty, N. et al. (2015). Points to consider in the development of seed stocks of pluripotent stem cells for clinical applications: International Stem Cell Banking Initiative (ISCBI). *Regen. Med.* 10 (2 Suppl): 1–44. https://doi.org/10.2217/rme.14.93.

Arayne, M.S., Sultana, N., and Zaman, M.K. (2008). Historical incidents leading to the evolution of good manufacturing practice. *Accredit. Qual. Assur.* 13: 431–432. https://doi.org/10.1007/s00769-008-0363-0.

ATCC SDO Workgroup ASN-0002 (2010). Cell line misidentification: the beginning of the end. *Nat. Rev. Cancer* 10 (6): 441–448. https://doi.org/10.1038/nrc2852.

Bairoch, A. (2018). The Cellosaurus, a cell-line knowledge resource. *J. Biomol. Tech.* 29 (2): 25–38. https://doi.org/10.7171/jbt.18-2902-002.

Bandrowski, A., Brush, M., Grethe, J.S. et al. (2016). The Resource Identification Initiative: a cultural shift in publishing. *Brain Behav.* 6 (1): e00417. https://doi.org/10.1002/brb3.417.

Begley, C.G. and Ellis, L.M. (2012). Drug development: raise standards for preclinical cancer research. *Nature* 483 (7391): 531–533. https://doi.org/10.1038/483531a.

Begley, C.G. and Ioannidis, J.P. (2015). Reproducibility in science: improving the standard for basic and preclinical research. *Circ. Res.* 116 (1): 116–126. https://doi.org/10.1161/CIRCRESAHA.114.303819.

Begley, C.G., Buchan, A.M., and Dirnagl, U. (2015). Robust research: institutions must do their part for reproducibility. *Nature* 525 (7567): 25–27. https://doi.org/10.1038/525025a.

Ben-David, U., Siranosian, B., Ha, G. et al. (2018). Genetic and transcriptional evolution alters cancer cell line drug response. *Nature* 560: 325–330. https://doi.org/10.1038/s41586-018-0409-3.

Bissell, M. (2013). Reproducibility: the risks of the replication drive. *Nature* 503 (7476): 333–334.

British Standards Institution (2012). *PAS 83:2012. Developing Human Cells for Clinical Applications in the European Union and the United States of America – Guide.* London: BSI.

Britten, C.M., Janetzki, S., van der Burg, S.H. et al. (2011). Minimal information about T cell assays: the process of reaching the community of T cell immunologists in cancer and beyond. *Cancer Immunol. Immunother.* 60 (1): 15–22. https://doi.org/10.1007/s00262-010-0940-z.

Byrne, J.A. and Labbe, C. (2017). Striking similarities between publications from China describing single gene knockdown experiments in human cancer cell lines. *Scientometrics* 110 (3): 1471–1493. https://doi.org/10.1007/s11192-016-2209-6.

Cancer Research UK (2017). Grant conditions. https://www.cancerresearchuk.org/sites/default/files/funding_grant_conditions.pdf (accessed 4 July 2018).

Casadevall, A., Steen, R.G., and Fang, F.C. (2014). Sources of error in the retracted scientific literature. *FASEB J.* 28 (9): 3847–3855. https://doi.org/10.1096/fj.14-256735.

Chalmers, I. and Glasziou, P. (2009). Avoidable waste in the production and reporting of research evidence. *Lancet* 374 (9683): 86–89. https://doi.org/10.1016/S0140-6736(09)60329-9.

Chambers, K., Collings, A., Graf, C. et al. (2019). Towards minimum reporting standards for life scientists. *MetaArXiv* 30 April https://doi.org/10.31222/osf.io/9sm4x.

Coecke, S., Balls, M., Bowe, G. et al. (2005). Guidance on Good Cell Culture Practice. A report of the second ECVAM Task Force on Good Cell Culture Practice. *Altern. Lab. Anim* 33 (3): 261–287.

De Los Angeles, A., Ferrari, F., Fujiwara, Y. et al. (2015). Failure to replicate the STAP cell phenomenon. *Nature* 525 (7570): E6–E9. https://doi.org/10.1038/nature15513.

Drucker, D.J. (2016). Never waste a good crisis: confronting reproducibility in translational research. *Cell Metab.* 24 (3): 348–360. https://doi.org/10.1016/j.cmet.2016.08.006.

EDQM (2020). European Pharmacopoeia (Ph.Eur.). https://www.edqm.eu/en/european-pharmacopoeia-ph-eur-10th-edition (accessed 30 July 2020).

Eskes, C., Bostrom, A.C., Bowe, G. et al. (2017). Good Cell Culture Practices & in vitro toxicology. *Toxicol. in Vitro* 45 (Pt 3): 272–277. https://doi.org/10.1016/j.tiv.2017.04.022.

Fang, F.C., Steen, R.G., and Casadevall, A. (2012). Misconduct accounts for the majority of retracted scientific publications. *Proc. Natl Acad. Sci. U.S.A.* 109 (42): 17028–17033. https://doi.org/10.1073/pnas.1212247109.

FDA (2002). Science and the regulation of biological products: from a rich history to a challenging future. https://www.fda.gov/downloads/aboutfda/whatwedo/history/forgshistory/historyoffdascentersandoffices/ucm582569.pdf (accessed 9 July 2018).

FDA (2018). FDA regulation of human cells, tissues, and cellular and tissue-based products (HCT/P's) product list. https://www.fda.gov/biologicsbloodvaccines/tissuetissueproducts/regulationoftissues/ucm150485.htm (accessed 9 July 2018).

FDA (2019). Electronic Code of Federal Regulations: Title 21 Food and Drugs. https://www.ecfr.gov/cgi-bin/ecfr?page=browse (accessed 9 April 2018).

Fennrich, S., Hennig, U., Toliashvili, L. et al. (2016). More than 70 years of pyrogen detection: current state and future perspectives. *Altern. Lab. Anim* 44 (3): 239–253.

Freedman, L.P., Gibson, M.C., Wisman, R. et al. (2015). The culture of cell culture practices and authentication – results from a 2015 survey. *BioTechniques* 59 (4): 189–190, 192. doi: https://doi.org/10.2144/000114344.

Freshney, R.I. (2002). Cell line provenance. *Cytotechnology* 39 (2): 55–67. https://doi.org/10.1023/A:1022949730029.

Fusenig, N.E., Capes-Davis, A., Bianchini, F. et al. (2017). The need for a worldwide consensus for cell line authentication: experience implementing a mandatory requirement at the International Journal of Cancer. *PLoS Biol.* 15 (4): e2001438. https://doi.org/10.1371/journal.pbio.2001438.

Gazdar, A.F., Girard, L., Lockwood, W.W. et al. (2010). Lung cancer cell lines as tools for biomedical discovery and research. *J. Natl Cancer Inst.* 102 (17): 1310–1321. https://doi.org/10.1093/jnci/djq279.

Geraghty, R.J., Capes-Davis, A., Davis, J.M. et al. (2014). Guidelines for the use of cell lines in biomedical research. *Br. J. Cancer* 111 (6): 1021–1046. https://doi.org/10.1038/bjc.2014.166.

Giraldo, O., Garcia, A., and Corcho, O. (2018). A guideline for reporting experimental protocols in life sciences. *PeerJ* 6: e4795. https://doi.org/10.7717/peerj.4795.

Goodman, S.N., Fanelli, D., and Ioannidis, J.P. (2016). What does research reproducibility mean? *Sci. Transl. Med.* 8 (341): 341ps12. https://doi.org/10.1126/scitranslmed.aaf5027.

Grizzle, W.E., Sexton, K.C., and Bell, W.C. (2008). Quality assurance in tissue resources supporting biomedical research. *Cell Preserv. Technol.* 6 (2): 113–118. https://doi.org/10.1089/cpt.2008.9993.

Grizzle, W.E., Gunter, E.W., Sexton, K.C. et al. (2015). Quality management of biorepositories. *Biopreserv. Biobank* 13 (3): 183–194. https://doi.org/10.1089/bio.2014.0105.

Hall, E. (2017). The notorious Chang liver. http://iclac.org/case-studies/chang-liver (accessed 11 July 2018).

Hammond, M., Marghitoiu, L., Lee, H. et al. (2014). A cytotoxic leachable compound from single-use bioprocess equipment that causes poor cell growth performance. *Biotechnol. Progr.* 30 (2): 332–337. https://doi.org/10.1002/btpr.1869.

Harris, R. (2017). *Rigor Mortis: How Sloppy Science Creates Worthless Cures, Crushes Hope, and Wastes Billions*. New York: Basic Books.

Hartung, T., Balls, M., Bardouille, C. et al. (2002). Good Cell Culture Practice. ECVAM Good Cell Culture Practice Task Force Report 1. *Altern. Lab. Anim* 30 (4): 407–414.

Heathman, T.R., Nienow, A.W., McCall, M.J. et al. (2015). The translation of cell-based therapies: clinical landscape and manufacturing challenges. *Regen. Med.* 10 (1): 49–64. https://doi.org/10.2217/rme.14.73.

Horbach, S.P. and Halffman, W. (2017). The ghosts of HeLa: how cell line misidentification contaminates the scientific literature. *PLoS One* 12 (10): e0186281. https://doi.org/10.1371/journal.pone.0186281.

Hughes, P., Marshall, D., Reid, Y. et al. (2007). The costs of using unauthenticated, over-passaged cell lines: how much more data do we need? *BioTechniques* 43 (5):575, 577–8, 581–2 passim. doi: https://doi.org/10.2144/000112598.

ICLAC (2014). Cell line checklist for manuscripts and grant applications. http://iclac.org/resources/cell-line-checklist (accessed 27 July 2018).

ICLAC (2019). Cell line policy for research institutions. https://iclac.org/resources/cell-line-policy (accessed 13 November 2019).

IEST (2018). Standards and recommended practices. http://www.iest.org/standards-rps (accessed 16 July 2018).

International Stem Cell Banking Initiative (2009). Consensus guidance for banking and supply of human embryonic stem cell lines for research purposes. *Stem Cell Rev.* 5 (4): 301–314. https://doi.org/10.1007/s12015-009-9085-x.

Ioannidis, J.P. (2005). Why most published research findings are false. *PLoS Med.* 2 (8): e124. https://doi.org/10.1371/journal.pmed.0020124.

ISO (2015a). ISO 9001:2015. Quality Management Systems – Requirements. Geneva: International Organization for Standardization.

ISO (2015b). ISO 14644-1:2015. Cleanrooms and Associated Controlled Environments – Part 1: Classification of Air Cleanliness by Particle Concentration. Geneva: International Organization for Standardization.

Karp, N.A. (2018). Reproducible preclinical research – is embracing variability the answer? *PLoS Biol.* 16 (3): e2005413. https://doi.org/10.1371/journal.pbio.2005413.

Katsura, I., Igarashi, K., Ito, T., et al. (2014). Report on STAP cell research paper investigation. http://www3.riken.jp/stap/e/c13document52.pdf (accessed 27 June 2018).

Kniss, D.A. and Summerfield, T.L. (2014). Discovery of HeLa cell contamination in HES cells: call for cell line authentication in reproductive biology research. *Reprod. Sci.* 21 (8): 1015–1019. https://doi.org/10.1177/1933719114522518.

Knoepfler, P. (2015). Reviewing post-publication peer review. *Trends Genet.* 31 (5): 221–223. https://doi.org/10.1016/j.tig.2015.03.006.

Korch, C., Hall, E.M., Dirks, W.G. et al. (2018). Authentication of M14 melanoma cell line proves misidentification of MDA-MB-435 breast cancer cell line. *Int. J. Cancer* 142 (3): 561–572. https://doi.org/10.1002/ijc.31067.

Lazic, S.E., Clarke-Williams, C.J., and Munafo, M.R. (2018). What exactly is 'N' in cell culture and animal experiments? *PLoS Biol.* 16 (4): e2005282. https://doi.org/10.1371/journal.pbio.2005282.

Macleod, M.R., Michie, S., Roberts, I. et al. (2014). Biomedical research: increasing value, reducing waste. *Lancet* 383 (9912): 101–104. https://doi.org/10.1016/S0140-6736(13)62329-6.

Maley, C.C., Aktipis, A., Graham, T.A. et al. (2017). Classifying the evolutionary and ecological features of neoplasms. *Nat. Rev. Cancer* 17 (10): 605–619. https://doi.org/10.1038/nrc.2017.69.

Moher, D., Glasziou, P., Chalmers, I. et al. (2016). Increasing value and reducing waste in biomedical research: who's listening? *Lancet* 387 (10027): 1573–1586. https://doi.org/10.1016/S0140-6736(15)00307-4.

Nardone, R.M. (2007). Eradication of cross-contaminated cell lines: a call for action. *Cell Biol. Toxicol.* 23 (6): 367–372. https://doi.org/10.1007/s10565-007-9019-9.

National Science Foundation (2015). Social, behavioral, and economic sciences perspective on robust and reliable science. https://www.nsf.gov/sbe/ac_materials/sbe_robust_and_reliable_research_report.pdf (accessed 26 June 2018).

Nature Editors (2015). Announcement: time to tackle cells' mistaken identity. *Nature* 520: 264. https://doi.org/10.1038/520264a.

Nature Editors (2018). Checklists work to improve science. *Nature* 556: 273–274. https://doi.org/10.1038/d41586-018-04590-7.

Neimark, J. (2015). Line of attack. *Science* 347 (6225): 938–940. https://doi.org/10.1126/science.347.6225.938.

Niepel, M., Hafner, M., Mills, C.E. et al. (2019). A multi-center study on the reproducibility of drug-response assays in mammalian cell lines. *Cell Syst.* https://doi.org/10.1016/j.cels.2019.06.005.

NIH (2015). NOT-OD-16-011: implementing rigor and transparency in NIH and AHRQ research grant applications. https://grants.nih.gov/grants/guide/notice-files/not-od-16-011.html (accessed 4 July 2018).

NIH (2017). Principles and guidelines for reporting preclinical research. https://www.nih.gov/research-training/rigor-reproducibility/principles-guidelines-reporting-preclinical-research (accessed 4 July 2018).

OECD (1998). OECD principles on Good Laboratory Practice (as revised in 1997). http://www.oecd.org/officialdocuments/publicdisplaydocumentpdf/?cote=env/mc/chem(98)17&doclanguage=en (accessed 4 July 2018).

OECD (2018a). *Guidance Document on Good in Vitro Method Practices (GIVIMP)*. Paris: OECD Publishing.

OECD (2018b). OECD series on principles of Good Laboratory Practice (GLP) and compliance monitoring. http://www.oecd.org/chemicalsafety/testing/oecdseriesonprinciplesofgoodlaboratorypracticeglpandcompliancemonitoring.htm (accessed 4 July 2018).

Olivieri, A., Degenhardt, O.S., McDonald, G.R. et al. (2012). On the disruption of biochemical and biological assays by chemicals leaching from disposable laboratory plasticware. *Can. J. Physiol. Pharmacol.* 90 (6): 697–703. https://doi.org/10.1139/y2012-049.

Pamies, D., Bal-Price, A., Simeonov, A. et al. (2017). Good Cell Culture Practice for stem cells and stem-cell-derived models. *ALTEX* 34 (1): 95–132. https://doi.org/10.14573/altex.1607121.

Pamies, D., Bal-Price, A., Chesne, C. et al. (2018). Advanced Good Cell Culture Practice for human primary, stem cell-derived and organoid models as well as microphysiological systems. *ALTEX* 35 (3): 353–378. https://doi.org/10.14573/altex.1710081.

Plant, A.L., Locascio, L.E., May, W.E. et al. (2014). Improved reproducibility by assuring confidence in measurements in biomedical research. *Nat. Methods* 11 (9): 895–898. https://doi.org/10.1038/nmeth.3076.

Plant, A.L., Becker, C.A., Hanisch, R.J. et al. (2018). How measurement science can improve confidence in research results. *PLoS Biol.* 16 (4): e2004299. https://doi.org/10.1371/journal.pbio.2004299.

Prinz, F., Schlange, T., and Asadullah, K. (2011). Believe it or not: how much can we rely on published data on potential drug targets? *Nat. Rev. Drug Discov.* 10 (9): 712. https://doi.org/10.1038/nrd3439-c1.

Rayment, E.A. and Williams, D.J. (2010). Concise review: mind the gap: challenges in characterizing and quantifying cell- and tissue-based therapies for clinical translation. *Stem Cells* 28 (5): 996–1004. https://doi.org/10.1002/stem.416.

Sarntivijai, S., Ade, A.S., Athey, B.D. et al. (2008). A bioinformatics analysis of the cell line nomenclature. *Bioinformatics* 24 (23): 2760–2766. https://doi.org/10.1093/bioinformatics/btn502.

Stacey, G. (2017). Stem cell banking: a global view. *Methods Mol. Biol.* 1590: 3–10. https://doi.org/10.1007/978-1-4939-6921-0_1.

Stanley, T.D., Doucouliagos, H., and Ioannidis, J.P. (2017). Finding the power to reduce publication bias. *Stat. Med.* 36 (10): 1580–1598. https://doi.org/10.1002/sim.7228.

Teytelman, L., Stoliartchouk, A., Kindler, L. et al. (2016). Protocols.io: virtual communities for protocol development and discussion. *PLoS Biol.* 14 (8): e1002538. https://doi.org/10.1371/journal.pbio.1002538.

Vasilevsky, N.A., Brush, M.H., Paddock, H. et al. (2013). On the reproducibility of science: unique identification of research resources in the biomedical literature. *PeerJ* 1: e148. https://doi.org/10.7717/peerj.148.

WHO (2009). Handbook: Good Laboratory Practice (GLP): Quality Practices for Regulated Non-Clinical Research and Development. http://www.who.int/tdr/publications/documents/glp-handbook.pdf (accessed 4 July 2018).

WHO (2016). Good Manufacturing Practices. http://www.who.int/biologicals/vaccines/good_manufacturing_practice/en (accessed 9 July 2018).

Wood, J., Mahajan, E., and Shiratori, M. (2013). Strategy for selecting disposable bags for cell culture media applications based on a root-cause investigation. *Biotechnol. Progr.* 29 (6): 1535–1549. https://doi.org/10.1002/btpr.1802.

PART III

Medium and Substrate Requirements

After reading the following chapters in this part of the book, you will be able to:

(8) **Culture Vessels and Substrate:**

(a) List examples of materials that are used as culture substrates and describe how these materials can be altered to improve their performance.

(b) Apply coatings of extracellular matrix (ECM) to culture vessels (with additional reading from Chapter 12, which discusses aseptic technique).

(c) Explain why feeder layers are used and discuss their advantages and limitations.

(d) List examples of commonly used culture vessels and discuss why each example might be chosen for handling cultures, e.g., with reference to cell yield and cost.

(e) Discuss why imaging, suspension culture, and 3D culture may require specialized vessels and list examples of products that are used for these applications.

(9) **Defined Media and Supplements:**

(a) List examples of balanced salt solutions and discuss how they are used.

(b) Explain why a buffer is included in culture medium and discuss the relationship between bicarbonate (HCO_3^-) and CO_2 levels.

(c) Estimate the pH of culture medium, using phenol red as a pH indicator.

(d) Explain the difference between "basal" and "complete" media and provide examples of some basal media formulations.

(e) Explain why antibiotics should not be used for routine handling and comment on exceptional circumstances in which their use is justified.

(10) **Serum-Free Media:**

(a) Define "chemically defined medium," "animal product-free medium," and "xeno-free medium."

(b) Explain why culture in serum-free medium is desirable and list some of the advantages and disadvantages associated with this approach.

(c) Describe how to choose a serum-free medium for a particular cell line or cell type.

(d) Discuss how to perform adaptation when a cell line is converted from serum-containing to serum-free media.

(e) Explain how to "build" a new serum-free medium by adding supplements.

(11) **Preparation and Sterilization:**

(a) Define "disinfection," "decontamination," and "sterilization."

(b) Prepare and sterilize glassware, caps and closures, and other apparatus.

Freshney's Culture of Animal Cells: A Manual of Basic Technique and Specialized Applications, Eighth Edition. Amanda Capes-Davis and R. Ian Freshney.
© 2021 John Wiley & Sons Ltd. Published 2021 by John Wiley & Sons Ltd.
Companion website: www.wiley.com/go/freshney/cellculture8

(c) Explain the various grades of water purity and list examples of their uses.

(d) Prepare and sterilize ultrapure water (UPW; also known as Type 1 water), balanced salt solutions, and other heat-stable liquids by autoclaving.

(e) Prepare and sterilize culture media by sterile filtration (with additional reading from Chapter 12, which discusses aseptic technique).

(f) Test media samples for sterility and describe how media may be tested for performance if required prior to use.

CHAPTER 8

Culture Vessels and Substrates

Tissue culture was initially performed using coverslips (Harrison 1907). Glass coverslips were used for ease of observation and as a suitable substrate for hanging drop culture – a three-dimensional (3D) technique that was used in the earliest tissue culture experiments (see Section 27.4.1). However, tissue culture became cumbersome when cells were maintained using coverslips for extended periods of time, resulting in the design of new vessels specifically for tissue culture. One popular design was the Carrel D flask, which included a flat surface for microscopic observation and an opening with an angled neck for aseptic handling (see Figure 1.1b) (Carrel 1923). Carrel's original design was for a flat, round flask, but this format was not conducive to later techniques, where cells were grown on pieces of cellophane that were cut to shape (Evans et al. 1947). Wilton Earle and colleagues modified the shape of the flask to give a rectangular surface for cell culture (see Figure 1.1c) (Earle and Highhouse 1954). Earle referred to the new design as a "T-flask" because it was hand-blown from Pyrex glass tubing. Earle's rectangular T-flask has since become the most widely used design for tissue culture. However, the broad use of tissue culture in many different fields has resulted in a dizzying array of culture vessels and substrates. This chapter discusses the commonly used culture substrates and considers how to choose a culture vessel for your work.

8.1 ATTACHMENT AND GROWTH REQUIRE-MENTS

The earliest continuous cell lines grew successfully in tissue culture flasks and other vessels such as roller tubes (see Section 28.3.1) (Earle et al. 1943; Scherer et al. 1953). Following this early success, the majority of vertebrate cells were cultured as monolayers on glass and other artificial substrates. However, glass was a poor substrate for many cultures. Other materials were tested as substrates for tissue culture, including bacteriological grade plastic (polystyrene) dishes. Initial attempts were unsuccessful until the polystyrene was treated to improve adhesion (Curtis et al. 1983), resulting in our modern "tissue culture grade" plasticware.

Substrate treatments are usually designed to alter the surface charge or mimic the properties of the extracellular matrix (ECM). Cells grow *in vivo* in a close relationship with the ECM, which provides the architecture and external signals required for adhesion, growth, migration, and differentiation (see Section 2.2.4). It therefore follows that cells will attach and grow better *in vitro* on a similar substrate (see Section 8.3). Many substrate coatings and treatments rely on the presence of the tripeptide arginyl-glycyl-aspartic acid (RGD), which is a common motif in many ECM proteins. The RGD motif improves cell adhesion by binding to many integrins and possibly by altering surface charge (see Section 2.2.1) (Villard et al. 2006; Widhe et al. 2013).

The ability of a culture to attach and grow on a flat surface depends on the cells that are present as well as the substrate. Most normal cells need to spread out on a substrate to proliferate, and inadequate spreading due to poor adhesion or overcrowding will inhibit proliferation (Folkman et al. 1979; Ireland et al. 1989; Danen and Yamada 2001; Shen et al. 2012). Spontaneous growth in suspension is restricted to hematopoietic cell lines, rodent ascites tumors, and a few other selected cell lines, such as human small cell lung cancer

Freshney's Culture of Animal Cells: A Manual of Basic Technique and Specialized Applications, Eighth Edition. Amanda Capes-Davis and R. Ian Freshney.
© 2021 John Wiley & Sons Ltd. Published 2021 by John Wiley & Sons Ltd.
Companion website: www.wiley.com/go/freshney/cellculture8

(Carney et al. 1981). However, many transformed cell lines can be made to grow in suspension and become independent of the surface charge on the substrate. Cells shown to require attachment for growth are said to be anchorage dependent; cultures that have undergone transformation frequently become anchorage independent and can be propagated in suspension (see Sections 3.6.2, 14.7).

Growing cells on flat surfaces changes their appearance and behavior (Lovitt et al. 2014). Culture in three dimensions results in more physiological behavior; this was demonstrated as early as 1951, when tumor cells were grown on sponge matrix to form organized aggregates with structural features that resembled the original tissue (Leighton 1951). Cells can be grown in three dimensions using scaffold-based systems or can be grown as aggregates in suspension (see Sections 8.6.3–8.6.5). Cells may also require feeder layers or specific coatings (see Sections 8.3.2, 8.4); this is particularly important for stem cell culture. Stem cell and 3D culture are discussed in greater detail later in this book (see Chapters 23, 27).

8.2 SUBSTRATE MATERIALS

8.2.1 Common Substrate Materials

Plastic (polystyrene). Single-use sterile polystyrene flasks, Petri dishes, or multiwell plates provide a simple, reproducible substrate for culture. They are usually of good optical quality, and the growth surface is flat, providing uniformly distributed and reproducible monolayer cultures. As manufactured, polystyrene is hydrophobic and does not provide a suitable surface for cell attachment. The substrate must be correctly charged to allow cell adhesion, or at least to allow the adhesion of cell-derived attachment factors, which will, in turn, allow cell adhesion and spreading. Tissue culture plastics are therefore treated by corona discharge, γ-irradiation, or chemical agents to produce a charged, wettable surface. Treated polystyrene is usually referred to as "tissue culture grade" plasticware. Because the resulting product can vary in quality and cultures may vary in their sensitivity to substrate changes, you should always perform your own testing before you commit to new plasticware (see Section 7.5.3).

Glass. The move toward disposable plasticware (see Section 5.1.2) means that glass is no longer routinely used for tissue culture, but it continues to be used for some applications, e.g. glass slides or coverslips are commonly used for live cell imaging (Phelps et al. 2017). Glass is optically clear and carries a surface charge that makes it suitable for many cell types (see Section 2.2). It is also cheap, easily washed without losing its growth-supporting properties, and can be sterilized readily by dry or moist heat. However, not all cell types grow well on glass and culture behavior may vary with the type of glass used and the washing process. High optical quality glass is alkaline and often has a high lead content, which may be detrimental to cell adhesion and growth. An acid wash can be used to improve adhesion and ensures that the substrate is clean for experimental work, e.g. when preparing metaphase spreads using glass slides (see Protocol P17.2) (MacLeod et al. 2017). Treatment with strong alkali (e.g. NaOH or caustic detergents) during washing may render glass unsatisfactory for culture; this can also be remedied using an acid wash (Earle 1943). A number of other methods can be used to improve adhesion, such as an ethylene diamine tetra-acetic acid (EDTA) treatment followed by a sodium carbonate wash (Rappaport and Bishop 1960). However, in most modern tissue culture laboratories, poor adhesion and spreading on glass surfaces are usually addressed by adding coatings (see Section 8.3.2).

8.2.2 Alternative Substrate Materials

Polymers. Polystyrene is by far the most common and cheapest plastic substrate. However, cells can be grown on a variety of plastics; other polymers are becoming increasingly popular because of their use in microfluidic culture (see Section 21.5) (Tsao 2016). A large number of polymers have been used for tissue culture, including polycarbonate (PC) (Karamichos et al. 2014), polyethylene (PE) (Jurkiewicz et al. 2014), polyethylene glycol (PEG) (Cui et al. 2017), polyethylene terephthalate (PET) (Huang et al. 2013), polytetrafluorethylene (PTFE) (Krishna et al. 2011), polymethyl methacrylate (PMMA; Plexiglas, Perspex, Lucite) (Gottwald et al. 2008; Alvarez et al. 2014), and polyvinylchloride (PVC) (Gabriel et al. 2012).

Polydimethylsiloxane (PDMS) is increasingly used for the fabrication of microfluidic devices, including organ-on-chip devices (see Section 29.5.2) (Sia and Whitesides 2003). PDMS is non-toxic to cells, permeable to gases, and impermeable to water. It is also relatively cheap (compared to silicon and glass) and is readily fabricated. PDMS is easily pourable and can be cured by exposure to heat (originally using the flame of a Bunsen burner) (Harris et al. 1980). Thus, it can be poured into a mold for casting using equipment that is readily accessible to research laboratories, e.g. using an oven at 65 °C for 24 hours (Friend and Yeo 2010). Cells do not adhere well to PDMS, but the surface can be modified to improve adhesion (Wang et al. 2010). More specialized facilities can add layer-by-layer deposits to generate polyelectrolyte multilayer (PEM) film coatings, which can be used to control surface properties such as charge, wettability, stiffness, and roughness (Guo et al. 2017).

Fibers. Rayon, nylon, polylactic acid (PLA), polyglycolic acid (PGA), and silk are often used for 3D constructs in tissue engineering. These applications are discussed in a later chapter (see Minireview M27.1). Silk is particularly interesting because it is biocompatible and biodegradable, and can self-assemble to form solid structures and 3D matrices (Widhe et al. 2013; Jastrzebska et al. 2015). Natural silk can be genetically modified to improve cell adhesion, resulting in exciting new 3D substrates for clinical applications (Deptuch and Dams-Kozlowska 2017). Other fibers can be modified to improve their suitability as culture substrates, such as paper (Lantigua et al. 2017).

Metals. Cells may be grown on stainless-steel disks (Birnie and Simons 1967) or other metallic surfaces such as titanium (Bledsoe et al. 2004). Observation of the cells on an opaque substrate usually requires surface interference microscopy, unless very thin metallic films are used. Bengt Westermark developed a method for the growth of fibroblasts and glia on palladium (Westermark 1978). Using electron microscopy shadowing equipment, he produced islands of palladium on agarose, which does not allow cell attachment in fluid media. The size and shape of the islands were determined by masks made by photoetching (see Section 8.3.7),

and the palladium was applied by shadowing under vacuum, as used in electron microscopy. Because the layer was very thin, it remained transparent.

8.3 SUBSTRATE TREATMENTS

Substrates can be treated to alter cell adhesion, growth, and differentiation. The need for substrate treatments will vary with the culture. For example, LNCaP cells may adhere poorly to polystyrene and depend on the presence of ECM to develop a tumorigenic phenotype (Russell and Kingsley 2003; Liberio et al. 2014). Culture of LNCaP cells on fibronectin, poly-L-lysine, poly-L-ornithine, or laminin results in a fibroblast-like morphology, while collagen IV results in cell aggregation (see Figure 8.1). The choice of

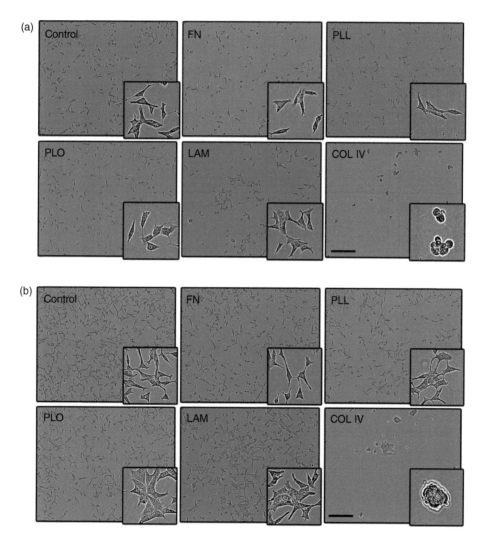

Fig. 8.1. LNCaP morphology on coated substrates. LNCaP cells grown on polystyrene (control), fibronectin (FN), poly-L-lysine (PLL), poly-L-ornithine (PLO), laminin (LAM), and collagen IV (COL IV). Images were taken at (a) 24 hours and (b) 96 hours after plating using the IncuCyte® system (Sartorius; see Figure 18.4). Scale bar represents 300 μm. *Source*: Liberio et al. (2014), https://doi.org/10.1371/journal.pone.0112122, licensed under CC BY 4.0.

surface treatment will also vary with your application; for example, LNCaP cells can be grown on laminin to improve performance in wound healing assays (Liberio et al. 2014). Finally, the choice of surface treatment may depend on its source. If you require xeno-free culture conditions (see Section 10.7), refer to the manufacturer's information to see if coatings or treatments are animal-derived and look for recombinant components or other synthetic products.

8.3.1 Substrate Conditioning

A well-established piece of tissue culture lore has it that used glassware supports growth better than new (Paul 1975). If that is true, it may be due to a change in surface charge, surface etching, or minute traces of residue left after culture. The growth of cells in a flask can be used to improve the surface for a second seeding. This approach has been used to grow endothelium on ECM derived from confluent monolayers of 3T3 cells, which were then removed with Triton™ X-100 (Gospodarowicz et al. 1980). Residual ECM has been used to promote differentiation in ovarian granulosa cells, to study tumor cell behavior, and for vascular morphogenesis (Vlodavsky et al. 1980; Du et al. 2014). The substrate can also be conditioned by pretreating it with spent medium from another culture or with serum (Stampfer et al. 1980; Hakala et al. 2009).

8.3.2 Extracellular Matrix (ECM) Coatings

A wide range of ECM products are available to coat glass or plastic surfaces (see Table 8.1). Some ECM products are partially purified, if not completely chemically defined; others are a mixture of matrix products that have been poorly characterized and may also contain bound growth factors. Undefined natural products may be the only route toward getting a culture established and may be appropriate for some applications. However, if ECM is required, the long-term objective should be to employ defined components. Collagen, fibronectin, laminin, and vitronectin are all commonly used as substrate coatings. Some of these ECM constituents are available as recombinant proteins and are therefore well defined and purified (e.g. recombinant laminin-521). ECM products are available precoated onto plates, dishes, and other substrates (BioCoat™ Cultureware, Corning; Millicoat, EMD Millipore). They may also be purchased in bottled form for the user to customize coatings for specific cell types, such as normal human bronchial epithelial (NHBE) cells, which are traditionally grown on a mixture of fibronectin, collagen, and bovine serum albumin (BSA) (see Supp. S24.4). In some cases, products can be added to the culture medium (Biological Industries 2018).

The Engelbreth-Holm-Swarm (EHS) mouse sarcoma is the most common source of naturally derived ECM products, with Matrigel® being the best-known example (Kleinman and Martin 2005). Matrigel consists of approximately 60%

laminin, 30% collagen IV, and 8% entactin; the latter is a bridging molecule that contributes to ECM structure (Corning 2013). Matrigel contains a number of factors that promote outgrowth of specific cell types from tissue explants, proliferation, differentiation, and angiogenesis (Corning 2013). Matrigel can be added by the user or obtained precoated on tissue culture plates, dishes, or filter well inserts (BioCoat Cultureware, Corning). Protocol P8.1 describes how to coat tissue culture plasticware using Matrigel.

PROTOCOL P8.1. APPLICATION OF MATRIGEL COATINGS

Contributed by Amanda Capes-Davis. This protocol has been adapted from Corning (2013).

Outline

Thaw Matrigel bottle and aliquot using precooled pipette tips and plasticware.

Materials (sterile)

☐ Matrigel matrix (Corning)
☐ Substrate for coating, e.g. dishes, multiwell plates
☐ Culture medium
☐ Pipette tips and plasticware (e.g. dishes or multiwell plates), prechilled
☐ Ice bath

Materials (non-sterile)

☐ Biological safety cabinet (BSC) and associated consumables (see Protocol P12.1)

Procedure

1. Place Matrigel bottle at 4 °C on ice to thaw slowly overnight.
2. Place pipette tips and plasticware at –20 °C to prechill.
3. Place thawed Matrigel, pipette tips, and plasticware on ice and keep cool throughout.
4. Prepare a work space for aseptic technique using a BSC.
5. Mix the Matrigel by gently swirling the bottle and dilute as per the recommended dilution factor (see Certificate of Analysis).
6. Aliquot diluted Matrigel onto dishes or multiwell plates. Depth will vary with your application:
   (a) For thin coat (cells grow on top): add 50 μl/cm² growth area.
   (b) For thick coat (cells grow within the layer): add 150–200 μl/cm² growth area.

7. Leave at room temperature for at least 1 hour before using (some laboratories prefer to incubate at 37 °C for 30 min). Unbound material can be gently aspirated just before use.
8. Coated dishes or plates may be stored (sealed) for up to seven days.

Notes

Step 1: Matrigel can take longer to thaw if the protein concentration is high and will start to form a gel at above 10 °C. Corning recommends that Matrigel is stored in a frost-free freezer; repeated freeze-thaw cycles can make it "clumpier."
Step 4: Aseptic technique is described in detail in a later chapter (see Sections 12.2–12.4).
Step 6: For information on the growth area of various plasticware see Section 8.5.

8.3.3 Collagen and Gelatin

Collagen has been used in tissue culture since the 1950s and is important for a number of *in vitro* model systems (Ehrmann and Gey 1956; Plant et al. 2009). Collagen may be used in its native form as an undenatured gel; this form may be required for expression of differentiated functions (see Sections 2.2.4, 24.2.5, 26.3.3). For example, undenatured collagen has been shown to support neurite outgrowth (Xie et al. 2013) and differentiation of breast cells (Nicosia and Ottinetti 1990; Berdichevsky et al. 1992) and hepatocytes (Sattler et al. 1978). Diluting the concentrated collagen 1:10 with culture medium and neutralizing to pH 7.4 causes the collagen to gel, so dilution and dispensing must be rapid. It is best to add growth medium to the collagen gel 4–24 hours before adding cells, to ensure that the gel equilibrates with the medium. At this stage, fibronectin (25–50 µg/ml), laminin (1–5 µg/ml), or both may also be added to the medium.

Collagen may be used in denatured form (gelatin). Heat denaturation exposes RGD motifs in collagen, which are important for adhesion (see Section 8.1) (Plant et al. 2009). Gelatin improves the attachment of many cell types, such as epithelial cells (Sarang et al. 2003; Patel et al. 2014) and is beneficial for the culture of muscle and endothelial cells (Richler and Yaffe 1970; Folkman et al. 1979). Coating with denatured collagen may be achieved by using rat tail collagen or commercially supplied alternatives (see Table 8.1). Simply pour the collagen solution over the surface of the dish, drain off the excess, and allow the residue to dry. Because this procedure sometimes leads to detachment of the collagen layer during culture, a protocol has been devised to ensure that the collagen remains firmly anchored by cross-linking to the plastic with carbodiimide (Macklis et al. 1985).

Collagen and gelatin, like silk (see Section 8.2.2), are biocompatible and biodegradable and thus have considerable potential for 3D culture and clinical applications. Collagen and gelatin have been used to produce thin films, 3D matrices, and tunable hydrogels for tissue engineering (Elsdale and Bard 1972; Davidenko et al. 2016; DeWitt and Rylander 2018). These 3D applications are examined in more detail later in this book (see Chapter 27).

8.3.4 ECM Mimetic Treatments

Substrates that are not naturally adhesive can be derivatized by covalent attachment of an RGD tripeptide (see Section 8.1) (Brandley et al. 1987; Gabriel et al. 2012). RGD is commonly synthesized as the pentapeptide glycine-arginine-glycine-aspartic acid-serine (GRGDS). Synthetic peptides may also contain flanking sequences from ECM components such as fibronectin and vitronectin (e.g. PureCoat™ ECM Mimetic and Synthemax® Surface, Corning). 1-Ethyl-3-(3-dimethylaminopropyl) carbodiimide hydrochloride (EDC) and N-hydroxysulfosuccinimide (NHS) have been used to derivatize processed silk to construct scaffolds for bone tissue engineering and this treatment is potentially applicable to a number of different substrates (Hofmann et al. 2006; Lao et al. 2008).

8.3.5 Polymer Coatings

Polymer coatings are typically used to alter the surface charge of the culture vessel (e.g. PureCoat Cultureware, Corning). The most commonly used polymer coating is polylysine, which improves adhesion of many different cell types to solid surfaces, most likely through its positive surface charge (see Section 2.2) (Mazia et al. 1975). The synthetic poly-D-lysine isomer is often used in preference to poly-L-lysine, as it is less readily digested by extracellular proteases, but both the D- and L-isomers have been used. Poly-D-lysine was initially used as a coating to stimulate clonal growth of normal fibroblasts in serum-free medium; the same technique can be used to promote neurite outgrowth (see Sections 20.6.3, 24.4.1). Higher molecular weights become more viscous to handle but have more binding sites; both 100 kD (MP Biomedicals) and 500 kD (Santa Cruz Biotechnology) are available.

Polylysine coatings are usually quite easy to prepare (McKeehan and Ham 1976):

(1) Make up 0.1 mg/ml of poly-D-lysine in ultrapure water (UPW) (see Section 11.5.1).
(2) Using aseptic technique (see Protocol P12.1), add poly-D-lysine to the culture vessel (~ 5 ml/25 cm²). Leave at room temperature for five minutes.
(3) Remove excess coating solution and wash the plates with Dulbecco's phosphate buffered saline without Ca²⁺ and Mg²⁺ (DPBS-A, ~ 5 ml/25 cm²). Remove the wash.
(4) Plates may be used immediately or stored for several weeks before use.

TABLE 8.1. Extracellular matrix (ECM) materials.

Product	Composition	Source	Supplier[a]
BioCoat Cultureware	Various ECM components including collagens, fibronectin, gelatin, and laminin; precoated plates, inserts, and dishes	Various	Corning
CELLstart™	Various ECM and serum proteins	Human origin	ThermoFisher Scientific
Cell-Tak™	Polyphenolic proteins	*Mytilus edulis*	Corning
Collagens (various)	Collagen I, II, III, IV	Human, bovine, murine, rat tail	Advanced BioMatrix, Corning, Sigma-Aldrich, STEMCELL Technologies, ThermoFisher Scientific
Cultrex® BME	Laminin, fibronectin, collagen IV, proteoglycans, growth factors (growth factor depleted available)	EHS sarcoma	Amsbio, Sigma-Aldrich, Trevigen
ECL Cell Attachment Matrix	Entactin, collagen IV, laminin	EHS sarcoma	Merck Millipore, Sigma-Aldrich
EHS Matrix Extract	Laminin, collagen IV, growth factors	EHS sarcoma	Sigma-Aldrich
Fibronectin	Attachment protein from ECM	Natural	Advanced BioMatrix, Biological Industries, Corning, Sigma-Aldrich, STEMCELL Technologies, ThermoFisher Scientific
Fibronectin	Attachment protein from ECM	Recombinant	Sigma-Aldrich, Takara Bio
Gelatin	Denatured collagen	Natural	Corning, STEMCELL Technologies, Sigma-Aldrich, ThermoFisher Scientific
Geltrex™ Basement Membrane Matrix	Laminin, collagen IV, entactin, heparin sulfate proteoglycan (growth factor depleted available)	EHS sarcoma	ThermoFisher Scientific
Heparan sulfate	Matrix proteoglycan	Natural	Sigma-Aldrich
Laminin	Attachment protein from basement membrane	Natural	Corning, Sigma-Aldrich, ThermoFisher Scientific
Laminin	Attachment protein from basement membrane, e.g. laminin-521	Recombinant	Amsbio, BioLamina, Biological Industries, STEMCELL Technologies
Matrigel	Laminin, fibronectin, collagen IV, proteoglycans, growth factors (growth factor depleted available)	EHS sarcoma	Corning
MaxGel™ ECM	Human collagens, laminin, fibronectin, tenascin, elastin, proteoglycans, glycosaminoglycans	Natural (cell culture derived)	Sigma-Aldrich
Millicoat™	Various ECM components including collagens, fibronectin, and laminin; precoated plates and microplate strips	Various	Merck Millipore
Osteopontin	Glycoprotein containing RGD motif	Natural	Corning
ProNectin® F (SmartPlastic®)	Fibronectin-like protein polymer with multiple copies of the RGD-containing epitope	Recombinant	MP Biomedicals, Sigma-Aldrich
Tropoelastin	Water-soluble precursor to elastin	Natural	Advanced BioMatrix, STEMCELL Technologies
Vitronectin	Attachment protein from ECM	Natural	Advanced BioMatrix, Corning, Merck Millipore, Sigma-Aldrich, STEMCELL Technologies, ThermoFisher Scientific
Vitronectin	Attachment protein from ECM	Recombinant	Advanced BioMatrix, Biological Industries, Merck Millipore, STEMCELL Technologies, Sigma-Aldrich, ThermoFisher Scientific

[a]Suppliers are drawn from product searches and may vary with your location.

8.3.6 Non-Adhesive Substrates and Patterning

Sometimes cell attachment is undesirable. Initially, non-adhesive substrates were used for selection of virally transformed colonies, which are anchorage independent (Macpherson and Montagnier 1964). Selection was achieved by plating cells in agar, as untransformed cells do not form colonies readily in agar (see Protocol P20.2). There are three principles involved in such a system: (i) prevention of attachment at the base of the dish, where spreading and anchorage-dependent growth would occur; (ii) immobilization of the cells such that daughter cells remain associated with the colony, even if they are non-adhesive; and (iii) the potential for alterations in gene expression in suspension cells (Feng et al. 2003). Soft agar colony forming assays continue to be used to measure the clonogenicity of tumor stem cells (Sant and Johnston 2017). Methocel (methylcellulose of viscosity 4000 cP) is also used. Because Methocel is a high-viscosity sol, cells will sediment slowly through it, so it is commonly used with an underlay of agar (see Protocol P20.3). Untreated polystyrene (see Section 8.2.1) can be used without an agar underlay, but some attachment and spreading may occur.

Non-adhesive substrates can be used in combination with adhesive substrates to produce striking patterns (Kleinfeld et al. 1988). Techniques for patterning of polystyrene include photoetching with ultraviolet (UV) irradiation (Welle et al. 2014), hot embossing (Brown et al. 2013), and use of coatings such as poly-L-lysine and myelin (Belkaid et al. 2013). Patterned cultures have a number of applications, including tissue engineering and high-throughput screening (see Sections 27.7.2, 28.6). As 3D bioprinting becomes more widely used, print-based patterning techniques will become more achievable for tissue culture laboratories (see Section 28.7.3) (Roth et al. 2004; Yamaguchi et al. 2012).

8.3.7 Other Surface Treatments

Several other surface treatments are available as commercial products, including:

(1) **CellBIND®.** This Corning product consists of polystyrene that has been treated using a patented plasma surface treatment. The process results in increased oxygen-containing functional groups, resulting in a more hydrophilic and wettable surface. This surface is used to improve cell attachment in roller bottles (see Section 28.3.1) and in other tissue culture plasticware. It also appears to improve ECM adsorption (Teare et al. 2001; van Kooten et al. 2004).

(2) **Primaria™.** This Corning product (applied to Falcon plasticware) consists of polystyrene that has been treated using a vacuum gas plasma process. The gases used contain both oxygen and ammonia, resulting in the presence of oxygen- and nitrogen-containing functional groups. The more common treatment used for tissue culture plasticware, corona discharge (see Section 8.2.1), does not produce such a high level of nitrogen-containing functional groups. This surface treatment is beneficial for some cell types including chondrocytes, mesenchymal stromal cells (MSCs), and ovarian carcinoma (see Section 25.3.1) (Petit et al. 2011; Rampersad et al. 2011).

(3) **Nunc UpCell™.** This ThermoFisher Scientific product consists of polystyrene with an additional layer of a temperature-responsive polymer. Reducing the temperature to 20 °C leads to cell detachment without the need for trypsinization. This concept is particularly useful for cell sheet-based engineering (Kobayashi and Okano 2010).

8.4 FEEDER LAYERS

Substrate treatments help to improve cell attachment, growth, and differentiation, but this may not be sufficient for some fastidious cells, particularly at low densities. The survival of low-density cultures and specific cell types can be improved by growing cells in the presence of feeder layers (Puck and Marcus 1955). Feeder cells are treated to induce growth arrest but provide a supportive *in vitro* environment for the culture of other cell types. A confluent feeder layer can prevent overgrowth from fibroblasts or other rapidly growing cells that are present alongside the cell type of interest (Pourreyron et al. 2011). Feeder layers are used for selection and long-term expansion of specific cell types, such as keratinocytes and tumor cells (see Sections 24.2.1, 25.4.2). Feeder layers are also used for conditional reprogramming, which is a state of unlimited cell proliferation that can be induced by specific culture conditions (see Section 22.3.4).

Mouse embryonic fibroblasts (MEFs) have been used as feeder cells for epidermal keratinocytes since the 1970s (Rheinwald and Green 1975). Howard Green and colleagues used 3T3 feeder layers to study long-term expansion of keratinocytes in culture, which relies on the presence of stem cells within the population (Barrandon and Green 1987). This early work led to the first stem cell therapy using cultured cells (Hynds et al. 2018). MEF feeder layers continue to be widely used for the culture of embryonic stem cells (ESCs) and induced pluripotent stem cells (iPSCs) (Llames et al. 2015). Initially, clones and feeder layer were kept separate and thus the effect was due to soluble factors. Most cultures are now grown directly on the feeder layer, resulting in a likely combination of medium conditioning (e.g. by metabolites and growth factors) and substrate conditioning (e.g. by collagen production) (Hauschka and Konigsberg 1966; Takahashi and Okada 1970). Cells grown on feeder layers often display changes in morphology; cells may be less well spread, show denser staining, and look more highly differentiated (see Figure 8.2; see also Figures 18.2c, 20.1d).

Other species and cell types can be used as feeder layers. The survival and extension of neurites by central and peripheral neurons can be enhanced by culturing the neurons on a

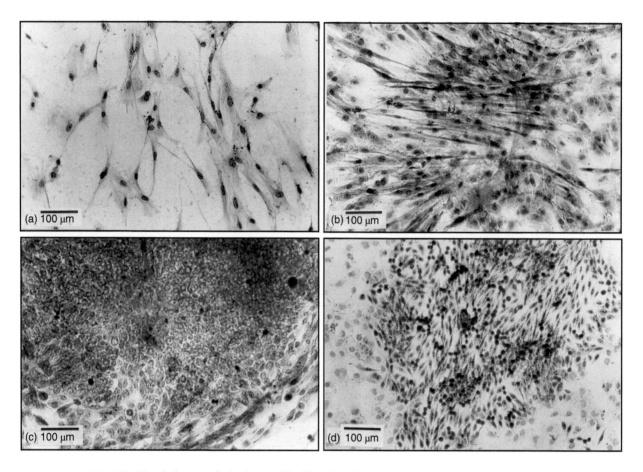

Fig. 8.2. Morphology on feeder layers. Fibroblasts from human breast carcinoma are grown on (a) plastic and (b) a confluent feeder layer of fetal human intestinal (FHI) cells. Epithelial cells from human breast carcinoma are grown on (c) plastic and (d) the same confluent feeder layer as in (b). *Source*: R. Ian Freshney.

monolayer of glial cells, although in this case the effect may be due to a diffusible factor rather than direct cell contact (Corbett et al. 2013; Xu et al. 2013; Ganz et al. 2014). Human feeder cells can be used for xeno-free culture (see Section 10.7), including fibroblasts (e.g. foreskin fibroblasts; see Supp. S24.1), amniocytes, and bone marrow-derived MSCs. Feeder-free substrates can also be used for ESC and iPSC culture (e.g. ECM components and artificial substrates) (Joddar and Ito 2013; Desai et al. 2015). However, feeder-based systems tend to be less expensive for long-term use and continue to be used by many laboratories.

How do you develop a feeder layer for your culture? MEFs can be isolated directly from mouse embryos (see Protocol P13.1) or purchased from suppliers (e.g. MEF (C57BL/6), ATCC SCRC-1008; STO, ECACC 86032003). A specific cell line or subclone may be used for some cell types – for example, the Swiss-3T3 J2 subclone is preferred for human epidermal keratinocytes (Pourreyron et al. 2011; Hynds et al. 2018). Feeder layers can vary in quality; it is important to familiarize yourself with the appearance of a "good-quality" feeder layer and determine the passage number and cell

density that are best for your application (Healy and Ruban 2015). Protocols for preparing feeder layers are provided in later chapters (see Protocols P20.7, P25.2).

8.5 CHOICE OF CULTURE VESSEL

Disposable plastic flasks, dishes, and multiwell plates are available from a wide range of suppliers (e.g. see Suppliers list at the end of this chapter). The common features of these culture vessels are listed in Table 8.2. The choice of culture vessel will depend on the purpose of the culture. If culture is for the production of cells or product, the major determinants will be (i) the cell yield required; (ii) whether cells grow in suspension or as a monolayer; (iii) whether cells require coatings or other treatments to display the desired phenotype; (iv) whether the culture should be vented to the atmosphere or sealed; (v) the need for good optical quality; (vi) if the purpose is analytical, the number of replicates and the number of cells required per assay sample (determined by the sensitivity of the assay); (vii) the frequency of sampling, e.g. simultaneous

TABLE 8.2. Culture vessel characteristics.

Culture vessel	Replicates per item	Culture volume[a,b]	Surface area $(cm^2)^b$	Approximate cell yield (HeLa)[b,c]
Microwell plates				
96-well plate	96	0.1–0.2 ml	0.32	6.4×10^4
384-well plate	384	25–50 μl	0.056	1.1×10^4
1536-well plate	1536	8 μl	0.025	5.0×10^3
Multiwell plates				
4-well plate	4	2	1.9	3.8×10^5
6-well plate	6	2	9.5	1.9×10^6
12-well plate	12	1	3.8	7.6×10^5
24-well plate	24	0.5	1.9	3.8×10^5
48-well plate	48	0.25	0.95	1.9×10^5
Petri dishes				
35-mm diameter	1	2	9.0	1.8×10^6
60-mm diameter	1	5	21	4.2×10^6
90-mm diameter	1	10	49	9.8×10^6
100-mm diameter	1	11	55	1.1×10^7
150-mm diameter	1	30	152	3.0×10^7
Flasks				
$10\,cm^2$ (T10)	1	2	10	2.0×10^6
$25\,cm^2$ (T25)	1	5	25	5.0×10^6
$75\,cm^2$ (T75)	1	20	75	1.5×10^7
$175\,cm^2$ (T175)	1	50	175	3.5×10^7
$225\,cm^2$ (T225)	1	75	225	4.5×10^7
Roller bottles				
$850\,cm^2$	1	200	850	1.7×10^8
$1750\,cm^2$	1	400	1750	3.5×10^8
Multisurface propagators				
Nunc Triple flask	1 (3 layers)	100–200	500	1.0×10^8
Falcon Multi-Flask (Corning)	1 (3–5 layers)	120–200	525–875	$1.1\text{–}1.8 \times 10^8$
HYPERFlask (Corning)	1 (10 layers)	560	1720	3.4×10^8
Nunc Cell Factory™ (ThermoFisher Scientific)	1 (1–40 layers)	200–8000	632–25 284	$1.2 \times 10^8\text{–}5.1 \times 10^9$
CellStack® (Corning)	1 (1–40 layers)	150–8000	636–25 440	$1.3 \times 10^8\text{–}5.1 \times 10^9$
HYPERStack (Corning)	1 (12–120 layers)	1300-13 000	6000-60 000	$1.2 \times 10^9\text{–}1.2 \times 10^{10}$
Spinner flasks				
500 ml (unsparged)	1	50		3.0×10^7
5000 ml (sparged)	1	4000		2.4×10^9

Source: Amanda Capes-Davis and R. Ian Freshney. Information for Corning culture vessels: Corning (2012). Information for Thermo Scientific microplates; Thermo Scientific (2014).

[a] The culture volume (used to estimate medium volume per replicate) is given per milliliter unless otherwise stated. If recommendations are not included for your culture vessel, use 0.2–0.3 ml medium per cm^2 growth surface area as a starting point. Volumes are usually increased for small spaces to account for evaporation and the meniscus effect.

[b] Volume, surface area, and yield will vary with the source, the cell line, the culture method, and the counting method. Always refer to supplier information and optimize culture conditions for your application.

[c] The cell yield that is used in this table is $2 \times 10^5/cm^2$, based on analysis of HeLa (see Figure 8.3). However, cell yield may vary between laboratories even when the same cell line is examined; analysis of HeLa elsewhere in this book gave a yield that was closer to $1 \times 10^5/cm^2$ (see Figure 14.1).

or sequential; (viii) the mode of analysis, e.g. measurement of fluorescence; and (ix) the cost.

8.5.1 Cell Yield

For monolayer cultures, the cell yield is proportional to the available surface area of the flask (see Figure 8.3). The yield from culture of HeLa is likely to be approximately 1–2×10^5 cells/cm^2 when fully confluent, but the yield from a finite cell line (e.g. diploid fibroblasts) could be one-fifth to one-tenth of the HeLa figure. The maximum cell yield is also regulated by the nutrient concentration, particularly the amino acid concentration (Amable and Butler 2008), the presence of the requisite growth factors, and the buildup of metabolites such as lactate and ammonia (see Section 9.4.4). Growth rate will be influenced by the uptake of oxygen, usually determined by the depth of medium (dependent on the ratio of the medium volume to culture surface area, usually $0.2 \, ml/cm^2$) and the production of CO_2, which depresses the pH.

Culture vessels come in a range of sizes, from microwell plates to cell factories (see Table 8.2). The middle of the size range embraces both Petri dishes and flasks ranging from 10 to 225 cm^2 (see Section 8.5.3). Culture at the large end of the size range becomes increasingly complicated, due to the requirement for a large surface area for attached cells and the need to maintain adequate oxygen and CO_2 exchange. Multilayer flasks are commonly used to increase yield in research laboratories and are included in this chapter (see Section 8.5.4). Roller bottles, multisurface propagators, and

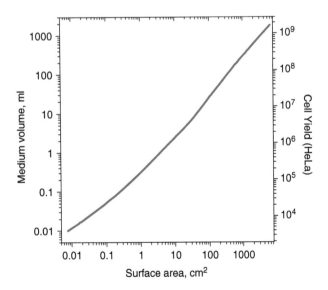

Fig. 8.3. Cell yield and surface area. The graph shows the approximate cell yield (for HeLa), based on the volume of the medium and the size of the vessel. This relationship is non-linear, as smaller vessels tend to be used with proportionally more medium than is used with larger vessels.

other high-yield vessels are discussed in a later chapter (see Section 28.3).

8.5.2 Multiwell Plates

Small volumes and multiple replicates are best handled using multiwell plates (e.g. 6- or 24-well plates) and microwell plates (e.g. 96-well plates; see Figure 8.4a). Multiwell plates are usually clear polystyrene, while microwell plates are available in clear or opaque formats (e.g. white, gray, or black). White or gray plates are preferred for luminescent assays, while black plates are preferred for fluorescent assays. The bottom of the well may be clear or opaque; clear wells are of course required for microscopic examination. Some microwell plates are designed to easily break apart so that wells can be removed for individual processing (e.g. Stripwell™ microplates, Corning).

Tissue culture plates traditionally come with loose-fitting lids but can be sealed with adhesive film to reduce evaporation, e.g. AeraSeal™ film (Alpha Laboratories; see Figure 8.4b) or SealMate™ sealing system (Excel Scientific). Evaporation affects the wells at the periphery of the plate more than it does the central wells, resulting in a phenomenon known as the edge effect, where cells behave differently in peripheral and central wells. A similar effect can arise due to thermal gradients (see Section 29.2.2). Edge effect is more obvious in smaller volumes (e.g. in 1536-well plates). This can result in a confounding effect in cytotoxicity assays, where evaporation results in toxicity due to increased salt concentration (Murray et al. 2016). Some plates allow liquid to be added to the outer moat or the inter-well space, which can improve insulation and reduce the edge effect (Wagener and Plennevaux 2014).

8.5.3 Flasks and Petri Dishes

Flasks are usually designated by their surface area (e.g. 25 or 175 cm^2, often abbreviated to T25 or T175, respectively), whereas Petri dishes are referred to by diameter (e.g. 35 or 90 mm). Both are now almost exclusively made of polystyrene (see Figures 8.5, 8.6); always check that dishes are tissue culture grade and not bacteriological grade (see Section 8.2.1). Flasks are typically used for routine handling and tend to be less prone to contamination compared to Petri dishes. Petri dishes are cheaper and quicker to handle and are typically used for clonogenic assays (see Section 19.4.1) and other experimental work. Flasks and dish designs may be customized for certain applications – for example, TPP® flasks (TPP Techno Plastic Products AG) have a peel-off cover for easy access to the growth surface of the flask.

8.5.4 Multilayer Flasks

If higher yields are required for adherent cells (e.g. $\sim 1 \times 10^9$ HeLa cells or 2×10^8 MRC-5 fibroblasts), increasing the size

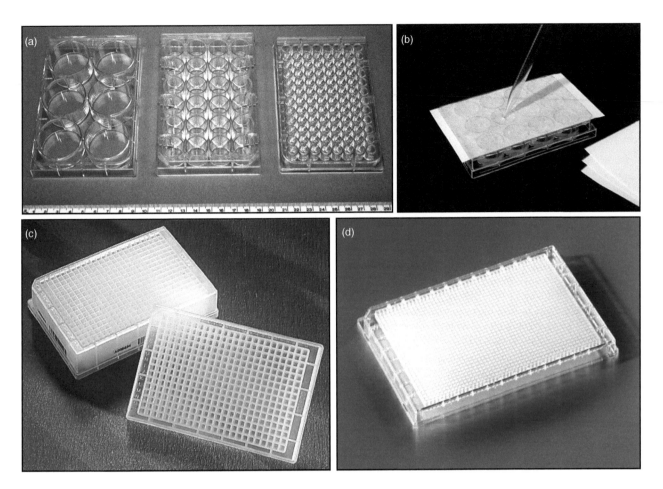

Fig. 8.4. Multiwell plates and microwell plates. Representative surface areas and culture volumes are listed elsewhere (see Table 8.2). (a) Nunc 6-well and 24-well multiwell plates and 96-well microwell plate. (b) AeraSeal gas-permeable sealing film. (c) Corning 384-well microwell plate with square wells with deep well plate at rear. (d) Corning 1536-well microwell plate with circular wells. Scale bar only applies to (a) although plate dimensions are the same. *Source*: (a, c, d) R. Ian Freshney; (b) Courtesy of Alpha Laboratories.

and number of conventional flasks becomes cumbersome and other, more specialized vessels are required. Multilayer flasks offer an intermediate step to increase surface area (see Figure 8.7). These vessels typically have a similar footprint to a 175-cm² flask; the user adds and removes cells and media using a similar process to the conventional flask, although care is needed to handle all layers equally and to avoid foaming during pipetting. If multilayer flasks do not provide sufficient yield, large multisurface propagators or roller bottles may be required (see Sections 28.3.1–28.3.2). You should also consider whether it is possible to grow the cells in suspension or to move to a different cell line that can be grown in suspension (see Section 8.6.2; see also Section 28.2). Growing cells in suspension is a good way to increase yield with a relatively modest investment in equipment.

Examples of multilayer flasks include the Nunc Triple-Flask™ (ThermoFisher Scientific), the Falcon® Multi-Flask (Corning), the HYPERFlask® (Corning), the Millicell®

HY Multilayer Culture Flask (Merck Millipore), and the Nest 5-layer flask (Wuxi NEST Biotechnology). Most multilayer flasks use the same substrate as conventional flasks, but the HYPERFlask is an interesting exception; it uses a gas-permeable film as a culture substrate, which allows more layers to be included in the same space. The HYPER-Flask is completely filled with medium instead of leaving a "headspace," with a narrow space provided between each layer for gas exchange (see Figure 8.7c; see also Table 8.2). The same technology is used in the HYPERStack®, which is a larger multisurface propagator (see Section 28.3.1).

8.5.5 Lids and Venting

Multiwell and Petri dishes, chosen for replicate sampling or cloning, have loose-fitting lids to give easy access to the dish. Consequently, they are not sealed and require a humidified atmosphere with control of CO_2 levels (see Section 9.2.3).

Fig. 8.5. Petri dishes. Dishes shown are 3.5-cm, 5-cm, and 9-cm diameter. Representative surface areas and culture volumes are listed elsewhere (see Table 8.2). Square Petri dishes are also available, with dimensions 9 × 9 cm. A grid pattern can be provided to help in scanning the dish – for example, in counting colonies – but can interfere with automatic colony counting. *Source*: R. Ian Freshney.

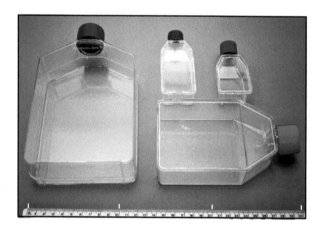

Fig. 8.6. Plastic flasks. Flasks shown are 10 and 25 cm² (Falcon, Corning), 75 cm² (Corning), and 185 cm² (Nalge Nunc). Markers above rule are at 10-cm intervals. Representative surface areas and culture volumes are listed elsewhere (see Table 8.2). *Source*: R. Ian Freshney.

Vented lids should be used with molded plastic supports inside (see Figure 8.8a, arrow) to prevent the thin film of liquid which can form around the inside of the lid and partially seal some dishes, due to liquid condensation or agitation. If a perfect seal is required, some multiwell dishes can be sealed with self-adhesive film (see Figure 8.4b).

Flasks should be vented when in a CO_2 incubator to allow CO_2 to enter or to allow excess CO_2 to escape in excessive acid-producing cell lines. A solid or "plug" cap should

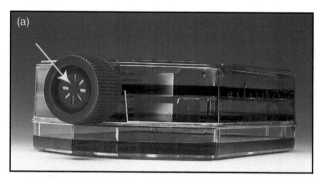

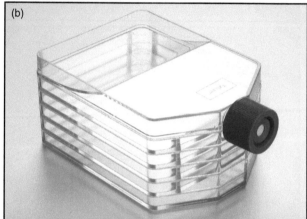

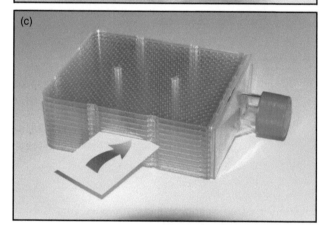

Fig. 8.7. Multilayer flasks. (a) The Nunc TripleFlask (ThermoFisher Scientific) has three growth surfaces that are seeded simultaneously, giving 500 cm² surface area for growth. As the headspace for the gas phase is smaller, this flask is best used with a filter cap (see arrow) in a CO_2 incubator. (b) The Falcon Multi-Flask (Corning) has up to five layers, giving up to 875 cm² surface area for growth. (c) The HYPERFlask (Corning) has a combined growth area of 1720 cm², using a gas-permeable membrane to remove the headspace above the medium. The arrow points to an air space between growth surfaces where gas exchange can occur. *Source*: (a) Courtesy of Nalge Nunc International; (b) Courtesy of Corning; (c) R. Ian Freshney, flask provided by TAP Biosystems.

Fig. 8.8. Venting petri dishes and flasks. (a) Vented Petri dish. Small ridges (see arrow) raise the lid from the base and prevent a thin film of liquid (e.g. condensate) from sealing the lid and reducing the rate of gas exchange. (b) Gas-permeable cap (see also Figure 8.7a) on a 175-cm² flask (Falcon); impermeable caps can be seen in the background. *Source*: (a) R. Ian Freshney; (b) Amanda Capes-Davis, courtesy of CMRI.

always be used in a non-CO_2 incubator or warmroom; these caps can be vented in a CO_2 incubator by slackening the caps one full turn. Flasks in a CO_2 incubator can also use caps with permeable filters that permit equilibration with the gas phase (see Figure 8.8b). Gas-permeable caps are preferable in a CO_2 incubator as they allow gas exchange without risk of contamination. The HYPERFlask is an exception to these general rules (see Section 8.5.4). This flask is made from gas-permeable film and is sealed with a solid cap, which should be kept closed in a CO_2 incubator.

8.5.6 Uneven Growth

Sometimes, uneven growth is observed in a dish or flask due to non-uniform distribution of cells across the growth surface.

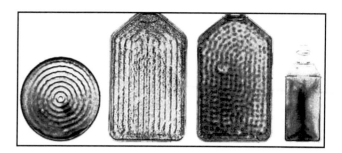

Fig. 8.9. Non-random growth. Examples of ridges seen in cultured monolayers in dishes and flasks, probably due to resonance in the incubator from fan motors or from opening and closing incubator doors. *Source*: Courtesy of Nunc.

This is commonly caused by vibration, which may occur due to opening and closing of the incubator, a faulty fan motor or other equipment, or personnel walking on floating floors. Vibration can perturb the medium, resulting in resonance or standing waves in the flask that, in turn, will produce a wave pattern in the monolayer and variations in cell density (see Figure 8.9). Eliminating vibration and minimizing entry into the incubator will help to reduce uneven growth. Placing a heavy weight in the tray or box with the plates and separating it from the shelf with plastic foam may also help alleviate the problem (Nielsen 1989). However, great care must be taken to wash and sterilize such foam pads, as they tend to harbor contamination. Further problem-solving for uneven growth is provided in a later chapter (see Section 31.8.3).

8.5.7 Cost

Most laboratories now use plastic because of its convenience, optical clarity, and quality. However, disposable plastic culture vessels are a major component of the cost of tissue culture. There is often a need to balance the cost of such plasticware against convenience. For example, Petri dishes are cheaper than flasks with an equivalent surface area, but they require humid, CO_2-controlled conditions and are more prone to contamination. Changing substrates can be challenging for some cell types. Always test a new substrate to decide if it is suitable for your cells and application (see Section 7.5.3). Moving to cheaper plasticware is false economy if your cells do not grow well on it.

Short-term work using multiple replicates is sometimes still performed using glass bottles and tubes. Glass vessels are more variable than plastic because they are usually drawn from standard pharmaceutical supplies, which are not designed for tissue culture. Glass vessels used for culture should have (i) one reasonably flat surface; (ii) a deep screw cap with a good seal and non-toxic liner; and (iii) shallow-sloping shoulders to facilitate harvesting of monolayer cells after trypsinization and to improve the efficiency of washing. Further suggestions for culture of replicate samples can be found in a previous chapter (see Section 7.6.2).

8.6 APPLICATION-SPECIFIC VESSELS

8.6.1 Imaging

Low-power microscopic observation is performed easily on flasks, Petri dishes, and multiwell plates with the use of an inverted microscope. However, difficulties may be encountered with microwell plates (and even 24-well plates) when using phase contrast because of the size of the meniscus relative to the diameter of the well. The meniscus effect may be reduced by limiting tissue culture treatment of the plastic to the bottom of the well. Large roller bottles are also difficult to examine; it is usually necessary to remove the condenser, in which case phase contrast will not be available. If microscopy plays a major part in your analysis, you will need to decide on the best substrate for the imaging equipment and procedure that you intend to use.

Vessels or substrates that are commonly used for imaging in research laboratories (apart from the plates, flasks, and Petri dishes described above) include:

(1) **Coverslips.** Sterile coverslips fit into multiwell dishes or Petri dishes (which need not be of tissue culture grade) for cells to attach before staining and processing.

(2) **Microscope slides.** Sterile slides fit into 90-mm Petri dishes for cells to attach before staining and processing. Cells in suspension may also be spun down onto a microscope slide using a cytocentrifuge (see Section 18.5.1).

(3) **Chambered slides and coverslips.** These microscope slides and coverslips have an attached chamber that can be removed when cells are fixed and processed (see Figure 8.10a, b; see also Supp. S18.3). Chambered slides and coverslips are available from a number of suppliers (e.g. Flask on cover glass II, Sarstedt; Millicell EZ Slides, Merck Millipore; Lab-Tek™ flasks and slides, ThermoFisher Scientific).

(4) **Imaging dishes and plates.** Some dishes and multiwell plates are available with glass or film at the bottom of the vessel, e.g. Cell Imaging Dishes and Plates, Eppendorf (see Figure 8.10c). Other examples include Glass Bottom Dishes (MatTek), High Content Screening Glass Bottom Microplates (Corning), and petriPERM dishes (Sartorius). Glass thickness may vary but usually 170 μm is available, corresponding to a coverglass thickness of #1.5. Some films may be gas-permeable, allowing optimal transfer of CO_2 and oxygen.

(5) **Closed culture chambers.** A frame is used to create one or more watertight chambers that contain observation areas made from glass, film, or gas-permeable membrane. Access ports are used to add or remove cells and reagents. Examples include the CoverWell™ imaging chamber (Grace Bio-Labs), Nunc OptiCell™ culture chamber (now discontinued), and the Sykes-Moore chamber (Bellco Glass). Some chambers are designed primarily for ease of

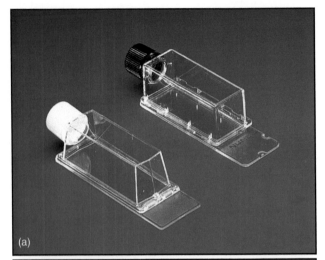

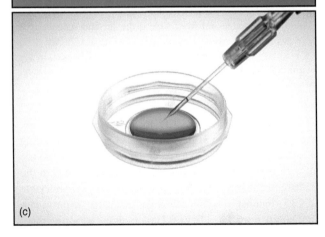

Fig. 8.10. Imaging vessels. Examples include (a) Lab-Tek Flask on Slide (ThermoFisher Scientific); (b) Lab-Tek chamber slides (ThermoFisher Scientific); and (c) Cell Imaging Dishes (Eppendorf). Slide-based systems (a, b) have detachable plastic chambers, while the bottom of the cell imaging dishes (c) is made of coverglass. *Source*: (a, b) Courtesy of Nunc; (c) Courtesy of Eppendorf, with permission.

imaging, while others provide a controlled environment for live-cell imaging (see Section 18.4.3).

(6) **Filter well inserts.** Some membranes used in filter well inserts (see Section 8.6.4) are sufficiently transparent to allow phase contrast microscopy. PC, PET, and collagen-coated PTFE have all been successfully used for immunofluorescent imaging (Gillespie et al. 2016).

How do you choose a vessel for imaging? Think about the substrates that best suit your cells, the microscope, and the procedure including any downstream processing steps. Some modern microscopes include autocorrection features that adjust for the thickness and optical properties of the substrate (e.g. Autocorr objectives, Zeiss). As a result, you may be able to use plastic (polystyrene) flasks, dishes, and multiwell plates for live-cell imaging on some microscopes. However, polystyrene is affected by heat and by a number of organic solvents including acetone, ethyl acetate, and xylene. Thus, you may need to use glass or solvent-resistant plastic for slides or coverslips.

Solvent-resistant plastics include Thermanox™ and Permanox™ (e.g. Nunc Thermanox coverslips, ThermoFisher Scientific; Permanox microscope slides, VWR). Thermanox coverslips are resistant to heat and many solvents. However, Thermanox is autofluorescent and is of poor optical quality. These coverslips should be mounted on slides with cells uppermost and a conventional glass coverslip on top. Permanox can be used for immunofluorescence and is of good optical quality, but slides may warp with some organic solvents (e.g. xylene). Aqueous mounting medium is recommended for Permanox slides and coverslips.

8.6.2 Suspension Culture

Cells that grow in suspension can be grown in any type of vessel including flasks, Petri dishes, and multiwell plates. Vessels must be sterile but do not need to be treated for cell attachment. Increasing the yield of cells growing in suspension requires only that the volume of the medium be increased, as long as cells in deep culture are kept agitated and sparged with 5% CO_2 in air. This can be done using shaker or spinner culture. Shaker culture can be performed using standard laboratory equipment, provided a suitable environment is provided for tissue culture, e.g. by installing a laboratory shaker inside a CO_2 incubator (see Figure 8.11b). Incubation shakers are also available, e.g. the Multitron Cell (Infors HT). Spinner culture requires access to a spinner flask, which uses a magnetic stirrer or a top-driven suspended paddle to gently agitate the culture. Spinner flasks are discussed later in this book, along with more specialized vessels that are now available for scale-up of suspension cultures (see Section 28.2).

Erlenmeyer flasks and Fernbach flasks are commonly used for shaker culture and are available in glass or polycarbonate from various manufacturers, including Corning, ThermoFisher Scientific, and TriForest. These classic shapes have

Fig. 8.11. Shaker flasks. (a) Optimum Growth flasks (Thomson Instrument Company) vary in capacity from 125 ml to 5 l. The vessel at left has an inversion transfer cap installed to assist with aseptic fluid transfer. (b) Optimum Growth flasks positioned on a shaker that has been installed in a CO_2 incubator. *Source*: Courtesy of Thomson Instrument Company, with permission.

been modified over time, resulting in new designs that are reported to reduce shearing and improve cell yield, e.g. the Optimum Growth™ flask (Thomson Instrument Company). Optimum Growth flasks are available in a range of sizes with useful accessories for shaker culture, such as inversion transfer caps to assist with aseptic transfer of cells (see Figure 8.11a). TubeSpin® bioreactors (TPP Techno Plastic Products) have been designed for culture in orbital shakers; CHO cells can be grown in volumes of 5–20 ml to give densities of 10^7 cells/ml (Stettler et al. 2007). Orbitally shaken bioreactors have been used successfully for scalable production of adeno-associated virus (AAV) vectors (Blessing et al. 2019). Further information on bioreactors is provided later in this book (see Sections 28.2, 28.3).

8.6.3 Scaffold-Free 3D Culture

Scaffold-free 3D culture is usually performed in suspension using a non-adhesive substrate, e.g. for spheroid culture (see Section 27.4.1). The cheapest non-adhesive substrate is untreated polystyrene; this is often used for hanging drop culture and may be successful in plates or dishes for some cell types (Mittler et al. 2017). However, attachment and spreading onto untreated polystyrene may occur in some assays (see Section 8.3.7). Coatings with agar or hydrophilic polymers are commonly used to minimize adhesion and promote spheroid formation. Commercially available spheroid microplates include AggreWell™ (STEM-CELL Technologies), GravityTRAP (InSphero), Nunclon Sphera™ (ThermoFisher Scientific), and Ultra-Low Attachment (ULA) plates (Corning). These microwell plates can be used to develop tumor spheroids for high-throughput drug discovery assays (see Protocol P27.1).

8.6.4 Permeable Supports

Vessels containing semipermeable membranes are used to mimic the basement membrane *in vivo* and to create chambers or compartments that allow cell separation or co-culture. These porous membranes act as gas-permeable substrates; they also allow the passage of water and (depending on the pore size) larger molecules or even cells. Growing cells on a water-permeable support increases diffusion of oxygen, CO_2, and nutrients. The permeability of the surface to which the cell is anchored may also induce polarity by simulating the basement membrane, especially if that membrane is coated with collagen, laminin, or fibronectin. Such polarity may be vital for full functional expression in secretory epithelia and many other types of cells (Chambard et al. 1987; Artursson and Magnusson 1990; Mullin et al. 1997; Babayeva et al. 2013).

Permeable supports are typically made of synthetic polymers, which may be treated to improve cell adhesion (see Section 8.3). "Native" semipermeable membranes have also been manufactured from ECM components (Mondrinos et al. 2017). Permeable supports are used in filter well inserts (see Section 27.5.2), where they have been used extensively to study cell–cell interaction, cell–matrix interaction, differentiation and polarity, transepithelial permeability, and tissue modeling. The resulting cell-based models are considered to be part of the movement toward 3D culture; cells grown on this substrate are able to interact with the environment in three dimensions, depending on the format used (Justice et al. 2009). Permeable supports are also used in the Corning HYPERFlask (see Section 8.5.4) and in some multisurface propagators and bioreactors (see Sections 28.2.4, 28.3.2, 28.3.4).

8.6.5 Scaffold-Based 3D Culture

Growth of cells on a flat surface results in the loss of many functional and morphological characteristics, due to the loss of tissue architecture and cell–cell interaction (see Sections 2.4.4, 26.3.3; see also Minireview M27.1). These deficiencies encouraged the exploration of matrices to provide a scaffold for 3D culture, which are now used extensively in tissue

engineering (Vunjak-Novakovic and Freshney 2006). Such matrices must enable cells to attach, proliferate, and differentiate in three dimensions; they must also degrade *in vivo* to be replaced by endogenous matrix.

Gels and sponges derived from animals and other "natural" sources were originally used as scaffolds or matrices for 3D culture. Fibrin clots were one of the first media to be used for primary culture and are still used, either as crude plasma clots or as purified fibrinogen mixed with thrombin (see Sections 1.2.1, 13.4). These systems generate a 3D gel in which cells may migrate and grow, either on the solid–gel interface or within the gel (Leighton 1991). Other materials include cellulose sponge (either alone or coated with collagen) (Leighton et al. 1968), collagen gel (Douglas et al. 1980), Matrigel (Kleinman and Martin 2005), AlgiMatrix (Justice et al. 2009), and Gelfoam® (Tome et al. 2014). Rationally designed hydrogels are also becoming available as matrices for 3D culture. The term "hydrogel" refers to a cross-linked network of polymers that contains a high water content (Tibbitt and Anseth 2009). Scaffolds and matrices for 3D culture are discussed in a later chapter (see Chapter 27).

Suppliers

Supplier	URL
Advanced BioMatrix	http://www.advancedbiomatrix.com
Alpha Laboratories	www.alphalabs.co.uk
Amsbio	http://www.amsbio.com/cells-and-cell-culture.aspx
Bellco Glass	http://www.bellcoglass.com
BioLamina	http://www.biolamina.com
Biological Industries	http://www.bioind.com/worldwide
Corning	http://www.corning.com/worldwide/en/products/life-sciences/products/cell-culture.html
Eppendorf	http://www.eppendorf.com/oc-en
Excel Scientific	http://www.excelscientific.com/index.html
Grace Bio-Labs	http://gracebio.com
Infors HT	http://www.infors-ht.com/en
InSphero	http://insphero.com
MatTek Corporation	http://www.mattek.com
Merck Millipore	http://www.merckmillipore.com
MP BioMedicals	http://www.mpbio.com
Santa Cruz Biotechnology	http://www.scbt.com/scbt/home
Sarstedt	http://www.sarstedt.com/en/home

Supplier	URL
Sartorius	http://www.sartorius.com/en
Sigma-Aldrich (Merck)	http://www.sigmaaldrich.com/life-science/cell-culture.html
STEMCELL Technologies	http://www.stemcell.com
Takara Bio	http://www.takarabio.com
ThermoFisher Scientific	http://www.thermofisher.com/us/en/home/life-science/cell-culture.html
Thomson Instrument Company	http://htslabs.com
TPP Techno Plastic Products	http://www.tpp.ch
Trevigen	http://trevigen.com/products-services
TriForest	http://www.triforest.com
VWR	http://si.vwr.com/store
Wuxi NEST Biotechnology	http://www.cell-nest.com/page1?_l=en

REFERENCES

Alvarez, Z., Sena, E., Mattotti, M. et al. (2014). An efficient and reproducible method to culture Bergmann and cortical radial glia using textured PMMA. *J. Neurosci. Methods* 232: 93–101. https://doi.org/10.1016/j.jneumeth.2014.05.011.

Amable, P. and Butler, M. (2008). Cell metabolism and its control in culture. In: *Animal Cell Technology: From Biopharmaceuticals to Gene Therapy* (eds. L. Castilho, A. Moraes, E. Augusto and M. Butler), 75–110. New York: Garland.

Artursson, P. and Magnusson, C. (1990). Epithelial transport of drugs in cell culture. II: Effect of extracellular calcium concentration on the paracellular transport of drugs of different lipophilicities across monolayers of intestinal epithelial (Caco-2) cells. *J. Pharm. Sci.* 79 (7): 595–600.

Babayeva, S., Rocque, B., Aoudjit, L. et al. (2013). Planar cell polarity pathway regulates nephrin endocytosis in developing podocytes. *J. Biol. Chem.* 288 (33): 24035–24048. https://doi.org/10.1074/jbc.M113.452904.

Barrandon, Y. and Green, H. (1987). Three clonal types of keratinocyte with different capacities for multiplication. *Proc. Natl Acad. Sci. U.S.A.* 84 (8): 2302–2306.

Belkaid, W., Thostrup, P., Yam, P.T. et al. (2013). Cellular response to micropatterned growth promoting and inhibitory substrates. *BMC Biotech.* 13: 86. https://doi.org/10.1186/1472-6750-13-86.

Berdichevsky, F., Gilbert, C., Shearer, M. et al. (1992). Collagen-induced rapid morphogenesis of human mammary epithelial cells: the role of the alpha 2 beta 1 integrin. *J. Cell Sci.* 102 (Pt 3): 437–446.

Biological Industries (2018). Application note: novel precoating-free protocol for culturing hPSC using vitronectin. https://www.bioind.com/media/wysiwyg/product/nutristem/precoating-free-protocol.pdf (accessed 11 March 2018).

Birnie, G.D. and Simons, P.J. (1967). The incorporation of 3H-thymidine and 3H-uridine into chick and mouse embryo cells cultured on stainless steel. *Exp. Cell. Res.* 46 (2): 355–366. https://doi.org/10.1016/0014-4827(67)90073-0.

Bledsoe, J.G., Slack, S.M., and Turitto, V.T. (2004). Cyclic mechanical strain alters tissue-factor activity in rat osteosarcoma cells cultured on a titanium substrate. *J. Biomed. Mater. Res. A* 70 (3): 490–496. https://doi.org/10.1002/jbm.a.30108.

Blessing, D., Vachey, G., Pythoud, C. et al. (2019). Scalable production of AAV vectors in orbitally shaken HEK293 cells. *Mol. Ther. Methods Clin. Dev.* 13: 14–26. https://doi.org/10.1016/j.omtm.2018.11.004.

Brandley, B.K., Weisz, O.A., and Schnaar, R.L. (1987). Cell attachment and long-term growth on derivatizable polyacrylamide surfaces. *J. Biol. Chem.* 262 (13): 6431–6437.

Brown, A., Burke, G.A., and Meenan, B.J. (2013). Patterned cell culture substrates created by hot embossing of tissue culture treated polystyrene. *J. Mater. Sci. Mater. Med.* 24 (12): 2797–2807. https://doi.org/10.1007/s10856-013-5011-5.

Carney, D.N., Bunn, P.A. Jr., Gazdar, A.F. et al. (1981). Selective growth in serum-free hormone-supplemented medium of tumor cells obtained by biopsy from patients with small cell carcinoma of the lung. *Proc. Natl Acad. Sci. U.S.A.* 78 (5): 3185–3189.

Carrel, A. (1923). A method for the physiological study of tissues in vitro. *J. Exp. Med.* 38 (4): 407–418.

Chambard, M., Mauchamp, J., and Chabaud, O. (1987). Synthesis and apical and basolateral secretion of thyroglobulin by thyroid cell monolayers on permeable substrate: modulation by thyrotropin. *J. Cell. Physiol.* 133 (1): 37–45. https://doi.org/10.1002/jcp.1041330105.

Corbett, G.T., Roy, A., and Pahan, K. (2013). Sodium phenylbutyrate enhances astrocytic neurotrophin synthesis via protein kinase C (PKC)-mediated activation of cAMP-response element-binding protein (CREB): implications for Alzheimer disease therapy. *J. Biol. Chem.* 288 (12): 8299–8312. https://doi.org/10.1074/jbc.M112.426536.

Corning (2012). Surface areas and recommended medium volumes for Corning cell culture vessels. https://www.corning.com/catalog/cls/documents/application-notes/cls-an-209.pdf (accessed 23 October 2018).

Corning (2013). Corning Matrigel Matrix frequently asked questions. https://www.corning.com/media/worldwide/cls/documents/cls-dl-cc-026%20dl.pdf (accessed 14 August 2018).

Cui, X., Hartanto, Y., and Zhang, H. (2017). Advances in multicellular spheroids formation. *J. R. Soc. Interface* 14 (127) https://doi.org/10.1098/rsif.2016.0877.

Curtis, A.S., Forrester, J.V., McInnes, C. et al. (1983). Adhesion of cells to polystyrene surfaces. *J. Cell Biol.* 97 (5 Pt 1): 1500–1506.

Danen, E.H. and Yamada, K.M. (2001). Fibronectin, integrins, and growth control. *J. Cell. Physiol.* 189 (1): 1–13. https://doi.org/10.1002/jcp.1137.

Davidenko, N., Schuster, C.F., Bax, D.V. et al. (2016). Evaluation of cell binding to collagen and gelatin: a study of the effect of 2D and 3D architecture and surface chemistry. *J. Mater. Sci. Mater. Med.* 27 (10): 148. https://doi.org/10.1007/s10856-016-5763-9.

Deptuch, T. and Dams-Kozlowska, H. (2017). Silk materials functionalized via genetic engineering for biomedical applications. *Mater. (Basel)* 10 (12) https://doi.org/10.3390/ma10121417.

Desai, N., Rambhia, P., and Gishto, A. (2015). Human embryonic stem cell cultivation: historical perspective and evolution of xeno-free culture systems. *Reprod. Biol. Endocrinol.* 13: 9. https://doi.org/10.1186/s12958-015-0005-4.

DeWitt, M.R. and Rylander, M.N. (2018). Tunable collagen microfluidic platform to study nanoparticle transport in the tumor microenvironment. *Methods Mol. Biol.* 1831: 159–178. https://doi.org/10.1007/978-1-4939-8661-3_12.

Douglas, W.H., McAteer, J.A., Dell'orco, R.T. et al. (1980). Visualization of cellular aggregates cultured on a three dimensional collagen sponge matrix. *In Vitro* 16 (4): 306–312.

Du, P., Subbiah, R., Park, J.H. et al. (2014). Vascular morphogenesis of human umbilical vein endothelial cells on cell-derived macromolecular matrix microenvironment. *Tissue Eng. Part A* 20 (17–18): 2365–2377. https://doi.org/10.1089/ten.TEA.2013.0693.

Earle, W.R. (1943). Production of malignancy in vitro. I. Method of cleaning glassware. *J. Natl Cancer Inst.* 4 (2): 131–133. https://doi.org/10.1093/jnci/4.2.131.

Earle, W.R. and Highhouse, F. (1954). Culture flasks for use with plane surface substrate tissue cultures. *J. Natl Cancer Inst.* 14 (4): 841–851. https://doi.org/10.1093/jnci/14.4.841.

Earle, W.R., Schilling, E.L., Stark, T.H. et al. (1943). Production of malignancy in vitro. IV. The mouse fibroblast cultures and changes seen in the living cells. *J. Natl Cancer Inst.* 4 (2): 165–212.

Ehrmann, R.L. and Gey, G.O. (1956). The growth of cells on a transparent gel of reconstituted rat-tail collagen. *J. Natl Cancer Inst.* 16 (6): 1375–1403.

Elsdale, T. and Bard, J. (1972). Collagen substrata for studies on cell behavior. *J. Cell Biol.* 54 (3): 626–637.

Evans, V.J., Earle, W.R., Schilling, E.L. et al. (1947). The use of perforated cellophane for the growth of cells in tissue culture. *J. Natl Cancer Inst.* 8 (3): 103–119.

Feng, G., Hicks, P., and Chang, P.L. (2003). Differential expression of mammalian or viral promoter-driven gene in adherent versus suspension cells. *In Vitro Cell. Dev. Biol. Anim.* 39 (10): 420–423. https://doi.org/10.1290/1543-706x(2003)039<0420:deomov>2.0.co;2.

Folkman, J., Haudenschild, C.C., and Zetter, B.R. (1979). Long-term culture of capillary endothelial cells. *Proc. Natl Acad. Sci. U.S.A.* 76 (10): 5217–5221.

Friend, J. and Yeo, L. (2010). Fabrication of microfluidic devices using polydimethylsiloxane. *Biomicrofluidics* 4 (2) https://doi.org/10.1063/1.3259624.

Gabriel, M., Strand, D., and Vahl, C.F. (2012). Cell adhesive and antifouling polyvinyl chloride surfaces via wet chemical modification. *Artif. Organs* 36 (9): 839–844. https://doi.org/10.1111/j.1525-1594.2012.01462.x.

Ganz, J., Arie, I., Ben-Zur, T. et al. (2014). Astrocyte-like cells derived from human oral mucosa stem cells provide neuroprotection in vitro and in vivo. *Stem Cells Transl. Med.* 3 (3): 375–386. https://doi.org/10.5966/sctm.2013-0074.

Gillespie, J.L., Anyah, A., Taylor, J.M. et al. (2016). A versatile method for Immunofluorescent staining of cells cultured on permeable membrane inserts. *Med. Sci. Monit. Basic Res.* 22: 91–94.

Gospodarowicz, D., Delgado, D., and Vlodavsky, I. (1980). Permissive effect of the extracellular matrix on cell proliferation in vitro. *Proc. Natl Acad. Sci. U.S.A.* 77 (7): 4094–4098.

Gottwald, E., Lahni, B., Thiele, D. et al. (2008). Chip-based three-dimensional cell culture in perfused micro-bioreactors. *J. Vis. Exp.* 2008 (15) https://doi.org/10.3791/564.

Guo, S., Zhu, X., and Loh, X.J. (2017). Controlling cell adhesion using layer-by-layer approaches for biomedical applications. *Mater. Sci. Eng. C Mater. Biol. Appl.* 70 (Pt 2): 1163–1175. https://doi.org/10.1016/j.msec.2016.03.074.

Hakala, H., Rajala, K., Ojala, M. et al. (2009). Comparison of biomaterials and extracellular matrices as a culture platform for multiple, independently derived human embryonic stem cell lines. *Tissue Eng. Part A* 15 (7): 1775–1785. https://doi.org/10.1089/ten.tea.2008.0316.

Harris, A.K., Wild, P., and Stopak, D. (1980). Silicone rubber substrata: a new wrinkle in the study of cell locomotion. *Science* 208 (4440): 177–179. https://doi.org/10.1126/science.6987736.

Harrison, R.G. (1907). Observations on the living developing nerve fiber. *Proc. Soc. Exp. Biol.* 4 (1): 140–143.

Hauschka, S.D. and Konigsberg, I.R. (1966). The influence of collagen on the development of muscle clones. *Proc. Natl Acad. Sci. U.S.A.* 55 (1): 119–126.

Healy, L. and Ruban, L. (2015). *Atlas of Human Pluripotent Stem Cells in Culture*. London: Springer.

Hofmann, S., Kaplan, D., Vunjak-Novakovic, G. et al. (2006). Tissue engineering of bone. In: *Culture of Cells for Tissue Engineering* (eds. G. Vunjak-Novakovic and R.I. Freshney), 325–373. Hoboken, NJ: Wiley.

Huang, Q., Liang, W., Xu, D. et al. (2013). Ultrastructural observations of human epidermal melanocytes cultured on polyethylene terephthalate film. *Micron* 48: 49–53. https://doi.org/10.1016/j.micron.2013.02.008.

Hynds, R.E., Bonfanti, P., and Janes, S.M. (2018). Regenerating human epithelia with cultured stem cells: feeder cells, organoids and beyond. *EMBO Mol. Med.* 10 (2): 139–150. https://doi.org/10.15252/emmm.201708213.

Ireland, G.W., Dopping-Hepenstal, P.J., Jordan, P.W. et al. (1989). Limitation of substratum size alters cytoskeletal organization and behaviour of Swiss 3T3 fibroblasts. *Cell Biol. Int. Rep.* 13 (9): 781–790.

Jastrzebska, K., Kucharczyk, K., Florczak, A. et al. (2015). Silk as an innovative biomaterial for cancer therapy. *Rep. Pract. Oncol. Radiother.* 20 (2): 87–98. https://doi.org/10.1016/j.rpor.2014.11.010.

Joddar, B. and Ito, Y. (2013). Artificial niche substrates for embryonic and induced pluripotent stem cell cultures. *J. Biotechnol.* 168 (2): 218–228. https://doi.org/10.1016/j.jbiotec.2013.04.021.

Jurkiewicz, E., Husemann, U., Greller, G. et al. (2014). Verification of a new biocompatible single-use film formulation with optimized additive content for multiple bioprocess applications. *Biotechnol. Progr.* 30 (5): 1171–1176. https://doi.org/10.1002/btpr.1934.

Justice, B.A., Badr, N.A., and Felder, R.A. (2009). 3D cell culture opens new dimensions in cell-based assays. *Drug Discov. Today* 14 (1–2): 102–107. https://doi.org/10.1016/j.drudis.2008.11.006.

Karamichos, D., Funderburgh, M.L., Hutcheon, A.E. et al. (2014). A role for topographic cues in the organization of collagenous matrix by corneal fibroblasts and stem cells. *PLoS One* 9 (1): e86260. https://doi.org/10.1371/journal.pone.0086260.

Kleinfeld, D., Kahler, K.H., and Hockberger, P.E. (1988). Controlled outgrowth of dissociated neurons on patterned substrates. *J. Neurosci.* 8 (11): 4098–4120.

Kleinman, H.K. and Martin, G.R. (2005). Matrigel: basement membrane matrix with biological activity. *Semin. Cancer Biol.* 15 (5): 378–386. https://doi.org/10.1016/j.semcancer.2005.05.004.

Kobayashi, J. and Okano, T. (2010). Fabrication of a thermoresponsive cell culture dish: a key technology for cell sheet tissue engineering. *Sci. Technol. Adv. Mater.* 11 (1): 014111. https://doi.org/10.1088/1468-6996/11/1/014111.

van Kooten, T.G., Spijker, H.T., and Busscher, H.J. (2004). Plasma-treated polystyrene surfaces: model surfaces for studying cell-biomaterial interactions. *Biomaterials* 25 (10): 1735–1747.

Krishna, Y., Sheridan, C., Kent, D. et al. (2011). Expanded polytetrafluoroethylene as a substrate for retinal pigment epithelial cell growth and transplantation in age-related macular degeneration. *Br. J. Ophthalmol.* 95 (4): 569–573. https://doi.org/10.1136/bjo.2009.169953.

Lantigua, D., Kelly, Y.N., Unal, B. et al. (2017). Engineered paper-based cell culture platforms. *Adv. Healthc. Mater.* 6 (22) https://doi.org/10.1002/adhm.201700619.

Lao, L., Tan, H., Wang, Y. et al. (2008). Chitosan modified poly(L-lactide) microspheres as cell microcarriers for cartilage tissue engineering. *Colloids Surf. B Biointerfaces* 66 (2): 218–225. https://doi.org/10.1016/j.colsurfb.2008.06.014.

Leighton, J. (1951). A sponge matrix method for tissue culture; formation of organized aggregates of cells in vitro. *J. Natl Cancer Inst.* 12 (3): 545–561.

Leighton, J. (1991). Radial histophysiologic gradient culture chamber: rationale and preparation. *In Vitro Cell. Dev. Biol.* 27A (10): 786–790.

Leighton, J., Mark, R., and Justh, G. (1968). Patterns of three-dimensional growth in vitro in collagen-coated cellulose sponge: carcinomas and embryonic tissues. *Cancer Res.* 28 (2): 286–296.

Liberio, M.S., Sadowski, M.C., Soekmadji, C. et al. (2014). Differential effects of tissue culture coating substrates on prostate cancer cell adherence, morphology and behavior. *PLoS One* 9 (11): e112122. https://doi.org/10.1371/journal.pone.0112122.

Llames, S., Garcia-Perez, E., Meana, A. et al. (2015). Feeder layer cell actions and applications. *Tissue Eng. Part B Rev.* 21 (4): 345–353. https://doi.org/10.1089/ten.TEB.2014.0547.

Lovitt, C.J., Shelper, T.B., and Avery, V.M. (2014). Advanced cell culture techniques for cancer drug discovery. *Biology (Basel)* 3 (2): 345–367. https://doi.org/10.3390/biology3020345.

Macklis, J.D., Sidman, R.L., and Shine, H.D. (1985). Cross-linked collagen surface for cell culture that is stable, uniform, and optically superior to conventional surfaces. *In Vitro Cell. Dev. Biol.* 21 (3 Pt 1): 189–194.

MacLeod, R.A., Kaufmann, M.E., and Drexler, H.G. (2017). Cytogenetic harvesting of cancer cells and cell lines. *Methods Mol. Biol.* 1541: 43–58. https://doi.org/10.1007/978-1-4939-6703-2_5.

Macpherson, I. and Montagnier, L. (1964). Agar suspension culture for the selective assay of cells transformed by polyoma virus. *Virology* 23: 291–294.

Mazia, D., Schatten, G., and Sale, W. (1975). Adhesion of cells to surfaces coated with polylysine. Applications to electron microscopy. *J. Cell Biol.* 66 (1): 198–200.

McKeehan, W.L. and Ham, R.G. (1976). Stimulation of clonal growth of normal fibroblasts with substrata coated with basic polymers. *J. Cell Biol.* 71 (3): 727–734.

Mittler, F., Obeid, P., Rulina, A.V. et al. (2017). High-content monitoring of drug effects in a 3D spheroid model. *Front. Oncol.* 7: 293. https://doi.org/10.3389/fonc.2017.00293.

Mondrinos, M.J., Yi, Y.S., Wu, N.K. et al. (2017). Native extracellular matrix-derived semipermeable, optically transparent, and inexpensive membrane inserts for microfluidic cell culture. *Lab Chip* 17 (18): 3146–3158. https://doi.org/10.1039/c7lc00317j.

Mullin, J.M., Marano, C.W., Laughlin, K.V. et al. (1997). Different size limitations for increased transepithelial paracellular solute flux across phorbol ester and tumor necrosis factor-treated epithelial cell sheets. *J. Cell. Physiol.* 171 (2): 226–233. https://doi.org/10.1002/(SICI)1097-4652(199705)171:2<226::AID-JCP14>3.0.CO;2-B.

Murray, D., McWilliams, L., and Wigglesworth, M. (2016). High-throughput cell toxicity assays. *Methods Mol. Biol.* 1439: 245–262. https://doi.org/10.1007/978-1-4939-3673-1_16.

Nicosia, R.F. and Ottinetti, A. (1990). Modulation of microvascular growth and morphogenesis by reconstituted basement membrane gel in three-dimensional cultures of rat aorta: a comparative study of angiogenesis in matrigel, collagen, fibrin, and plasma clot. *In Vitro Cell. Dev. Biol.* 26 (2): 119–128.

Nielsen, V. (1989). Vibration patterns in tissue culture vessels. Nunc Bulletin 2. Roskilde, Denmark: A/S Nunc.

Patel, D., Haque, A., Jones, C.N. et al. (2014). Local control of hepatic phenotype with growth factor-encoded surfaces. *Integr. Biol. (Camb.)* 6 (1): 44–52. https://doi.org/10.1039/c3ib40140e.

Paul, J. (1975). *Cell and Tissue Culture*, 5e. Edinburgh: Churchill Livingstone.

Petit, A., Demers, C.N., Girard-Lauriault, P.L. et al. (2011). Effect of nitrogen-rich cell culture surfaces on type X collagen expression by bovine growth plate chondrocytes. *Biomed. Eng. Online* 10: 4. https://doi.org/10.1186/1475-925X-10-4.

Phelps, E.A., Cianciaruso, C., Santo-Domingo, J. et al. (2017). Advances in pancreatic islet monolayer culture on glass surfaces enable super-resolution microscopy and insights into beta cell ciliogenesis and proliferation. *Sci. Rep.* 7: 45961. https://doi.org/10.1038/srep45961.

Plant, A.L., Bhadriraju, K., Spurlin, T.A. et al. (2009). Cell response to matrix mechanics: focus on collagen. *Biochim. Biophys. Acta* 1793 (5): 893–902. https://doi.org/10.1016/j.bbamcr.2008.10.012.

Pourreyron, C., Purdie, K.J., Watt, S.A. et al. (2011). Feeder layers: co-culture with nonneoplastic cells. *Methods Mol. Biol.* 731: 467–470. https://doi.org/10.1007/978-1-61779-080-5_37.

Puck, T.T. and Marcus, P.I. (1955). A rapid method for viable cell titration and clone production with HeLa cells in tissue culture: the use of X-irradiated cells to supply conditioning factors. *Proc. Natl Acad. Sci. U.S.A.* 41 (7): 432–437.

Rampersad, S., Ruiz, J.C., Petit, A. et al. (2011). Stem cells, nitrogen-rich plasma-polymerized culture surfaces, and type X collagen suppression. *Tissue Eng. Part A* 17 (19–20): 2551–2560. https://doi.org/10.1089/ten.TEA.2010.0723.

Rappaport, C. and Bishop, C.B. (1960). Improved method for treating glass to produce surfaces suitable for the growth of certain mammalian cells in synthetic medium. *Exp. Cell. Res.* 20: 580–584.

Rheinwald, J.G. and Green, H. (1975). Serial cultivation of strains of human epidermal keratinocytes: the formation of keratinizing colonies from single cells. *Cell* 6 (3): 331–343.

Richler, C. and Yaffe, D. (1970). The in vitro cultivation and differentiation capacities of myogenic cell lines. *Dev. Biol.* 23 (1): 1–22.

Roth, E.A., Xu, T., Das, M. et al. (2004). Inkjet printing for high-throughput cell patterning. *Biomaterials* 25 (17): 3707–3715. https://doi.org/10.1016/j.biomaterials.2003.10.052.

Russell, P.J. and Kingsley, E.A. (2003). Human prostate cancer cell lines. *Methods Mol. Med.* 81: 21–39. https://doi.org/10.1385/1-59259-372-0:21.

Sant, S. and Johnston, P.A. (2017). The production of 3D tumor spheroids for cancer drug discovery. *Drug Discov. Today Technol.* 23: 27–36. https://doi.org/10.1016/j.ddtec.2017.03.002.

Sarang, Z., Haig, Y., Hansson, A. et al. (2003). Microarray assessment of fibronectin, collagen and integrin expression and the role of fibronectin-collagen coating in the growth of normal, SV40 T-antigen-immortalised and malignant human oral keratinocytes. *Altern. Lab. Anim* 31 (6): 575–585.

Sattler, C.A., Michalopoulos, G., Sattler, G.L. et al. (1978). Ultrastructure of adult rat hepatocytes cultured on floating collagen membranes. *Cancer Res.* 38 (6): 1539–1549.

Scherer, W.F., Syverton, J.T., and Gey, G.O. (1953). Studies on the propagation in vitro of poliomyelitis viruses. IV. Viral multiplication in a stable strain of human malignant epithelial cells (strain HeLa) derived from an epidermoid carcinoma of the cervix. *J. Exp. Med.* 97 (5): 695–710.

Shen, B., Delaney, M.K., and Du, X. (2012). Inside-out, outside-in, and inside-outside-in: G protein signaling in integrin-mediated cell adhesion, spreading, and retraction. *Curr. Opin. Cell Biol.* 24 (5): 600–606. https://doi.org/10.1016/j.ceb.2012.08.011.

Sia, S.K. and Whitesides, G.M. (2003). Microfluidic devices fabricated in poly(dimethylsiloxane) for biological studies. *Electrophoresis* 24 (21): 3563–3576. https://doi.org/10.1002/elps.200305584.

Stampfer, M., Hallowes, R.C., and Hackett, A.J. (1980). Growth of normal human mammary cells in culture. *In Vitro* 16 (5): 415–425.

Stettler, M., Zhang, X., Hacker, D.L. et al. (2007). Novel orbital shake bioreactors for transient production of CHO derived IgGs. *Biotechnol. Progr.* 23 (6): 1340–1346. https://doi.org/10.1021/bp070219i.

Takahashi, K. and Okada, T.S. (1970). An analysis of the effect of "conditioned medium" upon the cell culture at low density. *Develop. Growth Differ.* 12 (2): 65–77.

Teare, D.O., Emmison, N., Ton-That, C. et al. (2001). Effects of serum on the kinetics of CHO attachment to ultraviolet-ozone modified polystyrene surfaces. *J. Colloid Interface Sci.* 234 (1): 84–89. https://doi.org/10.1006/jcis.2000.7282.

Thermo Scientific (2014). Thermo Scientific microplates guide. https://assets.thermofisher.com/tfs-assets/lcd/brochures/d10948.pdf (accessed 23 October 2018).

Tibbitt, M.W. and Anseth, K.S. (2009). Hydrogels as extracellular matrix mimics for 3D cell culture. *Biotechnol. Bioeng.* 103 (4): 655–663. https://doi.org/10.1002/bit.22361.

Tome, Y., Uehara, F., Mii, S. et al. (2014). 3-dimensional tissue is formed from cancer cells in vitro on Gelfoam®, but not on Matrigel. *J. Cell. Biochem.* 115 (8): 1362–1367. https://doi.org/10.1002/jcb.24780.

Tsao, C.W. (2016). Polymer microfluidics: simple, low-cost fabrication process bridging academic lab research to commercialized production. *Micromachines* 7 (12): 225. https://doi.org/10.3390/mi7120225.

Villard, V., Kalyuzhniy, O., Riccio, O. et al. (2006). Synthetic RGD-containing alpha-helical coiled coil peptides promote integrin-dependent cell adhesion. *J. Pept. Sci.* 12 (3): 206–212. https://doi.org/10.1002/psc.707.

Vlodavsky, I., Lui, G.M., and Gospodarowicz, D. (1980). Morphological appearance, growth behavior and migratory activity of human tumor cells maintained on extracellular matrix versus plastic. *Cell* 19 (3): 607–616.

Vunjak-Novakovic, G. and Freshney, R.I. (2006). *Culture of Cells for Tissue Engineering*. Hoboken, NJ: Wiley-Liss.

Wagener, J. and Plennevaux, C. (2014). Application note: Eppendorf 96-well cell culture plate – a simple method of minimizing the edge effect in cell-based assays. https://www.eppendorf.com/fileadmin/knowledgebase/asset/sp-en/105706.pdf (accessed 18 August 2018).

Wang, L., Sun, B., Ziemer, K.S. et al. (2010). Chemical and physical modifications to poly(dimethylsiloxane) surfaces affect adhesion of Caco-2 cells. *J. Biomed. Mater. Res. A* 93 (4): 1260–1271. https://doi.org/10.1002/jbm.a.32621.

Welle, A., Weigel, S., and Bulut, O.D. (2014). Patterning of polymeric cell culture substrates. *Methods Cell Biol.* 119: 35–53. https://doi.org/10.1016/B978-0-12-416742-1.00003-2.

Westermark, B. (1978). Growth control in miniclones of human glial cells. *Exp. Cell. Res.* 111 (2): 295–299.

Widhe, M., Johansson, U., Hillerdahl, C.O. et al. (2013). Recombinant spider silk with cell binding motifs for specific adherence of cells. *Biomaterials* 34 (33): 8223–8234. https://doi.org/10.1016/j.biomaterials.2013.07.058.

Xie, J., Pak, K., Evans, A. et al. (2013). Neurotrophins differentially stimulate the growth of cochlear neurites on collagen surfaces and in gels. *Neural Regen. Res.* 8 (17): 1541–1550. https://doi.org/10.3969/j.issn.1673-5374.2013.17.001.

Xu, S.L., Bi, C.W., Choi, R.C. et al. (2013). Flavonoids induce the synthesis and secretion of neurotrophic factors in cultured rat astrocytes: a signaling response mediated by estrogen receptor. *Evid. Based Complement. Alternat. Med.* 2013: 127075. https://doi.org/10.1155/2013/127075.

Yamaguchi, S., Ueno, A., Akiyama, Y. et al. (2012). Cell patterning through inkjet printing of one cell per droplet. *Biofabrication* 4 (4): 045005. https://doi.org/10.1088/1758-5082/4/4/045005.

CHAPTER 9

Defined Media and Supplements

Initial attempts to culture cells were performed in natural, largely undefined media based on body fluids and tissue extracts (see Section 1.2.1). With the establishment of continuous cell lines and their widespread use in many laboratories, there was a demand for larger amounts of medium and for greater consistency and quality. Greater demand led to the introduction of chemically defined media based on analyses of body fluids and nutritional biochemistry (Morgan et al. 1950; Morton 1970). This chapter aims to describe the general principles of medium composition, including its physicochemical properties and individual components, with a focus on widely used serum-supplemented media. Serum-free media are discussed in the following chapter (see Chapter 10).

9.1 MEDIUM DEVELOPMENT

The minimum requirements for growth of cells in culture were defined by Albert Fischer, Harry Eagle, and other tissue culture pioneers (Fischer 1925; Morton 1970; Yao and Asayama 2017). Early media were supplemented with various natural substances, including (i) serum derived from calves, horses, and humans; (ii) chick embryonic extract; and (iii) protein hydrolysates such as Bacto™ Peptone and tryptose phosphate broth. As more continuous cell lines became available, it became apparent that the majority were able to grow in a relatively simple formulation made up of amino acids, vitamins, inorganic salts, and glucose, supplemented with serum (Eagle 1955, 1959).

The early medium formulations contained the essential components for growth, but they were not intended to be optimal for all cell lines (Waymouth 1978). Later formulations were modified to improve the growth of specific cell types or for specific applications (see Section 9.4). Most commonly used formulations favor cell growth, which may occur at the expense of cell differentiation, and require further optimization to achieve truly physiological culture conditions. For example, further efforts to optimize medium for culture of human ovarian cancer cells have improved long-term growth, rates of cell line establishment, and resemblance to the original tumor (Ince et al. 2015).

9.2 PHYSICOCHEMICAL PROPERTIES

9.2.1 pH

Most cell lines grow well at pH 7.4. Although the optimal pH varies relatively little among many cultures, normal fibroblast cell lines often perform best at pH 7.4–7.7 and transformed cells may do better at pH 7.0–7.4 (Eagle 1973). Regardless of the exact value, however, it is important that the pH remains as close to optimum as possible throughout the growth cycle. Most cultures that grow well at pH 7.4 display diminished cell growth above pH 7.7 and below pH 6.5; viability is rapidly lost above pH 7.8 and below pH 6.0 (Eagle 1973). Transformed cells that are grown beyond confluence display a lowered pH due to lactate release and will quickly enter apoptosis. This programmed cell death is not necessarily due to pH depression; it is more likely to accompany nutrient and growth factor depletion and toxic metabolite accumulation. Typically, cultures are passaged before this problem arises (see Section 14.6). A few cell types have markedly different pH

Freshney's Culture of Animal Cells: A Manual of Basic Technique and Specialized Applications, Eighth Edition. Amanda Capes-Davis and R. Ian Freshney.
© 2021 John Wiley & Sons Ltd. Published 2021 by John Wiley & Sons Ltd.
Companion website: www.wiley.com/go/freshney/cellculture8

pH6.5   pH7.0   pH7.4   pH7.6

Fig. 9.1. pH standards. Phenol red pH indicator in a standard set of solutions. Far left and far right are unacceptable and need immediate action, i.e. medium change, subculture, or gassing. *Source*: R. Ian Freshney.

requirements; for example, insect cell lines are commonly maintained at pH 6.2 and epidermal cells have been maintained at pH 5.5 (Grace 1962; Eisinger et al. 1979). In special cases, the optimal pH can be determined by analysis of the culture's growth curve, cloning efficiency, or differentiation status (see Protocols P19.3, 19.4; see also Section 26.3).

Phenol red is commonly added to culture medium as a pH indicator. It is red at pH 7.4 and becomes orange at pH 7.0, yellow at pH 6.5, lemon-yellow below pH 6.5, more pink at pH 7.6, and purple at pH 7.8 (see Figure 9.1). Because the assessment of color is highly subjective, it is useful to make up a set of standards for comparison; a protocol for preparing pH standards is available as Supplementary Material online (see Supp. S9.1). However, phenol red may be toxic to some cell types, may increase background fluorescence in imaging studies, and is weakly estrogenic (Grady et al. 1991; Price 2017; OECD 2018). Culture media without phenol red are usually available, if required.

9.2.2 Buffering

A buffer is incorporated into culture medium to stabilize the pH level. Buffering is essential under two sets of conditions: (i) open dishes, where loss of CO_2 to the atmosphere causes the pH to rise; and (ii) overproduction of CO_2 and lactate in transformed cell lines that reach high cell concentrations, causing the pH to fall. Bicarbonate is used more frequently than any other buffer because of its low toxicity, low cost, and nutritional benefit to the culture. However, it requires a CO_2-enriched atmosphere and is a relatively poor buffer at physiological pH (see Figure 9.2; see also Section 9.2.3). Good's buffers (e.g. HEPES and Tricine) were developed as alternatives for biological research (Good et al. 1966). HEPES at 10–20 mM is a stronger buffer in the pH 7.2–7.6 range, but its concentration must be more than double that of bicarbonate for adequate buffering when it is used with exogenous CO_2 (see Table 9.1). One of the authors (R. Ian Freshney) has used a variation of Ham's F12 with 20 mM HEPES, 8 mM bicarbonate, and 2% CO_2 for the culture of a number of different cell lines. It allows the handling of microtitration and other multiwell plates outside the incubator without an excessive rise in pH. However, HEPES can be both toxic and

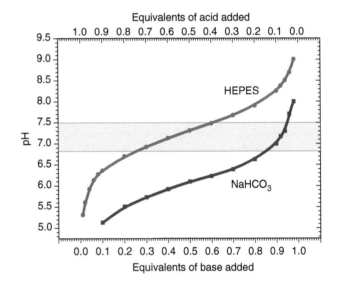

Fig. 9.2. Buffering by HEPES and bicarbonate. The graphs show the effect of titrating with acid (top axis) or base (bottom axis) on the change in pH in the presence of either HEPES or sodium bicarbonate. It can be seen from the upper graph that pH is buffered more effectively in the physiological range (shaded horizontal bar) by HEPES than by bicarbonate (lower graph). *Source*: Based on Shipman, C., Jr "Control of Culture pH with synthetic buffers." In Tissue culture: Methods and applications, edited by P. F. Kruse, Jr. and M. K. Patterson, 709-712. New York: Academic Press, 1973.

expensive; any changes should be assessed by specific testing (see Section 11.8.2).

9.2.3 Carbon Dioxide (CO_2) and Sodium Bicarbonate ($NaHCO_3$)

CO_2 is present in the atmosphere at around 0.04% and is elevated in gassed incubators to between 2% and 10% (usually 5%). CO_2 in the gas phase (whether at atmospheric or elevated levels) dissolves in culture medium, establishing an equilibrium with bicarbonate ions (HCO_3^-) and lowering the pH. The higher the CO_2 concentration, the more HCO_3^- will be required. Because dissolved CO_2, HCO_3^-, and pH

TABLE 9.1. Relationship between bicarbonate, CO_2, and HEPES.

Compound	Eagle's MEM, Hanks's salts	Low HCO3⁻ + buffer	Eagle's MEM, Earle's salts	DMEM
NaHCO₃	4 mM	10 mM	26 mM	44 mM
CO₂	Atmospheric and evolved from culture	2%	5%	10%
HEPES[a] (if required)	10 mM	20 mM	50 mM	–

[a] If HEPES is used, the equivalent molarity of NaCl must be omitted and osmolality must be checked.

are all interrelated, it is difficult to determine the major direct effect of CO_2 on the cells.

The atmospheric CO_2 tension regulates the concentration of dissolved CO_2 directly, as a function of temperature. This regulation in turn produces H_2CO_3, which dissociates according to the reaction:

$$H_2O + CO_2 \Longleftrightarrow H_2CO_3 \Longleftrightarrow H^+ + HCO_3^- \qquad (9.1)$$

HCO_3^- has a fairly low dissociation constant with most of the cations available in media, so it tends to re-associate, leaving the medium acidic. Thus, the net result of increasing atmospheric CO_2 is to lower the pH. The effect of elevated CO_2 tension can be neutralized by increasing the bicarbonate concentration:

$$NaHCO_3 \Longleftrightarrow Na^+ + HCO_3^- \qquad (9.2)$$

The increased HCO_3^- concentration pushes Eq. 9.1 to the left until equilibrium is reached at pH 7.4. If another alkali (e.g. NaOH) is used instead, the net result is the same:

$$NaOH + H_2CO_3 \Longleftrightarrow NaHCO_3 + H_2O \qquad (9.3)$$
$$\Longleftrightarrow Na^+ + HCO_3^- + H_2O \qquad (9.4)$$

The equivalent $NaHCO_3$ concentrations that are commonly used are listed here and as part of specific medium formulations (see Table 9.1; see also Appendix C). Intermediate values of CO_2 and HCO_3^- may be used, provided that the concentrations of both are varied proportionately. Because many media are made up in acidic solution and may incorporate a buffer, it is difficult to predict how much bicarbonate to use when other alkali may also end up as bicarbonate, as in Eq. 9.4. When preparing a new medium for the first time, add the specified amount of bicarbonate and then sufficient 1 N NaOH such that the medium equilibrates to the desired pH after incubation in a Petri dish at 37 °C, in the correct CO_2 concentration, overnight. When dealing with a medium that is already at working strength, vary the amount of HCO_3^-

to suit the gas phase (see Table 9.1), and leave the medium overnight to equilibrate at 37 °C. Each medium formulation has a recommended bicarbonate concentration and CO_2 tension for achieving the correct pH and osmolality, but minor variations will occur with different methods of preparation.

With the introduction of Good's buffers into tissue culture, there was some speculation that, as CO_2 was no longer necessary to stabilize the pH, it could be omitted. This proved to be untrue (Itagaki and Kimura 1974), at least for a large number of cell types, particularly at low cell concentrations. Although 20 mM HEPES can control pH within the physiological range, the absence of atmospheric CO_2 shifts the equilibrium in Eq. 9.1 to the left, eventually eliminating dissolved CO_2, and ultimately HCO_3^-, from the medium. This chain of events appears to limit cell growth, although whether the cells require the dissolved CO_2 or the HCO_3^- (or both) is not clear. HCO_3^-, CO_2, and HEPES concentrations are interrelated and must be controlled (see Table 9.1).

The addition of pyruvate to the medium enables cells to increase their endogenous production of CO_2, making them independent of exogenous CO_2 and HCO_3^-. Leibovitz's L-15 medium (Leibovitz 1963) contains a higher concentration of sodium pyruvate (550 mg/l) but lacks $NaHCO_3$ and does not require CO_2 in the gas phase. Buffering is achieved via the relatively high amino acid concentrations. Because it does not require CO_2, L-15 medium is sometimes recommended for the transportation of tissue samples. Sodium β-glycerophosphate can also be used to buffer autoclavable media lacking CO_2 and HCO_3^- (Waymouth 1978), and a non-HEPES proprietary medium is available (CO_2 Independent Medium, ThermoFisher Scientific). If elimination of CO_2 is important for cost saving, convenience, or other reasons, it might be worth considering one of these formulations, but only after appropriate testing with the cell line of interest.

In summary, cultures in open vessels need to be incubated in an atmosphere of CO_2, the concentration of which is in equilibrium with the sodium bicarbonate in the medium (see Table 9.1). It may not be necessary to add CO_2 to the gas phase if cells are at moderately high concentration ($\geq 1 \times 10^5$ cells/ml) and are grown in sealed flasks, provided the bicarbonate concentration is kept low (~ 4 mM), particularly if the cells are high acid producers. However, it will be necessary to add CO_2 to the gas phase at low cell concentrations (e.g. during initial seeding at lower cell densities for cloning or other applications) and with some primary cultures. When venting of the culture vessel is required to allow absorption of CO_2 (or its escape in the case of high acid producers), it is necessary to leave the cap slack or to use a CO_2-permeable cap (see Figure 8.8).

9.2.4 Oxygen

The other major constituent of the gas phase is oxygen. As with CO_2, the concentration of dissolved oxygen in culture medium will be determined by its concentration in the over-lying atmosphere. Because the depth of the culture medium can influence the rate of oxygen diffusion to the cells (Buck et al. 2014), it is advisable to keep the medium depth at 2–5 mm (0.2–0.5 ml/cm^2) in static culture. Cultures vary in their oxygen requirements. Some organ cultures (particularly from late-stage embryos, newborns, or adults) require up to 95% oxygen in the gas phase (Trowell 1959; De Ridder and Mareel 1978). This requirement may be a problem of diffusion, related to the geometry and gaseous penetration of organ cultures (see Section 27.8), or it may arise due to differences between differentiated and rapidly proliferating cells. Oxygen diffusion may also become limiting in porous microcarriers (see Sections 27.5.4, 28.3.3).

In the absence of an appropriate carrier, such as hemoglobin, reactive oxygen species will be present that are toxic to the cell. Providing the correct oxygen tension is always a compromise between satisfying respiratory requirements and avoiding toxicity. Whereas most cells require oxygen for respiration *in vivo*, cultured cells often rely mainly on glycolysis (see Section 9.4.4). In transformed cells, a high proportion of glycolysis is anaerobic, but oxygen is still required (Green et al. 1958; Fleischaker and Sinskey 1981). Oxygen toxicity may be reduced by adding free radical scavengers to the medium such as glutathione, 2-mercaptoethanol (β-mercaptoethanol), or dithiothreitol. It has been suggested that the requirement for selenium in medium is related to oxygen toxicity, as selenium is a cofactor in glutathione synthesis (McKeehan and Ham 1976). Oxygen tolerance – and selenium as well – may be provided by serum, so the control of oxygen tension is likely to be more critical in serum-free media.

Culture at decreased oxygen concentration is preferred by many laboratories because it is closer to physiological conditions. This requires specialized equipment for incubation and handling, as described in Minireview M9.1.

Minireview M9.1. **Minireview: Hypoxic Cell Culture** Contributed by Bob Geraghty, Cancer Research UK Cambridge Institute, University of Cambridge.
The *in vitro* laboratory culture of mammalian cells is still routinely performed in incubators at a temperature of 37 °C and in an atmospheric oxygen concentration of approximately 21%, supplemented with an enhanced CO_2 concentration of between 5% and 10%. The latter is used to facilitate the CO_2/bicarbonate buffering system commonly used in cell culture media to maintain a physiological pH of between 6.8 and 7.4 (see Sections 9.2.2, 9.2.3). However, dissolved oxygen concentrations measured in human tissues *in vivo* are never as high as 21%. Most mammalian tissues have a dissolved oxygen concentration of between 2% and 9%, depending on their distance from an oxygenated blood supply (Bertout et al. 2008; Haque et al. 2013). Standard cell culture conditions are therefore a poor representation of conditions found *in vivo* and it has been known for many

years that mammalian cell cultures often show an increased rate of growth in low oxygen conditions similar to those found *in vivo* (Cooper et al. 1958; Richter et al. 1972). Embryologists also routinely culture embryos in low oxygen environments (Thompson et al. 1990; Catt and Henman 2000), and human embryonic stem cells (hESCs), grown in low oxygen concentrations of around 2%, have been shown to exhibit improved clonal recovery, reduced chromosomal abnormalities, and less morphological differentiation than those grown in normoxic conditions (Forsyth et al. 2006).

Hypoxia, defined as <2% oxygen (Bertout et al. 2008), may occur within tissues if the demand for oxygen exceeds its supply and the physiological oxygen tension of the tissue cannot be maintained. In non-pathological cells the process of oxygen homeostasis is highly regulated with various cellular mechanisms which allow the cell to "sense" oxygen (Brunelle et al. 2005; Mansfield et al. 2005) and to respond and adapt to hypoxic conditions. This adaption to hypoxia is largely controlled by the hypoxia inducible factors (HIFs) which are able to control the transcription of specific genes in hypoxic conditions and drive or strongly influence many important cellular processes, including metabolism, senescence, autophagy, hematopoiesis, pH regulation, inflammation, vascular response to hypoxia, and angiogenesis (Koh et al. 2008; Majmundar et al. 2010).

Low oxygen conditions and the HIFs have a role in regulating pluripotency, proliferation, and differentiation of human embryonic and mesenchymal stem cells. This has generated significant research interest given their potential uses in tissue engineering, regenerative medicine, and stem cell therapies (Ma et al. 2009; Haque et al. 2013). The role of hypoxia in cancer is also generating significant research interest (Bertout et al. 2008). It is known that solid tumors are less oxygenated than their normal tissue counterparts (see Figure 9.3) and the extent of hypoxia and related HIF expression within tumors has a significant impact on tumor progression. An increase in tumor hypoxia is indicative of an aggressive tumor phenotype, increased metastatic potential, therapeutic resistance, and poor patient survival (Brahimi-Horn et al. 2007), and a better understanding of the pathways involved in the hypoxic environment during tumor progression could contribute to breakthroughs in cancer therapy (Jing et al. 2019).

Scientists researching any of the cellular processes described will therefore need to culture cells in hypoxic or reduced oxygen concentrations, and equipment designed to facilitate this is now readily available. The simplest option is to use a hypoxia chamber. These are small, self-contained, sealed "boxes" that are designed to fit inside any standard temperature-controlled laboratory cell culture incubator (see Figure 9.4). Cell cultures are placed inside the chambers which are then flushed and filled with the required low oxygen gas mixture. Hypoxia chambers are relatively cheap and no other specialist equipment, other than the correct gas mixture, is required. Because of their small size and volume, they use less gas and are virtually maintenance free and are a good option for laboratories

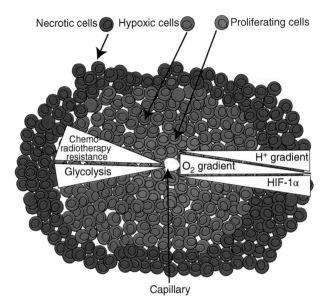

Necrotic cells ● Hypoxic cells ● Proliferating cells

Fig. 9.3. Schematic representation of a hypoxic tumor mass. Oxygen concentration decreases with distance from capillaries. Typically, rapidly dividing tumor cells will be found close to the capillaries, with non-dividing and then necrotic cells increasing with distance from the capillary. This gradient of cell viability will correspond to a decreasing oxygen concentration gradient, an increase in HIF-1α levels, a decrease in extracellular pH, and an increase in resistance to radio- and chemotherapies. *Source:* Courtesy of Bob Geraghty.

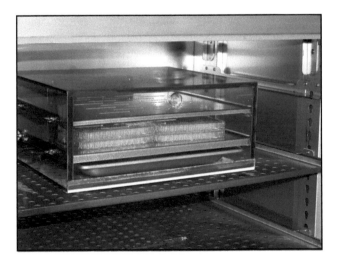

Fig. 9.4. Hypoxia chamber. This example is a custom-made sealed, gassable, plastic hypoxia chamber. *Source:* Courtesy of Reeve Irvine Institute.

that only occasionally perform hypoxic cell culture on a small scale.

The alternative is to use a tri-gas incubator. These are full sized incubators with similar dimensions to standard laboratory CO_2 incubators and are available from most major manufacturers and suppliers of laboratory CO_2 incubators.

They require a supply of CO_2 and N_2, with the N_2 being used to displace oxygen levels within the incubator chamber, allowing oxygen concentrations as low as 1–2% to be achieved and maintained. Tri-gas incubators often have a divided inner chamber with up to three compartments, each with its own inner door (see Figure 9.5). This allows one compartment to be opened without losing the atmosphere in the remaining compartments – a useful feature for larger-scale experiments – and means that cell culture vessels can be removed from one compartment without affecting the atmosphere in the others. A divided chamber also helps reduce the amount of gas used. Modern tri-gas incubators usually have inbuilt, automated decontamination functions. This is an important consideration, particularly where cells are being cultured for potential human therapeutic use and must be absolutely free of microbial contamination. The use of antimicrobial copper or high-content copper alloy interiors can also reduce the risk of contamination, as can in-chamber high-efficiency particulate air (HEPA) filtration.

A major disadvantage, also found with the small hypoxia chambers, is that each time cell cultures are removed for microscopic examination, drug additions, media changes, passage, or other manipulations, the low oxygen atmosphere will be lost and the cells exposed to an ambient atmosphere and temperature. Even very short exposures to ambient oxygen can adversely affect the results of hypoxic cell culture experiments (Olbryt et al. 2014).

These fluctuations in the oxygen concentration can be minimized by using a hypoxia workstation or "glove-box" (see Figure 9.6). These are similar to a Class III BSC (see Figure 6.7c) and are completely isolated from the ambient environment. They are available from a number of specialist manufacturers including Don Whitley Scientific, Baker Ruskinn, HypOxygen, Oxford Optronix, Plas-Labs, and BioSpherix. This type of workstation is by far the best option for hypoxic cell culture work. They are large enough to perform the entire cell culture process without having to remove the cell culture vessels at any stage and expose them to ambient conditions. Access to the workstation chamber is only possible through glove ports or a pass-through interlock, used to move media, reagents, and consumables into the workstation chamber. Hypoxia workstations vary in size from small bench-mounted units suitable for occasional small-scale hypoxic cell culture work, up to much larger modular workstations consisting of interconnected chambers capable of housing an entire cell culture laboratory, including incubators, microscopes, and live cell imaging equipment. However, hypoxia workstations are expensive, bulky pieces of equipment; they require a dedicated CO_2 and nitrogen gas supply and will consume large volumes of these gases when operating. Manipulation of cell cultures through glove ports can also be difficult and uncomfortable. Being sealed units, any microbial contamination can be very difficult to eliminate and will involve shutting the equipment down and either fumigating the chamber using a chemical fumigant, or taking

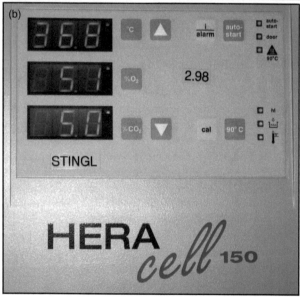

Fig. 9.5. Tri-gas incubator. Heraeus HERA cell 150 tri-gas cell culture incubator (ThermoFisher Scientific), operating with a 5% oxygen and 5% CO_2 atmosphere. (a) Incubator showing three-door access; (b) read-out of temperature and oxygen and CO_2 concentrations. *Source*: Bob Geraghty, courtesy of Cancer Research UK Cambridge Institute.

the workstation apart and cleaning internal surfaces with a suitable disinfectant. However, the ability of the hypoxia workstation to maintain precise, controlled, and continuous low oxygen or hypoxic conditions means that it is the best choice for all *in vitro* cell culture experiments that need to closely and continuously reproduce *in vivo* conditions.

A further advance in reproducing the *in vivo* microenvironment was made by Xcell Biosciences in 2017 with the introduction of the AVATAR™ Advanced Primary Cell Culture Incubator (see Figure 9.7). As well as allowing accurate fine control of oxygen concentrations within the incubator as low as 0.1%, the AVATAR allows control of pressure within the incubator up to a maximum of 5.0 psi. This allows cell culture conditions to be established that closely mimic the *in vivo* microenvironment, as the interstitial fluid pressure in mammalian tissues is higher than atmospheric pressure and in the case of solid tumors can be as high as 3.0 psi. The manufacturers claim that controlling oxygen concentration and pressure allows improved propagation of patient tumor cells, macrophage and induced pluripotent stem cell (iPSC) generation, and enhanced expansion of primary cells.

One other important consideration to be aware of is that, unless treated, the dissolved oxygen concentration in cell culture media and other reagents used for low oxygen or hypoxic cell culture work is likely to be close to ambient (Newby et al. 2005). Media and reagents must therefore be preconditioned or degassed before use. The simplest method involves placing the medium in a hypoxia chamber or incubator set to the required oxygen concentration and allowing it to equilibrate (Newby et al. 2005). To prepare larger volumes of preconditioned media there are dedicated automated conditioning units available such as the HypoxyCOOL™ (Baker Ruskinn). These are capable of reducing the dissolved oxygen concentration of up to 4.5 l of cell culture medium down to 2% in under three hours and will maintain media sterility, temperature, and pH. Cell culture media can also be conditioned by sparging with sterile filtered nitrogen gas, to displace dissolved oxygen. Using sparging, it is possible to reduce the oxygen concentration down to approximately 1.5% after 15 minutes, or down to 0% after 30 minutes, without affecting pH (Newby et al. 2005). Vacuum degassing can also be used. This method will remove up to 85% of the dissolved gasses from the medium and if kept under vacuum until required the medium will not regas (Newby et al. 2005). However, because all dissolved gasses (including CO_2) will be removed, HEPES buffered cell culture media should be used as these do not rely on a CO_2/bicarbonate buffering system to maintain pH. But be aware that HEPES can be toxic and some cells, particularly at low cell densities, may require CO_2/bicarbonate (see Section 9.2.3). One other method of lowering the dissolved oxygen concentration in cell culture media involves adding glucose oxidase and catalase. This two-enzyme system catalyzes the oxidation of glucose to gluconic acid utilizing

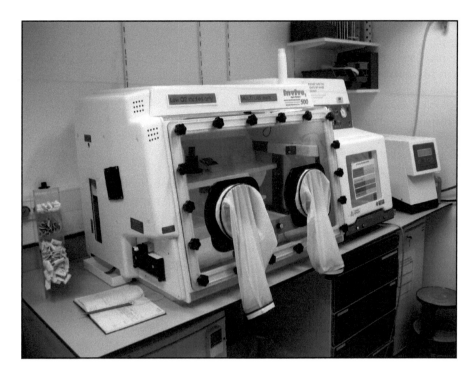

Fig. 9.6. Hypoxia workstation. This example is a Ruskinn InvivO$_2$ 500 Hypoxia workstation at the Cancer Research UK Cambridge Institute. *Source*: Bob Geraghty; image courtesy of Cathy Pauley, CRUK CI Laboratory Management.

Fig. 9.7. AVATAR Advanced Primary Cell Culture Incubator. (a) This incubator (Xcell Biosciences) allows fine control of temperature, CO$_2$, oxygen, and pressure levels, as shown by (b) a closer view of the front panel, to accurately reproduce the *in vivo* microenvironment. *Source*: Bob Geraghty; image courtesy of CRUK CI, Research Instrumentation and Cell Services Core Facility.

dissolved oxygen in the medium and can lower the dissolved oxygen to 1% in under 10 minutes (Baumann et al. 2008). For all of the methods described, a portable dissolved oxygen meter with a sterile electrode will be required to accurately measure dissolved oxygen concentrations.

9.2.5 Temperature

The optimal temperature for cell culture depends on (i) the body temperature of the animal from which the cells were obtained; (ii) any anatomic variation in temperature (e.g. the optimal temperature for cells derived from skin or testis may be lower than that for cells from the rest of the body); and (iii) the incorporation of a safety factor to allow for minor errors in regulating the incubator. Thus, the temperature recommended for most warm-blooded animal cell lines is 37 °C, close to body heat but set a little lower for safety, as overheating is a more serious problem than underheating. Cultured mammalian cells will tolerate considerable drops in temperature and can survive several days at 4 °C, but they cannot tolerate more than about 2 °C above normal (39.5 °C) for more than a few hours and will die quite rapidly at temperatures ≥40 °C.

The incubation temperature should be maintained within ±0.5 °C to ensure reproducible results. Doors of incubators or warmrooms must not be left open longer than necessary and large items or volumes of liquid should be prewarmed before exposing cells to them. The spatial distribution of temperature within the incubator or warmroom must also be uniform (see Sections 4.2.2, 5.3), with no cold spots and with free circulation of air within the chamber. This means that a large number of flasks should not be stacked together when first placed in the incubator or warmroom; space must be allowed between them for air to circulate. Because the gas phase within the flasks expands at 37 °C, flasks (particularly large flasks and when stacked) must be vented by briefly slackening the cap 30–60 minutes after placing the flasks in the incubator or warmroom. If vented caps are used in a CO_2 incubator, this is unnecessary.

Cells from poikilotherms (organisms whose body temperature changes with their environment; see Section 24.6) tolerate a wide temperature range, between 15 and 26 °C. Simulating *in vivo* conditions (e.g. for cold-water fish) may require an incubator with cooling as well as heating, to keep the incubator temperature below ambient. If necessary, poikilothermic animal cells can be maintained at room temperature, but the variability of the ambient temperature in laboratories makes this undesirable, and a cooled incubator is preferable. In contrast, birds have a higher body temperature. Avian cells should be maintained at 38.5 °C for maximum growth, but will grow quite satisfactorily (if more slowly) at 37 °C.

A number of temperature-sensitive (ts) mutant cell lines have been developed that allow the expression of specific genes below a set temperature, but not above it (Wyllie et al. 1992). These mutants facilitate studies on cell regulation, but also emphasize the narrow range within which one can operate; the two discriminating temperatures are usually only about 2–3 °C apart. The use of ts mutants may require an incubator with cooling as well as heating, to compensate for a warm ambient temperature.

Apart from its direct effect on cell growth, temperature will influence pH due to the increased solubility of CO_2 at lower temperatures and, possibly, because of changes in ionization and the pK_a of the buffer. The pH should be adjusted to 0.2 units lower at room temperature than at 37 °C, as it will rise during incubation. In preparing a medium for the first time, it is best to make up the complete medium and incubate a sample overnight at 37 °C under the correct gas tension, in order to check the pH (see Section 9.2.3).

9.2.6 Osmolality

Most cultured cells have a fairly wide tolerance for osmotic pressure (Waymouth 1970). As the osmolality of human plasma is about 290 mOsm/kg water, it is reasonable to assume that this level is the optimum for human cells *in vitro*, although it may vary for some species or cell types, e.g. 310 mOsm/kg is optimum for mice and 350 mOsm/kg for hESCs (Waymouth 1970; Ludwig et al. 2006). In practice, osmolalities between 260 and 320 mOsm/kg are quite acceptable for most cells but, once selected, should be kept consistent at ±10 mOsm/kg. Slightly hypotonic medium may be better for Petri dish or open-plate culture to compensate for evaporation during incubation. Dulbecco's modified Eagle's medium (DMEM) (see Section 9.4) is relatively hypertonic, with an osmolality of about 340 mOsm/l; this may be too high for some normal cells (Price 2017).

An osmometer is used to measure osmolality (see Figure 9.8). Osmometer measurements are based on depression of the freezing point, or elevation of the vapor pressure, of the medium. Changes in osmolality are generally achieved

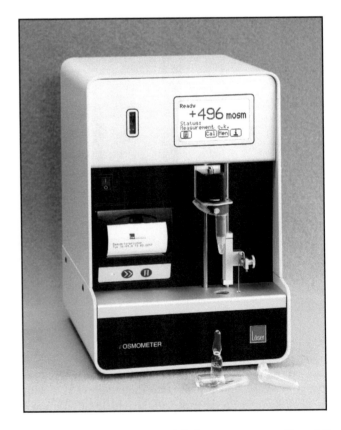

Fig. 9.8. Osmometer. This model (Löser osmometer, Type 16) accepts samples of 50–100 μl. *Source:* Image courtesy of Camlab Ltd with permission.

by altering the sodium chloride concentration, e.g. to compensate for different bicarbonate concentrations (see Table 9.1) or to allow for the addition of HEPES. The measurement of osmolality is a useful quality control test if you are making up the medium yourself, as it helps to guard against errors in weighing, dilution, etc. It is particularly important to monitor osmolality if alterations are made in the constitution of the medium. The addition of HEPES and drugs dissolved in strong acids and bases (with subsequent neutralization) can all markedly affect osmolality.

9.2.7 Viscosity

The viscosity of a culture medium is influenced mainly by the serum content. Viscosity may play a role in some processes due to macromolecular crowding (Rashid et al. 2014), but in most cases has little effect on cell growth. Viscosity becomes more important when a cell suspension is agitated (e.g. when a suspension culture is stirred) or when cells are dissociated after trypsinization. This is particularly relevant for low-serum or serum-free media and for stirred bioreactors, where agitation can result in the development of eddies (which dissipate power) and cell damage due to localized shearing (see Sections 10.4.3, 28.2.4) (Cherry and Papoutsakis 1988; Sen et al. 2002). Cell damage may be reduced in these conditions by adding polymers to increase the viscosity of the medium. Carboxymethylcellulose (CMC, Sigma-Aldrich) has been used successfully as a shear protectant at concentrations of 0.5–1.0% when growing human pluripotent stem cells (hPSCs) in a stirred bioreactor (Abbasalizadeh et al. 2012). Other agents that have been used to increase viscosity include polyvinylpyrrolidone (PVP) and various dextran preparations (Mizrahi and Moore 1970; Sen et al. 2002).

9.2.8 Surface Tension and Foaming

Foaming can occur when suspension cultures are agitated in spinner flasks or bioreactors, or when 5% CO_2 in air is bubbled through medium containing serum. Foam formation results in cell damage and reduced productivity during scale-up (see Minireview M28.5). The mechanisms for cell damage are not entirely clear. Cells may be trapped within the foam layer or damaged when bubbles rupture (Hu et al. 2011). Foaming may limit gaseous diffusion if a film from a foam or spillage gets into the capillary space between the lid and the base of a Petri dish or between a slack cap and the neck of a flask. The rate of protein denaturation may also increase, as may the risk of contamination if the foam reaches the neck of the culture vessel.

Although cells may be less sensitive to hydrodynamic stress than first thought (Nienow 2006; Godoy-Silva et al. 2009), CO_2 should not be bubbled through the medium when gassing a flask as this may generate foam and can also spread aerosols from the flask. Foaming in spinner flasks or bioreactors can be prevented by adding an antifoam agent. Various agents are available including Pluronic™ F-68 (various suppliers, used at 0.01–0.1%), Antifoam 204, Antifoam SE-15, EX-CELL® Antifoam (all from Sigma-Aldrich), and FoamAway™ (ThermoFisher Scientific). These agents alter the surface tension of the culture medium and may also protect cells against shear stress; Pluronic F-68 results in improved viability even in a bubble-free environment, which may relate to changes in the cells' hydrophobicity (Tharmalingam et al. 2008). However, antifoam agents can also have adverse effects and their impact should be assessed before selecting a specific agent for a cell culture application (Velugula-Yellela et al. 2018).

9.3 BALANCED SALT SOLUTIONS

A balanced salt solution is composed of inorganic salts and may include sodium bicarbonate and glucose (see Table 9.2). Balanced salt solutions were developed to imitate the natural cellular environment, minimize cellular trauma, and maximize maintenance of function (Waymouth 1970). Commonly used solutions include Earle's balanced salt solution (EBSS), Hanks's balanced salt solution (HBSS), and Dulbecco's phosphate buffered saline (DPBS) (Earle et al. 1943; Hanks and Wallace 1949; Dulbecco and Vogt 1954). "DPBS" is used to distinguish Dulbecco's formulation from simple phosphate-buffered saline (PBS), which lacks potassium chloride and is simply isotonic sodium chloride with phosphate buffer. Balanced salt solution recipes are often modified – for instance, by omitting glucose or phenol red from HBSS or by leaving out Ca^{2+} or Mg^{2+} ions from DPBS. DPBS without Ca^{2+} and Mg^{2+} is traditionally known as PBS Solution A or "DPBS-A," as used throughout this book. Always check for modifications when purchasing a balanced salt solution and quote any modifications to the published formula in publications.

Balanced salt solutions are used as the basis of many complete media (see Section 9.4). For example, commercial suppliers will provide Minimal Essential Medium (MEM) (see Section 9.4) with Hanks's salts or with Earle's salts, indicating which formulation was used. Inclusion of Hanks's salts requires sealed flasks with a gas phase of air, whereas Earle's salts imply a higher bicarbonate concentration compatible with growth in 5% CO_2. Balanced salt solutions are also used as diluents for concentrates of amino acids and vitamins to make complete media, as isotonic washes or dissection media, and for short incubations (up to about four hours, usually with glucose present).

The choice of balanced salt solution is dependent on both the CO_2 tension (see Section 9.2.3) and the intended use of the solution. Ca^{2+} and Mg^{2+} are usually omitted when balanced salt solutions are used for tissue disaggregation, monolayer dispersal, or suspension culture. Cells in suspension generally need reduced levels of Ca^{2+} and will grow in a concentration of about 20 mg/l $CaCl_2$ with less clumping (Price

TABLE 9.2. Balanced salt solutions.

Component	MW	EBSS g/l	EBSS mM	Without Ca²⁺ and Mg²⁺ (DPBS-A) g/L	Without Ca²⁺ and Mg²⁺ (DPBS-A) mM	With Ca²⁺ and Mg²⁺ g/L	With Ca²⁺ and Mg²⁺ mM	HBSS g/l	HBSS mM	Spinner salts (as in S-MEM) g/l	Spinner salts (as in S-MEM) mM
Reference		(Earle et al. 1943)		(Dulbecco and Vogt 1954)				(Hanks and Wallace 1949)		(Eagle 1959)	
Inorganic salts											
CaCl₂ (anhydrous)	111	0.20	1.8	–	–	0.10	0.90	0.14	1.3	–	–
KCl	74.55	0.40	5.4	0.20	2.7	0.20	2.7	0.40	5.4	0.40	5.4
KH₂PO₄	136.1			0.20	1.5	0.20	1.5	0.06	0.44		
MgCl₂.6H₂O	203.3			–	–	0.10	0.49	0.10	0.49	0.2	1.0
MgSO₄.7H₂O	246.5	0.20	0.83					0.10	0.41		
NaCl	58.44	6.8	116	8.0	137	8.0	137	8.0	137	6.8	116
NaHCO₃	84.01	2.2	26					0.35	4.2	2.0	24
Na₂HPO₄.7H₂O	268.1			2.2	8.1	2.2	8.1	0.09	0.34		
NaH₂PO₄.H₂O	138	0.14	1.0							1.4	10
Other components											
D-glucose	180.2	1.0	5.5					1.0	5.5	1.0	5.5
Phenol red[b]	354.4	0.05	0.14					0.02	0.06	0.02	0.06
Gas phase		5% CO₂				Air		Air		5% CO₂	

[a]DPBS was originally made up from three solutions. Solution (a) consisted of NaCl, KCl, Na₂HPO₄, and KH₂PO₄; solution (b) was CaCl₂; and solution (c) was MgCl₂ (Dulbecco and Vogt 1954).

[b]The amounts of phenol red in this table vary from some other sources and are derived from Paul (1975).

2017). Spinner's MEM (SMEM), based on Eagle's Spinner salt solution, is deficient in Ca²⁺ in order to reduce cell aggregation and attachment (see Table 9.2). Ca²⁺ can be added to SMEM to reach the optimal concentration for any intended use (Price 2017).

EBSS and HBSS contain bicarbonate; these and DPBS also rely on phosphate buffering, which is not at its most effective at physiological pH. Tris-buffered balanced salt solution is more effective (Paul 1975), but cells sometimes require a period of adaptation and/or selection. HEPES (10–20 mM) is currently the most effective buffer in the pH 7.2–7.8 range (see Figure 9.2), and Tricine in the pH 7.4–8.0 range, although both tend to be expensive if used in large quantities. If HEPES is used, the equivalent amount of NaCl should be omitted to maintain the correct osmolality.

9.4 MEDIA FORMULATIONS

The term "complete medium" implies a medium that is sufficient for tissue culture. It usually consists of a defined basal medium component and supplements such as serum; unstable constituents may be added just before use (e.g. glutamine; see Section 9.4.1). The most commonly used basal media were developed to support particular cell lines or conditions but have now become classic formulations with general applications (Morton 1970). Widely used, classic media include (see Appendix C for media formulations):

(1) **Medium 199 (M199)** (Morgan et al. 1950). Developed for nutritional studies of chick embryonic fibroblasts without serum. It continues to be used for organ culture.

(2) **Eagle's Basal Medium (BME)** (Eagle 1955). Developed for culture of L-929 and HeLa cells.

(3) **CMRL 1066** (Parker et al. 1957). Modified from Medium 199 to grow L-929 cells without serum. It has since been used alone or in combination with other media for more demanding applications (e.g. cloning of monkey kidney cells).

(4) **MB 752/1** (Waymouth 1959). Developed to study the nutritional requirements of L-929 cells using a minimal number of components in serum-free conditions. It continues to be used for a variety of cell types and for organ culture.

(5) **McCoy's 5A** (McCoy et al. 1959). Developed to study the nutritional requirements of Walker 256 carcinoma cells. It is a complex medium that has been used for many types of primary cultures (including explant culture) and cell lines.

(6) **Eagle's Minimum Essential Medium (MEM)** (Eagle 1959). Modified from BME by increasing the range and concentration of its constituents.

TABLE 9.3. Essential and non-essential amino acids.

Amino acid (L-form)	In vivo	In vitro MEM EE[a]	MEM NEAA[b] Supplement	Ham's F12 complete
Alanine	NE	NE	No	Yes
Arginine	NE	E	No	Yes
Asparagine	NE	NE	Yes	Yes
Aspartic acid	NE	NE	Yes	Yes
Cystine	NE	E	No	No
Cysteine	NE	NE	No	Yes
Isoleucine	E	E	No	Yes
Glutamic acid	NE	NE	Yes	Yes
Glutamine	NE	E	Yes	Yes
Glycine	NE	NE	Yes	Yes
Histidine	E	E	No	Yes
Leucine	E	E	No	Yes
Lysine	E	E	No	Yes
Methionine	E	E	No	Yes
Phenylalanine	E	E	No	Yes
Proline	NE	NE	Yes	Yes
Serine	NE	NE	Yes	Yes
Threonine	E	E	No	Yes
Tryptophan	E	E	No	Yes
Tyrosine	NE	E	No	Yes
Valine	E	E	No	Yes

[a] Essential amino acids as defined in Eagle's MEM.
[b] Non-essential amino acids available as a supplement from suppliers (e.g. Sigma-Aldrich) and found in αMEM.
Abbreviations not found in main text: E, essential; NE, non-essential.

(7) **DMEM** (Dulbecco and Freeman 1959). Modified from MEM for culture of mouse fibroblasts in transformation and virus propagation studies. It has twice the amino acid concentrations and four times the vitamin concentrations of MEM, and uses twice the HCO_3^- and CO_2 concentrations to achieve better buffering.

(8) **Ham's F12** (Ham 1965). Developed to clone Chinese hamster ovary (CHO) cells in low-serum or serum-free medium. The "F" series (including F12 and its predecessor F10) was developed from NCTC 109 (see Section 10.3) for culture of fibroblast-like cells. F12 is now widely used for clonogenic assays and primary culture (see Protocols P19.4, P20.1; see also Section 13.2.3).

(9) **DMEM/F12** (Barnes and Sato 1980). A 1:1 mixture of DMEM and Ham's F12 that combines the richness of F12 with the higher nutrient concentration of DMEM. Although not necessarily optimal for all cell lines, this combination has provided an empirical formula that is effective for serum-free formulations and for supplementation with special additives for many different cell types.

(10) **RPMI 1640** (Moore et al. 1967). Modified from McCoy's 5A for suspension culture and lacks calcium. Although

originally used for lymphoid cells, it is widely used for adherent cultures.

(11) **αMEM** (Stanners et al. 1971). Modified from MEM for somatic cell hybridization studies. It has additional amino acids and vitamins, as well as nucleosides and lipoic acid, and has been used for a wide range of cell types.

(12) **L-15** (Leibovitz 1963). Developed to provide buffering in the absence of HCO_3^- and CO_2. It is often used for transport and primary culture for this reason, but its value was diminished by the introduction of HEPES and the demonstration that HCO_3^- and CO_2 are often essential for optimal cell growth, regardless of the requirement for buffering. It continues to be used for culture of poikilothermic cells, e.g. fish cell lines (see Section 24.6).

Which media formulations are the most popular? Eagle's MEM was the most widely used formulation for many years. The most popular basal media in modern tissue culture laboratories are DMEM, MEM, and RPMI 1640, making up about 70% of total sales; blended DMEM/F12 adds a further 7% of sales (Arora 2013). Other formulations seldom account for more than 5% of the total.

9.4.1 Amino Acids

Amino acids are defined as essential or non-essential, depending on whether they can be synthesized in the body. The terms are also used in culture media to define what is essential or non-essential for growth *in vitro* (see Table 9.3) (Eagle 1955). Requirements are different for tissue culture because some amino acids are synthesized in the liver *in vivo*. Essential amino acids *in vitro* include arginine, cystine and/or cysteine, glutamine, and tyrosine. Non-essential amino acids (NEAA) are often added to culture medium as well, either because a particular cell type is unable to make them or because they are lost by leakage into the medium. NEAA are available as supplements or included in the formulation, as in αMEM (Stanners et al. 1971).

Individual requirements for amino acids will vary from one cell type to another. Eagle's BME was developed using mouse fibroblasts (L-929) and human cervical carcinoma cells (HeLa). Different cell types may have different requirements – for example, aspartic acid and glutamic acid are toxic to postnatal neurons (Price 2017). Requirements will also vary with the cell yield; aspartic acid, glutamic acid, and cysteine are self-limiting for most CHO cell lines grown to high density (Price 2017). Nutrient concentrations are, on the whole, low in Ham's F12 (which was optimized by cloning) and high in DMEM (which was optimized for growth at higher cell densities for viral propagation) (Dulbecco and Freeman 1959; Ham 1965).

Glutamine is required by most cells, although some cell lines will utilize glutamate. Evidence suggests that glutamine is mainly used by cultured cells as a source of energy and carbon (see Section 9.4.4) (Butler and Christie 1994). Glutamine is unstable in culture medium and generates toxic ammonia as a by-product (Hassell et al. 1991). The rate of decomposition is temperature-dependent; glutamine has a half-life of three to five days at 37 °C. Glutamine is normally added just before use and its lability may then determine the shelf life of the medium to which it has been added (see Section 9.8). Glutamine dipeptides are more stable at 37 °C; the most popular, GlutaMAX™ (ThermoFisher Scientific), is an alanyl-glutamine dipeptide that is bioavailable due to the action of dipeptidase.

9.4.2 Vitamins

Like the amino acids, vitamin requirements have been determined empirically (Eagle 1955). Eagle's MEM contains only water-soluble vitamins (the B group, plus choline, folic acid, inositol, and nicotinamide, but excluding biotin; see Appendix C). Other required vitamins presumably are provided by serum. Biotin is present in most of the more complex media, including the serum-free recipes; *p*-aminobenzoic acid (PABA) is present in Medium 199, CMRL 1066 (which was derived from Medium 199), and RPMI 1640. The fat-soluble vitamins (A, D, E, and K) are present only in Medium 199; vitamin A is present in LHC-9

and vitamin E in MCDB 110 (see Appendix C). Vitamin selection often relates to the cell line used in medium development, e.g. Fischer's medium has a high folate concentration because of the folate dependence of L5178Y, which was used in the development of the medium (Fischer and Sartorelli 1964). Vitamin selection is also influenced by practicality. For example, the presence of pyridoxal in DMEM resulted in a precipitate that shortened its shelf life, leading to its substitution by pyridoxine (Price 2017).

9.4.3 Inorganic Salts

The inorganic ions in culture medium are primarily based on the composition of balanced salt solution (see Section 9.3). Most media derive their salt concentrations from EBSS (high bicarbonate; gas phase, 5% CO_2) or HBSS (low bicarbonate; gas phase, air). Na^+, K^+, Mg^{2+}, Ca^{2+}, Cl^-, SO_4^{2-}, PO_4^{3-}, and HCO_3^- are the major components that contribute to medium osmolality (see Section 9.2.6). Divalent cations, particularly Ca^{2+}, are required by some cell adhesion molecules, such as the cadherins (see Section 2.2.2). The concentration of Ca^{2+} in the medium can also determine whether cells proliferate or differentiate (see Section 26.3.1) (Yuspa et al. 1981; Boyce and Ham 1983). Na^+, K^+, and Cl^- regulate membrane potential, whereas SO_4^{2-}, PO_4^{3-}, and HCO_3^- have roles as anions required by the matrix and nutritional precursors for macromolecules, as well as regulators of intracellular charge.

9.4.4 Glucose

Glucose (4–20 mM) is included in most media as an energy source. Glucose is converted to pyruvate by glycolysis, which may be converted to lactate or acetoacetate and may enter the citric acid cycle and be oxidized to form CO_2 and water (Alberts et al. 2015). Glutamine and pyruvate are also used as energy sources *in vitro* (Price 2017). Glutamine can be utilized as a carbon source by oxidation to glutamate by glutaminase and entry into the citric acid cycle by transamination to 2-oxoglutarate. Continuous cell lines tend to engage in "wasteful" glucose and glutamine metabolism that leads to accumulation of inhibitory byproducts such as lactate and ammonia (Young 2013). This behavior is particularly associated with their rapid growth; many cell lines are able to "rewire" their metabolism depending on their proliferation rate and access to nutrients. In the plateau phase of the growth cycle (see Figure 14.1), cells may decrease their lactate production or even switch to lactate consumption (Templeton et al. 2013).

9.4.5 Other Components

A variety of other compounds appear in complex media, including proteins, peptides, nucleosides, citric acid cycle intermediates, pyruvate, and lipids. These constituents have been found to be necessary when the serum concentration is reduced, and they may help in cloning and in maintaining certain specialized cells, even in the presence of serum. Phenol red is commonly added as a pH indicator (see Section 9.2.1).

9.5 SERUM

Serum is a complex mixture of constituents, many of which are not well defined. It is a popular supplement because of its antitrypsin activity, its positive effects on cellular proliferation and attachment, and its protective, detoxifying actions (e.g. serum viscosity can protect against shear damage) (Price 2017). Serum is a source of growth factors, lipids, inorganic salts, and hormones, many of which may be bound to protein (see Table 9.4). However, serum has many disadvantages and there are an increasing number of options for serum-free culture systems (see Sections 10.1, 10.5). It is important to consciously decide whether serum should be used or can be replaced.

Bovine sera are derived from fetal (unborn calves), newborn (< 3 weeks old), calf (3 weeks–12 months), or adult (> 12 months) animals (Nims and Harbell 2017). Sera should be accompanied by information on the country of origin, ideally with traceability certification from an industry body such as the International Serum Industry Association (ISIA) (OECD 2018). Approximately 45% of serum is sourced from three countries (New Zealand, Australia, and the United States) because of their low risk of infectious diseases and robust infrastructure for animal management (ISIA 2018). Fetal bovine serum (FBS) is the most popular serum choice, particularly for more demanding cell lines and for cloning. Calf serum has weaker growth-promoting activity, which can be useful for studying contact inhibition or differentiation (Rubin and Xu 1989; Yao and Asayama 2017). Calf and adult sera have higher lipid levels compared to FBS, which may influence the fatty acid composition of continuous cell lines (Stoll and Spector 1984; Yao and Asayama 2017).

Non-bovine sera are also available commercially. Horse serum is preferred to calf serum by some laboratories, as it can be obtained from a closed donor herd and is often more consistent from batch to batch. Cultures grown in horse serum may also be less likely to metabolize polyamines, due to lower levels of polyamine oxidase; polyamines are mitogenic for some cells (Hyvonen et al. 1988; Kaminska et al. 1990). Human serum is used for culture of some human stem cell lines (see Section 10.7). This may prove to be a safer approach for biological products or cell-based therapies because testing can be done in accordance with blood bank standards (Dos Santos et al. 2017).

9.5.1 Protein

Although proteins are a major component of serum, the functions of many of these proteins *in vitro* remain obscure. It

TABLE 9.4. Constituents of serum.

Constituent	Range of concentration[a]
Proteins and polypeptides	40–80 mg/ml
Albumin	20–50 mg/ml
Fetuin[b]	10–20 mg/ml
Fibronectin	1–10 µg/ml
Globulins	1–15 mg/ml
Protease inhibitors: α_1-antitrypsin, α_2-macroglobulin	0.5–2.5 mg/ml
Transferrin	2–4 mg/ml
Growth factors	
EGF, PDGF, IGF-I and -II, FGF, IL-1, IL-6	1–100 ng/ml
Amino acids	0.01–1.0 µM
Lipids	2–10 mg/ml
Cholesterol	10 µM
Fatty acids	0.1–1.0 µM
Linoleic acid	0.01–0.1 µM
Phospholipids	0.7–3.0 mg/ml
Carbohydrates	1.0–2.0 mg/ml
Glucose	0.6–1.2 mg/ml
Hexosamine[c]	1.2–6 mg/ml
Lactic acid[d]	0.5–2.0 mg/ml
Pyruvic acid	2–10 µg/ml
Polyamines	
Putrescine, spermidine	0.1–1.0 µM
Urea	170–300 µg/m
Inorganics	0.14–0.16 M
Calcium	4–7 mM
Chlorides	100 µM
Iron	10–50 µM
Potassium	5–15 mM
Phosphate	2–5 mM
Selenium	0.01 µM
Sodium	135–155 mM
Zinc	0.1–1.0 µM
Hormones	0.1–200 nM
Hydrocortisone	10–200 nM
Insulin	1–100 ng/ml
Triiodothyronine	20 nM
Thyroxine	100 nM

(continued)

TABLE 9.4. *(continued)*

Constituent	Range of concentration[a]
Vitamins	10 ng-10 µg/ml
Vitamin A	10–100 ng/ml
Folate	5–20 ng/ml

Source: Evans, V. J. and Sanford, K. K. (1978). "Development of defined media for studies on malignant transformation in culture." In Nutritional requirements of cultured cells, edited by H. Katsuta, 149–194. Tokyo and Baltimore: Japan Scientific Societies Press and University Park Press; Cartwright, T. and Shah, G.P. (1994). "Culture media." In Cell culture: A practical approach, edited by J. M. Davis, 57–89. New York: Oxford University Press.

[a] The range of concentrations is very approximate and is intended to convey only the order of magnitude.

[b] In fetal serum only.

[c] Highest in human serum.

[d] Highest in fetal serum.

is possible that relatively few proteins are required, other than as carriers for minerals, fatty acids, and hormones. Serum proteins where requirements have been demonstrated include (i) albumin, which has a number of roles including as a carrier protein and antioxidant (Francis 2010); (ii) serpins, such as α_1-antitrypsin; (iii) globulins, such as α_2-macroglobulin, which are trypsin inhibitors (de Vonne and Mouray 1978); (iv) fibronectin, which promotes cell attachment, although probably not as effectively as cell-derived fibronectin (Hielscher et al. 2016); (v) fetuin, which enhances cell attachment (Nie 1992); and (vi) transferrin, which binds iron to make it bioavailable and less toxic (Guilbert and Iscove 1976). Serum proteins also increase the viscosity of culture media, reducing shear stress during pipetting and stirring, and may add to their buffering capacity.

9.5.2 Hormones and Growth Factors

A number of growth factors and hormones are present in serum (Gospodarowicz and Moran 1976). Many are now added as supplements in serum-free media (see Section 10.4.1) and are available commercially as recombinant proteins. Growth factors and hormones that are recognized as important for tissue culture include:

(1) *Platelet-derived growth factor (PDGF).* It was noted some years ago that serum obtained by clotting whole blood stimulated cell proliferation more than serum from which the cells were removed by physical means (e.g. centrifugation) (Currie 1981). This observation appears to be due to the release of growth factors, particularly PDGF, from platelets during clotting. PDGF stimulates growth in fibroblasts and glia, but other platelet-derived factors, such as transforming growth factor-β (TGF-β), may inhibit growth and/or promote differentiation in epithelial cells (Lechner et al. 1981).

(2) *Insulin.* This hormone promotes the uptake of glucose and amino acids (Dimitriadis et al. 2011; Karim et al. 2012). It may owe its mitogenic effect to this property or to activity via the insulin-like growth factor (IGF)-I receptor. IGF-I and IGF-II bind to the insulin receptor, but also have their own specific receptors, to which insulin may bind with lower affinity. Insulin requires zinc to be bioactive and is unstable at 37 °C. If used as a supplement, it is added to the medium at a comparatively high concentration in the presence of zinc (Price 2017).

(3) *Hydrocortisone.* This is present in serum – particularly FBS – in varying amounts and can promote cell attachment and proliferation (Guner et al. 1977; Ballard 1979; McLean et al. 1986). However, under certain conditions (e.g. at high cell density) hydrocortisone can be cytostatic (Freshney et al. 1980) and can induce differentiation (Moscona and Piddington 1966; Speirs et al. 1991; McCormick et al. 1995).

(4) *Other growth factors.* Growth hormone, epidermal growth factor (EGF), fibroblast growth factors (FGFs), IGFs, nerve growth factor (NGF), vascular endothelial growth factor (VEGF), and angiogenin may be present in serum and may be required for specific cell types (Gospodarowicz and Moran 1976).

(5) *Other hormones.* Triiodothyronine, estrogens, androgens, and progesterone are present in serum and may be required *in vitro* (see Section 10.4.1) (Price 2017).

9.5.3 Nutrients and Metabolites

Serum may also contain amino acids, glucose, oxo(keto)acids, nucleosides, and a number of other nutrients and intermediary metabolites. These may be important in simple media but are less so in complex media, particularly those with higher amino acid concentrations and other defined supplements.

9.5.4 Lipids

Lipids are important as structural components for membranes, as a source of energy, and in a variety of signaling, transport, and biosynthesis pathways (Price 2017). Linoleic acid, oleic acid, ethanolamine, and phosphoethanolamine are present in serum in small amounts, usually bound to proteins such as albumin. Lipid levels in serum usually increase with the age of the animal, which may be a factor when selecting serum for cultures with high lipid requirements (see Section 9.5).

9.5.5 Trace Elements

Serum replacement experiments have suggested that copper, iron, zinc, and other trace elements may be bound to serum protein, probably albumin (Ham and McKeehan 1978). Copper, iron, and zinc are included in some medium formulations (e.g. DMEM, Ham's F12). Selenium (Na_2SeO_3) is a particularly important trace element; it is a cofactor for glutathione peroxidase, which helps to detoxify free radicals, and can act as an iron carrier in culture (McKeehan et al. 1976; Zhang et al. 2006). Selenium is found in most serum-free formulations or is added as a supplement, often in combination with insulin and transferrin (see Section 10.4.3; see also Appendix C).

9.5.6 Inhibitors

Serum may contain substances that inhibit cell proliferation (Harrington and Godman 1980; Liu et al. 1992). Some of these may be artifacts of preparation (e.g. bacterial toxins from contamination before filtration, or antibodies that cross-react with surface epitopes on the cultured cells). Others may be physiological negative growth regulators, such as TGF-β (Massague 2012). Heat inactivation is sometimes used to reduce the inhibitory effects of serum (see Section 9.7.3).

9.6 OTHER MEDIA SUPPLEMENTS

9.6.1 Conditioned Medium

Medium conditioning refers to its alteration by metabolically active cells, which release a variety of factors that are important for growth, differentiation, invasion, and angiogenesis (Dowling and Clynes 2011). Conditioned medium can be used to improve the survival of low-density clones or normal cells (Puck and Marcus 1955; Price 2017). Conditioned medium can be generated from the same cell line (e.g. by keeping half of the spent medium when refeeding) or from a different cell line. Conditioning the medium with embryonic fibroblasts or other cell lines remains a valuable method to culture difficult cells, along with feeder layers (see Section 8.4) (Stampfer et al. 1980). However, conditioning medium adds undefined components and should be eliminated as soon as the active constituents are determined. Conditioned medium has also been associated with cross-contamination (see Section 17.1) (Patel et al. 2003).

9.6.2 Antibiotics

Antibiotics were originally introduced into culture media to reduce the frequency of microbial contamination (Coriell 1962). However, the use of laminar airflow equipment and good aseptic technique has made the routine use of antibiotics unnecessary (see Sections 5.1.1, 12.4). Chronic antibiotic use has a number of disadvantages, including:

(1) **Relaxed aseptic technique.** Paradoxically, the early tissue culture pioneers recognized that removal of antibiotics was the best way to prevent contamination (Coriell 1962). Absence of antibiotics enforces meticulous aseptic technique and makes any lapses in technique more obvious at an earlier stage.

(2) **Inappropriate antibiotic selection.** Commonly used antibiotics are ineffective against many microbial contaminants. For example, mycoplasmas do not have a rigid cell wall, which makes penicillin ineffective against them (see Section 16.4.2).

(3) **Suppression of microbial contamination.** Many antibiotics are cytostatic rather than being cytotoxic, resulting in slower growth but not eradication. Thus, some microorganisms can persist in the presence of antibiotics as low-level, cryptic contaminants that only become evident when antibiotics are removed (Coriell 1962).

(4) **Antibiotic resistance.** Even if antibiotics are initially effective, they encourage the development of antibiotic-resistant organisms (Taylor-Robinson and Bebear 1997).

(5) **Effects on cell metabolism.** The presence of antibiotics can result in toxicity and widespread disruption of gene expression and regulation (Ryu et al. 2017). Toxicity is particularly evident in serum-free media and with antifungal agents (Price 2017; OECD 2018).

Hence it is strongly recommended that routine culture be performed in the absence of antibiotics and that their use be restricted to cultures where microbial contamination is a higher risk (e.g. primary cultures), for selection of genetically modified cultures that express antibiotic-resistant genes, and for treatment of microbial contamination if the culture is unique and irreplaceable (see Protocol P16.2). If conditions demand the broader use of antibiotics, they should be removed as soon as possible. If long-term use of antibiotics is considered desirable, parallel cultures should be maintained free of antibiotics as a quality control measure (see Figure 14.5).

9.7 CHOICE OF COMPLETE MEDIUM

How do you choose a complete medium for your cells? Common choices are listed here for various cell types and cell lines (see Table 9.5) and further suggestions can be found in the literature (Mather 1998; Price 2017; Yao and Asayama 2017). Look for information on the medium formulation (see Section 9.4), your cell type, the origin of your cell line, and its handling at a cell repository. A cell repository will provide information on media used for currently available cell lines, usually on data sheets that are sent to recipients and can be accessed on its website (see Table 15.3).

Many continuous cell lines (e.g. L-929, HeLa), primary cultures of fibroblasts from various species, and cell lines derived from them can be maintained on a relatively simple medium such as Eagle's MEM, supplemented with calf serum or FBS. More complex media may be required when a specialized function is being expressed or when cells are subcultured at a low seeding concentration ($< 1 \times 10^3$/ml), as in cloning (see Sections 20.7, 26.4). The more demanding culture conditions that require complex media frequently require FBS rather than calf or horse serum, unless the formulation specifically allows for the omission of serum.

TABLE 9.5. Selecting a suitable complete medium.

Cell line/cell type	Medium[a]	Serum
3T3 cells	MEM, DMEM	CS
Chick embryo fibroblasts	Eagle's MEM	CS
Chinese hamster ovary (CHO)	Eagle's MEM, Ham's F12	CS
Chondrocytes	Ham's F12	FBS
Continuous cell lines	Eagle's MEM, DMEM	CS
Endothelium	DMEM, Medium 199, MEM	CS
Fibroblasts	Eagle's MEM	CS
Fish cells (see Section 24.6.1)	L-15	FBS
Glial cells	MEM, DMEM/F12	FBS
Glioma	MEM, DMEM/F12	FBS
HeLa cells	Eagle's MEM	CS
Hematopoietic cells	RPMI 1640, αMEM	FBS
Human diploid fibroblasts	Eagle's MEM	CS
Human leukemia	RPMI 1640	FBS
Human tumors	DMEM/F12, L-15, RPMI 1640	FBS
Insect cells (see Section 24.6.2)	Grace's medium	FBS
Keratinocytes	αMEM	FBS
L-929 cells (NCTC clone 929)	Eagle's MEM	CS
Lymphoblastoid cell lines (human)	RPMI 1640	FBS
Mammary epithelium	DMEM/F12, RPMI 1640	FBS
MDCK dog kidney epithelium	DMEM, DMEM/F12	FBS
Melanocytes	Medium 199	FBS
Melanoma	DMEM, MEM, DMEM/F12	FBS
Mouse embryo fibroblasts	Eagle's MEM	CS
Mouse leukemia	Fischer's, RPMI 1640	FBS, HoS
Mouse erythroleukemia	DMEM/F12, RPMI 1640	FBS, HoS
Mouse myeloma	DMEM, RPMI 1640	FBS
Mouse neuroblastoma	DMEM, DMEM/F12	FBS
Neurons	DMEM	FBS
NRK rat kidney fibroblasts	Eagle's MEM, DMEM	CS
Rat minimal-deviation hepatoma (e.g. HTC, MDH)	DMEM/F12, Swim's S77	FBS
Skeletal muscle	DMEM, F12	FBS, HoS
Syrian hamster fibroblasts (e.g. BHK 21)	Eagle's MEM, DMEM, GMEM	CS

[a]These recommendations act as a starting point; more information can be found in source references and cell repository catalogs. See Table 10.5 for suggestions when selecting serum-free media. More information on early classic formulations is available elsewhere (Morton 1970; Sigma-Aldrich 2018).
Abbreviations not found in main text: CS, calf serum; GMEM, Glasgow Minimum Essential Medium; HoS, horse serum.

You may need to compromise in your choice of medium or serum because of cost. Most common media formulations can be purchased in powdered form and are relatively cheap and simple to prepare (see Protocol P11.5). However, serum can be an expensive reagent. If cell yield is not important, the cost of serum should be calculated on the basis of the volume of the medium. If the objective is to produce large quantities of cells, the cost of serum should be calculated on a per-cell basis. Thus, if a culture grows to 1×10^6 cells/ml in serum A and 2×10^6 cells/ml in serum B, serum B is less expensive by a factor of two, given that product formation or some other specialized function is the same. If FBS is essential but too costly, try mixing it with calf serum to reduce the concentration of FBS. You may also wish to reduce the concentration or leave out serum altogether and move toward serum-free media (see Section 10.6). Always test your medium and serum choices in the cell culture application of interest before you make a final decision.

9.7.1 Serum Testing

Broadly speaking, serum testing is performed to detect microbial contaminants and assess performance in culture (Nims and Harbell 2017). Testing is usually performed by the manufacturer as part of quality assurance and the results are made available in the Certificate of Analysis (see Section 7.5). Contaminant testing includes sterility testing (see Section 11.8.1) and specific assays for mycoplasma and viruses (e.g. bovine viral diarrhea virus). Other assays may be performed, depending on the incidence of viruses and other pathogens in donor herds, which will vary from country to country (see Section 9.5). Performance testing is also usually performed by the manufacturer and focuses on viability, growth, or plating of a small number of cell lines.

Testing of serum by the manufacturer does not guarantee that the serum is free from contamination – for example, prions are usually not tested. Similarly, performance testing by the manufacturer does not guarantee that serum will give good results for your specific cell line or application. It is important to perform your own testing in addition to the manufacturer's efforts. User testing typically focuses on:

(1) **Cloning efficiency.** Cells are plated at low density and incubated until colonies form (see Supp. S11.7). During cloning, the cells are at a low density and hence are at their most sensitive, making this a very stringent test. Each batch should be tested at a range of concentrations, from 2% to 20%. This approach reveals whether one serum is equally effective at lower concentrations, saving money and prolonging the life of the batch, and whether higher concentrations result in toxicity.

(2) **Growth analysis.** A growth curve should be plotted for each batch (see Supp. S11.8) to determine the lag period, population doubling time (PDT), and saturation density (cell density at "plateau"). A long lag period implies that the culture has to adapt to the serum; short doubling times are preferable if you want a lot of cells quickly; and a high saturation density provides more cells for a given amount of serum and will be more economical.

(3) **Other characterization.** Clearly, the cells must do what you require of them in the new serum, whether they are acting as substrate for a given virus, secreting a specific cell product, differentiating, or expressing a characteristic sensitivity to a given drug. Hence batch comparison may also need specific functional assays.

(4) **Sterility testing.** Serum from a reputable supplier will have been tested for microbial contaminants (e.g. mycoplasma). However, contamination may occur during preparation of complete medium or may have been missed. It is advisable to include serum and complete medium in routine screening before it is finally accepted.

Although this takes time, your efforts will be repaid if your cultures adjust to the new serum with minimal impact. A simple cell growth experiment using commercially available media and multiwell plates can be performed in about two weeks. Assays for cloning efficiency and expression of specialized functions will narrow the choice further. You may find that the best conditions do not agree with those mentioned in the literature (see Section 9.7). Reproducing the conditions found in another laboratory may be difficult because of variations in preparation, the impurities present in reagents and water, and differences between batches of serum. It is to be hoped that, as the need for serum decreases and reagent purity increases, the standardization of media will improve.

9.7.2 Serum Batch Reservation

Serum performance may vary between different batches or lots. Such variation results from differing methods of preparation and sterilization, different ages and storage conditions, and variations in animal stocks from which the serum was derived, including different strains and disparities in environmental conditions. It is important to select a batch, use it for as long as possible, and replace it, eventually, with one as similar to it as possible. Most serum suppliers will reserve a batch for a period of time to allow customers time to test a sample in their application; suppliers may also be willing to hold the batch once it has been purchased and to supply bottles to the customer on request. Several laboratories may choose to share a batch to reduce the impact on their budget.

9.7.3 Serum Treatment

Serum can be heat inactivated by incubation at 56 °C for 30 minutes before dispensing aliquots for storage. Heat inactivation was originally performed to inactivate complement and to reduce the cytotoxic action of immunoglobulins without damaging polypeptide growth factors. However, heat inactivation of serum may reduce its capacity to promote cell attachment, and it may not significantly change immune responses (Giard 1987; Leshem et al. 1999). It is often used because of the adoption of a previous protocol, without any concrete evidence that it is beneficial for the desired cell culture application.

Other serum treatments are required for specific applications and may be performed by the serum supplier. Treatments available from the supplier include:

(1) **Dialysis.** Used in early medium development to remove the low molecular weight components of serum; it is still used for metabolic assays (Yao and Asayama 2017).

(2) **Charcoal stripping.** Used to remove non-polar constituents of serum such as hormones, growth factors, cytokines, and some viruses (Nims and Harbell 2017). It is particularly used for experiments that focus on hormone or lipid function.

(3) *Immunoglobulin G (IgG) stripping.* Used to reduce and hopefully remove undesirable antibodies from the serum (Nims and Harbell 2017).

(4) *γ-Irradiation.* Used to inactivate microbial contaminants that may not be detected using standard testing. Irradiation is carried out by commercial irradiators commissioned by serum manufacturers and appears to be effective for mycoplasmas and many viruses; other viruses such as circoviruses, parvoviruses, and polyomaviruses appear to be relatively resistant to the doses of radiation that are normally applied to serum (Nims et al. 2011).

9.8 STORAGE OF MEDIUM AND SERUM

Opinions differ as to the shelf life of different media. As a rough guide, media made up without glutamine should last six to nine months at 4 °C. Once glutamine, serum, or antibiotics are added, the shelf life is reduced to that of the most sensitive essential component (Price 2017). Substitution of glutamine with a glutamyl dipeptide will improve its longevity (see Section 9.4.1). Complete media should typically be used within two to four weeks, but always refer to the manufacturer's instructions and monitor the bottle for changes. Stored media can show precipitation. If this does not redissolve on heating, it is likely to be a chemical alteration of one of the constituents (e.g. metal ions combining with phosphate) and may not be reversible. Medium may be stored frozen, but do not refreeze it, as components may degrade or precipitate (Price 2017).

Serum should be stored frozen at less than −15 °C (Nims and Harbell 2017). Autodefrost freezers have been raised as a concern by some laboratories, but in the authors' experience, autodefrost cycles have a minor impact on serum performance. Many of the essential constituents of serum are small proteins, polypeptides, and simpler organic and inorganic compounds that may be insensitive to cryogenic damage, particularly if solutions are stored in volumes > 100 ml. Serum bottles should be thawed in a refrigerator overnight before use. If faster thawing is required, serum may be placed at room temperature for around 10 minutes and then thawed at 37 °C, with gentle mixing (e.g. swirling) to improve heat transfer (Nims and Harbell 2017).

All medium and serum should be stored shielded from light. Fluorescent lighting results in deterioration of riboflavin, tryptophan, and tyrosine and the production of cytotoxic products, such as hydrogen peroxide (Wang and Nixon 1978; Edwards and Silva 2001). The addition of HEPES increases the production of cytotoxic products (Zigler et al. 1985). While part of the toxic effect on cells may be due to photochemical deterioration, there is also a direct mutagenic effect on cells, particularly in transformed cultures (Bradley and Sharkey 1977; Parshad et al. 1980).

Suppliers

Supplier	URL
Baker Ruskinn	http://bakerco.com/products/grow
BioSpherix	http://www.biospherix.com
Camlab	www.camlab.co.uk
Don Whitley Scientific	www.dwscientific.co.uk
HypOxygen	http://hypoxygen.com
Oxford Optronix	http://www.oxford-optronix.com
Plas-Labs, Inc.	http://www.plas-labs.com
Sigma-Aldrich (Merck)	http://www.sigmaaldrich.com/life-science/cell-culture.html
ThermoFisher Scientific	http://www.thermofisher.com/us/en/home/life-science/cell-culture.html
Xcell Biosciences	www.xcellbio.com

Supp. S9.1 Preparation of pH Standards.

REFERENCES

Abbasalizadeh, S., Larijani, M.R., Samadian, A. et al. (2012). Bioprocess development for mass production of size-controlled human pluripotent stem cell aggregates in stirred suspension bioreactor. *Tissue Eng. Part C Methods* 18 (11): 831–851. https://doi.org/10.1089/ten.TEC.2012.0161.

Alberts, B., Johnson, A., Lewis, J. et al. (2015). *Molecular Biology of the Cell*, 6e. New York: Garland Science.

Arora, M. (2013). Cell culture media: a review. *Mater. Methods* 3: 175. https://doi.org/10.13070/mm.en.3.175.

Ballard, P.L. (1979). Glucocorticoids and differentiation. *Monogr. Endocrinol.* 12: 493–515.

Barnes, D. and Sato, G. (1980). Methods for growth of cultured cells in serum-free medium. *Anal. Biochem.* 102 (2): 255–270.

Baumann, R.P., Penketh, P.G., Seow, H.A. et al. (2008). Generation of oxygen deficiency in cell culture using a two-enzyme system to evaluate agents targeting hypoxic tumor cells. *Radiat. Res.* 170 (5): 651–660. https://doi.org/10.1667/RR1431.1.

Bertout, J.A., Patel, S.A., and Simon, M.C. (2008). The impact of O_2 availability on human cancer. *Nat. Rev. Cancer* 8 (12): 967–975. https://doi.org/10.1038/nrc2540.

Boyce, S.T. and Ham, R.G. (1983). Calcium-regulated differentiation of normal human epidermal keratinocytes in chemically defined clonal culture and serum-free serial culture. *J. Invest. Dermatol.* 81 (1 Suppl): 33s–40s.

Bradley, M.O. and Sharkey, N.A. (1977). Mutagenicity and toxicity of visible fluorescent light to cultured mammalian cells. *Nature* 266 (5604): 724–726.

Brahimi-Horn, M.C., Chiche, J., and Pouyssegur, J. (2007). Hypoxia and cancer. *J. Mol. Med. (Berl.)* 85 (12): 1301–1307. https://doi.org/10.1007/s00109-007-0281-3.

Brunelle, J.K., Bell, E.L., Quesada, N.M. et al. (2005). Oxygen sensing requires mitochondrial ROS but not oxidative phosphorylation. *Cell Metab.* 1 (6): 409–414. https://doi.org/10.1016/j.cmet.2005.05.002.

Buck, L.D., Inman, S.W., Rusyn, I. et al. (2014). Co-regulation of primary mouse hepatocyte viability and function by oxygen and matrix. *Biotechnol. Bioeng.* 111 (5): 1018–1027. https://doi.org/10.1002/bit.25152.

Butler, M. and Christie, A. (1994). Adaptation of mammalian cells to non-ammoniagenic media. *Cytotechnology* 15 (1–3): 87–94.

Cartwright, T. and Shah, G.P. (1994). Culture media. In: *Cell Culture: A Practical Approach* (ed. J.M. Davis), 57–89. New York: Oxford University Press.

Catt, J.W. and Henman, M. (2000). Toxic effects of oxygen on human embryo development. *Hum. Reprod.* 15 (Suppl 2): 199–206.

Cherry, R.S. and Papoutsakis, E.T. (1988). Physical mechanisms of cell damage in microcarrier cell culture bioreactors. *Biotechnol. Bioeng.* 32 (8): 1001–1014. https://doi.org/10.1002/bit.260320808.

Cooper, P.D., Burt, A.M., and Wilson, J.N. (1958). Critical effect of oxygen tension on rate of growth of animal cells in continuous suspended culture. *Nature* 182 (4648): 1508–1509.

Coriell, L.L. (1962). Detection and elimination of contaminating organisms. *Natl Cancer Inst. Monogr.* 7: 33–53.

Currie, G.A. (1981). Platelet-derived growth-factor requirements for in vitro proliferation of normal and malignant mesenchymal cells. *Br. J. Cancer* 43 (3): 335–343.

De Ridder, L. and Mareel, M. (1978). Morphology and 125I-concentration of embryonic chick thyroids cultured in an atmosphere of oxygen. *Cell Biol. Int. Rep.* 2 (2): 189–194.

Dimitriadis, G., Mitrou, P., Lambadiari, V. et al. (2011). Insulin effects in muscle and adipose tissue. *Diabetes Res. Clin. Pract.* 93 (Suppl 1): S52–S59. https://doi.org/10.1016/S0168-8227(11)70014-6.

Dos Santos, V.T., Mizukami, A., Orellana, M.D. et al. (2017). Characterization of human AB serum for mesenchymal stromal cell expansion. *Transfus Med. Hemother.* 44 (1): 11–21. https://doi.org/10.1159/000448196.

Dowling, P. and Clynes, M. (2011). Conditioned media from cell lines: a complementary model to clinical specimens for the discovery of disease-specific biomarkers. *Proteomics* 11 (4): 794–804. https://doi.org/10.1002/pmic.201000530.

Dulbecco, R. and Freeman, G. (1959). Plaque production by the polyoma virus. *Virology* 8 (3): 396–397.

Dulbecco, R. and Vogt, M. (1954). Plaque formation and isolation of pure lines with poliomyelitis viruses. *J. Exp. Med.* 99 (2): 167–182.

Eagle, H. (1955). The specific amino acid requirements of a mammalian cell (strain L) in tissue culture. *J. Biol. Chem.* 214 (2): 839–852.

Eagle, H. (1959). Amino acid metabolism in mammalian cell cultures. *Science* 130 (3373): 432–437.

Eagle, H. (1973). The effect of environmental pH on the growth of normal and malignant cells. *J. Cell. Physiol.* 82 (1): 1–8. https://doi.org/10.1002/jcp.1040820102.

Earle, W.R., Schilling, E.L., Stark, T.H. et al. (1943). Production of malignancy in vitro. IV. The mouse fibroblast cultures and changes seen in the living cells. *J. Natl. Cancer Inst.* 4 (2): 165–212.

Edwards, A.M. and Silva, E. (2001). Effect of visible light on selected enzymes, vitamins and amino acids. *J. Photochem. Photobiol. B* 63 (1–3): 126–131.

Eisinger, M., Lee, J.S., Hefton, J.M. et al. (1979). Human epidermal cell cultures: growth and differentiation in the absence of differentiation in the absence of dermal components or medium supplements. *Proc. Natl Acad. Sci. U.S.A.* 76 (10): 5340–5344.

Evans, V.J. and Sanford, K.K. (1978). Development of defined media for studies on malignant transformation in culture. In: *Nutritional Requirements of Cultured Cells* (ed. H. Katsuta), 149–194. Tokyo and Baltimore: Japan Scientific Societies Press and University Park Press.

Fischer, A. (1925). *Tissue Culture: Studies in Experimental Morphology and General Physiology of Tissue Cells In Vitro*. London: William Heinemann.

Fischer, G.A. and Sartorelli, A.C. (1964). Development, maintenance and assay of drug resistance. *Methods Med. Res.* 10: 247–262.

Fleischaker, R.J. and Sinskey, A.J. (1981). Oxygen demand and supply in cell culture. *Eur. J. Appl. Microbiol. Biotechnol.* 12 (4): 193–197.

Forsyth, N.R., Musio, A., Vezzoni, P. et al. (2006). Physiologic oxygen enhances human embryonic stem cell clonal recovery and reduces chromosomal abnormalities. *Cloning Stem Cells* 8 (1): 16–23. https://doi.org/10.1089/clo.2006.8.16.

Francis, G.L. (2010). Albumin and mammalian cell culture: implications for biotechnology applications. *Cytotechnology* 62 (1): 1–16. https://doi.org/10.1007/s10616-010-9263-3.

Freshney, R.I., Sherry, A., Hassanzadah, M. et al. (1980). Control of cell proliferation in human glioma by glucocorticoids. *Br. J. Cancer* 41 (6): 857–866.

Giard, D.J. (1987). Routine heat inactivation of serum reduces its capacity to promote cell attachment. *In Vitro Cell Dev. Biol.* 23 (10): 691–697.

Godoy-Silva, R., Chalmers, J.J., Casnocha, S.A. et al. (2009). Physiological responses of CHO cells to repetitive hydrodynamic stress. *Biotechnol. Bioeng.* 103 (6): 1103–1117. https://doi.org/10.1002/bit.22339.

Good, N.E., Winget, G.D., Winter, W. et al. (1966). Hydrogen ion buffers for biological research. *Biochemistry* 5 (2): 467–477.

Gospodarowicz, D. and Moran, J.S. (1976). Growth factors in mammalian cell culture. *Annu. Rev. Biochem.* 45: 531–558. https://doi.org/10.1146/annurev.bi.45.070176.002531.

Grace, T.D. (1962). Establishment of four strains of cells from insect tissues grown in vitro. *Nature* 195: 788–789.

Grady, L.H., Nonneman, D.J., Rottinghaus, G.E. et al. (1991). pH-dependent cytotoxicity of contaminants of phenol red for MCF-7 breast cancer cells. *Endocrinology* 129 (6): 3321–3330. https://doi.org/10.1210/endo-129-6-3321.

Green, M., Henle, G., and Deinhardt, F. (1958). Respiration and glycolysis of human cells grown in tissue culture. *Virology* 5 (2): 206–219.

Guilbert, L.J. and Iscove, N.N. (1976). Partial replacement of serum by selenite, transferrin, albumin and lecithin in haemopoietic cell cultures. *Nature* 263 (5578): 594–595.

Guner, M., Freshney, R.I., Morgan, D. et al. (1977). Effects of dexamethasone and betamethasone on in vitro cultures from human astrocytoma. *Br. J. Cancer* 35 (4): 439–447.

Ham, R.G. (1965). Clonal growth of mammalian cells in a chemically defined, synthetic medium. *Proc. Natl Acad. Sci. U.S.A.* 53: 288–293.

Ham, R.G. and McKeehan, W.L. (1978). Development of improved media and culture conditions for clonal growth of normal diploid cells. *In Vitro* 14 (1): 11–22.

Hanks, J.H. and Wallace, R.E. (1949). Relation of oxygen and temperature in the preservation of tissues by refrigeration. *Proc. Soc. Exp. Biol. Med.* 71 (2): 196–200.

Haque, N., Rahman, M.T., Abu Kasim, N.H. et al. (2013). Hypoxic culture conditions as a solution for mesenchymal stem cell based regenerative therapy. *Sci. World J.* 2013: 632972. https://doi.org/10.1155/2013/632972.

Harrington, W.N. and Godman, G.C. (1980). A selective inhibitor of cell proliferation from normal serum. *Proc. Natl Acad. Sci. U.S.A.* 77 (1): 423–427.

Hassell, T., Gleave, S., and Butler, M. (1991). Growth inhibition in animal cell culture. The effect of lactate and ammonia. *Appl. Biochem. Biotechnol.* 30 (1): 29–41.

Hielscher, A., Ellis, K., Qiu, C. et al. (2016). Fibronectin deposition participates in extracellular matrix assembly and vascular morphogenesis. *PLoS One* 11 (1): e0147600. https://doi.org/10.1371/journal.pone.0147600.

Hu, W., Berdugo, C., and Chalmers, J.J. (2011). The potential of hydrodynamic damage to animal cells of industrial relevance: current understanding. *Cytotechnology* 63 (5): 445–460. https://doi.org/10.1007/s10616-011-9368-3.

Hyvonen, T., Alakuijala, L., Andersson, L. et al. (1988). 1-Aminooxy-3-aminopropane reversibly prevents the proliferation of cultured baby hamster kidney cells by interfering with polyamine synthesis. *J. Biol. Chem.* 263 (23): 11138–11144.

Ince, T.A., Sousa, A.D., Jones, M.A. et al. (2015). Characterization of twenty-five ovarian tumour cell lines that phenocopy primary tumours. *Nat. Commun.* 6: 7419. https://doi.org/10.1038/ncomms8419.

ISIA (2018). Geographic testing. https://www.serumindustry.org/traceability/geographic-testing (accessed 8 November 2018).

Itagaki, A. and Kimura, G. (1974). Tes and HEPES buffers in mammalian cell cultures and viral studies: problem of carbon dioxide requirement. *Exp. Cell. Res.* 83 (2): 351–361.

Jing, X., Yang, F., Shao, C. et al. (2019). Role of hypoxia in cancer therapy by regulating the tumor microenvironment. *Mol. Cancer* 18 (1): 157. https://doi.org/10.1186/s12943-019-1089-9.

Kaminska, B., Kaczmarek, L., and Grzelakowska-Sztabert, B. (1990). The regulation of G0-S transition in mouse T lymphocytes by polyamines. *Exp. Cell. Res.* 191 (2): 239–245.

Karim, S., Adams, D.H., and Lalor, P.F. (2012). Hepatic expression and cellular distribution of the glucose transporter family. *World J. Gastroenterol.* 18 (46): 6771–6781. https://doi.org/10.3748/wjg.v18.i46.6771.

Koh, M.Y., Spivak-Kroizman, T.R., and Powis, G. (2008). HIF-1 regulation: not so easy come, easy go. *Trends Biochem. Sci* 33 (11): 526–534. https://doi.org/10.1016/j.tibs.2008.08.002.

Lechner, J.F., Haugen, A., Autrup, H. et al. (1981). Clonal growth of epithelial cells from normal adult human bronchus. *Cancer Res.* 41 (6): 2294–2304.

Leibovitz, A. (1963). The growth and maintenance of tissue-cell cultures in free gas exchange with the atmosphere. *Am. J. Hyg.* 78: 173–180.

Leshem, B., Yogev, D., and Fiorentini, D. (1999). Heat inactivation of fetal calf serum is not required for in vitro measurement of lymphocyte functions. *J. Immunol. Methods* 223 (2): 249–254.

Liu, L., Delbe, J., Blat, C. et al. (1992). Insulin-like growth factor binding protein (IGFBP-3), an inhibitor of serum growth factors other than IGF-I and -II. *J. Cell. Physiol.* 153 (1): 15–21. https://doi.org/10.1002/jcp.1041530104.

Ludwig, T.E., Levenstein, M.E., Jones, J.M. et al. (2006). Derivation of human embryonic stem cells in defined conditions. *Nat. Biotechnol.* 24 (2): 185–187. https://doi.org/10.1038/nbt1177.

Ma, T., Grayson, W.L., Frohlich, M. et al. (2009). Hypoxia and stem cell-based engineering of mesenchymal tissues. *Biotechnol. Progr.* 25 (1): 32–42. https://doi.org/10.1002/btpr.128.

Majmundar, A.J., Wong, W.J., and Simon, M.C. (2010). Hypoxia-inducible factors and the response to hypoxic stress. *Mol. Cell* 40 (2): 294–309. https://doi.org/10.1016/j.molcel.2010.09.022.

Mansfield, K.D., Guzy, R.D., Pan, Y. et al. (2005). Mitochondrial dysfunction resulting from loss of cytochrome c impairs cellular oxygen sensing and hypoxic HIF-alpha activation. *Cell Metab.* 1 (6): 393–399. https://doi.org/10.1016/j.cmet.2005.05.003.

Massague, J. (2012). TGFbeta signalling in context. *Nat. Rev. Mol. Cell Biol.* 13 (10): 616–630. https://doi.org/10.1038/nrm3434.

Mather, J.P. (1998). Making informed choices: medium, serum, and serum-free medium. How to choose the appropriate medium and culture system for the model you wish to create. *Methods Cell Biol.* 57: 19–30.

McCormick, C., Freshney, R.I., and Speirs, V. (1995). Activity of interferon alpha, interleukin 6 and insulin in the regulation of differentiation in A549 alveolar carcinoma cells. *Br. J. Cancer* 71 (2): 232–239.

McCoy, T.A., Maxwell, M., and Kruse, P.F. Jr. (1959). Amino acid requirements of the Novikoff hepatoma in vitro. *Proc. Soc. Exp. Biol. Med.* 100 (1): 115–118.

McKeehan, W.L. and Ham, R.G. (1976). Methods for reducing the serum requirement for growth in vitro of nontransformed diploid fibroblasts. *Dev. Biol. Stand.* 37: 97–98.

McKeehan, W.L., Hamilton, W.G., and Ham, R.G. (1976). Selenium is an essential trace nutrient for growth of WI-38 diploid human fibroblasts. *Proc. Natl Acad. Sci. U.S.A.* 73 (6): 2023–2027.

McLean, J.S., Frame, M.C., Freshney, R.I. et al. (1986). Phenotypic modification of human glioma and non-small cell lung carcinoma by glucocorticoids and other agents. *Anticancer Res.* 6 (5): 1101–1106.

Mizrahi, A. and Moore, G.E. (1970). Partial substitution of serum in hematopoietic cell line media by synthetic polymers. *Appl. Microbiol.* 19 (6): 906–910.

Moore, G.E., Gerner, R.E., and Franklin, H.A. (1967). Culture of normal human leukocytes. *JAMA* 199 (8): 519–524.

Morgan, J.F., Morton, H.J., and Parker, R.C. (1950). Nutrition of animal cells in tissue culture; initial studies on a synthetic medium. *Proc. Soc. Exp. Biol. Med.* 73 (1): 1–8.

Morton, H.J. (1970). A survey of commercially available tissue culture media. *In Vitro* 6 (2): 89–108.

Moscona, A.A. and Piddington, R. (1966). Stimulation by hydrocortisone of premature changes in the developmental pattern of glutamine synthetase in embryonic retina. *Biochim. Biophys. Acta* 121 (2): 409–411.

Newby, D., Marks, L., and Lyall, F. (2005). Dissolved oxygen concentration in culture medium: assumptions and pitfalls. *Placenta* 26 (4): 353–357. https://doi.org/10.1016/j.placenta.2004.07.002.

Nie, Z. (1992). Fetuin: its enigmatic property of growth promotion. *Am. J. Phys.* 263 (3 Pt 1): C551–C562. https://doi.org/10.1152/ajpcell.1992.263.3.C551.

Nienow, A.W. (2006). Reactor engineering in large scale animal cell culture. *Cytotechnology* 50 (1–3): 9–33. https://doi.org/10.1007/s10616-006-9005-8.

Nims, R.W. and Harbell, J.W. (2017). Best practices for the use and evaluation of animal serum as a component of cell culture medium. *In Vitro Cell. Dev. Biol. Anim.* 53 (8): 682–690. https://doi.org/10.1007/s11626-017-0184-8.

Nims, R.W., Gauvin, G., and Plavsic, M. (2011). Gamma irradiation of animal sera for inactivation of viruses and mollicutes – a review. *Biologicals* 39 (6): 370–377. https://doi.org/10.1016/j.biologicals.2011.05.003.

OECD (2018). *Guidance Document on Good In Vitro Method Practices (GIVIMP)*. Paris: OECD Publishing.

Olbryt, M., Habryka, A., Student, S. et al. (2014). Global gene expression profiling in three tumor cell lines subjected to experimental cycling and chronic hypoxia. *PLoS One* 9 (8): e105104. https://doi.org/10.1371/journal.pone.0105104.

Parker, R.C., Castor, L.N., and McCulloch, E.A. (1957). Altered cell strains in continuous culture. Special publications. *N. Y. Acad. Sci.* 5: 303–313.

Parshad, R., Sanford, K.K., Jones, G.M. et al. (1980). Susceptibility to fluorescent light-induced chromatid breaks associated with DNA repair deficiency and malignant transformation in culture. *Cancer Res.* 40 (12): 4415–4419.

Patel, V.A., Logan, A., Watkinson, J.C. et al. (2003). Isolation and characterization of human thyroid endothelial cells. *Am. J. Physiol. Endocrinol. Metab.* 284 (1): E168–E176. https://doi.org/10.1152/ajpendo.00096.2002.

Paul, J. (1975). *Cell and Tissue Culture*, 5e. Edinburgh: Churchill Livingstone.

Price, P.J. (2017). Best practices for media selection for mammalian cells. *In Vitro Cell. Dev. Biol. Anim.* 53 (8): 673–681. https://doi.org/10.1007/s11626-017-0186-6.

Puck, T.T. and Marcus, P.I. (1955). A rapid method for viable cell titration and clone production with Hela cells in tissue culture: the use of X-irradiated cells to supply conditioning factors. *Proc. Natl Acad. Sci. U.S.A.* 41 (7): 432–437.

Rashid, R., Lim, N.S., Chee, S.M. et al. (2014). Novel use for polyvinylpyrrolidone as a macromolecular crowder for enhanced extracellular matrix deposition and cell proliferation. *Tissue Eng. Part C Methods* 20 (12): 994–1002. https://doi.org/10.1089/ten.TEC.2013.0733.

Richter, A., Sanford, K.K., and Evans, V.J. (1972). Influence of oxygen and culture media on plating efficiency of some mammalian tissue cells. *J. Natl Cancer Inst.* 49 (6): 1705–1712.

Rubin, H. and Xu, K. (1989). Evidence for the progressive and adaptive nature of spontaneous transformation in the NIH 3T3 cell line. *Proc. Natl Acad. Sci. U.S.A.* 86 (6): 1860–1864.

Ryu, A.H., Eckalbar, W.L., Kreimer, A. et al. (2017). Use antibiotics in cell culture with caution: genome-wide identification of antibiotic-induced changes in gene expression and regulation. *Sci. Rep.* 7 (1): 7533. https://doi.org/10.1038/s41598-017-07757-w.

Sen, A., Kallos, M.S., and Behie, L.A. (2002). Expansion of mammalian neural stem cells in bioreactors: effect of power input and medium viscosity. *Brain Res. Dev. Brain Res.* 134 (1–2): 103–113.

Shipman, C. Jr. (1973). Control of culture pH with synthetic buffers. In: *Tissue Culture: Methods and Applications* (eds. P.F. Kruse Jr. and M.K. Patterson), 709–712. New York: Academic Press.

Sigma-Aldrich (2018). Classic media and salts. https://www.sigmaaldrich.com/life-science/cell-culture/cell-culture-products.html?tablepage=9630413 (accessed 25 October 2018).

Speirs, V., Ray, K.P., and Freshney, R.I. (1991). Paracrine control of differentiation in the alveolar carcinoma, A549, by human foetal lung fibroblasts. *Br. J. Cancer* 64 (4): 693–699.

Stampfer, M., Hallowes, R.C., and Hackett, A.J. (1980). Growth of normal human mammary cells in culture. *In Vitro* 16 (5): 415–425.

Stanners, C.P., Eliceiri, G.L., and Green, H. (1971). Two types of ribosome in mouse-hamster hybrid cells. *Nat. New Biol.* 230 (10): 52–54.

Stoll, L.L. and Spector, A.A. (1984). Changes in serum influence the fatty acid composition of established cell lines. *In Vitro* 20 (9): 732–738.

Taylor-Robinson, D. and Bebear, C. (1997). Antibiotic susceptibilities of mycoplasmas and treatment of mycoplasmal infections. *J. Antimicrob. Chemother.* 40 (5): 622–630. https://doi.org/10.1093/jac/40.5.622.

Templeton, N., Dean, J., Reddy, P. et al. (2013). Peak antibody production is associated with increased oxidative metabolism in an industrially relevant fed-batch CHO cell culture. *Biotechnol. Bioeng.* 110 (7): 2013–2024. https://doi.org/10.1002/bit.24858.

Tharmalingam, T., Ghebeh, H., Wuerz, T. et al. (2008). Pluronic enhances the robustness and reduces the cell attachment of mammalian cells. *Mol. Biotechnol.* 39 (2): 167–177. https://doi.org/10.1007/s12033-008-9045-8.

Thompson, J.G., Simpson, A.C., Pugh, P.A. et al. (1990). Effect of oxygen concentration on in-vitro development of preimplantation sheep and cattle embryos. *J. Reprod. Fertil.* 89 (2): 573–578.

Trowell, O.A. (1959). The culture of mature organs in a synthetic medium. *Exp. Cell. Res.* 16 (1): 118–147.

Velugula-Yellela, S.R., Williams, A., Trunfio, N. et al. (2018). Impact of media and antifoam selection on monoclonal antibody production and quality using a high throughput micro-bioreactor system. *Biotechnol. Progr.* 34 (1): 262–270. https://doi.org/10.1002/btpr.2575.

de Vonne, T.L. and Mouray, H. (1978). Human alpha2-macroglobulin and its antitrypsic and antithrombin activities in serum and plasma. *Clin. Chim. Acta* 90 (1): 83–85.

Wang, R.J. and Nixon, B.R. (1978). Identification of hydrogen peroxide as a photoproduct toxic to human cells in tissue-culture medium irradiated with "daylight" fluorescent light. *In Vitro* 14 (8): 715–722.

Waymouth, C. (1959). Rapid proliferation of sublines of NCTC clone 929 (strain L) mouse cells in a simple chemically defined medium (MB 752/1). *J. Natl Cancer Inst.* 22 (5): 1003–1017.

Waymouth, C. (1970). Osmolality of mammalian blood and of media for culture of mammalian cells. *In Vitro* 6 (2): 109–127.

Waymouth, C. (1978). Studies on chemically defined media and the nutritional requirements of cultures of epithelial cells. In: *Nutritional Requirements of Cultured Cells* (ed. H. Katsuta), 39–61. Tokyo and Baltimore: Japan Scientific Societies Press and University Park Press.

Wyllie, F.S., Bond, J.A., Dawson, T. et al. (1992). A phenotypically and karyotypically stable human thyroid epithelial line conditionally immortalized by SV40 large T antigen. *Cancer Res.* 52 (10): 2938–2945.

Yao, T. and Asayama, Y. (2017). Animal-cell culture media: history, characteristics, and current issues. *Reprod. Med. Biol.* 16 (2): 99–117. https://doi.org/10.1002/rmb2.12024.

Young, J.D. (2013). Metabolic flux rewiring in mammalian cell cultures. *Curr. Opin. Biotechnol.* 24 (6): 1108–1115. https://doi.org/10.1016/j.copbio.2013.04.016.

Yuspa, S.H., Koehler, B., Kulesz-Martin, M. et al. (1981). Clonal growth of mouse epidermal cells in medium with reduced calcium concentration. *J. Invest. Dermatol.* 76 (2): 144–146.

Zhang, J., Robinson, D., and Salmon, P. (2006). A novel function for selenium in biological system: selenite as a highly effective iron carrier for Chinese hamster ovary cell growth and monoclonal antibody production. *Biotechnol. Bioeng.* 95 (6): 1188–1197. https://doi.org/10.1002/bit.21081.

Zigler, J.S. Jr., Lepe-Zuniga, J.L., Vistica, B. et al. (1985). Analysis of the cytotoxic effects of light-exposed HEPES-containing culture medium. *In Vitro Cell Dev. Biol.* 21 (5): 282–287.

CHAPTER 10

Serum-Free Media

Serum has been described as "the last undefined component in cell culture media" (Hayashi et al. 1978). Early studies of cellular nutritional requirements resulted in the development of numerous synthetic media, including some that were entirely serum-free. However, once Eagle's Basal Medium (BME) and Minimum Essential Medium (MEM) were developed, they were used successfully to grow many different cell types with the addition of serum (see Section 9.1). As a result, many researchers moved away from the field of cell nutrition (Ham 1974). It is ironic that this neglected field is now seen as essential for today's specialized applications. Media that are free of animal products are required to minimize the use of such products for ethical, safety, and regulatory reasons (OECD 2018; van der Valk et al. 2018). Defined culture conditions are needed for an increasing number of applications including stem cell culture, three-dimensional (3D) culture, and cell-based therapies. This chapter examines the rationale for serum-free media, the development of serum-free formulations, and important points to consider when using serum-free media, supplements, or replacements.

For this book, the term "serum-free medium" is used to refer to medium that does not contain serum or unprocessed plasma; it may contain individual proteins or protein fractions (Karnieli et al. 2017). "Xeno-free medium" may contain human-derived components (e.g. human serum or plasma) but no other animal-derived components. "Animal product-free medium" has no human- or animal-derived components. "Chemically defined" medium is made up of purified recombinant proteins or other components whose chemical compositions or structures are known. More

guidance on terminology is available elsewhere (Karnieli et al. 2017; OECD 2018).

10.1 RATIONALE FOR SERUM-FREE MEDIUM

10.1.1 Disadvantages of Serum

Using serum as a medium supplement has a number of disadvantages:

(1) **Ethical concerns.** Concerns have been raised regarding the ethics of harvesting fetal bovine serum (FBS), which is pooled from a large number of donor animals (Jochems et al. 2002). Safeguards can be applied to minimize pain and distress, but there is still a need for replacement, reduction, and refinement (the 3Rs) when animals are used for scientific purposes (see Section 6.5.1) (van der Valk et al. 2004). Development of serum-free medium will support these important goals.

(2) **Availability.** The supply of serum may be restricted by drought in cattle-rearing areas, the spread of disease among herds, or economic or political reasons. This can create problems at any time, restricting the amount of serum available and the number of batches to choose from. It is estimated that approximately 500 000 l of FBS are sold per year (Gstraunthaler et al. 2013). Cases of fraud have been documented, e.g. where FBS appeared to be blended with bovine serum albumin (BSA) and other additives (Gstraunthaler et al. 2014). Demand continues to increase and will probably exceed supply, unless

Freshney's Culture of Animal Cells: A Manual of Basic Technique and Specialized Applications, Eighth Edition. Amanda Capes-Davis and R. Ian Freshney.
© 2021 John Wiley & Sons Ltd. Published 2021 by John Wiley & Sons Ltd.
Companion website: www.wiley.com/go/freshney/cellculture8

the majority of commercial users can adopt serum-free media. Although an average research laboratory may reserve 100–2001 of serum per year, a commercial biotechnology laboratory can use that amount or more in a week.

(3) *Contamination risk.* Serum may be contaminated by microorganisms that are present in donor animals or introduced during the harvesting process. These may be harmless to cell culture but represent a potential biohazard risk and an additional unknown factor outside the operator's control (Merten 1999). Most manufacturers now filter FBS, with some offering triple filtration using pore sizes down to 0.1 nm (Nims and Harbell 2017). These measures and serum testing have virtually eliminated the risk of mycoplasma contamination from most reputable suppliers. However, viral contamination is an ongoing risk that is managed by specific testing and γ-irradiation (see Sections 9.7.1, 9.7.3) (Hawkes 2015). Chemical contamination is also a possibility. For example, the continued use of antibiotics in cattle rearing results in a risk of contamination from tetracycline (personal communication, L. Steeb), which can affect experiments that use tetracycline-inducible expression systems.

(4) *Physiological variability.* The major constituents of serum are known (see Section 9.5), but serum also contains a wide range of minor components that may have a considerable effect on cell growth and complicate the response to test substances (see Table 9.4; see also Figure 29.10). These components include nutrients (amino acids, nucleosides, sugars, etc.), proteins, peptide growth factors, hormones, minerals, and lipids, the concentrations and actions of which have not been fully determined.

(5) *Growth inhibition.* As well as its growth-promoting activity, serum contains growth-inhibiting activity (see Section 9.5.6). Although stimulation usually predominates, the net effect of the serum is an unpredictable combination of both inhibition and stimulation of cell proliferation. This effect may vary with serum composition and culture conditions. For example, hydrocortisone is cytostatic to many cell types at high cell densities, such as glia and lung epithelium (Guner et al. 1977; McLean et al. 1986), but may be mitogenic at low cell densities (see Section 9.5.2).

(6) *Differentiation.* FBS causes squamous differentiation of human bronchial epithelial cells; several serum-free formulations have been developed for this cell type, including LHC-9 (see Section 10.6.1) (Ke et al. 1988). Human embryonic stem cells (hESCs), induced pluripotent stem cells (iPSCs), and other stem cell populations may also spontaneously differentiate in the presence of serum (Skottman and Hovatta 2006; Furue et al. 2010).

(7) *Standardization.* Standardization of experimental protocols is difficult because of serum batch-to-batch variations over time and between laboratories. For example, variable androgen levels in serum may affect experiments where there is a need to maintain androgen-free culture conditions (Fiandalo et al. 2018). Although charcoal stripping may be used to reduce androgen levels in serum, serum-free medium offers a more defined and standardized alternative for experimental work.

(8) *Downstream applications.* The presence of serum creates a major obstacle for cell-based therapies. Although risk assessment processes exist for the use of serum in biological production (Nims and Harbell 2017), replacement of serum-containing media with chemically defined media is optimal for good manufacturing practice (GMP) (see Section 7.2.3) (Unger et al. 2008).

(9) *Quality control.* Because of its physiological variability and unpredictable effects, changing serum batches requires extensive testing to ensure that the replacement is as close as possible to the previous batch. Testing is performed to detect microbial contamination and to assess performance in culture (see Sections 9.7.1 and 11.8).

(10) *Specificity.* If more than one cell type is being cultured, each type may require a different type or batch of serum (e.g. fetal or calf serum). If this is the case, several batches must be held on reserve simultaneously. Co-culturing different cell types will present a further dimension of complexity in testing.

(11) *Shelf life and consistency.* Typically, a batch of serum will last for at least one year, but some of its desirable characteristics may deteriorate during that time. It must then be replaced with another batch that may be selected as similar, but will never be identical, to the first batch.

(12) *Cost.* Cost is often cited as a disadvantage of serum. Certainly, serum constitutes the major part of the cost of a bottle of medium (more than 10 times the cost of the chemical constituents), but if it is replaced by defined constituents, the cost of these may be as high as that of the serum. As the demand for defined constituents rises, the increased market size should lower the price and serum-free media should become relatively cheaper. Recent estimates suggest that the difference in cost between defined culture media (or xeno-free media) and culture on feeder layers is negligible if the full cost of implementation is considered (Vecchi and Wakatsuki 2015).

10.1.2 Advantages of Serum-Free Media

All of the above problems can be eliminated by removal of serum and other natural, undefined products. Serum-free media have additional benefits, including:

(1) *Achievement of chemically defined medium.* Given that pure constituents are used, a defined medium formulation can be standardized regardless of where it is used and by whom. Not only does this allow easier validation of

industrial processes, but also it means that research labs can replicate conditions to repeat and confirm experimental data.

(2) ***Selection of cell types.*** One of the major advantages of serum-free media is the ability to develop formulations that are selective for a particular cell type. Fibroblastic overgrowth can be inhibited in breast and skin cultures by using MCDB 170 or MCDB 153 (Peehl and Ham 1980; Hammond et al. 1984), and melanocytes can be cultivated in the absence of fibroblasts and keratinocytes (Naeyaert et al. 1991). Many of these selective media are available commercially, often with the cell type of interest (see Section 10.6.1; see also Table 24.1).

(3) ***Control of proliferation and differentiation.*** Almost every cell type is likely to have unique nutritional requirements for survival, proliferation, and differentiation (van der Valk et al. 2010). Culture conditions can thus be modified to promote proliferation or induce differentiation as required.

(4) ***Control of pluripotency and lineage specificity.*** As noted previously, hESCs and iPSCs tend to differentiate in the presence of serum (see Section 10.1.1). This tendency depends on a number of variables, including the type of feeder layer and the method used for cell dissociation (Furue et al. 2010). Clearly, it is important to use defined conditions to induce pluripotency or lineage differentiation in a reproducible manner, particularly for cell-based therapies that require GMP (see Section 7.2.3).

10.1.3 Disadvantages of Serum-Free Media

Serum-free media are not without disadvantages:

(1) ***Reagent purity.*** The removal of serum also removes its protective, detoxifying action (see Section 9.5). Although removing this action is no doubt desirable, it may not always be achievable. In the absence of serum, the purity of reagents and water and the cleanliness of all apparatus should be extremely high (see Section 10.6.2).

(2) ***Multiplicity of media.*** Each cell type appears to require a different recipe, and cultures from malignant tumors may vary in requirements from tumor to tumor, even within one class of tumors. Although this degree of specificity may be an advantage to laboratories that need to isolate specific cell types, it presents a problem if you need to initiate or maintain cell lines of several different origins.

(3) ***Availability.*** The increasing number of serum-free media may lead to the decision to discontinue some product lines that are less commercially viable. Some suppliers will manufacture media as a custom service, but this option is not always cost-effective, and the exact formulation may not be available (e.g. for proprietary media).

(4) ***Selectivity.*** Unfortunately, the transition to serum-free conditions, however desirable, is not always as straightforward as it seems. Some media may select a sublineage that

is not typical of the whole population; even in continuous cell lines, some degree of selection may still occur.

(5) ***Reduced proliferation or pluripotency.*** Growth is often slower in serum-free media (Chowdhury et al. 2012; Gottipamula et al. 2014). Fewer generations are achieved with finite cell lines (Fu et al. 2003). For stem cell cultures, pluripotency may be lost with passaging (Yamasaki et al. 2014).

10.2 DEVELOPMENT OF SERUM-FREE MEDIUM

Serum-free media have been used since the early days of tissue culture to standardize media among laboratories, provide specialized media for specific cell types, and eliminate variable natural products and the risk of microbial contamination (Morton 1970). Serum-free media formulations were initially developed for L-929, HeLa, and other continuous cell lines (Evans et al. 1964; Blaker et al. 1971). Although a degree of selection may have been involved in their adaptation to serum-free conditions, the early investigators discovered that serum could be reduced or omitted without apparent cell selection if appropriate nutritional and hormonal modifications were made (Barnes and Sato 1980a,b; Mather 1998). Media formulations were subsequently developed for normal diploid cells and specific cell types (Ham 1974; Carney et al. 1981).

Historically, there are three different approaches to the development of serum-free medium. The first approach was adopted by Richard Ham and coworkers to successfully develop a series of defined media for culture of normal diploid cells (Ham and McKeehan 1978). Ham would take a known recipe (usually with serum protein or other supplements), reduce the supplements until growth was impaired, and alter the medium constituents individually or in groups. Clonal growth experiments would be performed to detect improved growth; once detected, the supplements would be reduced until growth was again impaired and the process would be repeated. If a group of compounds was found to be effective in reducing serum supplementation, the active constituents would be identified by the systematic omission of single components. The optimum concentration of each constituent would be determined at the minimum concentration of serum protein that was required for satisfactory clonal growth. This approach results in a serum-free medium formulation that is optimized for the cell type of interest (see Section 10.3). Although effective, this is a time-consuming and laborious process, involving growth curves and clonal growth assays at each stage; it is not unreasonable to spend at least three years developing a new medium for a new cell type.

The time-consuming nature of the first approach led to a second approach, used by Gordon Sato and other investigators (Barnes and Sato 1980a; Carney et al. 1981). A known recipe would be used as a base (e.g. a 1 : 1 mixture of Dulbecco's modified Eagle's medium (DMEM) and Ham's F12; see Appendix C) and a shorter list of substances would be added as supplements, with the optimal concentration of each substance assessed at limiting serum concentration. This approach results in a serum-free medium that is supplemented with various growth factors and other individual serum constituents. It is often used by investigators who wish to "build" a serum-free medium that is sufficient for their needs without a major investment in time and materials (see Section 10.4).

A third "design of experiments" approach is now used to develop or modify media formulations, based on a statistical experimental strategy (Xiao et al. 2014; Yao and Asayama 2017). The investigator applies a series of defined input factors (e.g. multiple growth factors at multiple concentrations) and measures a series of defined output responses (e.g. cell attachment, morphology, growth, and differentiation). Statistical analysis of input factors and output responses allows the investigator to look for non-linear relationships between growth factors and minimizes the number of experiments required for medium design (Yao and Asayama 2017).

10.3 SERUM-FREE MEDIA FORMULATIONS

More than 100 serum-free media formulations have been developed to date (van der Valk et al. 2010; see also the fetal calf serum (FCS)-free database [www.fcs-free.org]). Some early formulations were described previously, including Medium 199, CMRL 1066, MB 752/1, and Ham's F12 (see Section 9.4). Although these media are commonly used with serum, they were originally developed for low-serum or serum-free culture. Other classic serum-free media include (see Appendix C for formulations):

(1) *The NCTC series* (Evans et al. 1964; Morton 1970) was developed for L-929 (see Figure 1.1) and other continuous cell lines. The prefix NCTC indicates that media were developed at the National Cancer Institute (NCI) Tissue Culture Section.

(2) *Iscove's modified Dulbecco's medium (IMDM)* (Iscove and Melchers 1978) was developed for serum-free culture of lymphoid cells, allowing maturation of B-cells to produce IgM and IgG. IMDM was originally modified from DMEM. IMDM has been further modified to give WCM5, which was developed for large-scale manufacture of biological products in Chinese hamster ovary (CHO) cells (Keen and Rapson 1995).

(3) *The MCDB 100 series* (Ham and McKeehan 1978) was developed for culture of normal human diploid cells. Media in this series included MCDB 110 (for lung fibroblasts), MCDB 131 (for vascular endothelium), MCDB 153 (for keratinocytes), and MCDB 170 (for mammary epithelium) (Peehl and Ham 1980; Bettger et al. 1981; Hammond et al. 1984; Knedler and Ham 1987). The prefix MCDB indicates that media were

developed at the Department of Molecular, Cellular, and Developmental Biology at the University of Colorado in Boulder.

(4) *The MCDB 200 series* (McKeehan and Ham 1976a) was developed for culture of chicken cells, including MCDB 202 (for chick embryo fibroblasts).

(5) *The MCDB 300 series* (Hamilton and Ham 1977) was developed for culture of Chinese hamster cells, including MCDB 302 (for CHO cells).

(6) *The MCDB 400 series* (Shipley and Ham 1983) was developed for culture of mouse cells, including MCDB 402 (for NIH 3T3 cells, see Section 3.6.1).

(7) *WAJC 404* (McKeehan et al. 1982) was developed for culture of prostate epithelial cells and can be used to isolate normal epithelial cells from fibroblasts in primary cultures. The WAJC 400 series was originally modified from MCDB 151. The prefix WAJC indicates that it was developed at the W. Alton Jones Cell Science Center (McKeehan et al. 1984).

(8) *LHC-9* (Lechner and LaVeck 1985) was developed for culture of human bronchial epithelial cells. The basal medium formulation (LHC Basal, used for LHC-8 and LHC-9) was modified from MCDB 151. The prefix LHC indicates that it was developed at the Laboratory of Human Carcinogenesis, NCI.

Many proprietary formulations are also available, particularly for CHO, HEK293, or hybridoma culture (see Section 10.6.1). Some of these formulations were specifically developed for large scale production of biologicals, where there are safety concerns and a clear need for a product that is free of serum proteins (Merten 1999). Others are applicable to specific cell types. Although evidence has been published for the effectiveness of proprietary media, commercial recipes are often a trade secret, and you can only rely on the supplier's advice or, better, screen a number of media over several subcultures with your own cells. The latter can be a time-consuming exercise but is justified if you are planning long-term work with the cells or if optimal cellular performance is required (e.g. for scale-up culture).

It is important to carefully review the quality control testing that has been performed by commercial suppliers of serum-free media (see Sections 7.5.2, 9.7.1). Ideally, medium would be tested against the cells that you wish to grow (e.g. keratinocyte growth medium should be tested on keratinocytes). Some suppliers provide the appropriate cells with the medium, which should ensure the correct quality control testing. Examples include some cell repositories (see Table 15.3), Lonza, Sigma-Aldrich, and ThermoFisher Scientific. Other suppliers may perform quality control testing using common continuous cell lines, in which case there is no guarantee that media will work for your cell type and you will have to do your own testing.

10.4 SERUM-FREE SUPPLEMENTS

The essential factors in serum have been described (see Section 9.5) and include proteins, hormones and growth factors, essential nutrients and metabolites, lipids, and minerals. Some essential factors are included in serum-free media formulations, but others are not and must be added to basal media for optimal growth. Supplemented media may be prepared in the laboratory from basal media and supplements or purchased from suppliers – for example, the MegaCell™ media (Sigma-Aldrich) and Advanced™ media (ThermoFisher Scientific) are classic media formulations with additional supplements, resulting in a decreased need for serum.

Supplemented media are available for serum-free culture of specific cell types (see Section 10.6). For example, TeSR1 was modified from DMEM/F12 for culture of human pluripotent stem cells (hPSCs) (Ludwig et al. 2006b). Essential 8 (E8) medium is a streamlined version of TeSR1, consisting of DMEM/F12 and an additional seven constituents (see Section 23.4.2) (Chen et al. 2011). E8 and other examples of serum-free supplemented media are listed in Appendix C, along with their source references. If existing serum-free formulations or supplemented media are not suitable for your application, you may need to "build" your own medium by adding additional supplements (see Section 10.2). Additions can be divided into a series of steps or modules (see Figure 10.1), which can be summarized as (van der Valk et al. 2010):

(1) **Select basal medium.** A 1 : 1 mixture of Ham's F12 and DMEM is used by many laboratories as a suitable basal medium. Supplementation with insulin, transferrin, and selenium (ITS) is recommended as a starting point (Gstraunthaler 2003).

(2) **Pre-coat culture vessels.** Serum-free culture may result in reduced adhesion and an increased dependence on attachment factors (van der Valk et al. 2010). Poly-D-lysine has been used for many years to improve attachment and growth in serum-free media (see Section 8.3.5) (McKeehan and Ham 1976b). A variety of extracellular matrix (ECM) coatings are also used (see Section 8.3.2), including fibronectin (25–50 μg/ml) and laminin (1–5 μg/ml).

(3) **Add hormones and growth factors.** A number of mitogenic factors may be used; some have broad utility while others are cell type-specific (see Section 10.4.1).

(4) **Add antioxidants, vitamins, and lipids.** Selenium has antioxidant activity and is already recommended for basal medium as noted above (1). Additional antioxidants, vitamins, and lipids may be used (see Section 10.4.2).

(5) **Add other supplements.** Additional supplements will depend on your cell type and application (see Section 10.4.3).

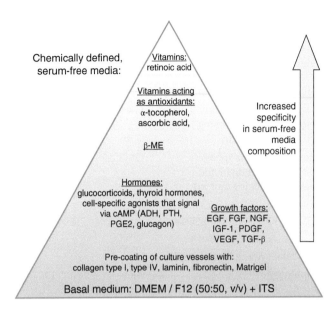

Fig. 10.1. Serum-free medium development. A modular approach that can be used to "build" serum-free media. Modules are underlined within the pyramid, commencing with basal medium. Each additional module is associated with increased specificity in serum-free medium composition. Abbreviations not included in the text: ADH, antidiuretic hormone; β-ME, 2-mercaptoethanol; PGE2, prostaglandin E$_2$; PTH, parathyroid hormone. *Source*: Van der Valk, J., Brunner, D., De Smet, K., et al. (2010). Optimization of chemically defined cell culture media – replacing fetal bovine serum in mammalian in vitro methods. Toxicol In Vitro 24 (4):1053–63. © 2010 Elsevier.

10.4.1 Hormones and Growth Factors

Hormones are physiological constituents in circulating blood and were therefore amongst the first substances to be added when serum-free media were developed (van der Valk et al. 2010). Early work on cellular nutrition resulted in the discovery of novel mitogens and growth factors, later evolving to give therapeutic targets for cancer (Sato et al. 2018). Many peptide hormones and growth factors are used as supplements in serum-free media (see Table 10.1). Most are available in recombinant form (e.g. from Sigma-Aldrich, ThermoFisher Scientific, or the National Hormone and Peptide Program) (Parlow 2006).

Hormones that are commonly used to supplement serum-free media include insulin (1–10 U/ml), which acts as a mitogenic factor and enhances cloning efficiency in a number of different cell types, and hydrocortisone (10–100 nM), which improves the cloning efficiency of glia and fibroblasts and is needed for the maintenance of epidermal keratinocytes and some other epithelial cell types (see Sections 9.5.2, 20.6.3, and 24.2.1). Triiodothyronine (1–100 pM) was initially used for serum-free culture of GH3 cells (rat pituitary) and MDCK cells (dog kidney) (Hayashi et al. 1978; Barnes and Sato 1980a); it has since been used for lung epithelium,

TABLE 10.1. Growth factors and mitogens.

Name; synonyms[a]	Abbreviation	Molecular. mass (kD)[b]	Source[c]	Function
Colony stimulating factor family				
Granulocyte colony-stimulating factor; colony-stimulating factor-3, pluripoietin	G-CSF	18–23 g	Conditioned medium, mouse lung	Granulocyte progenitor proliferation and differentiation
Granulocyte-macrophage colony-stimulating factor; colony-stimulating factor-2	GM-CSF	14–30 g	Conditioned medium, mouse lung	Granulocyte/macrophage progenitor proliferation and differentiation
Macrophage colony-stimulating factor; colony-stimulating factor-1	M-CSF	47–74	B- and T-cells, monocytes, mast cells, fibroblasts	Macrophage progenitor proliferation and differentiation
Epidermal growth factor family				
Amphiregulin	AR	11	Conditioned medium, MCF-7	Autocrine growth factor for keratinocytes
Epidermal growth factor; urogastrone	EGF	6.2	Mouse submaxillary salivary gland, human urine	Active transport; DNA, RNA, protein, synthesis; mitogen for epithelial and fibroblastic cells; synergizes with IGF-1 and TGF-β
Heparin-binding EGF-like factor	HB-EGF	9.5	Conditioned medium, U937	Mitogen for fibroblasts, smooth muscle cells; chemotaxis
Heregulin (product of *NRG1*, *erb*B2 ligand)	HRG	70	Breast cancer cells	Mitogen for mammary and other epithelial cells
Transforming growth factor-α	TGF-α	5.5	Conditioned medium, transformed fibroblasts	Anchorage-independent growth, loss of contact inhibition
Fibroblast growth factor families				
Fibroblast growth factor-1; acidic fibroblast gf, endothelial cell gf, eye-derived gf-2, heparin binding gf-1, myoblast gf	FGF-1 (aFGF)	19 h	Bovine brain, pituitary	Mitogen for endothelial cells and other cell types
Fibroblast growth factor-2; basic fibroblast gf, eye-derived gf-1, heparin binding gf-2, prostatropin	FGF-2 (bFGF)	17 h	Bovine brain, pituitary	Mitogen; adipocyte and ovarian granulosa cell differentiation; proliferation of hESCs
Fibroblast growth factor-3 (product of *int*-2 oncogene)	FGF-3	21 h	Mammary tumors	Mitogen; morphogen; angiogenic
Fibroblast growth factor-4 (product of *hst*/KS3 oncogene)	FGF-4	20 h	Embryo, tumors	Mitogen; morphogen; angiogenic
Fibroblast growth factor-5	FGF-5	27 h	Fibroblasts, epithelial cells, tumors	Mitogen; morphogen; angiogenic
Fibroblast growth factor-6 (product of *hst*-2 oncogene)	FGF-6	19 h	Testis, heart, muscle	Mitogenic for fibroblasts; morphogen
Fibroblast growth factor-8 (including a, b, e, f isoforms); androgen-induced gf	FGF-8	22 h	Conditioned medium, SC-3	Mitogen; epithelial-mesenchymal transition; epithelial differentiation
Fibroblast growth factor-9; glia activating factor, heparin binding gf	FGF-9	23 h	Conditioned medium, NMC-G1	Mitogen; neuronal viability
Fibroblast growth factor-10; keratinocyte gf-2	FGF-10	19 h	Embryo	Mitogen for alveolar epithelium; trophoblast invasion; stimulates uPA and PAI-1

(continued)

TABLE 10.1. (*continued*)

Name; synonyms[a]	Abbreviation	Molecular. mass (kD)[b]	Source[c]	Function
Keratinocyte growth factor; fibroblast gf-7	KGF (FGF-7)	19 h	Lung fibroblasts	Keratinocyte proliferation and differentiation; prostate epithelial proliferation and differentiation
Insulin family				
Insulin	Ins	5.8	Pancreatic β islet cells	Glucose uptake and oxidation; amino acid uptake; glyconeogenesis
Insulin-like growth factor-1; somatomedin-C, sulfation factor, non-suppressible insulin-like activity	IGF-1	7.6	Liver	Mediates effect of growth hormone on cartilage sulfation; insulin-like activity
Insulin-like growth factor-2; multiplication stimulating activity (rat)	IGF-2	7.5	Conditioned medium, BRL-3A	Mediates effect of growth hormone on cartilage sulfation; insulin-like activity
Interferon families				
Interferon-α (various subtypes); leukocyte interferon	IFN-α	19–26	Macrophages	Antiviral; differentiation inducer; anticancer
Interferon-β (various subtypes); fibroblast interferon	IFN-β	22–27 g	Fibroblasts	Antiviral; differentiation inducer; anticancer
Interferon-γ; immune interferon, macrophage-activating factor	IFN-γ	21–24 dimer	Activated lymphocytes	Antiviral, macrophage activator; antiproliferative on transformed cells
Interleukin families				
Ciliary neurotrophic factor	CNTF	23	Eye	Neuronal viability
Interleukin-1; B-cell-activating factor, lymphocyte-activating factor, hematopoietin-1	IL-1	12–18	Activated macrophages	Induces IL-2 release
Interleukin-2; T-cell gf	IL-2	15	Conditioned medium, CD4+ lymphocytes (NK)	Supports growth of activated T-cells; stimulates LAK cells
Interleukin-3; mast cell gf, multipotential colony-stimulating factor	IL-3	14–28 g	Activated T-lymphocytes, myelomonocytic cell lines	Granulocyte/macrophage proliferation, differentiation
Interleukin-4; B-cell gf-1, B-cell stimulating factor-1	IL-4	12–20	Activated CD4+ lymphocytes	Competence factor for resulting B-cells; mast cell maturation (with IL-3)
Interleukin-5; B-cell gf-2, T-cell-replacing factor, eosinophil differentiating factor	IL-5	40–60 g dimer	T-lymphocytes	Eosinophil differentiation; progression factor for competent B-cells
Interleukin-6; B-cell stimulating factor-2, hepatocyte-stimulating factor, hybridoma-plasmacytoma gf, interferon-β2	IL-6	22–27 g	Activated T-lymphocytes, macrophages/monocytes, fibroblasts	Acute phase response; B-cell differentiation; keratinocyte differentiation; PC12 differentiation
Interleukin-7; hematopoietic growth factor, lymphopoietin 1	IL-7	15–17 g	Bone marrow stroma	Pre- and pro-B-cell growth factor
Interleukin-8; granulocyte chemotactic protein-1, monocyte-derived neutrophil chemotactic factor, neutrophil-activating protein, T-cell chemotactic factor	IL-8	11 h	LPS-activated monocytes, PHA lymphocytes, endothelial cells	Chemotactic factor for neutrophils, basophils, and T-cells

(*continued*)

TABLE 10.1. (*continued*)

Name; synonyms[a]	Abbreviation	Molecular. mass (kD)[b]	Source[c]	Function
Interleukin-9; human P-40, mast cell growth-enhancing activity, mouse T-helper gf	IL-9	25–40 g	Activated CD4+ T-lymphocytes	Growth factor for T-helper, megakaryocytes, mast cells (with IL-3)
Interleukin-10; cytokine synthesis inhibitory factor	IL-10	20	T-lymphocytes	Immune suppressor
Interleukin-11; adipogenesis inhibitory factor	IL-11	23	Conditioned medium, PU-34	Stimulates plasmacytoma proliferation and T-cell-dependent development of Ig-producing B-cells
Interleukin-12; cytotoxic lymphocyte maturation factor	IL-12	75 dimer	B-lymphocytes	Activated T-cell and NK cell growth factor; induces IFN-γ
Leukemia inhibitory factor	LIF	32–62 g	SCO cells	Maintains pluripotency in mESCs
Oncostatin M	OSM	28	Activated T-cells and PMA-treated monocytes	Differentiation inducer (with glucocorticoid); fibroblast mitogen
Neurotrophin family				
Brain-derived neurotrophic factor	BDNF	27	Brain	Neuronal viability
Nerve growth factor (β subunit)	NGF	26	Male mouse submaxillary salivary gland	Trophic factor; chemotactic factor; differentiation factor; neurite outgrowth in peripheral nerve
Neurotrophin-3	NT-3	29	Brain	Stimulation of neurite outgrowth
PDGF family				
Platelet-derived growth factor	PDGF	35–45 g dimer	Blood platelets	Mitogen for mesodermal and neuroectodermal cells; wound repair; synergizes with EGF and IGF-1
Stem cell factor (*c-kit* ligand); mast cell growth factor, steel factor	SCF	31 g dimer	Endothelial cells, fibroblasts, bone marrow, Sertoli cells	Promotes first maturation division of pluripotent hematopoietic stem cells
Vascular endothelial growth factor; vascular permeability factor	VEGF	42 g dimer	Bovine pituitary	Angiogenesis; vascular endothelial cell proliferation
Transforming growth factor-β family				
Activin		26 dimer	Gonads, pituitary, placenta	Morphogen; induces FSH release; maintains pluripotency in hESCs
Inhibin		27–34 dimer	Ovary	Morphogen; inhibits FSH secretion
Müllerian inhibition factor	MIF	70	Testis	Inhibition of Müllerian duct; inhibition of ovarian carcinoma
Nodal		39	Recombinant	Morphogen; maintains pluripotency in hESCs
Transforming growth factor β (six species)	TGF-β1–6	23–25 dimer	Blood platelets	Inhibition of epithelial cell proliferation; squamous differentiation; proliferation of hESCs

(continued)

TABLE 10.1. (*continued*)

Name; synonyms[a]	Abbreviation	Molecular. mass (kD)[b]	Source[c]	Function
Tumor necrosis factor family				
Tumor necrosis factor α; cachectin, cachexin	TNF-α	17	Recombinant	Catabolic; cachexia; shock
Tumor necrosis factor-β; lymphotoxin	TNF-β	20–25	Lymphocytes	Cytotoxic for tumor cells
Other growth factors and mitogens: human				
Angiogenin	ANG	14	Fibroblasts, lymphocytes, colonic epithelium	Angiogenic; endothelial mitogen
Connective tissue growth factor	CTGF	38 p	Peritoneal mesothelium, mesangial cells	Fibroblast mitogen; angiogenic; matrix production
Erythropoietin	EPO	30–34 g	Renal juxtaglomerular cells	Erythroid progenitor proliferation and differentiation
Hepatocyte growth factor; scatter factor	HGF	78 h	Fibroblasts, platelets	Epithelial morphogenesis; hepatocyte proliferation
Macrophage inflammatory protein-1α; C-C motif chemokine-3	MIP-1α	8	Macrophages	Hematopoietic stem cell inhibitor
Platelet-derived endothelial cell growth factor; gliostatin, thymidine phosphorylase	PD-ECGF	100 dimer	Blood platelets, fibroblasts, smooth muscle	Angiogenesis; endothelial cell mitogen; neuronal viability; glial cytostasis
Transferrin	TF	78	Liver	Iron transport; mitogen
Other growth factors and mitogens: non-human				
Cholera toxin	CTX	80–90	Cholera bacillus	Mitogen for some normal epithelia
Endothelial cell growth supplement (mixture of endothelial mitogens)	ECGS	Various proteins	Bovine pituitary	Endothelial mitogenesis
Endotoxin (mixture of bacterial molecules, toxic to cells)		Various proteins	Bacteria	Stimulates TNF production
Lipopolysaccharide	LPS	10	Gram-positive bacteria	Lymphocyte activation
Phorbol 12-myristate 13-acetate (12-*O*-tetradecanoylphorbol-13-acetate)	PMA (TPA)	0.617	Croton oil	Tumor promoter; mitogen for some epithelial cells and melanocytes; inducing agent for differentiation (see Table 26.3)
Phytohemagglutinin	PHA	30	Red kidney bean (*Phaseolus vulgaris*)	Lymphocyte activation
Pokeweed mitogen	PWM		Roots of pokeweed (*Phytolacca americana*)	Monocyte activation

Source: Based on Barnes, W. D., Sirbasku, D. A. and Sato, G. H. (1984). Methods for preparation of media, supplements, and substrata for serum-free animal cell culture. Vol. 1, Cell culture methods for molecular and cell biology. New York: Alan R. Liss; Lange, W., Brugger, W., Rosenthal, F. M., et al. (1991). The role of cytokines in oncology. Int J Cell Cloning 9 (4):252–73; Jenkins, N. (1992). Growth factors, a practical approach. Oxford: IRL Press at Oxford University Press; Bafico, A. and Aaronson, S. A. (2003). "Classification of growth factors and their receptors." In Holland-Frei cancer medicine, edited by D. W. Kufe, R. E. Pollock, R. R. Weichselbaum, R. C. Bast, T. S. Gansler, J. F. Holland and E. Frei. Hamilton, ON: BC Decker.
[a]Synonyms containing "gf" are abbreviated from "growth factor."
[b]Molecular mass is approximate and will vary depending on post-translational modification. The addition of "g" indicates glycosylation, "h" indicates heparin binding, and "p" indicates perlecan binding.
[c]Sources described are some of the original tissues from which the natural product was isolated. In almost all cases, the natural product has been replaced by cloned recombinant material that is available commercially.

chondrocytes, and other cell types (Lechner and LaVeck 1985; Alini et al. 1996). Various combinations of estrogen, androgen, or progesterone with hydrocortisone and prolactin at around 10 nM have been used for the maintenance of mammary epithelium (Klevjer-Anderson and Buehring 1980; Hammond et al. 1984; Strange et al. 1991).

Some hormones promote growth for unexpected cell types that would be considered "off target" for those hormones *in vivo* (Sato et al. 2018). For example, follicle-stimulating hormone (FSH) can stimulate the growth of B16 cells (mouse melanoma) and gastrin-releasing peptide (GRP) can stimulate the growth of bronchial epithelium and small cell lung carcinoma (Barnes and Sato 1980a; DeMichele et al. 1994; Hohla et al. 2007). Optimization of growth conditions for different cell types has demonstrated that specific cell types can be selected and others inhibited through the careful selection of supplements (Barnes and Sato 1980b).

Cytokines and other locally-acting growth factors also tend to have a wide-ranging specificity beyond those tissues in which activity was first demonstrated. For example, keratinocyte growth factor (KGF), besides showing activity with epidermal keratinocytes, induces proliferation, and differentiation in prostatic epithelium (Aaronson et al. 1991; Thomson et al. 1997; Planz et al. 1998). Hepatocyte growth factor (HGF) is mitogenic for hepatocytes but is also morphogenic for kidney tubules (see Section 26.3.4) (Balkovetz and Lipschutz 1999). Growth factors and cytokines acquire their specificity by virtue of the fact that their production is localized and that they have a limited range. Most act as paracrine factors (i.e. they act on adjacent cells from different lineages) and not by systemic distribution in the blood.

Growth factors and hormones may act synergistically or additively with each other. For example, the action of interleukin 6 (IL-6) and oncostatin M on A549 cells is dependent on dexamethasone, a synthetic hydrocortisone analogue (McCormick et al. 1995; McCormick and Freshney 2000). The action is due to the production of a heparan sulfate proteoglycan (HSPG) by A549 cells (Yevdokimova and Freshney 1997). The requirement for heparin or HSPG was first observed with fibroblast growth factor-2 (FGF-2) (Klagsbrun and Baird 1991), but it may be a more general phenomenon; β-glycan is involved in the cellular response to transforming growth factor-β (TGF-β) (Lopez-Casillas et al. 1993). Some growth factors are dependent on the activity of a second growth factor before they act (Phillips and Cristofalo 1988). For example, bombesin alone is not mitogenic in normal cells but requires the simultaneous or prior action of insulin or one of the insulin-like growth factors (IGFs) (Santiskulvong et al. 2004). Growth factors and cytokines may also converge on common signaling pathways. For example, prostaglandin E_1, vasopressin, and cholera toxin are effective for culture of epithelial cells due to their elevation of intracellular cAMP (Gstraunthaler 2003).

10.4.2 Antioxidants, Vitamins, and Lipids

Selenium (Na_2SeO_3), at around 20 nM, is found in most serum-free media formulations. Selenium is known to have antioxidant activity, which is important to minimize cellular oxidative stress (see Section 9.2.4). Antioxidants that are normally supplied to cells from the bloodstream may be unsuitable as culture supplements – for example, vitamin C (L-ascorbic acid) is unstable and generates toxic degradation products, which can cause artifacts in cell-based assays (Halliwell 2014). More stable forms of L-ascorbic acid are available (e.g. L-ascorbic acid 2-phosphate magnesium). Other supplements with antioxidant activity include albumin (see Section 10.4.3), dithiothreitol, glutathione, 2-mercaptoethanol (β-mercaptoethanol), and vitamin E (α-tocopherol).

At least seven vitamins are believed to be essential for tissue culture: choline, folic acid, nicotinamide, pantothenate, pyridoxal, riboflavin, and thiamine (van der Valk et al. 2010). Lipids or lipid precursors such as choline, linoleic acid, ethanolamine, or phosphoethanolamine are also required, although this may vary depending on the cell type (Roberson and Robertson 1995). Lipid mixtures are easily oxidized, and should be stored at single-use concentration with minimal exposure to air and light (Price 2017).

10.4.3 Other Supplements for Serum-Free Medium

Other supplements that are used in serum-free medium include:

(1) *Trace elements.* A number of trace elements are present in serum and may be added to serum-free media, including copper, iron, zinc, and selenium (see Sections 9.5.5, 10.4.2). Lithium chloride is added to TeSR1 medium for undifferentiated proliferation of stem cells, as an activator of the Wnt/β-catenin signaling pathway (Ludwig et al. 2006b; Silva et al. 2010). Some minerals may be pro-oxidant due to the presence of "free" ions, so it is important to test for detrimental effects (Halliwell 2014). Trace elements may be subject to unexpected fluctuations, even between batches from the same supplier; it may be necessary to test for variations (Keenan et al. 2018).

(2) *Proteins.* Transferrin (5–300 ng/ml) is required as a carrier for iron and may also have a mitogenic role. Recombinant transferrin may be used (e.g. Optiferrin™, InVitria) or transferrin may be substituted with other iron chelators. Residual transferrin may be present for over 30 passages after substitution due to the cyclic nature of iron trafficking (Keenan et al. 2006). BSA (fatty acid free, 0.5–10 mg/ml) is often added to improve cell growth and survival. However, BSA adds undefined constituents to the medium and microbial contamination is an ongoing risk. BSA can be substituted with human

serum albumin (HSA) or recombinant albumin (e.g. Cellastim, InVitria) (Francis 2010).

(3) **Tissue extracts.** Bovine pituitary extract (BPE) may be used as a protein supplement in conjunction with some keratinocyte serum-free media and can protect against oxidative stress (Hammond et al. 1984; Kent and Bomser 2003). Like BSA and transferrin, BPE adds undefined constituents to the medium and carries a risk of microbial contamination.

(4) **Amino acid hydrolysates.** These may be used to reduce or eliminate serum in large scale cultures for biotechnology applications (Ikonomou et al. 2001). Amino acid hydrolysates are relatively cheap and may be derived from plants (e.g. soy protein hydrolysate), reducing the risk of microbial contamination compared to animal products. However, hydrolysate mixtures may not be optimal for some applications and can vary from batch to batch (Hazeltine et al. 2016; McGillicuddy et al. 2018).

(5) **Polyamines.** Spermidine ($0.1–1.0\,\mu M$), spermine, and putrescine are often added to serum-free media and appear to be important for cell growth and epithelial stem cell function (Ramot et al. 2011; Mandal et al. 2013).

(6) **Protective additives and detergents.** One of the actions of serum is to increase medium viscosity (see Section 9.2.7). This is particularly important in stirred suspension culture as it helps to minimize shear stress. Carboxymethylcellulose (CMC, $1.2–30\,mg/ml$) (Telling and Ellsworth 1965; Moreira et al. 1995), Pluronic F-68 (polyoxyethylene and polyoxypropylene, $1\,mg/ml$) (Tharmalingam et al. 2008), polyvinylpyrrolidone (PVP, $1\,mg/ml$) (Rashid et al. 2014), and poly-γ-glutamic acid (γPGA) (Chun et al. 2012) have all been used to minimize mechanical damage.

10.5 SERUM REPLACEMENTS

Serum replacement products can be used to substitute for serum in conventional media (see Table 10.2). In some cases, the formulation is defined e.g. ITS Premix (Corning), Nutridoma (Sigma-Aldrich), and Serum Replacement 1–3 (Sigma-Aldrich). In other cases, the formulation may be partly defined or regarded as proprietary information. Serum replacements may be optimized for particular cell types or applications; for example, KnockOut™ Serum Replacement and KnockOut DMEM (ThermoFisher Scientific) were optimized for growth of mouse embryonic stem cells (mESCs) and can give higher rates of cloning and undifferentiated proliferation for hESC cultures compared to serum (Price et al. 1998; Amit et al. 2000; Skottman and Hovatta 2006).

Although using a single product as a serum replacement may seem relatively simple, such products do have disadvantages. Serum replacements may only be partially defined and some contain animal-derived components. As a result, serum

replacements may have an ongoing risk of microbial contamination and may vary in their performance from batch to batch (Chaudhry et al. 2008). It is important to compare a selection of serum-free media before deciding on the best option for your cell type or application. Try to avoid a serum replacement if you cannot obtain details of its composition.

10.6 USE OF SERUM-FREE MEDIUM

10.6.1 Choice of Serum-Free Media

The range of serum-free media, supplements, and serum replacements is now so large and diverse that it is not possible to make individual recommendations for specific tasks. Some suggestions are provided here as a starting point (see Table 10.3). The FCS-free database provides a more comprehensive approach, with the ability to search and filter results based on species, cell line or type, and other parameters. Check this database and the scientific literature, discuss with suppliers, obtain samples of relevant products, and screen those samples in your own assays (e.g. for growth, survival, or special functions).

When a cell line is obtained from the originator or a reputable cell repository, they will recommend the appropriate medium based on practical experience with that cell line and cell type. If possible, it is best to stay with the originator's recommendations for serum-free media, as this may be the only way to ensure that the cell line exhibits its specific properties. If the culture was previously grown in serum, it will usually need a period of adaptation to serum-free conditions (see Section 10.6.3). Serum-free conditions will affect the handling of the culture, e.g. its subculture and cryopreservation (see Sections 14.8, 15.3.2). Always perform culture testing when changes are made to the culture's medium to look for changes in growth and other behavior.

If serum-free medium is needed to promote selective growth of a particular cell type, that requirement will determine the choice of medium. For example, MCDB 153 was optimized for epidermal keratinocytes, LHC-9 for bronchial epithelium, HITES for small cell lung cancer, and MCDB 131 for vascular endothelium (see Appendix C for formulations). If serum-free medium is required for growth of continuous cell lines to reduce the risk of microbial contamination or the presence of serum proteins in the cell product, the choice will be wider and there will be several commercial sources to choose from (see Table 10.3). It may still be necessary to screen a number of different suppliers to ensure that (i) cells grow well in the new medium; (ii) product formation is maintained; (iii) a long period of adaptation is not required; and (iv) costs and delivery times are acceptable.

Serum-free media formulations are increasingly used for stem cell culture. Some are available commercially, while others can be prepared in the laboratory based on source references (see Sections 10.6.2, 23.4.3). Supplements are selected with reference to the basal medium used, the presence or

TABLE 10.2. Serum replacements.

Brand name	Supplier	Contains	Recommended for[a]
BIT 9500	STEMCELL Technologies	BSA, rh-Insulin, human transferrin (iron-saturated), IMDM	Human and mouse hematopoietic progenitor cells
CDM-HD	FiberCell Systems	Vitamins, co-factors, trace elements, micronutrients, recombinant growth factors, carbohydrates including glucose	High cell densities in hollow fiber bioreactors
Chemically defined lipid concentrate	ThermoFisher Scientific (similar products available from other suppliers)	Arachidonic acid, cholesterol, DL-α-tocopherol acetate, ethyl alcohol, linoleic acid, linolenic acid, myristic acid, oleic acid, palmitic acid, palmitoleic acid, Pluronic F-68, stearic acid, Tween-80	Insect cells, CHO, hybridomas
EX-CYTE	Merck Millipore	Cholesterol, phospholipids, fatty acids	Mammalian cell lines including hybridomas, myelomas, CHO, HEK-293
FreeAdd	Biowest	Not stated; chemically defined	Most cell lines including stem cells, primary cells, insect cells
ITSE AF	InVitria (similar products available from other suppliers)	Recombinant insulin and transferrin, selenium, ethanolamine	CHO, hybridomas, HEK293, Vero, rat crypt, HT-29, fibroblasts, MDCK, hamster, PER.C6, stem cells
ITS supplement	STEMCELL Technologies (similar products available from other suppliers)	Insulin, holo-transferrin (iron-saturated), selenium, PBS	mESCs, iPSCs
KnockOut	ThermoFisher Scientific	Defined serum-free formulation containing growth factors (xeno-free product available)	hESCs, mESCs, iPSCs
Nu-Serum™	Corning	Not stated	MAb and virus production, screening, transfection
Nutridoma™	Sigma-Aldrich	Albumin, insulin, transferrin, other defined organic and inorganic compounds	Mouse myelomas and hybridomas
PeproGrow™-1	PeproTech	Not stated	Adherent mammalian cell lines
PluriQ™	Amsbio (GlobalStem)	Not stated	hESCs, iPSCs
Sheff-CHO ACF	Kerry (Sheffield Bioscience)	Wheat hydrolysate, recombinant human albumin	CHO
Sheff-Vax ACF	Kerry (Sheffield Bioscience)	Recombinant growth factors	MDCK, BHK-21, HEK-293
Serum replacement 1/2	Sigma-Aldrich	Heat-treated BSA, heat-treated bovine transferrin, bovine insulin	General cell culture, stem cells
Serum replacement 3	Sigma-Aldrich	HSA, human transferrin, rh-Insulin	Anchorage-dependent and suspension cells, production of cell-secreted proteins
Serum replacement solution (SR100)	PeproTech	Not stated	HEK-293, HeLa, A549
SSS™	Irvine Scientific	84% HSA, 16% alpha and beta globulins, saline	Mouse embryo tested
TCH™	MP Biomedicals	Not stated	Human cells, production of cell-secreted proteins
TCM™	MP Biomedicals	Not stated	Primary cells, cell lines, stem cells
Ultroser™ G	Pall Corporation	Semi-defined composition	Amniocentesis, karyotyping. Also tested for many other cells

[a]Suppliers' recommendations.
Abbreviations not included in the text: MAb, monoclonal antibody; rh-Insulin, recombinant human insulin.

TABLE 10.3. Selecting a serum-free medium.

Cell line/cell type	Serum-free medium suggestions[a]	Commercial suppliers[b]
Adipocytes	PBM™-2, PRIME-SV® adipogenic differentiation SFM	Irvine Scientific, Lonza
BHK-21	CD BHK-21 production, Cellvento® BHK, MP-BHK	Merck Millipore, MP Biomedicals, ThermoFisher Scientific
Bronchial epithelial cells (see Supp. S24.4)	BEGM™, LHC-9, PneumaCult™	Lonza, STEMCELL Technologies, ThermoFisher Scientific
Chick embryo fibroblasts	Medium 199, MCDB 201, MCDB 202	Biological Industries, Sigma-Aldrich, US Biological
CHO	CCM5 CHO, CD CHO, Cellvento CHO, EX-CELL® 302 and 325, MCDB 302, ProCHO™, WCM5	GE Healthcare Life Sciences, Lonza, Sigma-Aldrich, Merck Millipore, ThermoFisher Scientific
Chondrocytes	CDM chondrocyte differentiation, PRIME-XV chondrogenic differentiation	Irvine Scientific, Lonza
Continuous cell lines	CMRL 1066, DMEM/F12 + supplements, Medium 199, MB 752/1, MCDB media, PC-1™, UltraCULTURE™	Biological Industries, Lonza, Sigma-Aldrich, ThermoFisher Scientific, US Biological
Corneal epithelial cells (see Supp. S24.2)	EpiLife™, Keratinocyte-SFM, MCDB 153	ThermoFisher Scientific, US Biological
Corneal endothelial cells	Human endothelial-SFM	ThermoFisher Scientific
Endothelial cells	EBM™Plus, human endothelial-SFM, MCDB 130, MCDB 131, Medium 131, PeproGrow EPC	Lonza, PeproTech, Sigma-Aldrich, ThermoFisher Scientific, US Biological
Fibroblasts (including human diploid fibroblasts)	FGM™, MCDB 110, MCDB 202, MCDB 402, Medium 106, PC-1, UltraCULTURE	Biomol, Lonza, Sigma-Aldrich, ThermoFisher Scientific, US Biological
Glioma cells (see also Freshney et al. 1980; Ledur et al. 2017)	Neurobasal + supplements, SF12 (Ham's F12 with extra essential and non-essential amino acids)	ThermoFisher Scientific
HEK293	293 SFM II, CD 293, CDM4HEK293, EX-CELL 293, SFM4HEK293	GE Healthcare Life Sciences, Sigma-Aldrich, ThermoFisher Scientific
HeLa (see also Hutchings and Sato 1978)	EX-CELL HeLa, GEM-HeLa, Ham's F12 + supplements, PC-1	Lonza, MP Biomedicals, Sigma-Aldrich
Hematopoietic cells	AIM V®, ImmunoCult™, Macrophage-SFM, MarrowMAX™, Stemline® II, StemPro®-34 SFM, StemSpan™ H3000, X-VIVO™ 10	Lonza, Sigma-Aldrich, STEMCELL Technologies, ThermoFisher Scientific
Hepatocytes	HBM™, hepatozyme-SFM	Lonza, ThermoFisher Scientific
Human cells in spheroid culture	3dGRO™ Spheroid, 3D tumorsphere medium XF, MammoCult™, PRIME-XV tumorsphere	Irvine Scientific, PromoCell, Sigma-Aldrich, STEMCELL Technologies
Human mesenchymal stromal cells (MSCs)	CTS StemPro MSC, PeproGrow hMSC, PRIME-XV MSC expansion, stemline mesenchymal stem cell expansion	Irvine Scientific, PeproTech, Sigma-Aldrich, ThermoFisher Scientific
Human pluripotent stem cells (hESCs, human iPSCs)	Essential 8™, KnockOut DMEM, mTeSR™1, PluriSTEM™, StemFlex™, StemPro hESC SFM, TeSR™-E8™	Merck Millipore, Sigma-Aldrich, STEMCELL Technologies, ThermoFisher Scientific
Human tumor cells (see also Carney et al. 1981)	Cancer cell line medium XF, HITES	PromoCell
Hepatocytes (see also Williams and Gunn 1974; Mitaka et al. 1991)	L-15, Williams' E	Sigma-Aldrich, ThermoFisher Scientific
Hybridomas	CCM1, CD hybridoma, ClonaCell™-HY, EX-CELL hybridoma, UltraDOMA™	Biological Industries, GE Healthcare Life Sciences, Lonza, Sigma-Aldrich, STEMCELL Technologies, ThermoFisher Scientific

(continued)

TABLE 10.3. (*continued*)

Cell line/cell type	Serum-free medium suggestions[a]	Commercial suppliers[b]
Insect cells (see also Sf9)	Drosophila-SFM, ESF 921™, Express Five® SFM, EX-CELL 405, EX-CELL 420, SFM4Insect	GE Healthcare Life Sciences, Oxford Expression Technologies, Sigma-Aldrich, ThermoFisher Scientific
Keratinocytes (see Supp. S24.1 and S24.3)	EpiLife, KGM™-2, Keratinocyte-SFM, KGM-CD, MCDB 153, stemline keratinocyte medium II	Biological Industries, Lonza, Merck Millipore, PromoCell, Sigma-Aldrich, ThermoFisher Scientific, US Biological
L-929 cells (NCTC clone 929)	CMRL 1066, MB 752/1, NCTC 109, NCTC 135	Biological Industries, Lonza, MP Biomedicals, Sigma-Aldrich, ThermoFisher Scientific, US Biological
Leukemia and normal leukocytes	HPGM™, LGM™-3, PRIME-XV hematopoietic cell basal, PRIME-XV mouse hematopoietic cell basal, UltraCULTURE	Biological Industries, Irvine Scientific, Lonza, Sigma-Aldrich, STEMCELL Technologies, ThermoFisher Scientific
Mammary epithelium (see also Supp. S24.8)	HuMEC, MCDB 170, MEBM™	Lonza, ThermoFisher Scientific, US Biological
MDCK	EX-CELL MDCK, K-1, PC-1, Plus MDCK, SFM4MegaVir, UltraMDCK™	Cesco Bioengineering, GE Healthcare Life Sciences, Lonza, Sigma-Aldrich
Melanocytes (see also Supp. S24.19)	HSM (see Appendix C), Medium 254, MGM™-4	Lonza, ThermoFisher Scientific
Mouse cells in spheroid culture	HepatiCult™, IntestiCult™, PancreaCult™	STEMCELL Technologies
Mouse embryonic fibroblasts (MEFs)	MCDB 402, PC-1	Lonza
Mouse neuroblastoma	MCDB 411, DMEM/F12 + N1	Sigma-Aldrich
Mouse pluripotent stem cells (mESCs, mouse iPSCs)	ESGRO®, iSTEM, KnockOut DMEM	Merck Millipore, Takara Bio, ThermoFisher Scientific
Neurons	BrainPhys™, NeuroCult™, Neurobasal + B27/G5/N2, PRIME-XV neural basal	Irvine Scientific, STEMCELL Technologies, Takara Bio, ThermoFisher Scientific
Osteoblasts	OBM™, PRIME-XV osteogenic differentiation	Irvine Scientific, Lonza
Prostate (see also Peehl 2002)	PEC (see Appendix C), PrEGM™, WJAC 404	Lonza
Renal cells (see also HEK293 and MDCK)	MsBM™, OptiPRO™ SFM, REGM™, VP-SFM	Lonza, ThermoFisher Scientific
Sf9	CCM3, EX-CELL 420, ExpiSf™ CD, Insectagro™, Sf-900 II SFM	Corning, GE Healthcare Life Sciences, Sigma-Aldrich, ThermoFisher Scientific
Skeletal myoblasts (see also Das et al. 2009)	L-15/Medium 199 (3 : 1) + supplements	Sigma-Aldrich, ThermoFisher Scientific
Smooth muscle cells	Medium 231	ThermoFisher Scientific
Urothelial cells	MCDB 153, KGM-2, Keratinocyte-SFM	Lonza, ThermoFisher Scientific, US Biological
Vero	EX-CELL Vero, MP-Vero, OptiPRO SFM, Plus Vero, SFM4MegaVir, VP-SFM	Cesco Bioengineering, GE Healthcare Life Sciences, MP Biomedicals, Sigma-Aldrich, ThermoFisher Scientific

Source: Based on Desai, N., Rambhia, P. and Gishto, A. (2015). Human embryonic stem cell cultivation: historical perspective and evolution of xeno-free culture systems. Reprod Biol Endocrinol 13:9; Mather, J. P. (1998). Making informed choices: medium, serum, and serum-free medium. How to choose the appropriate medium and culture system for the model you wish to create. Methods Cell Biol 57:19–30; Yao, T. and Asayama, Y. (2017). Animal-cell culture media: History, characteristics, and current issues. Reprod Med Biol 16 (2):99–117.

[a]Suggestions act as a starting point and are listed in alphabetical order. More suggestions are available from suppliers and in publications that describe serum-free media and their use (see Appendix C).

[b]Media may be available from one or multiple suppliers; supplier names are provided here as a starting point in alphabetical order. To find a medium supplier, use your favorite search engine to search for media names and fine-tune with additional key words, e.g. "serum free medium."

absence of a feeder layer, the species of origin, and the desired characteristics, e.g. whether pluripotency or differentiation is required. Culture of hPSC lines using serum-free media and defined culture systems is discussed in a later chapter (see Section 23.4).

10.6.2 Preparation of Serum-Free Media

Reagent quality is of particular importance when preparing serum-free media. Ultrapure water (UPW) and reagents should be used, with quality control testing to ensure that performance is consistent. Variations in quality may be traced back to quite minor variations in procedure – for example, whether UPW is stored or used fresh (Ludwig et al. 2006a). A recent comparison of serum-free media found that many laboratory-made media were unable to support hESC proliferation in the absence of feeder layers. Media obtained from the original laboratory resulted in better performance, suggesting that problems traced back to subtle differences in preparation or reagent quality (International Stem Cell Initiative Consortium et al. 2010).

The procedure for making up serum-free recipes is similar to that for preparing regular media (see Section 11.6.2) (Waymouth 1984). Constituents are generally made up as a series of stock solutions: (i) minerals and vitamins at 1000×; (ii) tyrosine, tryptophan, and phenylalanine in 0.1 N HCl at 50×; (iii) essential amino acids at 100× in water; (iv) salts at 10× in water; and (v) any other special cofactors, lipids, etc. at 1000× in the appropriate solvents. These are combined in the correct proportions and diluted to the final concentration, and then the pH and osmolality are checked. Care should be taken with solutions of Ca^{2+} and Fe^{2+} or Fe^{3+} to avoid precipitation. Metal salts tend to precipitate in alkaline pH in the presence of phosphate, particularly when the medium or salt solution is autoclaved, so cations in stock solutions should be kept at a low pH (below 6.5) and maintained phosphate free. Divalent cations are often added last, immediately before using the medium. Growth factors, hormones, and cell adhesion factors are also best added separately just before the medium is used, as they may need to be adjusted to suit your experimental conditions.

10.6.3 Adaptation to Serum-Free Media

Many continuous cell lines have been adapted to serum-free media, including HeLa, BHK-21, CHO-K1, and mouse myelomas (Xiao et al. 2014). This often involves a prolonged period of selection of what may be a minority component of the cell population, so it is important to ensure that the properties of the cell line are not lost during selection (Zander and Bemark 2008; Rodrigues et al. 2013). If a myeloma is to be used to generate a hybridoma, or a CHO cell is to be used for transfection, the selection of a serum-free line should be done before fusion or transfection to minimize the risk of loss of properties during selection (Ozturk et al. 2003).

Adaptation to serum-free medium is often challenging and requires culture over multiple passages (see Section 14.8). Various approaches may be used, including (i) reduction of serum content with a shift to serum-free, hormone-supplemented medium; (ii) sequential adaptation, using a mixture of serum-containing and serum-free medium; (iii) conditional medium adaptation, using a mixture of conditioned and serum-free medium; and (iv) inside adaptation, where confluent cells are fed with serum-free medium and then maintained at higher density (van der Valk et al. 2010). Always freeze down stocks before you start the adaptation process in case it is unsuccessful (see Protocol P15.1). At each step, wait until stable cell proliferation is established before moving to a lower concentration. In suspension cultures, this is done by monitoring the viable cell count at each subculture, keeping the minimal cell concentration higher than for normal subculture (see Protocol P14.3). If problems are observed, go back to the previous step for two to three passages before trying again (van der Valk et al. 2010).

10.7 XENO-FREE MEDIA

There is increasing pressure from regulatory authorities to remove all animal products from cultured cells that may be used for the manufacture of biological products or other cell-based therapies (Unger et al. 2008; van der Valk et al. 2018). Replacement of animal products reduces the risk of undetected pathogens and provides a higher level of standardization and reproducibility, which is important for GMP (see Section 7.2.3). Numerous alternatives to animal products have been explored, particularly for hPSCs that may be used for cell-based therapies (Karnieli et al. 2017). For example, hESCs can be successfully grown in defined culture systems in the absence of a feeder layer (Ludwig et al. 2006a,[b]). However, culture of hESCs in defined media without feeder cells is challenging even for experienced laboratories (International Stem Cell Initiative Consortium et al. 2010). To address this problem, many laboratories are working to develop xeno-free culture systems that contain human-derived supplements (Karnieli et al. 2017; OECD 2018).

Xeno-free culture systems substitute animal-derived products with more defined components or with human-derived products such as platelet lysate, serum, or serum constituents (e.g. human albumin). These are likely to carry a lower risk of contamination if they are sourced from certified blood donation centers; screening of human blood products is well established and consistent with GMP procedures. Human platelet lysate contains many of the growth factors and cytokines that are present in serum and can be prepared from blood products that are not suitable for transfusion, e.g. post-expiry samples (OECD 2018). Human AB serum is routinely tested for viral contamination and supports the culture of a number of cell types (Karnieli et al. 2017). Feeder layers must also be xeno-free, using human feeder cells or xeno-free coatings (see Sections 8.3, 8.4) (Skottman and Hovatta 2006). All materials should be prepared using xeno-free procedures and screened for human pathogens and other contaminants (e.g. leachable chemicals; see Section 7.5.3).

10.8 ANIMAL PRODUCT-FREE MEDIA

Although xeno-free culture systems are an important step forward, they may still contain undefined components (e.g. human serum) and carry a risk of contamination with human pathogens. It is important to develop more fully defined culture systems and to remove all animal-derived components, including human products. Serum, trypsin, and other animal-based materials may be replaced with non-vertebrate products (e.g. plant-based hydrolysates; see Section 10.4.3) or with recombinant proteins. Albumin is available in recombinant form (e.g. Cellastim, InVitria) or may be replaced by those factors that are normally bound to albumin in serum, including lipids, hormones, minerals, and growth factors (see Section 9.5). The exact mechanism for the beneficial effect of albumin is still unclear and is likely to include its role as a carrier protein (Francis 2010). Transferrin, insulin, and many other peptide hormone and growth factors are also available in recombinant form (see Section 10.4.1).

Adaptation to animal product-free medium may involve further selection and regular checks must be made to ensure satisfactory growth and product formation or retention of the correct phenotype. As culture systems become more defined, they tend to become more cell- and process-specific and require higher levels of reagent purity for consistent results (Karnieli et al. 2017). Feeder-free culture of hESCs can result in spontaneous differentiation or chromosomal abnormalities, raising concerns regarding chromosomal instability due to culture stress (Ludwig et al. 2006b; Valamehr et al. 2011). The entire culture system will need optimization, including the substrate on which cells are grown (Karnieli et al. 2017).

As an alternative strategy, it may be possible to expand cultures in serum-free medium and then switch to animal product-free medium when they reach high density, e.g. using hollow fiber perfusion systems (see Sections 27.5.3, 28.3.5). This may allow for expression of the correct phenotype and product formation, which may even be enhanced in the absence of cell proliferation. However, for a biological product to be accepted as a therapeutic agent, it may be necessary for the cells to avoid all exposure to animal components. Ideally, cells that are used for manufacture of biological products or cell-based therapies should be derived in GMP-compatible conditions (Unger et al. 2008).

10.9 CONCLUSIONS: SERUM-FREE MEDIA

Serum-free culture is desirable but there are challenges to overcome. The relative simplicity of serum-containing media, the specialized techniques required for the use of some serum-free media, the considerable investment in time, effort, and resources that go into preparing new recipes or even adapting existing ones, and the multiplicity of media required if more than one cell type is being handled all act as deterrents to many laboratories. The best way to overcome these challenges is to investigate the physiological processes that govern growth and differentiation, to understand what our cells need and how best to supply it. The pressure from biotechnology to make the purification of products easier and the need to eliminate all sources of contamination will eventually force the adoption of serum-free media on a more general scale. Any new *in vitro* method should be developed without the use of serum or other undefined products (OECD 2018). Using a culture system that is serum-free (and preferably chemically defined) will avoid potential sources of uncertainty in future cell culture applications.

Suppliers

Supplier	URL
Amsbio	www.amsbio.com/cells-and-cell-culture.aspx
Biological Industries	www.bioind.com/worldwide
Biomol GmbH	www.biomol.com
Biowest	www.biowest.net
Cesco Bioengineering	www.cescobio.com.tw
Corning	www.corning.com/worldwide/en/products/life-sciences/products/cell-culture.html
FiberCellSystems	www.fibercellsystems.com
GE Healthcare Life Sciences	www.gelifesciences.com/en/it
InVitria (Ventria Bioscience)	invitria.com
Irvine Scientific	www.irvinesci.com
Kerry (Sheffield BioScience)	www.sheffieldbioscience.com
Lonza	www.lonza.com
Merck Millipore	www.merckmillipore.com
MP BioMedicals	www.mpbio.com
Oxford Expression Technologies	oetltd.com
Pall Corporation	www.pall.com
PeproTech	www.peprotech.com
PromoCell	www.promocell.com
Sigma-Aldrich (Merck)	www.sigmaaldrich.com/life-science/cell-culture.html
STEMCELL Technologies	www.stemcell.com

Supplier	URL
Takara Bio	www.takarabio.com
ThermoFisher Scientific	www.thermofisher.com/us/en/home/life-science/cell-culture.html
US Biological	www.usbio.net

REFERENCES

Aaronson, S.A., Bottaro, D.P., Miki, T. et al. (1991). Keratinocyte growth factor. A fibroblast growth factor family member with unusual target cell specificity. *Ann. N.Y. Acad. Sci.* 638: 62–77.

Alini, M., Kofsky, Y., Wu, W. et al. (1996). In serum-free culture thyroid hormones can induce full expression of chondrocyte hypertrophy leading to matrix calcification. *J. Bone Miner. Res.* 11 (1): 105–113. https://doi.org/10.1002/jbmr.5650110115.

Amit, M., Carpenter, M.K., Inokuma, M.S. et al. (2000). Clonally derived human embryonic stem cell lines maintain pluripotency and proliferative potential for prolonged periods of culture. *Dev. Biol.* 227 (2): 271–278. https://doi.org/10.1006/dbio.2000.9912.

Bafico, A. and Aaronson, S.A. (2003). Classification of growth factors and their receptors. In: *Holland-Frei Cancer Medicine* (eds. D.W. Kufe, R.E. Pollock, R.R. Weichselbaum, et al.). Hamilton, ON: BC Decker.

Balkovetz, D.F. and Lipschutz, J.H. (1999). Hepatocyte growth factor and the kidney: it is not just for the liver. *Int. Rev. Cytol.* 186: 225–260.

Barnes, D. and Sato, G. (1980a). Methods for growth of cultured cells in serum-free medium. *Anal. Biochem.* 102 (2): 255–270.

Barnes, D. and Sato, G. (1980b). Serum-free cell culture: a unifying approach. *Cell* 22 (3): 649–655.

Barnes, W.D., Sirbasku, D.A., and Sato, G.H. (1984). *Methods for Preparation of Media, Supplements, and Substrata for Serum-Free Animal Cell Culture*. Vol. 1, Cell Culture Methods for Molecular and Cell Biology. New York: Alan R. Liss.

Bettger, W.J., Boyce, S.T., Walthall, B.J. et al. (1981). Rapid clonal growth and serial passage of human diploid fibroblasts in a lipid-enriched synthetic medium supplemented with epidermal growth factor, insulin, and dexamethasone. *Proc. Natl Acad. Sci. U.S.A.* 78 (9): 5588–5592.

Blaker, G.J., Birch, J.R., and Pirt, S.J. (1971). The glucose, insulin and glutamine requirements of suspension cultures of HeLa cells in a defined culture medium. *J. Cell Sci.* 9 (2): 529–537.

Carney, D.N., Bunn, P.A. Jr., Gazdar, A.F. et al. (1981). Selective growth in serum-free hormone-supplemented medium of tumor cells obtained by biopsy from patients with small cell carcinoma of the lung. *Proc. Natl Acad. Sci. U.S.A.* 78 (5): 3185–3189.

Chaudhry, M.A., Vitalis, T.Z., Bowen, B.D. et al. (2008). Basal medium composition and serum or serum replacement concentration influences on the maintenance of murine embryonic stem cells. *Cytotechnology* 58 (3): 173–179. https://doi.org/10.1007/s10616-008-9177-5.

Chen, G., Gulbranson, D.R., Hou, Z. et al. (2011). Chemically defined conditions for human iPSC derivation and culture. *Nat. Methods* 8 (5): 424–429. https://doi.org/10.1038/nmeth.1593.

Chowdhury, S.R., Aminuddin, B.S., and Ruszymah, B.H. (2012). Effect of supplementation of dermal fibroblasts conditioned medium on expansion of keratinocytes through enhancing attachment. *Indian J. Exp. Biol.* 50 (5): 332–339.

Chun, B.H., Lee, Y.K., and Chung, N. (2012). Poly-gamma-glutamic acid enhances the growth and viability of Chinese hamster ovary cells in serum-free medium. *Biotechnol. Lett.* 34 (10): 1807–1810. https://doi.org/10.1007/s10529-012-0982-8.

Das, M., Rumsey, J.W., Bhargava, N. et al. (2009). Developing a novel serum-free cell culture model of skeletal muscle differentiation by systematically studying the role of different growth factors in myotube formation. *In Vitro Cell Dev. Biol. Anim.* 45 (7): 378–387. https://doi.org/10.1007/s11626-009-9192-7.

DeMichele, M.A., Davis, A.L., Hunt, J.D. et al. (1994). Expression of mRNA for three bombesin receptor subtypes in human bronchial epithelial cells. *Am. J. Respir. Cell Mol. Biol.* 11 (1): 66–74. https://doi.org/10.1165/ajrcmb.11.1.8018339.

Desai, N., Rambhia, P., and Gishto, A. (2015). Human embryonic stem cell cultivation: historical perspective and evolution of xeno-free culture systems. *Reprod. Biol. Endocrinol.* 13: 9. https://doi.org/10.1186/s12958-015-0005-4.

Evans, V.J., Bryant, J.C., Kerr, H.A. et al. (1964). Chemically defined media for cultivation of long-term cell strains from four mammalian species. *Exp. Cell. Res.* 36: 439–474.

Fiandalo, M.V., Wilton, J.H., Mantione, K.M. et al. (2018). Serum-free complete medium, an alternative medium to mimic androgen deprivation in human prostate cancer cell line models. *Prostate* 78 (3): 213–221. https://doi.org/10.1002/pros.23459.

Francis, G.L. (2010). Albumin and mammalian cell culture: implications for biotechnology applications. *Cytotechnology* 62 (1): 1–16. https://doi.org/10.1007/s10616-010-9263-3.

Freshney, R.I., Sherry, A., Hassanzadah, M. et al. (1980). Control of cell proliferation in human glioma by glucocorticoids. *Br. J. Cancer* 41 (6): 857–866.

Fu, B., Quintero, J., and Baker, C.C. (2003). Keratinocyte growth conditions modulate telomerase expression, senescence, and immortalization by human papillomavirus type 16 E6 and E7 oncogenes. *Cancer Res.* 63 (22): 7815–7824.

Furue, M.K., Tateyama, D., Kinehara, M. et al. (2010). Advantages and difficulties in culturing human pluripotent stem cells in growth factor-defined serum-free medium. *In Vitro Cell Dev. Biol. Anim.* 46 (7): 573–576. https://doi.org/10.1007/s11626-010-9317-z.

Gottipamula, S., Ashwin, K.M., Muttigi, M.S. et al. (2014). Isolation, expansion and characterization of bone marrow-derived mesenchymal stromal cells in serum-free conditions. *Cell Tissue Res.* 356 (1): 123–135. https://doi.org/10.1007/s00441-013-1783-7.

Gstraunthaler, G. (2003). Alternatives to the use of fetal bovine serum: serum-free cell culture. *ALTEX* 20 (4): 275–281.

Gstraunthaler, G., Lindl, T., and van der Valk, J. (2013). A plea to reduce or replace fetal bovine serum in cell culture media. *Cytotechnology* 65 (5): 791–793. https://doi.org/10.1007/s10616-013-9633-8.

Gstraunthaler, G., Lindl, T., and van der Valk, J. (2014). A severe case of fraudulent blending of fetal bovine serum strengthens the case for serum-free cell and tissue culture applications. *Altern. Lab. Anim.* 42 (3): 207–209. https://doi.org/10.1177/026119291404200308.

Guner, M., Freshney, R.I., Morgan, D. et al. (1977). Effects of dexamethasone and betamethasone on in vitro cultures from human astrocytoma. *Br. J. Cancer* 35 (4): 439–447.

Halliwell, B. (2014). Cell culture, oxidative stress, and antioxidants: avoiding pitfalls. *Biomed. J.* 37 (3): 99–105. https://doi.org/10.4103/2319-4170.128725.

Ham, R.G. (1974). Nutritional requirements of primary cultures. A neglected problem of modern biology. *In Vitro* 10: 119–129.

Ham, R.G. and McKeehan, W.L. (1978). Development of improved media and culture conditions for clonal growth of normal diploid cells. *In Vitro* 14 (1): 11–22.

Hamilton, W.G. and Ham, R.G. (1977). Clonal growth of Chinese hamster cell lines in protein-free media. *In Vitro* 13 (9): 537–547.

Hammond, S.L., Ham, R.G., and Stampfer, M.R. (1984). Serum-free growth of human mammary epithelial cells: rapid clonal growth in defined medium and extended serial passage with pituitary extract. *Proc. Natl Acad. Sci. U.S.A.* 81 (17): 5435–5439.

Hawkes, P.W. (2015). Fetal bovine serum: geographic origin and regulatory relevance of viral contamination. *Bioresour. Bioprocess.* 2 (1): 34. https://doi.org/10.1186/s40643-015-0063-7.

Hayashi, I., Larner, J., and Sato, G. (1978). Hormonal growth control of cells in culture. *In Vitro* 14 (1): 23–30.

Hazeltine, L.B., Knueven, K.M., Zhang, Y. et al. (2016). Chemically defined media modifications to lower tryptophan oxidation of biopharmaceuticals. *Biotechnol. Progr.* 32 (1): 178–188. https://doi.org/10.1002/btpr.2195.

Hohla, F., Schally, A.V., Kanashiro, C.A. et al. (2007). Growth inhibition of non-small-cell lung carcinoma by BN/GRP antagonist is linked with suppression of K-Ras, COX-2, and pAkt. *Proc. Natl Acad. Sci. U.S.A.* 104 (47): 18671–18676. https://doi.org/10.1073/pnas.0709455104.

Hutchings, S.E. and Sato, G.H. (1978). Growth and maintenance of HeLa cells in serum-free medium supplemented with hormones. *Proc. Natl Acad. Sci. U.S.A.* 75 (2): 901–904.

Ikonomou, L., Bastin, G., Schneider, Y.J. et al. (2001). Design of an efficient medium for insect cell growth and recombinant protein production. *In Vitro Cell Dev. Biol. Anim.* 37 (9): 549–559.

International Stem Cell Initiative Consortium, Akopian, V., Andrews, P.W. et al. (2010). Comparison of defined culture systems for feeder cell free propagation of human embryonic stem cells. *In Vitro Cell Dev. Biol. Anim.* 46 (3–4): 247–258. https://doi.org/10.1007/s11626-010-9297-z.

Iscove, N.N. and Melchers, F. (1978). Complete replacement of serum by albumin, transferrin, and soybean lipid in cultures of lipopolysaccharide-reactive B lymphocytes. *J. Exp. Med.* 147 (3): 923–933.

Jenkins, N. (1992). *Growth Factors, a Practical Approach*. Oxford: IRL Press at Oxford University Press.

Jochems, C.E., van der Valk, J.B., Stafleu, F.R. et al. (2002). The use of fetal bovine serum: ethical or scientific problem? *Altern. Lab. Anim.* 30 (2): 219–227.

Karnieli, O., Friedner, O.M., Allickson, J.G. et al. (2017). A consensus introduction to serum replacements and serum-free media for cellular therapies. *Cytotherapy* 19 (2): 155–169. https://doi.org/10.1016/j.jcyt.2016.11.011.

Ke, Y., Reddel, R.R., Gerwin, B.I. et al. (1988). Human bronchial epithelial cells with integrated SV40 virus T antigen genes retain the ability to undergo squamous differentiation. *Differentiation* 38 (1): 60–66.

Keen, M.J. and Rapson, N.T. (1995). Development of a serum-free culture medium for the large scale production of recombinant protein from a Chinese hamster ovary cell line. *Cytotechnology* 17 (3): 153–163. https://doi.org/10.1007/BF00749653.

Keenan, J., Pearson, D., and Clynes, M. (2006). The role of recombinant proteins in the development of serum-free media. *Cytotechnology* 50 (1–3): 49–56. https://doi.org/10.1007/s10616-006-9002-y.

Keenan, J., Horgan, K., Clynes, M. et al. (2018). Unexpected fluctuations of trace element levels in cell culture medium in vitro: caveat emptor. *In Vitro Cell Dev. Biol. Anim.* 54 (8): 555–558. https://doi.org/10.1007/s11626-018-0285-z.

Kent, K.D. and Bomser, J.A. (2003). Bovine pituitary extract provides remarkable protection against oxidative stress in human prostate epithelial cells. *In Vitro Cell Dev. Biol. Anim.* 39 (8–9): 388–394. https://doi.org/10.1290/1543-706X(2003)039<0388:BPEPRP>2.0.CO;2.

Klagsbrun, M. and Baird, A. (1991). A dual receptor system is required for basic fibroblast growth factor activity. *Cell* 67 (2): 229–231.

Klevjer-Anderson, P. and Buehring, G.C. (1980). Effect of hormones on growth rates of malignant and nonmalignant human mammary epithelia in cell culture. *In Vitro* 16 (6): 491–501.

Knedler, A. and Ham, R.G. (1987). Optimized medium for clonal growth of human microvascular endothelial cells with minimal serum. *In Vitro Cell Dev. Biol. Anim.* 23 (7): 481–491.

Lange, W., Brugger, W., Rosenthal, F.M. et al. (1991). The role of cytokines in oncology. *Int. J. Cell Cloning* 9 (4): 252–273. https://doi.org/10.1002/stem.5530090403.

Lechner, J.F. and LaVeck, M.A. (1985). A serum-free method for culturing normal human bronchial epithelial cells at clonal density. *J. Tissue Cult. Methods* 9 (2): 43–48.

Ledur, P.F., Onzi, G.R., Zong, H. et al. (2017). Culture conditions defining glioblastoma cells behavior: what is the impact for novel discoveries? *Oncotarget* 8 (40): 69185–69197. https://doi.org/10.18632/oncotarget.20193.

Lopez-Casillas, F., Wrana, J.L., and Massague, J. (1993). Betaglycan presents ligand to the TGF beta signaling receptor. *Cell* 73 (7): 1435–1444.

Ludwig, T.E., Bergendahl, V., Levenstein, M.E. et al. (2006a). Feeder-independent culture of human embryonic stem cells. *Nat. Methods* 3 (8): 637–646. https://doi.org/10.1038/nmeth902.

Ludwig, T.E., Levenstein, M.E., Jones, J.M. et al. (2006b). Derivation of human embryonic stem cells in defined conditions. *Nat. Biotechnol.* 24 (2): 185–187. https://doi.org/10.1038/nbt1177.

Mandal, S., Mandal, A., Johansson, H.E. et al. (2013). Depletion of cellular polyamines, spermidine and spermine, causes a total arrest in translation and growth in mammalian cells. *Proc. Natl. Acad. Sci. U.S.A.* 110 (6): 2169–2174. https://doi.org/10.1073/pnas.1219002110.

Mather, J.P. (1998). Making informed choices: medium, serum, and serum-free medium. How to choose the appropriate medium and culture system for the model you wish to create. *Methods Cell Biol.* 57: 19–30.

McCormick, C. and Freshney, R.I. (2000). Activity of growth factors in the IL-6 group in the differentiation of human lung adenocarcinoma. *Br. J. Cancer* 82 (4): 881–890. https://doi.org/10.1054/bjoc.1999.1015.

McCormick, C., Freshney, R.I., and Speirs, V. (1995). Activity of interferon alpha, interleukin 6 and insulin in the regulation of differentiation in A549 alveolar carcinoma cells. *Br. J. Cancer* 71 (2): 232–239.

McGillicuddy, N., Floris, P., Albrecht, S. et al. (2018). Examining the sources of variability in cell culture media used for biopharmaceutical production. *Biotechnol. Lett.* 40 (1): 5–21. https://doi.org/10.1007/s10529-017-2437-8.

McKeehan, W.L. and Ham, R.G. (1976a). Methods for reducing the serum requirement for growth in vitro of nontransformed diploid fibroblasts. *Dev. Biol. Stand.* 37: 97–98.

McKeehan, W.L. and Ham, R.G. (1976b). Stimulation of clonal growth of normal fibroblasts with substrata coated with basic polymers. *J. Cell Biol.* 71 (3): 727–734.

McKeehan, W.L., Adams, P.S., and Rosser, M.P. (1982). Modified nutrient medium MCDB 151, defined growth factors, cholera toxin, pituitary factors, and horse serum support epithelial cell and suppress fibroblast proliferation in primary cultures of rat ventral prostate cells. *In Vitro* 18 (2): 87–91.

McKeehan, W.L., Adams, P.S., and Rosser, M.P. (1984). Direct mitogenic effects of insulin, epidermal growth factor, glucocorticoid, cholera toxin, unknown pituitary factors and possibly prolactin, but not androgen, on normal rat prostate epithelial cells in serum-free, primary cell culture. *Cancer Res.* 44 (5): 1998–2010.

McLean, J.S., Frame, M.C., Freshney, R.I. et al. (1986). Phenotypic modification of human glioma and non-small cell lung carcinoma by glucocorticoids and other agents. *Anticancer Res.* 6 (5): 1101–1106.

Merten, O.W. (1999). Safety issues of animal products used in serum-free media. *Dev. Biol. Stand.* 99: 167–180.

Mitaka, T., Sattler, G.L., and Pitot, H.C. (1991). Amino acid-rich medium (Leibovitz L-15) enhances and prolongs proliferation of primary cultured rat hepatocytes in the absence of serum. *J. Cell. Physiol.* 147 (3): 495–504. https://doi.org/10.1002/jcp.1041470316.

Moreira, J.L., Santana, P.C., Feliciano, A.S. et al. (1995). Effect of viscosity upon hydrodynamically controlled natural aggregates of animal cells grown in stirred vessels. *Biotechnol. Progr.* 11 (5): 575–583. https://doi.org/10.1021/bp00035a012.

Morton, H.J. (1970). A survey of commercially available tissue culture media. *In Vitro* 6 (2): 89–108.

Naeyaert, J.M., Eller, M., Gordon, P.R. et al. (1991). Pigment content of cultured human melanocytes does not correlate with tyrosinase message level. *Br. J. Dermatol.* 125 (4): 297–303.

Nims, R.W. and Harbell, J.W. (2017). Best practices for the use and evaluation of animal serum as a component of cell culture medium. *In Vitro Cell Dev. Biol. Anim.* 53 (8): 682–690. https://doi.org/10.1007/s11626-017-0184-8.

OECD (2018). *Guidance Document on Good in Vitro Method Practices (GIVIMP)*. Paris: OECD Publishing.

Ozturk, S., Kaseko, G., Mahaworasilpa, T. et al. (2003). Adaptation of cell lines to serum-free culture medium. *Hybrid. Hybridomics* 22 (4): 267–272. https://doi.org/10.1089/153685903322329009.

Parlow, A.F. (2006). National hormone and peptide program: peptide hormones, antisera and other reagents available. *J. Clin. Endocrinol. Metab.* 91 (4): 1608–1610. https://doi.org/10.1210/jcem.91.4.9997.

Peehl, D.M. (2002). Human prostatic epithelial cells. In: *Culture of Epithelial Cells* (eds. R.I. Freshney and M. Freshney), 171–194. Hoboken, NJ: Wiley-Liss.

Peehl, D.M. and Ham, R.G. (1980). Clonal growth of human keratinocytes with small amounts of dialyzed serum. *In Vitro* 16 (6): 526–540.

Phillips, P.D. and Cristofalo, V.J. (1988). Classification system based on the functional equivalency of mitogens that regulate WI-38 cell proliferation. *Exp. Cell. Res.* 175 (2): 396–403.

Planz, B., Wang, Q., Kirley, S.D. et al. (1998). Androgen responsiveness of stromal cells of the human prostate: regulation of cell proliferation and keratinocyte growth factor by androgen. *J. Urol.* 160 (5): 1850–1855.

Price, P.J. (2017). Best practices for media selection for mammalian cells. *In Vitro Cell Dev. Biol. Anim.* 53 (8): 673–681. https://doi.org/10.1007/s11626-017-0186-6.

Price, P.J., Goldsborough, M.D., and Tilkins, M.L. (1998). Patent WO/1998/030679: embryonic stem cell serum replacement. WIPO. https://patentscope.wipo.int/search/en/detail.jsf?docid=wo1998030679.

Ramot, Y., Tiede, S., Biro, T. et al. (2011). Spermidine promotes human hair growth and is a novel modulator of human epithelial stem cell functions. *PLoS One* 6 (7): e22564. https://doi.org/10.1371/journal.pone.0022564.

Rashid, R., Lim, N.S., Chee, S.M. et al. (2014). Novel use for polyvinylpyrrolidone as a macromolecular crowder for enhanced extracellular matrix deposition and cell proliferation. *Tissue Eng. Part C Methods* 20 (12): 994–1002. https://doi.org/10.1089/ten.TEC.2013.0733.

Roberson, K.M. and Robertson, C.N. (1995). Isolation and growth of human primary prostate epithelial cultures. *Methods Cell Sci.* 17 (3): 177–185.

Rodrigues, M.E., Costa, A.R., Henriques, M. et al. (2013). Advances and drawbacks of the adaptation to serum-free culture of CHO-K1 cells for monoclonal antibody production. *Appl. Biochem. Biotechnol.* 169 (4): 1279–1291. https://doi.org/10.1007/s12010-012-0068-z.

Santiskulvong, C., Sinnett-Smith, J., and Rozengurt, E. (2004). Insulin reduces the requirement for EGFR transactivation in bombesin-induced DNA synthesis. *Biochem. Biophys. Res. Commun.* 318 (4): 826–832. https://doi.org/10.1016/j.bbrc.2004.04.100.

Sato, J.D., Okamoto, T., Barnes, D. et al. (2018). A tribute to Dr. Gordon Hisashi Sato (December 24, 1927–March 31, 2017). *In Vitro Cell Dev. Biol. Anim.* 54 (3): 177–193. https://doi.org/10.1007/s11626-018-0230-1.

Shipley, G.D. and Ham, R.G. (1983). Multiplication of Swiss 3T3 cells in a serum-free medium. *Exp. Cell. Res.* 146 (2): 249–260.

Silva, A.K., Yi, H., Hayes, S.H. et al. (2010). Lithium chloride regulates the proliferation of stem-like cells in retinoblastoma cell lines: a potential role for the canonical Wnt signaling pathway. *Mol. Vis.* 16: 36–45.

Skottman, H. and Hovatta, O. (2006). Culture conditions for human embryonic stem cells. *Reproduction* 132 (5): 691–698. https://doi.org/10.1530/rep.1.01079.

Strange, R., Li, F., Friis, R.R. et al. (1991). Mammary epithelial differentiation in vitro: minimum requirements for a functional response to hormonal stimulation. *Cell Growth Differ.* 2 (11): 549–559.

Telling, R.C. and Ellsworth, R. (1965). Submerged culture of hamster kidney cells in a stainless steel vessel. *Biotechnol. Bioeng.* 7: 417–434.

Tharmalingam, T., Ghebeh, H., Wuerz, T. et al. (2008). Pluronic enhances the robustness and reduces the cell attachment of mammalian cells. *Mol. Biotechnol.* 39 (2): 167–177. https://doi.org/10.1007/s12033-008-9045-8.

Thomson, A.A., Foster, B.A., and Cunha, G.R. (1997). Analysis of growth factor and receptor mRNA levels during development of the rat seminal vesicle and prostate. *Development* 124 (12): 2431–2439.

Unger, C., Skottman, H., Blomberg, P. et al. (2008). Good manufacturing practice and clinical-grade human embryonic stem cell lines. *Hum. Mol. Genet.* 17 (R1): R48–R53. https://doi.org/10.1093/hmg/ddn079.

Valamehr, B., Tsutsui, H., Ho, C.M. et al. (2011). Developing defined culture systems for human pluripotent stem cells. *Regen. Med.* 6 (5): 623–634. https://doi.org/10.2217/rme.11.54.

van der Valk, J., Mellor, D., Brands, R. et al. (2004). The humane collection of fetal bovine serum and possibilities for serum-free cell and tissue culture. *Toxicol. In Vitro* 18 (1): 1–12.

van der Valk, J., Brunner, D., De Smet, K. et al. (2010). Optimization of chemically defined cell culture media--replacing fetal bovine serum in mammalian in vitro methods. *Toxicol. In Vitro* 24 (4): 1053–1063. https://doi.org/10.1016/j.tiv.2010.03.016.

van der Valk, J., Bieback, K., Buta, C. et al. (2018). Fetal bovine serum (FBS): past – present – future. *ALTEX* 35 (1): 99–118. https://doi.org/10.14573/altex.1705101.

Vecchi, J.T. and Wakatsuki, T. (2015). The stagnant adaptation of defined and xeno-free culture of iPSCs in academia. *Arch. Stem Cell Res.* 3 (1).

Waymouth, C. (1984). Preparation and use of serum-free culture media. In: *Cell Culture Methods for Molecular and Cell Biology; Vol. 1: Methods for Preparation of Media, Supplements, and Substrata for Serum-Free Animal Cell Culture* (eds. W.D. Barnes, D.A. Sirbasku and G.H. Sato), 23–68. New York: Alan R. Liss.

Williams, G.M. and Gunn, J.M. (1974). Long-term cell culture of adult rat liver epithelial cells. *Exp. Cell. Res.* 89 (1): 139–142.

Xiao, Z., Sabourin, M., Piras, G. et al. (2014). Screening and optimization of chemically defined media and feeds with integrated and statistical approaches. *Methods Mol. Biol.* 1104: 117–135. https://doi.org/10.1007/978-1-62703-733-4_9.

Yamasaki, S., Taguchi, Y., Shimamoto, A. et al. (2014). Generation of human induced pluripotent stem (Ips) cells in serum- and feeder-free defined culture and TGF-Beta1 regulation of pluripotency. *PLoS One* 9 (1): e87151. https://doi.org/10.1371/journal.pone.0087151.

Yao, T. and Asayama, Y. (2017). Animal-cell culture media: history, characteristics, and current issues. *Reprod. Med. Biol.* 16 (2): 99–117. https://doi.org/10.1002/rmb2.12024.

Yevdokimova, N. and Freshney, R.I. (1997). Activation of paracrine growth factors by heparan sulphate induced by glucocorticoid in A549 lung carcinoma cells. *Br. J. Cancer* 76 (3): 281–289.

Zander, L. and Bemark, M. (2008). Identification of genes deregulated during serum-free medium adaptation of a Burkitt's lymphoma cell line. *Cell Prolif.* 41 (1): 136–155. https://doi.org/10.1111/j.1365-2184.2007.00500.x.

CHAPTER 11

Preparation and Sterilization

In the early days of tissue culture, most media and reagents were made up in the laboratory in which they were used. Cells were usually grown on glass substrates, using "medical flats" or Carrel flasks (the forerunners of our modern tissue culture flasks; see Figure 1.1b, c), which required careful washing and sterilization. The time taken to prepare and sterilize media, reagents, and glassware was a significant component of the technical support required for tissue culture. The introduction of commercially made media and plastic flasks transformed tissue culture by eliminating much of the preparation time previously required. As a result, most people entering the field will not need to prepare their own media or culture vessels. However, some larger establishments still prepare their own media, and even smaller laboratories may need to prepare special media, reagents, or substrates. It is important to appreciate some basic concepts to understand how to prepare glassware, equipment, and reagents appropriately.

11.1 TERMINOLOGY: PREPARATION

To describe how to prepare cell culture equipment and materials, it is necessary to agree on the meaning of commonly used terms such as "disinfection" and "sterilization." For this book, the following definitions are used:

(1) **Cleaning** is the removal of dirt, organic matter, and stains (WHO 2004). Cleaning reduces the amount of organic material that may be present on surfaces, resulting in a more consistent environment for disinfection or sterilization.

(2) **Disinfection** is a physical or chemical means of killing microorganisms, although not necessarily spores (WHO 2004; Yoo 2018). Its effectiveness may vary based on the material to be disinfected, the amount of organic matter, the environmental conditions (e.g. temperature), the process used, and other variables (CDC and NIH 2009).

(3) **Sterilization** is a process that kills or removes all classes of microorganisms and spores (WHO 2004). Following the process of sterilization, the probability of a microorganism surviving on an object or within a liquid is less than one in 10^6, which is referred to as the sterility assurance level (CDC and NIH 2009).

(4) **Decontamination** is a physical or chemical process that renders an area, device, item, or material safe to handle (CDC and NIH 2009). In this context, decontamination usually refers to the removal or killing of any microorganisms that may be present. Decontamination is an important requirement for biosafety and is discussed elsewhere (see Section 6.3.5).

(5) **Detergents** are chemicals or mixtures of chemicals that are used for cleaning (e.g. to wash glassware or mop floors). Detergents usually work by making dirt, organic matter, and other impurities more soluble. Some detergents may kill microorganisms, but this is not a requirement for items to be clean, as defined here.

(6) **Disinfectants** are chemicals or mixtures of chemicals that are used to kill microorganisms, although not necessarily

Freshney's Culture of Animal Cells: A Manual of Basic Technique and Specialized Applications, Eighth Edition. Amanda Capes-Davis and R. Ian Freshney.
© 2021 John Wiley & Sons Ltd. Published 2021 by John Wiley & Sons Ltd.
Companion website: www.wiley.com/go/freshney/cellculture8

spores (WHO 2004). Some disinfectants may result in sterilization, but this will depend on the starting material and the process used – for example, whether a disinfectant is used at sufficient concentration for a sufficient length of time.

11.2 STERILIZATION METHODS

A number of sterilization methods are used in tissue culture laboratories, with the choice of method largely depending on the stability of the material at high temperatures (see Table 11.1). In general, solid items with a high resistance to heat are best sterilized using dry heat, e.g. metals, glass, and thermostable plastics such as polytetrafluorethylene (PTFE). Heat stable liquids are best sterilized using pressurized steam sterilization (autoclaving). Heat-sensitive plastics will require irradiation and heat-labile liquids are usually filtered. Sterilization methods that are used for various apparatus and liquid reagents are summarized in Tables 11.2 and 11.3.

Sterilization is only possible if items are clean and suitably prepared (McDonnell et al. 2013). Usually, laboratories develop Standard Operating Procedures (SOPs) that describe how shared items should be prepared and sterilized (e.g. glassware). Each SOP should include protocols for cleaning, disinfection, and sterilization as required. Protocols for glassware, other apparatus, water, and other liquids are included in this chapter as starting points. If items are sterilized, the SOP should explain how to confirm that sterilization is successful; this may involve temperature monitoring or sterility indicators (see Section 11.2.7). A risk assessment must also be performed so that any safety concerns can be identified and duly managed (see Section 6.1.1).

11.2.1 Dry Heat Sterilization

Dry heat is one of the simplest and most effective methods of sterilization, provided all parts of the load reach the correct temperature for the required period. The temperature should be maintained at 160 °C (320 °F) for a minimum of

TABLE 11.1. Methods of sterilization.

Method	Conditions	Materials	Limitations
Dry heat	160 °C (320 °F), minimum 1 h	Heat stable: metals, glass, PTFE	Some charring may occur, e.g. of indicating tape and cotton plugs
Pressurized steam (autoclaving)	121 °C (250 °F), 100 kPa (15 lb/in.^{2}), minimum 20 min	Heat-stable liquids: water, salt solutions, autoclavable media. Moderately heat-stable plastics: silicones, polycarbonate, nylon, polypropylene	Steam penetration requires steam-permeable packaging and a drying cycle after sterilization is complete. Large fluid loads need time to heat up
Irradiation			
γ-Irradiation	25 kGy	Plastics, organic scaffolds, heat-sensitive reagents and pharmaceuticals	Chemical alteration of plastics can occur. Macromolecular degradation is possible
Electron beam	25 kGy	Plastics, organic scaffolds, heat-sensitive reagents and pharmaceuticals	Needs high-energy source. Not suitable for average laboratory installation
Plasma	Ionized gas, e.g. hydrogen peroxide, argon, oxygen	Polymers (e.g. PDMS), biodegradable scaffolds	Needs specialized equipment (plasma sterilizer). May alter the properties of some substrates
Chemical			
Ethylene oxide	1 h	Heat-labile plastics	Hazardous gas. Leaves toxic residue so items must be ventilated for 24–48 h after treatment
Hypochlorite	300–2500 ppm, 30 min	Contaminated solutions, plastics	Hazardous chemical. Needs extensive washing. May leave residue
Filtration	0.1- to 0.2-μm porosity filters	All aqueous solutions; particularly suitable for heat -labile reagents and media. Specify low protein binding for growth factors, etc.	Not suitable for some chemicals. Slow with viscous solutions

PDMS, Polydimethylsiloxane; PTFE, polytetrafluorethylene.

TABLE 11.2. Sterilization methods for apparatus.

Item	Sterilization[a]
Ampoules for freezer, glass	Dry heat
Ampoules for freezer, plastic	Autoclave (usually bought sterile)
Apparatus containing glass and silicone tubing	Autoclave
Disposable tips for micropipettes	Autoclave in autoclavable trays or nylon bags
Filters, reusable	Autoclave; do not use prevacuum or postvacuum cycles; remove air by displacement
Glassware	Dry heat
Glass bottles with screw caps	Autoclave with cap slack
Glass coverslips	Dry heat
Glass slides	Dry heat
Glass syringes	Autoclave (separate piston if PTFE)
Instruments for dissection	Dry heat
Magnetic stirrer bars	Autoclave
Pasteur pipettes, glass	Dry heat
Pipettes, glass	Dry heat
Polycarbonate	Autoclave
Repeating pipettes or syringes	Autoclave (separate PTFE pistons from glass barrels)
Screw caps	Autoclave
Silicone grease (for isolating clones)	Autoclave in glass Petri dish
Silicone tubing	Autoclave
Stoppers, rubber, and silicone	Autoclave
Test tubes	Dry heat

[a]Conditions required for sterilization are summarized in Table 11.1.

one hour. This restricts the use of dry heat sterilization to glass, metals, and heat-resistant plastics such as polycarbonate and PTFE. Temperature monitoring should be performed and some heat-resistant indicators may be added to the load (see Section 11.2.7). However, paper and temperature-sensitive tape may char and produce volatiles and are best avoided.

⬥ *Safety Note. Items should be allowed to cool before they are handled and personal protective equipment (PPE) should be worn when handling hot items.*

11.2.2 Pressurized Steam Sterilization (Autoclaving)

Sterilization by pressurized steam (autoclaving) is the most widely used method for laboratories and healthcare facilities (CDC 2008). Autoclaving is usually performed at 121°C (250°F) and 100 kPa (15 lb/in.^{2}) for a minimum of 20 minutes; the use of humid air under pressure results in rapid sterilization compared to dry heat (see Figure 11.1). Temperature monitoring should be performed from the center of the load to confirm that the correct conditions were reached for sterilization. Temperature-sensitive tape (autoclave tape) is usually added to individual items; other indicators of sterility may also be used (see Section 11.2.7).

This method of sterilization is used for heat-stable liquids such as water, balanced salt solutions, and a small number of media formulations (Waymouth 1978). If carried out in sealed containers, such as borosilicate glass (e.g. Pyrex), there will be no evaporation, but there is a risk of bottles exploding; usually caps are left slack and need to be tightened after completion of the cycle with the requisite cooling period. Evaporation should be monitored and, if necessary, made up with sterile ultrapure water (UPW) under sterile handling conditions (see Section 12.3). The autoclave should be supplied with distilled, deionized, or reverse osmosis (RO) water (see Section 11.5.1) to minimize contamination from condensate within the autoclave.

Autoclaving is also used to sterilize dry materials such as tubing, swabs, or instruments. Penetration of steam through the packaging is essential in order that the correct conditions of temperature and humidity are achieved for effective sterilization (see Figure 11.1). This can be done by including a pre-evacuation ("prevac") period, where the air in the chamber is evacuated, or by purging the chamber with steam before starting sterilization. Because steam penetration is essential, items sterilized by autoclaving will be wet after sterilization and need to be dried. Most automated autoclaves will include a post-evacuation ("postvac") period where the chamber is

TABLE 11.3. Sterilization methods for liquids.

Solution	Sterilization[a]	Storage
Agar	Autoclave	Room temperature
Amino acids	Filter	4°C
Antibiotics	Filter	−20°C
Bacto-peptone	Autoclave	Room temperature
Bovine serum albumin	Filter (use stacked filters)	4°C
Carboxymethyl cellulose	Autoclave with reduced cycle, filter[b]	4°C
Collagenase	Filter	−20°C
DMSO	Self-sterilizing; dispense into aliquots in sterile tubes. If filtered, use nylon	Room temperature; keep dark, avoid contact with rubber or plastics (except polypropylene)
Drugs	Filter (check for binding; use low-binding filter if necessary)	−20°C
EDTA	Autoclave	Room temperature
Glucose, 20%	Autoclave	Room temperature
Glucose, 1–2%	Filter (low concentrations; caramelizes if autoclaved)	Room temperature
Glutamine	Filter	−20°C
Glycerol	Autoclave	Room temperature
Growth factors	Filter (low protein binding)	−20°C
HEPES	Autoclave	Room temperature
HCl, 1 M	Filter	Room temperature
Lactalbumin hydrolysate	Autoclave	Room temperature
Methocel (methylcellulose)	Autoclave	4°C
NaHCO$_3$	Filter	Room temperature
NaOH, 1 M	Filter	Room temperature
Phenol red	Autoclave	Room temperature
Salt solutions (without glucose)	Autoclave	Room temperature
Serum	Filter (use stacked filters)	−20°C
Sodium pyruvate, 100 mM	Filter	−20°C
Transferrin	Filter	−20°C
Tryptose	Autoclave	Room temperature
Trypsin	Filter	−20°C
Vitamins	Filter	−20°C
Water	Autoclave	Room temperature

[a]Conditions required for sterilization are summarized in Table 11.1.
[b]Difficult to sterilize as it may degrade with autoclaving. A reduced temperature (e.g. 100°C for 30 minutes) may be suitable, provided sterility indicators confirm successful sterilization; lower viscosity solutions may be filtered.

evacuated after sterilization is complete to enable items to dry off. Prevac and postvac should not be used with sterilization of liquids as it is unnecessary and would increase evaporation significantly. Laboratories typically have separate runs for "wet" and "dry" items to provide the correct conditions for both groups (see Section 5.4.4).

If a small bench-top autoclave or pressure cooker is used for sterilization (see Figure 5.20a), take care that enough water is put in at the start to allow for evaporation. Make sure that the autoclave boils vigorously for 10–15 minutes before pressurizing to displace all the air. After autoclaving, items should be placed in a dry oven or on a drying rack to dry off.

◆ *Safety Note. Suitable PPE should be worn when operating the autoclave, including elbow-length insulated gloves. Take care to avoid burns when releasing steam and handling hot items; keep your face well clear of escaping steam when you open doors, lids, etc. Liquids may be superheated after*

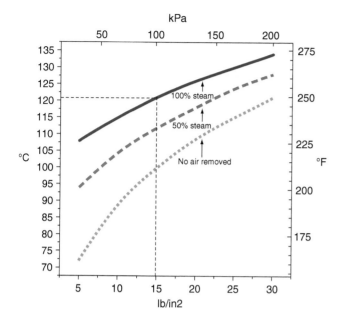

Fig. 11.1. Effect of humidity on temperature in autoclave. The top curve shows that 121 °C at 100 kPa (15 lb/in.²) can only be achieved when all the air is removed. Even when 50% of air is removed, it still requires 150 kPa (23 lb/in.²) to reach 121 °C. *Source*: R. Ian Freshney, based on data from Breach (1968).

autoclaving and should be allowed to cool before moving **(see Section 6.2.4).**

11.2.3 Irradiation

Irradiation can be used to sterilize heat-labile materials (e.g. plastics), heat-labile liquids (e.g. serum), or biomaterials (e.g. bone samples). This may involve γ-irradiation (usually from an isotope source such as ^{60}Co) or electron beam sterilization (Dai et al. 2016). Because the dose required is quite high (≥25 kGy), this is usually done at a central facility. Items should be packaged and sealed; polythene may be used and sealed by heat welding. Irradiation may affect performance of some items and can generate volatiles from some plastics (Nims et al. 2011; Wood et al. 2013; Harrell et al. 2018). Performance of irradiated items should be assessed as part of quality control testing (see Sections 7.5.3 and 9.7.1). Electron beam sterilization may be preferable to γ-irradiation for biomaterials (Hoburg et al. 2015).

11.2.4 Plasma Sterilization

Plasma is generated by subjecting gases to pulsed discharges of direct current, radio frequency, or microwaves. Plasma treatment can be used to kill fungi, bacteria, spores, viruses, and even prions; the mechanism of action may vary with the gas, but is likely to involve oxidation from the plasma (which is essentially ionized gas) and from generation of additional

free radicals (Shintani et al. 2010). Plasma sterilization is increasingly used for specialized applications, including biodegradable scaffolds for tissue engineering and microfluidic organ-on-chip systems (Dai et al. 2016; Novak et al. 2018).

11.2.5 Chemical Sterilization

Many plastics cannot be exposed to the temperature required for autoclaving or dry-heat sterilization. These items may be sterilized by exposure to chemicals (gases or liquids) under conditions that completely kill or remove all microorganisms. Ethylene oxide gas is commonly used to sterilize medical devices and heat-labile plastics (Shintani 2017). However, two to three weeks are required to clear the gas from the plastic surface after sterilization. Other commonly used chemicals include hypochlorite solution ("bleach"), quaternary ammonium compounds, and formaldehyde gas. These chemicals are primarily used for decontamination (see Sections 6.3.5, 6.3.6) (Coté 2001; CDC 2008).

Alcohol (ethanol or isopropanol) is used to disinfect laboratory items and surfaces. For example, plastic items can be immersed in 70% alcohol for 30 minutes and then dried under ultraviolet (UV) light in a biological safety cabinet (BSC) or laminar flow hood. However, ethanol and isopropanol are not sufficient for sterilization; some microorganisms are resistant to alcohols, including hydrophilic viruses and spores (CDC 2008; Dai et al. 2016). Care must be taken with some plastics (e.g. Plexiglas, Perspex, Lucite), as they will depolymerize in alcohol or when exposed to UV light.

11.2.6 Filter Sterilization

Micropore filtration can be used to sterilize heat-labile liquids, including culture media. Whether sterilization is achieved will depend on the pore size and process that are used. A pore size of 0.45 μm was used for sterile filtration until the 1960s (Belgaid et al. 2014). Microbial contaminants were then identified that could pass through 0.45-μm filters, leading to the adoption of 0.2- or 0.22-μm filters for sterile filtration. These pore sizes are able to remove most bacteria, including *Brevundimonas diminuta*, which is often used for validation of sterile filtration processes (Belgaid et al. 2014). However, it is important to consider the microorganisms that are likely to be found in tissue culture reagents. Mycoplasma can pass through 0.2-μm filters, leading to the adoption of 0.1-μm filters for filtration of culture media (Folmsbee et al. 2010) (see Section 11.7.1).

Sterilization of biological products requires a specialized approach. Virus clearance can be achieved using nanofiltration and this approach is now used in tandem with other methods (Burnouf and Radosevich 2003; Cipriano et al. 2012). Advances in filter design have allowed protection of bioreactor cultures from virus contamination, including parvoviruses, which are challenging to eliminate because of their small size (about 20 nm diameter) and resistance to many methods of sterilization (Mann et al. 2015). Comprehensive quality control testing is required for all filter sterilization processes to demonstrate filter integrity and sterility (see Sections 11.7.4, 11.8).

11.2.7 Sterility Indicators

Sterility indicators are used to show that (i) an item has been through the sterilization procedure; (ii) an item or load has been at the requisite temperature, pressure, and humidity to ensure sterility; and (iii) the sterilization process has been successful as shown by killing of a test organism. The first type of indicator is commonly used and includes temperature-sensitive tape ("autoclave tape") or tags. Always remember that these indicators only show that the item has been through the process and do not prove that it is sterile. Their main function is to enable distinction between items awaiting sterilization and items that have been sterilized.

The second type of indicator, if carefully selected, will show that items have been exposed to the correct conditions for sterilization. Indicators are available from multiple suppliers including 3M, STERIS, and Merck Millipore. Browne's tubes contain heat-sensitive dyes that change color after a certain time period at the correct temperature and can be used for dry heat sterilization or autoclaving (e.g. Browne Autoclave Indicator Valves, STERIS). Indicator strips can be added to individually wrapped items before autoclaving to confirm time, temperature, and steam exposure (e.g. Comply™ Thermalog™ Steam Chemical Integrators, 3M).

The third type of indicator uses a biological sample as a test for the sterilization process. Biological indicators include spore-containing strips (e.g. Mesa Spore Strips, Mesa Labs) and ampoules (e.g. Sterikon® plus Bioindicator, Merck Millipore). Spore strips must be incubated in broth after autoclaving to confirm that spores have been inactivated. Ampoules contain spores and broth, which typically changes color after incubation if the spores survive. Biological indicators are commonly used in industrial and clinical settings. They are less common in research organizations but may be used to comply with safety regulations. For example, Australian laboratories that use autoclaves to decontaminate genetically modified organisms (GMOs) must perform monthly testing using a biological indicator or other suitable method (OGTR 2011).

To reflect the true outcome of sterilization, sterility indicators must be placed in the part of the load that is the most difficult to sterilize, i.e. the middle of a packed autoclave or oven. Temperature monitoring should also be performed from the middle of the load, with the temperature probe located in a replica of the items being sterilized (see Figure 5.20b). The temperature of the chamber effluent is often used, but this does not accurately reflect the temperature of the load. Temperature monitoring can provide a permanent record that is archived digitally or as a paper copy; sterility indicators provide a visual confirmation of sterilization and can be used to monitor several parts of the load simultaneously. These approaches are complementary and are both recommended.

11.3 GLASSWARE

All glassware used in tissue culture should be reserved for that purpose alone. When most culture work was done on glass, the quality (charge, chemical residue) of the glass surface was critical. The almost universal adoption of single-use disposable plasticware has removed the problem of maintaining glass flasks suitable for use as culture vessels. Nevertheless, whenever glass is used (either for storage or for culture) the problem of chemical contaminants leaching out into media or reagents remains; absolute cleanliness is essential for all glassware. Traces of heavy metals or other toxic substances can be difficult to remove and may be detectable only by a gradual deterioration in the culture. It follows that separate stocks must also be washed separately.

The requirements of tissue culture washing are higher than for general glassware. Although the level of soil may be less, a special detergent may be necessary (see Section 11.3.1). A disinfectant should also be added while soaking glassware to remove any potential biohazard and prevent microbial growth. The overall process for washing and sterilizing glassware is shown in Figure 11.2 and described in Protocol P11.1. Additional supplementary information is available online for preparing, sterilizing, and using glass pipettes (see Supp. S11.1). More guidance on glassware preparation is available from suppliers, particularly for specialized items (Corning 2009).

PROTOCOL P11.1. PREPARATION AND STERILIZATION OF GLASSWARE

Outline

Collect glassware for washing, rinse, and dry for storage. Sterilize using dry heat. See Figure 11.2 for an overview of the procedure.

❖ *Safety Note. Disinfectants and detergents may be hazardous (check Safety Data Sheets). Eye protection and suitable gloves should be used when washing glassware; additional PPE may be needed. Discard any broken or chipped glassware and do not reuse.*

Materials (non-sterile)

☐ Hypochlorite disinfectant or suitable alternative (e.g. Presept, Medisave; Haz-Tab, Guest Medical). Use hypochlorite-based solutions at 0.1% (1 g/l)

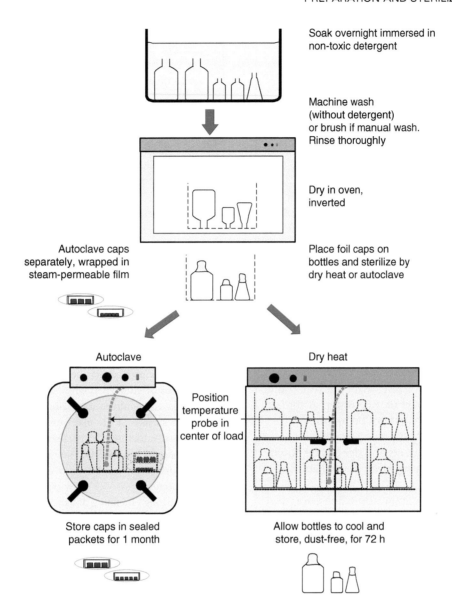

Soak overnight immersed in non-toxic detergent

Machine wash (without detergent) or brush if manual wash. Rinse thoroughly

Dry in oven, inverted

Autoclave caps separately, wrapped in steam-permeable film

Place foil caps on bottles and sterilize by dry heat or autoclave

Autoclave

Dry heat

Position temperature probe in center of load

Store caps in sealed packets for 1 month

Allow bottles to cool and store, dust-free, for 72 h

Fig. 11.2. Washing and sterilizing glassware. Sterilization conditions: autoclave at 121 °C (250 °F) for 15 minutes; place in oven at minimum of 160 °C (320 °F) for one hour (see Protocol P11.1). Caps are sterilized separately from bottles to avoid condensation forming in bottles if autoclaved with caps in place (see Figure 11.4).

of available chlorine; heavily soiled material may require up to 0.5% (5 g/l) (WHO 2004).
☐ Detergent (e.g. 7X or Decon)
☐ Soaking baths
☐ Bottle brushes
☐ Stainless-steel baskets (to collect washed and rinsed glassware for drying)
☐ Aluminum foil
☐ Sterility indicators for dry heat (e.g. Steraffirm Control Tubes, STERIS)

☐ Sterile-indicating tape or tabs (e.g. 3M Comply). This is different from the sterile-indicating tape used in autoclaves, as the sterilizing temperature is higher in an oven. Most autoclave tapes tend to char and release traces of volatile material from the adhesive, which can leave a deposit on the oven or even the glassware.
☐ Sterilizing oven, fan assisted, capable of reaching 160 °C (320 °F), and preferably with recording thermometer and flexible probe

Procedure

A. Collection and washing of glassware

1. Immediately after use, collect glassware into detergent containing disinfectant. It is important that glassware does not dry before soaking, or cleaning will be much more difficult.
2. Soak overnight in detergent.
3. Clean and rinse:
   (a) *Manual:* brush glassware the following morning, and rinse thoroughly in four complete changes of tap water followed by three changes of deionized, distilled, or RO water (see Section 11.5.2). A sink-rinsing spray is a useful accessory; otherwise bottles must be emptied and filled completely each time. Clipping bottles in a basket will help to speed up this stage.
   (b) *Glassware washing machine:* rinses should be done without detergent. If done on a spigot header (see Section 5.4.3), this can be reduced to two rinses with tap water and one with deionized or RO water. A further detergent wash should not be required.
4. After rinsing thoroughly, invert bottles, etc. in stainless-steel wire baskets and dry upside down. Use a glassware drying oven if available (see Section 5.4.3).
5. Cap bottles with aluminum foil when cool, and store.

B. Sterilization

6. Attach a small square of sterile-indicating tape or other indicator label to glassware, and date.
7. Place glassware in an oven with fan-circulated air and temperature set to 160 °C (320 °F).
8. To ensure that the center of the load reaches 160 °C (320 °F):
   (a) Place a sterility indicator in a bottle or typical item in the middle of the load.
   (b) If using a recording thermometer, place the sensor in a bottle or typical item in the middle of the load.
   (c) Do not pack the load too tightly; leave room for circulation of hot air.
9. Close the oven, check that the temperature returns to 160 °C (320 °F), seal the oven with a strip of tape with the time recorded on it (or use automatic locking and recorder), and leave for one hour.
10. After one hour, switch off the oven and allow it to cool with the door closed. It is convenient to put the oven on an automatic timer so that it can be left to switch off on its own overnight and be accessed in the morning. This precaution allows for cooling in a sterile environment and also minimizes the heat generated during the day, when it is hardest to deal with.
11. Use glassware within 24–48 hours.

11.3.1 Detergent Selection

As most cell culture is now carried out on disposable plastic, the major requirements for cleaning glassware are that (i) the detergent be effective in removing residue from the glass; and (ii) no toxic residue be left behind to leach out into the medium or other reagents. Some tissue culture suppliers provide a suitable detergent that has been tested with tissue culture, e.g. 7X Cleaning Solution (MP Biomedicals) or Decon 90 (Medisave) for manual washing. However, machine detergents will often come from the equipment supplier or a general laboratory supplier. Select a detergent that is effective in the water of your area, rinses off easily, and is non-toxic. If using glass as a culture substrate, caustic alkaline detergents render the surface of the glass unsuitable for cell attachment and require subsequent neutralization with 0.1 M HCl or H_2SO_4 (see Section 8.2.1). Neutral detergents do not alter the glass surface and can be removed with deionized, distilled, or RO water.

The washing efficiency of various detergents can be determined by washing heavily soiled glassware (e.g. a bottle of serum or medium containing serum that has been autoclaved). A visual check following the normal washup procedure will show which detergents have been effective. The presence of a toxic residue is best determined by examining cloning efficiency (see Sections 11.8.2, 19.4.1). For example, a glass Petri dish can be washed in the detergent, rinsed as indicated above (see Protocol P11.1), and then sterilized by dry heat before setting up the assay. A plastic Petri dish should be used as a control. This technique can also be used if you are anxious about residue left on glassware after autoclaving.

11.3.2 Glassware Sterilization

Glassware can be sterilized using two different approaches. The first is to use dry heat, which is preferred for most items; autoclaving is faster and more effective but carries a risk of leaving residue. Dry heat sterilization requires a minimum temperature of 160 °C (320 °F) for at least one hour, with additional time for cooling. Keep organic matter out of the oven. Do not use paper tape or packaging material unless you are sure that it will not release volatile products on heating. Such products will eventually build up on the inside of the oven (see Figure 11.3), making it smell when hot, and some deposition may occur inside the glassware being sterilized.

Fig. 11.3. Sterilizing oven. Pipette cans are stacked with spaces between to allow circulation of hot air. Brown staining on front of oven shows evidence of volatile material from sterile-indicating tape, a problem when using tape in a hot oven. *Source*: R. Ian Freshney.

Dry heat sterilization is not suitable for some bottle caps. Laboratories can autoclave the caps separately (see Section 11.3.3), which allows the bottles to cool down in the oven before they are removed. Bottles should be covered with foil and the caps added under aseptic conditions. Alternatively, bottles may be loosely capped with screw caps and foil, tagged with autoclave tape, and autoclaved for 20 minutes at 121 °C (250 °F) with a prevac and postvac included in the cycle (see Sections 5.4.4, 11.2.2). Caps must be very slack (loosened one complete turn) during autoclaving, to allow steam to enter the bottle and to prevent the liner (if one is used) from being sucked out of the cap and sealing the bottle. If a bottle becomes sealed during sterilization in an autoclave, sterilization will not be complete (see Figure 11.4). Caps may be tightened when the bottles have cooled down. Unfortunately, misting often occurs when bottles are autoclaved and a slight residue may be left when the mist evaporates. There is also a risk that bottles become contaminated as they cool by drawing in non-sterile air before they are sealed. Dry heat sterilization is therefore preferred.

11.3.3 Caps and Closures

There are two main types of caps that are in common use for glass bottles: (i) aluminum or phenolic plastic caps with synthetic rubber or silicone liners; and (ii) polypropylene caps that are wadless and reusable (e.g. high-temperature screw caps from Corning, Duran). The latter are deeply shrouded and have ring inserts for better sealing and to improve pouring (although pouring is not recommended in sterile work). Stoppers and similar closures may also need to be sterilized, although most are now disposable.

The following precautions should be observed:

(1) Wadless caps are preferable; if using caps with liners, the liners must be removed during washing.
(2) Polypropylene caps will seal only if screwed down tightly on a bottle with no chips or imperfections on the lip of the opening. Discard bottles with chipped necks.
(3) Shrouded caps are preferred to stoppers, but if the latter are required, use silicone or heavy metal-free white rubber stoppers in preference to those made of natural rubber.
(4) Do not leave aluminum caps or any other aluminum items in alkaline detergents for more than 30 minutes, as they will corrode.
(5) Do not put glassware together with caps in the same detergent bath, or the aluminum may contaminate the glass.
(6) Avoid detergents that are made for machine washing, as they are highly caustic.

PROTOCOL P11.2. PREPARATION AND STERILIZATION OF SCREW CAPS

Outline

Wash screw caps or closures, rinse, and dry for storage. Sterilize by autoclaving.

❖ *Safety Note. Disinfectants and detergents may be hazardous (check Safety Data Sheets). Eye protection and suitable gloves should be used when washing laboratory items; additional PPE may be needed.*

Materials (non-sterile)

☐ Detergent (e.g. 7X or Decon)
☐ Soaking baths
☐ Stainless-steel baskets (to collect caps for washing and drying)
☐ Sterility indicators (e.g. Thermalog, see Section 11.2.7)
☐ Glass Petri dishes (VWR; for packaging)
☐ Autoclavable nylon film (e.g. Portex) or bags (e.g. Cardinal Health; VWR)
☐ Sterile-indicating autoclave tape
☐ Autoclave with recording thermometer with flexible probe that can be inserted in load (see Figure 5.20b)

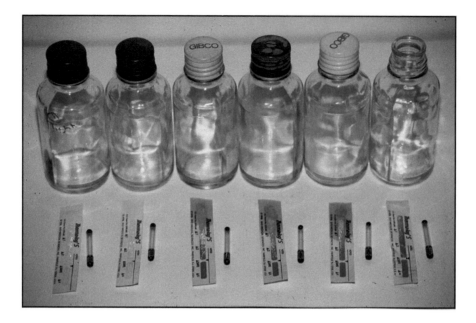

Fig. 11.4. Sterilizing capped bottles. These bottles were autoclaved with Thermalog sterility indicators inside (see Section 11.2.7), which turn blue with high temperature and steam; the blue area moves along the strip with time under the required sterilization conditions. The cap on the leftmost bottle was tight, and each succeeding cap was gradually slacker, until, finally, no cap was used on the bottle furthest to the right. The farthest left bottle is not sterile because no steam entered it. The second bottle is not sterile either, because the liner drew back onto the neck and sealed it. The next three bottles are all sterile, but the brown stain on the indicator shows that there was fluid in them at the end of the cycle. Only the bottle at the far right is sterile and dry. The glass indicators (Browne's tubes) all implied that their respective bottles were sterile. *Source*: R. Ian Freshney.

Procedure

A. Washing

1. Soak 30 minutes (maximum) in detergent.
2. Rinse thoroughly for two hours.
3. Metal or phenolic caps with liners: make sure all caps are submerged. Liners should be removed and replaced after rinsing, which may be carried out in either of two ways:
   (a) In a beaker (or stainless-steel pail) with running tap water led by a tube to the bottom. Stir the caps by hand every 15 minutes.
   (b) In a basket or, better, in a pipette-washing attachment. Rinse in an automatic washing machine, but do not use detergent in the machine.
4. Polypropylene caps:
   (a) Wash and rinse by hand as just described (extending the detergent soak if necessary). Because these caps may float, they must be weighted down during soaking and rinsing.

   (b) For automatic washers, after soaking in detergent, use pipette-washing attachment and normal cycle without machine detergent.
5. Stoppers: wash and sterilize as for caps. (There will be no problem with flotation in washing and rinsing.)

B. Sterilization

6. Prepare small caps for autoclaving: place in a glass Petri dish, caps with the open side down. Wrap Petri dish containing caps in cartridge paper or steam-permeable nylon film, and seal with autoclave tape (see Figure 11.5).
7. Prepare large caps for autoclaving: place in an autoclave paper bag and seal with autoclave tape. Various sizes are usually available.
8. Prepare a similar package with sterility indicator enclosed and insert in the middle of the load.
9. Autoclave for 20 minutes at 121 °C (250 °F) and 100 kPa (15 lb/in.^{2}) (see Figure 11.1).

Fig. 11.5. Packaging screw caps for sterilization. The caps are enclosed in a glass Petri dish, which is then sealed in an autoclavable nylon bag with autoclave sterile-indicating tape. *Source*: R. Ian Freshney.

11.4 OTHER LABORATORY APPARATUS

11.4.1 Cleaning and Packaging

All new apparatus and materials (tubing, filter holders, instruments, etc.) should be soaked in detergent overnight, thoroughly rinsed, and dried. Anything that will corrode in the detergent – mild steel, aluminum, copper, brass, etc. – should be washed directly by hand without soaking (or with soaking for 30 minutes only, using detergent if necessary), brushed, and then rinsed and dried. Aluminum centrifuge buckets and rotors must never be allowed to soak in detergent. Always check the manufacturer's instructions when selecting a suitable detergent and method of cleaning for equipment items.

Used items should be rinsed in tap water and immersed in detergent immediately after use. Allow them to soak overnight, brush if required, and then rinse and dry in the oven. Again, do not expose materials that might corrode to detergent for longer than 30 minutes. Particular care must be taken with items treated with silicone grease or silicone fluids. These items are best discarded, but, if kept, must be treated separately to remove the silicone. Silicones are very difficult to remove if they are allowed to spread to other apparatus, particularly glassware. Similarly, items used with agar or agarose are hard to clean and best discarded, but can be cleaned in boiling water if necessary.

Ideally, all apparatus being sterilized should be wrapped in a covering that allows steam to penetrate (see Figure 11.5) but is impermeable to dust, microorganisms, and mites. Autoclave bags are available with sterile-indicating marks that show up after sterilization (e.g. Cardinal Health). Semipermeable transparent nylon film (e.g. Portex), in rolls of flat tubes of different diameters, can be made up into bags with sterile-indicating tape (e.g. products from Camlab, STERIS, and VWR). Although expensive, such film can be reused several times before becoming brittle. Tubes and orifices

should be covered with tape and paper or nylon film before packaging, and needles or other sharp points should be shrouded with a glass test tube or other appropriate guard.

11.4.2 Sterilization or Disinfection

The need for sterilization will depend on what the apparatus is used for. Items that come into direct contact with tissue or cells (or that are used to prepare culture media or other reagents) must be sterilized, with the method determined by the material (see Table 11.2). Metallic items are best sterilized by dry heat. Silicone rubber (which should be used in preference to natural rubber), PTFE, polycarbonate, cellulose acetate, and cellulose nitrate filters, etc. should be autoclaved for 20 minutes at 121 °C (250 °F) and 100 kPa (15 lb/in.^{2}), with prevac and postvac steps in the cycle (see Section 11.2.2). Protocols may need to be modified for some items. For example, reusable filter assemblies are typically autoclaved using "wet" cycle conditions as described in Supplementary material online (see Supp. S11.2). If you intend to sterilize equipment, always check the manufacturer's instructions, e.g. you may need to separate centrifuge rotor components before autoclaving (Thermo Scientific 2008).

Items that do not have direct contact with tissue or cells may not need to be sterilized, but they should always be disinfected as an important part of aseptic technique. Work surfaces that are used for tissue culture should be wiped down using 70% alcohol (see Section 12.3.1). For equipment, always check the manufacturer's instructions when selecting a disinfection method.

11.5 WATER

11.5.1 Water Purity

Water is the simplest, but probably the most critical, constituent of all media and reagents, particularly for serum-free media (see Section 10.6.2). Water is often referred to as the "universal solvent"; this means that many different contaminants are present in the water supply (NIH 2013). Although potable "drinking" water is treated to remove or inactivate microorganisms, cellular fragments remain that will affect water quality when used for laboratory applications. Other contaminants that may be present in potable water include ions, organic molecules, hard and soft particulates, and colloids.

Water used in tissue culture must be of a very high purity. The purity of water is assessed using various parameters including the concentration of ions (measured by resistivity and conductivity), the pH, and the presence of organics, colloids, total particulates, and bacteria (NIH 2013; Merck Millipore 2018). Reference documents are available that set out detailed specifications for water quality (ISO 3696:1987 [International Organization for Standardization 1987]; CLSI GP40-A4-AMD [Clinical and Laboratory Standards Institute

TABLE 11.4. Grades of water purity.

Contaminant	Parameter	Type 3[a]	Type 2[a]	Type 1[a]
Ions	Resistivity at 25 °C, MΩ·cm	>0.05	>1.0	>18.0
Organics	TOC, ppb (µg/l)	<200	<50	<10
Pyrogens	EU/ml	NA	NA	<0.03
Particulates	Particulates >0.2 µm, units/ml	NA	NA	<1
Colloids	Silica, ppb	<1000	<10	<10
Bacteria	cfu/ml	<1000	<100	<1

Source: based on data in Merck Millipore, Water in the laboratory, 2018.
[a]These values are used as a general guide; the exact specifications will vary based on your application and the relevant technical specification (ISO 3696:1987 [International Organization for Standardization 1987]; ASTM D1193-06 [ASTM International 2018]).
cfu, Colony-forming units; EU, endotoxin unit; ppb, parts per billion; TOC, total organic carbon.

2006]; ASTM D1193-06 [ASTM International 2018]). Most of these documents divide laboratory water into three types or grades (see Table 11.4). Ultrapure water (UPW, also known as Type 1 water) is used for all tissue culture reagents and media. Type 2 water is used for some laboratory applications (e.g. preparation of buffers). Type 3 water is used for non-critical applications (e.g. filling water baths).

Always check that you are using the correct water quality for your application. Some applications have specialized water requirements – for example, next generation sequencing and other molecular biology applications may require UPW that is RNase- and DNase-free (Merck Millipore 2018). UPW should not be used to fill incubator pans and water baths because it leaches metal ions from equipment and can result in corrosion. Sterile water for pharmaceutical applications (e.g. water for injection) is manufactured to different specifications and should not be substituted for UPW unless it is known to be suitable for a particular application (Ludwig et al. 2006). UPW should be used immediately as water quality will decline with storage.

11.5.2 Purification Methods

Water purification systems use a combination of methods to achieve UPW (see Figure 11.6; see also Figure 5.18). The first stage is usually RO or distillation. The product of this first stage is semi-purified water (SPW), which is roughly equivalent to Type 3 water. A second stage of purification uses carbon filtration to remove both organic and inorganic colloids. The third stage uses high-grade mixed-bed deionization to remove ionized inorganic material. The product of the combined second and third stages is purified water (PW), which is roughly equivalent to Type 2 water. The final stage is micropore filtration to remove any microorganisms acquired from the system and to trap any resin that may have escaped from the deionizer (see Figure 11.6). If the water is recycled continuously from the micropore filter to the reservoir with the supply from the first stage turned off (e.g. overnight), the stored water gradually "polishes" (i.e. increases in

purity). This results in UPW. Water should be collected directly from the final stage micropore filter without being stored.

The choice of purification methods will depend on several factors including location, budget, and application. Because water supplies vary greatly, the degree of purification required may vary. Hard water will need a conventional ion-exchange water softener on the supply line before entering the purification system, but this will not be necessary with soft water. Distillation has the advantage that the water is heat sterilized, but it is more expensive because of power consumption and the need to clean out the boiler regularly. If glass distillation is used for the first stage, the still should be electric and automatically controlled, and the heating elements should be made of borosilicate glass or silica sheathed. RO purification depends on the integrity of the filtration membrane, which means that the conductivity of the effluent is also monitored. If RO is used for the first stage, the type of cartridge should be chosen to suit the pH of the water supply, as some membranes can become porous in extreme pH conditions.

Although water is purified, it may not necessarily be sterile. Sterilization requires a high degree of assurance (see Section 11.1), which may not be met using a water purification system. UPW can be sterilized by autoclaving as described in Protocol P11.3, which can also be adapted for use in training (see Supp. S30.2).

PROTOCOL P11.3. PREPARATION AND STERILIZATION OF ULTRAPURE WATER (UPW)

Outline

Sterilize water in suitable aliquots (e.g. 450 ml for media preparation from 10× concentrates) by autoclaving at 121 °C (250 °F) and 100 kPa (15 lb/in.²) for 20 minutes in bottles with the caps slack. Tighten caps when bottles are cool.

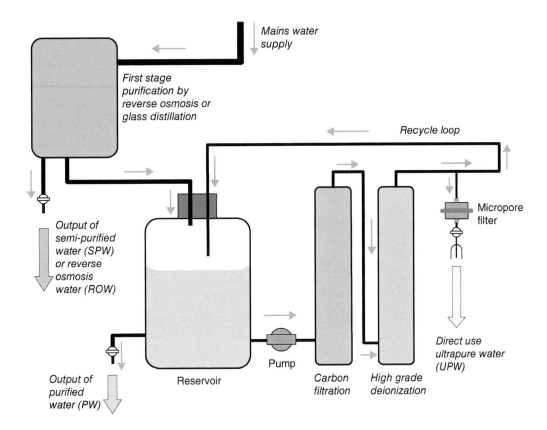

Fig. 11.6. Water purification. Tap water is fed to a storage container via RO or glass distillation to give SPW. This is then recycled back to the storage container via carbon filtration, deionization, and micropore filtration. PW for reagents is available at all times from the storage reservoir; UPW for media preparation is available from the micropore filter supply (at right of diagram). If the apparatus recycles continuously, water of the highest purity will be collected first thing in the morning (see also Figure 5.18).

❖ *Safety Note. Autoclaves can cause severe burns and are pressurized during operation, resulting in the risk of explosion. Autoclaves should only be operated by trained and competent personnel.*

Materials (non-sterile)

- ☐ Autoclavable bottles (e.g. Pyrex, polycarbonate); choose graduated bottles and ensure there is sufficient head space for any later additions (e.g. see Protocol P11.6)
- ☐ Screw caps to fit
- ☐ Sterility indicator (see Section 11.2.7)
- ☐ Sterile-indicating tape (e.g. Camlab; VWR)
- ☐ Marker pen or preprinted labels
- ☐ Water purification system (e.g. Elga-Veolia; Merck Millipore)
- ☐ Autoclave
- ☐ Reagent Log (see Table 11.5); record of reagent preparation and sterilization

Procedure

1. Create entry in Reagent Log (see Table 11.5); label bottles with date, contents, and batch number. (A label printer attached to the computer will generate labels automatically.)
2. Early in the morning after overnight recycling, run about 50 ml water to waste from purifier, check resistivity (or conductivity) and total organic carbon (TOC), and enter in log.
3. If water is within specified limits (resistivity $\geq 18\,M\Omega/cm$ at 25 °C, TOC $\leq 10\,ppb$), collect UPW directly into labeled bottles.
4. Fill to the specified mark, e.g. 430–450 ml if to be used for diluting 10× concentrated medium (see Protocol P11.6).
5. Place sterility indicator in one bottle (to be discarded when checked after autoclaving).
6. Close bottles with screw caps but leave caps slack by one complete turn.

TABLE 11.5. Layout for Reagent Log.

Batch number[a]		1	2	3	4	5, etc.
Date		16/2/18	17/2/18			
Solution		HBSS	Water			
Meter readings	$M\Omega \cdot cm$[b]	–	10			
	pH	6.5	6.8			
	mOsmol/kg	290	–			
	TOC (ppb)	–	10			
Sterilization	Minutes[c]	20	20			
	°C	121	121			
	Tape[d]	+ve	+ve			
	Indicator[e]	+ve	+ve			
Storage	Location	Fridge	Laboratory			
	°C	4	20			
	Date	17/2/14	18/2/14			
Quality control	Sterility	37 °C OK	37 °C OK			
	Cell growth	–	–			
	PE	–	–			
	Other	–	–			

[a]This is a suggested layout, but requirements will vary between laboratories, e.g. sterilization time and temperature may not be required if the autoclave records this information separately.
[b]Resistivity; conductivity, μS/cm, can also be used.
[c]Time starts when temperature reaches 121 °C.
[d]Sterile-indicating tape; +ve confirms that appropriate marking has appeared.
[e]Indicator placed inside vessel at center of load; +ve confirms that conditions are met, e.g. based on color change.

7. Attach sterile-indicating tape to each bottle (tape may be used to lightly hold caps in position).
8. Place bottles in autoclave with bottle containing sterility indicator in center of load.
9. Close autoclave and select a "wet" run at 121 °C (250 °F), 100 kPa (15 lb/in.²), for 20 minutes with prevac and postvac deselected.
10. Start sterilization cycle. On completion of cycle, check printout to confirm that correct conditions have been attained for the correct duration and enter in log.
11. Allow load to cool to below 50 °C (120 °F).
12. Open autoclave and retrieve bottles.
13. Tighten caps when bottles reach room temperature.
14. Check sterility indicator to confirm that sterilization conditions have been achieved and enter in log (see Table 11.5).
15. Replenish culture room stocks as required, confirming by examination of sterile-indicating tape that the bottles have been through sterilization cycle.
16. Add UPW to allow for evaporation before use.

Notes

Step 13: Sealing autoclaved bottles while they are still warm can cause the cap liner to be drawn into the bottle as it cools and the contents contract. There will also be a rapid intake of air into the bottle when it is first opened, which can cause contamination. Bottles that have been open during autoclaving should be allowed to cool to room temperature in a sterile atmosphere (e.g. in the autoclave or a horizontal laminar flow hood) before caps are tightened.

11.5.3 Monitoring and Maintenance

The quality of deionized water should be monitored by its resistivity (or the inverse, conductivity) and the system cartridge should be changed when a decrease in resistivity is observed. Resistivity of Type 1 water should be $\geq 18\,M\Omega/cm$ at 25 °C and the TOC should be ≤ 10 parts per billion (ppb; see Table 11.4). Resistivity meters and TOC monitors are usually supplied with water purification systems (see Figure 5.18) but can be purchased separately (e.g. M9 TOC Analyzers, GE Healthcare Life Sciences).

Many water purification systems now include continuous monitoring, alerts, or compliance software that allow users to save records and meet audit criteria (e.g. Millitrack® Compliance software, Merck Millipore). If they do not, laboratories should monitor their water purification systems regularly by checking the record of resistivity and TOC of water at each outlet weekly to determine that they conform to the set limits. A log should be kept for all reagents, including water, to record critical data during preparation and sterilization (see Table 11.5). Patterns will emerge that indicate the expected useful life of the RO, deionization, and charcoal cartridges, allowing laboratories to develop replacement programs that preempt these deteriorations. Similarly, laboratories can determine by observation when maintenance is required, e.g. how quickly the distillation boiler (if used) scales up. The frequency with which these operations must be carried out will vary from one laboratory to another, depending on the quality of the main water supply, and will emerge from keeping a record. Once a pattern emerges, it should be written into a SOP.

UPW should recycle through the purification system continually to minimize growth of microbial contamination. Tubing and connections can become contaminated with algae or other microorganisms. All tubing and reservoirs should be checked every three months or so and disinfection should be performed on a regular basis. A typical cleaning protocol usually involves the following:

(1) Disconnect tubing and other connections from the water purification system.
(2) Soak in hot water with a suitable detergent, e.g. hypochlorite. Rinse thoroughly.
(3) Soak in 1 M HCl. Rinse thoroughly in SPW.
(4) Reinstall tubing and connections and restart operation.
(5) Discard first batch of water.
(6) Test subsequent batches of water. Quality control can be performed by checking the cloning efficiency of a sensitive cell line at regular intervals in medium made up with the water (see Sections 11.8.2, 19.4.1).

11.6 MEDIA AND OTHER REAGENTS

The ultimate objective in preparing culture media and reagents is to produce them in a pure form to (i) avoid the accidental inclusion of toxic substances; (ii) enable the reagent to be totally defined, its preparation reproducible, and the functions of its constituents to be fully understood; and (iii) reduce the risk of microbial contamination. Most reagents are sterilized by filtration if they are heat labile and by autoclaving if they are heat stable (see Table 11.3). Protocols and instructions are given here for balanced salt solutions, basal and complete media, and serum. Additional information on recipes and procedures can be found in Appendix B.

11.6.1 Balanced Salt Solutions

The formulations of the balanced salt solutions are described elsewhere (see Section 9.3). Hanks's basal salt solution (HBSS) contains magnesium chloride in place of some of the magnesium sulfate originally recommended to reduce precipitation (see Table 9.2). It should be autoclaved below pH 6.5 and neutralized just before use for the same reason. Similarly, Dulbecco's phosphate buffered saline (DPBS) is often used without the addition of Ca^{2+} and Mg^{2+}. In that form, it should be referred to as DPBS without Ca^{2+} and Mg^{2+}, PBS Solution A, or DPBS-A; DPBS-A is used throughout this book. A calcium and magnesium supplement can be prepared as a separate solution (DPBS-B) and added just before use, but, in practice, it is simpler to purchase DPBS complete with calcium and magnesium or choose an alternative such as HBSS.

Many balanced salt solutions contain glucose (see Table 9.2). Because glucose may caramelize on autoclaving, it is best omitted during preparation and added after autoclaving. If glucose is prepared as a 100× concentrate (200 g/l), caramelization during autoclaving is reduced, and it can be used at 5–25 ml/l to give 1–5 g/l. DPBS-A does not contain glucose and is commonly used as a rinsing solution and solvent for trypsin or ethylene diamine tetra-acetic acid (EDTA). Protocol P11.4 can be used to prepare DPBS-A or any other balanced salt solution without glucose; it can also be used in training (see Supp. S30.3). Complete balanced salt solutions with glucose can be prepared as for complete media (see Protocols P11.5–P11.7) and sterilized by filtration (see Protocols P11.8, P11.9).

PROTOCOL P11.4. PREPARATION AND STERILIZATION OF DPBS-A

Outline

Dissolve powder with constant mixing, make up to final volume, check pH and conductivity, dispense into aliquots, and autoclave.

❖ *Safety Note. Autoclaves can cause severe burns and are pressurized during operation, resulting in the risk of explosion. Autoclaves should only be operated by trained and competent personnel.*

Materials (non-sterile)

☐ Dulbecco's phosphate buffered saline without Ca^{2+} and Mg^{2+} (DPBS-A) as powder (e.g. Sigma-Aldrich #D5652) or tablets (e.g. Oxoid #BR0014)
☐ UPW (see Section 11.5.1)
☐ Container: clear glass or clear plastic aspirator with tap outlet at base, Erlenmeyer flask or bottle with peristaltic metering pump and tubing (see Figures 5.6, 5.7)

- ☐ Magnetic stirrer and PTFE-coated follower
- ☐ Bottles for storage; borosilicate glass, graduated, with screw caps
- ☐ pH meter
- ☐ Conductivity meter or osmometer
- ☐ Sterile-indicating autoclave tape
- ☐ Autoclave
- ☐ Reagent Log (see Table 11.5); record of reagent preparation and sterilization

Procedure

1. Add UPW and magnetic follower to container.
2. Place container on magnetic stirrer and set to around 200 rpm.
3. Open packet of DPBS-A powder, or count out appropriate number of tablets, and add slowly to container while mixing.
4. Stir until completely dissolved.
5. Check pH of a sample and enter in log. pH should not vary more than 0.1 pH unit from the required value (the required pH value may vary with different formulations).
6. Check conductivity of a sample and enter in log. Conductivity should not vary more than 5% from 150 µS/cm.
7. Discard test samples; do not add back to main stock.
8. Dispense contents of container into graduated bottles.
9. Cap and seal bottles. Attach a small piece of autoclave tape or sterile-indicating tab and date.
10. Sterilize by autoclaving for 20 minutes at 121 °C (250 °F) and 100 kPa (15 lb/in.²) in sealed bottles.
11. Store at room temperature.
12. Complete record in Reagent Log.

Notes

Steps 5 and 6: It is important to check pH and conductivity as a quality control measure, to ensure that there has been no mistake in preparation. Osmolality can be used as an alternative (or addition) to conductivity and should show similar consistency between batches. Any adjustments, e.g. to osmolality, should be made after quality control checks have been made.

11.6.2 Basal and Complete Media

The term "complete medium" refers to medium that is sufficient for tissue culture (see Section 9.4). It consists of a defined basal component with the addition of unstable constituents (e.g. glutamine) and supplements (e.g. serum).

Basal media are usually available from commercial suppliers in three different formats: (i) powdered media, with or without $NaHCO_3$ and glutamine; (ii) concentrates (10×), usually without $NaHCO_3$ and glutamine, which are available separately; and (iii) working-strength stock solutions (1×), with or without glutamine and with a specified $NaHCO_3$ concentration depending on the CO_2 concentration to be used in the gas phase. Powdered media is cheapest but costs more in preparation time and resources. Concentrates allow for some alterations, e.g. inclusion of other supplements like non-essential amino acids (NEAA). Working-strength solutions are more expensive, but because the supplier has done the majority of the preparation, this option is the fastest and may have fewer quality concerns. Protocols P11.5, P11.6, and P11.7 are specific to each format; these protocols can also be used in training (see Supp. S30.5 and S30.7).

Always refer to the manufacturer's instructions and laboratory SOPs to ensure that you prepare media correctly and consistently. General points that apply whenever you prepare medium, regardless of the format, include:

(1) *Record keeping.* Label each bottle of complete medium to show the date and preparer. As soon as the medium is used for a particular cell line, label the bottle accordingly and do not use for any other cell line. Record the lot numbers of any reagents used in case of later problems. Supplements (e.g. serum) are not included in the original formulation and should be included in publications. Other additions (e.g. glutamine or $NaHCO_3$) are part of the formulation and are not supplements. They need not be indicated in publications unless their concentrations are changed.

(2) *Frozen constituents.* Many constituents are stored frozen. To minimize damage during thawing, single-use frozen aliquots may be prepared. Any unused material is usually discarded after thawing; select a suitable aliquot size to minimize waste. Antibiotics are not included as constituents in the protocols below because they should not be used for routine culture (see Section 9.6.2).

(3) *Bicarbonate adjustment.* Some media are designed for use with a high bicarbonate concentration and elevated CO_2 in the atmosphere (e.g. Eagle's Minimal Essential Medium (MEM) with Earle's salts), whereas others have a low bicarbonate concentration for use with a gas phase of air (e.g. Eagle's MEM with Hanks's salts; see Tables 9.1, 9.2). If a medium is changed and its bicarbonate concentration altered, it is important to make sure that the osmolality is still within an acceptable range. The osmolality should always be checked (see Section 9.2.6) when any significant alterations are made to a medium that are not in the original formulation.

(4) *pH adjustment.* Always equilibrate and check the pH at 37 °C, as the solubility of CO_2 decreases with increased

temperature and the pK_a of the HEPES will change. If using a new medium for the first time, pipette an aliquot into a flask or Petri dish, and incubate for at least one hour (preferably overnight) under your standard conditions, to ensure that the pH equilibrates at the correct value. If it does not, readjust the pH of the medium with 1 M HCl or 1 M NaOH and repeat incubation.

(5) **Precipitation.** Take care while preparing complex solutions to ensure that all constituents dissolve, are not removed during sterilization, and remain in solution after sterilization and storage. Concentrated media are often prepared at a low pH (3.5–5.0) to keep all the constituents in solution, but even then, some precipitation may occur. If the constituents are properly resuspended, they will usually redissolve on dilution. However, if the precipitate has been formed by degradation of some of the constituents, the quality of the medium may be reduced. It should be tested for performance (see Section 11.8.2) or replaced.

(6) **Method of sterilization.** Most media are sterilized using membrane filtration (see Section 11.7). However, some basal media can be autoclaved. Autoclavable media were initially developed in the 1960s due to concerns about the sterilizing filters that were used at that time (Waymouth 1978). Some classic basal formulations have also been modified to allow autoclaving (e.g. MEM AutoMod™, Sigma-Aldrich). The medium is buffered to pH 4.25 with succinate, in order to stabilize the B vitamins during autoclaving, and is subsequently neutralized. The procedure to follow is supplied in the manufacturer's instructions and is similar to that described earlier for DPBS-A (see Protocol P11.4).

(7) **Shelf life.** Perform quality control testing before releasing medium for use (see Section 11.8). Set the volume so that complete medium is completely used within three weeks. Glutamine is often added last, as it is unstable and will affect the shelf life. The half-life of glutamine in medium at 4 °C is about three weeks and at 37 °C about 3–5 days. It is best to buy glutamine separately and store it frozen or use a more stable glutamine dipeptide (see Section 9.4.1). All medium and serum should be stored shielded from light (see Section 9.8).

(8) **Scale-up.** If your consumption of medium is fairly high (>200 l/year), it may be better to get extra constituents included in a ready-made formulation, as this practice will work out to be cheaper. HEPES in particular is very expensive to buy separately. For large-scale requirements (>10 l/week), medium can be prepared in a pressure vessel, checked by sampling at intervals with a large pipette to determine whether solution is complete, and sterilized by positive pressure through an in-line disposable or reusable filter into a receiver vessel (see Section 11.7.3).

Powdered media.

PROTOCOL P11.5. PREPARATION OF MEDIUM FROM POWDER

Outline

Dissolve the entire contents of the pack in the correct volume of UPW, using a magnetic stirrer and adding the powder gradually with constant mixing. Filter the medium once all of the constituents have dissolved completely, without allowing to stand, in case any of the constituents precipitate or microbial contamination occurs. Add any supplements and unstable constituents and adjust the pH just before use. Note: materials and instructions will vary with the formulation and intended use. The procedure here is provided as an example only.

Materials (sterile)

□ Graduated bottles for medium, 500 ml size (20)
□ Caps for bottles (one per bottle)
□ Supplements and unstable constituents, e.g. glutamine, $NaHCO_3$, or serum

Materials (non-sterile)

□ BSC and associated consumables (see Protocol P12.1)
□ UPW (see Section 11.5.1; 10 l)
□ Powdered medium, e.g. Eagle's MEM with Earle's salts, without glutamine (10-l pack)
□ Sodium bicarbonate or other additives (see supplier instructions for specific formulation)
□ Container: Graduated Erlenmeyer flask or bottle, 10-l capacity plus head space (one)
□ Magnetic stirrer and PTFE-coated follower
□ pH meter
□ Conductivity meter or osmometer
□ Reagent Log (see Table 11.5); record of reagent preparation and sterilization

Procedure

1. Add UPW (less than final volume, e.g. 9.5 l) and magnetic follower to container.
2. Place container on magnetic stirrer and set to around 200 rpm.
3. Open packet of powder and add contents slowly to container while mixing; rinse out with UPW to make sure all contents recovered.
4. Stir until powder is completely dissolved (do not heat).
5. Add sodium bicarbonate or any other constituents that should be added prior to filtration; refer to the manufacturer's instructions for the specific formulation.

6. Adjust the pH to 7.4 or other level as desired for the cell type.
7. Make up to 10 l with UPW. Use aseptic technique in a BSC from this point onwards.
8. Sterilize by filtration (see Protocol P11.9), dispensing into 500-ml aliquots or 450-ml aliquots if serum is to be added later.
9. Collect samples for sterility testing (see Supp. S11.6).
10. Cap and seal bottles.
11. Store at 4 °C shielded from light (see Section 9.8).
12. Add glutamine, serum, and any other supplements just prior to use.
13. Complete record in Reagent Log, e.g. lot number for pack. Update with sterility result.

Notes

Step 6: If it is likely that the pH will be altered by later additions or procedures, the pH should be rechecked prior to use using an aliquot that is removed under sterile conditions.

Powdered media are prepared in specialized facilities and tested by the manufacturer for sterility and performance in culture; of course, the manufacturer's sterility testing only applies to the powder and not to its subsequent handling. Choose a pack size that you can make up all at once and use within its designated shelf life. Powdered media are mixed very efficiently, so, in theory, a pack may be subdivided for use at different times. However, in practice, it is better to match the size of the pack to the volume that you intend to prepare; once the pack is opened, the contents may deteriorate and some of the constituents may settle.

Concentrated (10×) media.

PROTOCOL P11.6. PREPARATION OF MEDIUM FROM 10× CONCENTRATE

Outline

Add concentrated medium and other constituents to an aliquot of UPW (see Section 11.5.1), giving a total volume of 500 ml. Adjust the pH and use the solution or return it to the refrigerator. Note: materials and instructions will vary with the formulation and intended use. The procedure here is provided as an example only, based on a volume of 500 ml.

Materials (sterile)

☐ Bottle containing premeasured aliquot of sterile UPW (see Protocol P11.3). For this example, an ~ 390-ml aliquot is used.

☐ Medium, 10× concentrate, e.g. Eagle's MEM with Hanks's salts or Earle's salts
☐ Glutamine, 200 mM stock
☐ NaHCO₃, 7.5% (0.89 M) stock
☐ HEPES (if required; only suitable for low NaHCO₃ medium), 1.0 M stock
☐ NaOH, 1 M (as required)
☐ Serum, e.g. fetal bovine serum (FBS)

Materials (non-sterile)

☐ BSC and associated consumables (see Protocol P12.1)

Procedure

1. Prepare a work space for aseptic technique using a BSC.
2. Thaw any frozen constituents.
3. Bring UPW bottle and other constituents to the BSC. Swab any bottles that have been in a water bath or refrigerator before placing in the BSC.
4. Uncap bottles. Prepare to transfer the appropriate volume of each addition to the UPW bottle, using a different pipette for each addition.
5. Add 10× concentrated medium, 1 : 10 (50 ml).
6. Add glutamine, 1 : 100 (5 ml).
7. Add NaHCO₃, with the amount depending on the gas mixture used. For MEM with Hanks's salts (as specified in Materials), add 4.5 ml. For MEM with Earle's salts for culture in a gas phase of 5% CO_2, add 13 ml.
8. Add HEPES if required to increase the buffering capacity. For MEM with Hanks's salts and a final concentration of 10 mM (1 : 100), add 5 ml. This allows the flask to be vented to atmosphere for some cell lines at a high cell density, resulting in acid production. For MEM with Earle's salts and a final concentration of 20 mM (1 : 50), but with only 4.5 ml NaHCO₃ at step 6 and 2% CO_2 in the gas phase, add 10 ml of HEPES.
9. Add 1 M NaOH to give pH 7.2 at 20 °C. When incubated, the medium will rise to pH 7.4 at 37 °C, but this figure may need to be checked by a trial titration the first time the recipe is used (see Protocol P11.7).
10. Add serum (50 ml) and any other supplements required.
11. Recap bottles and label to record additions, date, and user.
12. Collect samples for sterility testing (see Supp. S11.6).

Notes

Step 8: If HEPES is added, check the osmolality; it may be necessary to add additional water and accept the minor dilution of nutrients that this will cause.

Step 9: The amount of alkali needed to neutralize 10× concentrated medium (which is made up in acid to maintain the solubility of the constituents) may vary from batch to batch and from one medium to another. When making up a new medium for the first time, add the stipulated amount of $NaHCO_3$ and allow samples with varying amounts of alkali to equilibrate overnight at 37 °C in the appropriate gas phase. Check the pH the following morning, select the correct amount of alkali, and prepare the rest of the medium accordingly.

Preparing medium from 10× concentrate represents a good compromise between the economy of preparing medium from powder and the ease of using a 1× preparation. It is also a useful format when adding various supplements. If additional constituents are required and are made up in water, remove an equivalent volume from the amount of water already present in the bottle before adding the main constituents. If the additional constituents are in isotonic salt, they can be added to the final volume. Because it is isotonic, serum can be added to the final volume, although doing so will dilute the nutrients from the medium. Whatever procedure is followed, always be consistent.

The following formula can be used to calculate dilution when adding a component to medium:

$$\frac{\text{Required concentration}}{\text{Starting concentration}} \times \text{Required volume}$$

For a 4-mM final concentration of $NaHCO_3$ from 7.5% stock (0.89 M or 890 mM) this becomes:

$$\frac{4}{890} \times 100 = 0.45 \text{ ml}$$

and for 23 mM:

$$\frac{23}{890} \times 100 = 2.6 \text{ ml}$$

Working strength (1×) media.

PROTOCOL P11.7. PREPARATION OF MEDIUM FROM 1× STOCK

Outline

Check the formulation to see if additional constituents are needed. If the formulation lacks glutamine or $NaHCO_3$, add these constituents from the appropriate stock concentrates. Add any other supplements required. Note: materials and instructions will vary with the formulation and intended use. The procedure here is provided as an example only, based on a volume of 100 ml.

Materials (sterile)

☐ Working strength 1× medium, e.g. Eagle's MEM with Hanks's salts (100 ml)
☐ Glutamine, 200 mM stock (if not already included)
☐ $NaHCO_3$, 7.5% (0.89 M) stock (if not already included)
☐ Serum, e.g. FBS

Materials (non-sterile)

☐ BSC and associated consumables (see Protocol P12.1)

Procedure

1. Prepare a work space for aseptic technique using a BSC.
2. Thaw any frozen constituents.
3. Take medium to BSC with any other constituents that are required. Swab any bottles that have been in a water bath or refrigerator before placing in the BSC.
4. Uncap bottles. Prepare to transfer the appropriate volume of each addition to the stock bottle to make the correct dilution, using a different pipette for each addition.
5. Add glutamine, 1 : 100 (1 ml).
6. Add $NaHCO_3$ if not already included. For MEM with Hanks's salts (as specified in Materials), add 0.45 ml. If MEM is used with Earle's salts (see Note below), add 2.6 ml.
7. Add serum to give final concentration required, e.g. add 11 ml for 10% serum.
8. Recap bottles and label to record additions, date, and user. Move each new stock to the opposite side of the hood after use, so that you will know that it has been added.
9. Return complete medium to 4 °C or use directly.

Notes

Step 6: If the final $NaHCO_3$ concentration is 23 mM, as with Earle's salts, the medium should be used with 5% CO_2 (either by gassing the flask with CO_2 or by using a CO_2 incubator). Hanks's salts are designed for use in a sealed flask with a gas phase of air.

11.6.3 Serm

Serum was originally prepared by tissue culture laboratories; personnel would collect blood, allow it to clot, and separate the serum before passing it through filters of gradually reducing porosity (Paul 1975). Preparing serum is one of the more difficult procedures in tissue culture, because of variations in the quality and consistency of the raw materials and because of the difficulties encountered in sterile filtration due to particulate material, colloids, and viscosity. The move to commercial manufacture has greatly improved serum preparation and testing (see Section 9.7.1). Various serum treatments are also available commercially – for example, γ-irradiation of serum can be used to reduce the risk of microbial contamination (see Section 9.7.3). A protocol is provided in the Supplementary Material online for the rare occasion when serum may need to be prepared in the tissue culture laboratory (see Supp. S11.3).

11.7 STERILE FILTRATION

11.7.1 Filter Selection

Heat-labile liquids can be sterilized by filtration through a membrane of the appropriate pore size; 0.2- or 0.22-μm filters are widely used, with 0.1-μm filters recommended as an effective size for removal of mycoplasma (see Section 11.2.6). Membranes are constructed from various materials including polyethersulfone (PES), polyvinylidene difluoride (PVDF), polycarbonate (PC), nylon, cellulose acetate, cellulose nitrate, and ceramics. The choice of material usually depends on

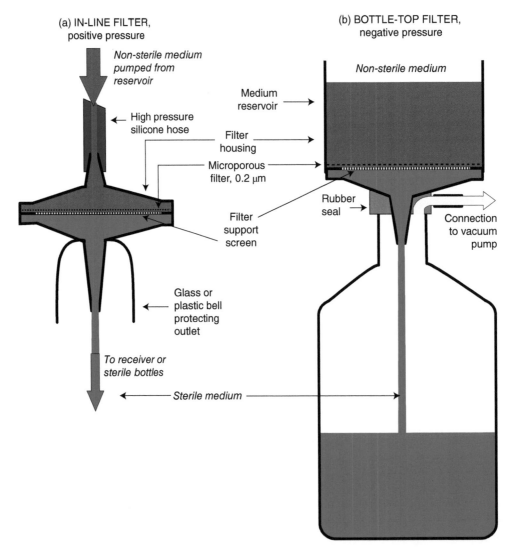

Fig. 11.7. Sterile filtration. (a) In-line filter. Non-sterile medium from peristaltic pump or pressure vessel. (b) Bottle-top filter or filter flask (designs are similar) for connection to vacuum pump. Medium is added to upper chamber and collected in lower. Lower chamber can be used for storage.

TABLE 11.6. Filter size and volume.

Filter size or designation[a]	Filtration area, cm^2 (in.^{2})	Process volume[b]
4 mm, Millex	0.10 (0.016)	1 ml
13 mm, Millex	0.65 (0.10)	10 ml
25 mm, Millex	3.9 (0.60)	100 ml
33 mm, Millex	4.52 (0.70)	100–200 ml
Sterivex	10 (1.6)	1–2 l
50 mm, Millex	19.6 (3.0)	4 l
Stericap PLUS	40 (6.2)	2–10 l
Steripak	100 (15.5) or 200 (31)	10–20 l
Millipak-20	100 (15.5)	10 l
Millipak-40	200 (31)	20 l
Millipak-60	300 (46.5)	30 l
Millipak-100	500 (77.5)	50 l
Millipak-200	1000 (155)	100 l

[a]Product names and information were collated from the Millipore catalog and are provided here as examples of disposable products. Similar products are available from other suppliers, e.g. Pall, Sartorius.
[b]Process volume will vary with the membrane, pore size, and the substance being filtered, e.g. colloidal solutions have a lower process volume compared to crystalloid solutions. Refer to the supplier for information on each product.

what is being filtered. Media and other water-based solutions require hydrophilic membranes with low levels of protein retention. Low-protein-binding filters are available from numerous suppliers and are commonly made from PES and PVDF (e.g. Merck Millipore, Pall, Sartorius). Polymer-based membranes are absolute filters with an array of holes of a uniform porosity; the number of holes per unit area increases as the size diminishes, to maintain a uniform flow rate. Most other filters are of the mesh variety and filter by entrapment. They generally have a faster flow rate and reduced clogging but will compress at high pressures.

Membrane filtration is performed using either positive pressure or negative pressure (see Figure 11.7). Positive pressure is applied manually (e.g. a syringe-tip filter) or using a peristaltic pump or pressure vessel that is in line with the filter. Negative pressure is applied using a vacuum pump that is positioned below a filter flask or bottle-top filter. The choice is based on the volume being processed and the equipment that is available (see Table 11.6; see also Figure 11.8). Positive-pressure filtration with a peristaltic pump was initially preferred because it reduced foaming and the pump could be switched on and off with instantaneous effect during collection. However, advances in the design of vacuum filter units have led to increased use of negative-pressure filtration in tissue culture laboratories (Coté 2001). It is often the simplest approach, particularly for small-scale operations, and will collect directly into storage vessels. However, it may cause an increase in pH because of release of CO_2.

Medium can become contaminated with toxic substances during filtration. Some filters are treated with wetting agents, which may leach out into the medium as it is being filtered (e.g. cellulose acetate, cellulose nitrate). Such filters should be washed by passing a balanced salt solution through them before use or by discarding the first aliquot of filtrate. Surfactant-free cellulose acetate is available, or an alternative material may be used that does not require a wetting agent (e.g. PES). Chemical resistance may also be important for some reagents; e.g. nylon (prewetted with water) is preferred for dimethyl sulfoxide (DMSO) (Coté 2001).

11.7.2 Disposable Filters

Most sterilizing filters used in tissue culture laboratories are disposable. Although disposable filters are more expensive than reusable filters, they are less time-consuming to use and give fewer failures. Disposable filter holder designs include simple disk filters, hollow-fiber units, and cartridges (see Figure 11.9). Syringe-tip filters are generally used for low-volume filtration (2–100 ml) and vary in size from 13- to 50-mm diameter (see Figure 11.9a). Protocol P11.8 describes how to use a syringe-tip filter for sterile filtration.

PROTOCOL P11.8. STERILE FILTRATION WITH SYRINGE-TIP FILTER

Outline

Fill a syringe with solution, attach filter, and expel solution through filter into sterile container. Note: the procedure here is provided as an example only. Various syringe-tip filters may be used.

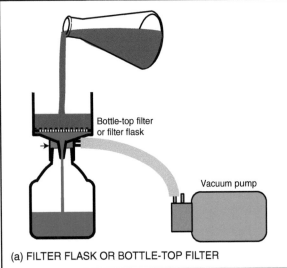

(a) FILTER FLASK OR BOTTLE-TOP FILTER

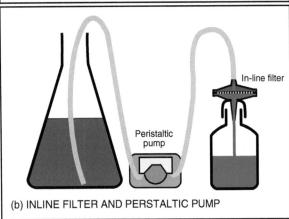

(b) INLINE FILTER AND PERSTALTIC PUMP

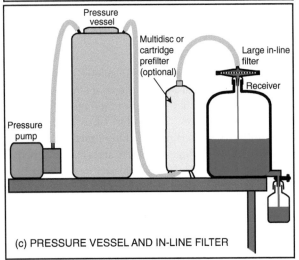

(c) PRESSURE VESSEL AND IN-LINE FILTER

Fig. 11.8. Options for sterile filtration. (a) Filter flask or bottle-top filter connected to vacuum pump; popular laboratory scale setup for 1- to 10-l medium. (b) In-line filter fed from large reservoir by peristaltic pump (see also Figure S11.5 in the Supplementary material); suitable for volumes up to 100 l, depending on size of filter. (c) Large-scale filtration with positive pressure pump, pressure vessel, prefilter (cartridge or multidisc, optional), and a final large-capacity sterilizing filter (see Figure 11.10, but without the prefilter). Suitable for 100–10 000 l depending on sizes of components. Semi-industrial to industrial scale. Not to scale.

Materials (sterile)

- ☐ Plastic syringe, 10- to 50-ml capacity
- ☐ Syringe-tip filter (e.g. Minisart® High Flow, Sartorius)
- ☐ Receiver vessel (e.g. a universal container)

Materials (non-sterile)

- ☐ BSC and associated consumables (see Protocol P12.1)
- ☐ Solution for sterilization (5–100 ml)

Procedure

1. Prepare a work space for aseptic technique using a BSC.
2. Uncap receiver vessel.
3. Unpack filter and attach to tip of syringe, holding the sterile filter within the bottom half of the packaging while attaching to syringe.
4. Fill syringe with solution to be sterilized.
5. Expel solution through filter into receiver vessel. Only moderate pressure is required.
6. Cap receiver vessel.
7. Discard syringe and filter.

Notes

Steps 4 and 5: The syringe may be refilled several times by returning the filter to the lower half of the sterile packaging, detaching it from the syringe, refilling the syringe, and reattaching the filter. If the back-pressure increases, change to a new filter.

Intermediate-sized filters (from 150 ml to 1 l) can be purchased as complete filter units for attaching to a vacuum line, with an upper chamber for the non-sterile solution and a lower chamber to receive the sterile liquid and use for storage (see Figure 11.9e). Intermediate-sized filters can also be purchased as bottle-top filters for use with a vacuum line and a regular medium bottle (see Figure 11.9f). Protocol P11.9 describes how to use a vacuum filter unit and can be adapted for use in training (see Supp. S30.5). A protocol for using a peristaltic pump with in-line filter is provided as Supplementary Material online (see Supp. S11.4).

PROTOCOL P11.9. STERILE FILTRATION WITH VACUUM FILTER UNIT

Outline

Attach vacuum pump to outlet, pour medium into top chamber of filter unit, switch on pump, and draw solution through to lower chamber; cap and

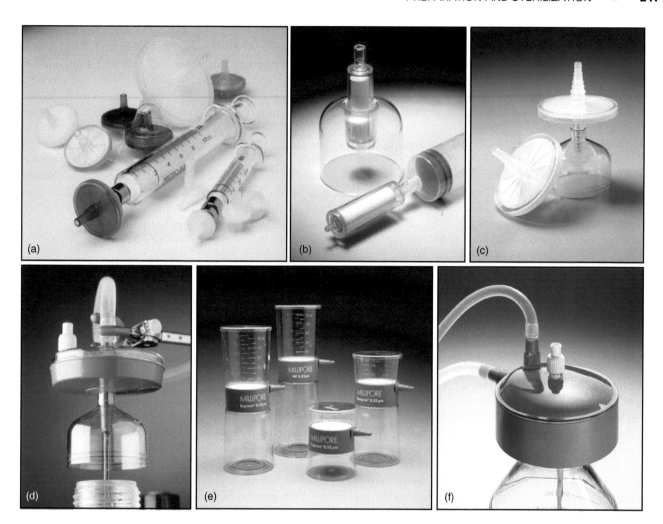

Fig. 11.9. Disposable sterilizing filters. Examples shown here (all from Millipore): (a) Millex® 25-mm disk syringe filter; (b) Sterivex high capacity (Luer fitting but can be attached via hose); (c) Millex 50-mm, in-line filter, with and without bell; (d) Steripak large in-line filter with bell (lowered in use to cover neck of bottle); (e) Stericup with storage vessels, also available as Stericap bottle-top filter; (f) bottle-top filter drawing from separate reservoir (see also Figure 11.7 with built-in reservoir). (a–d) Positive pressure; (e, f) negative pressure. *Source*: Courtesy of Millipore (UK) Ltd.

store. Note: the procedure here is provided as an example only. Various filter units or bottle-top filters may be used.

Materials (sterile)

- Vacuum filter unit (e.g. Stericup® filtration and storage unit, Merck Millipore)
- Cap for lower chamber (if chamber to be used for storage)

Materials (non-sterile)

- BSC and associated consumables (see Protocol P12.1)

- Medium for sterilization
- Vacuum pump or vacuum line
- Thick-walled connector tubing from pump or line, to fit filter flask connection tube
- Reagent Log (see Table 11.5); record of reagent preparation and sterilization

Procedure

1. Prepare a work space for aseptic technique using a BSC.
2. Connect side outlet of filter flask to vacuum pump (located outside BSC, e.g. on floor).
3. Remove cap from bottle and lid from top chamber of filter flask.

4. Pour non-sterile medium into top chamber.
5. Switch on pump.
6. Unpack cap for lower chamber, ready for use.
7. When liquid has all been drawn into lower chamber, switch off pump and detach filter housing and top chamber.
8. Collect samples for sterility testing (see Supp. S11.6).
9. Gas the head space in the lower chamber with 5% CO_2 (if required, see Note below).
10. Cap all containers and label with name of medium, date, and your initials.
11. Store lower chamber at 4 °C. Do not release for use until shown to be free of contamination.
12. Update Reagent Log with sterility test result.

Notes

Step 1: The BSC should be completely dedicated to media and supplement production (Geraghty et al. 2014). If this is not possible, the BSC should not be used for handling cultures for at least one hour beforehand and should be thoroughly cleaned and all non-essential equipment removed.

Step 9: Negative-pressure filtration systems may lead to loss of some dissolved CO_2 during filtration and an increased pH. It may be necessary to purge the head space of the bottle with 5% CO_2. Provided that the correct amount of $NaHCO_3$ is in the medium to suit the gas phase (see Section 9.2.3), the medium will re-equilibrate in the incubator, but this should be confirmed the first time the medium is used.

Filter flasks are available with a receiver capacity of 150–1000 ml. If a larger volume is to be filtered by negative pressure and dispensed into aliquots, it may be better to use a bottle-top filter, which can be set up to filter material directly into standard medium bottles (see Figure 11.9f, also available with upper reservoir). Alternatively, solutions may be filtered by positive pressure using a peristaltic pump, passing through a sterile in-line membrane or cartridge filter equipped with a bell to protect the receiver vessel from contamination (see Supp. S11.4; see also Figure 11.9b,c).

11.7.3 Reusable Filter Assemblies

Positive pressure is recommended for optimum performance of the filter and to avoid the removal of CO_2, which results from negative-pressure filtration. Filtration with positive pressure can also be scaled up by increasing the filter size and the size of the medium reservoir. The peristaltic pump is typically replaced by adding positive pressure to the medium reservoir in a pressure vessel (see Figure 11.10a). Large volumes of media (20–500 l) may be filtered using a large-scale in-line

filter assembly containing a disposable membrane or cartridge (see Figure 11.10c). Reusable filters may also be employed for large-scale filtration procedures (see Figure 11.11). Basic protocols for sterilizing reusable filter assemblies and for sterile filtration using a large in-line filter are provided as Supplementary material online (see Supp. S11.2, S11.5). The protocol in Supp. S11.5 is suitable for up to 50 l, but the volume can be increased to industrial scale by selecting the appropriate pressure vessel and filter, often a pleated cartridge or other multisurface filter contained in a cylindrical housing (see Figure 11.11b). Sterile filtration at this scale requires comprehensive quality control testing as part of an overarching quality assurance program (see Section 7.5).

11.7.4 Filter Testing

Sterile filtration depends on the integrity of the filter; any breaches will result in loss of sterility. Disposable filters from reputable suppliers will be tested before they are sold, but breaches may still occur during the process of filtration, particularly if you are filtering large volumes of liquid. Tests of filter integrity or performance include:

(1) **Bubble point testing.** When positive-pressure filtration is complete, raise the pump pressure until bubbles form in the effluent from the filter. This is the bubble point; the expected pressure level for your filter should be supplied by the manufacturer. A lower than expected bubble point can indicate a problem with the filter or sterile filtration process. In that case, any filtrate that has been collected should be regarded as non-sterile and refiltered. Single-use, disposable filters rarely fail the bubble point test, but reusable filters can fail, so they should be checked after every filtration run.

(2) **Downstream secondary filtration.** Place a demountable 0.45-μm sterile filter in the effluent line from the main sterilizing filter. Any contamination that passes because of failure in the first filter will be trapped in the second. At the end of the run, remove the second filter and place the filter on nutrient agar. If colonies grow, discard or refilter the medium. This method has the advantage that it monitors the entire filtrate, and not just a small fraction of it, although it does not detect contamination during other parts of the process such as bottling or capping.

11.8 MEDIUM TESTING

A medium that is prepared in the laboratory needs to be tested before use. Quality control testing for media and reagents can be broadly divided into:

(1) **Sterilization testing.** This focuses on the process of sterilization to ensure that the sterility assurance level has been achieved (see Section 11.1). This type of testing includes

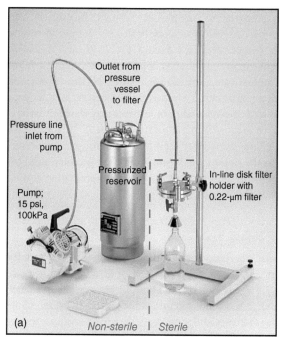
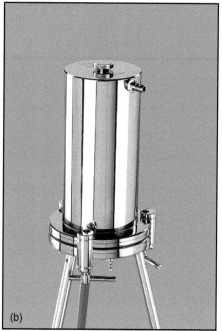
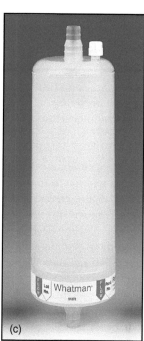

Fig. 11.10. Large-scale in-line filter assembly. (a) Filter assembly supplied from a pressurized reservoir (center) and connected via a sterile reusable 142-mm filter holder to a receiver flask (right). Substituting a larger receiver with a tap outlet would allow collection of the entire contents of the pressure vessel for later dispensing into sterile containers (see Figure 11.8c). Only the filter assembly and the receiver flask need be sterilized. (b) Filter holder for cartridge filter for large-scale sterile filtration in place of disc filter in (a) or for use as a prefilter (see Figure 11.8c). (c) Disposable cartridge filter, for use as a prefilter or sterilizing filter, depending on porosity. *Source*: (a, b) Courtesy of Sartorius Stedim; (c) Courtesy of Whatman.

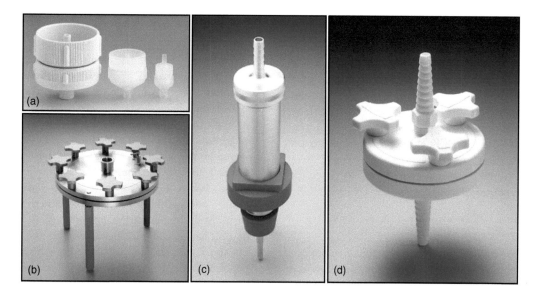

Fig. 11.11. Reusable filters. (a) Swinnex polypropylene, in-line, Luer fitting; (b) stainless-steel housing, 293-mm high-capacity disk-type filter; (c) 47-mm filter with reservoir; (d) 47-mm in-line filter with hose connections. *Source*: Courtesy of Millipore (UK) Ltd.

inclusion of sterility indicators in autoclave runs and filter testing to ensure that there are no breaches during sterile filtration (see Sections 11.2.7, 11.7.4)

(2) **Sterility testing**. This focuses on the detection of microorganisms in the final product after sterilization (see Section 11.8.1).

(3) **Culture testing**. This focuses on the performance of medium or serum and particularly its growth-promoting activity (see Section 11.8.2).

If the medium is purchased ready-made as a 1× working strength solution, it should be possible to rely on the quality control testing carried out by the supplier, other than any special requirements that you have of the medium. Likewise, if a 10× concentrate is used, quality control testing of the concentrate (and hopefully other supplements) will have been done, and the only variable will be the water used for dilution. Provided that the conductivity and TOC levels fall within specifications (see Section 11.5.3) and no major changes have been made in the supply of water, most research laboratories will accept this compliance as adequate quality control.

Sterility testing becomes essential if medium is prepared from powder and is thus sterilized in the laboratory, although, given that all the constituents have dissolved, you may be prepared to accept the quality control of the supplier regarding the medium's growth-promoting activity. Media made up from basic constituents will require complete quality control, involving both sterility testing and culture testing. Even large-scale manufacturers may struggle to exclude microorganisms from their preparation processes; for example, mycoplasma species can be difficult to eliminate due to their small size, flexible shape, and prolonged survival in reagent powders (Windsor et al. 2010).

All medium testing should be initiated well in advance of the exhaustion of the current stock of medium, so that proper comparisons may be made and there is time to have fresh medium prepared if samples fail quality control testing. All quality control tests should be recorded in a logbook or computer database, along with the other details relating to the preparation of each batch (see Table 11.5). The person supervising preparation and sterilization should review these records and determine failure rates and trends, to check for the need to alter procedures.

11.8.1 Sterility Testing

The term "sterility testing" usually refers to the microbiological culture of samples under conditions that will allow the growth of various bacteria and fungi. Additional testing may be needed for microorganisms that do not grow under standard conditions, such as mycoplasma (Folmsbee et al. 2014). Test samples may be collected at the beginning, middle, and end of a sterile filtration process or at the end of complete medium preparation. Samples are usually inoculated into several nutrient broths for microbiological culture at several different temperatures (see Figure 11.12) (ECACC 2016). A protocol for sterility testing is provided as Supplementary material online (see Supp. S11.6). Although this is provided as a starting point, sterility testing will be determined by your application. Some research laboratories may decide that it is sufficient to incubate samples of medium at 37 °C for up to a week as a "quick and dirty" test for microbiological contamination. At the opposite extreme, biological products used for therapeutic applications must undergo comprehensive testing (see Section 7.2.3).

11.8.2 Culture Testing

Media that are produced commercially will be tested for their performance in culture as part of quality assurance (see Section 7.5.2). If this quality control testing has not been performed, you should change your supplier! However, tissue culture laboratories still carry out their own culture testing under certain circumstances. Such testing is particularly useful if (i) the laboratory intends to make a bulk purchase (e.g. serum; see Section 9.7.1); (ii) the medium is for a special purpose that the commercial supplier is not able to test (e.g. culture of a cell type that is difficult to grow); (iii) medium has been made up in the laboratory from basic constituents; (iv) any additions or alterations are made to the medium; and (v) the medium is made up from powder and there is a risk of losing constituents or adding contaminants during filtration (see Section 11.7.1). Culture testing typically involves three different methodologies. All tests should be performed on the new batch of medium with your regular medium as a control.

Cloning efficiency. A clonogenic assay (see Section 19.4.1) can be used to detect minor deficiencies and low concentrations of toxins that are not apparent at higher cell densities. Cell lines should be selected for these assays that can support colony formation at low density, as determined through baseline assessment (see Protocol P19.4). Ideally, two or three suitable cell lines should be selected for clonogenic assays and frozen vials should be reserved specifically for this purpose. When testing medium, serum should be used at limiting concentration, which could otherwise mask deficiencies in the medium. For example, A549 cells grown in 10% FBS have a cloning efficiency of around 50%; growing A549 in 5% FBS should give a cloning efficiency of around 10%. If dropping the serum concentration to 5% has little or no effect, repeat with 2% or even 1% serum and adopt this as the correct limiting concentration for future assays.

The experimental design used to test medium by cloning efficiency is shown in Figure 11.13; a protocol with

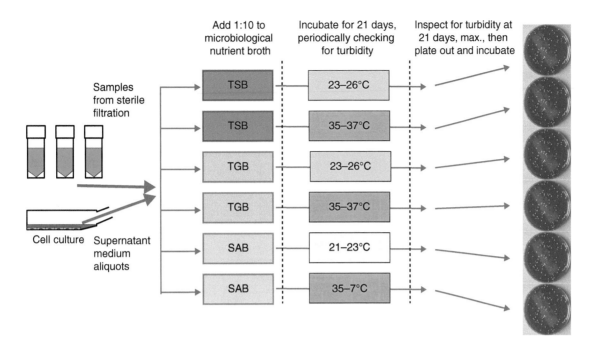

Fig. 11.12. Microbiological culture. Samples come from sterile filtration or from supernatant medium. Microbiological culture is performed for sterility testing using trypticase soya broth (TSB) for aerobic and facultative aerobic organisms; thioglycollate broth (TGB) for anerobic and microaerophilic organisms; and Sabouraud's broth (SAB) for fungi. Alternative broths include Todd-Hewitt broth (instead of TSB), brain–heart infusion broth (instead of TGB), and yeast malt broth (instead of SAB). *Source*: Based on Stacey, G. N. and Auerbach, J. M. (2007). "Quality control procedures for stem cell lines." In Culture of human stem cells, edited by R. I. Freshney, G. N. Stacey and J. M. Auerbach. Hoboken, NJ. 2007 John Wiley & Sons.

step-by-step instructions is provided in the Supplementary material online (see Supp. S11.7). Observational skills are important for this procedure. A clonal growth assay will not always detect insufficiencies in the amounts of particular constituents unless colony size is measured. For example, if the concentration of one or more amino acids is low, it may not affect the cloning efficiency but could influence the mean colony size.

Growth analysis. A growth curve (see Figures 14.1, 19.7) can be used to calculate a number of parameters, including (i) the lag phase before cell proliferation commences after subculture, indicating whether the cells are having to adapt to different conditions; (ii) the population doubling time (PDT) in the middle of the exponential growth phase, indicating the growth-promoting capacity of the medium; and (iii) the saturation density, indicating whether there are limiting concentrations of certain nutrients. In cell lines whose growth is not sensitive to density (see Figure 3.11), the saturation density indicates the total yield possible and usually reflects the total amino acid or glucose concentration. Remember that a medium that gives half the yield costs twice as much per cell produced.

Growth curves can be generated by counting cells at intervals, by image analysis of growing cultures, or using a microwell plate assay that uses a surrogate variable for cell number (see Protocol P19.3; see also Figure 18.5, Tables 19.1, 29.1). Cell lines used for medium testing should be sensitive to changes in conditions; as with cloning efficiency, frozen vials from two or three different cell lines should be reserved for future testing. Multiwell plates are often used to generate growth data (see Figure 11.14). A protocol for testing medium by growth is provided in the Supplementary material online (see Supp. S11.8).

Other characterization. The behavior of cells in culture can be highly dependent on their growth conditions. For example, A549 cells can be maintained at increased density when grown in Dulbecco modified Eagle medium (DMEM), but display more differentiated behavior in Ham's F12, particularly when grown in long-term culture (Cooper et al. 2016). If you are testing special functions, a standard test from the experimental system you are using should be performed on the new medium alongside the old one (e.g. differentiation in the presence of an inducer, viral propagation, the formation of a specific product, or the expression of a specific antigen).

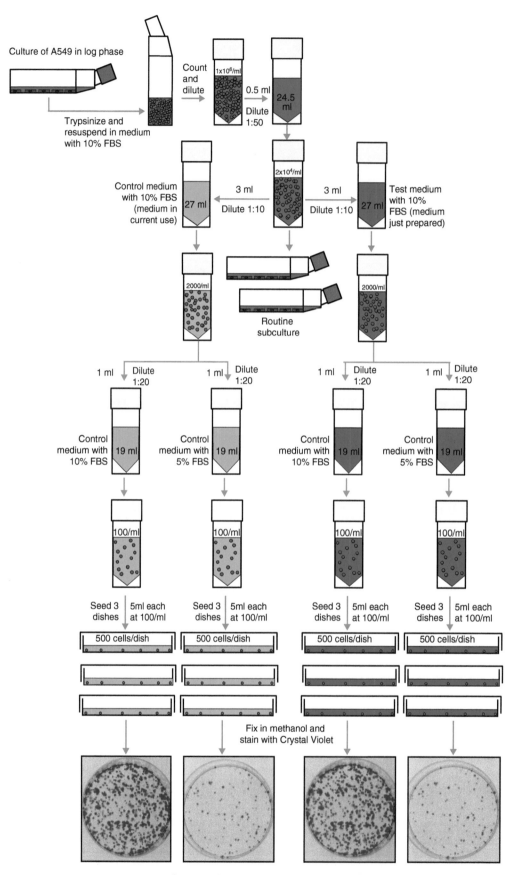

Fig. 11.13. Testing medium by cloning efficiency. Exponentially growing cells are trypsinized, counted, diluted in control and test media with 5% and 10% serum, and then plated out at 100 cells/ml. The colonies formed are stained and counted. *Source*: R. Ian Freshney.

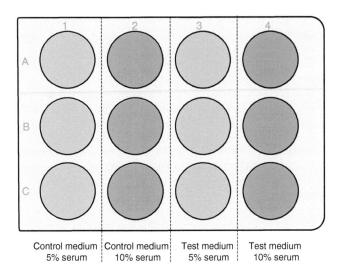

Fig. 11.14. Growth analysis in multiwell plates. Replicate cultures are set up in 12-well plates and harvested daily for cell counts. The first column has control medium with 5% serum; the second column has control medium with 10% serum; the third column has test medium with 5% serum; and the last column has test medium with 10% serum. One plate is harvested per day.

Suppliers

Supplier	URL
3 M	www.3m.com
Camlab	www.camlab.co.uk
Cardinal Health	http://www.cardinalhealth.com/en.html
Corning	http://www.corning.com/worldwide/en/products/life-sciences/products/cell-culture.html
DWK Life Sciences (Duran)	www.duran-group.com/en/home.html
Elga-Veolia	www.elgalabwater.com
GE Healthcare Life Sciences	http://www.gelifesciences.com/en/it
Guest Medical	guest-medical.co.uk/home/
Medisave	www.medisave.co.uk
Merck Millipore	www.merckmillipore.com
Mesa Labs	mesalabs.com/
MP BioMedicals	www.mpbio.com
Oxoid	http://www.oxoid.com/uk/blue/index.asp
Pall Corporation	www.pall.com
Sartorius	http://www.sartorius.com/en
Sigma-Aldrich (Merck)	http://www.sigmaaldrich.com/life-science/cell-culture.html
STERIS	www.steris-healthcare.com/products/ipt
VWR (Jencons)	https://uk.vwr.com/store/

Supp. S11.1 Preparation, Sterilization, and Use of Glass Pipettes.

Supp. S11.2 Sterilization of Reusable Filter Assemblies.

Supp. S11.3 Collection and Sterilization of Serum.

Supp. S11.4 Sterile Filtration Using Peristaltic Pump.

Supp. S11.5 Sterile Filtration with Large In-line Filter.

Supp. S11.6 Sterility Testing Using Microbiological Culture.

Supp. S11.7 Clonogenic Assay for Testing Medium.

Supp. S11.8 Growth Curve Analysis for Testing Medium.

REFERENCES

ASTM International (2018). ASTM D1193-06. Standard specification for reagent water. West Conshohocken, PA: ASTM International.

Belgaid, A., Benaji, B., Aadil, N. et al. (2014). Sterilisation of aseptic drug by sterile filtration: microbiology validation by microbiology challenge test. *J. Chem. Pharm. Res.* 6 (12): 760–770.

Breach, M.R. (1968). *Sterilization: Methods and Control*. London: Butterworths.

Burnouf, T. and Radosevich, M. (2003). Nanofiltration of plasma-derived biopharmaceutical products. *Haemophilia* 9 (1): 24–37.

CDC (2008). Guideline for disinfection and sterilization in healthcare facilities. https://www.cdc.gov/infectioncontrol/guidelines/disinfection/index.html (accessed 8 October 2018).

CDC and NIH (2009). Biosafety in microbiological and biomedical laboratories. https://www.cdc.gov/labs/bmbl.html (accessed 11 March 2019).

Cipriano, D., Burnham, M., and Hughes, J.V. (2012). Effectiveness of various processing steps for viral clearance of therapeutic proteins: database analyses of commonly used steps. *Methods Mol. Biol.* 899: 277–292. https://doi.org/10.1007/978-1-61779-921-1_18.

Clinical and Laboratory Standards Institute (2006). *CLSI GP40-A4-AMD. Preparation and Testing of Reagent Water in the Clinical Laboratory*, 4e. Annapolis Junction, MD: Clinical and Laboratory Standards Institute.

Cooper, J.R., Abdullatif, M.B., Burnett, E.C. et al. (2016). Long term culture of the A549 cancer cell line promotes multilamellar body formation and differentiation towards an alveolar type II Pneumocyte phenotype. *PLoS One* 11 (10): e0164438. https://doi.org/10.1371/journal.pone.0164438.

Corning (2009). Suggestions for cleaning glassware. http://www.corning.com/catalog/cls/documents/application-notes/cls_an_112_cleaningglassware.pdf (accessed 6 May 2016).

Coté, R.J. (2001). Sterilization and filtration. *Curr. Protoc. Cell Biol.* Chapter 1:Unit 1.4. doi: https://doi.org/10.1002/0471143030.cb0104s01.

Dai, Z., Ronholm, J., Tian, Y. et al. (2016). Sterilization techniques for biodegradable scaffolds in tissue engineering applications. *J. Tissue Eng.* 7 https://doi.org/10.1177/2041731416648810.

ECACC (2016). *Fundamental Techniques in Cell Culture: Laboratory Handbook*, 3e. Salisbury, UK: Public Health England, Sigma-Aldrich.

Folmsbee, M., Howard, G., and McAlister, M. (2010). Nutritional effects of culture media on mycoplasma cell size and removal by filtration. *Biologicals* 38 (2): 214–217. https://doi.org/10.1016/j .biologicals.2009.11.001.

Folmsbee, M., Lentine, K.R., Wright, C. et al. (2014). The development of a microbial challenge test with *Acholeplasma laidlawii* to rate mycoplasma-retentive filters by filter manufacturers. *PDA J. Pharm. Sci. Technol.* 68 (3): 281–296. https://doi.org/10.5731/ pdajpst.2014.00976.

Geraghty, R.J., Capes-Davis, A., Davis, J.M. et al. (2014). Guidelines for the use of cell lines in biomedical research. *Br. J. Cancer* 111 (6): 1021–1046. https://doi.org/10.1038/bjc.2014.166.

Harrell, C.R., Djonov, V., Fellabaum, C. et al. (2018). Risks of using sterilization by gamma radiation: the other side of the coin. *Int. J. Med. Sci.* 15 (3): 274–279. https://doi.org/10.7150/ijms.22644.

Hoburg, A., Keshlaf, S., Schmidt, T. et al. (2015). High-dose electron beam sterilization of soft-tissue grafts maintains significantly improved biomechanical properties compared to standard gamma treatment. *Cell Tissue Bank* 16 (2): 219–226. https://doi.org/10 .1007/s10561-014-9461-x.

International Organization for Standardization (1987). *ISO 3696: 1987. Water for Analytical Laboratory Use: Specification and Test Methods*. Geneva: International Organization for Standardization.

Ludwig, T.E., Bergendahl, V., Levenstein, M.E. et al. (2006). Feeder-independent culture of human embryonic stem cells. *Nat. Methods* 3 (8): 637–646. https://doi.org/10.1038/nmeth902.

Mann, K., Royce, J., Carbrello, C. et al. (2015). Protection of bioreactor culture from virus contamination by use of a virus barrier filter. *BMC Proc.* 9: P22. https://doi.org/10.1186/1753-6561- 9-S9-P22.

McDonnell, G., Dehen, C., Perrin, A. et al. (2013). Cleaning, disinfection and sterilization of surface prion contamination. *J. Hosp. Infect.* 85 (4): 268–273. https://doi.org/10.1016/j.jhin.2013.08 .003.

Merck Millipore (2018). Water in the laboratory. https://www .emdmillipore.com/us/en/water-purification/learning-centers/ tutorial/opab.qb.ixuaaae_mkorhe3j.nav (accessed 15 October 2018).

NIH (2013). Laboratory water: its importance and application. https://www.orf.od.nih.gov/policiesandguidelines/documents /dtr%20white%20papers/laboratory%20water-its%20importance %20and%20application-march-2013_508.pdf (accessed 15 May 2018).

Nims, R.W., Gauvin, G., and Plavsic, M. (2011). Gamma irradiation of animal sera for inactivation of viruses and mollicutes – a review. *Biologicals* 39 (6): 370–377. https://doi.org/10.1016/j.biologicals .2011.05.003.

Novak, R., Didier, M., Calamari, E. et al. (2018). Scalable fabrication of stretchable, dual channel, microfluidic organ chips. *J. Vis. Exp.* 140 https://doi.org/10.3791/58151.

OGTR (2011). Guidelines for the transport, storage and disposal of GMOs. www.ogtr.gov.au/internet/ogtr/publishing.nsf/ content/tsd-guidelines-toc (accessed 27 July 2012).

Paul, J. (1975). *Cell and Tissue Culture*, 5e. Edinburgh: Churchill Livingstone.

Shintani, H. (2017). Ethylene oxide gas sterilization of medical devices. *Biocontrol Sci.* 22 (1): 1–16. https://doi.org/10.4265/ bio.22.1.

Shintani, H., Sakudo, A., Burke, P. et al. (2010). Gas plasma sterilization of microorganisms and mechanisms of action. *Exp. Ther. Med.* 1 (5): 731–738. https://doi.org/10.3892/etm.2010.136.

Stacey, G.N. and Auerbach, J.M. (2007). Quality control procedures for stem cell lines. In: *Culture of Human Stem Cells* (eds. R.I. Freshney, G.N. Stacey and J.M. Auerbach). Hoboken, NJ: Wiley.

Thermo Scientific (2008). Thermo Scientific rotor care guide. www .thermofisher.com.au/uploads/file/scientific/applications/ equipment-furniture/thermo-scientific-rotor-care-guide.pdf (accessed 12 October 2018).

Waymouth, C. (1978). Studies on chemically defined media and the nutritional requirements of cultures of epithelial cells. In: *Nutritional Requirements of Cultured Cells* (ed. H. Katsuta), 39–61. Tokyo and Baltimore: Japan Scientific Societies Press and University Park Press.

WHO (2004). Laboratory biosafety manual. http://www.who.int/ csr/resources/publications/biosafety/who_cds_csr_lyo_2004_ 11/en (accessed 9 May 2018).

Windsor, H.M., Windsor, G.D., and Noordergraaf, J.H. (2010). The growth and long term survival of *Acholeplasma laidlawii* in media products used in biopharmaceutical manufacturing. *Biologicals* 38 (2): 204–210. https://doi.org/10.1016/j.biologicals .2009.11.009.

Wood, J., Mahajan, E., and Shiratori, M. (2013). Strategy for selecting disposable bags for cell culture media applications based on a root-cause investigation. *Biotechnol. Progr.* 29 (6): 1535–1549. https://doi.org/10.1002/btpr.1802.

Yoo, J.H. (2018). Review of disinfection and sterilization – back to the basics. *Infect. Chemother.* 50 (2): 101–109. https://doi.org/10 .3947/ic.2018.50.2.101.

PART IV

Handling Cultures

After reading the following chapters in this part of the book, you will be able to:

(12) *Aseptic Technique:*

(a) Define "aseptic technique" and explain its purposes when used for tissue culture.

(b) Select suitable equipment to provide asepsis and containment when handling cultures, including suitable personal protective equipment (PPE).

(c) Explain how to operate and maintain a biological safety cabinet (BSC) and how to minimize contamination in a CO_2 incubator.

(d) Discuss the general rules that should be observed when setting up any work area for tissue culture.

(e) Handle flasks, dishes, and plates using aseptic technique in a BSC and on the open bench.

(13) *Primary Culture:*

(a) Define "primary culture" and discuss options for its initiation and the various stages that must be considered during planning.

(b) Obtain tissue for primary culture by dissection (e.g. of mouse embryos) or by obtaining biopsy samples, with due reference to safety and ethical requirements.

(c) Perform explant culture using tissue fragments.

(d) Perform disaggregation of tissue fragments using trypsin (warm or cold), collagenase, or mechanical methods.

(e) Collect the necessary records regarding the culture's origins and handling.

(14) *Subculture and Cell Lines:*

(a) Discuss how to name a new cell line and explain why additional identifiers, such as Research Resource Identifiers (RRIDs), should also be used.

(b) Define "passage number" and "population doubling level" (PDL), with an explanation of how each variable is determined.

(c) Perform routine daily observation of a cell line to determine its phase of growth and detect problems.

(d) Feed an adherent culture and fill out a data record for feeding.

(e) Passage an adherent or a suspension culture and fill out a data record for subculture.

(15) *Cryopreservation and Banking:*

(a) Define "cryoprotectant" and "freezing medium" and discuss options when choosing a suitable freezing medium for cryopreservation.

(b) State the cooling rate that typically gives the best results for cultured cells and provide examples of controlled cooling devices that can be used to achieve it.

(c) Freeze cells in cryovials and record the necessary information for long-term storage in a cryofreezer.

(d) Thaw a frozen cryovial for ongoing culture and assess its viability.

(e) Explain the principles of cell banking and the purpose of a cell repository.

Freshney's Culture of Animal Cells: A Manual of Basic Technique and Specialized Applications, Eighth Edition. Amanda Capes-Davis and R. Ian Freshney.
© 2021 John Wiley & Sons Ltd. Published 2021 by John Wiley & Sons Ltd.
Companion website: www.wiley.com/go/freshney/cellculture8

CHAPTER 12

Aseptic Technique

Aseptic technique refers to a set of practices that are used to maintain asepsis, or the absence of microorganisms. Originally developed by Louis Pasteur to handle microbial cultures, aseptic technique was adapted by Robert Koch to control infection following surgical procedures (Schlich 2012). Alexis Carrel and other early tissue culture pioneers were practitioners of aseptic technique and invented flasks and instruments that allowed aseptic handling of cultures (Carrel 1923). However, many of the early techniques were considered overelaborate and slowed down handling of cultures. A better understanding of tissue culture has enabled us to identify the key elements of an aseptic environment and to develop a simple set of handling practices. The resulting procedures act as a foundation upon which all other tissue culture techniques are based.

12.1 OBJECTIVES OF ASEPTIC TECHNIQUE

12.1.1 Managing Contamination Risk

Contamination is a major problem in tissue culture and a constant threat to good cell culture practice (GCCP) (see Section 7.2.1). Bacteria, yeast, fungal spores, and viruses may be introduced via tissue culture personnel, the atmosphere, work surfaces, reagents, and other sources (see Section 16.1). Cells from other cultures may also be introduced, resulting in cross-contamination (see Section 17.2.2). Contamination can be minor and confined to one or two cultures, can spread among several cultures and compromise a whole experiment, or can be widespread and wipe out your (or even the whole laboratory's) entire stock of cells.

Aseptic technique aims to exclude all forms of contamination by establishing a standardized set of procedures when working in a tissue culture laboratory and by ensuring that everyone using the laboratory adheres to them (McGarrity and Coriell 1971). These procedures reduce the risk of contamination but cannot eliminate it entirely. Contamination must be further managed by (i) maintaining cultures without antibiotics (see Section 9.6.2); (ii) checking cultures carefully by eye and on a microscope, preferably by phase contrast, every time that they are handled; (iii) testing for contamination that is difficult to detect by eye, such as mycoplasma or cross-contamination (see Sections 16.4.1, 17.3.2, 17.3.3); (iv) checking reagents for sterility before use (including testing by laboratory personnel or by the supplier); and (v) separating bottles of media or other reagents by usage, ensuring that they are not shared with other people and that each bottle is dedicated to a single cell line.

12.1.2 Maintaining Sterility

Correct aseptic technique should provide a barrier between microorganisms in the environment and the pure, uncontaminated culture within its flask, dish, or other vessel. This sterility barrier cannot be absolute without working under conditions that would severely hamper most routine manipulations. An absolute barrier is required for some applications, e.g. biological production using Good Manufacturing Practice (GMP) (see Section 7.2.3). However, an absolute barrier is not required for most tissue culture laboratories, provided regular checks are performed for contamination. In this environment, aseptic technique consists of a number of

Freshney's Culture of Animal Cells: A Manual of Basic Technique and Specialized Applications, Eighth Edition. Amanda Capes-Davis and R. Ian Freshney.
© 2021 John Wiley & Sons Ltd. Published 2021 by John Wiley & Sons Ltd.
Companion website: www.wiley.com/go/freshney/cellculture8

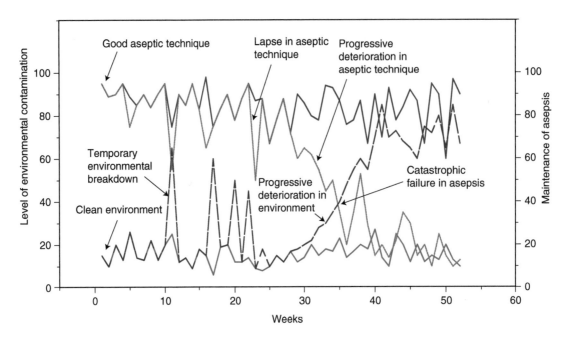

Fig. 12.1. Probability of contamination. The solid line in the top graph represents variability in technique against a scale of 100 (right-hand axis), which represents perfect aseptic technique. The solid line in the bottom graph represents fluctuations in environmental contamination, with zero being perfect asepsis (left-hand axis). The top line represents lapses in technique (e.g. forgetting to swab the work surface, handling a pipette too far down the body of the pipette, touching non-sterile surfaces with the tip of a pipette). The bottom line represents crises in environmental contamination (e.g. a high spore count, a contaminated incubator, contaminated reagents). As long as the lapses are minimal in degree and duration, the two lines do not overlap. When particularly bad lapses in technique (dotted line) coincide with severe environmental crises (dashed line), the lines overlap and the probability of infection increases (e.g. 10–11 weeks). If the breakdown in technique is progressive (dotted line, e.g. 28–36 weeks) and the deterioration in the environment is also progressive (dashed line, e.g. 25–42 weeks), the lines will overlap for prolonged periods, resulting in frequent, multispecific, and multifactorial contamination.

procedures that are designed to reduce the probability of infection, adopted largely on the basis of common sense and experience. There may not be an obvious correlation between omission of a particular step and subsequent contamination. The operator may abandon several precautions before the probability rises sufficiently that a contamination is likely to occur (see Figure 12.1). By then, the cause is often multifactorial, and, consequently, no simple single solution is obvious. If all precautions are maintained consistently, breakdowns in the sterility barrier will be rarer and more easily detected.

The sterility barrier must extend to the laboratory environment, the worker, and all materials that come into direct contact with the culture. Laboratory conditions have improved in some respects, thanks to the widespread use of biological safety cabinets (BSCs) and laminar flow hoods to provide sterile handling areas (see Section 4.2.1). However, the modern laboratory is often crowded and workers are likely to share tissue culture facilities and equipment. Each tissue culture laboratory must develop a set of Standard Operating Procedures (SOPs) (see Section 7.5.1), ensuring that all materials coming into direct contact with a culture are sterile and that all manipulations maintain the appropriate sterility barrier between the culture and its non-sterile surroundings. All laboratory personnel must adhere to these procedures and appropriate equipment checks and quality control measures must be in place (see Sections 5.1.1, 11.8).

12.2 ELEMENTS OF ASEPTIC ENVIRONMENT

Conditions for achieving a clean area for cell culture have changed over the years (see Figure 12.2), mainly due to the introduction of antibiotics, laminar flow cabinets, and filtered air conditioning. Clean room air, combined with laminar flow, has made the creation of an aseptic environment much simpler to attain and more reliable. As a result, it is no longer necessary or advisable to use antibiotics, which can mask lapses in aseptic technique (see Section 9.6.2).

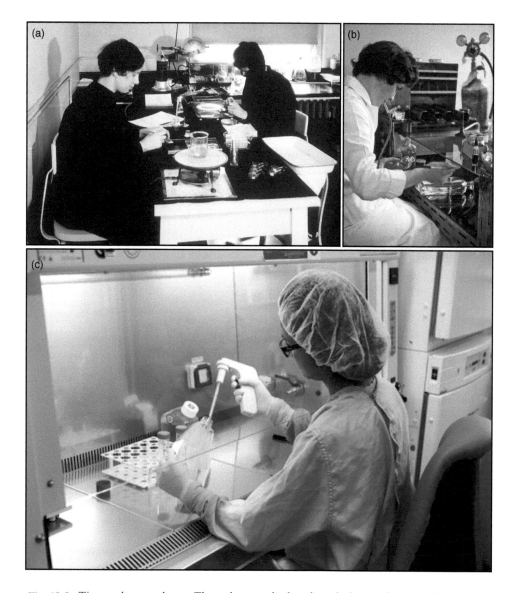

Fig. 12.2. Tissue culture work area. These photographs show how the layout of such areas has changed over the years. (a) Working on the open bench in 1931 at the Central Cancer Research Laboratory, Graduate School of Medicine, University of Pennsylvania. (b) Using a glass-topped table in 1961 in John Paul's tissue culture room at the Biochemistry Department, University of Glasgow. (c) Using a BSC in 2013 at CellBank Australia, CMRI. *Source*: (a) Courtesy of the National Cancer Institute, image AV-3100-4313, public domain; (b) R. Ian Freshney; (c) Courtesy of CellBank Australia.

12.2.1 Quiet Area

Tissue culture was originally performed on the open bench in a separate room or a quiet corner of the laboratory with little or no traffic and no other activity, using a Bunsen burner to provide a sterile field (see Figure 12.2a) (Sanders 2012). Although it is still possible to carry out sterile work on the open bench (see Section 12.4.2), it does not provide a stringent barrier for sterility. Most tissue culture laboratories perform aseptic work using a laminar flow hood (for asepsis) or a BSC (for asepsis and containment) (see Figures 12.2c, 12.3). Some applications may require a more specialized cleanroom

environment, e.g. for manufacture of biological products (see Section 7.5.5). However, these rooms require specific expertise and are costly to operate (Kruse et al. 1991).

12.2.2 Laminar Airflow

Equipment selection. Laminar airflow results in a constant, stable flow of filtered air passing over the work surface, which is used to protect materials from dust and contamination. There are two main types of airflow: (i) horizontal, where the air blows from the side facing you, parallel

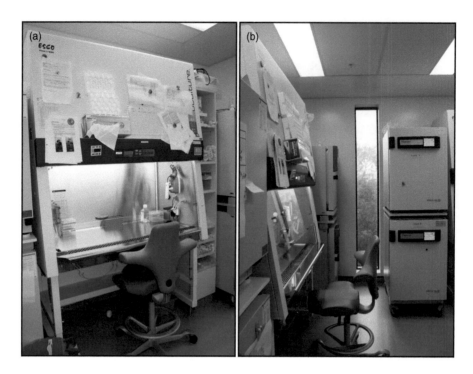

Fig. 12.3. BSC and surroundings. (a) Front view of an open BSC; (b) side view of the same BSC. Note the ergonomic design of the BSC and chair, the narrow cupboard next to the BSC for storage, and the mobile castors on which the incubators are mounted, allowing equipment to be moved for maintenance or cleaning. *Source*: Amanda Capes-Davis, courtesy of Leszek Lisowski and Vector and Genome Engineering Facility, CMRI.

to the work surface, and is not recirculated; and (ii) vertical, where the air blows down from the top of the hood onto the work surface and is drawn through the work surface and either recirculated or vented (see Figure 4.6). This is an important distinction when selecting equipment for tissue culture-related tasks. Always consider what the cabinet is designed to protect against and how that relates to your samples and activities.

Cabinets that are widely used in tissue culture laboratories include:

(1) *Horizontal laminar flow hoods* (also known as "clean workstations"). These provide stable airflow and good sterile protection but will blow aerosols into your face and are unsuitable for handling biohazardous or toxic material. Laminar flow hoods are typically used to prepare sterile reagents.

(2) *Biological safety cabinets (BSCs).* These use vertical airflow, recirculation of air, and filtration to direct air away from the user, minimize overspill from the work area of the cabinet, and remove aerosols from exhaust air. BSCs are typically used to handle cultures that require sterile protection but carry biohazard risk (see Section 6.3.4). Various classes and types of BSC have been developed (see Table 5.2). The Class II A2 BSC is commonly used in tissue culture laboratories, but the best cabinet for your application will depend on risk assessment (see Section 6.1.1).

(3) *Cytotoxic drug safety cabinets.* These are essentially BSCs with additional safety features to (i) provide a clean, controlled environment for aseptic handling of cytotoxic drugs; (ii) protect the user from toxic chemicals; and (iii) protect service engineers and other personnel who would otherwise be exposed to toxic chemicals on filters and internal surfaces.

(4) *Fume hoods.* These provide protection from volatile toxic chemicals but do not provide laminar airflow and cannot be used for asepsis or biological containment.

Purchase and positioning of the BSC has been discussed previously (see Sections 4.2.1, 5.1.1). Because you are likely to spend many hours sitting at the BSC, it is important to set up the space so that it is comfortable and place frequently used consumables near the working area (see Figure 12.3). The area should be kept clean and free of dust and should not contain equipment other than that connected with tissue culture. Activity in the area should be restricted to tissue culture work, and animals and microbiological cultures should be excluded. Non-sterile activities, such as sample processing, staining, or extractions, should be carried out elsewhere.

Equipment operation. The level of protection from any BSC depends on its mechanical performance and on the user's good laboratory practices (Kruse et al. 1991). Cabinets must undergo regular maintenance by competent service personnel, who will check high efficiency particulate air (HEPA) filters and perform certification testing (see Section 5.1.1). Always pay attention to pressure readings while the cabinet is operational. BSCs depend for their efficiency on a minimum pressure drop across the filter. When resistance builds up in the filter, the pressure drop will increase and the flow rate of air in the cabinet will decrease. Below 0.4 m/s (80 ft/min), the stability of the laminar airflow is compromised, and sterility can no longer be maintained. Most modern BSCs monitor pressure levels during operation. Report any alarms or error messages for repair, and do not use a faulty cabinet until the service engineer is confident that it is fully operational.

BSCs are designed for continuous operation, which may also help to control dust and other airborne particulates in the laboratory (CDC and NIH 2009). They are often a component of the ventilation system and may need to run continuously for this reason (see Section 4.2.1). However, continuous operation may not be feasible in some laboratories, e.g. due to increased power consumption. Always run the BSC for at least four minutes after turning it on to "purge" the cabinet, and do the same between cell lines and after work is complete before turning the cabinet off (CDC and NIH 2009).

Regular weekly checks should be made below the work surface. Typically, BSCs have easily removable (often sectional) work surfaces, which are designed to facilitate cleaning. Spillages should be mopped up when they occur, but occasionally they go unnoticed, so a regular check is imperative. Any spillage should be mopped up, the tray washed, and the area swabbed with a suitable disinfectant (e.g. ClearKlens Tego® 2001, VWR) followed by 70% alcohol (see Section 12.3.1). If a potentially corrosive disinfectant is used, care should be taken to remove all residue after the appropriate exposure time to prevent damage to the surfaces of the BSC. Swabs, tissue wipes, or gloves, if dropped below the work surface during cleaning, can end up on the primary filter and restrict airflow. Look for dropped items during cleaning and check the primary filter periodically. Many modern BSCs have a foam pre-filter under the work surface to prevent objects being carried into (and potentially damaging) the primary filter or fan. If fitted, the pre-filter should be checked and cleaned or replaced at regular intervals.

Ultraviolet (UV) lamps may be installed to assist with decontamination of the interior of the BSC between uses and after cleaning. UV light presents a radiation hazard, particularly to the eyes, and will lead to crazing of some clear plastic panels (e.g. Perspex) after six months to a year, particularly if used in conjunction with alcohol. The effectiveness of UV light is limited by its inability to reach into crevices; alcohol and other chemical disinfectants are more effective as they will run into crevices by capillarity. Despite these concerns, UV lamps continue to be used by many laboratories as

an adjunct to chemical disinfectants (Meechan and Wilson 2006). Thirty minutes of UV irradiation (on a timer) after swabbing when work is finished at the end of the day is probably a good practice. If UV lamps are used, they must be replaced regularly, as their output diminishes with time. They should also be cleaned regularly to maximize output.

❖ *Safety Note. BSCs should be closed and not used when UV irradiation is in progress. If UV exposure is a risk, protective goggles must be worn, and all exposed skin must be covered.*

12.2.3 Work Surface

Most experienced tissue culture practitioners arrange their work area for tissue culture using a consistent layout (see Figure 12.4a). Different layouts are used when working in a laminar flow hood and when working on the open bench (see Figure 12.5; see also Protocols P12.2, P12.3). The following

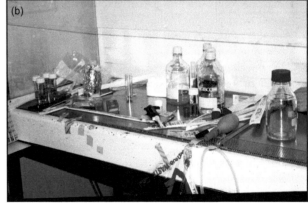

Fig. 12.4. Layout of BSC work area. (a) Correct layout with pipettes on the left at the back (propped up for easy access and to allow air flow to the rear grill), medium to the left of the work area, culture flasks central and well back from the front edge, and pipette controller on the right. Positions may be reversed for left-handed workers. (b) BSC used incorrectly. The BSC is too full, and many items encroach on the air intake at the front, destroying the laminar airflow and compromising both containment and sterility. *Source*: R. Ian Freshney.

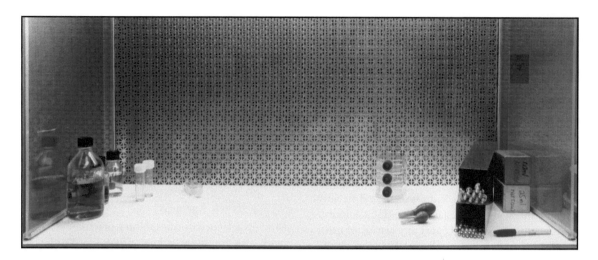

Fig. 12.5. Laminar flow hood work area. This layout is suitable when working in a horizontal laminar-flow hood. Positions may be reversed for left-handed workers. *Source:* R. Ian Freshney.

general rules should be observed when setting up any work area for tissue culture:

(1) Start with a completely clear surface.

(2) Swab the surface liberally with 70% alcohol (see Section 12.3.1).

(3) Bring onto the surface only those items you require for a particular procedure. If you have too many items inside the cabinet, you will inevitably brush the tip of a sterile pipette against a non-sterile surface (see Section 12.3.6). Furthermore, the laminar airflow will fail in a hood that is crowded with equipment (see Figure 12.4b).

(4) Arrange your work area so that you have (i) easy access to all items without having to reach over one to get at another; and (ii) a wide, clear space in the center of the bench to work on.

(5) Do not allow your hands or any other non-sterile items (even the outside of a flask is non-sterile) to pass over an open flask or dish. Even when using laminar flow, you should still work in a clear space with no obstructions between the central work area and the HEPA filter (see Figure 12.5).

(6) Keep items well back from the front edge. Working near the front of the cabinet can impair laminar airflow and threaten both asepsis and containment.

(7) Work within your range of vision, e.g. insert a pipette in a bulb or pipette controller with the tip of the pipette pointing away from you so that it is in your line of sight continuously and not hidden by your arm (see Section 12.3.6).

(8) Keep the work surface clean and tidy as you work.

(9) Mop up any spillage immediately and swab the area with 70% alcohol.

(10) Between procedures, remove everything that is no longer required and swab the surface down.

(11) When you have finished, remove all items and swab the work surface down again.

(12) Ensure that any space below the work surface is cleaned out regularly (see Section 12.2.2).

12.2.4 Personal Protective Equipment (PPE)

Personal protective equipment (PPE) is broadly defined as equipment that is worn to minimize exposure to hazards that cause workplace injuries and illnesses. In tissue culture laboratories, PPE is also used to protect cultures from the operator and the external environment. All humans carry a unique microbial "cloud" that enters the laboratory with them (Meadow et al. 2015). Skin, hair, clothing, and footwear are all potential sources of microbial contamination, which can be minimized by the use of appropriate PPE and aseptic technique. Most laboratories require personnel to wear a lab coat or gown, disposable gloves, and covered shoes when handling cultures or other biological samples (see Table 12.1). Face masks, caps, shoe covers, and protective suits may be required for laboratories that have more stringent requirements, but they are not necessary under normal conditions, particularly when working in a BSC. Additional PPE may be used for specific tasks, e.g. for thawing frozen vials (see Protocol P15.1).

Tissue culture can be performed aseptically using bare hands, but gloves are necessary to reduce the risk of exposure through cuts or abrasions that may be present on the hands but not noticed. They are also easier to keep clean than bare skin. Gloves should be worn with the cuff of the glove overlapping the cuff of the sleeve belonging to the gown or lab coat (see Figure 12.2c). Unlike in an operating theater, gloves used in tissue culture are not sterile and should be swabbed with 70% alcohol at regular intervals. If gloves are worn when touching and using communal equipment, their

TABLE 12.1. Personal protective equipment (PPE) for handling cultures.

Item	Description	Purpose
Commonly used PPE[a]		
Eye protection, e.g. safety goggles	Protective glasses or similar eyewear; should fit over prescription glasses.	Protect the eyes from splashes and projectiles (e.g. broken glass).
Lab coat	Made of impermeable, fire retardant, and easily laundered material. Long sleeves and closely fitting neck fastening. Wear buttoned up.	Protect the body and clothing from spills or fire.
Gown	Made of impermeable, fire retardant, and easily laundered material. Long sleeves and close-fitting cuffs. Fastening from the side or back.	Protect the body and clothing from spills or fire. Alternative to lab coat, preferred at BSL-3 / CL 3 (see Tables 6.3, 6.4).
Disposable gloves	Made of nitrile, latex, or vinyl. Choice determined by the substance being handled and any allergic history.	Protect the hands from exposure through broken skin or abrasions. Keep clean by swabbing.
Covered shoes	Made of impermeable material, covering the toes and as much of the feet as possible.	Protect the feet from spills.
Increased containment or asepsis[b]		
Head covering, e.g. disposable cap	Made of impermeable material.	Minimize transfer of microorganisms from hair.
Face mask	Disposable, made of breathable fabric. Covers the nose and mouth.	Minimize aerosols from coughing and sneezing.
Foot coverings	Disposable, made of impermeable material. Worn over shoes.	Minimize transfer of microorganisms to floors.
Laboratory clothing	Surgical attire ("scrubs") or similar.	Minimize transfer of microorganisms to clothing. Determined by risk assessment.
Protective suits	Worn over laboratory clothing.	Minimize exposure to hazardous substances. Determined by risk assessment.

[a]Used in many tissue culture laboratories while handling cultures in a BSC or handling biospecimens on the laboratory bench. Eye protection is usually provided by the glass panel of the BSC but will be needed at the bench.
[b]Used in some tissue culture laboratories with increased levels of containment or asepsis. PPE requirements will vary depending on your location and the nature of the risk (see Tables 6.3, 6.4).

sterility is compromised. Their use should be restricted to the culture suite and removed when leaving the suite or using communal equipment such as telephones. Lab coats and gowns should be frequently laundered and restricted to use in the tissue culture laboratory.

Always wash your hands when you enter the tissue culture laboratory and cover any obvious cuts or abrasions before donning gloves. If you have long hair, tie it back. Keep the temperature at a comfortable level to avoid excessive sweating, which may increase bacterial shedding (Coté 2001). When working aseptically on an open bench, do not talk; talking leads to increased risk of aerosol contamination. Talking is permissible when you are working in a BSC, with a barrier between you and the culture, but should still be kept to a minimum. If you have a cold, wear a face mask, or (better still) do not do any tissue culture during the height of the infection.

12.2.5 Reagents and Media

Reagents and media obtained from reputable suppliers will already have undergone strict quality control to ensure that they are sterile, but the outside surface of the bottle they come in is not. Some manufacturers supply bottles wrapped in polyethylene, which keeps them clean and allows them to be placed in a water bath to be warmed or thawed. The wrapping should be removed outside the BSC. Unwrapped bottles should be swabbed in 70% alcohol when they come from the refrigerator or from a water bath. The cap and neck areas will be particularly vulnerable as potential sources of contamination.

12.2.6 Cultures

The greatest contamination risk in the modern tissue culture laboratory is likely to be from other cultures. Cultures obtained from other laboratories carry a higher risk, because they may have been contaminated either at the source or in transit. Validation testing should be performed whenever a culture arrives in a laboratory (see Section 7.4). Even if certified as being contamination free, arriving cell lines should always be quarantined, i.e. they should be handled separately from the rest of your stocks until they are shown to be free from contamination (see Sections 4.2.3, 16.1.5). They may then be brought into the main tissue culture area. Antibiotics should not be used as they may suppress but not eliminate some contaminations and encourage poor aseptic technique (see Section 9.6.2).

12.3 STERILE HANDLING

12.3.1 Swabbing

A number of substances can be used to disinfect surfaces (McDonnell and Russell 1999; CDC 2008). Most tissue culture laboratories use 70% alcohol (ethanol or isopropanol) to swab down the work surface. Ethanol and isopropanol have broad bactericidal activity, act rapidly, and leave no toxic or corrosive residues behind. Ethanol is often bought in bulk, while isopropanol ("rubbing alcohol," IPA) is often purchased as a prepacked spray or as swabs. A spray is usually more effective (e.g. DECON-AHOL®, Veltek Associates, Inc.) and will cover a larger area more easily and penetrate crevices. Both are typically used at 70% (v/v); this concentration leads to a slower evaporation rate, making the treatment more effective, and alcohol is still active at this concentration. Methanol should be avoided as it is more toxic and has weaker bactericidal activity.

Swab down the work surface before and during work, particularly after any spillage, and swab it down again when you have finished. Swab bottles when you bring them to the work area, especially those coming from cold storage or a water bath, and swab flasks or boxes from the incubator. Swabbing sometimes removes labels, so use an alcohol-resistant marker. Swab gloves frequently as you handle cultures. Regular use of alcohol makes the skin more porous, giving another reason why gloves should be worn.

12.3.2 Flaming

Flaming is used when working on the open bench, where a Bunsen burner provides a sterile field for handling open bottles and dishes (Sanders 2012). Flaming is intended to remove any particles of dust or lint or, at least, to fix them to the object being flamed such that they do not drop off into a sterile area. It should be done briefly and systematically without allowing the object being flamed to become hot – it is not being used to sterilize the object! When working on an open bench (see Section 12.4.2), flame glass pipettes and the necks of bottles and screw caps before and after opening and closing. Work close to the flame, where there is an updraft due to convection, and do not leave bottles open.

Flaming should not be used when you are working in a laminar flow hood or BSC. An open flame can be a fire hazard in laminar airflow and can damage the HEPA filter or melt some of the plastic interior fittings. It also disrupts the laminar flow, which, in turn, compromises both the sterility of the BSC and its containment. If flaming is required (e.g. to burn off alcohol used to sterilize instruments), it should be performed outside the BSC or an alternative approach used (e.g. air drying within the BSC).

❖ *Safety Note. To avoid a potentially serious fire, care must be taken not to return the flamed instruments to the alcohol until they are fully extinguished.*

12.3.3 Capping

Deep screw caps are preferred for bottles that are used during aseptic procedures (see Section 11.3.3). Care must be taken during washing to ensure that all detergent is rinsed from behind rubber liners; wadless caps are preferred for this reason. The screw cap should be covered with aluminum foil to protect the neck of the bottle from sedimentary dust, although this foil shrouding is less necessary for deep polypropylene caps (e.g. from Corning, Duran®).

Caps should be held in the hand during pipetting if possible, avoiding the need to lay them down (see Figure 12.6). Placing the cap face-down may lead to the introduction of contaminants from the surface (or to the surface from the cap), while placing the cap face-up may lead to airborne contaminants if there are disturbances in laminar airflow. If it is necessary when working on the open bench, place the cap face-down on a clean surface and flame briefly before replacing it on the bottle. When working in a BSC or laminar flow hood, place the cap face-up toward the back of the work area to minimize exposure. Some caps are designed to reduce rolling, allowing them to be placed on the side.

12.3.4 Handling Bottles and Flasks

When you are working on the open bench, do not keep bottles vertical when open. Instead, keep them at an angle as shallow as possible without risking spillage (see Figure 5.6a). A bottle rack can be used to keep the bottles or flasks tilted. When you are working in a BSC, you can keep open bottles in a vertical position, but do not let your hands or any other items come between an open vessel or sterile pipette and the HEPA filter. Culture flasks should be laid down horizontally when open and, like bottles, held at an angle during manipulations. Flasks with angled necks are designed to facilitate pipetting when the flask is lying flat.

12.3.5 Pouring

Do not pour from one sterile container into another, unless the bottle or flask you are pouring from is to be used once

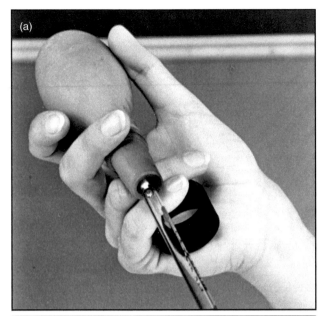

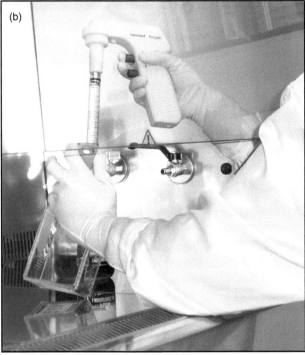

Fig. 12.6. Holding a cap while pipetting: (a) with a bulb; (b) with a pipette controller. The cap may be held in the crook of the little finger of the hand holding the bulb or pipette controller. Alternatively, the cap may be held by the thumb and index finger of the opposite hand and the remainder of the hand used to control the flask. *Source:* (a) R. Ian Freshney; (b) Amanda Capes-Davis.

only and then discarded – for example, if the bottle is premeasured to deliver all its contents in one delivery. The major risk in pouring lies in the generation of a bridge of liquid between the outside of the bottle and the inside, permitting contamination to enter the bottle during storage or incubation.

12.3.6 Pipetting

Serological pipettes and Pasteur pipettes are the easiest way to manipulate liquids when handling cultures (see Section 5.1.3). Pipetting in tissue culture is often a compromise between speed and precision. Speed is required to minimize deterioration during handling procedures, and precision is required for reproducibility during experimental work. An error of ±5% is usually acceptable, except under experimental conditions where greater precision may be required. Generally, using the smallest pipette compatible with the maneuver and volumes being dispensed will give the greater precision required of most quantitative experimental work; a larger pipette will allow quicker serial dispensing but with less accuracy. Whichever pipette size is used, always control the flow of liquid to avoid damaging the cells. This is usually done by tilting the culture vessel so that fluid is added and removed from the side of the vessel, away from the cell monolayer.

Pipettes of a convenient size range should be selected – 1, 2, 5, 10, and 25 ml cover most requirements, although 50- and 100-ml disposable pipettes are useful for preparing and aliquoting media (see Figure 5.2). Using fast-flow pipettes reduces accuracy slightly but gives considerable benefit in speed. Double-wrapped plastic pipettes should have the outer wrapping removed before being placed in the BSC. Bulk-wrapped plastic pipettes, which are not double wrapped, are not recommended. Although the unit cost will be cheaper, withdrawing pipettes from bulk wrapping while maintaining sterility is more difficult compared to individually wrapped pipettes. Also, the surplus of unused pipettes is difficult to store and keep sterile; opened items should not be passed on to another operator and may need to be discarded, adding to the unit cost.

A pipette controller must be used for all aseptic procedures (see Section 5.1.3). Mouth pipetting has been shown to be a contributory factor in mycoplasma contamination, is hazardous to the operator, and is now almost universally banned in tissue culture laboratories due to biohazard risk (see Section 6.3.1) (McGarrity 1976). Motorized pipette controllers are widely used (see Figures 5.3a, 12.6b); a rubber bulb is cheaper but will need more frequent replacement and cleaning (see Figure 12.6a). Try a selection of devices to find one that feels comfortable in your hand. The instrument you choose should accept all sizes of pipette that are available in the laboratory. You should be able to insert each pipette using correct technique (see Figure 12.7), without force and without the pipette falling out. Regulation of flow should be easy and rapid but at the same time capable of fine adjustment. You should be able to draw liquid up and down repeatedly (e.g. to disperse cells), and there should be no fear of carryover. The device should be easy to operate with one hand without fatigue.

All pipettes should have a cotton plug at the top to keep the pipette sterile during use. The plug is already in place for most plastic pipettes but must be inserted into glass pipettes before sterilization (see Supp. S11.1). The plug should prevent

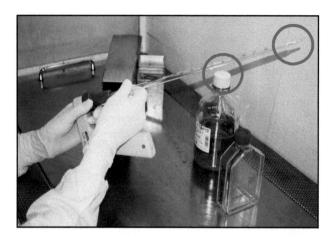

Fig. 12.7. Inserting a pipette in a pipette controller. To do so correctly, grip the pipette at a high point (above the graduations) and point it away from the user. Circled areas mark potential risks, i.e. inadvertently touching the bottle or the back of the cabinet. *Source*: R. Ian Freshney.

contamination from the bulb or pipette controller entering the pipette and reduce the risk of cross-contamination from pipette contents inadvertently entering the pipette controller. If the plug becomes wet, discard the pipette. If a pipette controller is contaminated with fluid, it must be exchanged for a clean one and the device must be cleaned before re-use. Most motorized pipette controllers have a protective filter which will need to be changed. Bulbs do not, so they should be washed out, rinsed with 70% alcohol, and dried.

12.3.7 Small-Volume Dispensing

Pipettors are useful for dispensing small volumes (≤ 1 ml), but the length of the sterile tip will limit the size of vessel used. If a sterile fluid is withdrawn from a container with a pipettor, the non-sterile stem must not touch the sides of the container. Reagent volumes of 10–20 ml may be sampled in 5-µl to 1-ml aliquots from a universal container or similar small vial, but withdrawing liquids from larger containers will risk contamination. Longer tips may be used with larger volumes. Pipettors should not be used for serial propagation unless filter tips are used. Care should be taken to distinguish between "PCR Clean" tips, which are RNAse-free but not sterile, and sterile tips that are suitable for tissue culture.

Pipettors are particularly useful when dealing with multiwell plates and are available with multiple tips (see Figure 5.3b). Automated or semi-automated plate handling systems are also widely available (see Figure 5.8). Care must be taken to avoid contamination when setting up automatic pipetting devices and repeating dispensers (see Section 5.1.3), but the increased speed in handling can cut down on fatigue and on the time that vessels are open to contamination. Automation of cell culture procedures is discussed in a later chapter (see Section 28.7).

Syringes are sometimes used for small volumes but should be discouraged as regular needles are too short to reach into most bottles. Syringing can also produce high shearing forces when you are dispensing cells, and the practice increases the risk of needlestick injuries (see Section 6.2.1). Wide-bore cannulae are preferable to needles but are usually slower to use, except with multiple-stepping or repeating dispensers (see Figure 5.4).

12.3.8 Large-Volume Dispensing

A different approach to fluid delivery must be adopted with culture vessels that exceed 100 ml of medium volume. If only a few flasks are involved, a 100-ml pipette, graduated bottle dispenser, or media bag may be quite adequate. However, if larger volumes (> 500 ml) or a large number of high-volume replicates are required, a peristaltic pump is preferable. Single fluid transfers of very large volumes (10–10 000 l) are usually achieved by preparing the medium in a sealed pressure vessel, sterilizing it by autoclaving or another suitable method, and then displacing it by positive pressure into the culture vessel. It is possible to dispense large volumes by pouring, but this should be restricted to a single action with a premeasured volume (see Section 12.3.5).

12.4 GOOD ASEPTIC TECHNIQUE

The essence of good aseptic technique embodies many of the principles of standard good laboratory practice (see Table 12.2). Keep a clean, clear space in which to work, and have on it only what you require for the immediate procedure being performed. Prepare as much as possible in advance, so that cultures are out of the incubator for the shortest possible time and the various manipulations are carried out quickly, easily, and smoothly. Keep everything in direct line of sight and develop an awareness of accidental contacts between sterile and non-sterile surfaces. Leave the area clean and tidy when you finish.

12.4.1 Aseptic Technique Using Laminar Airflow

Protocols P12.1 and P12.2 describe good aseptic technique when using a BSC; they can also be used in training (see Supp. S30.5, S30.6). Preparation of media and other reagents is discussed in more detail elsewhere (see Sections 11.5, 11.6).

PROTOCOL P12.1. ASEPTIC TECHNIQUE: HANDLING FLASKS IN A BSC

Background

The following protocol assumes that plastic disposable pipettes are used. Comments on reusable glass

TABLE 12.2. Good aseptic technique.

Subject	Do	Don't
BSCs	Swab down before and after use. Keep minimum amount of apparatus and materials in BSC. Work in direct line of sight.	Clutter up the BSC. Leave the BSC in a mess.
Contamination	Work without antibiotics. Check cultures regularly, by eye and microscope. Box Petri dishes and multiwell plates.	Open contaminated flasks in tissue culture. Carry infected cells, even with antibiotics. Leave contaminations unclaimed; dispose of them safely (see Protocol P16.1).
Mycoplasma	Test cells routinely.	Carry infected cells. Try to decontaminate cultures.
Importing cell lines	Get from a reliable source, e.g. a cell repository (see Table 15.3). Quarantine incoming cell lines. Perform validation testing, i.e. mycoplasma and authentication testing. Keep records.	Get from a source far removed from originator other than from a reputable cell repository.
Exporting cell lines	Perform validation testing, i.e. mycoplasma and authentication testing. Send data sheet. Package correctly for shipment.	Send contaminated cell lines. Pass on non-validated stock.
Glassware	Keep stocks separate.	Use for regular laboratory procedures.
Flasks	Pipette with flask sloped. Use gas-permeable caps in CO_2 incubator. Vent briefly at 37 °C if sealed and stacked.	Have too many open at once. Stack too high (see Figure 12.12).
Media and reagents	Swab bottles before placing them in BSC. Open only in BSC.	Share among cell lines. Share with others. Pour.
Pipettes	Use plugged pipettes. Change if contaminated or plug wetted. Use plastic for agar.	Use the same pipette for different cell lines. Share with other people. Overfill disposal cylinders.

pipettes can be found in the Supplementary material online (see Supp. S11.1).

Outline

Clean and swab down work area. Bring in bottles, pipettes, and other equipment. Carry out preparative procedures first (preparation of media and other reagents), followed by culture work. When finished, tidy up and wipe over surface with 70% alcohol.

❖ *Safety Note. Pipettes can break, resulting in sharps injuries (see Section 6.2.1). Take care to insert the pipette carefully into a bulb or pipette controller and in a direction to prevent injury to the hand should the pipette slip or break (see Figure 12.7). Dispose of pipettes separately into a pipette cylinder or similar container.*

Materials (sterile)

- ☐ Culture medium
- ☐ Disposable plastic pipettes in an assortment of sizes, e.g. 1, 5, 10, and 25 ml. Pipettes should be individually wrapped within an outer wrapper. Remove the required number of pipettes from the outer wrapper before bringing into the BSC. A pipette can or similar container may be used for storage.
- ☐ Culture flasks
- ☐ Other containers and tubes (as needed, will vary with the task)

Materials (non-sterile)

- ☐ BSC
- ☐ Pipette controller or bulb (see Figure 12.6)

□ Racks for tubes (as needed)

□ Marker pen with alcohol-insoluble ink

□ 70% alcohol in spray bottle

□ Lint-free swabs or wipes

□ Pipette cylinder, large sharps bin, or similar impermeable container (for used pipettes; on floor beside BSC), containing disinfectant (for decontamination), or a double-thickness biohazard waste bag (for autoclaving or incineration)

□ Vacuum pump or line connected to a waste receiver below BSC or waste beaker in BSC, both containing a suitable approved disinfectant (see Figures 5.6, 12.8; see also Section 6.3.6)

□ Sharps container (for glass Pasteur pipettes or other glass items)

□ Waste bin (for paper waste, swabs, and packaging; on floor beside BSC on opposite side from pipette cylinder or biohazard waste bag)

□ Notebook, pen, protocols

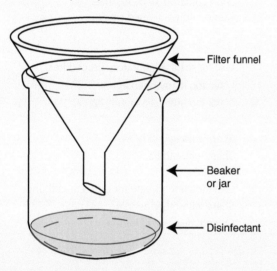

Filter funnel

Beaker
or jar

Disinfectant

Fig. 12.8. Waste beaker. The filter funnel is used to prevent the contents of the waste beaker from splashing back.

Procedure

1. Turn on the BSC if not run continuously (see Section 12.2.2). Allow cabinet to run for at least 4 minutes for the air to "purge" of contaminants.

2. Swab down the work surface and all other inside surfaces of BSC, including inside of front screen, with 70% alcohol and a lint-free swab or tissue.

3. Bring non-sterile materials that are required for handling procedures into the BSC, e.g. pipette controller, racks, marker pen, and waste beaker (if used). Leave other non-sterile materials outside the BSC but within reach, e.g. lint-free wipes and absorbent paper tissues.

4. Bring bottles of medium and reagents from cold storage or water bath, swab bottles with alcohol, and place those that you will need first in the BSC.

5. Collect pipettes and place at one side of the back of work surface in an accessible position. If using pipette cans, open and place lids on top or alongside, with the open side down. Arrange individually wrapped pipettes, sorted by size, ready for use.

6. Collect any other glassware, plastics, instruments, etc. that you will need, and place them close by (e.g. on a cart or adjacent bench).

7. Slacken, but do not remove, caps of all bottles about to be used.

8. Remove the cap of the flask into which you are about to pipette and the bottles that you wish to pipette from. Place the caps open side uppermost on the work surface, at the back of the BSC and behind the bottle, so that your hand will not pass over them. Alternatively, if you are handling only one cap at a time, keep hold of the cap in your hand while pipetting (see Figure 12.6) and replace it when you have finished pipetting.

9. Select a pipette. If pipettes are individually wrapped in plastic:
   (a) Open the pack at the top.
   (b) Peel the ends back, turning them outside in.
   (c) Insert the end of the pipette into the bulb or pipette controller.
   (d) Withdraw the pipette from the wrapping without it touching any part of the outside of the wrapping, or the pipette touching any non-sterile surface.
   (e) Discard the wrapping into the waste bin.

10. The pipette in the bulb or pipette controller will now be at right angles to your arm. Take care that the tip of the pipette does not touch the outside of a bottle or the inner surface of the BSC (see Figure 12.7, circled areas); always be aware of where the pipette is. Following this procedure is not easy when you are learning aseptic technique, but it is an essential requirement for success and will come with experience. If the pipette tip does touch a non-sterile surface, discard it immediately and continue with a new sterile pipette.

11. Tilt the medium bottle toward the pipette so that your hand does not come over the open neck, withdraw 5 ml medium (or whatever you require), and transfer it to a flask, also tilted.

12. Discard the pipette into the pipette cylinder or similar impermeable container for decontamination and disposal.
13. Recap the flask.
14. Replace the cap on the medium bottle and flasks. Bottles may be left open while you complete a particular maneuver but should always be closed whenever you leave the BSC.
15. On completion of the operation, tighten all caps, and place flasks in incubator.
16. Remove all solutions and materials no longer required from the work surface, and swab down. If you have used plastic, individually wrapped pipettes in sterile cans, you may recap the can and use the pipettes at your next session.
17. Turn the BSC off if not run continuously. The cabinet should have purged for at least 4 minutes after cultures are removed before turning the cabinet off.

Notes

Step 14: In vertical laminar flow, do not work immediately above an open vessel. In horizontal laminar flow, do not work behind an open vessel.

Petri dishes and multiwell plates are particularly prone to contamination because:

(1) A larger surface area is exposed when the dish is open.
(2) There is a risk of touching the rim of the dish when handling an open dish.
(3) There is a risk of carrying contamination from the work surface to the plate via the lid if the lid is laid down.
(4) Medium may enter the gap between the lid and the dish due to capillarity if the dish is tilted or shaken in transit to the incubator.
(5) There is a higher risk of contamination in the humid atmosphere of a CO_2 incubator.

The following practices will minimize the risk of contamination:

(1) Do not leave dishes open for an extended period.
(2) Do not work over an open dish or lid.
(3) When moving dishes (including to or from the incubator), take care not to tilt, swirl, or shake them. These actions may lead to uneven growth or allow the medium to enter the capillary space between the lid and the base. Other precautions include:
  a) Use "vented" dishes (see Figure 8.8).
  b) If medium still lodges in this space, discard the lid, blot any medium carefully from the outside of the rim

with a sterile tissue dampened with 70% alcohol, and replace the lid with a fresh one. (Make sure that the labeling is on the base!)
(4) Enclose dishes and plates in a transparent plastic box for incubation and swab the box with alcohol when it is retrieved from the incubator (see Section 12.5.2).

Protocol P12.2 provides specific instructions for handling dishes or plates.

PROTOCOL P12.2. ASEPTIC TECHNIQUE: HANDLING DISHES OR PLATES

Outline

See Protocol P12.1 for working in a BSC. The following additional instructions apply to Petri dishes or multiwell plates. "Dish" is used for ease of explanation.

Materials

☐ As per Protocol P12.1
☐ Sterile Petri dishes or multiwell plates

Procedure

1. Place dish or plate on one side of work area.
2. Position medium bottle and slacken the cap.
3. Bring dish to center of work area.
4. Remove bottle cap and fill pipette from bottle.
5. Remove lid from dish.
6. Add medium to dish, directing the stream gently low down on the side of the base of the dish.
7. Replace lid.
8. Return dish to side of BSC, taking care not to let the medium enter the capillary space between the lid and the base.
9. Discard pipette.
10. Tighten cap on medium bottle.

12.4.2 Aseptic Technique on the Open Bench

Protocol P12.3 emphasizes good aseptic technique when working on the open bench. Although tissue culture is now typically performed using laminar airflow, this procedure continues to be useful for some procedures, e.g. for handling microbiological broth or agar during sterility testing (see Supp. S11.6). It is also useful for laboratories that, for whatever reason, are not equipped with BSCs and laminar flow cabinets and therefore have no other option but to use the open bench.

Fig. 12.9. Open bench work area. Items are arranged in a crescent around the clear work space in the center. The Bunsen burner is located centrally, to be close by for flaming and to create an updraft over the work area. *Source*: R. Ian Freshney.

PROTOCOL P12.3. WORKING ON THE OPEN BENCH

Background

The following procedure uses pipetting of microbiological broth as an example of work that is still performed on the open bench. Plastic disposable pipettes are used; comments on reusable glass pipettes can be found in the Supplementary material online (see Supp. S11.1).

Outline

Clean and swab down work area. Set out bottles, pipettes, and other equipment (see Figure 12.9). Carry out preparative procedures first. Flame articles as necessary and keep the work surface clean and clear. When finished, tidy up and wipe over surface with 70% alcohol.

❖ *Safety Note. Pipettes can break, resulting in sharps injuries (see Section 6.2.1). Bunsen burners represent a fire hazard. Be careful when working with alcohol, e.g. swab the bench and then place the alcohol bottle well away from the naked flame.*

Materials (sterile)

☐ Microbiological broth
☐ Disposable plastic pipettes in an assortment of sizes, e.g. 1, 5, 10, and 25 ml. Pipettes should be individually wrapped within an outer wrapper. Remove the outer wrapper and place the pipettes in sterile cans or on a rack.
☐ Sterile containers, plates, dishes, etc. as required

Materials (non-sterile)

☐ Bunsen burner (or equivalent) and lighter
☐ Pipette controller or bulb (see Figure 12.6)
☐ Marker pen with alcohol-insoluble ink
☐ 70% alcohol in spray bottle
☐ Lint-free swabs or wipes
☐ Absorbent paper tissues
☐ Pipette cylinder, large sharps bin, or impermeable container holding double-thickness autoclavable biohazard waste bag (for used pipettes; on floor)
☐ Waste bin (for paper waste, swabs, and packaging; on floor on opposite side from pipette cylinder or biohazard waste bag)
☐ Suction line to aspirator (see Figure 5.6) or waste beaker (see Figure 12.8), both with disinfectant
☐ Notebook, pen, protocols

Procedure

1. Swab down bench surface with 70% alcohol.
2. Bring broth and other reagents from cold storage or water bath. Swab bottles with alcohol; place those that you will need first on the bench, leaving the others at the side.
3. Collect pipettes and any other glassware, plastics, or instruments. Place them close by, with pipettes at the side of the work surface in an accessible position.
4. Arrange your work space such that all handling procedures will be done on an area of bench close to the Bunsen burner so that airflow will be predominantly upwards due to convection.

Work as close to the burner as possible to be located within the area of updraft.

5. Flame necks of bottles, briefly rotating neck in flame, and slacken caps.

6. Select pipette. If pipettes are individually wrapped in plastic:
   (a) Open the pack at the top.
   (b) Peel the ends back, turning them outside in.
   (c) Insert the end of the pipette into the bulb or pipette controller.
   (d) Withdraw the pipette from the wrapping without it touching any part of the outside of the wrapping or the pipette touching any non-sterile surface.
   (e) Discard the wrapping into the waste bin.

7. Insert pipette in a bulb or pipette controller, pointing pipette way from you and holding it well above the graduations, so that the part of the pipette entering the bottle or flask will not be contaminated (see Figure 12.7).

8. The pipette in the bulb or pipette controller will now be at right angles to your arm. Take care that the tip of the pipette does not touch any non-sterile items. Always be aware of where the pipette is. Following this procedure is not easy when you are learning aseptic technique, but it is an essential requirement for success and will come with experience. If the pipette tip does touch a non-sterile surface, discard it immediately and continue with a new sterile pipette.

9. Holding the pipette still pointing away from you, remove the cap of the bottle and hold it in your hand (see Figure 12.6). If you have difficulty holding the cap in your hand while you pipette, place the cap on the bench, open side down.

10. Flame the neck of the bottle.

11. Tilt the bottle toward the pipette so that your hand does not come over the open neck.

12. Withdraw the requisite amount of fluid and hold.

13. Flame the neck or the bottle and recap.

14. Remove cap of receiving bottle, flame neck, insert fluid, reflame neck, and replace cap.

15. When finished, tighten caps.

16. Remove from the work surface all solutions and materials no longer required and swab down the work surface. If pipette cans have been used with individually wrapped plastic pipettes, these may be recapped and reused later.

Notes

> **Step 5:** Flame glass items only. Do not flame plastic pipettes or other plastic items. Do not flame bottles with an inserted plastic pouring ring.
>
> **Steps 10–12:** If bottles are to be left open, they should be sloped as close to horizontal as possible in laying them on the bench or on a bottle rest.

12.5 CONTROLLING EQUIPMENT CONTAMINATION

All apparatus and equipment used in the tissue culture area should be cleaned regularly to avoid the accumulation of dust and to prevent microbial growth in accidental spillages. Many laboratories have a roster that requires all personnel who share the space to contribute to cleaning activities and to document when cleaning is performed. Cleaning rosters should include the laboratory space itself as well as incubators, refrigerators, coldrooms, water baths, centrifuges, and microscopes. No major movement of equipment should take place while people are working aseptically; always schedule equipment maintenance for "quiet" periods with minimal culture work. Clean all items before they enter the laboratory, including replacement items (e.g. gas cylinders).

12.5.1 Incubators

Humidified incubators are a major source of microbial contamination (see Section 16.1.3). A number of features have been incorporated into their design to reduce microbial contamination, but even so, it can still be a problem (see Section 5.3.1). Fungal contamination is particularly likely in areas of condensation, which may be hidden behind the water pan or other accessories (see Figure 12.10). Any spillage or obvious condensation should be mopped up and the surface wiped down with a suitable disinfectant. Incubators should be cleaned at regular intervals; this may be weekly or monthly, depending on the level of atmospheric contamination and frequency of access. Monthly may suffice for a clean area with filtered room air, but a shorter interval will be required for a rural site, where the spore count is higher, or during construction work or renovation. Always refer to the manufacturer's instructions for your particular model. Some incubators offer built-in features to reduce contamination, such as high temperature cycles or UV light. High temperature cycles are useful to manage contamination but often mean that the incubator is out of use for extended periods during the sterilization cycle. UV light will not reach crevices or underneath the water pan.

Fig. 12.10. Incubator condensation. A zone of condensation (indicated by the rectangle) is present on the floor below the water pan, which was removed for this photograph. The surface temperature in that area was 2 °C cooler than the remainder of the floor. *Source:* Amanda Capes-Davis.

PROTOCOL P12.4. CLEANING CO₂ INCUBATORS

Background

These instructions are written to suit a variety of incubators. However, the choice of detergent and method will vary with the specific model. Check the manufacturer's recommendations.

Outline

Remove cultures to an alternative incubator, switch off the empty incubator, and clean with detergent and disinfectant. Autoclave accessories. Switch on and return to full operation.

Materials (non-sterile)

□ Alternative incubator or sealable container, e.g. plastic box with sealing tape
□ Neutral detergent, e.g. mild dish detergent
□ Disinfectant, e.g. 70% alcohol in spray bottle (see Notes below)
□ Sterile distilled water for rinsing and to refill the water pan
□ Lint-free swabs or wipes

Procedure

1. Remove all cultures to another CO_2 incubator, or enclose the cultures in a sealable container, gas them with CO_2, and place them in a regular incubator (see Section 12.5.3).
2. Switch off the incubator that is to be cleaned and any associated alarm systems.
3. Remove the water pan and any demountable shelves, racks, or panels. Set aside.
4. Wash the interior of the incubator with detergent; try to reach all corners and crevices.
5. Rinse the interior of the incubator with water.
6. Wipe the interior of the incubator with disinfectant.
7. Turn the incubator on and wait for it to dry. Leave the CO_2 off (e.g. set to zero) so it does not inject while the door is open. Shut the door once the interior is dry.
8. Wash the water pan and demountable accessories in detergent, rinse them in water, and wipe them with disinfectant. Autoclave if possible.
9. Return the water pan and demountable accessories to the incubator.
10. Run sterilization cycle if available and allow incubator to cool to normal working temperature.
11. Fill the water pan with sterile distilled water. Return CO_2 to normal and add water treatment if desired (see Section 16.1.3, "Humid Incubators").
12. When the conditions have stabilized, return the cultures to the incubator.

Notes

Step 6: 70% ethanol is not active against fungal spores, so it may be necessary to replace with a quaternary ammonium disinfectant (Bates et al. 2014). However, copper surfaces tend to reduce fungal contamination; the stronger disinfectant will not be necessary for some surfaces.

Step 10: Fungal spores may be resistant to some disinfectants and temperatures achieved during decontamination cycles. For this reason, a double decontamination cycle is recommended (with a minimum 12 hours gap between cycles) if a fungal contamination occurs in a CO_2 incubator. The first cycle will kill the vegetative cells. Any resistant spores will germinate in the gap between cycles, and the resulting vegetative cells will be killed in the second cycle.

Step 11: Typically, sterile distilled water at pH 7–9 should be used to refill the water pan (Bates et al. 2014). Ultrapure water (UPW) may lead to corrosion and should be avoided. Water treatment agents are sometimes used, but some agents release volatile organic chemicals and are harmful to the environment (see

Section 16.1.3). It is sensible to avoid routine use of water treatment agents and reserve them for limited "treatment" periods if required.

12.5.2 Boxed Cultures

When problems with contamination in humidified incubators recur frequently, it may be helpful to enclose culture vessels in a closed chamber (see Figure 9.4) or a plastic box. This is particularly useful for dishes and plates, which are at increased risk of contamination, and during persistent episodes of contamination (see Section 16.7). Boxes can also be used to culture cells under different atmospheric conditions or to protect from accidental spillage if cultures have increased biohazard risk. Such boxes should be swabbed with 70% alcohol before use, inside and outside, and allowed to dry in sterile air. When the box is removed from the incubator, it should be swabbed before being opened or introduced into your work area. Dishes can then be carefully removed and the interior of the box swabbed before reuse.

12.5.3 Gassing Cultures

Humid CO_2 incubators are now used to perform most tissue culture work. However, there may be circumstances where CO_2 incubators are not optimal or available. Different environmental conditions may be required for some cultures (Ahmed et al. 2016), or the incubator may break down and cannot be replaced. A sterile, premixed gas supply can be used to purge an individual flask or a culture box in these circumstances (see Figure 12.11). Gassing the flask avoids the need for a gassed incubator and gives the most uniform and rapid equilibration. After gassing, the flask or box can be sealed and placed in a dry incubator or warmroom. You

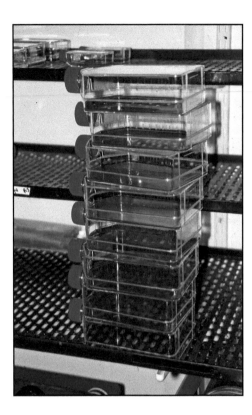

Fig. 12.12. Tilting flasks. The air space inside a flask expands in the incubator. In large flasks, this causes the flask to bulge and will tilt the flasks, increasing the tilt with the height of the stack. *Source*: R. Ian Freshney.

may need to release the pressure in the flasks after they have been in the incubator for about 30 minutes, as the flasks may distort due to expansion of the gas phase. This is particularly noticeable with larger flasks, leading to tilting of stacked flasks (see Figure 12.12).

Suppliers

Supplier	URL
Corning	http://www.corning.com/worldwide/en/products/life-sciences/products/cell-culture.html
DWK Life Sciences (Duran)	http://www.duran-group.com/en/home.html
Veltek Associates, Inc.	http://sterile.com/product-category/disinfectants-and-sporicides
VWR (Jencons)	https://uk.vwr.com/store/

REFERENCES

Ahmed, N.E., Murakami, M., Kaneko, S. et al. (2016). The effects of hypoxia on the stemness properties of human dental pulp stem cells (DPSCs). *Sci. Rep.* 6: 35476. https://doi.org/10.1038/srep35476.

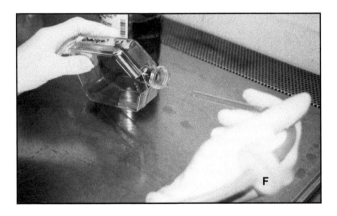

Fig. 12.11. Gassing a flask. A pipette is inserted into the supply line from the CO_2 source, and 5% CO_2 is used to flush the air out of the flask without bubbling through the medium. The letter "F" indicates a micropore filter inserted in the CO_2 line. *Source*: R. Ian Freshney.

Bates, M. K., D'Onofrio, J., Parrucci, M. L., et al. (2014). Back to basics: proper care and maintenance for your cell culture incubator. ThermoScientific Technical Notes. https://beta-static.fishersci.com/content/dam/fishersci/en_us/documents/programs/scientific/technical-documents/instruction-sheets/thermo-scientific-incubator-care-instruction.pdf (accessed 20 November 2018).

Carrel, A. (1923). A method for the physiological study of tissues in vitro. *J. Exp. Med.* 38 (4): 407–418.

CDC (2008). Guideline for disinfection and sterilization in healthcare facilities. https://www.cdc.gov/infectioncontrol/guidelines/disinfection/index.html (accessed 8 October 2018).

CDC and NIH (2009). Biosafety in microbiological and biomedical laboratories. https://www.cdc.gov/labs/bmbl.html (accessed 11 March 2019).

Coté, R.J. (2001). Aseptic technique for cell culture. *Curr. Protoc. Cell Biol.* Chapter 1:Unit 1.3. doi: https://doi.org/10.1002/0471143030.cb0103s00.

Kruse, R.H., Puckett, W.H., and Richardson, J.H. (1991). Biological safety cabinetry. *Clin. Microbiol. Rev.* 4 (2): 207–241.

McDonnell, G. and Russell, A.D. (1999). Antiseptics and disinfectants: activity, action, and resistance. *Clin. Microbiol. Rev.* 12 (1): 147–179.

McGarrity, G.J. (1976). Spread and control of mycoplasmal infection of cell cultures. *In Vitro* 12 (9): 643–648.

McGarrity, G.J. and Coriell, L.L. (1971). Procedures to reduce contamination of cell cultures. *In Vitro* 6 (4): 257–265. https://doi.org/10.1007/bf02625938.

Meadow, J.F., Altrichter, A.E., Bateman, A.C. et al. (2015). Humans differ in their personal microbial cloud. *Peer J.* 3: e1258. https://doi.org/10.7717/peerj.1258.

Meechan, P.J. and Wilson, C. (2006). Use of ultraviolet lights in biological safety cabinets: a contrarian view. *Appl. Biosafety* 11 (4): 222–227. https://doi.org/10.1177/153567600601100412.

Sanders, E.R. (2012). Aseptic laboratory techniques: volume transfers with serological pipettes and micropipettors. *J. Vis. Exp.* 31 (63): 2754. https://doi.org/10.3791/2754.

Schlich, T. (2012). Asepsis and bacteriology: a realignment of surgery and laboratory science. *Med. Hist.* 56 (3): 308–334. https://doi.org/10.1017/mdh.2012.22.

CHAPTER 13

Primary Culture

Primary culture refers to the first stage of the culture after isolation of the cells but before subculture. After the first subculture (passage), the culture ceases to be a primary culture and becomes a cell line (see Section 14.1). These definitions were formally accepted more than 25 years ago and continue to be widely used (Schaeffer 1990; Geraghty et al. 2014). Related terms such as "primary cell lines" or "primary cells" are also used, particularly by companies that supply early passage cultures of specialized cells. It is important to be clear about their terminology before you proceed with a purchase. In some cases, they may be dissociated cells from a primary culture, which will become a cell line when the recipient seeds them into culture. In other cases, they may be cells that are already passaged from a primary culture; technically, such a culture is a cell line rather than a primary culture. If subcultured from the primary culture, the correct term is "early passage cell line" and not "primary cell line," which is a contradiction in terms. "Primary cells" should refer to cells that are derived from the primary culture and ready for reseeding. This distinction may seem a little pedantic, but it is important as the stage of the culture will affect its uniformity and remaining lifespan, i.e. the number of generations left before senescence (see Section 3.2.1).

13.1 RATIONALE FOR PRIMARY CULTURE

Primary culture is essential for the establishment of a new cell line. However, primary cultures are also important models in their own right. Some laboratories prefer to work with primary cultures instead of cell lines for various reasons, including:

(1) **Difficulty establishing cell lines.** Some species and cell types are virtually impossible to culture for prolonged periods. For example, very few cell lines have been established from marine invertebrates (Bayne 1998; Yoshino et al. 2013). Only one mollusk cell line, Bge, has been established (Yoshino et al. 2013); the only crustacean cell line, OLGA-PH-J/92, is now known to be misidentified and actually comes from a different species (Lee et al. 2011).

(2) **Similarity to tissue of origin.** Longstanding concerns have been expressed about cell lines and their similarity (or lack of similarity) to the tissues from which they were established (see Sections 3.2–3.4). Many of these concerns relate to the duration of the culture; prolonged culture results in clonal evolution and other changes that are not apparent in the cell line's tissue of origin. A primary culture will avoid many of these changes and, therefore, is more likely to reflect its tissue of origin.

(3) **Maintenance of heterogeneity and microenvironment.** Tissue samples are highly heterogeneous, with varying cell populations and microenvironments. Depending on how primary culture is performed, it is possible to retain many different cell populations and preserve or recapitulate the microenvironment using *ex vivo* or three-dimensional (3D) culture (Miller et al. 2017; Miserocchi et al. 2017).

(4) **Access to rare biological material.** If biospecimens are derived from individuals with rare diseases or mutations, primary culture can be used to maximize their yield. For example, serial transfer of explant tissue can be used to maximize fibroblast outgrowth and improve cellular yield for genomic analysis (see Section 13.4).

Freshney's Culture of Animal Cells: A Manual of Basic Technique and Specialized Applications, Eighth Edition. Amanda Capes-Davis and R. Ian Freshney.
© 2021 John Wiley & Sons Ltd. Published 2021 by John Wiley & Sons Ltd.
Companion website: www.wiley.com/go/freshney/cellculture8

(5) ***Personalized applications.*** Primary cultures are increasingly important for donor-specific testing or as the starting material for cell-based therapy. For example, primary cancer cells can be used to establish patient-derived xenografts (PDXs), 3D culture systems, and other useful models (Miserocchi et al. 2017). Use of tissue and primary cells can avoid many of the problems that arise from cell line establishment, including low success rates and prolonged time in culture (Kodack et al. 2017).

Primary culture has limitations, as do all forms of tissue culture (see Section 1.5). Tissue samples may be difficult to access and limited material may be available. Primary culture is a transient state (ending at the first subculture) and most cells in the culture will have a limited lifespan, restricting the number of experiments that can be performed per sample. The behavior of primary cultures can vary considerably due to differences between individuals or culture conditions. For example, primary glioma cultures show different drug responses when cultured in different media (Ledur et al. 2017). Good cell culture practice (GCCP) (see Section 7.2.1) is essential for primary cultures, just as it is for all forms of tissue culture (Pamies et al. 2018). Good record keeping is particularly important and is discussed later in this chapter (see Section 13.8).

13.2 INITIATION OF PRIMARY CULTURE

There are four stages to consider when initiating a new primary culture: (i) acquisition of the sample; (ii) isolation of the tissue; (iii) dissection and/or disaggregation; and (iv) culture after seeding into the culture vessel. After isolation of the tissue, several options are available to generate a primary culture (see Figure 13.1). Fine dissection can be performed to generate primary explants, which are fragments of tissue that are encouraged to adhere to the culture vessel or other substrate (see Section 13.4). Over time, cells will migrate from the explants onto the substrate. Alternatively, disaggregation can be performed to produce a cell suspension (see Sections 13.5, 13.6). Cells will either attach to the substrate or continue to grow in suspension (e.g. as spheroids; see Section 27.4.1). Most normal untransformed cells will attach to a flat surface in order to survive and proliferate with maximum efficiency, although there are some exceptions, including hematopoietic cells and stem cells. Transformed cells, on the other hand, will often proliferate in suspension, particularly cells from transplantable animal tumors (see Section 3.6.2).

Several techniques can be used to disaggregate tissue for primary culture (see Figure 13.1). These techniques can be divided into (i) purely mechanical techniques, involving dissection with or without some form of maceration; and (ii) enzymatic techniques, utilizing proteases or other enzymes. Enzymatic disaggregation gives a better yield when more tissue is available. Mechanical disaggregation works well with soft tissues and some firmer tissues when the size of the viable yield is not important, or where loosely adherent cells are removed from a more fibrous stroma.

Disaggregation can damage primary cells, which in turn affects their ability to adhere to the culture substrate (Waymouth 1974). Cellular damage can be reduced by optimizing the procedure, e.g. the temperature and duration of enzymatic treatment (see Section 13.5.1). However, it is important to balance the need for a high yield against any damage that may result. You should also consider what you wish to do with the cells after initiating primary culture. For example, it may be necessary to retain cell–cell contact for culture of human pluripotent stem cells (hPSCs), where single cell suspensions can result in abnormal karyotypes, or for 3D culture (Kondo et al. 2011; Beers et al. 2012).

13.2.1 Proteases Used in Disaggregation

Trypsin is the most widely used enzyme for disaggregation and has been used in tissue culture for more than a hundred years. However, many other enzymes can be used for disaggregation (see Table 13.1). Most of these enzymes have protease activity, resulting in disruption of collagen and other components of the extracellular matrix (ECM). Fibronectin and laminin are protease sensitive, while proteoglycans (which are less sensitive) can sometimes be degraded by glycanases, e.g. hyaluronidase or heparinase. Elastin, which is a major component in arterial tissue and ligaments, is not sensitive to trypsin but can be degraded by elastase (Waymouth 1974).

The enzymes used most frequently for tissue disaggregation are crude preparations that contain varying amounts of trypsin, collagenase, elastase, hyaluronidase, and deoxyribonuclease (DNase), alone or in various combinations. Crude preparations are often more successful than purified enzyme preparations, because they contain other proteases as contaminants. An increasing number of enzymes are available from non-mammalian sources or in purified recombinant form (see Table 13.1). Purified enzyme preparations are generally less toxic and more specific in their action.

How do you decide on the best enzyme or enzyme mixture for your application? The easiest approach is to proceed from a simple disaggregation solution to a more complex solution (see Table 13.1), with trypsin alone or trypsin/ethylene diamine tetra-acetic acid (EDTA) as a starting point. Add or substitute other proteases to improve disaggregation and remove trypsin to increase viability if necessary. Proceed with caution when adding proteases; there is a risk that additions can result in proteolysis and inactivate some of the other enzymes.

Trypsin and Pronase give the most complete disaggregation but are more likely to cause damage. Collagenase and Dispase give incomplete disaggregation but are less harmful. Hyaluronidase and elastase are usually used in combination with other enzymes, e.g. collagenase (Brieno-Enriquez et al. 2010; Chi et al. 2017). Specific mixtures may be advisable for certain cell types. For example, elastase and DNase are

TABLE 13.1. Enzymes used in disaggregation of tissue.

Enzyme	Source[a]	Concentration	Medium	Recommended for[b]
Crude trypsin	Porcine pancreas	0.25%	DPBS-A or CMF ± 1 mM EDTA	Many tissues (see Sections 13.5.1, 13.5.2)
Purified trypsin	Porcine pancreas	25 µg/ml	DPBS-A or CMF ± 1 mM EDTA	Many tissues, e.g. lung (Finkelstein and Shapiro 1982)
TrypZean™	Recombinant trypsin, expressed in corn (Sigma-Aldrich)	As per instructions	DPBS-A or CMF	Many tissues
rTrypsin	Recombinant trypsin, expressed in bacteria (Novozymes)	As per instructions	DPBS-A or CMF ± 1 mM EDTA	Large-scale culture
TrypLE™ express, select	Recombinant trypsin-like protease, expressed in bacteria (ThermoFisher Scientific)	As per instructions	DPBS-A or CMF	Many tissues
Trypsin + collagenase	Porcine/bacterial	0.25% crude trypsin; 200 U/ml crude collagenase	DPBS-A or CMF ± 1 mM EDTA	Fibrous tissues
Crude collagenase	Bacterial	200 U/ml	Culture medium	Many tissues, e.g. breast epithelium (see Section 13.5.3)
Liberase™	Purified collagenase I + II (also Dispase or thermolysin)	As per instructions	Water or tissue dissociation buffer	Many tissues, e.g. liver and pancreatic islets (see Supp. S24.6, S24.7)
Accumax	Molluscan (Innovative Cell Technologies)	As per instructions	DPBS-A, CMF, culture medium	Many tissues
Papain	Papaya	20 U/ml	EBSS with 1 mM cysteine, 0.5 mM EDTA	Many tissues, e.g. neural cells (Grozdanov et al. 2010)
Dispase	Bacterial	0.1–1.0 mg/ml	DPBS-A or culture medium	Removal of epithelium in sheets (does not dissociate epithelium)
Pronase	Bacterial	0.1–1.0 mg/ml	Culture medium	Provision of good single-cell suspensions (may be harmful to some cells)
DNase I	Various	2–20 µg/m	Culture medium	Decreased clumping and presence of DNA fragments
Hyaluronidase (usually with collagenase)	Bovine testes	100 U/ml	Culture medium	Breast epithelium (Stampfer et al. 2002)
Elastase (usually with collagenase)	Porcine pancreas	100 µg/ml	DPBS-A or culture medium	Fibrous tissues, e.g. rat aorta (Chi et al. 2017)

[a]Products may come from multiple suppliers; examples are included here in parentheses.
[b]Suppliers' recommendations, with additions from publications where cited. Additional information is available from suppliers and other references (Worthington Biochemical Corporation 2018).
[c]CMF, Calcium- and magnesium-free saline; DPBS-A, Dulbecco's phosphate buffered saline without Ca^{2+} and Mg^{2+}.

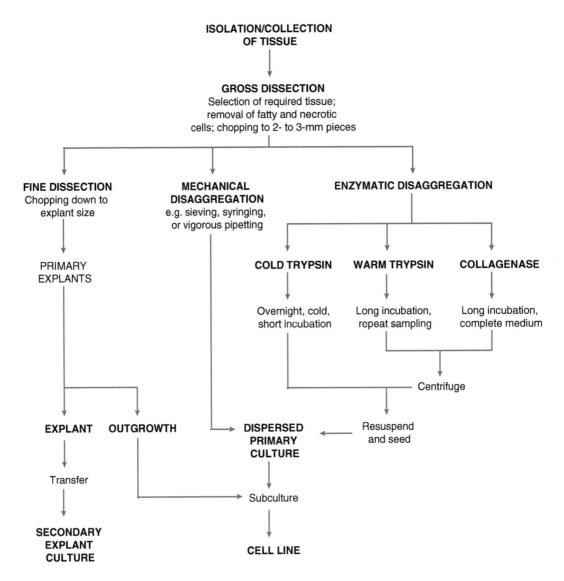

Fig. 13.1. Options for primary culture. Multiple paths may be successful, including: left, explant culture; center, mechanical disaggregation; right, enzymatic disaggregation. An explant may be transferred to allow further outgrowth to form ("secondary explant culture"), while the outgrowth from the explant may be subcultured to form a cell line.

effective for primary culture of type II alveolar cells, and collagenase and Dispase for intestinal epithelial cells (Booth and O'Shea 2002; Dobbs and Gonzalez 2002). Always check for recommendations from scientists with expertise in that particular tissue (Masters 1991).

Some disaggregation methods result in clusters or aggregates of epithelial cells, e.g. collagenase or the cold trypsin method (see Section 13.5.3; see also Figure 13.2). Such aggregates may be desirable if you wish to perform spheroid culture or isolate particular structures, such as ovarian follicles or pancreatic islets (Dolmans et al. 2006; Kondo et al. 2011). Aggregates can be selected under a dissection microscope and transferred to individual wells in a microtitration plate, either alone or with a feeder layer (see Section 8.4; see also Protocols P20.7, P25.2).

Enzymes used in disaggregation can be removed by washing the cells by centrifugation after disaggregation is complete. If the viable cell yield is high enough, diluting the cell suspension to an appropriate seeding concentration may be sufficient to dilute out the enzyme. Washing is usually insufficient for trypsin, which continues to bind cellular structures (Waymouth 1974). Trypsin should be neutralized by adding serum-containing medium (effective because serum has antiprotease activity; see Table 9.4) or a trypsin inhibitor (e.g. soybean trypsin inhibitor) (Morgan and Parker 1949).

13.2.2 Other Agents Used in Disaggregation

Chelating agents are often added to trypsin, such as EDTA or ethylene glycol tetra-acetic acid (EGTA). These agents

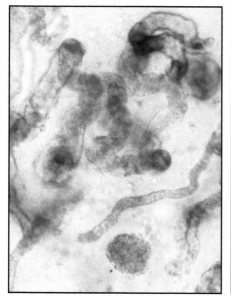

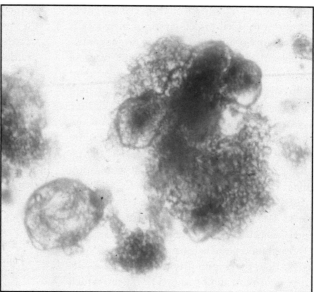

(a) Cold Trypsin, Mouse Kidney. Disaggregation of newborn mouse kidney by cold trypsin method. The connective tissue has dissociated but fragments of tubules and glomeruli remain intact. Bright field illumination; 4x objective. Source: R. Ian Freshney.

(b) Collagenase, Human Colon Carcinoma. As with the cold trypsin method, the connective tissue has dissociated following collagenase treatment but colonic crypts are preserved intact. Bright field illumination; 4x objective. Source: R. Ian Freshney.

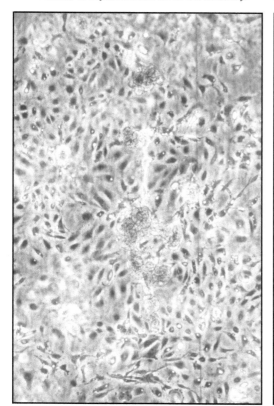

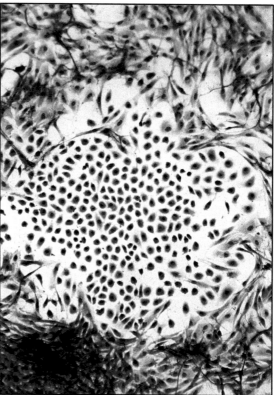

(c) Primary Culture, Kidney Tubules. Outgrowth of cells from attached fragments following disaggregation by cold trypsin method. Phase contrast; 10x objective. Source: R. Ian Freshney.

(d) Primary Culture, Newborn Rat Kidney. Primary culture from newborn rat kidney following disaggregation by cold trypsin method. Giemsa stained; 10x objective. Source: Courtesy of Mary G. Freshney.

Fig. 13.2. Disaggregation using cold trypsin or collagenase.

are used to chelate divalent cations involved in cell adhesion, particularly calcium ions. Cell to cell adhesion in tissues is mediated by a variety of homotypic interacting glycopeptides (cell adhesion molecules; see Section 2.2.2). Some of these molecules (particularly cadherins) are calcium-dependent and hence are sensitive to chelating agents. Integrins, which bind to the arginyl-glycyl-aspartic acid (RGD) motif (see Section 8.1), also have calcium-binding domains and are affected by calcium depletion. EGTA can be used at a higher concentration as it is less toxic than EDTA, e.g. 1 mM EDTA and 6.5 mM EGTA are used in TEGPED trypsinization solution (Stampfer et al. 2002).

DNase is used to disperse DNA released from lysed cells (see Table 13.1). DNA tends to impair proteolysis and promote reaggregation, resulting in clumping or a gelatinous appearance. DNase should be added sequentially after trypsin has been deactivated, as the trypsin may degrade the DNase. A gelatinous appearance can also be caused by mucin, which may be released during primary culture of intestinal epithelium. Mucin can be dispersed by washing cells in a solution of 0.05% dithiothreitol (DTT) and 3 mM EDTA (Whitehead et al. 1999).

13.2.3 Common Features of Disaggregation

Although each tissue may require a different set of conditions, certain requirements are shared by most of them:

(1) Fat and necrotic tissues are best removed during dissection.
(2) The tissue should be chopped finely with sharp scalpels to cause minimum cellular damage.
(3) Enzymes used for disaggregation should be removed subsequently by gentle centrifugation. Some enzymes require inactivation (e.g. trypsin; see Section 13.5.1).
(4) The concentration of cells in the primary culture should be much higher than that normally used for subculture, because the proportion of cells from the tissue that survives in primary culture may be quite low.
(5) Embryonic tissue disaggregates more readily, yields more viable cells, and proliferates more rapidly in primary culture than does adult tissue.
(6) A rich medium, such as Ham's F12, is preferable to a simple medium, such as Eagle's Minimum Essential Medium (MEM). If serum is required, fetal bovine serum (FBS) often gives better survival than does calf or horse. Isolation of specific cell types will probably require selective serum-free media (see Sections 10.6.1, 20.7.1, 25.4.1).

13.3 TISSUE ACQUISITION AND ISOLATION

Acquisition and isolation of tissue are important steps in primary culture. The nature of the tissue sample and its handling and storage conditions all contribute to the success or failure of the primary culture. Success rates will vary depending on the amount of trauma during tissue collection, the duration of ischemia, and the procedure used for tissue isolation. For example, tissue biopsies and surgical samples can result in different success rates for primary culture of human gastric mucous epithelial cells (Rutten et al. 1996). Removal of the underlying muscle layer from surgical samples is thought to reduce subsequent tissue necrosis and improve success rates in primary culture.

Correct processing and storage of tissue are essential if you are to have the correct information available at publication (see Section 7.3.2) (Geraghty et al. 2014). The following steps provide a good set of samples and information (Sachs et al. 2018):

(1) Photograph the sample before cutting tissue into small pieces (e.g. 1–3 mm^3).
(2) Process the sample for primary culture as described in this chapter.
(3) Fix two randomly chosen pieces in formalin for histopathological analysis and immunohistochemistry.
(4) Freeze the remaining tissue for later use. Always snap-freeze some pieces for later authentication testing (see Protocol P13.2; see also Section 17.2.3). Tissue fragments can also be subjected to cryopreservation for later culture (see Protocol P25.1).

13.3.1 Non-Human Tissue Samples

Before you work with animal tissue, make sure that you comply with any relevant ethical requirements as an essential part of GCCP (see Section 6.4.2) (Geraghty et al. 2014; Pamies et al. 2018). The need for bioethics will vary with the animal's age and species, the nature of the work, and your location. Always seek advice from your local compliance office or animal ethics committee. In the United Kingdom, live animals and fetuses that have passed two-thirds of their gestation or incubation period are protected under the Animals (Scientific Procedures) Act (UK Home Office 2018). In the United States, live vertebrates are covered by the Public Health Service (PHS) Policy on Humane Care and Use of Laboratory Animals (OLAW 2018).

Animal tissue must be obtained without causing undue pain or suffering and specified methods of humane killing may be required. Usually, separating the embryo from the fetal membranes or placenta followed by decapitation is regarded as humane, particularly if the gravid uteri or eggs are placed at 4 °C beforehand. An attempt should be made to sterilize the site of the resection with 70% alcohol if the site is likely to be contaminated (e.g. skin). Remove the tissue aseptically and transfer it to dissection balanced salt solution (DBSS) (see Appendix B) as soon as possible. Do not dissect animals in the tissue culture laboratory, as the animals may carry microbial contamination. If a delay in transferring the tissue is unavoidable, it can be held at 4 °C for up to 72 hours, although a better yield will usually result from a quicker transfer.

13.3.2 Mouse Embryo

Mouse embryos are a convenient source of undifferentiated mesenchymal cells. These cultures are often referred to as mouse embryonic fibroblasts (MEFs) and are used as feeder layers (see Section 8.4; see also Protocols P20.7, P25.2). Full term is about 19–21 days, depending on the strain. The optimal age for preparing cultures from a whole disaggregated embryo is around 13 days, when the embryo is relatively large and the cell yield is sufficiently high (see Figures 13.3, 13.4). This age contains a high proportion of undifferentiated mesenchyme. However, isolation and handling of later embryos may require a license, so embryos at 9–10 days may be preferable. Although the amount of tissue recovered from younger embryos will be substantially less, a higher proportion of the cells will grow. Once a given age is selected and shown to be satisfactory, this embryonic age should be used for all further studies. The phenotype of the cells in culture, the matrix they produce, and the conditioning of the medium that they provide may be different at other ages.

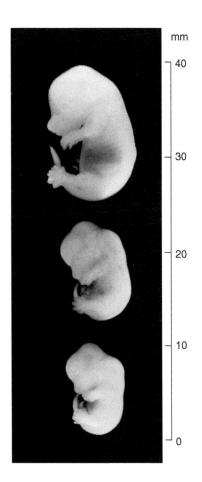

Fig. 13.3. Mouse embryos. Animals are shown from the twelfth, thirteenth, and fourteenth days of gestation. The 12-day embryo (bottom) came from a small litter (three) and is larger than would normally be found at this stage. *Source*: R. Ian Freshney.

Protocol P13.1 describes the isolation of whole mouse embryos and does not apply to individual organs. Most individual organs, with the exception of the brain and the heart, begin to form at about the ninth day of gestation, but are difficult to isolate until about the eleventh day. Dissection of individual organs is easier at 13–14 days, and most of the organs are completely formed by the eighteenth day. Stages in mouse embryo development are described in the eMouse Atlas Project (EMAP) (Armit et al. 2017).

PROTOCOL P13.1. ISOLATION OF MOUSE EMBRYOS

Background

Before you start work with animal tissue, check that you comply with any relevant requirements for biosafety and bioethics. Always perform a risk assessment to determine the hazards associated with your work (see Section 6.1.1) and ensure that animals are handled humanely in accordance with ethical requirements (see Sections 6.4.1, 13.3.1).

Outline

Remove uterus aseptically from a timed pregnant mouse and dissect out embryos.

- *Safety Note. The Bunsen burner is a fire hazard. Do not use in the laminar flow hood. When sterilizing instruments by dipping them in alcohol and flaming them, take care not to return the instruments to alcohol while they are still alight!*

Animals

- ☐ Timed pregnant mice (see step 1 of this protocol)

Materials (sterile)

- ☐ DBSS in 30- to 50-ml screw-capped tube or universal container (used for tissue samples, see Appendix B)
- ☐ Balanced salt solution, 50 ml in a sterile beaker (used to cool instruments after flaming)
- ☐ Petri dishes, 90 mm
- ☐ Pointed forceps
- ☐ Pointed scissors

Materials (non-sterile)

- ☐ Laminar flow hood
- ☐ Alcohol, 70%, in wash bottle (used to disinfect surfaces and mouse skin)
- ☐ Alcohol, 70% (to sterilize instruments; see Figure 6.6)

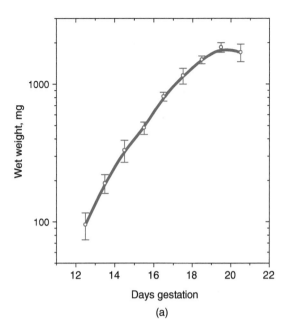
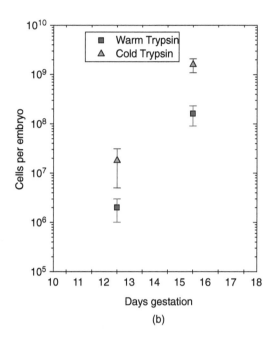

Fig. 13.4. Total wet weight and cell yield per mouse embryo. (a) Wet weight is given per embryo without placenta or membranes, mean ± standard deviation. *Source*: adapted from Paul et al. (1969). (b) Viable cell yield is given per embryo after incubation in 0.25% trypsin at 37 °C for four hours with no intermediate harvesting (squares) or after soaking in 0.25% trypsin at 4 °C for five hours and incubation at 37 °C for 30 minutes (triangles) (see Protocol P13.4). *Source*: adapted from Paul, J., Conkie, D. and Freshney, R. I. (1969). Erythropoietic cell population changes during the hepatic phase of erythropoiesis in the foetal mouse. Cell Tissue Kinet 2 (4):283–294.

☐ Bunsen burner (positioned away from laminar flow)

Procedure

A. Animal Preparation and Dissection

1. *Induction of estrus.* If males and females are housed separately, put together for mating. Estrus will be induced in the female three days later, when the maximum number of successful matings will occur. This process enables the planned production of embryos at the appropriate time. The timing of successful matings may be determined by examining the vagina each morning for a hard, mucous plug.
2. *Dating the embryos.* The day of detection of a vaginal plug, or the "plug date," is noted as day zero, and the development of the embryos is timed from this date. Day 13 is optimal for primary culture.
3. Kill the mouse using a humane method (e.g. AVMA 2013) and swab the ventral surface liberally with 70% alcohol (see Figure 13.5a).
4. Tear the ventral skin transversely at the median line just over the diaphragm (see Figure 13.5b). Grasping the skin on both sides of the tear, pull

in opposite directions to expose the untouched ventral surface of the abdominal wall (see Figure 13.5c).
5. Cut longitudinally along the median line of the exposed abdomen with sterile scissors, revealing the viscera (see Figure 13.5d). At this stage, the two horns of the uterus, filled with embryos, are obvious in the posterior abdominal cavity (see Figure 13.5e).
6. Dissect out the uterus into a 30-ml or 50-ml screw-capped tube or universal container containing 10 ml or 20 ml DBSS (see Figure 13.5f).
7. Take the intact uteri to the tissue culture laboratory and transfer them to a fresh Petri dish of sterile DBSS (see Figure 13.5g).

B. Embryo Dissection

8. Tear the uterus with two pairs of sterile forceps to reveal the embryos. Keep the points of the forceps close together to avoid distorting the uterus and bringing too much pressure to bear on the embryos (see Figure 13.5g, h).
9. Free the embryos from the membranes (see Figure 13.5i) and placenta and place them to one side of the dish to bleed.

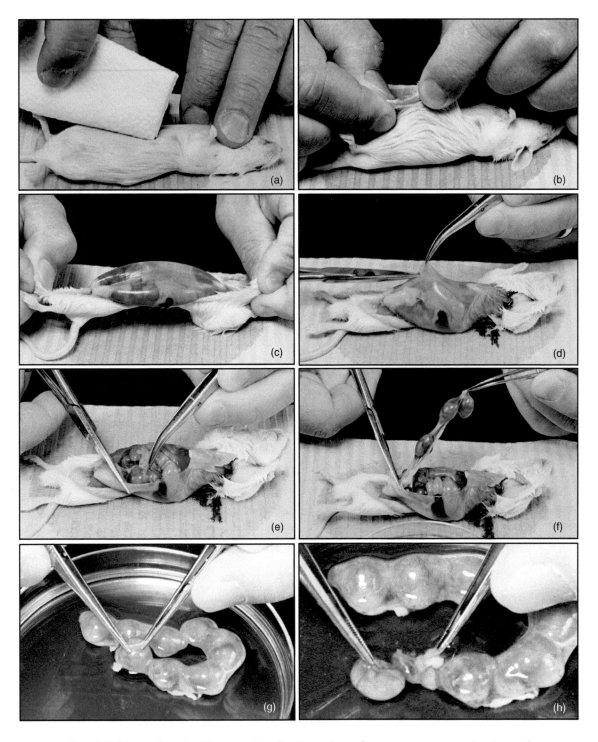

Fig. 13.5. Mouse dissection. Key steps in collecting embryos from a pregnant mouse (see Protocol P13.1) include (a) swabbing the abdomen; (b, c) tearing the skin to expose the abdominal wall; (d) opening the abdomen; (e) revealing the uterus *in situ*; (f) removing the uterus; (g, h) dissecting the embryos from the uterus; (i) removing the membranes; (j) removing the head (optional); (k) chopping the embryos; (l) transferring pieces to a trypsinization flask (for warm trypsinization, see Protocol P13.4); (m) transferring pieces to a small Erlenmeyer flask (for cold trypsinization, see Protocol P13.5); (n) flask on ice. *Source*: R. Ian Freshney.

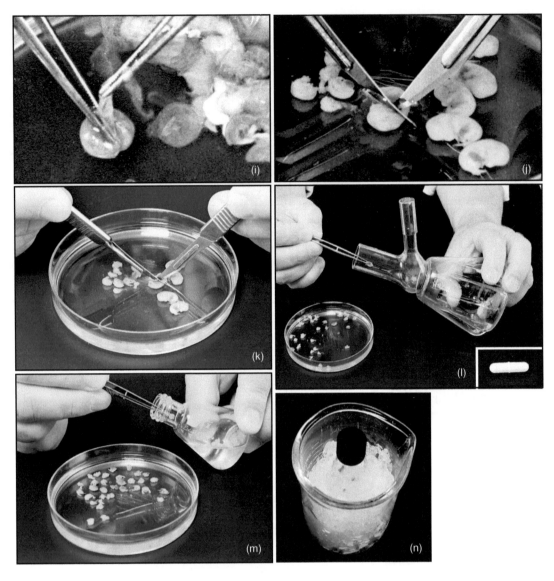

Fig. 13.5. (*Continued*)

10. Transfer the embryos to a fresh Petri dish. If a large number of embryos is required (i.e. more than 4–5 l), it may be helpful to place the dish on ice.

11. Remove any parts of the embryo that are not required and chop the tissue into smaller pieces (see Figure 13.5j, k).

12. If warm trypsinization will be used, transfer pieces to a trypsinization flask (see Figure 13.5l; see also Protocol P13.4). If cold trypsinization will be used, transfer pieces to a small Erlenmeyer flask and place on ice.

13. Make sure you record key information (e.g. the collection date and animal species and strain) and put aside tissue samples for later use (e.g. authentication; see Section 13.3).

Notes

Step 3: Dissection is performed using a laminar flow hood. A laminar flow hood does not protect from pathogens; select equipment based on the level of risk (see Section 6.3.4).

Step 7: All of the preceding steps should be done outside the tissue culture laboratory, using the laminar flow hood and rapid technique to help maintain sterility. Do not take live animals into the tissue culture laboratory, as the animals may carry contamination. If an animal carcass must be handled in the tissue culture area, make sure that the carcass is immersed in alcohol briefly, or thoroughly swabbed, and disposed of quickly after use.

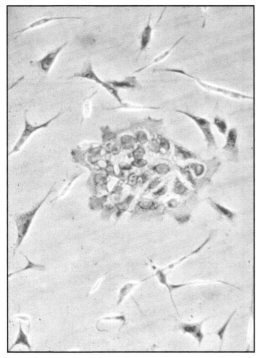

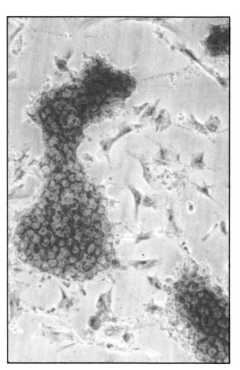

(a) Chick Embryo Lung. Primary culture of 13-day chick embryo lung, 48 h after disaggregation by the cold trypsin method. Light Giemsa stain and phase contrast; 10x objective.

(b) Chick Embryo Liver. Culture details as for (a).

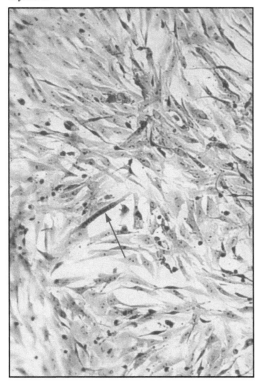

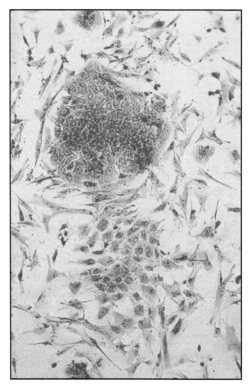

(c) Chick Embryo Thigh Muscle. Culture details as for (a). Note multinucleate myotube (red arrow) resulting from fusion of myocytes (small dark-stained spindles). Giemsa stain, bright field; 10x objective.

(d) Chick Embryo Kidney. Culture details as for (a). Giemsa stain, bright field; 10x objective. Source for all images: R. Ian Freshney.

Fig. 13.6. Culture of chick embryo organ rudiments.

13.3.3 Chick Embryo

The chick embryo is a major model system for developmental biology and for viral propagation (Stern 2005; Brauer and Chen 2015). Chick embryos, like mouse embryos, can provide predominantly mesenchymal cell primary cultures for feeder layers and for cell proliferation analysis. Chick embryos are also helpful for teaching; they are larger than mouse embryos at the equivalent stage of development and thus are easier to dissect. Because of their larger size, it is easier to generate specific cell types from individual organs, e.g. lung epithelium, hepatocytes, muscle, and kidney cells (see Figure 13.6).

Protocols to isolate chick embryos and to disaggregate chick embryo organ rudiments are provided in the Supplementary Material online (see Supp. S13.1, S13.2). A training exercise is available for the primary culture of the chick embryo, which may be useful for advanced trainees (see Supp. S30.15). An atlas of chick embryo development is available as part of the eChick Atlas Project (ECAP) (Wong et al. 2013).

13.3.4 Human Biopsy Samples

Human tissue has additional requirements and challenges that must be addressed, including the need for ethical review and informed consent (see Sections 6.4.2, 6.4.3). For human biopsy material, it is usually necessary to obtain approval or consent from (i) the local ethics committee; (ii) the attending physician or surgeon; and (iii) the individual donor or their next of kin if the donor is unable to give consent (see Section 6.4.3). Biopsy sampling is usually performed for diagnostic purposes, and hence the needs of the pathologist must be met first. This factor is less of a problem if extensive surgical resection or non-pathological tissue (e.g. placenta or umbilical cord) is involved.

The biopsy is often performed at a time that is not always convenient to the tissue culture laboratory, so some formal collection or storage system must be employed for times when you or someone on your team cannot be there. If delivery to your lab is arranged, there must be a system for receiving specimens, recording key information (see Section 13.8), and alerting the person who will perform the culture that the specimens have arrived. Otherwise, unique material may be lost or spoiled, and subsequent work will be severely limited.

PROTOCOL P13.2. HANDLING HUMAN BIOPSIES

Background

Before you start work with human tissue, check that you comply with ethical, safety, and transport requirements. Always perform a risk assessment to determine the appropriate biosafety level for your work (see Sections 6.1.1, 6.3.3) and ensure that your project has passed through ethical review and the donor has given informed consent (see Sections 6.4.2, 6.4.3). The sample should be suitably packaged for transport (see Section 15.8).

Although aseptic technique will be used during collection, human biopsies and tissue samples may still carry pathogens and other microbial contaminants. Superficial specimens (skin biopsies, melanomas, etc.) and gastrointestinal tract specimens are particularly prone to contamination. Contamination occurs even when a disinfectant wash is given before skin biopsy and a parenteral antibiotic is given before gastrointestinal surgery. It may be advantageous to consult a medical microbiologist to determine which flora to expect in a specific tissue and then choose your antibiotics for collection and dissection accordingly.

This protocol focuses primarily on steps before culture commences. Further handling will depend on the amount of material available and the method of primary culture to be used (see Protocols P13.3–P13.7). Remaining material should be frozen for future use (see Protocol P25.1).

Outline

Consult with hospital staff, provide labeled container(s) of medium, and arrange for collection of samples from operating room or pathologist.

- **Safety Note. Pathogens may be present in biopsy material. Human tissue should be handled in a biological safety cabinet (BSC) (see Section 6.3.4). Equipment and waste must be decontaminated after use or before disposal (see Sections 6.3.5, 6.3.6). A risk assessment must be performed to decide if other control measures are needed, e.g. testing for specific pathogens (see Section 6.3.1).**

Materials (sterile)

☐ Specimen tubes (15–30 ml) with leak-proof caps. Tubes should contain Collection Medium (about half full; see Appendix B) and should be clearly labeled with your name, laboratory address, and telephone number.

Procedure

1. Discuss with the donor's clinical care team to ensure that you understand their requirements when collecting and processing the sample, and that they understand your requirements

for collection, storage, and transport to the laboratory.

2. Provide containers of collection medium to the anteroom of the operating theater or to the pathology laboratory. Make sure that all specimen tubes and packaging are clearly labeled (see sterile Materials above).

3. Make arrangements to be alerted when the material is ready for collection.

4. Collect the containers after surgery or have someone send them to you immediately after collection and inform you when they have been dispatched.

5. On arrival, record the key information for each sample and assign a unique sample identifier number (see Notes below and Section 13.8).

6. Transfer the sample to the tissue culture laboratory for dissection and culture. Usually, if kept at 4 °C, biopsy samples survive for at least 24 hours and even up to three or four days, although the longer the time from surgery to culture, the more the samples are likely to deteriorate.

7. If microbial contamination is a significant risk, decontamination may be helpful. If the surgical sample is large enough (> 200 mg), a brief dip in 70% alcohol (30–60 seconds) will help to reduce superficial contamination without harming the center of the tissue sample.

Notes

Step 5: Key information usually includes the date and time of collection and arrival, the number of samples, the tissue of origin, the disease diagnosis, and the age and sex of the donor. A unique sample identifier number should be assigned on arrival and incorporated into the subsequent culture designation. This enables cultures to be de-identified in subsequent laboratory records and publications (see Supp. S6.4). Information should be recorded in a hand-written log book or electronic database. If hand-written, the information should be copied to an electronic database for improved access and long-term storage.

13.4 PRIMARY EXPLANTATION

The primary explant technique was the original method used to initiate tissue culture (Harrison 1907; Carrel 1912). As originally performed, a fragment of tissue was embedded in blood plasma or lymph, mixed with heterologous serum and embryo extract, and placed on a coverslip that was inverted over a concavity slide. The clotted plasma held the tissue in

place and the explant could be examined with a conventional microscope. The heterologous serum induced clotting of the plasma and, together with the embryo extract and plasma, supported cell proliferation and stimulated migration from the explant. This technique has largely been replaced by the simplified method described below (see Protocol P13.3), which is also useful for training (see Supp. 30.15).

Primary explants are particularly valuable for small or precious tissue samples, where there is a risk of losing cells during disaggregation. Primary explants can also be used to avoid the damage caused by disaggregation and to reduce the need for proteases, e.g. to isolate mesenchymal stromal cells (MSCs) from human adipose tissue (Priya et al. 2014). Disadvantages of primary explantation include difficulty with explant adhesion and selection of more migratory cells within the outgrowth. However, most cells (particularly embryonic cells) can migrate out successfully, provided sufficient time is given for their outgrowth and suitable culture conditions are used for their selection.

Adhesion and migration can be stimulated by placing a glass coverslip on top of the explant, with the explant near the edge of the coverslip. The plastic dish may be scratched through the explant to attach the tissue to the flask (see Supp. S24.4) (Elliget and Lechner 1992). Substrate treatments can also be used to promote attachment, e.g. coating with ECM components or polylysine (see Sections 8.3.2, 8.3.5). Historically, plasma clots have been used to promote attachment. To use a plasma clot, place a drop of plasma on the plastic surface and embed the explant in the drop. The plasma should clot after a few minutes, whereupon medium can be added. Purified fibrinogen and thrombin can be used to give a similar result (Nicosiea and Ottinetti 1990).

Historically, tissue explants were often passaged to a new plasma clot to encourage further outgrowth. A similar approach can be used to increase the yield from small amounts of tissue (Huschtscha et al. 2012). Explants can be transferred between dishes up to 80 times, increasing the cellular yield by up to two orders of magnitude, and can be cryopreserved for later use. The procedure used for maximal serial transfer (MST) of skin explants is provided as Supplementary Material online (see Supp. S13.3).

PROTOCOL P13.3. CULTURE OF PRIMARY EXPLANTS

Background

Explants are encouraged to adhere to the surface by seeding pieces in a small volume of medium with a high concentration (i.e. 40–50%) of serum. Surface tension should hold the pieces in place until they adhere spontaneously to the surface (see Figure 13.7a). Once this is achieved, outgrowth of

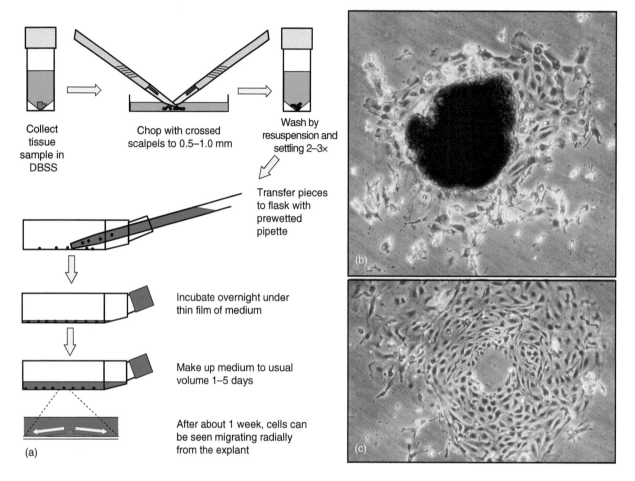

Fig. 13.7. Primary explant culture. (a) Schematic diagram of stages in dissection and seeding primary explants; (b) primary explant culture from mouse squamous skin carcinoma, showing explant and early stage of outgrowth about three days after explantation; (c) outgrowth after removal of explant, about seven days after explantation. 10× objective. *Source*: R. Ian Freshney.

cells usually follows (see Figure 13.7b, c; see also Figure 13.2c).

Outline

Chop the tissue finely and rinse. Seed the pieces onto the surface of a culture flask or Petri dish.

Materials (sterile or aseptically prepared)

☐ Tissue sample in Collection Medium or DBSS (see Appendix B)
☐ DBSS, 100 ml
☐ Growth medium (e.g. DMEM:F12 with 20% FBS), 100 ml
☐ Petri dishes, 90 mm (three; non-tissue culture grade is suitable)
☐ Centrifuge tubes, 15 ml or 50 ml, or universal containers (two)

☐ Culture flasks, 25 cm² (number determined by tissue size, 5 per 100 mg of tissue)
☐ Forceps, straight, and curved, for dissection and transfer of tissue
☐ Scalpels, #11 blade (two)
☐ Pipettes, 10 ml or 25 ml, with wide tips

Materials (non-sterile)

☐ BSC and associated consumables (see Protocol P12.1)

Procedure

A. Preparation of Tissue Pieces:

1. Using a BSC, prepare a work space for aseptic technique and bring in the tissue sample in a Petri dish or similar container.

2. Transfer the tissue to fresh, sterile DBSS in a 90-mm Petri dish, and rinse. Be careful not to lose the tissue.

3. Transfer the tissue to a second Petri dish. Dissect off unwanted tissue, such as fat or necrotic material, and transfer to a third dish.

4. Chop finely with crossed scalpels (see Figure 13.7a, top) into 1- to 3-mm^3 cubes. Embryonic organs, if they do not exceed this size, are better left whole.

5. Transfer the tissue pieces to a centrifuge tube or universal container. Be careful not to lose the pieces – use a prewetted pipette with a wide tip. Allow the pieces to settle.

6. Wash the tissue by resuspending the pieces in DBSS, allowing the pieces to settle, and removing the supernatant.

7. Repeat this step two more times and resuspend the pieces in the last wash.

B. Explant Culture:

8. Transfer the pieces (remember to wet the pipette) to a culture flask, with about 20–30 pieces per 25 cm^2 flask.

9. Remove most of the fluid and add 1 ml growth medium per 25 cm^2 growth surface. Tilt the flask gently to spread the pieces and medium evenly over the growth surface.

10. Cap the flask and incubate in a humidified incubator at 37 °C for 18–24 hours.

11. Add 1 ml medium the following day, allowing the medium volume to increase as the pieces adhere to the growth surface.

12. Make up the medium volume gradually to 5 ml per 25 cm^2 over the next three to five days.

13. Change the medium weekly until a substantial outgrowth of cells is observed (see Figure 13.7b).

14. Once an outgrowth has formed, the remaining explant may be picked off with a scalpel (see Figure 13.7c) and transferred to a fresh culture vessel. Repeat from step 8.

15. Replace the medium in the first flask until the outgrowth has spread to cover at least 50% of the growth surface, at which point the cells may be subcultured (see Protocol P14.2).

Notes

Steps 2 and 3: Clean healthy tissue with little blood may not need two transfers and can be dissected in the first dish, after transfer from the transport medium, which will act as the first wash.

Step 5: Tissue pieces tend to stick where they are not wanted, e.g. the inside of a pipette. When pipetting tissue samples, always prewet pipettes by pipetting liquid up and down before use.

Steps 6 and 7: These washes may be omitted if there is little blood or necrotic tissue.

13.5 ENZYMATIC DISAGGREGATION

Enzymatic disaggregation of the tissue avoids selection by rate of migration and yields a higher number of cells that are more representative of the whole tissue in a shorter time. However, just as the primary explant technique selects on the basis of cell migration, disaggregation techniques will select protease- and mechanical stress-resistant cells. Embryonic tissue disperses more readily and gives a higher yield of proliferating cells than newborn or adult tissue. The increasing difficulty in obtaining viable proliferating cells with increasing age is due to several factors, including the onset of differentiation, an increase in ECM and fibrous connective tissue, and a reduction of the undifferentiated proliferating cell pool. When procedures of greater severity are required to disaggregate the tissue (e.g. longer trypsinization or increased agitation), the more fragile components of the tissue may be destroyed. For example, it is very difficult to obtain complete dissociation in fibrous tumors using trypsin while still retaining viable carcinoma cells.

13.5.1 Trypsin

Crude trypsin is by far the most common enzyme used in tissue disaggregation, as it is tolerated quite well by many cells and is effective for many tissues (Waymouth 1974). Crude trypsin (often with chelating agents; see Section 13.2.2) is available in powdered form or in solution (typically 0.25%, although higher and lower concentrations are available). Trypsin solutions are mostly stored frozen and thawed just before use to prevent autodigestion. Some trypsin solutions include stabilizing agents that allow storage at 2–8 °C (e.g. StableCell™ trypsin, Sigma-Aldrich). The optimum pH for trypsin is between 7.8 and 8.5, but a pH around 7.6 is normally used to avoid excessive alkalinity.

The choice of which trypsin grade to use has always been difficult, as there are two opposing trends: (i) the purer the trypsin, the less toxic it becomes, and the more predictable its action; (ii) the cruder the trypsin, the more effective it may be, because of the presence of other proteases. Crude trypsin was traditionally supplied at 1 : 250 grade, which is a specific activity of 250 USP U/mg and is equivalent to 750 BAEE U/mg (the abbreviation BAEE refers to Nα-benzoyl-L-arginine ethyl ester, which acts as a substrate to measure trypsin activity). Some current trypsin products are close to this; Sigma-Aldrich quotes a specific activity of 1000–2000 BAEE U/mg for # T4799. However, the activity of purified trypsin will vary with the product. For example, trypsin type IX-S is 13 000–20 000 BAEE U/mg (#T0303, Sigma-Aldrich), while chromatographically purified trypsin is around 10 000 BAEE U/mg (#TRL3, Worthington Biochemical Corporation). Both of these products are more than 10 times as active as the original 1 : 250 grade. Recombinant trypsin is typically formulated to have similar activity to the crude trypsin that is used for cell culture (see Table 13.1). Specific activity is often not quoted, although there are exceptions, e.g. Prospec quotes a specific activity of 10 000 BAEE U/mg for recombinant human trypsin, and Roche quotes 10 800 BAEE U/ml for recombinant porcine trypsin. Recombinant trypsin products are assumed to be purer and are typically animal product free.

In practice, the balance between disaggregation ability and sensitivity to toxic effects may be difficult to predict. A series of preliminary test experiments is often useful to determine the optimum (i) enzyme grade and purity; (ii) concentration; (iii) duration; and (iv) temperature for good cellular yield. As trypsin batches may vary in specific activity, you may need to test a series of different concentrations with a particular tissue before selecting the correct one. Crude trypsin is often used at 0.25%; purer grades of trypsin typically allow lower concentrations, e.g. 0.01–0.05%. It is important to minimize the exposure of cells to active trypsin in order to preserve maximum viability. Hence, when whole tissue is being trypsinized at 37 °C, dissociated cells should be collected every half hour, and the trypsin should be removed by centrifugation and neutralized (see Section 13.2.1). If reaggregation occurs after cells are resuspended, incubate the cells in DNase (10–20 μg/ml) for 10–20 minutes and recentrifuge (see Section 13.2.2).

The following procedure (Protocol P13.4) is useful as a starting point for warm trypsinization. A related procedure (Protocol P13.5) follows for cold trypsinization (see Section 13.5.2). Both protocols can be used for training (see Supp. 30.11).

PROTOCOL P13.4. WARM TRYPSIN DISAGGREGATION

Background

The warm trypsin method is useful for the disaggregation of large amounts of tissue in a relatively short time, particularly for chopped whole mouse embryos or chick embryos (see Protocol P13.1; see also Supp. S13.1). It does not work as well with adult tissue, in which there is a lot of fibrous connective tissue, and mechanical agitation can be damaging to some of the more sensitive cell types, such as epithelium.

Outline

Chop the tissue finely and stir in trypsin for a few hours. Collect dissociated cells every half hour until disaggregation is complete. Centrifuge, resuspend, and pool the cell suspensions. Seed into flasks in growth medium (see Figure 13.8).

Materials (sterile or aseptically prepared)

- Tissue sample in Collection Medium or DBSS (see Appendix B)
- DBSS (see Appendix B), 50 ml
- Dulbecco's phosphate buffered saline without Ca^{2+} and Mg^{2+} (DPBS-A), 200 ml
- Trypsin (1 : 250 crude trypsin or equivalent, 2.5% solution), 100 ml
- Growth medium with serum (e.g. DMEM/F12 with 10% FBS), 200 ml
- Petri dishes, 90 mm (3)
- Centrifuge tubes, 15 ml or 50 ml, or universal containers (2)
- Trypsinization flask: spinner flask (see Figure 28.1) or shaker flask (preferably indented, see Figure 13.5l), 250 ml (1/g tissue)
- Culture flasks, 25 cm^2 or 75 cm^2 (number and size determined by tissue sample)
- Forceps, straight and curved, for dissection and transfer of tissue
- Scalpels, #11 blade (2)
- Pipettes, 10 ml or 25 ml, with wide tips
- Magnetic follower, autoclaved in a test tube (1)

Materials (non-sterile)

- BSC and associated consumables (see Protocol P12.1)
- Viability stain, e.g. trypan blue (0.4%), 5 ml
- Cell counting equipment (hemocytometer or automated counter; see Section 19.1)
- Magnetic stirrer (for stirrer flask) or laboratory shaker (for shaker flask) in incubator or warmroom

Procedure

1. Using a BSC, prepare tissue pieces as described in Protocol P13.3 (steps 1–7; see also Figure 13.8, top).
2. Transfer tissue pieces to the empty trypsinization flask. Use a prewetted, wide-bore pipette or pour the resuspended pieces in one single action. Allow the pieces to settle.

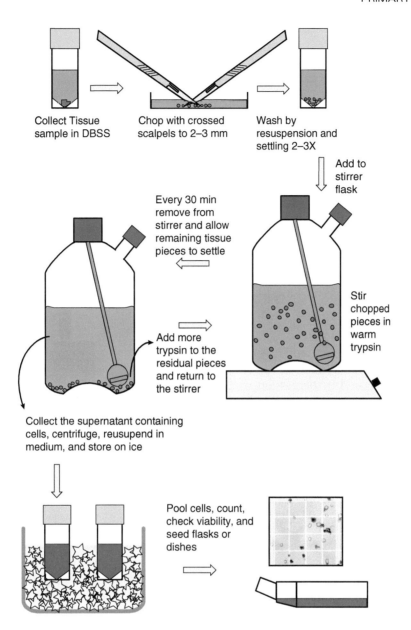

Collect Tissue
sample in DBSS

Chop with crossed
scalpels to 2–3 mm

Wash by
resuspension and
settling 2–3X

Add to
stirrer
flask

Stir
chopped
pieces in
warm
trypsin

Every 30 min
remove from
stirrer and allow
remaining tissue
pieces to settle

Add more
trypsin to the
residual pieces
and return to
the stirrer

Collect the supernatant containing
cells, centrifuge, reusupend in
medium, and store on ice

Pool cells, count,
check viability, and
seed flasks or
dishes

Fig. 13.8. Warm trypsin disaggregation. Tissue is coarsely chopped (top) and stirred in trypsin (middle), and dissociated cells are collected at intervals and stored on ice to be pooled later (bottom). *Source*: R. Ian Freshney.

3. Carefully remove most of the residual fluid in the trypsinization flask. Add 20 ml trypsin (2.5%) and 180 ml of DPBS-A (final concentration 0.25% in 200 ml total volume).

4. Add the sterile magnetic follower to the flask.

5. Cap the flask and place it on the magnetic stirrer at 37 °C.

6. Stir at about 100 rpm for 30 minutes at 37 °C.

7. After 30 minutes, collect disaggregated cells as follows:

 (a) Allow the pieces to settle.

 (b) Pour off the supernatant into a centrifuge tube and place it on ice. Carefully wipe off any medium running down the outside of the side arm from which you have poured, using a lint-free swab and 70% alcohol.

 (c) Add fresh 0.25% trypsin to the pieces remaining in the flask, and continue to stir and incubate for a further 30 minutes.

 (d) Centrifuge the harvested cells from step 14(b) at approximately 500 *g* for five minutes.

(e) Resuspend the resulting pellet in 10 ml growth medium and store the suspension on ice.

8. Repeat steps 4–7 until complete disaggregation occurs or until no further disaggregation is apparent (usually three to four hours).

9. Collect and pool the chilled cell suspensions.

10. Remove any large remaining aggregates by filtering through sterile muslin or a cell strainer (see Section 13.6) or by allowing the aggregates to settle.

11. Count the cells by hemocytometer or automated cell counter. As the cell population will be very heterogeneous, assess cell viability using trypan blue (see Section 19.2.1) and observe the degree of aggregation, e.g. using a hemocytometer.

12. Dilute the cell suspension to 1×10^5 to 1×10^6 cells/ml in growth medium. Seed cells at approximately 2×10^4 to 2×10^5 cells/cm^2 into as many flasks as are required.

13. Change the medium at regular intervals (every two to four days as dictated by depression of pH). Check the supernatant for viable cells before discarding it, as some cells can be slow to attach or may prefer to proliferate in suspension.

Notes

Step 7: All centrifuge speeds are quoted as g, the radial acceleration relative to gravity. For the conversion from rpm to g, see Appendix B.

Step 12: When the survival rate is unknown or unpredictable, a cell count is of little value (e.g. in tumor biopsies, in which the proportion of necrotic cells may be quite high). In this case, set up a range of concentrations from about 5–25 mg of tissue per milliliter of medium.

13.5.2 Trypsin with Cold Pre-Exposure

One of the disadvantages of using trypsin for tissue disaggregation is the damage that may result from exposure to trypsin at 37 °C (Waymouth 1974). This damage is progressive, hence the need to harvest disaggregated cells after 30 minutes of incubation in warm trypsin rather than exposing cells to the full period of incubation (i.e. three to four hours). A simple method of minimizing damage to the cells is to soak tissue in trypsin at 4 °C for 6–18 hours to allow penetration of the enzyme with little tryptic activity. Trypsin is taken up into the cell by pinocytosis, which is inhibited at 4 °C, resulting in less entry into the cell to cause damage (Ham and McKeehan 1978). After soaking at 4 °C, the tissue will only require 20–30 minutes at 37 °C for disaggregation (Cole and Paul 1966).

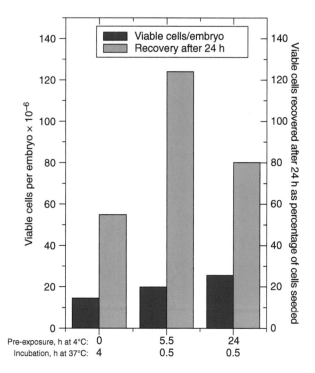

Fig. 13.9. Warm and cold trypsinization. Yield of viable cells per 12.5-day embryo increases by cold trypsinization up to 24 hours at 4 °C, but recovery after 24 hours culture is greatest with shorter cold trypsinization (>100%, implying cell proliferation), perhaps because some of the cells released by longer cold trypsinization are not proliferative (see also Table 13.2).

The cold trypsin method usually gives a higher yield of viable cells, with improved survival after 24 hours' culture (see Table 13.2; see also Figures 13.4, 13.9). It can preserve more different cell types than the warm method (see Figures 13.2, 13.6). It is also convenient; no stirring or centrifugation is required and the incubation at 4 °C may be done unattended overnight. It is, however, limited by the amount of tissue that can be processed at one time. This method is particularly useful for embryonic material. Cultures from mouse embryos contain more epithelial cells when prepared by the cold trypsin method, and erythroid cultures from 13-day fetal mouse liver respond to erythropoietin (EPO) after this treatment, but not after the warm trypsin method or mechanical disaggregation (personal communication, D. Conkie) (Cole and Paul 1966). Cultures from 10- to 13-day chick embryos show evidence of several different cell types characteristic of the tissue of origin. After three to five days, contracting cells may be seen in the heart cultures, colonies of pigmented cells in the pigmented retina culture, and the beginning of myotube fusion in skeletal muscle cultures. A general protocol for cold trypsin is provided below (see Protocol P13.5); a separate protocol for chick embryonic organ rudiments is provided as Supplementary Material online (see Supp. S13.2).

TABLE 13.2. Cell yield from 12.5-day embryo by warm and cold trypsinization.

Duration and temperature of trypsinization		After trypsinization			After 24 h in culture	
		Cells recovered per embryo $\times 10^{-7}$	Percentage viability by dye exclusion (trypan blue)	Total no. of viable cells $\times 10^{-7}$	Recovered (% of total seeded)	Viability (% of viable cells seeded)
4°C	37°C					
0 h	4 h	1.69	86	1.45	47.2	54.9
5.5 h	0.5 h	3.32	60	1.99	74.5	124[a]
24 h	0.5 h	3.40	75	2.55	60.3	80.2

[a]A figure greater than 100% implies that there has been some cell proliferation, exceeding the number of cells lost.

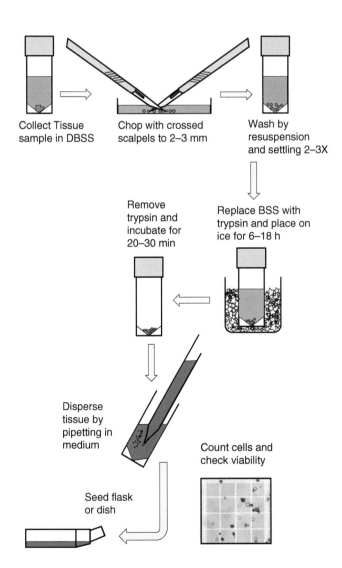

Collect Tissue sample in DBSS

Chop with crossed scalpels to 2–3 mm

Wash by resuspension and settling 2–3X

Replace BSS with trypsin and place on ice for 6–18 h

Remove trypsin and incubate for 20–30 min

Disperse tissue by pipetting in medium

Count cells and check viability

Seed flask or dish

Fig. 13.10. Cold trypsin disaggregation. Tissue is coarsely chopped (top), placed in trypsin at 4 °C, incubated briefly at 37 °C (middle), and dispersed by pipetting in medium (bottom). Images of cultures following cold trypsinization are shown elsewhere (see Figures 13.2, 13.6).

PROTOCOL P13.5. COLD TRYPSIN DISAGGREGATION

Background

The cold trypsin method is particularly suitable for small amounts of tissue, such as embryonic organs. This procedure gives good reproducible cultures from 10- to 13-day chick embryos.

Outline

Chop tissue and place in trypsin at 4 °C for 6–18 hours. Incubate after removing the trypsin and disperse the cells in warm medium (see Figure 13.10).

Materials (sterile or aseptically prepared)

☐ Tissue sample in Collection Medium or DBSS (see Appendix B)
☐ DBSS (see Appendix B), 100 ml
☐ Trypsin (1 : 250 crude trypsin or equivalent, 0.25% solution), 10–50 ml
☐ Growth medium with serum (e.g. DMEM/F12 with 10% FBS), 100–500 ml
☐ Petri dishes, 90 mm (3)
☐ Centrifuge tubes, 15 ml or 50 ml, or universal containers (2)
☐ Trypsinization container, e.g. glass Erlenmeyer flask, 25 ml or 50 ml (1/g tissue)
☐ Culture flasks, 25 cm² or 75 cm² (number and size determined by tissue sample)
☐ Forceps, straight and curved, for dissection and transfer of tissue
☐ Scalpels, #11 blade (2)
☐ Pipettes, 10 ml or 25 ml, with wide tips

Materials (non-sterile)

☐ BSC and associated consumables (see Protocol P12.1)
☐ Viability stain, e.g. trypan blue (0.4%), 1 ml
☐ Cell counting equipment (hemocytometer or automated counter; see Section 19.1)

Procedure

1. Using a BSC, prepare tissue pieces as described in Protocol P13.3 (steps 1–7; see also Figure 13.10, top).
2. Transfer the tissue pieces to the empty trypsinization container. Use a prewetted, wide-bore pipette or pour the resuspended pieces in one single action. Allow the pieces to settle.
3. Carefully remove the residual fluid and add trypsin solution (0.25% trypsin, 10 ml/g of tissue, prechilled at 4 °C).
4. Place the mixture at 4 °C for 6–18 hours (time will vary depending on tissue).
5. Remove and discard the trypsin carefully, leaving the tissue with only the residual trypsin. Place the tube at 37 °C for 20–30 minutes.
6. Add warm medium, approximately 1 ml for every 100 mg of original tissue, and gently pipette the mixture up and down until the tissue is completely dispersed.
7. If some tissue does not disperse, the cell suspension may be filtered through sterile muslin, stainless-steel mesh (100–200 μm), or a cell strainer (see Section 13.6), or the larger pieces may simply be allowed to settle.
8. Count the cells by hemocytometer or electronic cell counter. As the cell population will be very heterogeneous, assess cell viability using trypan blue (see Section 19.2.1) and observe the degree of aggregation, e.g. using a hemocytometer.
9. Dilute the cell suspension to 1×10^5 to 1×10^6 cells/ml in growth medium. Seed cells at approximately 2×10^4 to 2×10^5 cells/cm^2 into as many flasks as are required.
10. Change the medium at regular intervals (every two to four days as dictated by depression of pH). Check the supernatant for viable cells before discarding it, as some cells can be slow to attach or may prefer to proliferate in suspension.

Notes

Step 5: Some tissue types may depress the pH when a large amount of tissue is used, e.g. brain tissue. In those cases, replace the trypsin with a fresh solution before 37 °C incubation.

Step 7: When there is a lot of tissue, increasing the volume of medium in step 6 to 2 ml for every 100 mg of original tissue will facilitate settling and subsequent collection of cells in the supernatant. Most larger pieces should be dispersed in two to three minutes. If there are strands of tissue that do not sediment easily, it is possible that these are aggregated with DNA. DNase (2–20 μg/ml) may be added for 10–20 minutes to disperse these (see Section 13.2.2).

Step 9: When the survival rate is unknown or unpredictable, a cell count is of little value (e.g. in tumor biopsies, for which the proportion of necrotic cells may be quite high). In this case, set up a range of concentrations from about 5–25 mg of tissue per milliliter of medium.

13.5.3 Other Enzymatic Procedures

Trypsin may not be suitable for disaggregation because it can be damaging (e.g. to pancreatic islet cells) or ineffective (e.g. for very fibrous tissue, such as fibrous connective tissue). Many other enzymes can be used for disaggregation (see Table 13.1). Because ECM often contains collagen, collagenase is an obvious choice. Disaggregation in collagenase has proved suitable for the culture of human tumors, breast, liver, pancreas, and many other tissue types, particularly epithelium (e.g. see Supp. S24.6,S24.7) (Lasfargues 1957; Freshney 1972; Speirs 2004; Kin et al. 2007). Collagenase was originally used as a component of culture media and has relatively low toxicity; it is not inhibited by serum and does not require neutralization (Waymouth 1974). However, with more than 1 g of tissue it can become expensive because of the amount of collagenase required. Collagenase will also release most of the connective tissue cells, accentuating the problem of fibroblastic outgrowth, so it may need to be followed by selective culture or cell separation (see Section 20.7, see also Chapter 21).

PROTOCOL P13.6. COLLAGENASE DISAGGREGATION

Background

Disaggregation in collagenase is simple and effective for many tissues: embryonic, adult, normal, and malignant. It requires no mechanical agitation or special equipment. Crude collagenase is often used and may depend, for some of its action, on contamination with other proteases. More highly purified grades are available, but they may not be as effective as crude collagenase.

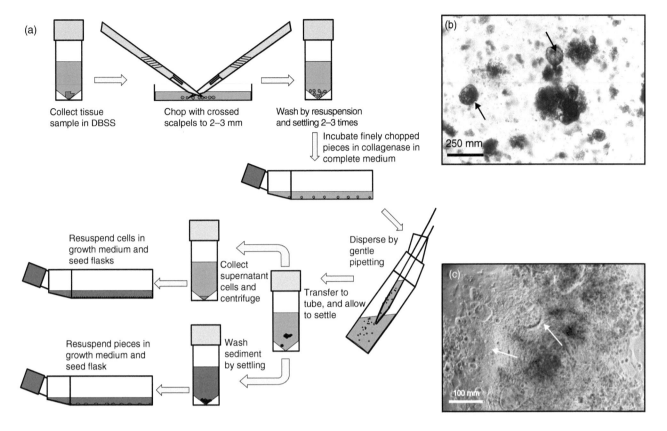

Fig. 13.11. Collagenase disaggregation. (a) Schematic diagram of dissection and collagenase disaggregation. (b) Cell clusters from human colonic carcinoma after 48 hours incubation in collagenase. Arrows point to probable epithelial clusters. (c) Cells after removal of collagenase, further disaggregation by pipetting, and 48 hours culture. The clearly defined rounded clusters (black arrows) in (b) have formed epithelium-like sheets (white arrows); some are still 3D, some are spreading as sheets, and the more irregularly shaped clusters are producing fibroblasts. *Source*: (b, c) R. Ian Freshney.

Outline

Place finely chopped tissue in complete medium containing collagenase and incubate. When tissue is disaggregated, remove collagenase by centrifugation and seed cells at a high concentration (see Figure 13.11).

Materials (sterile or aseptically prepared)

- ☐ Tissue sample in Collection Medium or DBSS (see Appendix B)
- ☐ DBSS (see Appendix B), 100 ml
- ☐ Collagenase, 2000 U/ml (e.g. Type IA, #C2674 Sigma-Aldrich), 5 ml
- ☐ Growth medium (e.g. DMEM/F12 with 20% FBS), 100 ml
- ☐ Petri dishes, 90 mm (3)
- ☐ Centrifuge tubes, 15 ml or 50 ml, or universal containers (2)
- ☐ Culture flasks, 25 cm² (number determined by tissue size)

- ☐ Forceps, straight, and curved, for dissection and transfer of tissue
- ☐ Scalpels, #11 blade (2)
- ☐ Pipettes, 10 ml or 25 ml, with wide tips

Materials (non-sterile)

- ☐ BSC and associated consumables (see Protocol P12.1)

Procedure

1. Using a BSC, prepare tissue pieces as described in Protocol P13.3 (steps 1–7; see also Figure 13.11a, top).
2. Transfer 20–30 pieces to one 25 cm² flask and 100–200 pieces to a second flask.
3. Drain off the DBSS and add 4.5 ml of complete growth medium to each flask.
4. Add 0.5 ml of crude collagenase (2000 U/ml, final concentration 200 U/ml in 5 ml total volume).

5. Incubate at 37 °C for 4–48 hours without agitation. Tumor tissue may be left up to five days or more if disaggregation is slow (e.g. in scirrhous carcinomas of the breast or the colon). It may be necessary to centrifuge the tissue and resuspend it in fresh medium and collagenase before that amount of time has passed if an excessive drop in pH is observed (i.e. pH < 6.5).

6. Check for effective disaggregation by gently moving the flask. The pieces of tissue will "smear" on the bottom of the flask and, with gentle pipetting, will break up into single cells and small clusters or aggregates (see Figure 13.11b).

7. Small clusters of epithelial cells may resist collagenase in some tissues (e.g. lung, kidney, colon, or breast carcinoma). These clusters may be separated from the rest by allowing them to settle for about two minutes. Clusters may form islands of epithelial cells if they are washed with DBSS, allowed to settle, and the sediment is resuspended in medium and seeded. Epithelial cells generally survive better if they are not completely dissociated.

8. When complete disaggregation has occurred, or when the supernatant cells are collected after removing clusters by settling, centrifuge the cell suspension from the disaggregate and any washings at 50–100 g for three minutes.

9. Discard the supernatant DBSS or medium, resuspend and combine the pellets in 5 ml of medium, and seed in a 25-cm² flask. If the pH fell during collagenase treatment (pH ≤ 6.5 by 48 hours), dilute the suspension two- to threefold in medium after removing the collagenase.

10. Replace the medium after 48 hours.

Notes

Step 7: This step is not required for other tissue types, such as glioma (see Section 25.5.5).

Steps 8 and 9: Some cells, particularly macrophages, may adhere to the first flask during the collagenase incubation. Transferring the cells to a fresh flask after collagenase treatment (and subsequent removal of the collagenase) removes many of the macrophages from the culture. The first flask may be cultured as well, if required. Light trypsinization will remove any adherent cells other than macrophages.

The increasing use of tissue culture in cell-based therapy has resulted in a move towards xeno-free and/or chemically defined reagents (see Sections 10.7, 10.8). Many proteases are derived from bacteria, including crude collagenase (*Clostridium histolyticum*), Pronase (*Streptomyces griseus*), and Dispase (*Bacillus polymyxa*). Although bacterial proteases are not derived from animals, they may carry high endotoxin levels and their activities may vary from batch to batch (Dolmans et al. 2006). Liberase (Roche CustomBiotech) is a blend of purified collagenase and neutral proteases (e.g. Dispase, thermolysin) that has been endotoxin tested. It can be used to successfully isolate human ovarian follicles and pancreatic islets (Dolmans et al. 2006; Brandhorst et al. 2017). Accumax (Innovative Cell Technologies) and Papain (e.g. Sigma-Aldrich, # P4762) are non-mammalian, non-bacterial alternatives to trypsin and collagenase that can be used for tissue disaggregation. However, a consistent and standardized enzyme blend is not yet available; this will require a better understanding of how the various enzymes function at the molecular level (Brandhorst et al. 2017).

13.6 MECHANICAL DISAGGREGATION

The outgrowth of cells from primary explants is a relatively slow process, can be highly selective, and is really only suitable for small amounts of tissue. Enzymatic digestion is more labor intensive and there is a risk of proteolytic damage to cells during enzymatic digestion. For these reasons, mechanical disaggregation is preferred by many laboratories. Methods include (i) spillage, collecting cells that spill out when the tissue is carefully sliced and the slices scraped (Lasfargues 1973); (ii) sieving, pressing tissue through a series of cell strainers with progressively finer mesh; (iii) syringing, pushing tissue through a syringe (Zaroff et al. 1961); and (iv) pipetting, using a wide-bore pipette to triturate the tissue (see Figure 13.12). Spillage and sieving are probably the gentlest mechanical methods, while pipetting and, particularly, syringing are most likely to generate shear stress and mechanical damage (Faillei et al. 2009).

Only soft tissues respond well to this technique, including spleen, embryonic liver, embryonic and adult brain, and some human and animal soft tumors. Even with brain, where fairly complete disaggregation can be obtained, the viability of the resulting suspension is lower than that achieved with enzymatic digestion. When the availability of tissue is not a limitation and the efficiency of the yield is unimportant, mechanical and enzymatic disaggregation can both be used to produce an equivalent number of viable cells. The mechanical method will take less time but at the expense of much more tissue.

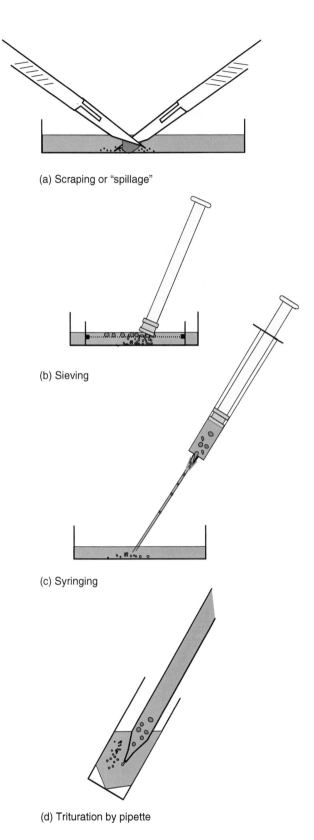

(a) Scraping or "spillage"

(b) Sieving

(c) Syringing

(d) Trituration by pipette

Fig. 13.12. Mechanical disaggregation. Methods may include (a) scraping or "spillage," where cells are released through the cutting action, or abrasion of the cut surface; (b) sieving, where tissue is forced through a sieve using a syringe piston; (c) syringing, where tissue is drawn into syringe through a wide-bore needle or cannula; and (d) trituration by pipette, where tissue is pipetted up and down through a wide-bore pipette.

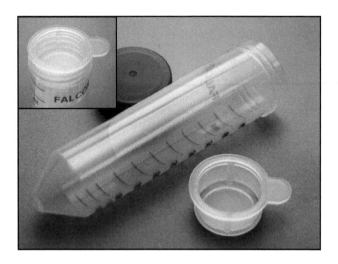

Fig. 13.13. Cell strainer. This example is a disposable polypropylene filter and tube that can be used to strain aggregates from primary suspensions (Corning). It can also be used for disaggregating soft tissues (see also Figure 13.12b). *Source*: R. Ian Freshney.

PROTOCOL P13.7. MECHANICAL DISAGGREGATION BY SIEVING

Background

This method of disaggregation has been found to be moderately successful with soft tissues, such as brain. Dead cells may be eliminated by centrifugation (see Section 13.7).

Outline

Push the tissue through a series of sieves for which the mesh is gradually reduced in size until a reasonable suspension of single cells and small aggregates is obtained. The suspension is then diluted and cultured directly.

Materials (sterile or aseptically prepared)

- ☐ Tissue sample in Collection Medium or DBSS (see Appendix B)
- ☐ Growth medium (e.g. DMEM/F12 with 10% FBS), 100 ml
- ☐ Petri dishes, 90 mm (3)
- ☐ Cell strainers (see Figure 13.13) with different mesh sizes, e.g. 20, 100, and 500 μm
- ☐ Centrifuge tubes, 50 ml (2)
- ☐ Disposable plastic syringes, 2 ml or 5 ml (2)
- ☐ Forceps (1 pair)
- ☐ Scalpels, #11 blade (2)

Materials (non-sterile)

☐ BSC and associated consumables (see Protocol P12.1)
☐ Viability stain, e.g. trypan blue (0.4%), 5 ml
☐ Cell counting equipment (hemocytometer or automated counter; see Section 19.1)

Procedure

1. Using a BSC, prepare tissue pieces as described in Protocol P13.3 but chop the tissue into larger pieces, about 3–5 mm^3.
2. Place a few pieces at a time into a cell strainer with a large mesh size, e.g. 500 μm. Push the tissue through the mesh into growth medium by applying gentle pressure with the piston of a disposable plastic syringe. Pipette more medium through the sieve to wash the cells through it.
3. Pipette the partially disaggregated tissue into a second cell strainer with finer porosity, perhaps 100 μm mesh, and repeat step 2.
4. If a single cell suspension is necessary, pipette the remaining tissue into a third cell strainer, perhaps 20 μm mesh, and repeat step 2. In general, the more highly dispersed the cell suspension, the higher the sheer stress required, and the lower the resulting viability.
5. Count the cells by hemocytometer or automated cell counter. Look for cell aggregates and assess viability by trypan blue (see Section 19.2.1).
6. Dilute the cell suspension to 1×10^5 to 1×10^6 cells/ml in growth medium. Seed the culture flasks at 2×10^5, 1×10^6, and 2×10^6 cells/ml.
7. Change the medium at regular intervals (every two to four days as dictated by depression of pH) and check the supernatant for viable cells before discarding it.

Notes

Step 2: Cell strainers of various mesh sizes are available from Corning, pluriSelect, and some other manufacturers. A stainless-steel or polypropylene sieve of 1 mm mesh may also be used.

13.7 ENRICHMENT OF VIABLE CELLS

When tissue is disaggregated and seeded into primary culture, only a proportion of the cells will survive and generate a primary culture (see Sections 3.1.1, 3.2.1) (Zaroff et al. 1961). Some cells will be viable but not ready for attachment; others will be non-viable because of conditions within the tissue or damage during disaggregation. Cell viability can be assessed by dye exclusion, flow cytometry, or other methods (see Section 19.2). Non-viable cells are normally removed at the first change of medium in adherent cultures. Where viable cells remain in suspension, non-viable cells are gradually diluted out when cell proliferation starts. Non-viable cells and debris can also be removed by centrifuging the cells on a mixture of Ficoll™ and sodium metrizoate (e.g. Hypaque™) (de Vries et al. 1973). Viable cells collect at the interface between the medium and the Ficoll/metrizoate, and the dead cells form a pellet at the bottom of the tube. This technique is similar to the preparation of lymphocytes from peripheral blood using density gradient centrifugation (see Section 21.1.1; see also Supp. S24.20).

PROTOCOL P13.8. ENRICHMENT OF VIABLE CELLS

Background

This enrichment procedure can be scaled up or down and works with lower ratios of density medium to cell suspension (e.g. 5 ml of cell suspension over 1 ml of density medium).

Outline

Up to 2×10^7 cells in 9 ml of medium may be layered on top of 6 ml of Ficoll-Paque in a 25- to 50-ml screw-capped centrifuge tube. The mixture is then centrifuged and viable cells are collected from the interface.

Materials (sterile or aseptically prepared)

☐ Cell suspension with as few aggregates as possible, 10–20 ml
☐ Clear centrifuge tubes or universal containers (2)
☐ Culture flasks, 25 cm^2 or 75 cm^2
☐ Density medium: Ficoll/metrizoate (e.g. Ficoll-Paque, GE Healthcare Life Sciences), 1.077 g/ml density, 10–20 ml
☐ Growth medium (e.g. DMEM/F12 with 10% FBS), 100 ml
☐ Pasteur pipette

Materials (non-sterile)

☐ BSC and associated consumables (see Protocol P12.1)
☐ Cell counting equipment (hemocytometer or automated counter; see Section 19.1)

Procedure

1. Allow major aggregates in the cell suspension to settle.
2. Layer 9 ml of the cell suspension onto 6 ml of the density medium. This step should be done in a wide, transparent centrifuge tube with a cap, such as a 30-ml universal container. If a 50-ml centrifuge tube is used, double the aforementioned volumes.
3. Centrifuge the mixture for 15 minutes at 400 g.
4. Carefully remove the top layer without disturbing the interface using a Pasteur pipette.
5. Collect the interface carefully using a Pasteur pipette.
6. Dilute the mixture to 20 ml in growth medium.
7. Centrifuge the mixture at 70 g for 10 minutes.
8. Discard the supernatant and resuspend the pellet in 5 ml of growth medium.
9. Repeat steps 7 and 8 in order to wash cells free of density medium.
10. Count the cells by hemocytometer or automated cell counter.
11. Seed the culture flask(s).

Primary cultures are initially heterogeneous, but any early equilibrium is likely to be short-lived due to the rapid proliferation of the fibroblasts within the culture (Lasfargues 1957). Active measures must be taken to select for the desired cell type or separate it from other cells within the population. Selection and separation techniques are discussed later in this book (see Section 20.7; see also Chapter 21).

13.8 RECORD KEEPING FOR PRIMARY CULTURE

It is essential to keep good records of the culture's origin and derivation, including the origin of the tissue, any relevant pathology, and the procedures used for disaggregation and primary culture (see Table 13.3). This is important for GCCP even if the cells do not proceed beyond the primary culture stage (Pamies et al. 2018). Any relevant ethical approvals should be included; these will be required when you publish the work later on. The location of any stored samples (including other tissue and DNA from the same donor) should also be recorded and can be used as reference material for later authentication testing (see Sections 13.3, 17.2.3). Records can be kept in a notebook or another hardcopy file of record sheets, but it is best at this stage to initiate a record in a computer database; this record then becomes the first step in maintaining the provenance of any subsequent cell lines (see Section 7.3). Negative outcomes should be retained, with the reason for failure if apparent, for future analysis of success rates.

TABLE 13.3. Data record for primary culture.

Category of information	Information	For each sample
Arrival	Date	
	Time	
	Person responsible for sample	
Identifier	Unique ID for sample[a]	
Origin of tissue	Species	
	Strain	
	Age	
	Sex	
	Pathology or animal identifier	
	Tissue	
	Site	
	Stored tissue/DNA location	
	Other comments	
Pathology	Normal/benign/malignant	
	Histology	
	Other comments	
Disaggregation agent	Trypsin, collagenase, etc.	
	Concentration	
	Duration	
	Diluent	
	Other comments	
Cell count[b]	Concentration after resuspension (C_I)	
	Volume (V_I)	
	Yield ($Y = C_I \times V_I$)	
	Yield per gram (wet weight of tissue)	
	Other comments	
Seeding	Number (N) and type of vessel (flask, dish, or multiwell plate)	
	Final concentration (C_F)	
	Volume per flask, dish, or well (V_F)	
	Other comments	
Medium	Basal medium	
	Supplier and batch no.	
	Serum type and concentration	
	Supplier and batch no.	
	Other additives	
	CO_2 concentration	
	Other comments	

TABLE 13.3. (*continued*)

Category of information	Information	For each sample
Coating or treatment	E.g. fibronectin, Matrigel®, collagen	
	Supplier and batch no.	
	Other comments	
Subculture	Recovery at first subculture, cells/flask	
	Percentage (cells recovered ÷ cells seeded)	
	Cell line name (see Section 14.2.1)	
	Other comments	

[a]A unique sample identifier should be used to identify the original tissue sample and retained within the culture's designation. The patient's name, initials, or other personal information should not be used for this purpose. Many laboratories keep a log of tissue samples and generate identifying numbers from a consecutive list of sample numbers in the log book or electronic database.
[b]Additional material is available elsewhere that demonstrates how to calculate additional parameters following cell counting (see Supp. S30.15).

13.9 CONCLUSIONS: PRIMARY CULTURE

The disaggregation of tissue and preparation of the primary culture make up the first, and perhaps most vital, stage in the culture of cells with specific functions. If the required cells are lost at this stage, the loss is irrevocable. Many different cell types may be cultured successfully if the correct techniques are chosen at this stage. In general, trypsin is more severe than collagenase, but is sometimes more effective in creating a single-cell suspension. Collagenase does not dissociate epithelial cells readily, but this characteristic can be an advantage for separating the epithelial cells from stromal cells and maintaining viability in epithelial clusters. Mechanical disaggregation is much quicker than enzymatic procedures but damages more cells. The best approach is to try out the techniques described in this chapter (see Protocols P13.3–P13.7) and select the one that works best in your system. If none of those methods is successful, try using additional or alternative enzymes at different concentrations (see Table 13.1), and consult later chapters in this book and the scientific literature (see Chapters 24 and 25).

Suppliers

Supplier	URL
GE Healthcare Life Sciences	http://www.gelifesciences.com/en/it
Innovative Cell Technologies, Inc.	http://www.accutase.com
Novozymes	http://www.novozymes.com/en/biology
pluriSelect	http://www.pluriselect.com/row/?___store=row
ProSpec	http://www.prospecbio.com
Roche CustomBiotech	http://custombiotech.roche.com/home.html
Sigma-Aldrich (Merck)	http://www.sigmaaldrich.com/life-science/cell-culture.html
ThermoFisher Scientific	http://www.thermofisher.com/us/en/home/life-science/cell-culture.html

Supp. S13.1 Isolation of Chick Embryos

Supp. S13.2 Disaggregation of Chick Embryo Organ Rudiments

Supp. S13.3 Maximal Serial Transfer (MST) of Human Fibroblasts from Skin Explants

REFERENCES

Armit, C., Richardson, L., Venkataraman, S. et al. (2017). eMouse-Atlas: an atlas-based resource for understanding mammalian embryogenesis. *Dev. Biol.* 423 (1): 1–11. https://doi.org/10.1016/j.ydbio.2017.01.023.

AVMA (2013). Guidelines for the euthanasia of animals. https://www.avma.org/kb/policies/pages/euthanasia-guidelines.aspx (accessed 28 November 2018).

Bayne, C.J. (1998). Invertebrate cell culture considerations: insects, ticks, shellfish, and worms. *Methods Cell Biol.* 57: 187–201.

Beers, J., Gulbranson, D.R., George, N. et al. (2012). Passaging and colony expansion of human pluripotent stem cells by enzyme-free dissociation in chemically defined culture conditions. *Nat. Protoc.* 7 (11): 2029–2040. https://doi.org/10.1038/nprot.2012.130.

Booth, C. and O'Shea, J.A. (2002). Isolation and culture of intestinal epithelial cells. In: *Culture of Epithelial Cells* (eds. R.I. Freshney and M.G. Freshney), 303–335. Hoboken, NJ: Wiley-Liss.

Brandhorst, D., Brandhorst, H., and Johnson, P.R.V. (2017). Enzyme development for human islet isolation: five decades of progress or stagnation? *Rev. Diabet. Stud.* 14 (1): 22–38. https://doi.org/10.1900/RDS.2017.14.22.

Brauer, R. and Chen, P. (2015). Influenza virus propagation in embryonated chicken eggs. *J. Vis. Exp.* 97: 52421. https://doi.org/10.3791/52421.

Brieno-Enriquez, M.A., Robles, P., Garcia-Cruz, R. et al. (2010). A new culture technique that allows in vitro meiotic prophase development of fetal human oocytes. *Hum. Reprod.* 25 (1): 74–84. https://doi.org/10.1093/humrep/dep351.

Carrel, A. (1912). On the permanent life of tissues outside of the organism. *J. Exp. Med.* 15 (5): 516–528.

Chi, J., Meng, L., Pan, S. et al. (2017). Primary culture of rat aortic vascular smooth muscle cells: a new method. *Med. Sci. Monit.* 23: 4014–4020.

Cole, R.J. and Paul, J. (1966). The effects of erythropoietin on haem synthesis in mouse yolk sac and cultured foetal liver cells. *J. Embryol. Exp. Morphol.* 15 (2): 245–260.

de Vries, J.E., van Benthem, M., and Rumke, P. (1973). Separation of viable from nonviable tumor cells by flotation on a Ficoll-Triosil mixture. *Transplantation* 15 (4): 409–410.

Dobbs, L.G. and Gonzalez, R.F. (2002). Isolation and culture of pulmonary alveolar epithelial type II cells. In: *Culture of Epithelial Cells* (eds. R.I. Freshney and M.G. Freshney), 278–301. Hoboken, NJ: Wiley-Liss.

Dolmans, M.M., Michaux, N., Camboni, A. et al. (2006). Evaluation of Liberase, a purified enzyme blend, for the isolation of human primordial and primary ovarian follicles. *Hum. Reprod.* 21 (2): 413–420. https://doi.org/10.1093/humrep/dei320.

Elliget, K.A. and Lechner, J.F. (1992). Normal human bronchial epithelial cell cultures. In: *Culture of Epithelial Cells* (ed. R.I. Freshney), 181–196. New York: Wiley-Liss.

Faillei, A., Consolini, R., Legitimo, A. et al. (2009). The challenge of culturing human colorectal tumor cells: establishment of a cell culture model by the comparison of different methodological approaches. *Tumori* 95 (3): 343–347.

Finkelstein, J.N. and Shapiro, D.L. (1982). Isolation of type II alveolar epithelial cells using low protease concentrations. *Lung* 160 (2): 85–98.

Freshney, R.I. (1972). Tumour cells disaggregated in collagenase. *Lancet* 2 (7775): 488–489.

Geraghty, R.J., Capes-Davis, A., Davis, J.M. et al. (2014). Guidelines for the use of cell lines in biomedical research. *Br. J. Cancer* 111 (6): 1021–1046. https://doi.org/10.1038/bjc.2014.166.

Grozdanov, V., Muller, A., Sengottuvel, V. et al. (2010). A method for preparing primary retinal cell cultures for evaluating the neuroprotective and neuritogenic effect of factors on axotomized mature CNS neurons. *Curr. Protoc. Neurosci.* 53 (1): 3.22.1–3.22.10, Chapter 3: Unit 3.22. doi: https://doi.org/10.1002/0471142301.ns0322s53.

Ham, R.G. and McKeehan, W.L. (1978). Nutritional requirements for clonal growth of nontransformed cells. In: *Nutritional Requirements of Cultured Cells* (ed. H. Katsuta), 63–111. Tokyo and Baltimore: Japan Scientific Societies Press and University Park Press.

Harrison, R.G. (1907). Observations on the living developing nerve fiber. *Proc. Soc. Exp. Biol.* 4 (1): 140–143.

Huschtscha, L.I., Napier, C.E., Noble, J.R. et al. (2012). Enhanced isolation of fibroblasts from human skin explants. *Biotechniques* 53 (4): 239–244. https://doi.org/10.2144/0000113939.

Kin, T., Johnson, P.R., Shapiro, A.M. et al. (2007). Factors influencing the collagenase digestion phase of human islet isolation. *Transplantation* 83 (1): 7–12. https://doi.org/10.1097/01.tp.0000243169.09644.e6.

Kodack, D.P., Farago, A.F., Dastur, A. et al. (2017). Primary patient-derived cancer cells and their potential for personalized cancer patient care. *Cell Rep.* 21 (11): 3298–3309. https://doi.org/10.1016/j.celrep.2017.11.051.

Kondo, J., Endo, H., Okuyama, H. et al. (2011). Retaining cell–cell contact enables preparation and culture of spheroids composed of pure primary cancer cells from colorectal cancer. *Proc. Natl Acad. Sci. U.S.A.* 108 (15): 6235–6240. https://doi.org/10.1073/pnas.1015938108.

Lasfargues, E.Y. (1957). Cultivation and behavior in vitro of the normal mammary epithelium of the adult mouse. *Anat. Rec.* 127 (1): 117–129.

Lasfargues, E.Y. (1973). Human mammary tumors. In: *Tissue Culture: Methods and Applications* (eds. P.F. Kruse Jr. and M.K. Patterson), 45–50. New York: Academic Press.

Ledur, P.F., Onzi, G.R., Zong, H. et al. (2017). Culture conditions defining glioblastoma cells behavior: what is the impact for novel discoveries? *Oncotarget* 8 (40): 69185–69197. https://doi.org/10.18632/oncotarget.20193.

Lee, L.E., Bufalino, M.R., Christie, A.E. et al. (2011). Misidentification of OLGA-PH-J/92, believed to be the only crustacean cell line. *In Vitro Cell Dev. Biol. Anim.* 47 (9): 665–674. https://doi.org/10.1007/s11626-011-9447-y.

Masters, J.R.W. (1991). *Human Cancer in Primary Culture, a Handbook*. Dordrecht: Kluwer Academic Publishers.

Miller, D.H., Sokol, E.S., and Gupta, P.B. (2017). 3D primary culture model to study human mammary development. *Methods Mol. Biol.* 1612: 139–147. https://doi.org/10.1007/978-1-4939-7021-6_10.

Miserocchi, G., Mercatali, L., Liverani, C. et al. (2017). Management and potentialities of primary cancer cultures in preclinical and translational studies. *J. Transl. Med.* 15 (1): 229. https://doi.org/10.1186/s12967-017-1328-z.

Morgan, J.F. and Parker, R.C. (1949). Use of antitryptic agents in tissue culture; crude soybean trypsin-inhibitor. *Proc. Soc. Exp. Biol. Med.* 71 (4): 665–668.

Nicosiea, R.F. and Ottinetti, A. (1990). Modulation of microvascular growth and morphogenesis by reconstituted basement membrane gel in three-dimensional cultures of rat aorta: a comparative study of angiogenesis in matrigel, collagen, fibrin, and plasma clot. *In Vitro Cell Dev. Biol.* 26 (2): 119–128.

OLAW (2018). PHS policy on humane care and use of laboratory animals. https://olaw.nih.gov/policies-laws/phs-policy.htm (accessed 28 November 2018).

Pamies, D., Bal-Price, A., Chesne, C. et al. (2018). Advanced good cell culture practice for human primary, stem cell-derived and organoid models as well as microphysiological systems. *ALTEX* 35 (3): 353–378. https://doi.org/10.14573/altex.1710081.

Paul, J., Conkie, D., and Freshney, R.I. (1969). Erythropoietic cell population changes during the hepatic phase of erythropoiesis in the foetal mouse. *Cell Tissue Kinet.* 2 (4): 283–294. https://doi.org/10.1111/j.1365-2184.1969.tb00238.x.

Priya, N., Sarcar, S., Majumdar, A.S. et al. (2014). Explant culture: a simple, reproducible, efficient and economic technique for isolation of mesenchymal stromal cells from human adipose tissue and lipoaspirate. *J. Tissue Eng. Regener. Med.* 8 (9): 706–716. https://doi.org/10.1002/term.1569.

Rutten, M.J., Campbell, D.R., Luttropp, C.A. et al. (1996). A method for the isolation of human gastric mucous epithelial cells for primary culture: a comparison of biopsy vs surgical tissue. *Methods Cell Sci.* 18: 269–281.

Sachs, N., de Ligt, J., Kopper, O. et al. (2018). A living biobank of breast cancer organoids captures disease heterogeneity. *Cell* 172 (1–2): 373–386. e10. doi: https://doi.org/10.1016/j.cell.2017.11.010.

Schaeffer, W.I. (1990). Terminology associated with cell, tissue, and organ culture, molecular biology, and molecular genetics. Tissue Culture Association Terminology Committee. *In Vitro Cell Dev. Biol.* 26 (1): 97–101.

Speirs, V. (2004). Primary culture of human mammary tumor cells. In: *Culture of Human Tumor Cells* (eds. R. Pfragner and R.I. Freshney), 205–219. Hoboken, NJ: Wiley-Liss.

Stampfer, M.R., Yaswen, P., and Taylor-Papadimitriou, J. (2002). Culture of human mammary epithelial cells. In: *Culture of Epithelial Cells* (eds. R.I. Freshney and M.G. Freshney), 95–135. Hoboken, NJ: Wiley-Liss.

Stern, C.D. (2005). The chick; a great model system becomes even greater. *Dev. Cell* 8 (1): 9–17. https://doi.org/10.1016/j.devcel.2004.11.018.

UK Home Office (2018). Guidance: animal testing and research. https://www.gov.uk/guidance/research-and-testing-using-animals (accessed 14 June 2018).

Waymouth, C. (1974). To disaggregate or not to disaggregate: injury and cell disaggregation, transient or permanent? *In Vitro* 10: 97–111.

Whitehead, R.H., Demmler, K., Rockman, S.P. et al. (1999). Clonogenic growth of epithelial cells from normal colonic mucosa from both mice and humans. *Gastroenterology* 117 (4): 858–865.

Wong, F., Welten, M.C., Anderson, C. et al. (2013). eChickAtlas: an introduction to the database. *Genesis* 51 (5): 365–371. https://doi.org/10.1002/dvg.22374.

Worthington Biochemical Corporation (2018). Tissue dissociation guide. http://www.worthington-biochem.com:8080/tissue-guide (accessed 27 November 2018).

Yoshino, T.P., Bickham, U., and Bayne, C.J. (2013). Molluscan cells in culture: primary cell cultures and cell lines. *Can. J. Zool.* 91 (6) https://doi.org/10.1139/cjz-2012-0258.

Zaroff, L., Sato, G., and Mills, S.E. (1961). Single-cell platings from freshly isolated mammalian tissue. *Exp. Cell Res.* 23 (3): 565–575. https://doi.org/10.1016/0014-4827(61)90016-7.

CHAPTER 14

Subculture and Cell Lines

Early research on cell growth and tumorigenesis resulted in cells spending an extended time in culture, leading to the development of the first cell lines. The mouse "L" strain was initiated in 1941 by exposing cells to 20-methylcholanthrene; this culture was successfully cloned several years later to give the L-929 cell line (see Figure 1.1) (Earle et al. 1943; Sanford et al. 1948). The human HeLa cell line was initiated 10 years later from a tumor biopsy (Gey et al. 1952). It was quickly recognized that these cell lines were easily cultivated and could be used in other fields of research. L-929 and HeLa continue to be widely used and have been joined by many other cell lines; more than 100 000 cell lines are now reported in the Cellosaurus knowledge resource (Bairoch 2018). This chapter focuses on what to consider when you initiate or choose a cell line and how to maintain a cell line over time. Always remember that cell lines have limitations as model systems (see Sections 1.5, 3.2–3.4). Many are derived from malignant tumors or transformed cells, and most will display clonal evolution and genetic drift with prolonged handling. Validation and characterization are essential parts of Good Cell Culture Practice (GCCP) and are discussed later in this book (see Chapters 16–18).

14.1 TERMINOLOGY: CELL LINE AND SUBCULTURE

The term "cell line" can mean different things to different people. The consensus view (from the Tissue Culture Association, now the Society for In Vitro Biology) is that a cell line arises from a primary culture when it is transferred into the next culture vessel (Schaeffer 1990; Masters 2000). Cell lines may be finite or continuous. A "finite cell line" has a limited lifespan and will undergo senescence after a certain number of population doublings, while a "continuous cell line" is essentially immortal (see Section 3.5). These definitions have proven to be controversial. Some tissue culture experts reserve the term "cell line" for immortal cultures and use the term "cell strain" for finite cultures (Hayflick and Moorhead 1961; Hayflick 1990). Unfortunately, many publications do not provide sufficient detail to determine whether a cell line is immortal or not, making this a difficult distinction in many cases. The consensus terminology is used throughout this book for consistency (for a full glossary of terms see Appendix A).

The terms "subculture" or "passage" (used interchangeably throughout this book) refer to the transfer of cells or tissue from one culture vessel to another. As part of this process, the material can be subdivided to allow increased space for proliferation and fresh medium can be added (Schaeffer 1990). Some other related terms are used in this chapter. "Passage number" refers to the total number of times the culture has been passaged. "Population doubling level" (PDL) refers to the total number of population doublings that the cells have undergone in culture; the term "generation number" is also commonly used, with one generation being equivalent to one PDL. Passage number and PDL can be used to estimate the age of the culture (see Section 14.2.2). "Split ratio" refers to the extent to which a culture is subdivided during passaging; for example, a split ratio of 1 : 4 means that cells in

Freshney's Culture of Animal Cells: A Manual of Basic Technique and Specialized Applications, Eighth Edition. Amanda Capes-Davis and R. Ian Freshney.
© 2021 John Wiley & Sons Ltd. Published 2021 by John Wiley & Sons Ltd.
Companion website: www.wiley.com/go/freshney/cellculture8

one 25-cm² flask are passaged into four 25-cm² flasks. The split ratio can be used to provide an estimate of the PDL (see Sections 14.2.2, 14.6.4).

A cell line may contain numerous clonal populations with similar or distinct phenotypes. Using consensus terminology, "cell strain" refers to a culture where cloning or selection has been performed to isolate individual cells and/or select for specific properties or markers (see Sections 20.2–20.4) (Schaeffer 1990). A cell strain may be a "primary strain" (from primary culture), "finite cell strain" (from a finite cell line), or "continuous cell strain" (from a continuous cell line). Despite best efforts to clarify, these terms are confusing and are difficult to find in online searches; the term "strain" is also used to describe laboratory animals and is a common word in research publications. These days, many tissue culture personnel prefer to use the term "derivative," accompanied by a description of how each derivative was established. Provided the description is sufficient, this is probably the simplest and clearest approach.

14.2 INITIATING A CELL LINE

The first subculture represents an important transition for a culture. The need to subculture implies that the primary culture has increased to occupy all of the available substrate. Hence, cell proliferation has become an important feature. The primary culture may have a variable growth fraction (see Table 19.4), but the growth fraction is usually high (> 80%) after the first subculture, although this will depend on the types of cells present in the culture. From a very heterogeneous primary culture, containing many of the cell types present in the original tissue, a more homogeneous cell line emerges. In addition to its biological significance, this process has considerable practical importance (see Section 3.2.1). More cells will become available for propagation, characterization, and cryopreservation. The potential increase in cell number and uniformity will also open up a much wider range of potential applications (see Section 1.3).

14.2.1 Cell Line Names and Identifiers

Confusion can arise in naming cell lines, particularly when short names are used. For example, the name "EJ-1" refers to a bladder cancer cell line (which is misidentified and is actually T24), a large B-cell lymphoma cell line, and a bacteriophage of *Streptococcus pneumonia* (Masters et al. 2001; Goy et al. 2003; Goh et al. 2007). A similar name, "EJ," refers to a different bladder cancer cell line and an endometrial cancer cell line (Lin et al. 1985; Isaka et al. 2002). Confusion can also arise from designations such as 3T3, which is meant to identify the regimen of maintenance (3×10^5 cells per 50 mm Petri dish every three days) rather than the actual name of the cells. Several cell lines have the designation "3T3" (NIH 3T3, 3T3-Swiss albino, and BALB/3T3 being the most popular), which were all maintained using the same

regime. The derivation of cloned strains, such as 3T3-L1 and 3T3-J2, has added uniqueness to the names of these valuable cultures, but the lesson to be learned is that each cell line should have a unique identifier. This need has become more evident with the increasing use of online searches for cell line information.

Guidelines for naming new cell lines are available from the International Cell Line Authentication Committee (ICLAC) (ICLAC 2015). Rules of confidentiality preclude the use of a donor's name or initials (see Supp. S6.4). Instead, use a sample log to record new tissue samples and add a unique identifier for each sample (see Protocol P13.2). The sample identifier, perhaps linked to a tissue identifier letter code, can then be used to establish the cell line designation, e.g. LT156 would be lung tumor biopsy number 156. This method is less likely to generate ambiguities and points to the record associated with the tissue sample. When reported in publications, it is helpful to prefix the cell line designation with a code indicating the laboratory in which it was derived (e.g. WI for Wistar Institute, NCI for National Cancer Institute, SK for Sloan-Kettering). Check to see if your cell line name is unique in online searches and avoid other names and commonly used abbreviations. For example, a search for "KB" brings up the KB cell line but also NF-κB and kilobase (kb) (Vaughan et al. 2017). Punctuation and unusual characters can give rise to problems when searching for a cell line in a database. Avoid using spaces, asterisks, sub- or superscripts, slashes, query or exclamation marks, periods, commas, colons or semi-colons, Greek letters, or other symbols. Any sublines or clones can be given a suffix as a number or alphanumeric binomial after a hyphen, e.g. LT156-D3, where "D3" is the numerical coordinate from a microwell plate.

Despite these efforts, the research community needs a process to generate a unique cell line identifier for publication (see Section 7.3.2). Cell repositories have traditionally dealt with this problem by giving each cell line an accession number when it is deposited. When you obtain a cell line from a cell repository, always record the accession number in your records, which identifies the source for publication (see Section 14.9). However, the repository accession number does not fully address the problem, particularly for cell lines that can be obtained from multiple sources. It is important to utilize (i) guidelines for naming a cell line; (ii) a central registry where its name can be registered, if available; and (iii) genomic evidence for its authenticity. A central registry is available for human pluripotent stem cell (hPSC) lines (Kurtz et al. 2018). Registration with the human pluripotent stem cell registry (hPSCreg) results in the generation of a standardized name, locking that name against further use. This approach represents best practice but is difficult to implement for all cell lines, because of the large number of existing cell lines and the ease with which new cell lines are generated. To improve reporting in publications, Research Resource Identifiers (RRIDs) have been developed to unambiguously identify materials, including cell lines (Bandrowski et al. 2016; Bairoch 2018). For example, bladder cell line "EJ-1" in the example above should be reported as "EJ-1 (RRID:CVCL_2893)," while endometrial cell line "EJ" should be reported as "EJ (RRID:CVCL_7039)." If you generate a new cell line, you can contact Cellosaurus to request its inclusion, which will generate an RRID for publication (Bairoch 2018).

14.2.2 Culture Age

Finite cell lines proliferate for a limited number of cell generations before reaching senescence, usually at 20–80 PDL (see Sections 3.2.1, 3.5). The PDL at which senescence occurs *in vitro* is known as the Hayflick limit. As senescence approaches, finite cell lines undergo biochemical and morphological changes that are suggestive of aging (see Figure 3.7) (Reddel 2000; Shay and Wright 2000). The behavior of a continuous cell line that has overcome senescence is markedly different to that of a finite cell line (see Table 14.1). Its behavior will also change with handling, e.g. due to selection of subpopulations within the culture. For all these reasons, the age of a culture is important in order to understand its behavior.

Passage number is used to estimate the time that a cell line has spent in culture, particularly with continuous cell lines. The passage number is usually indicated by the prefix "p" and increases by one with each subculture. PDL (see Section 14.1) is used to estimate the number of cell generations, particularly with finite cell lines, where this will determine the number of cell generations that remain before senescence. Passage number and PDL are not equivalent (unless the culture is subcultured with a 1 : 2 split ratio, in which case the two are roughly the same). PDL is calculated using a formula, based on the exponential increase in cell number during proliferation (Hayflick 1973):

$$N_{finish} = N_{start} \times 2^n \tag{14.1}$$

"N_{finish}" refers to the total number of cells at subculture, "N_{start}" to the number in the starting inoculum, and "n" to the number of cell generations. Rearranging the equation and using logarithms to the base 10, this becomes:

$$n = \frac{\log N_{finish} - \log N_{start}}{\log 2} \tag{14.2}$$

Log 2 can be included as a numerical value (0.301), which gives:

$$n = (\log N_{finish} - \log N_{start}) \times 3.32 \tag{14.3}$$

The final 14.3 can be used to determine the PDL for each subculture; each number should be added to the previous PDL, resulting in a cumulative total. If cells are not counted at each subculture, the PDL can be estimated from the split ratio

TABLE 14.1. Properties of finite and continuous cell lines.

Properties	Finite	Continuous (transformed)
Ploidy	Euploid, diploid	Aneuploid, heteroploid
Transformation	Normal	Immortal, altered growth control, tumorigenic
Anchorage dependence	Yes	No
Contact inhibition	Yes	No
Density limitation of cell proliferation	Yes	Reduced or lost
Mode of growth	Usually monolayer	Monolayer or suspension
Maintenance	Cyclic	Steady state possible
Serum requirement	High	Low
Yield	Low	High
Cloning efficiency	Low	High
Markers	Usually tissue-specific	Variable due to dedifferentiation or transdifferentiation
Special functions (e.g. virus susceptibility, differentiation)	May be retained	Often lost
Doubling time[a]	Slow (PDT > 36 h)	Rapid (PDT 12–36 h)
Culture lifespan	< 100 PDL, depends on cell type	≥ 100 PDL

[a]More information on PDT is available elsewhere (see Section 19.3.3).

(see Section 14.6.4). PDL increases by one for a split ratio of 1 : 2, by two for a split ratio of 1 : 4, by three for a split ratio of 1 : 8, and so on. These numbers are all approximate and do not allow for loss through cell death or differentiation, or premature aging and withdrawal from the cell cycle, which probably take place between each subculture. It is also difficult to estimate the number of cell generations in the primary culture before passaging commences.

14.2.3 Cell Line Validation

While continued subculture extends the lifetime of a culture and its availability, it also increases the risk of cross-contamination (see Section 17.1). Whenever more than one cell line is maintained in a laboratory there is always a risk that cells from one cell line will be accidentally introduced into the other, resulting in cross-contamination and (if the contaminating cells overgrow the authentic material) a misidentified cell line. Misidentified cell lines can arise due to lapses in aseptic technique, experimental errors (e.g. mislabeling or seeding the wrong flask), or poor inventory control in the cryofreezer. A recent study demonstrated that 14% of all cell lines received at a cell repository were misidentified over a 25-year period (Drexler et al. 2017). The incidence from primary sources declined to 6% in the last five years, suggesting that recent measures to reduce misidentification have reduced the size of the problem in source laboratories. However, cross-contamination continues to be a concern for all laboratories.

Mycoplasma contamination is another common risk that must be considered when generating a new cell line. While infection of a primary culture or an early passage cell line often leads to rapid degeneration and loss of the culture, mycoplasma contamination seems to be better tolerated in continuous cell lines and often goes undetected. A recent study demonstrated that 22% of all cell lines received at a cell repository were positive for mycoplasma over a 25-year period, although the incidence decreased dramatically in the last five years (Drexler et al. 2017). Mycoplasma arises from a number of sources, including the operator, the environment, contamination of equipment or reagents, and other cell lines (see Section 16.1). The major risk for today's laboratories is likely to be contamination of other cell lines. All laboratories must have a testing program in place to ensure that cell lines remain free of mycoplasma contamination (see Section 16.4.1). The consequences of a major outbreak far exceed the costs of performing regular tests.

Cell culture risks can be avoided through GCCP and validation testing (see Sections 7.2.1, 7.4). To summarize requirements when generating a new cell line:

(1) Use good aseptic technique whenever you handle cultures (see Section 12.4).
(2) Put aside tissue samples or DNA from the donor as reference samples for later authentication testing (see Section 13.3).
(3) Perform cryopreservation to store the culture at low passage; these "banks" of vials act as seed stock for later use (see Section 15.5).
(4) Perform authentication testing on the cell line and tissue sample, using a suitable consensus test method (see Sections 17.3.2, 17.3.3). If the cell line is misidentified, look for earlier passage seed stock and repeat the testing. Discard any misidentified stocks.
(5) Perform mycoplasma testing on the cell line (see Section 16.4.1). If mycoplasma is detected, look for earlier passage seed stock and repeat the testing. Discard or treat any contaminated stocks (see Protocols P16.1, P16.5).

(6) Perform additional characterization on the cell line as required for your experiments and the cell type (e.g. Drexler and Matsuo 1999).
(7) Publish testing and characterization data when you first publish the cell line and deposit authentication data in a database (e.g. BioSample) (Barrett et al. 2012).
(8) Deposit the cell line at a cell repository (see Section 15.6). Cell repositories provide validation testing, act as backup storage for your frozen stocks, and reduce the burden of distributing the cell line on request. Cell lines that are deposited at a cell repository can remain useful models well after the originating laboratory has closed or its stocks are discarded or lost.

14.3 CHOOSING A CELL LINE

Tissue culture laboratories have generated a large number of cell lines, many of which are available through cell repositories. Over time, many cell lines have become accepted as models for specific tissue types or disease states. Some commonly used examples are listed in Table 14.2; additional information is available in publications that describe these cell lines and their validation, many of which are cited in Cellosaurus (Bairoch 2018). Always remember that commonly used cell lines can be misidentified. For example, the KB cell line was originally reported as an epidermoid (squamous cell carcinoma [SCC]) cell line from the larynx; the cell line was shown to be misidentified more than 50 years ago and is actually HeLa, from cervical adenocarcinoma (Gartler 1968). Unfortunately, KB continues to be used as a popular research model for oral or laryngeal cancer. It was used in more than 600 journal articles between 2000 and 2014; 90% of articles described it as an oral, laryngeal, or SCC cell line (Vaughan et al. 2017). Misidentified cell lines should be avoided unless there is strong scientific justification for their use. For example, in the case of KB, some derivatives possess multidrug resistance and are correctly used for this research field, provided they are correctly described as HeLa derivatives (Gillet et al. 2013). A register of misidentified cell lines is available to check before you start work with a new cell line (ICLAC 2019).

How do you choose a cell line for your work? Usually, you will have a particular hypothesis or application that will determine the best cell line for your needs. Apart from specific functional requirements, general parameters to consider include:

(1) *Finite or continuous.* Is there a continuous cell line that expresses the right functions? A continuous cell line generally is easier to maintain, grows faster, clones more easily, produces a higher cell yield per flask, and is more readily adapted to serum-free medium (see Table 14.1).
(2) *Normal or transformed.* Many cell lines show signs of transformation, which usually implies that they have

TABLE 14.2. Commonly used cell lines.

Cell line	RRID[a]	Tissue origin	Species (strain)	Age	Morphology[b]	Ploidy	Characteristics
Finite, from normal tissue							
IMR-90	CVCL_0347	Lung	Human	Embryonic	Fibroblast	Diploid	Susceptible to human viral infection; contact inhibited
MRC-5	CVCL_0440	Lung	Human	Embryonic	Fibroblast	Diploid	Susceptible to human viral infection; contact inhibited
MRC-9	CVCL_2629	Lung	Human	Embryonic	Fibroblast	Diploid	Susceptible to human viral infection; contact inhibited
WI-38	CVCL_0579	Lung	Human	Embryonic	Fibroblast	Diploid	Susceptible to human viral infection
Continuous, from normal tissue							
3T3-L1	CVCL_0123	Whole animal	Mouse (Swiss albino)	Embryonic	Fibroblast	Aneuploid	Adipose differentiation
BALB/3T3 clone A31	CVCL_0184	Whole animal	Mouse (BALB/c)	Embryonic	Fibroblast	Aneuploid	Contact inhibited; readily transformed
BEAS-2B	CVCL_0168	Lung	Human	Adult	Epithelial		Pulmonary epithelium
BHK-21 clone 13	CVCL_1915	Kidney	Syrian hamster	Newborn	Fibroblast	Aneuploid	Transformable by polyoma
BRL-3A	CVCL_0606	Liver	Rat (Buffalo)	Newborn	Epithelial		IGF-2 production
C2	CVCL_6812	Skeletal muscle	Mouse (C3H)	Embryonic	Fibroblastoid		Myotubes
CHO-K1	CVCL_0214	Ovary	Chinese hamster	Adult	Fibroblast	Diploid	Simple karyotype
CV-1	CVCL_0223	Kidney	Green Monkey	Adult	Epithelioid		Viral substrate and assay
COS-7	CVCL_0224	CV-1 derivative	Green Monkey	Adult	Epithelioid		Good host for DNA transfection
HaCaT	CVCL_0038	Keratinocytes	Human	Adult	Epithelial	Diploid	Keratinization
HEK293	CVCL_0045	Kidney	Human	Embryonic	Epithelial	Aneuploid	Readily transfected
L6	CVCL_0385	Skeletal muscle	Rat	Embryonic	Fibroblastoid		Myotubes
LLC-PK1	CVCL_0391	Kidney	Pig	Adult	Epithelial	Diploid	Na+-dependent glucose uptake
MDCK	CVCL_0422	Kidney	Dog (Cocker spaniel)	Adult	Epithelial	Diploid	Dome formation, transport
NIH 3T3	CVCL_0594	Whole animal	Mouse (NIH Swiss)	Embryonic	Fibroblast	Aneuploid	Contact inhibited; feeder layer for hPSCs
NRK-49F	CVCL_2144	Kidney	Rat (Osborne-Mendel)	Adult	Fibroblast	Aneuploid	Induction of suspension growth by TGF-α,β
STO	CVCL_3420	Whole animal	Mouse (SIM)	Embryonic	Fibroblast	Aneuploid	Feeder layer for hPSCs
Vero	CVCL_0059	Kidney	Green Monkey	Adult	Fibroblast	Aneuploid	Viral substrate and assay

(*continued*)

TABLE 14.2. (*continued*)

Cell line	RRID[a]	Tissue origin	Species (strain)	Age	Morphology[b]	Ploidy	Characteristics
Continuous, from neoplastic tissue							
A2780	CVCL_0134	Ovarian adenocarcinoma	Human	Adult	Epithelial	Aneuploid	Chemosensitive with resistant variants
A-549	CVCL_0023	Lung adenocarcinoma	Human	Adult	Epithelial	Aneuploid	Synthesizes surfactant
B16	CVCL_F936	Melanoma	Mouse (C57BL/6)	Adult	Fibroblastoid	Aneuploid	Melanin
B35	CVCL_1951	Neuroblastoma	Rat (BDIX)	Juvenile	Neuronal	Aneuploid	Neurites; outgrowth induced by cAMP
C-1300	CVCL_4343	Neuroblastoma	Mouse (A/J)	Adult	Neuronal	Aneuploid	Neurites
C6	CVCL_0194	Malignant glioma	Rat (Wistar Furth)	Newborn	Fibroblastoid	Aneuploid	GFAP, GPDH
Caco-2	CVCL_0025	Colon adenocarcinoma	Human	Adult	Epithelial	Aneuploid	Transports ions and amino acids
EB3	CVCL_1185	Burkitt lymphoma	Human	Juvenile	Suspension	Diploid	EBV+
GH1	CVCL_0610	Pituitary tumor	Rat (Wistar Furth)	Adult	Epithelioid	Aneuploid	Growth hormone
HeLa	CVCL_0030	Cervical adenocarcinoma	Human	Adult	Epithelial	Aneuploid	Rapid growth, viral substrate
HeLa S3	CVCL_0058	HeLa derivative	Human	Adult	Epithelial	Aneuploid	High cloning efficiency; will grow well in suspension
Hep-G2	CVCL_0027	Hepatoblastoma	Human	Adult	Epithelioid	Aneuploid	Retains some microsomal metabolizing enzymes
HL-60	CVCL_0002	Acute myeloid leukemia	Human	Adult	Suspension	Aneuploid	Phagocytosis; neotetrazolium blue reduction
HT-29	CVCL_0320	Rectosigmoid adenocarcinoma	Human	Adult	Epithelial	Aneuploid	Differentiation inducible with NaBt
K-562	CVCL_0004	Chronic myeloid leukemia	Human	Adult	Suspension	Aneuploid	Hemoglobin
L1210	CVCL_0382	Leukemia	Mouse (DBA/2)	Adult	Suspension	Aneuploid	Rapidly growing; suspension

L-929 (NCTC clone 929)	CVCL_0462	Subcutaneous (carcinogen exposure)	Mouse (C3H/An)	Adult	Fibroblast	Aneuploid	First cloned cell strain (see Figure 1.1)
MCF-7	CVCL_0031	Pleural effusion from breast ductal carcinoma	Human	Adult	Epithelial	Aneuploid	ER+, dome formation, α-lactalbumin
MCF-10[c]	CVCL_0598 / CVCL_5553	Fibrocystic breast tissue	Human	Adult	Epithelial + suspension	Near diploid	Dome formation
P388D1	CVCL_0477	Lymphoma	Mouse (DBA/2)	Adult	Suspension	Aneuploid	Macrophage-like
CCRF S-180	CVCL_2874	Sarcoma	Mouse (CFW)	Adult	Fibroblast	Aneuploid	Cancer chemotherapy screening
SK-HEP-1	CVCL_0525	Hepatoma	Human	Adult	Endothelial	Aneuploid	Endothelial-like; Factor VIII
U-251MG	CVCL_0021	Malignant glioma	Human	Adult	Glial	Aneuploid	GFAP
WEHI-3B D+	CVCL_4357	Leukemia	Mouse (BALB/c)	Adult	Suspension	Aneuploid	IL-3 production
ZR-75-1	CVCL_0588	Ascites from breast ductal carcinoma	Human	Adult	Epithelial	Aneuploid	ER−, EGFR+

[a]RRIDs can be used to find additional information in the Cellosaurus knowledge resource, including source publications (https://web.expasy.org/cellosaurus) (Bairoch 2018).
[b]For further comments on morphology see Section 18.1.2.
[c]MCF-10 has two RRIDs, which identify its adherent and floating populations (MCF-10A and MCF-10F).
ER, Estrogen receptor; GFAP, glial fibrillary acidic protein; GPDH, glycerol phosphate dehydrogenase.

increased tumorigenicity (see Table 3.2). If this is undesirable, it may be possible to obtain an immortalized cell line that is not tumorigenic, e.g. BHK-21 clone 13 (see Table 14.2).

(3) **Species.** A non-human cell line may have a lower biohazard risk (see Section 6.3.1) and the tissue from which it was derived may be more accessible.

(4) **Cell type.** Some cell types are difficult to maintain in culture, while others have multiple cell lines available. Look for advice from laboratories that have worked with various cell lines from your tissue or disease of interest, e.g. breast cancer (Holliday and Speirs 2011). Omics data (genome, transcriptome, proteome, etc.) are also helpful, e.g. when selecting breast cancer cell lines as tumor models (Vincent et al. 2015).

(5) **Source.** If you intend to use a finite cell line, are there sufficient stocks available, or will you have to generate your own line(s)? If you intend to use a continuous cell line, are authenticated stocks available? If such stocks are not available in your laboratory, is the cell line available from a reputable source (see Section 15.6)?

(6) **Validation.** Is the cell line known to be misidentified? If your source has not performed validation testing, always do so on arrival and quarantine the cell line until the results are known (see Sections 14.2.3, 16.2). Testing should be repeated at intervals to confirm that the cell line remains authentic and free of mycoplasma. This is best done as part of cell banking (see Section 15.5).

(7) **Two-dimensional (2D) or three-dimensional (3D).** Most commonly used cell lines grow as adherent monolayers, but some grow as 3D spheroids in suspension (see Sections 14.7, 27.4.1). Some cell lines can be adapted to 3D culture systems; these systems tend to be relatively complex compared to 2D systems, but provide many benefits (see Minireview M27.1).

(8) **Characterization.** Does the cell line express the correct characteristics? Because its behavior can change with culture conditions and handling, it is important to demonstrate it has the correct phenotype before starting work (see Section 18.1).

(9) **Growth characteristics.** How quickly do the cells grow and what yield do you require? You will need to consider various parameters, including the culture's population doubling time (PDT), saturation density, and cloning efficiency (see Sections 19.3.3, 19.4).

(10) **Stability.** How stable is the cell line's genotype and phenotype (see Sections 3.3, 3.4; see also Figure 20.2)? Has it been cloned? If not, can you clone it, and how long would the cloning process take to generate sufficient frozen and usable stocks?

(11) **Controls.** If you are using a mutant, transfected, transformed, or otherwise abnormal cell line, is there a normal equivalent available for comparison?

14.4 MAINTAINING A CELL LINE

Once a culture is initiated, it will need periodic feeding (replacement of medium), followed by subculture if the cells are proliferating (also known as passaging; see Section 14.1). Intervals between medium changes and between subcultures will vary between cell lines. Rapidly growing continuous cell lines, such as HeLa, are usually subcultured once per week and the medium changed four days later. Slower growing cell lines (particularly finite cell lines) may need to be subcultured only every two, three, or even four weeks; the medium should be changed weekly between subcultures.

The need for subculture is determined by the growth curve of the cell line (see Figure 14.1; see also Section 19.3). Following subculture and reseeding at a lower concentration, growth usually follows a standard pattern. A lag period is followed by a period of exponential growth, known as the log phase. Cells should be fed during this phase to ensure that their nutritional requirements are met. Following the log phase, cells enter the plateau phase, where proliferation stops (in normal cells) or where any remaining cell proliferation is matched by cell loss (often the case for transformed cells). Cells should be subcultured at the top end of the log phase before entering the plateau phase. This will ensure that the seeding efficiency (the number of cells that attach and grow after subculture) is high and the lag period is minimized.

14.4.1 Routine Observation

Routine observation is essential to determine the phase of growth and to detect problems. Observation should be macroscopic (e.g. looking for medium turbidity and color changes that indicate high or low pH) and microscopic (e.g. looking at cell morphology and density). The cells should also be checked for any signs of deterioration, such as granularity around the nucleus, cytoplasmic vacuolation, and rounding up of the cells with detachment from the substrate (see Figure 14.2). Such signs may imply that the culture requires a medium change or may indicate a more serious problem, such as microbial or chemical contamination, inadequate or toxic medium or serum, or senescence (see Sections 31.1, 31.3–31.5). For example, BEAS-2B cells are meant to be grown serum-free and the addition of serum can cause deterioration (see Figure 14.2). Medium deficiencies can also initiate apoptosis (see Sections 2.5, 29.2.5). Medium changes and subculture frequency should aim to prevent signs of deterioration, as it is often difficult to reverse. Think about any actions that may be required as you perform routine observation of each culture (see Table 14.3).

Familiarity with the morphology of the cell line is essential (see Section 18.4; see also Figures 14.3, 14.4). Subtle changes in morphology may give the first indication of microbial contamination or cross-contamination. Always save images for any cell line that is handled in the laboratory to act as references, particularly when a new member of staff is being

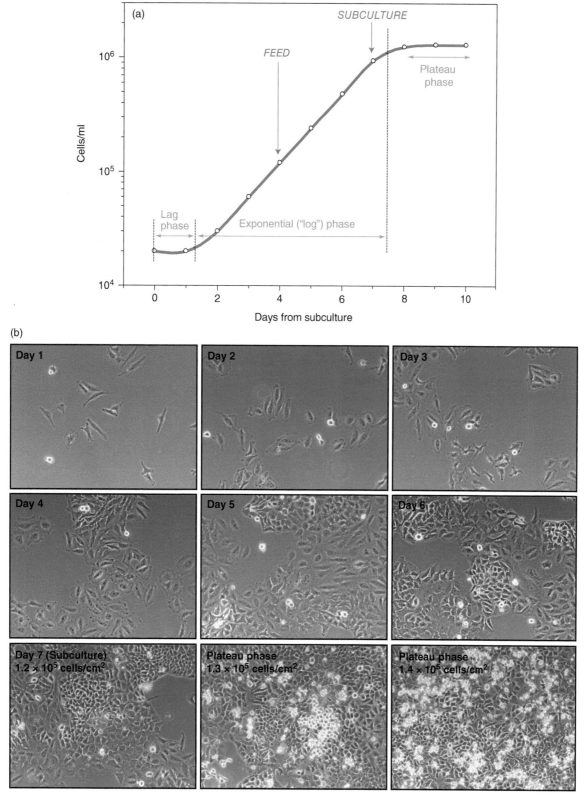

Fig. 14.1. Growth curve and morphology. (a) Semilog plot of cell concentration versus time, showing the lag phase, exponential "log" phase, and plateau, and indicating times at which feeding and subculture should be performed. The parameters that can be derived from the growth curve are shown elsewhere (see Figure 19.7). (b) Representative images of HeLa cells over the same time course. By day 7, cells have reached 1.2×10^5 cells/cm^2. After the culture reaches plateau phase, the cells increasingly detach from the monolayer, which limits any further increase in the yield of adherent cells. *Source*: (b) Images and data courtesy of Elsa Moy, CellBank Australia.

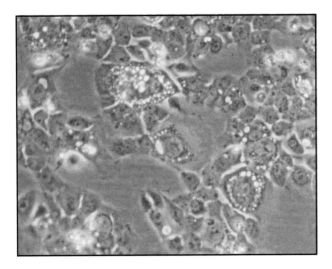

Fig. 14.2. Unhealthy cells. BEAS-2B cells showing signs of poor health due, in this case, to medium inadequacy. The cytoplasm of the cells becomes granular, particularly around the nucleus, and vacuolation occurs. Cells may become more refractile at the edge if spreading is impaired. *Source*: R. Ian Freshney.

introduced to culture work. Images should be saved at different cell densities (low and high), ideally with a cell count to calculate the number of cells per centimeter squared. A training exercise in cell culture observation is provided in the Supplementary material online, with a record that can be filled out during culture observation (see Supp. S30.8; see also Table S30.1).

14.4.2 Standardization of Culture Conditions

Standardization of culture conditions is an important part of routine maintenance. Although some conditions may alter because of the demands of experimentation, development, and production, routine maintenance should adhere to consistent, defined conditions. Parameters that need standardization include:

(1) *Culture age.* Some cell lines gain or lose markers or characteristics with ongoing handling. For example, the BEAS-2B cell line is used as a model for human airway epithelium but its behavior is known to change at later passage (Noah et al. 1995). Passage number and/or PDL should be recorded at each subculture and cells should be replaced from frozen stocks at regular intervals (see Section 15.5.3).

(2) *Subculture conditions.* Different cell lines will vary in their proliferation rates and behavior at low and high densities. It is important to understand the growth cycle of the cell line, which will help to determine the best conditions for subculture (see Section 14.6.4).

(3) *Medium.* The culture medium will influence the selection of different cell types and alter their phenotypic

TABLE 14.3. Examination of cells during routine maintenance.

Criterion	Action indicated[a]
Contamination (e.g. cloudy medium; see Figure 16.2)	Discard culture and medium and reagents used with it; identify and eliminate source if repeated or widespread (see Sections 16.2, 31.1)
pH of medium (see Figure 9.1)	If pH low, feed or subculture, depending on cell density; if pH high, check flasks for leaks and incubator for problems, e.g. CO_2 level; if pH high for all cultures, check medium preparation
Confluence (density)	If high (e.g. > 70% confluent), subculture; if moderate (50–70% confluent), check pH and feed if necessary
Morphology	Confirm against archival images at same cell density
Mitoses	Indicates proliferation; no action required unless absent, then check growth conditions
Deterioration	Indicates possible contamination or incorrect medium; check growth conditions and correct if necessary; test for contamination; discard and replace from frozen stock
Altered growth pattern	Indicates possible cross-contamination or transformation; check authenticity (see Sections 17.3.2, 17.3.3); replace from frozen stock
Heterogeneity (evidence of mixed cell types)	Confirm against archival images at same cell density; check authenticity (see Sections 17.3.2, 17.3.3); replace from frozen stock

[a]This table focuses on actions that should be taken based on observation. A separate Culture Observation Record is available to help trainees perform routine observation (see Supp. S30.1).

expression. Once you have selected a suitable medium for your cell line (see Section 9.7), continue to use it and avoid changing suppliers. If you must alter the medium, perform culture testing to understand its impact (see Section 11.8.2). Moving toward serum-free medium will help to eliminate variation between serum batches (see Section 10.6.1). If this is not feasible, serum replacements may offer greater consistency and are generally cheaper than growth factor supplementation, although they do not offer the control over the physiological environment afforded by defined serum-free medium (see Section 10.5).

(4) *Substrate.* Most flasks and dishes from major suppliers will give similar results, but there may be minor

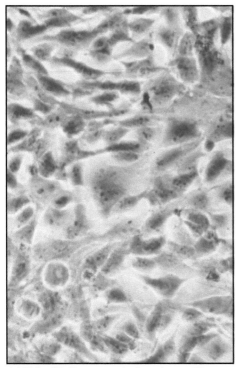

(a) **HeLa-S3.** *Human cervical carcinoma. Phase contrast; 40x objective.*

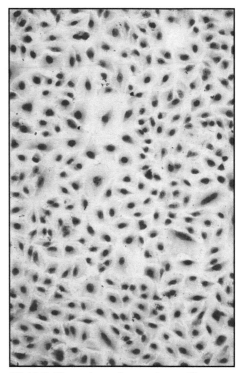

(b) **A549.** *Human lung adenocarcinoma. Giemsa stained; 10x objective.*

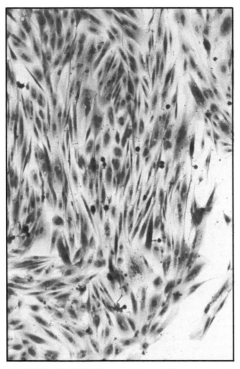

(c) **MOG-G-UVW.** *Human glioma. Giemsa stained; 10x objective.*

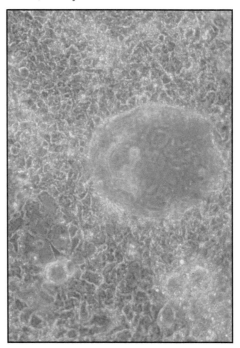

(d) **Caco-2.** *Human colorectal carcinoma showing dome formation. This cell line should be subcultured before it reaches this stage and while it is still subconfluent. Phase contrast; 10x objective. Source for all images: R. Ian Freshney.*

Fig. 14.3. Continuous cell lines from human tumors.

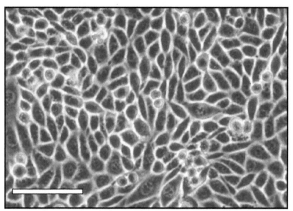

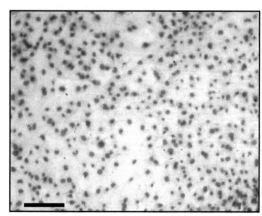

(a) CHO-K1. Clone of CHO, continuous cell line from Chinese hamster ovary. The scale bar indicates 100 µm. Phase contrast. Source: R. Ian Freshney.

(b) 3T3-Swiss Albino. Fibroblast-like cells from Swiss mouse embryo. The scale bar indicates 100 µm. Giemsa stained. Source: R. Ian Freshney.

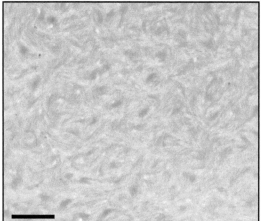

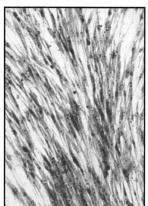

(c) BHK-21 Clone 13. Baby hamster kidney fibroblasts after confluence, showing typical swirling pattern formed by parallel arrays of cells. The scale bar indicates 2 mm. Giemsa stained. Source: R. Ian Freshney.

(d) BHK-21 Clone 13. Left panel: cells are approaching confluence and assuming parallel arrays. Right panel: cells are post-confluent and forming a second monolayer of more densely stained cells. The scale bars indicate 200 µm. Giemsa stained. Source: R. Ian Freshney.

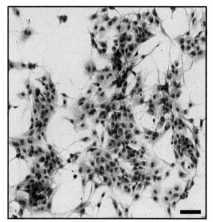

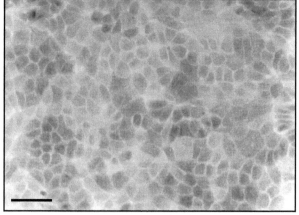

(e) MDCK. Madin Darby canine kidney cells. Epithelial-like cell line from dog kidney. The scale bar indicates 100 µm. Source: R. Ian Freshney.

(f) Mv1Lu. Mink lung epithelial cell line. Stained by immunoperoxidase for cytokeratin (AE3 primary antibody). The scale bar indicates 200 µm. Source: Courtesy of M.Z. Khan.

Fig. 14.4. Continuous cell lines from normal, non–human animal tissue.

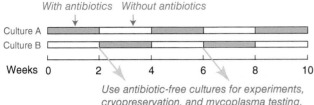

With antibiotics   Without antibiotics

Use antibiotic-free cultures for experiments, cryopreservation, and mycoplasma testing. Experimental culture can contain antibiotics as long as they are known not to interfere with processes under investigation.

Fig. 14.5. Parallel cultures and antibiotics. A suggested scheme for maintaining parallel cell cultures with and without antibiotics, such that each culture always spends part of the time out of antibiotics.

variations due to the treatment of the plastic for tissue culture (see Section 8.2.1). Hence it is preferable to use one type of flask or dish and avoid changing suppliers. If the cell line is normally grown on a treated surface or coating (e.g. Matrigel®), it should continue to be used (see Section 8.3).

14.4.3 Use of Antibiotics

Antibiotics should be avoided during routine maintenance of cell lines (see Sections 9.6.2, 16.1.5). The protocols provided in this chapter for refeeding and subculture all use antibiotic-free media. However, there may be some circumstances where antibiotics are warranted, e.g. primary culture of tissue types at increased risk of contamination (see Protocol P13.2). The choice of antibiotics will depend on the microorganisms that are likely to be present (see Table 16.2). If antibiotics are used, they should be ceased after a limited time or some antibiotic-free stocks should be maintained to reveal any cryptic contaminations. These stocks can be maintained in parallel, alternating with and without antibiotics, until antibiotic-free culture is possible (see Figure 14.5). It is not advisable to adopt this procedure as a permanent regime. If chronic contamination is suspected, cultures should be discarded or the contamination treated before reverting to antibiotic-free maintenance (see Sections 16.2, 16.3.2).

14.5 REPLACING MEDIUM (FEEDING)

14.5.1 Criteria for Replacing Medium

Five factors indicate the need for the replacement of culture medium:

(1) **Changes in pH.** Most cells stop growing as the pH falls from pH 7.0 to pH 6.5 and start to lose viability below pH 6.5, so if the medium goes from red through orange to yellow (see Figure 9.1), the medium should be changed.

Both the rate of fall and the absolute level should be considered. A culture at pH 7.0 that falls 0.1 pH units in one day will not come to harm if left a day or two longer before feeding, but a culture that falls 0.4 pH units in one day will need to be fed within 24–48 hours and cannot be left over a weekend without feeding.

(2) **Time since subculture.** The medium should be changed periodically even if cells are not proliferating, e.g. senescent cultures (see Section 14.2.2). The cells will take up nutrients and excrete products, some constituents of the medium will become exhausted or will degrade spontaneously, and the pH may change.

(3) **Cell density.** Cultures at a high density (number of cells per centimeter squared of substrate) or concentration (number of cells per milliliter of medium) exhaust the medium faster than those at a low density. Increasing cell density is usually evident in the rate of change of pH, but not always.

(4) **Cell type.** Normal cells (e.g. diploid fibroblasts) usually stop dividing at a high cell density because of cell crowding, shape change, growth factor depletion, and other factors (see Figure 3.1a). The cells block in the G1 phase of the cell cycle and deteriorate very little, even if left for two to three weeks or longer. However, many continuous cell lines and some hPSC lines deteriorate rapidly at high cell densities, unless the medium is changed daily or they are subcultured.

(5) **Morphological deterioration.** This factor must be anticipated by regular examination and familiarity with the cell line (see Section 14.4.1). If deterioration is allowed to progress too far, it will be irreversible, as the cells may enter apoptosis (see Sections 2.5, 29.2.3).

14.5.2 Holding Medium

A holding medium may be used for feeding when stimulation of mitosis, which usually accompanies a medium change, even at high cell densities, is undesirable. Holding media are usually regular media with the serum concentration reduced to 0.5% or 2% (or eliminated completely), or serum-free media without growth factors. Removal of serum or growth factors is used to inhibit mitosis and thus reduce cell growth. Holding media are sometimes used to maintain cell lines with a finite lifespan in a non-proliferative state so that the limited number of cell generations available to them are not used up (see Section 14.2.2). Reduction of serum and cessation of cell proliferation can also promote expression of the differentiated phenotype in some cells (Maltese and Volpe 1979; Schousboe et al. 1979). However, many continuous cell lines are unsuitable for this procedure. These cell lines will either continue to divide successfully or deteriorate, because transformed cells do not block in a regulated fashion in G$_1$ of the cell cycle.

Medium used for the collection of biopsy samples can also be referred to as holding medium. To avoid confusion, it is referred to as Collection Medium throughout this book (see Appendices A and B).

14.5.3 Standard Procedure for Feeding

Feeding a culture implies removal of spent medium and replenishment with fresh medium; if the cells are growing slowly, it may be preferable to remove half of the medium, allowing the remainder to act as conditioned medium (see Section 9.6.1). Cell lines should always be fed separately and the biological safety cabinet (BSC) should be wiped down and allowed to "purge" before introducing the next cell line (see Section 12.2.2). Always use separate bottles or aliquots of medium for each cell line; this is an important rule to manage the risk of contamination (see Tables 16.1, 17.2). Handling procedures generate airborne droplets, which are readily transferred to other vessels (Coriell 1962; McGarrity 1976). Droplet spread is reduced when using a BSC or laminar flow hood but cannot be eliminated entirely. Feeding cultures in Petri dishes or multiwell plates requires additional care as they are more prone to contamination, due to the exposed surface of the culture and manipulation of the lids.

How much medium should you add during feeding? The usual ratio of medium volume to surface area is $0.2–0.5\,ml/cm^2$ (see Section 8.5.1; see also Table 8.2). The upper limit is set by gaseous diffusion through the liquid layer, and the optimum ratio depends on the oxygen requirement of the cells. Cells with a high oxygen requirement do better in shallow medium (e.g. 2 mm) and those with a low requirement may do better in deep medium (e.g. 5 mm). If the depth of the medium is greater than 5 mm, gaseous diffusion may become limiting. For suspension culture, stirring the medium improves oxygenation up to a depth of around 5 cm, after which it will be necessary to sparge the medium (see Section 28.2). For adherent culture, this problem can be addressed by growing cells in roller bottles, multisurface propagators, or bioreactors that provide a supportive scaffold or permit the use of microcarriers (see Section 28.3).

Protocol P14.1 can be used as a standard procedure for feeding adherent cultures. It can also be used for training (see Supp. S30.9). A worksheet for handling cultures is available as part of the training exercise (see Table S30.3).

PROTOCOL P14.1. FEEDING ADHERENT CULTURES

Background

This procedure can be used to feed adherent cultures in flasks, Petri dishes, or multiwell plates. Separate cell lines should be fed separately. Flasks, dishes, and plates that contain the same cell line can be fed together, but the techniques used are slightly different.

Outline

Examine the culture by eye and using an inverted microscope. Decide if feeding is required. If required, remove the spent medium and add fresh medium. Return the culture to the incubator.

- *Safety Note. Pipettes can break, resulting in sharps injuries (see Section 6.2.1). Dispose of pipettes into a pipette cylinder or similar container. Dispose of glass Pasteur pipettes separately into a sharps container. Perform decontamination as part of safe disposal practices.*

Materials (sterile or aseptically prepared)

- ☐ Culture(s) in flask, Petri dish, or multiwell plate
- ☐ Complete medium (see Section 9.7)
- ☐ Pipettes in an assortment of sizes, usually 1, 5, 10, and 25 ml (plugged to prevent aspiration into the pipette controller or bulb)
- ☐ Unplugged pipettes, e.g. Pasteur pipettes (if using a pump or vacuum line)

Materials (non-sterile)

- ☐ BSC and associated consumables (see Protocol P12.1)

Procedure

A. Before Feeding

1. Prewarm bottles of complete medium and any other reagents required, e.g. in a water bath.
2. Prepare a work space for aseptic technique using a BSC.
3. Bring medium and any other reagents necessary for the procedure to the BSC. Swab each item with 70% alcohol and place in the BSC (see Protocol P12.1).
4. Take the culture out of the incubator and bring it to the microscope. Observe the culture carefully for signs of contamination or deterioration (see Section 14.4.1).
5. Check the previously described criteria (see Section 14.5.1) and, based on your knowledge of its behavior, decide whether to feed the culture.
6. If feeding is not required, return the culture to the incubator. If feeding is required, take the culture to the BSC, swab it with 70% alcohol, and place it in the BSC. If Petri dishes or multiwell plates have been enclosed in a protective box (see Section 12.5.2), remove from the box and swab with 70% alcohol before bringing dishes or plates to the BSC.

B. Feeding Flasks

7. Take a sterile pipette and insert it into the pipette controller. If using a pump or vacuum line, select a Pasteur pipette instead and connect it to the system's tubing.
8. Uncap the flask and tilt it so that the medium moves to the bottom of the flask. Withdraw the medium and discard it into a waste beaker. If using a pump or vacuum line, use the Pasteur pipette to aspirate the medium instead. Recap the flask and discard the pipette.
9. Take a fresh pipette and uncap the medium bottle. Take an aliquot of fresh medium, using the same volume that was removed from the flask. Recap the medium bottle.
10. Uncap the flask and transfer the medium in the pipette to the bottom of the flask. Avoid pipetting directly onto the cell monolayer, which might lead to loss of adhesion. Recap the flask and discard the pipette.
11. Lay the culture flat so that the fresh medium covers the adherent monolayer and return it to the incubator. If you do not have access to an incubator with the correct gas mixture, gas the flask before you return it to the incubator (see Section 12.5.3).

C. Feeding Dishes and Plates

12. Stack all dishes or plates carefully on one side of work area, ready to pick up.
13. Move the dish or plate that you intend to handle to the center of the work area.
14. Take a sterile pipette and insert it into the pipette controller. If using a pump or vacuum line, select a Pasteur pipette instead and connect it to the system's tubing.
15. Remove the lid and place it behind the dish, open side up. Grasp the dish as low down on the base as you can, taking care not to touch the rim of the dish or to let your hand come over the open area of the dish or lid. Tilt slightly so the medium becomes deeper in one area.
16. Place the pipette tip close to the adherent monolayer in the area where the medium is deepest and withdraw the medium. If working with a multiwell plate, withdraw medium from each well. Discard the spent medium and replace the lid.
17. Move the dish or plate to the other side of the work area.
18. Repeat steps 13–17 for the remaining dishes or plates. Once complete, discard the pipette. One

pipette may be used for several dishes or plates if it is a large pipette and serial deliveries are made to each dish or well. Replace if asepsis is breached accidentally.

19. Repeat step 12 so the dishes or plates are positioned ready for a second round of handling.
20. Move the dish or plate that you intend to handle to the center of the work area.
21. Take a fresh pipette and uncap the medium bottle. Take an aliquot of fresh medium, using the same volume that was removed from the flask. Recap the medium bottle.
22. Remove the lid and place it behind the dish, open side up. Add the medium to the dish, directing the stream gently low down on the side of the base. If working with a multiwell plate, add medium to each well. Replace the lid and discard the pipette. Do not return a used pipette to the medium bottle.
23. Move the dish or plate to the other side of the work area. Do not let the medium enter the capillary space between the lid and the base; wipe away any spills.
24. Repeat steps 22 and 23 for the remaining dishes or plates. Once complete, return dishes or plates to the incubator.

D. After Feeding

25. Clear away all pipettes, glassware, and other movable items. Return unused media and reagents to cold storage.
26. Swab down all surfaces with 70% alcohol and allow the BSC to purge for 4 minutes.
27. Record observations and actions using a record sheet, lab book, or electronic database.

Notes

Steps 1–6: The order and timing of these steps will vary from day to day. Many tissue culture personnel prefer to observe all cultures at the beginning of the working day and decide on any required actions at that point, before warming medium and reagents. On other days, it will be clear what actions are required based on the previous day's observations and actions.
Steps 7–11: This procedure assumes that you can use one hand to remove the cap, hold it while pipetting, and replace the cap afterwards. Using one hand may not be possible, e.g. when feeding multiple flasks. If you need both hands, remove the cap first, place it to one side while pipetting, and then replace it after you have discarded the pipette (see Section 12.3.3).

Steps 12–24: This procedure assumes that you cannot use one hand to hold the lid of the dish or plate while pipetting; these lids are larger than flask or bottle caps. However, with practice, you may be able to open the lid sufficiently and tilt the dish to transfer medium without removing the lid completely. This is quicker and safer.

14.6 SUBCULTURE (PASSAGING)

14.6.1 Criteria for Subculture

When increasing cell density results in all the available substrate becoming occupied, or when the cell concentration exceeds the capacity of the medium, cells will reach a plateau phase where growth ceases or is greatly reduced (see Figure 14.1). Subculture should be performed as determined by the following criteria:

(1) *Cell density.* Finite cell lines should be subcultured as soon as they reach confluence (see Figure 14.6d; see also Figure 18.1, confluent images). If left more than 24 hours beyond this point, normal cells will withdraw from the cycle and take longer to recover when reseeded. Continuous cell lines should also be subcultured on reaching confluence or shortly after. Although they will continue to proliferate beyond confluence, they will start to deteriorate after about two doublings and reseeding efficiency will decline. Some epithelial cell lines should be subcultured before they reach confluence as they become too difficult to detach after confluence, e.g. Caco-2.

(2) *Medium exhaustion and pH.* Exhaustion of the medium usually indicates that the medium requires replacement (see Section 14.5.1), but if a fall in pH occurs so rapidly that the medium must be changed more frequently, subculture may be required. Usually, a drop in pH is accompanied by an increase in cell density, which is the prime indicator of the need to subculture.

(3) *Time since feeding or subculture.* Routine subculture is best performed according to a strict schedule, so that reproducible behavior is achieved and monitored. If cells are still not confluent by the appropriate time, increase the seeding density; if they reach confluence too soon, reduce the seeding density. Determination of the correct seeding density and subculture interval is best done by plotting a growth curve (see Section 19.3). Once this routine is established, recurrent growth should be consistent in duration and cell yield from a given seeding density. Deviations from this pattern signify a departure from normal conditions or indicate deterioration of the cells. Ideally, a cell concentration should be found that allows for the cells to be subcultured after seven days, with the medium being changed after four days.

(4) *Procedural requirements.* When cells are required for other procedures, they must be subcultured to increase the cell yield or to change the type of culture vessel or medium. Ideally, this procedure should be done at the regular subculture time, when it will be known that the culture is performing routinely, resulting in more predictable behavior. Demands for cells do not always fit the established routine for maintenance, and compromises must be made. Whatever the circumstances, cells should not be subcultured while still within the lag period, and they should be taken between the middle of the log phase and the time at which they will enter the plateau phase. If the procedure requires cells at plateau phase, they will need frequent feeding or continuous perfusion before harvesting.

14.6.2 Dissociation Agents

For adherent cultures, subculture usually involves removal of the medium followed by dissociation of the cell monolayer. Various procedures can be used for cell dissociation (see Table 14.4). Some loosely adherent cells may be subcultured by shaking the bottle, collecting the cells in the medium, and diluting as appropriate in fresh medium (e.g. HeLa-S3). Mechanical dissociation can be performed using a cell scraper or cell lifter (e.g. available from Corning, STEMCELL Technologies, or ThermoFisher Scientific). Cell scrapers are used in flasks and roller bottles, while cell lifters are often used with plates and cellular aggregates. However, most cell lines are subcultured using enzymatic dissociation. A similar approach is used to enzymatic tissue disaggregation (see Section 13.5), but with a lower enzymatic concentration and treatment time.

Trypsin is the most commonly used enzyme for cell dissociation and is usually combined with ethylene diamine tetra-acetic acid (EDTA). The attachment of cells to each other and to the culture substrate is mediated by cell surface glycoproteins and is often mediated by Ca^{2+} (see Section 2.2.2). Other proteins and proteoglycans (derived from the cells and from serum) become associated with the cell surface and the substrate and will facilitate cell adhesion. Subculture usually requires chelation of Ca^{2+} and degradation of the extracellular matrix (ECM) and, potentially, the extracellular domains of some cell adhesion molecules. Hence, a protease, such as trypsin, is often combined with EDTA to bind divalent cations.

Make sure that each batch of trypsin that you use has the same activity. The specific activity of trypsin may vary between suppliers or between batches, requiring a different concentration to achieve the same activity (see Section 13.5.1). The pH for optimal proteolytic activity of trypsin is between 7.8 and 8.5, which is too high for cultured cells. An alkaline pH should be selected, usually around pH 7.6, as a compromise between maximal activity and minimal cell damage. Using a prewash will remove residual medium

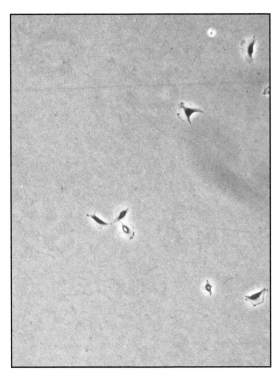

(a) Newly subcultured monolayer. NRK cells 24 h after subculture. Phase contrast; 10x objective.

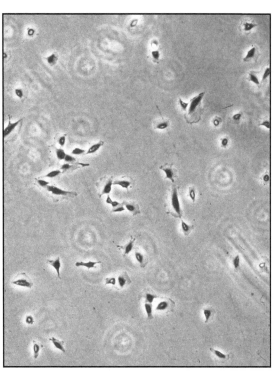

(b) Entering log phase. NRK cells 48 h after subculture. Phase contrast; 10x objective.

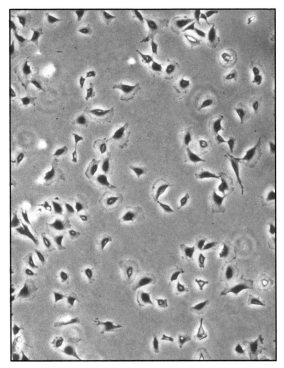

(c) Mid-log phase cells. NRK cells 3 d after subculture; ready for medium change. Phase contrast; 10x objective.

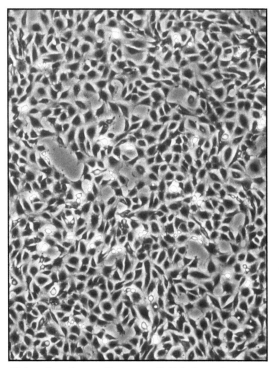

(d) Late log phase cells. NRK cells 7 d after subculture and ready for subculture again. Phase contrast; 10x objective. Source for all images: R. Ian Freshney.

Fig. 14.6. Phases of the growth cycle.

TABLE 14.4. Cell dissociation procedures.

Procedure or enzyme[a]	Pretreatment	Dissociation agent	Dissociation medium	Application
Shake-off	None	Gentle mechanical shaking, rocking, or vigorous pipetting	Culture medium	Mitotic or other loosely adherent cells
Scraping	None	Cell scraper (e.g. Falcon)	Culture medium	Cell lines for which proteases are to be avoided; rarely gives a single-cell suspension, can damage some cells
Trypsin	DPBS-A	Crude trypsin, 0.01–0.5%, usually 0.25%	DPBS-A, CMF	Most continuous cell lines, many early passage cell lines
Cold trypsin	DPBS-A, 4 °C	Crude trypsin, 0.25%, 4 °C	DPBS-A, CMF	Cell lines that are sensitive to trypsin damage
Trypsin + EDTA prewash	DPBS-A + 1 mM EDTA	Crude trypsin, 0.25%	DPBS-A	Cell lines that are strongly adherent
Trypsin + EDTA/EGTA	DPBS-A + 1 mM EDTA	Crude trypsin, 0.25%; EDTA, 1 mM	DPBS-A	Many epithelial cells, but some can be sensitive to EDTA; 1–5 mM EGTA can be used instead
Trypsin + collagenase	DPBS-A + 1 mM EDTA	Crude trypsin, 0.25%; crude collagenase, 200 U/ml	DPBS-A + 1 mM EDTA	Dense cultures and multilayers, particularly with fibroblasts
Purified trypsin	DPBS-A	10–25 μg/ml, 4 °C	DPBS-A	Serum-free subculture
Purified collagenase IV	None	50–200 U/ml	Culture medium	Serum-free subculture; PSC culture
Recombinant trypsin, e.g. TrypZean®	DPBS-A	As per instructions	As provided	Serum-free and routine subculture
TrypLE™ Express	DPBS-A	As per instructions	As provided	Serum-free and routine subculture
Liberase™	None	As per instructions	As provided	Serum-free and routine subculture
Accutase®	None	As per instructions	As provided	Serum-free subculture; PSC culture
Papain	None	Papain, 20 U/ml	EBSS +1 mM cysteine, 0.5 mM EDTA	Neural progenitors and oligodendrocytes (Young and Levison 1997); more commonly used for tissue disaggregation
Dispase	None	Dispase, 0.1–1.0 mg/ml	Culture medium	Epithelial cells (removes monolayer without dissociation); serum-free subculture; PSC culture
Pronase	None	Pronase, 0.1–1.0 mg/ml	Culture medium	Provision of good single-cell suspensions; may be harmful to some cells
EDTA (Versene)	DPBS-A + 0.5 mM EDTA	EDTA, 0.5 mM	DPBS-A	PSC culture

[a]Enzymes used for dissociation are available in varying degrees of purity. Crude preparations contain other proteases that may be helpful in dissociating some cells but may be toxic to other cells. Purer grades are often used at a lower concentration (micrograms per milliliter), as their specific activities (enzyme units per gram) are higher (see Section 13.5.1). Enzyme suppliers are listed in a previous table (see Table 13.1). CMF, Calcium- and magnesium-free saline; DPBS-A, Dulbecco's phosphate buffered saline without Ca^{2+} and Mg^{2+}.

from a culture and will also restore the pH to around pH 7.4. Subsequent addition of trypsin will make it sufficiently alkaline, but care should be taken with highly acidic cultures to ensure that the pH of the trypsin is alkaline.

Some cell monolayers cannot be dissociated in trypsin and require alternative proteases such as Pronase, Dispase, or collagenase (see Table 14.4). Of these proteases, Pronase is the most effective but can be harmful to some cells. Dispase and collagenase are generally less toxic than trypsin but may not give complete dissociation of epithelial cells. Purified or recombinant enzymes are increasingly used, particularly for serum-free culture (see Section 14.8; see also Table 13.1). Examples include Accutase (Innovative Cell Technologies), Liberase (Roche CustomBiotech), and TrypLE Express (ThermoFisher Scientific). These enzymes may be substituted for trypsin, but always assess their efficacy for your cell line or cell type. The severity of the treatment required will depend on the cell type, as does the sensitivity of the cells to proteolysis. A protocol should be selected with the least severity that is compatible with the generation of a cell suspension with high viability.

14.6.3 Standard Procedure for Subculture

Protocol P14.2 describes the subculture of an adherent monolayer using trypsin (see Figures 14.7, 14.8). The protocol can also be modified for use in training (see Supp. S30.11). As noted in this chapter and elsewhere in this book, it is important to handle each cell line separately and use separate bottles or aliquots of reagents. Otherwise there is a significant risk of contamination by microorganisms or by another cell line (see Sections 16.1.1, 17.2.3).

PROTOCOL P14.2. TRYPSINIZATION OF ADHERENT CELLS

Background

This protocol uses the NRK cell line as an example; one 75-cm^2 flask is passaged to give four 75-cm^2 flasks. Various procedures can be used for trypsinization. An EDTA prewash is used here to remove traces of medium, divalent cations, and serum. This procedure can be carried out without the prewash, or with 1 mM EDTA in the trypsin and a prewash of Dulbecco's phosphate buffered saline without Ca^{2+} and Mg^{2+} (DPBS-A). The choice of enzyme and method will depend on the cell line and cell type (see Table 14.4).

Outline

Remove the medium and rinse the monolayer. Expose the cells briefly to trypsin, then remove the trypsin. Incubate the cells, then disperse in medium, count, dilute, and reseed the subculture.

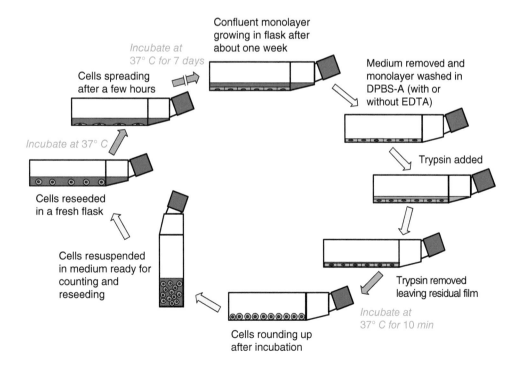

Fig. 14.7. Subculture of adherent monolayer. Stages in the subculture and growth of adherent cells, using trypsin for cell dissociation (see also Figures 14.6, 14.8).

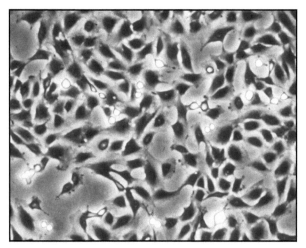

(a) NRK monolayer before trypsinization. Phase contrast; 20x objective.

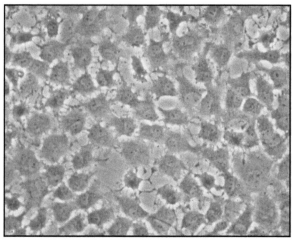

(b) NRK monolayer after DPBS-A/EDTA prewash. Phase contrast; 20x objective.

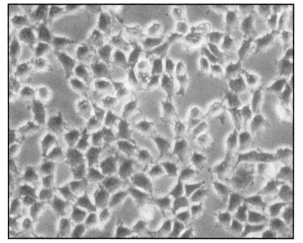

(c) NRK monolayer immediately after trypsin removal. Phase contrast; 20x objective.

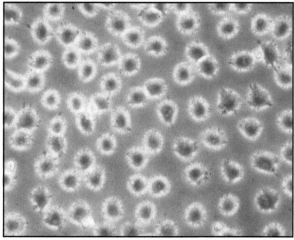

(d) NRK monolayer 1 min after trypsin removal. Phase contrast; 20x objective.

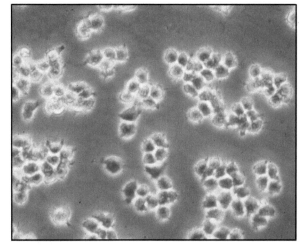

(e) NRK monolayer 5 min after trypsin removal. Phase contrast; 20x objective.

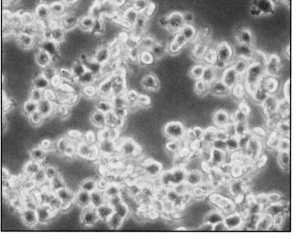

(f) Fully disaggregated monolayer. 10 min after removal of trypsin and ready for dispersing and counting. Phase contrast; 20x objective. Source for all images: R. Ian Freshney.

Fig. 14.8. Changes in morphology during trypsinization.

• *Safety Note. Pipettes can break, resulting in sharps injuries (see Section 6.2.1). Dispose of pipettes into a pipette cylinder or similar container. Dispose of glass Pasteur pipettes separately into a sharps container.*

Materials (sterile or aseptically prepared)

☐ NRK cells, 75-cm² flask at late log phase (just reaching confluence)
☐ Prewash: DPBS-A with 1 mM EDTA, 20 ml
☐ Trypsin: 0.25% in DPBS-A (see Table 14.4), 20 ml at 4 °C
☐ Complete medium: Eagle's Minimum Essential Medium (MEM) with Earle's salts and 10% fetal bovine serum (FBS), 200 ml
☐ Pipettes in an assortment of sizes, usually 1, 5, 10, and 25 ml (plugged to prevent aspiration into the pipette controller or bulb)
☐ Unplugged pipettes, e.g. Pasteur pipettes (if using a pump or vacuum line)
☐ Universal containers or centrifuge tubes (2)
☐ Culture flasks, 75 cm² (4)

Materials (non-sterile)

☐ BSC and associated consumables (see Protocol P12.1)
☐ Cell counting equipment (hemocytometer or automated counter; see Section 19.1)

Procedure

A. Before Subculture

1. Prewarm bottles of culture medium and DPBS-A, e.g. in a water bath.
2. Prepare a work space for aseptic technique using a BSC.
3. Bring growth medium and other reagents to the BSC. Swab each item with 70% alcohol and place in the BSC (see Protocol P12.1).
4. Take the culture out of the incubator and bring it to the microscope. Observe it carefully for signs of contamination or deterioration (see Section 14.4.1).
5. Check the previously described criteria (see Section 14.6.1) and, based on your knowledge of its behavior, decide whether to passage the culture.
6. If subculture is not required, return the flask to the incubator. If subculture is required, take the flask

to the BSC, swab it with 70% alcohol, and place it in the BSC.

B. Trypsinization

7. Take a sterile pipette and insert it into the pipette controller. If using a pump or vacuum line, select a Pasteur pipette instead and connect it to the system's tubing.
8. Uncap the flask and tilt it so that the medium moves to the bottom of the flask. Withdraw the medium and discard it into a waste beaker. If using a pump or vacuum line, use the Pasteur pipette to aspirate the medium instead. Recap the flask and discard the pipette.
9. Pipette 15 ml prewash (DPBS-A with 1 mM EDTA, 0.2 ml/cm²) onto the side of the flask opposite the cells (avoid pipetting over the monolayer). Rinse the prewash over the cells and remove the prewash by pipette from the side opposite the cells. This step is designed to remove traces of serum that would inhibit the action of the trypsin and to deplete the divalent cations, which contribute to cell adhesion (see Section 14.6.2).
10. Pipette 7.5 ml cold trypsin (0.25%, 0.1 ml/cm²) onto the side of the flask opposite the cells. Turn the flask over and lay it down. Ensure that the monolayer is completely covered by gently rocking the flask. Leave the flask stationary for 15–30 seconds. (A shorter exposure may be necessary if the trypsin is at room temperature.)
11. Tilt the flask gently to remove the trypsin from the monolayer and quickly check that the monolayer is not detaching. (Using trypsin at 4 °C helps to prevent premature detachment if this turns out to be a problem.)
12. Withdraw all but a few drops of the trypsin and incubate the flask lying flat at 37 °C. The flask should be incubated until the cells round up. When the bottle is tilted, the monolayer should slide down the surface. (In this example, cells detach after 10 minutes, but the process can take 5–15 minutes at 37 °C, depending on the cell type and growth phase.)
13. Pipette 10 ml medium (0.1–0.2 ml/cm²) onto the monolayer and disperse the cells by repeated pipetting over the surface.
14. Pipette the cell suspension up and down a few times, with the tip of the pipette resting on the bottom corner of the flask. Avoid introducing air bubbles or foaming.

15. Count the cells with a hemocytometer or an automated cell counter.

C. Seeding New Flasks

16. Calculate the volume of cell suspension (v) for the required cell concentration and medium volume using the formula:

$$\frac{Required\ concentration}{Starting\ concentration} \times Required\ volume$$

17. Dilute the cells to the required medium volume using one of two approaches:
    (a) Add the appropriate volume of cell suspension to a premeasured volume of medium. For example, the volume of cell suspension required to dilute 2.36×10^6 cells/ml to 2×10^4 cells/ml in 20 ml medium would be:

    $$\frac{2 \times 10^4}{2.36 \times 10^6} \times 20 = 0.17\ ml$$

    i.e. dilute 0.17 ml up to 20 ml in each flask.
    (b) Add the appropriate volume of cell suspension to the total volume required and distribute that volume among several flasks. For example, the volume of cell suspension required to dilute 2.36×10^6 cells/ml to 2×10^4 cells/ml in 100 ml would be:

    $$\frac{2 \times 10^4}{2.36 \times 10^6} \times 100 = 0.85\ ml$$

    i.e. dilute 0.85 ml up to 100 ml total and aliquot 20 ml into each flask.
18. Label four 75-cm² flasks with the cell line designation, the date, and your initials. If following step 17a above, seed the new flasks by adding 19.15 ml medium to each flask followed by 0.17 ml of the trypsinized cell suspension. If following step 17b above, dilute 0.85 ml cell suspension into 100 ml medium, mix, and dispense 20 ml into each 75 cm² flask.
19. Lay the culture flat so that the fresh medium covers the adherent monolayer and return it to the incubator. If you do not have access to an incubator with the correct gas mixture, gas the flask before you return it to the incubator (see Section 12.5.3).

D. After Subculture

20. Clear away all pipettes, glassware, and other movable items. Return unused media and reagents to cold storage.
21. Swab down all surfaces with 70% alcohol and allow the BSC to purge for 4 minutes.
22. Record observations and actions using a record sheet, lab book, or electronic database.

Notes

Steps 1–6: The order and timing of these steps will vary from day to day. Many tissue culture personnel prefer to observe all cultures at the beginning of the working day and decide on any required actions at that point, before warming medium and reagents. On other days, it will be clear what actions are required based on the previous day's observations and actions.

Step 12: In each case, the main dissociating agent, be it trypsin or EDTA, is present only briefly, and the incubation is performed in the residue after most of the dissociating agent has been removed. Do not leave the flasks longer than necessary, but, on the other hand, do not force the cells to detach before they are ready to do so, or else damage and clumping may result.

Step 14: The outcome should be a single cell suspension, resulting in an accurate cell count and uniform growth on reseeding, without mechanical damage. Some cell lines disperse easily, whereas others require more vigorous pipetting. Primary cell suspensions and early passage cultures are particularly prone to damage because of their greater fragility and larger size. Continuous cell lines are usually more resilient and require more vigorous pipetting. It may be helpful to centrifuge the cells (e.g. for 5 minutes at 250 *g*), remove the supernatant, and resuspend the pellet in a small volume of medium. This approach can give a more uniform cell suspension.

Step 17: Procedure (a) is useful for routine subculture when only a few flasks are used and precise cell counts and reproducibility are not critical. Procedure (b) is preferable when setting up several replicates, because the total number of manipulations is reduced and the concentrations of cells in each flask will be identical.

Step 18: Always label flasks and other culture vessels before cells are added to help prevent misidentification. Labeling should be on the side of a flask so that viewing on the microscope is not impaired, and on the edge of the base of a Petri dish or multiwell plate (in case the lid gets separated from the base).

Step 19: As the expansion of air inside plastic flasks causes larger flasks to swell, the pressure should be released by briefly slackening the cap 30 minutes after placing the flask in the incubator. Incubation restores the correct shape as the gas phase expands. If flasks are cultured in a humidified CO_2 incubator, using permeable caps will avoid this problem.

14.6.4 Growth Cycle and Split Ratios

It is essential to establish a growth curve for each cell line that is handled in the laboratory (see Figure 14.1). Routine subculture leads to the repetition of a standard growth cycle (see Figure 14.9a). If you understand the growth curve, you can determine the parameters that will give the most consistent growth cycle, including the seeding concentration, the subculture interval (the duration of growth before subculture), and the cell yield at the next subculture. These parameters will also determine the duration of experiments and the appropriate times for sampling to give experimental consistency. Cells at different phases of the growth cycle behave differently with respect to cell proliferation, enzyme activity, glycolysis and respiration, synthesis of specialized products, and many other properties. More information on growth curves and their analysis are provided in a later chapter (see Section 19.3).

Once standardized parameters have been determined, each subculture should yield approximately the same cell count from a given seeding concentration. At this point a split ratio can be selected for routine subculture (see Sections 14.1, 14.2.2). Split ratios are best chosen in multiples of two (1:2, 1:4, 1:8, etc.). This makes the calculation of PDL easier; for example, a culture divided eightfold requires three doublings to return to the same cell density (see Figure 14.9b). A fragile or slowly growing line should be split 1:2, whereas a faster growing cell line can be split 1:8. Rapidly growing continuous cell lines are split at higher split ratios ($\geq 1:16$). For such cell lines, it is usually easier to seed a specific cell count. As a general rule, most continuous cell lines subculture satisfactorily at a seeding concentration of between 1×10^4 and 5×10^4 cells/ml, finite fibroblast cell lines subculture at about the same concentration, and more fragile cultures (such as endothelium and some early passage epithelia) subculture at around 1×10^5 cells/ml.

How do you select a suitable split ratio? When handling a cell line for the first time, or when using an early passage culture with which you have little experience, it is good practice to subculture the cell line with a split ratio of 1:2 or 1:4. It may be helpful to seed two or three flasks at different split ratios and use the best split ratio at the next subculture. As

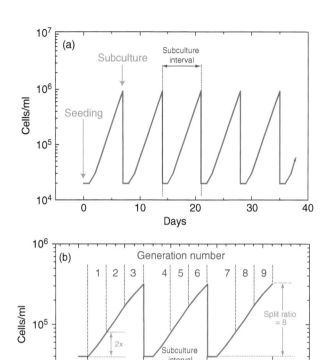

Fig. 14.9. Serial subculture. (a) Repetition of the standard growth cycle during propagation of a cell line. If cells are growing correctly, they should reach the same concentration (peaks) after the same time in each cycle, given that the seeding concentration (troughs) and subculture interval remain constant. (b) Generation number and passage. Each subculture represents one passage, but the generation number depends on the split ratio (here a split ratio of 1:8 is used, resulting in three generations per passage). Generation numbers are equivalent to population doublings and are used to estimate culture age (see Sections 14.1, 14.2.2).

you gain experience with that cell line, it may be possible to increase the split ratio (i.e. to reduce the cell concentration after subculture), but always keep one flask at a low split ratio when attempting to increase the split ratio of the rest. A growth curve should be plotted as soon as sufficient cells have been accumulated and stored.

Always record the cell count at each subculture. The cell count results in a better estimation of the cell yield and allows you to monitor ongoing growth, which is an important part of the provenance of the cell line (see Section 7.3.1). Otherwise, minor alterations will not be detected for several passages, e.g. if the cells are diluted too much or not diluted enough (see Figure 14.10). Split errors will affect the behavior of the culture. For example, if cells are diluted too much, the lag period may increase beyond a reasonable period (> 24 hours) or faster growing subpopulations may be selected and dominate the culture in later passages.

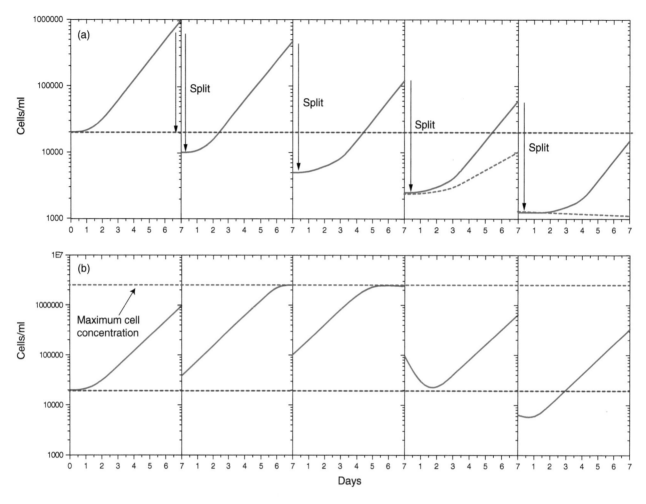

Fig. 14.10. Split errors. Cells must be seeded at the correct concentration at subculture. (a) Cells that are diluted too much, repeatedly, will show a poorer yield and may eventually die out; (b) cells that become too dense before subculture will reseed at too high a concentration, reach plateau earlier, and may deteriorate. Solid graph lines indicate the outcome without cell loss, while dotted lines indicate where the incorrect concentration is likely to result in cell loss.

14.7 MAINTAINING SUSPENSION CULTURES

Most primary cultures and cell lines grow as adherent monolayers. However, some cell types prefer to grow in suspension, either as single cells or as aggregates (referred to here as spheroids; see Section 27.4.1). Cells that grow in suspension include murine ascites tumors, many leukemia and lymphoma cell lines, and some colorectal and breast tumors (see Figure 14.11). Other cells can be kept in suspension mechanically by shaking the culture (see Figure 8.11) or using a magnetic stirrer or paddle to agitate the culture (spinner culture; see Section 28.2.1). Suspension culture has numerous advantages (see Table 14.5). Dissociation agents such as trypsin are not required, so subculture is quicker and less traumatic for the cells. Large quantities of cells may be produced and harvested without increasing the surface area of the substrate. If dilution of the culture is continuous and the cell concentration is kept constant, a steady state can be reached that is not readily achievable in monolayer culture (see Section 28.2.3).

14.7.1 Standard Procedure for Suspension Culture

Cells that grow spontaneously in suspension can be maintained in regular culture flasks or dishes, which need not be tissue culture treated (although they must be sterile, of course). The rules regarding the depth of medium in static cultures are as for monolayers, i.e. 2–5 mm to allow for gas exchange. Replacement of the medium (feeding) is not usually carried out with suspension cultures. Instead, the culture is (i) expanded by dilution into fresh culture vessels; (ii) diluted into one fresh culture vessel and the surplus discarded; or (iii) the bulk of the cell suspension is withdrawn and the residue is diluted back to an appropriate seeding

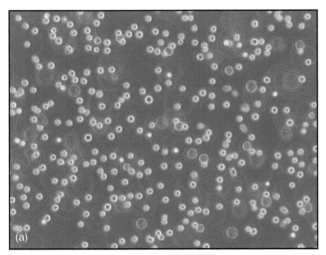

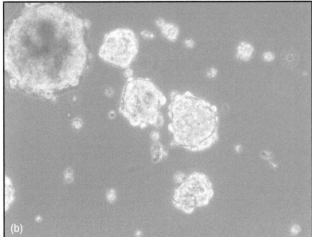

Fig. 14.11. Cell lines in suspension. (a) HL-60 leukemia cells, growing as a single cell suspension; (b) LIM1863 colorectal carcinoma cells, growing as spheroids. Phase contrast; 10x objective. *Source:* Courtesy of Elsa Moy, CellBank Australia.

TABLE 14.5. Monolayer versus suspension culture.

Monolayer	Suspension
Culture requirements	
Cyclic maintenance	Steady state
Trypsin passage	Dilution
Limited by surface area	Limited by volume (gas exchange)
Growth properties	
Contact inhibition	Homogeneous suspension
Cell interaction in 2D	Cell interaction in 3D
Diffusion boundary	
Useful for	
Broad research use	Biological production and harvesting
Cell staining	Culture scale-up
In situ treatments and extractions	3D applications, e.g. spheroids
Applicable to	
Most cell types, including primary cultures	Transformed cells, hematopoietic cells, some solid tumors

Protocol P14.3 describes subculture of a cell line in suspension. The procedure can be adapted for use in training (see Supp. S30.12).

concentration in the same culture vessel. The third option is usually best avoided; the probability of contamination gradually increases with any buildup of minor spillage on the neck of the flask during dilution.

The criteria for subculture are similar to those for adherent cells (see Section 14.6.1), including:

(1) *Cell concentration.* This should not exceed 1×10^6 cells/ml for most suspension-growing cells under standard conditions.

(2) *Medium exhaustion and pH.* The latter is linked to cell concentration and declines as the cell concentration rises.

(3) *Time since subculture.* A growth curve should be plotted and is usually similar to that for monolayer cells (see Figure 14.1), with a shorter lag period. Subculture should follow a regular schedule as determined by the growth curve.

PROTOCOL P14.3. SUBCULTURE OF SUSPENSION CELLS

Background

This protocol is based on use of a stirrer flask with a magnetic stirrer platform, but a shaker flask or spinner flask could be used instead (see Figures 8.11, 28.1).

Outline

Withdraw a sample of the cell suspension, count the cells, and seed an appropriate volume of the cell suspension into fresh medium in a new flask, restoring the cell concentration to the initial seeding level.

Materials (sterile or aseptically prepared)

☐ Culture in tissue culture flask, shaker flask, or stirrer flask (1)
☐ Growth medium, 500 ml

□ Pipettes in an assortment of sizes, usually 1, 5, 10, and 25 ml (plugged to prevent aspiration into the pipette controller or bulb)

□ Unplugged pipettes, e.g. Pasteur pipettes (if using a pump or vacuum line)

□ Universal containers or centrifuge tubes (2)

□ Stirrer flask (1), e.g. Techne 1-l flask (Cole-Parmer)

□ Small tissue culture flask or stirrer flask (1)

Materials (non-sterile)

□ BSC and associated consumables (see Protocol P12.1)

□ Cell counting equipment (hemocytometer or automated counter; see Section 19.1)

□ Magnetic stirrer platform

Procedure

1. Prepare the BSC, perform routine observation, and decide whether subculture is required as previously described for adherent cells (see steps 1–7, Protocol P14.2).

2. Open the flask in the BSC. If necessary, gently mix the cell suspension and disperse any clumps by pipetting the suspension up and down. Remove a sample to count the cells.

3. Add medium to a new stirrer flask to a maximum depth of 5 cm. If a greater volume of medium is required, CO_2 must be supplied via a sparging tube (see Figure 28.1a).

4. Add a sufficient number of cells to give a final concentration of 1×10^5 cells/ml for slow-growing cells (36–48 hours PDT) or 2×10^4 cells/ml for rapidly growing cells (12–24 hours PDT).

5. Cap the stirrer flask and place on magnetic stirrers set at 60–100 rpm, in an incubator or warm room at 37 °C. If required, gas the air space of the stirrer flask with 5% CO_2.

6. If the cells grow spontaneously in suspension, you may wish to add medium to another small flask and seed cells at the same concentration. This can be used as a backup culture or to seed the next stirrer culture.

7. Tidy and swab down the sterile work area and record your actions in a log book or database.

Notes

Step 1: Routine observation is more difficult with suspension cultures than with adherent cells, as the medium may appear cloudy due to the culture itself. Look for the presence of debris and for cells that are in poor condition as indicated by an irregular outline, shrinkage, and/or granularity. Healthy cells should appear clear, with a visible nucleus, and are often found in small aggregates in static culture.

Step 2: Cell aggregates require gentle dissociation if they are to be counted as a single cell suspension. It may be helpful to centrifuge the cells and resuspend the pellet or use a gentle form of enzymatic dissociation. However, single cell suspensions will result in damage or differentiation with some cell types (e.g. hPSCs) and should be avoided where this is the case.

Step 4: PDT is calculated from the growth curve (see Section 19.3.3).

Step 5: Take care that the stirrer motor does not overheat the culture. Insert a mat under the bottle if necessary. Induction-driven stirrers generate less heat and have no moving parts.

14.8 SERUM-FREE SUBCULTURE

Use of serum-free media (see Section 10.6) results in some challenges to overcome during subculture. Serum acts to inhibit residual proteolytic activity after trypsinization and has a protective, detoxifying effect (see Section 9.5). Serum-free subculture is more likely to produce cellular damage, due to the agent that is used for dissociation or to stress from suboptimal culture conditions (Ham and McKeehan 1978). Stem cells are particularly vulnerable to damage during subculture. For example, subculture of human embryonic stem cells (hESCs) can lead to loss of viability and chromosomal rearrangements, ranging from small subtelomeric deletions to aneuploidy (Thomson et al. 2008). Generation of a single cell suspension results in hESCs becoming more sensitive to treatments and more prone to cell death (Beers et al. 2012).

Factors to consider when choosing a serum-free subculture protocol include:

(1) *Cell type and subculture timing.* Some cell types that are grown in serum-free media may not attach if subcultured too early or may differentiate if allowed to overgrow. For example, hESCs that are grown in TeSR1 medium should be passaged within a 24-hour window for optimal results (Ludwig et al. 2006).

(2) *Method of dissociation.* hESC cultures were originally passaged using mechanical methods (Ohnuma et al. 2014). Enzymatic dissociation can be used, but methods that produce cell aggregates tend to increase survival

rates (e.g. Dispase; see Table 14.4). Gentle enzyme-free methods can be used for hPSCs that are grown in serum- and feeder-free conditions. These methods are usually based on manipulation of divalent cations, e.g. using EDTA (Beers et al. 2012).

(3) *Trypsin.* Special care is needed when performing trypsinization in serum-free culture. Trypsin is damaging and tends to produce a single cell suspension, which may affect viability. Cellular damage can be reduced by diluting trypsin, using the smallest volume possible, and performing trypsinization at 4 °C (Ham and McKeehan 1978). Because crude trypsin is a complex mixture of proteases, some of which may require different inhibitors, it is preferable to use purified or recombinant trypsin (e.g. TrypZean, Sigma-Aldrich).

(4) *Alternative proteases.* Other proteases may be used for tissue disaggregation and cell dissociation (see Tables 13.1, 14.4). Proteases that can be used in serum-free culture include Pronase, Dispase, Accutase (Innovative Cell Technologies), Liberase (Roche CustomBiotech), and TrypLE Express (ThermoFisher Scientific). Pronase is very effective but can be toxic to some cells. Pronase and Dispase are less effective for epithelial cells compared to fibroblasts and may not give a single cell suspension if there are epithelial cells present.

(5) *Protease inhibitors.* In the absence of serum, other inhibitors must be used to inhibit residual trypsin

activity, e.g. soybean trypsin inhibitor. Cells may also be washed by centrifugation, although it may still be advisable to include a trypsin inhibitor in the wash after trypsinization. Bacterial proteases (e.g. Dispase and Pronase) are typically not neutralized by trypsin inhibitors and should be removed by centrifugation. It is possible that Pronase can be inactivated by dilution without subsequent neutralization in serum-free conditions (personal communication, W. McKeehan).

14.9 RECORD KEEPING FOR CELL LINES

It is important to keep records of routine maintenance, including feeding and subculture (see Tables 14.6, 14.7). A written notebook, electronic spreadsheet, or electronic database can all be used; the most important requirement is that records are regularly updated. Such records are required for GCCP as part of the provenance of the cell line, and may be called on during publication, e.g. if reviewers ask at which passage authentication testing was performed (see Sections 7.3.2, 7.4.2). If Standard Operating Procedures (SOPs) are used (see Section 7.5.1), records need say only "Fed" or "Subcultured," with the date and a note of the cell numbers and reference to the SOP. Comments on visual assessment, changes to reagents, deviations from the SOP, or additional actions should also be recorded (e.g. changes in medium supplier or cryopreservation of token vials).

TABLE 14.6. Data record, feeding.

Cell line . . . . Source . . . . Operator . . . .

Parameter	Date:
Culture	Identifier, e.g. cell line name, barcode on flask
	Passage number
	Population doubling level (PDL)
Observations	Cell appearance
	Cell density (confluence)
	Medium pH (approx.)
	Medium clarity / turbidity
Reagents/conditions	Medium type
	Batch no.
	Serum type and concentration
	Batch no.
	Other reagents
	CO_2 concentration
Other actions	E.g. aliquot of medium stored for testing
Comments	

TABLE 14.7. Data record, subculture.

Cell line .... Source .... Operator ....

Parameter	Date:
Culture	Identifier, e.g. cell line name, barcode on flask
	Passage number
	Population doubling level (PDL)
Observations	Cell appearance
	Cell density (confluence)
	Medium pH (approx.)
	Medium clarity/turbidity
Dissociation	Prewash
	Agent, e.g. trypsin/EDTA
	Duration
	Temperature
	Other method, e.g. scraping
Cell count (see Section 19.1)	Concentration after resuspension (C_I)
	Volume after resuspension (V_I)
	Yield ($Y = C_I \times V_I$)
	Yield per flask/dish, etc.
Seeding (see Protocol P14.2)	Number (N) and type of vessel (flask, dish, microwell plate)
	Final concentration (C_F)
	Volume per flask, dish, or well (V_F)
	Split ratio ($Y \div C_F \times V_F \times N$), or number of flasks seeded ÷ number of flasks trypsinized, where the flasks are of same size
Reagents	Medium type
	Batch no.
	Serum type and concentration
	Batch no.
	Other reagents
	CO_2 concentration
Substrate	E.g. Matrigel, collagen
Other actions	E.g. pellet frozen for DNA/RNA
Comments	

Suppliers

Supplier	URL
Cole-Parmer (includes Argos Technologies, Techne)	www.coleparmer.com
Corning	http://www.corning.com/worldwide/en/products/life-sciences/products/cell-culture.html
Innovative Cell Technologies, Inc.	www.accutase.com
Roche CustomBiotech	custombiotech.roche.com/home.html
Sigma-Aldrich (Merck)	http://www.sigmaaldrich.com/life-science/cell-culture.html
STEMCELL Technologies	www.stemcell.com
ThermoFisher Scientific	http://www.thermofisher.com/us/en/home/life-science/cell-culture.html

REFERENCES

Bairoch, A. (2018). The Cellosaurus, a cell-line knowledge resource. *J. Biomol. Tech.* 29 (2): 25–38. https://doi.org/10.7171/jbt.18-2902-002.

Bandrowski, A., Brush, M., Grethe, J.S. et al. (2016). The resource identification initiative: a cultural shift in publishing. *Brain Behav.* 6 (1): e00417. https://doi.org/10.1002/brb3.417.

Barrett, T., Clark, K., Gevorgyan, R. et al. (2012). BioProject and BioSample databases at NCBI: facilitating capture and organization of metadata. *Nucleic Acids Res.* 40 (Database issue): D57–D63. https://doi.org/10.1093/nar/gkr1163.

Beers, J., Gulbranson, D.R., George, N. et al. (2012). Passaging and colony expansion of human pluripotent stem cells by enzyme-free dissociation in chemically defined culture conditions. *Nat. Protoc.* 7 (11): 2029–2040. https://doi.org/10.1038/nprot.2012.130.

Coriell, L.L. (1962). Detection and elimination of contaminating organisms. *Natl Cancer Inst. Monogr.* 7: 33–53.

Drexler, H.G. and Matsuo, Y. (1999). Guidelines for the characterization and publication of human malignant hematopoietic cell lines. *Leukemia* 13 (6): 835–842.

Drexler, H.G., Dirks, W.G., MacLeod, R.A. et al. (2017). False and mycoplasma-contaminated leukemia-lymphoma cell lines: time for a reappraisal. *Int. J. Cancer* 140 (5): 1209–1214. https://doi.org/10.1002/ijc.30530.

Earle, W.R., Schilling, E.L., Stark, T.H. et al. (1943). Production of malignancy in vitro. IV. The mouse fibroblast cultures and changes seen in the living cells. *J. Natl Cancer Inst.* 4 (2): 165–212.

Gartler, S.M. (1968). Apparent HeLa cell contamination of human heteroploid cell lines. *Nature* 217 (5130): 750–751.

Gey, G.O., Coffman, W.D., and Kubicek, M.T. (1952). Tissue culture studies of the proliferative capacity of cervical carcinoma and normal epithelium. *Cancer Res.* 12: 264–265.

Gillet, J.P., Varma, S., and Gottesman, M.M. (2013). The clinical relevance of cancer cell lines. *J. Natl Cancer Inst.* 105 (7): 452–458. https://doi.org/10.1093/jnci/djt007.

Goh, S., Ong, P.F., Song, K.P. et al. (2007). The complete genome sequence of *Clostridium difficile* phage phiC2 and comparisons to phiCD119 and inducible prophages of CD630. *Microbiology* 153 (Pt 3): 676–685. https://doi.org/10.1099/mic.0.2006/002436-0.

Goy, A., Ramdas, L., Remache, Y.K. et al. (2003). Establishment and characterization by gene expression profiling of a new diffuse large B-cell lymphoma cell line, EJ-1, carrying t(14;18) and t(8;14) translocations. *Lab. Invest.* 83 (6): 913–916.

Ham, R.G. and McKeehan, W.L. (1978). Development of improved media and culture conditions for clonal growth of normal diploid cells. *In Vitro* 14 (1): 11–22.

Hayflick, L. (1973). Subculturing human diploid fibroblast cultures. In: *Tissue Culture: Methods and Applications* (eds. P.F. Kruse Jr. and M.K. Patterson), 220–223. New York: Academic Press.

Hayflick, L. (1990). In the interest of clearer communication. *In Vitro Cell Dev. Biol.* 26 (1): 1–6.

Hayflick, L. and Moorhead, P.S. (1961). The serial cultivation of human diploid cell strains. *Exp. Cell. Res.* 25: 585–621.

Holliday, D.L. and Speirs, V. (2011). Choosing the right cell line for breast cancer research. *Breast Cancer Res.* 13 (4): 215. https://doi.org/10.1186/bcr2889.

ICLAC (2015). Naming a cell line. http://iclac.org/resources/cell-line-names (accessed 9 December 2018).

ICLAC (2019). Register of misidentified cell lines. http://iclac.org/databases/cross-contaminations (accessed 11 November 2019).

Isaka, K., Nishi, H., Nakada, T. et al. (2002). Establishment and characterization of a new human cell line (EJ) derived from endometrial carcinoma. *Hum. Cell* 15 (4): 200–206.

Kurtz, A., Seltmann, S., Bairoch, A. et al. (2018). A standard nomenclature for referencing and authentication of pluripotent stem cells. *Stem Cell Rep.* 10 (1): 1–6. https://doi.org/10.1016/j.stemcr.2017.12.002.

Lin, C.W., Lin, J.C., and Prout, G.R. Jr. (1985). Establishment and characterization of four human bladder tumor cell lines and sublines with different degrees of malignancy. *Cancer Res.* 45 (10): 5070–5079.

Ludwig, T.E., Bergendahl, V., Levenstein, M.E. et al. (2006). Feeder-independent culture of human embryonic stem cells. *Nat. Methods* 3 (8): 637–646. https://doi.org/10.1038/nmeth902.

Maltese, W.A. and Volpe, J.J. (1979). Induction of an oligodendroglial enzyme in C-6 glioma cells maintained at high density or in serum-free medium. *J. Cell. Physiol.* 101 (3): 459–469. https://doi.org/10.1002/jcp.1041010312.

Masters, J.R. (2000). Human cancer cell lines: fact and fantasy. *Nat. Rev. Mol. Cell Biol.* 1 (3): 233–236. https://doi.org/10.1038/35043102.

Masters, J.R., Thomson, J.A., Daly-Burns, B. et al. (2001). Short tandem repeat profiling provides an international reference standard for human cell lines. *Proc. Natl Acad. Sci. U.S.A.* 98 (14): 8012–8017. https://doi.org/10.1073/pnas.121616198.

McGarrity, G.J. (1976). Spread and control of mycoplasmal infection of cell cultures. *In Vitro* 12 (9): 643–648.

Noah, T.L., Yankaskas, J.R., Carson, J.L. et al. (1995). Tight junctions and mucin mRNA in BEAS-2B cells. *In Vitro Cell Dev. Biol. Anim.* 31 (10): 738–740. https://doi.org/10.1007/BF02634112.

Ohnuma, K., Fujiki, A., Yanagihara, K. et al. (2014). Enzyme-free passage of human pluripotent stem cells by controlling divalent cations. *Sci. Rep.* 4: 4646. https://doi.org/10.1038/srep04646.

Reddel, R.R. (2000). The role of senescence and immortalization in carcinogenesis. *Carcinogenesis* 21 (3): 477–484.

Sanford, K.K., Earle, W.R., and Likely, G.D. (1948). The growth in vitro of single isolated tissue cells. *J. Natl Cancer Inst.* 9 (3): 229–246.

Schaeffer, W.I. (1990). Terminology associated with cell, tissue, and organ culture, molecular biology, and molecular genetics. Tissue Culture Association Terminology Committee. *In Vitro Cell Dev. Biol.* 26 (1): 97–101.

Schousboe, A., Thorbek, P., Hertz, L. et al. (1979). Effects of GABA analogues of restricted conformation on GABA transport in astrocytes and brain cortex slices and on GABA receptor binding. *J. Neurochem.* 33 (1): 181–189. https://doi.org/10.1111/j.1471-4159.1979.tb11720.x.

Shay, J.W. and Wright, W.E. (2000). Hayflick, his limit, and cellular ageing. *Nat. Rev. Mol. Cell Biol.* 1 (1): 72–76. https://doi.org/10.1038/35036093.

Thomson, A., Wojtacha, D., Hewitt, Z. et al. (2008). Human embryonic stem cells passaged using enzymatic methods retain a normal karyotype and express CD30. *Cloning Stem Cells* 10 (1): 89–106. https://doi.org/10.1089/clo.2007.0072.

Vaughan, L., Glanzel, W., Korch, C. et al. (2017). Widespread use of misidentified cell line KB (HeLa): incorrect attribution and its impact revealed through mining the scientific literature. *Cancer Res.* 77 (11): 2784–2788. https://doi.org/10.1158/0008-5472 .CAN-16-2258.

Vincent, K.M., Findlay, S.D., and Postovit, L.M. (2015). Assessing breast cancer cell lines as tumour models by comparison of mRNA expression profiles. *Breast Cancer Res.* 17: 114. https://doi.org/10.1186/s13058-015-0613-0.

Young, G.M. and Levison, S.W. (1997). An improved method for propagating oligodendrocyte progenitors in vitro. *J. Neurosci. Methods* 77 (2): 163–168.

CHAPTER 15

Cryopreservation and Banking

Cell lines have been used by research laboratories since the 1940s (see Table 1.1). Efforts were made to collect cell lines at early passage and freeze stocks for later use. The early cell lines were frozen in glycerol, since it was shown as early as 1949 that glycerol prevents cells from dying while being frozen (Polge et al. 1949). Cells frozen in glycerol were later found to retain viability after prolonged storage in liquid nitrogen (Coriell et al. 1964; Greene et al. 1967). Using this approach, the Cell Culture Collection Committee collected an initial panel of 23 "certified cell lines" for deposit at the American Type Culture Collection (ATCC) under accession numbers CCL-1 to -25 (Coriell et al. 1964; Ledley 1964). Participating laboratories tested all cell lines for microbial contamination and species of origin. Unfortunately, methods to detect intraspecies cross-contamination were limited at that time; 14 of these cell lines are now known to be misidentified (see Chapter 17). However, the overall approach was successful in preserving early passage cultures and managing the risk of contamination through specific testing. This chapter focuses on how to perform cryopreservation, which is defined as the application of low temperatures to preserve the structural and functional integrity of cells and tissues (Pegg 2007; Hunt 2019), and how to prepare seed stocks or "cell banks" for ongoing use (Hay 1988). Although cell repositories specialize in these procedures, the general principles can be applied in any laboratory to ensure that unique and irreplaceable cell lines are preserved at low passage for future use.

15.1 PRINCIPLES OF CRYOPRESERVATION

Unprotected freezing is normally lethal to cells. Freezing of water and the formation of crystalline ice results in cellular damage, primarily due to the mechanical action of the ice crystals and secondary changes in the composition of the liquid phase (Pegg 2007; Hunt 2019). To preserve cell viability, conditions and procedures must minimize intracellular ice crystal formation and reduce cryogenic damage from foci of high-concentration solutes that form when intracellular water freezes. These aims can be achieved by (i) using a hydrophilic cryoprotectant to sequester water; (ii) freezing slowly at a controlled rate to allow water to leave the cell, but not so slowly that ice crystal growth is encouraged; (iii) storing the cells at the lowest possible temperature to minimize the effects of high salt concentrations; and (iv) thawing rapidly to minimize ice crystal growth and generation of solute gradients as the residual intracellular ice melts. Alternatively, cell viability can be preserved by inducing a vitreous or glassy state which prevents ice crystal formation (vitrification; see Section 15.1.4).

15.1.1 Cryoprotectants

Cryoprotective agents (referred to throughout this book as cryoprotectants) act to reduce the amount of ice formed at any given temperature and also protect against other adverse chemical and physical phenomena during the

Freshney's Culture of Animal Cells: A Manual of Basic Technique and Specialized Applications, Eighth Edition. Amanda Capes-Davis and R. Ian Freshney.
© 2021 John Wiley & Sons Ltd. Published 2021 by John Wiley & Sons Ltd.
Companion website: www.wiley.com/go/freshney/cellculture8

freezing process. For living cells there are two broad groups of cryoprotectants: penetrating and non-penetrating. Penetrating cryoprotectants work by lowering the freezing point of the cells, protecting against solute toxicity, stabilizing the cell membrane, and protecting the cytoskeleton. They must be soluble in water and able to penetrate cells with low toxicity (Pegg 2007). Non-penetrating cryoprotectants work by enhancing dehydration, increasing the glass-forming potential, and protecting the cell membranes. Examples of penetrating cryoprotectants include dimethyl sulfoxide (DMSO), glycerol, propanediol, ethanediol, and methanol. Examples of non-penetrating cryoprotectants include polyethylene glycol (PEG), sucrose, trehalose, Ficoll, serum, and albumin.

Glycerol and DMSO are the most widely used cryoprotectants and are typically added to the medium in which cells are suspended before freezing (see Section 15.3.2) (Polge et al. 1949; Lovelock and Bishop 1959). Concentrations of 5–10% are common, although a wide range of concentrations (2–20%) can be used. DMSO should be colorless; it must be stored in glass or polypropylene, as it is a powerful solvent and will leach impurities out of rubber and some plastics. DMSO is assumed to be self-sterilizing but can be filter sterilized using a nylon filter, preferably after dilution in medium. Glycerol should be not more than one year old; it may become toxic after prolonged storage, due to light-induced conversion to acrolein. It can be sterilized by autoclaving.

DMSO is typically more effective than glycerol, possibly because it diffuses more rapidly through cell membranes (Coriell et al. 1964). However, its use can cause problems for some cultures. DMSO is known to induce differentiation in hematopoietic cell lines (Collins et al. 1979; Young et al. 2004; Jiang et al. 2006). It may also be toxic at the concentrations used in freezing media (Matsumura et al. 2010). Toxicity is likely to be a cell-type-specific event, perhaps relating to intracellular penetration. Some laboratories find that DMSO is less toxic if freezing medium is used at 4 °C; once added, all subsequent steps should be performed at the same temperature until freezing commences (see Protocol P15.1) (Wewetzer and Dilmaghani 2001). There are additional concerns about its use for biological products with potential therapeutic use *in vivo*. DMSO has well-recognized side effects when administered to patients, many of which are associated with induction of histamine release (Morris et al. 2014). Laboratory personnel should also be careful when handling DMSO, which can penetrate many synthetic and natural membranes (see Section 6.2.2). Consequently, DMSO may carry hazardous substances into the circulation through skin and even through gloves.

- *Safety Note. Always wear suitable personal protective equipment (PPE) when handling DMSO and check chemical resistance when selecting gloves. Nitrile gloves are suitable for brief contact but will degrade if exposed to DMSO for a few minutes, so gloves should be changed promptly after contact.*

These concerns have led to the adoption of other cryoprotectants, including PEG, polyvinylpyrrolidone (PVP), hydroxyethyl starch (HES), or trehalose (Suzuki et al. 1995; Monroy et al. 1997; Pasch et al. 2000). Trehalose is a sugar that is used by some organisms to withstand extreme cold or desiccation, making it a "natural" cryoprotectant with less toxicity compared to other choices (Crowe and Crowe 2000). Trehalose normally does not enter cultured cells, but it can be introduced when the cell membrane is more permeable, e.g. following genetic modification or the use of specific buffers (Beattie et al. 1997; Eroglu et al. 2000; Buchanan et al. 2010).

15.1.2 Cooling Rate

Most cultured cells survive best if they are cooled at −1 °C/min (Coriell et al. 1964; Leibo and Mazur 1971; Hunt 2019). This is probably a compromise between fast freezing (minimizing ice crystal growth) and slow cooling (encouraging the extracellular migration of water). However, maintaining a consistent cooling rate is not a straightforward process, as can be seen by the temperature curve that is generated during freezing (see Figure 15.1a). Formation of ice crystals is an exothermic process, resulting in a spike in temperature and associated loss of viability; this is referred to as the latent heat of fusion (see Figure 15.1b) (Morris 2007). Latent heat must be absorbed in some way, either by dispersal through the insulating material that surrounds the sample or by accelerated cooling just before ice crystals begin to form (see Figure 15.1c).

The shape of the freezing curve will vary based on a number of factors, including (i) the ambient temperature; (ii) any insulation surrounding the cells, including the cryovial; (iii) the specific heat and volume of the cryovial contents; (iv) the onset of freezing (nucleation); and (v) latent heat absorption once nucleation occurs. The actual cooling rate in Figure 15.1c is faster than the recommended −1 °C/min, but this may be less important than the control around the eutectic point, which is defined as the temperature at which a homogenous mixture of substances melts or solidifies. Seeding ice crystal formation as the cell suspension reaches the eutectic point can initiate freezing and improve survival. These factors are explored elsewhere in more depth and are important for optimization of cryopreservation procedures (Mazur 1984; Morris and Acton 2013; Hubel 2017; Hunt 2019).

15.1.3 Storage Temperature

What temperature should be used for long-term storage of frozen cells? Storage at −80 °C may be sufficient for some cell types but results in loss of viability and degradation of

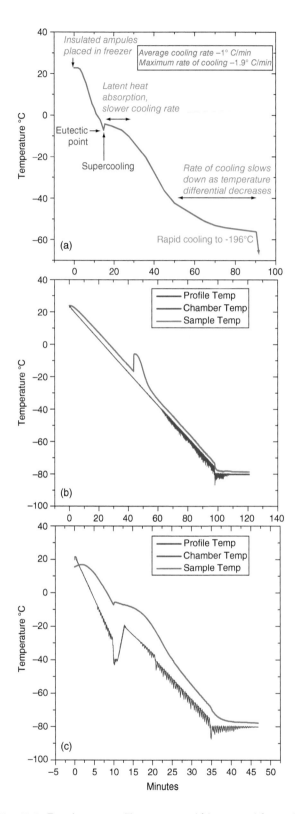

Fig. 15.1. Freezing curves. Temperature within a cryovial containing medium during freezing under different conditions. (a) Cryovial clipped with five other cryovials on an aluminum cane, enclosed in a cardboard tube, placed within a polyurea-foam tube, and placed in a freezer at −70 °C. (b) Cryovial placed in a rate-controlled freezer set to cool at −1 °C/min with no attempt to control for cooling. (c) Cryovial placed in a rate-controlled freezer with a multistep program to control cooling conditions. *Source*: (a) R. Ian Freshney; (b, c) courtesy of Elsa Moy and Amanda Capes-Davis, CellBank Australia.

phenotype for most cryopreserved cells (Hubel et al. 2014; Simione and Sharp 2017). Once frozen, cells should be stored at a temperature below the glass transition temperature (T_g), which for a solution of water and 10% DMSO is calculated to be −132.58 °C (Murthy 1998; Hubel et al. 2014). Below this temperature, the solution is highly viscous and the mobility of molecules within the sample is greatly reduced, which essentially means that metabolic processes are placed in stasis. For long-term storage, frozen cells should be kept in cryofreezers that maintain temperatures at less than −150 °C throughout the storage area (see Section 15.2.3) (Simione and Sharp 2017).

15.1.4 Vitrification

Vitrification refers to the solidification of a liquid into a glass, which is defined as a non-crystalline solid with the molecular structure of a liquid (strictly, an extremely viscous liquid with the mechanical properties of a solid) (Fahy and Wowk 2015). Rapid cooling of cells below the T_g induces vitrification and can be used to avoid ice crystal formation. Vitrification is particularly useful for cryopreservation of embryos. Their three-dimensional (3D) structure, although very small, is likely to limit diffusion during a slow cooling process. Vitrification is also commonly used for human pluripotent stem cells (hPSCs) (Richards et al. 2004; Kaindl et al. 2018).

During vitrification, cryoprotectants are typically added at stepwise increasing concentrations to minimize osmotic effects. Suitable agents include DMSO, glycerol, ethylene glycol, and sucrose. Embryos or cell suspensions are placed in a plastic capillary tube ("straw") or loaded onto a vitrification device before being placed directly into liquid nitrogen. Although vitrification can be very effective, it is difficult to perform in bulk and samples cannot be transported afterwards in dry ice, which (at a temperature of −78.5 °C) is above the T_g of the sample and will therefore affect its viability. Over time, techniques have been developed to allow stem cell cryopreservation at −1 °C/min, e.g. using Rho kinase (ROCK) inhibitors to promote survival. A protocol for stem cell cryopreservation is included in a later chapter (see Protocol P23.3).

15.2 APPARATUS FOR CRYOPRESERVATION

15.2.1 Cryovials

Glass ampoules or plastic cryovials can be used for cryopreservation. Cell repositories have traditionally preferred heat-sealed glass ampoules for seed stocks because the long-term storage properties of glass are well characterized and, when correctly performed, sealing is absolute. However, glass ampoules carry a risk of explosion during thawing (see Section 6.2.3). Glass ampoules have now largely been replaced by plastic cryovials, which are safer than glass (provided they are rated for cryogenic storage) and are more convenient and easier to label. Cryovials are usually made

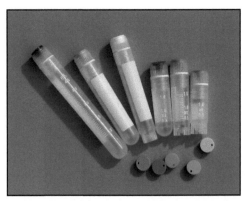

(d) Color-coded Cryovials. Polypropylene cryovials of various sizes, ranging from 1.0 to 4.6 mL. Colored inserts can be added to the caps to help identify cryovials. Source: Courtesy of Alpha Laboratories.

Fig. 15.2. Color-coded cryovials. Polypropylene cryovials of various sizes are shown, ranging from 1.0 to 4.6 ml. Colored inserts can be added to the caps to help identify cryovials. *Source*: courtesy of Alpha Laboratories.

from polypropylene and are available in a range of sizes (e.g. 1–5 ml; see Figure 15.2).

• *Safety Note. Inexperienced users should avoid using glass ampoules due to the risk of explosion during thawing. If glass ampoules are used, they must be perfectly and quickly sealed in a gas-oxygen flame. If sealing takes too long, the cells will heat up and die, and the air in the cryovial will expand and blow a hole in the top of the ampoule. If the ampoule is not perfectly sealed, nitrogen may enter during storage and cause explosion of the vial during thawing. If glass ampoules are used, they should be stored in vapor phase or in an isothermal freezer (see Section 15.2.3).*

Plastic cryovials have a screw cap and may have an external or internal thread; externally threaded vials are designed to reduce contamination, while the internally threaded vials usually have an "O" ring to improve their seal (Ryan 2004). Both require the correct torsion for closing, as they will leak if too slack or too tight (due to distortion of the "O" ring). It is worth practicing with a new batch to make sure that they seal correctly. Cryovials can be checked for leakage by placing them in a dish of stain, e.g. 1% methylene blue in 70% alcohol, at 4 °C for 10 minutes before freezing (Ryan 2004).

Cryovials must be clearly labeled using a method that will survive prolonged periods of time in cryogenic conditions. Remember, cultures that are stored in liquid nitrogen may well outlive you! At the very least, they will outlive your stay in a particular laboratory. The label should show the cell line's name or numerical identifier and, preferably, the date, passage number, and user's initials, although the latter is not always feasible in the available space. Use a fine-tipped marker that is alcohol-resistant or a cryogenic

label that is reported to withstand liquid nitrogen storage (e.g. Cryo-Babies, Sigma-Aldrich; FreezerBondz™ labels, Brady; LabTAG cryogenic labels, GA International). Always test to see if the printing remains legible and the label remains firmly adherent after handling. This can be done by wiping the labeled cryovial with alcohol, storing in the cryofreezer, and thawing in a water bath.

Some tissue culture laboratories use additional measures to ensure that labeling remains intact for the life of the cryovial and its records are rapidly accessible. Such measures include adding printed barcodes to the cryovial or using vials with pre-existing labels that were added during manufacture. Commercial sample management systems are available with barcoded cryovials and compatible barcode readers, sample racks, and software (e.g. FluidX, Brooks Life Sciences). Laboratories may also develop their own customized systems using standalone electronic databases, barcode software, and printable labels. Customized systems are likely to be cheaper and specific to your needs but will take time and expertise to make the various components work together. Simple additions are also helpful; for example, colored caps or cap inserts can be used to identify stocks that are reserved for cell banking procedures (see Section 15.5).

Always check that your cryovials are compatible with the cryofreezer and its storage system (see Section 15.2.3). Most current systems are designed to accept 1.2-ml cryovials, which are commonly used for storage. It may be difficult to fit cryovials where additional material has been added, such as Cryoflex™ (ThermoFisher Scientific). Cryoflex is recommended when cryovials are stored in liquid phase to minimize the risks of explosion or contamination (see Section 15.2.3). If Cryoflex is used, it is heat sealed in position over the vial before freezing, with the cryovials placed on ice to prevent overheating (Wewetzer and Dilmaghani 2001). Cryoflex must be removed before thawing; if left in position, it will slow thawing and may allow contaminated water from the water bath to be trapped between the Cryoflex and the vial.

15.2.2 Controlled Cooling Devices
A controlled cooling rate (see Figure 15.1) can be achieved using a controlled cooling device. Tissue culture laboratories use an assortment of cooling devices, including:

(1) *Foam insulation.* A polystyrene foam box (∼ 15 mm wall thickness) can be used with the cryovials placed in a rack or lying on cotton wool. Alternatively, a foam rack can be used with a second rack taped over the top (e.g. racks supplied with Falcon 15-ml conical tubes #C352099, Corning). The insulation should be sufficient for cryovials to cool at −1 °C/min when placed at −70 to −90 °C but also absorb the latent heat of fusion. Foam boxes can be surprisingly effective, but this effectiveness may vary between boxes and between different locations inside the box.

(2) ***A commercial freezing container*** (see Figure 15.3). Various formats are available to suit different cryovial sizes and sample numbers. Freezing containers are placed in a freezer at −70 °C or − 90 °C; devices may need isopropanol (e.g. Mr. Frosty™, ThermoFisher Scientific) or may be alcohol-free (e.g. CoolCell®, Corning). Commercial freezing containers usually produce consistent freezing profiles but have some limitations, e.g. space is limited to a set number of cryovials.

(3) ***A programmable rate-controlled freezer*** (see Figure 15.4). This allows increased reproducibility, active control of the freezing process, and handling of large volumes, e.g. >100 cryovials or cryobags containing > 100 ml (Massie et al. 2014). Users can monitor temperature data using a sample probe and adjust the freezing curve by editing the freezing program (see Figure 15.1c). Rate-controlled freezers are available from several suppliers (e.g. Custom Biogenic Systems, Planer PLC, or ThermoFisher Scientific). These devices typically work by injecting liquid nitrogen into the sample chamber at variable rates, in response to a preset freezing program. Liquid nitrogen-free devices are also available that utilize Stirling closed-cycle, regenerative heat engines as a completely sealed heat transfer system (e.g. via Freeze™ controlled-rate freezers, GE Healthcare Life Sciences) (Walker 1983).

Always perform viability testing to ensure that the device is satisfactory, particularly when freezing a new cell type or a new cell line (see Section 15.4.4). Poor viability is addressed by looking at the entire cryopreservation procedure (see Section 15.4). Even small details are important, such as the length of time spent in the freezing container. If cryovials are placed in a freezing container at −70 °C, they will cool rapidly to around −50 °C, but the cooling rate falls off significantly after that (see Figure 15.1a). Hence, the time that they spend in the −70 °C freezer needs to be longer than the amount of time projected by a − 1 °C/min cooling rate, as the bottom of the curve is asymptotic. Freezing containers should be kept in the freezer overnight and cryovials transferred to the cryofreezer on the following day. Cryovials will heat up at a rate of −10 °C/min when removed from the freezing device. It is critical that they do not warm up above −50 °C, as they will start to deteriorate, so the transfer to the cryofreezer must take significantly less than two minutes.

Further optimization of the cooling process is discussed elsewhere (Hubel 2017). A programmable rate-controlled freezer is useful, if available, to analyze temperature data and adjust the freezing program (see Figure 15.1). Freezing programs typically consist of a series of steps, including (i) a holding period to allow the samples and chamber to equilibrate; (ii) slow cooling until the sample temperature is close to the eutectic point; (iii) a brief period of rapid cooling to induce nucleation in the sample and absorb the latent heat of fusion; (iv) a brief period of rewarming to −20 °C to

Fig. 15.3. Freezing containers. Insulated blocks with spaces for cryovials. (a) Mr. Frosty (ThermoFisher Scientific), which relies on the specific heat of the coolant (isopropanol) added to the base to insulate the container and give a cooling rate of approximately −1 °C/min. (b) CoolCell (Corning), which relies on an insulating outer housing and thermoconductive solid core. *Source:* R. Ian Freshney.

correct for supercooling in the chamber; and (v) a return to slow cooling until the sample reaches −60 °C. Each step can be edited to optimize cryopreservation for various cell types.

15.2.3 Cryofreezers

Mechanical freezers can be used for storage at ultralow temperatures, but access to a mechanical freezer may result in temperature fluctuations as high as −30 °C (Simione and Sharp 2017). Liquid nitrogen Dewars or freezers (referred to here as "cryofreezers") are preferred for storage of cells and tissue, ensuring that correct temperature conditions are maintained with maximum stability and security (Coecke et al. 2005; Simione and Sharp 2017). Various cryofreezer designs may be used (see Figures 15.5, 15.6). Tissue culture laboratories often have more than one design; a new cryofreezer may be purchased to add to existing storage capacity rather than replace the previous model (see Figure 15.6e). The various designs affect how samples are stored and how cryofreezers are maintained.

Fig. 15.4. Rate-controlled freezer (CryoMed™, ThermoFisher Scientific) with door open, showing the freezing chamber with cryovials in position. Racks are modular, depending on the sample number and format, and can be added to fill the chamber. *Source*: Amanda Capes-Davis, courtesy of CMRI.

Narrow versus wide neck. Liquid nitrogen evaporation is an important concern for all cryofreezers. Some models are designed with narrow necks to reduce evaporation. Access to samples can become a little awkward, but these systems can typically be used with a range of racks and storage boxes (see Figure 15.6a, b, d). Larger systems may have a rotating carousel installed that allows the user to select a specific rack for removal. Cryofreezers with wide necks are easier to access but tend to have a faster evaporation rate (see Figure 15.6c).

Liquid versus vapor phase. Early cryofreezer designs added liquid nitrogen to the storage chamber, completely covering the samples ("liquid phase"). Storage in liquid phase means that liquid nitrogen is easily topped up and there is potentially less risk to sample integrity during disasters (see Section 4.3). However, liquid phase storage has a number of inherent problems and important safety concerns. Racks can be heavy and slower to remove due to drain time and there is a greater risk of splashing liquid nitrogen causing potential injury. Liquid nitrogen can be taken up by leaky cryovials, resulting in explosion due to rapid gas expansion on thawing. There is also a greater likelihood of transfer of contamination between cryovials or between the tank and its contents (Bielanski 2014). For example, an outbreak of hepatitis B in six patients was traced back to contamination of a cryofreezer from a bone marrow sample (Tedder et al. 1995; Hawkins et al. 1996). Mycoplasma and some other microorganisms can survive in liquid nitrogen without cryopreservation and may become concentrated over time, due to the constant evaporation of liquid nitrogen and ice buildup (Morris 2005; Bielanski 2014). All cryofreezers must be regarded as potentially contaminated, although the risk is likely to be extremely low

if samples are properly sealed (Pomeroy et al. 2010; Bielanski 2014).

• *Safety Note. Biohazardous material must not be stored in the liquid phase and teaching and demonstrating should also not be done with liquid phase storage.*

Improvements in cryofreezer design have led to storage in nitrogen vapor, which is generated by adding liquid nitrogen to the base or within the wall of the chamber ("vapor phase"). Storage in vapor phase means that there are fewer safety concerns when handling vials; the risks of explosion and transfer of contamination are reduced, although contamination cannot be eliminated, particularly if ice buildup occurs (Grout and Morris 2009). However, vapor phase storage does have limitations. Liquid nitrogen must be continually supplied, resulting in the need to auto-fill the cryofreezer from a supply tank (see Figure 15.6f). Vapor systems require electrical supply and are potentially more vulnerable to disturbances in power supply. A temperature gradient may also be present within the chamber from the bottom to the top of the racking system. Temperature gradients may span up to 80 °C, from −190 °C to around −110 °C, although they can be minimized through improved design and use of aluminum racking systems (Rowley and Byrne 1992). Despite some limitations, design improvements have led to the broad uptake of vapor phase storage and these systems are available from multiple suppliers (e.g. Chart MVE, Custom Biogenic Systems, Taylor-Wharton, Worthington Industries).

Some vapor phase systems have the liquid nitrogen located within the wall of the freezer and not in the storage chamber (see Figure 15.5d). These cryofreezers are referred to here as isothermal freezers and are available from some manufacturers (e.g. Custom Biogenic Systems). Isothermal freezers reduce evaporation of liquid nitrogen and can eliminate the temperature gradient. However, the nitrogen level is not visible and is difficult to measure independently, resulting in greater reliance on the freezer's internal monitoring. Problems with the auto-fill have occurred with some isothermal freezers due to accumulation of ice or other particulate material.

Cryofreezer storage systems. Vertical storage systems are usually employed in cryofreezers, with samples added from the bottom up to minimize the risk of thawing or temperature fluctuations. Cylindrical canisters were originally used for storage of straws and canes, followed by "pie racks" with triangular drawers and square racks with cardboard or plastic boxes. All of these formats can be used for 1.2-ml cryovials, with the choice depending on the user's preference and the shape of the cryofreezer (see Figure 15.6b–d).

Cryovials can be clipped on to an aluminum cane, inserted into a sleeve, and placed within canisters. Sleeves were originally cardboard, but transparent sleeves are now available, such

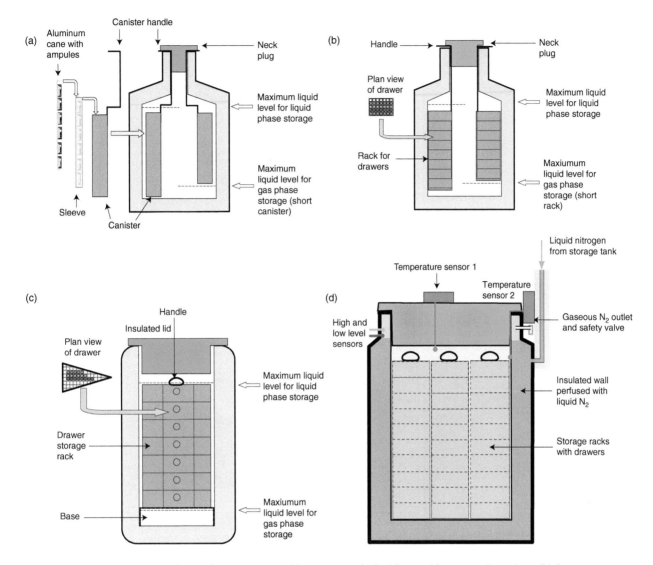

Fig. 15.5. Cryofreezer design variations. (a) Narrow-necked with cryovials on canes in canisters (high capacity, low boil-off rate); (b) narrow-necked with cryovials in square racks (moderate capacity, low boil-off rate); (c) wide-necked with cryovials in triangular racks (high capacity, high boil-off rate); (d) wide-necked with storage in drawers, with piped liquid nitrogen perfused through freezer wall (isothermal tank) and level controlled automatically by high- and low-level sensors (high capacity, even temperature throughout chamber).

as the CryoSleeve™ (ThermoFisher Scientific). The cane system has the advantage that cryovials can be handled in multiples of six at a time, with all the cryovials on one cane being from the same cell line, making the transfer from the cooling device to the cryofreezer easier and quicker. Canes can be labeled with colored tabs and one cane can be partially withdrawn to retrieve a cryovial without exposing other vials to the warm atmosphere and without the risk of replacing vials in the wrong location.

Racks and boxes are commonly used for cryovial storage. Many users feel that retrieval from boxes is easier; individual cryovials can be identified by the drawer number and the coordinates within the drawer, and the location recorded in a database. It does mean, however, that the total contents of the drawer (up to 100 cryovials) are exposed to warm temperatures whenever a vial is added or removed. The whole rack must be lifted out if you are accessing one of the lower drawers and the risk of returning a cryovial to the wrong location is higher. Boxes are held in place with a pin, which if forgotten can lead to cryovials or whole boxes floating away in liquid phase storage. Racks can also be quite heavy, particularly in liquid phase where liquid nitrogen is present in each box. A lifting device can be used when handling racks (e.g. Cryo-lift, Custom Biogenic Systems). Large numbers of cryovials should be handled on dry ice and the rack may need to be returned to the chamber to cool periodically during handling.

Fig. 15.6. Cryofreezers and storage systems. (a) Narrow-necked cryofreezer with storage on canes in canisters. (b) Interior of cryofreezer in (a), looking down on canes in canister, positioned in center as it would be for withdrawal. Normal storage position is under shoulder of freezer, just visible top right. (c) Wide-necked cryofreezer with storage in triangular drawers. (d) Narrow-necked cryofreezer with storage in square drawers. (e) Various small-capacity cryofreezers used by a research laboratory. (f) Medium-capacity cryofreezers used for vapor phase storage, with a single liquid nitrogen tank supplying two devices. (g) High-capacity cryofreezer with offset access port open, revealing canes in canisters. (h) High-capacity cryofreezers in a cell repository. Nearest freezer shows connections for monitoring and automatic filling. *Source*: (a–d) R. Ian Freshney; (e, f) Amanda Capes-Davis, courtesy of CMRI; (g, h) Courtesy of ATCC.

Many larger laboratories, biobanks, and cell repositories have converted to using a consensus format developed by the Society for Biomedical Screening (SBS) – now the Society for Laboratory Automation and Screening (SLAS) – for cryovial boxes, racks, and racking systems (SLAS 2019). This format will typically hold 96 cryovials in an SBS microtiter format box and has the advantage of allowing more samples to be stored in a given space. It is also compatible with most of the robotic liquid handling platforms that are used in larger tissue culture laboratories and facilities (see Sections 5.1.3, 28.7.1, 28.7.2). These cryostorage boxes and racks are often used with cryovials that have permanent unique barcodes on their base and sides, which can be read by many automated systems. SBS format cryovials and boxes are available from several specialist suppliers (e.g. FluidX, Greiner Bio-One, Micronic).

Cryofreezer maintenance. The investment in the contents of a cryofreezer may be considerable and must be protected by ongoing maintenance and automated monitoring and alarm systems. Maintenance should include regular removal of ice crystals on the lid and other accessible locations. Ice is the enemy of all frozen storage and will increase the risk of equipment breakdown and contamination over time. Always report cryofreezer faults promptly for repair by trained and competent personnel. Never adjust liquid nitrogen connections or carry out repairs yourself unless you are trained and competent to do so safely. Otherwise, "repairs" can lead to catastrophic accidents (Lowe 2006).

Monitoring and alarms are best carried out using an equipment alarm system with independent temperature sensors (see Section 4.3.2). At least one sensor should be positioned high in the chamber to act as an early warning system for rising chamber temperatures. Temperature and alarm information should be accessible remotely so that laboratory members are alerted to problems at an early stage, e.g. if temperatures rise above $-150\,°C$. Cryofreezers that are currently on the market typically include monitoring of temperature and liquid nitrogen levels; these systems are designed to minimize the risk of faults or failures and may have auto-fill cycles that are triggered if liquid nitrogen levels reach the set low point. However, monitoring systems may still fail, and failures can be difficult to detect if cryofreezers are infrequently accessed or positioned away from the main laboratory. Contingency planning should always cover loss of sample integrity due to cryofreezer failure, e.g. loss of liquid nitrogen supply. Simple measures can be adopted to manage the associated risks, including storing samples across multiple cryofreezers and arranging offsite storage (see Section 4.3.1).

15.3 REQUIREMENTS FOR CRYOPRESERVATION

15.3.1 When to Freeze

There are certain requirements that should be met before cell lines are considered for cryopreservation (see Table 15.1)

(Stacey and Dowall 2007; Parker 2011). The most important considerations relate to the culture itself. Cells should appear healthy on routine observation (see Table 14.3). The culture should be in an exponential "log" phase of growth (see Figure 14.1) (Terasima and Yasukawa 1977). Do not allow the culture to become confluent and enter its plateau phase of growth, where viability is likely to be impaired (Stacey and Dowall 2007). The cell yield should be sufficient for a high cell concentration in each cryovial (see Section 15.3.3).

How soon should a culture be frozen after its inception? In theory, you should optimize the culture conditions and wait for cells to proliferate before you start to freeze down samples. Finite cell lines are grown to around the fifth population doubling (see Section 14.2.2) in order to generate sufficient cells for freezing. However, in practice, you may have too many cells to handle in the time available or you may be concerned that the culture or its properties will be lost with further handling. Many tissue culture personnel choose to freeze down "token" vials whenever additional cells become available at subculture, e.g. plating half of the culture in a new flask and freezing the other half in one or two cryovials. Token vials are particularly useful if you discover a problem with the culture, e.g. if later authentication testing shows that it is misidentified. Preserving early passage material gives the best possible opportunity to understand the cause of the problem and hopefully retrieve authentic material.

Continuous cell lines usually grow rapidly, giving no shortage of cells available for freezing. It may be tempting to reserve cryopreservation for "backup" of the cell line after experimental work is complete. However, experienced practitioners have found that it is far better to freeze cells first and perform validation testing, resulting in a reliable source of cells for experimental work (Stacey and Dowall 2007). Cell banking procedures are discussed later in this chapter (see Section 15.5).

15.3.2 Freezing Medium

Freezing medium refers to the solution containing cryoprotectant (see Section 15.1.1) in which cells are suspended during cryopreservation. The choice of an appropriate freezing medium will depend on whether the cells require serum (including many continuous mammalian cell lines) or can be grown serum-free (including normal/cancer stem cell cultures, certain types of suspension cultures, and cells used in human *in vivo* applications). If serum can be used, freezing medium will typically contain cryoprotectant, basal medium, and serum at an increased concentration (20–50%). The addition of DMSO to basal medium normally increases the pH; it may be necessary to gas the medium with CO_2 or adjust the pH prior to use if basal medium is present (Morris 2007). Alternatively, 90% serum can be used with cryoprotectant and no basal medium. Serum provides better protection and pH control, and serum-based freezing

TABLE 15.1. Requirements before freezing.

Status	Criterion	Action indicated
Culture	Primary culture	Freeze excess for later use but viability may be poor
	Finite cell line	Freeze token number of cryovials as early as possible; increase number as cells proliferate
	Continuous cell line	Use cell banking approach; repeat with derivative cell lines, e.g. clonal populations
Standardization	Protocol	Develop Standardized Operating Procedure (SOP) for cryopreservation and optimize for viability
	Freezing medium	Select optimal freezing medium (see Section 15.3.2)
	Medium	Select optimal medium and adhere to this selection
	Serum (if used)	Select a batch for ongoing use (see Section 9.7.2)
	Substrate	Select one type and supplier, although not necessarily one size or configuration
Validation	Provenance	Keep records on cell line, cryopreservation, and thawing (see Sections 7.3.1, 15.7)
	Viability testing	Test viability during thawing and/or 24 hours after plating (see Section 15.4.4)
	Other	Test for authenticity and microbial contamination, e.g. mycoplasma, sterility testing (see Sections 11.8.1, 16.2, 17.2.3)
Characterization	Transformation	Determine transformed status (see Table 3.2)
	Other	Test for other characteristics depending on the cell type and intended use

medium can be dispensed into aliquots and stored at −20 °C for later use.

Selection of freezing medium becomes more challenging if cells are grown in serum-free media or cannot be individualized to form a single-cell suspension. For example, hPSC aggregates are more sensitive to supercooling and vary in their permeability to cryoprotectant, making cells more vulnerable to damage (Li et al. 2018). Various reagents may be added to improve hPSC viability, including ROCK inhibitors and trehalose (Claassen et al. 2009; Hanna and Hubel 2009). Serum may be substituted with various reagents, including human serum albumin (5%), methylcellulose (0.1%), Pluronic F68 (1–5%), and PVP (3%) (Merten et al. 1995; Gonzalez Hernandez and Fischer 2007; Hanna and Hubel 2009). Optimization of the freezing process will be required and should extend to the freezing curve and the subculture conditions used (Liu and Chen 2014; Li et al. 2018). Adding 50–90% conditioned medium (serum-free medium in which the cells were grown) to both the freezing and recovery medium may improve the post freezing recovery and survival.

15.3.3 Cell Concentration

Cells appear to survive freezing best when cell suspensions are frozen at a high concentration. This is largely an empirical observation but may relate to improved survival if cells are leaky because of cryogenic damage. A high concentration at freezing also allows for loss of some cells without impairing

cloning efficiency (which tends to decline at lower density) and increases dilution of the cryoprotectant when medium is added after thawing (which reduces toxicity and may render centrifugation unnecessary). The number of cells frozen per cryovial should be sufficient to allow for 1 : 20 dilution on thawing to dilute out the cryoprotectant. It is also important to keep the cell concentration higher than at normal passage (Morris 2007). For example, for cells that are normally passaged at 1×10^5/ml, 1×10^7 cells may be frozen in 1 ml of freezing medium and the cells diluted to 20 ml after thawing. This would give 5×10^5 cells/ml (five times the normal seeding concentration) and dilutes the cryoprotectant from 10% to 0.5%, at which concentration it is less likely to be toxic. Cells normally seeded at 2×10^4/ml can be diluted 1 : 100 to give 1×10^5/ml and a cryoprotectant concentration of 0.1%.

15.4 CRYOPRESERVATION PROCEDURES

15.4.1 Cryopreservation in Cryovials

Protocol 15.1 is suitable for most continuous cell lines, but some cell types will need optimization of the freezing curve (see Figure 15.1) or other conditions. Most cells are frozen in DMSO but there are some cells for which glycerol is preferred (see Section 15.1.1). Protocol P15.1 is suitable for either cryoprotectant and may be adapted for use in training (see Supp. S30.13).

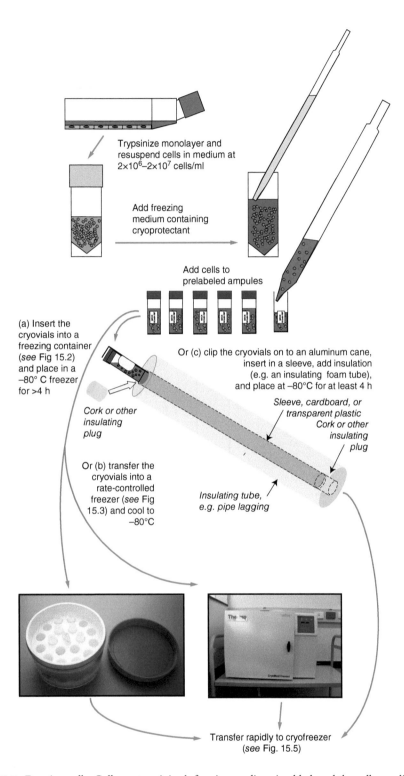

Trypsinize monolayer and
resuspend cells in medium at
2×10^6–2×10^7 cells/ml

Add freezing
medium containing
cryoprotectant

Add cells to
prelabeled ampules

(a) Insert the
cryovials into a
freezing container
(*see* Fig 15.2)
and place in a
−80° C freezer
for >4 h

Cork or other
insulating
plug

Or (c) clip the cryovials on to an aluminum cane,
insert in a sleeve, add insulation
(e.g. an insulating foam tube),
and place at −80°C for at least 4 h

*Sleeve, cardboard, or
transparent plastic
Cork or other
insulating
plug*

Or (b) transfer the
cryovials into a
rate-controlled
freezer (*see* Fig
15.3) and cool to
−80°C

*Insulating tube,
e.g. pipe lagging*

Transfer rapidly to cryofreezer
(*see* Fig. 15.5)

Fig. 15.7. Freezing cells. Cells are trypsinized, freezing medium is added, and the cells are aliquoted into cryovials and frozen using a controlled cooling device. Cells may be slowly cooled: (a) in a freezing container; (b) in a rate-controlled freezer; or (c) using foam insulation (e.g. pipe lagging surrounding a cane to which samples are clipped). Once cells have completed the freezing process (see Figure 15.1), cryovials should be promptly transferred to the cryofreezer for long-term storage.

PROTOCOL P15.1. FREEZING CELLS IN CRYOVIALS

Background

Cells are frozen at subculture, where they are suspended in freezing medium instead of plating into new culture vessels. Otherwise, subculture should be performed as per usual for that cell type (see Protocols P14.2,14.3). The passage number will increase by one to indicate that subculture is performed, but stays the same when cells are thawed (Parker 2011).

Outline

Grow the culture to late log phase, prepare a high-concentration cell suspension in medium with a cryoprotectant, aliquot into cryovials, and freeze slowly (see Figure 15.7).

❖ *Safety Note. All personnel working with liquid nitrogen should be trained to do so safely and wear suitable PPE. Wear a closed lab coat, face shield, and insulated gloves when handling frozen cryovials (see Sections 6.2.3,6.2.4). Wear chemically resistant gloves when handling DMSO (see Section 6.2.2).*

Materials (sterile or aseptically prepared)

☐ Culture to be frozen
☐ Freezing medium (see Section 15.3.2), stored in aliquots at −20°C or made up fresh
☐ Universal containers or centrifuge tubes
☐ Pipettes in an assortment of sizes, usually 1, 5, 10, and 25 ml
☐ Syringe, 1–5 ml, for dispensing glycerol if used (because it is viscous)
☐ Cryovials, 1.2 ml

Materials (non-sterile)

☐ PPE (see Safety Note)
☐ Biological safety cabinet (BSC) and associated consumables (see Protocol P12.1)
☐ Cell counting equipment (hemocytometer or automated counter; see Section 19.1)
☐ Controlled cooling device (see Section 15.2.2)
☐ Forceps (×1), for moving small items in the cryofreezer
☐ Dry ice
☐ Cryofreezer database (see Section 15.7)

Procedure

A. Before Cryopreservation

1. Examine the culture for cell morphology, confluence, and freedom from contamination. Cells should be healthy and subconfluent (at late log phase before plateau; see Figure 14.1). Proceed if these and other requirements for cryopreservation are met (see Table 15.1).
2. Prewarm reagents for subculture and thaw freezing medium (if stored at −20°C).
3. Prepare a work space for aseptic technique using a BSC.
4. Perform subculture as per usual for that cell type to give a cell suspension.
5. Count the cells with a hemocytometer or an automated cell counter. If cells have been subcultured from multiple flasks, combine them at this point to give a total cell count that will be used to generate a consistent batch of cryovials.

B. Cryopreservation

6. Calculate the cell yield and the volume of freezing medium that should be added for the required cell concentration, usually between 2×10^6 and 2×10^7 cells/ml. This volume will determine the number of cryovials that can be frozen, e.g. at 1 ml per vial.
7. Label the cryovials with the cell line's designation, passage number, cell count, date, etc.
8. Bring premade aliquots of freezing medium to the BSC or make up fresh, using either DMSO or glycerol (usually at 10%) with medium and/or serum (see Section 15.3.2).
9. Add freezing medium to the cells and mix gently to resuspend, e.g. by inverting the tube.
10. Aliquot 1 ml cell suspension into each cryovial. Cap each cryovial with sufficient torsion to seal the cryovial without distorting the gasket.
11. Once the cell suspension has been fully transferred or you have reached the required number of cryovials, transfer all cryovials to the controlled cooling device for freezing.
12. If you have any remaining cells, consider how these may be used, e.g. for validation testing.

C. Transfer to Cryofreezer

13. Check the cryofreezer database or other records to look for available spaces.
14. Remove cryovials from the controlled cooling device and place on dry ice. Bring to the cryofreezer. Put on PPE for working with liquid nitrogen.

15. Open the cryofreezer and navigate to the empty spaces that were previously identified. Place the cryovials in the correct locations and record the coordinates of each cryovial.

16. Return the box, drawer, or cane to its correct location. If you are working with racks that use a pin to keep boxes or drawers in place, do not forget to return the pin!

17. Update the records in the cryofreezer database (see Section 15.7).

Notes

Step 5: Some tissue culture personnel do not count cells at each passage. If you do not, always record the size of vessel that was used; if you freeze cells from a 25-cm^2 flask, the next person is likely to get good results from thawing cells into the same sized vessel (Parker 2011).

Step 9: If freezing medium is added at 4 °C (see Section 15.1.1), place cells on ice from this point onwards. Otherwise, you do not normally need to place cryovials on ice. A delay of up to 30 minutes at room temperature is not harmful when using DMSO and is beneficial when using glycerol.

Step 10: The volume suggested here is for 1.2-ml cryovials and will vary with the vial size. If multiple cryovials are used, take care to keep the cell suspension consistent, e.g. by gently mixing at regular intervals during aliquoting. Do not use a large pipette, as the cells will fall with gravity and later cryovials will have fewer cells.

Step 11: If a foam box or commercial freezing container is used, leave it overnight in a freezer between −70 and −90 °C and transfer cryovials to liquid nitrogen on the following day.

Step 12: Leftover cells are representative of that batch of cryovials and would indicate if any microbial contaminants have been introduced, e.g. during aliquoting. These cells are particularly useful for sterility and mycoplasma testing (see Sections 11.8.1, 16.4.1).

Step 15: This transfer must be done quickly (< 2 minutes), as the cryovials will reheat at ~10 °C/min, and the cells will deteriorate rapidly if the temperature rises above −50 °C. When transferring multiple samples, it is helpful to have two people at work – one to handle the cryovials and the other to record their coordinates.

15.4.2 Cryopreservation in Other Vessels

Cryopreservation may be performed in other vessels, provided they can be used safely for frozen storage (some vessels may crack or explode at low temperatures). Flasks have been frozen by growing the cells to late log phase, adding 5–10% DMSO to the smallest volume of medium that will effectively cover the monolayer, and putting the flask in a polystyrene container of 15-mm wall thickness (Ohno et al. 1991). The insulated container is placed in a freezer at −70 to −90 °C and will cool at approximately −1 °C/min. Cells may survive for several months if the flask in its container is not removed from the freezer. Multiwell plates may also be frozen in the same manner, with about 150 µl freezing medium per well for a 24-well plate (Ure et al. 1992). This approach can be used to store large numbers of clones during evaluation procedures.

15.4.3 Thawing Stored Cryovials

The procedure used for thawing is an important part of preserving viability. The cryovial should be thawed rapidly at the cell line's normal growth temperature (typically 37 °C), to minimize intracellular ice crystal growth during the warming process. This can be done using a water bath or a dry heat block, provided heat is rapidly transferred from the dry block to the cryovial. After thawing, the cell suspension should be diluted slowly to avoid osmotic damage that may reduce viability. Some tissue culture personnel centrifuge the cell suspension after dilution, particularly if cells are sensitive to cryoprotectant (e.g. some suspension-growing cells) or if toxicity is a concern for later applications (see Section 15.1.1). However, if the cell concentration has been selected to incorporate a suitable dilution factor (see Section 15.3.3), centrifugation can be omitted and the medium replaced once cells have attached. Cells should be reseeded at a relatively high concentration to optimize recovery.

PROTOCOL P15.2. THAWING FROZEN CRYOVIALS

Outline

Thaw the cells in the cryovial rapidly, dilute them slowly, and reseed at a high cell concentration (see Figure 15.8).

❖ *Safety Note. All personnel working with liquid nitrogen should be trained to do so safely and wear suitable PPE. Wear a closed lab coat, face shield, and insulated gloves when handling frozen cryovials (see Sections 6.2.3, 6.2.4).*

Materials (sterile or aseptically prepared)

☐ Frozen cryovial
☐ Complete culture medium
☐ Centrifuge tubes, 15 ml (if centrifugation is required)
☐ Pipettes in an assortment of sizes, usually 1, 5, 10, and 25 ml

□ Culture flask, e.g. 25 cm^2

Materials (non-sterile)

□ PPE (see Safety Note)
□ BSC and associated consumables (see Protocol P12.1)
□ Water bath, dry heat block with correct cryovial inserts, or bucket containing clean (sterilized) water at 37 °C; cover if cryovial was stored in liquid phase (see Section 15.2.3)
□ Dry ice, for transferring cryovial from cryofreezer to water bath (if required; see Notes)
□ Cell counting equipment (hemocytometer or automated counter; see Section 19.1)
□ Trypan blue (0.4%) or alternative dye (see Section 19.2.1)
□ Forceps (×1), for moving small items in the cryofreezer
□ Cryofreezer database (see Section 15.7)

Procedure

A. Transfer from Cryofreezer

1. Check the cryofreezer database or other records to identify the location of the cryovial.
2. Prewarm culture medium and prepare the water bath for thawing (see Figure 15.8).
3. Prepare a work space for aseptic technique using a BSC. Bring in culture medium and check the water bath is ready. Label the culture flask.
4. Put on PPE for working with liquid nitrogen. Open the cryofreezer and navigate to the coordinates of the cryovial you identified earlier.
5. Remove the cryovial and check that you have the correct vial, e.g. check the label and any color coding on the cap or cane.
6. Return the box, drawer, or cane to its correct location. If you are working with racks that use a pin to keep boxes or drawers in place, do not forget to return the pin! Shut the cryofreezer.

B. Thawing

7. Place the cryovial in sterile water (see Figure 15.8) or a dry heat block at 37 °C until thawed. Keep the cap and neck above the level of the water to reduce the risk of contamination.
8. Check the label again to confirm its identity. Swab the cryovial thoroughly with 70% alcohol and bring it to the BSC.
9. For cells that do not require centrifugation to remove the cryoprotectant:

(a) Transfer the contents of the cryovial to a culture flask using a 1-ml pipette.
(b) Add 10 ml medium to the cell suspension over about two minutes. Start by adding drops and then gradually increase the speed, slowly diluting the cells and cryoprotectant. Lay the flask down to spread the cells and medium over the surface of the flask.
(c) Place the culture flask in an incubator at 37 °C and 5% CO_2 (or as required).
(d) There should be a small volume of liquid remaining in the cryovial. Use this to count the cells, adding trypan blue to look for non-viable cells (see Protocols P19.1, 19.2).

10. For cells that require centrifugation:
(a) Transfer the contents of the cryovial to a 15-ml centrifuge tube using a 1-ml pipette or pipettor with sterile filter tip.
(b) Add 10 ml medium to the cell suspension over about two minutes. Start by adding drops and then gradually increase the speed, slowly diluting the cells and cryoprotectant.
(c) Centrifuge the cells at a low speed, e.g. 100 g for two minutes. Check that a pellet has formed.
(d) Remove the supernatant and discard. Resuspend the cells in fresh culture medium. Transfer to the culture flask and lay flat to spread the cells and medium over the surface.
(e) Place the culture flask in an incubator at 37 °C and 5% CO_2 (or as required).
(f) There should be a small volume of liquid remaining in the centrifuge tube. Use this to count the cells, adding trypan blue to look for non-viable cells (see Protocol P19.2).

C. After Thawing

11. Update the records in the cryofreezer database to show that the cryovial has been removed and any additional information, e.g. viability (see Section 15.4.4).
12. Check the culture flask after 24 hours.

Notes

Step 3: The flask retains the same passage number as the cryovial, since subculture is not performed as part of thawing.
Step 5: If you are not ready to thaw the cells, place the vial on dry ice until you reach step 7. If cryovials were stored in liquid phase, use a screw-top aluminum canister (vented) or similar container to guard against explosion during thawing (Morris 2007).

Step 7: If the cryovial has been submerged in liquid nitrogen during storage, the warming bath or heat block must be covered to guard against explosion during thawing. If Cryoflex is used, it should be removed prior to this step.

Step 12: For adherent cultures, confirm attachment and estimate the cell density. This can be used to estimate the percentage survival, based on photographs of cells at the expected density (cells per square centimeter; see Figure 14.6). If cells have attached, the medium may also be replaced. For suspension cultures, check the appearance (clear cytoplasm, lack of granularity) and dilute to the regular seeding concentration. This is best done by counting an aliquot of the cell suspension and staining with trypan blue or other dye to assess viability (see Sections 19.1, 19.2).

15.4.4 Viability Testing

Viability should be assessed following thawing and will guide how any remaining vials are handled. Many tissue culture laboratories assess viability by adding a dye such as trypan blue during cell counting to identify non-viable cells. Trypan blue is known to be toxic, with potential teratogenic effects; erythrosin B is becoming a more widely used, less toxic alternative for use in viability measurements (see Section 19.2.1). Analysis of viability at a cell repository found that most results exceeded 85% when cells were counted in the presence of trypan blue; cryovials were rejected if viability results were less than 75% (Morris 2007). Although the cell count after thawing is useful as a rapid estimate of viability, it does not detect cells that are damaged and will subsequently undergo cell death. For a more precise estimate of recovery, a single vial from the batch can be thawed, plated in a flask or dish, and counted after 24 hours. The cell count after 24 hours should exceed 80% and may exceed 100% if the culture recovers rapidly and proliferates beyond the number that was originally frozen. Although this culture is not grown beyond 24 hours, the cells may be used for other purposes, e.g. to generate a cell pellet for DNA extraction.

15.5 CELL BANKING PROCEDURES

15.5.1 Rationale for Cell Banking

Even the best-run laboratory is prone to equipment failure and contamination. Mycoplasma contamination and misidentification continue to occur with alarming frequency, particularly if cultures are obtained from secondary sources (e.g. colleagues) rather than primary sources where validation testing has been performed (see Section 14.2.3) (Drexler et al. 2017). Cell lines in continuous culture are also prone to heterogeneity and evolution with further passaging (see Sections

3.2–3.4). Reasons to freeze down validated stocks of cells are compelling, including:

(1) Senescence and the resulting extinction of the cell line.
(2) Transformation resulting in acquisition of malignancy-associated properties.
(3) Genotypic drift due to genetic instability.
(4) Phenotypic instability due to selection and dedifferentiation.
(5) Contamination by microorganisms.
(6) Cross-contamination by other cell lines.
(7) Misidentification due to cross-contamination or other mechanisms, e.g. mislabeling.
(8) Incubator failure.
(9) Saving time and materials by not maintaining lines other than those in current use.
(10) Reducing the need to address problems before publication by performing validation before experimental work commences.
(11) Need for distribution of validated stocks to other users.

15.5.2 Principles of Cell Banking

Cell banking is based on the concept that a seed stock must be stored for each cell line to preserve its unique characteristics at early passage (Hay 1988; Reid 2017). A tiered approach is used to ensure that early passage material can be preserved (see Figure 15.9). A token freeze is performed on surplus material to guard against future problems. Seed stock (also known as a master stock or master cell bank) is frozen once sufficient cells are available. A distribution stock (also known as working stock or a working cell bank) is then prepared from one of the seed stock vials and used for experimental work (in a laboratory) or for distribution to other laboratories (in a cell repository). Validation testing is required at each level to ensure that the cell line remains viable and contamination-free (see Figure 15.9). The process should be repeated for any derivatives, including cloned cell strains (see Section 20.1).

The overall process of cell banking can be scaled up or down, depending on the size of the laboratory and the popularity of the cell line. A small laboratory will probably ask each person to prepare their own seed stocks, with 12 cryovials often sufficient for this purpose (see Table 15.2a). A research organization or large laboratory may have a core facility or a dedicated individual who takes responsibility for preparing seed stocks, with the user requesting one or more vials to expand for their own work (see Table 15.2b). User stocks should never be passed on to other laboratories and should be discarded after work is complete. In contrast, a large cell repository will prepare master cell banks and working cell banks that may contain hundreds of vials (see Table 15.2c). For example, the master cell bank for the finite cell line IMR-90 consisted of 662 ampoules (Nichols et al. 1977). Laboratories that use these cell repositories will request single vials that become the basis for their own cell banking procedures.

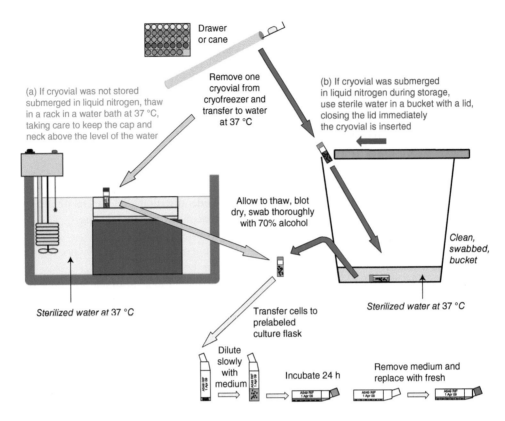

Fig. 15.8. Thawing cells. Cryovials are removed from the freezer and thawed rapidly at 37 °C. Cells may be thawed (a) in a water bath, keeping the water level below the cap and neck, if stored in vapor phase; or (b) at least 10 cm deep in a clean, alcohol-swabbed bucket with a lid, if the cryovials have been stored in liquid phase (in case of explosion; see Section 6.2.3).

Whatever the size of the laboratory, validation testing must be performed on all cell lines (see Figure 15.9). Characterization is also important, e.g. generation of a growth curve (Reid 2017). This will vary with the cell type and application; for example, stem cell lines will need to be assessed for genetic stability and for expression of the expected phenotype, e.g. pluripotency (see Section 23.7). The timing of testing and the method used may also vary. For example, a laboratory that can perform single nucleotide polymorphism (SNP) genotyping may generate a short tandem repeat (STR) profile initially to check for unexpected matches and then use SNP genotyping for subsequent comparison to their seed stocks (Yu et al. 2015).

15.5.3 Replacement of Culture Stocks

Seed stocks (master cell banks) are a high priority for ongoing storage and contingency planning (see Section 4.3.1). Seed stocks should be stored in at least two cryofreezers and at least two locations; unique and irreplaceable cell lines should be stored offsite, preferably at a cell repository where they can be shared with the research community. Care must be taken to ensure that seed stocks are not depleted. Further stocks should be generated when a certain number of cryovials are left from

the previous batch, e.g. five cryovials for a small laboratory. Distribution stocks (working cell banks) should be replenished from seed stocks using a similar approach.

Continuous cell lines that are used for experimental work should be discarded at regular intervals and replaced with frozen stocks. The time period will vary with the cell line and the changes that are reported with ongoing handling. As a rough guide, cultures that have a population doubling time of approximately 24 hours should be discarded once they have been in culture for three months. An earlier passage cryovial should be thawed at two-and-a-half months and expanded to replace the later passage material (see Figure 15.10). Cell lines with shorter or longer population doubling times may need shorter or longer replacement intervals, respectively.

15.6 CELL REPOSITORIES

Cell repositories carry out cell banking procedures for cell lines that are likely to be of interest to the scientific community. Repository staff perform culture, cryopreservation, validation testing, and then (if no problems are detected) distribute the cell line on behalf of the depositor. Cell repositories vary in size, from small core facilities to large

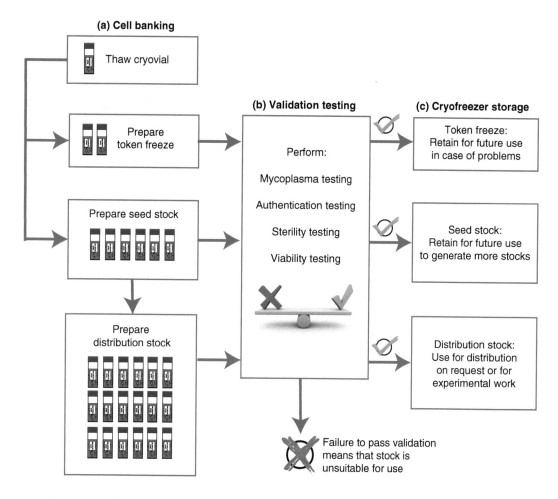

(a) Cell banking

Thaw cryovial

Prepare token freeze

Prepare seed stock

Prepare distribution stock

(b) Validation testing

Perform:

Mycoplasma testing

Authentication testing

Sterility testing

Viability testing

(c) Cryofreezer storage

Token freeze: Retain for future use in case of problems

Seed stock: Retain for future use to generate more stocks

Distribution stock: Use for distribution on request or for experimental work

Failure to pass validation means that stock is unsuitable for use

Fig. 15.9. Cell banking. (a) A cryovial is thawed and used to generate a token freeze and a seed stock (also known as a master stock or master cell bank). A single seed stock vial is used to generate distribution stock (also known as working stock or a working cell bank). The number of cryovials can be scaled up or down, depending on the scale of operation (see Table 15.2). (b) Validation testing is performed on representative samples from each stock; validation typically includes mycoplasma, authentication, sterility, and viability testing. (c) Vials are stored in the cryofreezer for various purposes, depending on the stock. *Source*: courtesy of Amanda Capes-Davis and CMRI.

organizations; examples of widely used cell repositories are listed in Table 15.3. Some facilities accept all cell types, while others focus on specific diseases or cell types, e.g. hPSC lines (Kim et al. 2017). A subset of cell repositories is accredited to receive deposits of cell lines as patented material; these organizations are listed as International Depository Authorities (IDAs) under the Budapest Treaty (WIPO 2019).

Cell repositories are essential sources of cell line information as well as early passage material. Most cell repositories make their catalogs available online, which contain recommendations for culture conditions and handling procedures. Many cell repositories share their validation testing data and contribute to searchable databases of STR profiles (Dirks and

Drexler 2013). Comparison of testing data between repositories has uncovered a number of cases of misidentification, demonstrating the importance of a consensus method for authentication (see Section 17.3.1). Cell repositories may also perform broader testing, e.g. of pluripotency and genetic stability in stem cell banks (Kim et al. 2017).

All cell lines should be obtained from primary sources, which include the originator, a cell repository that has performed cell banking and validation testing prior to distribution, or a supplier that is authorized to distribute the cell line on their behalf. Secondary sources (including colleagues in other laboratories) have an increased risk of contamination, misidentification, and other quality concerns (Drexler et al. 2017). When obtaining a cell line from a cell repository or

TABLE 15.2. Acquisition and storage of cell lines.

Stage	Source	No. of cryovials[a]	Use[b]
(a) Small laboratory or individual scientist			
Token freeze	Vial from originator or cell repository	1–3	Retained in case of problems
Seed stock	As for token freeze	12 (stored in two cryofreezers/locations)	Replenishment of stocks; restore when depleted; do not use for other purposes
Working stock	Vial from seed stock	20	Experimental work; do not pass on to other users
(b) Large laboratory or research organization			
Token freeze	Vial from originator or cell repository	1–3	Retained in case of problems
Seed stock	As for token freeze	36 (stored in three cryofreezers/locations)	Replenishment of stocks; restore when depleted; do not use for other purposes
Distribution stock	Vial from seed stock	50–100	Distribution to users within the same laboratory or organization
User stock (prepared by recipient)	Vial from distribution stock	20	Experimental work; do not pass on to other users
(c) Cell repository or industrial concern			
Token freeze	Vial from originator or cell repository	1–3	Retained in case of problems
Master cell bank	As for token freeze	60 (stored in multiple cryofreezers/locations)	Replenishment of stocks; restore when depleted; do not use for other purposes
Working cell bank	Vial from master cell bank	600 (stored in multiple cryofreezers/locations)	Distribution on request, e.g. to research groups, production facilities
User cell bank (prepared by recipient)	Vial from working cell bank	See (a, b) above	Generation of cell banks as described in (a, b) above

[a]Numbers are suggestions and will depend on the cell line, e.g. a finite cell line is usually frozen in a greater number of cryovials because it has a limited lifespan (see Section 14.2.2). The suggestion of 20 cryovials for the final stock is based on a five-year project where stock is replaced every three months (four per year).
[b]Use may be restricted by the terms of the MTA if obtained from another laboratory or cell repository.

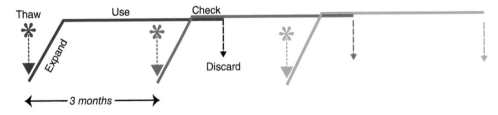

Fig. 15.10. Serial culture replacement. Cultured cells are discarded and replaced from frozen stocks at regular intervals determined by the cell line and the application. On each occasion, a frozen cryovial is thawed, expanded, and used to replace the current stock, which is then discarded. A period of checking is advisable to ensure that cells continue to express the desired phenotype.

TABLE 15.3. Commonly used cell repositories.

Repository name[a]	Website[b]
American Type Culture Collection (ATCC)	www.atcc.org
CellBank Australia	www.cellbankaustralia.com
Cell Resource Center, Institute of Basic Medical Sciences, Chinese Academy of Medical Sciences/Peking Union Medical College	www.cellresource.cn/
Childhood Cancer Repository, Children's Oncology Group	www.cccells.org/cellreqs-leuk.php
China Center for Type Culture Collection (CCTCC), Wuhan University	www.cctcc.org/
Common Access to Biological Resources and Information (CABRI) Collections	www.cabri.org/cabri/srs-doc/index.html
Coriell Institute Biorepositories	catalog.coriell.org/
European Bank for Induced Pluripotent Stem Cells (EBiSC)	www.EBiSC.org/
European Collection of Authenticated Cell Cultures (ECACC), Culture Collections Public Health England	www.phe-culturecollections.org.uk/products/celllines/index.aspx
Japanese Collection of Research Bioresources (JCRB), National Institute of Biomedical Innovation	cellbank.nibiohn.go.jp/english/
Korean Cell Line Bank	cellbank.snu.ac.kr/main/index.html
Leibniz-Institut Deutsche Sammlung von Mikroorganismen und Zellkulturen (DSMZ)	www.dsmz.de/catalogues.html
National Cell Bank of Iran, Pasteur Institute of Iran	en.pasteur.ac.ir/pages.aspx?id=823
RIKEN Bioresource Center Cell Bank	cell.brc.riken.jp/en/
The Tick Cell Biobank, Institute of Infection and Global Health	www.liverpool.ac.uk/infection-and-global-health/research/tick-cell-biobank/
UK Stem Cell Bank, National Institute for Biological Standards and Control (NIBSC)	www.nibsc.org/ukstemcellbank
WiCell Stem Cell Bank	www.wicell.org/home/stem-cells/stem-cells.cmsx

[a]This table primarily focuses on not-for-profit cell repositories that are known to perform mycoplasma and authentication testing; information has been collated from various websites and publications, e.g. ICLAC (2014); Stacey (2017). To look for holdings of a specific cell line, search for that cell line in the Cellosaurus knowledge resource and look for entries under "Cell line collections" (https://web.expasy.org/cellosaurus).
[b]Additional resources for cell line genotypic data are discussed in a later chapter (see Table 18.1).

supplier, it is important to be confident that the source is a reputable one. "Pirate" cell repositories and suppliers exist; these organizations habitually distribute cell lines where they are not authorized to do so and without concern for their quality. Cell lines from these "pirate" organizations have passed through an uncertain chain of custody that has been concealed from the customer's view. Such concerns can be addressed by (i) contacting the originator or their institution to check if a source is known to them; (ii) obtaining cell lines only from cell repositories or suppliers that are known to be reputable; and (iii) using the Cellosaurus knowledge resource to identify sources of a particular cell line (Bairoch 2018). Because Cellosaurus is a curated knowledge resource, cell line sources can be removed if evidence is provided to show that a particular supplier is not a reputable one.

15.7 RECORD KEEPING FOR FROZEN STOCKS

Records for frozen stocks may be kept in a variety of formats. The traditional approach was to keep paper-based records using a card index or folder system. For smaller laboratories, this approach can be effective (if all personnel take the time to update information), but it is prone to data loss and other difficulties, e.g. unclear handwriting. If you are currently setting up a record system, an electronic database is preferable and is essential for large laboratories where multiple users may require frozen stocks. Electronic databases should have regular backup arrangements (including local and offsite or cloud backup storage) and should be accessible to all laboratory personnel, preferably with different logins for each user for traceability. Larger systems usually include user access

TABLE 15.4. Cell line record.

Cell line	Freeze date	
	Passage/PDL	
	No. of vials	
	Location	
Species	Strain	Age/stage of development
Tissue	Diagnosis	Site
Source	Catalog no.	Reference
Mycoplasma Method		
Date of test	Result	
Authentication Method		
Date of test	Result	
Normal maintenance		
Subculture frequency		Seeding concentration
Medium change frequency		Medium composition — Agent, Serum (%)
Gas phase	Buffer	pH
Special characteristics		
Freeze instructions	Freezing medium	
Thaw instructions		
Dilute to		
Centrifuge to remove preservative?	Yes/No	
Special requirements		
Any other special conditions		
Biohazard precautions		
Person completing card	Date	

levels; for example, the database manager might be the only person who can edit certain entries or release cryovials for distribution. At this level, a specific individual or team may take responsibility for all cryostorage and respond to requests for cryovials from system users, resulting in more consistent data entry and sample handling.

Regardless of the format used, all cryofreezer records should provide:

(1) An index of cell line names and other designations, e.g. accession numbers (for cell repositories) or Research Resource Identifiers (RRIDs) (see Section 7.3.2).
(2) For each cell line, information on:
   (a) Source, e.g. originator or cell repository (including catalog number).
   (b) Culture conditions, e.g. temperature and gas levels, media, and substrates.
   (c) Handling conditions, e.g. subculture and cryopreservation procedures.
   (d) Validation testing, e.g. viability testing, mycoplasma testing, and STR profiles.
   (e) Number of cryovials stored and their locations.
   (f) Cell count in the cryovials or size of flask to be seeded.
   (g) Any restrictions that may apply to cryovials, e.g. seed stocks.
(3) An inventory of each cryofreezer with information on:
   (a) Storage system components, e.g. a diagram showing rack and box locations.
   (b) Cryovials, e.g. rack number, box number, and coordinates within the box.
   (c) Available storage spaces.

Examples of cell line and freezer records are provided as a starting point in Tables 15.4 and 15.5. Commercial inventory management systems are available, e.g. Freezerworks (Dataworks Development, Inc.), ItemTracker (ItemTracker Software), and Pro-curo (Pro-curo Software). Commercial biobanking systems may not include cell line information fields, but additional modules or customization may be available. Some large laboratories have developed customized databases that include other cell line provenance information and allow printing of cryovial labels (see Section 15.2.1). Laboratories can also use an electronic laboratory notebook (ELN) (e.g. LabArchives; https://www.labarchives.com) or management system (e.g. Quartzy) to record some information and look for ways to supply missing functionality, e.g. LabArchives offers a freezer box widget.

15.8 TRANSPORTING CELLS

Safety concerns must be addressed whenever cells are transported between laboratories, whether transport is local (e.g. flasks that are hand-carried to another laboratory in the same building) or distant (e.g. frozen cryovials that are shipped to an international laboratory). Generally speaking, if you are carrying cells to a nearby laboratory you should use a sealed, unbreakable secondary container to prevent accidental spillage (WHO 2004). The secondary container should be regularly decontaminated; choose a container that is autoclavable or resistant to chemical disinfectants. If dry ice or liquid nitrogen are used, the container should include a vent for CO_2 or nitrogen gas. Never travel with dry ice or liquid nitrogen in a confined space such as a lift because of the risk of asphyxia (see

TABLE 15.5. Freezer record.

Position Freezer no.		Rack/canister no.			Box/drawer no.			
Cell line name	Freeze date				Frozen by			
No. of cryovials frozen	No. of cells/cryovial				in .... ml			
Growth medium	Serum		Conc.		Freeze medium			
Method of cooling					Cooling rate			
Thawing record								
Thaw date	No. of cryovials	Seeding			Viability			Notes
		Conc.	Vol.	Medium	Dye exclusion	% attached by 24 h	Cloning efficiency	

Section 6.2.3). Additional requirements may apply, depending on the level of risk; if in doubt, always discuss with your supervisor or local safety officers.

Shipping cells to a distant laboratory is subject to strict national and international regulations (Simione and Sharp 2017). Generally speaking, shipping requires a triple packaging system (WHO 2004). The choice of packaging will depend on the hazards associated with the shipment, which should be clearly indicated by labeling the package with the appropriate hazard classification. Shipping should only be performed by personnel who are trained and competent to do so. More information on shipping cells is provided in the supplementary material online and can be used as a starting point when planning such a shipment (see Supp. S15.1).

Suppliers

Supplier	URL
Brady	www.brady.co.uk
Brooks Life Sciences (FluidX)	www.brookslifesciences.com
Chart MVE	www.chartindustries.com/life-sciences
Corning	www.corning.com/worldwide/en/products/life-sciences/products/cell-culture.html
Custom Biogenic Systems	www.custombiogenics.com
Dataworks Development, Inc.	www.freezerworks.com
GA International (LabTag)	www.labtag.com
GE Healthcare Life Sciences	www.gelifesciences.com/en/it
Greiner Bio-One	www.gbo.com/en_int.html
ItemTracker Software Ltd.	www.itemtracker.com
Micronic	www.micronic.com
Planer PLC	www.planer.com
Pro-curo Software Inc.	www.pro-curo.com
Sigma-Aldrich (Merck)	www.sigmaaldrich.com/life-science/cell-culture.html
Taylor-Wharton	www.twcryo.com
Tempshield	www.tempshield.com
ThermoFisher Scientific	www.thermofisher.com/us/en/home/life-science/cell-culture.html
Worthington Industries	worthingtonindustries.com/products/life-sciences-cryogenic-equipment

Supp. S15.1. Shipping Cells.

REFERENCES

Bairoch, A. (2018). The Cellosaurus, a cell-line knowledge resource. *J. Biomol. Tech.* 29 (2): 25–38. https://doi.org/10.7171/jbt.18-2902-002.

Beattie, G.M., Crowe, J.H., Lopez, A.D. et al. (1997). Trehalose: a cryoprotectant that enhances recovery and preserves function of human pancreatic islets after long-term storage. *Diabetes* 46 (3): 519–523.

Bielanski, A. (2014). Biosafety in embryos and semen cryopreservation, storage, management and transport. *Adv. Exp. Med. Biol.* 753: 429–465. https://doi.org/10.1007/978-1-4939-0820-2_17.

Buchanan, S.S., Pyatt, D.W., and Carpenter, J.F. (2010). Preservation of differentiation and clonogenic potential of human hematopoietic stem and progenitor cells during lyophilization and ambient storage. *PLoS One* 5 (9) https://doi.org/10.1371/journal.pone.0012518.

Claassen, D.A., Desler, M.M., and Rizzino, A. (2009). ROCK inhibition enhances the recovery and growth of cryopreserved human embryonic stem cells and human induced pluripotent stem cells. *Mol. Reprod. Dev.* 76 (8): 722–732. https://doi.org/10.1002/mrd.21021.

Coecke, S., Balls, M., Bowe, G. et al. (2005). Guidance on good cell culture practice: a report of the second ECVAM task force on good cell culture practice. *Altern. Lab. Anim.* 33 (3): 261–287.

Collins, S.J., Ruscetti, F.W., Gallagher, R.E. et al. (1979). Normal functional characteristics of cultured human promyelocytic leukemia cells (HL-60) after induction of differentiation by dimethylsulfoxide. *J. Exp. Med.* 149 (4): 969–974.

Coriell, L.L., Greene, A.E., and Silver, R.K. (1964). Historical development of cell and tissue culture freezing. *Cryobiology* 51: 72–79.

Crowe, J.H. and Crowe, L.M. (2000). Preservation of mammalian cells – learning nature's tricks. *Nat. Biotechnol.* 18 (2): 145–146. https://doi.org/10.1038/72580.

Dirks, W.G. and Drexler, H.G. (2013). STR DNA typing of human cell lines: detection of intra- and interspecies cross-contamination. *Methods Mol. Biol.* 946: 27–38. https://doi.org/10.1007/978-1-62703-128-8_3.

Drexler, H.G., Dirks, W.G., MacLeod, R.A. et al. (2017). False and mycoplasma-contaminated leukemia-lymphoma cell lines: time for a reappraisal. *Int. J. Cancer* 140 (5): 1209–1214. https://doi.org/10.1002/ijc.30530.

Eroglu, A., Russo, M.J., Bieganski, R. et al. (2000). Intracellular trehalose improves the survival of cryopreserved mammalian cells. *Nat. Biotechnol.* 18 (2): 163–167. https://doi.org/10.1038/72608.

Fahy, G.M. and Wowk, B. (2015). Principles of cryopreservation by vitrification. *Methods Mol. Biol.* 1257: 21–82. https://doi.org/10.1007/978-1-4939-2193-5_2.

Gonzalez Hernandez, Y. and Fischer, R.W. (2007). Serum-free culturing of mammalian cells – adaptation to and cryopreservation in fully defined media. *ALTEX* 24 (2): 110–116.

Greene, A.E., Athreya, B., Lehr, H.B. et al. (1967). Viability of cell cultures following extended preservation in liquid nitrogen. *Proc. Soc. Exp. Biol. Med.* 124 (4): 1302–1307.

Grout, B.W. and Morris, G.J. (2009). Contaminated liquid nitrogen vapour as a risk factor in pathogen transfer. *Theriogenology* 71

(7): 1079–1082. https://doi.org/10.1016/j.theriogenology.2008.12.011.

Hanna, J. and Hubel, A. (2009). Preservation of stem cells. *Organogenesis* 5 (3): 134–137.

Hawkins, A.E., Zuckerman, M.A., Briggs, M. et al. (1996). Hepatitis B nucleotide sequence analysis: linking an outbreak of acute hepatitis B to contamination of a cryopreservation tank. *J. Virol. Methods* 60 (1): 81–88.

Hay, R.J. (1988). The seed stock concept and quality control for cell lines. *Anal. Biochem.* 171 (2): 225–237.

Hubel, A. (2017). *Preservation of Cells: A Practical Manual*. Hoboken, NJ: Wiley.

Hubel, A., Spindler, R., and Skubitz, A.P. (2014). Storage of human biospecimens: selection of the optimal storage temperature. *Biopreserv. Biobanking* 12 (3): 165–175. https://doi.org/10.1089/bio.2013.0084.

Hunt, C.J. (2019). Technical considerations in the freezing, low-temperature storage and thawing of stem cells for cellular therapies. *Transfus. Med. Hemother.* 46 (3): 134–150. https://doi.org/10.1159/000497289.

ICLAC (2014). Choose wisely: obtaining cell lines from reliable sources. *GIT Lab. J.* 3–4: 22–24.

Jiang, G., Bi, K., Tang, T. et al. (2006). Down-regulation of TRRAP-dependent hTERT and TRRAP-independent CAD activation by Myc/Max contributes to the differentiation of HL60 cells after exposure to DMSO. *Int. Immunopharmacol.* 6 (7): 1204–1213. https://doi.org/10.1016/j.intimp.2006.02.014.

Kaindl, J., Meiser, I., Majer, J. et al. (2018). Zooming in on cryopreservation of hi PSCs and neural derivatives: a dual-center study using adherent vitrification. *Stem Cells Transl. Med.* 8 (3): 247–259. https://doi.org/10.1002/sctm.18-0121.

Kim, J.H., Kurtz, A., Yuan, B.Z. et al. (2017). Report of the International Stem Cell Banking Initiative workshop activity: current hurdles and progress in seed-stock banking of human pluripotent stem cells. *Stem Cells Transl. Med.* 6 (11): 1956–1962. https://doi.org/10.1002/sctm.17-0144.

Ledley, R.S. (1964). Animal cell strains. The Cell Culture Collection Committee has assembled and certified 23 strains of animal cells. *Science* 146 (3641): 241–243.

Leibo, S.P. and Mazur, P. (1971). The role of cooling rates in low-temperature preservation. *Cryobiology* 8 (5): 447–452.

Li, R., Yu, G., Azarin, S.M. et al. (2018). Freezing responses in DMSO-based cryopreservation of human iPS cells: aggregates versus single cells. *Tissue Eng. Part C Methods* 24 (5): 289–299. https://doi.org/10.1089/ten.TEC.2017.0531.

Liu, W. and Chen, G. (2014). Cryopreservation of human pluripotent stem cells in defined medium. *Curr. Protoc. Stem Cell Biol.* 31 1C 17.1–13. https://doi.org/10.1002/9780470151808.sc01c17s31.

Lovelock, J.E. and Bishop, M.W. (1959). Prevention of freezing damage to living cells by dimethyl sulphoxide. *Nature* 183 (4672): 1394–1395.

Lowe, D. (2006). In the pipeline. How not to do it: liquid nitrogen tanks. https://blogs.sciencemag.org/pipeline/archives/2006/03/08/how_not_to_do_it_liquid_nitrogen_tanks (accessed 3 January 2019).

Massie, I., Selden, C., Hodgson, H. et al. (2014). GMP cryopreservation of large volumes of cells for regenerative medicine: active control of the freezing process. *Tissue Eng. Part C Methods* 20 (9): 693–702. https://doi.org/10.1089/ten.TEC.2013.0571.

Matsumura, K., Bae, J.Y., and Hyon, S.H. (2010). Polyampholytes as cryoprotective agents for mammalian cell cryopreservation. *Cell Transplant.* 19 (6): 691–699. https://doi.org/10.3727/096368910X508780.

Mazur, P. (1984). Freezing of living cells: mechanisms and implications. *Am. J. Physiol.* 247 (3 Pt 1): C125–C142. https://doi.org/10.1152/ajpcell.1984.247.3.C125.

Merten, O.W., Petres, S., and Couve, E. (1995). A simple serum-free freezing medium for serum-free cultured cells. *Biologicals* 23 (2): 185–189. https://doi.org/10.1006/biol.1995.0030.

Monroy, B., Honiger, J., Darquy, S. et al. (1997). Use of polyethyleneglycol for porcine islet cryopreservation. *Cell Transplant.* 6 (6): 613–621.

Morris, G.J. (2005). The origin, ultrastructure, and microbiology of the sediment accumulating in liquid nitrogen storage vessels. *Cryobiology* 50 (3): 231–238. https://doi.org/10.1016/j.cryobiol.2005.01.005.

Morris, C.B. (2007). Cryopreservation of animal and human cell lines. *Methods Mol. Biol.* 368: 227–236. https://doi.org/10.1007/978-1-59745-362-2_16.

Morris, G.J. and Acton, E. (2013). Controlled ice nucleation in cryopreservation – a review. *Cryobiology* 66 (2): 85–92. https://doi.org/10.1016/j.cryobiol.2012.11.007.

Morris, C., de Wreede, L., Scholten, M. et al. (2014). Should the standard dimethyl sulfoxide concentration be reduced? Results of a European Group for Blood and Marrow Transplantation prospective noninterventional study on usage and side effects of dimethyl sulfoxide. *Transfusion* 54 (10): 2514–2522. https://doi.org/10.1111/trf.12759.

Murthy, S.S. (1998). Some insight into the physical basis of the cryoprotective action of dimethyl sulfoxide and ethylene glycol. *Cryobiology* 36 (2): 84–96. https://doi.org/10.1006/cryo.1997.2064.

Nichols, W.W., Murphy, D.G., Cristofalo, V.J. et al. (1977). Characterization of a new human diploid cell strain, IMR-90. *Science* 196 (4285): 60–63.

Ohno, T., Saijo-Kurita, K., Miyamoto-Eimori, N. et al. (1991). A simple method for in situ freezing of anchorage-dependent cells including rat liver parenchymal cells. *Cytotechnology* 5 (3): 273–277.

Parker, K.A. (2011). Storage of cell lines. *Methods Mol. Biol.* 731: 27–34. https://doi.org/10.1007/978-1-61779-080-5_3.

Pasch, J., Schiefer, A., Heschel, I. et al. (2000). Variation of the HES concentration for the cryopreservation of keratinocytes in suspensions and in monolayers. *Cryobiology* 41 (2): 89–96. https://doi.org/10.1006/cryo.2000.2270.

Pegg, D.E. (2007). Principles of cryopreservation. *Methods Mol. Biol.* 368: 39–57. https://doi.org/10.1007/978-1-59745-362-2_3.

Polge, C., Smith, A.U., and Parkes, A.S. (1949). Revival of spermatozoa after vitrification and dehydration at low temperatures. *Nature* 164 (4172): 666.

Pomeroy, K.O., Harris, S., Conaghan, J. et al. (2010). Storage of cryopreserved reproductive tissues: evidence that cross-contamination of infectious agents is a negligible risk. *Fertil. Steril.* 94 (4): 1181–1188. https://doi.org/10.1016/j.fertnstert.2009.04.031.

Reid, Y.A. (2017). Best practices for naming, receiving, and managing cells in culture. *In Vitro Cell Dev. Biol. Anim.* 53 (9): 761–774. https://doi.org/10.1007/s11626-017-0199-1.

Richards, M., Fong, C.Y., Tan, S. et al. (2004). An efficient and safe xeno-free cryopreservation method for the storage of human embryonic stem cells. *Stem Cells* 22 (5): 779–789. https://doi .org/10.1634/stemcells.22-5-779.

Rowley, S.D. and Byrne, D.V. (1992). Low-temperature storage of bone marrow in nitrogen vapor-phase refrigerators: decreased temperature gradients with an aluminum racking system. *Transfusion* 32 (8): 750–754.

Ryan, J. (2004). General guide for cryogenically storing animal cell cultures. https://www.corning.com/media/worldwide/cls/ documents/t_cryoanimalcc.pdf (accessed 19 August 2008).

Simione, F. and Sharp, T. (2017). Best practices for storing and shipping cryopreserved cells. *In Vitro Cell Dev. Biol. Anim.* 53 (10): 888–895. https://doi.org/10.1007/s11626-017-0214-6.

SLAS (2019). Microplate standards. https://slas.org/resources/ information/industry-standards (accessed 15 January 2020).

Stacey, G. (2017). Stem cell banking: a global view. *Methods Mol. Biol.* 1590: 3–10. https://doi.org/10.1007/978-1-4939-6921-0_1.

Stacey, G.N. and Dowall, S. (2007). Cryopreservation of primary animal cell cultures. *Methods Mol. Biol.* 368: 271–281. https:// doi.org/10.1007/978-1-59745-362-2_19.

Suzuki, T., Saha, S., Sumantri, C. et al. (1995). The influence of polyvinylpyrrolidone on freezing of bovine IVF blastocysts following biopsy. *Cryobiology* 32 (6): 505–510. https://doi.org/10 .1006/cryo.1995.1051.

Tedder, R.S., Zuckerman, M.A., Goldstone, A.H. et al. (1995). Hepatitis B transmission from contaminated cryopreservation tank. *Lancet* 346 (8968): 137–140.

Terasima, T. and Yasukawa, M. (1977). Dependence of freeze-thaw damage on growth phase and cell cycle of cultured mammalian cells. *Cryobiology* 14 (3): 379–381.

Ure, J.M., Fiering, S., and Smith, A.G. (1992). A rapid and efficient method for freezing and recovering clones of embryonic stem cells. *Trends Genet.* 8 (1): 6.

Walker, G. (1983). Stirling cryocoolers. In: *Cryocoolers* (ed. G. Walker), 95–184. Boston, MA: Springer.

Wewetzer, K. and Dilmaghani, K. (2001). Exposure to dimethyl sulfoxide at 37 degrees C prior to freezing significantly improves the recovery of cryopreserved hybridoma cells. *Cryobiology* 43 (3): 288–292. https://doi.org/10.1006/cryo.2001.2352.

WHO (2004). Laboratory biosafety manual. http://www.who.int/ csr/resources/publications/biosafety/who_cds_csr_lyo_2004_ 11/en/ (accessed 9 May 2018).

WIPO (2019). Depository institutions having acquired the status of International Depository Authority under the Budapest Treaty. https://www.wipo.int/export/sites/www/treaties/en/ registration/budapest/pdf/ida.pdf (accessed 27 October 2019).

Young, D.A., Gavrilov, S., Pennington, C.J. et al. (2004). Expression of metalloproteinases and inhibitors in the differentiation of P19CL6 cells into cardiac myocytes. *Biochem. Biophys. Res. Commun.* 322 (3): 759–765. https://doi.org/10.1016/j.bbrc.2004.07 .178.

Yu, M., Selvaraj, S.K., Liang-Chu, M.M. et al. (2015). A resource for cell line authentication, annotation and quality control. *Nature* 520 (7547): 307–311. https://doi.org/10.1038/nature14397.

PART V

Validation and Characterization

After reading the following chapters in this part of the book, you will be able to:

(16) *Microbial Contamination:*

  (a) List the major sources of contamination and discuss the overall approach to management of contamination (including persistent contamination).

  (b) Detect microbial contamination with bacteria, yeasts, or fungi by observation and comment on further testing that may be employed for problematic cases.

  (c) Discuss how to select a suitable method for mycoplasma detection and perform testing based on polymerase chain reaction (PCR) detection of 16S RNA or staining with Hoechst 33258.

  (d) Discuss how to detect viral contamination.

  (e) Dispose of contaminated cultures safely and discuss when eradication of microbial contamination may be attempted.

(17) *Cell Line Misidentification and Authentication:*

  (a) Define "cross-contamination" and "misidentification" and discuss why usage of misidentified cell lines has increased over time.

  (b) Define cell line "authentication" and provide examples of suitable techniques.

  (c) Interpret a short tandem repeat (STR) profile from a human cell line, with reference to the cell line authentication using STR (CLASTR) search tool from Cellosaurus.

  (d) Perform species detection using cytochrome c oxidase I (CO1) barcoding.

  (e) Provide examples of samples that are challenging to authenticate and discuss how to proceed if authentication testing does not give a clear result.

(18) *Cell Line Characterization:*

  (a) Discuss why laboratories should perform their own cell line characterization and list key requirements to consider when setting priorities for characterization.

  (b) Provide examples of cell line databases that act as resources for genotypic and phenotypic data, particularly those that are known to use authenticated cell lines.

  (c) Use an inverted microscope to observe the morphology of a culture and take representative photographs.

  (d) Fix and stain cultures using Giemsa or crystal violet staining solutions (to assess morphology), or using indirect immunofluorescence (to detect specific markers).

  (e) Discuss requirements for specialized microscopy if you intend to perform live-cell imaging or high-resolution imaging.

(19) *Quantitation and Growth Kinetics:*

  (a) Comment on the differences between direct and indirect counting methods.

  (b) Count cells manually using a hemocytometer and assess their viability.

Freshney's Culture of Animal Cells: A Manual of Basic Technique and Specialized Applications, Eighth Edition. Amanda Capes-Davis and R. Ian Freshney.
© 2021 John Wiley & Sons Ltd. Published 2021 by John Wiley & Sons Ltd.
Companion website: www.wiley.com/go/freshney/cellculture8

(c) Explain why automated devices should be considered for cell counting and comment on the major principles of operation that are used in such devices.

(d) Plot a growth curve and describe the three phases of growth that are evident and the parameters that can be derived from the growth curve.

(e) Explain how to perform a clonogenic assay and the unique information that comes from analysis of clonal populations (colonies).

CHAPTER 16

Microbial Contamination

In the early days of tissue culture, microbial contamination was so frequent that it was virtually impossible to store large batches of cells that were completely free of contamination (Barile et al. 1962; McGarrity and Coriell 1971). Approximately 50% of cell lines in use at that time were contaminated (Barile et al. 1962). Gerard McGarrity, who was a pioneer in the detection and control of cell culture contamination, performed a classic experiment to determine how cultures became contaminated, using a cell line that carried *Acholeplasma laidlawii* (see Section 16.5) (McGarrity 1976). During the course of everyday handling, *A. laidlawii* appeared throughout the local environment including the lip of the culture flask (20/20 samples) and the work surface (8/12 samples). As McGarrity noted when reporting this work, the droplets that are generated by cell handling procedures are relatively large and will sediment within the immediate area. Thus, although it is important to handle cultures in a biological safety cabinet (BSC) to eliminate airborne routes of contamination, microorganisms can spread from one culture to another through droplet spread or other forms of direct contact (McGarrity and Coriell 1971; McGarrity 1976). This chapter looks broadly at how contamination arises and how to manage it, before focusing on specific microorganisms such as mycoplasmas that, unfortunately, continue to be widespread in many tissue culture laboratories.

16.1 SOURCES OF CONTAMINATION

In this book, the term "contamination" generally refers to the presence of a foreign organism, which may be a microorganism or a different cell line; the latter is usually referred to as cross-contamination and is discussed separately (see Section 17.1). Chemical contamination may also occur due to leaching of chemicals from disposable plasticware, inadequate cleaning of glassware, or other causes (see Sections 7.5.2, 7.5.3, 11.3, 31.3) (Nims and Price 2017). However, this usually leads to a failure of the culture to thrive, followed by its disposal. By contrast, microbial contamination can persist in a culture for lengthy periods of time and may even contribute to its survival; some mycoplasma species can increase cloning efficiency and promote *in vitro* transformation through suppression of p53 activation and activation of NF-κB (see Section 20.6) (Tsai et al. 1995; Logunov et al. 2008).

Microbial contamination is an ongoing risk for any form of tissue culture that can be managed, but cannot be eliminated entirely. Aseptic technique is required to provide a barrier of sterility between the culture in its flask or dish and the external environment (see Section 12.1.2). If the skill and level of care of the operator are high and the atmosphere is clean, free of dust, and undisturbed, contamination as a result of manipulation will be rare. If the environment deteriorates (e.g. as a result of construction work or a seasonal increase in humidity), or if the operator's technique declines (through the omission of one or more apparently unnecessary precautions), the probability of contamination increases. If both happen simultaneously or sequentially, the results can be catastrophic (see Figure 12.1). These multiple sources often make it difficult to understand why a particular episode of contamination has occurred, leading to extensive problem solving to identify the various contributing factors (see Section 31.1).

Freshney's Culture of Animal Cells: A Manual of Basic Technique and Specialized Applications, Eighth Edition. Amanda Capes-Davis and R. Ian Freshney.
© 2021 John Wiley & Sons Ltd. Published 2021 by John Wiley & Sons Ltd.
Companion website: www.wiley.com/go/freshney/cellculture8

TABLE 16.1. Causes and prevention of contamination.

Route or cause	Action to prevent contamination[a]
Operator problems	
Operator hair, hands, breath, clothing	
Dust from skin, hair, or clothing dropped or blown into the culture	Wash hands thoroughly when entering the room. Wear suitable PPE (see Table 12.1). Rinse gloves frequently with 70% alcohol while handling cultures. Tie back long hair or wear a cap.
Aerosols from talking, coughing, sneezing, etc.	Keep talking to a minimum, and face away from the work surface when you talk. Avoid working with a cold or throat infection, or wear a mask.
Transmission of microbial contaminants from other tasks	Wear a lab coat or gown that is different from those in the general lab area or animal house.
Work surface	
Non-sterile surfaces and equipment	Clear work area of items not in immediate use. Swab all items with 70% alcohol before bringing to the sterile handling area.
Dust and spillage	Allow the air in the BSC to "purge" before starting work and between cell lines. Swab the surface with 70% alcohol before, during, and after work. Mop up any spillages immediately.
Manipulations, pipetting, dispensing, etc.	
Droplets, dust, skin particulates, etc. settling on the culture or bottle	Do not work over (vertical laminar flow and open bench) or behind and over (horizontal laminar flow) an open bottle or dish. Replace lids promptly once dishes or plates have been handled.
Spillage on necks and outside of bottles and on work surface	Do not pour liquids. Dispense or transfer by pipette or other transfer device. If pouring is unavoidable: (i) do so in one smooth movement; (ii) discard the bottle that you pour from; and (iii) wipe up any spillage.
Touching or holding pipettes too low down, touching necks of bottles, inside screw caps	Hold pipettes above graduations.
Splashback from waste beaker or pipetting	Discard waste into a beaker with a funnel or, preferably, by drawing off the waste into a reservoir by means of a vacuum pump. Avoid generating droplets during pipetting or other handling procedures.
Environmental problems	
Poor airflow or air quality	
Drafts, eddies, turbulence	Reduce traffic and extraneous activity.
High airborne particulate level, dust	Filter incoming air using a HEPA filter. Use air handling to generate positive pressure relative to the corridor. Clean the room regularly.
Mold on surfaces (including bench tops, walls, and floors)	
High temperature or humidity	Use air handling systems to control air temperature and humidity.
Dust, spillage on surfaces	Wipe down all surfaces, including walls, with disinfectant and mop the floors on a regular basis.
High spore count within room	Request servicing of air handling system. Look for a focus within the room. Consider fumigating the room (see Section 6.3.5).
Mites, insects, and other infestations in wooden furniture, on benches, in incubators, and on mice, etc. taken from the animal house	
Entry of mites, etc., into sterile packages	Avoid wooden furniture; use plastic laminate or stainless-steel bench tops. If wooden furniture is used, seal it with polyurethane varnish or wax polish and wash it regularly with disinfectant. Keep animals out of the tissue culture lab. Seal all sterile packs.

TABLE 16.1. (*continued*)

Route or cause	Action to prevent contamination[a]
Equipment problems	
BSCs	
Perforated filter	Check filters regularly for holes and leaks. Refer any alarms for a service engineer to address.
Change of filter needed	Check the pressure drop across the filter. Service and repair regularly.
Spillages, particularly in crevices or below a work surface	Clear around and below the work surface regularly. Wipe all surfaces with 70% alcohol and let alcohol run into crevices. Consider using a UV lamp to support chemical disinfection.
CO$_2$ humidified incubators	
Growth of molds and bacteria on walls and shelves in a humid atmosphere	Clean incubators regularly (see Protocol P12.4). Wipe shelves regularly with 70% alcohol or autoclave and replace. Remove any obvious spillages or condensation.
Growth of molds and bacteria in the water pan	Replace water on a regular basis and clean the water pan. Consider using a water treatment agent to retard fungal growth.
Growth of molds and bacteria in Petri dishes or multiwell plates	Enclose open dishes in plastic boxes with close-fitting lids (but do not seal the lids).
Dry incubators	
Growth of molds and bacteria on spillages	Wipe up any spillage using 70% alcohol. Clean incubators regularly.
Autoclave, oven	
Faulty sterilization equipment	Monitor performance using temperature monitoring and sterility indicators (see Section 11.2.7). Perform sterility testing of reagents (see Supp. S11.6). Perform regular service and repair.
Faulty sterilization procedures, e.g. due to an overfilled oven or sealed bottles, preventing the ingress of steam	Use sterility indicators throughout the load. In the autoclave, keep caps slack on empty bottles. Stack oven and autoclave correctly (see Protocols P11.1, P11.2). Refresh user training.
Other equipment	
Dust on cylinders, pumps, etc.	Wipe with 70% alcohol before bringing into the tissue culture lab.
Fungal spores in water baths or areas of condensation	Clean water baths on a regular basis. Remove condensation from equipment and surfaces.
Reagent problems	
Media and other solutions	
Non-sterile reagents and media	Filter or autoclave solutions before using them.
Poor commercial supplier	Ask about quality control testing. Test solutions; change suppliers.
Dirty storage conditions	Clean up storage areas and disinfect regularly.
Glassware and screw caps	
Dust and spores from storage	Shroud caps with foil for storage. Wipe bottles with 70% alcohol before taking them into the BSC. Clean out stores regularly. Replace stocks from the back of the shelf.
Plasticware	
Non-sterile plasticware	Purchase individually packaged items. Open within the BSC. Rewrap sealed items (e.g. unused flasks from a sleeve) and store in a dust-free area near the BSC.
Instruments	
Contact with a non-sterile surface or some other material	Resterilize instruments, e.g. use 70% alcohol, flame, and cool off the instruments. Do not grasp any part of an instrument or pipette that will pass into a culture vessel.

(*continued*)

TABLE 16.1. (*continued*)

Route or cause	Action to prevent contamination[a]
Media bottles and other glassware in use	
Dust and spores from incubator or refrigerator	Clean equipment regularly. Swab bottles before placing in BSC. Use screw caps instead of stoppers. Box plates and dishes (see Section 12.5.2).
Media under the cap and spreading to the outside of the bottle	Discard all bottles that show spillage on the outside of the neck. Do not pour.
Cell line problems	
Tissue samples	
Infected at source or during dissection	Keep animals out of the tissue culture lab. Incorporate antibiotics into the Collection Medium (see Appendix B). Dip potentially infected large-tissue samples in 70% alcohol for 30 s. Dip intestinal samples into sodium hypochlorite solution (see Supp. S24.5).
Other cell lines	
Contaminated on arrival	Quarantine incoming cell lines (see Section 16.2). Swab down the bench or BSC after use and do not use it until the next morning. Wait until cultures have been tested before bringing into general area.
Contaminated from other cell lines in lab	Maintain cells in antibiotic-free media. Check for microbial contamination by regular inspection and by testing for mycoplasma (see Sections 16.4, 16.5). Consider testing for viral contaminants, depending on risk assessment (see Section 16.6). Discard contaminated cultures (see Protocol P16.1).

[a]No one-to-one relationship between prevention and cause is intended throughout this table; preventative measures are interactive and may relate to more than one cause.

Sources of microbial contamination can be broadly grouped into (i) operator problems, e.g. lapses in aseptic technique; (ii) environmental problems, e.g. poor quality room air or inadequate cleaning of the facility; (iii) equipment problems, e.g. BSC faults or incubator contamination; (iv) reagent problems, e.g. contamination of a batch of medium; and (v) cell line problems, e.g. a contaminated culture that is brought into the laboratory (see Figure 16.1). Specific actions that can be taken to address each problem are listed in Table 16.1.

16.1.1 Operator Problems

Maintaining asepsis is one of the most difficult challenges that face the newcomer to tissue culture. Training in and adherence to good aseptic technique make up the single most effective way to manage the risk of contamination in a tissue culture laboratory (see Sections 12.3, 12.4). Awkwardness during early training can be overcome by experience, but even experienced workers can suffer from contamination, particularly if familiarity results in lapses in aseptic technique. Lapses that increase the probability of contamination (including both microbial contamination and cross-contamination) include:

(1) ***Inadequate hygiene and personal protective equipment (PPE).*** Most laboratories require personnel to wear a lab coat or gown, disposable gloves, and covered shoes when handling cultures or other biological samples. Failure to comply with basic hygiene requirements such as hand washing, or to wear suitable PPE, means that cultures are more likely to be exposed to external contaminants and personnel are exposed to biohazard risk (see Sections 6.3, 12.2.4).

(2) ***Incorrect use of the BSC.*** Common errors include working too near the front edge or failing to clear the BSC of non-essential equipment. If the work surface becomes overcrowded with bottles and equipment (see Figure 12.4b), the laminar airflow is disrupted. This in turn leads to the entry of non-sterile air into the BSC and the release of potentially biohazardous materials into the room. The risk of collision between sterile pipettes and non-sterile surfaces of bottles, etc. will also increase. You should bring into the BSC only those items that are directly involved in the current operation. Equipment operation and aseptic technique when using laminar airflow are discussed elsewhere (see Sections 12.2.2, 12.4.1).

(3) ***Working with two cell lines simultaneously.*** While this may be tempting to save time, it greatly increases the probability of microbial or cross-contamination.

(4) ***Sharing of reagents.*** Using the same bottle of medium or other reagent for two different cell lines increases the risk that contaminants will pass from one to the other

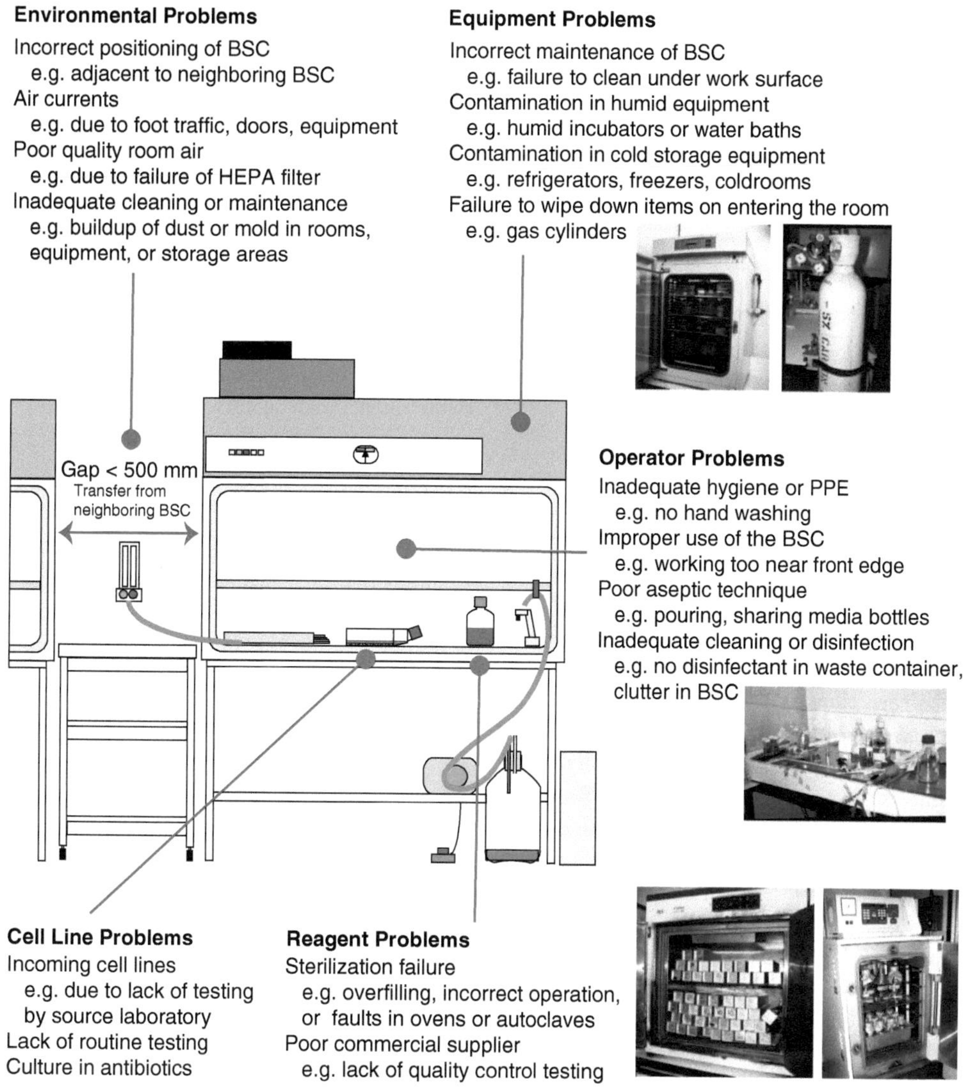

Environmental Problems

Incorrect positioning of BSC
  e.g. adjacent to neighboring BSC
Air currents
  e.g. due to foot traffic, doors, equipment
Poor quality room air
  e.g. due to failure of HEPA filter
Inadequate cleaning or maintenance
  e.g. buildup of dust or mold in rooms,
  equipment, or storage areas

Equipment Problems

Incorrect maintenance of BSC
  e.g. failure to clean under work surface
Contamination in humid equipment
  e.g. humid incubators or water baths
Contamination in cold storage equipment
  e.g. refrigerators, freezers, coldrooms
Failure to wipe down items on entering the room
  e.g. gas cylinders

Gap < 500 mm
Transfer from
neighboring BSC

Operator Problems

Inadequate hygiene or PPE
  e.g. no hand washing
Improper use of the BSC
  e.g. working too near front edge
Poor aseptic technique
  e.g. pouring, sharing media bottles
Inadequate cleaning or disinfection
  e.g. no disinfectant in waste container,
  clutter in BSC

Cell Line Problems

Incoming cell lines
  e.g. due to lack of testing
  by source laboratory
Lack of routine testing
Culture in antibiotics

Reagent Problems

Sterilization failure
  e.g. overfilling, incorrect operation,
  or faults in ovens or autoclaves
Poor commercial supplier
  e.g. lack of quality control testing

Fig. 16.1. Sources of contamination. Problems can arise due to operator technique, the environment within the facility, equipment, reagents, or other cell lines. Examples are given here showing how each source can give rise to contamination; for a more comprehensive list see Table 16.1. *Source*: R. Ian Freshney and Amanda Capes-Davis.

through airborne transmission or direct contact, e.g. by returning a used pipette to a medium bottle. This risk can be managed by dedicating a single bottle to a specific cell line or by aliquoting smaller volumes into separate bottles prior to use.

(5) *Pouring and pipetting problems.* Traditionally, mouth pipetting was an important cause of microbial contamination; tissue culture personnel carry mycoplasma species in their throats, which can be detected after speaking or sneezing (Kundsin and Praznik 1967; McGarrity 1976). Mouth pipetting can be eliminated by using a pipette controller (see Section 5.1.3). However, contamination can still occur through pouring liquids or using contaminated pipettes (see Sections 12.3.5, 12.3.6).

16.1.2 Environmental Problems

The environment in which tissue culture is carried out must be as clean as possible and free from disturbance and through traffic. Conducting tissue culture in the regular laboratory area should be avoided; a BSC will not give sufficient protection from the busy environment of the average laboratory. A clean, traffic-free area should be set aside for tissue culture, preferably as an isolated room or suite of rooms (see Sections 4.1.1, 4.2.1). Air currents from doors, refrigerators, centrifuges, and the movement of operators all increase the risk of contamination and should be minimized. Ideally, tissue culture areas should be supplied by air that has passed through a high-efficiency particulate air (HEPA) filter and is at positive pressure compared to the outer corridor, resulting in less entry of dust and fewer airborne contaminants (see Section 4.1.4). However, cleaning and maintenance are equally important and can be achieved in any tissue culture space, regardless of design constraints. A regular cleaning and maintenance program should be carried out that extends to all surfaces and equipment, including air handling units. Any items that harbor water can act as reservoirs for bacterial or fungal growth and should be cleaned as a high priority. Other areas that may become contaminated include water supply taps (faucets), hand lotion dispensers, and foot mats (McGarrity and Coriell 1971).

16.1.3 Equipment Problems

BSC operation and maintenance. In addition to operating a BSC correctly, the equipment must be maintained regularly by a competent engineer who can check the HEPA filters and perform certification testing for biohazard containment (see Sections 5.1.1, 6.3.4). Any spills should be promptly cleaned up and the area beneath the work surface (where spills accumulate) should be regularly checked and cleaned (see Section 12.2.2). Always allow the cabinet to "purge" for at least four minutes and wipe it down using a chemical disinfectant such as 70% alcohol (ethanol or isopropanol) before and after each cell line (see Sections 12.2.2, 12.3.1). Some laboratories use ultraviolet (UV) light as an adjunct to the chemical disinfectant, but its effectiveness will vary and should not be relied on for disinfection (see Section 12.2.2). Chemical disinfectants and laminar airflow are the key requirements to minimize microbial contamination when using a BSC.

Humid incubators. Frequent contamination of open plate or dish cultures can often be traced back to the incubator; the same is true for contamination events that affect multiple people in a shared facility. Historically, incubator contamination was managed by keeping flasks in a warm-room or a dry incubator, which did not require high levels of humidity (see Sections 4.2.2, 5.3.1). Cultures were grown in sealed flasks using low-CO_2 medium (e.g. based on Hanks's salts) or were gassed with CO_2 and then sealed (see Sections 9.3, 12.5.3). Currently, most laboratories use humid incubators and manage the risk of contamination by using flasks with gas-permeable caps and purchasing incubators with contamination control features (see Section 5.3.1). All incubators must be regularly cleaned to manage the risk of contamination, even if such features are present; a protocol for cleaning humid CO_2 incubators is available in another chapter (see Protocol P12.4).

Fungal growth is particularly likely to occur in the water pan and in areas of condensation on incubator surfaces (see Figure 12.10). Any areas of condensation should be wiped down regularly and the water pan should be cleaned on a weekly basis. Water in the pan should be replaced with sterile distilled water at pH 7–9 or as instructed by the manufacturer. Some laboratories add water treatment agents to retard fungal growth; examples include copper sulfate (1.0 g/l) and a wide range of quaternary ammonium disinfectants (2%) (Bates et al. 2014; Geraghty et al. 2014). Commercially available products include Roccal-D (now discontinued), AquaClean (WAK-Chemie Medical GmbH), Aquaguard-1 (Biological Industries, PromoCell), and SigmaClean® (Sigma-Aldrich). However, copper sulfate is harmful to the aquatic environment, while some other treatment agents release volatile organic chemicals, which may be harmful to cultures (Bates et al. 2014). As with antibiotic use (see Section 9.6.2), the most sensible approach is to avoid the routine use of such agents and reserve them for limited "treatment" periods. Remember, such agents will only protect the water pan; there is no substitute for regular cleaning!

Other equipment. Cold storage equipment and water baths can be major sources of contamination if they are not cleaned regularly. Coldrooms are particularly well recognized as sources of microbial contamination, due to contamination of bottle necks by molds (Fogh et al. 1971). Contamination may affect the interior or the exterior of some equipment; for example, refrigerators and freezers often have drip trays that are hidden from view and can become reservoirs of fungal spores. Equipment contamination can be managed by locating such equipment away from the sterile handling area, by performing regular maintenance and cleaning, and by swabbing down bottles and other reagents before they are placed in the BSC (see Sections 4.2.6, 5.5, 12.3.1). Any new equipment items (including gas cylinders) should be wiped down before entering the room.

16.1.4 Reagent Problems

There should be no risk of contamination from glassware, plasticware, media, and other reagents if the appropriate quality assurance (QA) is in place (see Section 7.5). Commercial suppliers should clearly state the quality control (QC) testing that has been performed to ensure that reagents are sterile or free from specific microorganisms (see Sections 7.5.2, 7.5.3) (Nims and Price 2017). Reagents that are prepared in-house should also pass through a process of QC testing. In-house QC testing typically focuses on the completeness of sterilization (e.g. using sterility indicators), the quality of water used for reagent preparation, and the sterility of the final product (see Sections 11.2.7, 11.5.3, 11.8). A procedure for sterility testing is available as Supplementary material online (see Supp. S11.6). Culture testing is not routinely performed to detect microbial contamination, but can provide essential information; for example, a historical case of yeast contamination (*Torula* species) was only detected when contaminated serum was used for insect culture at 28 °C (McGarrity and Coriell 1971). Serum can also be a source of viral contamination. For example, fetal bovine serum (FBS) can contain bovine viral diarrhea viruses (BVDV) and bovine polyomaviruses (BPyV) which can at least infect cultures that originate from cattle.

16.1.5 Cell Line Problems

The effectiveness of contamination control in modern tissue culture laboratories – particularly the use of laminar airflow, elimination of mouth pipetting, and the QA programs adopted by manufacturers of commercial products – means that the major contamination risk now comes from other cultures. This risk became evident as early as the 1960s, when it was observed that the arrival of a mycoplasma-contaminated culture led to the appearance of contamination in other cultures; if one culture contained mycoplasma, most (if not all) of the cultures in that laboratory were contaminated with the same species (O'Connell et al. 1964; McGarrity 1976).

Reasons for this relatively high risk of contamination include:

(1) **Incoming cell lines.** Any cell line that is brought into the tissue culture laboratory may carry microbial contamination. Whenever possible, all cell lines should be acquired via a reputable cell repository that performs testing for microbial and other forms of contamination (see Section 15.6). Cell lines from any other source, as well as other materials such as biopsies and primary cultures, should be regarded as contaminated until shown to be otherwise (see Section 16.2).

(2) **High titers of microorganisms.** Mycoplasma contamination typically results in a titer of 10^6–10^8 colony forming units (cfu)/ml in the culture supernatant (McGarrity et al. 1978). The original estimate was accompanied by a wry comment that such cultures could be viewed as mycoplasma cultures contaminated with cells rather than the other way around! High titers lead to an increased risk of contamination via aerosols, droplets, or other modes of transmission (McGarrity 1976).

(3) **Prolonged survival in dried form.** Fungi and some bacteria (e.g. *Bacillus* species) produce spores that are resistant to disinfection; a historical case of *Aspergillus fumigatus* persisted even after fumigation of the incubators and room in which the cultures were kept (Clarke et al. 1989). Some of these species grow slowly, which means that they may be difficult to detect in rapidly growing cultures (Coriell 1962; Fogh et al. 1971). Mycoplasma species do not produce spores, but they remain viable for prolonged periods (Kundsin 1968). *A. laidlawii* can be recovered from dry paper disks that are left for at least 168 days at 4 °C or 30 °C (Nagatomo et al. 2001).

(4) **Difficulties in detecting microorganisms.** Some viruses and intracellular bacteria do not cause major changes in the appearance of the culture even when high titers are present. Subtle changes may occur (e.g. changes in growth rate), but these may only be recognized if you are familiar with the cell line or have access to characterization data from before the contamination occurred. This is true for mycoplasma and for intracellular bacteria such as *Mycobacterium avium-intracellulare* complex (MAC). These bacteria are predominantly found in patients with acquired immunodeficiency syndrome (AIDS) and grow slowly *in vitro*; they are best detected using microbiological culture with a rich growth medium (Lelong-Rebel et al. 2009).

(5) **Chronic antibiotic use.** Many tissue culture laboratories add antibiotics to their media, under the assumption that this will prevent microbial contamination. However, chronic antibiotic use has a number of disadvantages (see Section 9.6.2). Perhaps the most obvious problem is an increase in antibiotic resistance; antibiotics tend to result in contamination by a smaller number of organisms with altered antibiotic susceptibility (McGarrity and Coriell 1971; Taylor-Robinson and Bebear 1997). Antibiotics may also suppress the growth of microorganisms, resulting in lower titers that are more difficult to detect (Coriell 1962; Fogh et al. 1971).

16.2 MANAGEMENT OF CONTAMINATION

A recent survey asked why researchers "are not authenticating cell lines and performing frequent and detailed quality controls on a regular basis." Almost a quarter of researchers chose the response, "I do not see the necessity; I am careful" (Freedman et al. 2015). Unfortunately, this response does not allow for the various sources of contamination that may arise and the occasional lapses in technique that can affect anyone, despite best efforts. McGarrity estimated that if a tissue culture worker experienced microbial contamination at a rate of 1%, this would lead to contamination of all their cell lines in one year (based on the estimate that cultures are handled twice a week) (McGarrity et al. 1978). With this in mind, it seems reasonable to conclude that contamination will affect everyone at some stage and take preventative steps to detect and eliminate microbial contaminants.

Prevention of contamination rests on good aseptic technique, cleaning and maintenance of the laboratory and its equipment, and screening of reagents to ensure that adventitious contaminants are not introduced during handling. Even with this foundation in place, three further steps must be taken to manage microbial contamination:

(1) **Quarantine incoming cell lines.** Ideally, quarantine should take place in a separate room with its own BSC and incubator (see Section 4.2.3). If this is not feasible, cells should be handled at the end of the day and the BSC should be cleaned and withdrawn from use until the following day. Once testing and a period of monitoring have been completed without problems, cell lines can join other stocks in general use.

(2) **Monitor for visible contamination.** As part of examining cells during routine maintenance, always check for contamination by eye and with a microscope (see Table 14.3). Microbial contamination may be immediately obvious through changes in pH, turbidity, or the appearance of

visible microorganisms when viewed using phase contrast microscopy (see Section 16.4).

(3) *Test for contaminants that are difficult to detect by eye.* All tissue culture laboratories should routinely test for mycoplasma (see Section 16.5). The need for additional testing will vary between laboratories, e.g. cells that have been passaged in mice are more likely to require testing for γ-retroviruses (see Section 16.6).

If microbial contamination is suspected but not proven, the culture should be quarantined while further investigation is performed. Once microbial contamination is clearly evident or has been confirmed through sterility testing (see Supp. S11.6) or other assays, the culture should be discarded. Do not attempt to treat cultures for microbial contamination unless they are unique and irreplaceable.

PROTOCOL P16.1. DISPOSAL OF CONTAMINATED CULTURES

Background

Microbial contamination may arise due to pathogenic organisms. Examples include *Mycoplasma pneumoniae*, which causes atypical pneumonia (and was originally discovered in tissue culture), and *A. fumigatus*, which may cause allergic reactions or infections in immunocompromised individuals (Hayflick 1965; Clarke et al. 1989). All contaminated cultures should be disposed of as biological waste in accordance with local safety regulations (see Sections 6.1.2, 6.3.6).

Solid biological waste is typically incinerated or autoclaved, while liquid biological waste is typically decontaminated using a chemical disinfectant prior to disposal. The choice of decontamination method may vary with the type of microbial contamination. Hypochlorite solutions ("bleach") are effective against a wide range of microorganisms and can kill spores if used at a suitable concentration (Russell 1990; WHO 2004). However, because these solutions are hazardous and can cause corrosion, some laboratories prefer to use low-level disinfectants such as Virkon™ (LANXESS Deutschland GmbH). Virkon can be effective against vegetative bacteria, yeasts, and viruses, but does not kill spores (Hernandez et al. 2000). Always perform a risk assessment before selecting a suitable disinfectant, considering the likely microorganisms that may be present, the conditions of disinfection, and the hazards associated with that agent.

Outline

Make a record of the cultures affected. Perform decontamination prior to disposal as solid or liquid biological waste. Perform additional problem solving as required (see Section 31.1).

- *Safety Note. Hypochlorite solutions can cause burns to skin or eyes, lung damage, and bleaching of clothing. PPE should be worn including eye protection, gloves, and a lab coat or apron. Always read the Safety Data Sheet (SDS) and perform a risk assessment before proceeding.*

Materials (non-sterile)

☐ Contaminated culture(s) to be discarded (flasks, dishes, plates, etc.)
☐ BSC and associated consumables (see Protocol P12.1)
☐ Suitable disinfectant, e.g. hypochlorite solution (see Background above)
☐ Waste receiver containing disinfectant (see Figures 5.6, 12.8)
☐ Soaking bath for disinfection of open cultures, e.g. polypropylene container

Procedure

1. Check other cultures for contamination. Decide whether the problem is a new or ongoing problem, and whether it is widespread, i.e. affecting two or more different cultures.
2. For any affected cultures, make a record of the nature of the contamination (e.g. bacterial rods or cocci, yeast, fungi), the operator, the incubator in which cells were grown, and any other information that may be useful for problem solving.
3. For any contaminated cultures:
   (a) If the culture is sealed and only has a small amount of medium, leave it unopened and transfer to solid biological waste for autoclaving or incineration prior to disposal.
   (b) If the culture is sealed but has sufficient medium that decontamination is unlikely to be successful, open the culture vessel in a BSC and transfer the medium into a waste receiver that contains an appropriate disinfectant. Transfer the vessel to solid biological waste and dispose of the treated liquid appropriately.
   (c) If the culture is open and there is a risk of airborne spread or spillage, soak the vessel in a disinfectant bath; discard appropriately after treatment is complete.
4. Wipe down the BSC and any other areas that may have been exposed to the affected cultures. Clean the incubator in which the cultures were grown (see Protocol P12.4).

5. If the contamination is new and is not widespread, discard the medium bottle and any other reagents (e.g. unfinished aliquots) that were opened in the presence of that culture.
6. If the contamination is new and widespread beyond a single incubator, discard all media, trypsin, and other reagents and stock solutions in current use.
7. If the same kind of contamination has occurred before, check stock solutions for contamination by sterility testing (see Supp. S11.6).
8. If the contamination is widespread and repeated:

   (a) Alert users of the room to a problem and discuss any recent changes that might lead to increased contamination risk. Bring together records of all contaminated cultures.
   (b) Alert facility staff to a possible problem in the room. Ask them to check air handling systems and other sources of contamination that might affect the room as a whole.
   (c) Check and clean any equipment that is likely to act as a focus of contamination such as water baths, coldrooms, and other cold storage equipment (see Section 16.1.3).
   (d) Check sterilization procedures, e.g. autoclave and oven temperatures, particularly in the center of the load. Use sterility indicators for problem solving (see Section 11.2.7).
   (e) Consider performing species testing to identify the specific microorganisms involved (see Section 16.4.1).
   (f) Consider performing environmental monitoring using air monitoring and microbial monitoring of surfaces (e.g. contact plates) to identify a further source of contamination.

Notes

Step 3: If using hypochlorite solution, it may be necessary to inactivate the hypochlorite before disposal, e.g. using sodium thiosulfate (Hegde et al. 2012).

Step 4: Typically, 70% alcohol is used to wipe down the BSC, but this will not be effective against spores. If fungal contamination is present, consider using a high-level disinfectant such as hypochlorite and then wipe down with 70% alcohol, which will help to remove the hypochlorite. Because such disinfectants may cause corrosion, their use should be based on risk assessment.

Step 7: If testing is negative, but contamination is still suspected, incubate 100 ml of the solution, filter it through a 0.2-μm filter, and plate out the filter on nutrient agar with an uninoculated control.

16.3 VISIBLE MICROBIAL CONTAMINATION

The major microbial contaminants consist of bacteria (including mycoplasmas), yeasts, fungi, and viruses (Fogh et al. 1971; McGarrity et al. 1978; Langdon 2004). Many (although not all) bacteria, fungi, and yeasts give rise to visible signs of contamination. Characteristic signs of microbial contamination include:

(1) *A sudden change in pH.* This is usually a decrease with most bacterial infections (see Figure 16.2a), very little change with yeast until the contamination is heavy, and sometimes an increase in pH with fungal contamination.
(2) *Cloudiness in the medium.* A slight film or scum may also be present on the surface and spots may be visible on the growth surface that dissipate when the flask is moved (see Figure 16.2b).
(3) *Visible microorganisms under the microscope.* Under a 10× objective, spaces between cells will appear granular and may shimmer with bacterial contamination (see Figure 16.3a). Some bacteria form clumps or associate with the cultured cells. Yeasts will often appear as separate round or ovoid particles that may bud off smaller particles (see Figure 16.3b). Fungi produce thin filamentous mycelia (see Figure 16.3c) and, sometimes, denser clumps of spores which may be blue or green. Under higher magnification, it may be possible to resolve individual bacteria, distinguish between rods and cocci, and observe motility.

16.3.1 Testing of Bacteria, Fungi, and Yeasts

If microbial contamination is clearly visible and is an isolated incident, it may only be necessary to record the general kind of contaminant and any other relevant information (see Protocol P16.1). Further investigation is required if it forms part of a recurrent pattern of contamination or if there is uncertainty as to the nature of the problem. For example, a cell line may show increased debris but no changes in pH or medium turbidity. Microbial contaminants can be distinguished from cell debris or precipitates of media constituents (particularly protein) by their motility, regular morphology, and the dispersal of clumps to form single organisms or "strings" when the culture is shaken. Precipitates may be crystalline or globular and irregular and are not usually as uniform in size. However, the "sick cell line" may be difficult to diagnose. A cryptic contaminant may be present that is difficult to detect because it is too small to be visible (e.g. mycoplasma) or is present at low titer due to slow growth or chronic antibiotic use.

Species testing may be considered in order to understand the nature of the contaminant and its origins, particularly if the contaminant recurs or is difficult to detect. Samples can be collected from culture supernatant (ideally spent medium from an antibiotic-free culture), by inoculating the supernatant into nutrient broth for expansion, or by streaking onto nutrient agar to obtain colonies. Environmental samples may

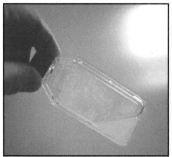

(a) Contaminated Flask. *Reduction in pH and cloudiness of medium.*

(b) Flocculated Contamination. *Bacterial, but with no drop in pH.*

(c) Contaminated Medium Bottle. *Yeast, showing cloudiness but no drop in pH.*

(d) Samples of Broth. *L-broth with test samples of medium. Control (left) is clear, as is the test sample (center); compare to the positive control at right.*

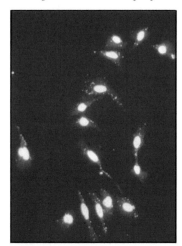

(e) Mycoplasma, Low Power. *Mycoplasma-contaminated culture as revealed by Hoechst 33258 staining using a 50x water-immersion objective.*

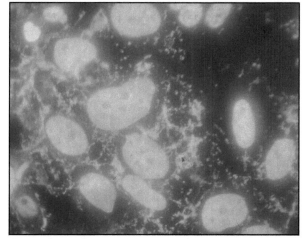

(f) Mycoplasma, High Power. *Hoechst 33258 staining demonstrates staining in the cytoplasm and on the plasma membrane using a 100x oil-immersion objective.*
Source for all images: R. Ian Freshney.

Fig. 16.2. Examples of contamination.

also be collected using contact plates and airborne monitoring (Cobo and Concha 2007; IEST 2018). Bacterial species can be identified by sequencing the 16S ribosomal RNA (rRNA) gene (Woo et al. 2008). Bacterial testing is performed as a service by some clinical microbiology laboratories and other providers, e.g. Charles River Laboratories.

There are a number of benefits from identifying the microbial contaminant. Unusual and novel causes of microbial contamination have been identified, including some parasites (e.g. *Hartmannella* and *Naegleria* species), *Mycobacterium avium*, and *Achromobacter* species (Fogh et al. 1971; Buehring et al. 1995; Gray et al. 2010). The safety risk of a microbial contaminant can also be assessed and antibiotic sensitivity testing performed using suitable nutrient agar, which can guide the selection of antibiotics for treatment (Clarke et al. 1989; Gray et al. 2010).

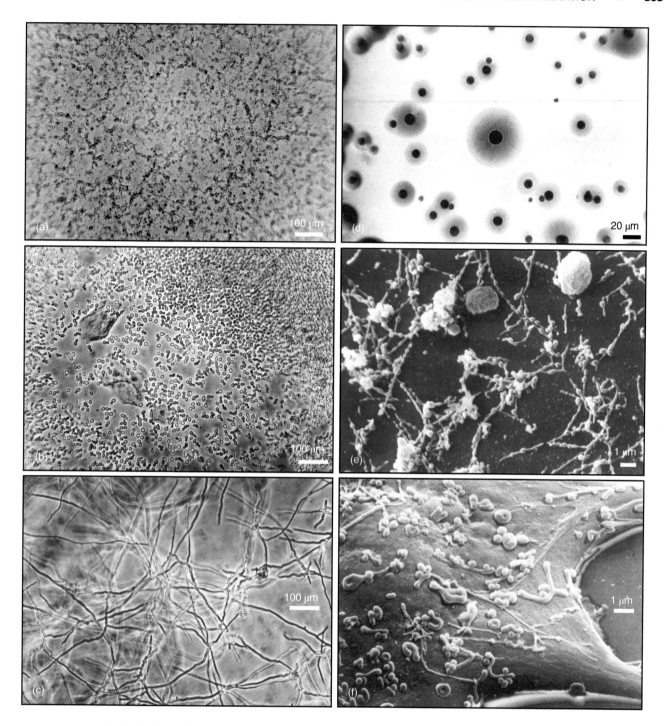

Fig. 16.3. Types of contamination. Various microorganisms can be identified in cell cultures, including (a) bacteria; (b) yeast; (c) mold; (d) mycoplasma colonies growing on special nutrient agar; (e, f) mycoplasma growing on the surface of cultured cells, visible using scanning electron microscopy. Scale bars are as indicated. *Source*: (a–c) R. Ian Freshney; (d–f) Courtesy of Dr M. Gabridge.

16.3.2 Eradication of Bacteria, Fungi, and Yeasts

The most reliable method of eradicating a microbial contamination is to discard the culture and the medium and reagents used with it (see Protocol P16.1). Treatment should be attempted only if the culture is unique and irreplaceable,

as there is a risk that the contaminant will be transmitted to other cell lines while treatment is in progress. There is also a risk that the effort will be unsuccessful and will produce hardier, antibiotic-resistant strains. If the decision is made to treat the culture with antibiotics, the culture should be quarantined throughout the process (see Section 16.2). If treatment is successful, continued testing should be performed to pick up any recurrences. If unsuccessful, the culture should be discarded as soon as failure becomes obvious.

PROTOCOL P16.2. TREATMENT OF MICROBIAL CONTAMINATION

Background

A number of antibiotics have been used in tissue culture that may be effective against bacteria, fungi, and yeasts (see Table 16.2). However, a significant number of bacterial strains are resistant to antibiotics, either naturally or by selection, so the control that they provide is never absolute. Fungal and yeast contaminations are particularly hard to control with antibiotics; they may be held in check but are seldom eliminated.

Outline

Wash the culture several times in a high concentration of antibiotics by rinsing the monolayer or by centrifugation and resuspension of non-adherent cells. Then grow the culture for three subcultures with, and three without, antibiotics. Monitor for recurrence of contamination.

- *Safety Note. Cultures that are contaminated with microorganisms have an associated biohazard risk (see Protocol P16.1).*

Materials (sterile or aseptically prepared)

☐ High-antibiotic wash solution, e.g. Dissection Balanced Salt Solution (DBSS; see Appendix B)
☐ High-antibiotic medium, e.g. Collection Medium (see Appendix B)
☐ Materials for trypsinization (see Protocol P14.2)

Materials (non-sterile)

☐ BSC and associated consumables (see Protocol P12.1)
☐ Inverted microscope with phase contrast optics, preferably with 40x and 100x objectives
☐ Disinfectant for surface decontamination and disposal (see Protocol P16.1)

Procedure

1. Place the contaminated culture in quarantine – either by location (e.g. a separate room or incubator) or by handling (e.g. handling the culture at the end of the day).

2. Working in the BSC, collect the contaminated medium carefully. If possible, the organism should be tested for species and for sensitivity to a range of individual antibiotics. If this is not possible, decontaminate and discard (see Protocol P16.1).
3. Wash the cells in DBSS (to dilute the contaminant):
   (a) For monolayers, rinse the culture three times with DBSS, trypsinize, and wash the cells twice more in DBSS by centrifugation and resuspension.
   (b) For suspension cultures, wash the culture five times in DBSS by centrifugation and resuspension.
4. Reseed a fresh flask at the lowest reasonable seeding density.
5. Add high-antibiotic medium and change every two days until the cells are confluent. Check the cultures by phase contrast microscopy for visible microbial contamination.
6. Subculture in a high-antibiotic medium.
7. Repeat steps 1–5 for three subcultures.
8. Remove the antibiotics, and culture the cells without them for a further three subcultures.
9. Culture the cells for a further two months without antibiotics.
10. Perform sterility testing to confirm that the contaminant has been eliminated (see Supp. S11.6).

16.4 MYCOPLASMA CONTAMINATION

In 1956, Lucille Robinson and colleagues were studying "pleuropneumonia-like organisms" (PPLOs), which were similar to the microorganism responsible for bovine pleuropneumonia (Robinson et al. 1956). They inoculated PPLO into HeLa cells, only to find that their control cells also showed evidence of PPLO. Cells that were frozen prior to October 1955 did not contain PPLOs, indicating that a contamination event had occurred. This initial finding opened the floodgates to widespread reports of contamination with these organisms (now commonly known as "mycoplasmas"), which continue to the present day. A large 25-year survey from 1990 to 2014 showed that 22% (169/762) of cell lines showed evidence of mycoplasma contamination (Drexler et al. 2017). While prevalence declined over that time, 6% (4/63) of cell lines from 2010 to 2014 remained mycoplasma-positive. It is important to stress that these figures come from cell lines that were deposited at a cell repository and thus may represent a "best case" scenario.

The term "mycoplasma" refers to the *Mycoplasma* genus to which many of these organisms belong (Stanbridge 1971). Over time, changes in taxonomy and the discovery of new

TABLE 16.2. Antibiotics used in tissue culture.

Antibiotic	Concentration, µg/ml (unless otherwise stated)		Activity against:[a]
	Working	Cytotoxic	
Amphotericin B (Fungizone)	2.5	30	Fungi, yeasts
Ampicillin	100		Bacteria, gram positive and gram negative
Ciprofloxacin	100		Bacteria, gram positive and gram negative
Erythromycin	100	300	Bacteria, gram positive and gram negative
Gentamicin sulfate	50	> 300	Bacteria, gram positive and gram negative
Kanamycin monosulfate	100	10 mg/ml	Bacteria, gram positive and gram negative
Neomycin sulfate	50	3000	Bacteria, gram positive and gram negative
Nystatin	50		Fungi, yeasts
Penicillin-G	100 U/ml	10 000 U/ml	Bacteria, gram positive
Polymixin B sulfate	50	1 mg/ml	Bacteria, gram negative
Streptomycin sulfate	10	20 mg/ml	Bacteria, gram positive and gram negative
Tetracycline hydrochloride	10	35	Bacteria, gram positive and gram negative

Source: Paul, J. (1975). Cell and tissue culture. 5th ed. Edinburgh: Churchill Livingstone; Sigma-Aldrich (2018). Cell culture antibiotic selection guide.
[a]There are no guarantees that any of these antibiotics will eradicate a specific microorganism. Always follow the manufacturer's instructions and continue to monitor after treatment is complete for recurrences. A list of antibiotics that can be used for treatment of mycoplasma is provided separately (see Table 16.6).

species have resulted in a broader family tree with at least 180 species (Razin et al. 1998). Six species are known to be common in tissue culture (see Table 16.3) (Drexler and Uphoff 2002). However, incidence data vary between different studies, with multiple species uncovered since the 1970s, e.g. *M. yeatsii* and *M. bovis* (Barile et al. 1978; Del Giudice 1998; Calcutt et al. 2015). All of these species belong to the class *Mollicutes* (derived from the Latin for "soft skin"), which can be distinguished from other bacteria by their small size (0.3–0.8 µm) and the absence of a cell wall. These characteristics help to explain why they are such a problem in tissue culture; organisms can pass through a 0.45-µm filter due to their size and ability to change shape (Drexler and Uphoff 2002).

Mycoplasma species grow best in the presence of a host cell and do not grow in standard microbiological broth. Special media have been developed for their culture (see Figure 16.3d), although even these are ineffective for some species (Volokhov et al. 2008). The reasons for their dependence on a host became clear when their full genomes were examined. Members of the class *Mollicutes* appear to have evolved toward a reduced genome that lacks certain key pathways, such as those for amino acid biosynthesis and

energy metabolism (Razin et al. 1998). These requirements are met by the host cell or the medium, e.g. leading to arginine depletion (Stanbridge 1971). Organisms grow on the surface of the cell (see Figure 16.3e, f); depending on the species and the host cell type, they can also invade or fuse with the host cell to take up residence within the cytoplasm (Tarshis et al. 2004; Yavlovich et al. 2004). This partly explains why mycoplasmas can be resistant to treatment with antibiotics; an intracellular reservoir of organisms may be present that permits chronic contamination (Razin et al. 1998).

Many cases of mycoplasma contamination result in reduced growth rates but do not destroy the host cell. However, cell behavior may be altered in many different ways, ranging from acute effects (e.g. lysis of the whole cell population) to chronic effects (e.g. chromosomal abnormalities, increased cloning efficiency, or increased rates of transformation) (Fogh et al. 1971). Mycoplasma contamination also leads to secondary effects when data from contaminated cells are used in other studies. Mycoplasma sequence has been identified on human microarrays and in at least 7% of samples from the Human Genome Project (Aldecoa-Otalora et al. 2009; Langdon 2014). A survey of RNA sequencing (RNA-seq) data found that 11% (52/484) of series with cultured samples mapped to mycoplasma (Olarerin-George and Hogenesch 2015).

16.4.1 Mycoplasma Detection

Because high titers do not necessarily cause changes in morphology, specific testing must be performed to detect mycoplasma contamination. Multiple methods have been used for mycoplasma detection; for example, uptake of radioisotopes by these organisms is evident using microautoradiography (see Figure 19.14). Many laboratories now use polymerase chain reaction (PCR) and DNA fluorescence, which can be performed in-house using laboratory reagents or commercial kits (see Table 16.4). Mycoplasma testing is also offered as a service by cell repositories and other testing providers. Unfortunately, there is no perfect method for mycoplasma detection, due to the large number of species that may be present and the risk that contamination may be present at low titer. Cell repositories often use two complementary methods in an effort to detect unusual species or low-level contaminants, or to resolve problematic samples, e.g. the "sick cell line" that appears to be mycoplasma-negative. Other approaches that may help to resolve problematic samples include retesting cultures after further passaging in the absence of antibiotics or after enrichment.

Factors that should be considered when assessing a method, kit, or provider include (i) the number of species detected and whether these include common tissue culture contaminants (see Table 16.3); (ii) the limit of detection (in cfu or genomic copies); and (iii) the control samples that will help to assess its reliability in practice. Some mycoplasma species are more difficult to detect than others; *M. fermentans* is particularly challenging, which may be due to its slow recovery after thawing or changing the culture medium (Volokhov et al. 2008). Control samples can be built up slowly over time from contaminated cultures, using aliquots of spent medium that are stored at −80 °C for later use in validation testing, or (ideally) purchased as reference stocks from microbial culture collections (Dabrazhynetskaya et al. 2014).

TABLE 16.4. Mycoplasma detection kits.

Product[a]	Principle	Supplier
PCR kits		
PCR Mycoplasma Test Kit	PCR of 16S rRNA	AppliChem
MycoSensor kits	PCR and real-time PCR, primer sites undisclosed	Agilent
EZ-PCR Mycoplasma Test Kit	PCR of 16S rRNA	Biological Industries
Milliprobe® Real-time Detection System	Real-time PCR of ribosomal RNA	Merck Millipore
Venor™ GeM kits	PCR and real-time PCR of 16S rRNA	Minerva Biolabs
PCR Mycoplasma Test Kits	PCR and real-time PCR of 16S rRNA	PromoCell
MycoTOOL kits	PCR and real-time PCR of 16S rRNA	Roche CustomBiotech
Microsart® kits	Real-time PCR of 16S rRNA	Sartorius (manufactured by Minerva Biolabs)
LookOut® kits	PCR and real-time PCR of 16S rRNA	Sigma-Aldrich
MycoSEQ™ kits	Real-time PCR, primer sites undisclosed	ThermoFisher Scientific
Other kits		
MycoAlert	Luminescence detection of ATP from mycoplasma-specific enzymes	Lonza
MTC-NI	DNA–RNA hybridization with acridinium ester labeled DNA and subsequent determination of the DNA–RNA hybrids by chemiluminescence	Merck Millipore
MycoFluor™	Fluorescent detection of mycoplasma using a nucleic acid stain	ThermoFisher Scientific
MycoProbe	Colorimetric detection of biotin-labeled probe for 16S rRNA in *M. hyorhinis, M. arginini, M. fermentans, M. orale, M. pirum, M. hominis, M. salivarium, Acholeplasma laidlawii*	R&D Systems
PCR ELISA	PCR of 16S rRNA, subsequent detection of amplification products by ELISA	Roche CustomBiotech

Source: courtesy of Cord Uphoff.
[a] These are representative examples; a complete list is impossible, due to the large number of suppliers and kits.

TABLE 16.3. Common mycoplasma species in tissue culture.

Species[a]	Frequency (%)	Natural host
M. orale	20–40	Human
M. hyorhinis	10–40	Swine
M. arginini	20–30	Bovine
M. fermentans	10–20	Human
M. hominis	10–20	Human
A. laidlawii	5–20	Bovine

Source: Drexler, H.G., Uphoff, C.C. Mycoplasma contamination of cell cultures: Incidence, sources, effects, detection, elimination, prevention. Cytotechnology 39, 75–90. © 2002 Springer Nature.

[a]These species are referred to as "mycoplasma" in the text because the term is familiar to tissue culture laboratories. The term "mollicute" may also be used (because all species belong to the class *Mollicutes*) and is more technically correct. As noted in the text, these six species are relatively common, but others are also found in tissue culture laboratories.

Direct culture and enrichment. Many mycoplasma species can be grown in Mycoplasma Broth Base or Agar with Mycoplasma Supplement (BD Biosciences, Oxoid, Sigma-Aldrich). Cultured cells are seeded into the broth, which is plated out onto mycoplasma agar; other approaches can also be used, such as direct plating onto agar or inoculation of cells into semi-solid medium (Taylor-Robinson 1978). Colonies are easily recognized by their size (approximately 200 μm diameter) and their "fried egg" appearance, with a dense center and lighter periphery (see Figure 16.3d). Direct microbiological culture is a highly sensitive method and is employed for regulatory purposes, e.g. to test cell lines used for vaccine production (Volokhov et al. 2011). However, direct culture should only be used with the appropriate facilities for separate handling and suitable microbiological expertise; these organisms are quite fastidious, and it is necessary to culture mycoplasma as a positive control. Some species or strains cannot be cultivated using this approach, including strains of *M. hyorhinis*, which is a common contaminant in tissue culture.

As a complementary approach, an indicator cell line can be used that allows the "uncultivated" species or strains to grow successfully (Volokhov et al. 2011). A suitable cell line must be chosen that is permissive for mycoplasma growth; the MDCK cell line is a particularly good choice for the species that commonly occur in tissue culture (Volokhov et al. 2008). Culture using an indicator cell line can be used as an enrichment technique to increase the sensitivity of detection for other test methods (Kong et al. 2007). A sample of medium can be used for enrichment, which avoids losing a valuable culture if it is your only flask. It is also helpful with fluorescent assays, to avoid false-positive results arising from autofluorescence or debris.

Nucleic acid technologies.

PROTOCOL P16.3. DETECTION OF MYCOPLASMA BY PCR

Contributed by C. C. Uphoff and H. G. Drexler, Leibniz Institute DSMZ, Braunschweig, Germany.

Background

The polymerase chain reaction (PCR) provides a very sensitive and specific assay for the direct detection of mycoplasmas in cell cultures with low expenditure of labor, time, and cost, and with simplicity, objectivity of interpretation, reproducibility, and easy documentation of results. Several primer sequences are published for both single and nested PCR and with narrow or broad specificity for mycoplasma or eubacteria species. In most cases, the 16S rDNA sequences are used as target sequences, because this gene contains regions with more and less conserved sequences. This gene also offers the opportunity to perform a PCR with the 16S rDNA or a reverse transcriptase-PCR (RT-PCR) with the cDNA of the 16S rRNA.

Here, we describe the use of a mixture of oligonucleotides for the specific detection of mycoplasmas. This approach reduces significantly the generation of false-positive results due to possible contamination of the solutions used for sample preparation and the PCR run, and of other materials with airborne bacteria. One of the main problems concerning PCRs with samples from cell cultures is the inhibition of the *Taq* polymerase by unspecified substances. To eliminate those inhibitors, we strongly recommend that the sample DNA be extracted and purified by conventional phenol-chloroform extraction, by salt-precipitation protocol, or by the more convenient column or matrix binding extraction methods.

To confirm the error-free preparation of the sample and PCR run, appropriate control reactions have to be included in the PCR. These comprise internal control DNA for every sample reaction and, in parallel, positive, negative, and water control reactions. The internal control consists of a DNA fragment with the same primer sequences for amplification, but it is of a different size than the amplicon of mycoplasma-contaminated samples. This control DNA is added to the PCR mixture in a previously determined limiting dilution to demonstrate the sensitivity of the PCR. The internal control DNA can be obtained from the authors. The performance of the PCR run highly depends on the employed *Taq* polymerase, buffer, and apparatus. Thus, the protocol might be needed to be individually adjusted for optimal conditions. This can be performed with dilution series of the positive or the internal control DNA.

Outline

Supernatant medium of adherent cells or of suspension cultures with settled cells is collected and the cellular particles are isolated by centrifugation. After washing steps, the DNA is extracted and an aliquot is used for nucleic acid amplification with mycoplasma-specific oligonucleotides. Appropriate positive and negative control reactions confirm the error-free run and verify the sensitivity of the PCR. The amplification products are visualized in an ethidium bromide stained agarose gel and subsequently documented.

Materials

- Phosphate-buffered saline (PBS): 137 mM NaCl, 2.7 mM KCl, 8.1 mM Na_2HPO_4, 1.5 mM KH_2PO_4, pH 7.2. Autoclave 20 minutes at 121 °C to sterilize the solution
- Tris acetic acid ethylene diamine tetra-acetic acid (EDTA) (Tris base, acetic acid, EDTA [TAE]), 50x: 2 M Tris base, 5.71% glacial acetic (v/v), 100 mM EDTA. Adjust to about pH 8.5
- DNA extraction and purification system, e.g. phenol/chloroform extraction and ethanol precipitation, or DNA extraction kits applying salt-precipitation or DNA binding matrices
- *Taq* DNA polymerase with the appropriate 10x buffer (Platinum hot start *Taq* polymerase; Invitrogen (Sigma-Aldrich), 10x buffer without $MgCl_2$)
- Loading buffer, 6x:10 mM Tris-HCl (pH 7.6), 0.03% (w/v) bromophenol blue, 0.03% (w/v) xylene cyanol FF, 60% glycerol (v/v), 60 mM EDTA
- Primers (see Table 16.5 for sequence):
  - Primer stock solutions: 100 μM in dH_2O, stored frozen at −20 °C
  - Working solutions: mix of forward primers at 5 μM each (Myco-5′) and mix of reverse primers at 5 μM each (Myco-3′) in dH_2O, aliquoted in small amounts (i.e. 25–50 μl aliquots), and stored frozen at −20 °C
- Internal control DNA: This may be prepared as published elsewhere (Uphoff et al. 1992) or can be obtained from the authors. A limiting dilution should be determined experimentally by performing PCR with a dilution series of the internal control DNA

- Positive control DNA: a tenfold dilution of any mycoplasma-positive sample prepared as described below or can be obtained from the authors
- Deoxy-nucleotide triphosphate mixture (dNTP-mix) Mixture contains 5 mM each of deoxyadenosine triphosphate (dATP), deoxycytidine triphosphate (dCTP), deoxyguanosine triphosphate (dGTP), and deoxythymidine triphosphate (dTTP) at 5 mM in H_2O, and stored as 50-μl aliquots at −20 °C
- Magnesium chloride (50 mM)
- Agarose, 1.3%–TAE gel
- Thermal cycler

TABLE 16.5. Primers for mycoplasma detection.

Direction	Sequence[a]
Forward primers (Myco-5′)	cgc ctg agt agt acg **twc** gc
	tgc ctg **r**gt agt aca ttc gc
	c**r**c ctg agt agt atg ctc gc
	cgc ctg ggt agt aca ttc gc
Reverse primers (Myco-3′)	gcg gtg tgt aca **ar**a ccc ga
	gcg gtg tgt aca aac ccc ga

Source: courtesy of Cord Uphoff.
[a] **r** = mixture of g and a; **w** = mixture of t and a.

Procedure

A. Sample Collection and Preparation of DNA

1. Before collecting the samples, the cell line to be tested for mycoplasma contamination should be cultured without any antibiotics for several days, or for at least two weeks after thawing. This should ensure that the titer of the mycoplasmas in the supernatant medium is within the detection limits of the PCR assay.
2. Collect 1 ml of the supernatant medium of adherent cells, or of cultures with settled suspension cells. Collecting the samples in this way will include some viable and/or dead cells, an advantage as some mycoplasma strains predominantly adhere to the eukaryotic cells or even invade them. The supernatant medium can be stored at 4 °C for a few days or at −20 °C for several weeks. After thawing, the samples should be processed immediately.
3. Centrifuge the supernatant medium at 13 000 g for 5 minutes and resuspend the pellet in 1 ml PBS by vortexing.
4. Centrifuge the suspension again and wash one more time with PBS as described in step 3.

5. After centrifugation, resuspend the pellet in 100 µl PBS by vortexing and then heat to 95 °C for 15 minutes.

6. Immediately after lysing the cells, the DNA is extracted and purified by standard phenol/chloroform extraction and ethanol precipitation or other DNA isolation methods.

B. PCR

Established rules to avoid DNA carry-over should be strictly followed: (i) the places where the DNA is extracted, the PCR is set up, and the gel is run after the PCR should be separated from each other; (ii) all reagents should be stored in small aliquots to provide a constant source of uncontaminated reagents; (iii) avoid reamplifications; (iv) reserve pipettes, tips, and tubes for their use in PCR only and irradiate pipettes frequently by UV light; (v) the succession of the PCR setup described below should be followed strictly; (vi) wear gloves during the whole sample preparation and PCR setup; and (vii) include the appropriate control reactions, such as internal, positive, negative, and the water control reaction.

1. Set up two reaction premixtures per sample to be tested with the following solutions:
   (a) Sample only: 1 µl dNTPs, 0.5 µl Myco-5′, 0.5 µl Myco-3′, 2.5 µl 10× PCR buffer, 1 µl MgCl$_2$, 0.2 µl hot start *Taq* DNA polymerase, 17.3 µl dH$_2$O.
   (b) Sample and DNA internal standard: 1 µl dNTPs, 0.5 µl Myco-5′, 0.5 µl Myco-3′, 2.5 µl 10× PCR buffer, 0.2 µl hot start *Taq* DNA polymerase, 16.3 µl dH$_2$O, 1 µl internal control DNA.

2. Prepare sufficient volumes of premixtures for multiple samples, including three additional control reactions without internal control DNA (for the positive, negative, and water controls) and two additional with the internal control DNA (for the positive and negative controls), plus a surplus for pipetting.

3. Transfer 23 µl of each premixed stock to 0.2 ml PCR tubes and add 2 µl dH$_2$O to the water control reactions.

4. Set aside all reagents used for the preparation of the premixed stocks.

5. Take out the samples of DNA to be tested and the positive control DNA. Do not handle the reagents and samples simultaneously.

6. Add 2 µl per DNA preparation to one reaction tube that contains no internal control DNA and to one tube containing the internal control DNA.

7. Transfer the reaction mixtures to the thermal cycler and start the PCR with the following parameters:
   (a) For 1 cycle (to activate the hot start *Taq* polymerase): 7 minutes at 95 °C; 3 minutes at 72 °C; 2 minutes at 65 °C; 5 minutes at 72 °C.
   (b) For 32 cycles: 4 seconds at 95 °C; 8 seconds at 65 °C; 16 seconds at 72 °C plus 1 s/cycle (to increase the extension time throughout the PCR).
   (c) Finish the reaction with a final amplification step at 72 °C for 10 minutes and then cool the samples to room temperature.

8. Prepare a 1.3% agarose-TAE gel containing ethidium bromide, 0.3 µg/ml. Submerge the gel in 1x TAE and add 12 µl of the amplification product (10 µl reaction mixtures plus 2 µl of 6x loading buffer) to each well and run the gel at 10 V/cm.

9. Visualize the specific products on a suitable UV light screen and document the results.

C. Interpretation of Results

1. All samples containing the internal control DNA should show a band at 986 bp (see Figure 16.4). A repeat of the PCR run may be performed. If the second run also shows no band for sample and the internal control, the whole procedure should be repeated with new cell culture supernatant.

2. Mycoplasma-positive samples should show a band at 502–520 bp, depending on the mycoplasma species (see Figure 16.4).

3. Contaminations of reagents with mycoplasma-specific DNA or PCR product are revealed by a band in the water control and/or in the negative control sample. Weak mycoplasma-specific bands can occur after treatment of infected cell cultures with antimycoplasma reagents for the elimination of mycoplasmas or when other antibiotics are applied routinely. In these cases, the positive reaction might be due to residual DNA in the culture medium derived from dead mycoplasma cells or to viable mycoplasma cells that are present at a very low titer.

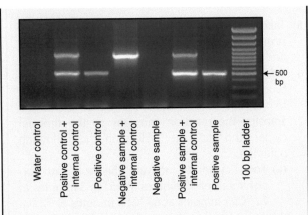

Fig. 16.4. Mycoplasma detection by PCR. PCR products from infected, uninfected, and control cells, electrophoresed on 1.3% agarose and stained with ethidium bromide. The 100-bp ladder consists of the following bands: 100, 200, 300, 400, 500 (strongly stained band), 600, 700, 800, 900, 1031, 1200, 1500, 2000, and 3000 bp. The wild-type mycoplasma bands are about 510 bp and the internal control band is almost 1000 bp. *Source*: Courtesy of Cord Uphoff.

Notes

The highly conserved regions of the 16S rRNA genes enable the selection of primers of wide specificity ("universal primers") which will react with DNA of any mycoplasma and even with the DNA of other prokaryotes, but also of primers with a higher mycoplasmal specificity (Uphoff and Drexler 1999). It is important that no cross-reactions with the DNA of cell lines occur. The amplification is usually performed as a single-step PCR, but for higher sensitivity and specificity a nested PCR can be accomplished. Usually a very high sensitivity level is not required in routine diagnosis as a large number of organisms are present in the sample. The high sensitivity of PCR may cause problems in producing false-positive results due to contamination with target DNA. Another possible problem is false-negative data caused by the inhibition of the *Taq* polymerase by components in the samples. This problem can be overcome by performing a DNA preparation of the sample and by spiking an internal control DNA into the sample. Using a limiting dilution of the internal control DNA, the PCR can easily be standardized. Ideally, the internal control is amplified with the same primers as the wild-type mycoplasma DNA fragment. A further development of the conventional PCR method is real-time PCR. This technique utilizes either SYBR Green for the detection of amplified DNA or a third quenched fluorescently labeled oligonucleotide, which can hybridize specifically to the amplified product

emanating from mycoplasmal DNA. If SYBR Green is used, a melting point determination is required to distinguish mycoplasma-specific amplification products from unspecific amplicons. PCR protocols for the establishment of the tests in any laboratory (Uphoff and Drexler 1999) as well as several PCR kits are commercially available (see Table 16.4).

Molecular hybridization may also be employed for mycoplasma detection, using probes that are complementary to mycoplasma DNA sequence. This approach has been used for microarray analysis using species-specific probes (Kong et al. 2007); it is also employed in some testing kits. The MycoProbe® Mycoplasma Detection Kit (R&D Systems) uses biotin-labeled probes and streptavidin and enzymic amplification steps to produce a colorimetric endpoint, while the Mycoplasma Tissue Culture Non-isotopic Rapid Detection (MTC-NI) System (originally Gen-Probe, later distributed by Merck Millipore) uses a single-stranded DNA probe (labeled with acridinium ester) that is homologous to mycoplasma 16S and 23S rRNA sequences.

DNA fluorescence. Mycoplasma DNA can be stained with Hoechst 33258 or 4′,6-diamidine-2′-phenylindole dihydrochloride (DAPI) for visual detection using fluorescence (Chen 1977). Staining of a contaminated culture results in a fine particulate or filamentous staining pattern in the cytoplasm, on the cell surface, and (with high titers) in surrounding areas when cells are examined under a 40× or 100× objective (see Figure 16.2e, f). The nuclei of the cultured cells are also brightly stained by this method and thereby act as a positive control for the staining procedure. Low levels of contamination from various microorganisms may be evident, making this method a useful general approach for organisms that are difficult to see under the microscope. However, it requires a degree of technical expertise to perform and is less sensitive than some other methods, e.g. PCR (Molla Kazemiha et al. 2016).

PROTOCOL P16.4. DETECTION OF MYCOPLASMA USING HOECHST 33258

Background

This procedure uses an indicator cell line, which is essential for suspension cultures but is also desirable for monolayer cultures. An indicator cell line results in increased sensitivity, can help to avoid problems with false-positive results, and allows enrichment of the contaminant (Volokhov et al. 2008). For this protocol, the indicator cell line should be a good host for mycoplasma but should also spread well in culture, have adequate cytoplasm to reveal any mycoplasma, and demonstrate minimal autofluorescence and debris.

Outline

Culture cells in the absence of antibiotics for at least one week, transfer the supernatant medium to an early log phase indicator culture, and incubate for three to five days. Fix and stain the cells. Look for extranuclear fluorescence (see Figure 16.2e, f).

Materials (sterile or aseptically prepared)

☐ Test culture, e.g. a seven-day monolayer or suspension culture
☐ Indicator culture, e.g. NRK, A549, Vero, MDCK (see also Volokhov et al. (2008))
☐ Culture medium for indicator cells, antibiotic-free
☐ Hanks's balanced salt solution without phenol red (HBSS-PR)
☐ Dulbecco's phosphate buffered saline without Ca^{2+} and Mg^{2+} (DPBS-A)
☐ Chambered slides (see Section 8.6.1) or Petri dishes, 60 mm
☐ Centrifuge tubes or universal containers

Materials (non-sterile)

☐ Hoechst 33258 staining solution, 50 ng/ml (see Appendix B)
☐ Deionized water
☐ Fixative: freshly prepared acetic acid: methanol (see Appendix B)
☐ Glycerol-based mounting medium with fade retardant if required, e.g. Fluoromount-G® or Fluoromount-G Anti-Fade (SouthernBiotech)
☐ Glass coverslips

Procedure

1. Seed indicator cells into chambered slides or Petri dishes at a seeding density that will give 50–60% confluence in four to five days.
2. Remove a sample of medium from the test culture and transfer to a centrifuge tube or universal container. If the cells grow in suspension or there is a large amount of debris, centrifuge the medium and remove the supernatant.
3. One day after seeding the indicator cells, inoculate with supernatant from the test culture, aiming for 150–300 µl supernatant per milliliter of total volume.
4. Incubate the indicator cells for a further three to four days until they reach 50–60% confluence.

At this point, the culture is ready for fixation and staining.

5. Remove the medium and discard it.
6. Rinse the monolayer with HBSS-PR or DPBS-A and discard the rinse.
7. Add fresh HBSS-PR or DPBS-A, diluted 50:50 with fixative. Rinse the monolayer and discard the rinse.
8. Add pure fixative, rinse, and discard the rinse.
9. Add more fixative (approximately 0.5 ml/cm²) and fix the cells for 10 minutes. Remove and discard the fixative.
10. Wash off the fixative with deionized water and discard the wash.
11. Add Hoechst 33258 staining solution for 10 minutes at room temperature. Remove and discard the stain.
12. Rinse the monolayer with deionized water and discard the rinse.
13. Mount a coverslip in a drop of mounting medium and blot off any surplus from the edges.
14. Examine the monolayer by epifluorescence using a 330- to 380-nm excitation filter and an LP 440-nm barrier filter. Mycoplasmas give pinpoints or filaments of fluorescence over the cytoplasm and, sometimes, in intercellular spaces (see Figure 16.2e, f). The pinpoints are close to the limits of resolution with a 50x objective (0.1–1.0 µm) and are usually regular in size and shape. Not all of the cells will necessarily be infected, so as much as possible of the preparation should be scanned before the culture is declared to be mycoplasma-free.

Notes

Step 1: Negative controls should be included, i.e. cells that are not inoculated with supernatant.

Step 2: Test samples may be stored briefly at 4 °C if indicator cells are not ready for inoculation.

Step 4: Cultures must not reach confluence by the end of the assay or staining will be inhibited and the subsequent visualization of mycoplasma will be impaired.

Step 10: Samples may be accumulated at this stage and stained later. If the sample is to be stored for later processing, dry the monolayer completely after removing the fixative.

Step 14: Sometimes a light, uniform staining of the cytoplasm is observed, probably due to RNA. This fluorescence tends to fade on storage of the preparation, and examination the next day (after storing dry and in the

> dark) usually gives clearer results. This artifact never has the sharp punctuate or filamentous appearance of mycoplasma and can be distinguished fairly readily with further experience in observation and comparison to negative control samples.

Fluorescence *in situ* hybridization (FISH) can be used as an alternative staining method. This procedure can be performed directly on fixed cells from the culture to be tested. A fluorescently labeled oligonucleotide hybridizes to a highly conserved region of eubacterial rRNA, which is present in the bacteria in high amounts, and leads to a signal detectable with a fluorescence microscope (Uphoff and Drexler 2014a). The specificity of the probe for bacterial rRNA impedes fluorescence that is not attributed to the contaminants. Thus, the time-consuming cultivation of an indicator cell line can be omitted. The assay can be performed in less than three hours, including some incubation steps.

Biochemical assays. Several assays have been developed to detect mycoplasma-specific enzymes such as arginine deiminase or nucleoside phosphorylase (Schneider and Stanbridge 1975; Levine and Becker 1978). The MycoAlert® Mycoplasma Detection Kit (Lonza) detects adenosine triphosphate (ATP) generated by carbamate kinase and/or acetate kinase, which are active in arginine-hydrolyzing and carbohydrate-fermenting mycoplasmas, respectively. Viable organisms are lyzed and the activity of the enzymes in the culture supernatant is measured by luminescence with luciferase. These ATP-generating enzymes are also found in several other bacteria, which means that the assay is not completely mycoplasma-specific, but this is unlikely to be a concern and may even be an advantage in the context of detecting broader microbial contamination. The MycoAlert assay is less sensitive than some other assay methods (e.g. PCR) (Molla Kazemiha et al. 2016), but it is fast and easy to perform, making it a suitable choice for routine screening. Another biochemical assay that detected toxicity with 6-methylpurine deoxyriboside (MycoTect) was shown to produce false-negative results (Uphoff et al. 1992) and has since been discontinued.

Immunological assays. Some assays have been developed that use mycoplasma-specific polyclonal or monoclonal antibodies; binding is detected using immunofluorescence or enzyme-linked immunosorbent assay (ELISA). While this approach can be effective for individual species (e.g. *M. pneumoniae*), it produces false-negative results *in vitro* when only certain species are addressed (Uphoff et al. 1992). A monoclonal antibody, CCM-2, has been produced that recognizes a common antigen (elongation factor TU) shared by most mycoplasmas (Blazek et al. 1990). The staining is more specific than DNA fluorescence and can thus be applied

directly to the test culture instead of using an indicator cell line. However, the interpretation of the staining can be subjective and difficult to reproduce between users. Immunological methods have largely been superseded in tissue culture laboratories by PCR-based assays, which can detect a broad range of species.

16.4.2 Mycoplasma Eradication

PROTOCOL P16.5. ERADICATION OF MYCOPLASMA CONTAMINATION

Contributed by C. C. Uphoff and H. G. Drexler, Leibniz Institute DSMZ, Braunschweig, Germany.

Background

Several agents are active against mycoplasma, including kanamycin, gentamicin, tylosin (Friend et al. 1966), polyanethol sulfonate (Mardh 1975), and 5-bromouracil in combination with Hoechst 33258 and UV light (Marcus et al. 1980). As some mycoplasma species can penetrate eukaryotic cells (e.g. *M. fermentans*), the eradication process must include both intracellular and extracellular mycoplasmas. Thus, the most successful agents have been antimycoplasmal antibiotics from the class of macrolides (e.g. tiamulin), fluoroquinolones (e.g. ciprofloxacin, enrofloxacin, sparfloxacin), and tetracyclines (e.g. minocyclin, doxycyclin) (Uphoff and Drexler 2002). Several products are available directly for the use in cell culture, such as Mycoplasma Removal Agent (MRA; MP Biomedicals), BM-Cyclin (Sigma-Aldrich), and Plasmocin (InvivoGen); antibiotics are applied singly or in combination therapy. The more recently developed membrane-active peptides are also applied alone or in combination with other antibiotics. The activity of alamethicin, dermaseptin B2, gramicidin S, and surfactin lipopeptide antibiotics against mycoplasmas is reduced in the presence of serum and should be applied in cultures without serum or with reduced serum concentrations to up to 5%. Kits are available from Lonza (MycoZap™ Elimination Reagent) and Minerva Biolabs (Mynox® Gold).

The antibiotic treatment of mycoplasma contamination in cell cultures is highly efficient and about three-quarters of treated cell cultures are cured with one of the mentioned antibiotics. Resistant mycoplasma strains can be eliminated subsequently with an antibiotic from another category. Nevertheless, resistance of the mycoplasma strain and loss of the culture due to cytotoxic effects of the antibiotic can occur. Thus, it is advantageous to have several antibiotics from different classes at hand, which can be used as alternatives. The prophylactic use of antibiotics can lead to the selection of resistant mycoplasma strains and is not recommended. Contaminated cultures should be treated as described in this protocol, using the antibiotics mentioned in Table 16.6 at the recommended concentrations and treatment times. However, this operation should not be undertaken unless it is absolutely essential, and even then it must be performed by experienced hands and in isolation (Uphoff and Drexler 2011, 2014b). It is far safer to discard infected cultures.

Outline

Wash the culture several times in culture medium or DPBS-A by rinsing the monolayer or by centrifugation and resuspension of non-adherent cells. Then grow the culture for one to three weeks with, and at least two weeks without, antibiotics. Test for contamination after post-treatment cultivation.

Materials (sterile or aseptically prepared)

- ☐ Cell culture for mycoplasma eradication
- ☐ Cell culture specific medium
- ☐ DPBS-A
- ☐ Antibiotics (see Table 16.6 and Uphoff and Drexler [2014b] for concentrations)
- ☐ Materials for subculture (see Protocol P14.2)

Materials (non-sterile)

- ☐ Microscope, preferably with 40x and 100x phase-contrast objectives
- ☐ Materials for mycoplasma detection (see Protocols P16.3, P16.4)

Procedure

1. Collect the contaminated medium carefully and dispose of as liquid biological waste (see Protocol P16.1).
2. Wash the cells in culture medium or DPBS-A (dilution can reduce the number of contaminants by two logs with each wash, unless they are adherent to the cells):
   (a) For monolayers, rinse the culture three times with DPBS-A, trypsinize, and wash the cells twice more in DPBS-A by centrifugation and resuspension.
   (b) For suspension cultures, wash the culture in DPBS-A by centrifugation and resuspension.
3. Reseed one or more fresh flasks at high seeding densities.

4. Add antibiotic solution at the recommended concentration (see Table 16.6) (Uphoff and Drexler 2014b). Change the culture medium every two days.
5. Subculture in antibiotic medium for the recommended time (see Table 16.6) (Uphoff and Drexler 2014b). If antibiotics from different classes are applied alternately with BM-Cyclin, cells should be washed before a new antibiotic is added.
6. Remove the antibiotics and culture the cells without them for at least a further two weeks.
7. Check the cultures by PCR or another detection assay (see Protocols P16.3, P16.4).
8. Culture the cells for a further two months without antibiotics, and check to make sure that all contamination has been eliminated.

16.5 VIRAL CONTAMINATION

Experienced tissue culture practitioners have expressed concerns for many years regarding the risk of undetected viruses in cell lines (Fogh et al. 1971; Todaro et al. 1973). Viral material may come from the original tissue sample; for example, the HeLa cell line is known to carry sequence from human papillomavirus-18 (HPV-18), which is a causative agent for cervical carcinoma (Boshart et al. 1984). About two-thirds of the HPV-18 genome is integrated into chromosome 8 near the *myc* proto-oncogene, resulting in its activation (Adey et al. 2013). As with all microbial contaminants, viruses may be transmitted from many other sources including the operator, reagents (e.g. serum or trypsin), and other cell lines (see Section 16.1) (Hay 1991; Merten 2002). Screening of reagents is performed by reputable suppliers to detect the viruses that represent the greatest risk (Marcus-Sekura et al. 2011). Unfortunately, it is impossible to screen reagents for all viruses and the risks from other sources are more difficult to quantify.

Over the last decade, it has become possible to examine large cell line panels for undetected viruses using molecular methods (Uphoff et al. 2010, 2015; Shioda et al. 2018). These studies have detected viral contamination in 3–5% of human cell lines, with the most common single contaminant being Epstein-Barr virus (EBV) in hematopoietic cell lines. This finding is only to be expected, considering that EBV is a causative agent for many lymphomas and is used as an immortalizing agent for lymphoblastoid cell lines (LCLs) (see Section 22.3.1). Other human pathogens were detected in a small number of cell lines (see Table 16.7).

Unexpectedly, about 3% of human cell lines carry γ-retroviruses, with murine leukemia viruses (MLVs) being particularly common (see Table 16.7) (Uphoff et al. 2015).

TABLE 16.6. Antibiotic treatments suitable for eradication of mycoplasmas from cell cultures.

Brand name	Generic names of constituents	Antibiotic category	Typical final concentration (µg/ml)	Duration of treatment
Baytril	Enrofloxacin	Fluoroquinolone	25	1 week
Ciprobay	Ciprofloxacin	Fluoroquinolone	10	2 weeks
MRA	Undisclosed	Fluoroquinolone	0.5	1 week
Zagam	Sparfloxacin	Fluoroquinolone	10	1 week
Plasmocin	1. Undisclosed	Fluoroquinolone	25	2 weeks
	2. Undisclosed	Protein synthesis inhibition		
BM-Cyclin	1. Tiamulin	Pleuromutilin	10	3 weeks
	2. Minocycline	Tetracycline	5	
MycoZap	1. Undisclosed	Membrane-active peptide	—	3 passages
	2. Undisclosed	—	—	
Mynox Gold	1. Surfactin	Membrane-active peptide	—	1 passage
	2. Undisclosed		—	3 passages

Source: courtesy of Cord Uphoff.

Many of these viruses are replication-competent and are believed to come from xenograft passaging in immunocompromised mice (Takeuchi et al. 2008; Sfanos et al. 2011; Zhang et al. 2011). Laboratories with access to xenograft facilities have increased rates of viral contamination; a study of seven independent laboratories showed that 17% (13/78) of samples from groups with xenograft facilities were positive for MLV, compared to 0% (0/50) from groups without xenograft facilities (Zhang et al. 2011).

The biohazard risks that are associated with these viruses may vary considerably. Some cell lines, such as HeLa, carry an incomplete viral genome and are thus unable to produce infectious virions; other cell lines carry full genomic copies that have been integrated into the host genome or persist as episomes (Adey et al. 2013; Cao et al. 2015). Cell lines that carry replication-competent virus may shed infectious virions, leading to a risk of further contamination or transmission to the operator. However, it may be unclear whether the virus is pathogenic in humans or, indeed, in any other species. For example, the xenotropic murine leukemia virus-related virus (XMRV) is produced at high titer by the 22Rv1 cell line, resulting in transmission to both mice and primates and contamination of other cell lines (Arias and Fan 2014). XMRV is now believed to have arisen from a laboratory artifact; it is likely that two viruses were co-packaged during xenografting and underwent recombination (Delviks-Frankenberry et al. 2012).

16.5.1 Detection of Viral Contamination

Multiple methods can be used to detect viral contamination, including (i) immunological assays, e.g. ELISA to detect viral proteins; (ii) enzymatic assays, e.g. reverse transcriptase assays to identify active γ-retrovirus; (iii) cytogenetic analysis, e.g. FISH to detect chromosomal integration; (iv) PCR-based assays to amplify viral sequence; (v) genomic analysis, e.g. screening whole exome sequence or RNA-seq data to detect expressed viral sequence; (vi) proteomic analysis to detect viral proteins; and (vii) electron microscopy to visualize intracellular viruses. Many cell lines have publicly available genomic and proteomic datasets that can be analyzed to detect viral contaminants (Chernobrovkin and Zubarev 2014; Cao et al. 2015; Uphoff et al. 2019). PCR-based assays can be highly effective for specific viruses, e.g. to screen for EBV in hematopoietic cell lines. Comparison of PCR-based assays and RNA-seq analysis has shown a high degree of concordance (Uphoff et al. 2019).

Although molecular methods such as PCR are widely accessible, their results must be interpreted with caution. Nucleic acid contamination can affect molecular procedures, leading to the appearance of false-positive results (Uphoff et al. 2015). A positive result should always be confirmed independently; for example, a positive result for MLV from PCR might be confirmed by performing a reverse transcriptase assay. Contamination with mouse DNA is known to affect many common laboratory reagents, including DNA binding columns (Erlwein et al. 2011).

16.5.2 Eradication of Viral Contamination

Theoretically, viral contamination can be eliminated using antiviral drugs. Studies of murine retroviruses have shown that many are intrinsically resistant to the antiretroviral drugs that are used in the clinic, but a small subset of agents appears to be effective (Powell et al. 1999; Smith et al. 2010). In practice, however, it is highly unlikely that a cell line can be completely cleared of virus, particularly if genomic integration has occurred. This leaves disposal (see Protocol P16.1) or tolerance as the only options. At the very least, the presence of

TABLE 16.7. Viral contamination of cell lines.

Viral contamination (including partial or complete viral genome)[a]	Cell lines carrying viral contamination[b]	References
B19V	FPC5JTO, XP2SA	Shioda et al. (2018)
BLV	FLK-BLV	Unpublished
BPyV	SK-BR-3	Uphoff et al. (2019)
BVDV	EBL, GM-7373	Unpublished
EBV	B95-8, BC-1, BC-2, BDCM, BONNA-12, CCRF-SB, CI-1, CRO-AP2, CRO-AP5, Daudi, DOHH-2 (EBV+ and EBV- exist), EB-1, EHEB, GRANTA-519, HAIR-M, HBL-6, HC-1, Hs 611.T, HS-Sultan, HuNS1, IM-9, ITSM, Jiyoye, JVM-2, JVM-3, JVM-13, KAI13, KE-97, MEC-1, MEC-2, Namalwa, NK-92, OCI-LY-19, P32/ISH, P3HR-1, RAFI, Raji, RM-P1, RPMI 1788, SLVL, SNU-719, T2, VAL, WIL2-NS, YT	Cesarman et al. (1995), Drexler et al. (1998), Uphoff et al. (2010), Cao et al. (2015), Shioda et al. (2018), and unpublished
β-Retrovirus (Mason-Pfizer monkey virus)	OC 314	Unpublished
γ-Retroviruses (including FeLV, MLVs, GALV, SMRV, XMRV)	1065met, 22RV1, 253J-BV, A3.01/F7, AC-1M46, BHY, BICR 18, BT-B, CAK1, COLO-677, DEL, DG-75, EKVX, EVSA-T, GB-1, Jurkat J6, KARPAS-1106P, KELLY, Ki-JK, KMM-1, KYSE-70, L3.3, LAMA-87, LAPC-4, LCL-HO, Li-7, LMSU, LX44, LX47, LX48, LXF-289, MES-SA, ML-1, Namalwa, NB-1, NCEB-1, NCI-N417, QGP-1, S-117, SCLC-21H, SK-GT-2, SK-MEL-1, SNU-C4, TYK-nu, TZM-bl, VCaP, XCL1, XCL3, YD-10B, YD-15, YD-38, YD-8	Raisch et al. (2003), Takeuchi et al. (2008), Uphoff et al. (2010), Sfanos et al. (2011), Zhang et al. (2011), Cao et al. (2015, Uphoff et al. (2015), and unpublished
HBV	Alexander (PLC/PRF/5), HEP-3B, huH-1, JHH-7, SNU-739, SNU-761, SNU-878, SNU-886	Lee et al. (1999), Uphoff et al. (2010), Shioda et al. (2018)
HHV-6	HUV-EC-C	Shioda et al. (2018)
HHV-7	AT(L)7KY	Shioda et al. (2018)
HHV-8 (KSHV)	BC-1, BC-2, BCP-1, CRO-AP2, CRO-AP3, CRO-AP5, HBL-6	Cesarman et al. (1995), Uphoff et al. (1998), Cao et al. (2015)
HPV-16	HeLa	Boshart et al. (1984), Adey et al. (2013)
HPV-18	NCI-H1341	Unpublished
HTLV-1	HuT 102, KHM-3S, MJ, MT-1, MT-2, MT-3, MT-4	Cao et al. (2015), Shioda et al. (2018)
Parainfluenza virus 5	AGS, JHH-6, KMRC-3	Unpublished

Source: Amanda Capes-Davisand Cord Uphoff; all unpublished data courtesy of Cord Uphoff.

[a]It is impossible to include all cases of viral contamination here. Many cell lines are untested; viral contaminants are also reported as isolated cases, which are difficult to find in the scientific literature. Always perform your own literature search to find out if a specific cell line is reported to carry viral contamination. Other resources that provide information on viral contamination include cell repository catalogs (see Table 15.3) and Cellosaurus (https://web.expasy.org/cellosaurus).

[b]Cell lines that have been transduced with viruses or viral sequence are not listed here; it is expected that those cell lines will continue to carry viral genetic sequence. Derivatives of the cell lines listed here are not included, due to space constraints, but may also carry viral contamination (including misidentified cell lines).

BLV, Bovine leukemia virus; BPyV, bovine polyomavirus; BVDV, bovine viral diarrhea virus; EBV, Epstein-Barr virus; FeLV, feline leukemia/sarcoma virus; GALV, gibbon ape leukemia virus; HBV, hepatitis B virus; HHV, human herpesvirus; HPV, human papillomavirus; HTLV, human T-cell lymphotropic virus; KSHV, Kaposi's sarcoma herpesvirus; MLV, murine leukemia virus; SMRV, squirrel monkey retrovirus; XMRV, xenotropic murine leukemia virus-related virus.

viral contamination should be considered during risk assessments and the question should be asked whether the cell line is worth using as a research model – particularly if it is known to produce infectious virions, resulting in potential risks for the operator and for other cell lines (Arias and Fan 2014).

16.6 DEALING WITH PERSISTENT CONTAMINATION

Many laboratories have suffered from periods of contamination that appear to be refractory to any, or to all, of the recommended remedies (see Table 16.1; see also Section 31.1). An increase in the contamination rate typically stems from deterioration in aseptic technique, but other factors are also likely to play a part. The constant use of antibiotics favors the development of chronic contamination (see Section 9.6.2). Many microorganisms are inhibited, but not killed, by antibiotics; such organisms will persist, undetected for most of the time but resurfacing when conditions change. Other contributing factors include an increased spore count in the atmosphere, equipment contamination (particularly in incubators, the coldroom, or the refrigerator), or a fault in sterilization procedures. Sterilization faults relate not only to the sterilizing equipment but also to the way that it is packed or the monitoring of the sterilization cycle.

There is no easy resolution to persistent contamination, other than to follow the previous recommendations in a logical and analytical fashion. Always gather data on the episodes of contamination and look for patterns, paying particular attention to changes in routine handling or maintenance and to new workers, new suppliers, and new equipment. A slight change in practices, the introduction of new personnel, or an increase in activity as more people use a facility can all contribute to an increase in the rate of contamination.

In some cases, the fault will come back to lack of training. It is often assumed that experienced practitioners can perform basic procedures such as sterilization, but in today's laboratories, these "basic" tasks are often performed by other staff. All new tissue culture personnel should be made familiar with sterilization procedures (see Chapter 11), even if they will not be called upon to carry out these procedures themselves. They should also be aware of the location of, and distinction between, sterile and non-sterile stocks (see Section 4.2.6). A simple error by a newcomer can cause severe problems that last for days or even weeks before the cause is discovered. Procedures must remain stringent, even if the reason is not always obvious to the operator, and alterations in routine should not be made casually. If correct practices are maintained, contamination may not be eliminated entirely, but it will be reduced considerably and detected early.

Suppliers

Supplier	URL
Agilent	www.agilent.com
AppliChem	http://www.itwreagents.com/countryselector
BD Biosciences	www.bdbiosciences.com
Biological Industries	http://www.bioind.com/worldwide
Charles River Laboratories	www.criver.com
InvivoGen	www.invivogen.com
LANXESS Deutschland GmbH	www.virkon.com
Lonza	www.lonza.com
Merck Millipore	www.merckmillipore.com
Minerva Biolabs	www.minerva-biolabs.com/en
MP BioMedicals	www.mpbio.com
Oxoid	http://www.oxoid.com/uk/blue/index.asp
PromoCell	www.promocell.com
R&D Systems	www.rndsystems.com
Roche CustomBiotech	custombiotech.roche.com/home.html
Sartorius	http://www.sartorius.com/en
Sigma-Aldrich (Merck)	http://www.sigmaaldrich.com/life-science/cell-culture.html
SouthernBiotech	www.southernbiotech.com
ThermoFisher Scientific	http://www.thermofisher.com/us/en/home/life-science/cell-culture.html
WAK-Chemie Medical GmbH	www.wak-chemie.net

REFERENCES

Adey, A., Burton, J.N., Kitzman, J.O. et al. (2013). The haplotype-resolved genome and epigenome of the aneuploid HeLa cancer cell line. *Nature* 500 (7461): 207–211. https://doi.org/10.1038/nature12064.

Aldecoa-Otalora, E., Langdon, W., Cunningham, P. et al. (2009). Unexpected presence of mycoplasma probes on human microarrays. *Biotechniques* 47 (6): 1013–1015. https://doi.org/10.2144/000113271.

Arias, M. and Fan, H. (2014). The saga of XMRV: a virus that infects human cells but is not a human virus. *Emerg. Microbes Infect.* 3 (4): e. https://doi.org/10.1038/emi.2014.25.

Barile, M.F., Malizia, W.F., and Riggs, D.B. (1962). Incidence and detection of pleuropneumonia-like organisms in cell cultures by fluorescent antibody and cultural procedures. *J. Bacteriol.* 84: 130–136.

Barile, M.F., Hopps, H.E., and Grabowski, M. (1978). Incidence and sources of mycoplasma contamination: a brief review. In: *Mycoplasma Infection of Cell Cultures* (eds. G.J. McGarrity, D.G. Murphy and W.W. Nichols), 35–45. New York: Plenum Press.

Bates, M. K., D'Onofrio, J., Parrucci, M. L., et al. (2014). Back to basics: proper care and maintenance for your cell culture incubator. ThermoScientific Technical Notes. https://beta-static.fishersci.com/content/dam/fishersci/en_us/documents/programs/scientific/technical-documents/instruction-sheets/thermo-scientific-incubator-care-instruction.pdf (accessed 20 November 2018).

Blazek, R., Schmitt, K., Krafft, U. et al. (1990). Fast and simple procedure for the detection of cell culture mycoplasmas using a single monoclonal antibody. *J. Immunol. Methods* 131 (2): 203–212. https://doi.org/10.1016/0022-1759(90)90191-w.

Boshart, M., Gissmann, L., Ikenberg, H. et al. (1984). A new type of papillomavirus DNA: its presence in genital cancer biopsies and in cell lines derived from cervical cancer. *EMBO J.* 3 (5): 1151–1157.

Buehring, G.C., Valesco, M., and Pan, C.Y. (1995). Cell culture contamination by mycobacteria. *In Vitro Cell Dev. Biol. Anim.* 31 (10): 735–737. https://doi.org/10.1007/BF02634111.

Calcutt, M.J., Szikriszt, B., Poti, A. et al. (2015). Genome sequence analysis of *Mycoplasma* sp. HU2014, isolated from tissue culture. *Genome Announc.* 3 (5) https://doi.org/10.1128/genomeA.01086-15.

Cao, S., Strong, M.J., Wang, X. et al. (2015). High-throughput RNA sequencing-based virome analysis of 50 lymphoma cell lines from the Cancer Cell Line Encyclopedia project. *J. Virol.* 89 (1): 713–729. https://doi.org/10.1128/JVI.02570-14.

Cesarman, E., Moore, P.S., Rao, P.H. et al. (1995). In vitro establishment and characterization of two acquired immunodeficiency syndrome-related lymphoma cell lines (BC-1 and BC-2) containing Kaposi's sarcoma-associated herpesvirus-like (KSHV) DNA sequences. *Blood* 86 (7): 2708–2714.

Chen, T.R. (1977). In situ detection of mycoplasma contamination in cell cultures by fluorescent Hoechst 33258 stain. *Exp. Cell. Res.* 104 (2): 255–262. https://doi.org/10.1016/0014-4827(77)90089-1.

Chernobrovkin, A.L. and Zubarev, R.A. (2014). Detection of viral proteins in human cells lines by xeno-proteomics: elimination of the last valid excuse for not testing every cellular proteome dataset for viral proteins. *PLoS One* 9 (3): e91433. https://doi.org/10.1371/journal.pone.0091433.

Clarke, J.H., Norman, J.A., and Lavery, E. (1989). Some observations on contamination of animal cell cultures by the fungus *Aspergillus fumigatus* and suggested control measures. *Cell Biol. Int. Rep.* 13 (9): 773–779. https://doi.org/10.1016/0309-1651(89)90054-4.

Cobo, F. and Concha, A. (2007). Environmental microbial contamination in a stem cell bank. *Lett. Appl. Microbiol.* 44 (4): 379–386. https://doi.org/10.1111/j.1472-765X.2006.02095.x.

Coriell, L.L. (1962). Detection and elimination of contaminating organisms. *Natl Cancer Inst. Monogr.* 7: 33–53.

Dabrazhynetskaya, A., Furtak, V., Volokhov, D. et al. (2014). Preparation of reference stocks suitable for evaluation of alternative NAT-based mycoplasma detection methods. *J. Appl. Microbiol.* 116 (1): 100–108. https://doi.org/10.1111/jam.12352.

Del Giudice, R.A. (1998). M-CMRL, a new axenic medium to replace indicator cell cultures for the isolation of all strains of mycoplasma hyorhinis. *In Vitro Cell Dev. Biol. Anim.* 34 (2): 88–89. https://doi.org/10.1007/s11626-998-0087-9.

Delviks-Frankenberry, K., Cingoz, O., Coffin, J.M. et al. (2012). Recombinant origin, contamination, and de-discovery of XMRV. *Curr. Opin. Virol.* 2 (4): 499–507. https://doi.org/10.1016/j.coviro.2012.06.009.

Drexler, H.G. and Uphoff, C.C. (2002). Mycoplasma contamination of cell cultures: incidence, sources, effects, detection, elimination, prevention. *Cytotechnology* 39 (2): 75–90. https://doi.org/10.1023/A:1022913015916.

Drexler, H.G., Uphoff, C.C., Gaidano, G. et al. (1998). Lymphoma cell lines: in vitro models for the study of HHV-8+ primary effusion lymphomas (body cavity-based lymphomas). *Leukemia* 12 (10): 1507–1517. https://doi.org/10.1038/sj.leu.2401160.

Drexler, H.G., Dirks, W.G., MacLeod, R.A. et al. (2017). False and mycoplasma-contaminated leukemia-lymphoma cell lines: time for a reappraisal. *Int. J. Cancer* 140 (5): 1209–1214. https://doi.org/10.1002/ijc.30530.

Erlwein, O., Robinson, M.J., Dustan, S. et al. (2011). DNA extraction columns contaminated with murine sequences. *PLoS One* 6 (8): e23484. https://doi.org/10.1371/journal.pone.0023484.

Fogh, J., Holmgren, N.B., and Ludovici, P.P. (1971). A review of cell culture contaminations. *In Vitro* 7 (1): 26–41. https://doi.org/10.1007/bf02619002.

Freedman, L.P., Gibson, M.C., Wisman, R. et al. (2015). The culture of cell culture practices and authentication – results from a 2015 survey. *Biotechniques* 59 (4): 189–190, 192. doi: https://doi.org/10.2144/000114344.

Friend, C., Patuleia, M.C., and Nelson, J.B. (1966). Antibiotic effect of tylosin on a mycoplasma contaminant in a tissue culture leukemia cell line. *Proc. Soc. Exp. Biol. Med.* 121 (4): 1009–1010. https://doi.org/10.3181/00379727-121-30950.

Geraghty, R.J., Capes-Davis, A., Davis, J.M. et al. (2014). Guidelines for the use of cell lines in biomedical research. *Br. J. Cancer* 111 (6): 1021–1046. https://doi.org/10.1038/bjc.2014.166.

Gray, J.S., Birmingham, J.M., and Fenton, J.I. (2010). Got black swimming dots in your cell culture? Identification of *Achromobacter* as a novel cell culture contaminant. *Biologicals* 38 (2): 273–277. https://doi.org/10.1016/j.biologicals.2009.09.006.

Hay, R.J. (1991). Operator-induced contamination in cell culture systems. *Dev. Biol. Stand.* 75: 193–204.

Hayflick, L. (1965). Tissue cultures and mycoplasmas. *Tex. Rep. Biol. Med.* 23 (Suppl 1): 285.

Hegde, J., Bashetty, K., and Krishnakumar, U.G. (2012). Quantity of sodium thiosulfate required to neutralize various concentrations of sodium hypochlorite. *Asian J. Pharm. Health Sci.* 2 (3): 390–393.

Hernandez, A., Martro, E., Matas, L. et al. (2000). Assessment of in-vitro efficacy of 1% Virkon against bacteria, fungi, viruses and spores by means of AFNOR guidelines. *J. Hosp. Infect.* 46 (3): 203–209. https://doi.org/10.1053/jhin.2000.0818.

IEST (2018). IEST-RP-CC018 Cleanroom housekeeping: operating and monitoring procedures. https://www.iest.org/standards-rps/recommended-practices/iest-rp-cc018 (accessed 30 October 2019).

Kong, H., Volokhov, D.V., George, J. et al. (2007). Application of cell culture enrichment for improving the sensitivity of mycoplasma detection methods based on nucleic acid amplification technology (NAT). *Appl. Microbiol. Biotechnol.* 77 (1): 223–232. https://doi.org/10.1007/s00253-007-1135-1.

Kundsin, R.B. (1968). Aerosols of mycoplasmas, L forms, and bacteria: comparison of particle size, viability, and lethality of ultraviolet radiation. *Appl. Microbiol.* 16 (1): 143–146.

Kundsin, R.B. and Praznik, J. (1967). Pharyngeal carriage of mycoplasma species in healthy young adults. *Am. J. Epidemiol.* 86 (3): 579–583. https://doi.org/10.1093/oxfordjournals.aje.a120767.

Langdon, S.P. (2004). Cell culture contamination. In: *Cancer Cell Culture: Methods and Protocols* (ed. S.P. Langdon), 309–317. Totowa, NJ: Humana Press.

Langdon, W.B. (2014). Mycoplasma contamination in the 1000 Genomes Project. *BioData Min.* 7: 3. https://doi.org/10.1186/1756-0381-7-3.

Lee, J.H., Ku, J.L., Park, Y.J. et al. (1999). Establishment and characterization of four human hepatocellular carcinoma cell lines containing hepatitis B virus DNA. *World J. Gastroenterol.* 5 (4): 289–295. https://doi.org/10.3748/wjg.v5.i4.289.

Lelong-Rebel, I.H., Piemont, Y., Fabre, M. et al. (2009). *Mycobacterium avium-intracellulare* contamination of mammalian cell cultures. *In Vitro Cell Dev. Biol. Anim.* 45 (1–2): 75–90. https://doi.org/10.1007/s11626-008-9143-8.

Levine, E.M. and Becker, B.G. (1978). Biochemical methods for detecting mycoplasma contamination. In: *Mycoplasma Infection of Cell Cultures* (eds. G.J. McGarrity, D.G. Murphy and W.W. Nichols), 87–104. New York: Plenum Press.

Logunov, D.Y., Scheblyakov, D.V., Zubkova, O.V. et al. (2008). Mycoplasma infection suppresses p53, activates NF-kappaB and cooperates with oncogenic Ras in rodent fibroblast transformation. *Oncogene* 27 (33): 4521–4531. https://doi.org/10.1038/onc.2008.103.

Marcus, M., Lavi, U., Nattenberg, A. et al. (1980). Selective killing of mycoplasmas from contaminated mammalian cells in cell cultures. *Nature* 285 (5767): 659–661. https://doi.org/10.1038/285659a0.

Marcus-Sekura, C., Richardson, J.C., Harston, R.K. et al. (2011). Evaluation of the human host range of bovine and porcine viruses that may contaminate bovine serum and porcine trypsin used in the manufacture of biological products. *Biologicals* 39 (6): 359–369. https://doi.org/10.1016/j.biologicals.2011.08.003.

Mardh, P.A. (1975). Elimination of mycoplasmas from cell cultures with sodium polyanethol sulphonate. *Nature* 254 (5500): 515–516. https://doi.org/10.1038/254515a0.

McGarrity, G.J. (1976). Spread and control of mycoplasmal infection of cell cultures. *In Vitro* 12 (9): 643–648.

McGarrity, G.J. and Coriell, L.L. (1971). Procedures to reduce contamination of cell cultures. *In Vitro* 6 (4): 257–265. https://doi.org/10.1007/bf02625938.

McGarrity, G.J., Vanaman, V., and Sarama, J. (1978). Methods of prevention, control and elimination of mycoplasma infection. In: *Mycoplasma Infection of Cell Cultures* (eds. G.J. McGarrity, D.G. Murphy and W.W. Nichols), 213–241. New York: Plenum Press.

Merten, O.W. (2002). Virus contaminations of cell cultures – a biotechnological view. *Cytotechnology* 39 (2): 91–116. https://doi.org/10.1023/A:1022969101804.

Molla Kazemiha, V., Bonakdar, S., Amanzadeh, A. et al. (2016). Real-time PCR assay is superior to other methods for the detection of mycoplasma contamination in the cell lines of the National Cell Bank of Iran. *Cytotechnology* 68 (4): 1063–1080. https://doi.org/10.1007/s10616-015-9862-0.

Nagatomo, H., Takegahara, Y., Sonoda, T. et al. (2001). Comparative studies of the persistence of animal mycoplasmas under different environmental conditions. *Vet. Microbiol.* 82 (3): 223–232. https://doi.org/10.1016/s0378-1135(01)00385-6.

Nims, R.W. and Price, P.J. (2017). Best practices for detecting and mitigating the risk of cell culture contaminants. *In Vitro Cell Dev. Biol. Anim.* 53 (10): 872–879. https://doi.org/10.1007/s11626-017-0203-9.

O'Connell, R.C., Wittler, R.G., and Faber, J.E. (1964). Aerosols as a source of widespread mycoplasma contamination of tissue cultures. *Appl. Microbiol.* 12: 337–342.

Olarerin-George, A.O. and Hogenesch, J.B. (2015). Assessing the prevalence of mycoplasma contamination in cell culture via a survey of NCBI's RNA-seq archive. *Nucleic Acids Res.* 43 (5): 2535–2542. https://doi.org/10.1093/nar/gkv136.

Paul, J. (1975). *Cell and Tissue Culture*, 5e. Edinburgh: Churchill Livingstone.

Powell, S.K., Artlip, M., Kaloss, M. et al. (1999). Efficacy of antiretroviral agents against murine replication-competent retrovirus infection in human cells. *J. Virol.* 73 (10): 8813–8816.

Raisch, K.P., Pizzato, M., Sun, H.Y. et al. (2003). Molecular cloning, complete sequence, and biological characterization of a xenotropic murine leukemia virus constitutively released from the human B-lymphoblastoid cell line DG-75. *Virology* 308 (1): 83–91. https://doi.org/10.1016/s0042-6822(02)00074-0.

Razin, S., Yogev, D., and Naot, Y. (1998). Molecular biology and pathogenicity of mycoplasmas. *Microbiol. Mol. Biol. Rev.* 62 (4): 1094–1156.

Robinson, L.B., Wichelhausen, R.H., and Roizman, B. (1956). Contamination of human cell cultures by pleuropneumonia-like organisms. *Science* 124 (3232): 1147–1148. https://doi.org/10.1126/science.124.3232.1147.

Russell, A.D. (1990). Bacterial spores and chemical sporicidal agents. *Clin. Microbiol. Rev.* 3 (2): 99–119. https://doi.org/10.1128/cmr.3.2.99.

Schneider, E.L. and Stanbridge, E.J. (1975). A simple biochemical technique for the detection of mycoplasma contamination of cultured cells. *Methods Cell Biol.* 10: 277–290. https://doi.org/10.1016/s0091-679x(08)60742-6.

Sfanos, K.S., Aloia, A.L., Hicks, J.L. et al. (2011). Identification of replication competent murine gammaretroviruses in commonly used prostate cancer cell lines. *PLoS One* 6 (6): e20874. https://doi.org/10.1371/journal.pone.0020874.

Shioda, S., Kasai, F., Watanabe, K. et al. (2018). Screening for 15 pathogenic viruses in human cell lines registered at the JCRB Cell Bank: characterization of in vitro human cells by viral infection. *R. Soc. Open Sci.* 5 (5): 172472. https://doi.org/10.1098/rsos.172472.

Sigma-Aldrich (2018). Cell culture antibiotic selection guide. https://www.sigmaaldrich.com/life-science/cell-culture/learning-center/antibiotic-selector.html (accessed 25 October 2018).

Smith, R.A., Gottlieb, G.S., and Miller, A.D. (2010). Susceptibility of the human retrovirus XMRV to antiretroviral inhibitors. *Retrovirology* 7: 70. https://doi.org/10.1186/1742-4690-7-70.

Stanbridge, E. (1971). Mycoplasmas and cell cultures. *Bacteriol. Rev.* 35 (2): 206–227.

Takeuchi, Y., McClure, M.O., and Pizzato, M. (2008). Identification of gammaretroviruses constitutively released from cell lines used for human immunodeficiency virus research. *J. Virol.* 82 (24): 12585–12588. https://doi.org/10.1128/JVI.01726-08.

Tarshis, M., Yavlovich, A., Katzenell, A. et al. (2004). Intracellular location and survival of *Mycoplasma penetrans* within HeLa cells. *Curr. Microbiol.* 49 (2): 136–140. https://doi.org/10.1007/s00284-004-4298-3.

Taylor-Robinson, D. (1978). Cultural and serologic procedures for mycoplasmas in tissue culture. In: *Mycoplasma Infection of Cell Cultures* (eds. G.J. McGarrity, D.G. Murphy and W.W. Nichols), 47–56. New York: Plenum Press.

Taylor-Robinson, D. and Bebear, C. (1997). Antibiotic susceptibilities of mycoplasmas and treatment of mycoplasmal infections. *J. Antimicrob. Chemother.* 40 (5): 622–630. https://doi.org/10.1093/jac/40.5.622.

Todaro, G.J., Arnstein, P., Parks, W.P. et al. (1973). A type-C virus in human rhabdomyosarcoma cells after inoculation into NIH Swiss mice treated with antithymocyte serum. *Proc. Natl Acad. Sci. U.S.A.* 70 (3): 859–862. https://doi.org/10.1073/pnas.70.3.859.

Tsai, S., Wear, D.J., Shih, J.W. et al. (1995). Mycoplasmas and oncogenesis: persistent infection and multistage malignant transformation. *Proc. Natl Acad. Sci. U.S.A.* 92 (22): 10197–10201.

Uphoff, C.C. and Drexler, H.G. (1999). Detection of mycoplasma contaminations in cell cultures by PCR analysis. *Hum. Cell* 12 (4): 229–236.

Uphoff, C.C. and Drexler, H.G. (2002). Comparative antibiotic eradication of mycoplasma infections from continuous cell lines. *In Vitro Cell. Dev. Biol. Anim.* 38 (2): 86–89. https://doi.org/10.1290/1071-2690(2002)038<0086:CAEOMI>2.0.CO;2.

Uphoff, C.C. and Drexler, H.G. (2011). Elimination of mycoplasmas from infected cell lines using antibiotics. *Methods Mol. Biol.* 731: 105–114. https://doi.org/10.1007/978-1-61779-080-5_9.

Uphoff, C.C. and Drexler, H.G. (2014a). Detection of *Mycoplasma* contamination in cell cultures. *Curr. Protoc. Mol. Biol.* 106: 28.4.1–28.4.4. https://doi.org/10.1002/0471142727.mb2804s106.

Uphoff, C.C. and Drexler, H.G. (2014b). Eradication of *Mycoplasma* contaminations from cell cultures. *Curr. Protoc. Mol. Biol.* 106: 28.5.1–28.5.12. https://doi.org/10.1002/0471142727.mb2805s106.

Uphoff, C.C., Gignac, S.M., and Drexler, H.G. (1992). Mycoplasma contamination in human leukemia cell lines. I. Comparison of various detection methods. *J. Immunol. Methods* 149 (1): 43–53. https://doi.org/10.1016/s0022-1759(12)80047-0.

Uphoff, C.C., Carbone, A., Gaidano, G. et al. (1998). HHV-8 infection is specific for cell lines derived from primary effusion (body cavity-based) lymphomas. *Leukemia* 12 (11): 1806–1809. https://doi.org/10.1038/sj.leu.2401194.

Uphoff, C.C., Denkmann, S.A., Steube, K.G. et al. (2010). Detection of EBV, HBV, HCV, HIV-1, HTLV-I and -II, and SMRV in human and other primate cell lines. *J. Biomed. Biotechnol.* 2010: 904767. https://doi.org/10.1155/2010/904767.

Uphoff, C.C., Lange, S., Denkmann, S.A. et al. (2015). Prevalence and characterization of murine leukemia virus contamination in human cell lines. *PLoS One* 10 (4): e0125622. https://doi.org/10.1371/journal.pone.0125622.

Uphoff, C.C., Pommerenke, C., Denkmann, S.A. et al. (2019). Screening human cell lines for viral infections applying RNA-Seq data analysis. *PLoS One* 14 (1): e0210404. https://doi.org/10.1371/journal.pone.0210404.

Volokhov, D.V., Kong, H., George, J. et al. (2008). Biological enrichment of *Mycoplasma* agents by cocultivation with permissive cell cultures. *Appl. Environ. Microbiol.* 74 (17): 5383–5391. https://doi.org/10.1128/AEM.00720-08.

Volokhov, D.V., Graham, L.J., Brorson, K.A. et al. (2011). Mycoplasma testing of cell substrates and biologics: review of alternative non-microbiological techniques. *Mol. Cell. Probes* 25 (2–3): 69–77. https://doi.org/10.1016/j.mcp.2011.01.002.

WHO (2004). Laboratory biosafety manual. http://www.who.int/csr/resources/publications/biosafety/who_cds_csr_lyo_2004_11/en (accessed 9 May 2018).

Woo, P.C., Lau, S.K., Teng, J.L. et al. (2008). Then and now: use of 16S rDNA gene sequencing for bacterial identification and discovery of novel bacteria in clinical microbiology laboratories. *Clin. Microbiol. Infect.* 14 (10): 908–934. https://doi.org/10.1111/j.1469-0691.2008.02070.x.

Yavlovich, A., Tarshis, M., and Rottem, S. (2004). Internalization and intracellular survival of *Mycoplasma pneumoniae* by non-phagocytic cells. *FEMS Microbiol. Lett.* 233 (2): 241–246. https://doi.org/10.1016/j.femsle.2004.02.016.

Zhang, Y.A., Maitra, A., Hsieh, J.T. et al. (2011). Frequent detection of infectious xenotropic murine leukemia virus (XMLV) in human cultures established from mouse xenografts. *Cancer Biol. Ther.* 12 (7): 617–628.

CHAPTER 17

Cell Line Misidentification and Authentication

In the 1960s, a committee of tissue culture experts set out to collect a set of "certified" cell lines for use by the research community (Ledley 1964). The committee developed comprehensive procedures for cell banking, resulting in the cryopreservation of seed stocks that were extensively tested for contamination (see Sections 15.5, 16.4). Although the committee members were aware of the risk of contamination from other species (which was initially reported in the 1950s), they were unable to test for contamination arising from the same species (Rothfels et al. 1959; ATCC SDO Workgroup ASN-0002 2010). That inability would prove to be disastrous. Only a few years after the committee announced that 23 "certified" cell lines were available, Stanley Gartler used 14 of its cell lines to study isoenzyme variation, looking at glucose-6-phosphate dehydrogenase (G6PD) and phosphoglucomutase (PGM) (Gartler 1967, 1968). All of the cell lines in Gartler's panel carried the same isoenzyme variants, including G6PD type A, which was not found in Caucasian individuals. Gartler concluded that all of these cell lines were contaminated with HeLa, which was known to come from an African-American woman (Gartler 1968; Skloot 2010). Gartler's findings were confirmed and extended by other researchers, including Walter Nelson-Rees, who pioneered the use of chromosomal markers to detect intraspecies contamination (Nelson-Rees et al. 1981). Many of the cell lines in Gartler's original panel continue to be widely used more than 50 years later, alongside more than 400 other misidentified cell lines (ICLAC 2019c). This chapter looks at the problem of cell line misidentification and its detection using authentication testing.

17.1 TERMINOLOGY: CROSS-CONTAMINATION, MISIDENTIFICATION, AND AUTHENTICATION

The term "cross-contamination" refers to the introduction of cells from another culture (Coriell 1962). The common assumption is that cross-contamination results in a mixed culture, where the original, authentic material persists alongside the contaminant. This may occur with a small number of cell types; a mixed culture containing mesenchymal stromal cells (MSCs) and HT-1080 fibrosarcoma cells can persist for three to four months, which may be caused by an antitumor effect from the MSCs that persist until they undergo senescence (Garcia et al. 2010). In practice, however, the mixed culture is a transient phenomenon. Most studies show that one population is overgrown by the other within a limited number of passages, particularly if a high split ratio is used (Nims et al. 1998; Jordan et al. 2012; Yu et al. 2015). One experiment, which introduced 100 CHO-K1 cells into a confluent flask of MRC-5 cells, demonstrated that the original culture was overgrown by the contaminant within four passages (25 days) (Nims et al. 1998).

In this book, the term "misidentification" refers to a cell line that no longer corresponds to the donor or species from which it was established (ICLAC 2019b). While this may occur due to cross-contamination, the cause of misidentification is often unknown and may arise due to mislabeling or other mechanisms. The definition of "misidentification" used here may seem overly narrow, as it does not extend to other characteristics of the cell line, e.g. tissue or disease of origin. A narrowly defined term is appropriate because

Freshney's Culture of Animal Cells: A Manual of Basic Technique and Specialized Applications, Eighth Edition. Amanda Capes-Davis and R. Ian Freshney.
© 2021 John Wiley & Sons Ltd. Published 2021 by John Wiley & Sons Ltd.
Companion website: www.wiley.com/go/freshney/cellculture8

of the likelihood that a culture's tissue- or disease-specific markers will change *in vitro*, leading to incorrect conclusions regarding its origins (see Sections 3.4, 17.3.1, 18.3.2). A cell line that comes from the correct donor but does not come from the reported tissue or disease (e.g. due to an incorrect diagnosis at the time the cell line was established) is referred to here as a "misclassified" cell line.

The term "authentication" refers to a process of testing that is used to confirm or verify the identity of a cell line, demonstrating that it is derived from the correct species and donor (ICLAC 2019b). Again, this is a relatively narrow definition that differs from some other publications, where authentication is taken to represent the entirety of cell line characterization (Nims and Reid 2017). A narrow definition is particularly relevant to cross-contamination, where cells are overgrown by HeLa or other rapidly growing tumor cell lines; testing for cells from a different species or donor is feasible as a mandatory step prior to publication, which is essential to prevent the widespread use of misidentified cell lines within the scientific literature (Fusenig et al. 2017). Broader cell line characterization is also important and is discussed separately (see Chapter 18).

17.2 MISIDENTIFIED CELL LINES

17.2.1 Impact

How common are misidentified cell lines in today's laboratories? A recent landmark publication examined the prevalence of misidentified and mycoplasma-contaminated cell lines in submissions to a cell repository between 1990 and 2014 (Drexler et al. 2017). Misidentified cell lines represented 14% (120/848) of all submissions. If obtained from primary sources (originators or cell repositories), the risk of encountering a misidentified cell line decreased across the 25-year period, but it remained high from secondary sources, at approximately one in six (Drexler et al. 2017). As with mycoplasma contamination (see Section 16.4), it is important to stress that deposit at a cell repository is usually a "best case" scenario. The prevalence of misidentified cell lines is considerably higher in some other studies; estimates in China range from 25% to 46%, based on data from Chinese cell repositories and testing services (Ye et al. 2015; Huang et al. 2017; Bian et al. 2017). These studies also show that 73–85% of cell lines that were established in China are misidentified. Many of these cell lines were established in the 1980s, when authentication testing was not widely available. This factor may partly explain the high prevalence of misidentified cell lines in China, as well as lack of information and language barriers (Ye et al. 2015; Bian et al. 2017).

The problem is compounded by the ongoing usage of known misidentified cell lines. The International Cell Line Authentication Committee (ICLAC) keeps a register of known misidentified cell lines, based on its review of publications and cell line provenance (see Section 7.3)

(Capes-Davis et al. 2010; Capes-Davis 2018). The ICLAC Register currently lists 486 cell lines that are known to be misidentified, with no known authentic material (ICLAC 2019c). Analysis of publication data (conducted in August 2017) showed that the cell lines in the ICLAC Register were used in 32 755 journal articles (Horbach and Halffman 2017). Far from declining over time, the usage of misidentified cell lines has increased in recent years and represents a major cause of irreproducible research (see Figure 7.1). Commonly used misidentified cell lines include HEp-2, KB, and MDA-MB-435 (see Table 17.1); all three have been highlighted as known misidentified cell lines but continue to be used despite these warnings (Vaughan et al. 2017; Korch et al. 2018; Gorphe 2019).

Some aspects of publications using misidentified cell lines may remain perfectly valid, such as the study of a molecular process. Usage of a misidentified cell line may therefore be justified under certain circumstances, provided their origins are correctly stated (although most studies do not do so) (Vaughan et al. 2017). However, any attempt to correlate cell behavior with the reported tissue of origin, its pathology, or other cell lines from the same donor will be totally invalidated by cross-contamination. If the mistaken identity is not acknowledged then this work may be cited by others, who may be entirely unaware of the problem and its impact on their work.

17.2.2 Causes

While propagation and cryopreservation extend the lifetime of a culture and its availability, they also increase the risk of cross-contamination. Whenever more than one cell line is maintained in a laboratory there is always a risk that cells from one cell line will be accidentally introduced into another and, if the growth rate of the contaminating cells is faster, that they will overgrow and eventually replace the original cell line. During the development of tissue culture, a number of cell strains evolved with very short doubling times and high plating efficiencies. Although these properties make such cell lines valuable experimental material, they also make them potentially hazardous for the cross-contamination of other cell lines. Testing of misidentified cell lines revealed that many arise from well-known, classic human tumor cell lines; HeLa is the most common contaminant (23% of cases), followed by T24 bladder carcinoma (4%), M14 melanoma, and HT-29 colon carcinoma (both 3%) (MacLeod et al. 1999; ICLAC 2019c).

Cross-contamination may arise at any time, due to the same sources of contamination already described for microbial contamination (see Section 16.1). This problem is particularly significant if it occurs in the original laboratory where a cell line was established, as is often the case (MacLeod et al. 1999). If the cell line is affected early, there will be very few samples from before the contamination occurred, making recovery of authentic material difficult or impossible. It also means that any subsequent experimental work is performed on the

TABLE 17.1. Commonly used misidentified cell lines.

Misidentified cell line[a]	RRID[b]	Claimed species	Claimed cell type	Contaminating cell line	Actual species	Actual cell type
ARO81–1 (ARO)	CVCL_0144	Human	Thyroid, anaplastic carcinoma	HT-29	Human	Colon carcinoma
CaOV	CVCL_M091	Human	Ovarian carcinoma	HeLa	Human	Cervical adenocarcinoma
CH1	CVCL_4992	Human	Ovarian carcinoma	PA1	Human	Teratocarcinoma
Chang liver	CVCL_0238	Human	Liver, normal hepatic cells	HeLa	Human	Cervical adenocarcinoma
CNE-2	CVCL_6889	Human	Nasal pharynx, carcinoma	HeLa	Human	Cervical adenocarcinoma
DAMI	CVCL_4360	Human	Leukemia, acute myeloid, M7	HEL	Human	Leukemia, acute myeloid, M6
ECV-304	CVCL_2029	Human	Endothelium, normal cells	T24	Human	Bladder carcinoma
EPC	CVCL_4361	Carp (Cyprinus carpio)	Epithelial papilloma	Unknown	Fathead minnow (Pimephales promelas)	Unknown
FL	CVCL_1905	Human	Amnion	HeLa	Human	Cervical adenocarcinoma
G-11 (HBT-3 derivative)	CVCL_U962	Human	Breast carcinoma	HeLa	Human	Cervical adenocarcinoma
HBL-100	CVCL_4362	Human	Breast carcinoma	Unknown	Human	Unknown
HCE	CVCL_M619	Human	Cervical carcinoma	HeLa	Human	Cervical adenocarcinoma
HEp-2 (H.Ep.-2)	CVCL_2940	Human	Laryngeal carcinoma	HeLa	Human	Cervical adenocarcinoma
HEp-2 (clone 2B)	CVCL_2817	Human	Laryngeal carcinoma	HeLa	Human	Cervical adenocarcinoma
HSG	CVCL_2517	Human	Salivary gland, submandibular	HeLa	Human	Cervical adenocarcinoma
Intestine 407 (Int-407, HEI)	CVCL_1907	Human	Intestinal cells (jejunum/ileum), embryonic	HeLa	Human	Cervical adenocarcinoma
KB	CVCL_0372	Human	Oral carcinoma	HeLa	Human	Cervical adenocarcinoma
KB-3-1 (KB derivative)	CVCL_2088	Human	Oral carcinoma	HeLa	Human	Cervical adenocarcinoma
McCoy	CVCL_3742	Human	Not specified	Strain L	Mouse	Connective tissue
MDA-MB-435	CVCL_0417	Human	Breast carcinoma	M14	Human	Melanoma
MKN28	CVCL_1416	Human	Gastric carcinoma	MKN74	Human	Gastric carcinoma
RGC-5	CVCL_4059	Rat	Retinal ganglion	661W	Mouse	Retina, photoreceptor cells
SH-2	CVCL_M622	Human	Breast carcinoma	HeLa	Human	Cervical adenocarcinoma
SK-N-MC	CVCL_0530	Human	Neuroblastoma	Unknown	Human	Sarcoma (Ewing's)
TE671	CVCL_1756	Human	Medulloblastoma	RD	Human	Sarcoma (rhabdomyosarcoma)
WiDR	CVCL_2760	Human	Colon carcinoma	HT-29	Human	Colon carcinoma
WISH	CVCL_1909	Human	Amnion, normal cells	HeLa	Human	Cervical adenocarcinoma

[a] Cell lines come from the ICLAC Register of Misidentified Cell Lines and represent a small subset of the full dataset (Capes-Davis et al. 2010; ICLAC 2019c). This subset was selected based on a search using the Web of Science database for all articles containing names from the ICLAC Register and either "cell(s)" or "cell line(s)" (Horbach and Halffman 2017). Cell lines that appeared in more than 200 publications in that search are listed here.

[b] RRIDs can be used to find additional information in the Cellosaurus knowledge resource, including source publications (https://web.expasy.org/cellosaurus) (Bairoch 2018).

misidentified cell line, leading to an extensive trail of damage within the scientific literature. If the cell line is affected late (e.g. in a secondary laboratory), some stocks will be misidentified, but earlier, authentic material is likely to be available from primary sources. The ICLAC Register of Misidentified Cell Lines includes a separate table for cell lines that were believed to be misidentified at one time but where authentic material has since been found (ICLAC 2019c). Other factors may play a role in the development of a misidentified cell line, such as mislabeling (see Table 17.2).

17.2.3 Eradication

Once misidentification occurs, it is impossible to treat; the emphasis must shift to detection and prevention (Coriell 1962). Because it may arise due to multiple factors, the problem must be addressed across multiple levels. Roland Nardone, who was a cell culture pioneer and passionate advocate for authentication testing, urged that (i) funding agencies should require cell line authentication as a condition for the award of grant and contract funds; (ii) journals should require authentication as a condition for publication; (iii) professional societies should endorse these policies and sponsor training activities to facilitate their adoption; and (iv) laboratory and department heads should ensure that staff members are aware of the problem and the methods required to detect it (Nardone 2007).

Nardone's call for action was heard by some journal editors, who incorporated authentication testing into their author guidelines (see Table 7.2) (Fusenig et al. 2017). It was also heard by several funding bodies, including the National Institutes of Health (NIH), who adopted requirements for enhancing reproducibility that include authentication of key resources (NIH 2015). A small number of professional societies and research organizations developed policy and training initiatives to promote authentication; ICLAC has a model policy that organizations can use as a starting point in this area (ICLAC 2019a). However, many scientists do not see the need for action until they experience the problem firsthand in their own laboratories.

What can be done by individuals who wish to eliminate misidentified cell lines from their work, but who work in organizations where such resources are not available? A number of simple, affordable steps can be taken to address the problem, including:

(1) **Before experimental work commences:**

 (a) If you intend to establish a new cell line, keep good records of the culture's origins and retain part of the original tissue sample to act as a reference for later authentication (see Section 13.8).

 (b) If you are working with an existing cell line, check that it is not already known to be misidentified. This can be done using the Cellosaurus knowledge resource or the ICLAC Register of Misidentified Cell Lines (Bairoch 2018; ICLAC 2019c).

 (c) Obtain the cell line from a primary source – either the originator or a cell repository – if at all possible.

 (d) If suitable authentication testing data are not available, test the cell line on arrival (see Section 17.3). This also applies to primary sources, since many originators do not perform authentication testing.

 (e) Perform cell banking to prepare frozen stocks for experimental work. Once these stocks are tested, you can return to them at regular intervals throughout experimental work (see Sections 15.5.2, 15.5.3).

(2) **During experimental work:**

 (a) Wipe down the biological safety cabinet (BSC), equipment, and supplies before and after working with a culture. Allow the cabinet to purge between cell lines.

 (b) Handle only one cell line at a time.

 (c) Handle rapidly growing lines after other cultures.

 (d) Dedicate one set of medium and other reagents to each cell line.

 (e) Label culture vessels clearly, distinctly, and unambiguously.

 (f) Add medium and any other reagents to the flask first and then add the cells last.

 (g) Use only plugged pipettes or filtered pipette tips for routine maintenance.

 (h) Never put a pipette back into a bottle of medium, trypsin, etc. after it has been in a culture flask containing cells. Never use the same pipette for different cell lines.

 (i) Check the characteristics of the culture regularly and call into question any sudden change in morphology, growth rate, or other phenotypic properties.

 (j) Store samples for authentication testing, particularly after sudden changes have occurred (see Sections 17.3.2, 17.3.3).

(3) **After experimental work:**

 (a) Check the author requirements for the journal where you submit your work to ensure that you comply with any specific instructions for authentication.

 (b) In the Materials and Methods of the manuscript, list the Research Resource Identifier (RRID) for the cell line and other necessary information, including the method used for authentication testing and the result (see Table 7.3).

 (c) Provide authentication testing data with the manuscript. Sharing of testing data is the best way to ensure that misidentified cell lines are detected across different locations. However, the level of data access may vary with the nature of the data and requirements for donor privacy (see Supp. S6.4).

TABLE 17.2. Cause and prevention of misidentification and cross-contamination.

Cause	Prevention
Sourcing cell lines	
Using known misidentified cell lines	Check Cellosaurus or the ICLAC Register of Misidentified Cell Lines[a] to see if a cell line is known to be misidentified
Obtaining cells from non-validated sources	Obtain a cell line from primary sources, i.e. the originator or a reputable cell repository (see Section 15.6); ask for authentication results; if untested, authenticate on arrival in the laboratory
Passing unauthenticated cell lines on to others	Do not pass on untested cell lines; refer requests to primary sources; if primary sources are no longer available, authenticate before the cell line is dispatched and make the results available to the recipient
Handling cultures	
Poor aseptic technique (see also Table 16.1)	Avoid splashing and other causes of aerosols; keep the work space clean and free from clutter; wipe up any spills promptly
Working with more than one cell line simultaneously, leading to transfer of aerosols between cultures	Never handle two different cell lines simultaneously; when using a BSC, always allow the cabinet to purge, and wipe down surfaces with 70% alcohol between each cell line (see Protocol P12.1)
Handling rapidly growing cell lines (e.g. HeLa) along with other slower growing lines	Always handle rapidly growing cell lines last, when other cultures and their media have been put away
Using the same medium and reagents for different cell lines	Always keep a separate set of medium and other reagents (trypsin, DPBS-A, EDTA, etc.) for each cell line; do not share these sets of reagents between cell lines or between users
Putting a pipette back into a medium or reagent bottle after being in contact with cells	Never put a pipette back into a medium or reagent bottle after it has been in a culture flask with cells; always put medium into culture vessel before cells
Using unplugged pipettes	Always use plugged pipettes when handling cells for further propagation
Using pipettors with unplugged tips	Use plugged pipette tips when handling cells for further propagation
Labeling, record keeping, and inventory	
Mislabeling at subculture	Label flasks, plates, and dishes before seeding cells
Mislabeling at freezing	Label cryovials before adding cells; only handle one cell line at a time
Poor record keeping leading to cryovials being listed in the wrong locations	Record locations promptly after adding to the cryofreezer; do not remove a vial unless you are sure of its identity and do not replace it without double-checking the correct location; perform episodic checks of random locations within the cryofreezer
Accidental transfer of cryovials to the wrong locations in the cryofreezer	Secure vials correctly in boxes or canes (e.g. with the "pin" in place that secures boxes into the rack); back up records with color coding of cryovial caps or other means
Thawing the wrong cryovial	Use permanent labeling that will not deteriorate in cryostorage (e.g. printed cryogenic labels or alcohol-resistant pens; see Section 15.2.1); check labels carefully before thawing
Seeding the wrong flask on thawing	Check that the label on the flask matches the cryovial before discarding the vial; do not thaw more than one cell line at a time
Validation and characterization	
Not carrying out authentication checks	Authenticate cell lines on arrival in the laboratory and after preparing cell stocks; repeat testing after experimental work (although this can be avoided by returning to frozen stocks; see Figure 15.10)
Not observing changes that may indicate misidentification	Always check cultures under the microscope before handling; refer to images of the culture at earlier passages; authenticate cultures if a change in morphology or behavior is observed

[a] For more information on these resources see Bairoch (2018); ICLAC (2019c).

17.3 CELL LINE AUTHENTICATION

17.3.1 Evolution of Authentication Techniques

Early work on cell line misidentification relied on isoenzyme and cytogenetic analysis and had quite astonishing success, allowing Walter Nelson-Rees to uncover more than 80 misidentified cell lines (Gartler and Farber 1973; Nelson-Rees et al. 1981; Capes-Davis et al. 2010). Both methods were highly specific for HeLa, which carried "HeLa marker" chromosomes and expressed an isoenzyme variant that was absent from Caucasian populations. Cell line panels were screened for the absence of these markers, but there was a risk that other contaminants would not be detected (Fogh et al. 1977). For example, a set of eight cell lines was uncovered that carried the same isoenzyme variants but did not carry HeLa markers; six were found to be misidentified, while two, SW-480 and SW-620, were legitimately derived from the same individual (Leibovitz et al. 1979).

The solution to this problem, as proposed by Jørgen Fogh and colleagues, was to use multiple polymorphic markers (initially 16 isoenzyme variants) to create a "genetic signature" that was unique to each cell line (Wright et al. 1981; Dracopoli and Fogh 1983b). Fogh determined that 38% (52/137) of cell lines in his testing panel had genetic signatures that could be distinguished from one another, which was a significant improvement on Gartler's two loci (G6PD and PGM) (Dracopoli and Fogh 1983b). However, it was still insufficient to discriminate between the hundreds of cell lines that were available at that time – much less the thousands that are in use today (Bairoch 2018). The need for new markers was particularly evident when Fogh reported that many tumor cell lines displayed loss of heterozygosity (LOH) with extended passage, resulting in a reduced dataset for authentication (Dracopoli and Fogh 1983a). Although isoenzyme analysis persisted for species identification (see Figure 17.1), the kit used for species detection (AuthentiKit) is no longer available, making this an historical technique.

In 1984, Alec Jeffreys discovered that tandemly repeated DNA sequences could be used to generate a "DNA fingerprint" to distinguish between individuals (Jeffreys et al. 1985; Jeffreys 2013). Jeffreys's discovery was quickly used for forensic analysis, but scientists also recognized that DNA

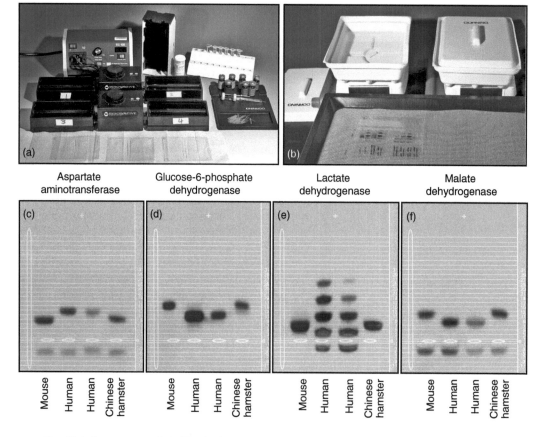

Fig. 17.1. Isoenzyme analysis. The AuthentiKit gel electrophoresis system (Innovative Chemistry; now discontinued) is used here to examine four isoenzyme variants. (a) A four-tank setup with power pack at rear, reagents on right, and three precast gels in the foreground; (b) staining and washing trays (Corning) with developed gel in foreground; (c–f) images from developed electropherograms. *Source:* Courtesy of Greg Sykes, ATCC.

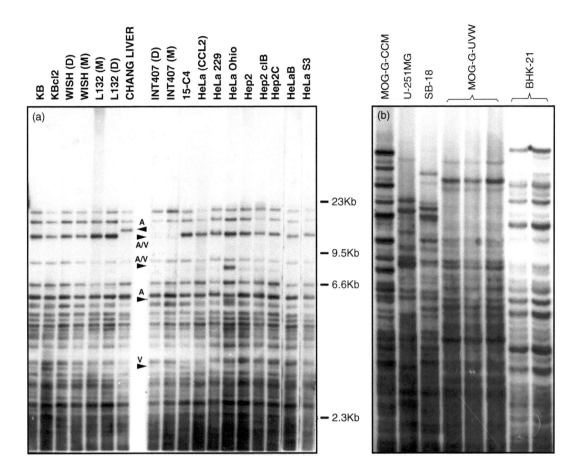

Fig. 17.2. DNA fingerprinting. Southern blots of DNA digested with the *Hinf* I restriction enzyme and hybridized with the minisatellite probe 33.15 (Jeffreys et al. 1985). (a) HeLa-contaminated cell lines and subclones of the HeLa cell line. Banding patterns are identical in cases when master banks (M) and their derivative working or distribution cell banks (D) were analyzed. DNA fingerprints were generally consistent between the cell lines, but some cases showed additional (A) or variable (V) bands (e.g. HeLa Ohio, INT 407, and Chang Liver). (b) Four human glioma cell lines (MOG-G-CCM, U-251MG, SB-18, and MOG-G-UVW at three separate freezings) and duplicate lanes of BHK-21 controls showing distinct fingerprints. *Source*: Courtesy of G. Stacey, NIBSC, UK.

fingerprinting could be employed for cell line authentication (Gill et al. 1985; Stacey et al. 1992). Cell line genomic DNA was digested with restriction enzymes and hybridized to one of the "Jeffreys probes," giving banding patterns specific to a single individual and any biosamples derived from them, including cell lines (see Figure 17.2). Although this method had a number of advantages (including the ability to test multiple species), standardization of the technique for comparison between laboratories was extremely difficult (National Research Council 1992). Typically, misidentified cell lines are detected when results are compared between different laboratories, bringing up unexpected matches.

The need for standardization led to the adoption of short tandem repeat (STR) loci for forensic analysis and other human identity testing (Butler 2006). STR loci can be amplified using a multiplex PCR; the resulting fragments are analyzed using capillary electrophoresis, allowing the number

of repeats in each allele to be determined (see Figure 17.3) (Robin et al. 2019). The allele calls are then recorded in a tabular format, which is easily shared between laboratories. Commercial kits have been developed for STR profiling that are supported by comprehensive population data, allowing the selection of suitable STR loci and allelic ladders for interlaboratory comparison (Butler 2006). STR profiling has been used to authenticate human cell lines since the late 1990s, resulting in publicly available data for more than 6400 cell lines (Tanabe et al. 1999; Masters et al. 2001; Robin et al. 2019). STR loci are also available for some other species, including African green monkey, dog, and mouse (Almeida et al. 2011, 2019; O'Donoghue et al. 2011). Based on more than 20 years of analysis, we can conclude that STR profiling is a robust and reliable method for detecting misidentified cell lines, with some caveats that will be discussed shortly (see Section 17.3.2).

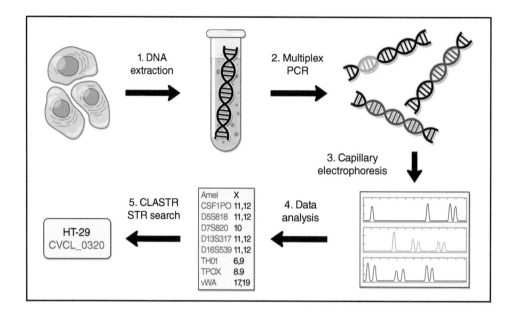

Fig. 17.3. STR profiling. The overall workflow of authentication testing using STR profiling is shown here; each step is described within the text. *Source*: Robin, T., Capes-Davis, A., & Bairoch, A. (2019). CLASTR: the Cellosaurus STR Similarity Search Tool A Precious Help for Cell Line Authentication. International Journal of Cancer. © 2019 John Wiley & Sons.

Although a standardized method is now available for cell line authentication, evolution of techniques is essential (Korch and Varella-Garcia 2018). Single nucleotide polymorphism (SNP) loci are particularly attractive as the basis for new technologies because of their compatibility with sequencing platforms (in contrast to STR loci, which are usually analyzed using capillary electrophoresis). SNP loci are diallelic, which gives them poor biostatistical power compared to STR loci, but their lower power can be compensated by increasing the number of loci used for analysis. The development of massively parallel sequencing (MPS) technologies allowed a large number of SNP loci to be analyzed simultaneously, resulting in new opportunities for cell line authentication (Demichelis et al. 2008; Schneider 2012). A suitable panel of 48 SNP loci can be used to authenticate human cell lines with a similar confidence level to a panel of 16 STR loci; some commercially available SNP panels can be used for this purpose, such as the SNP Trace™ panel (Fluidigm) (Liang-Chu et al. 2015; Yu et al. 2015). SNP analysis can also provide additional information on ancestry or other characteristics. SNP data have been used to estimate the genetic ancestral origin of 1393 human tumor cell lines, allowing assessment of their diversity (Dutil et al. 2019). While the data may have implications for donor privacy (see Supp. S6.4), such information is important to understand the relevance of cell line models. SNP analysis is rapid and cost-effective, particularly for laboratories that have access to next-generation sequencing platforms. However, a standardized set of SNP loci has not been developed for cell line authentication, which means that STR profiling is still required for interlaboratory comparison (Yu et al. 2015). SNP profiling also requires validation to ensure that complex genotypes (particularly in tumor cell lines) do not result in incorrect allele calls (Korch et al. 2018).

New technologies are needed for authentication testing, particularly for species where STR or SNP loci are unavailable for authentication (Almeida et al. 2016). Always remember, however, that methods for cell line authentication must be genotype-based. Phenotype-based methods may be misleading, as shown by analysis of the Chang liver cell line, which is actually HeLa but may express "liver-specific" markers (see Section 18.3.2) (Hall 2017; Korch et al. 2018). While there is some evidence that large-scale proteomic analysis using mass spectrometry may be effective for authentication (Lokhov et al. 2009; Karger et al. 2010), validation must be performed to ensure that such "proteomic footprints" do not vary with culture conditions or passaging.

Whatever the technology, the provenance of the cell line is absolutely essential for authentication (see Section 7.3) (Capes-Davis 2018). For example, the HBL-100 cell line has a unique STR profile, which might suggest that it is authentic. This cell line was reported to come from milk derived from a lactating female (Gaffney 1982). Unfortunately, STR profiling and fluorescence *in situ* hybridization (FISH) analysis demonstrated that HBL-100 carries a Y chromosome (Yoshino et al. 2006). The presence of a Y chromosome in a female cell line is an important indicator of misidentification, although its absence in a male cell line may simply indicate LOH (Yu et al. 2015).

17.3.2 Short Tandem Repeat (STR) Profiling

In 2008, John Masters and Roland Nardone proposed that a Standard should be written for human cell line authentication, under the auspices of the American National Standards Institute (ANSI) (Barallon et al. 2010). A working group was formed by the American Type Culture Collection Standards Development Organization (ATCC SDO) to draft the document, which aims to standardize the performance and interpretation of STR profiling. ANSI Standards pass through a process of stakeholder review, allowing the research community to provide input on the draft and thus develop a consensus approach. The ASN-0002 Standard successfully passed through this process and is now available from ANSI (ANSI 2012). The ASN-0002 working group also published extensively on the underlying principles set out in the Standard and their application (ATCC SDO Workgroup ASN-0002 2010; Capes-Davis et al. 2013; Reid et al. 2013). The Standard is currently undergoing revision and will be updated in the near future (ATCC SDO 2017).

Most laboratories outsource their STR profiling to an external testing provider. Cell repositories, forensic testing laboratories, and an increasing number of other providers offer cell line authentication as a service. If you are working with a testing provider, always ask questions about the service that they offer to assure yourself that they will provide what you need. Questions may include:

(1) *Are the genetic markers used linked with those used by professional bodies or other expert centers and does the assay conform to the relevant standard?* This is particularly relevant for human cell lines, where a Standard is available (ANSI 2012).

(2) *Does the provider have accreditation by an appropriate professional or government body or do they have formal affiliation with an expert group or organization?*

(3) *Does the provider have in-house expertise in the methods to assist in interpretation of results?* Or do they outsource their testing elsewhere?

(4) *What samples are suitable to ship for analysis?* Some testing providers prefer cells so that they can perform their own DNA extraction, while others will allow you to ship cell pellets, DNA, or samples spotted on collection cards (see Protocol P17.1).

(5) *What results are included in the test report that is returned to the customer?* Some testing providers only return the STR profile and you are expected to complete the data analysis, while others will perform data interpretation. A few laboratories provide multiple report options.

STR profiling can be performed in-house, but it requires an investment in time and infrastructure. Broadly speaking, the procedure can be divided into (i) DNA extraction; (ii) multiplex PCR of STR loci; (iii) capillary electrophoresis; (iv) data analysis; and (v) data interpretation, e.g. by searching STR profile databases for matches (see Figure 17.3) (Robin

et al. 2019). Unfortunately, space constraints mean that detailed instructions for STR profiling cannot be provided here. Some practical advice is given as a starting point for analysis of human cell lines, with particular emphasis on data interpretation. Refer to the ASN-0002 Standard and other resources for detailed guidance on assay development (ANSI 2012; Reid et al. 2013).

DNA extraction. Cells can be put aside for DNA extraction at a later date. Simply centrifuge the cells to generate a pellet, remove the supernatant, and store the pellet at −80 °C. Various DNA extraction methods can be successful, but removal of RNA is crucial (unless you use a quantitation method that is RNA-independent); the presence of RNA may lead to incorrect quantitation, resulting in failure of amplification or the generation of artifacts. Alternatively, cells can be spotted onto a sample collection card (see Protocol P17.1). These cards can be kept at room temperature and mailed economically to a distant testing provider. When spotting cells onto a sample collection card, always count the cells and record the approximate number in the pellet; the cell number can be used to estimate the amount of DNA present. Once the spot dries, a micro-punch (e.g. Harris Uni-core, Agar Scientific) can be used to remove a small disk from the center of this area for use in PCR. The device may be cleaned between samples to reduce cross-contamination by taking a punch from a blank card or by wiping the device with 70% alcohol (Fujii et al. 2011).

Multiplex PCR. For amplification of multiple STR loci, a commercial kit should be used that is optimized for multiplex PCR and includes the "core" set of STR loci required for cell line comparison. The current version of the ASN-0002 Standard requires eight core STR loci: CSF1PO, D5S818, D7S820, D13S317, D16S539, TH01, TPOX, and vWA (ANSI 2012). An additional locus, amelogenin, is added for sex determination. Kits that include the eight core STR loci are available from various suppliers, including Promega (e.g. PowerPlex® 18D System [Reid et al. 2013]), QIAGEN (e.g. Investigator IDplex Plus Kit), and ThermoFisher Scientific (e.g. AmpFlSTR™ Identifiler™ PCR Amplification Kit). A number of experts now recommend that the number of "core" STR loci is expanded to improve the discriminatory power of the assay (Bady et al. 2012; Yu et al. 2015; Huang et al. 2017). Such recommendations are likely to be incorporated into future versions of the ASN-0002 Standard. The kits listed above contain additional loci and are likely to remain current even after the Standard is revised. Always follow the supplier's instructions when using a specific kit; further advice and resources are available in the ASN-0002 Standard and from suppliers (ANSI 2012; Promega 2019).

Capillary electrophoresis. After performing the multiplex PCR, the fragments are separated for analysis using capillary electrophoresis. Within each primer pair, a dye labels one direction while the complementary primer remains unlabeled. These fluorescent dyes allow different loci to be identified even when the fragments overlap in size. Check that the fluorescent dyes in your kit are compatible with the specific genetic analyzer that will be used. Suitable instruments may include the 3500 Series Genetic Analyzers, the SeqStudio Genetic Analyzer (both from ThermoFisher Scientific), and the Spectrum Compact CE System (Promega). Kits also include internal size standards, allelic ladders, and controls; always follow the supplier's instructions regarding their use when preparing each sample run.

Data analysis. Analytical software is used to convert the raw data from capillary electrophoresis into an electropherogram (see Figure 17.4) and to "call" the alleles at each STR locus. This step is particularly important for data quality. A number of biological and instrumental artifacts can affect STR profiles, making their interpretation challenging and potentially prone to error (Butler 2011; Goor et al. 2011). Biological artifacts include stutter peaks (due to slippage of the DNA polymerase), split peaks (due to incomplete adenylation), and variant or "off-ladder" peaks (usually due to mutations in repeat or flanking regions of the STR locus). Instrumental artifacts include bleed-through between fluorescent dyes, dye blobs, and voltage spikes (Butler 2011). Analytical software must be able to reject stutter peaks from the analysis and to flag other quality concerns. Cell line samples tend to produce increased quality flags, due to their complex genotypes, which result in variable peak heights and the appearance of variant alleles or multiple peaks. The ASN-0002 Standard has specific guidance when performing analysis of human cell line data (ANSI 2012). Software packages that can be used for cell line authentication include GeneMapper™ *ID-X* (ThermoFisher Scientific), GeneMarker® HID (SoftGenetics), and the open source OSIRIS software program, which was developed by the National Center for Biotechnology Information (NCBI) (Goor et al. 2011).

Data interpretation. After analyzing the raw data, the allele calls are converted to a tabular format for comparison to other STR profiles (see Table 17.3) (Korch et al. 2018). Two key questions are asked during this process. Does the cell line correctly correspond to other samples from the original donor? Or does it incorrectly correspond to cell lines from different donors? The first question can be answered by comparison to reference samples from the original donor, including blood, tumor tissue, "sister" cell lines from the same donor, or early passages of the same cell line. The second question can be answered by comparison to a database of STR profiles from other cell lines (Dirks and Drexler 2013). Multiple databases have been developed over time, leading to a situation where numerous comparisons must be performed. Cellosaurus addressed this situation by collating STR profiles from cell repositories, publications, and other publicly available sources to give a single dataset, which can be searched with reference to the cell line authentication

using STR (CLASTR) similarity search tool (Robin et al. 2019). CLASTR is easy to use, provides multiple options for data comparison, and allows simultaneous searches using multiple STR profiles.

A match algorithm is used to give a numerical value when comparing different cell lines. Several match algorithms are effective for cell line comparison (Romano et al. 2008; Lorenzi et al. 2009; Capes-Davis et al. 2013). However, the most commonly used algorithms are the "Masters" and "Tanabe" algorithms (Tanabe et al. 1999; Masters et al. 2001). If using a similarity search tool such as CLASTR, one of these algorithms can be selected for pairwise comparison across the dataset (see Figure 17.5). Both algorithms use a tally of the total number of alleles at each locus and the number of shared alleles between two STR profiles (the "Reference" and "Test" sample; see Table 17.3).

The "Masters" algorithm, which is recommended in the current version of the ASN-0002 Standard, divides the number of shared alleles by the total number of alleles in either the test or the reference sample (ANSI 2012). While this algorithm is simple and easy to apply, it produces different results depending on whether the test or reference sample is used in the denominator of the equation. The "Tanabe" algorithm avoids this problem and produces comparable results to the Masters algorithm (Capes–Davis et al. 2013). The version now used for cell line comparison is as follows (Yu et al. 2015):

$$\text{Identity score} = 2 \times (\text{number of distinct alleles shared}$$
$$\text{by both samples})/((\text{total number of}$$
$$\text{distinct alleles in Reference sample})$$
$$+ (\text{total number of distinct alleles}$$
$$\text{in Test sample}))$$

The use of "distinct alleles" within the algorithm arises from different methods of counting single alleles in an STR profile. For example, the STR profile for the MDA-MB-435S cell line (see Table 17.3, Test sample 3) lists a single distinct allele, "12," at the STR locus D5S818. Some testing providers would list this as "12, 12" because a single allele in a diploid sample is likely to be homozygous. However, cell lines can have complex genotypes. In the case of MDA-MB-435S, other samples from the same donor exhibit two alleles at this locus ("11, 12") and thus the single allele call is likely to be due to LOH. For this reason, such alleles are best counted as one allele rather than two.

Cell lines from the same individual may display differences between their STR profiles, resulting in a decreased identity score (Capes-Davis et al. 2013). Loss of heterozygous alleles is common, as seen when early- and late-passage samples from the same donor are compared (see Table 17.3, Reference and Test samples 2 and 3). Some cell lines may also gain additional alleles (see Table 17.3, Test sample 1). Cell lines with microsatellite instability (MSI) are particularly likely to show

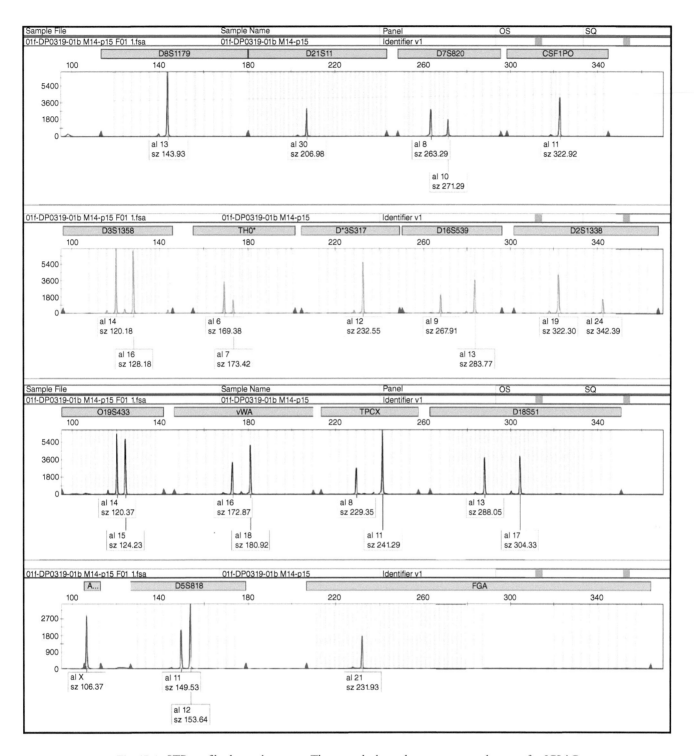

Fig. 17.4. STR profile electropherogram. The example shown here was generated as part of an ICLAC investigation into the origins of the M14 and MDA-MB-435 cell lines. The electropherogram shown comes from the M14 cell line; it is also used in Figure 17.5 and Table 17.3 (Test sample 2). *Source:* National Library of Medicine, BioSample accession SAMN06129842. Public domain. Data originally from [Korch et al. 2018], https://doi.org/10.1002/ijc.31067, licensed under CC BY 4.0.

TABLE 17.3. STR profile comparison.

Locus	Donor serum[a]	ML14 cell line[a]	M14 cell line[a]	MDA-MB-435S cell line[b]
STR profile				
Amelogenin	X, Y	X, Y	X	X
CSF1PO	11	11	11	11, 12
D2S1338	19, 24	19,24	19, 24	19, 24
D3S1358	14, 16	14, 16	14, 16	14
D5S818	11, 12	11, 12	11, 12	12
D7S820	8, 10	8, 10	8, 10	8
D8S1179	13, 14	13, 14	13	13, 14
D13S317	11, 12	11, 12, 13	12	12
D16S539	9, 13	13	9, 13	13
D18S51	13, 17	13, 17	13, 17	13, 17
D19S433	14, 15	14, 15	14, 15	14, 15
D21S11	30	30	30	30
FGA	21, 26	21, 25, 26	21	21
TH01	6, 7	6, 7	6, 7	6, 7
TPOX	8, 11	8, 11	8, 11	8, 11
vWA	16, 18	16, 18	16, 18	16, 18
Comparison using a match algorithm (Tanabe et al. 1999; Yu et al. 2015)				
Role in comparison	Reference sample	Test sample 1	Test sample 2	Test sample 3
Total number of alleles (including amelogenin)	30	31	26	24
Number of shared alleles with the reference sample	—	29	26	23
Percent match	—	95%	93%	85%

Source: data from Korch, C., Hall, E. M., Dirks, W. G., Ewing, M., Faries, M., Varella-Garcia, M., ... Capes-Davis, A. (2017). Authentication of M14 melanoma cell line proves misidentification of MDA-MB-435 breast cancer cell line. International Journal of Cancer, 142(3), 561–572.
[a]Obtained from the original institution where the M14 cell line was established, as part of an ICLAC investigation into the origins of the M14 and MDA-MB-435 cell lines. DNA was successfully extracted from donor serum and from two cell lines: a lymphoblastoid cell line (LCL) from the same donor, ML14, and the melanoma cell line M14.
[b]Obtained from ATCC (catalog number HTB-129D); deposited by the originator of the MDA-MB-435 cell line.

allelic gains, shifts, and losses during extended subculture (see Section 17.4.2). Unfortunately, MSI-positive cell lines display genetic instability when assessed across multiple assays. For example, the MSI-positive cell line CCRF-CEM has an unstable STR profile and also displays an unstable karyotype (MacLeod et al. 1999; Parson et al. 2005).

Despite these challenges, large-scale statistical analysis using 16 STR loci has shown that the majority of cell line STR profiles can be successfully grouped into synonymous samples (derived from the same donor with a high identity score) and non-synonymous samples (derived from different donors with a low identity score) (Yu et al. 2015). Match criteria are typically applied to interpret individual identity scores. The current version of the ASN-0002 Standard recommends that a percent match value of greater than 80% is interpreted as coming from the same donor, while a value of less than 55% indicates that samples come from different donors (ANSI 2012). Percent match values of

55–80% are indeterminant, which means that further analysis is required to reach a definitive conclusion. However, cutoff values are highly dependent on the number of STR loci used and the application. Some investigators recommend that a more rigorous cutoff value of 90% is used (Yu et al. 2015); unfortunately, this approach would miss a number of misidentified cell lines (see Table 17.3, Test sample 3). To ensure that misidentified cell lines are detected, the number of loci should be increased, which improves the discriminatory power of the assay, and cell line provenance and other evidence should be considered as part of data interpretation (Capes-Davis 2018).

17.3.3 CO1 DNA Barcoding

Interspecies misidentification occurs in about 10% of reported cases (ICLAC 2019c) but is impossible to detect with many of the existing STR profiling kits, which are human-specific.

Fig. 17.5. Cell line authentication using STR (CLASTR) similarity search tool. The example shown here is the STR profile from M14 (see Figure 17.4), entered into CLASTR for comparison with other human cell line STR profiles. The five highest results from the search are shown. The highest result is to M14, i.e. to itself. The next four results are to known misidentified cell lines (Korch et al. 2018). *Source*: Screen capture courtesy of Cellosaurus, with permission.

This issue can be addressed by (i) spiking human STR loci with primers specific for other species, e.g. mouse; (ii) developing new assays that are specific to other species; and (iii) performing species detection using an assay that detects a large number of species. All three approaches have been used for cell line authentication. For example, novel STR assays were developed to authenticate African green monkey, dog, and mouse cell lines (Almeida et al. 2011; O'Donoghue et al. 2011; Almeida et al. 2019). Strain-level testing was also performed for some rodent species using SNP or simple sequence length polymorphism (SSLP) analysis (Didion et al. 2014; Uchio-Yamada et al. 2017). However, the sheer number of *in vitro* species (see Section 1.5.3) (Bairoch 2018) means that species-specific methods cannot apply to all cell lines. An assay for species detection is essential to ensure that the correct species is present and to detect interspecies cross-contamination, e.g. following xenograft or feeder layer culture.

Species detection was performed for many years using isoenzyme analysis (O'Brien et al. 1980; Nims et al. 1998), but this approach became impractical when the kit that was used for species testing (AuthentiKit) was discontinued. Cytogenetic analysis can be used for species detection but requires specific expertise (see Section 17.3.4). Over time, these approaches were supplemented by PCR- and sequence-based methods, which are accessible, relatively easy to use, and (in the case of sequence-based assays) allow comparison of results using publicly available databases. A new sequence-based assay became possible in 2003, when Paul Hebert and colleagues proposed that a short, standardized region of sequence from the mitochondrial cytochrome *c* oxidase I (*CO1*) gene could be used as a "DNA barcode" for species identity (Hebert et al. 2003). This approach was enthusiastically adopted by the research community, in what has been described as the largest research collaboration in biodiversity science (Hebert et al. 2016). Resources are now available that allow any laboratory to perform CO1 DNA barcoding, using basic molecular biology equipment (Ivanova et al. 2012).

The following CO1 barcoding protocol (Protocol P17.1) was developed specifically for species detection using animal cell line samples. It is the basis for an ANSI Standard, which provides a comprehensive approach to this topic (ANSI 2015).

PROTOCOL P17.1. CO1 BARCODING OF ANIMAL CELLS

Contributed by G.R. Sykes, American Type Culture Collection, 10 801 University Boulevard, Manassas, VA 20110, USA.

Background

DNA barcoding is a species identification method achieved by examining a small, specific DNA segment (Hebert et al. 2003). Using this technique, a conserved region of the genome is amplified and sequenced, then compared to the equivalent sequences of other known species. Close similarities indicate a high likelihood of that sample belonging to that species. For animal cell lines, analyzing a section of the CO1 gene provides reliable species identification. This semi-conserved mitochondrial region has no introns, is maternally inherited, and usually lacks substantial variation within animal populations yet differs between species. Moreover, mitochondrial DNA recombination events rarely occur. The query sequence data are compared to thousands of high-quality, curated animal CO1 reference sequences within the publicly accessible Barcode of Life Data (BOLD) System (Ratnasingham and Hebert 2007).

Although CO1 barcoding is a powerful tool for cell line authentication, it has some limitations. The identification process employs Sanger sequencing, which recognizes the strongest base calls; samples evenly mixed with two animal species may produce data too disordered to resolve. Furthermore, low cross-contamination levels can slip by unnoticed; a solution is to use PCR-based approaches that can detect at least 1% contamination with primers specific to common laboratory animal models (mouse, rat, rabbit, etc.) (Cooper et al. 2007). With hybrid animals, multispecies hybridomas, transgenic cells, nuclear transfer, and other forms of genetic manipulation, CO1 barcodes reflect only the maternal mitochondrial origins – not necessarily all of the genetic sources within that cell. Additionally, CO1 barcoding only targets animal DNA; revealing microbial contamination requires other detection techniques (see Section 16.3.1).

The nearly universal primers within this protocol detect thousands of different animal species. However, some primer optimization may be needed for certain animal groups. Most species have little CO1 barcode variation, with some possible exceptions. Finally, one must be aware of one potentially confusing taxonomy aspect: species may undergo reclassification and updated nomenclature resulting in historic and current names describing the same organism.

Outline

DNA is purified from animal cells and amplified with CO1 barcode primers. The amplicon is then confirmed, purified, and sequenced. A sample is then loaded onto a genetic analyzer and the data are analyzed.

A. DNA Preparation

Note: Many genomic DNA purification procedures are compatible with CO1 barcoding, ranging from commercially available kits and robotic extractions to in-house preparations. Numerous methods can also be used for DNA quantitation (e.g. absorbance readings or PicoGreen® analysis). A laboratory's established procedures may be employed to obtain purified DNA at 10–300 ng/μl. The following protocol describes one DNA preparation method, which uses Whatman FTA™ cards (GE Healthcare Life Sciences).

Materials and Reagents

- BSC and associated consumables (see Protocol P12.1)
- Dulbecco's phosphate buffered saline without Ca^{2+} and Mg^{2+} (DPBS-A), pH 7.4
- TE: Tris-HCl, 10 mM, with 0.1 mM EDTA, pH 8.0
- FTA cards: Whatman Indicating FTA Cards (GE Healthcare Life Sciences)
- Whatman FTA Purification Reagent (GE Healthcare Life Sciences)
- Harris punch, 1.2 mm (GE Healthcare Life Sciences)
- Micropipettes and filtered, sterile tips
- Vortex mixer
- PCR tubes, 200 μl
- Microfuge tubes, 0.65–2.0 ml
- Tube racks

Procedure

Note: The optimum cell spot is 20 μl at 1 000 000 cells/ml. Less than 20 000 cells may produce data, but the signal strength might be low.

1. In a BSC, start with a cell pellet having 30 000–5 000 000 cells (see Section 14.6).
2. Calculate the volume of DPBS-A to resuspend the cell pellet: (total cells) $\div 10^3 =$ DPBS-A volume (μl).
3. Add DPBS-A to cell pellet and resuspend cells via vortexing.
4. Place 20 μl of cell suspension onto the prelabeled FTA card and air-dry the spot. The sample is stable at room temperature for at least 10 years (Fomovskaia et al. 2004; GE Healthcare 2015).
5. Prior to PCR, use the 1.2-mm Harris punch to transfer a center portion of the spot to a labeled 200-μl PCR tube. Positive controls may be previously spotted with authenticated Biosafety Level (BSL)-1 cell lines (see Table 6.3); no-template controls are blank sections of any FTA card.
6. Add 190 μl of FTA Purification Reagent to the PCR tube and incubate for five minutes at room temperature.
7. Remove the liquid carefully with a 200-μl pipette.
8. Repeat steps 6 and 7 twice more.
9. Dispense 190 μl of TE into each tube and incubate for five minutes.
10. Remove all of the TE with a 200-μl micropipette and discard.
11. Air-dry the punched section of card *in situ* in the tube, taking care not to lose the sample during drying due to airflow or static electricity. It is ready for PCR or may be stored at $-20\,°C$ for several weeks.

B. CO1 Barcode Amplification and Detection

Materials and Reagents

- Pre-PCR and post-PCR laboratories (see Appendix A for terminology)
- Samples: aqueous DNA (normalized to 5 ng/μl) or FTA punched samples
- Positive control(s) from authenticated cell lines; suggestions involving the following groups are:
  - Mammalian: SIRC (Statens Seruminstitut Rabbit Cornea) (ATCC® CCL-60™) – *Oryctolagus cuniculus* (rabbit)
  - Insect: Sf9 (ATCC CRL-1711™) – *Spodoptera frugiperda* (fall armyworm)
  - Fish: SJD.1 (ATCC CRL-2296™) – *Danio rerio* (zebrafish)
  - Amphibian: A6 (ATCC CCL-102™) – *Xenopus laevis* (African clawed frog)
  - Reptile: Gekko lung-1 (ATCC CCL-111™) – *Gekko gecko* (Tokay gecko)

- PCR tubes, 200 μl
- Microfuge tubes, 0.65–2.0 ml
- Tube racks
- Micropipettes and filtered, sterile tips
- Molecular-grade water
- Primers, 10 μM (see Table 17.4 for primer sequence)
- PCR buffer, 10× (provided with HotStart *Taq* DNA polymerase)
- MgCl$_2$, 50 mM
- dNTP, 10 mM
- HotStart *Taq* DNA polymerase, 5 U/μl (e.g. Platinum™ Taq, ThermoFisher Scientific)
- 100 bp ladder, 100 ng/μl (ThermoFisher Scientific)
- Vortex mixer
- Small centrifuge
- Thermocycler (e.g. Bio-Rad)
- Precast agarose gel and base: E-Gel® and iBase™ (2%, double comb, SYBR® Safe or ethidium bromide, ThermoFisher Scientific)
  - ***Safety Note. Ethidium bromide is a known mutagen and suspected carcinogen.***
- Transilluminator and camera (e.g. Bio-Rad Gel Doc XR+)

Procedure

1. Determine the total number of samples and controls. To compensate for pipetting error, multiply that number by 1.05 and round up to the nearest whole number for the "Factor."
2. In a pre-PCR laboratory, mix 100 μl each of primers VF1d_t1, VR1d_t1, LepF1_t1, and LepR1_t1 (see Table 17.4) in a clean, labeled microfuge tube.
3. Use the following table for master mix preparation, where "Master Mix volume" is the product of the "Factor" and "1× Reaction volume:"

Reagent	1× Reaction vol. (μl)	Factor	Master Mix vol. (μl)
Molecular-grade water	19.15[a]		
10× PCR buffer	2.5		
50 mM MgCl$_2$	1.25		
10 mM dNTP	0.5		
10 μM primer mix	0.5		
HotStart Taq	0.1		
Total volume	24.0[a]		

*For FTA samples, the water volume is 20.15 μl; the final reaction volume is 25.0 μl.

4. Thaw, vortex, and briefly spin PCR reagents.
5. Assemble the Master Mix in a microfuge tube, vortex, and briefly spin.
6. Add Master Mix and sample:
   (a) For aqueous DNA samples: aliquot 24 μl of Master Mix into each labeled 200-μl tube, then add 1 μl of sample, positive control DNA, or no-template water control.
   (b) For FTA samples: aliquot 25 μl of Master Mix into the tubes containing washed FTA card punches.
7. Seal, vortex, and briefly spin the tubes.
8. Bring the tubes to a post-PCR laboratory and place in a thermocycler.
9. Run the PCR using the following program:

Cycle	Temperature (°C)	Time
1x	94	1 min
5x	94	30 s
	50	40 s
	72	1 min
35x	94	30 s
	54	40 s
	72	1 min
1x	72	10 min
1x	4	Hold

10. After the run, place the tubes in a rack.
11. Insert E-Gel(s) into the iBase.
12. Load 10 μl of 100 bp ladder in the marker lanes.
13. Load 10 μl of amplicon (PCR product from sample) into appropriate wells.
14. Run the E-Gel(s) for 13 minutes.
15. With the correct filter for either SYBR Safe or ethidium bromide, photograph the gel(s) on a transilluminator. Afterwards, properly dispose of the gel(s).
16. Set PCR sample tubes with a strong band at ~750 bp aside for purification.

C. Amplicon Purification

Note: Different purification procedures are available. Some simple and effective methods include enzymatic (ExoSAP-IT®, ThermoFisher Scientific) and column (QIAquick® PCR Purification Kits, QIAGEN) techniques. Select a method and follow the manufacturer's protocol or use an amplicon purification procedure already established in your laboratory. Store purified samples at −20 °C.

D. CO1 Barcode Sequencing

Materials and Reagents

□ Post-PCR laboratory
□ BigDye® Terminator v3.1 Cycle Sequencing Kit (ThermoFisher Scientific)
□ Ice bucket
□ Primers, 10 μM (see Table 17.4 for primer sequence)
□ Microplate, 96-well, and plate sealer
□ Micropipettes and/or multichannel pipettes and filtered, sterile tips
□ Microfuge tubes, 0.65–2.0 ml
□ Tube racks
□ Vortex mixer
□ Large centrifuge (e.g. Eppendorf 5810R) with plate swing-buckets
□ Thermocycler (e.g. Bio-Rad)

Procedure

1. Create a plate map, which determines each reaction's placement in a 96-well plate (see Figure 17.6).
2. Determine the total reaction number including sequencing controls (e.g. pGEM®). To compensate for pipetting error, multiply that number by 1.125 (higher multiple than PCR since the volumes are smaller and solution more viscous) and round up to the nearest whole number for the "Factor."
3. Calculate a sequencing Master Mix for each primer. Sequencing is single directional; **never mix** forward (M13F) and reverse (M13R) primers in the same tube. "Master Mix volume" is the product of the "Factor" and "1x Reaction volume:"

Reagent	1x Reaction vol. (μl)	Factor	Master Mix vol. (μl)
BigDye3.1 mix	4.0		
Primer	1.0		
Total volume	5.0		

4. In a post-PCR laboratory, thaw, vortex, and briefly spin the above reagents.
5. Keep BigDye reagents on ice.
6. Assemble the Master Mixes in separate microfuge tubes, vortex, and briefly spin.
7. Dispense 5.0 μl of Master Mix into the appropriate wells of the 96-well plate.

8. Add 5.0 μl of purified amplicon template to the appropriate wells.
9. Seal and vortex the plate.
10. Briefly spin the plate in a large centrifuge at 400 *g* to draw droplets toward the bottom.
11. Load the plate in a thermocycler. Run the sequence reaction using the following program:

Cycle	Temperature (°C)	Time
1x	96	1 min
25x	96	10 s
	50	5 s
	60	4 min
1x	4	Hold

12. After the run, purify the samples.

E. Purification and Capillary Electrophoresis

Reagents and Materials

☐ Large centrifuge (e.g. Eppendorf 5810R) with plate swing-buckets
☐ Micropipettes and/or multichannel pipettes and filtered, sterile tips
☐ Microplates: 96-well, flat bottomed
☐ Performa® DTR Ultra 96-well plates (EdgeBio)
☐ MicroAmp™ Optical 96-Well Reaction Plate (ThermoFisher Scientific)
☐ Plate Septa 96-Well (ThermoFisher Scientific)
☐ 3500xL Genetic Analyzer (ThermoFisher Scientific) or equivalent
☐ Items for use with genetic analyzer (from same supplier as analyzer):
 • Capillary array, 50 cm
 • POP-7 polymer
 • Run buffers
 • Plate cassette

Procedure

1. Purify the samples using Performa DTR Ultra 96-well plates per the manufacturer's instructions. Capture purified samples in a MicroAmp Optical 96-Well Reaction Plate (MicroAmp Plate).
2. After purification, cover the samples in the MicroAmp Plate with plate septa.
3. Assemble the genetic analyzer cassette and MicroAmp Plate.
4. Place the loaded cassette onto the genetic analyzer, set time and voltage to collect at least 720 bases, and start the run.

F. Sequence Analysis

Materials

☐ CodonCode Aligner (CodonCode) or equivalent DNA sequence analysis software
☐ PC with internet access

Procedure

1. Transfer the electropherograms (.ab1 files) to the analysis PC.
2. In the CodonCode Aligner "File" menu, select "Import" and then "Add Samples" from the collected data into a new project.
3. Highlight the samples, enable the software to remove the PCR primer sequences, and trim the ends. Document any failed reactions and move them to the Trash.
4. Confirm high sequence quality of any controls.
5. Highlight the forward and reverse directions of a sample. In the "Contig" menu, select "Assemble."
6. Rename the new contig with the sample name. Open the contig and trim any remaining low-quality end sequences. If the forward primer sequence is reading right to left, press [Ctrl] + R to reverse the direction. Note: for the most reliable analysis, sequences should have at least 500 bases with quality scores > 20.
7. Repeat steps 5 and 6 for all samples.
8. Highlight all contigs. Under "File" menu, select "Export" and then "Consensus Sequences" (or "Samples" for any samples with only one sequence) as "One FASTA per Sample."
9. Open a FASTA (FAST-All – single-letter coding for nucleotides and amino acids) and copy the contents.
10. To compare the data, go to the BOLD website (http://boldsystems.org/index.php/ids_openidengine).
11. Select the database to search. The options enable the stringency of reference specimen queried. Paste the FASTA information in the "Enter sequences in FASTA format" field. Click "Submit."
12. The next screen provides the highest likelihood of species identification. Depending on the initial database queried, display options may include matches to individual reference samples, neighbor-joining trees, and closest match data clusters using barcode index numbers. See the BOLD website for the most current selections.
13. Repeat steps 11 and 12 for all samples.

TABLE 17.4. Primers for CO1 barcoding.

Primer name	Primer sequence
For CO1 barcode amplification (M13 tail is in bold):[a]	
VF1d_t1	**TGTAAAACGACGGCCAGT**TCTCAACCAACCACAARGAYATYGG
VR1d_t1	**CAGGAAACAGCTATGAC**TAGACTTCTGGGTGGCCRAARAAYCA
LepF1_t1	**TGTAAAACGACGGCCAGT**ATTCAACCAATCATAAAGATATTGG
LepR1_t1	**CAGGAAACAGCTATGAC**TAAACTTCTGGATGTCCAAAAAATCA
For CO1 barcode sequencing:	
M13F (-21) Universal	TGTAAAACGACGGCCAGT
M13R (-27) Universal	CAGGAAACAGCTATGAC

Source: courtesy of Greg Sykes, ATCC.
[a] Primer sequence has been modified from Ivanova et al. (2007). Note: this primer cocktail works on many species. Some primer sequence and PCR condition optimization may be necessary for specialized groups.

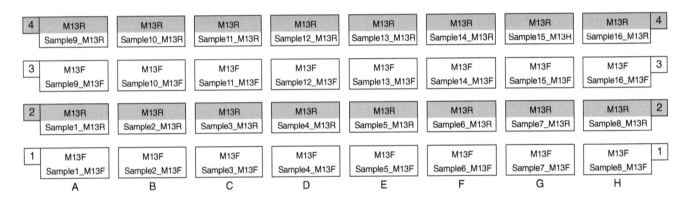

Fig. 17.6. DNA barcoding plate map. The plate map shown determines each reaction's placement in a 96-well plate (see Protocol P17.1, Part D). *Source*: courtesy of Greg Sykes, ATCC.

17.3.4 Cytogenetic Analysis

The chromosome content or karyotype of a cell line is one of the most characteristic and best-defined criteria for its identification and for determining the species and sex from which it was derived. Cytogenetic analysis was the first technique that definitively proved the existence of interspecies cross-contamination, more than five years before Gartler's discovery of cross-contamination with HeLa (see Section 17.3.1) (Rothfels et al. 1959; Defendi et al. 1960; Clausen and Syverton 1962). After Gartler's discovery, Nelson-Rees and colleagues used Giemsa banding to demonstrate intraspecies cross-contamination alongside isoenzyme analysis (Nelson-Rees et al. 1974, 1981). Other authentication techniques have continued to emerge, but cytogenetic analysis remains essential as a complementary approach to STR profiling (MacLeod et al. 2011; Korch and Varella-Garcia 2018). It is also a valuable part of cell line characterization (see Section 18.2.2). For example, cytogenetic analysis can distinguish between normal and transformed cells (see Section 3.6), because the chromosome number is more stable in normal cells (except in mice, where

the chromosome complement of normal cells can change quite rapidly after explantation).

Conventional cytogenetics relies on chromosome banding, which enables individual chromosome pairs to be identified when there is little morphological difference between them (Wang and Fedoroff 1972). For Giemsa banding, the chromosomal proteins are partially digested using crude trypsin (or trypsin and EDTA), producing a banded appearance on subsequent staining that is characteristic for each chromosome pair (see Figure 17.7a). An alternative technique, Quinacrine banding (Q-banding), does not require trypsinization; cells are stained in 5% (w/v) quinacrine dihydrochloride in 45% acetic acid, followed by rinsing the slide and mounting it in deionized water at pH 4.5 (Caspersson et al. 1968). Conventional techniques can be modified to discriminate between human and mouse chromosomes, principally to aid the karyotypic analysis of human/mouse hybrids (see Section 17.4.3). These methods include fluorescent staining with Hoechst 33258, which causes mouse centromeres to fluoresce more brightly than human centromeres, and staining with Giemsa at pH 11 (see Figure 17.7b, c) (Bobrow et al. 1972; Hilwig and Gropp 1972; Friend et al. 1976).

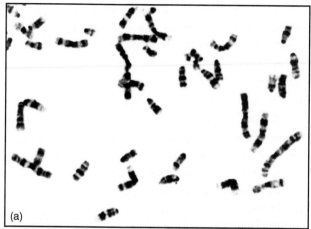

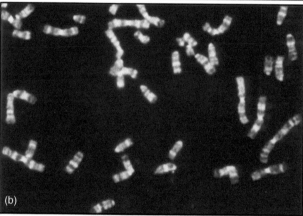

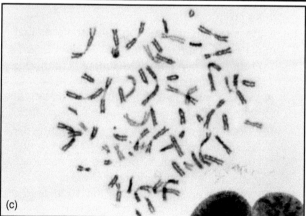

Fig. 17.7. Chromosome staining. (a) Human chromosomes banded by the standard trypsin-Giemsa technique. (b) The same preparation as in (a) but stained with Hoechst 33258. (c) Human–mouse hybrid stained with Giemsa at pH 11. Human chromosomes are less intensely stained than mouse chromosomes. Several human–mouse chromosomal translocations can be seen. *Source*: Courtesy of R. L. Church.

The karyotypes of human tumor cell lines can be difficult to study due to their complexity, which can cause confusion when investigating problematic cell lines (MacLeod et al. 2011). Fortunately, a number of molecular cytogenetic techniques have been developed that allow the complexity of tumor cell lines to be comprehensively examined (MacLeod et al. 2017; Wan 2017). Various molecular cytogenetic techniques can be used such as multiplex-fluorescence *in situ* hybridization (M-FISH), array comparative genomic hybridization (aCGH), and spectral karyotyping (SKY). STR profiling should also be performed; while cytogenetic techniques provide valuable and potentially unique information, a consensus method remains important for interlaboratory comparison.

The following procedure (Protocol P17.2) acts as a starting point for conventional cytogenetic analysis using Giemsa staining. Molecular cytogenetic techniques require specialized expertise and are outside the scope of this book; more information can be found in published protocols, books, and other resources (MacLeod et al. 2011; Heim and Mitelman 2015; De Braekeleer et al. 2017).

PROTOCOL P17.2. CHROMOSOME PREPARATION AND GIEMSA STAINING

Background

Dividing cells must be enriched for the number of metaphases (usually by treatment with Colcemid™, which inhibits the formation of the mitotic spindle) and then exposed to hypotonic conditions to release the chromosomal complement from the nucleus (Rothfels and Siminovitch 1958; Dracopoli et al. 2004). Optimization of the technique is important, particularly to match the hypotonic treatment to the requirements of the individual culture (MacLeod et al. 2017). A method for Giemsa banding of amniocytes is provided in the Supplementary material online as a further resource (see Supp. S23.4).

Outline

Arrest cells in metaphase, swell in hypotonic medium, and fix in acetic alcohol. Drop fixed cells on a slide, stain with Giemsa, and examine the results (see Figure 17.8).

❖ *Safety Note. Colcemid is less toxic than colchicine or vinblastine, which were initially used for chromosome preparations. However, it is still a hazardous substance. Perform a risk assessment and refer to the Safety Data Sheet (SDS) for more information on any hazardous substance (see Sections 6.1.1, 6.2.2).*

Materials (sterile or aseptically prepared)

☐ Culture of cells in log phase
☐ Colcemid (e.g. Democolcine, Sigma-Aldrich; KaryoMAX® Colcemid, ThermoFisher Scientific), 10 µM in balanced salt solution, e.g. DPBS-A
☐ Complete culture medium
☐ Trypsin, 0.25% crude
☐ Flask, 75 cm²
☐ Centrifuge tubes
☐ Pasteur pipettes

Materials (non-sterile)

☐ BSC and associated consumables (see Protocol P12.1)
☐ Hypotonic solution: 40 mM KCl, 25 mM sodium citrate
☐ Acetic acid: methanol fixative (see Appendix B)
☐ Giemsa stain solution (Sigma-Aldrich)
☐ Slide mounting medium (e.g. DPX, Sigma-Aldrich; Permount, Electron Microscopy Sciences)
☐ Glass microscope slides, precleaned (see Notes, Step 11)
☐ Coverslips (thickness #00) (e.g. Agar Scientific)
☐ Slide dishes
☐ Low-speed centrifuge
☐ Vortex mixer

Procedure

A. Chromosome Preparation

1. Subculture cells into a 75-cm² flask at $2-5 \times 10^4$ cells/ml ($4 \times 10^3-1 \times 10^4$ cells/cm²) in 20 ml culture medium.
2. Incubate the flask for approximately three to five days until the cells are in the log phase of growth.
3. Add 0.2 ml Colcemid solution to the medium already in the flask to give a final concentration of 0.1 µM. Incubate the culture for 4–6 h at 37 °C.
4. Remove the medium gently (without disturbing the cell monolayer), add 5 ml 0.25% trypsin, and incubate the culture for a further 10 minutes at 37 °C.
5. Remove the cells in the trypsin-containing medium and transfer to a centrifuge tube. Centrifuge the cells for two minutes at 100 g to give a cell pellet. Remove and discard the supernatant.
6. Resuspend the cells in 5 ml of hypotonic solution and leave them for 20 minutes at 37 °C.

7. Add an equal volume of freshly prepared, ice-cold acetic methanol, mixing constantly, and then centrifuge the cells at 100 g for two minutes. Remove and discard the supernatant.
8. "Buzz" the pellet on a vortex mixer (e.g. holding the bottom of the tube against the edge of the rotating cup) and slowly add 5 ml fresh acetic methanol with constant mixing. Leave the cells for 10 minutes on ice.
9. Centrifuge the cells for two minutes at 100 g. Remove and discard the supernatant.
10. Resuspend the pellet by "buzzing" in 0.2 ml acetic methanol to give a finely dispersed cell suspension.
11. Draw one drop of the suspension into the tip of a Pasteur pipette and drop from around 30 cm (12 in.) onto a cold slide. Tilt the slide and let the drop run down the slide as it spreads.
12. Allow the slide to dry and examine it on the microscope with phase contrast. If the cells are evenly spread and not touching, prepare more slides at the same cell concentration. If the cells are piled up and overlapping, dilute the suspension two- to fourfold and make a further drop preparation. If the cells from the diluted suspension are satisfactory, prepare more slides. If not, dilute the suspension further and repeat this step.
13. If the spread is still suboptimal, variations that may help to improve its quality include:
    (a) Altering hypotonic treatment conditions (see Notes, step 6).
    (b) Altering the height at which the cells are dropped.
    (c) Breathing on the slide prior to dropping the cell suspension (see Supp. S23.4). A fine layer of moisture is believed to encourage spreading (Henegariu et al. 2001).
    (d) Blowing the drop across the slide as the drop spreads.
    (e) Chilling the slide at -20 °C or on dry ice before dropping the cells onto it.
    (f) Chilling the fixed-cell suspension at 4 °C overnight before dropping the cells onto a slide.
    (g) Drying the slide more rapidly after dropping the cells (e.g. by heating over a flame).
    (h) Dropping the cells on a chilled slide (e.g. steep the slide in cold alcohol and dry it off) and then placing the slide over a beaker of boiling water to dry it.

(i) Burning off the fixative by flaming to ignite the residual fixative (this may make subsequent banding more difficult).

B. Giemsa Staining

14. Place the slides horizontally on a rack positioned over the sink.
15. Cover the cells completely with a few drops of neat Giemsa staining solution. Stain for two minutes.
16. Flood the slides with approximately 10 volumes of water.
17. Leave the slides for a further two minutes.
18. Displace the diluted stain with running water.
19. Finish by running the slides individually under tap water to remove any precipitated stain.
20. Check the staining under the microscope. If it is satisfactory, dry the slide thoroughly and mount with a coverslip using DPX or Permount.
21. Once staining is complete, assess the chromosomal complement of the culture:
    (a) Count the chromosome number per spread for between 50 and 100 spreads. Plot the results as a histogram (see Figure 3.4).
    (b) Photograph about 10–20 good spreads of banded chromosomes and use a specialized karyotyping program to visualize the karyotype. If such software is not available, a generic image analysis program may be used to cut out the individual chromosomes and paste them into a file where they can be rotated, trimmed, aligned, and sorted. Open source software is also available, e.g. ChromaWizard (Auer et al. 2018).

Notes

Step 1: Increased serum is commonly used, e.g. 20% fetal bovine serum (FBS). The quality of the preparation may vary depending on the batch of FBS that is used (MacLeod et al. 2017).

Step 3: Duration of the metaphase block may be increased to give more metaphases for chromosome counting or shortened to reduce chromosome condensation and improve banding.

Step 4: Some cell lines (e.g. CHO and HeLa) detach readily when in metaphase if the flask is shaken, eliminating trypsinization; this is known as "mitotic shake-off" (see Section 19.6). If using this technique, remove the Colcemid carefully after step 3 and replace it with the hypotonic solution. Shake the flask to dislodge cells in metaphase either before or after incubation in hypotonic solution and fix the cells as described in step 7 onwards.

Step 6: Optimization of hypotonic treatment is particularly important for tumor cell preparations. Options include (a) varying the proportion of KCl and sodium citrate in the hypotonic solution; (b) varying the duration of treatment from one to five minutes and from five to thirty minutes; and (c) varying the temperature, e.g. performing the treatment at room temperature or on ice (MacLeod et al. 2017).

Step 11: Additional slide cleaning is advisable if inconsistent spreading or quality is observed. An acid wash can be performed in 1 M HCl at 55 °C overnight, followed by washing in double-distilled water, sonication, and storage in 95% ethanol (MacLeod et al. 2017).

Step 18: Do not pour the stain off the slides as it will leave a scum of oxidized stain behind; always displace the stain with water. Even when staining in a dish, the stain is never poured off but must be displaced from the bottom with water (see Protocol P18.3).

17.4 AUTHENTICATION OF CHALLENGING SAMPLES

Cell line authentication is usually a straightforward process. If a cell line is authentic, STR profiling will show that it corresponds to other samples from the same donor and that it does not correspond to other cell lines such as HeLa; species testing will confirm that the correct species is present. If the cell line is misidentified, the STR profile or species data will clearly indicate that the cell line corresponds to a different donor or a different species. While misidentification may be an unexpected finding, the data usually allow a definitive conclusion to be reached and the only uncertainty relates to how the cell line came to be misidentified. However, a minority of cases are more difficult and require additional data to reach a definitive conclusion.

The commonest reasons for difficulties in authentication are lack of reference material and lack of information regarding the provenance of the cell line (see Section 7.3). Storage of reference material (e.g. tissue from the donor) is best practice and can be used to prove that a cell line is authentic. Unfortunately, however, most cell lines do not have reference material available, particularly when established more than 20 years ago. There are some notable exceptions to this finding; early donor material was tested to reach definitive conclusions on the origins of the U-87 MG cell line (established in 1968) and the M14 cell line (established in 1973) (Allen et al. 2016; Korch et al. 2018). However, original material is often discarded when cryostorage space becomes limited, particularly after the original laboratory is closed.

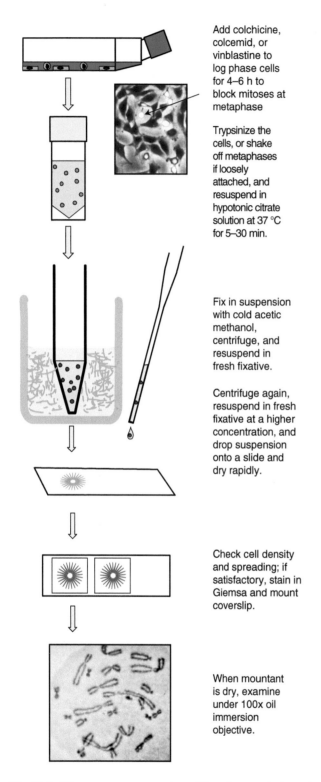

Add colchicine, colcemid, or vinblastine to log phase cells for 4–6 h to block mitoses at metaphase

Trypsinize the cells, or shake off metaphases if loosely attached, and resuspend in hypotonic citrate solution at 37 °C for 5–30 min.

Fix in suspension with cold acetic methanol, centrifuge, and resuspend in fresh fixative.

Centrifuge again, resuspend in fresh fixative at a higher concentration, and drop suspension onto a slide and dry rapidly.

Check cell density and spreading; if satisfactory, stain in Giemsa and mount coverslip.

When mountant is dry, examine under 100x oil immersion objective.

Fig. 17.8. Chromosome preparation. The drop technique is employed here to prepare chromosome spreads from monolayer cultures. *Source*: R. Ian Freshney.

There are also biological and technical reasons for authentication difficulties. Three types of samples are particularly difficult to authenticate for these reasons and are likely to require retesting or additional techniques to reach a definitive conclusion.

17.4.1 Cell Line Mixtures

A cell line that contains a mixture of cells from different donors will display an increased number of alleles in its STR profile. The current version of the ASN-0002 Standard states that a mixture is likely when more than two alleles are visible at three or more STR loci (ANSI 2012). However, the limit of detection will vary depending on the test method. Analysis of spiked mixture samples suggests that STR profiling can reliably detect a mixture at 10% of the total cell number; although small peaks may be evident at 5% or even 2%, they are often not "called" by the assay software, which is configured to exclude biological artifacts such as stutter peaks (Masters et al. 2001; Yu et al. 2015). SNP analysis may be more effective for low-level mixtures, but this will vary with the cell line (e.g. whether loci are present at high copy number) (Liang-Chu et al. 2015; Yu et al. 2015). If a mixture is suspected, the sample can be passaged further and then retested to see if a low-level mixture becomes more abundant and thus more obvious. Cloning by limiting dilution may also be attempted (see Section 20.3), followed by authentication testing of the resulting clonal populations.

A mixture between two different species will not be detected using STR profiling unless additional species-specific markers are present. Many cell repositories use a second method to detect interspecies mixtures that is independent of STR profiling (Yu et al. 2015; Bian et al. 2017). PCR-based methods were developed for species detection that generate different fragment sizes for different species; these methods can reportedly detect interspecies mixtures down to 1% (Cooper et al. 2007; Yu et al. 2015).

17.4.2 Cell Lines with Microsatellite Instability (MSI)

A small number of cell lines are likely to yield "indeterminant" results (55–80% match) when their STR profiles are compared over multiple passages (see Section 17.3.2). These cell lines, which are usually derived from tumors and are MSI-positive (see Section 25.3.4), display marked heterogeneity when assessed using cytogenetic analysis or STR profiling, with gain of heterozygosity at some STR loci (Hane et al. 1992; Masters et al. 2001; Parson et al. 2005). Distinguishing between an MSI-positive cell line and a mixture can be very difficult. The best approach is usually to retest the cell line over several passages and look for trends over time. If an MSI-positive cell line is tested over successive passages, genetic "drift" may be observed in the STR profile; for example, the peak height of one allele may gradually decrease as an adjacent peak emerges. Other STR loci remain relatively constant and appear to be less prone to drift, particularly TH01, TPOX, Penta D, and Penta E (Parson et al. 2005). These trends are different to those observed with a mixture, where an increasing number of alleles are seen across all loci. Adding additional STR loci to the core set is important to improve the level of discrimination of the assay (Bady et al. 2012). Another test method may also be informative; SNP analysis is reported to give higher percent match values with MSI-positive cell lines compared to STR profiling (Castro et al. 2013; Yu et al. 2015).

17.4.3 Hybrid Cell Lines

A hybrid cell line (see Section 22.3.1) may carry genomic material from either or both of the parental cells that were fused to establish it. STR profiling may therefore give a complex result, with some loci failing to amplify and others generating three or more alleles. Detection of interspecies hybrids can be difficult; mitochondrial-based methods such as CO1 DNA barcoding may only detect a single species, due to preferential retention of mitochondria from one species over the other (Nims et al. 2018). Cytogenetic analysis is likely to be more successful when analyzing such cases, but even then, the genetic composition of the cell line may continue to change with passage, e.g. human chromosomes are frequently lost from human–mouse hybrids (Weiss and Green 1967). Always perform authentication early (ideally when the hybrid cell line is newly formed) and compare to parental STR profiles if available, allowing confirmation of the cell line's identity and a baseline for further work (ANSI 2012).

Misidentified hybrid cell lines are challenging to detect, since the presence of a hybrid cell line is not immediately evident. This is best demonstrated by a set of seven nasopharyngeal cell lines that demonstrate a degree of similarity to HeLa but do not fit the typical match criteria (Chan et al. 2008; Strong et al. 2014). AdAH, CNE-1, CNE-2, HNE-1, HNE-2, HONE-1, and NPC-KT give indeterminant results when compared to HeLa, but they carry a unique allelic variant (13.3 at D13S317) that has only been seen in HeLa and its derivatives (ANSI 2012). Such cases usually require a process of detective work to resolve, which is often triggered by discrepancies in characterization that do not fit the reported donor, tissue, or disease (Capes-Davis 2018). For these nasopharyngeal cell lines, investigators were initially concerned by the absence of Epstein–Barr virus (EBV) and the presence of human papillomavirus-18 (HPV-18), which are unusual for nasopharyngeal carcinoma samples (Chan et al. 2008; Strong et al. 2014). RNA-seq analysis subsequently showed that the HPV-18 genomic integration patterns seen in these cell lines matched HeLa (Strong et al. 2014).

17.5 CONCLUSIONS: AUTHENTICATION

Lewis Coriell once observed that "like death and taxes, contamination or the threat of contamination is always with us" (Coriell 1962). Coriell would be appalled (but probably not surprised) at the extent of the problem in today's research endeavors. Microbial contamination and cross-contamination led to the widespread use of mycoplasma-contaminated and misidentified cell lines; other factors, such as the ongoing use of known misidentified cell lines, compounded these problems. While mycoplasma and authentication testing may seem daunting to laboratories that must tackle these problems for the first time, doing nothing is not an option. As a starting point, a search should be conducted using Cellosaurus or with comparison to the ICLAC Register of Misidentified Cell Lines to determine if cell lines are known to be misidentified before starting work; at minimum, cell lines should be authenticated as they enter and leave the laboratory (including as part of publications) (Kniss and Summerfield 2014). These two simple steps would minimize the impact of cell line misidentification on publications. Authentication testing should then be incorporated into routine laboratory procedures, detecting and eliminating problematic cell lines before they are used for experimental work.

Suppliers

Supplier	URL
Agar Scientific	http://www.agarscientific.com/fr
Bio-Rad Laboratories, Inc.	http://www.bio-rad.com
EdgeBio	http://www.edgebio.com
Electron Microscopy Sciences	http://www.emsdiasum.com/microscopy/default.aspx
Eppendorf	http://www.eppendorf.com/oc-en
Fluidigm	http://www.fluidigm.com
GE Healthcare Life Sciences	http://www.gelifesciences.com/en/it
Promega	http://www.promega.com
QIAGEN	http://www.qiagen.com/us
Sigma-Aldrich (Merck)	http://www.sigmaaldrich.com/life-science/cell-culture.html
SoftGenetics	http://softgenetics.com/index.php
ThermoFisher Scientific	http://www.thermofisher.com/us/en/home/life-science/cell-culture.html

REFERENCES

Allen, M., Bjerke, M., Edlund, H. et al. (2016). Origin of the U87MG glioma cell line: good news and bad news. *Sci. Transl. Med.* 8 (354): 354re3. https://doi.org/10.1126/scitranslmed .aaf6853.

Almeida, J.L., Hill, C.R., and Cole, K.D. (2011). Authentication of African green monkey cell lines using human short tandem repeat markers. *BMC Biotechnol.* 11: 102. https://doi.org/10 .1186/1472-6750-11-102.

Almeida, J.L., Cole, K.D., and Plant, A.L. (2016). Standards for cell line authentication and beyond. *PLoS Biol.* 14 (6): e1002476. https://doi.org/10.1371/journal.pbio.1002476.

Almeida, J.L., Dakic, A., Kindig, K. et al. (2019). Interlaboratory study to validate a STR profiling method for intraspecies identification of mouse cell lines. *PLoS One* 14 (6): e0218412. https://doi.org/10.1371/journal.pone.0218412.

ANSI (2012). ANSI/ATCC ASN-0002–2011: authentication of human cell lines: standardization of STR profiling. https://webstore.ansi.org/standards/atcc/ansiatccasn00022011.

ANSI (2015). ANSI/ATCC ASN-0003-2015: species-level identification of animal cells through mitochondrial cytochrome c oxidase subunit 1 (CO1) DNA barcodes. https://webstore.ansi.org/standards/atcc/ansiatccasn00032015.

ATCC SDO (2017). ANSI/ATCC ASN-0002 revision and reaffirmation. ATCC Standards Development Organization Newsletter, 9. Teddington: ATCC.

ATCC SDO Workgroup ASN-0002 (2010). Cell line misidentification: the beginning of the end. *Nat. Rev. Cancer* 10 (6): 441–448. https://doi.org/10.1038/nrc2852.

Auer, N., Hrdina, A., Hiremath, C. et al. (2018). ChromaWizard: an open source image analysis software for multicolor fluorescence in situ hybridization analysis. *Cytometry A* 93 (7): 749–754. https://doi.org/10.1002/cyto.a.23505.

Bady, P., Diserens, A.C., Castella, V. et al. (2012). DNA fingerprinting of glioma cell lines and considerations on similarity measurements. *Neuro-Oncology* 14 (6): 701–711. https://doi.org/10.1093/neuonc/nos072.

Bairoch, A. (2018). The Cellosaurus, a cell-line knowledge resource. *J. Biomol. Tech.* 29 (2): 25–38. https://doi.org/10.7171/jbt.18-2902-002.

Barallon, R., Bauer, S.R., Butler, J. et al. (2010). Recommendation of short tandem repeat profiling for authenticating human cell lines, stem cells, and tissues. *In Vitro Cell. Dev. Biol. Anim.* 46 (9): 727–732. https://doi.org/10.1007/s11626-010-9333-z.

Bian, X., Yang, Z., Feng, H. et al. (2017). A combination of species identification and STR profiling identifies cross-contaminated cells from 482 human tumor cell lines. *Sci. Rep.* 7 (1): 9774. https://doi.org/10.1038/s41598-017-09660-w.

Bobrow, M., Madan, K., and Pearson, P.L. (1972). Staining of some specific regions of human chromosomes, particularly the secondary constriction of No. 9. *Nat. New Biol.* 238 (82): 122–124. https://doi.org/10.1038/newbio238122a0.

Butler, J.M. (2006). Genetics and genomics of core short tandem repeat loci used in human identity testing. *J. Forensic Sci.* 51 (2): 253–265. https://doi.org/10.1111/j.1556-4029.2006.00046.x.

Butler, J.M. (2011). Forensic DNA testing. *Cold Spring Harb. Protoc.* 2011 (12): 1438–1450. https://doi.org/10.1101/pdb.top066928.

Capes-Davis, A. (2018). Cell line detective work: basic principles and molecular applications. *Adv. Mol. Pathol.* 1 (1): 229–238. https://doi.org/10.1016/j.yamp.2018.06.011.

Capes-Davis, A., Theodosopoulos, G., Atkin, I. et al. (2010). Check your cultures! A list of cross-contaminated or misidentified cell lines. *Int. J. Cancer* 127 (1): 1–8. https://doi.org/10.1002/ijc.25242.

Capes-Davis, A., Reid, Y.A., Kline, M.C. et al. (2013). Match criteria for human cell line authentication: where do we draw the line? *Int. J. Cancer* 132 (11): 2510–2519. https://doi.org/10.1002/ijc.27931.

Caspersson, T., Farber, S., Foley, G.E. et al. (1968). Chemical differentiation along metaphase chromosomes. *Exp. Cell Res.* 49 (1): 219–222. https://doi.org/10.1016/0014-4827(68)90538-7.

Castro, F., Dirks, W.G., Fahnrich, S. et al. (2013). High-throughput SNP-based authentication of human cell lines. *Int. J. Cancer* 132 (2): 308–314. https://doi.org/10.1002/ijc.27675.

Chan, S.Y., Choy, K.W., Tsao, S.W. et al. (2008). Authentication of nasopharyngeal carcinoma tumor lines. *Int. J. Cancer* 122 (9): 2169–2171. https://doi.org/10.1002/ijc.23374.

Clausen, J.J. and Syverton, J.T. (1962). Comparative chromosomal study of 31 cultured mammalian cell lines. *J. Natl Cancer Inst.* 28: 117–145.

Cooper, J.K., Sykes, G., King, S. et al. (2007). Species identification in cell culture: a two-pronged molecular approach. *In Vitro Cell. Dev. Biol. Anim.* 43 (10): 344–351. https://doi.org/10.1007/s11626-007-9060-2.

Coriell, L.L. (1962). Detection and elimination of contaminating organisms. *Natl Cancer Inst. Monogr.* 7: 33–53.

De Braekeleer, E., Huret, J.L., Mossafa, H. et al. (2017). Cytogenetic resources and information. In: *Cancer Cytogenetics: Methods and Protocols* (ed. T.S.K. Wan), 311–331. New York: Springer.

Defendi, V., Billingham, R.E., Silvers, W.K. et al. (1960). Immunological and karyological criteria for identification of cell lines. *J. Natl Cancer Inst.* 25: 359–385.

Demichelis, F., Greulich, H., Macoska, J.A. et al. (2008). SNP panel identification assay (SPIA): a genetic-based assay for the identification of cell lines. *Nucleic Acids Res.* 36 (7): 2446–2456. https://doi.org/10.1093/nar/gkn089.

Didion, J.P., Buus, R.J., Naghashfar, Z. et al. (2014). SNP array profiling of mouse cell lines identifies their strains of origin and reveals cross-contamination and widespread aneuploidy. *BMC Genomics* 15: 847. https://doi.org/10.1186/1471-2164-15-847.

Dirks, W.G. and Drexler, H.G. (2013). STR DNA typing of human cell lines: detection of intra- and interspecies cross-contamination. *Methods Mol. Biol.* 946: 27–38. https://doi.org/10.1007/978-1-62703-128-8_3.

Dracopoli, N.C. and Fogh, J. (1983a). Loss of heterozygosity in cultured human tumor cell lines. *J. Natl Cancer Inst.* 70 (1): 83–87.

Dracopoli, N.C. and Fogh, J. (1983b). Polymorphic enzyme analysis of cultured human tumor cell lines. *J. Natl Cancer Inst.* 70 (3): 469–476.

Dracopoli, N.C., Haines, J.L., Korf, B.R. et al. (2004). *Short Protocols in Human Genetics*. Hoboken, NJ: Wiley.

Drexler, H.G., Dirks, W.G., MacLeod, R.A. et al. (2017). False and mycoplasma-contaminated leukemia-lymphoma cell lines: time for a reappraisal. *Int. J. Cancer* 140 (5): 1209–1214. https://doi.org/10.1002/ijc.30530.

Dutil, J., Chen, Z., Monteiro, A.N. et al. (2019). An interactive resource to probe genetic diversity and estimated ancestry in cancer cell lines. *Cancer Res.* 79 (7): 1263–1273. https://doi.org/10.1158/0008-5472.CAN-18-2747.

Fogh, J., Wright, W.C., and Loveless, J.D. (1977). Absence of HeLa cell contamination in 169 cell lines derived from human tumors. *J. Natl Cancer Inst.* 58 (2): 209–214.

Fomovskaia, G., Smith, M. A., Davis, J. C., et al. (2004). Patent US6746841B1: FTA-coated media for use as a molecular diagnostic tool. https://patents.google.com/patent/wo2000062023a1/en.

Friend, K.K., Dorman, B.P., Kucherlapati, R.S. et al. (1976). Detection of interspecific translocations in mouse–human hybrids by alkaline Giemsa staining. *Exp. Cell Res.* 99 (1): 31–36. https://doi.org/10.1016/0014-4827(76)90676-5.

Fujii, K., Kitayama, T., Nakahara, H. et al. (2011). Degree of cross-contamination during the punching of FTA cards for STR typing. *Jpn J. Forens. Sci. Technol.* 16 (1): 67–72. https://doi.org/10.3408/jafst.16.67.

Fusenig, N.E., Capes-Davis, A., Bianchini, F. et al. (2017). The need for a worldwide consensus for cell line authentication: experience implementing a mandatory requirement at the International Journal of Cancer. *PLoS Biol.* 15 (4): e2001438. https://doi.org/10.1371/journal.pbio.2001438.

Gaffney, E.V. (1982). A cell line (HBL-100) established from human breast milk. *Cell Tissue Res.* 227 (3): 563–568. https://doi.org/10.1007/bf00204786.

Garcia, S., Bernad, A., Martin, M.C. et al. (2010). Pitfalls in spontaneous in vitro transformation of human mesenchymal stem cells. *Exp. Cell Res.* 316 (9): 1648–1650. https://doi.org/10.1016/j.yexcr.2010.02.016.

Gartler, S.M. (1967). Genetic markers as tracers in cell culture. *Natl Cancer Inst. Monogr.* 26: 167–195.

Gartler, S.M. (1968). Apparent HeLa cell contamination of human heteroploid cell lines. *Nature* 217 (5130): 750–751.

Gartler, S.M. and Farber, R.A. (1973). Biochemical identification of cells in culture. A. Human cell lines by enzyme polymorphism. In: *Tissue Culture: Methods and Applications* (eds. P.F. Kruse Jr. and M.K. Patterson), 797–804. New York: Academic Press.

Gill, P., Jeffreys, A.J., and Werrett, D.J. (1985). Forensic application of DNA "fingerprints". *Nature* 318 (6046): 577–579. https://doi.org/10.1038/318577a0.

Goor, R.M., Forman Neall, L., Hoffman, D. et al. (2011). A mathematical approach to the analysis of multiplex DNA profiles. *Bull. Math. Biol.* 73 (8): 1909–1931. https://doi.org/10.1007/s11538-010-9598-0.

Gorphe, P. (2019). A comprehensive review of Hep-2 cell line in translational research for laryngeal cancer. *Am. J. Cancer Res.* 9 (4): 644–649.

Hall, E. (2017). The notorious Chang liver. http://iclac.org/case-studies/chang-liver (accessed 11 July 2018).

Hane, B., Tummler, M., Jager, K. et al. (1992). Differences in DNA fingerprints of continuous leukemia-lymphoma cell lines from different sources. *Leukemia* 6 (11): 1129–1133.

GE Healthcare (2015). FTA cards: high-quality media for storage and transport of DNA. www.gelifesciences.co.jp/catalog/pdf/29156954aa_fta_whitepaper.pdf (accessed 21 november 2019).

Hebert, P.D., Cywinska, A., Ball, S.L. et al. (2003). Biological identifications through DNA barcodes. *Proc. Biol. Sci.* 270 (1512): 313–321. https://doi.org/10.1098/rspb.2002.2218.

Hebert, P.D., Hollingsworth, P.M., and Hajibabaei, M. (2016). From writing to reading the encyclopedia of life. *Philos. Trans. R. Soc. Lond. Ser. B Biol. Sci.* 371 (1702) https://doi.org/10.1098/rstb.2015.0321.

Heim, S. and Mitelman, F. (2015). *Cancer Cytogenetics: Chromosomal and Molecular Genetic Aberrations of Tumor Cells*, 4e. Hoboken, NJ: Wiley.

Henegariu, O., Heerema, N.A., Lowe Wright, L. et al. (2001). Improvements in cytogenetic slide preparation: controlled chromosome spreading, chemical aging and gradual denaturing. *Cytometry* 43 (2): 101–109.

Hilwig, I. and Gropp, A. (1972). Staining of constitutive heterochromatin in mammalian chromosomes with a new fluorochrome.

Exp. Cell Res. 75 (1): 122–126. https://doi.org/10.1016/0014-4827(72)90527-7.

Horbach, S.P. and Halffman, W. (2017). The ghosts of HeLa: how cell line misidentification contaminates the scientific literature. *PLoS One* 12 (10): e0186281. https://doi.org/10.1371/journal.pone.0186281.

Huang, Y., Liu, Y., Zheng, C. et al. (2017). Investigation of cross-contamination and misidentification of 278 widely used tumor cell lines. *PLoS One* 12 (1): e0170384. https://doi.org/10.1371/journal.pone.0170384.

ICLAC (2019a). Cell line policy for research institutions. https://iclac.org/resources/cell-line-policy (accessed 13 November 2019).

ICLAC (2019b). Definitions. https://iclac.org/resources/definitions (accessed 12 November 2019).

ICLAC (2019c). Register of misidentified cell lines. http://iclac.org/databases/cross-contaminations (accessed 11 November 2019).

Ivanova, N.V., Zemlak, T.S., Hanner, R.H. et al. (2007). Universal primer cocktails for fish DNA barcoding. *Mol. Ecol. Notes* 7 (4): 544–548. https://doi.org/10.1111/j.1471-8286.2007.01748.x.

Ivanova, N.V., Clare, E.L., and Borisenko, A.V. (2012). DNA barcoding in mammals. *Methods Mol. Biol.* 858: 153–182. https://doi.org/10.1007/978-1-61779-591-6_8.

Jeffreys, A.J. (2013). The man behind the DNA fingerprints: an interview with Professor Sir Alec Jeffreys. *Investig. Genet.* 4 (1): 21. https://doi.org/10.1186/2041-2223-4-21.

Jeffreys, A.J., Wilson, V., and Thein, S.L. (1985). Individual-specific "fingerprints" of human DNA. *Nature* 316 (6023): 76–79. https://doi.org/10.1038/316076a0.

Jordan, I., Munster, V.J., and Sandig, V. (2012). Authentication of the R06E fruit bat cell line. *Viruses* 4 (5): 889–900. https://doi.org/10.3390/v4050889.

Karger, A., Bettin, B., Lenk, M. et al. (2010). Rapid characterisation of cell cultures by matrix-assisted laser desorption/ionisation mass spectrometric typing. *J. Virol. Methods* 164 (1–2): 116–121. https://doi.org/10.1016/j.jviromet.2009.11.022.

Kniss, D.A. and Summerfield, T.L. (2014). Discovery of HeLa cell contamination in HES cells: call for cell line authentication in reproductive biology research. *Reprod. Sci.* 21 (8): 1015–1019. https://doi.org/10.1177/1933719114522518.

Korch, C. and Varella-Garcia, M. (2018). Tackling the human cell line and tissue misidentification problem is needed for reproducible biomedical research. *Adv. Mol. Pathol.* 1 (1): 209–228.e36. doi: https://doi.org/10.1016/j.yamp.2018.07.003.

Korch, C., Hall, E.M., Dirks, W.G. et al. (2018). Authentication of M14 melanoma cell line proves misidentification of MDA-MB-435 breast cancer cell line. *Int. J. Cancer* 142 (3): 561–572. https://doi.org/10.1002/ijc.31067.

Ledley, R.S. (1964). Animal cell strains. The Cell Culture Collection Committee has assembled and certified 23 strains of animal cells. *Science* 146 (3641): 241–243.

Leibovitz, A., Wright, W.C., Pathak, S. et al. (1979). Detection and analysis of a glucose 6-phosphate dehydrogenase phenotype B cell line contamination. *J. Natl Cancer Inst.* 63 (3): 635–645.

Liang-Chu, M.M., Yu, M., Haverty, P.M. et al. (2015). Human biosample authentication using the high-throughput, cost-effective SNPtrace(TM) system. *PLoS One* 10 (2): e0116218. https://doi.org/10.1371/journal.pone.0116218.

Lokhov, P., Balashova, E., and Dashtiev, M. (2009). Cell proteomic footprint. *Rapid Commun. Mass Spectrom.* 23 (5): 680–682. https://doi.org/10.1002/rcm.3928.

Lorenzi, P.L., Reinhold, W.C., Varma, S. et al. (2009). DNA fingerprinting of the NCI-60 cell line panel. *Mol. Cancer Ther.* 8 (4): 713–724. https://doi.org/10.1158/1535-7163.MCT-08-0921.

MacLeod, R.A., Dirks, W.G., Matsuo, Y. et al. (1999). Widespread intraspecies cross-contamination of human tumor cell lines arising at source. *Int. J. Cancer* 83 (4): 555–563. https://doi.org/10.1002/(sici)1097-0215(19991112)83:4<555::aid-ijc19>3.0.co;2-2.

MacLeod, R.A., Kaufmann, M., and Drexler, H.G. (2011). Cytogenetic analysis of cancer cell lines. *Methods Mol. Biol.* 731: 57–78. https://doi.org/10.1007/978-1-61779-080-5_6.

MacLeod, R.A., Kaufmann, M.E., and Drexler, H.G. (2017). Cytogenetic harvesting of cancer cells and cell lines. *Methods Mol. Biol.* 1541: 43–58. https://doi.org/10.1007/978-1-4939-6703-2_5.

Masters, J.R., Thomson, J.A., Daly-Burns, B. et al. (2001). Short tandem repeat profiling provides an international reference standard for human cell lines. *Proc. Natl Acad. Sci. U.S.A.* 98 (14): 8012–8017. https://doi.org/10.1073/pnas.121616198.

Nardone, R.M. (2007). Eradication of cross-contaminated cell lines: a call for action. *Cell Biol. Toxicol.* 23 (6): 367–372. https://doi.org/10.1007/s10565-007-9019-9.

National Research Council (1992). *DNA Technology in Forensic Science*. Washington, DC: The National Academies Press.

Nelson-Rees, W.A., Flandermeyer, R.R., and Hawthorne, P.K. (1974). Banded marker chromosomes as indicators of intraspecies cellular contamination. *Science* 184 (4141): 1093–1096. https://doi.org/10.1126/science.184.4141.1093.

Nelson-Rees, W.A., Daniels, D.W., and Flandermeyer, R.R. (1981). Cross-contamination of cells in culture. *Science* 212 (4493): 446–452.

NIH (2015). NOT-OD-16-011: implementing rigor and transparency in NIH and AHRQ research grant applications. https://grants.nih.gov/grants/guide/notice-files/not-od-16-011.html (accessed 4 July 2018).

Nims, R.W. and Reid, Y. (2017). Best practices for authenticating cell lines. *In Vitro Cell. Dev. Biol. Anim.* 53 (10): 880–887. https://doi.org/10.1007/s11626-017-0212-8.

Nims, R.W., Shoemaker, A.P., Bauernschub, M.A. et al. (1998). Sensitivity of isoenzyme analysis for the detection of interspecies cell line cross-contamination. *In Vitro Cell. Dev. Biol. Anim.* 34 (1): 35–39. https://doi.org/10.1007/s11626-998-0050-9.

Nims, R.W., Capes-Davis, A., Korch, C. et al. (2018). Authenticating hybrid cell lines. In: *Cell Culture* (ed. R.A. Mehanna). London: IntechOpen.

O'Brien, S.J., Shannon, J.E., and Gail, M.H. (1980). A molecular approach to the identification and individualization of human and animal cells in culture: isozyme and allozyme genetic signatures. *In Vitro* 16 (2): 119–135. https://doi.org/10.1007/bf02831503.

O'Donoghue, L.E., Rivest, J.P., and Duval, D.L. (2011). Polymerase chain reaction-based species verification and microsatellite analysis for canine cell line validation. *J. Vet. Diagn. Investig.* 23 (4): 780–785. https://doi.org/10.1177/1040638711408064.

Parson, W., Kirchebner, R., Muhlmann, R. et al. (2005). Cancer cell line identification by short tandem repeat profiling: power and limitations. *FASEB J.* 19 (3): 434–436. https://doi.org/10.1096/fj.04-3062fje.

Promega (2019). Cell line authentication testing. https://www.promega.com/resources/pubhub/cell-line-authentication-testing (accessed 19 November 2019).

Ratnasingham, S. and Hebert, P.D. (2007). Bold: the barcode of life data system (www.barcodinglife.org). *Mol. Ecol. Notes* 7 (3): 355–364. https://doi.org/10.1111/j.1471-8286.2007.01678.x.

Reid, Y., Storts, D., Riss, T. et al. (2013). Authentication of human cell lines by STR DNA profiling analysis. In: *Assay Guidance Manual [Internet]* (eds. G.S. Sittampalam, A. Grossman and K. Brimacombe). Bethesda, MD: Eli Lilly and Company and the National Center for Advancing Translational Sciences.

Robin, T., Capes-Davis, A., and Bairoch, A. (2019). CLASTR: the Cellosaurus STR similarity search tool – a precious help for cell line authentication. *Int. J. Cancer* 146 (5): 1299–1306. https://doi.org/10.1002/ijc.32639.

Romano, P., Manniello, A., Aresu, O. et al. (2008). Cell line data base: structure and recent improvements towards molecular authentication of human cell lines. *Nucleic Acids Res.* 37 (Suppl. 1): D925–D932. https://doi.org/10.1093/nar/gkn730.

Rothfels, K.H. and Siminovitch, L. (1958). An air-drying technique for flattening chromosomes in mammalian cells grown in vitro. *Stain. Technol.* 33 (2): 73–77. https://doi.org/10.3109/10520295809111827.

Rothfels, K.H., Axelrad, A.A., Siminovitch, L. et al. (1959). The origin of altered cell line from mouse, monkey, and man as indicated by chromosome and transplantation studies. *Proc. Can. Cancer Conf.* 3: 189–214.

Schneider, P.M. (2012). Beyond STRs: the role of diallelic markers in forensic genetics. *Transfus. Med. Hemother.* 39 (3): 176–180. https://doi.org/10.1159/000339139.

Skloot, R. (2010). *The Immortal Life of Henrietta Lacks*. New York: Crown Publishers.

Stacey, G., Bolton, B., Doyle, A. et al. (1992). DNA fingerprinting – a valuable new technique for the characterisation of cell lines. *Cytotechnology* 9 (1–3): 211–216. https://doi.org/10.1007/bf02521748.

Strong, M.J., Baddoo, M., Nanbo, A. et al. (2014). Comprehensive high-throughput RNA sequencing analysis reveals contamination of multiple nasopharyngeal carcinoma cell lines with HeLa cell genomes. *J. Virol.* 88 (18): 10696–10704. https://doi.org/10.1128/JVI.01457-14.

Tanabe, H., Takada, Y., Minegishi, D. et al. (1999). Cell line individualization by STR multiplex system in the cell bank found cross-contamination between ECV304 and EJ-1/T24. *Tissue Cult. Res. Commun.* 18 (4): 329–338. https://doi.org/10.11418/jtca1981.18.4_329.

Uchio-Yamada, K., Kasai, F., Ozawa, M. et al. (2017). Incorrect strain information for mouse cell lines: sequential influence of misidentification on sublines. *In Vitro Cell. Dev. Biol. Anim.* 53 (3): 225–230. https://doi.org/10.1007/s11626-016-0104-3.

Vaughan, L., Glanzel, W., Korch, C. et al. (2017). Widespread use of misidentified cell line KB (HeLa): incorrect attribution and its impact revealed through mining the scientific literature. *Cancer Res.* 77 (11): 2784–2788. https://doi.org/10.1158/0008-5472.CAN-16-2258.

Wan, T.S. (2017). Cancer cytogenetics: an introduction. *Methods Mol. Biol.* 1541: 1–10. https://doi.org/10.1007/978-1-4939-6703-2_1.

Wang, H.C. and Fedoroff, S. (1972). Banding in human chromosomes treated with trypsin. *Nat. New Biol.* 235 (54): 52–54. https://doi.org/10.1038/newbio235052a0.

Weiss, M.C. and Green, H. (1967). Human-mouse hybrid cell lines containing partial complements of human chromosomes and functioning human genes. *Proc. Natl Acad. Sci. U.S.A.* 58 (3): 1104–1111.

Wright, W.C., Daniels, W.P., and Fogh, J. (1981). Distinction of seventy-one cultured human tumor cell lines by polymorphic enzyme analysis. *J. Natl Cancer Inst.* 66 (2): 239–247.

Ye, F., Chen, C., Qin, J. et al. (2015). Genetic profiling reveals an alarming rate of cross-contamination among human cell lines used in China. *FASEB J.* 29 (10): 4268–4272. https://doi.org/10.1096/fj.14-266718.

Yoshino, K., Iimura, E., Saijo, K. et al. (2006). Essential role for gene profiling analysis in the authentication of human cell lines. *Hum. Cell* 19 (1): 43–48. https://doi.org/10.1111/j.1749-0774.2005.00007.x.

Yu, M., Selvaraj, S.K., Liang-Chu, M.M. et al. (2015). A resource for cell line authentication, annotation and quality control. *Nature* 520 (7547): 307–311. https://doi.org/10.1038/nature14397.

CHAPTER 18

Cell Line Characterization

When Wilton Earle, Katherine Sanford, and colleagues established the first two continuous cell lines in the 1940s, their work gave rise to a substantial increase in the number of cells that were available for characterization (see Figure 1.1) (Earle et al. 1943; Sanford and Evans 1982). The L strain and its clonal derivative, L-929, were established from normal fibroblasts through a process of transformation. This led to detailed analysis of their tumor-producing capacity, in the hope that it would lead to a better understanding of cancer (Sanford et al. 1956). Earle's laboratory also used photography, cinematography, and other techniques to characterize their cells, aiming to (i) identify the original cell type and tissue of origin; (ii) determine whether the cells maintained differentiated function and structure; and (iii) establish whether their biochemical behavior was affected by the medium or growth conditions (Sanford and Evans 1982). As more cell lines were established, this level of characterization became more difficult to achieve. Cell repositories came to hold hundreds of cell lines from multiple tissues, which often required quite different methods of characterization. Investigators were also constrained by a lack of suitable test methods, e.g. for tissue of origin (Fogh 1986). As a result, characterization testing tended to focus on common requirements for all cell lines, such as documenting cell morphology and growth (Reid 2011). By contrast, today's tissue culture laboratories have access to a huge number of assays for cell line characterization, and it is necessary to decide which ones are most important. This chapter starts by asking how to determine priorities for characterization and then describes a number of commonly used techniques, with a particular focus on imaging and staining techniques. Characterization

of stem cells, tumor cells, and 3D cultures are discussed in later chapters (see Sections 23.7, 25.3.4, 27.9).

18.1 PRIORITIES AND ESSENTIAL CHARACTERIZATION

Priorities for characterization testing can be determined based on six key requirements:

(1) *Validity, i.e. is the cell line worth using?* If a cell line is misidentified or contaminated, then any work that is performed using that cell line may be inaccurate or irreproducible (see Section 7.1.2). For this book, the term "validation testing" is used to refer to a subset of characterization methods that detect quality concerns such as contamination and misidentification (see Section 7.4). Validation testing is an essential priority for any tissue culture laboratory and should be performed for all cell lines, as part of their provenance (see Sections 7.3, 18.1.1).

(2) *Safety, i.e. does the cell line carry a known pathogen?* Pathogens may be present in the original donor sample or may arise due to contamination from elsewhere (e.g. from another cell line) or following genetic modification. Such pathogens are an important source of biohazard risk (see Sections 6.3.1, 6.3.2). The need for pathogen testing is primarily determined by the risk control measures that are used when handling cultures. For example, if a cell line is handled in a biological safety cabinet (BSC) (see Section 6.3.4), the need to test for pathogens may

Freshney's Culture of Animal Cells: A Manual of Basic Technique and Specialized Applications, Eighth Edition. Amanda Capes-Davis and R. Ian Freshney.
© 2021 John Wiley & Sons Ltd. Published 2021 by John Wiley & Sons Ltd.
Companion website: www.wiley.com/go/freshney/cellculture8

be reduced because the risk to the operator can be managed through containment. The application or outcome is also important; for example, extensive pathogen testing is required if cells are used to manufacture a biological product.

(3) *Culture behavior, i.e. what are the key properties of the cell line that enable you to handle it effectively and to monitor it during handling?* Typically, this question is answered by studying the morphology of the culture and by measuring its growth, e.g. by plotting a growth curve (see Sections 18.1.2, 18.1.3). These forms of characterization are essential for any tissue culture laboratory to optimize handling procedures and to detect changes in cell line behavior.

(4) *Functionality, i.e. which tissue and cell type was the cell line derived from, and does it display the typical properties that you would expect of that tissue or cell?* This is best determined by comparing the cell line to a panel of primary samples (original tissue or primary cultures). Many classic cell lines have already passed through such a comparison as part of a cell line panel, e.g. the NCI-60 panel. If a comparison has already been performed, the question becomes whether your sample of the cell line continues to express the same functionality over time and with further passaging. Testing will vary with the cell type and the application.

(5) *Abnormality, i.e. which properties of the cell line are not what you would expect for that tissue or cell type?* Many continuous cell lines (see Section 14.1) display evidence of transformation, which is associated with *in vivo* tumor formation. Because this change is relatively common, transformation assays are traditionally performed as part of the characterization of a new cell line (see Section 18.1.4). Other abnormalities include the presence of disease-specific mutations, the loss of tissue- or lineage-specific markers, or the gain of unexpected markers, e.g. due to epithelial–mesenchymal transition (EMT) (see Section 26.2.5). Apart from transformation assays, testing will usually vary with the cell type.

(6) *Usage, i.e. what is the application for which the cell line will be employed?* As a general rule, the more you rely on a cell line, the more you need to perform characterization testing. This will give a better understanding of the cell line's unique properties, its value as a research model, and the ways in which it resembles – or fails to resemble – the cell type, tissue, or disease of origin.

18.1.1 Validation Testing

Characterization of a cell line is vital, not only in determining its functionality but also in proving its authenticity. Special attention must be paid to the possibility that the cell line has become misidentified, due to cross-contamination with another cell line, mislabeling, or confusion in handling. This is such an important concern that a full chapter is dedicated to the topic of misidentified cell lines and authentication testing (see Chapter 17). Authentication testing must be incorporated into any work with cell lines, particularly those that are newly acquired or that are used in publications.

Broad testing for microbial contamination can be performed using sterility testing, where culture supernatant is inoculated into microbiological broth to look for growth (see Supp. S11.6). While this method can be effective for some bacteria and fungi, some microorganisms do not grow in standard microbiological media and require specific testing in order to be detected, such as mycoplasmas and viruses. Detection of mycoplasma and viral contamination is discussed elsewhere (see Sections 16.4, 16.5).

18.1.2 Morphology

Observation of morphology is the simplest and most direct technique for characterization of cells. Most tissue culture laboratories have a basic inverted microscope that can be used for morphological observation (see Section 5.2.1). Frequent brief observations of living cultures, preferably with phase-contrast optics, are more valuable than infrequent stained preparations that are studied at length. The former give a more general impression of the cell's morphology and its plasticity, and reveal differences in granularity and vacuolation that bear on the health of the culture. Unhealthy cells often become granular and then display vacuolation around the nucleus (see Figure 14.2).

It is important to keep a record of cell line morphology, using a camera to capture representative images (see Figure 18.1). Photographs should be taken shortly after acquisition of the cell line and at intervals thereafter, in case changes in morphology occur with handling. Images of cell lines in regular use should be displayed near the microscope or accessible on an adjacent computer. Such records can be supplemented with photographs of stained preparations and authentication testing data, stored either with the cell line record in a database or separately with links to the cell line database. Images should be compared to reference images from source publications, cell repository catalogs, or cell line atlases (see Table 15.3) (Lindl and Steubing 2013). Procedures for cell imaging are provided later in this chapter (see Section 18.4).

Descriptions of culture morphology can be helpful, but the terminology used can be misleading. The terms "fibroblastic" and "epithelial" are used rather loosely in tissue culture and often describe the appearance rather than the origin of the cells. "Fibroblast" is likely to refer to a bipolar or multipolar migratory cell, the length of which is usually more than twice its width. "Epithelial" is likely to refer to a cell that is polygonal, with more regular dimensions, and that grows in a discrete monolayer along with other cells. The difference in morphology may relate to the original cell type, but it also depends on whether cells require cell–cell adhesion for their survival and whether they are actively spreading

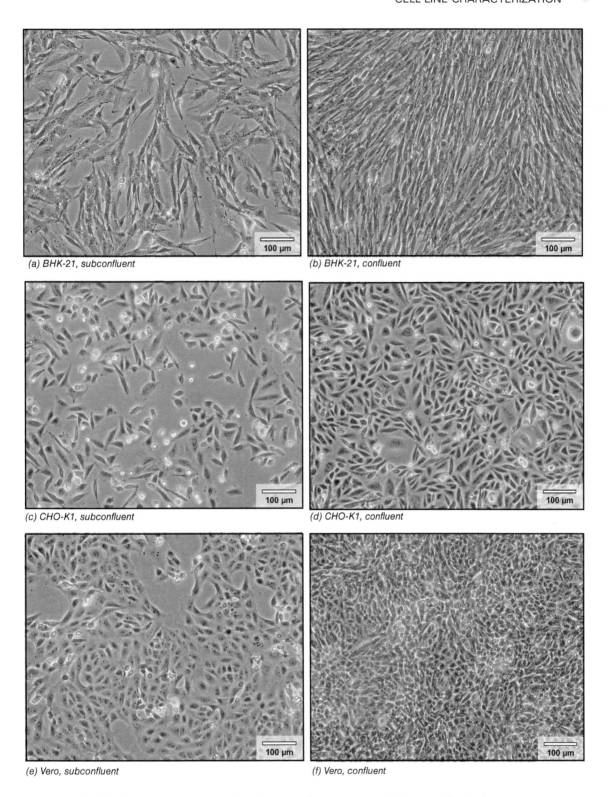

Fig. 18.1. Examples of cell morphology in culture. (a) BHK-21 (Syrian hamster kidney) at low density; the cells are well spread and randomly oriented, although some non-random orientation is beginning to appear. (b) BHK-21 at high density. (c) CHO-K1 (cloned line of Chinese hamster ovary) at low density; some cells are fibroblast-like, others more epithelioid. (d) CHO-K1 at high density; cells are refractile and more elliptical or epithelioid with fewer spindle-shaped cells. (e) Vero (African green monkey kidney) at low density; cells appear epithelial and are forming sheets. (f) Vero at high density; cells form a dense sheet with smaller cell diameters.

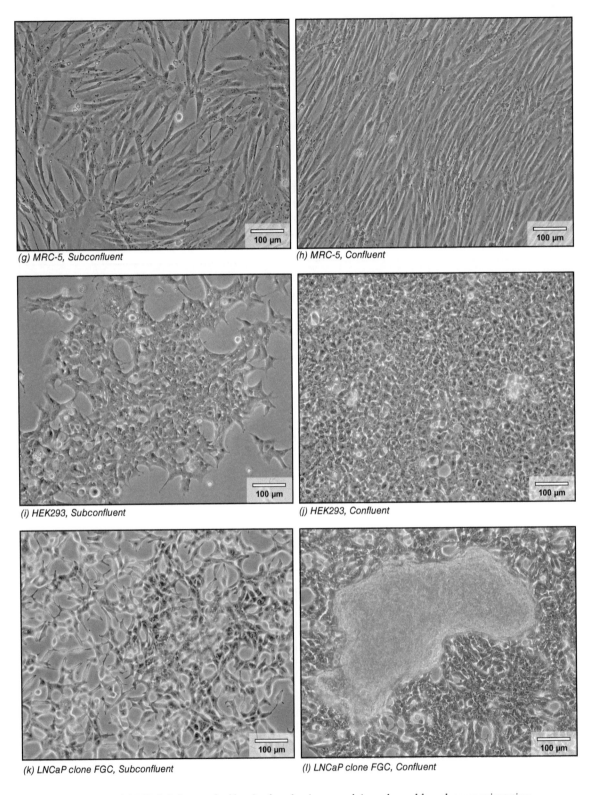

(g) MRC-5, Subconfluent

(h) MRC-5, Confluent

(i) HEK293, Subconfluent

(j) HEK293, Confluent

(k) LNCaP clone FGC, Subconfluent

(l) LNCaP clone FGC, Confluent

Fig. 18.1. (g) MRC-5 (human fetal lung) at low density; growth is random, although some orientation is beginning to appear. (h) MRC-5 at high density; parallel orientation is clearly displayed. (i) HEK293 (human embryonic kidney) at low density; cells are growing in sheets at mid–log phase. (j) HEK293 at high density; cells are now densely packed. (k) LNCaP clone FGC (cloned line of human prostate carcinoma) at medium to high density. (l) LNCaP clone FGC at high density, forming an aggregate.

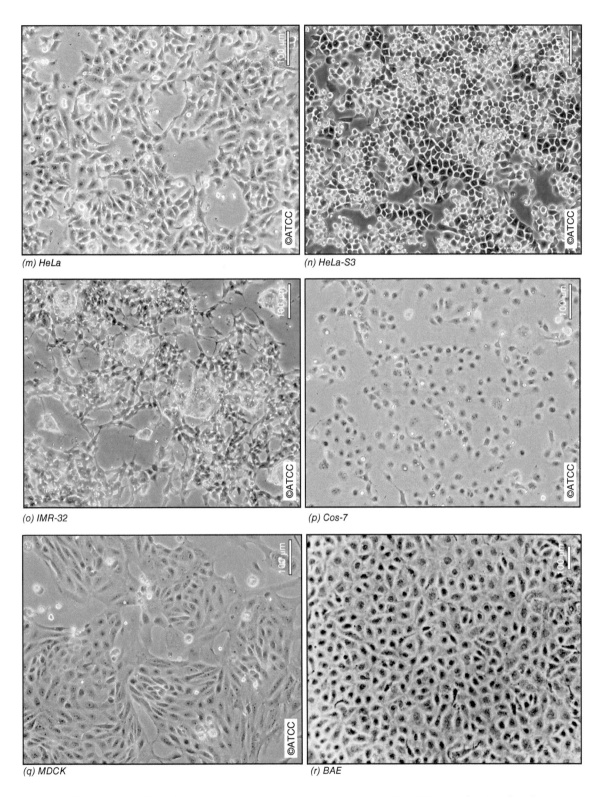

(m) HeLa

(n) HeLa-S3

(o) IMR-32

(p) Cos-7

(q) MDCK

(r) BAE

Fig. 18.1. (m) HeLa (human cervical carcinoma) at low density. (n) HeLa S3 clone showing altered appearance compared to its parent. (o) IMR–32 (human neuroblastoma). (p) COS-7 (Green monkey kidney). (q) MDCK (canine kidney). (r) BAE (bovine arterial endothelium).

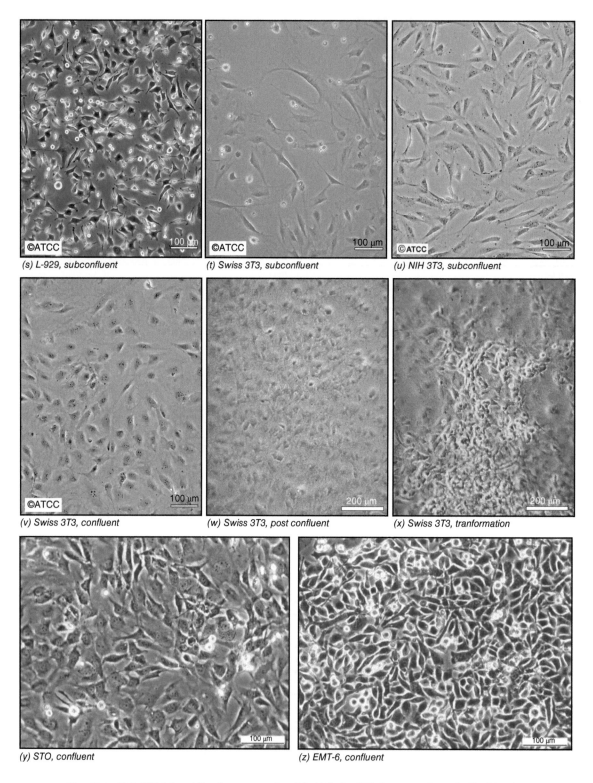

(s) L-929, subconfluent

(t) Swiss 3T3, subconfluent

(u) NIH 3T3, subconfluent

(v) Swiss 3T3, confluent

(w) Swiss 3T3, post confluent

(x) Swiss 3T3, tranformation

(y) STO, confluent

(z) EMT-6, confluent

Fig. 18.1. (s) L-929 (cloned line from mouse L-cells). (t) Swiss-3T3 (mouse embryonic fibroblasts) at low density. (t) NIH 3T3 at low density. (v) Swiss-3T3 at high density; note low-contrast appearance implying very flat monolayer. (w) Swiss-3T3 at very high density. (x) Swiss-3T3 showing a transformation focus of refractile cells overgrowing the monolayer, demonstrating why these cells should be subcultured well before they reach confluence. (y) STO (mouse embryo fibroblasts) at high density. (z) EMT-6 (mouse breast carcinoma) at high density; this cell line forms spheroids readily.

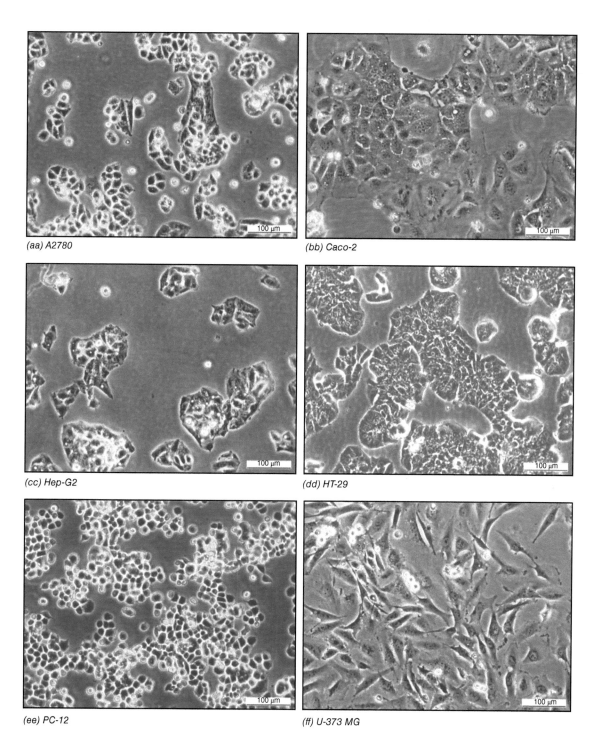

(aa) A2780

(bb) Caco-2

(cc) Hep-G2

(dd) HT-29

(ee) PC-12

(ff) U-373 MG

Fig. 18.1. (aa) A2780 (human ovarian carcinoma). (bb) Caco-2 (human colorectal carcinoma). (cc) Hep-G2 (human hepatoblastoma). (dd) HT-29 (human colorectal carcinoma). (ee) PC12 (rat adrenal pheochromocytoma). (ff) U-373MG (human malignant glioma). *Source*: (a–q, s–v) courtesy of ATCC; (r) courtesy of Peter Del Vecchio; (w, x) R. Ian Freshney; (y, z) courtesy of R. Ian Freshney; (aa–ff) courtesy of ECACC.

(see Section 2.2). Carcinoma-derived cells are often epithelial-like but more variable in morphology, e.g. LNCaP or HeLa (see Figure 18.1k–n). Some mesenchymal cells can assume an epithelium-like morphology at confluence, e.g. CHO-K1 or BAE (see Figure 18.1d, r). Strictly speaking, the terms "fibroblast-like" or "epithelial-like" (or similar terms such as "fibroblastoid" or "fibroblastic") should be used if the cell type of origin has not been confirmed (Reid 2017). The use of "blast" (e.g. in fibroblast, myoblast, or lymphoblast) denotes that cells are precursor cells which are still capable of dividing, but as with "fibroblast" and "epithelial," terminology is usually chosen based on convention rather than actual characterization.

Although very useful, culture morphology has limitations that should be recognized when used as a characterization technique. Most limitations relate to the plasticity of cellular morphology in response to different culture conditions. For example, epithelial cells growing in the center of a confluent sheet are usually regular, polygonal, and with a clearly defined birefringent edge, whereas the same cells growing at the edge of a patch may be more irregular and distended and, if transformed, may break away from the patch and become fibroblast-like in shape. Subconfluent fibroblasts from hamster kidney or human lung or skin assume multipolar or bipolar shapes and are well spread on the culture surface, but at confluence they are bipolar and less well spread (see Figure 18.1a, b, g, h; see also Figures 14.4d, 24.4a, b). Mouse 3T3 cells grow like multipolar fibroblasts at low cell density but become polygonal or pavement-like at confluence (see Figure 18.1s, t, v, w).

A change in cell morphology may indicate that cells are undergoing senescence, transformation, or other forms of cell line "evolution" (see Sections 3.5, 3.6). It may also indicate that cross-contamination or microbial contamination has occurred (see Chapters 16, 17). However, changes in morphology can also be traced back to changes in the substrate or the composition of the medium (Gospodarowicz et al. 1978; Freshney 1980). Comparative observations of cells should always be made at several different densities, using the same medium and substrate.

18.1.3 Growth Curve Analysis

A growth curve is normally plotted by recording the number of cells within the culture vessel over time, from seeding to subculture (see Figures 14.1, 19.7). This should be performed for every new cell line early in the culture process. Although plotting a growth curve for each cell line may seem somewhat excessive, it will help to determine the optimum seeding concentration and subculture timing, which makes the whole process more efficient thereafter. More information on calculating and interpreting growth curves can be found in the following chapter (see Section 19.3).

Another approach to growth analysis is to record the population doubling level (PDL) of the cell line (see Section 14.1).

The PDL is a cumulative value that is calculated at each passage, based on the number of cells that were initially seeded and the final cell count (see Section 14.2.2). The cumulative PDL acts as a rough estimate of the age of the cell line and is particularly important for finite cell lines such as normal human fibroblasts, where there is a limited number of population doublings before senescence occurs (see Figure 3.1e). It is also important to demonstrate that a cell line has become immortalized (see Sections 3.5, 22.3).

18.1.4 Transformation Assays

The term transformation refers to an *in vitro* phenotype that is associated with *in vivo* tumor formation and malignant behavior (see Section 3.6). The most direct assay for transformation is to inject a cell line into the "nude" mouse (or a similar rodent model) to generate a tumor (see Section 3.6.3). This approach can be used to assess tumorigenicity, but it also provides supporting data for the cell line's tissue of origin. There is a strong correlation between the histology of the xenograft tumor and that of the original tumor; the xenograft may even display increased differentiation, which can be used to help determine the primary site of the original tumor (Fogh 1986; Mohseny et al. 2011). This form of characterization is increasingly uncommon, due to the need to reduce the use of animals in research, although it continues to be used for patient-derived xenografts (PDXs) (see Section 25.4.5). Transformed cells also display aberrant growth control *in vitro*, including loss of contact inhibition, increased cloning efficiency, and anchorage-independent growth (see Section 3.6.2). Any changes in growth control can be used as the basis of an *in vitro* transformation assay (see Table 3.2). Cloning in soft agar or Methocel is particularly helpful, since it requires anchorage-independent growth and can be used to assess other characteristics such as cloning efficiency (see Protocols P20.2, P20.3).

18.2 GENOTYPE-BASED CHARACTERIZATION

The term "genotype" was originally defined as the sum total of all the genes in a gamete or zygote or (more briefly) as its genetic composition (Johannsen 2014; Fisch 2017). Wilhelm Johannsen's concept of "genotype" was focused on the importance of genes as the mechanism through which hereditary characteristics are transmitted. Johannsen rejected the proposal that chromosomes might also have special organizing functions, so the concept of "genotype" was initially quite a narrow one. However, it is now clear that hereditary characteristics are determined by other factors such as DNA methylation or chromatin modification, which control the way in which the genotype is expressed (Romanowska and Joshi 2019). For this book, genotype-based characterization is primarily focused on DNA sequence but may include other forms of analysis such as copy number variation (CNV), epigenetic analysis, or RNA sequencing (RNA-seq).

18.2.1 Sequence Analysis

Most widely used tumor cell lines have mutation data or other genotype-based information available via online databases (see Table 18.1). Many of these datasets arose from exploration of the cancer genome, focusing on somatic mutations that were acquired during cancer development (Greenman et al. 2007; Stratton et al. 2009). Human tumor cell lines are particularly useful as models to explore the cancer genome. This can be seen from an early study that generated genomic data from the melanoma cell line, COLO-829, and a lymphoblastoid cell line from the same donor, COLO-829BL (Pleasance et al. 2010). Comparison of the two genomes revealed a mutation signature in COLO-829 that was associated with ultraviolet (UV) light exposure and that hinted at resulting DNA repair

processes. More recent work has generated genomic data from hundreds of cell lines, exploring germline and somatic variants to understand how they contribute to tumorigenesis and drug sensitivity (Menden et al. 2018; Petljak et al. 2019). Although human cell lines tend to attract the most attention, genomic data have been published from non-human cell lines such as CHO-K1 and Vero (Xu et al. 2011; Osada et al. 2014).

When using an online database, always check that cell lines have been tested for authenticity (see Sections 17.3.2, 17.3.3). Early studies of p53 mutation status demonstrated that results were discordant for 23% of cell lines, which was probably caused by misidentification in many publications (Berglind et al. 2008). Cell line databases that are known to use authenticated cell lines include:

TABLE 18.1. Cell line databases with genomic, transcriptomic, or epigenomic data.

Database[a]	Organization	Reference	Website
Cancer Cell Line Encyclopedia (CCLE)	Broad Institute	Barretina et al. (2012)	https://portals.broadinstitute.org/ccle
Cancer Therapeutics Response Portal (CTRP)	Broad Institute	Basu et al. (2013)	http://portals.broadinstitute.org/ctrp.v2.1
Catalogue of Somatic Mutations in Cancer (COSMIC) Cell Lines Project	Wellcome Sanger Institute	Forbes et al. (2015)	https://cancer.sanger.ac.uk/cell_lines
Cell Model Passport	Wellcome Sanger Institute	van der Meer et al. (2019)	https://cellmodelpassports.sanger.ac.uk
Cellosaurus	SIB Swiss Institute of Bioinformatics	Bairoch (2018)	https://web.expasy.org/cellosaurus
Dependency Map (DepMap) Portal	Broad Institute	Ghandi et al. (2019)	https://depmap.org/portal
Genomics of Drug Sensitivity in Cancer (GDSC)	Wellcome Sanger Institute; Massachusetts General Hospital Cancer Center	Garnett et al. (2012)	https://www.cancerrxgene.org
Human Glioblastoma Cell Culture (HGCC) Resource	Uppsala University	Xie et al. (2015)	http://hgcc.se
IGRhCellID	Institute of Biomedical Sciences, Taiwan	Shiau et al. (2011)	http://igrcid.ibms.sinica.edu.tw/cgi-bin/index.cgi
Immuno Polymorphism Database – European Searchable Tumor Cell Line Database (IPD-ESTDAB)	EMBL-EBI	Robinson et al. (2013)	www.ebi.ac.uk/ipd/estdab/directory.html
Library of Integrated Network-Based Cellular Signatures (LINCS)	LINCS Consortium	Niepel et al. (2019)	http://www.lincsproject.org/lincs/tools
SelTar*base*	DKFZ	Woerner et al. (2010)	http://www.seltarbase.org
UMD TP53 Mutation Database		Leroy et al. (2013)	http://p53.fr/tp53-database

Abbreviations not found in the main text: DKFZ, Deutsches Krebsforschungszentrum; EMBL-EBI, European Molecular Biology Laboratory–European Bioinformatics Institute.

[a]The databases listed here include cell line information that will help the user decide on the validity of the cell line sample used to generate the data, e.g. authentication data, an index of confidence, or information on its source. However, cell line data are available through numerous databases, such as those hosted by the National Center for Biotechnology Information (NCBI). If you are searching for a specific cell line, it is a good idea to start by searching its name using Cellosaurus and then scroll down to look at its database listings.

(1) ***The Catalogue of Somatic Mutations in Cancer (COSMIC) Cell Line Project.*** Originally launched in 2004, the COSMIC Cell Line Project holds data from 1025 cell lines, but this number is expected to increase toward 1500 (Forbes et al. 2015). Authentication testing was performed using short tandem repeat (STR) and single nucleotide polymorphism (SNP) analysis; the results are available in a "QC" spreadsheet on the project website (Wellcome Sanger Institute 2019).

(2) ***The Cancer Cell Line Encyclopedia (CCLE).*** Originally published in 2012, the CCLE holds data from 1072 cell lines (Barretina et al. 2012; Ghandi et al. 2019). Authentication testing was performed for all cell lines using SNP analysis. Although the results are not publicly available, the authors highlight synonymous cell lines (i.e. cell lines that were established from the same person) in their dataset. These include both legitimate derivatives and misidentified cell lines.

(3) ***The Cell Model Passports database.*** As described in 2019, the Cell Model Passports site holds data from more than 1200 human tumor cell lines and organoid models (van der Meer et al. 2019). All of the cell lines included in the Genomics of Drug Sensitivity in Cancer (GDSC) panel are listed in this database (Garnett et al. 2012). Authentication testing was performed internally and during curation of the Cell Model Passports using STR profiling (van der Meer et al. 2019).

If you are working with a cell line that has online data available, how much effort should you invest in characterization of your own sample? Some degree of characterization is necessary because of the genomic variation that occurs in many continuous cell lines (see Section 3.3.2). Ideally, a "profiling" approach should be employed with direct comparison to reference genomes, similar to authentication testing (Hynds et al. 2018). A tool is available on the DepMap portal ("Cell STRAINER") to assess the degree of difference between a given cell line sample and a reference sample, which can be used for this purpose (Ben-David et al. 2019). However, many laboratories do not have the resources to generate large genomic datasets. At minimum, always confirm the presence of key driver mutations or other events that are relevant to your model or application. Never assume that a mutation will be present in your sample, particularly if that mutation is essential for your work. Other characterization testing that can be performed on a limited budget includes analysis of microsatellite instability (MSI) (see Section 25.3.4) and some forms of cytogenetic analysis.

18.2.2 Cytogenetic Analysis

Cytogenetic analysis has been used for cell line authentication since the 1950s and was one of the key methods used to detect HeLa cross-contamination in the 1970s (Rothfels et al. 1959; Nelson-Rees et al. 1981). It continues to be used for authentication alongside STR profiling (see Section 17.3.4). Chromosomal analysis can provide information on species and donor sex, and can be used to unequivocally identify individual cell lines, provided unique "marker" chromosomes are present (MacLeod et al. 2011; Korch and Varella-Garcia 2018). However, cytogenetic analysis has other benefits for cell line characterization. It can be used to study chromosomal aberrations and genetic fusion events that are associated with certain diseases, such as the Philadelphia chromosome. It is also an important method to assess genetic stability in stem cell lines, used for basic research (to ensure that genetic changes do not affect experimental results) and clinical applications (see Sections 3.3.1, 23.7) (Kyriakides et al. 2018).

Cytogenetic techniques that are particularly useful for cell lines include Giemsa banding, multiplex-fluorescence *in situ* hybridization (M-FISH), and spectral karyotyping (SKY) (see Section 17.3.4) (MacLeod et al. 2011, 2017; Korch and Varella-Garcia 2018). Higher resolution techniques are also available, such as array comparative genomic hybridization (aCGH) (Torsvik et al. 2014). All of these techniques rely on suitable methods for harvesting and slide preparation; it may be necessary to optimize harvesting protocols, since a method that works for one cell line may give suboptimal results for another (MacLeod et al. 2017). Other techniques may also provide information on the chromosomal composition of the cell. The variations in chromosome number that are found in most tumor cultures are often reflected in abnormal DNA content by flow cytometry (Krueger and Wilson 2011). The incidence of genetic instability and frequency of chromosomal rearrangement can be determined using the sister chromatid exchange assay (see Figure 29.2e, f) (Bryant et al. 2004).

18.2.3 Epigenetic Analysis

The epigenetic status of a cell is important for control of gene expression and plays a central role in lineage development and differentiation (Thiagarajan et al. 2014; Romanowska and Joshi 2019). Key epigenetic events include X chromosome inactivation and genomic imprinting, where one of the two alleles at a given locus is silenced in a parent-of-origin-specific manner, e.g. through DNA methylation. These events occur during embryonic development, which has led to a great deal of interest in the epigenetic changes that occur during stem cell culture, induction of pluripotency, and differentiation (Thiagarajan et al. 2014). The epigenetic status of a cell may also vary *in vitro* (Kyriakides et al. 2018). Early studies demonstrated that mouse embryonic stem cells (mESCs) displayed a marked degree of epigenetic instability (Dean et al. 1998; Humpherys et al. 2001). Broader analysis of human embryonic stem cell (hESC) lines gave a more complex picture; hESCs were relatively stable compared to their mouse counterparts, but variations in allele-specific expression were observed between different samples of the same cell line, suggesting that culture conditions can affect epigenetic status (Rugg-Gunn et al. 2007). This finding was confirmed by a study that assessed DNA methylation in an hESC line over more than 100 passages (Garitaonandia et al. 2015). By contrast, induced pluripotent stem cells (iPSCs) retained residual DNA methylation patterns from the somatic cells that were used as a starting point for reprogramming (Ma et al. 2014).

Clearly, epigenetic analysis is important, but it typically requires specialized techniques and is not routinely performed as part of cell line characterization in most laboratories. Methods that can be used for epigenetic analysis are discussed elsewhere (Tollefsbol 2011; Kagohara et al. 2018; Cazaly et al. 2019).

18.3 PHENOTYPE-BASED CHARACTERIZATION

The term "phenotype" refers to the cell's physical appearance along with its other observable characteristics, including finer measures of assessment (see Section 3.4). Phenotype-based techniques can assess various physical properties, including (i) cell morphology, e.g. through low and high power microscopy of stained and unstained cells; (ii) cell behavior, e.g. through live cell imaging to assess migration or contact inhibition; (iii) expression of markers that are specific to the cell type, lineage, tissue, or disease; (iv) the transcriptome, i.e. the full set of RNA molecules transcribed by the cell; (v) the proteome, i.e. the proteins synthesized by the cell; (vi) the metabolome, i.e. the active metabolic pathways or metabolites that are present in the cell; and (vii) the secretome, i.e. the factors secreted by the cell. Most tissue culture laboratories focus on the first four of these groups, although multi-omic analysis is becoming more widespread, leading to a rise in publicly available datasets and the need for reporting standards (Hayton et al. 2017; Robin et al. 2018).

Although it is often said that genotype determines phenotype, this saying is only partially correct for cells *in vitro*. The genotype of the cell may alter in culture, resulting in the development of a phenotype that was not observed in the original tissue, e.g. *in vitro* transformation (see Sections 3.6, 18.1.4). The cell may lose differentiated ("tissue-specific") markers or gain markers that are characteristic of a different cell type through changes in differentiation status, e.g. epithelial–mesenchymal transition (EMT) (see Section 26.2.5). The phenotype of the cell will also be determined by its culture conditions; this is particularly obvious when cells are grown in three-dimensional (3D) culture (see Chapter 27). For all of these reasons, it is important to approach the expression of "specific markers" with some caution. This topic will be discussed in isolation before considering how to perform phenotype-based characterization.

18.3.1 Cell Line-Specific Markers

Markers that uniquely identify a particular cell line are relatively rare and tend to be genotype- rather than phenotype-based. Examples include genetic mutations and chromosomal aberrations (e.g. translocations that result in unique fusion events). Many of these alterations occur in association with a specific tumor type (i.e. they are part of the cancer genome), which means that they are helpful to identify the tissue or disease of origin but may not be specific to a particular cell line. An alternative approach uses a number of highly polymorphic markers to generate a unique "fingerprint." This approach is used for authentication testing using STR profiling and some other genotype-based techniques (see Section 17.3.1). Phenotype-based properties are usually not truly unique, but they may be used to distinguish between cell lines from the same tissues that arise from different donors, particularly after cells are exposed to cytotoxins. Examples include enzymatic deficiencies (e.g. thymidine kinase-deficient cells) or drug resistance (e.g. vinblastine resistance, which is usually coupled to excess levels of P-glycoproteins) (Ouar et al. 1999).

18.3.2 Tissue- or Lineage-Specific Markers

During the process of embryonic development, cells become committed to a particular embryonic lineage and express more specialized properties (see Sections 2.4.1, 2.4.5). This progression is associated with the appearance of markers that are characteristic of that lineage and are distinct from markers that were expressed by the stem cells (see Table 23.2). Examples of lineage- or tissue-specific markers include myosin or tropomyosin for muscle, melanin for melanocytes, and hemoglobin for erythroid cells. These markers are highly specific but depend on the complete expression of the differentiated phenotype. Other markers may be expressed more broadly; for example, melanoma cells that do not express melanin are likely to express vimentin, which is expressed by a broad range of cell types (Quentmeier et al. 2001). Lineage markers that can help to establish the relationship of a cell line to its tissue of origin include cell surface antigens, intermediate filament proteins, and some enzymes. Although these markers may be described as lineage-specific, they are often more characteristic of the function of the cell than its origin from a particular germ layer.

Cell surface antigens. Cell surface antigens are particularly useful for characterization because they are accessible for antibody binding. Many of these molecules have been assigned "CD" names, based on the antibodies that recognize them. The term "CD" refers to "Clusters of Differentiation" and has been developed to identify groups of monoclonal antibodies that bind to common cell surface antigens (Clark et al. 2016; HCDM 2019). Cell surface antigens have been extensively used for characterization and affinity-based separation of hematopoietic cell lines (Drexler 2010). They are also important for the affinity-based separation of stem cell populations, including hematopoietic stem cells (HSCs), human pluripotent stem cells (hPSCs), and mesenchymal stromal cells (MSCs) (see Section 21.4) (Diogo et al. 2012).

As with genotype-based techniques such as STR profiling, it is essential to use multiple antigens to build up a cell type-specific "fingerprint;" for example, MSCs can be identified based on the presence of CD105, CD73, and CD90, combined with the absence of other surface antigens (see Table 23.5).

Intermediate filament proteins. Intermediate filament proteins are among the most widely used lineage or tissue markers (see Figure 18.2a–c) (Lane 1982; Ramaekers et al. 1982). Highly specific examples include desmin for muscle and glial fibrillary acidic protein (GFAP) for astrocytes (Costa et al. 2004; Sofroniew and Vinters 2010). However, many other intermediate filament proteins are less specific. Cytokeratins are commonly used as markers of epithelial cells but are also expressed in mesothelium (Moll et al. 1982; Wu et al. 1982; Moll 1994). Neurofilament protein marks neurons but also stains some neuroendocrine cells (Bishop et al. 1988; Kondo and Raff 2000). Vimentin, although usually restricted to mesodermally derived cells *in vivo*, can appear in other cell types *in vitro* (Quentmeier et al. 2001). Intermediate filament proteins can be useful markers in cell lines, but their expression may vary compared to original tissue samples; results must be interpreted with caution.

Enzymes. A number of enzymatic markers have been used to study specific cell types (see Table 18.2). The differentiated phenotype may not be constitutively expressed – and is quite likely not to be, as the cells are proliferating – so these markers may require induction in order to be detected. For example, alkaline phosphatase may be induced in lung adenocarcinoma cells following treatment with dexamethasone, oncostatin M, or interleukin 6 (IL-6) (see Section 26.3) (McCormick and Freshney 2000). Enzymatic analysis is likely to be performed in today's laboratories by measuring gene expression or performing an enzyme-linked immunosorbent assay (ELISA). Nevertheless, full functionality can only be confirmed by direct measurement of enzymatic activity.

Parameters that can be used for characterization of enzymatic activity include (i) the constitutive level (i.e. in the absence of inducers or repressors); (ii) the induced or adaptive level (i.e. in response to inducers and repressors); and (iii) the presence of isoenzyme polymorphisms. The latter have traditionally been used for authentication testing (see Section 17.3.1) but can also be tissue-specific. Creatine kinase (CK) MM isoenzyme is found in muscle, while the CK BB isoenzyme is characteristic of neuronal and neuroendocrine cells, as is neuron-specific enolase. Lactic dehydrogenase is present in most tissues, but as different isoenzymes; a high level of tyrosine aminotransferase, inducible by dexamethasone, is generally regarded as specific to hepatocytes (see Table 18.2) (Granner et al. 1968).

18.3.3 Transcriptomic Analysis

Although tissue- or lineage-specific markers can be helpful, they can also be confusing (see Section 26.5.1). Phenotypic variation may occur, leading to loss of differentiated markers or gain of markers that are not typical of the original tumor. Such variations may arise due to changes in differentiation status *in vitro* or within the original tumor (see Section 26.2). Moreover, studies of phenotypic variation are often limited by the small number of markers that are used for characterization. Large-scale analysis of the cell line transcriptome provides a more balanced view, particularly for cell lines that are widely used models and thus require more comprehensive characterization (see Section 18.1).

Analysis of the cell line transcriptome was initially performed using cDNA microarrays (Ross et al. 2000). Although various microarray formats continue to be used, RNA-seq is becoming more widely accessible and provides an important complementary approach (Malone and Oliver 2011; Lowe et al. 2017). Transcriptomic data are available for most widely used cell lines (see Table 18.1) and can be used for cell line characterization in many ways, including:

(1) *Identification of cell type or tissue of origin.* Early analysis using the NCI-60 panel demonstrated that cell lines from the same tissues typically clustered together; this clustering could be used to confirm tissue origin in most cases (Ross et al. 2000; Wang et al. 2006). Larger panels have since been used to confirm this finding, identify similarities to specific cell types or tumor subtypes, and uncover new therapeutic targets (Neve et al. 2006; Domcke et al. 2013; Quentmeier et al. 2019).

(2) *Selection of representative cell line models.* Although cell lines from the same tissues cluster together, they vary in their degree of similarity to the original tissue or tumor (Sandberg and Ernberg 2005). Transcriptomic analysis or multi-omic analysis can be used to determine which cell lines are the most appropriate models for specific tumors or subtypes. For example, multi-omic data have been used to rank 47 ovarian carcinoma cell lines, based on their suitability as models for high-grade serous ovarian cancer (Domcke et al. 2013). The two most popular ovarian cell lines, SK-OV-3 and A2780, were close to the bottom of the list.

(3) *Provision of complementary data for authentication testing.* Transcriptomic analysis can uncover cell lines that fail to cluster with the reported tissue of origin or display other unexpected behavior. For some cell lines, this is the first indication that the cell line has become misidentified, e.g. MDA-MB-435 (Ross et al. 2000). Tools are becoming available to compare data from different platforms (e.g. microarray and next-generation sequencing data) and to analyze variations in RNA-seq for authentication purposes (Blayney et al. 2016; Mohammad et al. 2019).

(4) *Understanding of common culture processes.* Transcriptomic analysis shows that many cell lines display a mesenchymal phenotype (Ross et al. 2000; Klijn et al. 2015). The mesenchymal gene signature is very similar to that

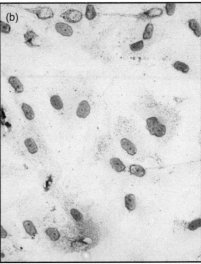

(a,b) Immunoperoxidase Staining.
(a) MOG-G-CCM glial cells stained for glial fibrillary acidic protein (GFAP); (b) human umbilical vein endothelial cells (HUVECs) stained for Factor VIII (granular staining). Source: R. Ian Freshney.

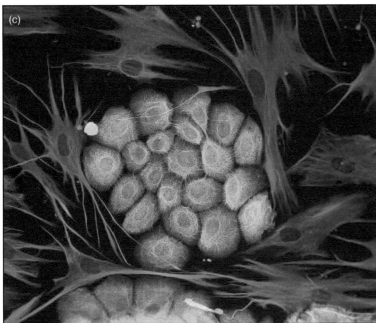

(c) Normal Human Keratinocytes on 3T3 Feeder Layer. *Keratinocytes are stained green using pancytokeratin, and 3T3 cells have been stained red using vimentin. Source: Courtesy of Hans-Jürgen Stark.*

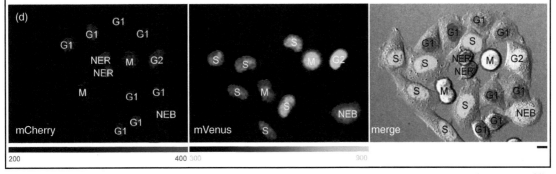

(d) HeLa Cells Expressing Fucci(CA). *Cells express Fucci(CA) constructs containing mCherry (left panel) and mVenus (middle panel); fluorescent data are then merged with a phase contrast image obtained using differential interference contrast (DIC; right panel). The Fucci(CA) constructs produce a color-distinct separation as cells progress through G1, S, and G2 of the cell cycle. M, mitosis; NEB, nuclear envelope breakdown; NER, nuclear envelope reformation. The scale bar represents 10 μm. Source: [Sakaue-Sawano et al. 2017], DOI 10.1016/j.molcel.2017.10.001, reproduced with permission of Elsevier.*

Fig. 18.2. Expression of tissue- or cell-specific markers.

TABLE 18.2. Enzymatic markers.

Enzyme	Cell type	Inducer	Repressor	References
Alkaline phosphatase	Type II pneumocyte (in lung alveolus)	Dexamethasone, oncostatin, IL-6	TGF-β	Edelson et al. (1988); McCormick and Freshney (2000)
Alkaline phosphatase	Enterocytes	Dexamethasone, sodium butyrate	—	Vachon et al. (1996)
Angiotensin-converting enzyme	Endothelium	Collagen, Matrigel	—	Del Vecchio and Smith (1981)
Creatine kinase BB	Neurons, neuroendocrine cells, SCLC	—	—	Gazdar et al. (1981)
Creatine kinase MM	Muscle cells	IGF-II	FGF-1,2,7	Stewart et al. (1996)
DOPA-decarboxylase	Neurons, SCLC	—	—	Gazdar et al. (1980); Chung et al. (2006)
Glutamyl synthetase	Astroglia (brain)	Hydrocortisone	Glutamine	Hallermayer and Hamprecht (1984)
Neuron-specific enolase	Neurons, neuro-endocrine cells	—	—	Hansson et al. (1984)
Non-specific esterase	Macrophages	PMA, Vitamin D_3	—	Murao et al. (1983)
Proline hydroxylase	Fibroblasts	Vitamin C	—	Pinnel et al. (1987)
Sucrase	Enterocytes	Sodium butyrate	—	Pignata et al. (1994); Vachon et al. (1996)
Tyrosinase	Melanocytes	cAMP	—	Park et al. (1993, 1999)
Tyrosine aminotransferase	Hepatocytes	Hydrocortisone	—	Granner et al. (1968)

Abbreviations not found in main text: FGF, fibroblast growth factor; IGF, insulin-like growth factor; PMA, phorbol 12-myristate 13-acetate; SCLC, small cell lung carcinoma; TGF-β, transforming growth factor-β.

seen after induction of EMT (see Section 26.2.5), suggesting that EMT is a common *in vitro* process regardless of tissue origin. Similarly, genes associated with multiple drug resistance are upregulated in many cell lines, including members of the NCI-60 panel (Gillet et al. 2011). Although these common gene expression signatures have caused concerns regarding the suitability of cell lines as research models (Gillet et al. 2013), they also provide opportunities to study these important processes in greater depth.

(5) ***Single-cell analysis.*** RNA-seq can be used for single-cell analysis, resulting in exciting new opportunities to investigate cancer genomics, stem cell biology, and the dynamics of clonal evolution (Navin 2015; Wen and Tang 2016; Lawson et al. 2018).

Always remember that transcriptomic analysis provides a "snapshot" of cell behavior that will depend on the culture conditions at that point in time. Cultures that are used for transcriptomic studies are often grown in antibiotics (which may explain upregulation of drug resistance genes) and harvested during the exponential phase of cell growth, when they are least likely to display differentiated behavior (Cooper et al.

2016; Ryu et al. 2017). It is important to optimize the culture conditions first, based on the behavior of the cell line, before you commence any large-scale characterization.

18.3.4 Behavioral Assays

Some cell types display certain behaviors that indicate their cell type or tissue of origin, such as muscle contraction or neuronal depolarization. Perhaps the most widely studied *in vitro* behavior relates to polarized transport across a cell monolayer, which can be used to predict the absorption of orally administered drugs (Hubatsch et al. 2007). Transport of inorganic ions, and the resultant transfer of water, is characteristic of absorptive and secretory epithelia (Abaza et al. 1974). The formation of "domes" in monolayer culture may indicate that such transport is occurring (see Figure 18.3; see also Figure 26.6a, b). These structures are caused by accumulation of water on the underside of the monolayer, which requires active fluid transport (Rabito et al. 1980; Lever 1981; Ramond et al. 1985). Domes have been observed in a handful of cell lines, including MDCK kidney cells, Caco-2 colon cells, and WIL lung cells. The Caco-2 cell line is a particularly good model of polarized transport; cells

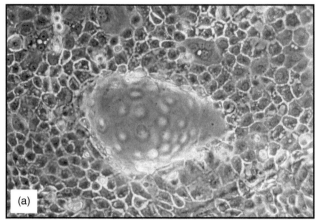

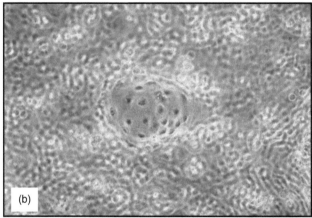

Fig. 18.3. Domes. This example was photographed in an epithelial monolayer, showing (a) lower focus on the monolayer; (b) upper focus on the top of the dome. *Source*: R. Ian Freshney.

can spontaneously differentiate to form polarized, confluent monolayers after 15–20 days on a filter well insert (see Section 27.5.2) (Borchardt 2011). Protocols are available elsewhere for production of polarized monolayers of Caco-2 cells (Hubatsch et al. 2007; Natoli et al. 2012).

18.4 CELL IMAGING

18.4.1 Microscopy

Routine observation using an inverted microscope is an essential part of cell line characterization (see Sections 5.2.1, 14.4.1). An inverted microscope is necessary because the thickness of the culture vessel makes observation difficult from above. Instead, the objective is positioned below the microscope stage, pointing upwards, and the light source is positioned above the culture vessel. As the thickness of the wall of the culture vessel still limits the working distance, the objective magnification is usually limited to 40×. Phase contrast optics are used in most inverted microscopes to visualize living cells, which are transparent using standard bright field illumination. In phase contrast microscopy, an

annular light path is masked by a corresponding dark ring in the objective and only diffracted light is visible, which enables unstained cells to be viewed with higher contrast. Dense regions appear darker and a characteristic "halo" of light can be seen surrounding the cell. Some inverted microscopes may use differential interference contrast (DIC) to provide higher contrast. In DIC microscopy, a prism is used to generate separate beams of light that have a difference in phase due to their different paths through the sample. The beams are then recombined by passing through a second prism, resulting in interference between the two beams. DIC results in a pseudo-3D appearance and is commonly used in confocal microscopes (see Figure 18.2d).

The inverted microscope is one of the most important tools in the tissue culture laboratory, but it is often used incorrectly. It is important to seek out training if you are a newcomer to microscopy; excellent resources are also available from Leica, Nikon, Olympus, Zeiss, and other microscope suppliers. The following protocol (Protocol P18.1) is provided as a starting point when using an inverted microscope.

PROTOCOL P18.1. USING AN INVERTED MICROSCOPE

Outline

Place the culture on the microscope stage, switch on the light, and select the correct optics. Focus on the specimen and center the condenser and phase ring, if necessary.

❖ *Safety Note. Prolonged microscope use can result in eyestrain or musculoskeletal issues. Be aware of your posture, adjust the seat and microscope to suit your needs, and take frequent breaks to stretch your muscles and rest your eyes.*

Materials (sterile or aseptically prepared)

☐ Cultures in flasks, Petri dishes, multiwell plates, or other vessels (see Section 8.6.1)

Materials (non-sterile)

☐ Inverted microscope (see Section 5.2.1) using phase contrast with 4x, 10x, and 20x objectives
☐ Lens tissue
☐ 70% alcohol

Procedure

1. Check that the microscope is clean and ready for use. If necessary, wipe the stage with 70% alcohol and clean the objectives using lens tissue.

2. Adjust the seat and microscope settings so that you are comfortable:

   (a) Set the seat to the correct height and distance from the bench such that your eyes are in line with the eyepieces without straining and your back maintains its natural curve without hunching. This will not be the same for all users, so take time to adjust the seat.

   (b) Position yourself at the bench such that your elbows are close to your body, your forearms do not rest on any hard edges, and your hands are free for focusing.

   (c) Adjust the eyepieces so that they are the correct distance apart for your eyes.

3. Switch on the power, bringing the lamp intensity up from its lowest to the correct intensity with the rheostat, instead of switching the lamp straight to bright illumination.

4. Check the alignment of the condenser and the light source by setting Köhler illumination:

   (a) Deselect the phase contrast condenser if there is one.

   (b) Place a stained slide, flask, or dish on the stage.

   (c) Close down the field aperture.

   (d) Focus on the image of the iris diaphragm.

   (e) Center the image of the iris in the field of view.

   (f) Open the field aperture until the edges are just invisible in the field of view.

5. Center the phase ring:

   (a) Reselect the correct phase contrast condenser.

   (b) If a phase telescope is provided, insert it in place of the standard eyepiece. Then focus on the phase ring, and check that the condenser ring (white on black) is concentric with the objective ring (black on white).

   (c) If no phase telescope is provided, replace the stained specimen with an unstained culture (living or fixed), refocus, and move the phase ring adjustment until optimum contrast is obtained.

6. Bring the culture vessel from the incubator to the microscope. Wipe down the flask, dish, or plate with 70% alcohol and place on the stage.

7. Select an objective to examine the cells. A 4x phase-contrast objective will not give sufficient detail but may be used to perceive patterns and select areas for investigation. A 10x phase-contrast objective will give sufficient detail for routine examination. Higher magnifications than 10x restrict the field of view but are helpful to assess intracellular features.

8. Record your observations (see Tables 14.3, S30.1). The following questions may be helpful:

   (a) Does the medium appear clear? Look for debris, floating granules, or other signs of contamination (see Figures 16.2, 16.3). Contamination is often more obvious as you change the focal plane, e.g. to focus above the cell monolayer.

   (b) Do the cells appear normal compared to reference photographs? Look for evidence of deterioration (e.g. granular, vacuolated cells) or changes in behavior.

   (c) What is the approximate level of confluence? This may be estimated using a percent value (e.g. 70–80%) or a description (e.g. sparse, subconfluent, confluent, dense).

9. Wipe down the flask, dish, or plate with 70% alcohol and return the culture to the incubator.

10. Repeat steps 6–9 until all vessels have been examined.

11. Once viewing is complete, turn the rheostat down and switch off the power to the microscope.

Notes

Steps 4 and 5: General information and tutorials on Köhler illumination and phase contrast condensers are available elsewhere (Murphy et al. 2019; Parry-Hill et al. 2019). For specific recommendations, always refer to the operating manual for that microscope model.

Step 7: With some inverted microscopes, the phase ring should be changed whenever the objective is changed. If you are not sure, check the operating manual for your model.

18.4.2 Photomicrography

Digital photography has greatly simplified the study of cells in culture. Film photography gave beautiful, high-resolution images of cultured cells, but was challenging for novices, who needed to wait until the film was developed before discovering if their images were successful. Digital images can be viewed immediately, and adjustments made to microscope and camera settings until suitable image quality is achieved. Most modern tissue culture laboratories invest in a digital camera, computer, and monitor to perform routine photomicrography (see Sections 5.2.2, 5.2.3). Always check that suitable image software is installed. Image acquisition software should be available from the microscope supplier, but some laboratories prefer to use their own software, particularly if image processing or analysis are required. Examples include Image-Pro® (Media Cybernetics), Imaris (Oxford Instruments) and MetaMorph® (Molecular Devices) software packages.

Any project that generates large numbers of images will require planning to ensure that sufficient data storage is available and files are suitably organized. It is important to estimate how much space you will need for digital images and discuss your requirements with information technology (IT) staff; many organizations formalize this process using a Research Data Management Plan. Each image should be accompanied by metadata that identify the sample, the project and experiment, the date and time, the microscope, and the objective. Always keep an unedited, high-resolution version of the original image and perform any editing on a separate copy. If images are stored in a shared database, it is essential to comply with any database rules for vocabulary (e.g. file names) and to fill out all required fields (Plant et al. 2011; Eliceiri et al. 2012).

PROTOCOL P18.2. DIGITAL PHOTOGRAPHY ON A MICROSCOPE

Outline

Using an inverted microscope (see Protocol P18.1), focus on a suitable area. Open imaging software, divert the light beam to the camera, capture the image, and save the file.

Materials (sterile or aseptically prepared)

☐ Cultures in flasks, Petri dishes, multiwell plates, or other vessels (see Section 8.6.1)

Materials (non-sterile)

☐ Inverted microscope (see Section 5.2.1) using phase contrast with 4x, 10x, and 20x objectives
☐ Digital camera and computer (see Section 5.2.2) with image acquisition software
☐ Lens tissue
☐ 70% alcohol
☐ Micrometer slide for size comparison (image software may supply this requirement)

Procedure

A. Preparation of Cultures for Imaging

1. Bring the culture vessel from the incubator to the microscope. Wipe down the flask, dish, or plate with 70% alcohol and place on the stage.

2. Examine the culture to make sure it is suitable for imaging. Make sure that it is free of debris; if necessary, passage the cells or change the culture medium before proceeding. If cells are passaged to a new flask, label the flask on the side to avoid obscuring the field of view.
3. Check for condensation that may affect image quality. If condensation is present:
   (a) For tissue culture flasks, gently turn the flask so that a film of medium runs over the roof, removing the condensation (do not allow the medium to run into the neck of the flask where contamination may occur). Allow the film of medium to drain by standing the flask vertically for 10–20 seconds.
   (b) For dishes or plates, open in a BSC and replace the lid (make sure that the dish or plate is labeled on the base).

B. Digital Imaging

4. Move the flask on the stage until a suitable area is found for imaging. Avoid any obvious imperfections or marks on the vessel.
5. Select the best objective for imaging. Typically, 4x is best for clones or patterning of a monolayer, 10x for a representative shot of cells, and 20x or greater for cellular detail.
6. Adjust the light intensity and the focus to give the best results as seen with the eyepiece.
7. Open the image acquisition software and (if necessary) divert the light beam from the eyepiece to the camera. Bring up the live image in the software.
8. Use the software to perform auto-adjustments, e.g. exposure time, white balance, or shading. Check to see if any manual adjustments are required, e.g. focus or light intensity.
9. Use the software to take the photograph. Check the resulting image; if it is not optimal, continue to adjust microscope and computer settings and take more photographs.
10. Once you have a selection of images you are satisfied with, wipe down the flask, dish, or plate with 70% alcohol and return the culture to the incubator.
11. Save the image files at a resolution that is suitable to the intended use. Add any additional metadata to identify the images for later use.
12. Use a micrometer slide or the image software to give the scale of magnification. Use this information to add a scale bar to the image or keep it for later reference.

Notes

Step 1: If flasks are not vented, it may be necessary to loosen the lid and allow the culture to cool, in order to avoid any distortion of the flask. Tighten the cap when the contents of the flask are at room temperature.

Step 4: In addition to looking for representative areas, think about the composition of the image. This may include placing key features at the center of the field of view and avoiding debris.

Step 8: When adjusting focus, it may help to increase the size of the field of view on the computer screen. If it looks sharp and clear, the final image should also look good.

Step 9: If the culture is left on the stage for a long period of time, turn down the light whenever possible to avoid overheating. An infrared filter may be incorporated to minimize overheating.

Step 11: Always retain the original, high-resolution file in case of later problems. Remember that you can always reduce the resolution later, but you cannot increase it! Publications usually require images at 300–600 dots per inch (dpi).

18.4.3 Live-Cell Imaging

Live-cell microscopy was initially performed in the 1920s by Ronald Canti and other tissue culture pioneers (Canti 1927; Stramer and Dunn 2015). However, uptake of "microcinematography" was relatively slow until the 1980s; the advent of computers made it possible to perform video enhancement and digital processing, which increased sensitivity and resolving power beyond the limits of photographic film (Dunn and Jones 2004). Scientists working in the 1990s developed a "paintbox" of fluorescent proteins to label living cells – mostly related to green fluorescent protein (GFP), which was isolated from the bioluminescent jellyfish *Aequorea victoria* (Matz et al. 2002; Tsien 2010). Genes encoding fluorescent proteins can now be introduced into the living cell and visualized using laser light at specific wavelengths. These advances have resulted in some beautiful studies of the living cell. For example, HeLa cells can be genetically modified to carry the Fucci (CA) sensor, which varies in color as cells progress through G1, S, and G2 phases of the cell cycle (see Figure 18.2d) (Sakaue-Sawano et al. 2017). Individual cells with fluorescent markers can be followed over time to study migration, changes in phenotype, or the fate of individual stem cells (Schroeder 2011; Piltti et al. 2018).

What equipment do you need to perform live-cell imaging? Many of the cameras that are available in tissue culture laboratories (see Section 5.2.2) can be used to capture changes in morphology or migration, using time-lapse or continuous video recording. These cameras are often connected to inverted microscopes that can perform both phase contrast and epifluorescence microscopy. However, live-cell imaging under these circumstances is limited by a lack of environmental control. Prolonged imaging is stressful for cells, resulting in phototoxicity, photobleaching of fluorescent signals, and changes in temperature, pH, and other variables (Dailey et al. 2006; Frigault et al. 2009). It is important to minimize light exposure to reduce phototoxicity and photobleaching and to use some form of environmental control for long experiments.

Suitable environmental control can be provided by (i) installing a camera within a CO_2 incubator; (ii) providing some form of chamber or enclosure for the microscope stage; or (iii) growing cells in a culture vessel where the environment can be controlled through perfusion. Live-cell imaging platforms that can be installed in a CO_2 incubator include the CytoSMART™ 2 (Lonza) and the IncuCyte® platform (Sartorius; see Figure 18.4). Such incubator-based imaging systems are particularly useful to improve the speed and convenience of cell-based assays. Images can be used to generate growth curves (see Figure 18.5) or to perform migration-based assays, e.g. scratch wound assays (Guy et al. 2017). If your imaging requirements cannot be met inside an incubator, then it will be necessary to use a microscope with environmental control. Environmental equipment may range from a simple heated stage to a complex stage-top incubator that can be used for high-resolution imaging (see Section 18.4.4). A protocol for time-lapse analysis is provided in the Supplementary Material online, using a stage-top chamber and MetaMorph software (see Supp. S18.1).

The choice of imaging equipment will depend on your expertise, budget, and application. Low-cost imaging platforms can be constructed if you have the necessary skills; it is possible to assemble an imaging system (including an incubator chamber and motorized stage) for less than €1250 (Walzik et al. 2015). However, many laboratories do not have this expertise and will either invest in commercial imaging systems or use a core facility that specializes in microscopy and imaging. Always discuss your requirements with local core facilities and arrange for a demonstration (ideally using your cells or assay) before you invest in any live-cell system. Further information on live-cell imaging is available from commercial suppliers of these systems; some excellent reviews are also available (Frigault et al. 2009; Jensen 2013; Ettinger and Wittmann 2014).

18.4.4 High-Resolution Imaging

Confocal microscopy. Confocal microscopy allows the "optical sectioning" of living or fixed cells and tissue samples (Petran et al. 1986; Dailey et al. 2006). It is widely used for subcellular imaging; tight junctions in polarized epithelia can be studied using immunofluorescent staining, or migrating

Fig. 18.4. IncuCyte Live-cell Analysis System (Sartorius). (a) Incubator with IncuCyte S3 in place; (b) closer view of open compartment showing flasks and multiwell plates. Reagents and consumables are also available for use with the system, including ClearView plates and Live-cell Imaging reagents (in foreground). *Source*: Courtesy of Sartorius, with permission.

cells can be labeled with fluorescent phalloidin to study actin filaments (Wakabayashi et al. 2007; Burnette et al. 2014). Confocal microscopy is also effective for non-transparent samples, making it useful for characterization of 3D cultures (Graf and Boppart 2010). Confocal data are stored digitally and can be processed in a number of ways, including the creation of a vertical section through the sample (a so-called Z-section). Several different technologies are available, including the confocal laser-scanning microscope and the spinning-disk confocal microscope (Jonkman and Brown 2015). However, there are some limitations when using these technologies. Penetration depth is limited to about 100 μm and prolonged light exposure can result in marked phototoxicity and photobleaching (Dailey et al. 2006; Graf and Boppart 2010). For 3D spheroids, it may be necessary to perform optical clearing to reduce light scattering, which will improve imaging depth and contrast (Costa et al. 2019; Dekkers et al. 2019).

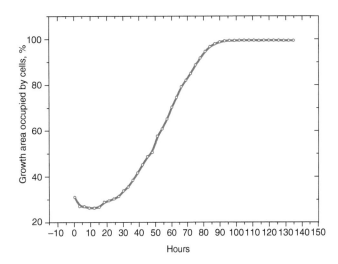

Fig. 18.5. IncuCyte growth curve. The human colon carcinoma cell line HCA7 colony 29 has been monitored by the IncuCyte for five to six days, resulting in a graph of cell confluence (i.e. the percentage of the growth area that is occupied by cells) over time. *Source*: Courtesy of J. Cooper, ECACC.

Two-photon microscopy. Two-photon microscopy was developed as an alternative to conventional confocal microscopy that offers high-resolution imaging of fluorescent samples (Denk et al. 1990). Like confocal microscopy, this technique provides thin optical sections, but it does so by restricting fluorescent light emission to the focal plane of the microscope, which gives greater penetration depth (up to 1 mm) with less phototoxicity and photobleaching (Rubart 2004; Costa et al. 2019). Two-photon microscopes are less widely available compared to confocal microscopes, but if accessible, they can be used for the characterization of 3D cultures that give rise to a high degree of light scattering (Graf and Boppart 2010). For example, two-photon imaging has been used to visualize the development of intestinal villi in live organ cultures and the infiltration of tumor cells within cerebral organoids (Walton and Kolterud 2014; Linkous et al. 2019).

Light-sheet fluorescence microscopy. Light-sheet microscopy was developed to combine optical sectioning with multiple-view imaging (Reynaud et al. 2008). The sample is illuminated from the side, using a cylindrical lens to focus a laser beam into a thin sheet of light. The microscope can acquire a stack of images from illumination of a single focal plane, which leads to much less phototoxicity and photobleaching. As with two-photon microscopy, light-sheet microscopes are expensive and difficult to access, but if your local core facility has such a microscope, it will allow live-cell imaging to be performed at high resolution for extended periods of time. For example, light-sheet microscopy can be used to study microtubule growth trajectories in HeLa cells as they undergo mitosis and to perform single-cell tracking in kidney organoids (Yamashita et al. 2015; Held et al. 2018).

18.5 CELL STAINING

18.5.1 Preparation of Cultures for Staining

Staining of fixed cells continues to be an important part of cell line characterization and may be combined with live-cell imaging, with the initial live-cell images matched to later staining data (Gomez-Villafuertes et al. 2017). Although staining and live-cell imaging may be complementary, there are some differences that must be considered. Staining usually requires a vessel or substrate that has been designed specifically for imaging, e.g. coverslips or chambered slides (see Figure 8.10a, b). Plastic vessels tend to warp in organic solvents (which are commonly used during staining procedures), so glass vessels are typically used, although some coverslips and slides are available that use solvent-resistant plastics (see Section 8.6.1). Non-adherent cells must be deposited on the surface for cytological observations. The conventional technique for blood cells – the preparation of a blood smear – does not work well because cultured cells tend to rupture during smearing, although this can be reduced by coating the slide with serum and then spreading the cells in serum. Most laboratories working with suspension cultures will spin cells onto slides using a cytocentrifuge (see Figure 18.6). A procedure for using a cytocentrifuge is provided as Supplementary Material online (see Supp. S18.2).

Fixation is usually performed prior to staining to preserve cellular architecture, although some stains can be used on live cells, e.g. Hoechst 33342 (Chazotte 2011). Methanol or paraformaldehyde are commonly used as fixatives for cell staining; glutaraldehyde, although used for some procedures, results in autofluorescence and is best avoided when examining fluorescent markers (Callis 2010). Methanol acts to precipitate proteins; it is commonly used as a 3 : 1 mixture with acetic acid (see Appendix B) or as a 1 : 1 mixture with acetone, which helps to permeabilize cell membranes (Held 2015). Methanol is best avoided when using GFP and other expressed fluorescent proteins, as precipitation may alter their function. Paraformaldehyde acts to cross-link proteins but does not permeabilize the membranes, so a separate permeabilization step must be performed, e.g. using Triton™ X-100. It is usually made up as a 4% solution in phosphate-buffered saline (PBS) and should be used fresh (see Appendix B). The fixation time will vary with the fixative and the sample. If 4% paraformaldehyde is used as a fixative, 10 minutes may be sufficient for some samples, but 45 minutes is recommended for organoids and a longer fixation time (up to four hours) may be required for organoids with fluorescent markers (Dekkers et al. 2019).

> • *Safety Note. Many fixatives are hazardous chemicals. Always read the Safety Data Sheet (SDS) and use suitable risk control measures (see Section 6.2.2).*

18.5.2 Histological Stains

Giemsa staining. Giemsa stain is commonly used in clinical laboratories to stain chromosomes and some pathogens (e.g. *Plasmodium vivax*). It also provides an easy way to stain fixed cells for bright field imaging or to count colonies, e.g. when testing medium for cloning efficiency (see Supp. S11.7). Giemsa stain is usually combined with May-Grünwald stain when staining blood, but not when staining cultured cells. It is a polychromatic stain that, when used alone, turns the nucleus pink or magenta, the nucleoli dark blue, and the cytoplasm pale gray-blue (see Figures 14.3b,c, 26.4e,f). It stains cells that are fixed in alcohol or formaldehyde but will not work correctly unless the preparation is completely anhydrous.

Giemsa staining is a simple procedure that gives a good high-contrast polychromatic stain, but precipitated stain may give a spotted appearance to the cells. This precipitate forms as a scum at the surface of the staining solution as well as throughout the solution and is often deposited on the surface of the slide, when water is added. Protocol P18.3 washes off the stain by upward displacement, rather than pouring it off or removing slides from a dish, which is designed to prevent the cells from coming in contact with the scum. Extensive washing at the end of the procedure is designed to remove any precipitate left on the preparation.

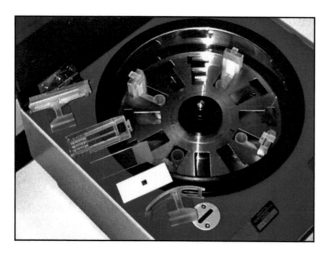

Fig. 18.6. Cytocentrifuge. The interior of the Cyto-Tek® centrifuge is shown here, with rotor and slide carriers in place; one carrier is disassembled on the left-hand side. *Source*: Courtesy of Sakura Finetek, with permission.

PROTOCOL P18.3. STAINING WITH GIEMSA

Outline

Fix the culture in methanol, stain it directly with undiluted Giemsa, and then dilute the stain 1 : 5. Wash the culture and examine it wet.

❖ *Safety Note. Methanol and Giemsa stains are hazardous. Refer to the SDS from the supplier for information on associated hazards and safety measures.*

Materials (sterile or aseptically prepared)

☐ Cultures in Petri dishes or other culture vessels

Materials (non-sterile)

☐ Dulbecco's phosphate buffered saline without Ca^{2+} and Mg^{2+} (DPBS-A)
☐ Anhydrous methanol
☐ DPBS-A: methanol, 1 : 1
☐ Giemsa stain solution, undiluted (e.g. #GS500, Sigma-Aldrich)
☐ Deionized water

Procedure

1. Remove and discard the medium.
2. Rinse the monolayer with DPBS-A and discard the rinse.
3. Add 5 ml DPBS-A: methanol per 25 cm^2 surface area and leave for two minutes. Remove and discard the DPBS-A: methanol.
4. Add 5 ml fresh methanol and leave for 10 minutes. Remove and discard the methanol.
5. Rinse with fresh anhydrous methanol and discard the methanol.
6. At this point, the flask may be dried and stored or stained directly. It is important that staining be done directly from fresh anhydrous methanol, even with a dry flask. If the methanol is poured off and the flask is left for some time, water will be absorbed by the residual methanol and will inhibit subsequent staining. Even "dry" monolayers can absorb moisture from the air and should be rinsed with fresh anhydrous methanol before staining.
7. Add neat Giemsa stain, 2 ml per 25 cm^2, making sure that the entire monolayer is covered and remains covered. Leave it for two minutes.
8. Dilute the stain with 8 ml water and agitate it gently for a further two minutes.

9. Displace the stain with water so that the scum that forms is floated off and not left behind to coat the cells. Wash the cells gently in running tap water until any pink cloudy background stain (precipitate) is removed, but stain is not leached out of cells (usually about 10–20 seconds).
10. Pour off the water, rinse the monolayer in deionized water, and examine the cells on the microscope while the monolayer is still wet.
11. Store the cells dry and rewet them to re-examine.

Notes

Step 1: This procedure assumes that a cell monolayer is being used, but fixed cell suspensions (see Supp. S18.2) can also be used, starting at step 6.

Crystal violet staining. Crystal violet can be used to stain fixed-cell preparations as an alternative to Giemsa. It is not as good as Giemsa for morphological observations, but it is easier to use and can be reused. It is a convenient stain to use for staining clones for counting, as it does not have the precipitation problems of Giemsa and is easier to wash off, giving a clearer background and making automated colony counting easier. Crystal violet is a monochromatic stain that turns the nucleus dark blue and the cytoplasm light blue.

PROTOCOL P18.4. STAINING WITH CRYSTAL VIOLET

Outline

Fix the culture in methanol, stain it directly with crystal violet, wash the culture, and examine.

❖ *Safety Note. Methanol and crystal violet stains are hazardous. Refer to the SDS from the supplier for information on associated hazards and safety measures.*

Materials (sterile or aseptically prepared)

☐ Cultures in Petri dishes or other culture vessels

Materials (non-sterile)

☐ DPBS-A
☐ Anhydrous methanol
☐ DPBS-A: methanol, 1 : 1
☐ Crystal violet solution, 0.1% (see Appendix B)
☐ Filter funnel and filter paper of a size appropriate to take 5 ml of stain from each dish

□ Bottle for recycled stain

Procedure

1. Remove and discard the medium.
2. Rinse the monolayer with DPBS-A and discard the rinse.
3. Add 5 ml DPBS-A: methanol per 25 cm² surface area and leave for two minutes. Remove and discard the DPBS-A: methanol.
4. Add 5 ml fresh methanol and leave for 10 minutes. Remove and discard the methanol.
5. Drain the dishes and allow them to dry.
6. Add 5 ml of crystal violet per 25 cm² (making sure that the whole of the growth surface is covered) and leave for 10 minutes.
7. Place the filter funnel in the recycle bottle and pour off the stain into the filter funnel.
8. Rinse the dish in tap water and then in deionized water and allow the dish to dry.

18.5.3 Immunocytochemistry

Tissue- or lineage-specific markers are commonly detected by the binding of antibodies to specific antigens. There are some caveats to this approach; the presence of a particular protein does not necessary mean that it is active, and many proteins that are expressed in fully differentiated cells are not observed in proliferating cultures (see Section 18.3.2). However, immunocytochemical analysis is easily performed if suitable antibodies are available. It can be a useful technique for cell line characterization, provided its limitations are clearly recognized (Quentmeier et al. 2001).

Immunological staining of cells or tissue may be direct or indirect. In direct staining, the specific antibody is conjugated to a fluorochrome or enzyme and used to stain the specimen directly. In indirect staining (which is more common), two antibodies are used: a primary antibody that recognizes the antigen in the specimen and a secondary antibody that binds to the primary antibody. The secondary antibody is conjugated to an enzyme or fluorochrome, which is used for visualization. The indirect approach may seem complex in theory, but in practice it is easily performed and allows the same method of visualization to be used with various primary antibodies. Enzymes that are used for immunocytochemistry include horseradish peroxidase and alkaline phosphatase, resulting in a colored product that can be seen using bright field microscopy (see Figure 18.2a, b). Fluorochrome-labeled antibodies result in fluorescent staining in a range of colors; while green is the most popular, due to the widespread use of fluorescein isothiocyanate (FITC), a large array of other colors are available (Novus Biologicals 2019). A protocol for indirect immunofluorescence using chambered slides is provided in the Supplementary Material online (see Supp. S18.3).

The specificity of the reaction must be assessed using appropriate controls. This is true for monoclonal antibodies and polyclonal antisera alike; a monoclonal antibody is highly specific for a particular epitope, but the specificity of the expression of the epitope to a particular cell type must still be demonstrated. For immunofluorescent staining, controls should be included to detect the presence of autofluorescence, which may occur naturally in some cell types or relate to the fixation method. Exposure to light results in photobleaching, which means that slides cannot be archived for long-term storage and must be shielded from light until imaging has been completed.

Suppliers

Supplier	URL
Leica	http://www.leica-microsystems.com
Lonza	http://www.lonza.com
Media Cybernetics	http://www.mediacy.com
Molecular Devices	http://www.moleculardevices.com
Nikon	http://www.nikoninstruments.com/applications/life-sciences
Olympus	http://www.olympus-lifescience.com/en
Oxford Instruments	http://imaris.oxinst.com
Sakura Finetek	http://www.sakuraus.com
Sartorius	http://www.sartorius.com/en
Sigma-Aldrich (Merck)	http://www.sigmaaldrich.com/life-science/cell-culture.html
Zeiss	http://www.zeiss.com/microscopy/int/home.html

Supp. S18.1 Time-Lapse Video Recording.

Supp. S18.2 Preparation of Suspension Cultures for Cytology by Cytocentrifuge.

Supp. S18.3 Immunofluorescence Using Chambered Slides.

REFERENCES

Abaza, N.A., Leighton, J., and Schultz, S.G. (1974). Effects of ouabain on the function and structure of a cell line (MDCK) derived from canine kidney. I. Light microscopic observations of monolayer growth. *In Vitro* 10 (3–4): 72–183.

Bairoch, A. (2018). The Cellosaurus, a cell-line knowledge resource. *J. Biomol. Tech.* 29 (2): 25–38. https://doi.org/10.7171/jbt.18-2902-002.

Barretina, J., Caponigro, G., Stransky, N. et al. (2012). The Cancer Cell Line Encyclopedia enables predictive modelling of anticancer drug sensitivity. *Nature* 483 (7391): 603–607. https://doi.org/10.1038/nature11003.

Basu, A., Bodycombe, N.E., Cheah, J.H. et al. (2013). An interactive resource to identify cancer genetic and lineage dependencies targeted by small molecules. *Cell* 154 (5): 1151–1161. https://doi.org/10.1016/j.cell.2013.08.003.

Ben-David, U., Beroukhim, R., and Golub, T.R. (2019). Genomic evolution of cancer models: perils and opportunities. *Nat. Rev. Cancer* 19 (2): 97–109. https://doi.org/10.1038/s41568-018-0095-3.

Berglind, H., Pawitan, Y., Kato, S. et al. (2008). Analysis of p53 mutation status in human cancer cell lines: a paradigm for cell line cross-contamination. *Cancer Biol. Ther.* 7 (5): 699–708. https://doi.org/10.4161/cbt.7.5.5712.

Bishop, A.E., Power, R.F., and Polak, J.M. (1988). Markers for neuroendocrine differentiation. *Pathol. Res. Pract.* 183 (2): 119–128.

Blayney, J.K., Davison, T., McCabe, N. et al. (2016). Prior knowledge transfer across transcriptional data sets and technologies using compositional statistics yields new mislabelled ovarian cell line. *Nucleic Acids Res.* 44 (17): e137. https://doi.org/10.1093/nar/gkw578.

Borchardt, R.T. (2011). Hidalgo, I. J., Raub, T. J., and Borchardt, R. T.: characterization of the human colon carcinoma cell line (Caco-2) as a model system for intestinal epithelial permeability, *Gastroenterology*, 96, 736–749, 1989 – the backstory. *AAPS J.* 13 (3): 323–327. https://doi.org/10.1208/s12248-011-9283-8.

Bryant, P.E., Gray, L.J., and Peresse, N. (2004). Progress towards understanding the nature of chromatid breakage. *Cytogenet. Genome Res.* 104 (1–4): 65–71. https://doi.org/10.1159/000077467.

Burnette, D.T., Shao, L., Ott, C. et al. (2014). A contractile and counterbalancing adhesion system controls the 3D shape of crawling cells. *J. Cell Biol.* 205 (1): 83–96. https://doi.org/10.1083/jcb.201311104.

Callis, G. (2010). Glutaraldehyde-induced autofluorescence. *Biotech. Histochem.* 85 (4): 269. https://doi.org/10.3109/10520290903472415.

Canti, R. G. (1927). Cells in tissue culture (normal and abnormal). https://wellcomecollection.org/works/yygn7b8s (accessed 18 September 2019).

Cazaly, E., Saad, J., Wang, W. et al. (2019). Making sense of the epigenome using data integration approaches. *Front. Pharmacol.* 10: 126. https://doi.org/10.3389/fphar.2019.00126.

Chazotte, B. (2011). Labeling nuclear DNA with Hoechst 33342. *Cold Spring Harb. Protoc.* 2011 (1): pdb prot5557. https://doi.org/10.1101/pdb.prot5557.

Chung, S., Shin, B.S., Hwang, M. et al. (2006). Neural precursors derived from embryonic stem cells, but not those from fetal ventral mesencephalon, maintain the potential to differentiate into dopaminergic neurons after expansion in vitro. *Stem Cells* 24 (6): 1583–1593. https://doi.org/10.1634/stemcells.2005-0558.

Clark, G., Stockinger, H., Balderas, R. et al. (2016). Nomenclature of CD molecules from the tenth human leucocyte differentiation antigen workshop. *Clin. Transl. Immunol.* 5 (1): e57. https://doi.org/10.1038/cti.2015.38.

Cooper, J.R., Abdullatif, M.B., Burnett, E.C. et al. (2016). Long term culture of the A549 cancer cell line promotes multilamellar body formation and differentiation towards an alveolar type II pneumocyte phenotype. *PLoS One* 11 (10): e0164438. https://doi.org/10.1371/journal.pone.0164438.

Costa, M.L., Escaleira, R., Cataldo, A. et al. (2004). Desmin: molecular interactions and putative functions of the muscle intermediate filament protein. *Braz. J. Med. Biol. Res.* 37 (12): 1819–1830. https://doi.org/10.1590/s0100-879x2004001200007.

Costa, E.C., Silva, D.N., Moreira, A.F. et al. (2019). Optical clearing methods: an overview of the techniques used for the imaging of 3D spheroids. *Biotechnol. Bioeng.* 116 (10): 2742–2763. https://doi.org/10.1002/bit.27105.

Dailey, M.E., Manders, E., Soll, D.R. et al. (2006). Confocal microscopy of living cells. In: *Handbook of Biological Confocal Microscopy* (ed. J.B. Pawley), 381–403. Boston, MA: Springer.

Dean, W., Bowden, L., Aitchison, A. et al. (1998). Altered imprinted gene methylation and expression in completely ES cell-derived mouse fetuses: association with aberrant phenotypes. *Development* 125 (12): 2273–2282.

Dekkers, J.F., Alieva, M., Wellens, L.M. et al. (2019). High-resolution 3D imaging of fixed and cleared organoids. *Nat. Protoc.* 14 (6): 1756–1771. https://doi.org/10.1038/s41596-019-0160-8.

Del Vecchio, P.J. and Smith, J.R. (1981). Expression of angiotensin-converting enzyme activity in cultured pulmonary artery endothelial cells. *J. Cell. Physiol.* 108 (3): 337–345. https://doi.org/10.1002/jcp.1041080307.

Denk, W., Strickler, J.H., and Webb, W.W. (1990). Two-photon laser scanning fluorescence microscopy. *Science* 248 (4951): 73–76. https://doi.org/10.1126/science.2321027.

Diogo, M.M., da Silva, C.L., and Cabral, J.M. (2012). Separation technologies for stem cell bioprocessing. *Biotechnol. Bioeng.* 109 (11): 2699–2709. https://doi.org/10.1002/bit.24706.

Domcke, S., Sinha, R., Levine, D.A. et al. (2013). Evaluating cell lines as tumour models by comparison of genomic profiles. *Nat. Commun.* 4: 2126. https://doi.org/10.1038/ncomms3126.

Drexler, H. G. (2010). Guide to leukemia-lymphoma cell lines. https://www.dsmz.de/research/human-and-animal-cell-lines/guide-to-leukemia-lymphoma-cell-lines.html (accessed 24 February 2011).

Dunn, G.A. and Jones, G.E. (2004). Cell motility under the microscope: Vorsprung durch Technik. *Nat. Rev. Mol. Cell Biol.* 5 (8): 667–672. https://doi.org/10.1038/nrm1439.

Earle, W.R., Schilling, E.L., Stark, T.H. et al. (1943). Production of malignancy in vitro. IV. The mouse fibroblast cultures and changes seen in the living cells. *J. Natl Cancer Inst.* 4 (2): 165–212.

Edelson, J.D., Shannon, J.M., and Mason, R.J. (1988). Alkaline phosphatase: a marker of alveolar type II cell differentiation. *Am. Rev. Respir. Dis.* 138 (5): 1268–1275. https://doi.org/10.1164/ajrccm/138.5.1268.

Eliceiri, K.W., Berthold, M.R., Goldberg, I.G. et al. (2012). Biological imaging software tools. *Nat. Methods* 9 (7): 697–710. https://doi.org/10.1038/nmeth.2084.

Ettinger, A. and Wittmann, T. (2014). Fluorescence live cell imaging. *Methods Cell Biol.* 123: 77–94. https://doi.org/10.1016/B978-0-12-420138-5.00005-7.

Fisch, G.S. (2017). Whither the genotype–phenotype relationship? An historical and methodological appraisal. *Am. J. Med. Genet. C Semin. Med. Genet.* 175 (3): 343–353. https://doi.org/10.1002/ajmg.c.31571.

Fogh, J. (1986). Human tumor lines for cancer research. *Cancer Investig.* 4 (2): 157–184.

Forbes, S.A., Beare, D., Gunasekaran, P. et al. (2015). COSMIC: exploring the world's knowledge of somatic mutations in human cancer. *Nucleic Acids Res.* 43 (Database issue): D805–D811. https://doi.org/10.1093/nar/gku1075.

Freshney, R.I. (1980). Culture of glioma of the brain. In: *Brain Tumours: Scientific Basis, Clinical Investigation and Current Therapy* (eds. D.G.T. Thomas and D.I. Graham), 21–50. London: Butterworths.

Frigault, M.M., Lacoste, J., Swift, J.L. et al. (2009). Live-cell microscopy – tips and tools. *J. Cell Sci.* 122 (Pt 6): 753–767. https://doi.org/10.1242/jcs.033837.

Garitaonandia, I., Amir, H., Boscolo, F.S. et al. (2015). Increased risk of genetic and epigenetic instability in human embryonic stem cells associated with specific culture conditions. *PLoS One* 10 (2): e0118307. https://doi.org/10.1371/journal.pone.0118307.

Garnett, M.J., Edelman, E.J., Heidorn, S.J. et al. (2012). Systematic identification of genomic markers of drug sensitivity in cancer cells. *Nature* 483 (7391): 570–575. https://doi.org/10.1038/nature11005.

Gazdar, A.F., Carney, D.N., Russell, E.K. et al. (1980). Establishment of continuous, clonable cultures of small-cell carcinoma of lung which have amine precursor uptake and decarboxylation cell properties. *Cancer Res.* 40 (10): 3502–3507.

Gazdar, A.F., Zweig, M.H., Carney, D.N. et al. (1981). Levels of creatine kinase and its BB isoenzyme in lung cancer specimens and cultures. *Cancer Res.* 41 (7): 2773–2777.

Ghandi, M., Huang, F.W., Jane-Valbuena, J. et al. (2019). Next-generation characterization of the Cancer Cell Line Encyclopedia. *Nature* 569 (7757): 503–508. https://doi.org/10.1038/s41586-019-1186-3.

Gillet, J.P., Calcagno, A.M., Varma, S. et al. (2011). Redefining the relevance of established cancer cell lines to the study of mechanisms of clinical anti-cancer drug resistance. *Proc. Natl Acad. Sci. U.S.A.* 108 (46): 18708–18713. https://doi.org/10.1073/pnas.1111840108.

Gillet, J.P., Varma, S., and Gottesman, M.M. (2013). The clinical relevance of cancer cell lines. *J. Natl Cancer Inst.* 105 (7): 452–458. https://doi.org/10.1093/jnci/djt007.

Gomez-Villafuertes, R., Paniagua-Herranz, L., Gascon, S. et al. (2017). Live imaging followed by single cell tracking to monitor cell biology and the lineage progression of multiple neural populations. *J. Vis. Exp.* 16 (130): 56291. https://doi.org/10.3791/56291.

Gospodarowicz, D., Greenburg, G., and Birdwell, C.R. (1978). Determination of cellular shape by the extracellular matrix and its correlation with the control of cellular growth. *Cancer Res.* 38 (11 Pt 2): 4155–4171.

Graf, B.W. and Boppart, S.A. (2010). Imaging and analysis of three-dimensional cell culture models. *Methods Mol. Biol.* 591: 211–227. https://doi.org/10.1007/978-1-60761-404-3_13.

Granner, D.K., Hayashi, S., Thompson, E.B. et al. (1968). Stimulation of tyrosine aminotransferase synthesis by dexamethasone phosphate in cell culture. *J. Mol. Biol.* 35 (2): 291–301.

Greenman, C., Stephens, P., Smith, R. et al. (2007). Patterns of somatic mutation in human cancer genomes. *Nature* 446 (7132): 153–158. https://doi.org/10.1038/nature05610.

Guy, J.B., Espenel, S., Vallard, A. et al. (2017). Evaluation of the cell invasion and migration process: a comparison of the video microscope-based scratch wound assay and the Boyden chamber assay. *J. Vis. Exp.* 17 (129): 56337. https://doi.org/10.3791/56337.

Hallermayer, K. and Hamprecht, B. (1984). Cellular heterogeneity in primary cultures of brain cells revealed by immunocytochemical localization of glutamine synthetase. *Brain Res.* 295 (1): 1–11. https://doi.org/10.1016/0006-8993(84)90810-2.

Hansson, E., Ronnback, L., Persson, L.I. et al. (1984). Cellular composition of primary cultures from cerebral cortex, striatum, hippocampus, brainstem and cerebellum. *Brain Res.* 300 (1): 9–18. https://doi.org/10.1016/0006-8993(84)91335-0.

Hayton, S., Maker, G.L., Mullaney, I. et al. (2017). Experimental design and reporting standards for metabolomics studies of mammalian cell lines. *Cell. Mol. Life Sci.* 74 (24): 4421–4441. https://doi.org/10.1007/s00018-017-2582-1.

HCDM (2019). Molecule information. http://www.hcdm.org/index.php/molecule-information (accessed 13 September 2019).

Held, P. (2015). Sample preparation for fluorescence microscopy: an introduction. https://www.biotek.com/assets/tech_resources/cell%20fixation%20white%20paper.pdf (accessed 19 September 2019).

Held, M., Santeramo, I., Wilm, B. et al. (2018). Ex vivo live cell tracking in kidney organoids using light sheet fluorescence microscopy. *PLoS One* 13 (7): e0199918. https://doi.org/10.1371/journal.pone.0199918.

Hubatsch, I., Ragnarsson, E.G., and Artursson, P. (2007). Determination of drug permeability and prediction of drug absorption in Caco-2 monolayers. *Nat. Protoc.* 2 (9): 2111–2119. https://doi.org/10.1038/nprot.2007.303.

Humpherys, D., Eggan, K., Akutsu, H. et al. (2001). Epigenetic instability in ES cells and cloned mice. *Science* 293 (5527): 95–97. https://doi.org/10.1126/science.1061402.

Hynds, R.E., Vladimirou, E., and Janes, S.M. (2018). The secret lives of cancer cell lines. *Dis. Model. Mech.* 11 (11) https://doi.org/10.1242/dmm.037366.

Jensen, E.C. (2013). Overview of live-cell imaging: requirements and methods used. *Anat. Rec. (Hoboken)* 296 (1): 1–8. https://doi.org/10.1002/ar.22554.

Johannsen, W. (2014). The genotype conception of heredity. 1911. *Int. J. Epidemiol.* 43 (4): 989–1000. https://doi.org/10.1093/ije/dyu063.

Jonkman, J. and Brown, C.M. (2015). Any way you slice it – a comparison of confocal microscopy techniques. *J. Biomol. Tech.* 26 (2): 54–65. https://doi.org/10.7171/jbt.15-2602-003.

Kagohara, L.T., Stein-O'Brien, G.L., Kelley, D. et al. (2018). Epigenetic regulation of gene expression in cancer: techniques, resources and analysis. *Brief. Funct. Genomics* 17 (1): 49–63. https://doi.org/10.1093/bfgp/elx018.

Klijn, C., Durinck, S., Stawiski, E.W. et al. (2015). A comprehensive transcriptional portrait of human cancer cell lines. *Nat. Biotechnol.* 33 (3): 306–312. https://doi.org/10.1038/nbt.3080.

Kondo, T. and Raff, M. (2000). Oligodendrocyte precursor cells reprogrammed to become multipotential CNS stem cells. *Science* 289 (5485): 1754–1757. https://doi.org/10.1126/science.289.5485.1754.

Korch, C. and Varella-Garcia, M. (2018). Tackling the human cell line and tissue misidentification problem is needed for reproducible biomedical research. *Adv. Mol. Pathol.* 1 (1): 209–228.e36. doi: https://doi.org/10.1016/j.yamp.2018.07.003.

Krueger, S.A. and Wilson, G.D. (2011). Flow cytometric DNA analysis of human cancers and cell lines. *Methods Mol. Biol.* 731: 359–370. https://doi.org/10.1007/978-1-61779-080-5_29.

Kyriakides, O., Halliwell, J.A., and Andrews, P.W. (2018). Acquired genetic and epigenetic variation in human pluripotent stem cells. *Adv. Biochem. Eng. Biotechnol.* 163: 187–206. https://doi.org/10.1007/10_2017_22.

Lane, E.B. (1982). Monoclonal antibodies provide specific intramolecular markers for the study of epithelial tonofilament organization. *J. Cell Biol.* 92 (3): 665–673. https://doi.org/10.1083/jcb.92.3.665.

Lawson, D.A., Kessenbrock, K., Davis, R.T. et al. (2018). Tumour heterogeneity and metastasis at single-cell resolution. *Nat. Cell Biol.* 20 (12): 1349–1360. https://doi.org/10.1038/s41556-018-0236-7.

Leroy, B., Fournier, J.L., Ishioka, C. et al. (2013). The TP53 website: an integrative resource centre for the TP53 mutation database and TP53 mutant analysis. *Nucleic Acids Res.* 41 (Database issue): D962–D969. https://doi.org/10.1093/nar/gks1033.

Lever, J.E. (1981). Regulation of dome formation in kidney epithelial cell cultures. *Ann. N. Y. Acad. Sci.* 372: 371–383. https://doi.org/10.1111/j.1749-6632.1981.tb15489.x.

Lindl, T. and Steubing, R. (2013). *Atlas of Living Cell Cultures*. Weinheim: Wiley-Blackwell.

Linkous, A., Balamatsias, D., Snuderl, M. et al. (2019). Modeling patient-derived glioblastoma with cerebral organoids. *Cell Rep.* 26 (12): 3203–3211. e5. doi: https://doi.org/10.1016/j.celrep.2019.02.063.

Lowe, R., Shirley, N., Bleackley, M. et al. (2017). Transcriptomics technologies. *PLoS Comput. Biol.* 13 (5): e1005457. https://doi.org/10.1371/journal.pcbi.1005457.

Ma, H., Morey, R., O'Neil, R.C. et al. (2014). Abnormalities in human pluripotent cells due to reprogramming mechanisms. *Nature* 511 (7508): 177–183. https://doi.org/10.1038/nature13551.

MacLeod, R.A., Kaufmann, M., and Drexler, H.G. (2011). Cytogenetic analysis of cancer cell lines. *Methods Mol. Biol.* 731: 57–78. https://doi.org/10.1007/978-1-61779-080-5_6.

MacLeod, R.A., Kaufmann, M.E., and Drexler, H.G. (2017). Cytogenetic harvesting of cancer cells and cell lines. *Methods Mol. Biol.* 1541: 43–58. https://doi.org/10.1007/978-1-4939-6703-2_5.

Malone, J.H. and Oliver, B. (2011). Microarrays, deep sequencing and the true measure of the transcriptome. *BMC Biol.* 9: 34. https://doi.org/10.1186/1741-7007-9-34.

Matz, M.V., Lukyanov, K.A., and Lukyanov, S.A. (2002). Family of the green fluorescent protein: journey to the end of the rainbow. *Bioessays* 24 (10): 953–959. https://doi.org/10.1002/bies.10154.

McCormick, C. and Freshney, R.I. (2000). Activity of growth factors in the IL-6 group in the differentiation of human lung adenocarcinoma. *Br. J. Cancer* 82 (4): 881–890. https://doi.org/10.1054/bjoc.1999.1015.

van der Meer, D., Barthorpe, S., Yang, W. et al. (2019). Cell model passports – a hub for clinical, genetic and functional datasets of preclinical cancer models. *Nucleic Acids Res.* 47 (D1): D923–D929. https://doi.org/10.1093/nar/gky872.

Menden, M.P., Casale, F.P., Stephan, J. et al. (2018). The germline genetic component of drug sensitivity in cancer cell lines. *Nat. Commun.* 9 (1): 3385. https://doi.org/10.1038/s41467-018-05811-3.

Mohammad, T.A., Tsai, Y.S., Ameer, S. et al. (2019). CeL-ID: cell line identification using RNA-seq data. *BMC Genomics* 20 (Suppl. 1): 81. https://doi.org/10.1186/s12864-018-5371-9.

Mohseny, A.B., Machado, I., Cai, Y. et al. (2011). Functional characterization of osteosarcoma cell lines provides representative models to study the human disease. *Lab. Investig.* 91 (8): 1195–1205. https://doi.org/10.1038/labinvest.2011.72.

Moll, R. (1994). Cytokeratins in the histological diagnosis of malignant tumors. *Int. J. Biol. Markers* 9 (2): 63–69.

Moll, R., Franke, W.W., Schiller, D.L. et al. (1982). The catalog of human cytokeratins: patterns of expression in normal epithelia, tumors and cultured cells. *Cell* 31 (1): 11–24. https://doi.org/10.1016/0092-8674(82)90400-7.

Murao, S., Gemmell, M.A., Callaham, M.F. et al. (1983). Control of macrophage cell differentiation in human promyelocytic HL-60 leukemia cells by 1,25-dihydroxyvitamin D3 and phorbol-12-myristate-13-acetate. *Cancer Res.* 43 (10): 4989–4996.

Murphy, D. B., Oldfield, R., Schwartz, S., et al. (2019). MicroscopyU: phase contrast microscope configuration. https://www.microscopyu.com/techniques/phase-contrast/phase-contrast-microscope-configuration (accessed 17 September 2019).

Natoli, M., Leoni, B.D., D'Agnano, I. et al. (2012). Good Caco-2 cell culture practices. *Toxicol. In Vitro* 26 (8): 1243–1246. https://doi.org/10.1016/j.tiv.2012.03.009.

Navin, N.E. (2015). The first five years of single-cell cancer genomics and beyond. *Genome Res.* 25 (10): 1499–1507. https://doi.org/10.1101/gr.191098.115.

Nelson-Rees, W.A., Daniels, D.W., and Flandermeyer, R.R. (1981). Cross-contamination of cells in culture. *Science* 212 (4493): 446–452.

Neve, R.M., Chin, K., Fridlyand, J. et al. (2006). A collection of breast cancer cell lines for the study of functionally distinct cancer subtypes. *Cancer Cell* 10 (6): 515–527. https://doi.org/10.1016/j.ccr.2006.10.008.

Niepel, M., Hafner, M., Mills, C.E. et al. (2019). A multi-center study on the reproducibility of drug-response assays in mammalian cell lines. *Cell Syst.* 9 (1): 35–48. https://doi.org/10.1016/j.cels.2019.06.005.

Novus Biologicals (2019). Immunocytochemistry (ICC) handbook. https://images.novusbio.com/design/br_iccguide.pdf (accessed 20 September 2019).

Osada, N., Kohara, A., Yamaji, T. et al. (2014). The genome landscape of the African green monkey kidney-derived Vero cell line. *DNA Res.* 21 (6): 673–683. https://doi.org/10.1093/dnares/dsu029.

Ouar, Z., Lacave, R., Bens, M. et al. (1999). Mechanisms of altered sequestration and efflux of chemotherapeutic drugs by multidrug-resistant cells. *Cell Biol. Toxicol.* 15 (2): 91–100.

Park, H.Y., Russakovsky, V., Ohno, S. et al. (1993). The beta isoform of protein kinase C stimulates human melanogenesis by activating tyrosinase in pigment cells. *J. Biol. Chem.* 268 (16): 11742–11749.

Park, H.Y., Perez, J.M., Laursen, R. et al. (1999). Protein kinase C-beta activates tyrosinase by phosphorylating serine residues in its cytoplasmic domain. *J. Biol. Chem.* 274 (23): 16470–16478. https://doi.org/10.1074/jbc.274.23.16470.

Parry-Hill, M., Sutter, R. T. and Davidson, M. W. (2019). MicroscopyU: microscope alignment for Kohler illumination. https://www.microscopyu.com/tutorials/kohler (accessed 17 September 2019).

Petljak, M., Alexandrov, L.B., Brammeld, J.S. et al. (2019). Characterizing mutational signatures in human cancer cell lines reveals episodic APOBEC mutagenesis. *Cell* 176 (6): 1282–1294. e20. doi: https://doi.org/10.1016/j.cell.2019.02.012.

Petran, M., Hadravsky, M., Benes, J. et al. (1986). In vivo microscopy using the tandem scanning microscope. *Ann. N. Y.*

Acad. Sci. 483: 440–447. https://doi.org/10.1111/j.1749-6632 .1986.tb34554.x.

Pignata, S., Maggini, L., Zarrilli, R. et al. (1994). The enterocyte-like differentiation of the Caco-2 tumor cell line strongly correlates with responsiveness to cAMP and activation of kinase a pathway. *Cell Growth Differ.* 5 (9): 967–973.

Piltti, K.M., Cummings, B.J., Carta, K. et al. (2018). Live-cell time-lapse imaging and single-cell tracking of in vitro cultured neural stem cells – tools for analyzing dynamics of cell cycle, migration, and lineage selection. *Methods* 133: 81–90. https:// doi.org/10.1016/j.ymeth.2017.10.003.

Pinnel, S.R., Murad, S., and Darr, D. (1987). Induction of collagen synthesis by ascorbic acid. A possible mechanism. *Arch. Dermatol.* 123 (12): 1684–1686. https://doi.org/10.1001/archderm .123.12.1684.

Plant, A.L., Elliott, J.T., and Bhat, T.N. (2011). New concepts for building vocabulary for cell image ontologies. *BMC Bioinform.* 12: 487. https://doi.org/10.1186/1471-2105-12-487.

Pleasance, E.D., Cheetham, R.K., Stephens, P.J. et al. (2010). A comprehensive catalogue of somatic mutations from a human cancer genome. *Nature* 463 (7278): 191–196. https://doi.org/ 10.1038/nature08658.

Quentmeier, H., Osborn, M., Reinhardt, J. et al. (2001). Immunocytochemical analysis of cell lines derived from solid tumors. *J. Histochem. Cytochem.* 49 (11): 1369–1378. https://doi.org/10 .1177/002215540104901105.

Quentmeier, H., Pommerenke, C., Dirks, W.G. et al. (2019). The LL-100 panel: 100 cell lines for blood cancer studies. *Sci. Rep.* 9 (1): 8218. https://doi.org/10.1038/s41598-019-44491-x.

Rabito, C.A., Tchao, R., Valentich, J. et al. (1980). Effect of cell–substratum interaction on hemicyst formation by MDCK cells. *In Vitro* 16 (6): 461–468.

Ramaekers, F.C., Puts, J.J., Kant, A. et al. (1982). Use of antibodies to intermediate filaments in the characterization of human tumors. *Cold Spring Harb. Symp. Quant. Biol.* 46 (Pt 1): 331–339. https:// doi.org/10.1101/sqb.1982.046.01.034.

Ramond, M.J., Martinot-Peignoux, M., and Erlinger, S. (1985). Dome formation in the human colon carcinoma cell line Caco-2 in culture. Influence of ouabain and permeable supports. *Biol. Cell.* 54 (1): 89–92.

Reid, Y.A. (2011). Characterization and authentication of cancer cell lines: an overview. *Methods Mol. Biol.* 731: 35–43. https:// doi.org/10.1007/978-1-61779-080-5_4.

Reid, Y.A. (2017). Best practices for naming, receiving, and managing cells in culture. *In Vitro Cell. Dev. Biol. Anim.* 53 (9): 761–774. https://doi.org/10.1007/s11626-017-0199-1.

Reynaud, E.G., Krzic, U., Greger, K. et al. (2008). Light sheet-based fluorescence microscopy: more dimensions, more photons, and less photodamage. *HFSP J.* 2 (5): 266–275. https://doi.org/10 .2976/1.2974980.

Robin, T., Bairoch, A., Muller, M. et al. (2018). Large-scale reanalysis of publicly available HeLa cell proteomics data in the context of the human proteome project. *J. Proteome Res.* 17 (12): 4160–4170. https://doi.org/10.1021/acs.jproteome.8b00392.

Robinson, J., Halliwell, J.A., McWilliam, H. et al. (2013). IPD – the Immuno Polymorphism Database. *Nucleic Acids Res.* 41 (Database issue): D1234–D1240. https://doi.org/10.1093/nar/gks1140.

Romanowska, J. and Joshi, A. (2019). From genotype to phenotype: through chromatin. *Genes (Basel)* 10 (2) https://doi.org/10 .3390/genes10020076.

Ross, D.T., Scherf, U., Eisen, M.B. et al. (2000). Systematic variation in gene expression patterns in human cancer cell lines. *Nat. Genet.* 24 (3): 227–235. https://doi.org/10.1038/73432.

Rothfels, K.H., Axelrad, A.A., Siminovitch, L. et al. (1959). The origin of altered cell line from mouse, monkey, and man as indicated by chromosome and transplantation studies. *Proc. Can. Cancer Conf.* 3: 189–214.

Rubart, M. (2004). Two-photon microscopy of cells and tissue. *Circ. Res.* 95 (12): 1154–1166. https://doi.org/10.1161/01.RES .0000150593.30324.42.

Rugg-Gunn, P.J., Ferguson-Smith, A.C., and Pedersen, R.A. (2007). Status of genomic imprinting in human embryonic stem cells as revealed by a large cohort of independently derived and maintained lines. *Hum. Mol. Genet.* 16 (R2): R243–R251. https://doi.org/10.1093/hmg/ddm245.

Ryu, A.H., Eckalbar, W.L., Kreimer, A. et al. (2017). Use antibiotics in cell culture with caution: genome-wide identification of antibiotic-induced changes in gene expression and regulation. *Sci. Rep.* 7 (1): 7533. https://doi.org/10.1038/s41598-017-07757-w.

Sakaue-Sawano, A., Yo, M., Komatsu, N. et al. (2017). Genetically encoded tools for optical dissection of the mammalian cell cycle. *Mol. Cell* 68 (3): 626–640, e5. doi: https://doi.org/10.1016/j .molcel.2017.10.001.

Sandberg, R. and Ernberg, I. (2005). Assessment of tumor characteristic gene expression in cell lines using a tissue similarity index (TSI). *Proc. Natl Acad. Sci. U.S.A.* 102 (6): 2052–2057. https:// doi.org/10.1073/pnas.0408105102.

Sanford, K.K. and Evans, V.J. (1982). A quest for the mechanism of "spontaneous" malignant transformation in culture with associated advances in culture technology. *J. Natl Cancer Inst.* 68 (6): 895–913.

Sanford, K.K., Hobbs, G.L., and Earle, W.R. (1956). The tumor-producing capacity of strain L mouse cells after 10 years in vitro. *Cancer Res.* 16 (2): 162–166.

Schroeder, T. (2011). Long-term single-cell imaging of mammalian stem cells. *Nat. Methods* 8 (Suppl. 4): S30–S35. https://doi.org/ 10.1038/nmeth.1577.

Shiau, C.K., Gu, D.L., Chen, C.F. et al. (2011). IGRhCellID: integrated genomic resources of human cell lines for identification. *Nucleic Acids Res.* 39 (Database issue): D520–D524. https://doi .org/10.1093/nar/gkq1075.

Sofroniew, M.V. and Vinters, H.V. (2010). Astrocytes: biology and pathology. *Acta Neuropathol.* 119 (1): 7–35. https://doi.org/10 .1007/s00401-009-0619-8.

Stewart, C.E., James, P.L., Fant, M.E. et al. (1996). Overexpression of insulin-like growth factor-II induces accelerated myoblast differentiation. *J. Cell. Physiol.* 169 (1): 23–32. https://doi.org/10.1002/(SICI)1097-4652(199610)169:1<23:: AID-JCP3>3.0.CO;2-G.

Stramer, B.M. and Dunn, G.A. (2015). Cells on film – the past and future of cinemicroscopy. *J. Cell Sci.* 128 (1): 9–13. https://doi .org/10.1242/jcs.165019.

Stratton, M.R., Campbell, P.J., and Futreal, P.A. (2009). The cancer genome. *Nature* 458 (7239): 719–724. https://doi.org/10.1038/ nature07943.

Thiagarajan, R.D., Morey, R., and Laurent, L.C. (2014). The epigenome in pluripotency and differentiation. *Epigenomics* 6 (1): 121–137. https://doi.org/10.2217/epi.13.80.

Tollefsbol, T.O. (2011). Advances in epigenetic technology. *Methods Mol. Biol.* 791: 1–10. https://doi.org/10.1007/978-1-61779-316-5_1.

Torsvik, A., Stieber, D., Enger, P.O. et al. (2014). U-251 revisited: genetic drift and phenotypic consequences of long-term cultures of glioblastoma cells. *Cancer Med.* 3 (4): 812–824. https://doi.org/10.1002/cam4.219.

Tsien, R.Y. (2010). Nobel lecture: constructing and exploiting the fluorescent protein paintbox. *Integr. Biol. (Camb.)* 2 (2–3): 77–93. https://doi.org/10.1039/b926500g.

Vachon, P.H., Perreault, N., Magny, P. et al. (1996). Uncoordinated, transient mosaic patterns of intestinal hydrolase expression in differentiating human enterocytes. *J. Cell. Physiol.* 166 (1): 198–207. https://doi.org/10.1002/(SICI)1097-4652(199601)166:1<198::AID-JCP21>3.0.CO;2-A.

Wakabayashi, Y., Chua, J., Larkin, J.M. et al. (2007). Four-dimensional imaging of filter-grown polarized epithelial cells. *Histochem. Cell Biol.* 127 (5): 463–472. https://doi.org/10.1007/s00418-007-0274-x.

Walton, K.D. and Kolterud, A. (2014). Mouse fetal whole intestine culture system for ex vivo manipulation of signaling pathways and three-dimensional live imaging of villus development. *J. Vis. Exp.* 91: e51817. https://doi.org/10.3791/51817.

Walzik, M.P., Vollmar, V., Lachnit, T. et al. (2015). A portable low-cost long-term live-cell imaging platform for biomedical research and education. *Biosens. Bioelectron.* 64: 639–649. https://doi.org/10.1016/j.bios.2014.09.061.

Wang, H., Huang, S., Shou, J. et al. (2006). Comparative analysis and integrative classification of NCI60 cell lines and primary tumors using gene expression profiling data. *BMC Genomics* 7: 166. https://doi.org/10.1186/1471-2164-7-166.

Wellcome Sanger Institute (2019). What is the Cell Lines Project? https://cancer.sanger.ac.uk/cell_lines/help/desc (accessed 11 September 2019).

Wen, L. and Tang, F. (2016). Single-cell sequencing in stem cell biology. *Genome Biol.* 17: 71. https://doi.org/10.1186/s13059-016-0941-0.

Woerner, S.M., Yuan, Y.P., Benner, A. et al. (2010). SelTarbase, a database of human mononucleotide-microsatellite mutations and their potential impact to tumorigenesis and immunology. *Nucleic Acids Res.* 38 (Database issue): D682–D689. https://doi.org/10.1093/nar/gkp839.

Wu, Y.J., Parker, L.M., Binder, N.E. et al. (1982). The mesothelial keratins: a new family of cytoskeletal proteins identified in cultured mesothelial cells and nonkeratinizing epithelia. *Cell* 31 (3 Pt 2): 693–703.

Xie, Y., Bergstrom, T., Jiang, Y. et al. (2015). The human glioblastoma cell culture resource: validated cell models representing all molecular subtypes. *EBioMedicine* 2 (10): 1351–1363. https://doi.org/10.1016/j.ebiom.2015.08.026.

Xu, X., Nagarajan, H., Lewis, N.E. et al. (2011). The genomic sequence of the Chinese hamster ovary (CHO)-K1 cell line. *Nat. Biotechnol.* 29 (8): 735–741. https://doi.org/10.1038/nbt.1932.

Yamashita, N., Morita, M., Legant, W.R. et al. (2015). Three-dimensional tracking of plus-tips by lattice light-sheet microscopy permits the quantification of microtubule growth trajectories within the mitotic apparatus. *J. Biomed. Opt.* 20 (10): 101206. https://doi.org/10.1117/1.JBO.20.10.101206.

CHAPTER 19

Quantitation and Growth Kinetics

Once the early tissue culture laboratories developed the technical skills to grow large numbers of cells, they required an accurate, easy, and rapid method for cell counting (Sanford et al. 1951). A number of methods were used to measure "growth" in tissue culture, including changes in mitotic index, labeling index, and cell mass (dry weight) (Cunningham and Kirk 1942). Although helpful, these methods were all thought to be inadequate, leading to an alternative solution that was based on direct observation. Cell proliferation was measured by taking representative samples from each culture, lysing the cells with citric acid, staining the nuclei with crystal violet, and counting the nuclei using a hemocytometer (Sanford et al. 1951). A similar (somewhat simplified) technique continues to be widely used today; a survey of counting methods showed that 71% of researchers counted their cells using a hemocytometer (Ongena et al. 2010). However, a greater variety of techniques and equipment are now available for quantitation, allowing a better understanding of the key attributes of each culture. This chapter describes cell counting equipment and procedures, with a particular emphasis on applying these techniques to the study of cell proliferation. Related topics such as cell viability are discussed here if shared assays are employed for their measurement, such as counting cells in the presence of trypan blue (see Section 19.2). However, the broader topic of cell survival and cytotoxicity is explored in a later chapter (see Chapter 29).

19.1 CELL COUNTING

Although estimates can be made of the stage of growth of a culture from its appearance under the microscope, standardization of culture conditions and proper quantitative experiments require that cells are counted at subculture. Some experienced tissue culture personnel may choose to omit cell counting when performing routine handling of some cell lines, using a suitable split ratio, but it is still possible for split errors to occur (see Section 14.6.4; see also Figure 14.10). Regardless of experience, a cell count is essential for any cell lines that grow rapidly, requiring a high split ratio, or for experiments where replicate cultures will be used (see Section 7.6).

Methods to determine cell number can be broadly divided into (i) direct methods, where cell number is measured through observation or other forms of direct analysis; and (ii) indirect methods, where a separate variable is measured as a surrogate for cell number (see Table 19.1). If you use an indirect method for cell counting, it is important to calibrate your assay against a direct method. The need for calibration was highlighted by a recent multicenter study of drug-response assays in mammalian cell lines (Niepel et al. 2019). Two of the drugs tested, etoposide and palbociclib, gave rise to irreproducible data across the various test centers. Differences were found to relate to an adenosine triphosphate (ATP)-based assay that was used by one center as a surrogate for cell number. Treatment with etoposide and palbociclib resulted in an increase in cell size, which altered the amount of ATP that was present in the cell. Thus, an ATP-based assay could not be used to give an accurate measurement of cell number for those two drugs, highlighting the importance of assay selection (Niepel et al. 2017; Niepel et al. 2019). Indirect methods are commonly used in cytotoxicity assays and are discussed in more detail in a later chapter (see Section 29.2).

Freshney's Culture of Animal Cells: A Manual of Basic Technique and Specialized Applications, Eighth Edition. Amanda Capes-Davis and R. Ian Freshney.
© 2021 John Wiley & Sons Ltd. Published 2021 by John Wiley & Sons Ltd.
Companion website: www.wiley.com/go/freshney/cellculture8

TABLE 19.1. Determination of cell number.

Test	Method	Advantages	Disadvantages
Direct			
Hemocytometer slide	Count number of cells in given area of marked slide under microscope and extrapolate back to stock cell suspension	Inexpensive; cells, clumps, etc. are visible; viability stain can be used	Needs single cell suspension; slow; high statistical error; low sensitivity
Resistance-based electronic particle counter	Dilute cell sample in isotonic saline and count at prescribed setting; based on resistance change when cells pass through narrow orifice	Quick; suitable for high-throughput analysis; low error rates; allows analysis of other properties, e.g. cell volume, aggregation, debris	Does not allow cells to be visualized; requires dilution to avoid coincident signals; orifice prone to blockage
Image analysis	Scan of microscope view of stained and unstained cells in a special counting chamber	Same principle as hemocytometer but faster and more accurate; cells, clumps, etc. are visible	Needs single cell suspension; may need disposable counting chambers
Flow cytometry	Count by light scatter, fixed or unfixed cell population, or by DNA fluorescent stain	Low error rates and high sensitivity; allows multiple parameters to be measured simultaneously	Needs single cell suspension; technically more complex
Indirect[a]			
Amido black staining (Schulz et al. 1994)	Stain fixed cells with amido black 10B (naphthalene black B); destain and elute; measure absorbance	Based on protein levels; economical; rapid; suitable for multiwell plates	Stains fixed cells; requires destaining and elution; requires a suitable microplate reader
Coomassie Blue staining (Baumgarten 1985)	Stain cells with Coomassie Blue solution; measure absorbance	Based on protein levels; easy; economical; rapid; suitable for multiwell plates; solubilization occurs during staining	Potential errors with stratified cells, high level of insoluble protein; requires a suitable microplate reader
Propidium iodide staining (Wan et al. 1994)	Permeabilize cells using Triton X-100 and EDTA; stain with propidium iodide; measure fluorescence	Based on DNA levels; economical; rapid; less autofluorescence compared to other DNA stains	Stains permeabilized cells; may bind to double-stranded RNA; requires a suitable microplate reader
Sulforhodamine B staining (Skehan et al. 1990)	Stain fixed cells with sulforhodamine; measure absorbance (increased sensitivity can be achieved by measuring fluorescence)	Based on protein levels; easy; economical; rapid; highly sensitive; suitable for multiwell plates	Stains fixed cells; requires a suitable microplate reader

[a] The indirect methods listed here are those that relate to DNA or protein levels, which offer a consistent relationship to cell number. Other methods are sometimes used to estimate cell number that are based on cell metabolism or viability; such methods have a more variable relationship to cell number and are listed separately (see Table 29.1).

Accuracy must also be considered when selecting a suitable method for cell counting. The need to work under sterile conditions and to minimize the time spent by cells outside the incubator often leads to a degree of compromise between speed and accuracy. Although primary culture and normal maintenance require a quantitative approach to ensure reproducibility, speed of handling is the key to good cell survival and growth, so in these situations reduced accuracy (in pipetting, for example) may be acceptable. However, accuracy should dominate under experimental conditions, even if operations become a little slower. An error of ±10%

may be acceptable in routine maintenance, but error rates should be much lower under experimental conditions.

19.1.1 Manual Cell Counting

The concentration of a cell suspension may be determined by placing the cells in a hemocytometer for direct visualization and manual counting (see Figure 19.1). The hemocytometer is an optically flat chamber slide that, as its name suggests, was originally designed to count blood cells (Verso 1964; Vembadi et al. 2019). Each hemocytometer chamber consists of a defined area of known depth (usually 0.1 mm), resulting in a

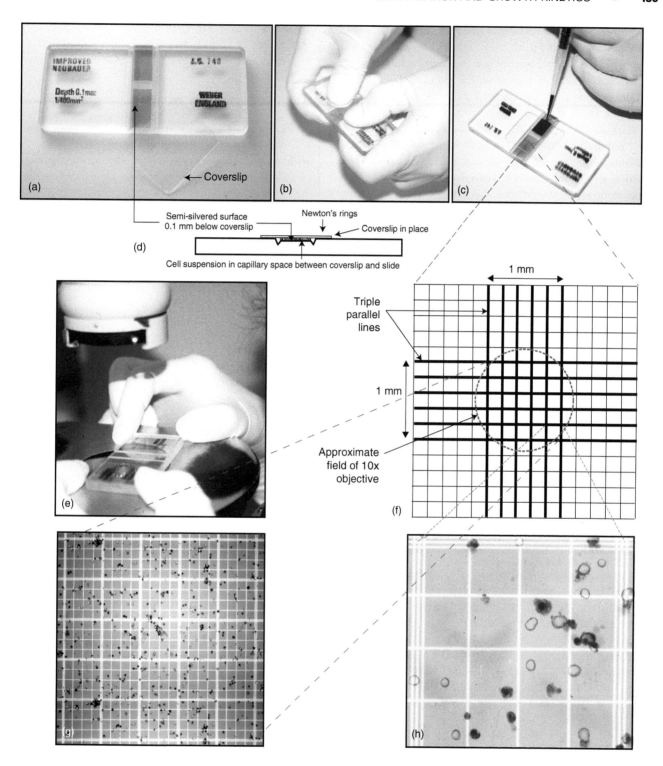

Fig. 19.1. Using a hemocytometer slide. (a) Slide (improved Neubauer type) and coverslip before use. (b) Pressing coverslip down onto slide. (c) Adding a cell suspension (with naphthalene black stain) to an assembled slide. (d) Longitudinal section of the slide, showing the position of the cell sample in a 0.1-mm-deep chamber. (e) Viewing slide on microscope. (f) Magnified view of the total area of the grid. The central area enclosed by the dotted circle is that area which would be covered by the average 10× objective. This area covers approximately the central 1 × 1 mm square of the grid. (g) View under a 10× objective, showing the 25 smaller (200 × 200 μm) squares of a slide, which make up the central (1 × 1 mm) square. (h) View under 40× objective, showing one of the smaller squares, bounded by three parallel lines and containing 16 of the smallest (50 × 50 μm) squares. Viable cells are unstained and clear, with a refractile ring around them; non-viable cells (stained with naphthalene black) are dark and have no refractile ring. *Source*: (a–c, e, g, h) R. Ian Freshney.

defined volume of cell suspension (see Figure 19.1a–d). The chamber has an etched grid that is visible using an inverted microscope with a 10× objective (see Figure 19.1e, f); the user counts the cells that are present within the grid. A viability stain can be used to distinguish between living and dead cells (see Figure 19.1g, h; see also Section 19.2).

A hemocytometer provides the most economical, compact, and portable method for cell counting. However, manual counting is a subjective process and can be extremely laborious if you are counting multiple samples. It requires a minimum of 2×10^5 cells/ml and is prone to error, both in the method of sampling and in the total number of cells counted. Formal assessment of counting accuracy has found that the error rate is typically less than 10%, but this will depend on the user's expertise, the choice of an optimum concentration for counting, and the presence or absence of clumping (Sanford et al. 1951; Cadena-Herrera et al. 2015). If counting is performed using a hemocytometer, it is important to use a standardized approach. The following protocol (Protocol P19.1) is useful for standardization and can be adapted for use in training (see Supp. S30.10).

PROTOCOL P19.1. CELL COUNTING BY HEMOCYTOMETER

Background

This procedure describes how to count cells without assessing their viability. A separate procedure is used to count cells with trypan blue for viability assessment (see Protocol P19.2).

Most of the errors in this procedure occur by incorrect sampling and transfer of cells to the chamber. Make sure that the cell suspension is properly mixed, break up any aggregates, and do not allow the cells time to settle or adhere in the tip of the pipette before transferring them to the chamber. If aggregation cannot be eliminated, lyse the cells in 0.1 M citric acid containing 0.1% crystal violet at 37 °C for one hour and then count the nuclei (Sanford et al. 1951).

Outline

Prepare a single-cell suspension and the hemocytometer slide. Add the cells to the counting chamber and transfer to an inverted microscope for counting.

Materials (sterile or aseptically prepared)

- Cell line or primary culture, e.g. during subculture (see Protocols P14.2, P14.3)
- Complete culture medium
- Small containers, e.g. microcentrifuge tubes or cryovials

- Pipettes in various sizes, including 1 ml (alternatively, a Pasteur pipette could be used)
- Pipettor tips

Materials (non-sterile)

- Biological safety cabinet (BSC) and associated consumables (see Protocol P12.1)
- Inverted microscope with phase contrast optics (see Protocol P18.1)
- Tally counter
- Hemocytometer (improved Neubauer, 0.1 mm depth; see Figure 19.1a)
- Rectangular glass coverslip (see Figure 19.1a)
- 70% alcohol in spray or squirt bottle
- Lint-free swabs or wipes (suitable to wipe the hemocytometer without scratching)
- Pipettors, including P200 size

Procedure

A. Prepare the Cell Sample for Counting

1. Prepare a work space for aseptic technique using a BSC.
2. For adherent cells: perform subculture using the standard procedure for that cell line (e.g. trypsinization; see Protocol P14.2). Resuspend in culture medium to give a single-cell suspension. Aim for approximately 1×10^6 cells/ml, based on an estimate of the pellet size.
3. For cells in suspension: gently mix the cell suspension and disperse any clumps by pipetting the suspension up and down (see Protocol P14.3).
4. Use a sterile pipette to transfer about 1 ml of cell suspension to a small sterile container.

B. Prepare the Hemocytometer Slide for Counting

5. Clean the surface of the slide with 70% alcohol and dry using a lint-free wipe, taking care not to scratch the semi-silvered coating.
6. Clean the coverslip with 70% alcohol and dry using a lint-free wipe.
7. Wetting the edges of the coverslip very slightly, press it down over the grooves and semi-silvered counting area (see Figure 19.1b), with the edges of the coverslip extending beyond the outermost grooves.
8. Check for the appearance of interference patterns ("Newton's rings" – rainbow colors between the coverslip and the slide, like the rings formed by oil on water). These indicate that the coverslip is properly attached, thereby determining the depth of the counting chamber. If Newton's rings are not evident, remove the coverslip and try again.

9. Use the P200 pipettor to mix the cell sample thoroughly, pipetting vigorously to disperse any clumps, and collect an aliquot of approximately 30–40 μl.

10. Transfer the cell suspension immediately to the edge of the hemocytometer chamber and expel the aliquot near the front of the coverslip (see Figure 19.1c). Let the liquid be drawn under the coverslip by capillarity. Do not overfill or underfill the chamber; the fluid should run only to the edges of the grooves.

11. Mix the cell suspension, reload the pipettor, and fill the second chamber if there is one.

12. Blot off any surplus fluid (without drawing from under the coverslip) and transfer the slide to the microscope stage.

C. Perform Cell Counting

13. Select a 10x objective and focus on the grid lines in the chamber (see Figure 19.1e, f).

14. Move the slide so that the field you see is the central area of the grid and is the largest area that you can see bounded by three parallel lines. The area of this large square is 1 mm^2. With a standard 10x objective, this area will almost fill the field, or the corners will be slightly outside the field, depending on the field of view.

15. Start to count the cells lying within this large square (see Figure 19.2a). Use the subdivisions (also bounded by three parallel lines) and single grid lines as an aid for counting. Count cells that lie on the top and left-hand lines of each square, but not those on the bottom or right-hand lines, to avoid counting the same cells twice.

16. As you count, use the tally counter to record the number. For routine subculture, 100–300 cells should be counted; the more cells that are counted, the more accurate the count becomes. For more precise quantitative experiments, 500–1000 cells should be counted.

17. If there are very few cells (< 100/mm^2), count one or more additional squares (each 1 mm^2) surrounding the central square millimeter (see Figure 19.2b).

18. If there are too many cells (> 1000/mm^2), count only five small squares (each 0.04 mm^2 and bounded by three parallel lines) across the diagonal of the larger (1-mm^2) square, giving a total of 0.2 mm^2 (see Figure 19.2c).

19. If the slide has two chambers, move to the second chamber and do a second count. If not, rinse the slide and repeat the count with a fresh sample.

20. Once counting is complete, rinse the hemocytometer and coverslip using 70% alcohol, allow to air-dry, and put away safely for next time.

D. Calculate the Cell Concentration

21. Using the cell counts that were recorded by the tally counter, determine the average number of cells in one large square (as indicated by the triple parallel lines; see Figure 19.1f–h). This is your cell count for calculating the concentration.

22. Use the following formula to determine the cell concentration:

$$Concentration = \frac{cell\ count}{volume}$$

23. If the improved Neubauer hemocytometer is used, the depth of the chamber is 0.1 mm and the area of the central square (as indicated by the triple parallel lines, see Figure 19.1f) is 1 mm^2. Thus, the volume counted is 0.1 mm^3 or 1×10^{-4} ml, giving the following formula:

$$Concentration = cell\ count \times 10^4$$

24. This result indicates the concentration (expressed in cells per milliliter) of the cell suspension from step 4. To determine the cell yield, multiply the concentration by the total volume of the cell suspension (expressed in milliliters). To determine the cell density in the original flask (for monolayer cultures), divide the cell yield by the surface area of the original flask. To determine the cell concentration in the original flask (for suspension cultures), divide the cell yield by the volume of medium in the original flask.

Notes

Step 2: Where samples are being counted in a growth experiment, the trypsin need not be removed and the cells can be dispersed in the trypsin and counted directly, or after diluting 50 : 50 with medium containing serum if the cells tend to reaggregate.

Step 3: A minimum of approximately 1×10^6 cells/ml is required for this method, so the cells may need to be concentrated by centrifugation (at 100 g for two minutes) and resuspended in a smaller volume. This step can also help to disperse clumps.

Step 7: Some tissue culture personnel prefer to wet the ridges on either side of the counting chamber – any method is OK, provided Newton's rings are evident and there is no liquid in the counting chamber before the cells are added.

Steps 7–10: It is important to seat the coverslip correctly and add the correct amount of fluid. If incorrectly done, the dimensions of the counting chamber will change, which will affect the accuracy of the final count.

Step 13: If phase contrast is not available and focusing is difficult because of poor contrast, close down the field iris, or make the lighting slightly oblique by offsetting the condenser.

Step 16: Hemocytometer counting has a high statistical error especially if < 100 cells are counted per mm^2 (Hsiung et al. 2013).

Step 19: Always record each cell count and the number of squares counted before zeroing the tally counter.

Step 21: Usually, the total cell count is added up and divided by the number of large squares that were counted. For example, if the cell concentration is relatively low and the nine squares that make up the grid are all counted (see Figure 19.2b), the cell count should be divided by 9. If the cell concentration is relatively high and only five small squares were counted (see Figure 19.2c), the cell count should be multiplied by 5.

Step 24: Cells per flask is the figure most useful in routine subculture, while cells per milliliter and cells per square centimeter are more appropriate for plotting a growth curve (see Protocol P19.3).

19.1.2 Automated Cell Counting

Automated cell counting is rapid, can be used for high-throughput applications, and has a low inherent statistical error because of the high number of cells counted. The question of whether to invest in such a device usually comes down to cost. It is important to consider the time saved if counting is no longer performed manually to help decide if the equipment cost represents a worthwhile investment. The costs associated with repairs and consumables should also be considered, as these can become quite expensive over time with some instruments. Today's automated cell counters usually rely on three major principles of operation, which help to explain some of the variations in cost and modes of operation (see Table 19.2). Always discuss with suppliers to determine how their instrument may differ from others on the market, arrange for a demonstration before proceeding, and follow the supplier's specific instructions for that device.

Using the Coulter principle. The Coulter counter, which was originally invented by Wallace Coulter for the analysis of blood samples, has been widely adopted by tissue culture laboratories for automated cell counting (Coulter 1949; Vembadi et al. 2019). Coulter's invention was based on the principle that living cells are poor electrical conductors; as the cells pass through a small aperture, they result in a transient change in electrical resistance that can be measured and used to assess cell number, viability, and other characteristics (Stewart and Crosland-Taylor 1959). Devices that use the Coulter principle for cell counting include the CASY Cell Counter (OMNI Life Science; see Figure 19.3a), the Multisizer 4e Coulter Counter (Beckman Coulter Life Science), and the Scepter™ Cell Counter (Sigma-Aldrich).

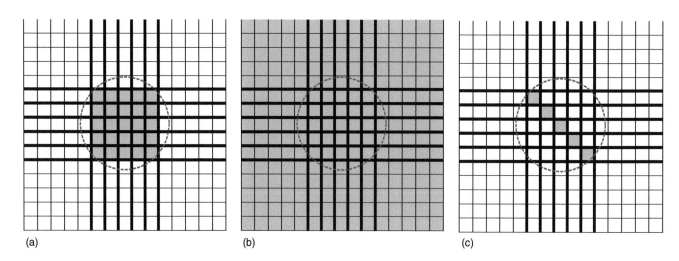

(a) (b) (c)

Fig. 19.2. Selecting hemocytometer fields for counting. Count the cells in an area sufficient to give between 200 and 500 cells (shaded here in gray). (a) For an anticipated count of 2×10^6–1×10^7 cells/ml, count the central 25 squares (1 mm^2) and multiply by 10^4. (b) For a lower cell concentration, count the whole grid (9 mm^2), divide by 9, and multiply by 10^4. (c) For a higher cell concentration, count five of the smallest squares (0.04 mm^2 each = 0.2 mm^2), preferably in a diagonal array, and multiply by 5×10^4.

TABLE 19.2. Automated cell counters.

Name	Supplier	Principle of action
Accuri C6	BD Biosciences	Flow cytometry of single cell stream; multiparametric
ADAM™ series	NanoEnTek	Image analysis using PI
Amnis FlowSight	Luminex	Flow cytometry of single cell stream; image analysis
CASY	OMNI Life Science (OLS)	Electrical resistance change as cell passes through orifice
Cedex HighRes	Roche CustomBiotech	Image analysis, viability assessment with trypan blue
Cellaca MX High-throughput Automated Cell Counter	Nexcelom Bioscience	Image analysis, viability assessment with trypan blue; some models allow fluorescence imaging
CellDrop Cell Counter	DeNovix	Image analysis, viability assessment with trypan blue; some models allow fluorescence imaging
Cellometer series	Nexcelom Bioscience	Image analysis, viability assessment with trypan blue; some models allow fluorescence imaging
Corning Cell Counter	Corning	Image analysis, viability assessment with trypan blue
Countess II	ThermoFisher Scientific	Image analysis, viability assessment with trypan blue; some models allow fluorescence imaging
CyFlow Counter	Sysmex Partec	Flow cytometry of single cell stream; multiparametric
EVE™ series	NanoEnTek	Image analysis, viability assessment with trypan blue
Guava easyCyte™	Luminex	Flow cytometry of single cell stream; multiparametric
Guava Muse®	Luminex	Flow cytometry of single cell stream; multiparametric
LUNA™ series	Logos Biosystems	Image analysis, viability assessment with trypan blue
Moxi Z Mini Automated Cell Counter	Orflo	Electrical resistance change as cell passes through orifice
Multisizer 4e Coulter Counter	Beckman Coulter Life Sciences	Electrical resistance change as cell passes through orifice
NucleoCounter series	ChemoMetec	Image analysis using trypan blue, PI, AO, or DAPI in whole and lysed cells (nuclei)
Scepter Cell Counter	Sigma-Aldrich	Electrical resistance change as cell passes through orifice
TC20™ Automated Cell Counter	Bio-Rad	Image analysis, viability assessment with trypan blue
Vi-CELL BLU	Beckman Coulter Life Sciences	Image analysis, viability assessment with trypan blue
Vi-CELL XR	Beckman Coulter Life Sciences	Image analysis, viability assessment with trypan blue

Abbreviations not found in main text: AO, acridine orange; PI, propidium iodide.

This type of electronic cell counter usually has three main components: (i) an orifice tube, with a 150-μm orifice, connected to a metering pump: (ii) an amplifier, pulse-height analyzer, and scaler connected to two electrodes, one in the orifice tube and one in the sample beaker; and (iii) an analog and a digital readout showing the cell count and a number of other parameters, such as cell volume and size distribution (see Figure 19.3b, c). When the count is initiated, a measured volume of cell suspension is drawn through the orifice. As each cell passes through the orifice, it changes the resistance to the current flowing through the orifice by an amount proportional to the volume of the cell. This change in resistance generates a pulse that is amplified and counted. Because the size of the pulse is proportional to the volume of the cell (or any other particle) passing through

the orifice, a series of signals of varying pulse height are generated.

Thresholds are set to determine which range of particle sizes is counted. In many modern counters, the lower threshold can be set manually on the readout, while the upper is either set to infinity or adjustable, depending on the model. Setting the lower threshold to 7.0 μm will include non-viable cells, whereas a higher setting (12 μm in Figure 19.3b) will exclude most of the non-viable cells. Non-viable cells have a smaller apparent diameter because the plasma membrane is leaky, the cytoplasm has the same resistance as the electrolyte, and the resistance signal is generated by the nucleus. Counts at and above 30 μm indicate aggregation; aggregates can be partially excluded by setting an appropriate upper threshold (e.g. 25 or 30 μm in Figure 19.3b), or can be allowed for using a

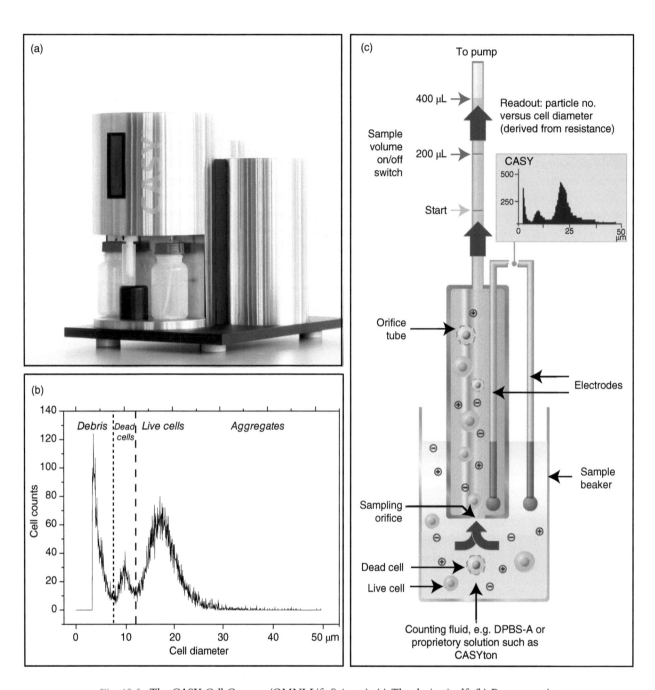

Fig. 19.3. The CASY Cell Counter (OMNI Life Science). (a) The device itself. (b) Representative sample data. Cell number plotted against cell diameter gives a size distribution analysis that enables the lower threshold (vertical dashed line) to be set. In this display, the upper threshold could be set to 30 μm or set to infinity. The peak between the vertical dashed line and the vertical dotted line represents non-viable cells that can be used with the viable cell count to determine percentage viability. (c) Principle of operation. A measured volume of the cell sample (in suitable buffer) is drawn into the orifice tube. A cell passing through the orifice alters the flow of current and generates a signal, the amplitude of which is proportional to the volume of the cell. *Source*: (a) Courtesy of OMNI Life Science; (b, c) Courtesy of Schärfe Systems, with permission.

statistical calculation available within the counter's operating program. In some cases, the software is capable of correcting for clumping of cells and coincidence (i.e. when two or more cells go through the aperture at once).

Using image analysis. Some automated cell counters can scan cells in an optical counting chamber, using visible light or fluorescence; an image is recorded using a digital camera for post-image processing and quantification. A number of cell counters now rely on image analysis (see Figure 19.4), including the CellDrop™ (DeNovix), the Countess II (ThermoFisher Scientific), the NucleoCounter® NC-200™ (ChemoMetec), and the Vi-CELL™ BLU (Beckman Coulter Life Sciences). Viable and non-viable cells can be distinguished by adding trypan blue or, in models that can detect fluorescence, by adding propidium iodide, acridine orange, or 4′,6-diamidine-2′-phenylindole dihydrochloride (DAPI) (see Section 19.2). Systems that are based on trypan blue can be used from 5×10^4 to 1×10^7 cells/ml, while those using propidium iodide can be used down to 5×10^3 cells/ml.

Some image-based counters use disposable chambered slides or reusable slides, with a few systems allowing hemocytometer slides to be used. Others do not require slides for counting; samples are either injected into the imaging chamber (e.g. the Vi-CELL BLU) or dropped onto a stage for counting (e.g. the CellDrop). An autofocus system is used in some devices to determine the best focal plane for analysis, while others rely on manual focusing. Many devices generate an analog plot of cell size distribution, which allows thresholds to be set for gating. Ease of use and cost will vary a great deal based on these different factors. Always ask for a demonstration and discuss with users to get a sense of the practical advantages and disadvantages when using these instruments.

If you have access to a live-cell imaging system, it should be possible to use it to perform cell counting. Provided the image is spread in one layer on a known area, automated cell counting can be performed using various image analysis software packages, e.g. ImageJ (Grishagin 2015). Some suppliers provide reagents and protocols for this purpose; for example, cells can be transiently transfected using the IncuCyte® NucLight™ range of labeling reagents and then counted using an IncuCyte live-cell system (Sartorius; see Figure 18.4) (Essen Bioscience 2019).

Using flow cytometry. Bench-top flow cytometers (see Section 21.4) can be used for cell counting and multiparameter assessment. Examples of such bench-top systems include the Accuri™ C6 Plus flow cytometer (BD Biosciences), the CyFlow Counter (Sysmex Partec), and the Guava® series of cell analyzers (Luminex; see Figure 19.5). These instruments can give a great deal of useful information about the characteristics of the culture in addition to an accurate cell count. Parameters such as viability (by uptake of propidium iodide) or apoptosis (by annexin V staining) can be readily

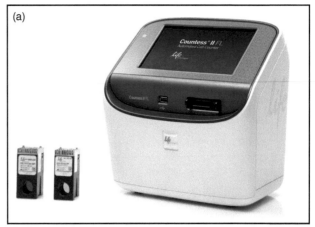

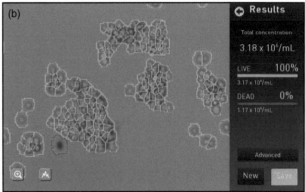

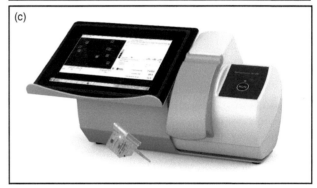

Fig. 19.4. Image analysis cell counters. (a) The Countess FL (ThermoFisher Scientific) with accessory fluorescence filters to the left; (b) representative sample data screen, including image of cells being counted (note that the 3T3 cells counted here are clumped but still resolved into individual cells); (c) the NucleoCounter NC-200 (ChemoMetec) with cell sample cassette in front. *Source:* (a, b) Courtesy of ThermoFisher; (c) Courtesy of ChemoMetec.

quantified using a compact bench-top flow cytometer (see Figure 19.6). Some instruments also allow simultaneous flow cytometry and imaging for direct visualization, e.g. the Amnis® FlowSight® (Luminex).

Bench-top flow cytometers allow the measurement of multiple parameters, making them valuable tools for analysis of cellular characteristics and staining. They are relatively

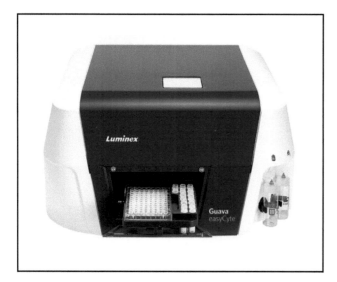

Fig. 19.5. Guava easyCyte flow cytometer (Luminex). Instrument with sample tray open; note that this system allows sampling from individual tubes or 96-well plates. Sampling is performed using a microcapillary flow cell, which allows direct sampling by aspiration. *Source*: Courtesy of Luminex, with permission.

easy to handle compared to the larger flow cytometers, but they are likely to be more expensive and complicated compared to devices that only perform cell counting. Always consider what you need from such a device, discuss your requirements with the supplier, and ask for a demonstration before investing in any new equipment.

19.1.3 Counting Adherent Cells

All of the previous methods are based on dispersal of cells into a single-cell suspension (although Coulter-based and image-based systems may be able to compensate for some degree of clumping). This may not be possible for some cultures, where cells cannot be harvested for counting or are too few to count in suspension, e.g. small numbers of cells in multiwell plates. Alternative approaches for adherent cells *in situ* include:

(1) **Live-cell image analysis.** If you have access to a live-cell imaging system, you should be able to use it to count the number of cells in each well or to monitor changes in cell number or confluence over time (see Section 18.4.3). Cells may also be stained with fluorescent dyes or conjugated antibodies to assist with counting, to distinguish between viable and non-viable cells, or to assess lineage- or tissue-specific markers (see Section 18.3.2). Live-cell imaging offers a direct, rapid, and non-invasive approach if the necessary equipment is available, e.g. in a local core facility.

(2) **Fixation and staining of cells for later analysis.** Cells can be fixed and stained with Giemsa or crystal violet (see

Protocols P18.3, P18.4) for manual counting or image analysis. Absorption can also be measured after crystal violet staining using a densitometer; this approach has been used to calculate the number of cells per colony in clonal growth assays (McKeehan et al. 1977). However, staining with crystal violet can also vary considerably from one cell line to another (Skehan et al. 1990).

(3) **Measurement of total protein.** The amount of protein in solubilized cells can be estimated directly by measuring the absorbance at 280 nm, with minimal interference from nucleic acids and other constituents. Absorbance at 280 nm can detect down to 100 µg of protein, or about 2×10^5 cells. Colorimetric assays are more sensitive; the Bradford assay (based on Coomassie brilliant blue G-250) can give a linear relationship between absorbance and cell number between 1×10^4 and 2×10^5 cells/well in a 96-well plate (Baumgarten 1985). Testing of dyes with the NCI-60 panel demonstrated that sulforhodamine B gave the best results at low cell densities, detecting $1–2 \times 10^3$ cells/well (Skehan et al. 1990; Orellana and Kasinski 2016).

(4) **Measurement of total DNA.** The amount of DNA in viable or non-viable cells may be measured using various fluorescent dyes, including DAPI, Hoechst 33258, and propidium iodide (Brunk et al. 1979; Labarca and Paigen 1980; Wan et al. 1994). The fluorescence emission of Hoechst 33258 at 458 nm is increased by interaction of the dye with DNA at pH 7.4, in the presence of high salt to dissociate the chromatin protein. This method gives a sensitivity of 10 ng/ml but requires intact double-stranded DNA and may be limited by autofluorescence (which is increased by Triton X-100 and some other agents used for cell lysis). Propidium iodide results in less autofluorescence and can give a linear relationship between cell number and fluorescence between 2×10^3 and 1×10^6 cells/well in a 96-well plate (Wan et al. 1994).

(5) **Uptake of reagents by living cells.** One of the most popular *in situ* assays for counting cells is based on uptake and metabolism of 3-(4,5-dimethylthiazol-2-yl)-2,5-diphenyltetrazolium bromide (MTT) (see Protocol P29.1). However, this does not necessarily result in a linear relationship with cell number (Keepers et al. 1991; van Tonder et al. 2015). The MTT assay (and others based on cell metabolism) is commonly used to assess cytotoxicity and is discussed elsewhere (see Section 29.2.2).

19.1.4 Cell Weight and Packed Cell Volume (PCV)

Wet weight is seldom used to estimate cell number unless very large numbers are involved, because the amount of adherent extracellular liquid gives a large error. As a rough guide, however, there are about 2.5×10^8 HeLa cells (14–16 µm in diameter) per gram wet weight, about $8–10 \times 10^8$ cells/g for

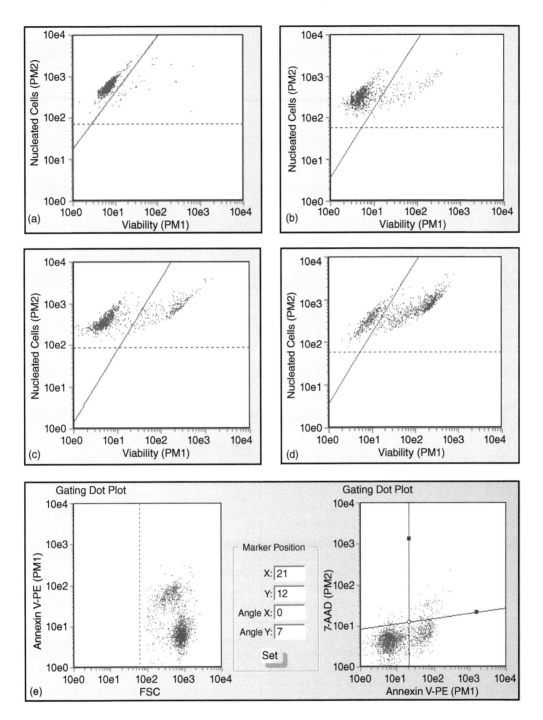

Fig. 19.6. Output from Guava flow cytometer. Dot plot from deteriorating cell culture. (a–d) Four panels show progressive accumulation of dead cells, stained with propidium iodide, in upper right quadrant. (e) Bottom two panels show apoptotic cells, staining with antibody to annexin V (bottom right quadrant in right-hand image). *Source*: Courtesy of Edward Burnett, ECACC, Public Health England.

TABLE 19.3. Comparison of cell size, volume, and mass.

Cell type	Diameter (μm)	Volume (μm³)	Cells/g × 10⁻⁶ Calculated	Cells/g × 10⁻⁶ Measured
Murine leukemia (e.g. L5178Y or Friend cells)	11–12	800	1250	1000
HeLa	14–16	1200	800	250
Human diploid fibroblasts	16–18	2500	400	180

murine leukemias, myelomas, and hybridomas (11–12 μm in diameter), and about 1.8×10^8 cells/g for human diploid fibroblasts (16–18 μm in diameter; see Table 19.3). Similarly, dry weight is seldom used; salt derived from the medium contributes to the weight of unfixed cells, and fixed cells lose some of their low molecular weight intracellular constituents and lipids. Packed cell volume (PCV) is more commonly employed, particularly for blood samples but also as a rapid estimate of cell number at subculture. PCV can be determined by centrifugation into a special calibrated tube, e.g. the PCV tube (TPP Techno Plastic Products).

19.2 CELL VIABILITY

19.2.1 Dye Exclusion Assays

Viable cells are impermeable to many colorimetric staining reagents such as trypan blue, naphthalene black (also known as amido black), and erythrosin B (see Figure 19.1g, h) (Kaltenbach et al. 1958; Schrek 1958; Krause et al. 1984). Although these dyes are excluded from viable cells, they are taken up by non-viable cells with ruptured plasma membranes and thus provide a rapid assay for the measurement of cell death (Crowley et al. 2016). Trypan blue is the most popular choice for bright field examination, although concerns have been raised about its use due to safety hazards (as a known carcinogen) and its toxic effect on some cell types (van Dooren et al. 2004; Kim et al. 2016). Some fluorescent dyes can also be used for dye exclusion assays, such as acridine orange (see Figure 29.2c, d) and propidium iodide (Hathaway et al. 1964; Darzynkiewicz et al. 1994). Acridine orange and propidium iodide give excellent results when used together for viability assessment, provided suitable equipment is available to detect fluorescence (Mascotti et al. 2000).

Dye exclusion assays are used primarily to measure viable cells after a potentially traumatic procedure, such as primary disaggregation, cell separation, or cryopreservation; some dyes are also used for cytotoxicity testing (see Section 29.2.3). These assays tend to overestimate viability and results should be interpreted with caution. For example, 90% of cells thawed from liquid nitrogen may exclude trypan blue, but only 60% may prove to be capable of attachment 24 hours later. Dye exclusion assays also tend to be uninformative during routine culture because most non-viable cells are lost in the discarded medium and prewash prior to trypsinization. However, these assays are useful in certain circumstances (e.g. to assess dead cells after cryopreservation and thawing), and can be readily combined with cell counting using manual or automated approaches. The following procedure (Protocol P19.2) uses trypan blue to assess cell viability during hemocytometer counting and can be adapted for use in training (see Supp. S30.10).

PROTOCOL P19.2. CELL COUNTING USING TRYPAN BLUE

Background

See Protocol P19.1 for counting cells using a hemocytometer. The following instructions apply to the addition of trypan blue, which requires minor adjustments to the counting protocol. If a dye is added for viability, you must compensate for the additional dilution to obtain an absolute count.

Outline

Prepare a single-cell suspension and the hemocytometer slide. Dilute the cell suspension 1:1 with a trypan blue solution and add the stained cells to the counting chamber. Count the cells using an inverted microscope and calculate the cell concentration with adjustment for the dilution.

- *Safety Note. Trypan blue is a hazardous substance (see Section 6.2.2). Refer to the Safety Data Sheet (SDS) from the supplier for more information.*

Materials (sterile or aseptically prepared)

- Cell line or primary culture for testing, e.g. during thawing after cryopreservation (see Protocol P15.2)
- Trypan blue solution, 0.4% (e.g. #T8154, Sigma-Aldrich)
- Otherwise as per Protocol P19.1

Materials (non-sterile)

- Differential tally counter (e.g. differential cell counter, Bal Supply)
- Otherwise as per Protocol P19.1

Procedure

1. Prepare a cell sample for counting as described previously (see Protocol P19.1, steps 1–4). When preparing the cell suspension, aim for approximately 2×10^6 cells/ml, based on an estimate of the pellet size. This concentration (twice that suggested in the previous protocol) is required because the addition of trypan blue will dilute the sample.

2. Prepare the hemocytometer slide for counting as described previously (see Protocol P19.1, steps 5–8).

3. Using a pipettor, transfer 50 µl of trypan blue solution to a fresh sample tube.

4. Gently mix the cell suspension and transfer 50 µl to the tube containing the trypan blue.

5. Gently mix the trypan blue and the cell suspension. Remove an aliquot of about 30–40 µl and use it to load one of the counting chambers of the hemocytometer slide (see Protocol P19.1, step 10). Repeat with the second chamber, if there is one.

6. Leave the slide for one to two minutes before starting to count (do not leave any longer, or viable cells will deteriorate and take up the stain).

7. Transfer the slide to the microscope stage and start to count the cells as described previously (see Protocol P19.1, steps 12–15).

8. The differential cell counter allows two separate tallies. Use one tally for viable (unstained) cells and the other for non-viable (stained) cells (see Figure 19.1g, h).

9. Continue to count until you have sufficient cells to be able to calculate the cell concentration (see Protocol P19.1, steps 17–20). Keep a record of your tallies and the number of squares counted.

10. Using the cell counts that were recorded by the tally counter, determine the average number of stained cells in one large square (as indicated by the triple parallel lines; see Figure 19.1f–h) and the average number of unstained cells in the same area. These are your cell counts for calculating cell concentration and the percentage of viable cells.

11. Calculate the viable cell concentration as described previously (see Protocol P19.1, steps

22–24), but use the following formula:

$$Concentration = unstained\ cells \times 2 \times 10^4$$

12. Calculate the percentage of viable cells using the following formula:

$$Percent\ viable\ cells = \frac{unstained\ cells}{unstained + stained\ cells} \times 100$$

Notes

Step 11: This formula is based on the one used previously (see Protocol P19.1, step 23) but multiplies the result by two to adjust for the 1 : 1 mixture of cells:trypan blue. The formula could be adjusted for different dilutions or to determine the total cell concentration, including both viable and non-viable populations.

19.2.2 Dye Uptake Assays

Neutral red can penetrate cell membranes through passive diffusion, where it is retained inside the lysosomes in viable cells. This property allows it to be used in a classic colorimetric assay for cell viability (Borenfreund and Puerner 1985). Neutral red tends to precipitate, so the medium with stain is usually incubated overnight and centrifuged before use. Uptake is then quantified by fixing the cells in formaldehyde, solubilizing the stain in acetic ethanol, and measuring absorbance at 570 nm. Uptake of neutral red depends on the cell's capacity to maintain pH gradients through the production of ATP (Repetto et al. 2008). Thus, depending on the experimental design, it is possible to distinguish between viable and non-viable cells and to assess lysosomal function (Repetto and Sanz 1993). Neutral red uptake can be used to assess cytotoxicity, e.g. for phototoxicity testing using BALB/c 3T3 cells (see Section 29.1.1).

Several fluorescent molecules can also be used to assess cell viability. These molecules are introduced to the cell as non-fluorescent dyes, which are modified to produce a fluorescent product that is retained by the living cell. Generation of a fluorescent signal thus depends on enzymatic activity and membrane integrity, making these reagents useful for simple fluorescence microscopy, flow cytometry, or cytotoxicity assays (see Sections 18.4, 21.4, 29.2.3). Commonly used dyes include calcein-acetoxymethyl (calcein-AM) and diacetyl fluorescein (DAF). Calcein-AM is a lipid-soluble molecule that passively crosses the cell membrane, where it is converted to calcein by esterase cleavage of the AM ester. Likewise, DAF can penetrate the cell membrane through passive means and is hydrolyzed to produce fluorescein (Rotman and Papermaster 1966). DAF is often combined with propidium iodide; viable cells will fluoresce green due to the presence of fluorescein, while non-viable cells will fluoresce red due to propidium iodide. A protocol is provided in the Supplementary Material online for staining with DAF and propidium iodide to estimate cell viability (see Supp. S19.1).

19.3 CELL PROLIFERATION

Measurement of cell proliferation is often performed during experimental work (e.g. in cytotoxicity assays; see Section 29.2), but it is also a central part of routine culture practice and its importance is frequently underestimated. Proliferation data can be used to predict the behavior of a cell line at different densities, to select the best time for subculture and the best seeding density, and to standardize the growth conditions for ongoing handling (see Sections 14.4.2, 14.6.4). Once proliferation data are available, the results can be used to monitor the cell line's behavior over time and for comparison when testing new media, sera, culture vessels, or substrates.

As with cell counting, cell proliferation can be assessed using various methods (see Table 19.4). The most important requirement is to establish a growth curve, in which the cell number is assessed at regular intervals between one passage and the next. The growth curve can be used to determine other key parameters such as the population doubling time (PDT) (see Section 19.3.3). Another important requirement is to determine the cloning efficiency, i.e. the number of clonal populations that arise after plating cells at low density (see Section 19.4). Depending on your application, additional measures of cell proliferation may be required to provide a baseline for experimental work (see Sections 19.5, 19.6).

TABLE 19.4. Measurement of cell proliferation.

Term	Definition	Measurement or method
Cell cycle time	Time from one point in the cell cycle until the cell reaches the same point again	Percentage of mitoses labeled with [³H]thymidine after 30 minutes' incubation and sampled during 24–48 hours' cold thymidine chase; percentage of mitoses with BrdU pulse labeling, followed by staining with anti-BrdU at intervals during cold thymidine chase
Cloning efficiency	Number of cells capable of forming clones (colonies derived from one cell) following subculture at a low cell density	Trypsinize and plate out cells as single cells; count colonies above a defined threshold (usually 50 cells) after 10 days to 3 weeks
Colony-forming efficiency	Number of cells capable of forming discrete colonies following subculture at a low cell density	Same as cloning efficiency, provided colonies are all formed from single clonal populations and have not merged during growth
Division index	Percentage of cells undergoing cell division at any one time	Stain with antibodies to PCNA or Ki-67
Growth curve	Semi-log plot of cell number against time from subculture	Count cells on a daily basis following subculture until fully confluent
Growth fraction	Percentage of cells in cycle during a set time period identified by labeling	Prolonged label (24–48 hours) with [³H]thymidine; staining of cells that have incorporated BrdU over a comparable time period
Labeling index	Percentage of cells labeled with radioactive precursor or fluorescent antibody Ki-67 or to BrdU or PCNA	[³H]-thymidine incorporation into DNA followed by autoradiography; staining with fluorescent antibody
Mitotic index	Percentage of cells in mitosis	Count mitoses as percentage of total population
Population doubling time (PDT)	Time for a given cultured cell population to double; product of cell division, cell death, and non-dividing cells	Parameter generated from the exponential phase of the growth curve
Saturation density	Cell density at plateau phase	Parameter generated from the plateau phase of the growth curve
Seeding efficiency	Number of cells that have attached by the end of the lag period	Plate out cells, trypsinize, and count after 18–24 hours
Viability index	Percentage of cells excluding viability stain, e.g. trypan blue	Microscopic observation or image analysis of unfixed cells stained with viability stain

19.3.1 The Growth Curve

All cell lines, regardless of their origin, display three distinct growth phases as they proliferate following subculture (see Figure 19.7) (Reid 2017). Cultures vary significantly in many of their properties when assessed during the lag phase, the period of exponential growth (log phase), and the stationary phase (plateau phase). This explains why single time points cannot be used to monitor growth if the shape of the growth curve is not known. A reduced cell count after, say, five days could be caused by (i) a reduced growth rate of some or all of the cells; (ii) a longer lag period, implying adaptation or cell loss (it is difficult to distinguish between the two); or (iii) a reduction in saturation density (see Section 19.3.3). Once the growth curve has been established, single time point observations can be made based on the predicted phases of growth for that cell line.

Lag phase. After subculture and reseeding, the cell number may remain constant (or even decrease) for a limited period of time, with little evidence of proliferation (Earle et al. 1954). This "lag phase" is a period of adaptation during which adherent cells attach to the substrate, replace elements of the cell surface and the extracellular matrix (ECM) that were lost during trypsinization, and spread out. Cell spreading requires an intact cytoskeleton that can interact with the ECM; F-actin filaments are particularly important for this process, and will rapidly increase in number before any significant change occurs in the cell shape (see Section 2.2.3) (Mooney et al. 1995). Cells in suspension may also experience a lag phase, during which they "condition" the medium to their needs (Earle et al. 1954). The activity of enzymes, such as DNA polymerase, increases, followed by the synthesis of new DNA and structural proteins. Some specialized cell products may disappear and not reappear until the cessation of cell proliferation at a high cell density.

Log phase. After the lag phase, most cells enter the cell cycle to actively undergo cell division, resulting in a growth fraction of between 90% and 100% for many continuous cell lines (see Section 2.3.1; see also Table 19.4). This results in a period of exponential or logarithmic growth that is usually referred to as the "log phase." It is the optimal time for sampling, because the population is at its most uniform and the viability is high. However, the cells are randomly distributed in the cell cycle and, for some purposes, may need to be synchronized (see Section 19.6). The duration of the log phase will depend on the growth rate of the cells, the initial seeding density, and the density that inhibits cell proliferation; it normally terminates one to two population doublings after confluence is reached. The slope of the growth curve during the log phase can be used to calculate the population doubling time (PDT) (see Section 19.3.3).

Plateau phase. Toward the end of the log phase, the culture becomes confluent, i.e. all of the available growth surface becomes occupied and all cells make contact with surrounding cells. At this stage, the growth fraction falls to between 0% and 10% and the culture enters the plateau, or stationary, phase. The cells may become less motile; some fibroblasts become oriented with respect to one another, forming a typical parallel array of cells (see Figure 3.1a). "Ruffling" of the plasma membrane is reduced, and the cell occupies less surface area of substrate and presents less of its own surface to the medium. Normal cells that have entered the plateau phase have low motility, a reduced growth fraction, may be more differentiated, and may become polarized. They generally tend to secrete more ECM and may be more difficult to disaggregate. Transformed cells (see Section 3.6) may behave similarly except that the plateau phase will have more cell motility, a higher growth fraction, and more apoptotic cells, and may tend to shed more cells into the supernatant medium (see Figure 14.1b).

Normal cells typically display contact inhibition at confluence, which can be defined as the inhibition of movement and proliferation when cells are in contact with one another (Ribatti 2017). This phenomenon may relate to increased cell density rather than direct contact (Stoker and Rubin 1967). Contributing factors include changes in cell shape, buildup of inhibitors, depletion of nutrients, and the presence of diffusion boundaries near the cell surface (Stoker 1973; Folkman and Moscona 1978; Holley et al. 1978). However, many cultures continue to proliferate (although at a reduced rate) at confluence, provided their media are regularly replenished. Human embryonic lung and adult skin fibroblasts, which express contact inhibition of movement,

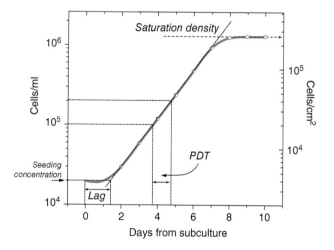

Fig. 19.7. Growth curve and parameters. Semilog plot of cell concentration (left axis) and cell density (right axis) against time after subculture, obtained by daily sampling of replicate 25-cm² flasks. It should be possible to draw a straight line through the part of the plot that represents the log phase. Calculate the population doubling time (PDT) from the middle region of this best-fit line, and the lag time from the point where it intersects with the original seeding concentration. Calculate the saturation density by reading the cell density at the top end of the curve, once the culture reaches plateau phase.

will continue to proliferate, laying down layers of collagen between the cell layers until multilayers of six or more cells can be reached under optimal conditions; the major limiting factor becomes exhaustion of medium components (Kruse et al. 1970; Dulbecco and Elkington 1973).

Thus, the terms "plateau" and "stationary" are not strictly accurate and should be used with caution. Rather than implying the complete cessation of cell proliferation, plateau phase represents a steady state in which cell division is balanced by cell loss. However, the concept of plateau phase has some practical uses, particularly to determine the saturation density of the culture (see Section 19.3.3).

19.3.2 Experimental Design

Although it is important to establish a growth curve to map out the three phases of cell growth (which can vary considerably between different cell lines), it takes some effort to generate the necessary data. It is important to design growth curve experiments to minimize unnecessary labor. Factors to consider include:

(1) *Availability of equipment.* A live-cell analysis system (see Section 18.4.3), if available, can be used to generate a growth curve without the need to harvest cells. For example, the IncuCyte platform (Sartorius) can be used to generate growth curves, based on the percentage of the growth surface that is occupied by cells (see Figure 18.5). An automated cell counter (see Section 19.1.3) can be used to reduce the manual labor involved in counting cells with a hemocytometer.

(2) *Direct versus indirect methods.* A direct method of measuring cell number gives a clear experimental readout; for example, saturation density can be calculated by simply looking for the number of cells per square centimeter at plateau phase on the growth curve (see Figure 19.7). Indirect methods are less likely to give a direct readout, since they measure a surrogate instead of cell number, but they are also less labor-intensive and allow multiple variables to be examined simultaneously on the same multiwell plate.

(3) *Choice of culture vessel.* Growth curve experiments can be performed in flasks, dishes, or multiwell plates. Flasks or dishes are more labor intensive; even limiting the number of replicates to two per day for eight days would require a minimum of 16 flasks. This approach is useful for comparing different reagents, but for a quantitative assay, it is preferable to use multiwell plates and to sample a whole plate at each timepoint. Growth curve experiments can be performed in microplates, but as the area per well is very small, it is difficult to perform cell counts and it may be necessary to pool four or eight wells together to give a reproducible result. Alternatively, the cell count may be estimated using an indirect method that provides an absorbance reading.

(4) *Choice of seeding density.* An initial cell concentration of about 2×10^4 cells/ml (in flasks) is suitable for a rapidly growing continuous cell line and 1×10^5 cells/ml for a slower-growing finite cell line. If the experiment is set up using multiwell plates, then at least three seeding densities can be used (see Figure 19.8), which will give a better understanding of the cell line's behavior at high and low density.

Generating data for the growth curve. Protocol P19.3 describes how to generate proliferation data for a growth curve using multiwell plates. An alternative protocol to generate a growth curve using flasks is provided in the Supplementary Material online (see Supp. S19.2).

PROTOCOL P19.3. GENERATING A GROWTH CURVE USING MULTIWELL PLATES

Background

This protocol is based on direct cell counting but can be adapted for use with indirect assays, e.g. the sulforhodamine B assay (Orellana and Kasinski 2016). Instructions are written for adherent cells but can easily be modified for suspension cultures (see Notes following the procedure).

Samples are plated here in 12-well plates at three seeding densities, giving four wells at each density per plate (see Figure 19.8a). Three of these wells are counted and the fourth well is stained using Giemsa or crystal violet, which allows plates to be set aside for imaging at a later date. Staining could be replaced by imaging prior to counting; this would allow the fourth well to be counted or would provide extra space for a fourth cell density to be examined.

Outline

Set up a series of multiwell plates with cultures at three different cell concentrations. Harvest one plate each day for cell counting until the culture reaches the plateau phase.

Materials (sterile or aseptically prepared)

- Cell line for growth assessment
- Dulbecco's phosphate buffered saline without Ca^{2+} and Mg^{2+} (DPBS-A)
- Trypsin (0.25%)/EDTA (10 mM) solution
- Complete culture medium
- Multiwell plates, 12-well (8)
- Bottles or tubes, 50–100 ml (3, for preparing different cell concentrations)
- Microcentrifuge tubes, 2 ml

Materials (non-sterile)

- Biological safety cabinet (BSC) and associated consumables (see Protocol P12.1)
- Plastic box or trays to hold the plates
- Cell counting equipment (hemocytometer or automated counter; see Section 19.1)
- Giemsa stain or crystal violet solution (see Protocols P18.3, P18.4)

Procedure

A. Preparation of Multiwell Plates

1. Grow the cells in one to two 75-cm^2 flasks until they are close to confluence (which should be equivalent to late log phase).
2. Perform subculture using the standard procedure for that cell line (e.g. trypsinization; see Protocol P14.2).
3. Resuspend in culture medium to give a single-cell suspension. Count the cells and calculate the cell concentration (see Protocol P19.1).
4. Dilute the cell suspension in culture medium to 1×10^5 cells/ml, 3×10^4 cells ml, and 1×10^4 cells/ml; you will need at least 64 ml at each concentration (8 ml per plate, eight plates).
5. Aliquot the following into each of eight 12-well plates (see Figure 19.8a for a guide):
   (a) Row A (top four wells): 2 ml 1×10^4 cells/ml.
   (b) Row B (middle four wells): 2 ml 3×10^4 cells/ml.
   (c) Row C (bottom four wells): 2 ml 1×10^5 cells/ml.
6. Place the plates in a humid CO_2 incubator or a sealed box gassed with 5% CO_2.

B. Harvesting Plates

7. After 24 hours, remove the first plate from the incubator and bring it to the BSC.
8. Remove the medium completely from the wells in the first three columns (see Figure 19.8a). The fourth column will be used for staining.
9. Wash the wells with DPBS-A and remove.
10. Add 0.5 ml of trypsin/EDTA to the wells and incubate the plate for 15 minutes.
11. Add 0.5 ml medium with serum to each well.
12. Disperse the cells in the trypsin/EDTA/medium and transfer the cell suspension to a prelabeled

tube, e.g. indicating the row and column number for that well.
13. This leaves one column of cells that have not been harvested. Stain these wells using Giemsa or crystal violet as a record of cell confluence on that day (see Protocols P18.3, P18.4).
14. Count the cells using an automated cell counter following the supplier's instructions. If an automated device is not available, use a hemocytometer (see Protocol P19.1).
15. Record the counts for that day.
16. Repeat steps 8–15 at 48 and 72 hours using the next two plates.
17. Change the medium at 72 hours, or sooner if indicated by a drop in pH (see Protocol P14.1). Keep changing the medium every one, two, or three days, as indicated by the fall in pH.
18. Continue harvesting daily for rapidly growing cells (i.e. cells with a PDT of 12–24 hours), but reduce the frequency of sampling to every two days for slowly growing cells (i.e. cells with a PDT > 24 h) until the plateau phase is reached.

Notes

Step 5: Add the cell suspension slowly from the center of the well, so that it does not swirl around the well. Similarly, do not shake the plate to mix the cells, as the circular movement of the medium will concentrate the cells in the middle of the well.

Steps 8–14: For suspension cultures, instead of performing trypsinization, simply transfer the cell suspension to the prelabeled tube. Make sure that the cell suspension is well mixed and completely disaggregated before counting. If cells are to be stained, you will need to spin the cells down onto a slide using a cytocentrifuge (see Supp. S18.2).

Step 14: A hemocytometer may be difficult to use for lower cell concentrations. Try reducing the volume of trypsin to 0.1 ml and disperse the cells carefully, using a pipettor without frothing the trypsin. Transfer the cells to the hemocytometer and count enough squares to give a count in excess of 200 cells (see Protocol P19.1).

Step 17: For suspension cultures, it may not be necessary to refeed each well, which would require centrifuging the cells to resuspend them in the fresh medium. Alternatively, the wells could be topped up with additional medium and the cell concentration adjusted accordingly.

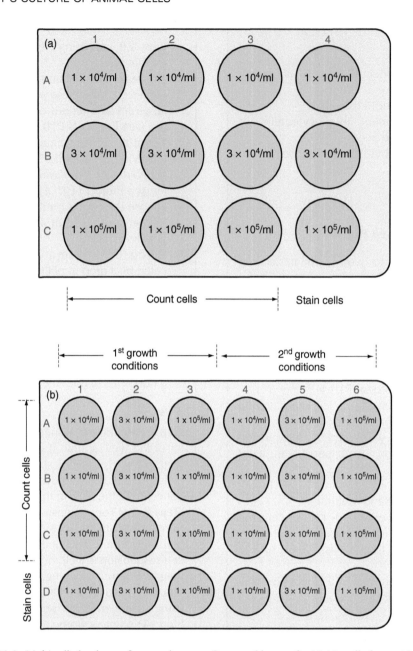

Fig. 19.8. Multiwell plate layout for growth curves. Suggested layouts for (a) 12-well plates with cells at three different concentrations and wells allocated for counting and staining; (b) 24-well plates with an additional variable, in this case a different set of growth conditions (e.g. a different batch of serum).

Plotting the growth curve. After counting is complete, the data can be used to plot a growth curve as follows:

(1) *Transfer the raw counts from each time point and seeding density to a table.* Calculate the mean from the wells that were counted in triplicate. The mean value is referred to here as the primary count – this is the number of cells per milliliter of trypsin/EDTA/medium (see Protocol P19.3).

(2) *Use the primary count to determine the total cell number.* If 1 ml trypsin/EDTA/medium was used for counting,

the primary count is the same as the total number of cells per well. If a different volume has been used, multiply the primary count by the volume (expressed in milliliters).

(3) *Use the total cell number to calculate the cell concentration.* Divide the total cell number by the volume of medium used during culture to give the number of cells/ml.

(4) *For adherent cells, use the total cell number to calculate the cell density.* Divide the total cell number by the surface area of the well that you used to give the number of cells per square centimeter. As there are small variations among manufacturers, it is best to refer to the supplier

for the surface area of your multiwell plate. As a rough guide, the surface area of each well in a 12-well plate is approximately 3.8 cm² (see Table 8.2).

(5) ***Plot the cell concentration and the cell density on a log scale against time on a linear scale.*** In the example shown previously (see Figure 19.7), the concentration is given on the left vertical axis and the cell density is on the right vertical axis. The scale on both vertical axes is the same but out of register by a factor that depends on the number of square centimeters per milliliter of medium. For example, in a 25-cm² flask, there is 5 ml medium covering 25 cm², which translates to 0.2 ml/cm² or 5 cm²/ml medium. Thus, the right-hand axis will be out of register with the left by a factor of 5, and 1×10^5 on the left axis will be equivalent to 2×10^4 on the right-hand axis.

(6) ***Plot the data for each seeding density and compare the results.*** The optimum seeding density should resemble the example shown previously (see Figure 19.7); some suboptimal results and their interpretation are shown in Figure 19.9.

(7) ***Examine the stained cells at each density.*** Staining can help to determine whether the distribution of cells in the flasks or wells is uniform, whether the cells are growing up the sides of the well, and whether cell morphology changes in each phase.

Medium volume, cell concentration, and cell density. When extrapolating results from one culture vessel to another, it is important to remember that the volume of medium used in a multiwell plate is often proportionately higher than in a flask for a given surface area. This means that the cell density in the multiwell plate will be higher for the same cell concentration. For example, if 2×10^4 cells/ml are seeded in 5 ml into a 25-cm² flask, the cell density at seeding will be 4000 cells/cm² ($20\,000 \times 5 \div 25$). If the same cell concentration is seeded in 2 ml in a 12-well plate, the cell density will be 10 500 cells/cm² ($20\,000 \times 2 \div 3.8$). This density is more than twice that of the flask, which means that the cells will reach plateau at least one day earlier. If an exact comparison is intended, the ratio of medium to culture surface area must be the same; this would translate to a volume of 0.75 ml in a 12-well plate in order to achieve the same density as a 25-cm² flask. Unfortunately, such a low volume would cause uneven cellular distribution due to the shape of the meniscus, and cells would tend to concentrate at the edges of the wells. This problem increases as the wells get smaller, because the relative effect of the meniscus increases with a decrease in diameter of the well. A reasonable compromise is to ensure that the cell density (cells per square centimeter) is the same, although the cell concentration will be less.

Ideally, if a growth curve is being used to establish conditions for routine maintenance, the growth curve should be performed in the same size of vessel used for routine subculture. In practice, however, the design of the experiment is usually constrained by the number of cells required to set up the assay and the associated labor. Always use a range of seeding densities in multiwell plates (including lower cell densities that are likely to translate to seeding densities in flasks) and use the number of cells per square centimeter when extrapolating results from one vessel to another.

19.3.3 Parameters Derived from the Growth Curve

Lag time. The duration of the lag phase, or "lag time," is obtained by extrapolating a line drawn through the points for the log phase until it intersects the seeding concentration (see Figure 19.7) and then reading off the elapsed time since seeding equivalent to that intercept. The duration of the lag phase will vary with the cell line, the seeding density, and choice of substrate, but for most continuous cell lines, it is usually no more than 24 hours. A longer lag time may indicate that the seeding density is too low, resulting in a longer time for the cells to recover from subculture and commence active proliferation.

Population doubling time (PDT). PDT is obtained by drawing a best-fit line through the points for the log phase and using the line to determine the time taken for the cells to double in number (see Figure 19.7). PDTs vary from 12 to 15 hours in rapidly growing mouse leukemia cell lines (e.g. L1210), to 24–36 hours in many adherent continuous cell lines, and up to 60–72 hours in slow-growing finite cell lines. An increase in PDT may indicate that transformation has occurred, particularly if combined with an increase in saturation density (see Figure 19.10). A decrease in PDT may

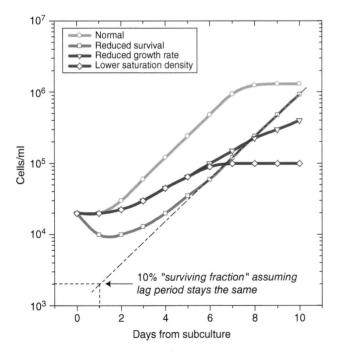

Fig. 19.9. Interpretation of growth curves. Changes in the shape of a growth curve can be interpreted in a number of different ways, but the labels in the key of this plot indicate what would normally be deduced from these curves.

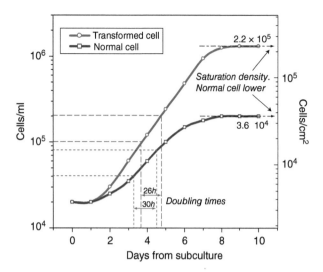

Fig. 19.10. Changes in growth curve dynamics. Transformation produces an increase in the saturation density of transformed cells, relative to that found in the equivalent normal cells. This increase is often accompanied by a shorter PDT. *Source*: R. Ian Freshney, based on data from normal human glia and glioma cell lines.

indicate suboptimal culture conditions or contamination; mycoplasma contamination commonly results in reduced cell proliferation, as evidenced by decreased rates of DNA synthesis (Nardone et al. 1965). Variations give an early warning that the conditions may have changed, requiring testing for a possible cause.

The PDT of a cell line should not be confused with its population doubling level (PDL) (see Sections 14.2.2, 14.6.4). The PDT is an average value that is obtained by studying growth during one subculture. It describes the net result of a wide range of division rates within the whole population, including a rate of zero for any cells that are not actively proliferating or that are undergoing cell death. By contrast, the PDL is a cumulative value that is adjusted at each subculture and aims to estimate the total number of cell generations that have occurred within the culture (see Figure 14.9).

Saturation density. Saturation density refers to the density of the cells (cells per square centimeter of growth surface) in the plateau phase (see Figure 19.7); a similar term, "plateau level," is used for the cell concentration (cells per milliliter of medium) in the same phase. Saturation density and plateau level are difficult to measure accurately, as a steady state is not easily achievable. Cells are likely to experience suboptimal conditions at the end of a growth curve experiment due to exhaustion of the medium. Ideally, the culture should be perfused to avoid nutrient limitation or growth factor depletion, but a reasonable compromise is to grow the cells on a restricted area (say a 15-mm-diameter coverslip or filter well) in a 90-mm Petri dish with 20 ml of medium that is replaced daily. Under these conditions, limitation of growth by the medium is minimal and the cell density exerts the major effect. A cell count under these conditions is a more accurate

and reproducible measurement of saturation density compared to a cell count in plateau under conventional conditions.

Although "saturation density" does not typically apply to suspension cultures, non-adherent cells can still enter plateau because of exhaustion of the medium. Frequently, however, suspension cells will enter apoptosis quite quickly, show a marked fall in cell concentration, and may not generate a stable plateau. With normal cells, a steady state may be achievable by not replenishing the growth factors in the medium; cells may be seeded and grown and the plateau reached without changing the medium. Clearly, the conditions used to attain the plateau phase must be carefully defined.

Cultures that have transformed spontaneously or have been transformed by virus or chemical carcinogens will usually reach a higher cell density in the plateau phase than their normal counterparts (see Figure 19.10) (Westermark 1973). This higher density is accompanied by a higher growth fraction and reduced density limitation of cell proliferation (see Figure 3.11). Transformed cells may also display other forms of aberrant growth control such as anchorage independence, i.e. they can easily be made to grow in suspension (see Section 3.6.1). However, transformed cells do not always form a stable plateau, particularly without regular medium replacement; many of the cells may enter apoptosis, resulting in a rapid decline in cell number.

19.4 CLONING EFFICIENCY

Cloning efficiency refers to the percentage of cells that form clones at low density (see Section 20.1). A number of related terms are used, which may vary between different research fields. "Colony-forming efficiency" refers to the percentage of cells that form colonies and is typically used in clonogenic assays, which aim to measure cloning efficiency. "Plating efficiency" is a broader term that encompasses these two terms and other concepts such as the efficiency of cell attachment (Schaeffer 1990). "Cloning efficiency" is used in preference to "plating efficiency" here because the term is more clearly defined, but for it to be strictly accurate, there must be some degree of confidence that "clones" originally came from single cells.

19.4.1 Clonogenic Assays

When cells are plated out as a single-cell suspension at low cell densities (2–50 cells/cm^2), the individual cells will give rise to discrete colonies (see Figure 20.1). Colony formation can be used as the endpoint of an assay to measure cloning efficiency; cells are diluted in dishes or plates, where adherent colonies can be counted, or are seeded into a gel or viscous solution for suspension cloning (see Sections 20.2, 20.3). Such assays are useful because they reveal differences in the growth rate within a population and distinguish between alterations in the growth rate (colony size) and cell survival (colony number). Protocol P19.4 describes how to perform a clonogenic assay; it can be adapted to test media and other reagents and perform cytotoxicity assays (see Supp. S11.7, S29.1).

PROTOCOL P19.4. CLONOGENIC ASSAY FOR ATTACHED CELLS

Outline

Perform serial dilutions of the cell line and plate the cells in Petri dishes at five different densities (see Figure 19.11). Incubate until colonies form; stain and count the colonies.

Materials (sterile or aseptically prepared)

- Adherent cell line for growth assessment
- Complete culture medium, e.g. Ham's F12
- Flasks, 25 cm^2 (3)
- Petri dishes, 60 mm (17)
- Centrifuge tubes or universal containers for dilution

Materials (non-sterile)

- BSC and associated consumables (see Protocol P12.1)
- Plastic box or trays to hold the dishes
- Cell counting equipment (hemocytometer or automated counter; see Section 19.1)
- Reagents for crystal violet staining (see Protocol P18.4)

Procedure

1. Grow the cells in a 25-cm^2 flask until close to confluence (late log phase).
2. Perform subculture using the standard procedure for that cell line (e.g. trypsinization; see Protocol P14.2).
3. Before or during subculture, prepare the dishes and tubes for the assay:
   (a) Label the dishes on the side of the base (to ensure the label is retained during counting).
   (b) Label the tubes that will be used for dilution of the cells (see Figure 19.11).
   (c) Measure out medium for the dilution steps into the labeled tubes (see Figure 19.11). There should be more than enough medium for three replicates at each dilution.
4. Resuspend in culture medium to give a single-cell suspension. Count the cells and calculate the cell concentration (see Protocol P19.1). Readjust the cell concentration to give a starting concentration of 1×10^6 cells/ml.
5. Perform serial dilutions (see Figure 19.11) to give a series of concentrations and then seed at various densities as follows:
   (a) 2×10^4 cells/ml: seed into two 25-cm^2 flasks for routine maintenance.

   (b) 2000 cells/ml: seed 5 ml into two dishes, giving 10 000 cells/dish. These two dishes will act as controls in case the cloning is unsuccessful (to prove that there were cells present in the top dilution, at least).
   (c) 200 cells/ml: seed 5 ml into three dishes to give 1000 cells/dish.
   (d) 100 cells/ml: seed 5 ml into three dishes to give 500 cells/dish.
   (e) 50 cells/ml: seed 5 ml into three dishes to give 250 cells/dish.
   (f) 20 cells/ml: seed 5 ml into three dishes to give 100 cells/dish.
   (g) 10 cells/ml: seed 5 ml into three dishes to give 50 cells/dish.
6. Put the Petri dishes in a transparent plastic box or on a tray and place in a humid CO_2 incubator. Incubate the dishes until colonies are visible to the naked eye (one to three weeks).
7. Once the colonies are visible, stain the colonies with crystal violet (see Protocol P18.4).
8. Count the colonies in each dish, excluding those below 50 cells per colony.
9. Calculate the mean of the three counts at each seeding density.
10. Calculate the colony-forming efficiency for each seeding density as follows:

Colony forming efficiency

$$= \frac{number\ of\ colonies\ formed}{number\ of\ cells\ seeded} \times 100$$

Notes

Step 7: Once stained, dishes can be left at room temperature for a prolonged period of time (up to at least 50 weeks) before they are counted (Franken et al. 2006).

Step 8: It is necessary to define a threshold above which colonies will be counted. If the majority of the colonies are between a hundred and a few thousands, set the threshold at 50 cells per colony. In practice, this is a fairly natural threshold when counting by eye. However, if the colonies are very small (< 100 cells), set the threshold at 16 cells per colony. Below 16 cells, equivalent to four cell consecutive divisions, it would be hard to presume continued cell proliferation.

Step 10: If it can be confirmed that each colony grew from a single cell, this becomes the cloning efficiency (see Section 19.4).

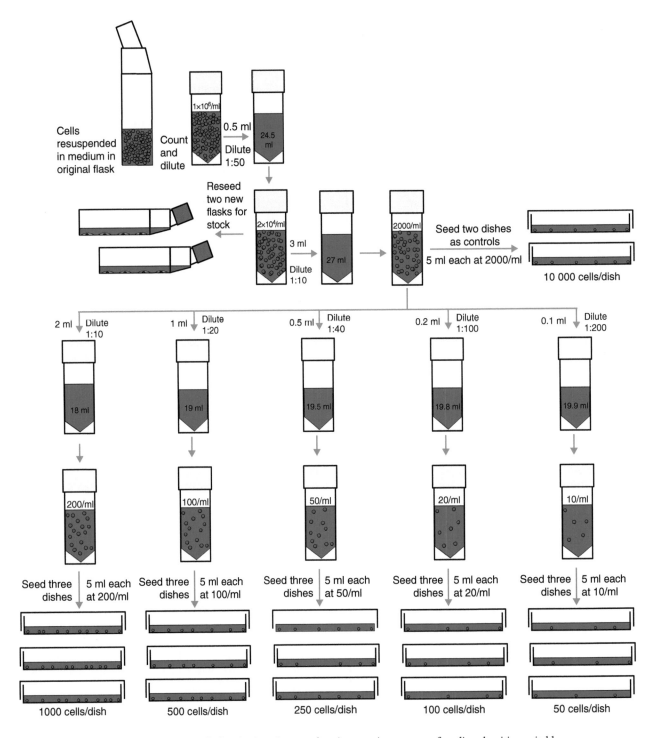

Fig. 19.11. Diluting cells for cloning. Suggested regime to give a range of seeding densities, suitable for cloning a cell line for the first time or establishing the linearity of cloning efficiency versus seeding concentration (see Figure 19.13). Once a suitable concentration has been selected, the cells may be diluted, with fewer steps, to the desired concentration, e.g. when testing culture conditions (see Figure 11.13).

19.4.2 Colony Counting

Colony counting can be performed using (i) a microscope to visualize and manually count the colonies; (ii) an image taken under the microscope, which is then analyzed to detect and count the colonies; or (iii) an automated colony counter that has been designed specifically for this purpose. The choice of approach will depend on the extent to which the assay is used, the degree of standardization required, and the availability of equipment. Colony morphology may also be a factor; colonies that overlap more than ~ 20% or have irregular outlines can be challenging to distinguish using automated approaches. However, automated methods can be used very successfully to reduce the variability that occurs with manual colony counting (Lumley et al. 1997).

Software applications and add-ons that can be used for colony recognition include OpenCFU, several ImageJ macros (e.g. "Cell Colony Edge" or "ColonyArea"), and a CellProfiler Pipeline ("Cell Colony Counting") (Geissmann 2013; Guzman et al. 2014; Choudhry 2016). These methods vary in their approaches and their overall performance (Choudhry 2016); always test against your own cells before deciding if they are suitable. Automated colony counters usually come with validation data (although it is still a good idea to test your own assay format) and may be able to count 3D structures, e.g. colonies in soft agar or Methocel and tumor spheroids (see Sections 20.3, 27.4.1). For example, the GelCount™ (Oxford Optronix; see Figure 19.12) can be used to measure diameter and optical density, which allows calculation of area and volume, in soft agar-based cytotoxicity assays and other clonogenic or 3D assays (Kajiwara et al. 2008).

19.4.3 Analysis of Colony Formation

Cells may grow differently as isolated colonies at different densities, resulting in changes in cloning efficiency. This change is particularly important for cytotoxicity assays, where cells may be plated at different densities in order to achieve similar counts after treatment with cytotoxic agents (Pomp et al. 1996). Cloning efficiency can also be influenced by other variables, including the oxygen and CO_2 concentration, the selection of medium and serum, and changes in substrate and subculture routines. Always start by standardizing culture conditions and use a range of seeding densities that are likely to be employed in your assay so that you can assess the linearity of cloning efficiency (see Figure 19.13). A decrease in cloning efficiency at low density may indicate that the cells do not survive at very low densities; this can sometimes be minimized using a feeder layer (see Section 8.4). If the cloning efficiency falls at higher concentrations, it implies that the cells are aggregating or colonies are coalescing.

Heterogeneity in clonal growth rates reflects differences in the capacity for cell proliferation between lineages within a population. These differences are not necessarily expressed in an interacting monolayer at higher densities, when cell communication becomes possible. Thus, cloning efficiency at low

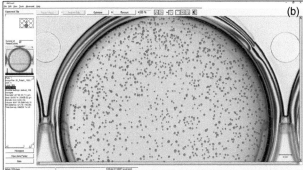

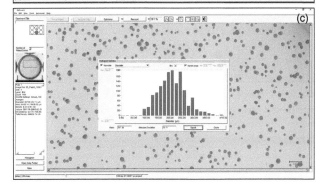

Fig. 19.12. GelCount Colony Counter (Oxford Optronix). (a) Counter with sample tray open, showing a range of sample formats that can be used; (b) screen image with colonies flagged; (c) histogram inset showing size distribution of selected colonies. *Source*: Courtesy of Oxford Optronix, with permission.

density does not equate to the recovery of adherent cells after seeding at higher densities. Survival at higher densities is more properly referred to as the seeding efficiency, which can be assessed by counting the number of cells that attach to the culture substrate before mitosis commences (e.g. at the end of the lag phase). This time is a difficult point to define, as the window between maximum cell attachment and the initiation of mitosis may be quite narrow, and the events may even overlap. The seeding efficiency is calculated by dividing the number of cells that have attached by the total number of cells that were seeded in the culture vessel, expressed as a percentage.

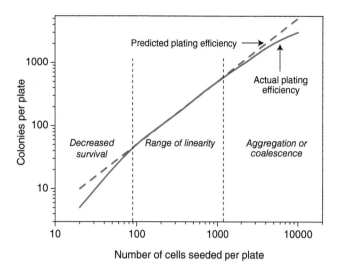

Fig. 19.13. Linearity of cloning efficiency. If plating efficiency remains constant over a wide range of cell concentrations, the curve is linear (dashed line), whereas if there is poor survival at low densities or aggregation or coalescence at high densities, plating efficiency decreases (solid line).

19.5 DNA SYNTHESIS

Measurements of DNA synthesis are considered representative of the amount of cell proliferation within a culture. Traditionally, DNA synthesis was assessed by exposing the cells to a radiolabeled DNA precursor such as [3H]thymidine, which is incorporated into replicated DNA strands. Exposure may be for short periods (0.5–1 hour) for rate estimations, or for longer periods (24 hours or more) to measure accumulated DNA synthesis when the basal rate is low (e.g. in plateau phase). [3H]thymidine should not be used for incubations longer than 24 hours or at high specific activities because radiolysis of DNA could occur, due to the short path length of β-emission ($\sim 1\,\mu m$) from decaying tritium. [3H]thymidine incorporation assays are commonly performed in multiwell plates; after incubation with the radioisotope, the cells are washed, precipitated using trichloroacetic acid, lyzed, and counted using a liquid scintillation counter or a suitable microplate reader. Methods for thymidine incorporation assays are described elsewhere (Griffiths and Sundaram 2011; PerkinElmer 2019). [3H]thymidine can also be visualized using microautoradiography (see Figure 19.14); a protocol for this technique is available in the Supplementary Material online (see Supp. S19.3).

Many research laboratories now minimize their use of ionizing radiation (see Supp. S6.2), leading to substitution with non-radioactive DNA precursors. The thymidine analog 5'-bromo-2'-deoxyuridine (BrdU) is widely used to assess DNA proliferation. BrdU uptake can be assessed using anti-BrdU antibodies, allowing visualization by immunofluorescence or measurement by enzyme-linked immunosorbent assay (ELISA). BrdU ELISA kits are available from various suppliers (e.g. Abcam, Roche CustomBiotech). Direct comparison of BrdU and [3H]thymidine assays showed that these methods were generally in agreement (see Figure 19.15) (Khan et al. 1991). The labeling index (percentage of cells labeled with [3H]thymidine or BrdU) was consistently low in mink lung cells (Mv1Lu) but increased in oncogene-transfected derivatives.

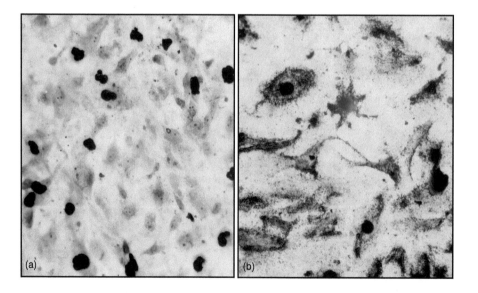

Fig. 19.14. Microautoradiographs. Normal glial cells were incubated with 0.1 μCi/ml (3.7 KBq/ml), 200 Ci/mmol (7.4 GBq/μmol), of [3H]thymidine for 24 hours, washed, and processed as described elsewhere (see Supp. S19.3). (a) Typical densely labeled nuclei, suitable for determining the labeling index (see Figure 19.15; see also Figure 3.9d). (b) A similar culture that is infected with mycoplasma, showing [3H]thymidine incorporation in the cytoplasm. *Source*: R. Ian Freshney.

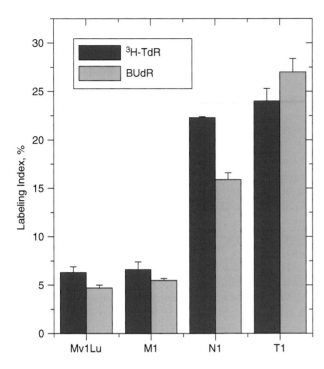

Fig. 19.15. Labeling index. Mink lung cell line Mv1Lu and oncogene-transfected derivatives were labeled with [^{3}H]thymidine for one hour before performing microautoradiography (see Supp. S19.3); replicate samples were labeled for one hour with BrdU, fixed, and stained by immunoperoxidase using an antibody directed against BrdU. Labeled nuclei were counted as percentage of the total in each case. Mv1Lu is the control cell line; M1 is Mv1Lu transfected with the *myc* oncogene; N1 is Mv1Lu cells transfected with normal human *ras* and T1 with mutant human *ras*. T1 cells were tumorigenic and had a statistically significant (p < 0.001) increase in labeling index compared to Mv1Lu using both methods. *Source*: Based on Khan, M. Z., Spandidos, D. A., Kerr, D. J., et al. (1991). Oncogene transfection of mink lung cells: effect on growth characteristics in vitro and in vivo. Anticancer Res 11 (3):1343–8.

19.6 CELL CYCLE ANALYSIS

A number of techniques have been developed in order to follow the progression of cells through the cell cycle. Techniques can be broadly divided into (i) fractionation or separation of actively dividing cells; (ii) blockade of mitosis, which leads to synchronization of the culture once the blockade is released; and (iii) single-cell analysis based on cell cycle-specific markers. A combination of approaches may be used; for example, cells can be labeled with a "pulse" of BrdU (with or without synchronization), stained with anti-BrdU antibody and propidium iodide (for total DNA), and analyzed using flow cytometry or confocal analysis (Cecchini et al. 2012). A number of proteins are upregulated during cell division, such as proliferating cell nuclear antigen (PCNA) and Ki-67, which can be used to identify dividing cells and potentially to discriminate between phases of the cell cycle (Schonenberger et al. 2015; Sun and Kaufman

2018). Cells can also be genetically modified to carry cell cycle-specific markers; for example, the Fucci(CA) fluorescent sensor gives a three-color readout that distinguishes between G_1, S, and G_2 during live-cell imaging (see Figure 18.2d) (Sakaue-Sawano et al. 2017).

Fractionation or separation of cells relies on changes in their physical properties during cell division. One of the simplest techniques for separating synchronized cells is mitotic shake-off. Monolayer cells tend to round up at metaphase and may detach when the flask in which they are growing is shaken. This method works well with some adherent cell lines such as HeLa and CHO (Terasima and Tolmach 1963; Tobey et al. 1967). Placing the cells at 4 °C for 30–60 minutes can also be used to synchronize cells in cycle and enhances the yield at mitotic shake-off (Miller et al. 1972; Rieder and Cole 2002). However, fractionation and cell cycle synchronization can be challenging, particularly to the newcomer to this area. Some excellent resources on physical fractionation and cell cycle synchronization are available elsewhere (Banfalvi 2017).

Suppliers

Supplier	URL
Abcam	www.abcam.com/
Bal Supply	balsupply.com/
BD Biosciences	http://www.bdbiosciences.com
Beckman Coulter Life Sciences	http://www.beckman.com
Bio-Rad Laboratories, Inc.	http://www.bio-rad.com
ChemoMetec	http://chemometec.com
Corning	http://www.corning.com/worldwide/en/products/life-sciences/products/cell-culture.html
DeNovix	http://www.denovix.com
Logos Biosystems	http://logosbio.com
Luminex	http://www.luminexcorp.com
NanoEnTek	http://www.nanoentek.com/?lang=en
Nexcelom Bioscience	http://www.nexcelom.com
OMNI Life Science (OLS)	cellcounting.de/?lang=en
Orflo	http://www.orflo.com/v/vspfiles/home.html
Oxford Optronix	http://www.oxford-optronix.com
Roche CustomBiotech	http://custombiotech.roche.com/home.html
Sartorius	http://www.sartorius.com/en
Sigma-Aldrich (Merck)	http://www.sigmaaldrich.com/life-science/cell-culture.html
Sysmex Partec	http://www.sysmex-partec.com
ThermoFisher Scientific	http://www.thermofisher.com/us/en/home/life-science/cell-culture.html
TPP Techno Plastic Products	http://www.tpp.ch

Supp. S19.1 Estimation of Viability by Dye Uptake.

Supp. S19.2 Generating a Growth Curve Using Flasks.

Supp. S19.3 Microautoradiography of Cultured Cells.

REFERENCES

Banfalvi, G. (2017). *Cell Cycle Synchronization*. New York: Humana Press.

Baumgarten, H. (1985). A simple microplate assay for the determination of cellular protein. *J. Immunol. Methods* 82 (1): 25–37. https://doi.org/10.1016/0022-1759(85)90221-2.

Borenfreund, E. and Puerner, J.A. (1985). A simple quantitative procedure using monolayer cultures for cytotoxicity assays (HTD/NR-90). *J. Tissue Cult. Methods* 9 (1): 7–9. https://doi.org/10.1007/bf01666038.

Brunk, C.F., Jones, K.C., and James, T.W. (1979). Assay for nanogram quantities of DNA in cellular homogenates. *Anal. Biochem.* 92 (2): 497–500. https://doi.org/10.1016/0003-2697(79)90690-0.

Cadena-Herrera, D., Esparza-De Lara, J.E., Ramirez-Ibanez, N.D. et al. (2015). Validation of three viable-cell counting methods: manual, semi-automated, and automated. *Biotechnol. Rep. (Amst.)* 7: 9–16. https://doi.org/10.1016/j.btre.2015.04.004.

Cecchini, M.J., Amiri, M., and Dick, F.A. (2012). Analysis of cell cycle position in mammalian cells. *J. Vis. Exp.* 59: 3491. https://doi.org/10.3791/3491.

Choudhry, P. (2016). High-throughput method for automated colony and cell counting by digital image analysis based on edge detection. *PLoS One* 11 (2): e0148469. https://doi.org/10.1371/journal.pone.0148469.

Coulter, W. H. (1949). Patent US2656508A: means for counting particles suspended in a fluid. https://patents.google.com/patent/us2656508a/en.

Crowley, L.C., Marfell, B.J., Scott, A.P. et al. (2016). Dead cert: measuring cell death. *Cold Spring Harb. Protoc.* 2016 (12) https://doi.org/10.1101/pdb.top070318.

Cunningham, B. and Kirk, P.L. (1942). Measure of "growth" in tissue culture. *J. Cell. Comp. Physiol.* 20 (3): 343–358. https://doi.org/10.1002/jcp.1030200309.

Darzynkiewicz, Z., Li, X., and Gong, J. (1994). Assays of cell viability: discrimination of cells dying by apoptosis. *Methods Cell Biol.* 41: 15–38. https://doi.org/10.1016/s0091-679x(08)61707-0.

van Dooren, B.T., Beekhuis, W.H., and Pels, E. (2004). Biocompatibility of trypan blue with human corneal cells. *Arch. Ophthalmol.* 122 (5): 736–742. https://doi.org/10.1001/archopht.122.5.736.

Dulbecco, R. and Elkington, J. (1973). Conditions limiting multiplication of fibroblastic and epithelial cells in dense cultures. *Nature* 246 (5430): 197–199. https://doi.org/10.1038/246197a0.

Earle, W.R., Schilling, E.L., Bryant, J.C. et al. (1954). The growth of pure strain L cells in fluid-suspension cultures. *J. Natl Cancer Inst.* 14 (5): 1159–1171.

Essen Bioscience (2019). IncuCyte cell count proliferation assay. https://www.essenbioscience.com/media/uploads/files/8000-0396-e00_cell_transduction_protocol_uqzyd9n.pdf (accessed 25 September 2019).

Folkman, J. and Moscona, A. (1978). Role of cell shape in growth control. *Nature* 273 (5661): 345–349. https://doi.org/10.1038/273345a0.

Franken, N.A., Rodermond, H.M., Stap, J. et al. (2006). Clonogenic assay of cells in vitro. *Nat. Protoc.* 1 (5): 2315–2319. https://doi.org/10.1038/nprot.2006.339.

Geissmann, Q. (2013). OpenCFU, a new free and open-source software to count cell colonies and other circular objects. *PLoS One* 8 (2): e54072. https://doi.org/10.1371/journal.pone.0054072.

Griffiths, M. and Sundaram, H. (2011). Drug design and testing: profiling of antiproliferative agents for cancer therapy using a cell-based methyl-[3H]-thymidine incorporation assay. In: *Cancer Cell Culture: Methods and Protocols* (ed. I.A. Cree), 451–465. Totowa, NJ: Humana Press.

Grishagin, I.V. (2015). Automatic cell counting with ImageJ. *Anal. Biochem.* 473: 63–65. https://doi.org/10.1016/j.ab.2014.12.007.

Guzman, C., Bagga, M., Kaur, A. et al. (2014). ColonyArea: an ImageJ plugin to automatically quantify colony formation in clonogenic assays. *PLoS One* 9 (3): e92444. https://doi.org/10.1371/journal.pone.0092444.

Hathaway, W.E., Newby, L.A., and Githens, J.H. (1964). The acridine orange viability test applied to bone marrow cells. I. Correlation with trypan blue and eosin dye exclusion and tissue culture transformation. *Blood* 23: 517–525.

Holley, R.W., Armour, R., and Baldwin, J.H. (1978). Density-dependent regulation of growth of BSC-1 cells in cell culture: growth inhibitors formed by the cells. *Proc. Natl Acad. Sci. U.S.A.* 75 (4): 1864–1866. https://doi.org/10.1073/pnas.75.4.1864.

Hsiung, F., McCollum, T., Hefner, E., et al. (2013). Comparison of count reproducibility, accuracy, and time to results between a hemocytometer and the TC20 automated cell counter. https://www.bio-rad.com/webroot/web/pdf/lsr/literature/bulletin_6003.pdf (accessed 24 September 2019).

Kajiwara, Y., Panchabhai, S., and Levin, V.A. (2008). A new preclinical 3-dimensional agarose colony formation assay. *Technol. Cancer Res. Treat.* 7 (4): 329–334. https://doi.org/10.1177/153303460800700407.

Kaltenbach, J.P., Kaltenbach, M.H., and Lyons, W.B. (1958). Nigrosin as a dye for differentiating live and dead ascites cells. *Exp. Cell Res.* 15 (1): 112–117. https://doi.org/10.1016/0014-4827(58)90067-3.

Keepers, Y.P., Pizao, P.E., Peters, G.J. et al. (1991). Comparison of the sulforhodamine B protein and tetrazolium (MTT) assays for in vitro chemosensitivity testing. *Eur. J. Cancer* 27 (7): 897–900. https://doi.org/10.1016/0277-5379(91)90142-z.

Khan, M.Z., Spandidos, D.A., Kerr, D.J. et al. (1991). Oncogene transfection of mink lung cells: effect on growth characteristics in vitro and in vivo. *Anticancer Res.* 11 (3): 1343–1348.

Kim, S.I., Kim, H.J., Lee, H.J. et al. (2016). Application of a non-hazardous vital dye for cell counting with automated cell counters. *Anal. Biochem.* 492: 8–12. https://doi.org/10.1016/j.ab.2015.09.010.

Krause, A.W., Carley, W.W., and Webb, W.W. (1984). Fluorescent erythrosin B is preferable to trypan blue as a vital exclusion dye for mammalian cells in monolayer culture. *J. Histochem. Cytochem.* 32 (10): 1084–1090. https://doi.org/10.1177/32.10.6090533.

Kruse, P.F. Jr., Keen, L.N., and Whittle, W.L. (1970). Some distinctive characteristics of high density perfusion cultures of diverse cell types. *In Vitro* 6 (1): 75–88. https://doi.org/10.1007/bf02616136.

Labarca, C. and Paigen, K. (1980). A simple, rapid, and sensitive DNA assay procedure. *Anal. Biochem.* 102 (2): 344–352. https://doi.org/10.1016/0003-2697(80)90165-7.

Lumley, M.A., Burgess, R., Billingham, L.J. et al. (1997). Colony counting is a major source of variation in CFU-GM results between centres. *Br. J. Haematol.* 97 (2): 481–484. https://doi.org/10.1046/j.1365-2141.1997.492695.x.

Mascotti, K., McCullough, J., and Burger, S.R. (2000). HPC viability measurement: trypan blue versus acridine orange and propidium iodide. *Transfusion* 40 (6): 693–696. https://doi.org/10.1046/j.1537-2995.2000.40060693.x.

McKeehan, W.L., McKeehan, K.A., Hammond, S.L. et al. (1977). Improved medium for clonal growth of human diploid fibroblasts at low concentrations of serum protein. *In Vitro* 13 (7): 399–416. https://doi.org/10.1007/bf02615100.

Miller, G.G., Walker, G.W., and Giblak, R.E. (1972). A rapid method to determine the mammalian cell cycle. *Exp. Cell Res.* 72 (2): 533–537. https://doi.org/10.1016/0014-4827(72)90023-7.

Mooney, D.J., Langer, R., and Ingber, D.E. (1995). Cytoskeletal filament assembly and the control of cell spreading and function by extracellular matrix. *J. Cell Sci.* 108 (Pt 6): 2311–2320.

Nardone, R.M., Todd, J., Gonzalez, P. et al. (1965). Nucleoside incorporation into strain L cells: inhibition by pleuropneumonia-like organisms. *Science* 149 (3688): 1100–1101. https://doi.org/10.1126/science.149.3688.1100.

Niepel, M., Hafner, M., Chung, M. et al. (2017). Measuring cancer drug sensitivity and resistance in cultured cells. *Curr. Protoc. Chem. Biol.* 9 (2): 55–74. https://doi.org/10.1002/cpch.21.

Niepel, M., Hafner, M., Mills, C.E. et al. (2019). A multi-center study on the reproducibility of drug-response assays in mammalian cell lines. *Cell Syst.* 9 (1): 35–48, e5. doi: https://doi.org/10.1016/j.cels.2019.06.005.

Ongena, K., Das, C., Smith, J.L. et al. (2010). Determining cell number during cell culture using the Scepter cell counter. *J. Vis. Exp.* 45: 2204. https://doi.org/10.3791/2204.

Orellana, E.A. and Kasinski, A.L. (2016). Sulforhodamine B (SRB) assay in cell culture to investigate cell proliferation. *Bio. Protoc.* 6 (21) https://doi.org/10.21769/BioProtoc.1984.

PerkinElmer (2019). Thymidine incorporation assays. https://www.perkinelmer.com/lab-products-and-services/application-support-knowledgebase/radiometric/thymidine-incorporation-assays.html (accessed 3 October 2019).

Pomp, J., Wike, J.L., Ouwerkerk, I.J. et al. (1996). Cell density dependent plating efficiency affects outcome and interpretation of colony forming assays. *Radiother. Oncol.* 40 (2): 121–125.

Reid, Y.A. (2017). Best practices for naming, receiving, and managing cells in culture. *In Vitro Cell. Dev. Biol. Anim.* 53 (9): 761–774. https://doi.org/10.1007/s11626-017-0199-1.

Repetto, G. and Sanz, P. (1993). Neutral red uptake, cellular growth and lysosomal function: in vitro effects of 24 metals. *Altern. Lab. Anim* 21: 501–507.

Repetto, G., del Peso, A., and Zurita, J.L. (2008). Neutral red uptake assay for the estimation of cell viability/cytotoxicity. *Nat. Protoc.* 3 (7): 1125–1131. https://doi.org/10.1038/nprot.2008.75.

Ribatti, D. (2017). A revisited concept: contact inhibition of growth. From cell biology to malignancy. *Exp. Cell Res.* 359 (1): 17–19. https://doi.org/10.1016/j.yexcr.2017.06.012.

Rieder, C.L. and Cole, R.W. (2002). Cold-shock and the mammalian cell cycle. *Cell Cycle* 1 (3): 169–175.

Rotman, B. and Papermaster, B.W. (1966). Membrane properties of living mammalian cells as studied by enzymatic hydrolysis of fluorogenic esters. *Proc. Natl Acad. Sci. U.S.A.* 55 (1): 134–141. https://doi.org/10.1073/pnas.55.1.134.

Sakaue-Sawano, A., Yo, M., Komatsu, N. et al. (2017). Genetically encoded tools for optical dissection of the mammalian cell cycle. *Mol. Cell* 68 (3): 626–640, e5. doi: https://doi.org/10.1016/j.molcel.2017.10.001.

Sanford, K.K., Earle, W.R., Evans, V.J. et al. (1951). The measurement of proliferation in tissue cultures by enumeration of cell nuclei. *J. Natl Cancer Inst.* 11 (4): 773–795.

Schaeffer, W.I. (1990). Terminology associated with cell, tissue, and organ culture, molecular biology, and molecular genetics. Tissue Culture Association Terminology Committee. *In Vitro Cell. Dev. Biol.* 26 (1): 97–101.

Schonenberger, F., Deutzmann, A., Ferrando-May, E. et al. (2015). Discrimination of cell cycle phases in PCNA-immunolabeled cells. *BMC Bioinform.* 16: 180. https://doi.org/10.1186/s12859-015-0618-9.

Schrek, R. (1958). Slide-chamber method to measure sensitivity of cells to toxic agents; application of the method to normal and leukemic human lymphocytes. *A.M.A. Arch. Pathol.* 66 (4): 569–576.

Schulz, J., Dettlaff, S., Fritzsche, U. et al. (1994). The amido black assay: a simple and quantitative multipurpose test of adhesion, proliferation, and cytotoxicity in microplate cultures of keratinocytes (HaCaT) and other cell types growing adherently or in suspension. *J. Immunol. Methods* 167 (1–2): 1–13. https://doi.org/10.1016/0022-1759(94)90069-8.

Skehan, P., Storeng, R., Scudiero, D. et al. (1990). New colorimetric cytotoxicity assay for anticancer-drug screening. *J. Natl Cancer Inst.* 82 (13): 1107–1112. https://doi.org/10.1093/jnci/82.13.1107.

Stewart, J.W. and Crosland-Taylor, P.J. (1959). Cell counts. *Postgrad. Med. J.* 35 (407): 502–513. https://doi.org/10.1136/pgmj.35.407.502.

Stoker, M.G. (1973). Role of diffusion boundary layer in contact inhibition of growth. *Nature* 246 (5430): 200–203. https://doi.org/10.1038/246200a0.

Stoker, M.G. and Rubin, H. (1967). Density dependent inhibition of cell growth in culture. *Nature* 215 (5097): 171–172. https://doi.org/10.1038/215171a0.

Sun, X. and Kaufman, P.D. (2018). Ki-67: more than a proliferation marker. *Chromosoma* 127 (2): 175–186. https://doi.org/10.1007/s00412-018-0659-8.

Terasima, T. and Tolmach, L.J. (1963). Growth and nucleic acid synthesis in synchronously dividing populations of HeLa cells. *Exp. Cell Res.* 30: 344–362. https://doi.org/10.1016/0014-4827(63)90306-9.

Tobey, R.A., Anderson, E.C., and Petersen, D.F. (1967). Properties of mitotic cells prepared by mechanically shaking monolayer cultures of Chinese hamster cells. *J. Cell. Physiol.* 70 (1): 63–68. https://doi.org/10.1002/jcp.1040700109.

van Tonder, A., Joubert, A.M., and Cromarty, A.D. (2015). Limitations of the 3-(4,5-dimethylthiazol-2-yl)-2,5-diphenyl-2H-tetrazolium bromide (MTT) assay when compared to three commonly used cell enumeration assays. *BMC. Res. Notes* 8: 47. https://doi.org/10.1186/s13104-015-1000-8.

Vembadi, A., Menachery, A., and Qasaimeh, M.A. (2019). Cell cytometry: review and perspective on biotechnological advances. *Front. Bioeng. Biotechnol.* 7: 147. https://doi.org/10.3389/fbioe.2019.00147.

Verso, M.L. (1964). The evolution of blood-counting techniques. *Med. Hist.* 8: 149–158. https://doi.org/10.1017/s0025727300029392.

Wan, C.P., Sigh, R.V., and Lau, B.H. (1994). A simple fluorometric assay for the determination of cell numbers. *J. Immunol. Methods* 173 (2): 265–272. https://doi.org/10.1016/0022-1759(94)90305-0.

Westermark, B. (1973). The deficient density-dependent growth control of human malignant glioma cells and virus-transformed glia-like cells in culture. *Int. J. Cancer* 12 (2): 438–451.

PART VI

Physical and Genetic Manipulation

After reading the following chapters in this part of the book, you will be able to:

(20) **Cell Cloning and Selection:**

(a) Perform dilution cloning of adherent cells in dishes or microwell plates.

(b) Perform suspension cloning of anchorage-independent cells using soft agar or methylcellulose (Methocel®).

(c) Select colonies for further growth following a successful cloning procedure.

(d) Discuss how to stimulate cloning efficiency if colonies fail to grow.

(e) Explain how selective media, substrates, or techniques can be used to select for a specific cell type and discourage overgrowth by other cells, e.g. fibroblasts.

(21) **Cell Separation and Sorting:**

(a) Discuss how physical properties such as cell density and cell size may be used to achieve cell separation.

(b) Perform cell separation by centrifugation through a density gradient and describe how to optimize a density gradient for a specific sample.

(c) Perform magnetic separation using magnetic beads.

(d) Explain the basic principles of fluorescence-activated cell sorting (FACS) using a flow cytometer.

(e) Define "microfluidics" and provide examples of its advantages and applications for cell culture and for cell-based assays.

(22) **Genetic Modification and Immortalization:**

(a) Provide examples of physical, chemical, and biological methods of gene delivery, and discuss the advantages and disadvantages of each system.

(b) Explain the basic principles of three widely used gene editing technologies, with a particular emphasis on clustered regularly interspaced short palindromic repeat (CRISPR) systems and conserved CRISPR-associated protein (Cas).

(c) Transfect a culture with a DNA vector using calcium phosphate coprecipitation.

(d) Transfect a culture with CRISPR/Cas9 ribonucleoprotein (RNP) using electroporation.

(e) Transfect actively dividing cells with human telomerase reverse transcriptase (hTERT) for immortalization using lipofection and discuss other methods that can be used for immortalization.

(f) Discuss how to select for the presence of the correct genetic modification and list examples of artifacts that may arise associated with that modification.

Freshney's Culture of Animal Cells: A Manual of Basic Technique and Specialized Applications, Eighth Edition. Amanda Capes-Davis and R. Ian Freshney.
© 2021 John Wiley & Sons Ltd. Published 2021 by John Wiley & Sons Ltd.
Companion website: www.wiley.com/go/freshney/cellculture8

CHAPTER 20

Cell Cloning and Selection

One of the major technical challenges of cell culture comes from the need to grow a single, isolated cell into a clone or pure culture (Sanford et al. 1948). Cultures originate from a mixed population of cells in a tissue sample, and their properties may change over time as the cells respond to the culture environment. There is an obvious need to select a specific cell type, expand it as a clonal population, and preserve its particular properties for further study. However, early efforts to do so were unsuccessful (Moen 1935). Wilton Earle, Katherine Sanford, and colleagues at the National Cancer Institute (NCI) speculated that the difficulty was caused by inadequate culture conditions, resulting in the need for cells to modify their environment before they could proliferate. Sanford established the first cloned cultures by growing single cells in glass capillary tubes, which reduced the volume of medium that cells were required to modify (see Figure 1.1) (Sanford et al. 1948). Capillaries were coated with a thin lining of plasma clot and the culture medium was "conditioned" by exposure to other, actively growing cultures. Although this work was performed many years ago, the basic principles still apply to cell cloning and selection. Many cell types can now be grown at low density and their clonal populations selected and expanded for further characterization. However, their conditions must be suitably optimized for growth at low density if individual cells are to survive and proliferate. This chapter focuses on basic techniques that can be performed in any tissue culture laboratory to provide a suitable environment for cell cloning and selection.

20.1 TERMINOLOGY: CLONING AND SELECTION

For this chapter, "clone" refers to a population of cells that is derived from a single cell by mitosis (Schaeffer 1990). "Cloning" refers to the process used to generate a clonal population, usually through limiting dilution or growth in suspension. The cloned cell population may be described as a "cell strain" (see Section 14.1) or a clonal derivative of the parental cell line. Cloning may also be used to assess the ability of cells to survive at low density; this is typically done by studying cloning efficiency, which is defined as the percentage of cells that form clones at low cell density (Schaeffer 1990). Such assays are referred to in this book as clonogenic assays (see Section 19.4.1), although various terms are used in the scientific literature, including colony-forming unit (CFU) or colony-forming cell (CFC) assays. It is important to distinguish between generation of a clonal derivative and the use of the technique in a clonogenic assay, because these approaches have different requirements, e.g. clonogenic assays do not require selection (Ham 1974).

In this book, "selection" refers to the process used to isolate a clonal population, including any associated procedures to discourage growth of unwanted cell populations. The term "selection" may be used elsewhere to refer to isolation or separation of cells that possess a particular characteristic or marker, e.g. using density gradients or antibody-based techniques. Separation-based techniques are discussed separately in the next chapter (see Chapter 21). "Cloning" and "selection" may also refer to isolation of cells that have been

Freshney's Culture of Animal Cells: A Manual of Basic Technique and Specialized Applications, Eighth Edition. Amanda Capes-Davis and R. Ian Freshney.
© 2021 John Wiley & Sons Ltd. Published 2021 by John Wiley & Sons Ltd.
Companion website: www.wiley.com/go/freshney/cellculture8

genetically modified to carry specific mutations or transcripts. Gene editing procedures are discussed in a later chapter (see Section 22.2).

20.2 CLONING BY LIMITING DILUTION

Limiting dilution was initially introduced as a rapid and reliable method for cloning cells, using a similar approach to bacterial cloning (Puck and Marcus 1955; Marcus et al. 2006). Cultures were serially diluted to a point where limiting numbers of cells were present (e.g. one to five cells) and left undisturbed for at least a week to see if discrete colonies would form (see Figures 20.1, 20.2) (Paul 1975). Cloning efficiency was originally quite poor, leading to the introduction of feeder cells to improve it (Marcus et al. 2006). Later improvements in medium design resulted in higher cloning efficiencies for many continuous cell lines (e.g. HeLa). Dilution cloning of these cell lines was then used as an assay to measure cloning efficiency and clonal growth (Ham 1974).

Primary cultures and finite cell lines tend to have poor cloning efficiencies, which reduces the effectiveness of limiting dilution as a technique. However, cloning of primary cultures and finite cell lines can succeed with some cell types. Cloning has been performed successfully on renal juxtaglomerular and glomerular cells and on testicular Sertoli cells (Muirhead et al. 1990; Troyer and Kreisberg 1990; Zwain et al. 1991). Other cell types that have been cloned successfully include satellite cells from skeletal muscle, oval cells from liver, and basal cells from immortalized airway epithelium (Zeng et al. 2002; Suh et al. 2003; Walters et al. 2013). Even if successful, the limited lifespan of primary cultures or finite cell lines is likely to become a major obstacle to further culture (see Section 14.2.2). By the time the clone has produced a usable number of cells, it may already be near to senescence (see Figure 20.3). In most cases, for cloning to be a viable option, a finite cell line will need to be immortalized (Walters et al. 2013).

Limiting dilution is used in a clonogenic assay to generate clonal populations (colonies), which are counted to determine the cloning efficiency. This approach is useful when performing toxicity testing; for example, it can be used to demonstrate delayed effects after exposure to cytotoxic drugs or ionizing radiation (see Section 29.2.4). Dilution cloning is also used in a number of assays that measure stem cell capacity, proliferation, and survival (see Sections 19.4, 23.6). Formation of a colony from a single cell is proof of its ability to undergo self-renewal, resulting in the use of such assays to study stem cells and their growth potential in the epidermis and in other tissues (Barrandon and Green 1987; Young et al. 2004). Although some use is historical, clonogenic assays are still used routinely; for example, these assays are useful tools to assess the quality of mesenchymal stromal cells (MSCs) following isolation procedures (Penfornis and Pochampally 2016). Many tissue culture laboratories use clonogenic assays to test new media and serum (see Sections 9.7.1, 11.8.2).

Dilution cloning may be used as a method to obtain a pure derivative or "strain" for further work. However, the "purity" of the culture may be debated. Cloned populations can display heterogeneity as they are expanded for use, due to inadequate cloning conditions or changes in phenotype with further handling (see Section 3.4; see also Figure 20.2). Cloning may help to reduce the heterogeneity of a culture, but it is important to perform ongoing characterization to ensure that the correct phenotype continues to be expressed.

20.2.1 Dilution Cloning in Dishes

Dilution cloning of adherent cells may be carried out in Petri dishes, multiwell plates, or flasks, depending on the application. It is relatively easy to discern individual colonies in Petri dishes, although there may be some doubt as to whether the colony is truly clonal in nature. Micromanipulation is the only conclusive method for determining genuine clonality (i.e. that a colony was derived from one cell). However, if a single-cell suspension produces symmetrical colonies, the colonies are likely to be clones, particularly if colony formation is monitored at the early stages.

The following protocol (Protocol P20.1) is useful to generate single colonies in Petri dishes, which can be selected for expansion in culture (see Section 20.4.1). This protocol can also be adapted for use in clonogenic assays (see Protocol P19.4; see also Supp. S11.7, S29.1).

PROTOCOL P20.1. DILUTION CLONING

Outline

Seed the cells at low density and incubate until colonies form (see Figure 20.4).

Materials (sterile or aseptically prepared)

- Cells, e.g. CHO-K1, 25-cm^2 flask, late log phase
- Complete culture medium, e.g. Ham's F12 with 10% fetal bovine serum (FBS)
- Trypsin, 0.25%, crude (250 USP U/ml or equivalent; see Section 13.5.1).
- Pipettes in an assortment of sizes, usually 1, 5, 10, and 25 ml
- Petri dishes, 6 cm, one pack
- Centrifuge tubes or universal containers, for dilution

Materials (non-sterile)

- Biological safety cabinet (BSC) and associated consumables (see Protocol P12.1)
- Plastic box or trays to hold the dishes
- Cell counting equipment (hemocytometer or automated counter; see Section 19.1)

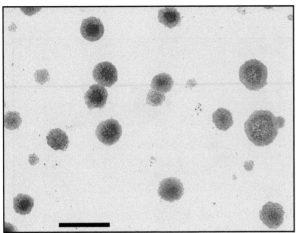

(a) HeLa Clones. *HeLa-S3 cloned by dilution cloning and stained with Giemsa after 3 weeks. The scale bar represents 5 mm. Source: R. Ian Freshney.*

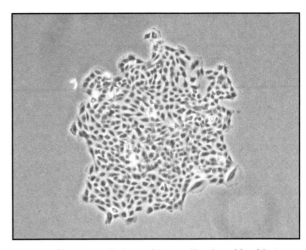

(b) NRK Clone. *Small clone of NRK cells, cloned by dilution cloning. Phase contrast, 4x objective. Source: R. Ian Freshney.*

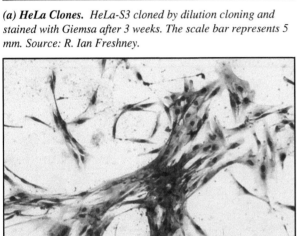

(c) Breast Carcinoma, Cloned on Plastic. *Microphotograph of secondary culture from human breast carcinoma (JUW), 4000 cells/cm², growing on plastic. Mainly fibroblasts. Giemsa stain; 10x objective. Source: R. Ian Freshney.*

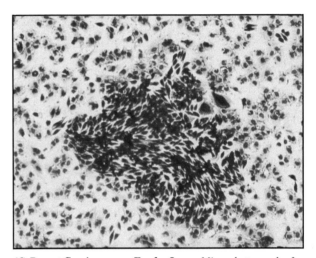

(d) Breast Carcinoma on Feeder Layer. *Microphotograph of epithelial colony from human breast carcinoma cells, 400 cells/cm², growing on confluent feeder layer of FHs 74 Int cells. Giemsa stain; 10x objective. Source: R. Ian Freshney.*

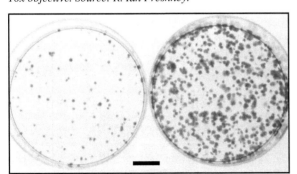

(e) Effect of Serum on Plating Efficiency. *Mv1Lu cells, transfected with myc oncogene, plated with 500 cells per 6 cm Petri dish, and fixed and stained 2 weeks later. Cells have been grown in 10% FBS (left) or 20% FBS (right). The scale bar represents 10 mm. Source: Courtesy of M.Z.Khan.*

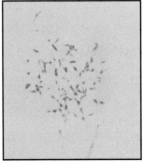

(f) Effect of Glucocorticoid on Plating Efficiency. *Early passage cell line from human glioma cloned in the absence (left) and presence (right) of 25 μM dexamethasone. Source: R. Ian Freshney.*

Fig. 20.1. Cell cloning.

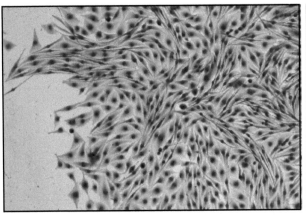

(a) Clone of Continuous Glioma Cell Line, MOG-G-CCM. *Elliptical and spindle-shaped morphology. Giemsa stained; 10x objective.*

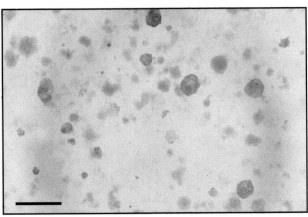

(d) Cloned Culture of Early Passage Cell Line from Human non-small cell lung carcinoma (NSCLC). *The scale bar represents 10 mm.*

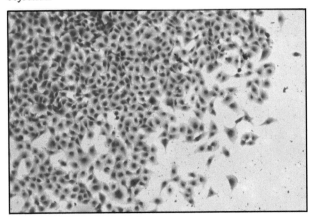

(b) Second Clone of Continuous Glioma Cell Line, MOG-G-CCM. *Epithelioid morphology. Giemsa stained; 10x objective.*

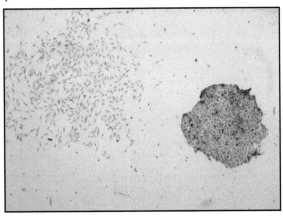

(e) Clones from Human NSCLC. *Detail from (d). Giemsa stained; 4x objective.*

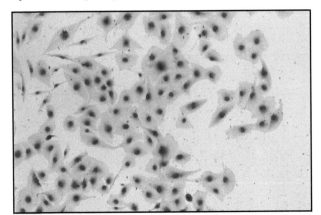

(c) Third Clone of Continuous Glioma Cell Line, MOG-G-CCM. *Squamous epithelioid morphology. Same magnification as (a) and (b). Giemsa stained; 10x objective.*

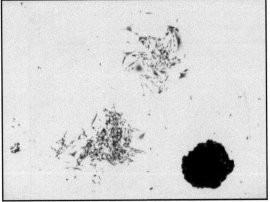

(f) Cloned Culture of A2780. *Continuous cell line from human ovarian carcinoma. Giemsa stained; 4x objective. Source for all images: R. Ian Freshney.*

Fig. 20.2. Morphological diversity in cell clones.

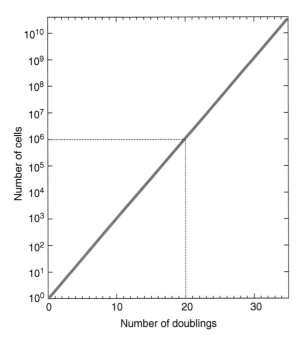

Fig. 20.3. Clonal cell yield. The relationship of the cell yield in a clone to the number of population doublings, e.g. 20 doublings are required to produce 10^6 cells.

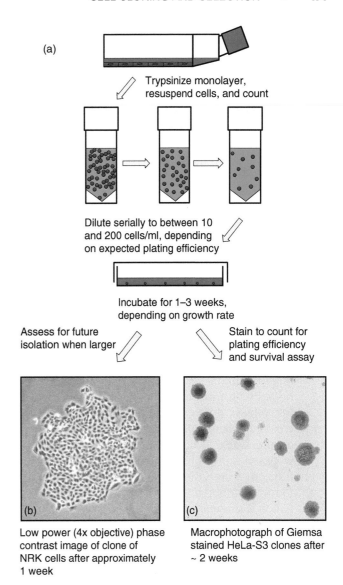

(a)

Trypsinize monolayer, resuspend cells, and count

Dilute serially to between 10 and 200 cells/ml, depending on expected plating efficiency

Incubate for 1–3 weeks, depending on growth rate

Assess for future isolation when larger

Stain to count for plating efficiency and survival assay

(b) Low power (4x objective) phase contrast image of clone of NRK cells after approximately 1 week

(c) Macrophotograph of Giemsa stained HeLa-S3 clones after ~ 2 weeks

Fig. 20.4. Dilution cloning. (a) Cells from a trypsinized monolayer culture are counted and diluted sufficiently to generate isolated colonies. (b) If isolation is required, colonies should be allowed to grow until they reach a suitable size for isolation (usually larger than the colony shown here). (c) If cloning is being performed for a clonogenic assay, the colonies should be fixed, stained, and counted (see Sections 19.4.1, 29.2.4). *Source*: (b, c) R. Ian Freshney.

Procedure

1. Prewarm bottles of growth medium and any other reagents required, e.g. in a water bath.
2. Prepare a work space for aseptic technique using a BSC.
3. Trypsinize the cells to produce a single-cell suspension (see Protocol P14.2).
4. While the cells are trypsinizing, number the dishes (on the side of the base) and measure out medium for the dilution steps. Up to four dilution steps may be necessary to reduce a regular monolayer accurately to a concentration suitable for cloning.
5. When the cells round up and start to detach, disperse the monolayer to a single cell suspension in 5 ml medium containing serum or trypsin inhibitor (see Section 13.2.1).
6. Count the cells with a hemocytometer or an automated cell counter.
7. Dilute the cell suspension to 1×10^5 cells/ml (for CHO-K1 this will be approximately 1 : 10 or 1 : 20, depending on the number of cells in the flask).
8. Prepare serial dilutions of the cell suspension, aiming for approximately 50 colonies per Petri dish. For CHO-K1 this will be 10 cells/ml (expected cloning efficiency 100%,

see Table 20.1) and the dilution steps will be as follows:
(a) Dilute 200 μl of the 1×10^5 cells/ml suspension to 20 ml (1 : 100), giving 1×10^3 cells/ml.
(b) Dilute 200 μl of the 1×10^3 cells/ml suspension to 20 ml (1 : 100), giving 10 cells/ml.

(c) If you wish to add a variable to the cloning conditions (e.g. a range of serum concentrations, different sera, or growth factors), prepare a range of tubes at this stage and add 200 µl of the 1×10^3 cells/ml suspension to each of them separately.

(d) If you are cloning cells for the first time, choose a range of 10–2000 cells/ml to determine the cloning efficiency (see Protocol P19.4).

9. Seed three Petri dishes each with 5 ml medium containing cells from the final dilution stage. It is also advisable to seed dishes from the 1×10^3 cells/ml suspension as controls, to confirm that cells were present in case no clones form at the lower concentration.

10. Put the Petri dishes in a transparent plastic box or on a tray and place in a humid CO_2 incubator. Leave the cultures untouched for one week for the colonies to form.

11. If no colonies are visible after one week, continue to culture for another week. Dishes may be fed with fresh medium, but this is optional (see Notes). If no colonies are visible after two weeks, feed the dishes and culture them for a third week. If no colonies appear by three weeks, it is unlikely that they will appear at all.

12. If colonies are visible, check that they are uniformly distributed (see Notes) and no contamination is present.

13. Proceed to select colonies for culture (see Protocol P20.4) or count colonies to measure cloning efficiency (see Notes).

Notes

Step 3: It is fundamental to the concept of cloning that a single cell suspension is produced. Under-trypsinizing will produce clumps and over-trypsinizing will reduce the viability of the cells. When cloning a new cell line for the first time, you may need to try different trypsin exposure times and different recipes to give optimum cloning efficiency from a good single-cell suspension. For cells other than CHO-K1, you may need to try different trypsin activities or use other proteases (see Table 14.4).

Step 11: As the density of cells during cloning is very low, the need to feed the dishes after one week is debatable. Feeding mainly counteracts the loss of unstable nutrients (such as glutamine), replaces growth factors that have degraded or been depleted, and compensates for evaporation. However, it also increases the risk of contamination, so it is reasonable to leave dishes for two weeks without feeding. If it is necessary to leave the dishes for a third week, the medium should be replaced, or at least half of it.

Step 12: Incorrect seeding may cause preferential formation of colonies at the center of the plate. This may occur because the cells were seeded into the center of a plate that already contained medium or because swirling of the medium in the plate led to a centralized location. It can also be caused by resonance in the incubator due to vibration, e.g. from an imbalanced fan motor or excessive opening and closing of the incubator (see Figure 8.9).

Step 13: If colonies are used to measure cloning efficiency, they are usually fixed and stained with crystal violet for later counting (see Protocols P18.4, P19.4).

20.2.2 Dilution Cloning in Microwell Plates

If the prime purpose of cloning is to isolate colonies, seeding into microwell plates can be an advantage (see Figure 20.5). When the clones grow up, isolation is easy, although the plates must be monitored at the early stages in order to mark which wells genuinely have single clones. The statistical probability of a well having a single clone can be increased by reducing the seeding density to a level such that only 1 in 5 or 10 wells would be expected to have a colony. For example, from Table 20.1, 100 cells/ml at 10% cloning efficiency would give 10 colonies/ml, or 1 colony/0.1 ml, as added to a microtitration plate – i.e. 1 colony/well. If the seeding concentration is reduced to 10 cells/ml, then, theoretically, only 1 in 10 wells will contain a colony, and the probability of wells containing more than one colony is very low.

TABLE 20.1. Relationship of seeding density to cloning efficiency.

Expected cloning efficiency (%)	Optimal cell number to be seeded			
	Per ml	Per cm^2	Per dish, 60 mm	Per dish, 90 mm
0.1	1×10^4	2×10^3	40 000	100 000
1.0	1×10^3	200	4000	10 000
10	100	20	400	1000
50	20	4	80	200
100	10	2	40	100

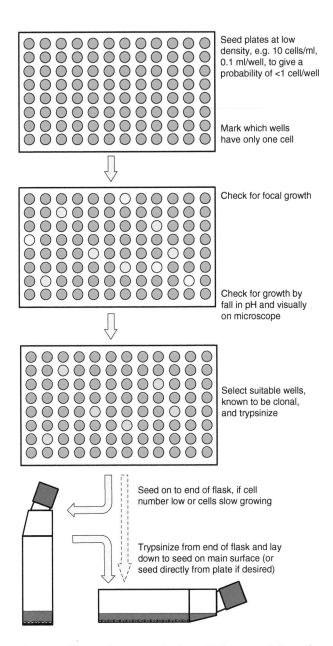

Seed plates at low density, e.g. 10 cells/ml, 0.1 ml/well, to give a probability of <1 cell/well

Mark which wells have only one cell

Check for focal growth

Check for growth by fall in pH and visually on microscope

Select suitable wells, known to be clonal, and trypsinize

Seed on to end of flask, if cell number low or cells slow growing

Trypsinize from end of flask and lay down to seed on main surface (or seed directly from plate if desired)

Fig. 20.5. Cloning in microwell plates. Cells are seeded at a low enough concentration to give a probability of <1 cell/well, resulting in one cell in some wells. These wells can be identified and marked during visual examination on the microscope a few hours after plating. When a fall in pH indicates growth in a marked well, this is confirmed by microscopic observation and the colony is trypsinized.

20.3 CLONING IN SUSPENSION

Cloning in suspension was initially used for the selective growth of cells infected with polyoma virus (Macpherson and Montagnier 1964). Cultures that were transduced with large quantities of virus developed "transformed" colonies, where cells grew in an anchorage-independent manner. Anchorage-independent growth was subsequently used as an indicator of *in vitro* transformation, where it is closely associated with malignant behavior (see Section 3.6.2). Cloning in suspension can be used to isolate anchorage-independent cell populations, using a similar approach to limiting dilution (see Section 19.4.1). Suspension-based clonogenic assays can be used to measure the clonogenicity of cancer stem cells (CSCs) (see Section 25.6) and to assess potential anticancer drugs (Hamburger and Salmon 1977; Sant and Johnston 2017). Hematopoietic stem cells (HSCs) from bone marrow, peripheral blood, or umbilical cord blood form colonies in suspension-based assays (Sant and Johnston 2017). Depending on the cells and growth factors used, colonies may generate undifferentiated cells with high repopulation efficiency, *in vivo* or *in vitro*, or may mature into colonies of differentiated hematopoietic cells with very little repopulation efficiency. Cloning then becomes an assay for self-renewal capacity and stem cell identity.

Suspension cloning is performed by seeding into a gel, such as agar or agarose, or into a viscous solution, such as methylcellulose (Methocel®, Dow Chemical Company). To hold the colony together and prevent mixing, cells are suspended in agar or Methocel and plated on an agar underlay or into dishes that are not treated for tissue culture (see Section 8.3.6). The stability of the gel, or viscosity of the Methocel, ensures that daughter cells do not break away from the colony as it forms. Detachment and formation of daughter colonies can occur even in monolayer cloning for some cell lines, such as HeLa S3 and CHO, and will give an erroneous cloning efficiency. Formation of daughter colonies can be minimized by cloning in Methocel without an underlay, and by allowing the cells to sediment onto the plastic surface. Most cell types clone in suspension with a lower efficiency than in monolayer, some cells by two or three orders of magnitude. The isolation of colonies is, however, much easier.

20.3.1 Soft Agar

The following procedure (Protocol P20.2) can be used for cloning in soft agar, followed by selection or counting of colonies. The procedure can also be adapted for high-throughput screening using microwell plates and simultaneous cloning and selection of hybridomas and transfected cell lines (Ke et al. 2004; Horman et al. 2013; Wognum and Lee 2013). It is important to select an agar that is non-toxic, e.g. Difco Noble agar (BD Biosciences) (Paul 1975). Agarose can be used instead of agar and may give better results for some cell types due to reduced levels of sulfated polysaccharides, which are believed to inhibit normal cell proliferation (Macpherson 1973). Some types of agarose have a lower gelling temperature and can be manipulated more easily at 37 °C; they can be gelled at 4 °C and then returned to 37 °C.

PROTOCOL P20.2. CLONING IN AGAR

Contributed by Mary Freshney when at Cancer Research UK Beatson Institute, Glasgow, G61 1BD, United Kingdom.

Outline

Agar is liquid at high temperatures, but gels at 37 °C. Cells are suspended in warm agar medium and, when incubated after the agar gels, form discrete colonies that may be isolated easily (see Figure 20.6; see also Protocol P20.5).

Materials (sterile)

- Noble agar (BD Biosciences)
- Medium at double strength (2×) for agar preparation (i.e. Ham's F12, RPMI 1640, Dulbecco's modified Eagle's medium [DMEM], or CMRL 1066)
- Serum (FBS) if required
- Other supplements if required, e.g. growth factors, hormones (to be added to the 0.6% agar underlay)
- Growth medium, 1×, for cell dilutions
- Sterile ultrapure water (UPW)
- Sterile conical flask
- Universal containers, bijoux containers, or centrifuge tubes for dilution
- Petri dishes, 3.5 cm, non-tissue-culture grade
- Pipettes, including sterile plastic disposable pipettes for agar solutions

Materials (non-sterile)

- BSC and associated consumables (see Protocol P12.1)
- Cell counting equipment (hemocytometer or automated counter; see Section 19.1)
- Bunsen burner and tripod (located away from the BSC)
- Water bath at 55 °C
- Water bath at 37 °C
- Tray

Procedure

1. Prepare a work space for aseptic technique using a BSC.
2. Calculate the cell dilutions required for the assay. For an assay to measure the cloning efficiency of a cell line, prepare to set up three dishes for each cell dilution. Convenient cell numbers per 3.5 cm dish are 1000, 333, 111, and 37, i.e. serial one-third dilutions of the cell suspension (see step 12 below).

3. Label the Petri dishes on the side of the base. It is convenient to place them on a tray.
4. Prepare 2× medium containing 40% FBS and keep it at 37 °C.
5. Weigh out 1.2 g agar.
6. Measure 100 ml sterile UPW into a sterile conical flask and another 100 ml into a sterile bottle. Add the 1.2 g agar to the flask. Cover the flask and boil the solution for two minutes. Alternatively, the agar may be sterilized in the autoclave in advance, but, if subsequently stored, it will still need to be boiled or microwaved, in order to melt it for use.
7. Transfer the boiled agar and the bottle of sterile UPW to a water bath at 55 °C.
8. Prepare a 0.6% agar underlay by combining an equal volume of 2× medium and 1.2% agar (see Figure 20.6a). Keep the underlay at 37 °C. If any growth factors, hormones or other supplements are being used, they should be added to the underlay medium at this point.
9. Add 1 ml 0.6% agar medium to each dish, mix, and ensure that the medium covers the base of the dish. Leave the dishes at room temperature to set.
10. Prepare the cell suspension and count the cells (see Figure 20.6c).
11. Prepare 0.3% agar medium and keep it at 37 °C. This medium may be prepared by diluting 2× medium at 37 °C with 1.2% agar at 55 °C and UPW at 55 °C in the respective proportions of 2 : 1 : 1 (see Figure 20.6b).
12. Prepare the following cell dilutions:
    (a) 1×10^5 cells/ml.
    (b) Dilute 1×10^5 cells/ml by 1/3 to give 3.3×10^4 cells/ml.
    (c) Dilute 3.3×10^4 cells/ml by 1/3 to give 1.1×10^4 cells/ml.
    (d) Dilute 1.1×10^4 cells/ml by 1/3 to give 3.7×10^3 cells/ml.
13. Label four bijoux bottles or tubes, one for each dilution, and pipette 40 μl of each cell dilution, including the 1×10^5/ml concentration, into the respective container. Add 4 ml of 0.3% agar medium at 37 °C to each container, mix, and pipette 1 ml from each container onto each of three Petri dishes (see Figure 20.6c). This will give final concentrations as follows:
    (a) 1×10^3 cells/ml per dish.
    (b) 330 cells/ml per dish.
    (c) 110 cells/ml per dish.
    (d) 37 cells/ml per dish.

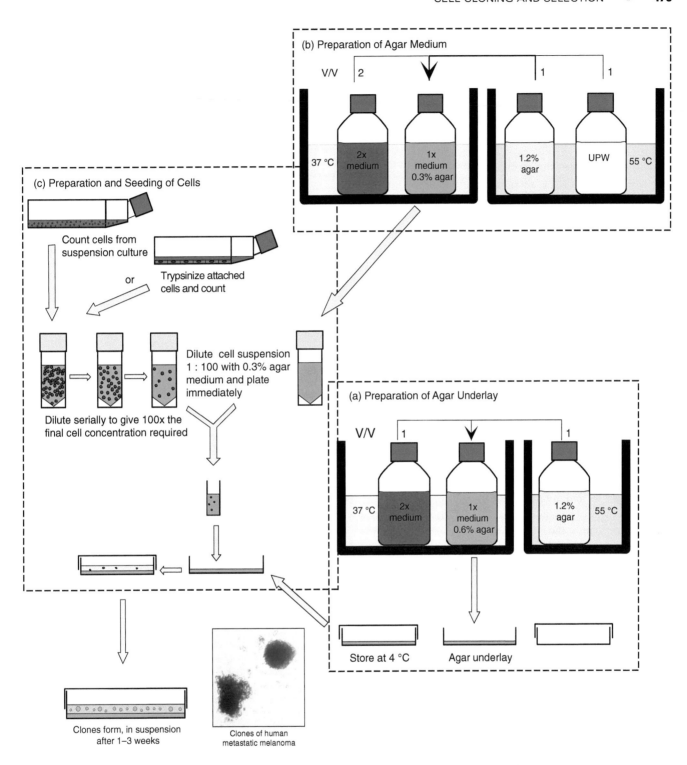

Fig. 20.6. Cloning in suspension in agar. Cultured cells or primary suspensions from bone marrow or tumors form colonies in suspension, using agar or low-melting-temperature agarose, which is then allowed to gel. An underlay prevents attachment to the base of the dish. (a) Preparation of agar underlay: agar, 1.2%, at 55 °C is mixed with 2× medium at 37 °C and dispensed immediately into dishes, where it is allowed to gel at room temperature or 4 °C. (b) Preparation of agar medium: agar, 1.2%, and UPW are maintained at 55 °C and mixed with 2× medium to give 0.3% agar for cloning. The use of low-melting-point agarose allows all solutions to be maintained at 37 °C, but this agarose can be more difficult to gel. (c) Cells grown in suspension (derived from suspension culture or trypsinized from an attached monolayer) are counted and diluted serially, and the final product is diluted with agar or agarose and seeded onto an agar underlay.

14. Allow the solution in the Petri dishes to gel at room temperature.

15. Put the Petri dishes into a clear plastic box with a lid and incubate them at 37 °C in a humid incubator for 10 days.

Notes

Step 4: Prepare the medium from 10× concentrate to half the recommended final volume and add twice the normal concentration of serum if required.

Step 8: If a titration of growth factors is being carried out or a selection of different factors is being used, add the required amount to the Petri dishes before the underlay is added.

Step 13: Always be sure that the agar medium for the top layer has had adequate time to cool to 37 °C before adding the cells to it.

20.3.2 Methylcellulose (Methocel)

Some laboratories prefer to use methylcellulose (Methocel) for cloning in suspension, because of the impurities that may be present in agar and the complexity of handling melted agar with cells (Buick et al. 1979). Methocel is a viscous solution and not a gel. It has a higher viscosity when warm and, because it is a sol and not a gel, cells will sediment through it slowly. It is therefore essential to use an underlay with Methocel. Colonies form at the interface between the Methocel and the agar (or agarose) underlay, placing themselves in the same focal plane and making analysis and photography easier.

PROTOCOL P20.3. CLONING IN METHOCEL

Outline

Suspend the cells in medium containing Methocel and seed the cells into dishes containing an agar or agarose underlay (see Figure 20.7).

Materials (sterile)

- Noble agar (BD Biosciences)
- Methocel (Dow Chemical Company), 4 Pa-s (4000 cP), 1.6% in UPW; place on ice
- Medium at double strength (2×) for agar and Methocel preparation (i.e. Ham's F12, RPMI 1640, DMEM, or CMRL 1066)
- Serum (FBS) if required
- Growth medium, 1×, for cell dilutions
- Sterile ultrapure water (UPW)
- Sterile conical flask

- Universal containers, bijoux containers, or centrifuge tubes for dilution
- 3.5-cm Petri dishes, non-tissue-culture grade
- Pipettes, including sterile plastic disposable pipettes for agar solutions
- Syringes without needles to dispense Methocel (because of its viscosity Methocel tends to cling to the inside of pipettes, making dispensing difficult and inaccurate)

Materials (non-sterile)

- BSC and associated consumables (see Protocol P12.1)
- Cell counting equipment (hemocytometer or automated counter; see Section 19.1)
- Bunsen burner and tripod (located away from the BSC)
- Water bath at 55 °C
- Water bath at 37 °C
- Tray

Procedure

1. Prepare a work space for aseptic technique using a BSC.

2. Prepare agar underlays as described previously (see Protocol P20.2, steps 2–9).

3. Dilute the Methocel to 0.8% with an equal volume of 2× medium. Mix it well and keep on ice.

4. Prepare the cell suspension and count the cells.

5. Prepare the following cell dilutions:
- **(a)** 1×10^5 cells/ml.
- **(b)** Dilute 1×10^5 cells/ml by 1/3 to give 3.3×10^4 cells/ml.
- **(c)** Dilute 3.3×10^4 cells/ml by 1/3 to give 1.1×10^4 cells/ml.
- **(d)** Dilute 1.1×10^4 cells/ml by 1/3 to give 3.7×10^3 cells/ml.

6. Label four bijoux bottles or tubes, one for each dilution, and pipette 40 μl of each cell dilution, including the 1×10^5/ml concentration, into the respective container. Add 4 ml of 0.8% Methocel medium to each container and mix well with a vortex (if the cells are known to be particularly fragile, mix by gently aspirating up and down using a syringe).

7. Use a syringe to add 1 ml from each container to each of four Petri dishes (see Figure 20.7). This will give final concentrations as follows:
- **(a)** 1×10^3 cells ml per dish.
- **(b)** 330 cells ml per dish.
- **(c)** 110 cells ml per dish.
- **(d)** 37 cells ml per dish.

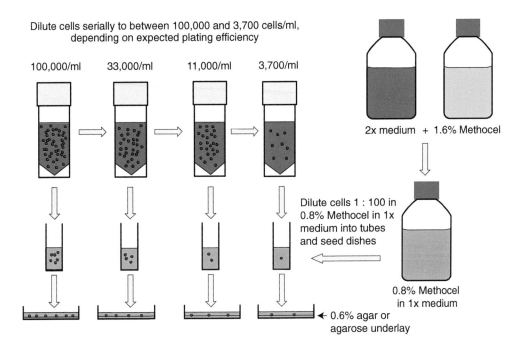

Dilute cells serially to between 100,000 and 3,700 cells/ml, depending on expected plating efficiency

100,000/ml   33,000/ml   11,000/ml   3,700/ml

2x medium + 1.6% Methocel

Dilute cells 1 : 100 in 0.8% Methocel in 1x medium into tubes and seed dishes

0.8% Methocel in 1x medium

← 0.6% agar or agarose underlay

Fig. 20.7. Cloning in suspension in Methocel. A series of cell dilutions is prepared, diluted 1 : 100 in Methocel medium, and plated into non-tissue culture-grade dishes or dishes with an agar underlay.

8. Incubate the dishes in a humid incubator until colonies form.

Notes

Step 6: Methocel is viscous, so manipulations are easier to perform with a syringe without a needle.

Step 8: Fresh medium may be added, 1 ml per dish or well, after one week and then removed and replaced with more fresh medium after two weeks. Because the colonies form at the interface between the agar and the Methocel, this can be done without disturbing the colonies.

20.4 SELECTION OF CLONES

20.4.1 Adherent Clones

When cloning is used to generate clonal populations, colonies must be isolated for further propagation. If dilution cloning in Petri dishes has been used (see Protocol P20.1), a barrier must be created between different colonies by placing a cloning ring or cylinder around each colony (see Figure 20.8a). Cloning rings are often made from glass, but other materials may be used such as stainless steel, ceramics, or plastic. Plastic rings may be purchased or cut from thick-walled tubing made from nylon, silicone, or Teflon. Whatever the material, it must have a smooth base (allowing it to seal onto the base of the Petri dish) and an internal diameter that is just wide enough to enclose one whole clone and exclude any adjacent clones. Cloning rings are usually anchored onto the dish using silicone grease; they can also be anchored using low melting point agarose to reduce leakage (see Figure 20.8b) (Mathupala and Sloan 2009).

PROTOCOL P20.4. ISOLATION OF ADHERENT CLONES WITH CLONING RINGS

Outline

Position a cloning ring around the colony, trypsinize the colony within the ring, and transfer the cells to one of the wells of a 24-well plate, or directly to a 25-cm² flask (see Figure 20.9).

Materials (sterile)

☐ Sterile cloning rings or cylinders (e.g. glass cloning cylinders, Sigma-Aldrich; Bel-Art™ polystyrene cloning cylinders, SP Scienceware); glass rings can be sterilized by autoclaving or dry heat in a glass Petri dish

☐ Silicone grease; sterilize by autoclaving or dry heat in the same dish as the cloning rings

☐ Trypsin 0.25% in Dulbecco's phosphate buffered saline without Ca²⁺ and Mg²⁺ (DPBS-A)

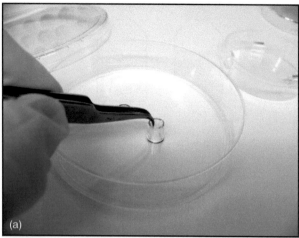

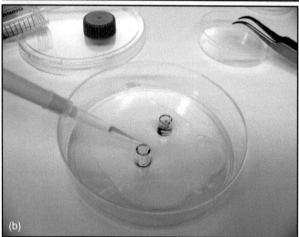

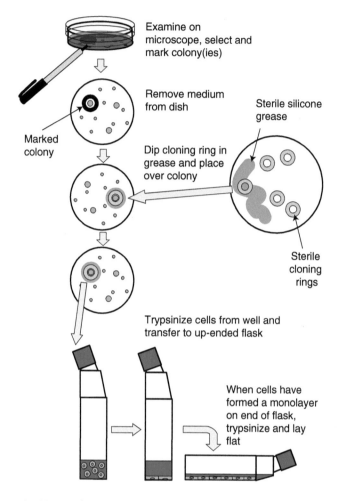

Fig. 20.8. Cloning rings. (a) Tapered cloning rings are placed over demarcated cell clones; (b) after positioning, low melting point agarose is added and allowed to harden to seal the rings in place. This allows trypsinization within the ring without risk of leakage. *Source*: Mathupala, S. and Sloan, A. A. (2009). An agarose-based cloning-ring anchoring method for isolation of viable cell clones. Biotechniques 46 (4):305–7. doi: 10.2144/000113079, reproduced with permission from BioTechniques as agreed by Future Science Ltd.

Fig. 20.9. Isolation of adherent clones. Colonies are examined under the microscope, and suitable colonies are selected and marked. The medium is removed and cloning rings, dipped in silicone grease, are placed around each colony, which is then trypsinized from within the ring.

□ Growth medium
□ 24-well plate (one per procedure) and/or 25-cm² flask (one per colony)
□ Sterile pipettor tips with filter, or plugged Pasteur pipettes with a bent end
□ Sterile forceps

Materials (non-sterile)

□ BSC and associated consumables (see Protocol P12.1)
□ Pipettor, 50–200 µl

□ Object marker (preferably fitting into the objective nosepiece of a microscope (e.g. Nikon Object Marker, SEO Enterprises)); a felt-tip pen can also be used

Procedure

1. Examine the Petri dish under the microscope and mark the colonies that you wish to isolate, using an object marker or a felt-tip pen on the underside of the dish.
2. Prepare a work space for aseptic technique using a BSC.
3. Remove the medium from the dish and rinse the clones gently with DPBS-A.
4. Using sterile forceps, take one cloning ring, dip it in silicone grease, and press it down on the

dish alongside the silicone grease to spread the grease around the base of the ring.

5. Position the ring (greased side down) around the desired colony.

6. Repeat steps 4 and 5 for other colonies in the dish.

7. Add sufficient 0.25% trypsin to fill the hole in each ring (100–400 µl, depending on the internal diameter of the ring). Leave the trypsin for 20 seconds and then remove it.

8. Close the dish and incubate for 15 minutes at 37 °C.

9. Add 100–400 µl medium to each ring.

10. Taking each clone in turn, pipette the medium up and down to disperse the cells, and transfer the medium to the well of a 24-well plate or to a 25-cm^2 flask standing on end (see Figure 20.9). Use a separate pipette, or a separate pipettor tip, for each clone.

11. Wash out each ring with another volume of medium (100–400 µl) and transfer the medium to the same well or flask.

12. Make up the medium to a suitable final volume (e.g. 0.5–1.0 ml per well in a 24-well plate), close the plate, and incubate it. If you are using flasks, add 1 ml of medium to each flask and incubate the flasks standing on end.

13. When the clone grows to fill the available space, passage into a larger volume. If cells were transferred to a 24-well plate, passage into a 25-cm^2 flask, incubated conventionally with 5 ml medium. If cells were transferred to an upended flask (see Figure 20.9), trypsinize when confluent, resuspend in 5 ml medium, and lay the flask down flat. Continue the incubation.

Notes

Steps 4 and 5: Instead of silicone grease, low melting point agarose may be used to anchor the cloning rings. A detailed protocol is provided elsewhere (Mathupala and Sloan 2009).

Step 6: Contamination is a risk following this procedure. Select the best colonies first and discard the plate after a limited number of manipulations (e.g. after lifting two to three colonies).

Steps 10 and 11: The dish will dry out if left open for too long. It is important to limit the number of clones isolated or cover the dish between manipulations.

Isolation of clones with cloning rings is time-consuming and can be challenging, e.g. if cloning rings leak or plates dry

out. A number of alternative methods have been developed for isolation of adherent clones, including:

(1) Automated colony screening and picking. The ClonePix™ II Mammalian Colony Picker (Molecular Devices) can be used to automate imaging and selection of high-value clones. Typically, this system is used for hybridomas or other cell lines that express a target product. Cells are seeded in semisolid agar (see Section 20.4.2) and clones are selected using a pin that transfers the clone to a well of a 96-well plate.

(2) Dilution cloning in multiwell plates instead of Petri dishes (see Section 20.2.2), followed by trypsinization of individual wells. Clonal origin of the colony must be confirmed during its formation by regular microscopic observation.

(3) Addition of small coverslips or coverslip fragments to the Petri dish before cells are added (Martin 1973). When plated at the correct density, some colonies will be singly distributed on a piece of glass and may be transferred to a fresh dish or plate.

(4) Addition of paper discs (e.g. 5- to 6-mm discs punched from Whatman #1 filter paper and sterilized by autoclaving) after colonies have formed (Domann and Martinez 1995). Each paper disc can be dipped in trypsin, placed on a colony, incubated while the trypsin takes effect, and then transferred to a 24-well plate containing medium.

(5) Isolation of cells from a specific colony by simultaneously scraping and aspirating cells into the tip of a pipettor; this technique is referred to as the "scratch and sniff" method (Karin 1999).

(6) Culture of a dilute cell suspension in a sterile glass capillary tube (e.g. a 50-µl Drummond Microcap, Sigma-Aldrich), allowing colonies to form inside the tube. The tube is then carefully broken on either side of a colony and transferred to a fresh plate. This procedure was first used for cloning in the 1940s and then simplified to make it easier to use (Sanford et al. 1948, 1961). It has since been adapted as a clonogenic assay using agar-containing glass capillaries (Maurer and Echarti 1990).

(7) Irradiation of the culture flask with shielding applied to protect a specific colony. A piece of lead is cut from a 2-mm-thick sheet to the appropriate size, e.g. 2–5 mm diameter. The flask is irradiated with 30 Gy and then trypsinized; cells are plated in the same flask, using the irradiated cells as a feeder layer for the protected colony. However, viability of the shielded colony may be impaired (Sjostedt et al. 2014).

20.4.2 Suspension Clones

The isolation of colonies growing in suspension is comparatively straightforward, provided you have access to a dissection microscope.

PROTOCOL P20.5. ISOLATION OF SUSPENSION CLONES

Outline

Draw the colony into a pipettor or Pasteur pipette, and transfer the colony to a flask or the well of a multiwell plate (see Figure 20.10).

Materials (sterile)

- Growth medium in universal container or centrifuge tube
- 24-well plate or 25-cm² flask
- Sterile pipettor tips with filter

Materials (non-sterile)

- BSC and associated consumables (see Protocol P12.1)

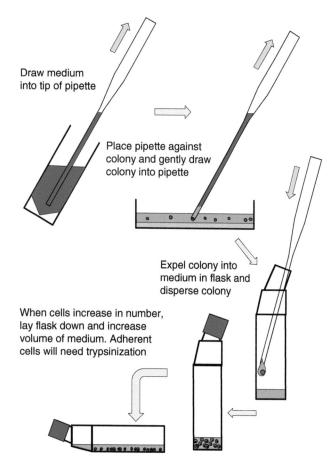

Draw medium into tip of pipette

Place pipette against colony and gently draw colony into pipette

Expel colony into medium in flask and disperse colony

When cells increase in number, lay flask down and increase volume of medium. Adherent cells will need trypsinization

Fig. 20.10. Isolation of suspension clones. Colonies are marked (see Figure 20.9) and drawn into a pipettor tip or Pasteur pipette. Each colony is transferred to a culture flask, dispersed in medium, and incubated. The volume of medium is increased when cells start to grow.

- Dissecting microscope, 20–50× magnification
- Pipettor, 50–200 μl
- Object marker (see Protocol P20.4) or felt-tip pen

Procedure

1. Examine the Petri dish under the microscope and mark the colonies that you wish to isolate, using an object marker or a felt-tip pen on the underside of the dish.
2. Prepare a work space for aseptic technique using a BSC.
3. Pipette 1 ml of medium into each well of a 24-well plate or into a 25-cm² flask that is standing on end (see Figure 20.10).
4. Find the colony you wish to isolate using a dissecting microscope. Set the pipettor to 100 μl and draw approximately 50 μl into the pipette tip. Place the tip of the pipette against the colony to be isolated and gently draw it in. Use a separate pipettor tip for each colony.
5. Transfer the contents of the pipettor tip to a well in a 24-well plate or a 25-cm² flask that is standing on end (see Figure 20.10). Flush out the colony with medium.

Notes

Step 5: If agar has been used, you may need to pipette the colony up and down a few times in the well to disperse the agar. If Methocel has been used, the colony will settle in the gel and adherent cells will attach and grow out. Cells that normally grow in suspension will settle but, of course, will not attach.

20.5 REPLICA PLATING

Bacterial colonies can be transferred from a primary to secondary plates by pressing a moist pad gently down onto colonies that are growing on a nutrient agar plate, which is then transferred to a fresh agar plate. Various attempts have been made to adapt this technique to cell culture (Robb 1973). For example, a filter or mesh screen can be placed over monolayer clones and transferred to a fresh dish after a few days (Hornsby et al. 1992). For clones that have been developed in microtitration plates, there are a number of transfer devices available, e.g. the Transtar (Corning; see Figure 5.8). Transfer devices can be used with suspension cultures directly or with adherent cultures after trypsinization and resuspension. Robotic systems may also enable high-throughput replica plating (see Section 28.6).

20.6 STIMULATION OF CLONING EFFICIENCY

One of the greatest challenges when cloning cultures is poor cloning efficiency, particularly for early passage cultures (see Section 20.2). While the cloning efficiency for continuous cell lines seldom drops below 10%, it may decline to 0.5–5% (or even to zero) for primary cultures and finite cell lines. Even for human tumor cells, which proliferate continuously *in vivo*, cloning efficiency is likely to be less than 1% (see Section 25.3.3). Studies of normal keratinocytes have shown that clonal populations have different capacities for self-renewal (Barrandon and Green 1987). Some clones have extensive proliferative capacity ("holoclones"), some have very limited growth ("paraclones"), while a third group develop colonies that may display either behavior ("meroclones"). The clonogenic cells that give rise to these colonies are likely to represent the stem cells of that population (Beaver et al. 2014). Hence the number may be quite low in normal adult tissues, which would limit cloning efficiency. It is also possible that the stem cells that give rise to these populations are in equilibrium with the rest of the population and that this equilibrium is re-established in a cloned derivative.

Attempts to improve cloning efficiency generally focus on improving cellular nutrition, based on the assumption that cells require a greater range of nutrients at low densities, because of loss by leakage, or that cell-derived diffusible signals or conditioning factors are absent or too dilute at low density culture. The intracellular metabolic pool of a leaky cell in a dense population will soon reach equilibrium with the surrounding medium, while that of an isolated cell never will. The broader culture environment is also important, particularly for stem cells, which are prone to spontaneous differentiation and apoptosis *in vitro* (see Section 23.4.1). It is likely that many of the stem cells that are initially present in cloning experiments do not survive and proliferate at low density because culture conditions are suboptimal. The development of optimized culture conditions for pluripotent stem cell (PSC) lines may help to improve cloning efficiency and analysis of low-density cultures (see Section 23.4.2).

Poor cloning efficiency may be less of a concern if cloning is used as an assay (e.g. for testing medium; see Section 11.8.2). However, fluctuations in cloning efficiency can still occur that will affect the interpretation of clonogenic assays. For example, cloning efficiency may vary when cells are plated at different densities, e.g. to assess the effects of different radiation dosages (Pomp et al. 1996). Mycoplasma contamination is known to increase cloning efficiency; some mycoplasma species induce cellular transformation, resulting in anchorage-independent growth (Macpherson 1973; Tsai et al. 1995). Always include controls in clonogenic assays to detect unexpected changes in cloning efficiency.

20.6.1 Cloning Using Conditioned Medium

Conditioned medium is perhaps the easiest way to stimulate cloning efficiency, as it is easily made by exposing culture medium to metabolically active cells (see Section

9.6.1). However, conditioned medium has some limitations, including the use of undefined components and the risk of cross-contamination from conditioning cells (Patel et al. 2003). Always perform authentication to ensure that the "cloned" population is not an artifact due to cross-contamination. The risk of cross-contamination can be reduced (although not eliminated) by freezing and thawing, centrifugation, or filtration to remove viable cells. The risk can also be reduced by using the same cells for conditioning as for cloning, but this may not be ideal for cloning efficiency. For example, culture of hematopoietic precursors is improved by using conditioned medium from various tumor cell lines, including 5637 (human bladder carcinoma), Mo (human leukemia), and WEHI-3 (mouse leukemia) (Drexler 2004). These cell lines are likely to secrete regulatory proteins and mitogenic factors that are required for the proliferation of hematopoietic cells (Drexler 2010).

Protocol P20.6 provides a general approach when preparing conditioned medium from various cell types. Some variations are suggested in the Protocol Notes.

PROTOCOL P20.6. PREPARATION OF CONDITIONED MEDIUM

Outline

Grow cells to the late log phase and harvest conditioned medium. Centrifuge, filter, freeze, and thaw to remove viable cells. Dilute with fresh medium before use.

Materials (sterile or aseptically prepared)

- ☐ Conditioning cells (see Notes below)
- ☐ Culture medium for cloning, e.g. Ham's F12 with 10% FBS
- ☐ Centrifuge tubes
- ☐ Sterilizing filter, e.g. 0.22- or 0.45-μm filter flask

Materials (non-sterile)

- ☐ BSC and associated consumables (see Protocol P12.1)

Procedure

1. Decide on the best cells for use to condition medium ("conditioning cells").
2. Grow conditioning cells to 50% confluence.
3. Change the medium and incubate for 48 hours.
4. Collect the conditioned medium into tubes and centrifuge at 1000 g for 10 minutes.
5. Filter the medium through a sterilizing filter.
6. Store the medium frozen at −20 °C and thaw it before use.
7. Add the conditioned medium to the cloning medium, using one part conditioned medium to two parts cloning medium.

Notes

Step 1: As a starting point, try the same cell line at higher density or mouse embryonic fibroblasts (MEFs) (see Protocol P13.1). Refer to the literature and other resources for your cell type. For example, 5637 bladder carcinoma cells may be used to condition medium for the establishment of leukemia-lymphoma cell lines (see Supp. S25.3).

Step 2: If conditioned medium is unsuccessful, wait until just before confluence and try conditioning the medium using cells at plateau rather than log phase.

Step 3: If conditioned medium is unsuccessful, try extending the conditioning period to four days.

Step 5: The medium may need to be clarified first by prefiltration through 5- and 1.2-μm filters (see Supp. S11.3).

Step 6: Medium may be stored as aliquots or multiple collections may be made from confluent cells and pooled.

20.6.2 Cloning on Feeder Layers

The reason that some cells do not clone well may be related to their inability to survive at low cell densities. One way to maintain cells at clonogenic densities, but mimic high cell densities at the same time, is to clone the cells onto a growth-arrested feeder layer (see Section 8.4). Although feeder layers are widely used for stem cell culture and primary culture of specific cell types (e.g. epidermal keratinocytes), they were originally developed to improve cloning efficiency (Puck and Marcus 1955).

Feeder cells typically overgrow any clonal populations unless they are treated to induce growth arrest. Commonly used treatments include γ-irradiation and mitomycin C (see Figure 20.11). The amount of mitomycin C used for growth arrest should be $\geq 2\,\mu g/1 \times 10^6$ cells; lower dosages may result in continued proliferation of some feeder cells (Macpherson and Bryden 1971). It is important that the cells enter the cell cycle after mitomycin C treatment, or the DNA damage may be repaired. Mitomycin C treated feeder cells can be stored at 4 °C for up to one week (personal communication, C. Wigley) or frozen for later use (see Protocols P15.1, P15.2). When conditions are correct, the feeder cells will remain viable for up to three weeks, but they will eventually die out and are not carried over if the colonies are isolated.

Feeder layers, like conditioned medium, have limitations when used for cloning. Continued proliferation of cells within the feeder layer can occur, resulting in cross-contamination of clonal populations (Schneider et al. 2008). Feeder

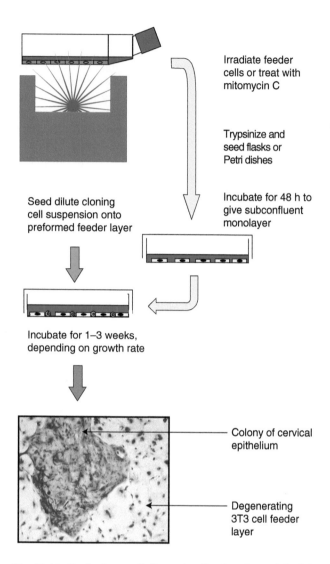

Irradiate feeder cells or treat with mitomycin C

Trypsinize and seed flasks or Petri dishes

Incubate for 48 h to give subconfluent monolayer

Seed dilute cloning cell suspension onto preformed feeder layer

Incubate for 1–3 weeks, depending on growth rate

Colony of cervical epithelium

Degenerating 3T3 cell feeder layer

Fig. 20.11. Feeder layers. Cells are irradiated and trypsinized (or treated with mitomycin C) and seeded at a low density to enhance cloning efficiency. *Source*: R. Ian Freshney, image courtesy of M. G. Freshney.

layers act as an undefined substrate and may carry microbial contaminants, resulting in safety concerns for later applications (Llames et al. 2015). Always test for microbial contamination and perform authentication testing on any cloned derivative that is isolated using this method. If feeder cells come from a different species, cross-contamination can be readily detected, provided the feeder cells make up a high enough proportion of the culture (see Sections 17.3.3, 17.3.4). Low levels of contamination (< 10%) are harder to detect (see Section 17.4.1).

Various species and cell types can be used to generate feeder layers, as described in a previous chapter (see Section 8.4). Early-passage MEFs probably produce more extracellular matrix (ECM) components than do established cell lines, but screening different cells is the only way to be sure which type of cell is best for your specific application. Cells may vary in their sensitivity to irradiation or mitomycin C; for example, 3T3 cells will become resistant if they undergo transformation (Pourreyron et al. 2011). A trial run should be carried out before cloning is attempted to ensure that none of the feeder cells survive. Even then, it is advisable to seed additional control plates with feeder cells alone when cloning on a feeder layer.

PROTOCOL P20.7. PREPARATION OF FEEDER LAYERS

Outline

Commence culture of feeder cells (e.g. MEFs). Induce growth arrest using irradiation or treatment with mitomycin C. Reseed at medium density before using for cloning.

❖ *Safety Note. Mitomycin C is a hazardous chemical. Refer to its Safety Data Sheet for recommended handling and Personal Protective Equipment (PPE). Irradiation is performed using a high-energy source. Refer to local safety regulations for radiation protection and monitoring procedures (see also Supp. S6.2).*

Materials (sterile or aseptically prepared)

☐ Primary culture of MEFs from 13-day mouse embryo (see Protocols P13.1, P13.4, P13.5) or other suitable feeder cells, e.g. 3T3 cell line (Pourreyron et al. 2011)
☐ Culture medium for MEFs
☐ Culture medium for cells to be cloned
☐ Mitomycin C (Sigma-Aldrich), 100 μg/ml stock, in Hanks's balanced salt solution (HBSS) or serum-free medium (store shielded from light)
☐ Culture flasks or Petri dishes
☐ Centrifuge tubes, universal containers

Materials (non-sterile)

☐ BSC and associated consumables (see Protocol P12.1)
☐ Irradiator or X-ray machine that can deliver 30 Gy in ≤ 30 minutes (alternative to mitomycin C)

Procedure

1. Trypsinize MEFs (see Protocol P14.2).
2. Reseed the cells in flasks at 1×10^5 cells/ml and incubate for three to five days. Observe to ensure that cells are subconfluent and still dividing.
3. Block further proliferation using one of the following methods:

(a) Mitomycin C treatment in flasks or dishes (Macpherson and Bryden 1971): make up mitomycin C in culture medium to give a final concentration of 0.25 μg/ml. Refeed subconfluent cells ($\sim 1 \times 10^5$ cells/ml) with medium containing mitomycin C. Incubate at 37 °C overnight ($\sim$ 18 hours) and replace the medium on the following day. Trypsinize flasks after 24–72 hours for use as a feeder layer.

(b) Mitomycin C treatment in suspension (Stanley 2002): make up mitomycin C in culture medium to give a final concentration of 20 μg/ml. Trypsinize the flask and resuspend at 1×10^7 cells/ml in medium containing mitomycin C (giving $2 \, \mu g/10^6$ cells). Incubate at 37 °C for one hour and centrifuge the cells, removing the mitomycin C after centrifugation. Reseed cells at 1×10^5 cells/ml or store the suspension at 4 °C for up to 48 hours (see Notes below).

(c) Irradiation in flasks: bring flasks to the irradiator or X-ray machine and expose to 60 Gy. Trypsinize flasks after 24–72 hours for use as a feeder layer.

(d) Irradiation in suspension: trypsinize the flask and resuspend cells in a container that can be used for irradiation. Bring the cell suspension to the irradiator or X-ray machine and expose to 60 Gy. Reseed cells at 1×10^5 cells/ml or store the suspension at 4 °C for up to 48 hours (see Notes below).

4. Plate feeder cells for use in cloning at 5×10^4 cells/ml (1×10^4 cells/cm^2).
5. Seed cells for cloning after a further 24–48 hours.

Notes

Step 1: A single batch of primary MEFs may provide feeder layers for an extended period of time. Freeze surplus cells for later use.

Step 3: It is usually possible to store feeder cells at 4 °C for up to 48 hours (Pourreyron et al. 2011). However, their effectiveness may be reduced.

Step 4: The cell density will vary depending on your application. Approximately 50% confluence is commonly used for human pluripotent stem cell (hPSC) culture (Healy and Ruban 2015), while confluent feeder layers are preferred for tumor cell culture (see Section 25.3.2). Try different densities in a trial run.

20.6.3 Optimization of Clonal Growth

Cloning efficiency may be optimized in a more defined fashion by focusing on specific aspects of the culture system, including:

(1) **Medium.** If possible, choose a rich medium that has been optimized for the specific cell type. For example, Ham's F12 medium was originally designed for the clonal growth of normal human fibroblasts (see Section 9.4; see also Appendix C).

(2) **Serum.** When serum is required, FBS is generally better than calf or horse serum. Select a batch for cloning experiments that gives a high cloning efficiency during tests (see Section 9.7.1; see also Figure 20.1e).

(3) **Trypsin.** Purified trypsin (0.05 μg/ml) may be preferable to crude trypsin, although opinions vary (Brown and Kiehn 1977). For example, the manufacturer of TrypLE™ Express has shown that it can improve cloning efficiency for A549 cells (with serum) and MDCK cells (serum-free) (Nestler et al. 2004). Remember that the specific activity of different trypsin preparations may vary, resulting in the need to adjust the concentration used (see Section 13.5.1). Cloning rates can be improved by using the smallest possible amount of trypsin at 4 °C (Ham and McKeehan 1978).

(4) **Substrate treatments.** Polylysine and other positively charged molecules improve the cloning efficiency of human fibroblasts in low serum concentrations (see Section 8.3.5) (McKeehan and Ham 1976; Ham and

McKeehan 1978). Rat tail collagen and fibronectin (5 μg/ml) can also improve cloning efficiency for some cell types (Ham 1974; Barnes and Sato 1980).

(5) **Hormones.** Insulin (1×10^{-10} IU/ml) increases the cloning efficiency of several cell types (Hamilton and Ham 1977). Dexamethasone (2.5×10^{-5} M or 10 μg/ml) improves the cloning efficiency of chick myoblasts and human normal glial cells, fibroblasts, glioma, and melanoma (see Figure 20.12a). It also gives increased clonal growth (colony size) if removed five days after plating (Freshney et al. 1980). Lower concentrations are preferable for epithelial cells, e.g. 1×10^{-7} M for lung carcinoma (see Figure 20.12b).

(6) **Intermediary metabolites.** Clonal growth requires special media constituents that may not be required for other cultures (Ham 1974). Oxo-acids (keto-acids) such as pyruvate or α-oxoglutarate (α-ketoglutarate) can increase cloning efficiency and are included in some rich media formulations, such as Ham's F12 and DMEM (see Appendix C) (McKeehan and McKeehan 1979). Nucleosides are present in αMEM (minimum essential medium) and may help to explain its higher cloning efficiency with human tumor cells (Baker et al. 1988). The addition of nucleosides may limit replication stress, at least in some cultures, e.g. during early tumorigenesis (Bester et al. 2011).

(7) **Carbon dioxide.** CO_2 is essential for maximum cloning efficiency in most cell types. Although 5% CO_2 is typically used, 2% is sufficient for many cells and may even

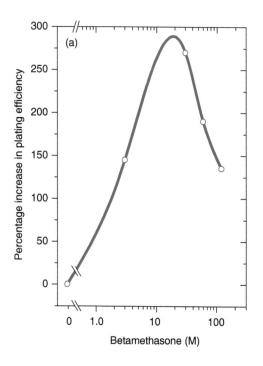

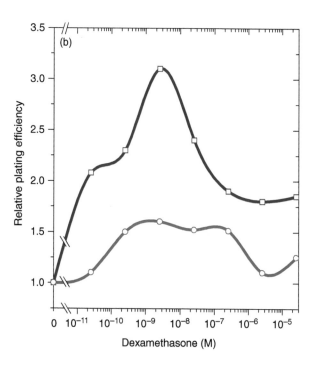

Fig. 20.12. Effect of glucocorticoids on cloning. Hydrocortisone analogs enhance the cloning efficiency of several cell types: (a) betamethasone and a human glioma cell line; (b) dexamethasone with SK-MES-1 (squares) and A549 (circles) human non-small cell lung carcinoma cells.

be better for human glia and fibroblasts. HEPES (20 mM) should be used with 2% CO_2 to protect the cells against pH fluctuations and CO_2 supply failure. At the other extreme, DMEM is designed for 10% CO_2 and is frequently used to clone myeloma hybrids for monoclonal antibody production (although perhaps not at the higher bicarbonate/CO_2 concentration). Always adjust the bicarbonate concentration if the CO_2 tension is altered, so that equilibrium is reached at pH 7.4 (see Table 9.1).

(8) *Oxygen.* Low oxygen tension (hypoxia; see Minireview M9.1) may improve cloning efficiency for some cell types (Richter et al. 1972). This effect is enhanced by incorporating rat red blood cells into the culture medium (Courtenay et al. 1978). The problem may lie with the toxicity of free oxygen; with red blood cells present it will be bound to hemoglobin. It may be possible to produce a similar effect by adding free radical scavengers such as 2-mercaptoethanol (50 µM), glutathione (1 mM), or α-thioglycerol (75 µM) (Iscove et al. 1980). Perfluorocarbons can also be used in suspension culture to enhance oxygen transfer (Lowe et al. 1998).

20.7 SELECTIVE CULTURE CONDITIONS

The success of cloning depends on achieving (i) a suitable cloning efficiency; and (ii) a truly clonal population, i.e. one derived from a single cell. In practice, these requirements are often not achieved. Some cell types are difficult or impossible to clone, or the process may not be feasible due to limited lifespan (see Section 20.2). Additional cells may be present in a colony for various reasons, including incomplete dispersal of cells, reaggregation before attachment, and migration or detachment of cells after adhesion (Ham 1974). As a result, overgrowth of cells from other lineages is a major problem. Theoretically, a 90% pure culture of cell line A will be 50% overgrown by a 10% contamination with cell line B in 10 days, given that B grows 50% faster than A (see Figure 20.13). Additional factors may encourage overgrowth; for example, fibroblasts may actively inhibit the growth of other cell types (Halaban 2004). Culture conditions must be employed that select for the desired lineage. In practice, there may be a series of selective conditions to discourage overgrowth by fibroblasts or other unwanted lineages, encourage proliferation of the desired cell type (e.g. tumor rather than normal cells), and induce expression of differentiated characteristics (see Sections 25.4, 26.3).

20.7.1 Selective Media and Inhibitors

Selective media are frequently used for microorganisms, but this approach is limited for animal cells because of the basic metabolic similarities of most cell types from any one species. The problem is accentuated by the effect of serum, which tends to mask the selective properties of different media. The most successful approach to selectively grow specific cell types

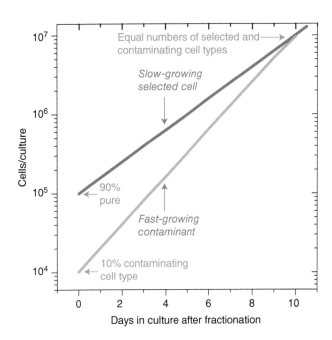

Fig. 20.13. Overgrowth in mixed culture. Overgrowth of a slow-growing cell line by a rapidly growing contaminant. This figure portrays a hypothetical example, but it demonstrates that a 10% contamination with a cell population that doubles every 24 hours will reach equal proportions with a cell population that doubles every 36 hours after only 10 days of growth.

at early passage is to use serum-free media (see Section 10.1.2). Many selective media are available commercially (e.g. fibroblast control kits, CHI Scientific) and are frequently supplied alongside the cell type of interest when purchasing primary or early passage cultures (see Table 24.1).

Positive selection focuses on selection of the cell type of interest through unique properties or markers. For example, selective HAT medium can be used to positively select for hybrid clones from somatic hybridization experiments. HAT medium contains hypoxanthine, aminopterin, and thymidine (see Appendix B) and selects hybrids that express hypoxanthine guanine phosphoribosyltransferase and thymidine kinase, generated from parental cells that are deficient in one or the other enzyme (see Supp. S24.21, S24.22) (Littlefield 1964). Similarly, antibiotics can be used to positively select for cells that carry an antibiotic resistance gene following transfection (see Section 22.4.1).

Negative selection of undesirable populations may be combined with positive approaches or used in isolation. Perhaps the best example of this approach was the development of a monoclonal antibody that targeted the stromal cells of a human breast carcinoma (Edwards et al. 1980). Used with complement, this antibody proved to be cytotoxic to fibroblasts from several tumors and helped to purify a number of malignant cell lines. Drug- or toxin-conjugated antibodies can also be used to kill certain cell types selectively (Paraskeva et al. 1985; Beattie et al. 1990). However, selective antibodies are typically used for magnetic separation rather than cell killing (see Section 21.3).

Nutritional selection has been achieved by substituting L-valine for D-valine in the culture medium, which selects for some epithelial cell types, e.g. kidney tubular epithelial cells (Gilbert and Migeon 1975, 1977). Cells must express D-amino acid oxidase to convert D-valine into L-valine, which is an essential amino acid (see Table 9.3). This approach has been used to isolate bovine mammary epithelial cells, rat Schwann cells, and endothelial cells from rat brain (Sordillo et al. 1988; Armati and Bonner 1990; Abbott et al. 1992). However, it is often ineffective against human fibroblasts (Masson et al. 1993). *Cis*-hydroxyproline can be used to inhibit fibroblast overgrowth, but may prove toxic to other cell types (Kao and Prockop 1977).

Different sensitivities to drugs or growth factors can be used for selection. Fibroblast overgrowth has been inhibited by adding sodium ethylmercurithiosalicylate, phenobarbitone, or carbamazepine (Braaten et al. 1974; Fry et al. 1979; Parada-Turska et al. 2013). Fibroblasts also tend to be more sensitive to G418 (Geneticin) at $100\,\mu g/ml$ (Halaban and Alfano 1984). When a mixture of cells shows different responses to growth factors, it is possible to stimulate one cell type with the appropriate growth factor and then, taking advantage of the increased sensitivity of the more rapidly growing cells, kill the cells selectively with cytarabine (cytosine arabinoside) or irradiation. Alternatively, an inhibitor may be added or a growth factor removed to halt proliferation in one population; the remaining cells can be killed with cytarabine or irradiation.

20.7.2 Selective Substrates

Feeder layers allow the selective growth of specific cell types, such as epidermal and cervical keratinocytes (see Sections 24.2.1, 24.2.9). Confluent feeder layers can repress stromal overgrowth during culture of tumor cells, including breast and colon carcinoma (see Section 25.4.2; see also Figure 20.1c, d; Supp. S25.1). The role of the feeder layer is probably quite complex. To illustrate this point, one of the authors (R. Ian Freshney) found that human glioma would grow on confluent feeder layers of normal glia, whereas cells derived from normal brain would not. Feeder cells provide not only ECM for adhesion of the epithelium, but also positively acting growth factors and negative regulators that inactivate transforming growth factor-β (TGF-β) (Maas-Szabowski and Fusenig 1996).

Substrate treatments can be used to select for specific cell types. Matrigel® and laminin have been shown to favor epithelial survival and proliferation (Bissell et al. 1987; Kibbey et al. 1992). Collagen and fibronectin can be used to enhance epithelial cell attachment and growth in some tissue types (see Supp. S24.4) and to support endothelial cell growth and function (Relou et al. 1998; Martin et al. 2004). Because

the constituents of the ECM are increasingly understood, it should be possible to create more selective substrates by mixing various collagens with laminin, proteoglycans, and other proteins. More specialized techniques can also be used to modulate the physical properties of the substrate, including its surface charge, stiffness, and topography (Kshitiz et al. 2012). These approaches are increasingly used to regulate stem cell lineage specification but could be used to select for specific cell types in other contexts.

20.7.3 Selection by Adhesion and Detachment

Different cell types have different affinities for the culture substrate and attach at different rates. If a primary cell suspension is seeded into one flask and transferred to a second flask after 30 minutes, a third flask after 1 hour, and so on for up to 24 hours, the most adhesive cells will be found in the first flask and the least adhesive in the last. Macrophages will tend to remain in the first flask, fibroblasts in the next few flasks, epithelial cells in the next few flasks, and, finally, hematopoietic cells in the last flask. A similar method has been used to isolate human chondroprogenitor cells from cartilage (Archer et al. 2007). This method can be modified using collagenase for disaggregation of tissue fragments (see Protocol P13.6). Most of the cells that are released will not attach within 48 hours unless the collagenase is removed. However, macrophages will migrate out of the tissue fragments and attach during this period; the remaining cells are transferred to a fresh flask after 48–72 hours of collagenase treatment.

Different cell types detach from the culture substrate at different rates. Periodic brief exposure to trypsin can remove fibroblasts from cultures of human fetal intestine and skin (Owens et al. 1974; Milo et al. 1980). Collagenase exposure for a few days at a time can be used to remove fibroblasts from cultures of breast tissue, leaving the epithelial cells in place (Lasfargues 1973). In contrast, ethylene diamine tetra-acetic acid (EDTA) appears to release epithelial cells more readily than it does fibroblasts (Paul 1975). Dispase II (Roche CustomBiotech) can be used to selectively dislodge sheets of epithelium from human cervical cultures grown on feeder layers of 3T3 cells, without dislodging the 3T3 layer (see Supp. S25.1). This technique may be used to select against stromal fibroblasts.

20.7.4 Selection by Anchorage-Independent Growth

Cloning in soft agar can be used to select for cells that display anchorage-independent growth and exclude most of the normal cells (see Sections 20.3.1, 25.4.3). This technique was originally used to select colonies of fibroblasts after viral transformation (Macpherson and Montagnier 1964). Normal cells will not form colonies in suspension with the high efficiency

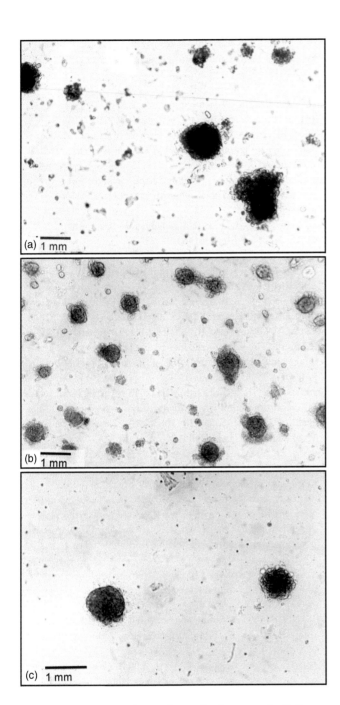

Fig. 20.14. Suspension clones of normal and tumor cells. Cells were plated out at 5×10^5 per 35 mm dish (2.5×10^5 cells/ml) in 1.5% Methocel over a 1.25% agar underlay. Colonies were photographed after three weeks. Cells are (a) melanoma; (b) human normal embryonic skin fibroblasts; (c) human normal adult glia. Scale bars as shown. *Source*: R. Ian Freshney.

of virally transformed cells, although they will often do so with low cloning efficiencies. The difference between transformed and untransformed cells is not as clear with early passage tumor cell lines, as cloning efficiencies can be quite low. Normal glia and fetal skin fibroblasts also form colonies in suspension with similar efficiencies (< 1%; see Figure 20.14).

20.8 CONCLUSIONS: CLONING AND SELECTION

The best advice on cloning comes from experienced practitioners such as Richard Ham. Ham advised beginners to (i) start with a procedure from the literature that has been empirically demonstrated to be successful; (ii) experimentally verify the efficacy of each modification that you introduce; and (iii) be as gentle as possible (Ham 1974). Successful cloning will mean that you have a pure population of cells with no need for selection. However, this is often not possible and further selection must be performed. In practice, you may need to combine several selection techniques or apply selective conditions for a prolonged period of time. For example, epithelial cells in your culture may be selected through selective adhesion and detachment techniques (see Section 20.7.3), but this advantage is rapidly lost if the remaining fibroblasts have a growth advantage (see Figure 20.13). Selective conditions are often combined with physical separation techniques, as described in the following chapter, to fully optimize procedures for a specific cell type.

Suppliers

Supplier	URL
BD Biosciences	http://www.bdbiosciences.com
Bel-Art (SP Scienceware)	http://www.belart.com
CHI Scientific	http://www.chiscientific.com/index1.aspx
Dow Chemical Company	http://www.dow.com/en-us/product-search/methocelpremium
Molecular Devices	http://www.moleculardevices.com
Roche CustomBiotech	http://custombiotech.roche.com/home.html
Sigma-Aldrich (Merck)	http://www.sigmaaldrich.com/life-science/cell-culture.html

REFERENCES

Abbott, N.J., Hughes, C.C., Revest, P.A. et al. (1992). Development and characterisation of a rat brain capillary endothelial culture: towards an in vitro blood–brain barrier. *J. Cell Sci.* 103 (Pt 1): 23–37.

Archer, C.W., Oldfield, S., Redman, S. et al. (2007). Isolation, characterization, and culture of soft connective tissue stem/progenitor cells. In: *Culture of Human Stem Cells* (eds. R.I. Freshney, G.N. Stacey and J.M. Auerbach), 233–248. Hoboken, NJ: Wiley.

Armati, P.J. and Bonner, J. (1990). A technique for promoting Schwann cell growth from fresh and frozen biopsy nerve utilizing D-valine medium. *In Vitro Cell Dev. Biol.* 26 (12): 1116–1118.

Baker, F.L., Ajani, J., Spitzer, G. et al. (1988). High colony-forming efficiency of primary human tumor cells cultured in the adhesive-tumor-cell culture system: improvements with medium and serum alterations. *Int. J. Cell Cloning* 6 (2): 95–105. https://doi.org/10.1002/stem.5530060203.

Barnes, D. and Sato, G. (1980). Methods for growth of cultured cells in serum-free medium. *Anal. Biochem.* 102 (2): 255–270.

Barrandon, Y. and Green, H. (1987). Three clonal types of keratinocyte with different capacities for multiplication. *Proc. Natl Acad. Sci. U.S.A.* 84 (8): 2302–2306.

Beattie, G.M., Lappi, D.A., Baird, A. et al. (1990). Selective elimination of fibroblasts from pancreatic islet monolayers by basic fibroblast growth factor-saporin mitotoxin. *Diabetes* 39 (8): 1002–1005.

Beaver, C.M., Ahmed, A., and Masters, J.R. (2014). Clonogenicity: holoclones and meroclones contain stem cells. *PLoS One* 9 (2): e89834. https://doi.org/10.1371/journal.pone.0089834.

Bester, A.C., Roniger, M., Oren, Y.S. et al. (2011). Nucleotide deficiency promotes genomic instability in early stages of cancer development. *Cell* 145 (3): 435–446. https://doi.org/10.1016/j.cell.2011.03.044.

Bissell, D.M., Arenson, D.M., Maher, J.J. et al. (1987). Support of cultured hepatocytes by a laminin-rich gel. Evidence for a functionally significant subendothelial matrix in normal rat liver. *J. Clin. Invest.* 79 (3): 801–812. https://doi.org/10.1172/JCI112887.

Braaten, J.T., Lee, M.J., Schenk, A. et al. (1974). Removal of fibroblastoid cells from primary monolayer cultures of rat neonatal endocrine pancreas by sodium ethylmercurithiosalicylate. *Biochem. Biophys. Res. Commun.* 61 (2): 476–482.

Brown, M. and Kiehn, D. (1977). Protease effects on specific growth properties of normal and transformed baby hamster kidney cells. *Proc. Natl Acad. Sci. U.S.A.* 74 (7): 2874–2878.

Buick, R.N., Stanisic, T.H., Fry, S.E. et al. (1979). Development of an agar-methyl cellulose clonogenic assay for cells in transitional cell carcinoma of the human bladder. *Cancer Res.* 39 (12): 5051–5056.

Courtenay, V.D., Selby, P.J., Smith, I.E. et al. (1978). Growth of human tumour cell colonies from biopsies using two soft-agar techniques. *Br. J. Cancer* 38 (1): 77–81.

Domann, R. and Martinez, J. (1995). Alternative to cloning cylinders for isolation of adherent cell clones. *Biotechniques* 18 (4): 594–595.

Drexler, H.G. (2004). Establishment and culture of human leukemia-lymphoma cell lines. In: *Culture of Human Tumor Cells* (eds. R. Pfragner and R.I. Freshney), 319–348. Hoboken, NJ: Wiley-Liss.

Drexler, H. G. (2010). Guide to leukemia-lymphoma cell lines. https://www.dsmz.de/research/human-and-animal-cell-lines/guide-to-leukemia-lymphoma-cell-lines.html (accessed 24 February 2011).

Edwards, P.A., Easty, D.M., and Foster, C.S. (1980). Selective culture of epithelioid cells from a human squamous carcinoma using a monoclonal antibody to kill fibroblasts. *Cell Biol. Int. Rep.* 4 (10): 917–922.

Freshney, R.I., Sherry, A., Hassanzadah, M. et al. (1980). Control of cell proliferation in human glioma by glucocorticoids. *Br. J. Cancer* 41 (6): 857–866.

Fry, J.R., Wiebkin, P., and Bridges, J.W. (1979). Phenobarbitone induction of microsomal mono-oxygenase activity in primary cultures of adult rat hepatocytes [proceedings]. *Biochem. Soc. Trans.* 7 (1): 119–121.

Gilbert, S.F. and Migeon, B.R. (1975). D-valine as a selective agent for normal human and rodent epithelial cells in culture. *Cell* 5 (1): 11–17.

Gilbert, S.F. and Migeon, B.R. (1977). Renal enzymes in kidney cells selected by D-valine medium. *J. Cell. Physiol.* 92 (2): 161–167. https://doi.org/10.1002/jcp.1040920204.

Halaban, R. (2004). Culture of melanocytes from normal, benign, and malignant lesions. In: *Culture of Human Tumor Cells* (eds. R. Pfragner and R.I. Freshney), 289–318. Hoboken, NJ: Wiley-Liss.

Halaban, R. and Alfano, F.D. (1984). Selective elimination of fibroblasts from cultures of normal human melanocytes. *In Vitro* 20 (5): 447–450.

Ham, R.G. (1974). Unique requirements for clonal growth. *J. Natl Cancer Inst.* 53 (5): 1459–1463.

Ham, R.G. and McKeehan, W.L. (1978). Development of improved media and culture conditions for clonal growth of normal diploid cells. *In Vitro* 14 (1): 11–22.

Hamburger, A.W. and Salmon, S.E. (1977). Primary bioassay of human tumor stem cells. *Science* 197 (4302): 461–463.

Hamilton, W.G. and Ham, R.G. (1977). Clonal growth of Chinese hamster cell lines in protein-free media. *In Vitro* 13 (9): 537–547.

Healy, L. and Ruban, L. (2015). *Atlas of Human Pluripotent Stem Cells in Culture*. London: Springer.

Horman, S.R., To, J., and Orth, A.P. (2013). An HTS-compatible 3D colony formation assay to identify tumor-specific chemotherapeutics. *J. Biomol. Screen.* 18 (10): 1298–1308. https://doi.org/10.1177/1087057113499405.

Hornsby, P.J., Yang, L., Lala, D.S. et al. (1992). A modified procedure for replica plating of mammalian cells allowing selection of clones based on gene expression. *Biotechniques* 12 (2): 244–251.

Iscove, N.N., Guilbert, L.J., and Weyman, C. (1980). Complete replacement of serum in primary cultures of erythropoietin-dependent red cell precursors (CFU-E) by albumin, transferrin, iron, unsaturated fatty acid, lecithin and cholesterol. *Exp. Cell. Res.* 126 (1): 121–126.

Kao, W.W. and Prockop, D.J. (1977). Proline analogue removes fibroblasts from cultured mixed cell populations. *Nature* 266 (5597): 63–64.

Karin, N.J. (1999). Cloning of transfected cells without cloning rings. *Biotechniques* 27 (4): 681–682. https://doi.org/10.2144/99274bm10.

Ke, N., Albers, A., Claassen, G. et al. (2004). One-week 96-well soft agar growth assay for cancer target validation. *Biotechniques* 36 (5): 826–828, 830, 832-3. doi: https://doi.org/10.2144/04365ST07.

Kibbey, M.C., Royce, L.S., Dym, M. et al. (1992). Glandular-like morphogenesis of the human submandibular tumor cell line A253 on basement membrane components. *Exp. Cell. Res.* 198 (2): 343–351.

Kshitiz, P.J., Kim, P., Helen, W. et al. (2012). Control of stem cell fate and function by engineering physical microenvironments. *Integr. Biol. (Camb.)* 4 (9): 1008–1018.

Lasfargues, E.Y. (1973). Human mammary tumors. In: *Tissue Culture: Methods and Applications* (eds. P.F. Kruse Jr. and M.K. Patterson), 45–50. New York: Academic Press.

Littlefield, J.W. (1964). Selection of hybrids from matings of fibroblasts in vitro and their presumed recombinants. *Science* 145 (3633): 709–710.

Llames, S., Garcia-Perez, E., Meana, A. et al. (2015). Feeder layer cell actions and applications. *Tissue Eng. Part B Rev.* 21 (4): 345–353. https://doi.org/10.1089/ten.TEB.2014.0547.

Lowe, K.C., Davey, M.R., and Power, J.B. (1998). Perfluorochemicals: their applications and benefits to cell culture. *Trends Biotechnol.* 16 (6): 272–277.

Maas-Szabowski, N. and Fusenig, N.E. (1996). Interleukin-1-induced growth factor expression in postmitotic and resting fibroblasts. *J. Invest. Dermatol.* 107 (6): 849–855.

Macpherson, I. (1973). Soft agar techniques. In: *Tissue Culture: Methods and Applications* (eds. P.F. Kruse Jr. and M.K. Patterson), 276–280. New York: Academic Press.

Macpherson, I. and Bryden, A. (1971). Mitomycin C treated cells as feeders. *Exp. Cell. Res.* 69 (1): 240–241.

Macpherson, I. and Montagnier, L. (1964). Agar suspension culture for the selective assay of cells transformed by polyoma virus. *Virology* 23: 291–294.

Marcus, P.I., Sato, G.H., Ham, R.G. et al. (2006). A tribute to Dr. Theodore T. Puck (September 24, 1916–November 6, 2005). *In Vitro Cell Dev. Biol. Anim.* 42 (8–9): 235–241. https://doi.org/10.1290/0606039A.1.

Martin, G.M. (1973). Dilution plating on coverslip fragments. In: *Tissue Culture: Methods and Applications* (eds. P.F. Kruse Jr. and M.K. Patterson), 264–266. New York: Academic Press.

Martin, S.M., Schwartz, J.L., Giachelli, C.M. et al. (2004). Enhancing the biological activity of immobilized osteopontin using a type-1 collagen affinity coating. *J. Biomed. Mater. Res. A* 70 (1): 10–19. https://doi.org/10.1002/jbm.a.30052.

Masson, E.A., Atkin, S.L., and White, M.C. (1993). D-valine selective medium does not inhibit human fibroblast growth in vitro. *In Vitro Cell Dev. Biol. Anim.* 29A (12): 912–913.

Mathupala, S. and Sloan, A.A. (2009). An agarose-based cloning-ring anchoring method for isolation of viable cell clones. *Biotechniques* 46 (4): 305–307. https://doi.org/10.2144/000113079.

Maurer, H.R. and Echarti, C. (1990). Clonogenic assays for hematopoietic and tumor cells using agar-containing capillaries. *Methods Mol. Biol.* 5: 379–394. https://doi.org/10.1385/0-89603-150-0:379.

McKeehan, W.L. and Ham, R.G. (1976). Stimulation of clonal growth of normal fibroblasts with substrata coated with basic polymers. *J. Cell Biol.* 71 (3): 727–734.

McKeehan, W.L. and McKeehan, K.A. (1979). Oxocarboxylic acids, pyridine nucleotide-linked oxidoreductases and serum factors in regulation of cell proliferation. *J. Cell. Physiol.* 101 (1): 9–16. https://doi.org/10.1002/jcp.1041010103.

Milo, G.E., Ackerman, G.A., and Noyes, I. (1980). Growth and ultrastructural characterization of proliferating human keratinocytes in vitro without added extrinsic factors. *In Vitro* 16 (1): 20–30.

Moen, J.K. (1935). The development of pure cultures of fibroblasts from single mononuclear cells. *J. Exp. Med.* 61 (2): 247–260.

Muirhead, E.E., Rightsel, W.A., Pitcock, J.A. et al. (1990). Isolation and culture of juxtaglomerular and renomedullary interstitial cells. *Methods Enzymol.* 191: 152–167.

Nestler, L., Evege, E., McLaughlin, J. et al. (2004). TrypLE express: a temperature stable replacement for animal trypsin in cell dissociation applications. *Quest* 1 (1): 42–47. https://www.thermofisher.com/document-connect/document-connect.html?url=https://assets.thermofisher.com/tfs-assets/lsg/brochures/o-13047_tryple.pdf (accessed 29 January 2019).

Owens, R.B., Smith, H.S., and Hackett, A.J. (1974). Epithelial cell cultures from normal glandular tissue of mice. *J. Natl Cancer Inst.* 53 (1): 261–269.

Parada-Turska, J., Nowicka-Stazka, P., Majdan, M. et al. (2013). Anti-epileptic drugs inhibit viability of synoviocytes in vitro. *Ann. Agric. Environ. Med.* 20 (3): 571–574.

Paraskeva, C., Buckle, B.G., and Thorpe, P.E. (1985). Selective killing of contaminating human fibroblasts in epithelial cultures derived from colorectal tumours using an anti Thy-1 antibody-ricin conjugate. *Br. J. Cancer* 51 (1): 131–134. https://doi.org/10.1038/bjc.1985.19.

Patel, V.A., Logan, A., Watkinson, J.C. et al. (2003). Isolation and characterization of human thyroid endothelial cells. *Am. J. Physiol. Endocrinol. Metab.* 284 (1): E168–E176. https://doi.org/10.1152/ajpendo.00096.2002.

Paul, J. (1975). *Cell and Tissue Culture*, 5e. Edinburgh: Churchill Livingstone.

Penfornis, P. and Pochampally, R. (2016). Colony forming unit assays. *Methods Mol. Biol.* 1416: 159–169. https://doi.org/10.1007/978-1-4939-3584-0_9.

Pomp, J., Wike, J.L., Ouwerkerk, I.J. et al. (1996). Cell density dependent plating efficiency affects outcome and interpretation of colony forming assays. *Radiother. Oncol.* 40 (2): 121–125.

Pourreyron, C., Purdie, K.J., Watt, S.A. et al. (2011). Feeder layers: co-culture with nonneoplastic cells. *Methods Mol. Biol.* 731: 467–470. https://doi.org/10.1007/978-1-61779-080-5_37.

Puck, T.T. and Marcus, P.I. (1955). A rapid method for viable cell titration and clone production with HeLa cells in tissue culture: the use of X-irradiated cells to supply conditioning factors. *Proc. Natl Acad. Sci. U.S.A.* 41 (7): 432–437.

Relou, I.A., Damen, C.A., van der Schaft, D.W. et al. (1998). Effect of culture conditions on endothelial cell growth and responsiveness. *Tissue Cell* 30 (5): 525–530.

Richter, A., Sanford, K.K., and Evans, V.J. (1972). Influence of oxygen and culture media on plating efficiency of some mammalian tissue cells. *J. Natl Cancer Inst.* 49 (6): 1705–1712.

Robb, J.A. (1973). Replica plating. In: *Tissue Culture: Methods and Applications* (eds. P.F. Kruse Jr. and M.K. Patterson), 270–274. New York: Academic Press.

Sanford, K.K., Earle, W.R., and Likely, G.D. (1948). The growth in vitro of single isolated tissue cells. *J. Natl. Cancer Inst.* 9 (3): 229–246.

Sanford, K.K., Covalesky, A.B., Dupree, L.T. et al. (1961). Cloning of mammalian cells by a simplified capillary technique. *Exp. Cell. Res.* 23: 361–372.

Sant, S. and Johnston, P.A. (2017). The production of 3D tumor spheroids for cancer drug discovery. *Drug Discov. Today Technol.* 23: 27–36. https://doi.org/10.1016/j.ddtec.2017.03.002.

Schaeffer, W.I. (1990). Terminology associated with cell, tissue, and organ culture, molecular biology, and molecular genetics. Tissue Culture Association Terminology Committee. *In Vitro Cell Dev. Biol.* 26 (1): 97–101.

Schneider, A., Spitkovsky, D., Riess, P. et al. (2008). "The good into the pot, the bad into the crop!" – a new technology to free stem cells from feeder cells. *PLoS One* 3 (11): e3788. https://doi.org/10.1371/journal.pone.0003788.

Sjostedt, S., Bezak, E., and Marcu, L. (2014). Experimental investigation of the cell survival in dose cold spot. *Acta Oncol.* 53 (1): 16–24. https://doi.org/10.3109/0284186X.2013.787165.

Sordillo, L.M., Oliver, S.P., and Akers, R.M. (1988). Culture of bovine mammary epithelial cells in D-valine modified medium: selective removal of contaminating fibroblasts. *Cell Biol. Int. Rep.* 12 (5): 355–364.

Stanley, M.A. (2002). Culture of human cervical epithelial cells. In: *Culture of Epithelial Cells* (eds. R.I. Freshney and M.G. Freshney), 138–169. Hoboken, NJ: Wiley-Liss.

Suh, H., Song, M.J., and Park, Y.N. (2003). Behavior of isolated rat oval cells in porous collagen scaffold. *Tissue Eng.* 9 (3): 411–420. https://doi.org/10.1089/107632703322066598.

Troyer, D.A. and Kreisberg, J.I. (1990). Isolation and study of glomerular cells. *Methods Enzymol.* 191: 141–152.

Tsai, S., Wear, D.J., Shih, J.W. et al. (1995). Mycoplasmas and oncogenesis: persistent infection and multistage malignant transformation. *Proc. Natl Acad. Sci. U.S.A.* 92 (22): 10197–10201.

Walters, M.S., Gomi, K., Ashbridge, B. et al. (2013). Generation of a human airway epithelium derived basal cell line with multipotent differentiation capacity. *Respir. Res.* 14: 135. https://doi.org/10.1186/1465-9921-14-135.

Wognum, B. and Lee, T. (2013). Simultaneous cloning and selection of hybridomas and transfected cell lines in semisolid media. *Methods Mol. Biol.* 946: 133–149. https://doi.org/10.1007/978-1-62703-128-8_9.

Young, H.E., Duplaa, C., Romero-Ramos, M. et al. (2004). Adult reserve stem cells and their potential for tissue engineering. *Cell Biochem. Biophys.* 40 (1): 1–80. https://doi.org/10.1385/CBB:40:1:1.

Zeng, C., Pesall, J.E., Gilkerson, K.K. et al. (2002). The effect of hepatocyte growth factor on Turkey satellite cell proliferation and differentiation. *Poult. Sci.* 81 (8): 1191–1198. https://doi.org/10.1093/ps/81.8.1191.

Zwain, I.H., Morris, P.L., and Cheng, C.Y. (1991). Identification of an inhibitory factor from a Sertoli clonal cell line (TM4) that modulates adult rat Leydig cell steroidogenesis. *Mol. Cell. Endocrinol.* 80 (1–3): 115–126.

CHAPTER 21

Cell Separation and Sorting

Cloning is an effective way to isolate a population of cells from a single cell type, but, as noted previously, some cell types are difficult or impossible to clone using standard procedures and culture conditions (see Section 20.7). Overgrowth of the desired population by more rapidly growing cells, such as fibroblasts, is a major concern. Even if cloning is successful, the clonal population may fail to express the desired behavior or may exhibit heterogeneous behavior. For example, colonies of cells from human fetal kidney and amniotic fluid can express at least two phenotypes, despite originating from single cells (Gilbert et al. 1981). There is a clear need for additional methods to isolate or enrich specific cell populations, based on their physical properties or behavior. This need is particularly urgent when working with rare cell populations, such as circulating tumor cells or hematopoietic stem cells (HSCs) (Shields et al. 2015). Circulating tumor cells are present at a frequency of approximately one cell per 10^6–10^7 nucleated blood or bone marrow cells (Pantel et al. 2008). Selection techniques must be used for such rare populations to positively select for the desired cell type. Negative selection techniques may also be needed to remove undesirable cell types, or a combination of positive and negative techniques may be used. Such techniques are typically based on (i) differences in physical properties such as cell density or size; (ii) the affinity of antibodies or other molecules for specific ligands that are present on the cell surface; or (iii) the presence of unique properties that identify the cell type of interest. These three principles can be applied by any tissue culture laboratory to achieve cell separation and sorting. They can also be used in specialized applications such as microfluidics (see Section 21.5).

21.1 CELL DENSITY AND ISOPYCNIC CENTRIFUGATION

21.1.1 Density Gradient Centrifugation

Typically, mammalian cells have a density (mass to volume ratio) between 1.055 and 1.110 g/ml (Pretlow and Pretlow 1989). Variations in density for a specific cell type are relatively low; the coefficient of variation for human erythrocytes is 11–15% for cell size, but only 0.5% for cell density (Norouzi et al. 2017). Cell density is therefore an attractive physical property for separation of different cell types. Isopycnic centrifugation is the most common technique used for density-based separation (see Figure 21.1). A sample is centrifuged in a density gradient, resulting in sedimentation of each component at an equilibrium position that is determined by its density. This approach has been used to separate multiple cell types from blood, bone marrow, liver, spleen, sperm, and other cell samples (Pertoft 2000; GE Healthcare 2014). For example, stem cell populations have been isolated from bone marrow, adipose tissue, and skeletal muscle using density gradient centrifugation (Juopperi et al. 2007; Insausti et al. 2012; Ceusters et al. 2017).

Cells should be centrifuged in a density medium that is non-toxic and non-viscous at high densities (1.10 g/ml), and that exerts little osmotic pressure in solution. Density media have been used since the 1960s and include Ficoll™ (GE Healthcare Life Sciences), Histopaque® (Sigma-Aldrich), Lymphoprep™ (Abbott Diagnostics Technologies AS), OptiPrep™ (Sigma-Aldrich, STEMCELL Technologies), and Percoll™ (GE Healthcare Life Sciences) (Boyum 1968a, b; Pertoft et al. 1977). Simple one-step separations

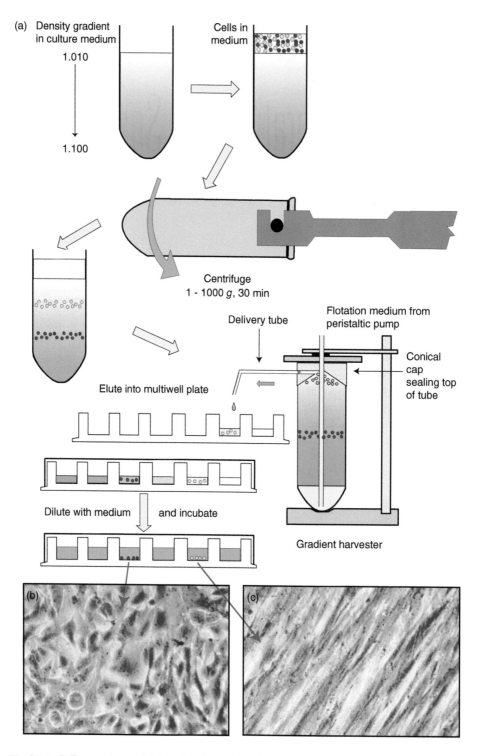

Fig. 21.1. Cell separation by density. (a) Cells are layered on to a preformed gradient (see Figures 21.2, 21.3) and the tube centrifuged. The tube is placed on a gradient harvester, and flotation medium (e.g. Fluorochemical FC43) is pumped down the inlet tube to the bottom of the gradient, displacing the gradient and cells upward and out through the delivery tube into a multiwell plate. The cells are diluted with medium (so that they will sink) and cultured. (b) Purified HeLa cells recovered from an artificial mixture of HeLa and MRC-5 fibroblasts. (c) Purified MRC-5 eluting at a lower density. *Source*: R. Ian Freshney; gradient harvester after an original design by G. D. Birnie.

can be performed by layering cultured cells in medium or blood cells in defibrinated or heparinized plasma over a layer of Ficoll-Paque at 1.077 g/ml. This technique can be used to separate viable from non-viable cells (see Protocol P13.8), separate lymphocytes from plasma and erythrocytes, or enrich the nucleated component of the bone marrow containing HSCs and mesenchymal stromal cells (MSCs) (see Supp. S23.5, 24.20). More sophisticated separations can be performed by layering cells on a density gradient (1.02–1.10 g/ml) (Pretlow and Pretlow 1989; Recktenwald 1997; Ito and Shinomiya 2001). Gradients can be generated by (i) layering different densities using a pipette, syringe, or pump; (ii) using a special gradient former (see Figure 21.2); or (iii) centrifuging Percoll at high speed for a self-forming gradient (see Figure 21.3). Because high g forces are not required, isopycnic sedimentation can be performed on any centrifuge without specialized equipment. Protocol P21.1 is useful as a starting point for cell separation by centrifugation using a density gradient.

PROTOCOL P21.1. CELL SEPARATION BY CENTRIFUGATION ON A DENSITY GRADIENT

Outline

Form a gradient, centrifuge cells through the gradient, collect fractions, dilute with medium, and culture. Use a gradient former and harvester if available.

- *Safety Note. Always balance samples carefully when performing centrifugation.*

Materials (Sterile)

- ☐ Percoll (GE Healthcare Life Sciences)
- ☐ Fluorinert™ FC-43 (Sigma-Aldrich)
- ☐ Culture medium
- ☐ Dulbecco's phosphate buffered saline without Ca^{2+} and Mg^{2+} (DPBS-A)
- ☐ Trypsin, 0.25%
- ☐ Centrifuge tubes, 15 or 50 ml
- ☐ Pipettes in an assortment of sizes, usually 1, 5, 10, and 25 ml
- ☐ Multiwell plates, 24- or 96-well
- ☐ Syringe with wide-bore needle (for preparing the gradient and harvesting cell fractions)
- ☐ Gradient former, e.g. SG 50 Gradient Maker (GE Healthcare Life Sciences)
- ☐ Gradient harvester, e.g. BR-188 Density Gradient Fractionation System (Brandel)

Materials (non-sterile)

- ☐ Biological safety cabinet (BSC) and associated consumables (see Protocol P12.1)

- ☐ Cell counting equipment (hemocytometer or automated counter; see Section 19.1)
- ☐ Centrifuge and rotor rated for the appropriate speed

Procedure

A. Gradient Preparation

1. Prepare a work space for aseptic technique using a BSC.
2. Prepare a stock solution of Percoll at the required density and osmolality, as per the manufacturer's instructions.
3. Prepare the gradient using a gradient former, if available. A continuous linear gradient may be produced by mixing, for example, 1.020 g/ml with 1.080 g/ml Percoll (see Figure 21.2).
4. If a gradient former is not available, the gradient can be prepared manually by layering Percoll in multiple steps in a 15- or 50-ml centrifuge tube:
   (a) Adjust the density of the Percoll medium to 1.10 g/ml and its osmotic strength to 290 mosmol/kg.
   (b) Mix the Percoll and regular media in varying proportions to give the desired density range (e.g. 1.020–1.100 g/ml) in 10 or 20 steps.
   (c) Layer one step over another, 1 or 2 ml per step, with a pipette, syringe, or peristaltic pump, starting with the densest solution and building up a stepwise density gradient in a centrifuge tube. Keep the tip of the instrument against the wall of the tube just above the surface to avoid splashing and mixing at the interface.
5. If preferred, the gradient can be prepared by centrifugation:
   (a) Place the medium containing Percoll at density 1.085 g/ml in a centrifuge tube.
   (b) Centrifuge at 20 000 g for one hour.
6. Gradients may be used immediately or left overnight (capped and left undisturbed).

B. Gradient Centrifugation

7. Perform subculture as per usual for that cell type (see Protocol P14.2, steps 1–16).
8. Count the cells and check to make sure there is a single cell suspension.
9. Using a syringe, pipettor, or fine-tipped pipette, layer up to 2 × 10^7 cells in 2 ml culture medium on top of the gradient.

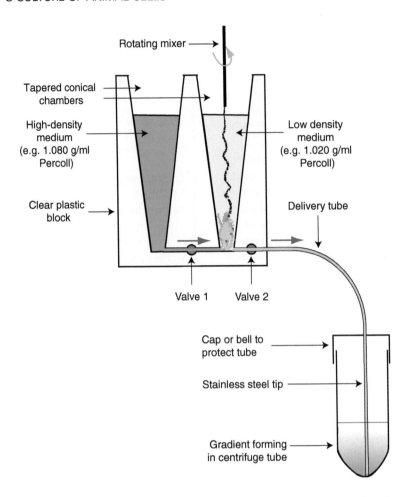

Fig. 21.2. Gradient former. Two chambers are cut in a solid transparent plastic block and connected by a thin canal across the bases of the chambers and exiting to the exterior. A delivery tube with a stainless steel tip, long enough to reach the bottom of the centrifuge tube, is inserted in the outlet. With the valves closed, the left-hand chamber is filled with high-density medium and the right-hand chamber with low-density medium. Valve 1 is opened, the stainless-steel tip is placed in the bottom of the tube, the stirrer is started, and then Valve 2 is opened. As the solution in the left chamber runs into the right, it mixes and gradually increases the density, while the mixture is running out into the tube.

10. Centrifuge the gradient at 100–1000 *g* for 20 minutes. Increase centrifuge speed gradually at the start of the run and do not apply the brake at the end of the run.

C. Fraction Collection and Culture

11. Collect fractions with a syringe or a gradient harvester (see Figure 21.1):
    (a) Collect 1-ml fractions into a 24-well plate or 0.1-ml fractions into a 96-well plate.
    (b) Take samples at intervals for cell counting and for determining the density (ρ) of the gradient medium during optimization (see Section 21.1.2).

12. Add an equal volume of medium to each well and mix to ensure that the cells settle to the bottom of the well. Incubate the plate at 37 °C and the required gas levels, e.g. 5% CO_2.

13. Change the medium to remove the Percoll after 24–48 hours incubation.

Notes

Step 2: Percoll is typically supplied as a 23% colloidal solution in water with a density of 1.130 ± 0.005 g/ml (GE Healthcare 2014). A stock solution should be made up at the correct osmolality as per the manufacturer's instructions.

Step 4: It is also possible, and may be preferable, to layer from the bottom, starting with the least dense

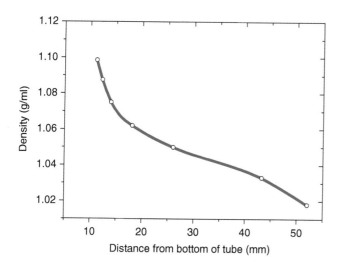

Fig. 21.3. Centrifuge-derived gradient. Gradient generated by spinning Percoll at 20 000 *g* for one hour.

solution and injecting each layer of progressively higher density below the previous one, using a syringe or peristaltic pump.

Step 5: Centrifugation generates a sigmoid gradient (see Figure 21.3). The shape of the curve is determined by the starting concentration of Percoll, the duration and centrifugal force of the centrifugation, the shape of the tube, and the type of rotor (Pertoft 2000).

Step 10: Alternatively, the tube may be allowed to stand on the bench for four hours and will sediment under 1 *g*.

Step 13: Density media can be removed after fractionation by washing. However, Percoll is non-toxic to cells and does not significantly adhere to membranes (Pertoft 2000).

21.1.2 Optimization of Density Gradients

Density-based centrifugation can be extremely effective when clear differences in density (≥ 0.02 g/ml) exist between cells. However, it requires a uniform cell suspension and a knowledge of the density of the cell type of interest (Grover et al. 2011). Cell density may be affected by the position of the cells in the growth cycle (plateau phase cells are denser), uptake of density medium, and the presence of serum (Freshney 1976; Pretlow and Pretlow 1989). The technique requires careful optimization when it is first performed for a new sample type. Points to consider during optimization include:

(1) **Gradient density and separation.** The density of collected fractions can be measured using a refractometer or density meter (e.g. Mettler-Toledo). Density marker beads (colored marker beads of known densities) can be used as controls to determine the density of regions within the gradient. Marker beads are increasingly difficult to source but continue to be manufactured by a few suppliers (e.g. Cospheric). It is important to assess the efficacy of the separation procedure by examining the number of cells that are harvested in each fraction and the recovery, purity, and viability of the desired cell type (Pretlow and Pretlow 1989; Tomlinson et al. 2013).

(2) **Density medium.** Ficoll and Percoll are commonly used for isopycnic centrifugation. Percoll may form aggregates on autoclaving or during prolonged storage; this can be minimized during autoclaving by reducing contact with air (GE Healthcare 2014). Ficoll is a little more viscous than Percoll at high densities and may cause some cells to agglutinate. Ficoll is often used in combination with iodinated compounds such as metrizoate (e.g. Ficoll-Paque, Hypaque, and Isopaque), which were originally developed for radiography. Metrizamide (Santa Cruz Biotechnology), a non-ionic derivative of metrizoate, is less viscous than Ficoll at high densities but may be taken up by some cells (Rickwood and Birnie 1975; Freshney 1976).

(3) **Gradient type.** Multistep gradients are useful when working with a limited number of cell types with known densities (Pertoft 2000). However, continuous gradients are more sensitive and selective (Pretlow and Pretlow 1989).

(4) **Centrifugal force and duration.** Centrifugal force can be damaging for cells, particularly after they have ceased sedimentation. The degree of force required may also vary with the density medium, e.g. Percoll is less viscous than Ficoll and thus requires less centrifugal force and time (Pretlow and Pretlow 1989).

(5) **Cell position.** Samples are usually layered on top of a density gradient; this reduces contamination by cell debris, which remains in the top layer (Pertoft 2000). Samples can be layered underneath the gradient, but this may increase the risk that the density medium is taken up by the cells (Freshney 1976). Cells may also be incorporated into the gradient during its formation by centrifugation. Although only one spin is required, spinning the cells at such a high *g* force can be damaging.

21.2 CELL SIZE AND SEDIMENTATION VELOCITY

21.2.1 Velocity Sedimentation at Unit Gravity

Cell size (cross-sectional area) is more variable compared to cell density, but it is still useful for separation. The size of a cell becomes the major determinant of sedimentation velocity at unit gravity (1 *g*) and a significant component at higher sedimentation rates at elevated *g*. The relationship between particle size and sedimentation rate at 1 *g*, although complex for submicron-sized particles, is fairly simple for cells and can be expressed approximately as (Miller and Phillips 1969):

$$\nu \approx \frac{r^2}{4} \qquad (21.1)$$

where ν is the sedimentation rate in millimeters per hour and r is the radius of the cell in micrometers (see Table 19.3).

Cells with a size difference of at least 0.5 μm (or a very large difference in density) can be separated using velocity sedimentation at unit gravity (Wells 1989). A shallow gradient is used to provide stability during loading and unloading and to prevent mixing from thermal effects. For example, cells may be layered over a serum gradient in medium and allowed to settle through the medium according to the above equation. Unit gravity sedimentation is useful for cell populations where there are major differences in cell size or when aggregates are being separated from single cells, e.g. after collagenase digestion (see Protocol P13.6). However, unit gravity sedimentation is unable to handle large numbers of cells (> 1×10^6 cells/cm^2 of surface area at the top of the gradient) and does not give particularly good separations unless the mean cell sizes are very different and the cell populations are homogeneous in size.

21.2.2 Centrifugal Elutriation

Velocity sedimentation is used to separate cells at increased g using centrifugal elutriation. Originally referred to as "counter-streaming centrifugation" by its inventor, the technique uses a counterflow of medium that opposes the centrifugal force as the basis for separation (Lindahl 1948; Sanders and Soll 1989). A centrifugal elutriator (Beckman Coulter Life Sciences) is required for this technique, consisting of a specially designed centrifuge and rotor (see Figure 21.4a). A cell suspension is pumped into the separation chamber in the rotor while it is spinning. Centrifugal force tends to push the cells to the outer edge of the rotor (see Figure 21.4b). Meanwhile, medium is pumped through the chamber such that the centripetal flow rate balances the sedimentation rate of the cells. If the cells were uniform, they would remain stationary in one position, but because they vary in size, density, and cell surface configuration, they tend to sediment at different rates. Because the sedimentation chamber is tapered, the flow rate increases toward the edge of the rotor, and a continuous range of flow rates is generated. Cells of differing sedimentation rates will therefore reach equilibrium at different positions in the chamber. The sedimentation chamber is illuminated by a stroboscopic light and can be observed through a viewing port. When the cells are seen to reach equilibrium, the flow rate is increased and the cells are pumped out into receiving vessels.

Centrifugal elutriation has a number of advantages. It is relatively fast and gentle for most cell types. Equilibrium is reached in a few minutes and the whole procedure may be completed within three hours and repeated as often as necessary (Delgado et al. 2017). Each run can process approximately 1×10^8 cells; further improvements in scale can be made through modifications to equipment design (Lutz et al. 1992). Separation can be performed in complete culture medium and the cells cultured directly afterward. However, the apparatus is fairly expensive, and a considerable amount of experience is required before effective separations may be made. Details of the procedure are provided in the operating manual for the elutriator (Beckman Coulter Life Sciences), but it usually involves (i) setting up and sterilizing the apparatus; (ii) calibration; (iii) loading the sample and establishing the equilibrium conditions; and (iv) harvesting fractions.

A number of cell types have been separated using this technique, including lung epithelial cells and MSCs from umbilical cord and bone marrow (Lag et al. 1996; Majore et al. 2009; Hall et al. 2013). It is particularly useful to separate cells at different phases of the cell cycle (Banfalvi 2008; Barradas et al. 2015; Delgado et al. 2017). The technique may be used in combination with fluorescence-activated cell sorting (FACS); while centrifugal elutriation gives moderate resolution with a high yield, FACS gives high resolution with a low yield (see Section 21.4).

21.3 MAGNETIC SEPARATION AND SORTING

Some cells have intrinsic magnetic properties that can be used for separation. For example, erythrocytes that have been infected with malaria species (e.g. *Plasmodium falciparum*) accumulate hemozoin pigment, which has paramagnetic properties (Zborowski and Chalmers 2011; Coronado et al. 2014). Apart from these rare instances, cell separation and sorting can be achieved using particles or "beads" with paramagnetic properties. Such beads consist of a magnetic core, a surface coating, and a ligand that can bind to the cell surface in a specific manner (Modh et al. 2018). The ligand is usually an antibody that has been raised against a specific cell surface epitope to select for the desired cell type (or for undesirable cell types, resulting in their depletion). Cells will bind to the antibody and thus to the bead, resulting in magnet-activated cell sorting (MACS) when exposed to a magnetic field (Antoine et al. 1978). Other ligands can be used such as peptides or aptamers, which are single-stranded nucleic acids that specifically bind to their target molecules with high affinity (Modh et al. 2018).

Magnetic separation techniques have many advantages. They are usually rapid, are cost-effective, and generate high yields compared to FACS (see Section 21.4) (Sutermaster and Darling 2019). Cells may be sorted using positive selection (where beads are bound to the desired cell type) or negative selection (where beads are bound to undesirable cell types, resulting in their depletion). Both approaches can be used together to collect different cell populations. The technique is effective for many samples and cell types, including stem cells and circulating tumor cells (Bertoncello et al. 1991;

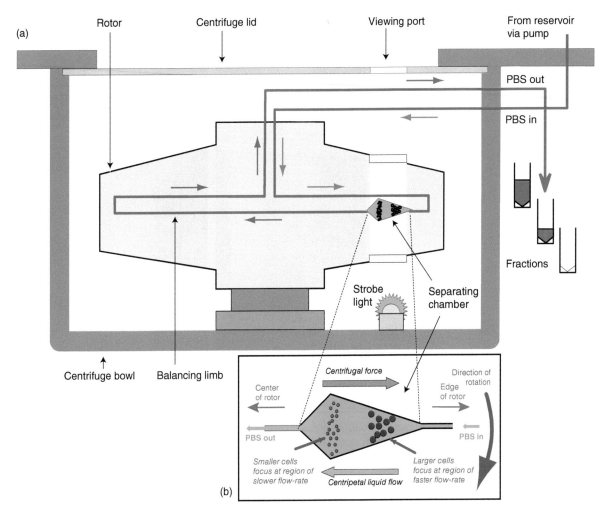

(a)

Rotor   Centrifuge lid   Viewing port   From reservoir via pump

PBS out

PBS in

Fractions

Strobe light   Separating chamber

Centrifuge bowl   Balancing limb

(b)

Center of rotor   *Centrifugal force*   Direction of rotation   Edge of rotor

PBS out   PBS in

Smaller cells focus at region of slower flow-rate   *Centripetal liquid flow*   *Larger cells focus at region of faster flow-rate*

Fig. 21.4. Centrifugal elutriator. (a) Section through the rotor of a centrifugal elutriator (Beckman Coulter Life Sciences). A cell suspension enters at the center of the rotor and is pumped to the periphery and then into the outer end of the separating chamber. The return loop is via the opposite side of the rotor, to maintain balance. (b) A closer view of the separating chamber.

Fong et al. 2009; Zamay et al. 2015). Magnetic separation can also be used for high-throughput and clinical applications using Good Manufacturing Practice (GMP) (Apel et al. 2013).

The most widely used magnetic beads come from ThermoFisher Scientific and Miltenyi Biotec (see Figure 21.5) (Plouffe et al. 2015). Dynabeads® (ThermoFisher Scientific) consist of uniform, micro-sized (> 1 μm diameter) polystyrene particles that contain magnetic iron (Kemshead and Ugelstad 1985). When a cell suspension is mixed with the beads and then placed in a magnetic field (see Figure 21.5a–c), the cells that have attached to the beads are drawn to the side of the separating chamber. The cells and beads are released when the current is switched off. Cells can be released and eluted from the beads, e.g. using FlowComp™ (ThermoFisher Scientific). Kits are available for positive or negative separation of various cell types. You

can also use your own antibody for separation by purchasing beads that are surface-activated or coated with streptavidin, anti-biotin antibody, or species-specific secondary antibodies.

MACS® MicroBeads (Miltenyi Biotec) consist of nano-sized beads (∼ 50 nm diameter) made from an iron oxide core with a dextran coating (Miltenyi et al. 1990; Apel et al. 2007). Cells are labeled with the beads and applied to a separation column containing a ferromagnetic matrix, placed in a magnetic field (see Figures 21.5d, e, 21.6). Labeled cells are retained, while unlabeled cells flow through the column. Once the column is removed from the magnet, the column matrix rapidly demagnetizes and the labeled cells are easily removed from the column. Cells are usually cultured with the beads attached, although removal of the beads can be achieved, e.g. with REAlease™ (Miltenyi Biotec). The technique can be used for semi- or fully automated processing if suitable equipment is available (e.g. autoMACS Pro Separator, Miltenyi Biotec).

(a-c) Magnetic Sorting with Dynabeads® (ThermoFisher Scientific). Negative Sort; *committed progenitor cells from bone marrow suspension bound to Dynabeads with antibodies to lineage markers. Lineage negative (stem) cells are not bound and remain in the suspension ready for sorting by flow cytometry. (a) Inserting the tube into the magnetic holder; (b) tube immediately after being placed in magnetic holder; (c) tube 30 s after placement in magnetic holder. Source: R. Ian Freshney.*

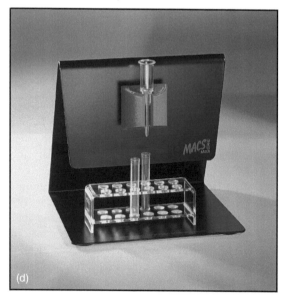

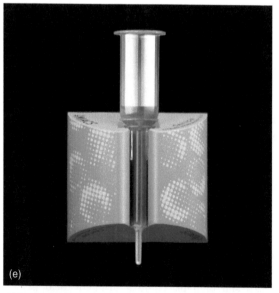

(d-e) Magnetic Cell Sorting with the Midi-MACS Separator (Miltenyi Biotec). (d) Column, magnet, rack, and stand; (e) Midi-MACS Separator with LD Column. Source: courtesy of Miltenyi Biotec.

Fig. 21.5. Magnetically activated cell sorting.

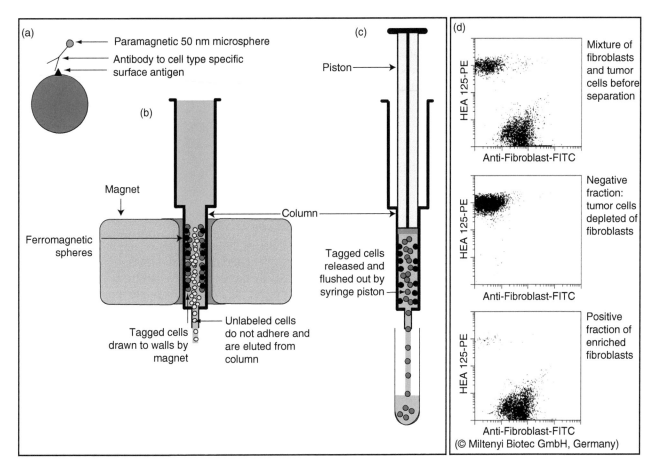

Fig. 21.6. Magnetic cell sorting using MACS technology. Separation of tumor cells from fibroblasts using MACS Microbeads for positive separation (see also Protocol P21.2). (a) Cells are preincubated with antibodies raised against a cell type-specific surface antigen and conjugated to MACS MicroBeads. (b) When the cells are introduced into the column, cells bound to MicroBeads are retained in the magnetic field, while unlabeled cells go straight through. (c) The magnetically bound cells are released when the column is removed from the magnet and flushed out with the piston. (d) Dot plot from flow cytometry of a culture containing tumor cells and fibroblasts, before MACS sorting (top) and after MACS sorting, looking at the flow-through fraction (middle, tumor cells), and the fraction released from the magnet (bottom, enriched fibroblasts). *Source*: © Miltenyi Biotec (used with permission).

It is important to remember that the label or the bead itself may alter cell surface receptor function or signal transduction (Apel et al. 2007). Micro-sized beads tend to remain on the surface of the cell, which may affect downstream applications; cells should be eluted from the beads if this is the case (Plouffe et al. 2015). Nano-sized beads may be internalized through endocytosis, resulting in the accumulation of synthetic nanoparticles within the cell itself and potential loss of viability (Plouffe et al. 2015). However, internalization can also be used in a beneficial sense; for example, nanoparticles can be exploited for three-dimensional (3D) culture by magnetic levitation (see Section 27.4.2) (Haisler et al. 2013). Always include controls to assess the impact of the label and the bead, and, if required, use a technique where labelling is

not required, where the bead can be removed, or where undesired cell populations are depleted through negative selection.

PROTOCOL P21.2. MAGNET-ACTIVATED CELL SORTING (MACS)

Contributed by Amanda Capes-Davis. The procedure has been adapted from the manufacturer's instructions for a specific kit (MACS Tumor Cell Isolation Kit, Miltenyi Biotec).

Background

This procedure is provided as an example of how to isolate human tumor cells from single cell suspensions using magnetic separation. In this example, negative separation is performed by targeting non-tumor cell types, resulting in their binding to the column. Unlabeled tumor cells pass through the column for collection. Always refer to the manufacturer for specific instructions that will apply to your kit, sample, and application.

Outline

Prepare a single cell suspension, e.g. by trypsinization. Label the cells with antibody-conjugated MicroBeads and perform magnetic separation using a MACS Column. Collect the unlabeled cells and elute the labeled cells for separate culture and analysis.

- *Safety Note. Commercial kits may contain hazardous substances. In this kit, reagents contain sodium azide and should be diluted with running water before discarding. Always refer to the manufacturer's warnings and instructions.*

Materials (sterile or aseptically prepared)

- ☐ Primary culture of human tumor, e.g. human breast carcinoma
- ☐ Pre-separation filter (70 µm) to remove clumps (e.g. Miltenyi Biotec, #130-095-823)
- ☐ MACS Tumor Cell Isolation Kit (Miltenyi Biotec, #130-108-339)
- ☐ MACS LS Column
- ☐ Buffer (from kit, prepared as per manufacturer's instructions and pre-cooled on ice)
- ☐ Collection tubes to match volumes being collected

Materials (non-sterile)

- ☐ BSC and associated consumables (see Protocol P12.1)
- ☐ Cell counting equipment (hemocytometer or automated counter; see Section 19.1)
- ☐ Suitable MACS Separator, e.g. MidiMACS™ Separator (see Figure 21.5d, e)

Procedure

1. Prepare a work space for aseptic technique using a BSC.
2. Generate a single-cell suspension using a method that is suitable for your sample and cell type, e.g. trypsinization (see Protocols P13.4, P14.2).
3. Count the cells and check that a single-cell suspension has been generated. If required, use a

pre-separation filter to remove clumps from the cell suspension.
4. Centrifuge the single-cell suspension at 300 g for 10 minutes.
5. Remove and discard the supernatant. Resuspend the cell pellet in 60 µl buffer.
6. Add the antibody-conjugated MicroBeads from the kit as per instructions (e.g. 20 µl Non-Tumor Depletion Cocktails A and B). Mix gently and incubate at 2–8 °C for 15 minutes.
7. Adjust the total volume up to 500 µl using buffer.
8. Place the LS column in the magnetic field of the MACS Separator.
9. Prepare the column by rinsing with 3 ml buffer. Collect and discard the rinse.
10. Apply the cell suspension onto the column and collect the flow-through into a tube.
11. Wash the column with 2 × 1 ml buffer and combine with the flow-through from step 10. These fractions contain the unlabeled, enriched tumor cells.
12. Remove the column from the MACS Separator. Add 3 ml buffer to the column and firmly push the plunger into the column to flush out and collect the remaining cells. This fraction contains labeled non-tumor cells, e.g. fibroblasts.
13. Add culture medium to the unlabeled tumor cells and either plate for culture (e.g. in a 25-cm² flask) or perform further analysis to determine their yield and purity.
14. If desired, repeat step 13 for the labeled non-tumor cells.

Notes

Step 2: A single-cell suspension is important for effective separation using magnetic beads. Clumps may also result in clogging of the column.

Step 3: The volumes provided in this protocol as an example are suitable for up to 2×10^6 tumor cells or up to 1×10^7 total cells.

Step 4: All steps from this point should be performed using pre-cooled solutions on ice to reduce non-specific labelling.

21.4 FLUORESCENCE-ACTIVATED CELL SORTING (FACS)

Flow cytometry is a technology that examines multiple individual cell parameters, allowing analysis and sorting based on physical properties and immunolabeling (Ibrahim and van den Engh 2007; Picot et al. 2012). The technology is based on the projection of a single stream of cells through a laser beam in

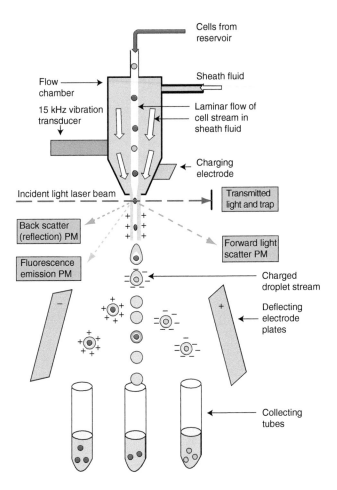

Fig. 21.7. Principles of flow cytometry. The cell stream enters at the top, and sheath liquid is injected around the cell stream to generate a laminar flow within the flow chamber. As the cell stream exits the chamber, it cuts a laser beam, and the resulting signal triggers the charging electrode, thereby charging the cell stream. The cell stream then breaks up into droplets, induced by the 15-kHz vibration transducer attached to the flow chamber. The droplets carry the charge briefly applied to the exiting cell stream and are deflected by the electrode plates below the flow chamber. The charge is applied to the plates with a sufficient delay to allow for the transit time from cutting the beam to entering the space between the plates.

such a way that the light scattered from the cells is detected at several angles and fluorescence emissions (see Figure 21.7). The optical collection system records scattered light, fluorescence signal emissions, and other variables (e.g. electrical impedance), resulting in signals that are processed by the instrument's electronic system. Data are usually displayed in a univariate histogram or bivariate scatter plot and should be published with sufficient information to ensure that the procedure is reproducible (Ibrahim and van den Engh 2007; Alvarez et al. 2010).

Flow cytometry is a powerful tool for cell analysis. Cells can be pretreated with a fluorescent stain (e.g. propidium iodide for DNA), a fluorescent antibody, or a fluorescent probe to analyze cells based on the presence or absence of specific labels. Flow cytometry can be used to examine many different cell properties, using light scatter (e.g. cell size and density), fluorescence (e.g. DNA, RNA, or protein content; enzyme activity; or specific antigens), and other variables. The technology has been used extensively for DNA analysis of human cancers and cell lines (Krueger and Wilson 2011). The commonest application is probably analysis of hematopoietic cells (Yeung and So 2009), where disaggregation into the obligatory single-cell suspension is relatively simple. The technology has also been widely used for characterization and separation of stem cell populations, including human pluripotent stem cells (hPSCs), HSCs, and MSCs (Diogo et al. 2012).

Flow cytometers with the capacity for FACS use the emission signals from each cell to sort various populations into sample collection tubes (see Figure 21.8) (Bonner et al. 1972). If specific coordinates are set to delineate sections of the display ("gating"), the cell sorter will divert those cells with properties that would place them within these coordinates (e.g. high or low light scatter, high or low fluorescence) into the appropriate receiver tube below the cell stream. Although the technology used for sorting may vary, most high-speed cell sorters use the "jet-in-air" method (Picot et al. 2012). The cell stream is broken up into droplets by a high-frequency vibration applied to the flow chamber and the droplets containing single cells with specific attributes are charged as they leave the chamber. These droplets are deflected, left or right according to the charge applied, as they pass between two oppositely charged plates. The charge is applied briefly and at a set time after the cell has cut the laser beam such that the droplet containing one specifically marked cell is deflected into the correct tube. The concentration in the cell stream must be low enough that the gap between cells is sufficient to prevent two cells from inhabiting one droplet.

FACS allows multiple cell populations to be isolated from a single sample with high levels of recovery and purity (Picot et al. 2012). For example, FACS was used for the initial purification of mouse HSCs and continues to be used for the isolation and enrichment of rare stem cell populations (Armstrong et al. 2012). However, the procedure can be time-consuming and tends to produce lower cell viability compared to gentler approaches (e.g. MACS; see Section 21.3). Some applications may be limited by the cell yield – about 1×10^7 cells is a reasonable maximum number of cells that can be processed at one time. Flow cytometers traditionally have a high capital cost and require specialized expertise; they are often operated by core facilities that provide the necessary maintenance and user training.

Bench-top flow cytometers are now available that are cheaper and easier to handle (Picot et al. 2012). These instruments may be used as an alternative to automated cell counters (e.g. the Guava® easyCyte flow cytometer, Luminex; see Figure 19.5). They offer greater accuracy when counting cells and additional parameters, such as viability

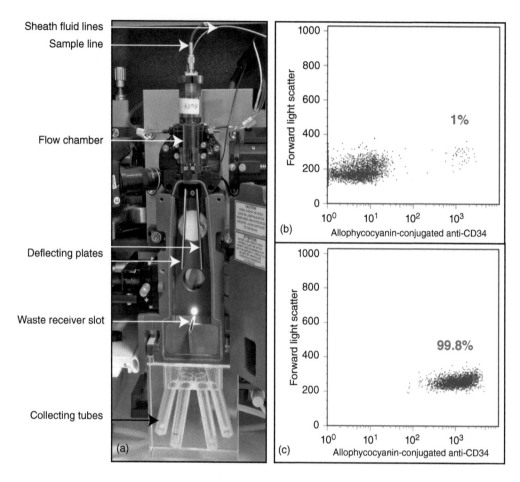

Fig. 21.8. Fluorescence-activated cell sorter (FACS). (a) Close-up of flow chamber and separation compartment of FACSAria (BD Biosciences). Although this model is no longer current, it demonstrates the principles of FACS, resulting in separation of fractions into the collecting tubes at the bottom. The instrument was used to isolate CD34+ cells from a peripheral blood sample, following magnetic cell separation using CliniMACS (Miltenyi Biotech). (b) Percentage of CD34+ cells, within a total mononuclear cell population, prior to enrichment. (c) Purified cells following CD34+ selection. *Source:* R. Ian Freshney; data courtesy of Paul O'Gorman Leukemia Research Centre, Glasgow.

by diacetyl fluorescein (DAF) uptake, apoptotic index, and DNA content. When looking at bench-top flow cytometers, remember that the cheaper cost usually comes with reduced capacity, e.g. fewer lasers and detectors. Always consider a range of options and arrange for a demonstration of your preferred choice prior to purchase.

21.5 MICROFLUIDIC SORTING

The techniques described so far can be very effective for cell separation, but they all have limitations. There is likely to be some overlap between cell populations if physical properties are used for separation, and individual cells may vary in their physical properties over time, e.g. during cell division (Pretlow and Pretlow 1989). Affinity-based techniques rely on the use of suitable ligands on the cell surface, which may

not always be available. Intracellular ligands are usually inaccessible in living cells, which is an important consideration if cells are to be cultured after sorting. Some of the techniques described here are effective for rare cell populations (e.g. magnetic sorting or FACS), but they are not effective for all such populations. For example, techniques that require single-cell suspensions cannot be used to isolate clusters of circulating tumor cells, which are important for cancer metastasis (Au et al. 2017). For all these reasons, biomedical engineers are developing microfluidic devices to provide rapid, high-throughput cell sorting and analysis on a miniature scale.

"Microfluidics" can be defined as the science and technology of systems that process or manipulate small amounts of fluids, from 10^{-9} to 10^{-18} l, using microchannels (Whitesides 2006). Microfluidic sorting relies on addition of cells to a specially fabricated biochip, where they can be manipulated in microscale volumes (Tasoglu et al. 2013). Such biochips have been used for bacterial cells for many years, as demonstrated by an early application using a microfabricated fluorescence-activated cell sorter (Fu et al. 1999). Over time, the principles of cell sorting and separation have been applied to produce biochips for specific cell types. For example, microfluidic devices can be developed to isolate circulating tumor cells in the peripheral blood of cancer patients and to detect CD4+ T-lymphocytes in patients with acquired immunodeficiency syndrome (AIDS) (Cheng et al. 2007; Nagrath et al. 2007). Microfluidic devices may utilize any principle that is used for separation at a larger scale, including magnetic nanoparticles, antibodies, or selective adhesion (Tasoglu et al. 2013). Cellular properties, such as density, can also be used for separation without the need for additional molecules (see Figure 21.9) (Norouzi et al. 2017). In many cases, cells perform better in microfluidic culture and their behavior can be more tightly controlled (Gagliano et al. 2016; Luni et al. 2016). Microfluidic technology is complementary to other recent advances, e.g. to achieve compartmentalization for single cell analysis (Hosic et al. 2016).

Publications on microfluidics in the life sciences have soared in recent years (Shields et al. 2017). Although only a handful of devices have actively progressed toward commercialization, the techniques have become more broadly accessible to many laboratories (Perkel 2015; Shields et al. 2017). Microfluidic cell culture is now an important application that can be used to generate models for various fields, such as organ-on-chip systems for toxicity testing (see Section 29.5.2).

Minireview M21.1. **Microfluidic Cell Culture**   Contributed by Roland Kaunas, Department of Biomedical Engineering, Texas A&M University, TX, USA.
Microfluidic cell culture provides a number of advantages over conventional culture in plates and flasks:

(1) The small scale of these systems greatly reduces the necessary volumes of expensive media and supplements.
(2) The cell local microenvironment can be precisely defined through spatial and temporal control over the delivery of soluble molecules and patterning of adhesive contacts (Folch and Toner 2000).
(3) Fluid flow overcomes diffusion-limited growth, thus facilitating the use of 3D matrices that require local convective transport of nutrient and waste products.
(4) Complex combinations of microfluidic features can be integrated to allow sample assays within the device and mechanical actuation to dynamically alter flows.

Several issues must be considered when using microfluidic systems, however, including the choice of construction materials, effects of fluid forces, and limited access to the internal compartments.

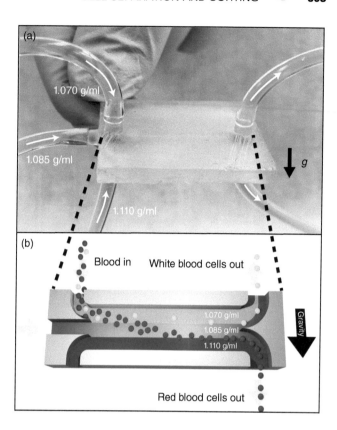

Fig. 21.9. Density sorter chip. (a) Photograph of a 3D-printed density sorter chip with; (b) cross-sectional illustration of its operating principles. The chip has three inlets for fluids of three different densities (1.070, 1.085, and 1.110 g/ml). Under laminar flow conditions, these fluids form a microscale density gradient flowing across the horizontal channel. The density gradient can be used to separate leukocytes in whole blood (average density $\rho = 1.080$ g/ml) from erythrocytes (average density $\rho = 1.110$ g/ml), resulting in a thousand-fold enrichment without centrifugation. *Source*: Norouzi, N., Bhakta, H. C. and Grover, W. H. (2017). Sorting cells by their density. PLoS One 12 (7):e0180520. doi: 10.1371/journal.pone.0180520., https://doi.org/10.1371/journal.pone.0180520, licensed under CC BY 4.0.

Applications of Microfluidic Cultures

Perfusion culture is particularly relevant to the culture of cells normally found in highly perfused tissues such as the liver and kidney (Leclerc et al. 2003; Jang and Suh 2010). For example, a device consisting of two polydimethylsiloxane (PDMS) layers separated by a porous membrane was used to culture primary rat kidney inner medullary collecting duct cells such that the apical sides were subjected to physiologically relevant levels of fluid shear stress, while the basolateral side faced a static fluid-containing chamber, resulting in monolayer architecture and transepithelial transport properties that were lost in cultures on glass (Jang and Suh 2010). Thus, multiple layers of PDMS can be used to create multifunctional devices to recapitulate features of particular tissues. In addition to fluid shear stress, cells can be subjected to tensile forces. Simultaneous

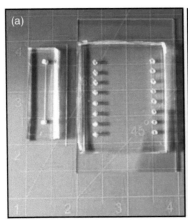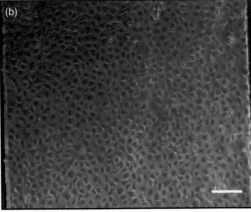

Fig. 21.10. Microfluidic culture. Microfluidic device to apply controlled fluid wall shear stress. (a) Single and multiple channel configurations of the device. In both configurations, each channel has a single inlet and single outlet port to which tubing is inserted. (b) Phase contrast image of a human lymphatic endothelial cell monolayer cultured within a channel. The edges of the channel can be seen on the left and right borders. Scale bar represents 100 μm. *Source*: Courtesy of Roland Kaunas.

application of cyclic stretch and fluid shear stress can be generated on a thin, elastic PDMS membrane bisecting a perfused microchannel, with stretch produced by applying a cyclic vacuum in flanking gas grooves (Zheng et al. 2012).

Microfluidic devices provide an excellent system for studying cell migration in 3D matrices. Collagen gels can be poured into a region between two parallel microchannels using closely spaced posts, with the collagen flow stopped by surface tension between posts (Shin et al. 2012). Culture media flowing through the adjacent channels at different pressures and concentrations of chemotactic factors result in interstitial flow and chemotactic gradients within the matrix that stimulate directional migration and vessel network formation from invading endothelial cells (Shin et al. 2012).

Device Construction and Operation

The typical microfluidic device is constructed from PDMS poured over a patterned silicon wafer and cured in an oven at 80 °C for one hour. While PDMS is inert and non-cytotoxic, surface treatments such as adsorption of extracellular matrix (ECM) protein are necessary to support cell adhesion directly on PDMS surfaces (Halldorsson et al. 2015). PDMS is often fused to glass, in which case glass is the surface for cell adhesion. PDMS is highly permeable to gases if sufficiently thin (~ 100 μm), which facilitates maintenance of media pH in CO_2 incubators. Since PDMS is also permeable to water vapor, changes in osmolarity of the medium may occur due to evaporation if the environment outside of the chamber is dry (Heo et al. 2007). A simple solution is to saturate the PDMS microfluidic devices with culture media for 24 hours prior to introducing cells. PDMS is also permeable to small, hydrophobic macromolecules, which may significantly reduce the concentrations of these molecules in cell culture.

Perfusion

Flow in microfluidic devices can be driven by simple gravity or by external pumps for more precise control. The flexible nature of PDMS also allows the incorporation of valves and on-chip pumps (Meyvantsson and Beebe 2008). Compared to conventional cultures, convective flow in microfluidic systems substantially reduces the amount of culture medium and expensive additives required for a given number of cells. Fluid shear stresses generated by flow can be detrimental to cultures in cases where weak cell adhesion is disrupted by fluid forces. This is particularly critical to initial cell adhesion, where flow is transiently stopped until the cells have sufficiently spread. A balance must be chosen between time for cell attachment and need for medium exchange for cell survival. For cells that are characterized by loose adhesions, such as HSCs, the regions of cell culture can be isolated from the bulk flow to minimize fluid shear stress with diffusion distances small enough to still provide adequate transport of nutrients and waste (Lecault et al. 2011).

Fluid shear stress is sometimes applied intentionally to provide physiologically relevant biomechanical cues (see Section 26.3.7). For example, human lymphatic endothelial cells were grown within channels to measure intracellular calcium signaling induced by fluid shear stress (see Figure 21.10) (Jafarnejad et al. 2015). In this simple configuration, the magnitude of shear stress is uniform throughout the cell culture. Regions with a range of fluid shear stress magnitudes within a single device can be generated by varying the channel cross-sectional area. Another significant feature of microfluidics is the laminar nature of the flow, which allows the formation of stable concentration gradients through convergence of multiple streams of differing concentrations of molecules such as chemotactic agents to study cell motility (Li Jeon et al. 2002).

Assays

The transparent properties of PDMS and continuous perfusion facilitate long-term live imaging of cell cultures. Thus, microfluidic systems are ideal for assays based on fluorescent probes or morphological measurements. The small number of cells and limited access to the cells in a closed system provide a challenge to performing endpoint assays often used with conventional plate cultures. Whole cells or cell lysates can be extracted by flowing trypsin or cell lysate buffer through the device, although a significant fraction of material is lost in the process. On-chip versions of macroscale assays, including reverse transcriptase-PCR (RT-PCR) and cytometry, have also been developed (Dittrich et al. 2006; Yi et al. 2006).

21.6 CONCLUSIONS: SORTING AND SEPARATION

Where do you start if you are a beginner to cell sorting and separation? First, you need to understand your cells. Do they have surface markers or other properties that could be exploited for separation and sorting? Look for publications from other laboratories that have succeeded in isolating the same cell type and discuss with colleagues to see what techniques and equipment are available in your location. Start with a relatively simple technique and work up toward more specialized procedures. Density gradient centrifugation and MACS are usually the easiest techniques for the beginner and use equipment that is readily available in most tissue culture laboratories or can be purchased at moderate expense. If the resolution or yield is insufficient, it may be necessary to perform FACS or centrifugal elutriation. Centrifugal elutriation is useful for the rapid sorting of large numbers of cells, but FACS will probably give the purest cell population, based on the combined application of two or more stringent criteria. When a purification of several logs is required, MACS is often used as a first step, with the final purification performed by FACS. This approach has been used successfully for isolation of many challenging cell types, e.g. human embryonic stem cells (hESCs) (Fong et al. 2009). The most effective sorting often employs two or more parameters to obtain pure populations of cells and requires a good understanding of that cell type and its unique properties.

Suppliers

Supplier	URL
Abbott Diagnostics Technologies AS	www.axis-shield-density-gradient-media.com
BD Biosciences	www.bdbiosciences.com
Beckman Coulter Life Sciences	www.beckman.com
Brandel	http://www.brandel.com/index.html
Cospheric	www.cospheric.com
GE Healthcare Life Sciences	http://www.gelifesciences.com/en/it
Luminex	www.luminexcorp.com
Mettler Toledo	http://www.mt.com/us/en/home.html
Miltenyi Biotec	https://www.miltenyibiotec.com/us-en/products/macs-cell-separation.html
Santa Cruz Biotechnology	http://www.scbt.com/scbt/home
Sigma-Aldrich (Merck)	http://www.sigmaaldrich.com/life-science/cell-culture.html
STEMCELL Technologies	www.stemcell.com
ThermoFisher Scientific	http://www.thermofisher.com/us/en/home/life-science/cell-culture.html

REFERENCES

Alvarez, D.F., Helm, K., Degregori, J. et al. (2010). Publishing flow cytometry data. *Am. J. Physiol. Lung Cell Mol. Physiol.* 298 (2): L127–L130. https://doi.org/10.1152/ajplung.00313.2009.

Antoine, J.C., Ternynck, T., Rodrigot, M. et al. (1978). Lymphoid cell fractionation on magnetic polyacrylamide – agarose beads. *Immunochemistry* 15 (7): 443–452.

Apel, M., Heinlein, U.A.O., Miltenyi, S. et al. (2007). Magnetic cell separation for research and clinical applications. In: *Magnetism in Medicine: A Handbook* (eds. W. Andra and H. Nowak), 571–595. Weinheim, Germany: Wiley-VCH.

Apel, M., Bruning, M., Granzin, M. et al. (2013). Integrated clinical scale manufacturing system for cellular products derived by magnetic cell separation, centrifugation and cell culture. *Chem. Ing. Tech.* 85 (1–2): 103–110.

Armstrong, L., Lako, M., Buckley, N. et al. (2012). Editorial: our top 10 developments in stem cell biology over the last 30 years. *Stem Cells* 30 (1): 2–9. https://doi.org/10.1002/stem.1007.

Au, S.H., Edd, J., Haber, D.A. et al. (2017). Clusters of circulating tumor cells: a biophysical and technological perspective. *Curr. Opin. Biomed. Eng.* 3: 13–19. https://doi.org/10.1016/j.cobme.2017.08.001.

Banfalvi, G. (2008). Cell cycle synchronization of animal cells and nuclei by centrifugal elutriation. *Nat. Protoc.* 3 (4): 663–673. https://doi.org/10.1038/nprot.2008.34.

Barradas, O.P., Jandt, U., Becker, M. et al. (2015). Synchronized mammalian cell culture: part I – a physical strategy for synchronized cultivation under physiological conditions. *Biotechnol. Progr.* 31 (1): 165–174. https://doi.org/10.1002/btpr.1944.

Bertoncello, I., Bradley, T.R., and Watt, S.M. (1991). An improved negative immunomagnetic selection strategy for the purification of primitive hemopoietic cells from normal bone marrow. *Exp. Hematol.* 19 (2): 95–100.

Bonner, W.A., Hulett, H.R., Sweet, R.G. et al. (1972). Fluorescence activated cell sorting. *Rev. Sci. Instrum.* 43 (3): 404–409.

Boyum, A. (1968a). Isolation of leucocytes from human blood. Further observations. Methylcellulose, dextran, and ficoll as erythrocyte aggregating agents. *Scand. J. Clin. Lab. Invest. Suppl.* 97: 31–50.

Boyum, A. (1968b). Isolation of mononuclear cells and granulocytes from human blood. Isolation of monuclear cells by one centrifugation, and of granulocytes by combining centrifugation and sedimentation at 1 g. *Scand. J. Clin. Lab. Invest. Suppl.* 97: 77–89.

Ceusters, J., Lejeune, J.P., Sandersen, C. et al. (2017). From skeletal muscle to stem cells: an innovative and minimally-invasive process for multiple species. *Sci. Rep.* 7 (1): 696. https://doi.org/10.1038/s41598-017-00803-7.

Cheng, X., Irimia, D., Dixon, M. et al. (2007). A microchip approach for practical label-free CD4+ T-cell counting of HIV-infected subjects in resource-poor settings. *J. Acquir. Immune Defic. Syndr.* 45 (3): 257–261. https://doi.org/10.1097/QAI.0b013e3180500303.

Coronado, L.M., Nadovich, C.T., and Spadafora, C. (2014). Malarial hemozoin: from target to tool. *Biochim. Biophys. Acta* 1840 (6): 2032–2041. https://doi.org/10.1016/j.bbagen.2014.02.009.

Delgado, M., Kothari, A., Hittelman, W. N., et al. (2017). Preparation of primary acute lymphoblastic leukemia cells in different cell cycle phases by centrifugal elutriation. *J. Vis. Exp.* 2017 (129):56418. doi: https://doi.org/10.3791/56418.

Diogo, M.M., da Silva, C.L., and Cabral, J.M. (2012). Separation technologies for stem cell bioprocessing. *Biotechnol. Bioeng.* 109 (11): 2699–2709. https://doi.org/10.1002/bit.24706.

Dittrich, P.S., Tachikawa, K., and Manz, A. (2006). Micro total analysis systems. Latest advancements and trends. *Anal. Chem.* 78 (12): 3887–3908. https://doi.org/10.1021/ac0605602.

Folch, A. and Toner, M. (2000). Microengineering of cellular interactions. *Annu. Rev. Biomed. Eng.* 2: 227–256. https://doi.org/10.1146/annurev.bioeng.2.1.227.

Fong, C.Y., Peh, G.S., Gauthaman, K. et al. (2009). Separation of SSEA-4 and TRA-1-60 labelled undifferentiated human embryonic stem cells from a heterogeneous cell population using magnetic-activated cell sorting (MACS) and fluorescence-activated cell sorting (FACS). *Stem Cell Rev.* 5 (1): 72–80. https://doi.org/10.1007/s12015-009-9054-4.

Freshney, R.I. (1976). Separation of cultured cells by isopycnic centrifugation in metrizamide gradients. In: *Biological Separations* (ed. D. Rickwood), 123–130. London: Information Retrieval.

Fu, A.Y., Spence, C., Scherer, A. et al. (1999). A microfabricated fluorescence-activated cell sorter. *Nat. Biotechnol.* 17 (11): 1109–1111. https://doi.org/10.1038/15095.

Gagliano, O., Elvassore, N., and Luni, C. (2016). Microfluidic technology enhances the potential of human pluripotent stem cells. *Biochem. Biophys. Res. Commun.* 473 (3): 683–687. https://doi.org/10.1016/j.bbrc.2015.12.058.

Gilbert, S.F., Axelman, J., and Migeon, B.R. (1981). Phenotypic heterogeneity within clones of fetal human cells. *Am. J. Hum. Genet.* 33 (6): 950–956.

Grover, W.H., Bryan, A.K., Diez-Silva, M. et al. (2011). Measuring single-cell density. *Proc. Natl Acad. Sci. U.S.A.* 108 (27): 10992–10996. https://doi.org/10.1073/pnas.1104651108.

Haisler, W.L., Timm, D.M., Gage, J.A. et al. (2013). Three-dimensional cell culturing by magnetic levitation. *Nat. Protoc.* 8 (10): 1940–1949. https://doi.org/10.1038/nprot.2013.125.

Hall, S.R., Jiang, Y., Leary, E. et al. (2013). Identification and isolation of small CD44-negative mesenchymal stem/progenitor cells from human bone marrow using elutriation and polychromatic flow cytometry. *Stem Cells Transl. Med.* 2 (8): 567–578. https://doi.org/10.5966/sctm.2012-0155.

Halldorsson, S., Lucumi, E., Gomez-Sjoberg, R. et al. (2015). Advantages and challenges of microfluidic cell culture in poly-dimethylsiloxane devices. *Biosens. Bioelectron.* 63: 218–231. https://doi.org/10.1016/j.bios.2014.07.029.

GE Healthcare (2014). Cell separation media: methodology and applications. www.gelifesciences.co.kr/wp-content/uploads/2016/07/023.6_cell-separation-media.pdf (accessed 4 February 2019).

Heo, Y.S., Cabrera, L.M., Song, J.W. et al. (2007). Characterization and resolution of evaporation-mediated osmolality shifts that constrain microfluidic cell culture in poly(dimethylsiloxane) devices. *Anal. Chem.* 79 (3): 1126–1134. https://doi.org/10.1021/ac061990v.

Hosic, S., Murthy, S.K., and Koppes, A.N. (2016). Microfluidic sample preparation for single cell analysis. *Anal. Chem.* 88 (1): 354–380. https://doi.org/10.1021/acs.analchem.5b04077.

Ibrahim, S.F. and van den Engh, G. (2007). Flow cytometry and cell sorting. *Adv. Biochem. Eng./Biotechnol.* 106: 19–39. https://doi.org/10.1007/10_2007_073.

Insausti, C.L., Blanquer, M.B., Olmo, L.M. et al. (2012). Isolation and characterization of mesenchymal stem cells from the fat layer on the density gradient separated bone marrow. *Stem Cells Dev.* 21 (2): 260–272. https://doi.org/10.1089/scd.2010.0572.

Ito, Y. and Shinomiya, K. (2001). A new continuous-flow cell separation method based on cell density: principle, apparatus, and preliminary application to separation of human buffy coat. *J. Clin. Apher.* 16 (4): 186–191.

Jafarnejad, M., Cromer, W.E., Kaunas, R.R. et al. (2015). Measurement of shear stress-mediated intracellular calcium dynamics in human dermal lymphatic endothelial cells. *Am. J. Physiol. Heart Circ. Physiol.* 308 (7): H697–H706. https://doi.org/10.1152/ajpheart.00744.2014.

Jang, K.J. and Suh, K.Y. (2010). A multi-layer microfluidic device for efficient culture and analysis of renal tubular cells. *Lab Chip* 10 (1): 36–42. https://doi.org/10.1039/b907515a.

Juopperi, T.A., Schuler, W., Yuan, X. et al. (2007). Isolation of bone marrow-derived stem cells using density-gradient separation. *Exp. Hematol.* 35 (2): 335–341. https://doi.org/10.1016/j.exphem.2006.09.014.

Kemshead, J.T. and Ugelstad, J. (1985). Magnetic separation techniques: their application to medicine. *Mol. Cell. Biochem.* 67 (1): 11–18.

Krueger, S.A. and Wilson, G.D. (2011). Flow cytometric DNA analysis of human cancers and cell lines. *Methods Mol. Biol.* 731: 359–370. https://doi.org/10.1007/978-1-61779-080-5_29.

Lag, M., Becher, R., Samuelsen, J.T. et al. (1996). Expression of CYP2B1 in freshly isolated and proliferating cultures of epithelial rat lung cells. *Exp. Lung Res.* 22 (6): 627–649.

Lecault, V., Vaninsberghe, M., Sekulovic, S. et al. (2011). High-throughput analysis of single hematopoietic stem cell proliferation in microfluidic cell culture arrays. *Nat. Methods* 8 (7): 581–586. https://doi.org/10.1038/nmeth.1614.

Leclerc, E., Sakai, Y., and Fujii, T. (2003). Cell culture in a three-dimensional network of PDMS (polydimethylsiloxane) microchannels. *Biomed. Microdevices* 5: 109–114.

Li Jeon, N., Baskaran, H., Dertinger, S.K. et al. (2002). Neutrophil chemotaxis in linear and complex gradients of interleukin-8 formed in a microfabricated device. *Nat. Biotechnol.* 20 (8): 826–830. https://doi.org/10.1038/nbt712.

Lindahl, P.E. (1948). Principle of a counter-streaming centrifuge for the separation of particles of different sizes. *Nature* 161 (4095): 648.

Luni, C., Giulitti, S., Serena, E. et al. (2016). High-efficiency cellular reprogramming with microfluidics. *Nat. Methods* 13 (5): 446–452. https://doi.org/10.1038/nmeth.3832.

Lutz, M.P., Gaedicke, G., and Hartmann, W. (1992). Large-scale cell separation by centrifugal elutriation. *Anal. Biochem.* 200 (2): 376–380.

Majore, I., Moretti, P., Hass, R. et al. (2009). Identification of subpopulations in mesenchymal stem cell-like cultures from human umbilical cord. *Cell Commun. Signal.* 7: 6. https://doi.org/10.1186/1478-811X-7-6.

Meyvantsson, I. and Beebe, D.J. (2008). Cell culture models in microfluidic systems. *Annu. Rev. Anal. Chem. (Palo Alto Calif.)* 1: 423–449. https://doi.org/10.1146/annurev.anchem.1.031207.113042.

Miller, R.G. and Phillips, R.A. (1969). Separation of cells by velocity sedimentation. *J. Cell. Physiol.* 73 (3): 191–201. https://doi.org/10.1002/jcp.1040730305.

Miltenyi, S., Muller, W., Weichel, W. et al. (1990). High gradient magnetic cell separation with MACS. *Cytometry* 11 (2): 231–238. https://doi.org/10.1002/cyto.990110203.

Modh, H., Scheper, T., and Walter, J.G. (2018). Aptamer-modified magnetic beads in biosensing. *Sensors (Basel)* 18 (4) https://doi.org/10.3390/s18041041.

Nagrath, S., Sequist, L.V., Maheswaran, S. et al. (2007). Isolation of rare circulating tumour cells in cancer patients by microchip technology. *Nature* 450 (7173): 1235–1239. https://doi.org/10.1038/nature06385.

Norouzi, N., Bhakta, H.C., and Grover, W.H. (2017). Sorting cells by their density. *PLoS One* 12 (7): e0180520. https://doi.org/10.1371/journal.pone.0180520.

Pantel, K., Brakenhoff, R.H., and Brandt, B. (2008). Detection, clinical relevance and specific biological properties of disseminating tumour cells. *Nat. Rev. Cancer* 8 (5): 329–340. https://doi.org/10.1038/nrc2375.

Perkel, J. (2015). Designing a microscale lab. *Biotechniques* 58 (3): 97–100. https://doi.org/10.2144/000114260.

Pertoft, H. (2000). Fractionation of cells and subcellular particles with Percoll. *J. Biochem. Bioph. Methods* 44 (1–2): 1–30.

Pertoft, H., Rubin, K., Kjellen, L. et al. (1977). The viability of cells grown or centrifuged in a new density gradient medium, Percoll(™). *Exp. Cell. Res.* 110 (2): 449–457.

Picot, J., Guerin, C.L., Le Van Kim, C. et al. (2012). Flow cytometry: retrospective, fundamentals and recent instrumentation. *Cytotechnology* 64 (2): 109–130. https://doi.org/10.1007/s10616-011-9415-0.

Plouffe, B.D., Murthy, S.K., and Lewis, L.H. (2015). Fundamentals and application of magnetic particles in cell isolation and enrichment: a review. *Rep. Prog. Phys.* 78 (1): 016601. https://doi.org/10.1088/0034-4885/78/1/016601.

Pretlow, T.G. and Pretlow, T.P. (1989). Cell separation by gradient centrifugation methods. *Methods Enzymol.* 171: 462–482.

Recktenwald, D. (1997). *Cell Separation Methods and Applications*. New York: Marcel Dekker.

Rickwood, D. and Birnie, G.D. (1975). Metrizamide, a new density-gradient medium. *FEBS Lett.* 50 (2): 102–110.

Sanders, M.J. and Soll, A.H. (1989). Cell separation by elutriation: major and minor cell types from complex tissues. *Methods Enzymol.* 171: 482–497.

Shields, C.W., Reyes, C.D., and Lopez, G.P. (2015). Microfluidic cell sorting: a review of the advances in the separation of cells from debulking to rare cell isolation. *Lab Chip* 15 (5): 1230–1249. https://doi.org/10.1039/c4lc01246a.

Shields, C.W., Ohiri, K.A., Szott, L.M. et al. (2017). Translating microfluidics: cell separation technologies and their barriers to commercialization. *Cytometry B Clin. Cytom.* 92 (2): 115–125. https://doi.org/10.1002/cyto.b.21388.

Shin, Y., Han, S., Jeon, J.S. et al. (2012). Microfluidic assay for simultaneous culture of multiple cell types on surfaces or within hydrogels. *Nat. Protoc.* 7 (7): 1247–1259. https://doi.org/10.1038/nprot.2012.051.

Sutermaster, B.A. and Darling, E.M. (2019). Considerations for high-yield, high-throughput cell enrichment: fluorescence versus magnetic sorting. *Sci. Rep.* 9 (1): 227. https://doi.org/10.1038/s41598-018-36698-1.

Tasoglu, S., Gurkan, U.A., Wang, S. et al. (2013). Manipulating biological agents and cells in micro-scale volumes for applications in medicine. *Chem. Soc. Rev.* 42 (13): 5788–5808. https://doi.org/10.1039/c3cs60042d.

Tomlinson, M.J., Tomlinson, S., Yang, X.B. et al. (2013). Cell separation: terminology and practical considerations. *J. Tissue Eng.* 4: 2041731412472690. https://doi.org/10.1177/2041731412472690.

Wells, J. (1989). Cell separation using velocity sedimentation at unit gravity and buoyant density centrifugation. *Methods Enzymol.* 171: 497–512.

Whitesides, G.M. (2006). The origins and the future of microfluidics. *Nature* 442 (7101): 368–373. https://doi.org/10.1038/nature05058.

Yeung, J. and So, C.W. (2009). Identification and characterization of hematopoietic stem and progenitor cell populations in mouse bone marrow by flow cytometry. *Methods Mol. Biol.* 538: 301–315. https://doi.org/10.1007/978-1-59745-418-6_15.

Yi, C.Q., Li, C.W., Jin, S.L. et al. (2006). Microfluidics technology for manipulation and analysis of biological cells. *Anal. Chim. Acta* 560: 1–23.

Zamay, G.S., Kolovskaya, O.S., Zamay, T.N. et al. (2015). Aptamers selected to postoperative lung adenocarcinoma detect circulating tumor cells in human blood. *Mol. Ther.* 23 (9): 1486–1496. https://doi.org/10.1038/mt.2015.108.

Zborowski, M. and Chalmers, J.J. (2011). Rare cell separation and analysis by magnetic sorting. *Anal. Chem.* 83 (21): 8050–8056. https://doi.org/10.1021/ac200550d.

Zheng, W., Jiang, B., Wang, D. et al. (2012). A microfluidic flow-stretch chip for investigating blood vessel biomechanics. *Lab Chip* 12 (18): 3441–3450. https://doi.org/10.1039/c2lc40173h.

CHAPTER 22

Genetic Modification and Immortalization

As cell lines became more widely used in the 1950s, scientists realized that they were particularly useful for the study of cancer. Many early cell lines developed the ability to proliferate indefinitely, which is now known as immortalization (see Minireview M3.1). Some cell lines formed tumors in nude mice and displayed other behavior consistent with malignancy; this phenotype is often referred to as transformation (see Section 3.6). In the 1960s, Hilary Koprowski and colleagues discovered that simian virus 40 (SV40) could be used to induce transformation (Koprowski et al. 1962). Purified viral DNA was also applied to cells directly, but the plasma membrane was a major obstacle to nucleic acids, resulting in a search for chemicals that could be used as vehicles for gene delivery (Brash et al. 1987). In the 1970s, it was discovered that cells could be transformed by purified viral genomic material using calcium phosphate coprecipitation (Graham and van der Eb 1973). A number of other techniques were developed for gene delivery, particularly by developmental biologists to study gain or loss of function in mouse models (Westphal 1984). Such historical efforts to improve gene delivery, and the development of recombinant DNA technology, paved the way for our modern gene editing methods. This chapter examines the commonly used techniques that are employed for gene delivery and gene editing *in vitro*. Practical examples focus on immortalization of cultures, which continues to be an important requirement in many tissue culture laboratories. Design and generation of recombinant vectors and nucleic acids are described elsewhere (Green and Sambrook 2012).

22.1 GENE DELIVERY

A variety of methods can be used to transfer genetic material into primary cultures, cell lines, and intact tissue samples (see Table 22.1). Methods can be grouped into (i) physical methods, which use electrical impulses or other physical factors to cross the plasma membrane; (ii) chemical methods, which typically use chemical agents to form a complex with the nucleic acid before passing into the cell; and (iii) biological methods, which use viruses or recombination-deficient viral vectors to achieve the same result (Kim and Eberwine 2010). Chemical agents and physical methods are the most popular methods in research laboratories (see Sections 22.1.1–22.1.3). However, advances in the design and availability of viral vectors have led to a recent resurgence in popularity for biological methods (see Section 22.1.4). Gene delivery using non-viral methods is often referred to as "transfection," while delivery using whole virus or viral vectors is referred to as "transduction." These terms have some practical relevance but there is increasing overlap between the various approaches; for example, viral genetic material may be transferred into a cell using a physical or chemical method.

Gene delivery techniques may result in transient or stable expression, depending on the material and method used. Transient transfection systems are typically used for short-term reporter assays or protein production. Expression levels are usually relatively high but persist for a shorter duration compared to stable systems, e.g. protein products are often harvested at 7–14 days post-transfection (Gutierrez-Granados

Freshney's Culture of Animal Cells: A Manual of Basic Technique and Specialized Applications, Eighth Edition. Amanda Capes-Davis and R. Ian Freshney.

TABLE 22.1. Methods of gene delivery.

Method	Advantages	Disadvantages
Physical		
Electroporation (see Protocol P22.3)	Successful for most cell types and some intact tissues; relatively easy to perform; highly reproducible	Requires specific equipment; may be toxic; requires optimization for each cell line
Microcell-mediated chromosome transfer	Allows a single chromosome to be introduced into a culture and maintained by selection	Requires expertise to perform the procedure and subsequent cytogenetic analysis
Microinjection	Allows precise delivery of a specific nucleic acid into a specific cell	Requires expertise and specialized equipment; limited utility for large numbers of cells
Microparticle or nanoparticle delivery, e.g. using Helios® Gene Gun (Bio-Rad)	Allows precise control over composition and delivery, e.g. using a pulse of helium gas; successful for large DNA fragments and some cell types that are hard to transfect	Requires specific equipment and gas supply; may be toxic; requires optimization for each cell line
Chemical		
Calcium phosphate coprecipitation (see Protocol P22.1; see also Supp. S22.1)	Successful for many cell types and some intact tissues; relatively cheap and easy to perform; suitable for uncloned DNA or high DNA concentration	Low efficiency and reproducibility; not successful for some cell types due to toxicity, differentiation, or aggregation; requires optimization for each cell line; DNA integrates as multiple copies in tandem
Cationic lipids, e.g. Lipofectamine (ThermoFisher Scientific), FuGENE 6 (Promega) (see Protocols P22.2, P22.4)	Successful for many cell types and some intact tissues; relatively high efficiency and low toxicity compared to other chemical agents; optimized for various target molecules including DNA, RNA, oligonucleotides, and some proteins	Relatively high cost compared to other chemical methods; may not be compatible with serum; ineffective for many terminally differentiated cells; limited utility for large DNA fragments or scale-up
Cationic polymers, e.g. DEAE-dextran, PEI	Commonly used for transient transfection, e.g. using PEI; relatively cheap; can be used in combination with other agents	Relatively low efficiency for many reagents, e.g. DEAE-dextran; may not be compatible with some media; further adaptation is needed to improve efficiency and biocompatibility
Strontium phosphate coprecipitation	Successful for some cell types where calcium phosphate is toxic or induces differentiation, e.g. normal human bronchial epithelial cells	As for calcium phosphate coprecipitation
Biological		
Whole virus, e.g. adenovirus, EBV, Moloney leukemia virus, SV40	Historical technique used to induce specific behavior, e.g. immortalization; high specificity and efficiency for transduction, depending on virus and tissue tropism	Biosafety risk; many cell lines continue to shed virus; superseded by viral vectors for most cell types
AAV vectors	Low biosafety risk (not known to be a human pathogen); successful for many cell types and non-dividing cells; typically maintained as a long-lasting episome; high specificity and efficiency, depending on serotype used	Traditionally requires a helper virus for replication; limited insert size (up to 4.9 kb); genomic integration may occur
Adenoviral vectors	Successful for many cell types and non-dividing cells; high efficiency; high titers during production; suitable for large DNA fragments (> 8 kb)	Biosafety risk due to origin from human pathogen and immunogenicity; transient expression with no genomic integration
Baculovirus vectors	Low biosafety risk (not pathogenic in vertebrates); high level of protein expression in insect cells (e.g. Sf cell lines); suitable for large DNA fragments	Normally used only for insect cells (although recombinant vectors may offer transient expression in mammalian cells)
Retroviral vectors (including lentiviral vectors)	Successful for many cell types; high efficiency; stable expression with genomic integration; lentiviral vectors are successful for non-dividing cells	Biosafety risk due to origin from human pathogen; random integration of transgene, resulting in insertional mutagenesis
Sendai virus (see Protocol P23.1)	Low biosafety risk for humans (not known to be a human pathogen); successful for many cell types and non-dividing cells; high efficiency	Biosafety risk for mice (known to be a mouse pathogen); transient expression with no genomic integration (this may be an advantage for some applications, e.g. iPSCs)

et al. 2018). Stable transfection systems are used when integration into the host genome is desirable, resulting in long-term expression. Viral transduction is often preferred for these applications, but stable transfection can be achieved using non-viral methods such as electroporation. The likelihood of stable transfection can be increased through a careful choice of method and materials, e.g. linearization of DNA will increase the number of stable transfectants with some methods of gene delivery (Stuchbury and Munch 2010).

Genetic modification of any living organism or material – including cells in culture – can have unexpected consequences. Safety risks and ethical implications are important for all genetically modified organisms (GMOs) and require reassessment with today's specialized *in vitro* applications (see Sections 6.3.7, 6.4.2) (Munsie and Gyngell 2018). As part of planning gene delivery experiments, always assess the risks and ensure that you comply with safety and ethical regulations (see Sections 6.1.1, 6.1.2).

22.1.1 Transfection with Calcium Phosphate

Calcium phosphate transfection relies on the precipitation of DNA in the presence of calcium and phosphate ions (Graham and van der Eb 1973). This results in formation of a visible sediment, which becomes closely associated with the cell membrane and facilitates uptake of DNA. Calcium phosphate coprecipitation can be used for transient or stable transfection of a number of cell types and samples, including primary cultures, cell lines, and intact tissue samples (Conn et al. 1998; James et al. 1998). It is the cheapest transfection method but has a relatively low transfection efficiency, often in the range of 10–30% (Somasundaram et al. 1992; Goetze et al. 2004). This may not be a disadvantage for some applications, particularly if a fluorescent marker is used to determine which cells have been successfully transfected. It can promote aggregation of cells in suspension and is typically used only for adherent cultures (Gutierrez-Granados et al. 2018).

The success of calcium phosphate transfection can vary considerably, depending on the cell type, DNA, and minor variations in procedure. The best transfection efficiencies are associated with formation of a fine sediment that does not include large crystals; a fine sediment is more likely to form when the pH, DNA concentration, incubation time, and incubation temperature have all been optimized (Conn et al. 1998). Protocol P22.1 has been used successfully for transfection of neuronal cultures and can be used as a starting point for other cell types.

PROTOCOL P22.1. CALCIUM PHOSPHATE CO-PRECIPITATION

Contributed by Ornella Tolhurst when at the Developmental Neurobiology Unit, CMRI.

Outline

Subculture cells and incubate for 24 hours. Mix DNA with $CaCl_2$ and HEPES to form a precipitate. Add the precipitate to cells and incubate overnight. Wash cells to remove precipitate.

Materials (sterile or aseptically prepared)

- ☐ Cells for transfection
- ☐ DNA for transfection, 10 μg (made up as 1 μg/μl sterile solution; see Notes)
- ☐ $CaCl_2$ 2 M solution
- ☐ HEPES ×2 strength (see Appendix B)
- ☐ Tris base, ethylene diamine tetra-acetic acid (EDTA) (TE) buffer (10 mM Tris–HCl, pH 8.0; EDTA 1 mM)
- ☐ Culture medium
- ☐ Dulbecco's phosphate buffered saline without Ca^{2+} and Mg^{2+} (DPBS-A)
- ☐ Culture dishes, 100 mm (1 per transfection)
- ☐ Centrifuge tubes, 15 ml (2 per transfection)
- ☐ Plastic serological pipettes in an assortment of sizes, including 1 ml if available
- ☐ Fine plastic pipette, e.g. plastic Pasteur pipette with bulb

Materials (non-sterile)

- ☐ Biological safety cabinet (BSC) and associated consumables (see Protocol P12.1)
- ☐ Cell counting equipment (hemocytometer or automated counter; see Section 19.1)

Procedure

A. Day Before Transfection

1. Subculture cells using customary procedure (e.g. see Protocol P14.2).
2. Count the cells with a hemocytometer or an automated cell counter.
3. Plate cells to give approximately 5×10^5 cells per plate on the day of transfection. Incubate plate for 24 hours under standard culture conditions, e.g. 37 °C and 5% CO_2.

B. Day of Transfection

4. Prewarm culture medium and prepare a work space for aseptic technique using a BSC.
5. Refeed cells with fresh culture medium (10 ml per plate) and return to the incubator.
6. Prepare two centrifuge tubes as follows:
   (a) Tube 1: add 10 μg DNA, 63 μl 2 M $CaCl_2$, and TE to a total volume of 500 μl.
   (b) Tube 2: add 500 μl 2× HEPES.

7. Use a small pipette (e.g. 1 ml) to bubble air through the HEPES buffer in Tube 2.
8. While bubbling air through the buffer, use a fine plastic pipette to add the contents of Tube 1. Do this slowly drop by drop. Gently flick the tube a few times to mix the solution.
9. A very fine precipitate should start to appear, making the solution look cloudy. Leave the precipitate to stand for 30 minutes at room temperature.
10. Remove the culture plate from the incubator. Gently flick the centrifuge tube a few times to mix the contents and pipette the precipitate drop by drop onto the cells in the culture dish.
11. Swirl the plate to evenly spread the precipitate over the cells.
12. Return the plate to the incubator and incubate with the precipitate for 24 hours. The precipitate should be visible when viewed under the microscope.

C. Day after Transfection

13. Prewarm culture medium and prepare a work space for aseptic technique using a BSC.
14. Aspirate and discard the precipitate and medium. Wash with DPBS-A (3 × 5 ml each) and then culture medium (1 × 5 ml).
15. Refeed cells, e.g. with medium plus G418 to detect cells expressing neomycin resistance.

Notes

Step 6: DNA should be sterile and of high purity (Conn et al. 1998). DNA can be sterilized by performing ethanol precipitation and the pellet air-dried before resuspending in sterile TE buffer. Some experts advise that water should be used instead of TE; buffers containing Tris may alter the pH of the precipitate and reduce transfection efficiency (Conn et al. 1998). Supercoiled DNA has a higher transfection efficiency compared to linear DNA.

Steps 6 and 7: This method is sometimes referred to as "calcium chloride transfection," due to the use of $CaCl_2$ as a source of calcium ions, but HEPES is equally important as a source of phosphate ions.

Step 9: The duration may affect the quality of the precipitate. Most procedures use 20–60 minutes.

Step 12: Overnight incubation may result in toxicity for some cultures; incubation may be reduced to four hours if required. Some experts recommend adding a glycerol or dimethyl sulfoxide (DMSO) shock after incubation to improve efficiency, but this may cause increased toxicity (Conn et al. 1998).

Altering the CO_2 level or using CO_2-independent media may also improve efficiency; this will alter the pH, which is an important variable when optimizing the procedure (Goetze et al. 2004).

22.1.2 Transfection with Cationic Lipids and Polymers

Cationic lipids have traditionally been widely used as transfection reagents in research laboratories because of their relatively high efficiency, low toxicity, and ease of use. Cationic lipids act by forming a positively charged complex or "liposome" with nucleic acids, resulting in the term "lipofection" to describe the technique (Felgner et al. 1987). Once formed, the lipid complex enters the cell by endocytosis or by fusing with the cell membrane, resulting in transient transfection. Complexes can also enter the nucleus, allowing stable transfection to occur through genomic integration (Friend et al. 1996). At least 12 companies provide cationic lipids as lipofection reagents (Hawley-Nelson et al. 2008). Most companies now offer multiple products that have been optimized for different applications, cell types, and target molecules (including DNA, mRNA, synthetic RNA, oligonucleotides, and some proteins). Lipofectamine (ThermoFisher Scientific) is probably the most widely used cationic lipid over time, but the original formulation has undergone extensive optimization since its initial release (Hawley-Nelson et al. 2008). Most of the currently available lipofection reagents use proprietary formulations.

Cationic polymers were used for gene delivery before the development of lipid-based reagents, starting with diethylaminoethyl (DEAE)-dextran, which was used to improve the efficiency of viral transduction (McCutchan and Pagano 1968). DEAE-dextran gave a relatively low efficiency, resulting in its replacement by calcium phosphate transfection. Polyethylenimine (PEI) has been more successful, particularly for transient transfection (Gutierrez-Granados et al. 2018). More recent generations of cationic polymers have been optimized to improve their efficiency and biocompatibility, and show considerable promise for gene therapy (Zhang and Wagner 2017). Some of these cationic polymers are now also marketed for research laboratories, e.g. TurboFect (ThermoFisher Scientific) and jetPEI® (Polyplus-transfection).

How do you choose a transfection reagent? Although the currently available reagents have been extensively optimized, performance still varies with the cell line and the target molecule. Review the literature as it relates to your specific cell line or cell type, discuss with local suppliers, and ask for a sample or a discount on your preferred reagent that will allow you to test its efficiency and toxicity on your own cultures. Protocol P22.2 is provided as a starting point for optimization using FuGENE® 6 (Promega) (Jacobsen et al. 2004).

PROTOCOL P22.2. OPTIMIZATION OF LIPOFECTION

Contributed by Amanda Capes-Davis; adapted from Jacobsen et al. (2004) and the FuGENE 6 Technical Manual (Promega).

Background

Protocols can vary considerably between transfection reagents. Always follow the manufacturer's instructions for the specific reagent you wish to use. You should also optimize the procedure for your specific cell line. Key variables include the number of cells plated, the amount of DNA used, and the ratio of DNA to reagent. This procedure focuses on the latter variable. Using a multiwell plate will minimize the amount of reagent used during optimization.

Outline

Subculture cells into a 96-well plate. Combine medium and FuGENE 6 Transfection Reagent, followed by plasmid DNA. Add to cells and incubate for 24–48 hours. Assess performance.

Materials (sterile or aseptically prepared)

☐ Cell line for optimization
☐ DNA for transfection (1 µg/µl), e.g. plasmid containing reporter transgene
☐ FuGENE 6 Transfection Reagent (Promega), stored in its original vial at 4 °C
☐ Serum-free culture medium, aliquot for mixing with transfection reagent
☐ 96-well plate (1)
☐ Microcentrifuge tubes (4)

Materials (non-sterile)

☐ BSC and associated consumables (see Protocol P12.1)
☐ Cell counting equipment (hemocytometer or automated counter; see Section 19.1)

Procedure

A. Day before Transfection

1. Plan the optimization procedure, including the number of variables to be tested, the number of wells required, and the plate layout.
2. Dispense an aliquot of serum-free medium that has not been exposed to cells.
3. Subculture cells using customary procedure (e.g. see Protocol P14.2).

4. Count the cells with a hemocytometer or an automated cell counter.
5. Plate cells so that they are approximately 50–80% confluent on the day of transfection. Incubate plate for 24 hours under standard culture conditions, e.g. 37 °C and 5% CO_2.

B. Day of Transfection

6. Prewarm an aliquot of culture medium (see step 2) to room temperature and prepare a work space for aseptic technique using a BSC.
7. Using microcentrifuge tubes, add FuGENE 6 to medium at four different volumes (to optimize the FuGENE 6:DNA ratio, with each tube containing sufficient reagent for 20 wells):
   (a) Tube 1 (6:1 ratio): add 12 µl FuGENE 6 to 86 µl medium.
   (b) Tube 2 (4:1 ratio): add 8 µl FuGENE 6 to 90 µl medium.
   (c) Tube 3 (3:1 ratio): add 6 µl FuGENE 6 to 92 µl medium.
   (d) Tube 4 (1.5:1 ratio): add 3 µl FuGENE 6 to 95 µl medium.
8. Mix each tube immediately after adding and incubate at room temperature for five minutes.
9. Add 2 µl DNA (1 µg/µl) to each tube, giving a total volume of 100 µl in all four tubes.
10. Mix each tube immediately after adding and incubate at room temperature for 15 minutes.
11. Bring the 96-well plate containing cells to the BSC. Add 5 µl FuGENE 6:DNA mixture to each well, in accordance with the experimental plan. Mix gently, e.g. by pipetting.
12. Return plates to the incubator and incubate under normal conditions for 24–48 hours before assessing transfection efficiency, toxicity, and other variables.

Notes

Step 2: The reagent manual states that FuGENE 6 should not be dispensed into aliquots. This creates a potential risk for cross-contamination. To minimize the risk, always handle FuGENE 6 and associated reagents separately and mix with an aliquot of medium that has not been exposed to cells in culture. FuGENE 6 can be used in the presence of serum, but the initial mixture should be prepared in serum-free medium (Jacobsen et al. 2004).

Step 3: The health of the culture is a major factor in transfection efficiency. If possible, passage cells when they are in an exponential phase of growth (see Figure 14.1).

Step 5: Medium should be antibiotic-free. In addition to the usual concerns regarding antibiotic use (see Section 9.6.2), they can reduce transfection efficiency (Jacobsen et al. 2004).

Step 7: Always add FuGENE 6 directly to the medium and do not drop down the side of the tube. Contact with plastic can dramatically reduce its transfection efficiency (Jacobsen et al. 2004).

Step 12: It is not necessary to remove the transfection reagent.

22.1.3 Electroporation

Electroporation refers to the delivery of genetic material using an electrical impulse. The technique is based on an early observation that applying short electrical impulses to mouse lyoma cells resulted in a transient increase in permeability, allowing DNA transfer (Neumann et al. 1982). The increase in permeability is associated with formation of pores within the plasma membrane, but it persists after the pores close, so additional mechanisms must play a part in gene delivery (Mir 2014). Electrical impulses can be used to induce transient or stable transfection, cell–cell fusion, or cell death, depending on the field strength and pulse duration. For stable transfection, the procedure can be used to optimize the number of copies that are introduced per cell; this compares favorably to some other methods such as calcium phosphate transfection, where DNA often integrates as multiple copies in tandem (Potter and Heller 2003).

Cells are usually placed in suspension for electroporation (see Protocols P14.2, P14.3). The cell suspension is washed and resuspended at a high concentration in ice-cold electroporation buffer, e.g. Gene Pulser® Electroporation Buffer (Bio-Rad) (Jordan et al. 2008). Aliquots of the cell suspension are transferred to sterile electroporation cuvettes and the DNA is added on ice. Each cuvette is placed in an electroporator to receive an electrical impulse (see Figure 22.1). Cuvettes are kept on ice for a further 10–15 minutes before removing the transfected cells and rinsing with culture medium. The number of cells, amount of DNA, and optimal procedure will vary, depending on the cell line and whether transient or stable transfection is required (Potter and Heller 2003; Jordan et al. 2008).

Electroporation requires optimization and access to the necessary equipment, but if these requirements can be met, it is efficient, easy to perform, and highly reproducible. Equipment may be purchased with programs and kits that have been optimized for specific cell types, such as the Nucleofector™ System (Lonza). Other systems provide a standard protocol but also allow the user to customize parameters, such as the Gene Pulser Xcell™ Electroporation System (Bio-Rad; see Figure 22.1) and the Neon Transfection System (ThermoFisher Scientific). Always ask for a demonstration and test the electroporator on your own cells (as recommended for any equipment that will be used for tissue culture!) before you commit to what may be an expensive purchase.

22.1.4 Viral Transduction

Whole viruses such as SV40 were originally used to immortalize cells *in vitro*, resulting in safety concerns for some long-standing cell lines (see Section 22.3.1). As viruses became increasingly used for gene delivery, tissue culture laboratories began to develop viral vectors from human adenovirus and γ-retroviruses such as Moloney murine leukemia virus (MMLV). These early vectors had some disadvantages. Adenovirus could be used for transient expression but did not allow genomic integration, while MMLV did not allow genomic integration in non-dividing cells (see Table 22.1). The lentivirus family of retroviruses was seen as an attractive alternative for gene delivery because of its ability to induce genomic integration with high efficiency in dividing and non-dividing cells (Amado and Chen 1999). However, many lentiviruses are pathogens that require a high level of biological containment (see Section 6.3.3). Viral vectors were modified from these species to improve their safety and efficacy, resulting in their increasing use for gene delivery in the laboratory and gene-based therapy in the clinic (Lundstrom 2018).

A number of viral vectors are now available for use in tissue culture laboratories, each with their own advantages and disadvantages (see Table 22.1) (Addgene 2019). In this chapter, we focus on four groups of vectors derived from baculovirus, lentivirus, adenovirus, and adenovirus-associated virus (AAV). Sendai viral vectors are commonly used to generate induced pluripotent stem cell (iPSC) lines and are discussed in the following chapter (see Section 23.3).

Baculovirus expression vectors (BEVs). Baculovirus refers to a family of viruses (Baculoviridae) that infect many invertebrate species *in vivo* but are non-pathogenic for vertebrate species (Rohrmann 2013; van Oers et al. 2015). Baculovirus expression vector (BEV) systems were initially developed from *Autographa californica* multiple nucleopolyhedrovirus (*Ac*NPV) (Smith et al. 1983; Pennock et al. 1984). Over time, the efficacy of this expression system, its low biosafety risk, and easy access for research use have resulted in broad uptake of the technology (Rohrmann 2013; van Oers et al. 2015). BEV systems are now widely used to express recombinant proteins and to produce vaccines against papillomavirus, influenza virus, and other pathogens (Madhan et al. 2010). Recombinant BEVs have also been developed that allow gene delivery into mammalian cells (Chambers et al. 2018).

Baculovirus expression is usually performed in insect cell lines (e.g. Sf9), maintained in insect-specific media at 20–28 °C (see Section 24.6.2; see also Tables 10.2, 10.3). Some portions of the baculovirus genome are not required for its replication *in vitro*, such as the polyhedron gene (*polh*), allowing replacement by a transgene whose expression is

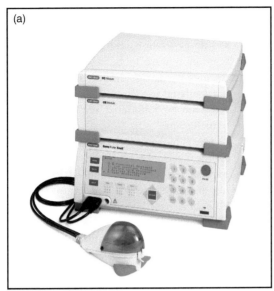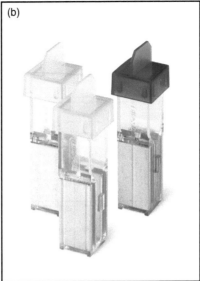

Fig. 22.1. Electroporation equipment. (a) The Gene Pulser Xcell Electroporation System. This device consists of a main unit and a choice of two stackable modules; the CE module is recommended for use with most eukaryotic cells. Cuvettes are placed into the ShockPod chamber, which is connected by leads to the front of the main unit. (b) A Gene Pulser electroporation cuvette. Cuvettes are supplied sterile and vary in gap size (0.1–0.4 cm); a higher gap size translates to a lower field strength. *Source*: Courtesy of Bio-Rad Laboratories, Inc. with permission.

driven by the *polh* promoter. *Polh*-negative viruses cannot replicate *in vivo*, which gives an additional biosafety feature. Usually, the transgene is cloned into a transfer vector that contains *polh* flanking sequences; this vector is co-transfected into insect cells along with the virus genome (Chambers et al. 2018). Co-transfection is required because of the size of the baculovirus genome, making it difficult to insert genes directly. Recombinant virus is shed into the culture medium, where it can be harvested and purified. Once viral stocks are available, they can be used for protein production, e.g. by transduction of large volumes of insect cells in suspension (see Section 8.6.2).

Co-transfection with a transfer vector originally gave very low success rates (typically 0.1–1%), requiring rounds of screening and purification using plaque assays (van Oers et al. 2015). Over time, a number of improvements have been made to increase success rates and simplify the procedure. For example, a "bacmid" system may be used that contains the baculovirus genome but can replicate as a plasmid in bacteria, allowing recombinant baculovirus to be produced before it is added to insect cells (Luckow et al. 1993). A number of different BEV systems are now available. Recent examples include the flashBAC™ System (Oxford Expression Technologies), which simplifies the production of recombinant virus and eliminates the need to purify recombinant virus, and the BaculoDirect™ system (ThermoFisher Scientific), which enables the transgene to be cloned directly into the baculovirus genome via a Gateway® entry vector. BEV

systems can be technically challenging for the beginner and it is impossible to give more than a brief overview here. Other resources are available that have further information on BEV systems (Murhammer 2016; Oxford Expression Technologies 2019).

Lentiviral vectors. Lentiviral vectors used in research laboratories are mostly derived from human immunodeficiency virus-1 (HIV-1). As such, they have obvious safety concerns, resulting in the need for specific guidance when lentiviral vectors are used in the laboratory or for gene-based therapy in the clinic (RAC 2006; Schlimgen et al. 2016; White et al. 2017). Risks associated with lentiviral vectors include (i) generation of replication-competent lentivirus (RCL); (ii) mutagenesis due to random insertion of the transgene; and (iii) development of malignant disease, e.g. from altered expression of oncogenes or tumor suppressor genes following random mutagenesis (Schlimgen et al. 2016). These risks must be assessed and managed appropriately whenever a lentiviral vector is used.

Lentiviral vectors have undergone successive rounds of development since their introduction. First-generation vectors used a split genome design to reduce the risk of RCL. Genomic material was divided across three plasmid vectors: a packaging plasmid, an envelope plasmid, and a transfer plasmid that encodes the bulk of the viral genome and the ψ packaging signal (Naldini et al. 1996). This approach has been refined over successive generations to reduce safety risks (see Figure 22.2) (Schlimgen et al. 2016).

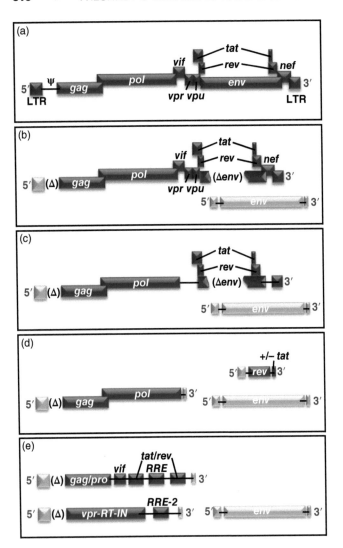

Fig. 22.2. Lentiviral vector development. (a) The wild-type HIV genome is used as the basis for lentiviral vectors. (b) First-generation vectors removed the envelope protein and the psi packaging signal and incorporated a heterologous promoter to reduce recombination potential. (c) Second-generation vectors removed accessory genes (*vif*, *vpr*, *vpu*, and *nef*) to reduce the virulence of any potential RCL. (d) Third-generation vectors eliminated the transactivator gene, *tat*, and split the vector into three plasmids to reduce further recombination potential, retaining only the three genes necessary for transgene expression (*gag*, *pol*, *rev*). (e) Fourth-generation vectors split the *gag* and *pol* onto separate plasmids to further reduce recombination potential. This generation added back some HIV genes to enhance transduction efficiency and transgene expression. *Source*: Schlimgen, R., Howard, J., Wooley, D., et al. (2016). Risks Associated With Lentiviral Vector Exposures and Prevention Strategies. J Occup Environ Med 58 (12):1159–1166. © 2016 Wolters Kluwer.

First- and second-generation vectors had a higher risk of generating RCL compared to later generations, although second-generation vectors should have a lower virulence if replication were to be restored. Testing for RCL may be required to manage the risk and must be performed for clinical lentiviral use (RAC 2006). The nature of the transgene must also be considered, e.g. oncogenes carry a higher risk of malignancy.

Once biohazard risks have been assessed and managed appropriately, the transgene can be cloned into a suitable transfer vector and introduced (together with the packaging and envelope plasmids) into a suitable cell line. The HEK293T cell line is typically used for viral production. Culture medium containing lentivirus is harvested, concentrated and purified, and titrated prior to use (Li et al. 2012). Lentiviral vectors and production services are available from Addgene and some commercial suppliers, e.g. Applied Biological Materials (abm), Sigma-Aldrich, Takara Bio, ThermoFisher Scientific, and Vigene Biosciences.

Adenoviral vectors. Adenoviral vectors are generally derived from human adenovirus-5. This virus is pathogenic in humans, where it causes respiratory disease, but has a lower risk compared to wild-type lentiviruses. Adenoviral vectors are particularly used for applications where cancer gene therapy is the intended outcome, resulting in more than 400 gene therapy trials that have commenced to date (Wold and Toth 2013). Adenoviral vectors typically carry deletions in the E1 region, which is essential for replication, and may lack E2, E3, and E4 regions (Wold and Toth 2013). Vectors carrying these deletions must be grown on complementary cell lines that express the relevant deleted genes. Other "helper-dependent" adenoviral vectors lack all non-viral coding sequence, allowing larger inserts to be used, but retain the origin of replication and sequence required for packaging of virions. These vectors must be grown in the presence of a helper adenovirus. Adenoviral vectors are available from Addgene; laboratories may choose to generate their own recombinant adenovirus, e.g. using the AdEasy system (Luo et al. 2007; Kamens 2015). Commercial suppliers also offer vectors and may provide production services, e.g. abm, Takara Bio, Vector Biolabs, and Vigene Biosciences.

Adenovirus-associated virus (AAV) vectors. AAVs were initially discovered as a contaminant of adenovirus preparations and are non-pathogenic in humans (Naso et al. 2017). These viruses require a helper virus such as adenovirus to be present for replication to occur. If a helper virus is not present, AAVs can persist as a circular episome or through genomic integration (Lisowski et al. 2015). AAV's potential for viral transduction was originally seen more than 30 years ago (Hermonat and Muzyczka 1984), but today's recombinant AAV vectors have been extensively modified for use in the laboratory and for gene-based therapy in the clinic. Several AAV-based products have been approved in Europe and the United States for treatment of genetic diseases (Naso et al. 2017; Rodrigues et al. 2018).

In its native state, the AAV genome includes coding sequences for Rep (Replication) and Cap (Capsid) proteins, flanked by two inverted terminal repeats (ITRs). At least 12 AAV serotypes exist, which vary in tissue tropism depending on their capsid sequence. In recombinant AAV vectors, the viral coding sequence is removed and the transgene of interest is cloned into an expression cassette between the ITRs, where it is driven by a mammalian promoter, e.g. the cytomegalovirus (CMV) promoter/enhancer. Pseudo-serotyping can be used to configure the AAV vector for specific tissue types (Lisowski et al. 2015; Naso et al. 2017). Rep and Cap proteins are supplied separately, along with a helper virus. Helper virus-free systems are also available that provide the necessary adenovirus genes using specific plasmids or packaging cell lines, e.g. HEK293 cells. AAV vectors and production services are available from Addgene and some commercial suppliers, e.g. abm, Takara Bio, Vector Biolabs, and Vigene Biosciences.

22.2 GENE EDITING

The term "gene editing" refers to a family of technologies that allow targeted changes in DNA sequence at high efficiency in multiple species (DeWitt et al. 2017). The idea of making targeted changes in DNA sequence is not a new one; gene targeting has been performed in mice and a small number of other species for many years, but the efficiency of gene targeting was low, requiring screening or enrichment to identify the rare cells where targeting was successful (Capecchi 2005). The field was transformed by the discovery that nucleases could be "programmed" to target specific DNA sequences in eukaryotic cells (Kim et al. 1996; Chandrasegaran and Carroll 2016). Three gene editing technologies are now widely used in research laboratories, based on (i) zinc-finger nucleases (ZFNs); (ii) transcription activator-like effector nucleases (TALENs); and (iii) RNA-guided nucleases that use the clustered regularly interspaced short palindromic repeat (CRISPR) and a conserved CRISPR-associated protein (Cas) (Carroll 2017).

The three gene editing technologies discussed here are distinctly different but share some core similarities (see Table 22.2). Each technology relies on a nuclease that can target a specific genetic sequence, resulting in the generation of a DNA double-strand break (DSB) at that location. Once the nuclease has acted, naturally occurring cellular mechanisms are used to repair the DSB. DNA repair pathways used in gene editing can be divided into (see Figure 22.3) (Carroll 2014; Danner et al. 2017):

(1) **Homology-directed repair (HDR).** These pathways use homologous sequence as a template for DNA repair. Removal of 5′ nucleotides at the cleavage site results in

a single strand with an exposed 3′ end, which invades the homologous sequence and is extended to repair the break point. The extended strand then pairs with the exposed sequence from the opposite side of the break. A template may be supplied for HDR, which will help to determine the specific pathway used. HDR has a low rate of error but is utilized only in the S and G_2 phases of the cell cycle (see Section 2.3.1).

(2) **Non-homologous end joining (NHEJ).** These pathways do not require a template for DNA repair. Instead, the DSB ends are held in close alignment by a protein complex, the Ku70-Ku80 heterodimer, resulting in direct ligation of the DSB. Classic NHEJ is faster than HDR and can occur throughout the cell cycle, making it the predominant DNA repair mechanism in many forms of gene editing. Unfortunately, NHEJ has a relatively high rate of error due to the random insertion or deletion of one or more nucleotides (indels) during the repair process (see Sections 22.4.3, 22.4.4).

22.2.1 Zinc Finger Nucleases (ZFNs)

ZFNs were originally developed as "hybrid restriction enzymes" in which the DNA cleavage domain of the *Fok* I endonuclease was linked to various zinc finger proteins (see Figure 22.4a) (Kim et al. 1996). Zinc finger motifs consist of approximately 30 amino acids that bind to the DNA double helix, based on recognition of a 3–4 base pair (bp) nucleotide sequence. Usually, ZFNs encode at least three zinc finger motifs positioned in tandem, which bind to a 9–18 bp target site and cleave the DNA if present as a dimer (Carroll 2014; Chandrasegaran and Carroll 2016). Binding preferences vary with the specific zinc fingers used – for example, the first two ZFNs bound to 5′-GAG GAG GCT-3′ and 5′-GAG GGA TGT-3′ (Kim et al. 1996). A large library of ZFNs has been developed by Sangamo Therapeutics, which licensed the technology for research use and is engaged in clinical trials for gene-based therapy (Chandrasegaran 2017).

ZFNs are currently used in research laboratories for a range of applications, including generation of targeted knockout animals and insertion of transgenes into human and rodent genomes. ZFN reagents, kits, and services are available for research use (e.g. Zinc Finger Consortium Reagents, Addgene; CompoZr® technology, Sigma-Aldrich) (Hansen et al. 2012; Kamens 2015). Affinity and specificity may vary, and it takes time and effort to optimize ZFNs for specific genes and cell lines (Carroll 2014). Artifacts may also occur, such as generation of indels through NHEJ at the target site and other off-target effects (see Sections 22.4.3, 22.4.4).

22.2.2 Transcription Activator-Like Effector Nucleases (TALENs)

TALENs, like ZFNs, are recombinant "hybrid" or fusion proteins that link the DNA cleavage domain of the *Fok* I

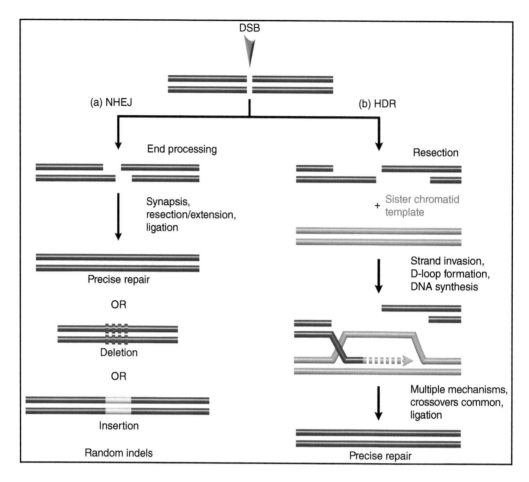

Fig. 22.3. DSB repair pathways. Endogenous DNA repair can occur following formation of a DSB by (a) NHEJ, resulting in random indels; (b) HDR, which uses a template DNA strand for precise repair. *Source*: Lino, C. A., Harper, J. C., Carney, J. P., & Timlin, J. A. (2018). Delivering CRISPR: a review of the challenges and approaches. Drug Delivery, 25(1), 1234–1257. Licensed under CCBY 4.0.

endonuclease to a DNA-binding motif (see Figure 22.4b). For TALENs, this motif comes from the world of plant biology (Baltes et al. 2017). Transcription activator-like effector proteins are produced by some members of the bacterial genus *Xanthomonas*, where they promote infection of plant cells by binding host DNA (Doyle et al. 2013). These proteins recognize DNA through a unique TALE domain, which consists of tandem repeats of a 33–35 amino acid sequence. Repeats differ at amino acids 12 and 13, which are known as the repeat variable diresidue (RVD). Each RVD recognizes one nucleotide in the target sequence (Boch et al. 2009; Moscou and Bogdanove 2009). Unlike ZFNs, where neighboring zinc fingers may alter specificity, RVDs have a one-to-one relationship with specific nucleotides, e.g. asparagine–isoleucine (NI) consistently targets adenine and thus base pair adenine:thymine (A:T) (Carroll 2014).

TALEN design has benefited from the experience gained through ZFNs and offers greater control over target specificity for the user. However, TALENs are larger than ZFNs (because their motifs recognize 1 bp, while zinc fingers recognize 3–4 bp) and their highly repetitive sequence can cause difficulties when translated *in vitro* (Chandrasegaran and Carroll 2016). For example, the TALE motif can be unstable when used with lentiviral vector systems, resulting in extensive rearrangement and loss of function (Holkers et al. 2013). Despite these caveats, TALEN constructs can be assembled by research laboratories with remarkable speed and efficiency, e.g. using Golden Gate cloning (Cermak et al. 2011; Engler and Marillonnet 2014). TALEN kits are available from Addgene and some commercial suppliers (e.g. ThermoFisher Scientific).

Although TALENs have been partly overshadowed by later developments, they are a powerful platform for gene editing in plants and animals (Baltes et al. 2017; Carroll 2017). Ongoing challenges for TALEN-based applications include the need to deliver functional nucleases at high efficiency, resulting in less screening for difficult cell types, while avoiding genomic integration and long-term expression, which is usually not required once gene editing is completed (Hackett and Somia 2014).

TABLE 22.2. Gene editing technologies.

Characteristic	ZFNs	TALENs	CRISPR/Cas9
Year of development	1996 (Kim et al. 1996)	2010–2011 (Christian et al. 2010; Miller et al. 2011; Mussolino et al. 2011)	2012–2013 (Jinek et al. 2012; Cho et al. 2013; Cong et al. 2013; Hwang et al. 2013; Mali et al. 2013)
Composition	Recombinant, programmable fusion protein	Recombinant, programmable fusion protein	Naturally occurring, programmable RNP containing CRISPR RNA and Cas9 protein
Source of nuclease activity	Cleavage domain of *Fok* I endonuclease; dimerization is required for nuclease activity	Cleavage domain of *Fok* I endonuclease; dimerization is required for nuclease activity	Cas9 protein
Source of DNA binding and specificity	Zinc fingers used in parallel; target sequence determined by choice of zinc fingers and neighboring modules; each zinc finger determines specificity for 3–4 bp	TALE domain repeats used in tandem; target sequence determined by choice of RVD in TALE repeats; each RVD determines specificity for 1 bp	CRISPR loci; target sequence determined by choice of spacer sequence
DSB repair pathway	NHEJ or HDR, depending on experimental design, e.g. availability of donor sequence	As for ZFNs	As for ZFNs
Mode of delivery	DNA/mRNA encoding a ZFN pair (for upstream and downstream sequence) is introduced using transfection, electroporation, or viral transduction	DNA/mRNA encoding a TALEN pair (for upstream and downstream sequence) is introduced using lipofection, electroporation, or viral transfection	CRISPR/Cas9 is introduced as DNA (in an expression vector), RNA, or a preassembled RNP complex using various methods (Lino et al. 2018)

See text for abbreviations. "Programmable" means that the specificity of the system is determined by the user.

22.2.3 CRISPR/Cas RNA-Guided Nucleases

CRISPR/Cas systems use a naturally occurring method of DNA integration that is present in many bacteria and archaea (Sorek et al. 2013; Hille et al. 2018). These prokaryotes carry a distinctive locus, known as a clustered regularly interspaced short palindromic repeat (CRISPR), which consists of a repeated sequence (typically 20–50 bp), interspersed with unique spacer sequences of a similar length. CRISPR loci were originally identified in *Escherichia coli* (Ishino et al. 1987). Over time, it was recognized that such loci were present in many prokaryotic species and were positioned close to CRISPR-associated (*cas*) genes (Mojica et al. 2000; Jansen et al. 2002). It is now understood that CRISPR loci and Cas proteins work together in prokaryotes as an adaptive immune mechanism against bacteriophages and plasmids (Sorek et al. 2013).

The unique spacer sequences in CRISPR loci are derived from foreign genetic material (Bolotin et al. 2005; Mojica et al. 2005; Pourcel et al. 2005). When a prokaryote with a CRISPR/Cas system is exposed to a bacteriophage, the host acquires new spacer sequences from the phage DNA, which are integrated into the CRISPR locus. CRISPR loci containing the new spacer material are transcribed into CRISPR RNA (crRNA), which is assembled into a ribonucleoprotein (RNP) complex with one or more Cas proteins. These RNPs can then bind to the corresponding foreign DNA, leading to its cleavage (Barrangou et al. 2007; Garneau et al. 2010). The spacer sequence is retained as part of the genome, allowing a rapid response if the bacteriophage reinfects the cell.

Once the biology of the CRISPR/Cas system was understood, its potential for use in gene editing became obvious to a number of laboratories (Jinek et al. 2012; Cong et al. 2013; Hwang et al. 2013; Cho et al. 2013; Mali et al. 2013). Various CRISPR/Cas systems exist *in vivo*. The Type II CRISPR/Cas system is particularly appealing for gene editing because only three components are needed for activity (Sorek et al. 2013). The first component, the Cas9 protein, has two nuclease domains, allowing cleavage of double-stranded DNA. The remaining two components are crRNA and a *trans*-activating crRNA (tracrRNA). These two RNA transcripts can be produced *in vitro* as a single guide RNA (Jinek et al. 2013; Carroll 2014). In practice, the

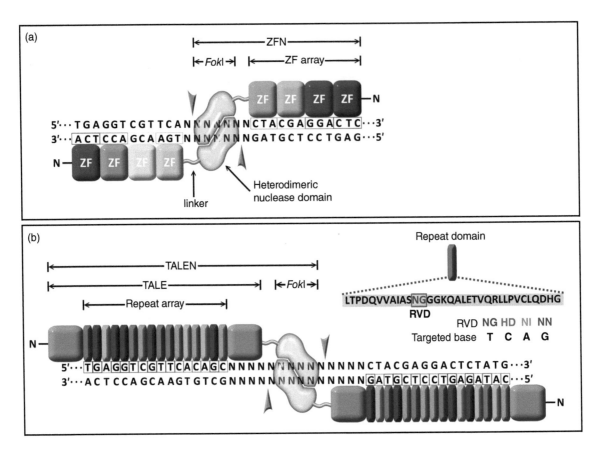

Fig. 22.4. ZFN and TALEN composition. (a) ZFN design, showing the *Fok* I endonuclease linked to a zinc finger (ZF) array and binding to DNA as a dimer. Each ZF recognizes 3–4 bp. (b) TALEN design, showing the *Fok* I endonuclease linked to a TALE motif and binding to DNA as a dimer. Each TALE motif consists of an array of tandem repeats; each repeat has an RVD that recognizes a single nucleotide. *Source:* Lino, C. A., Harper, J. C., Carney, J. P., & Timlin, J. A. (2018). Delivering CRISPR: a review of the challenges and approaches. Drug Delivery, 25(1), 1234–1257. Licensed under CCBY 4.0.

number of RNA transcripts may vary with the application, so the term "guide RNA" (gRNA) is used in this chapter.

CRISPR/Cas genome editing has now been used in over 200 vertebrate, invertebrate, plant, and microbial species (Farboud et al. 2018). A PubMed search for "CRISPR" and "Cas9" (conducted in January 2020) showed that this platform had accumulated more than 10 000 publications since its discovery in 2012. Applications include editing of genomic material, "knock in" or "knock out" of genetic material, and regulation of transcription (Hille and Charpentier 2016). CRISPR/Cas9-associated plasmids can be obtained from the Addgene plasmid repository, which reported in 2015 that it had already distributed more than 25 000 of these constructs (Kamens 2015). CRISPR/Cas9 products and services are also widely available from many commercial suppliers, e.g. abm, GenScript, Integrated DNA Technologies, OriGene, Oxford Genetics, Santa Cruz Biotechnology, Sigma-Aldrich, and ThermoFisher Scientific.

What do you need to develop CRISPR/Cas9 technology in your own laboratory? The beauty of the CRISPR/Cas9 system is that it can be employed by any laboratory with molecular biology skills (Farboud et al. 2018). Specific components will vary with your application, but broadly speaking, you will need (see Figure 22.5) (Jinek et al. 2013):

(1) **Cas9 protein.** Initially, Cas9 from *Streptococcus pyogenes* was optimized for use in human cells and expressed as a fusion protein with a nuclear localization signal and green fluorescent protein (GFP). Cas9 sequence can also be introduced as mRNA, or purified Cas9 protein may be introduced directly.

(2) **Guide RNA (gRNA).** gRNA was initially designed to carry a guide region that is complementary to the target sequence, a scaffold for Cas9 binding, and a termination signal for RNA transcription. Streamlined protocols and RNA expression vectors are now available that simplify this process considerably (Nageshwaran et al. 2018), but it is still necessary to choose a suitable target sequence. This sequence (also known as a protospacer, since it

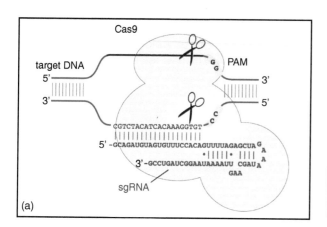

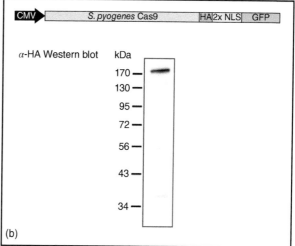

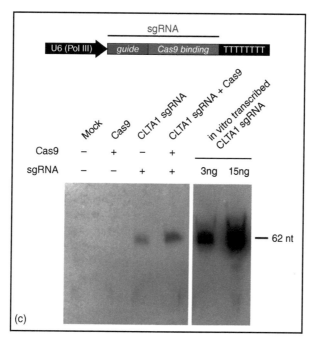

Fig. 22.5. CRISPR/Cas9 composition. (a) Schematic diagram of CRISPR/Cas9. Cas9 protein binds to a short guide RNA (sgRNA), forming an RNP complex. Guide sequence within the sgRNA is designed to bind to a specific sequence. Once bound, the Cas9 nuclease will cleave the target DNA. (b) Design of the DNA vector used to deliver Cas9 (upper panel). *S. pyogenes* Cas9 is expressed under the control of the CMV promoter; expression protein is confirmed using Western blotting (lower panel). (c) Design of the sgRNA used to deliver the guide region and the Cas9 binding region (upper panel). RNA is expressed under the control of the U6 polymerase III promoter; expression is confirmed using Northern blotting (lower panel). This set of examples focuses on a specific target (CLTA, human clathrin light chain gene) and method of delivery; various methods can now be used to deliver CRISPR/Cas9 to the cell. *Source*: Jinek, M., East, A., Cheng, A., Lin, S., Ma, E., & Doudna, J. (2013). RNA-programmed genome editing in human cells. eLife, 2. https://doi.org/10.7554/eLife .00471.003, Licensed under CC BY 3.0.

will be incorporated as a spacer in the CRISPR locus) consists of 20 nucleotides next to a protospacer-adjacent motif (PAM). For Cas9 from *S. pyogenes*, the optimal PAM sequence is 5′-NGG-3′, where N is any of the four DNA bases. Resources are available to help design gRNA, identify PAMs, and select target sequences (Graham and Root 2015).

(3) **Donor DNA.** As noted previously (see Section 22.2), a template may be used to promote HDR after cleavage by the Cas9 protein, resulting in a lower error rate. Careful design of the experiment can be used to aid selection, e.g. protecting donor DNA with phosphorothioate can improve HDR efficiency (Liang et al. 2017).

Cas9 and gRNA sequence may be delivered as DNA (cloned into expression plasmids), as RNA (Cas9 mRNA and separate gRNA), or as a preassembled RNP complex (Lino et al. 2018). Preassembly of the RNP complex is becoming popular; it is a highly efficient method that comes with fewer off-target effects, due to the RNP's limited lifespan (Kim et al. 2014; Lin et al. 2014). It is also effective for a broad range of species and cell types, even those that are hard to transfect (Farboud et al. 2018; Liu et al. 2018). Protocol P22.3 focuses on delivery of the CRISPR/Cas9 RNP into medaka (*Oryzias latipes*) fish cells, where gene editing can be challenging (Liu et al. 2018). The mammalian U6 polymerase III promoter, which is normally used to express sgRNA, does not work well in this species. To overcome this challenge, CRISPR/Cas9 can be preassembled from Cas9 protein, crRNA, and tracrRNA before delivery to the culture.

PROTOCOL P22.3. DELIVERY OF CRISPR/CAS9 RNP USING ELECTROPORATION

Adapted from Liu et al. (2018), courtesy of Ruowen Ge, Department of Biological Sciences, National University of Singapore. This protocol describes a portion of the experimental procedure. Refer to the source paper for the context and additional detail.

Outline

Assemble the CRISPR/Cas9 RNP from crRNA, tracr-RNA, and recombinant Cas9 protein. Trypsinize cells. Add the CRISPR/Cas9 RNP to the cell suspension and perform electroporation.

Materials (sterile or aseptically prepared)

- ☐ Medaka fish cells (grown in ESM4 medium at 37 °C in ambient air)
- ☐ crRNA (200 µM) as described separately (Liu et al. 2018)
- ☐ tracrRNA (200 µM) as described separately (Liu et al. 2018)
- ☐ Recombinant *S. pyogenes* Cas9 nuclease (61 µM; Integrated DNA Technologies)
- ☐ Carrier DNA (100 µM; Integrated DNA Technologies)
- ☐ DPBS
- ☐ Electroporation cuvettes (0.2-cm gap, Bio-Rad)
- ☐ Microcentrifuge tubes
- ☐ Cell culture dishes or flasks

Materials (non-sterile)

- ☐ BSC and associated consumables (see Protocol P12.1)
- ☐ Heating block
- ☐ Electroporator; Gene Pulser Xcell Electroporation System (Bio-Rad)

Procedure

1. Plate medaka fish cells, aiming for 1×10^6 cells at ~90% confluence on the day of the procedure.
2. On the day of electroporation, prewarm medium and prepare a work space for aseptic technique using a BSC.
3. Mix 5 µl crRNA and 5 µl tracrRNA and heat at 95 °C for 5 min. Incubate the mixture at room temperature for 30 minutes. This is the crRNA:tracrRNA complex.
4. In a separate tube, add 1.2 µl of the crRNA:tracrRNA complex to 2.1 µl DPBS and 1.7 µl Cas9 nuclease. Incubate at room temperature for 10–20 minutes. This is the RNP complex.
5. Subculture cells using trypsinization (see Protocol P14.2, steps 1–16).
6. Wash cells twice by centrifuging and resuspending in 1 ml DPBS.
7. Resuspend cells in 94 µl DPBS. Add 5 µl of the RNP complex and 1 µl carrier DNA. Pipette gently to mix. Transfer to a sterile electroporation cuvette.
8. Perform electroporation using 220 V and 5 ms.
9. After electroporation, plate cells into fresh dishes or flasks and incubate overnight.
10. Refeed cells 24 hours after electroporation using fresh medium.
11. Perform genotyping seven days after electroporation.

Notes

Step 3: For information on the design of crRNA and tracrRNA please refer to the source paper.

22.3 IMMORTALIZATION

22.3.1 Early Immortalization Strategies

Whole virus. Initially, SV40 and other tumor-associated viruses were used to extend culture lifespan and study *in vitro* transformation (see Section 3.6). Whole virus is still used to immortalize some cell types, particularly to generate lymphoblastoid cell lines (LCLs). Early attempts to grow lymphocytes in culture showed that immortalized lines could be derived from very dense cell pellets in a culture tube (Moore et al. 1967). These cells were from the B-lymphocyte lineage and were stimulated to proliferate by endogenous Epstein-Barr virus (EBV). Generation of LCLs using EBV is now a routine and widely used procedure (see Section 24.5.1). Multiple viral genes are required to immortalize an LCL, including *EBNA1* (which allows replication by the host's DNA polymerase) and *LMP1* (which prevents apoptosis) (Farrell 2019).

Many cell lines that were immortalized using whole virus continue to carry the viral genome and may produce infectious virions, resulting in increased biohazard risk (see Sections 6.3.1, 16.5). MRC-5V2 was immortalized using whole SV40, which continues to be detected even after 20 years in culture (Huschtscha and Holliday 1983; Morelli et al. 2004). Similarly, EBV can be detected in LCLs following *in vitro* immortalization and in other hemopoietic cell lines due to *in vivo* infection (Uphoff et al. 2010; Shioda et al. 2018). Always consider genetic modifications when performing a risk assessment and handle the culture using suitable containment (see Section 6.3.3).

Somatic cell hybridization. Somatic cell hybrid cell lines arise through fusion of somatic cells of different origins (Nims et al. 2018). Somatic cell hybridization is an essential process *in vivo* but can also occur *in vitro*, where it has been used to study senescence and immortalization (Ephrussi and Weiss 1965; Whitaker et al. 1992). Various fusogenic agents can be used to induce somatic cell hybridization *in vitro*, including polyethylene glycol (PEG), electrical pulses (electrofusion), and some viruses (e.g. Sendai virus) (Nims et al. 2018). Use of PEG is arguably the simplest approach; cells are covered with PEG solution (48% in serum-free medium) for exactly one minute and then the culture is rinsed and fed using complete growth medium (Whitaker et al. 1992). A proportion of the cells that fuse progress to nuclear fusion, and a proportion of those cells progress through mitosis, such that both sets of chromosomes replicate together and a hybrid is formed.

An increasing body of evidence has demonstrated that spontaneous cell–cell fusion occurs *in vivo*, where it plays an important role in horizontal transmission of malignancy and tumor evolution (Jacobsen et al. 2006; Goldenberg et al. 2012; Zhou et al. 2015). Spontaneous cell–cell fusion has also been reported *in vitro*; for example, mouse chromosomes were found in a human tumor cell line (NCEB-1) established

using mouse feeder cells (see Section 25.3.3). Such events can be detected through authentication testing, but the detection of a cell–cell hybrid is a challenging task (see Section 17.4.3) (Nims et al. 2018). Always perform authentication testing on a known hybrid early after its establishment, to gain a baseline for later comparison.

22.3.2 Immortalization Using Viral Genes and Oncogenes

Although whole virus is effective, single viral genes can be used for immortalization with less biohazard risk. The most commonly used viral gene is the SV40 large T antigen (SV40 TAg, abbreviated as LT in some studies). SV40 TAg is encoded by the early region of the SV40 genome and can be used to extend the lifespan of various adherent cells, including fibroblasts, keratinocytes, and endothelial cells (Mayne et al. 1996; Punchard et al. 1996; Steinberg 1996). A second, small t antigen is also transcribed from the early region and is implicated in transformation (see Section 3.6), but SV40 TAg is essential for extension of cellular lifespan (Hahn et al. 2002). Other viral genes that are used for immortalization include adenovirus E1A and human papillomavirus (HPV) E6 and E7 (Peters et al. 1996; Seigel 1996; Le Poole et al. 1997). Most of these genes act by inactivating p53 and pRB/p16^{INK4a} tumor suppressor pathways; some may activate telomerase and contribute to genomic instability (Yugawa and Kiyono 2009).

Typically, the immortalizing gene is introduced by transfection or viral transduction before cells enter senescence. This extends their proliferative lifespan for another 20–30 population doublings, whereupon the cells cease proliferation and enter crisis (see Minireview M3.1). After a period in crisis that may last for some months, clones begin to emerge with unlimited growth potential. The frequency of immortalization with SV40 TAg is about 3×10^{-7} when used with human diploid fibroblasts (Shay and Wright 1989), and may vary between 1×10^{-5} and 1×10^{-9}. A protocol for immortalization of human fibroblasts using SV40 TAg is provided in the Supplementary Material online (see Supp. S22.1).

The absence of proliferation during crisis implies that viral genes can extend the lifespan of the culture but may be insufficient for immortalization (De Silva et al. 1994). Oncogenes such as *HRAS* or *MYC* may be used to induce immortalization in some cell types (Boukamp et al. 1990; Gil et al. 2005). However, these genes are often associated with acquisition of a malignant phenotype and the timing of their delivery can be critical, e.g. activated *HRAS* may cause transformation in immortalized cells, but may induce premature senescence in normal cells (Newbold 2002). In many cases, oncogenes result in autonomous growth control and their gene products are permanently active and cannot be regulated. Such excessive proliferative signaling may trigger cell senescence, unless this response is impaired, e.g. by loss of p53 (Hanahan and Weinberg 2011).

Over time, some immortalizing genes have been modified to provide a conditional form of immortalization. This is perhaps best illustrated by the *H-2K^b-tsA58* transgenic mouse ("Immortomouse"), which carries a temperature-sensitive variant of SV40 TAg under the control of the *H-2K^b* promoter element (Jat et al. 1991; Kern and Flucher 2005). Cells express SV40 TAg when maintained under permissive conditions (33 °C with interferon (IFN)-γ), but not when transferred to non-permissive conditions (39.5 °C without IFN-γ). A number of cell lines have been established from challenging tissues, including normal adult colonic epithelium, using this mouse model (Whitehead and Robinson 2009).

22.3.3 Immortalization Using Telomerase

For a cell to have an unlimited lifespan, it must have a mechanism to maintain telomere length. Otherwise, progressive shortening of its telomeres will lead to senescence (see Minireview M3.1). Telomere length can be maintained by telomerase enzyme activity or by an alternative mechanism for lengthening telomeres (ALT), which depends on homologous recombination (Counter et al. 1992; Bryan et al. 1995; Sobinoff and Pickett 2017). Usually, such a mechanism will be active in cancer cells and continuous cell lines, but repressed in normal human tissues (Avilion et al. 1996). Telomerase can be introduced to bypass senescence and induce immortalization in normal human diploid cells and many other cell types (Bodnar et al. 1998; Vaziri and Benchimol 1998; Reddel 2000). Typically, cells that are immortalized with telomerase retain more properties of normal cells and are less likely to display other hallmarks of transformation, such as genomic instability (see Section 3.6.1) (Ouellette et al. 2000; Toouli et al. 2002).

Human telomerase consists of human telomerase reverse transcriptase (hTERT), which acts as a catalytic subunit, and telomerase RNA, which provides a template for telomere synthesis. hTERT is normally silenced by transcriptional repression, while telomerase RNA is constitutively expressed (Newbold 2002). hTERT can thus be used to immortalize normal human fibroblasts and various other primary cultures (Avilion et al. 1996; Counter et al. 1998). hTERT is unable to immortalize some cell types in isolation, including keratinocytes, mammary epithelial cells, mammary fibroblasts, and some endothelial cells (Kiyono et al. 1998; O'Hare et al. 2001). In these cases, several genes may be required, e.g. hTERT and SV40 TAg (O'Hare et al. 2001). Immortalization with hTERT may require previous inactivation of pRB/p16^INK4a and possibly p53; cells may also have limiting amounts of telomerase RNA, resulting in low levels of telomerase activity even when sufficient hTERT is present (Noble et al. 2004; Cao et al. 2008).

Protocol P22.4 has been used to successfully immortalize adipose tissue-derived mesenchymal stromal cells (MSCs) from patients with Parkinson's disease using hTERT (Moon et al. 2013). The protocol introduces a plasmid construct containing the hTERT open reading frame, expressed under the control of the myeloproliferative sarcoma virus promoter (pGRN145), using lipofection. It is suitable for actively dividing MSCs; non-dividing cells will require a different method, e.g. lentiviral transduction.

PROTOCOL P22.4. IMMORTALIZATION USING HTERT TRANSFECTION

Adapted from Moon et al. (2013), courtesy of Dr Hyo Eun Moon and Dr Sun Ha Paek, Seoul National University College of Medicine. This protocol describes a portion of the experimental procedure. Refer to the source paper for the context and additional detail.

Outline

Subculture cells and incubate for 24 hours. Dilute Lipofectamine® LTX reagent and hTERT plasmid separately in medium and then mix together. Add to cells and incubate for 24 hours.

Materials (sterile or aseptically prepared)

- MSCs (grown in Mesenchymal Stem Cell Expansion medium, Sigma-Aldrich)
- pGRN145 expression plasmid (ATCC #MBA-141; see Table 15.3)
- Lipofectamine LTX reagent (ThermoFisher Scientific)
- Opti-Minimum Essential Medium (MEM) I medium, serum-free (ThermoFisher Scientific)
- Hygromycin B (ThermoFisher Scientific)
- 24-well plate (1)
- Microcentrifuge tubes (2)

Materials (Non-sterile)

- BSC and associated consumables (see Protocol P12.1)

Procedure

1. On the day before transfection, subculture cells using the customary procedure for the cell type (e.g. see Protocol P14.2).
2. Plate cells in a 24-well plate, aiming for 90% confluence on the day of transfection. Incubate overnight at 37 °C.
3. On the day of transfection, prepare a work space for aseptic technique using a BSC.
4. In a microcentrifuge tube, add Lipofectamine LTX reagent (3.5–5.0 μl per well) to Opti-MEM I medium (50 μl per well). In a separate tube, add

pGRN145 plasmid (1 μg per well) to Opti-MEM I medium (50 μl per well).

5. Mix the contents of both tubes and incubate for 30–40 minutes at room temperature. Add Opti-MEM I medium to both tubes (500 μl).

6. Add the DNA : lipid : Opti-MEM I medium complexes to the wells in the 24-well plate (500 μl per well). Gently shake the plate to mix. Incubate for six to eight hours at 37 °C.

7. Remove the complexes and refeed the plate with fresh growth medium.

8. At 24 hours post-transfection, refeed the plate with fresh growth medium.

9. At 48 hours post-transfection, refeed the plate with medium containing hygromycin B (30 μg/ml). Continue culture with this antibiotic concentration for two to three weeks.

Notes

Step 2: Cells should be antibiotic-free for transfection procedures. As noted previously, antibiotics can reduce transfection efficiency (Jacobsen et al. 2004).

Step 4: The amount of reagent required for efficient transfection will vary with the cell type and the passage number (see Protocol P22.2 for a typical optimization experiment).

Step 5: A more recent formulation of Lipofectamine LTX is now available where PLUS™ Reagent is added to the tube containing DNA. See the supplier's instructions for more information.

Step 9: Cell death should occur rapidly for non-transfected cells. After transfected cells have been selected, the hygromycin B concentration can be reduced to 10 μg/ml.

22.3.4 Conditional Reprogramming

Cellular lifespan is affected by various factors, including hostile growth conditions, which can cause culture stress and induce growth arrest (see Section 3.5.2). Conversely, cellular lifespan can be extended by optimization of growth conditions. Early work in this field was performed by James Rheinwald and Howard Green, who used feeder layers and epidermal growth factor (EGF) to delay senescence of human epidermal keratinocytes (Rheinwald and Green 1975, 1977). Their technique has since been adapted to induce "conditional reprogramming" in a number of cell types (Chapman et al. 2010; Liu et al. 2012). Although the focus in this chapter is primarily on immortalization, it seems likely that cells also increase their level of potency, resulting in behavior that is more consistent with progenitor or stem cells (see Section 2.4.4).

Detailed protocols for conditional reprogramming (also known as "3T3 + Y") are available elsewhere (Chapman et al. 2014; Liu et al. 2017). Key requirements include:

(1) **Use irradiated 3T3 J2 mouse fibroblasts as a feeder layer.** The Swiss-3T3J2 subclone (Kerafast) was originally isolated for use as a keratinocyte feeder layer (Rheinwald and Green 1975). These cells support epithelial cell growth and inhibit growth of stromal cells at early passage. Although the J2 subclone is preferred, other 3T3 cell lines have been used as feeder layers (e.g. NIH 3T3) (Gao et al. 2017).

(2) **Supplement the culture medium with Rho kinase (ROCK) inhibitor.** The technique was initially developed using Y-27632 (10 μM), but other ROCK inhibitors can be used, including fasudil hydrochloride (20 μM), HA1000 hydrochloride (20 μM), and GSK 429286 (100 nM) (Chapman et al. 2014). ROCK inhibitors can also be used as "survival factors" for human pluripotent stem cells (hPSCs) (see Section 23.4.1).

(3) **Use Rheinwald-Green F-medium to replicate the original findings.** F-medium was originally developed for keratinocytes but can be used for other cell types (Liu et al. 2017). It consists of a 3 : 1 mixture of Ham's F-12 and Dulbecco's modified Eagle's medium (DMEM), supplemented with fetal bovine serum (FBS, 5%), hydrocortisone (0.4 μg/ml), insulin (5 μg/ml), cholera toxin (8.4 ng/ml), EGF (10 ng/ml), adenine (24 μg/ml), and antibiotics (Chapman et al. 2014).

Keratinocytes that have been studied under these conditions resemble early passage cultures (see Figure 22.6) and exhibit upregulation of telomerase, increased cell growth, and inhibition of differentiation. The feeder layer is probably responsible for upregulation of telomerase, while ROCK inhibitor causes increased cellular proliferation, inhibition of apoptosis, and inhibition of differentiation (Chapman et al. 2010, 2014). The resulting phenotype shares many characteristics with adult stem cells, although conditionally reprogrammed cells retain their initial lineage commitment and thus are not pluripotent (Suprynowicz et al. 2012). The phenotype is rapidly lost when reprogramming conditions are removed. Cells grow at a slower rate for several passages, followed by senescence (Chapman et al. 2014).

Conditional reprogramming conditions represent an exciting advance in cellular immortalization, but there are some cautions to consider alongside the excitement. Several laboratories have reported difficulties growing lung carcinoma cells due to preferential selection of normal lung epithelial cells (Gao et al. 2017; Hynds et al. 2018; Sette et al. 2018). It is likely that ROCK inhibitors affect different cell lineages in different ways. Lineage preferences have been demonstrated in pluripotent stem cell (PSC) cultures, where ROCK inhibitors are used to reduce apoptosis, but may induce markers of epithelial-mesenchymal transition (EMT)

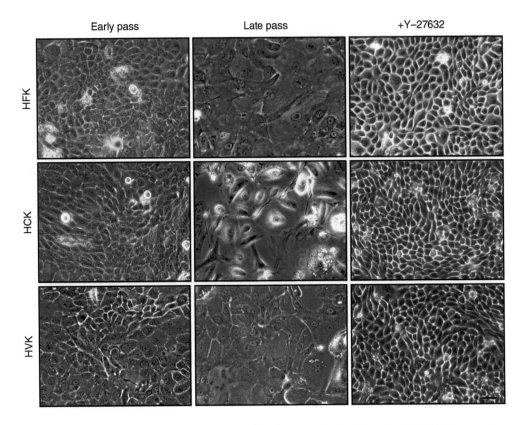

Fig. 22.6. Conditional reprogramming. Human keratinocytes obtained from foreskin (HFK, top row), ectocervix (HCK, middle row), and vagina (HVK), grown on 3T3J2 feeder cells in F-medium. Left column: cells at early passage (p. 1). Middle column: cells near senescence (HFK, p. 15; HCK, p. 9; HVK, p. 5). Right column: cells grown with the addition of Y-27632 (HFK, p. 100; HCK, p. 29; HVK, p. 26). Note that the morphology of cells grown under conditional reprogramming conditions in the right column is similar to the early passage cultures in the left column. *Source*: Chapman, S., Liu, X., Meyers, C., et al. (2010). Human keratinocytes are efficiently immortalized by a Rho kinase inhibitor. J Clin Invest 120 (7):2619–26. doi: 10.1172/JCI42297, reproduced with permission of American Society for Clinical Investigation.

(see Section 26.2.5) (Maldonado et al. 2016). Conditions may need to be optimized for specific cell types, e.g. using bronchial epithelial growth medium (BEGM) and 2% oxygen for human bronchial epithelial cells (Peters-Hall et al. 2018).

22.4 SCREENING AND ARTIFACTS

22.4.1 Selection of Modified Cells

Many genetic modifications are lost as cultures are passaged. Although loss is expected with a transient transfection system (see Section 22.1), it may also occur after stable transfection or gene editing, due to absence of genomic integration or the presence of clonal diversity within the culture. If the modification is detrimental to cell growth, it will be rapidly overtaken by any remaining cells that carry the wild-type sequence. Reporter genes may be used to identify cells where gene delivery has been successful, e.g. luciferase (*luc*) or GFP

(Stepanenko and Heng 2017). If long-term genetic modification is required, selection is usually required to isolate cells with the desired genetic modification and eliminate other populations. Plasmid vectors containing selectable markers are available from Addgene and many commercial suppliers (e.g. InvivoGen, Takara Bio, and ThermoFisher Scientific).

Cytotoxic antibiotics are commonly used for selection. These methods rely on the introduction of an antibiotic resistance gene alongside the transgene of interest. Various antibiotics can be used, including G418 (a derivative of neomycin), blasticidin S, histidinol, hygromycin B, puromycin, and zeocin (a derivative of bleomycin) (Stepanenko and Heng 2017). The antibiotic must be used at a concentration that will positively select for transfected cells against wild-type cells, as determined by titration of the dosage (Nakatake et al. 2013). Negative selection is also possible by using the herpes simplex virus thymidine kinase 1 gene (*HSVtk*), which activates ganciclovir (InvivoGen) into a cytotoxic product and thus sensitizes transfected cells to the drug.

Drug resistance may be induced experimentally without gene delivery. This is usually done to study the development of drug resistance during therapy. Cells that are exposed to gradually increasing concentrations of methotrexate, or other cytotoxic substances, will develop resistance to the toxic effects of the drug over a prolonged period of time (Biedler et al. 1972). For methotrexate, resistance due to amplification of the dihydrofolate reductase (*DHFR*) gene generally develops the most rapidly, although other mechanisms may confer part or all of the resistant phenotype, e.g. alteration in antifolate transport and/or mutations affecting enzyme structure or affinity. Techniques for developing drug-resistant cell lines are described elsewhere (McDermott et al. 2014).

22.4.2 Toxicity

Any form of genetic modification can have unintended consequences, which can act as artifacts in experimental procedures or risk the safety of individuals in a clinical setting. The most obvious unintended result is toxicity, resulting in cell death. All methods of gene delivery can be toxic to cells; toxicity will vary with the cell type, the choice of reagents, and the experimental procedure (Stepanenko and Heng 2017). Most methods can be optimized to reduce toxicity, but for some cell types toxicity will persist and a different method must be used. For example, calcium phosphate is consistently toxic to normal human bronchial epithelial cells, resulting in the development of a variant technique using strontium phosphate (see Table 22.1) (Reddel et al. 1988). Toxicity is also known to occur with gene editing, most likely due to off-target DNA cleavage (see Section 22.2) (Kim et al. 2009; Cho et al. 2014; Mussolino et al. 2014).

Genetic modification may cause more subtle changes due to cell stress or other causes (Stepanenko and Heng 2017). The process of transfection is associated with changes in gene transcription, which often go undetected (Jacobsen et al. 2009). Changes in transcription can have many causes, including (i) genetic compensation, in response to decreased function of the target gene; (ii) a response to the vector backbone; (iii) interactions between multiple vectors; (iv) expression of a reporter gene; (v) antibiotic selection; (vi) the transfection reagent; or (vii) the transfection method. While it is important to use empty vector as a control, this may itself cause artifacts (Stepanenko and Heng 2017). Always optimize transfection conditions (see Protocol P22.2) and consider how to best design experiments to detect subtle artifacts.

22.4.3 Indels and Rearrangements at the Target Site

Gene editing commonly causes unintended genetic changes at the target site. Small insertions and deletions (indels) can

occur when NHEJ is used for DNA repair, which has a relatively high error rate (see Section 22.2). New tools and methods are becoming available to reduce the frequency of indel formation. ZFNs, TALENs, and Cas9 proteins have all been redesigned as "nickases" that cut only one strand of the DNA, favoring the more accurate HDR pathways (albeit at lower efficiency) (Kim et al. 2012; Cho et al. 2014; Wu et al. 2014). However, indels can still occur even when nickases are used, suggesting that nicks may be converted to DSBs *in vitro* (Cho et al. 2014). Testing for indels can be performed using various methods, including the mismatch cleavage assay, which is relatively simple and cost-effective to use for screening (Zischewski et al. 2017).

Larger rearrangements may occur, e.g. during "knock-in" experiments. Insertion of large DNA fragments (> 1.5 kb) using CRISPR/Cas9 results in relatively high success rates compared to homologous recombination, but also leads to more rearrangements at the target site (Rezza et al. 2019). Larger rearrangements can be detected using Southern blotting of the target region, amplified fragment length polymorphism (AFLP) analysis, or by sequencing the target area (Zischewski et al. 2017; Rezza et al. 2019).

22.4.4 Off-Target Effects

The term "off-target effect" is used here to refer to unexpected consequences of genetic modification that occur away from the target region. Off-target effects are a concern for all forms of genetic modification. For example, antisense RNA or oligonucleotides may cause off-target effects due to interference with other genes that carry a similar sequence (Eisen and Smith 2008). All such experiments must include suitable controls to detect artifacts due to off-target effects, followed by validation; for example, immunocytochemistry using a specific antibody can be used to confirm loss of function (Eisen and Smith 2008; Zischewski et al. 2017; Kimberland et al. 2018).

Off-target mutations are a particular concern for applications of the CRISPR/Cas9 system (*Nature* Editors 2018). This is partly due to its widespread use and the risk of unsanctioned gene-based therapy, but CRISPR/Cas9 has some inherent design challenges that make it potentially more prone to off-target mutations. The target sequence required for CRISPR is shorter than for other gene editing technologies and must be located next to a PAM (see Section 22.2.3), which narrows the available design options. Off-target activity will depend on the target sequence and its homology elsewhere in the genome, the selection of gRNA and Cas9 nuclease, and the cell's DNA repair pathways, e.g. cancer cell lines may have high levels of off-target activity due to faulty DNA repair machinery (Zischewski et al. 2017). Activity levels can be reduced through good experimental design, e.g. using preassembled CRISPR/Cas9 to reduce exposure time (see Protocol P22.3) (Kimberland et al. 2018). However, off-target mutations remain an ongoing concern for CRISPR/Cas9 and other gene editing technologies. Detection of these mutations is challenging and is discussed elsewhere (Zischewski et al. 2017).

22.4.5 Oncogenesis

Any form of genetic modification may alter the cell's inherent risk of tumor development, e.g. due to random integration of a transgene near an oncogene or tumor suppressor, resulting in gain or loss of function. This risk is increased if an oncogenic virus, viral gene, or oncogene is introduced (see Sections 22.3.1, 22.3.2). Tumor development is part of a broader phenotype that is often referred to as *in vitro* transformation (see Section 3.6). Transformed cells will produce tumors in nude mice *in vivo* and display evidence of invasiveness and angiogenesis (see Section 3.6.3). Such cells also become immortalized *in vitro* and display genetic instability and aberrant growth control. Characteristics of transformation (see Table 3.2) should be assessed when a cell line is established, to determine whether it originates from neoplastic cells or has undergone transformation in culture. For tumor samples, this will confirm that cells are derived from the neoplastic component of the tumor, rather than from normal equivalent cells, infiltrating fibroblasts, blood vessel cells, or inflammatory cells.

The most widely used assay for malignant transformation is to inject tumor cells into a "nude" mouse or similar *in vivo* model (see Section 3.6.3). Transplantable tumor cells ($\sim 1 \times 10^6$) will produce invasive or metastasizing tumors in a high proportion of cases, whereas the same number of normal cells of similar origin will not. Tumorigenesis assays should always be accompanied by histology to confirm its similarity to the original tumor and to demonstrate that it is invasive. Such *in vivo* models provide a good indicator of malignancy and are a mainstay of cancer bioassays. However, these assays tend to use large numbers of animals and can be time consuming and expensive, and generate false negative results (Creton et al. 2012).

Various *in vitro* cell transformation assays have been developed that can replace or supplement animal models, based on the transformed phenotype (see Table 3.2). For example, colony formation in soft agar is known to increase after viral transformation (see Sections 3.6.2, 20.3). However, the applicability of the technique to spontaneous tumors is less clear. Although many human tumors contain a small percentage of cells (< 1.0%) that are clonogenic in agar, some normal cells will clone in suspension, including normal human fibroblasts (Hamburger and Salmon 1977; Peehl and Stanbridge 1981). Thus, the value of this technique for examining human tumor cells in short-term cultures is in doubt, although the question of whether these are stem cells is still open.

More than one criterion is necessary to confirm neoplastic status. Such characteristics are often expressed in normal cells at particular stages of development but expressed inappropriately in transformed cells. The exceptions are gross aneuploidy, heteroploidy, and tumorigenicity, which, taken together, are regarded as conclusive positive indicators of malignant transformation. With the advent of genomic analysis as a routine procedure, it is now easier to compare the genotype of cultured cells with the genotype of the original tumor. This not only confirms the origin of the cell line but also will show whether transformation has occurred.

Suppliers

Supplier	URLs
Applied Biological Materials (abm)	www.abmgood.com
Bio-Rad Laboratories, Inc.	www.bio-rad.com
GenScript	www.genscript.com
Integrated DNA Technologies	sg.idtdna.com/pages
InvivoGen	www.invivogen.com
Kerafast	www.kerafast.com
Lonza	www.lonza.com
OriGene	www.origene.com
Oxford Expression Technologies	oetltd.com/
Oxford Genetics	www.oxfordgenetics.com
Polyplus-transfection	www.polyplus-transfection.com
Promega	www.promega.com
Sangamo Therapeutics	www.sangamo.com
Santa Cruz Biotechnology	http://www.scbt.com/scbt/home
Sigma-Aldrich (Merck)	http://www.sigmaaldrich.com/life-science/cell-culture.html
Takara Bio	www.takarabio.com
ThermoFisher Scientific	http://www.thermofisher.com/us/en/home/life-science/cell-culture.html
Vector Biolabs	www.vectorbiolabs.com
Vigene Biosciences	www.vigenebio.com

Supp. S22.1 Fibroblast Immortalization Using SV40 TAg.

REFERENCES

Addgene (2019). Viral plasmids and resources. https://www.addgene.org/viral-vectors (accessed 11 April 2019).

Amado, R.G. and Chen, I.S. (1999). Lentiviral vectors – the promise of gene therapy within reach? *Science* 285 (5428): 674–676.

Avilion, A.A., Piatyszek, M.A., Gupta, J. et al. (1996). Human telomerase RNA and telomerase activity in immortal cell lines and tumor tissues. *Cancer Res.* 56 (3): 645–650.

Baltes, N.J., Gil-Humanes, J., and Voytas, D.F. (2017). Genome engineering and agriculture: opportunities and challenges. *Prog. Mol.*

Biol. Transl. Sci. 149: 1–26. https://doi.org/10.1016/bs.pmbts .2017.03.011.

Barrangou, R., Fremaux, C., Deveau, H. et al. (2007). CRISPR provides acquired resistance against viruses in prokaryotes. *Science* 315 (5819): 1709–1712. https://doi.org/10.1126/science .1138140.

Biedler, J.L., Albrecht, A.M., Hutchison, D.J. et al. (1972). Drug response, dihydrofolate reductase, and cytogenetics of amethopterin-resistant Chinese hamster cells in vitro. *Cancer Res.* 32 (1): 153–161.

Boch, J., Scholze, H., Schornack, S. et al. (2009). Breaking the code of DNA binding specificity of TAL-type III effectors. *Science* 326 (5959): 1509–1512. https://doi.org/10.1126/science.1178811.

Bodnar, A.G., Ouellette, M., Frolkis, M. et al. (1998). Extension of life-span by introduction of telomerase into normal human cells. *Science* 279 (5349): 349–352.

Bolotin, A., Quinquis, B., Sorokin, A. et al. (2005). Clustered regularly interspaced short palindrome repeats (CRISPRs) have spacers of extrachromosomal origin. *Microbiology* 151 (Pt 8): 2551–2561. https://doi.org/10.1099/mic.0.28048-0.

Boukamp, P., Stanbridge, E.J., Foo, D.Y. et al. (1990). c-Ha-ras oncogene expression in immortalized human keratinocytes (HaCaT) alters growth potential in vivo but lacks correlation with malignancy. *Cancer Res.* 50 (9): 2840–2847.

Brash, D.E., Mark, G.E., Farrell, M.P. et al. (1987). Overview of human cells in genetic research: altered phenotypes in human cells caused by transferred genes. *Somat. Cell Mol. Genet.* 13 (4): 429–440.

Bryan, T.M., Englezou, A., Gupta, J. et al. (1995). Telomere elongation in immortal human cells without detectable telomerase activity. *EMBO J.* 14 (17): 4240–4248.

Cao, Y., Huschtscha, L.I., Nouwens, A.S. et al. (2008). Amplification of telomerase reverse transcriptase gene in human mammary epithelial cells with limiting telomerase RNA expression levels. *Cancer Res.* 68 (9): 3115–3123. https://doi.org/10.1158/0008-5472.CAN-07-6377.

Capecchi, M.R. (2005). Gene targeting in mice: functional analysis of the mammalian genome for the twenty-first century. *Nat. Rev. Genet.* 6 (6): 507–512. https://doi.org/10.1038/nrg1619.

Carroll, D. (2014). Genome engineering with targetable nucleases. *Annu. Rev. Biochem.* 83: 409–439. https://doi.org/10.1146/annurev-biochem-060713-035418.

Carroll, D. (2017). Genome editing: past, present, and future. *Yale J. Biol. Med.* 90 (4): 653–659.

Cermak, T., Doyle, E.L., Christian, M. et al. (2011). Efficient design and assembly of custom TALEN and other TAL effector-based constructs for DNA targeting. *Nucleic Acids Res.* 39 (12): e82. https://doi.org/10.1093/nar/gkr218.

Chambers, A.C., Aksular, M., Graves, L.P. et al. (2018). Overview of the baculovirus expression system. *Curr. Protoc. Protein Sci.* 91: 5.4.1–5.4.6. https://doi.org/10.1002/cpps.47.

Chandrasegaran, S. (2017). Recent advances in the use of ZFN-mediated gene editing for human gene therapy. *Cell Gene Ther. Insights* 3 (1): 33–41. https://doi.org/10.18609/cgti .2017.005.

Chandrasegaran, S. and Carroll, D. (2016). Origins of programmable nucleases for genome engineering. *J. Mol. Biol.* 428 (5 Pt B): 963–989. https://doi.org/10.1016/j.jmb.2015.10.014.

Chapman, S., Liu, X., Meyers, C. et al. (2010). Human keratinocytes are efficiently immortalized by a Rho kinase inhibitor. *J. Clin. Invest.* 120 (7): 2619–2626. https://doi.org/10.1172/JCI42297.

Chapman, S., McDermott, D.H., Shen, K. et al. (2014). The effect of Rho kinase inhibition on long-term keratinocyte proliferation is rapid and conditional. *Stem Cell Res. Ther.* 5 (2): 60. https://doi.org/10.1186/scrt449.

Cho, S.W., Kim, S., Kim, J.M. et al. (2013). Targeted genome engineering in human cells with the Cas9 RNA-guided endonuclease. *Nat. Biotechnol.* 31 (3): 230–232. https://doi.org/10.1038/nbt.2507.

Cho, S.W., Kim, S., Kim, Y. et al. (2014). Analysis of off-target effects of CRISPR/Cas-derived RNA-guided endonucleases and nickases. *Genome Res.* 24 (1): 132–141. https://doi.org/10.1101/gr .162339.113.

Christian, M., Cermak, T., Doyle, E.L. et al. (2010). Targeting DNA double-strand breaks with TAL effector nucleases. *Genetics* 186 (2): 757–761. https://doi.org/10.1534/genetics.110.120717.

Cong, L., Ran, F.A., Cox, D. et al. (2013). Multiplex genome engineering using CRISPR/Cas systems. *Science* 339 (6121): 819–823. https://doi.org/10.1126/science.1231143.

Conn, K.J., Degterev, A., Fontanilla, M.R. et al. (1998). Calcium phosphate transfection. In: *DNA Transfer to Cultured Cells* (eds. K. Ravid and R.I. Freshney), 111–124. New York: Wiley-Liss.

Counter, C.M., Avilion, A.A., LeFeuvre, C.E. et al. (1992). Telomere shortening associated with chromosome instability is arrested in immortal cells which express telomerase activity. *EMBO J.* 11 (5): 1921–1929.

Counter, C.M., Hahn, W.C., Wei, W. et al. (1998). Dissociation among in vitro telomerase activity, telomere maintenance, and cellular immortalization. *Proc. Natl Acad. Sci. U.S.A.* 95 (25): 14723–14728.

Creton, S., Aardema, M.J., Carmichael, P.L. et al. (2012). Cell transformation assays for prediction of carcinogenic potential: state of the science and future research needs. *Mutagenesis* 27 (1): 93–101. https://doi.org/10.1093/mutage/ger053.

Danner, E., Bashir, S., Yumlu, S. et al. (2017). Control of gene editing by manipulation of DNA repair mechanisms. *Mamm. Genome* 28 (7–8): 262–274. https://doi.org/10.1007/s00335-017-9688-5.

De Silva, R., Whitaker, N.J., Rogan, E.M. et al. (1994). HPV-16 E6 and E7 genes, like SV40 early region genes, are insufficient for immortalization of human mesothelial and bronchial epithelial cells. *Exp. Cell. Res.* 213 (2): 418–427. https://doi.org/10.1006/excr.1994.1218.

DeWitt, M.A., Corn, J.E., and Carroll, D. (2017). Genome editing via delivery of Cas9 ribonucleoprotein. *Methods* 121–122: 9–15. https://doi.org/10.1016/j.ymeth.2017.04.003.

Doyle, E.L., Stoddard, B.L., Voytas, D.F. et al. (2013). TAL effectors: highly adaptable phytobacterial virulence factors and readily engineered DNA-targeting proteins. *Trends Cell Biol.* 23 (8): 390–398. https://doi.org/10.1016/j.tcb.2013.04.003.

Eisen, J.S. and Smith, J.C. (2008). Controlling morpholino experiments: don't stop making antisense. *Development* 135 (10): 1735–1743. https://doi.org/10.1242/dev.001115.

Engler, C. and Marillonnet, S. (2014). Golden Gate cloning. *Methods Mol. Biol.* 1116: 119–131. https://doi.org/10.1007/978-1-62703-764-8_9.

Ephrussi, B. and Weiss, M.C. (1965). Interspecific hybridization of somatic cells. *Proc. Natl Acad. Sci. U.S.A.* 53 (5): 1040–1042.

Farboud, B., Jarvis, E., Roth, T.L. et al. (2018). Enhanced genome editing with Cas9 ribonucleoprotein in diverse cells and organisms. *J. Vis. Exp.* 135: 57350. https://doi.org/10.3791/57350.

Farrell, P.J. (2019). Epstein-Barr virus and cancer. *Annu. Rev. Pathol.* 14: 29–53. https://doi.org/10.1146/annurev-pathmechdis-012418-013023.

Felgner, P.L., Gadek, T.R., Holm, M. et al. (1987). Lipofection: a highly efficient, lipid-mediated DNA-transfection procedure. *Proc. Natl Acad. Sci. U.S.A.* 84 (21): 7413–7417.

Friend, D.S., Papahadjopoulos, D., and Debs, R.J. (1996). Endocytosis and intracellular processing accompanying transfection mediated by cationic liposomes. *Biochim. Biophys. Acta* 1278 (1): 41–50.

Gao, B., Huang, C., Kernstine, K. et al. (2017). Non-malignant respiratory epithelial cells preferentially proliferate from resected non-small cell lung cancer specimens cultured under conditionally reprogrammed conditions. *Oncotarget* 8 (7): 11114–11126. https://doi.org/10.18632/oncotarget.14366.

Garneau, J.E., Dupuis, M.E., Villion, M. et al. (2010). The CRISPR/Cas bacterial immune system cleaves bacteriophage and plasmid DNA. *Nature* 468 (7320): 67–71. https://doi.org/10.1038/nature09523.

Gil, J., Kerai, P., Lleonart, M. et al. (2005). Immortalization of primary human prostate epithelial cells by c-Myc. *Cancer Res.* 65 (6): 2179–2185. https://doi.org/10.1158/0008-5472.CAN-03-4030.

Goetze, B., Grunewald, B., Baldassa, S. et al. (2004). Chemically controlled formation of a DNA/calcium phosphate coprecipitate: application for transfection of mature hippocampal neurons. *J. Neurobiol.* 60 (4): 517–525. https://doi.org/10.1002/neu.20073.

Goldenberg, D.M., Zagzag, D., Heselmeyer-Haddad, K.M. et al. (2012). Horizontal transmission and retention of malignancy, as well as functional human genes, after spontaneous fusion of human glioblastoma and hamster host cells in vivo. *Int. J. Cancer* 131 (1): 49–58. https://doi.org/10.1002/ijc.26327.

Graham, D.B. and Root, D.E. (2015). Resources for the design of CRISPR gene editing experiments. *Genome Biol.* 16: 260. https://doi.org/10.1186/s13059-015-0823-x.

Graham, F.L. and van der Eb, A.J. (1973). Transformation of rat cells by DNA of human adenovirus 5. *Virology* 54 (2): 536–539.

Green, M.R. and Sambrook, J. (2012). *Molecular Cloning: A Laboratory Manual*, 4e. Cold Spring Harbor, NY: Cold Spring Harbor Laboratory Press.

Gutierrez-Granados, S., Cervera, L., Kamen, A.A. et al. (2018). Advancements in mammalian cell transient gene expression (TGE) technology for accelerated production of biologics. *Crit. Rev. Biotechnol.* 38 (6): 918–940. https://doi.org/10.1080/07388551.2017.1419459.

Hackett, P.B. and Somia, N.V. (2014). Delivering the second revolution in site-specific nucleases. *Elife* 3: e02904. https://doi.org/10.7554/eLife.02904.

Hahn, W.C., Dessain, S.K., Brooks, M.W. et al. (2002). Enumeration of the simian virus 40 early region elements necessary for human cell transformation. *Mol. Cell. Biol.* 22 (7): 2111–2123.

Hamburger, A.W. and Salmon, S.E. (1977). Primary bioassay of human tumor stem cells. *Science* 197 (4302): 461–463.

Hanahan, D. and Weinberg, R.A. (2011). Hallmarks of cancer: the next generation. *Cell* 144 (5): 646–674. https://doi.org/10.1016/j.cell.2011.02.013.

Hansen, K., Coussens, M.J., Sago, J. et al. (2012). Genome editing with CompoZr custom zinc finger nucleases (ZFNs). *J. Vis. Exp.* 64: e3304. https://doi.org/10.3791/3304.

Hawley-Nelson, P., Ciccarone, V., and Moore, M.L. (2008). Transfection of cultured eukaryotic cells using cationic lipid reagents. *Curr. Protoc. Mol. Biol.* Chapter 9:Unit 9.4. doi: https://doi.org/10.1002/0471142727.mb0904s81.

Hermonat, P.L. and Muzyczka, N. (1984). Use of adeno-associated virus as a mammalian DNA cloning vector: transduction of neomycin resistance into mammalian tissue culture cells. *Proc. Natl Acad. Sci. U.S.A.* 81 (20): 6466–6470.

Hille, F. and Charpentier, E. (2016). CRISPR-Cas: biology, mechanisms and relevance. *Philos. Trans. R Soc. Lond. B Biol. Sci.* 371 (1707) https://doi.org/10.1098/rstb.2015.0496.

Hille, F., Richter, H., Wong, S.P. et al. (2018). The biology of CRISPR-Cas: backward and forward. *Cell* 172 (6): 1239–1259. https://doi.org/10.1016/j.cell.2017.11.032.

Holkers, M., Maggio, I., Liu, J. et al. (2013). Differential integrity of TALE nuclease genes following adenoviral and lentiviral vector gene transfer into human cells. *Nucleic Acids Res.* 41 (5): e63. https://doi.org/10.1093/nar/gks1446.

Huschtscha, L.I. and Holliday, R. (1983). Limited and unlimited growth of SV40-transformed cells from human diploid MRC-5 fibroblasts. *J. Cell Sci.* 63: 77–99.

Hwang, W.Y., Fu, Y., Reyon, D. et al. (2013). Efficient genome editing in zebrafish using a CRISPR-Cas system. *Nat. Biotechnol.* 31 (3): 227–229. https://doi.org/10.1038/nbt.2501.

Hynds, R.E., Ben Aissa, A., Gowers, K.H.C. et al. (2018). Expansion of airway basal epithelial cells from primary human non-small cell lung cancer tumors. *Int. J. Cancer* 143 (1): 160–166. https://doi.org/10.1002/ijc.31383.

Ishino, Y., Shinagawa, H., Makino, K. et al. (1987). Nucleotide sequence of the iap gene, responsible for alkaline phosphatase isozyme conversion in *Escherichia coli*, and identification of the gene product. *J. Bacteriol.* 169 (12): 5429–5433.

Jacobsen, L.B., Calvin, S.A., Colvin, K.E. et al. (2004). FuGENE 6 transfection reagent: the gentle power. *Methods* 33 (2): 104–112. https://doi.org/10.1016/j.ymeth.2003.11.002.

Jacobsen, B.M., Harrell, J.C., Jedlicka, P. et al. (2006). Spontaneous fusion with, and transformation of mouse stroma by, malignant human breast cancer epithelium. *Cancer Res.* 66 (16): 8274–8279. https://doi.org/10.1158/0008-5472.CAN-06-1456.

Jacobsen, L.B., Calvin, S.A., and Lobenhofer, E.K. (2009). Transcriptional effects of transfection: the potential for misinterpretation of gene expession data generated from transiently transfected cells. *Biotechniques* 47 (1): 617–624. https://doi.org/10.2144/000113132.

James, M.F., Rich, C.B., Trinkaus-Randall, V. et al. (1998). In vitro tissue transfection by calcium phosphate. In: *DNA Transfer to Cultured Cells* (eds. K. Ravid and R.I. Freshney), 157–177. New York: Wiley-Liss.

Jansen, R., Embden, J.D., Gaastra, W. et al. (2002). Identification of genes that are associated with DNA repeats in prokaryotes. *Mol. Microbiol.* 43 (6): 1565–1575.

Jat, P.S., Noble, M.D., Ataliotis, P. et al. (1991). Direct derivation of conditionally immortal cell lines from an H-2Kb-tsA58 transgenic mouse. *Proc. Natl Acad. Sci. U.S.A.* 88 (12): 5096–5100.

Jinek, M., Chylinski, K., Fonfara, I. et al. (2012). A programmable dual-RNA-guided DNA endonuclease in adaptive bacterial immunity. *Science* 337 (6096): 816–821. https://doi.org/10.1126/science.1225829.

Jinek, M., East, A., Cheng, A. et al. (2013). RNA-programmed genome editing in human cells. *Elife* 2: e00471. https://doi.org/10.7554/eLife.00471.

Jordan, E.T., Collins, M., Terefe, J. et al. (2008). Optimizing electroporation conditions in primary and other difficult-to-transfect cells. *J. Biomol. Tech.* 19 (5): 328–334.

Kamens, J. (2015). The Addgene repository: an international nonprofit plasmid and data resource. *Nucleic Acids Res.* 43 (Database issue): D1152–D1157. https://doi.org/10.1093/nar/gku893.

Kern, G. and Flucher, B.E. (2005). Localization of transgenes and genotyping of H-2kb-tsA58 transgenic mice. *Biotechniques* 38, 1: 38, 40, 42. doi: https://doi.org/10.2144/05381BM03.

Kim, T.K. and Eberwine, J.H. (2010). Mammalian cell transfection: the present and the future. *Anal. Bioanal.Chem.* 397 (8): 3173–3178. https://doi.org/10.1007/s00216-010-3821-6.

Kim, Y.G., Cha, J., and Chandrasegaran, S. (1996). Hybrid restriction enzymes: zinc finger fusions to Fok I cleavage domain. *Proc. Natl Acad. Sci. U.S.A.* 93 (3): 1156–1160.

Kim, H.J., Lee, H.J., Kim, H. et al. (2009). Targeted genome editing in human cells with zinc finger nucleases constructed via modular assembly. *Genome Res.* 19 (7): 1279–1288. https://doi.org/10.1101/gr.089417.108.

Kim, E., Kim, S., Kim, D.H. et al. (2012). Precision genome engineering with programmable DNA-nicking enzymes. *Genome Res.* 22 (7): 1327–1333. https://doi.org/10.1101/gr.138792.112.

Kim, S., Kim, D., Cho, S.W. et al. (2014). Highly efficient RNA-guided genome editing in human cells via delivery of purified Cas9 ribonucleoproteins. *Genome Res.* 24 (6): 1012–1019. https://doi.org/10.1101/gr.171322.113.

Kimberland, M.L., Hou, W., Alfonso-Pecchio, A. et al. (2018). Strategies for controlling CRISPR/Cas9 off-target effects and biological variations in mammalian genome editing experiments. *J. Biotechnol.* 284: 91–101. https://doi.org/10.1016/j.jbiotec.2018.08.007.

Kiyono, T., Foster, S.A., Koop, J.I. et al. (1998). Both Rb/p16INK4a inactivation and telomerase activity are required to immortalize human epithelial cells. *Nature* 396 (6706): 84–88. https://doi.org/10.1038/23962.

Koprowski, H., Ponten, J.A., Jensen, F.C. et al. (1962). Transformation of cultures of human tissue infected with simian virus SV40. *J. Cell Comp. Physiol.* 59 (3): 281–292. https://doi.org/10.1002/jcp.1030590308.

Le Poole, I.C., van den Berg, F.M., van den Wijngaard, R.M. et al. (1997). Generation of a human melanocyte cell line by introduction of HPV16 E6 and E7 genes. *In Vitro Cell. Dev. Biol. Anim.* 33 (1): 42–49. https://doi.org/10.1007/s11626-997-0021-6.

Li, M., Husic, N., Lin, Y. et al. (2012). Production of lentiviral vectors for transducing cells from the central nervous system. *J. Vis. Exp.* 63: e4031. https://doi.org/10.3791/4031.

Liang, X., Potter, J., Kumar, S. et al. (2017). Enhanced CRISPR/Cas9-mediated precise genome editing by improved design and delivery of gRNA, Cas9 nuclease, and donor DNA. *J. Biotechnol.* 241: 136–146. https://doi.org/10.1016/j.jbiotec.2016.11.011.

Lin, S., Staahl, B.T., Alla, R.K. et al. (2014). Enhanced homology-directed human genome engineering by controlled timing of CRISPR/Cas9 delivery. *Elife* 3: e04766. https://doi.org/10.7554/eLife.04766.

Lino, C.A., Harper, J.C., Carney, J.P. et al. (2018). Delivering CRISPR: a review of the challenges and approaches. *Drug Deliv.* 25 (1): 1234–1257. https://doi.org/10.1080/10717544.2018.1474964.

Lisowski, L., Tay, S.S., and Alexander, I.E. (2015). Adeno-associated virus serotypes for gene therapeutics. *Curr. Opin. Pharmacol.* 24: 59–67. https://doi.org/10.1016/j.coph.2015.07.006.

Liu, X., Ory, V., Chapman, S. et al. (2012). ROCK inhibitor and feeder cells induce the conditional reprogramming of epithelial cells. *Am. J. Pathol.* 180 (2): 599–607. https://doi.org/10.1016/j.ajpath.2011.10.036.

Liu, X., Krawczyk, E., Suprynowicz, F.A. et al. (2017). Conditional reprogramming and long-term expansion of normal and tumor cells from human biospecimens. *Nat. Protoc.* 12 (2): 439–451. https://doi.org/10.1038/nprot.2016.174.

Liu, Q., Yuan, Y., Zhu, F. et al. (2018). Efficient genome editing using CRISPR/Cas9 ribonucleoprotein approach in cultured medaka fish cells. *Biol. Open.* 7 (8) https://doi.org/10.1242/bio.035170.

Luckow, V.A., Lee, S.C., Barry, G.F. et al. (1993). Efficient generation of infectious recombinant baculoviruses by site-specific transposon-mediated insertion of foreign genes into a baculovirus genome propagated in *Escherichia coli. J. Virol.* 67 (8): 4566–4579.

Lundstrom, K. (2018). Viral vectors in gene therapy. *Diseases* 6 (2) https://doi.org/10.3390/diseases6020042.

Luo, J., Deng, Z.L., Luo, X. et al. (2007). A protocol for rapid generation of recombinant adenoviruses using the AdEasy system. *Nat. Protoc.* 2 (5): 1236–1247. https://doi.org/10.1038/nprot.2007.135.

Madhan, S., Prabakaran, M., and Kwang, J. (2010). Baculovirus as vaccine vectors. *Curr. Gene Ther.* 10 (3): 201–213.

Maldonado, M., Luu, R.J., Ramos, M.E. et al. (2016). ROCK inhibitor primes human induced pluripotent stem cells to selectively differentiate towards mesendodermal lineage via epithelial–mesenchymal transition-like modulation. *Stem Cell Res. Ther.* 17 (2): 222–227. https://doi.org/10.1016/j.scr.2016.07.009.

Mali, P., Yang, L., Esvelt, K.M. et al. (2013). RNA-guided human genome engineering via Cas9. *Science* 339 (6121): 823–826. https://doi.org/10.1126/science.1232033.

Mayne, L.V., Price, T.N.C., Morwood, K. et al. (1996). Development of immortal human fibroblast cell lines. In: *Culture of Immortalized Cells* (eds. R.I. Freshney and M.G. Freshney), 77–93. New York: Wiley-Liss.

McCutchan, J.H. and Pagano, J.S. (1968). Enhancement of the infectivity of simian virus 40 deoxyribonucleic acid with diethylaminoethyl-dextran. *J. Natl Cancer Inst.* 41 (2): 351–357.

McDermott, M., Eustace, A.J., Busschots, S. et al. (2014). In vitro development of chemotherapy and targeted therapy drug-resistant cancer cell lines: a practical guide with case studies. *Front. Oncol.* 4: 40. https://doi.org/10.3389/fonc.2014.00040.

Miller, J.C., Tan, S., Qiao, G. et al. (2011). A TALE nuclease architecture for efficient genome editing. *Nat. Biotechnol.* 29 (2): 143–148. https://doi.org/10.1038/nbt.1755.

Mir, L.M. (2014). Electroporation-based gene therapy: recent evolution in the mechanism description and technology developments. *Methods Mol. Biol.* 1121: 3–23. https://doi.org/10.1007/978-1-4614-9632-8_1.

Mojica, F.J., Diez-Villasenor, C., Soria, E. et al. (2000). Biological significance of a family of regularly spaced repeats in the genomes of archaea, bacteria and mitochondria. *Mol. Microbiol.* 36 (1): 244–246.

Mojica, F.J., Diez-Villasenor, C., Garcia-Martinez, J. et al. (2005). Intervening sequences of regularly spaced prokaryotic repeats derive from foreign genetic elements. *J. Mol. Evol.* 60 (2): 174–182. https://doi.org/10.1007/s00239-004-0046-3.

Moon, H.E., Yoon, S.H., Hur, Y.S. et al. (2013). Mitochondrial dysfunction of immortalized human adipose tissue-derived mesenchymal stromal cells from patients with Parkinson's disease. *Exp. Neurobiol.* 22 (4): 283–300. https://doi.org/10.5607/en.2013.22.4.283.

Moore, G.E., Gerner, R.E., and Franklin, H.A. (1967). Culture of normal human leukocytes. *JAMA* 199 (8): 519–524.

Morelli, C., Barbisan, F., Iaccheri, L. et al. (2004). SV40-immortalized human fibroblasts as a source of SV40 infectious virions. *Mol. Med.* 10 (7–12): 112–116. https://doi.org/10.2119/2004-00037.Morelli.

Moscou, M.J. and Bogdanove, A.J. (2009). A simple cipher governs DNA recognition by TAL effectors. *Science* 326 (5959): 1501. https://doi.org/10.1126/science.1178817.

Munsie, M. and Gyngell, C. (2018). Ethical issues in genetic modification and why application matters. *Curr. Opin. Genet. Dev.* 52: 7–12. https://doi.org/10.1016/j.gde.2018.05.002.

Murhammer, D.W. (2016). *Baculovirus and Insect Cell Expression Protocols.* New York: Humana Press.

Mussolino, C., Morbitzer, R., Lutge, F. et al. (2011). A novel TALE nuclease scaffold enables high genome editing activity in combination with low toxicity. *Nucleic Acids Res.* 39 (21): 9283–9293. https://doi.org/10.1093/nar/gkr597.

Mussolino, C., Alzubi, J., Fine, E.J. et al. (2014). TALENs facilitate targeted genome editing in human cells with high specificity and low cytotoxicity. *Nucleic Acids Res.* 42 (10): 6762–6773. https://doi.org/10.1093/nar/gku305.

Nageshwaran, S., Chavez, A., Cher Yeo, N. et al. (2018). CRISPR guide RNA cloning for mammalian systems. *J. Vis. Exp.* 140 https://doi.org/10.3791/57998.

Nakatake, Y., Fujii, S., Masui, S. et al. (2013). Kinetics of drug selection systems in mouse embryonic stem cells. *BMC Biotech.* 13: 64. https://doi.org/10.1186/1472-6750-13-64.

Naldini, L., Blomer, U., Gallay, P. et al. (1996). In vivo gene delivery and stable transduction of nondividing cells by a lentiviral vector. *Science* 272 (5259): 263–267.

Naso, M.F., Tomkowicz, B., Perry, W.L. 3rd et al. (2017). Adeno-associated virus (AAV) as a vector for gene therapy. *BioDrugs* 31 (4): 317–334. https://doi.org/10.1007/s40259-017-0234-5.

Nature Editors (2018). Keep off-target effects in focus. *Nat. Med.* 24 (8): 1081. https://doi.org/10.1038/s41591-018-0150-3.

Neumann, E., Schaefer-Ridder, M., Wang, Y. et al. (1982). Gene transfer into mouse lyoma cells by electroporation in high electric fields. *EMBO J.* 1 (7): 841–845.

Newbold, R.F. (2002). The significance of telomerase activation and cellular immortalization in human cancer. *Mutagenesis* 17 (6): 539–550.

Nims, R.W., Capes-Davis, A., Korch, C. et al. (2018). Authenticating hybrid cell lines. In: *Cell Culture* (ed. R.A. Mehanna), 151–169. London: IntechOpen.

Noble, J.R., Zhong, Z.H., Neumann, A.A. et al. (2004). Alterations in the p16(INK4a) and p53 tumor suppressor genes of hTERT-immortalized human fibroblasts. *Oncogene* 23 (17): 3116–3121. https://doi.org/10.1038/sj.onc.1207440.

van Oers, M.M., Pijlman, G.P., and Vlak, J.M. (2015). Thirty years of baculovirus-insect cell protein expression: from dark horse to mainstream technology. *J. Gen. Virol.* 96 (Pt 1): 6–23. https://doi.org/10.1099/vir.0.067108-0.

O'Hare, M.J., Bond, J., Clarke, C. et al. (2001). Conditional immortalization of freshly isolated human mammary fibroblasts and endothelial cells. *Proc. Natl Acad. Sci. U.S.A.* 98 (2): 646–651. https://doi.org/10.1073/pnas.98.2.646.

Ouellette, M.M., McDaniel, L.D., Wright, W.E. et al. (2000). The establishment of telomerase-immortalized cell lines representing human chromosome instability syndromes. *Hum. Mol. Genet.* 9 (3): 403–411.

Oxford Expression Technologies (2019). baculoCOMPLETE: a complete laboratory guide to the baculovirus expression system and insect cell culture. https://oetltd.com/wp-content/uploads/2019/01/baculocomplete-user-guide-2019-20-compressed.pdf (accessed 19 March 2019).

Peehl, D.M. and Stanbridge, E.J. (1981). Anchorage-independent growth of normal human fibroblasts. *Proc. Natl Acad. Sci. U.S.A.* 78 (5): 3053–3057.

Pennock, G.D., Shoemaker, C., and Miller, L.K. (1984). Strong and regulated expression of *Escherichia coli* beta-galactosidase in insect cells with a baculovirus vector. *Mol. Cell. Biol.* 4 (3): 399–406.

Peters, D.M., Dowd, N., Brandt, C. et al. (1996). Human papilloma virus E6/E7 genes can expand the lifespan of human corneal fibroblasts. *In Vitro Cell. Dev. Biol. Anim.* 32 (5): 279–284.

Peters-Hall, J.R., Coquelin, M.L., Torres, M.J. et al. (2018). Long-term culture and cloning of primary human bronchial basal cells that maintain multipotent differentiation capacity and CFTR channel function. *Am. J. Physiol. Lung Cell Mol. Physiol.* 315 (2): L313–L327. https://doi.org/10.1152/ajplung.00355.2017.

Potter, H. and Heller, R. (2003). Transfection by electroporation. *Curr. Protoc. Mol. Biol.* Chapter 9:Unit 9 3. doi: https://doi.org/10.1002/0471142727.mb0903s62.

Pourcel, C., Salvignol, G., and Vergnaud, G. (2005). CRISPR elements in *Yersinia pestis* acquire new repeats by preferential uptake of bacteriophage DNA, and provide additional tools for evolutionary studies. *Microbiology* 151 (Pt 3): 653–663. https://doi.org/10.1099/mic.0.27437-0.

Punchard, N., Watson, D., Thomson, R. et al. (1996). Production of immortal human umbilical vein endothelial cells. In: *Culture of Immortalized Cells* (eds. R.I. Freshney and M.G. Freshney), 203–238. New York: Wiley-Liss.

RAC (2006). Biosafety considerations for research with lentiviral vectors. https://osp.od.nih.gov/wp-content/uploads/lenti_containment_guidance.pdf (accessed 17 March 2019).

Reddel, R.R. (2000). The role of senescence and immortalization in carcinogenesis. *Carcinogenesis* 21 (3): 477–484.

Reddel, R.R., Ke, Y., Gerwin, B.I. et al. (1988). Transformation of human bronchial epithelial cells by infection with SV40 or adenovirus-12 SV40 hybrid virus, or transfection via strontium phosphate coprecipitation with a plasmid containing SV40 early region genes. *Cancer Res.* 48 (7): 1904–1909.

Rezza, A., Jacquet, C., Le Pillouer, A. et al. (2019). Unexpected genomic rearrangements at targeted loci associated with CRISPR/Cas9-mediated knock-in. *Sci. Rep.* 9 (1): 3486. https://doi.org/10.1038/s41598-019-40181-w.

Rheinwald, J.G. and Green, H. (1975). Serial cultivation of strains of human epidermal keratinocytes: the formation of keratinizing colonies from single cells. *Cell* 6 (3): 331–343.

Rheinwald, J.G. and Green, H. (1977). Epidermal growth factor and the multiplication of cultured human epidermal keratinocytes. *Nature* 265 (5593): 421–424.

Rodrigues, G.A., Shalaev, E., Karami, T.K. et al. (2018). Pharmaceutical development of AAV-based gene therapy products for the eye. *Pharm. Res.* 36 (2): 29. https://doi.org/10.1007/s11095-018-2554-7.

Rohrmann, G.F. (2013). *Baculovirus Molecular Biology*, 3e. Bethesda, MD: National Center for Biotechnology Information.

Schlimgen, R., Howard, J., Wooley, D. et al. (2016). Risks associated with lentiviral vector exposures and prevention strategies. *J. Occup. Environ. Med.* 58 (12): 1159–1166. https://doi.org/10.1097/JOM.0000000000000879.

Seigel, G.M. (1996). Establishment of an E1A-immortalized retinal cell culture. *In Vitro Cell. Dev. Biol. Anim.* 32 (2): 66–68.

Sette, G., Salvati, V., Giordani, I. et al. (2018). Conditionally reprogrammed cells (CRC) methodology does not allow the in vitro expansion of patient-derived primary and metastatic lung cancer cells. *Int. J. Cancer* 143 (1): 88–99. https://doi.org/10.1002/ijc.31260.

Shay, J.W. and Wright, W.E. (1989). Quantitation of the frequency of immortalization of normal human diploid fibroblasts by SV40 large T-antigen. *Exp. Cell. Res.* 184 (1): 109–118.

Shioda, S., Kasai, F., Watanabe, K. et al. (2018). Screening for 15 pathogenic viruses in human cell lines registered at the JCRB Cell Bank: characterization of in vitro human cells by viral infection. *R Soc. Open Sci.* 5 (5): 172472. https://doi.org/10.1098/rsos.172472.

Smith, G.E., Summers, M.D., and Fraser, M.J. (1983). Production of human beta interferon in insect cells infected with a baculovirus expression vector. *Mol. Cell. Biol.* 3 (12): 2156–2165.

Sobinoff, A.P. and Pickett, H.A. (2017). Alternative lengthening of telomeres: DNA repair pathways converge. *Trends Genet.* 33 (12): 921–932. https://doi.org/10.1016/j.tig.2017.09.003.

Somasundaram, C., Tournier, I., Feldmann, G. et al. (1992). Increased efficiency of gene transfection in primary cultures of adult rat hepatocytes stimulated to proliferate: a comparative study using the lipofection and the calcium phosphate precipitate methods. *Cell Biol. Int. Rep.* 16 (7): 653–662.

Sorek, R., Lawrence, C.M., and Wiedenheft, B. (2013). CRISPR-mediated adaptive immune systems in bacteria and archaea. *Annu. Rev. Biochem.* 82: 237–266. https://doi.org/10.1146/annurev-biochem-072911-172315.

Steinberg, M.L. (1996). Immortalization of human epidermal keratinocytes by SV40. In: *Culture of Immortalized Cells* (eds. R.I. Freshney and M.G. Freshney), 95–120. New York: Wiley-Liss.

Stepanenko, A.A. and Heng, H.H. (2017). Transient and stable vector transfection: pitfalls, off-target effects, artifacts. *Mutat. Res.* 773: 91–103. https://doi.org/10.1016/j.mrrev.2017.05.002.

Stuchbury, G. and Munch, G. (2010). Optimizing the generation of stable neuronal cell lines via pre-transfection restriction enzyme digestion of plasmid DNA. *Cytotechnology* 62 (3): 189–194. https://doi.org/10.1007/s10616-010-9273-1.

Suprynowicz, F.A., Upadhyay, G., Krawczyk, E. et al. (2012). Conditionally reprogrammed cells represent a stem-like state of adult epithelial cells. *Proc. Natl Acad. Sci. U.S.A.* 109 (49): 20035–20040. https://doi.org/10.1073/pnas.1213241109.

Toouli, C.D., Huschtscha, L.I., Neumann, A.A. et al. (2002). Comparison of human mammary epithelial cells immortalized by simian virus 40 T-antigen or by the telomerase catalytic subunit. *Oncogene* 21 (1): 128–139. https://doi.org/10.1038/sj.onc.1205014.

Uphoff, C.C., Denkmann, S.A., Steube, K.G. et al. (2010). Detection of EBV, HBV, HCV, HIV-1, HTLV-I and -II, and SMRV in human and other primate cell lines. *J. Biomed. Biotechnol.* 2010: 904767. https://doi.org/10.1155/2010/904767.

Vaziri, H. and Benchimol, S. (1998). Reconstitution of telomerase activity in normal human cells leads to elongation of telomeres and extended replicative life span. *Curr. Biol.* 8 (5): 279–282.

Westphal, H. (1984). Gene transfer into mammalian cells and embryos. *J. Natl Cancer Inst.* 72 (4): 777–782.

Whitaker, N.J., Kidston, E.L., and Reddel, R.R. (1992). Finite life span of hybrids formed by fusion of different simian virus 40-immortalized human cell lines. *J. Virol.* 66 (2): 1202–1206.

White, M., Whittaker, R., Gandara, C. et al. (2017). A guide to approaching regulatory considerations for lentiviral-mediated gene therapies. *Hum. Gene Ther. Methods* 28 (4): 163–176. https://doi.org/10.1089/hgtb.2017.096.

Whitehead, R.H. and Robinson, P.S. (2009). Establishment of conditionally immortalized epithelial cell lines from the intestinal tissue of adult normal and transgenic mice. *Am. J. Physiol. Gastrointest. Liver Physiol.* 296 (3): G455–G460. https://doi.org/10.1152/ajpgi.90381.2008.

Wold, W.S. and Toth, K. (2013). Adenovirus vectors for gene therapy, vaccination and cancer gene therapy. *Curr. Gene Ther.* 13 (6): 421–433.

Wu, Y., Gao, T., Wang, X. et al. (2014). TALE nickase mediates high efficient targeted transgene integration at the human multi-copy ribosomal DNA locus. *Biochem. Biophys. Res. Commun.* 446 (1): 261–266. https://doi.org/10.1016/j.bbrc.2014.02.099.

Yugawa, T. and Kiyono, T. (2009). Molecular mechanisms of cervical carcinogenesis by high-risk human papillomaviruses: novel functions of E6 and E7 oncoproteins. *Rev. Med. Virol.* 19 (2): 97–113. https://doi.org/10.1002/rmv.605.

Zhang, P. and Wagner, E. (2017). History of polymeric gene delivery systems. *Top. Curr. Chem. (Cham.)* 375 (2): 26. https://doi.org/10.1007/s41061-017-0112-0.

Zhou, X., Merchak, K., Lee, W. et al. (2015). Cell fusion connects oncogenesis with tumor evolution. *Am. J. Pathol.* 185 (7): 2049–2060. https://doi.org/10.1016/j.ajpath.2015.03.014.

Zischewski, J., Fischer, R., and Bortesi, L. (2017). Detection of on-target and off-target mutations generated by CRISPR/Cas9 and other sequence-specific nucleases. *Biotechnol. Adv.* 35 (1): 95–104. https://doi.org/10.1016/j.biotechadv.2016.12.003.

PART VII

Stem Cells and Differentiated Cells

After reading the following chapters in this part of the book, you will be able to:

(23) *Culture of Stem Cells:*

(a) List examples of tissues that are commonly used to establish stem cell cultures and discuss their advantages and disadvantages.

(b) Describe how to induce pluripotency in somatic cells.

(c) Passage a human pluripotent stem cell (hPSC) line using EDTA and chemically defined conditions (E8 medium and vitronectin coating).

(d) Perform cryopreservation of an hPSC line using chemically defined conditions in the presence of Rho kinase (ROCK) inhibitor.

(e) Discuss how to perform characterization of an undifferentiated hPSC line.

(24) *Culture of Specific Cell Types:*

(a) Discuss how to obtain "specialized" cells, i.e. specific cell types that continue to express their differentiated characteristics or can be induced to do so.

(b) Perform primary culture of various cell types, including epithelial, mesenchymal, neuroectodermal, and hematopoietic cells (with reference to the Supplementary material online, which contains all protocols for this chapter).

(c) Describe how to generate hybridomas for production of monoclonal antibodies.

(d) Explain how a chimeric antigen receptor (CAR) T-cell is generated.

(e) Discuss how to culture poikilothermic cells from fish, insects, and other species.

(25) *Culture of Tumor Cells:*

(a) Freeze tumor biopsies for later culture and snap freeze portions for other applications (e.g. to extract DNA for authentication testing).

(b) Provide examples of techniques that can be used for positive selection of tumor cells and for negative selection of other cell types.

(c) Culture tumor cells using a confluent feeder layer.

(d) Perform primary culture of colorectal tumor cells, mammary tumor cells, and leukemia-lymphoma cells (with reference to the Supplementary material online).

(e) Explain the meaning of "cancer stem cell" (CSC) and discuss why *in vitro* analysis of CSCs is challenging.

(26) *Differentiation:*

(a) Provide examples of continuous cell lines with differentiated properties and describe the advantages and disadvantages of using them to study differentiation.

Freshney's Culture of Animal Cells: A Manual of Basic Technique and Specialized Applications, Eighth Edition. Amanda Capes-Davis and R. Ian Freshney.
© 2021 John Wiley & Sons Ltd. Published 2021 by John Wiley & Sons Ltd.
Companion website: www.wiley.com/go/freshney/cellculture8

(b) Discuss why cultures change their differentiation status *in vitro* and the meanings of commonly used terms such as "dedifferentiation" and "transdifferentiation."

(c) Select soluble factors to induce differentiation of a particular cell type, including physiological or non-physiological factors.

(d) Describe how to change the culture environment to promote cell differentiation.

(e) Discuss how to achieve stem cell differentiation and why this is an ongoing challenge for many laboratories.

CHAPTER 23

Culture of Stem Cells

The term "stem cell" was used as early as 1868, when Ernst Haeckel described unicellular organisms as "Stammzellen" that gave rise to multicellular organisms, in an evolutionary tree of species development (Ramalho-Santos and Willenbring 2007). Haeckel used the term in a later work to refer to the fertilized egg as an ancestral "stem cell" that gives rise to all cell types in an organism. It is now used to describe cells that can both renew themselves and give rise to differentiated cells (see Section 2.4.2). However, despite a century of interest in this concept, it is technically challenging to cultivate stem cells *in vitro*. Hematopoietic stem cells (HSCs) were proven to exist in 1963, but these cells could only be enriched from the adult bone marrow following the development of flow cytometry 20 years later (Becker et al. 1963; Spangrude et al. 1988). The recent explosion in stem cell research can be traced back to advances in cell culture techniques and reagents (Armstrong et al. 2012). The development of feeder layers, substrate treatments, and serum-free media formulations – all discussed in previous chapters (see Part III) – have enabled today's stem cell research and tomorrow's clinical applications. This chapter provides a starting point for tissue culture laboratories that are new to stem cell culture. It is impossible to discuss all stem cell types in a single chapter, so we focus primarily on human pluripotent stem cells (hPSCs). Other books are available that explain the intricacies of stem cell biology and explore stem cell culture at a more specialized level (Freshney et al. 2007; Lanza and Atala 2014; Crook and Ludwig 2017). Guidelines are also available that set out the fundamental principles and standards required for stem cell research and clinical translation (ISSCR 2016).

23.1 TERMINOLOGY: STEM CELLS

Stem cell terminology often varies between publications and can be quite confusing. Logically, this is not surprising. Stem cells vary in their characteristics depending on their origin and the type of characterization that is performed. They also tend to be studied in different areas of specialization, resulting in inconsistent reporting. Some journals and organizations have developed glossaries or position statements on stem cell terms, which help to provide a consensus approach (Smith 2006; ISSCR 2016; NIH 2019b). Where a consensus has not been developed, the simplest approach is to describe stem cells in accordance with their source and potency (see Section 2.4.2) (Ilic and Polak 2011).

Stem cells can be isolated from many different sources, which all have advantages and disadvantages for research and clinical applications (see Table 23.1). Stem cells from various sources can be broadly divided into (i) prenatal stem cells, which include embryonic stem cells (ESCs) from the pre-implantation embryo; (ii) perinatal stem cells from tissues such as the amniotic fluid, placenta, and umbilical cord; (iii) postnatal stem cells (often described as adult stem cells), which include HSCs and other cell types that give rise to specific lineages; (iv) reprogrammed stem cells, which have been manipulated *in vitro* to increase their level of potency; and (v) disease-related stem cells, which are capable of self-renewal and differentiation but where regulation of these processes may be aberrant. The latter include teratoid tumors, which contain cells from all three germ layers (see Section 2.4.1), and cancer stem cells (CSCs) (see Minireview M25.1).

Freshney's Culture of Animal Cells: A Manual of Basic Technique and Specialized Applications, Eighth Edition. Amanda Capes-Davis and R. Ian Freshney.
© 2021 John Wiley & Sons Ltd. Published 2021 by John Wiley & Sons Ltd.
Companion website: www.wiley.com/go/freshney/cellculture8

TABLE 23.1. Sources of stem cells.

Source[a]	Potency[b]	Advantages	Disadvantages
Prenatal			
Zygote (fertilized egg)	Totipotent	Can be used for clinical applications, e.g. *in vitro* fertilization (IVF); may be used for research in some non-human species following ethical review and approval	Ethical concerns regarding use of embryos for research
Blastocyst (pre-implantation, inner cell mass)	Pluripotent (ESCs, EPL cells, FAB-SCs)	Natural source of PSCs; widely studied	Ethical concerns regarding use of embryos for research; ESCs from different species have different growth factor dependencies
Epiblast (post-implantation)	Pluripotent (EpiSC)	Natural source of PSCs; dependence of mouse epiSCs on Activin/Nodal signaling is similar to hESCs, allowing similar culture conditions	Ethical concerns regarding use of embryos for research; cells are prone to spontaneous differentiation, genetic instability, and apoptosis under suboptimal conditions
Embryonic germ cells (late embryo/early fetus)	Pluripotent	Cells display long-term self-renewal and relatively good genomic stability; historically easier to maintain before hPSC culture conditions	Ethical concerns regarding use of embryos for research
Trophoblast	Multipotent (differentiated cell types of the placenta)	Can be isolated in mouse and some other species, e.g. rhesus macaque	Difficult or impossible to isolate in many species, including human
Perinatal			
Amniotic fluid; stem cells from amniotic epithelium	Multipotent, some reports of pluripotency (AFSCs)	Can be used to examine the fetal karyotype	Requires invasive procedure; limited amount of material
Placenta; stem cells from amnion, placental villi, placental blood	Multipotent (MSCs)	Easily accessible without invasive techniques; sample is usually discarded	Discarded samples must be handled appropriately to preserve viability and function
Umbilical cord; stem cells from cord blood and Wharton's jelly	Multipotent (HSCs and MSCs), some reports of pluripotency (VSELs)	Easily accessible without invasive techniques; sample is usually discarded; HSCs and MSCs can be isolated and characterized using readily available equipment and techniques	Discarded samples must be handled appropriately to preserve viability and function
Postnatal: tissue-specific stem cells			
Adipose tissue	Multipotent (MSCs); some reports of pluripotency (ASCs, Muse cells)	Accessible with medical procedure; sample is usually discarded	Discarded samples must be handled appropriately to preserve viability and function
Blood	Multipotent (HSCs, MSCs)	Accessible with venipuncture; HSCs can be increased by mobilizing bone marrow stem cells into the circulation	Stem cells are rare cell populations (e.g. 1 : 100 000 blood cells for HSCs)

(Continued)

TABLE 23.1. (*continued*)

Source[a]	Potency[b]	Advantages	Disadvantages
Bone marrow	Multipotent (HSCs, MSCs, EPCs); some reports of pluripotency (MAPCs, MIAMI cells, Muse cells, VSELs)	Accessible with medical procedure; HSCs and MSCs can be isolated and characterized using readily available equipment and techniques	Requires invasive procedure; limited amount of material; stem cells are rare cell populations (e.g. 1 : 10 000 bone marrow cells for HSCs)
Teeth (dental pulp)	Multipotent (MSCs)	Accessible with dental procedure	Discarded samples must be handled appropriately to preserve viability and function
Reprogrammed stem cells			
Adult somatic cell (nuclear transfer)	Totipotent (cloning)	Used for research, e.g. to develop mammalian cloning (Dolly the sheep)	Ethical concerns regarding use of cloning for research or clinical use
Adult somatic cell (somatic cell hybridization)	Pluripotent	Used for research, e.g. historical method used to induce pluripotency	Ethical concerns regarding genetic manipulation of embryonic cells
Adult somatic cell (reprogramming using defined factors)	Pluripotent (iPSC)	Fewer ethical concerns compared to many other sources; potentially accessible from any donor or tissue type	Behavior may differ compared to ESCs; cells are prone to spontaneous differentiation, genetic instability, and apoptosis under suboptimal conditions; risk of oncogenesis due to activation of c-Myc or off-target effects
Disease-related stem cells			
Cancer stem cells	Disordered potency	Potential for cancer detection and treatment, e.g. of metastasis	Disease-related stem cells; abnormal regulation of potency and lineage
Teratoma, teratocarcinoma	Pluripotent (embryonal carcinoma cells)	Used as a model to study embryonic development and lineage selection	Disease-related stem cells; abnormal regulation of potency and lineage

Source: Based on Gao, L., Thilakavathy, K. and Nordin, N. (2013). A plethora of human pluripotent stem cells. Cell Biol Int 37 (9):875–87; Ng, Yuk Yin, Baert, Miranda R.M., de Haas, Edwin F.E., et al. (2009). "Isolation of Human and Mouse Hematopoietic Stem Cells." In Genetic Modification of Hematopoietic Stem Cells: Methods and Protocols, edited by Christopher Baum, 13–21. Totowa, NJ: Humana Press; Pera, M. F. and Tam, P. P. (2010). Extrinsic regulation of pluripotent stem cells. Nature 465 (7299):713–20.

[a]This table focuses on sources of stem cells that are commonly available to tissue culture laboratories. It is not intended to be comprehensive for all stem cell sources or types.

[b]Definitions of totipotent, pluripotent, and multipotent can be found elsewhere (see Section 2.4.2).

Potency refers to the range of commitment origins that are available to the stem cell (see Section 2.4.2), which determines the range of differentiated cell types that can arise from it. A pluripotent cell can give rise to any cell type from the three germ layers, while a multipotent cell can give rise to multiple cell types but cannot do so from all three germ layers. However, it is important to note that the potency of a culture is heavily reliant on the type of testing that has been used to demonstrate it (see Section 23.7) (Damjanov and Andrews

2016; International Stem Cell Initiative 2018). If a suitable method for pluripotency testing is not used, it is impossible to say whether a stem cell is pluripotent or multipotent – an important distinction for cell-based therapy.

Perhaps the most hotly debated term is that of the mesenchymal stromal cell (MSC) – also known as the mesenchymal stem cell, multipotent stromal cell, or medicinal signaling cell (Dominici et al. 2006; Caplan 2017). MSCs come from different sources (see Table 23.1) and are expanded and characterized using different methods, resulting in cells with varying characteristics. MSCs from the connective tissue of adult rat are capable of self-renewal and can differentiate into mesodermal lineages, including adipocytes, chondrocytes, and myoblasts (Seruya et al. 2004). By contrast, cell lines established from human mesenchymal tissue can differentiate into multiple lineages but have limited growth potential *in vitro* (Sudo et al. 2007); it is possible that only a proportion of these cultures are MSCs or that their capacity for self-renewal is lost *in vitro*. The term "mesenchymal stem cell" has also been misused in a clinical context, where patients may be told that they will receive direct medical benefit from unproven "stem cell" products (Caplan 2017). Such cultures are often not well characterized and their benefits (and risks) are unknown. For this book, the term "mesenchymal stromal cell" is used to be consistent with the recommendations of the International Society for Cellular Therapy (Dominici et al. 2006).

23.2 EMBRYONIC STEM CELLS (ESCS)

Early studies of pluripotency were performed using teratomas that occurred spontaneously in the mouse 129 strain (Stevens and Little 1954). Teratoid tumors may contain cell types from any of the three germ layers; they also contain an undifferentiated cell type, known as an embryonal carcinoma cell. Intraperitoneal injection of embryonal carcinoma cells into mice resulted in the formation of embryoid bodies, which are clusters of undifferentiated cells that can differentiate into the three germ layers, although they lack the normal axes of development that occur *in vivo* (Kleinsmith and Pierce 1964; Wesselschmidt and McDonald 2006). Clonal populations of these cells were established in xenograft models *in vivo*, where they differentiated into cell types from all three germ layers. Embryonal carcinoma cells were able to participate in normal development if they were transferred to a blastocyst and then allowed to develop to term, resulting in the formation of a chimeric mouse (Gokhale and Andrews 2012). These studies, while important in their own right, laid the foundation for culture of ESCs and for their characterization through teratoma or chimera formation (see Section 23.7).

23.2.1 Mouse (mESCs)

Further technical advances made it possible for scientists to isolate and establish cell lines from the inner cell mass of the mouse embryo (see Figure 23.1a, b). The early mESC lines were established by growing cells in serum-containing media in the presence of a feeder layer of STO fibroblasts (see Section 8.4), resulting in clusters of mESCs that grew on top of the feeder layer (see Figure 23.1c) (Evans and Kaufman 1981; Martin 1981). These mESC lines behaved in a similar manner to embryonal carcinoma cells and expressed many of the same pluripotency-associated markers, e.g.

stage-specific embryonic antigen-1 (SSEA-1) (Gokhale and Andrews 2012). However, unlike embryonal carcinoma cells, mESC lines displayed a normal karyotype (Evans and Kaufman 1981). Procedures for primary culture of mESCs are available in the Supplementary Material online (see Supp. S23.1).

Although mESC lines were successfully established in the 1980s, their ongoing culture presented a number of challenges. Cells grown on feeder layers would proliferate, but not differentiate, while removal of the feeder layer resulted in the formation of embryoid bodies that spontaneously differentiated into various cell types (Evans and Kaufman 1981). It was then discovered that leukemia inhibitory factor (LIF) could be used to maintain pluripotency, removing the requirement for feeder cells (Smith et al. 1988; Williams et al. 1988). LIF stimulates mESC proliferation and maintains an undifferentiated phenotype, as shown by expression of the transcription factor Pou5f1 (previously known as Oct-3/4), SSEA-1, and alkaline phosphatase (Furue et al. 2005). Bone morphogenetic proteins (BMPs) such as BMP-4 are often used in combination with LIF, especially in serum-free media. BMP-4 can have a synergistic effect with LIF (due to induction of Id gene expression via the Smad pathway), although it has induced epithelial differentiation in some studies (Ying et al. 2003; Furue et al. 2005). Culture conditions represent an equilibrium of many different factors, so it is unlikely that a single factor or signaling pathway will have the same effect under all circumstances. Procedures for propagation of mESC lines using various culture conditions are available in the Supplementary Material online (see Supp. S23.2; see also Figure 23.1). Cells may be grown entirely serum- and feeder-free using N2B27 medium (see Appendix B), supplemented with LIF and BMP-4 (see Supp. S23.1, S23.2).

Although mESC lines were the first to be established, other pluripotent cell types can be grown from the mouse embryo (see Table 23.2) (Pera and Tam 2010; Wu and Izpisua Belmonte 2015). Epiblast stem cells (EpiSCs) can be isolated from the post-implantation embryo, i.e. at a later stage of embryogenesis (Brons et al. 2007; Tesar et al. 2007). EpiSCs are pluripotent but display a different set of growth requirements compared to mESCs. Instead of relying on LIF for self-renewal, EpiSCs rely on (i) fibroblast growth factor-2 (FGF-2); (ii) Activin or Nodal; and (iii) activation of the Wnt signaling pathway (Chou et al. 2008). These requirements are very similar to those of human embryonic stem cells (hESCs). EpiSCs and hESCs are likely to represent an embryonic state that is "primed" for differentiation, whereas mESCs appear to represent an earlier, "naïve" state (Wu and Izpisua Belmonte 2015).

Refinement of culture conditions resulted in a further expansion in the pluripotent cell types that may be grown *in vitro*. Early primitive ectoderm-like (EPL) cells were generated by exposing mESCs to conditioned medium (Rathjen et al. 1999). More recently, mESCs were isolated using

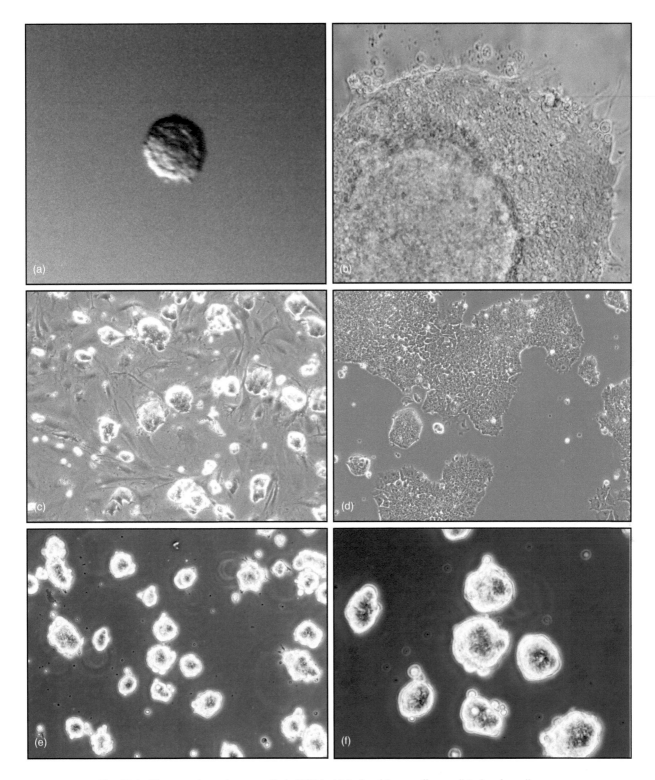

Fig. 23.1. Mouse embryonic stem cells (mESCs). (a) Isolated inner cell mass; "rind and core" structure of hypoblast growing round epiblast. (b) mESC-like outgrowths from cultured epiblast. (c) mESC colonies growing on feeder layers with serum and LIF. (d) Subcultured mESCs in serum and LIF. (e, f) mES colonies growing in 2i medium: (e) 10×; (f) 20× magnification. *Source:* Courtesy of J. P. Wray and A. Smith. (For associated protocols see Supp. S23.1, 23.2.)

TABLE 23.2. Properties of various pluripotent cell populations grown *in vitro*.

Property	mESCs	Mouse EPL cells	Mouse FAB-SCs	mEpiSCs	hPSCs
Stem cell genes[a]					
Pou5f1 (Oct-4)	Expressed	Expressed	Expressed	Expressed	Expressed
Nanog	Expressed	Expressed	Expressed	Expressed	Expressed
Sox2	Expressed	Expressed	Expressed	Expressed	Expressed
Klf4	Expressed	Not determined	Not determined	Not expressed	Expressed
Dppa3	Expressed	Not expressed	Not determined	Not expressed	Expressed
Zfp42 (Rex1)	Expressed	Not expressed	Not determined	Not expressed	Expressed
Gbx2	Expressed	Not expressed	Expressed	Not expressed	Expressed
Fgf5	Not expressed	Expressed	Not determined	Expressed	Not expressed
Cell-surface markers					
SSEA-1	Expressed	Expressed	Expressed	Expressed	Not expressed
SSEA-3, SSEA-4	Not expressed	Not expressed	Not expressed	Not expressed	Expressed
Alkaline phosphatase	Expressed	Expressed	Not determined	Not expressed	Expressed
Response to factors (requirement is for self-renewal)					
LIF	Required	Required[b]	Required[b]	Not required	Not required
Nodal and/or Activin	Not required[c]	Not determined	Required[d]	Required	Required
FGF-2	Not required	Not determined	Required[d]	Required[e]	Required
Developmental potential					
Teratoma	Yes	Yes	No	Yes	Yes
Chimera	Yes	No	No	No	Not determined

Source: Pera, M. F. and Tam, P. P. (2010). Extrinsic regulation of pluripotent stem cells. Nature 465 (7299):713–20. © 2010 Springer Nature.
[a]Since most data here relate to mouse cells, mouse gene nomenclature is used throughout. Human gene names are typically written in uppercase, not lowercase.
[b]Cells grown in LIF revert to an ESC-like state.
[c]One study shows long-term self-renewal of mESCs in Activin.
[d]Cells are derived and maintained in these factors, but dependence on these factors for self-renewal has not been rigorously examined.
[e]The requirement for FGF-2 has not been rigorously determined for these cells.
Abbreviations: see main text.

similar conditions to those employed for EpiSCs, resulting in the development of another stem cell type with different properties (see Table 23.2) (Chou et al. 2008). These cells are known as FAB-SCs because of their dependence on FGF-2, Activin, and BIO (an inhibitor of glycogen synthase kinase-3β). Stem cell populations can now be established with expanded potential that can contribute to extra-embryonic tissues in a chimera assay (Yang et al. 2017). Looking at these studies, it seems reasonable to conclude that pluripotency is a dynamic state that can be regulated by defined factors – but that there are still inherent differences between cells at different embryonic stages, e.g. between ESCs from the pre-implantation mouse embryo and EpiSCs from the post-implantation embryo (see Table 23.2).

23.2.2 Human (hESCs)

The first hESC lines were established 17 years after mESC lines were developed (Thomson et al. 1998). This delay was partly due to ethical reasons. The first five hESC lines were established from embryos that were produced by *in vitro* fertilization (IVF) for clinical purposes, and were donated after informed consent and ethical approval, but the topic remains a complex and controversial one (Sipp et al. 2018). There are also significant differences between mouse and human biology, resulting in the need for time to optimize the culture conditions. James Thomson and colleagues were initially successful in establishing an ESC line from rhesus monkey and used essentially the same conditions for hESC culture (Thomson et al. 1995, 1998).

There were some early similarities between the methods used to establish mESCs and hESCs (Evans and Kaufman 1981; Martin 1981; Thomson et al. 1998). All of the early approaches used cells from the inner cell mass of the pre-implantation embryo, which were grown in the presence of serum on feeder layers. A core set of transcription factors were expressed in both species, including POU5F1, NANOG, and SOX2 (see Table 23.2) (Pera and Tam 2010).

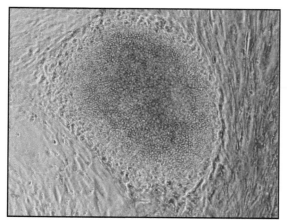

(a) *Human Embryonic Stem Cells (hESCs). Colony of hESCs on feeder layer (see also Fig. 23.2) Source: R. Ian Freshney, from [Cooke and Minger 2007].*

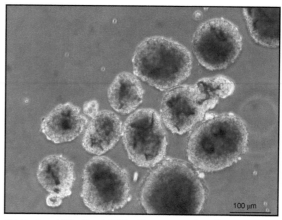

(b) *Embryoid Bodies. hESC cells grown as free-floating embryoid bodies. Source: R. Ian Freshney, from [Cooke and Minger 2007].*

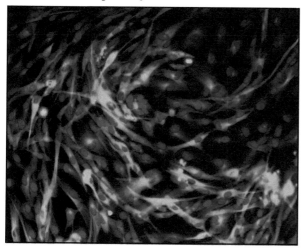

(c) *Differentiation in hESCs. Immunofluorescent staining for nestin (green); nuclei stained with Hoechst 33342 (blue) Source: R. Ian Freshney, from [Jackson et al. 2007].*

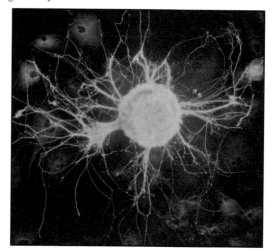

(d) *Differentiation in human embryonal carcinoma cells. Immunofluorescent staining for neural marker TUJ-1. Source: R. Ian Freshney, from [Przyborski 2007].*

Fig. 23.2. Morphology of human embryonic stem cells (hESCs).

However, there were also noticeable differences between mESCs and hESCs, including (Wu and Izpisua Belmonte 2015):

(1) *Morphology.* mESC colonies are usually raised or "dome" shaped, while hESC colonies tend to be flatter (see Figure 23.2a, b; see also Figure 23.3) (Cooke and Minger 2007). However, hESC morphology is liable to change; these cells are highly sensitive to the culture conditions and require frequent (daily) observation (Healy and Ruban 2015).

(2) *Sensitivity to dissociation.* hESCs are markedly sensitive to dissociation into a single cell suspension, resulting in rapid onset of apoptosis (Beers et al. 2012). Enzymatic methods that result in individualized cells (i.e. single cell suspensions) are more likely to give poor survival rates and abnormal karyotypes (Chen et al. 2010).

(3) *Expression of surface markers.* SSEA-1 is an important marker of pluripotency in mESC lines. However, early studies of human embryonal carcinoma cells showed that SSEA-1 was associated with a differentiated phenotype in human cells (Andrews et al. 1982). Human embryonal carcinoma cells and hESCs express SSEA-3 and SSEA-4 rather than SSEA-1 (Andrews et al. 1982; Gokhale and Andrews 2012).

(4) *Growth factor requirements.* Instead of LIF, hESCs require Nodal or Activin to maintain pluripotency (Vallier et al. 2005). Nodal and Activin are both members of the transforming growth factor-β (TGF-β) family (see Table 10.1); their effect on pluripotency is mediated by SMAD2/3 proteins, which bind to POU5F1 and NANOG (Pauklin and Vallier 2015). However, Nodal and Activin are not sufficient to sustain self-renewal in

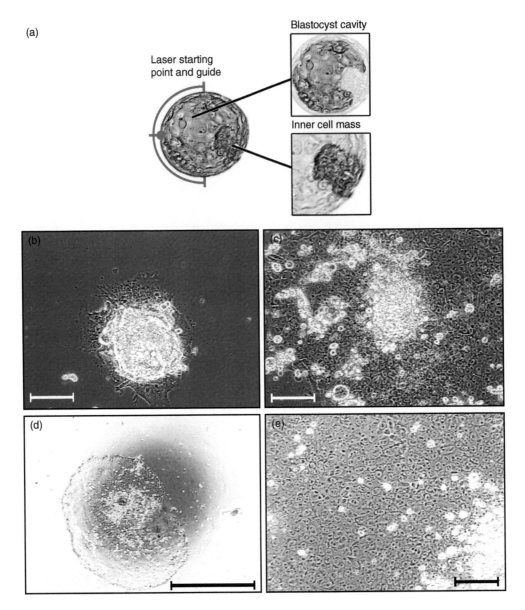

Fig. 23.3. Isolation of human embryonic stem cells (hESCs). (a) Strategy for isolating blastocyst by laser dissection. (b, c) Cultured, zona-free blastocyst at (b) two days and (c) seven days after plating. (d, e) Large colony ready for passage, visualized using (d) 4× objective and (e) 20× objective. Scale bars (b, c, e) 100 μm; (d) 1000 μm. *Source*: Courtesy of Z. Hewitt, R. Weightman, and H. Moore. (For associated protocol see Supp. S23.3.)

the absence of serum; FGF-2 must be added to maintain pluripotency under serum-free and feeder-free conditions. The effect of FGF-2 on pluripotency is dependent on Activin/Nodal signaling, where it appears to have a synergistic effect (Vallier et al. 2005; Pauklin and Vallier 2015).

Twenty years after the first five hESC lines were established, nearly 400 such cell lines have been approved for use with federal funding in the United States and more than 300 cell lines are listed as approved by the Medical Research Council in the United Kingdom (Ludwig et al. 2018). Registries of hESC lines are available from the Human Pluripotent Stem Cell Registry (hPSCreg), the International Stem Cell Forum (ISCF), and the National Institutes of Health (NIH) (Seltmann et al. 2016; ISCF 2019; NIH 2019a). The most frequently used hESC lines are typically older, well-characterized cultures (see Table 23.3), with an associated decline in the establishment of new hESC lines (Guhr et al. 2018; Ludwig et al. 2018). However, there is still a need to establish new cell lines from specific diseases or for

TABLE 23.3. Frequently used hESC lines.

hESC line name (commonly used)	hPSCreg name[a]	Number of publications (2008–2016)	Percentage of publications, total hESC research (2008–2016)
H9[b][c]	WAe009-A	2202	46.4
H1[b][c]	WAe001-A	1114	23.5
H7[b][c]	WAe007-A	351	7.4
HES-3[c]	ESIBIe003-A	287	6.0
HUES9	HVRDe009-A	178	3.8
BG01[c]	VIACe001-A	172	3.6
HES-2[c]	ESIBIe002-A	172	3.6
KhES-1	KUIMSe001-A	134	2.8
HUES7	HVRDe007-A	100	2.1
KhES-3	KUIMSe003-A	96	2.0
HSF-6[c]	UCSFe002-A	92	1.9
H14[b][c]	WAe014-A	85	1.8
HUES6	HVRDe006-A	75	1.6
HUES1	HVRDe001-A	67	1.4
HUES3	HVRDe003-A	66	1.4
HUES8	HVRDe008-A	65	1.4
HS181	KIe001-A	61	1.3
HSF-1[c]	UCSFe003-A	56	1.2
MEL-1	SCSe001-A	54	1.1
CA1	MSHRIe001-A	50	1.1
H13[b][c]	WAe013-A	48	1.0

Source: Guhr, A., Kobold, S., Seltmann, S., et al. (2018). Recent Trends in Research with Human Pluripotent Stem Cells: Impact of Research and Use of Cell Lines in Experimental Research and Clinical Trials. Stem Cell Reports 11 (2):485–496. © 2018 Elsevier.

[a]hPSCreg assigns a name to each hPSC line that acts as a unique identifier (see Section 23.7) (Kurtz et al. 2018). The hPSCreg name is also included in other knowledge resources, e.g. Cellosaurus.

[b]One of the first five hESC lines to be established (Thomson et al. 1998; Ludwig et al. 2018).

[c]Derived prior to 22 August 2001, when United States President George W. Bush allowed federal funding of research using a list of eligible hESC lines (NIH 2019a).

specific applications. Procedures to isolate the inner cell mass using laser dissection and to propagate hESC colonies are described in the Supplementary Material online (see Supp. S23.3; see also Figure 23.3).

Ethical regulations apply to all work using hESCs, whether a cell line is newly established or has been in use for some time (see Sections 6.4.2, 6.4.3). In the United Kingdom, the Human Fertilization and Embryology Authority (HFEA, www.hfea.gov.uk) licenses all research projects that derive hESCs, but not the cell lines themselves; these are regulated by the Human Tissue Authority (HTA) (HTA 2019). In the United States, guidance on hESCs and their use in

federally funded research is available from the Office for Human Research Protections (OHRP) and NIH (OHRP 2002; NIH 2016). Ethical regulations will also apply to other pluripotent cells that may be isolated from the embryo or fetus, such as embryonic germ cells (see Table 23.1).

23.2.3 Other Species

ESCs have been isolated and grown *in vitro* from a number of species including chickens, zebrafish, and rabbits (Pain et al. 1996; Fan and Collodi 2006; Honda et al. 2008). The number of species is difficult to assess because the methods used for their characterization may vary. Expression of pluripotency-associated markers may be used in some reports as proof of pluripotency, but as noted in relation to SSEA-1 (see Table 23.2), differences in biology exist that may cause confusion when comparing different species. *In vivo* assays for teratoma or chimera formation offer more stringent assessments of pluripotency (Hackett and Fortier 2011). Most ESC lines from other species have been established using feeder layers, although feeder-free conditions have been developed for some species (e.g. zebrafish) (Ho et al. 2014). Growth factor requirements need to be better understood to develop feeder-free conditions for other species, such as the role of chicken stem cell factor (SCF) (Farzaneh et al. 2017).

23.3 INDUCTION OF PLURIPOTENCY

The development of induced pluripotent stem cells (iPSCs) followed the discovery that differentiated somatic cells could be "reprogrammed" into other cell types using somatic cell nuclear transfer (SCNT) or somatic cell hybridization (Gurdon 1962; Tada et al. 2001; Cowan et al. 2005). Inspired by these studies, Shinya Yamanaka and colleagues speculated that overexpression of factors that are important for ESC lineages could convert somatic cells into an ESC fate, inducing pluripotency and immortality (Takahashi and Yamanaka 2016). They selected and screened 24 candidates before identifying four factors that were sufficient to induce pluripotency in human and mouse adult fibroblasts (Takahashi and Yamanaka 2006; Takahashi et al. 2007). These four transcription factors – POU5F1 (originally known as Oct-3/4), SOX2, KLF4, and MYC – are known as the "OSKM" or "Yamanaka" factors. Another set of factors – POU5F1, SOX2, NANOG, and LIN28A – are known as the "OSNL" factors and were developed by James Thomson's group to induce pluripotency in human somatic cells (Yu et al. 2007).

Thousands of iPSCs have now been generated using these reprogramming factors (Beers et al. 2015; Takahashi and Yamanaka 2016), but their delivery to the cell continues to be challenging. Retroviral vectors were used to generate the first iPSCs, resulting in high efficiency but some associated safety concerns (Okita and Yamanaka 2011). Retroviral vectors result in genomic integration, with associated off-target

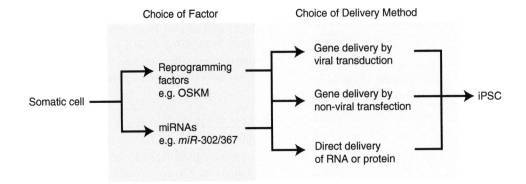

Fig. 23.4. Induced pluripotent stem cells (iPSCs). Overview of approaches to induce pluripotency in somatic human cells. *Source*: Rony, I. K., Baten, A., Bloomfield, J. A., et al. (2015). Inducing pluripotency in vitro: recent advances and highlights in induced pluripotent stem cells generation and pluripotency reprogramming. Cell Prolif 48 (2):140–56.© John Wiley & Sons.

effects (see Sections 22.4.4, 22.4.5). Early studies showed that as many as 20% of iPSC chimeras developed tumors, due to reactivation of the *Myc* transgene (Okita et al. 2007). This led to a search for delivery methods that could address safety concerns but also offer suitable levels of efficiency. Delivery methods can be broadly grouped into (see Figure 23.4):

(1) **Gene delivery by viral transduction.** Integrating viral vectors (e.g. γ-retrovirus or lentivirus) continue to be widely used for research applications. Cre-mediated excision can be performed to remove the reprogramming vector; this technique can be used to generate iPSCs from a wide variety of biospecimens, e.g. from 4-ml blood samples (Sommer et al. 2012). Alternatively, non-integrating vectors from adenovirus or Sendai virus can be used to reduce the risk of genomic integration (Stadtfeld et al. 2008; Fusaki et al. 2009; Ono et al. 2012). Sendai viral vectors are non-pathogenic in humans (although they are pathogenic in mice), have a relatively high efficiency, can be used with various sample types (including blood samples and dermal fibroblasts), and can be introduced using a minimal number of steps, making them a popular choice for high-throughput applications (Nakanishi and Otsu 2012; Beers et al. 2015).

(2) **Gene delivery by non-viral transfection.** Repeated plasmid transfections have been used to generate iPSCs from somatic mouse cells, but this approach results in low transfection efficiency, particularly in human cells (estimated at <0.0002%) (Okita et al. 2007; Okita and Yamanaka 2011). Other approaches can be used to improve efficiency and reduce off-target effects. For example, polycistronic vectors and transposon-based systems have been designed that give robust but transient expression of factors in the absence of virus (Kaji et al. 2009; Woltjen et al. 2009; Davis et al. 2013). Such molecular approaches are beyond the scope of this book and are reviewed elsewhere (Sommer and Mostoslavsky 2010; Rony et al. 2015).

(3) **RNA or protein delivery.** Reprogramming factors can be introduced directly as peptides (fused with a cell-penetrating peptide) or as synthetic mRNAs (with modifications to promote their half-life and efficient translation) (Kim et al. 2009; Warren et al. 2010). Intriguingly, this work has opened up new avenues to study the molecular pathways involved in pluripotency. Some microRNAs (miRNAs) – small, non-coding RNAs that can modulate mRNA expression – are highly expressed in ESCs. Delivery of *miR-302/367* or other miRNAs can induce pluripotency without the need for additional reprogramming factors, resulting in an alternative approach to iPSC generation (Sridharan and Plath 2011; Sandmaier and Telugu 2015).

Techniques to induce pluripotency can be highly specialized and the newcomer to stem cell culture may feel uncertain as to how to proceed. Start by looking for existing iPSC lines that may meet your needs within registries (e.g. hPSCreg) and cell repository catalogues (see Table 15.3) (Seltmann et al. 2016). Some core facilities generate iPSCs as a service, as do commercial suppliers (e.g. Cellaria and REPROCELL). Services may seem expensive, but remember that labor costs make up a large part of the overall cost when generating iPSCs (Beers et al. 2015). It is wise to estimate the overall cost of generating iPSCs in-house before deciding if a service is good value for money. If you decide to generate your own iPSCs, always check the literature to determine which approaches are reported to succeed for your cell type. Reprogramming vectors can be obtained from the Addgene plasmid repository (Addgene 2019), with reprogramming kits available from some commercial suppliers, e.g. the Simplicon™ RNA Reprogramming Kit (Merck Millipore) or the CytoTune™-iPS Sendai Reprogramming Kit (ThermoFisher Scientific). Protocol P23.1 uses a kit to generate iPSCs from BJ cells (human fibroblasts) and is used to illustrate the cell culture requirements for iPSC generation.

PROTOCOL P23.1. GENERATION OF IPSCS USING SENDAI VIRAL VECTORS

Contributed by Amanda Capes-Davis, adapted from the CytoTune-iPS 2.0 Sendai Reprogramming Kit User Guide (ThermoFisher Scientific) with modifications from other resources (Beers et al. 2015; Kasai-Brunswick et al. 2018).

Background

This kit introduces three Sendai viral vectors, carrying sequence for the four OSKM factors. BJ human fibroblasts are used here as a positive control for the procedure; the manufacturer states that the reprogramming efficiency is 0.02–1.2% for this kit and cell line. The procedure can be modified for other cell types, feeder-free culture, or high-throughput applications. Further information can be found in the manufacturer's instructions and the cited references.

Outline

Subculture human fibroblasts (BJ cells) into a 6-well plate. Refeed with medium containing the three Sendai viral-vectors. Incubate cells for seven days; while this is in progress, prepare iPSC medium and mouse embryonic fibroblast (MEF) feeder layers. Transfer cells to a fresh 6-well plate containing MEF feeder cells and wait for iPSC colonies to emerge. Select colonies for expansion and cryopreservation.

❖ *Safety Note. Use a biosafety cabinet (BSC) and personal protective equipment (PPE) for all procedures using cells or vectors. Sendai virus is a respiratory pathogen in mice. Vectors have been modified to address safety concerns, e.g. to remove their capacity to produce infectious virions. However, care should be taken to avoid any exposure, particularly through aerosols or mucosal membranes.*

Materials (sterile or aseptically prepared)

☐ CytoTune-iPS 2.0 Sendai Reprogramming Kit (ThermoFisher Scientific) containing KOS, hc-Myc, and hKlf4 vectors (stored at −80 °C)
☐ BJ human fibroblasts (ATCC #CRL-2522; see Table 15.3)
☐ MEFs, frozen vial of inactivated cells ready for use as feeder layer (see Protocol P20.7)
☐ Fibroblast medium (see step 2 for constituents)
☐ iPSC medium (see step 14 for constituents)
☐ Attachment Factor (ThermoFisher Scientific #S006100, containing 0.1% gelatin)

☐ 6-well plates
☐ Medium bottles (100 ml)
☐ Centrifuge tubes

Materials (non-sterile)

☐ BSC and associated consumables (see Protocol P12.1)
☐ Cell counting equipment (hemocytometer or automated counter; see Section 19.1)

Procedure

A. Before Transduction

1. Plan your experiment to ensure that you have the correct number of cells and all reagents are available. One kit is sufficient for a minimum of five wells in a 6-well plate, but the procedure can be modified to run a greater number of samples at lower volume (see Notes).
2. Prepare fibroblast medium (volumes here are for 100 ml total):
   (a) DMEM: 89 ml
   (b) Fetal bovine serum (FBS): 10 ml
   (c) MEM non-essential amino acid solution (100x): 1 ml
   (d) β-Mercaptoethanol (55 mM): 100 µl
3. Two days before transduction:
   (a) Subculture cells using trypsin/EDTA (see Protocol P14.2).
   (b) Count the cells with a hemocytometer or an automated cell counter.
   (c) Plate an equal number of cells into each well of a 6-well plate, so that they are approximately 50–80% confluent on the day of transfection.
4. Incubate for two days using standard culture conditions, e.g. 37 °C and 5% CO_2.

B. Day of Transduction

5. Check the cells on the plate and decide if they have reached the appropriate confluence to proceed with transduction.
6. Prewarm fibroblast medium and prepare a work space for aseptic technique using a BSC.
7. Trypsinize the cells from one well and count as per step 3b. This well is used to determine the cell count for each well in the plate; cells are discarded after counting.
8. Calculate the volume to be added for each viral vector, based on the desired multiplicity of infection (MOI) (see Notes), the viral titer, and

the cell count using the formula:

$$\text{Volume of virus } (\mu L) = \frac{\text{MOI (CIU per cell)}}{\text{Viral titer(CIU per } \mu L)} \\ \times \text{number of cells}$$

9. Thaw the tubes containing the three Sendai virus vectors by briefly immersing in a 37 °C water bath for 5–10 seconds, followed by thawing at room temperature. Place thawed tubes on ice.
10. Aliquot fibroblast medium (1 ml per well) into a centrifuge tube and add the required volume of each vector to the medium. Gently mix by pipetting.
11. Refeed the cells in each well using the medium containing the viral vectors.

C. After Transduction

12. One day after transduction, remove the medium containing the viral vectors and refeed with fresh fibroblast medium.
13. Continue to refeed transduced cells with fresh fibroblast medium every second day.
14. Prepare iPSC medium (volumes here are for 100 ml total):
    (a) DMEM/F12 with GlutaMAX™ supplement (see Section 9.4.1): 78 ml
    (b) KnockOut™ serum replacement (see Section 10.5): 20 ml
    (c) MEM non-essential amino acids (10 mM): 1 ml
    (d) β-Mercaptoethanol (55 mM): 100 μl
    (e) FGF-2 (10 μg/ml, added just before use): 40 μl
15. Prepare a feeder layer using a separate (fresh) 6-well plate:
    (a) Coat the wells in the plate using Attachment Factor and incubate for 30 minutes at 37 °C or for two hours at room temperature. Aspirate the remaining liquid. Plates may be used immediately or stored at room temperature for up to 24 hours.
    (b) Thaw a frozen vial of inactivated MEFs (e.g. see Protocol P15.2).
    (c) Count the cells with a hemocytometer or an automated counter.
    (d) Resuspend MEFs in fibroblast medium to give 2.5×10^5 cells/ml.
    (e) Add MEF suspension (1 ml per well) to the wells in the plate. Add additional medium to give a total volume of 2 ml per well.

(f) Incubate using standard culture conditions, e.g. 37 °C and 5% CO_2. Feeder layers should be used within three to four days after plating.
16. Seven days after transduction, transfer transduced cells onto MEF feeder layers:
    (a) Trypsinize the cells using trypsin/EDTA or an alternative procedure (see Notes).
    (b) Count the cells with a hemocytometer or an automated cell counter.
    (c) Resuspend the cells in fibroblast medium and plate onto MEF feeder cells at 2×10^4 to 1×10^5 cells per well.
17. Eight days after transduction, remove the fibroblast medium and refeed with iPSC medium.
18. On each day following, refeed with fresh iPSC medium and observe the plates, looking for the emergence of iPSC colonies.
19. Once colonies have emerged, allow them to increase in size until they are ready for manual selection. A marker pen under the plate can be used to label colonies as they emerge.
20. Select colonies for further expansion, cryopreservation, and characterization.

Notes

Step 1: For large-scale studies, a single kit may be used to reprogram 24–48 samples using a 48-well plate (see Table 8.2) (Beers et al. 2015).

Step 2: The choice of medium will vary with your cell type. iPSCs have been generated from peripheral blood mononuclear cells (PBMCs) using an erythroblast proliferation culture medium (QBSF-60 Stem Cell Medium, Quality Biological) supplemented with SCF (50 ng/ml), IL-3 (10 ng/ml), IGF-1 (40 ng/ml), ascorbic acid (50 μg/ml), erythropoietin (EPO; 2 U/ml), dexamethasone (1 μM), and antibiotics (Sommer et al. 2012; Kasai-Brunswick et al. 2018).

Steps 3–5: The manufacturer recommends using $2–3 \times 10^5$ cells per well on the day of transfection, which should translate to 50–80% confluence (although this will vary with the cell type). This may be modified for high-throughput applications to streamline workflows, e.g. plating 1×10^4 cells per well in a 48-well plate for transduction two hours after plating (Beers et al. 2015).

Step 8: The manufacturer recommends that a suitable starting MOI for each vector is 5 (for KOS), 5 (for hc-Myc), and 3 (for hKlf4). However, MOI can be optimized; a lower MOI will increase the possible number of experiments per kit. Viral titers will vary from lot to lot and will be listed on the Certificate of

Analysis. For this formula, the viral titer is expressed in CIU per microliter.

Steps 14 and 15: If feeder-free conditions are required, cells can be maintained in Matrigel®-coated dishes using Essential 8 medium and subcultured using EDTA/PBS (Beers et al. 2015). Sendai reprogramming is commonly performed feeder-free, particularly for high-throughput applications.

Step 16: Instead of trypsinization, colonies may be picked manually. From this point, cells are more sensitive to dissociation into single cells or to prolonged trypsinization. Gentler enzymatic techniques may be used instead of trypsin, e.g. Accutase (Innovative Cell Technologies).

Step 19: BJ cells normally produce iPSC colonies within two weeks of transduction, but this will vary with the cell type. Allow up to four weeks for colonies to form. Colonies tend to appear in waves. The first wave is likely to be karyologically abnormal and is typically ignored in favor of the second wave. It is a good idea to mark colonies to indicate their order of appearance, e.g. using a marker pen underneath the plate.

Step 20: Procedures for subculture, cryopreservation, and characterization are described elsewhere (see Section 23.4). Culture and screening of colonies are usually the most labor-intensive parts of iPSC generation.

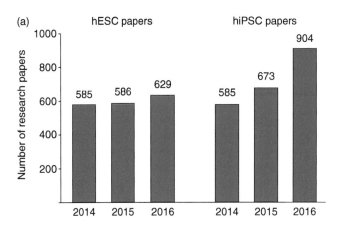

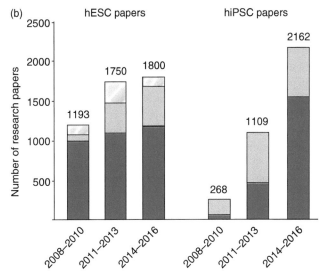

Fig. 23.5. Trends in human pluripotent stem cell (hPSC) research. Research papers that reported experimental use of hESCs (left) or human iPSCs (right): (a) from 2014 to 2016; (b) from 2008 to 2016, grouped into three-year periods. Papers that used both hESCs and human iPSCs in the same study are shown in light gray. Papers that used hESCs for comparison to verify certain characteristics of iPSCs ("gold standard" usage) are indicated by the gray-white striped lines. *Source*: Guhr, A., Kobold, S., Seltmann, S., et al. (2018). Recent Trends in Research with Human Pluripotent Stem Cells: Impact of Research and Use of Cell Lines in Experimental Research and Clinical Trials. Stem Cell Reports 11 (2):485–496. © 2018 Elsevier.

23.4 HUMAN PLURIPOTENT STEM CELL (HPSC) LINES

23.4.1 Evolution in Culture of hPSCs

The successful culture of hESCs – and, more recently, the reprogramming of iPSCs from adult human cells – led to an upsurge in popularity for hPSC lines, reagents, and techniques (see Figure 23.5) (Guhr et al. 2018). hESC and human iPSC lines display many similar features and can be grown using the same reagents and techniques (Kramer et al. 2016). However, these cell lines are also highly sensitive to their growth conditions, resulting in a number of problems for tissue culture laboratories that are new to their requirements. Problems can be broadly grouped into three different areas:

(1) *Spontaneous differentiation* (see Section 23.4.3). Maintenance of pluripotency is heavily dependent on maintaining the correct culture environment, including the substrate on which cells are grown, the medium used, the level of confluence, and the time in culture. Changes in morphology may indicate that the culture environment is suboptimal, resulting in differentiation or poor health. It is essential to monitor hPSC morphology

and assess the level of spontaneous differentiation within the culture (Healy and Ruban 2015; EBiSC 2017).

(2) *Genetic instability* (see Section 3.3). hPSC lines initially display a diploid karyotype but acquire genetic and epigenetic variation with further culture (Draper et al. 2004; Spits et al. 2008; Kyriakides et al. 2018). Analysis of more than 1700 hPSC lines found that 12–13% had an abnormal karyotype, with hESCs and iPSCs displaying similar abnormalities (Taapken et al. 2011). While some changes appeared to be random, others resulted in upregulation of genes associated with pluripotency

and the cell cycle, resulting in a growth advantage and the potential for clonal evolution (Mayshar et al. 2010; International Stem Cell Initiative et al. 2011).

(3) *Apoptosis* (see Section 2.5). As noted previously, hESCs are prone to apoptosis when dispersed into single cell suspensions (see Section 23.2.2); the same is true for human iPSCs. To improve survival, established hPSC lines are typically passaged using methods that retain cells in aggregate form. Mechanical methods or collagenase IV are commonly used for culture on feeder layers, while Dispase or EDTA are preferred for feeder-free culture (see Table 14.4; see also Section 23.4.3).

Why are hPSC lines so prone to genetic instability and apoptosis? Human embryonal carcinoma cells and hESCs appear to respond to replication stress by inducing cell apoptosis instead of DNA repair, due to a failure to block entry into S phase (see Section 2.3.2) (Desmarais et al. 2012). This results in an increased rate of cell death under stress and to the potential accumulation of genetic defects if apoptosis is disrupted. By contrast, cell survival is heavily dependent on cell–cell contact (Chen et al. 2010). Loss of direct contact results in actin–myosin contraction, membrane blebbing, and rapid onset of cell death. Interestingly, Rho kinase (ROCK) inhibitors are commonly used as a "survival factor" for dissociated iPSCs (see Section 23.4.4) (Watanabe et al. 2007; Kurosawa 2012). ROCK inhibitors directly affect myosin binding and disrupt actin–myosin contraction, which suggests that they act through the cytoskeleton to ameliorate the loss of cell contact under these circumstances (Chen et al. 2010).

23.4.2 Culture Conditions

Culture using serum and serum replacements. hESCs were originally isolated using serum-containing media on feeder layers (Thomson et al. 1998). Serum can induce differentiation in hPSC lines and has a number of other disadvantages (see Section 10.1), making it best avoided for today's hPSC cultures, although it continues to be used in some stem cell differentiation protocols (Lin and Talbot 2011; Pamies et al. 2017). Serum replacements are now widely available and can be directly substituted for serum (see Table 10.2). These products are likely to be more consistent in their performance and many have been optimized for stem cell culture, e.g. KnockOut Serum Replacement (ThermoFisher Scientific; see Section 10.5).

Culture on feeder layers. hPSC lines continue to be maintained on feeder layers in many research laboratories (Healy and Ruban 2015). Feeder layers were originally used to maintain pluripotency and were associated with less genomic instability compared to feeder-free systems (Draper et al. 2004; Skottman and Hovatta 2006). Feeder-free growth conditions have been greatly improved since these early studies, but their effectiveness can vary considerably (International Stem Cell Initiative Consortium et al. 2010). Research laboratories continue to use feeder layers because (i) they offer a reliable and effective way to maintain pluripotency; (ii) staff have already developed the necessary expertise; (iii) feeder cells have already been stored and are available for use; or (iv) feeder-based systems are cheaper compared to defined conditions, which may require expensive supplements. The latter reason is likely to become less compelling as the demand increases for defined reagents, resulting in lower costs (see Section 10.1.1). Procedures for thawing, subculture, and cryopreservation of hPSC lines on feeder layers are described elsewhere (Borowski et al. 2012).

Feeder layers commonly consist of MEFs that have been inactivated by irradiation or treatment with mitomycin C (see Protocol P20.7). Although effective, their effects are poorly defined and variable; MEF cultures may also contain undetected pathogens (see Sections 6.3.1, 10.7). Mouse feeder layers can be replaced by human cells (e.g. fibroblasts from human foreskin), but some human fibroblast cultures are less effective in supporting hPSC growth compared to MEFs (Richards et al. 2003; Skottman and Hovatta 2006). If hPSCs are to be used in clinical applications, feeder cells must be derived using Good Manufacturing Practice (GMP) (see Section 7.2.3) (Prathalingam et al. 2012).

Culture using defined conditions. Various defined media formulations have been optimized for hPSC culture by laboratories with this expertise, allowing removal of animal components and increased standardization. A defined substrate and method of dissociation is usually recommended for use with each formulation, giving a defined culture system (see Table 23.4). It is unwise to "mix and match" the components in various systems unless you have the expertise to do so effectively. For example, culture in NutriStem hPSC XF on a coating of laminin-521 is reported to provide protection against apoptosis when hPSCs are dissociated into individual cells (Rodin et al. 2014). Enzymatic methods can be used under those conditions that would result in a low cell survival for other substrates, where cells should be passaged in aggregate form.

To discuss hPSC handling procedures, we will focus on one example from a number of possible culture systems (see Table 23.4). E8 medium is used here for purposes of illustration because it is chemically defined, can be prepared in-house or purchased from commercial suppliers, and can support long-term maintenance of stem cell lines (International Stem Cell Initiative Consortium et al. 2010). E8 was developed from TeSR, which was in turn adapted from DMEM/F12 to optimize proliferation of undifferentiated hESCs (see Section 10.4) (Ludwig et al. 2006). TeSR medium contains DMEM/F12 and 18 additional supplements, including albumin, which is not chemically defined. E8 medium consists of DMEM/F12 and seven additional constituents (i.e. "Essential 8" or "E8"; see Appendix C), all of which are fully defined (Chen et al. 2011).

TABLE 23.4. Defined culture systems for hPSCs.

Complete medium[a]	Basal formulation[b]	Substrate treatment[c]	Dissociation method	References
E8 (Essential 8, ThermoFisher Scientific; TeSR-E8™, STEMCELL Technologies)	DMEM/F12	Vitronectin	EDTA	Chen et al. (2011); Beers et al. (2012)
hESF9	ESF	Collagen type I, fibronectin, laminin, vitronectin	PBS with Ca^{2+} and without Mg^{2+}	Furue et al. (2008); Ohnuma et al. (2014)
NutriStem hPSC XF (Biological Industries, Corning)	Proprietary information	Laminin-521	Scraping, trypsin/EDTA, TrypLE Select, Gentle Cell Dissociation Reagent	Rodin et al. (2014)
PluriSTEM (Merck Millipore)	DMEM/F12	Matrigel	Scraping, Dispase	Frank et al. (2012)
StemPro® (ThermoFisher Scientific)	DMEM/F12	Matrigel, Geltrex™, CELLstart™	Scraping, collagenase type IV	Wang et al. (2007); Swistowski et al. (2009)
TeSR (mTeSR™1 and mTeSR2, STEMCELL Technologies)	DMEM/F12	Matrigel, collagen IV + fibronectin + laminin + vitronectin	Dispase, EDTA	Ludwig et al. (2006); Beers et al. (2012)

Source: Amanda Capes-Davis.

[a] This table lists examples of defined culture systems that have been used to maintain undifferentiated growth of hPSCs. Due to the large number of publications and suppliers in this area, it is not possible to provide a comprehensive list. A more extensive list of serum-free media is provided in an earlier chapter (see Table 10.3).

[b] A basal formulation provides a starting point before supplements are added to give the complete medium. If you are making media in-house, please refer to the references in the final column for information on the complete formulation. Note that some commercial media are proprietary and thus those formulations are not available.

[c] Treatments are typically used as coatings prior to the cells being added. The treatments listed here are those reported in the cited references. The plus sign indicates that components are used as a mixture. A more extensive list of substrates and their suppliers is provided in an earlier chapter (see Table 8.1).

Always remember that hPSC lines may vary in their behavior under different culture conditions. If cells have adapted to a particular set of culture conditions, there is no guarantee that they will maintain an undifferentiated phenotype under new conditions or that they will remain responsive to differentiation using a particular procedure. Before any changes are made, always perform cryopreservation using the original culture conditions so that stocks are available in case the new conditions prove to be unsuitable. Cultures are likely to require a gradual process of adaptation over several passages.

23.4.3 Feeding and Subculture

hPSC cultures should be observed on a daily basis, comparing their appearance under the microscope to previous images or an atlas of hPSC morphology (see Figures 23.6, 23.7) (Healy and Ruban 2015; EBiSC 2017). A scoring system is useful to give a consistent approach when examining the health of an hPSC line and looking for differentiated cells (EBiSC 2017). When refeeding, medium should be mostly removed (leaving a thin film of medium covering the cell layer) and then replaced by gently adding medium to the side of the well (EBiSC 2017). Typically, hPSC cultures need to be refed on a daily basis due to a high sensitivity to acidification of the culture medium (Wilmes et al. 2017). Some protocols and media formulations have been developed that can reduce the need for weekend feeding, e.g. by refeeding cells with an increased volume of medium or growing cells in a restricted area such as a coated coverslip (EBiSC 2017; Wilmes et al. 2017).

PROTOCOL P23.2. SUBCULTURE IN CHEMICALLY DEFINED CONDITIONS

Contributed by European Bank for Induced Pluripotent Stem Cells (EBiSC).

Background

The protocol provided here has been adapted from EBiSC's Protocol for the Use of Induced Pluripotent Stem Cells (De Sousa et al. 2017; EBiSC 2017). The full Protocol describes several approaches using E8 (with vitronectin) or mTeSR (with Matrigel or Geltrex), with or without ROCK inhibitor. A single approach is used here for illustration; please refer to the cited protocol for more information on other approaches (EBiSC 2017). Training videos are available on the EBiSC website (www.EBiSC.org) that show the various procedures (EBiSC 2019).

Outline

Prepare vitronectin coating in a 6-well plate. Wash the cells and add EDTA. After incubation at room temperature, remove the EDTA and gently wash the cells with E8 medium to detach cell clusters. Transfer detached clusters to the fresh vitronectin-coated plate.

Materials (sterile or aseptically prepared)

☐ iPSC line
☐ E8 medium (e.g. Essential 8™, ThermoFisher Scientific; TeSR™-E8™, STEMCELL Technologies), prepared according to the manufacturer's instructions
☐ Dulbecco's phosphate buffered saline without Ca^{2+} and Mg^{2+} (DPBS-A)
☐ EDTA 0.5 mM working solution (see Notes)
☐ Recombinant human vitronectin 500 μg/ml (e.g. VTN-N #A14700, ThermoFisher Scientific; Vitronectin XF™, STEMCELL Technologies)
☐ 6-well plate
☐ Centrifuge or sample tubes

Materials (non-sterile)

☐ BSC and associated consumables (see Protocol P12.1)

Procedure

A. Before Subculture

1. Check the culture and decide if it is suitable to proceed, e.g. if cells look healthy, retain a characteristic hPSC morphology, and are at 70–80% confluence (see Figures 23.6, 23.7).
2. Prepare a work space for aseptic technique using a BSC and allow the medium, DPBS-A, and EDTA to equilibrate to room temperature.
3. Prepare vitronectin-coated plates:
   (a) Prepare aliquots of vitronectin by dispensing into 60-μl aliquots for storage at –80 °C.
   (b) Thaw a 60-μl aliquot of vitronectin (one aliquot per 6-well plate).
   (c) Dilute the vitronectin 1:100 by adding 60 μl to 6 ml of room temperature DPBS-A.
   (d) Add 1 ml of the diluted vitronectin to each well. Incubate at room temperature for one hour.
   (e) Remove the residual liquid and discard. Once coated, plates may be sealed and stored at 5 °C for up to three days.
   (f) Before use, add medium to the required wells.

B. Subculture

4. Remove the existing spent medium from each vessel or well requiring passage.

5. Add 1 ml DPBS-A per well and aspirate the liquid. Repeat this wash step.
6. Add 1 ml 0.5 mM EDTA per well and incubate at room temperature for approximately five minutes.
7. On completion of the incubation time, observe the colonies on the plate, looking for evidence of feathered edges and holes that indicate cells have detached from each other (but not from the culture vessel).
8. Gently tilt the plate containing EDTA so that the liquid collects at the bottom of the well and remove the EDTA.
9. Immediately after removal of EDTA, add 1 ml medium per well to neutralize the EDTA. Use this medium to gently wash the surface of the well to remove the cell clusters. Two or three gentle aspirations should be sufficient.
10. Transfer the cell suspension to a fresh vitronectin-coated 6-well plate at the required split ratio (see Section 14.2.2).
11. Gently move the plate from side to side and back and forth to distribute the cell clusters in each well. Incubate using standard culture conditions, e.g. 37 °C and 5% CO_2.

Notes

Step 1: If the cell line has come from a cell repository, refer to their instructions for handling and the optimal split ratio for that cell line. Established iPSC lines should be passaged during the exponential phase of growth (see Figure 14.1), using a split ratio that allows subculture every four to five days; typically this translates to a split ratio of 1:4 to 1:8 (EBiSC 2017).

Step 5: Wash steps are necessary to remove residual Ca^{2+} and Mg^{2+} from the medium.

Step 6: The working solution of EDTA is diluted from 0.5 M stock using DPBS-A and filtered; it may be stored at 4 °C for up to six months. The osmolarity of the EDTA solution can be adjusted to 340 mOsm using NaCl to give a similar osmolarity to E8 medium, which helps to reduce damage during dissociation (Liu and Chen 2014). The amount of time required for dissociation from the plate should be optimized for each cell line.

Step 9: Aim to dislodge the cell clusters but not generate a single cell suspension, which would reduce cell survival. It is better to be gentle and leave a few cells behind. Addition of Ca^{2+} and Mg^{2+} will result in the colonies starting to reattach, so proceed promptly to the next step.

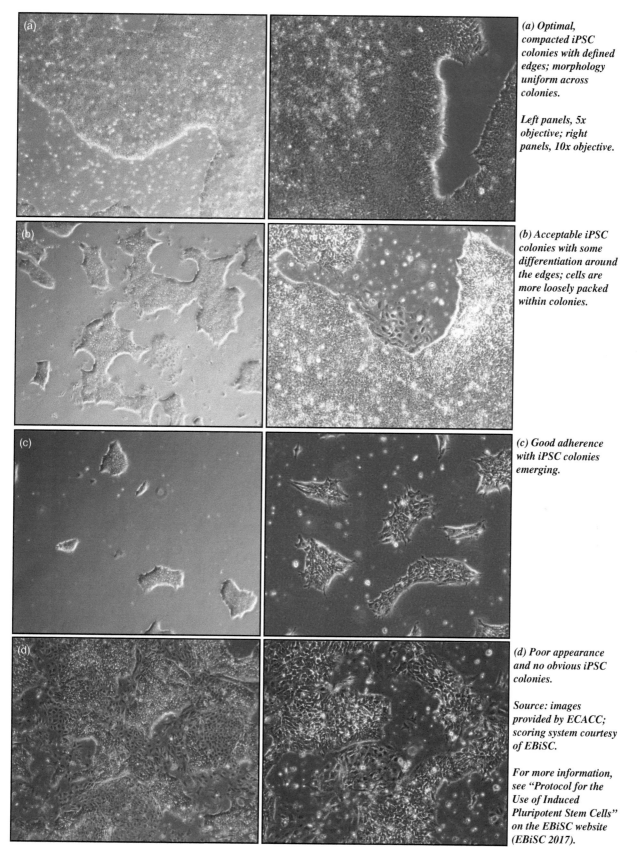

(a) Optimal, compacted iPSC colonies with defined edges; morphology uniform across colonies.

Left panels, 5x objective; right panels, 10x objective.

(b) Acceptable iPSC colonies with some differentiation around the edges; cells are more loosely packed within colonies.

(c) Good adherence with iPSC colonies emerging.

(d) Poor appearance and no obvious iPSC colonies.

Source: images provided by ECACC; scoring system courtesy of EBiSC.

For more information, see "Protocol for the Use of Induced Pluripotent Stem Cells" on the EBiSC website (EBiSC 2017).

Fig. 23.6. Morphology of induced pluripotent stem cells (iPSCs).

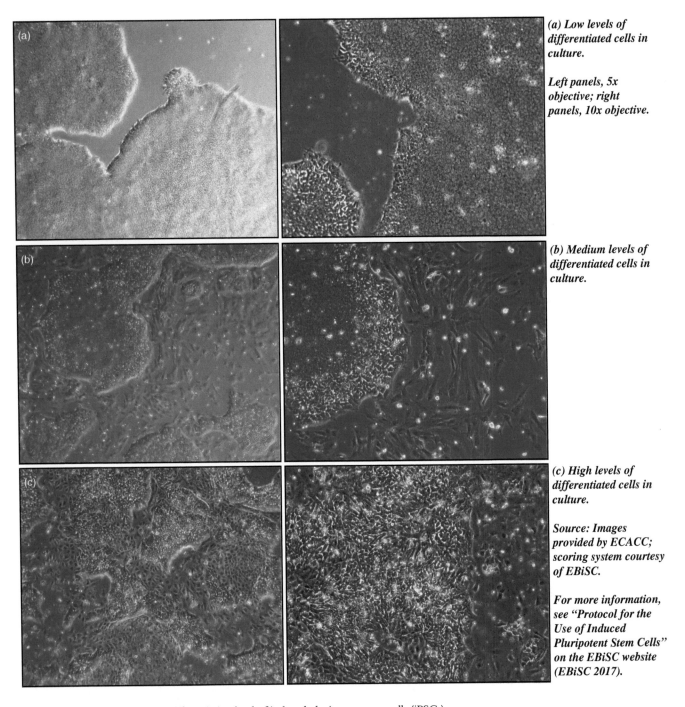

(a) Low levels of differentiated cells in culture.

Left panels, 5x objective; right panels, 10x objective.

(b) Medium levels of differentiated cells in culture.

(c) High levels of differentiated cells in culture.

Source: Images provided by ECACC; scoring system courtesy of EBiSC.

For more information, see "Protocol for the Use of Induced Pluripotent Stem Cells" on the EBiSC website (EBiSC 2017).

Fig. 23.7. Differentiation level of induced pluripotent stem cells (iPSCs).

23.4.4 Cryopreservation and Thawing

Early hESC lines were frozen using vitrification, which is widely used for embryos and oocytes (see Section 15.1.4). Vitrification results in good viability but has some disadvantages; the procedure is difficult to scale up and the freezing medium used can be harmful for hPSCs (Miyazaki and Suemori 2016). Cryopreservation of HSCs can be performed using a controlled cooling rate of −1 °C/minute (see Section 15.1.2), so a similar approach was adopted for hPSCs (Hanna and Hubel 2009). This initially resulted in poor viability results for many reasons, including (i) apoptosis when hPSCs are dissociated as individual cells; (ii) variable permeability to cryoprotectant when hPSCs are frozen as cellular aggregates; (iii) toxicity when dimethyl sulfoxide (DMSO) is used for hPSCs or for cell-based therapy (see Section 15.1.1); (iv) the presence of serum in freezing medium; and (v) lack of standardization when measuring hPSC viability.

Optimization of culture conditions and procedures will help to improve viability following cryopreservation. For example, cell survival can be improved if cells are passaged two to three days after the last subculture, compared to five to six days afterwards (Liu and Chen 2014). Other variables that may be optimized include the method of dissociation, the composition of the freezing medium (including the level of DMSO and choice of medium), the cooling rate, and the seeding temperature for ice crystal formation (see Section 15.1.2) (Liu and Chen 2014; Li et al. 2018). If toxicity from DMSO is a concern, it may be possible to substitute with another cryoprotectant such as ethylene glycol, propylene glycol, glycerol, or trehalose (Katkov et al. 2011; Ntai et al. 2018).

Cell viability and growth can also be improved through the addition of ROCK inhibitors (e.g. Y-27632), which reduce apoptosis when aggregates are dissociated into individual cells under serum-free and feeder-free conditions (Watanabe et al. 2007; Martin-Ibanez et al. 2008; Claassen et al. 2009). ROCK inhibitors can increase survival and plating efficiency of hPSCs when used in various handling procedures, including subculture and cloning (Kurosawa 2012). During cryopreservation, ROCK inhibitors can be added to the freezing medium as described in Protocol P23.3, and to the growth medium immediately after thawing. However, addition of ROCK inhibitors may have unexpected consequences. Treatment with Y-27632 has led to selective differentiation of iPSCs into mesendodermal lineages, in a process that appears similar to epithelial–mesenchymal transition (EMT) (see Section 26.2.5) (Maldonado et al. 2016). Use of ROCK inhibitors should be temporary (e.g. the first 24 hours after thawing or just before and after cell dissociation) and should be avoided for routine culture.

PROTOCOL P23.3. CRYOPRESERVATION USING ROCK INHIBITOR

Contributed by the European Bank for Induced Pluripotent Stem Cells (EBiSC).

Background

The protocol provided here has been adapted from EBiSC's Protocol for the Use of Induced Pluripotent Stem Cells (De Sousa et al. 2017; EBiSC 2017). The full Protocol describes several approaches using E8 (with vitronectin) or mTeSR (with Matrigel or Geltrex), with or without ROCK inhibitor. A single approach is used here for illustration; please refer to the cited protocol for more information on other approaches (EBiSC 2017). Training videos are

available on the EBiSC website (www.EBiSC.org) that show the various procedures (EBiSC 2019).

Outline

Subculture cells two to three days before cryopreservation to give 70–80% confluence. Detach cells from the plate using EDTA and wash with freezing medium to detach cell clusters. Transfer to a cryovial and freeze slowly using a controlled cooling device.

❖ *Safety Note. Y-27632 is a hazardous substance; freezing medium contains DMSO, which may act as a carrier for hazardous substances and is rapidly absorbed through skin (see Section 6.2.2). Refer to a Safety Data Sheet for PPE and other precautions. All personnel working with liquid nitrogen should be trained to do so safely and wear suitable PPE (see Sections 6.2.3, 6.2.4).*

Materials (sterile or aseptically prepared)

- iPSC line
- E8 medium (e.g. Essential 8, ThermoFisher Scientific; TeSR-E8™, STEMCELL Technologies), prepared according to the manufacturer's instructions
- CryoStor® CS10 cryopreservation medium (e.g. #C2874, Sigma-Aldrich), prechilled
- Y-27632 10 mM solution (e.g. #1254 Tocris, #72302 STEMCELL Technologies)
- DPBS-A
- EDTA 0.5 mM working solution (see Protocol P23.2)
- Cryovials
- Centrifuge tubes

Materials (non-sterile)

- BSC and associated consumables (see Protocol P12.1)
- PPE (see Safety Note)
- Controlled cooling device, e.g. Mr. Frosty™ (ThermoFisher Scientific; see Section 15.2.2)
- Cryofreezer and associated database (see Section 15.7)

Procedure

1. Passage hPSCs two to three days before cryopreservation using EDTA into a 6-well plate coated with vitronectin in accordance with the previous protocol (see Protocol P23.2).

2. On the day of cryopreservation, check the culture and decide if it is suitable to proceed, e.g. if cells look healthy, retain a characteristic hPSC morphology, and are at 70–80% confluence (see Figures 23.6, 23.7).

3. Prepare a work space for aseptic technique using a BSC and allow the medium, DPBS-A, and EDTA to equilibrate to room temperature. Other reagents and equipment (e.g. CryoStor and Mr. Frosty) should be kept chilled (5 °C ± 3 °C).

4. Prepare freezing medium by adding ROCK inhibitor to the CryoStor cryopreservation medium, giving a final concentration of 10 µM (1 µl 10 mM Y-27632 solution in 1 ml CryoStor). Prepare sufficient freezing medium to give 1 ml per cryovial.

5. Remove the existing medium from the 6-well plate containing the iPSC line.

6. Add 1 ml DPBS-A per well and aspirate the liquid. Repeat this wash step.

7. Add 1 ml 0.5 mM EDTA per well and incubate at room temperature for approximately five minutes, observing the cells for evidence that they have detached from each other (see Protocol P23.2).

8. Gently tilt the plate so that the liquid collects at the bottom of the well and remove the EDTA.

9. Immediately after removal of EDTA, add 1 ml freezing medium per well. Use the freezing medium to gently wash the surface of the well to remove the cell clusters. Do not aspirate more than three times to avoid breaking the cell clusters into individual cells.

10. Transfer 1 ml of freezing medium containing cell clusters into a labeled cryovial.

11. Transfer cryovials to the controlled cooling device and place at –80 °C overnight.

12. On the day after cryopreservation, transfer frozen vials to the cryofreezer and update the cryofreezer database (see Protocol P15.1, steps 13–17).

Notes

Step 1: If the cell line has come from a cell repository, refer to their instructions for timing and cell confluence at cryopreservation. Subculture procedures are important to achieve good viability and may require optimization (Liu and Chen 2014). Cells should be in the logarithmic phase of growth and at the appropriate confluence (usually 70–80%).

Step 4: The choice of freezing medium will depend on your requirements (e.g. ease of use) and later applications. Options include (i) a commercial freezing medium as used here; (ii) a serum-based freezing medium (e.g. 50% medium, 40% FBS, and 10% DMSO); or (iii) a defined freezing medium (e.g. 90% E8 medium and 10% DMSO) (Liu and Chen 2014; EBiSC 2017). ROCK inhibitor is optional and is used here for illustration. To prepare the stock solution, resuspend Y-27632 dihydrochloride in sterile DPBS-A or water to give a 10-mM concentration. Aliquots of Y-27632 may be stored shielded from light at –20 °C for up to six months.

Step 10: If multiple wells from the same hPSC line are handled together, they may be pooled in a centrifuge tube and an aliquot removed for counting (see Section 19.1). One well of a 6-well plate is estimated to give approximately $1–2 \times 10^6$ cells; aim to dispense this amount per cryovial (EBiSC 2017). If a cell count has not been performed, newcomers to the procedure are advised to freeze no more than two cryovials per well (Liu and Chen 2014).

The procedure used to thaw hPSCs is similar to that described previously for somatic cells (see Protocol P15.2). Cells should be thawed rapidly at 37 °C using a water bath or bucket containing sterile water or a dry heat block. After transfer to a larger tube or vessel, medium should be added slowly, drop by drop, to minimize osmotic shock (EBiSC 2017). Cells may be gently centrifuged (e.g. at 200 g for two minutes) and resuspended in E8 medium for plating onto vitronectin, where they should attach within the first 30 minutes. Cryopreservation following EDTA dissociation is expected to give 30–80% efficiency; recovery will vary with the age of the culture, with newly established lines tending to have lower recovery rates. Recovery can be improved by growing cells in the presence of 5% oxygen for the first two days after thawing (Liu and Chen 2014).

23.5 PERINATAL STEM CELLS

Amniotic fluid, placental tissue, and umbilical cord blood all contain cell types that exhibit self-renewal and the ability to differentiate into various lineages (Al Shammary and Moretti 2016). Human amniotic fluid can only be accessed if amniocentesis is performed as a diagnostic procedure (e.g. to assess the fetal karyotype); a procedure for culture of amniocytes is provided in the Supplementary Material online (see Supp. S23.4). Human placental tissue and umbilical cord blood are more readily available, as these samples are normally discarded after birth and may be collected and banked provided informed, written consent has been obtained (Harris 2016; Kusuma et al. 2018). Most stem cell populations from perinatal tissue are HSCs (in umbilical cord blood) or MSCs, which are identified through the presence of a shared phenotype,

TABLE 23.5. Minimal criteria to identify MSCs.

Characteristic	MSC criteria[a]
Adherence to plastic	Adheres to plastic when maintained in standard culture conditions using tissue culture flasks
Presence of surface antigens	Expression of CD105, CD73, and CD90 in ≥ 95% of the population as measured by flow cytometry
Absence of surface antigens	Expression of CD45, CD34, CD14 or CD11b, CD79α or CD19, and HLA-DR in ≤ 2% of the population as measured by flow cytometry
Induction of differentiation	Differentiates into osteoblasts, adipocytes, and chondroblasts as demonstrated by *in vitro* staining

Source: Dominici, M., Le Blanc, K., Mueller, I., et al. (2006). Minimal criteria for defining multipotent mesenchymal stromal cells. The International Society for Cellular Therapy position statement. Cytotherapy 8 (4):315–17.© 2006 Elsevier.
[a]Criteria are recommended by the ISCT for laboratory-based research and preclinical studies. They should not be confused with release specifications for clinical studies.
HLA-DR, Human leukocyte antigen-DR isotype.

e.g. expression of specific surface antigens. Criteria for MSCs have been developed by the International Society for Cellular Therapy (ISCT) (see Table 23.5) and include the ability to differentiate into three mesodermal lineages, indicating that these cells are multipotent. Pluripotent cells may be present in perinatal tissues, such as amniotic fluid stem cells (AFSCs) or very small embryonic-like stem cells (VSELs) in umbilical cord blood, but these cell types form a very small proportion of the total cell population (Gao et al. 2013).

23.6 ADULT STEM CELLS

Adult stem cells were originally discovered in the hematopoietic system, where scientists injected bone marrow cells into irradiated mice to identify colonies that were capable of self-renewal and differentiation (Becker et al. 1963; Laurenti and Gottgens 2018). HSCs are responsible for the continuous production and turnover of all mature lymphoid and myeloid blood cells, including red blood cells, white blood cells, and platelets. However, our understanding of HSC biology has grown in recent years; it is now clear that these multipotent cells vary in the degree to which they self-renew and in their capacity to give rise to all differentiated blood cell types (Laurenti and Gottgens 2018). HSCs can be isolated from various samples (including bone marrow, peripheral blood, and umbilical cord blood) using magnetic sorting or flow cytometry, based on expression of CD34, absence of CD38,

and other markers (see Figures 21.5 and 21.8) (Ng et al. 2009; Frisch and Calvi 2014). A number of assays may be used for the characterization of HSCs, such as the colony forming cell (CFC) assay, the cobblestone area-forming cell (CAFC) assay, and the long-term culture initiating cell (LTC-IC) assay (Frisch and Calvi 2014).

Bone marrow contains a number of stem cell populations that are not part of the hematopoietic system, although they may be necessary for its support. MSCs were identified in bone marrow through their ability to generate colonies in clonogenic assays and to differentiate into osteoblasts (Owen and Friedenstein 1988). Bone marrow MSCs can be isolated by plating mononuclear cells in plastic dishes, where the MSCs adhere to the dish while the hemopoietic cells are gradually washed away (see Figure 23.8) (Gregory and Prockop 2007). A protocol for generating MSCs from bone marrow is provided as Supplementary Material online (see Supp. S23.5). Other stem cell populations that have been described in the bone marrow include VSELs, Muse cells, marrow-isolated adult multilineage inducible (MIAMI) cells, multipotent adult stem cells (MACS), multipotent adult progenitor cells (MAPCs), and endothelial progenitor cells (EPCs) (Ratajczak et al. 2008; Gao et al. 2013; Simerman et al. 2016).

Although the bone marrow is a fruitful source of stem cell populations, it is only accessible through an invasive procedure that produces a limited amount of material. Alternative sources of MSCs include perinatal tissues (see Section 23.5) and an increasing number of adult tissues including adipose tissue, dental pulp, and muscle (Murray et al. 2014). MSCs appear to arise from cell populations that are associated with the blood vessel wall, such as pericytes and adventitial perivascular cells, which would explain why MSCs can be isolated from virtually any vascularized tissue (Murray et al. 2014; Caplan 2017). Adipose tissue and teeth are particularly promising sources of stem cells for therapeutic applications because they are normally discarded following surgical or dental procedures. These populations include MSCs and other cell types, such as adipose-derived stem cells (ASCs), that are reported to be pluripotent. MSCs can display quite different characteristics when derived from different sources, or when isolated and examined using different methods. Procedures for isolation of MSCs from adipose tissue and dental pulp are described elsewhere (Palumbo et al. 2018; Guirado et al. 2019).

MSCs from bone marrow and other sources have been utilized in cell-based therapies for more than 20 years (Lazarus et al. 1995; Robb et al. 2019). Although they may serve as progenitors for tissue repair, MSCs secrete bioactive factors that modulate the immune response and promote other therapeutic outcomes (Horwitz and Dominici 2008; Caplan 2017). Unfortunately, the widespread availability of source material has led to a rise in unproven cellular therapies, resulting in marketing of MSCs as "stem cell" treatments (Dominici et al. 2015). Such unproven products often lack

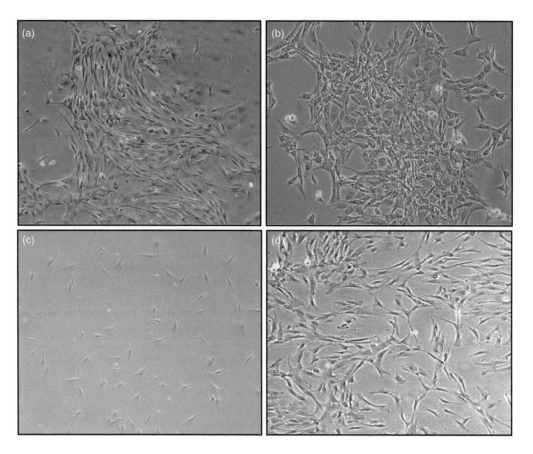

Fig. 23.8. Bone marrow-derived mesenchymal stromal cells (MSCs). (a, b) Colonies formed after plating of whole bone marrow mononuclear cells, showing morphology and optimal passaging density of MSCs. (c) An early passage culture of MSCs. (d) A monolayer at the appropriate density for passage. *Source*: Courtesy of C. Gregory and D. Prockop. (For associated protocol see Supp. S23.5.)

quality assurance or safety assessment, resulting in very poor outcomes for recipients – for example, a batch of products derived from umbilical cord blood caused infections in 12 patients in 2018 due to bacterial contamination (Perkins et al. 2018).

23.7 STEM CELL CHARACTERIZATION AND BANKING

The basic principles of Good Cell Culture Practice (GCCP) apply to all cell types in culture, including stem cells and their derivatives (see Section 7.2.1) (OECD 2018). However, stem cell lines have additional requirements for characterization and reporting that must be met (Pamies et al. 2017). These include obtaining ethical approvals that are suitable for stem cell-based work and using the correct nomenclature for any cell lines that are established (see Sections 6.4, 14.2.1). A standardized nomenclature is used for hPSC lines, which combines five descriptive elements to give a unique identifier (see Table 23.3) (Kurtz et al. 2018). A new identifier can be generated by filling out a form on the hPSCreg website

with the required information, including evidence that ethical approval and informed consent have been obtained (hPSCreg 2019).

Stem cell characterization and banking are specialized areas, but, fortunately, there are some excellent resources for further information (International Stem Cell Banking Initiative 2009; Stacey et al. 2016; Crook and Ludwig 2017). Broadly speaking, stem cell characterization should include assessment of (Healy et al. 2011):

(1) **Purity (freedom from microbial contamination).** Some bacteria and fungi can be detected through direct observation, but inspection does not detect all microorganisms and is ineffective for mycoplasma (see Sections 16.3, 16.4). Mycoplasma testing should be routinely performed on all cell lines, including stem cell lines (see Section 16.4.1). Testing for sterility and viral pathogens may also be required, depending on risk assessment (see Sections 11.8.1, 16.5). A recent study of 47 human iPSC lines screened by EBiSC showed that three (6.4%) returned a positive result from sterility testing and one (2%) carried a viral pathogen (De Sousa et al. 2017).

TABLE 23.6. Quality control (QC) for undifferentiated hPSC lines.

Quality control test[a]	Examples of test methods	Required characteristics
Authenticity	STR profiling; SNP genotyping	Test sample corresponds to reference, e.g. donor tissue or parental cell line
Bacteria/fungi	Sterility testing	Contamination not detected
Mycoplasma	Direct culture in broth or agar and indirect test using indicator culture or DNA stain	Contamination not detected
Viability post-thaw	Cell counting, colony morphology	Thawed vial generates viable colonies that are predominantly free of differentiated cells; efficiency of recovery should be reported for each bank/lot
Karyotype	G-banding; FISH	Normal diploid karyotype after examining a minimum number of metaphase spreads
Pluripotency	Teratoma formation; *in vitro* trilineage differentiation through spontaneous or directed differentiation (see Figure 23.9)	Test sample meets defined criteria using a specific protocol and method of data analysis
Antigen expression	Positive for a range of hPSC markers by flow cytometry (e.g. SSEA-3, SSEA-4, TRA-1-60, TRA-1-81, POU5F1); negative for SSEA-1	High proportion of cells (typically > 70%) positive for each marker
Growth characteristics	Growth curve; population doubling time (PDT); stable growth pattern, e.g. stable four-day growth to confluence	Report growth curve or PDT assessment using standard culture conditions
Genetic stability	SNP genotyping; CGH analysis	Report data available
Gene expression	Expression profiling using microarray, Q-PCR, or whole exome sequencing; expression of core hPSC genes as well as markers of differentiated cell types	Report data available

Source: International Stem Cell Banking Initiative (2009). Consensus guidance for banking and supply of human embryonic stem cell lines for research purposes. Stem Cell Rev 5 (4):301–14. © 2009 Springer Nature.
[a] These tests were originally proposed for an hESC master cell bank (for cell banking concept see Section 15.5). Testing requirements will vary with the intended use, e.g. a working cell bank may require less extensive testing.
Abbreviations not found within the main text: CGH, comparative genome hybridization; FISH, fluorescence *in situ* hybridization; Q-PCR, quantitative-polymerase chain reaction; SNP, single nucleotide polymorphism; STR, short tandem repeat.

(2) *Authenticity.* Cross-contamination and misidentification may affect any cell line, including stem cell lines. Authentication testing should be performed on all cell lines to exclude these possibilities (see Section 17.3.2). The same study of 47 human iPSC lines screened at EBiSC showed that 17% were misidentified (De Sousa et al. 2017); this was the most common cause of iPSC lines failing to pass quality control criteria.

(3) *Viability.* Recovery following cryopreservation should be addressed through viability testing (see Sections 15.4.4, 19.2). Cryopreservation of stem cell lines can be challenging, as noted previously for hPSC lines, which are often frozen as aggregates rather than individual cells to minimize apoptosis (see Section 23.4.4). For these cell lines, it is important to assess colony recovery during the first 48 hours after thawing (De Sousa et al. 2017). Markers of apoptosis may also be informative, e.g. expression of annexin IV (Healy et al. 2011).

(4) *Genotype.* Genetic instability is an important concern for stem cell lines, particularly hPSC lines that may be used for preclinical research or cell-based therapies. The karyotype of the cell line should be examined, using sufficient metaphase spreads to be confident that it is a normal diploid sample (a minimum of 20 metaphases for a chromosome count and eight metaphases for analysis of banding patterns) (International Stem Cell Banking Initiative 2009). Other methods for assessment of genetic stability may also be used (see Table 23.6).

(5) *Phenotype.* Stem cell lines should be actively growing, have a characteristic morphology, and express markers that are consistent with the stem cell type, e.g. MSCs will express CD105, CD73, and CD90 (see Table 23.5). The phenotype should be consistent with an undifferentiated state and will change when cells undergo differentiation, e.g. hPSC lines that undergo differentiation display reduced expression of POU5F1 and increased expression of SSEA-1 (see Table 23.2).

(6) *Potency.* Stem cell lines must be able to differentiate into all three germ layers if they are to be considered pluripotent (see Sections 2.4.1, 2.4.2). Pluripotency assays have

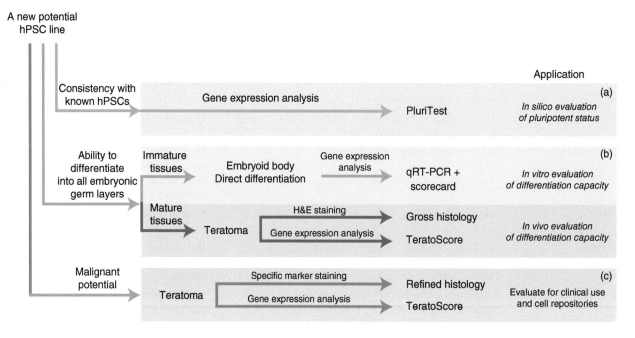

Fig. 23.9. The role of pluripotency assays. A new potential iPSC line can be assessed using: (a) gene expression analysis (PluriTest) to ensure it is consistent with known hPSCs; (b) embryoid body, teratoma formation, or other assays to ensure it can differentiate into all embryonic germ layers; and (c) teratoma formation to examine its malignant potential. *Source*: International Stem Cell Initiative (2018). Assessment of established techniques to determine developmental and malignant potential of human pluripotent stem cells. Nat Commun 9 (1):1925. Licensed under CCBY 4.0.

traditionally relied on teratoma formation (where cells are injected into nude mice) or chimera formation (where cells are transferred into early mouse embryos). However, chimera formation cannot be assessed with hPSCs due to ethical reasons, while teratoma formation requires a large number of mice and is time- and labor-intensive. *In vitro* methods may be suitable for some (research-based) applications. Such methods include (i) analysis of embryoid bodies (Pettinato et al. 2015); (ii) analysis of the cell line's pluripotent transcriptome in a bioinformatic assay (PluriTest) (Muller et al. 2011); or (iii) use of a kit to induce trilineage differentiation (STEMdiff™ kit, STEMCELL Technologies). These methods are useful to assess differentiation and to predict pluripotency, but the teratoma assay remains the "gold standard" to assess pluripotency and malignant potential in preclinical safety testing (see Figure 23.9) (International Stem Cell Initiative 2018).

Cell banking is an essential concept for any cell line, ensuring that a suitable number of vials is stored at early passage and at an appropriate level of quality to be useful for many years into the future. The general principles of cell banking have been described in a previous chapter (see Section 15.5). Further protocols and detailed requirements for stem cell banking are available for laboratories or cell repositories that wish to develop specialized expertise in this area (Crook

and Ludwig 2017). Points to consider when preparing hPSC stocks for clinical translation are discussed elsewhere (Andrews et al. 2015).

23.8 CONCLUSIONS: CULTURE OF STEM CELLS

Despite its technical challenges, stem cell culture is a rapidly expanding field. In the 20 years since the first hESC lines were established, governments have invested more than $2.2 billion in hPSC-related research; in 2018, 32 clinical trials were underway or had been completed using hPSC lines (Guhr et al. 2018; Ludwig et al. 2018). Most clinical trials to date have used research grade cell lines that were subsequently adapted for culture using GMP (see Section 7.2.3). A number of unproven "stem cell" products are also being offered to patients based on dubious claims of therapeutic efficacy, leading to potential adverse consequences for their recipients (see Section 23.6). The best way to address these issues is to require that all tissue culture laboratories meet GCCP requirements that relate to their particular application, including a good understanding of the *in vitro* system and the various factors that may affect it (Pamies et al. 2017). A better understanding of stem cell biology is essential to develop more physiological systems that can be adopted for use with all stem cell applications.

Suppliers

Supplier	URL
BioLife Solutions	www.biolifesolutions.com
Biological Industries	www.bioind.com/worldwide
Cellaria	www.cellariabio.com
Corning	www.corning.com/worldwide/en/products/life-sciences/products/cell-culture.html
Innovative Cell Technologies, Inc.	www.accutase.com
Merck Millipore	www.merckmillipore.com
Quality Biological, Inc. (QBI)	www.qualitybiological.com
REPROCELL	www.reprocell.com
Sigma-Aldrich (Merck)	www.sigmaaldrich.com/life-science/cell-culture.html
STEMCELL Technologies	www.stemcell.com
ThermoFisher Scientific	www.thermofisher.com/us/en/home/life-science/cell-culture.html
Tocris	www.tocris.com

Supp. S23.1 Derivation and Primary Culture of Mouse Embryonic Stem Cells (mESCs).

Supp. S23.2 Propagation of Mouse Embryonic Stem Cell (mESC) Lines.

Supp. S23.3 Derivation and Culture of Human Embryonic Stem Cells (hESCs).

Supp. S23.4 Culture of Amniocytes.

Supp. S23.5 Mesenchymal Stromal Cell (MSC) Production from Human Bone Marrow.

REFERENCES

Addgene (2019). Viral plasmids and resources. https://www.addgene.org/viral-vectors (accessed 11 April 2019).

Al Shammary, M. and Moretti, F.M. (2016). Placental stem cells and culture methods. In: *Fetal Stem Cells in Regenerative Medicine: Principles and Translational Strategies* (eds. D.O. Fauza and M. Bani), 277–292. New York: Springer.

Andrews, P.W., Goodfellow, P.N., Shevinsky, L.H. et al. (1982). Cell-surface antigens of a clonal human embryonal carcinoma cell line: morphological and antigenic differentiation in culture. *Int. J. Cancer* 29 (5): 523–531.

Andrews, P.W., Baker, D., Benvinisty, N. et al. (2015). Points to consider in the development of seed stocks of pluripotent stem cells for clinical applications: International Stem Cell Banking Initiative (ISCBI). *Regener. Med.* 10 (2 Suppl): 1–44. https://doi.org/10.2217/rme.14.93.

Armstrong, L., Lako, M., Buckley, N. et al. (2012). Editorial: our top 10 developments in stem cell biology over the last 30 years. *Stem Cells* 30 (1): 2–9. https://doi.org/10.1002/stem.1007.

Becker, A.J., McCulloch, E.A., and Till, J.E. (1963). Cytological demonstration of the clonal nature of spleen colonies derived from transplanted mouse marrow cells. *Nature* 197: 452–454.

Beers, J., Gulbranson, D.R., George, N. et al. (2012). Passaging and colony expansion of human pluripotent stem cells by enzyme-free dissociation in chemically defined culture conditions. *Nat. Protoc.* 7 (11): 2029–2040. https://doi.org/10.1038/nprot.2012.130.

Beers, J., Linask, K.L., Chen, J.A. et al. (2015). A cost-effective and efficient reprogramming platform for large-scale production of integration-free human induced pluripotent stem cells in chemically defined culture. *Sci. Rep.* 5: 11319. https://doi.org/10.1038/srep11319.

Borowski, M., Giovino-Doherty, M., Ji, L. et al. (2012). Basic pluripotent stem cell culture protocols. In: *StemBook* (ed. The Stem Cell Research Community). Cambridge, MA: Harvard Stem Cell Institute.

Brons, I.G., Smithers, L.E., Trotter, M.W. et al. (2007). Derivation of pluripotent epiblast stem cells from mammalian embryos. *Nature* 448 (7150): 191–195. https://doi.org/10.1038/nature05950.

Caplan, A.I. (2017). Mesenchymal stem cells: time to change the name! *Stem Cells Transl. Med.* 6 (6): 1445–1451. https://doi.org/10.1002/sctm.17-0051.

Chen, G., Hou, Z., Gulbranson, D.R. et al. (2010). Actin–myosin contractility is responsible for the reduced viability of dissociated human embryonic stem cells. *Cell Stem Cell* 7 (2): 240–248. https://doi.org/10.1016/j.stem.2010.06.017.

Chen, G., Gulbranson, D.R., Hou, Z. et al. (2011). Chemically defined conditions for human iPSC derivation and culture. *Nat. Methods* 8 (5): 424–429. https://doi.org/10.1038/nmeth.1593.

Chou, Y.F., Chen, H.H., Eijpe, M. et al. (2008). The growth factor environment defines distinct pluripotent ground states in novel blastocyst-derived stem cells. *Cell* 135 (3): 449–461. https://doi.org/10.1016/j.cell.2008.08.035.

Claassen, D.A., Desler, M.M., and Rizzino, A. (2009). ROCK inhibition enhances the recovery and growth of cryopreserved human embryonic stem cells and human induced pluripotent stem cells. *Mol. Reprod. Dev.* 76 (8): 722–732. https://doi.org/10.1002/mrd.21021.

Cooke, J.A. and Minger, S.L. (2007). Human embryonal stem cell lines: derivation and culture. In: *Culture of Human Stem Cells* (eds. R.I. Freshney, G.N. Stacey and J.M. Auerbach), 22–59. Hoboken, NJ: Wiley.

Cowan, C.A., Atienza, J., Melton, D.A. et al. (2005). Nuclear reprogramming of somatic cells after fusion with human embryonic stem cells. *Science* 309 (5739): 1369–1373. https://doi.org/10.1126/science.1116447.

Crook, J.M. and Ludwig, T.E. (2017). *Stem Cell Banking: Concepts and Protocols. Vol. 1590*, Methods in Molecular Biology. New York: Humana Press.

Damjanov, I. and Andrews, P.W. (2016). Teratomas produced from human pluripotent stem cells xenografted into immunodeficient mice – a histopathology atlas. *Int. J. Dev. Biol.* 60 (10-11-12): 337–419. https://doi.org/10.1387/ijdb.160274id.

Davis, R.P., Nemes, C., Varga, E. et al. (2013). Generation of induced pluripotent stem cells from human foetal fibroblasts using the sleeping beauty transposon gene delivery system. *Differentiation* 86 (1–2): 30–37. https://doi.org/10.1016/j.diff.2013.06.002.

De Sousa, P.A., Steeg, R., Wachter, E. et al. (2017). Rapid establishment of the European Bank for induced Pluripotent Stem Cells (EBiSC) – the hot start experience. *Stem Cell Res.* 20: 105–114. https://doi.org/10.1016/j.scr.2017.03.002.

Desmarais, J.A., Hoffmann, M.J., Bingham, G. et al. (2012). Human embryonic stem cells fail to activate CHK1 and commit to apoptosis in response to DNA replication stress. *Stem Cells* 30 (7): 1385–1393. https://doi.org/10.1002/stem.1117.

Dominici, M., Le Blanc, K., Mueller, I. et al. (2006). Minimal criteria for defining multipotent mesenchymal stromal cells. The International Society for Cellular Therapy position statement. *Cytotherapy* 8 (4): 315–317. https://doi.org/10.1080/14653240600855905.

Dominici, M., Nichols, K., Srivastava, A. et al. (2015). Positioning a scientific community on unproven cellular therapies: the 2015 International Society for Cellular Therapy Perspective. *Cytotherapy* 17 (12): 1663–1666. https://doi.org/10.1016/j.jcyt.2015.10.007.

Draper, J.S., Smith, K., Gokhale, P. et al. (2004). Recurrent gain of chromosomes 17q and 12 in cultured human embryonic stem cells. *Nat. Biotechnol.* 22 (1): 53–54. https://doi.org/10.1038/nbt922.

EBiSC (2017). Protocol for the use of induced pluripotent stem cells. http://www.ebisc.org/files/other-doc/ebisc-user_protocol-v2-2017-05.pdf (accessed 16 July 2017).

EBiSC (2019). Virtual training library. http://www.ebisc.org/trainings (accessed 10 May 2019).

Evans, M.J. and Kaufman, M.H. (1981). Establishment in culture of pluripotential cells from mouse embryos. *Nature* 292 (5819): 154–156.

Fan, L. and Collodi, P. (2006). Zebrafish embryonic stem cells. *Methods Enzymol.* 418: 64–77. https://doi.org/10.1016/S0076-6879(06)18004-0.

Farzaneh, M., Attari, F., Mozdziak, P.E. et al. (2017). The evolution of chicken stem cell culture methods. *Br. Poult. Sci.* 58 (6): 681–686. https://doi.org/10.1080/00071668.2017.1365354.

Frank, S., Zhang, M., Scholer, H.R. et al. (2012). Small molecule-assisted, line-independent maintenance of human pluripotent stem cells in defined conditions. *PLoS One* 7 (7): e41958. https://doi.org/10.1371/journal.pone.0041958.

Freshney, R.I., Stacey, G.N., and Auerbach, J.M. (2007). *Culture of Human Stem Cells*. Hoboken, NJ: Wiley.

Frisch, B.J. and Calvi, L.M. (2014). Hematopoietic stem cell cultures and assays. *Methods Mol. Biol.* 1130: 315–324. https://doi.org/10.1007/978-1-62703-989-5_24.

Furue, M., Okamoto, T., Hayashi, Y. et al. (2005). Leukemia inhibitory factor as an anti-apoptotic mitogen for pluripotent mouse embryonic stem cells in a serum-free medium without feeder cells. *In Vitro Cell. Dev. Biol. Anim.* 41 (1–2): 19–28. https://doi.org/10.1290/0502010.1.

Furue, M.K., Na, J., Jackson, J.P. et al. (2008). Heparin promotes the growth of human embryonic stem cells in a defined serum-free medium. *Proc. Natl Acad. Sci. U.S.A.* 105 (36): 13409–13414. https://doi.org/10.1073/pnas.0806136105.

Fusaki, N., Ban, H., Nishiyama, A. et al. (2009). Efficient induction of transgene-free human pluripotent stem cells using a vector based on Sendai virus, an RNA virus that does not integrate into the host genome. *Proc. Jpn Acad. Ser. B Phys. Biol. Sci.* 85 (8): 348–362.

Gao, L., Thilakavathy, K., and Nordin, N. (2013). A plethora of human pluripotent stem cells. *Cell Biol. Int.* 37 (9): 875–877. https://doi.org/10.1002/cbin.10120.

Gokhale, P.J. and Andrews, P.W. (2012). The development of pluripotent stem cells. *Curr. Opin. Genet. Dev.* 22 (5): 403–408. https://doi.org/10.1016/j.gde.2012.07.006.

Gregory, C.A. and Prockop, D.J. (2007). Fundamentals of culture and characterization of mesenchymal stem/progenitor cells (MSCs) from bone marrow stroma. In: *Culture of Human Stem Cells* (eds. R.I. Freshney, G.N. Stacey and J.M. Auerbach), 208–232. Hoboken, NJ: Wiley.

Guhr, A., Kobold, S., Seltmann, S. et al. (2018). Recent trends in research with human pluripotent stem cells: impact of research and use of cell lines in experimental research and clinical trials. *Stem Cell Rep.* 11 (2): 485–496. https://doi.org/10.1016/j.stemcr.2018.06.012.

Guirado, E., Zhang, Y., and George, A. (2019). Establishment of stable cell lines from primary human dental pulp stem cells. *Methods Mol. Biol.* 1922: 21–27. https://doi.org/10.1007/978-1-4939-9012-2_3.

Gurdon, J.B. (1962). The developmental capacity of nuclei taken from intestinal epithelium cells of feeding tadpoles. *J. Embryol. Exp. Morphol.* 10: 622–640.

Hackett, C.H. and Fortier, L.A. (2011). Embryonic stem cells and iPS cells: sources and characteristics. *Vet. Clin. North Am. Equine Pract.* 27 (2): 233–242. https://doi.org/10.1016/j.cveq.2011.04.003.

Hanna, J. and Hubel, A. (2009). Preservation of stem cells. *Organogenesis* 5 (3): 134–137.

Harris, D.T. (2016). Umbilical cord blood stem cell populations. In: *Fetal Stem Cells in Regenerative Medicine: Principles and Translational Strategies* (eds. D.O. Fauza and M. Bani), 241–255. New York: Springer.

Healy, L. and Ruban, L. (2015). *Atlas of Human Pluripotent Stem Cells in Culture*. London: Springer.

Healy, L., Young, L., and Stacey, G.N. (2011). Stem cell banks: preserving cell lines, maintaining genetic integrity, and advancing research. *Methods Mol. Biol.* 767: 15–27. https://doi.org/10.1007/978-1-61779-201-4_2.

Ho, S.Y., Goh, C.W., Gan, J.Y. et al. (2014). Derivation and long-term culture of an embryonic stem cell-like line from zebrafish blastomeres under feeder-free condition. *Zebrafish* 11 (5): 407–420. https://doi.org/10.1089/zeb.2013.0879.

Honda, A., Hirose, M., Inoue, K. et al. (2008). Stable embryonic stem cell lines in rabbits: potential small animal models for human research. *Reprod. Biomed. Online* 17 (5): 706–715.

Horwitz, E.M. and Dominici, M. (2008). How do mesenchymal stromal cells exert their therapeutic benefit? *Cytotherapy* 10 (8): 771–774. https://doi.org/10.1080/14653240802618085.

hPSCreg (2019). How to register. https://hpscreg.eu/about/registration (accessed 19 May 2019).

HTA (2019). Regulating human embryonic stem cell lines for human application. www.hta.gov.uk/policies/regulating-human-embryonic-stem-cell-lines-human-application (accessed 3 May 2019).

Ilic, D. and Polak, J.M. (2011). Stem cells in regenerative medicine: introduction. *Br. Med. Bull.* 98: 117–126. https://doi.org/10.1093/bmb/ldr012.

International Stem Cell Banking Initiative (2009). Consensus guidance for banking and supply of human embryonic stem cell lines for research purposes. *Stem Cell Rev.* 5 (4): 301–314. https://doi.org/10.1007/s12015-009-9085-x.

International Stem Cell Initiative (2018). Assessment of established techniques to determine developmental and malignant potential of human pluripotent stem cells. *Nat. Commun.* 9 (1): 1925. https://doi.org/10.1038/s41467-018-04011-3.

International Stem Cell Initiative Consortium, Akopian, V., Andrews, P.W. et al. (2010). Comparison of defined culture systems for feeder cell free propagation of human embryonic stem cells. *In Vitro Cell Dev. Biol. Anim.* 46 (3–4): 247–258. https://doi.org/10.1007/s11626-010-9297-z.

International Stem Cell Initiative, Amps, K., Andrews, P.W. et al. (2011). Screening ethnically diverse human embryonic stem cells identifies a chromosome 20 minimal amplicon conferring growth advantage. *Nat. Biotechnol.* 29 (12): 1132–1144. https://doi.org/10.1038/nbt.2051.

ISCF (2019). Stem cell registry. http://www.stem-cell-forum.net/initiatives/isci/stem-cell-registry (accessed 3 May 2019).

ISSCR (2016). Guidelines for stem cell research and clinical translation. http://www.isscr.org/membership/policy/2016-guidelines/guidelines-for-stem-cell-research-and-clinical-translation (accessed 30 April 2019).

Kaji, K., Norrby, K., Paca, A. et al. (2009). Virus-free induction of pluripotency and subsequent excision of reprogramming factors. *Nature* 458 (7239): 771–775. https://doi.org/10.1038/nature07864.

Kasai-Brunswick, T.H., Silva Dos Santos, D., Ferreira, R.P. et al. (2018). Generation of patient-specific induced pluripotent stem cell lines from one patient with Jervell and Lange-Nielsen syndrome, one with type 1 long QT syndrome and two healthy relatives. *Stem Cell Res.* 31: 174–180. https://doi.org/10.1016/j.scr.2018.07.016.

Katkov, I.I., Kan, N.G., Cimadamore, F. et al. (2011). DMSO-free programmed cryopreservation of fully dissociated and adherent human induced pluripotent stem cells. *Stem Cells Int.* 2011: 981606. https://doi.org/10.4061/2011/981606.

Kim, D., Kim, C.H., Moon, J.I. et al. (2009). Generation of human induced pluripotent stem cells by direct delivery of reprogramming proteins. *Cell Stem Cell* 4 (6): 472–476. https://doi.org/10.1016/j.stem.2009.05.005.

Kleinsmith, L.J. and Pierce, G.B. Jr. (1964). Multipotentiality of single embryonal carcinoma cells. *Cancer Res.* 24: 1544–1551.

Kramer, N., Rosner, M., Kovacic, B. et al. (2016). Full biological characterization of human pluripotent stem cells will open the door to translational research. *Arch. Toxicol.* 90 (9): 2173–2186. https://doi.org/10.1007/s00204-016-1763-2.

Kurosawa, H. (2012). Application of rho-associated protein kinase (ROCK) inhibitor to human pluripotent stem cells. *J. Biosci. Bioeng.* 114 (6): 577–581. https://doi.org/10.1016/j.jbiosc.2012.07.013.

Kurtz, A., Seltmann, S., Bairoch, A. et al. (2018). A standard nomenclature for referencing and authentication of pluripotent stem cells. *Stem Cell Rep.* 10 (1): 1–6. https://doi.org/10.1016/j.stemcr.2017.12.002.

Kusuma, G.D., Abumaree, M.H., Pertile, M.D. et al. (2018). Isolation and characterization of mesenchymal stem/stromal cells derived from human third trimester placental chorionic villi and decidua basalis. In: *Preeclampsia : Methods and Protocols* (eds. P. Murthi and C. Vaillancourt), 247–266. New York: Springer.

Kyriakides, O., Halliwell, J.A., and Andrews, P.W. (2018). Acquired genetic and epigenetic variation in human pluripotent stem cells. *Adv. Biochem. Eng. Biotechnol.* 163: 187–206. https://doi.org/10.1007/10_2017_22.

Lanza, R. and Atala, A. (2014). *Essentials of Stem Cell Biology*, 3e. San Diego, CA: Academic Press.

Laurenti, E. and Gottgens, B. (2018). From haematopoietic stem cells to complex differentiation landscapes. *Nature* 553 (7689): 418–426. https://doi.org/10.1038/nature25022.

Lazarus, H.M., Haynesworth, S.E., Gerson, S.L. et al. (1995). Ex vivo expansion and subsequent infusion of human bone marrow-derived stromal progenitor cells (mesenchymal progenitor cells): implications for therapeutic use. *Bone Marrow Transplant* 16 (4): 557–564.

Li, R., Yu, G., Azarin, S.M. et al. (2018). Freezing responses in DMSO-based cryopreservation of human iPS cells: aggregates versus single cells. *Tissue Eng. Part C Methods* 24 (5): 289–299. https://doi.org/10.1089/ten.TEC.2017.0531.

Lin, S. and Talbot, P. (2011). Methods for culturing mouse and human embryonic stem cells. *Methods Mol. Biol.* 690: 31–56. https://doi.org/10.1007/978-1-60761-962-8_2.

Liu, W. and Chen, G. (2014). Cryopreservation of human pluripotent stem cells in defined medium. *Curr. Protoc. Stem Cell Biol.* 31: 1C 17 1–1C 17 13. https://doi.org/10.1002/9780470151808.sc01c17s31.

Ludwig, T.E., Levenstein, M.E., Jones, J.M. et al. (2006). Derivation of human embryonic stem cells in defined conditions. *Nat. Biotechnol.* 24 (2): 185–187. https://doi.org/10.1038/nbt1177.

Ludwig, T.E., Kujak, A., Rauti, A. et al. (2018). 20 years of human pluripotent stem cell research: it all started with five lines. *Cell Stem Cell* 23 (5): 644–648. https://doi.org/10.1016/j.stem.2018.10.009.

Maldonado, M., Luu, R.J., Ramos, M.E. et al. (2016). ROCK inhibitor primes human induced pluripotent stem cells to selectively differentiate towards mesendodermal lineage via epithelial–mesenchymal transition-like modulation. *Stem Cell Res.* 17 (2): 222–227. https://doi.org/10.1016/j.scr.2016.07.009.

Martin, G.R. (1981). Isolation of a pluripotent cell line from early mouse embryos cultured in medium conditioned by teratocarcinoma stem cells. *Proc. Natl Acad. Sci. U.S.A.* 78 (12): 7634–7638.

Martin-Ibanez, R., Unger, C., Stromberg, A. et al. (2008). Novel cryopreservation method for dissociated human embryonic stem cells in the presence of a ROCK inhibitor. *Hum. Reprod.* 23 (12): 2744–2754. https://doi.org/10.1093/humrep/den316.

Mayshar, Y., Ben-David, U., Lavon, N. et al. (2010). Identification and classification of chromosomal aberrations in human induced pluripotent stem cells. *Cell Stem Cell* 7 (4): 521–531. https://doi.org/10.1016/j.stem.2010.07.017.

Miyazaki, T. and Suemori, H. (2016). Slow cooling cryopreservation optimized to human pluripotent stem cells. *Adv. Exp. Med. Biol.* 951: 57–65. https://doi.org/10.1007/978-3-319-45457-3_5.

Muller, F.J., Schuldt, B.M., Williams, R. et al. (2011). A bioinformatic assay for pluripotency in human cells. *Nat. Methods* 8 (4): 315–317. https://doi.org/10.1038/nmeth.1580.

Murray, I.R., West, C.C., Hardy, W.R. et al. (2014). Natural history of mesenchymal stem cells, from vessel walls to culture vessels. *Cell. Mol. Life Sci.* 71 (8): 1353–1374. https://doi.org/10.1007/s00018-013-1462-6.

Nakanishi, M. and Otsu, M. (2012). Development of Sendai virus vectors and their potential applications in gene therapy and regenerative medicine. *Curr. Gene Ther.* 12 (5): 410–416.

Ng, Y.Y., Baert, M.R.M., de Haas, E.F.E. et al. (2009). Isolation of human and mouse hematopoietic stem cells. In: *Genetic Modification of Hematopoietic Stem Cells: Methods and Protocols* (ed. C. Baum), 13–21. Totowa, NJ: Humana Press.

NIH (2016). Guidelines for human stem cell research policy: questions and answers. https://stemcells.nih.gov/research/newcell_qa.htm (accessed 3 May 2019).

NIH (2019a). Human embryonic stem cell registry. https://grants.nih.gov/stem_cells/registry/current.htm (accessed 3 May 2019).

NIH (2019b). Stem cell information: glossary. https://stemcells.nih.gov/glossary.htm (accessed 30 April 2019).

Ntai, A., La Spada, A., De Blasio, P. et al. (2018). Trehalose to cryopreserve human pluripotent stem cells. *Stem Cell Res* 31: 102–112. https://doi.org/10.1016/j.scr.2018.07.021.

OECD (2018). *Guidance Document on Good in Vitro Method Practices (GIVIMP)*. Paris: OECD Publishing.

Ohnuma, K., Fujiki, A., Yanagihara, K. et al. (2014). Enzyme-free passage of human pluripotent stem cells by controlling divalent cations. *Sci. Rep.* 4: 4646. https://doi.org/10.1038/srep04646.

OHRP (2002). Human embryonic stem cells, germ cells, and cell-derived test articles. https://www.hhs.gov/ohrp/regulations-and-policy/guidance/guidance-on-research-involving-stem-cells/index.html (accessed 15 June 2018).

Okita, K. and Yamanaka, S. (2011). Induced pluripotent stem cells: opportunities and challenges. *Philos. Trans. R. Soc. Lond. B Biol. Sci.* 366 (1575): 2198–2207. https://doi.org/10.1098/rstb.2011.0016.

Okita, K., Ichisaka, T., and Yamanaka, S. (2007). Generation of germline-competent induced pluripotent stem cells. *Nature* 448 (7151): 313–317. https://doi.org/10.1038/nature05934.

Ono, M., Hamada, Y., Horiuchi, Y. et al. (2012). Generation of induced pluripotent stem cells from human nasal epithelial cells using a Sendai virus vector. *PLoS One* 7 (8): e42855. https://doi.org/10.1371/journal.pone.0042855.

Owen, M. and Friedenstein, A.J. (1988). Stromal stem cells: marrow-derived osteogenic precursors. *Ciba Found. Symp.* 136: 42–60.

Pain, B., Clark, M.E., Shen, M. et al. (1996). Long-term in vitro culture and characterisation of avian embryonic stem cells with multiple morphogenetic potentialities. *Development* 122 (8): 2339–2348.

Palumbo, P., Lombardi, F., Siragusa, G. et al. (2018). Methods of isolation, characterization and expansion of human adipose-derived stem cells (ASCs): an overview. *Int. J. Mol. Sci.* 19 (7) https://doi.org/10.3390/ijms19071897.

Pamies, D., Bal-Price, A., Simeonov, A. et al. (2017). Good cell culture practice for stem cells and stem-cell-derived models. *ALTEX* 34 (1): 95–132. https://doi.org/10.14573/altex.1607121.

Pauklin, S. and Vallier, L. (2015). Activin/nodal signalling in stem cells. *Development* 142 (4): 607–619. https://doi.org/10.1242/dev.091769.

Pera, M.F. and Tam, P.P. (2010). Extrinsic regulation of pluripotent stem cells. *Nature* 465 (7299): 713–720. https://doi.org/10.1038/nature09228.

Perkins, K.M., Spoto, S., Rankin, D.A. et al. (2018). Notes from the field: infections after receipt of bacterially contaminated umbilical cord blood-derived stem cell products for other than hematopoietic or immunologic reconstitution – United States, 2018. *MMWR Morb. Mortal. Wkly Rep.* 67 (50): 1397–1399. https://doi.org/10.15585/mmwr.mm6750a5.

Pettinato, G., Wen, X., and Zhang, N. (2015). Engineering strategies for the formation of embryoid bodies from human pluripotent stem cells. *Stem Cells Dev.* 24 (14): 1595–1609. https://doi.org/10.1089/scd.2014.0427.

Prathalingam, N., Ferguson, L., Young, L. et al. (2012). Production and validation of a good manufacturing practice grade human fibroblast line for supporting human embryonic stem cell derivation and culture. *Stem Cell Res. Ther.* 3 (2): 12. https://doi.org/10.1186/scrt103.

Ramalho-Santos, M. and Willenbring, H. (2007). On the origin of the term "stem cell". *Cell Stem Cell* 1 (1): 35–38. https://doi.org/10.1016/j.stem.2007.05.013.

Ratajczak, M.Z., Zuba-Surma, E.K., Wojakowski, W. et al. (2008). Bone marrow – home of versatile stem cells. *Transfus. Med. Hemother.* 35 (3): 248–259. https://doi.org/10.1159/000125585.

Rathjen, J., Lake, J.A., Bettess, M.D. et al. (1999). Formation of a primitive ectoderm like cell population, EPL cells, from ES cells in response to biologically derived factors. *J. Cell Sci.* 112 (Pt 5): 601–612.

Richards, M., Tan, S., Fong, C.Y. et al. (2003). Comparative evaluation of various human feeders for prolonged undifferentiated growth of human embryonic stem cells. *Stem Cells* 21 (5): 546–556. https://doi.org/10.1634/stemcells.21-5-546.

Robb, K.P., Fitzgerald, J.C., Barry, F. et al. (2019). Mesenchymal stromal cell therapy: progress in manufacturing and assessments of potency. *Cytotherapy* 21 (3): 289–306. https://doi.org/10.1016/j.jcyt.2018.10.014.

Rodin, S., Antonsson, L., Hovatta, O. et al. (2014). Monolayer culturing and cloning of human pluripotent stem cells on laminin-521-based matrices under xeno-free and chemically defined conditions. *Nat. Protoc.* 9 (10): 2354–2368. https://doi.org/10.1038/nprot.2014.159.

Rony, I.K., Baten, A., Bloomfield, J.A. et al. (2015). Inducing pluripotency in vitro: recent advances and highlights in induced pluripotent stem cells generation and pluripotency reprogramming. *Cell Prolif.* 48 (2): 140–156. https://doi.org/10.1111/cpr.12162.

Sandmaier, S.E. and Telugu, B.P. (2015). MicroRNA-mediated reprogramming of somatic cells into induced pluripotent stem cells. *Methods Mol. Biol.* 1330: 29–36. https://doi.org/10.1007/978-1-4939-2848-4_3.

Seltmann, S., Lekschas, F., Muller, R. et al. (2016). hPSCreg – the human pluripotent stem cell registry. *Nucleic Acids Res.* 44 (D1): D757–D763. https://doi.org/10.1093/nar/gkv963.

Seruya, M., Shah, A., Pedrotty, D. et al. (2004). Clonal population of adult stem cells: life span and differentiation potential. *Cell Transplant.* 13 (2): 93–101.

Simerman, A.A., Phan, J.D., Dumesic, D.A. et al. (2016). Muse cells: nontumorigenic pluripotent stem cells present in adult tissues – a paradigm shift in tissue regeneration and evolution. *Stem Cells Int.* 2016: 1463258. https://doi.org/10.1155/2016/1463258.

Sipp, D., Munsie, M., and Sugarman, J. (2018). Emerging stem cell ethics. *Science* 360 (6395): 1275. https://doi.org/10.1126/science.aau4720.

Skottman, H. and Hovatta, O. (2006). Culture conditions for human embryonic stem cells. *Reproduction* 132 (5): 691–698. https://doi.org/10.1530/rep.1.01079.

Smith, A. (2006). A glossary for stem-cell biology. *Nature* 441 (7097): 1060–1060. https://doi.org/10.1038/nature04954.

Smith, A.G., Heath, J.K., Donaldson, D.D. et al. (1988). Inhibition of pluripotential embryonic stem cell differentiation by purified polypeptides. *Nature* 336 (6200): 688–690. https://doi.org/10.1038/336688a0.

Sommer, C.A. and Mostoslavsky, G. (2010). Experimental approaches for the generation of induced pluripotent stem cells. *Stem Cell Res. Ther.* 1 (3): 26. https://doi.org/10.1186/scrt26.

Sommer, A.G., Rozelle, S.S., Sullivan, S. et al. (2012). Generation of human induced pluripotent stem cells from peripheral blood using the STEMCCA lentiviral vector. *J. Vis. Exp.* (68): 4327. https://doi.org/10.3791/4327.

Spangrude, G.J., Heimfeld, S., and Weissman, I.L. (1988). Purification and characterization of mouse hematopoietic stem cells. *Science* 241 (4861): 58–62.

Spits, C., Mateizel, I., Geens, M. et al. (2008). Recurrent chromosomal abnormalities in human embryonic stem cells. *Nat. Biotechnol.* 26 (12): 1361–1363. https://doi.org/10.1038/nbt.1510.

Sridharan, R. and Plath, K. (2011). Small RNAs loom large during reprogramming. *Cell Stem Cell* 8 (6): 599–601. https://doi.org/10.1016/j.stem.2011.05.009.

Stacey, G.N., Coecke, S., Price, A.B. et al. (2016). Ensuring the quality of stem cell-derived in vitro models for toxicity testing. *Adv. Exp. Med. Biol.* 856: 259–297. https://doi.org/10.1007/978-3-319-33826-2_11.

Stadtfeld, M., Nagaya, M., Utikal, J. et al. (2008). Induced pluripotent stem cells generated without viral integration. *Science* 322 (5903): 945–949. https://doi.org/10.1126/science.1162494.

Stevens, L.C. and Little, C.C. (1954). Spontaneous testicular teratomas in an inbred strain of mice. *Proc. Natl Acad. Sci. U.S.A.* 40 (11): 1080–1087.

Sudo, K., Kanno, M., Miharada, K. et al. (2007). Mesenchymal progenitors able to differentiate into osteogenic, chondrogenic, and/or adipogenic cells in vitro are present in most primary fibroblast-like cell populations. *Stem Cells* 25 (7): 1610–1617. https://doi.org/10.1634/stemcells.2006-0504.

Swistowski, A., Peng, J., Han, Y. et al. (2009). Xeno-free defined conditions for culture of human embryonic stem cells, neural stem cells and dopaminergic neurons derived from them. *PLoS One* 4 (7): e6233. https://doi.org/10.1371/journal.pone.0006233.

Taapken, S.M., Nisler, B.S., Newton, M.A. et al. (2011). Karotypic abnormalities in human induced pluripotent stem cells and embryonic stem cells. *Nat. Biotechnol.* 29 (4): 313–314. https://doi.org/10.1038/nbt.1835.

Tada, M., Takahama, Y., Abe, K. et al. (2001). Nuclear reprogramming of somatic cells by in vitro hybridization with ES cells. *Curr Biol* 11 (19): 1553–1558.

Takahashi, K. and Yamanaka, S. (2006). Induction of pluripotent stem cells from mouse embryonic and adult fibroblast cultures by defined factors. *Cell* 126 (4): 663–676. https://doi.org/10.1016/j.cell.2006.07.024.

Takahashi, K. and Yamanaka, S. (2016). A decade of transcription factor-mediated reprogramming to pluripotency. *Nat. Rev. Mol. Cell Biol.* 17 (3): 183–193. https://doi.org/10.1038/nrm.2016.8.

Takahashi, K., Tanabe, K., Ohnuki, M. et al. (2007). Induction of pluripotent stem cells from adult human fibroblasts by defined factors. *Cell* 131 (5): 861–872. https://doi.org/10.1016/j.cell.2007.11.019.

Tesar, P.J., Chenoweth, J.G., Brook, F.A. et al. (2007). New cell lines from mouse epiblast share defining features with human embryonic stem cells. *Nature* 448 (7150): 196–199. https://doi.org/10.1038/nature05972.

Thomson, J.A., Kalishman, J., Golos, T.G. et al. (1995). Isolation of a primate embryonic stem cell line. *Proc. Natl Acad. Sci. U.S.A.* 92 (17): 7844–7848. https://doi.org/10.1073/pnas.92.17.7844.

Thomson, J.A., Itskovitz-Eldor, J., Shapiro, S.S. et al. (1998). Embryonic stem cell lines derived from human blastocysts. *Science* 282 (5391): 1145–1147.

Vallier, L., Alexander, M., and Pedersen, R.A. (2005). Activin/nodal and FGF pathways cooperate to maintain pluripotency of human embryonic stem cells. *J. Cell Sci.* 118 (Pt 19): 4495–4509. https://doi.org/10.1242/jcs.02553.

Wang, L., Schulz, T.C., Sherrer, E.S. et al. (2007). Self-renewal of human embryonic stem cells requires insulin-like growth factor-1 receptor and ERBB2 receptor signaling. *Blood* 110 (12): 4111–4119. https://doi.org/10.1182/blood-2007-03-082586.

Warren, L., Manos, P.D., Ahfeldt, T. et al. (2010). Highly efficient reprogramming to pluripotency and directed differentiation of human cells with synthetic modified mRNA. *Cell Stem Cell* 7 (5): 618–630. https://doi.org/10.1016/j.stem.2010.08.012.

Watanabe, K., Ueno, M., Kamiya, D. et al. (2007). A ROCK inhibitor permits survival of dissociated human embryonic stem cells. *Nat. Biotechnol.* 25 (6): 681–686. https://doi.org/10.1038/nbt1310.

Wesselschmidt, R.L. and McDonald, J.W. (2006). Embryonic stem cells and neurogenesis. In: *Neural Development and Stem Cells* (ed. M.S. Rao), 299–341. Totowa, NJ: Humana Press.

Williams, R.L., Hilton, D.J., Pease, S. et al. (1988). Myeloid leukaemia inhibitory factor maintains the developmental potential of embryonic stem cells. *Nature* 336 (6200): 684–687. https://doi.org/10.1038/336684a0.

Wilmes, A., Rauch, C., Carta, G. et al. (2017). Towards optimisation of induced pluripotent cell culture: extracellular acidification results in growth arrest of iPSC prior to nutrient exhaustion. *Toxicol. In Vitro* 45 (Pt 3): 445–454. https://doi.org/10.1016/j.tiv.2017.07.023.

Woltjen, K., Michael, I.P., Mohseni, P. et al. (2009). piggyBac transposition reprograms fibroblasts to induced pluripotent stem cells. *Nature* 458 (7239): 766–770. https://doi.org/10.1038/nature07863.

Wu, J. and Izpisua Belmonte, J.C. (2015). Dynamic pluripotent stem cell states and their applications. *Cell Stem Cell* 17 (5): 509–525. https://doi.org/10.1016/j.stem.2015.10.009.

Yang, J., Ryan, D.J., Wang, W. et al. (2017). Establishment of mouse expanded potential stem cells. *Nature* 550 (7676): 393–397. https://doi.org/10.1038/nature24052.

Ying, Q.L., Nichols, J., Chambers, I. et al. (2003). BMP induction of id proteins suppresses differentiation and sustains embryonic stem cell self-renewal in collaboration with STAT3. *Cell* 115 (3): 281–292.

Yu, J., Vodyanik, M.A., Smuga-Otto, K. et al. (2007). Induced pluripotent stem cell lines derived from human somatic cells. *Science* 318 (5858): 1917–1920. https://doi.org/10.1126/science.1151526.

CHAPTER 24

Culture of Specific Cell Types

Specific cell types from many different tissues have been cultured to examine developmental processes and their pathologies or for use in cell-based screening or therapies. Cultures that are established from specific cell types and that continue to express differentiated characteristics (or can be induced to do so) are referred to here as "specialized cells." Considering that these cells are often difficult to access, and that stem cell cultures are becoming more widely available, do we still need to work with cells from different tissue types? Although it is possible to induce differentiation of stem cells into a number of specific cell types, this process is still not completely understood or controllable. For example, three-dimensional (3D) organoid culture can be used to induce the formation of retinal structures that contain differentiated photoreceptors, but their development is often limited to early postnatal development and cells may not fully express differentiated characteristics (see Section 27.3). We still have a great deal to learn about differentiation and how to control it with greater precision *in vitro*.

Most of this chapter focuses on the culture of specialized cells from humans and rodents, with an emphasis on primary culture from normal (non-malignant) tissues (see Sections 24.2–24.5). The final section focuses on poikilothermic species that require different culture conditions (see Section 24.6). Because of space constraints, all protocols in this chapter are provided as Supplementary Material in the companion book website. For all protocols, it is assumed that you have access to a tissue culture laboratory with essential equipment, such as a biological safety cabinet (BSC), and that you are comfortable with basic handling procedures (see Parts II, IV).

24.1 SPECIALIZED CELLS AND THEIR AVAILABILITY

For anyone wishing to culture specialized cells, the question will always arise whether to attempt isolation and selective culture yourself or to purchase the cells from a supplier. The generation of a specialized cell culture can be achieved by (i) isolating differentiated cells or tissue for short-term culture, either as a primary culture or as a finite cell line (see Section 3.2.1); (ii) establishing a continuous cell line through immortalization (see Section 22.3); or (iii) isolating stem cells, expanding them, and inducing them to differentiate (see Sections 23.4–23.6, 26.5.2). Primary culture avoids the risk that specialized cells or behavior will be lost when the culture is passaged, but results in a limited amount of material (see Section 13.1). If primary cultures are not available, tissue culture laboratories must work with finite or continuous cell lines that have been demonstrated to retain differentiated characteristics under specific culture conditions. A list of commonly used cell lines is available in an earlier chapter (see Table 14.2); more information is available through the Cellosaurus knowledge resource, which aims to describe all cell lines used in biomedical research (Bairoch 2018).

An extensive range of specialized cells can be purchased from commercial suppliers (see Table 24.1). This list of commercial suppliers includes some cell repositories, which may not be "commercial" suppliers in a business sense (since many cell repositories operate under a not-for-profit business model) but are included here for convenience. All reputable suppliers should offer a catalog that lists the specific cultures or cell lines that are available; each listing should include

Freshney's Culture of Animal Cells: A Manual of Basic Technique and Specialized Applications, Eighth Edition. Amanda Capes-Davis and R. Ian Freshney.
© 2021 John Wiley & Sons Ltd. Published 2021 by John Wiley & Sons Ltd.
Companion website: www.wiley.com/go/freshney/cellculture8

TABLE 24.1. Types and sources of specialized cells.

Tissue or system[a]	Cell type[a]	Supplier[b]
Aorta	Endothelial cells, fibroblasts, smooth muscle cells	ATCC, Cell Applications, Cell Biologics, Lifeline Cell Technology, Lonza, PromoCell, ScienCell Research Laboratories, ThermoFisher Scientific
Bladder	Epithelial cells, fibroblasts, smooth muscle cells	ATCC, Cell Applications, Cell Biologics, Lifeline Cell Technology, Lonza, ScienCell Research Laboratories, ZenBio
Blood	PBMCs, umbilical cord blood	Multiple suppliers
Bone	Osteoblasts	Cell Applications, Lonza, Neuromics, PromoCell, ScienCell Research Laboratories
Bone marrow	Mononuclear cells, stromal cells	ATCC, Lonza, ReachBio Research Labs
Breast	Mammary epithelial cells (HMECs), fibroblasts	ATCC, Cell Applications, Cell Biologics, Lifeline Cell Technology, Lonza, ScienCell Research Laboratories, ThermoFisher Scientific, ZenBio
Ear	Middle ear epithelial cells, fibroblasts	ScienCell Research Laboratories
Eye	Corneal epithelial cells, iris pigment epithelial cells, lens epithelial cells, retinal pigment epithelial cells	ATCC, Cell Applications, Cell Biologics, Lifeline Cell Technology, Lonza, MatTek Corporation, Neuromics, ScienCell Research Laboratories
	Tissue equivalent using filter well insert (cornea)	EPISKIN, MatTek Corporation
Fat	Adipocytes, preadipocytes	ATCC, Biopredic International, Cell Applications, Lonza, PromoCell, ScienCell Research Laboratories, ZenBio
Heart	Cardiac fibroblasts, cardiomyocytes, pericardial fibroblasts	Cell Applications, Cell Biologics, Lifeline Cell Technology, Lonza, Neuromics, PromoCell, ScienCell Research Laboratories
Intestine	Epithelial cells, myofibroblasts, smooth muscle cells	Cell Applications, Cell Biologics, Lonza, ScienCell Research Laboratories
	Tissue equivalent using filter well insert	MatTek Corporation
Joint	Chondrocyte, synoviocytes, tenocytes	Cell Applications, Lonza, Neuromics, PromoCell, ScienCell Research Laboratories, ZenBio
Kidney	Cortical epithelial cells, mesangial cells, proximal tubular epithelial cells, endothelial cells	ATCC, Biopredic International, Cell Applications, Cell Biologics, Lifeline Cell Technology, Lonza, Neuromics, PromoCell, ScienCell Research Laboratories, ZenBio
Liver	Hepatocytes, fibroblasts, endothelial cells, Kupffer cells, stellate cells	Biopredic International, Cell Applications, Cell Biologics, LifeNet Health, Lonza, Neuromics, ScienCell Research Laboratories, ThermoFisher Scientific, ZenBio
Lung	Fibroblasts, smooth muscle cells, small airway epithelial cells	ATCC, Cell Applications, Cell Biologics, Lifeline Cell Technology, Lonza, PromoCell, ScienCell Research Laboratories
	Tissue equivalent using filter well insert	MatTek Corporation
Lymphatic system	Endothelial cells, fibroblasts, microvascular endothelial cells, mononuclear cells, thymic epithelial cells and fibroblasts, tonsillar endothelial cells	ABM, Cell Biologics, Lonza, ScienCell Research Laboratories
Nervous system	Astrocytes, cerebellar granule cells, cortical neurons, meningeal cells, microglia, neural progenitors, Schwann cells	Cell Applications, Lonza, Neuromics, ScienCell Research Laboratories, ThermoFisher Scientific

TABLE 24.1. (*continued*)

Tissue or system[a]	Cell type[a]	Supplier[b]
Oral cavity	Keratinocytes, fibroblasts	ATCC, Cell Biologics, Lifeline Cell Technology, ScienCell Research Laboratories
	Tissue equivalent using filter well insert	EPISKIN, MatTek Corporation
Ovary	Ovarian epithelial cells, fibroblasts	Cell Biologics, ScienCell Research Laboratories
Pancreas	Pancreatic islets	Cell Biologics, Lonza
Placenta	Epithelial cells, villous trophoblasts, villous fibroblasts	Cell Applications, ScienCell Research Laboratories
Prostate	Epithelial cells, fibroblasts, smooth muscle cells	ATCC, Cell Applications, Cell Biologics, Lifeline Cell Technology, Lonza, ScienCell Research Laboratories
Skeletal muscle	Skeletal muscle cells	Cell Applications, Lifeline Cell Technology, Lonza, ScienCell Research Laboratories, ThermoFisher Scientific, ZenBio
Skin	Epidermal keratinocytes, epidermal melanocytes, dermal fibroblasts, dermal papillary cells, sebocytes	ATCC, Biopredic International, Cell Applications, Cell Biologics, Cellworks, Lifeline Cell Technology, Lonza, MatTek Corporation, PromoCell, ScienCell Research Laboratories, ThermoFisher Scientific, ZenBio
	Tissue equivalent using filter well insert	EPISKIN, Genoskin, MatTek Corporation
Testis	Leydig cells, Sertoli cells	ScienCell Research Laboratories
Trachea, bronchus	Epithelial cells (including NHBE cells), fibroblasts, smooth muscle cells	ATCC, Cell Applications, Cell Biologics, Lifeline Cell Technology, Lonza, MatTek Corporation, PromoCell, ScienCell Research Laboratories, ZenBio
Umbilical cord	Endothelial cells (including HUVECs), smooth muscle cells	ATCC, Biopredic International, Cell Applications, Cell Biologics, Cellworks, Lonza, Neuromics, PromoCell, ScienCell Research Laboratories, ThermoFisher Scientific, ZenBio
Uterus, cervix, vagina	Epithelial cells, fibroblasts, smooth muscle cells	ATCC, Lifeline Cell Technology, Lonza, PromoCell, ScienCell Research Laboratories
	Tissue equivalent using filter well insert	EPISKIN, MatTek Corporation
Vascular system, e.g. coronary or pulmonary arteries	Endothelial cells, smooth muscle cells, pericytes	ATCC, Cell Applications, Cell Biologics, Lifeline Cell Technology, Lonza, Neuromics, PromoCell, ScienCell Research Laboratories, ThermoFisher Scientific, ZenBio

[a] Because of the large number of products, it is impossible to generate a comprehensive list of tissues or cell types. This table is a starting point for further investigation, focused on primary human cells from healthy individuals. Frequently used cell lines, including embryonic stem cell (ESC) lines, are listed in earlier chapters (see Tables 14.2, 23.3). Suitable media for various tissues are also listed elsewhere (see Tables 9.5, 10.3).

[b] Information in this table has been collated from supplier websites and is subject to change. Always discuss with suppliers before deciding which product is the best choice for your needs. See the Suppliers list at the end of the chapter for website information. Suppliers are listed here in alphabetical order.

HMEC, Human mammary epithelial cell; HUVEC, human umbilical vein endothelial cell; NHBE, normal human bronchial epithelial; PBMC, peripheral blood mononuclear cell.

information on the culture's characteristics, the testing that has been performed, and the growth conditions and protocols that are required for its ongoing maintenance. Different cell types may display very different morphology and the supplier should be able to supply typical images (see Figure 24.1). In some cases, the supplier's terminology may be unclear and it is important to clarify their meaning. The term "primary cells" is commonly used, implying that materials are derived from primary cultures. However, cells may have been passaged and, technically, have become cell lines (though not "primary cell lines" which is incorrect terminology; see Sections 3.2.1, 14.1).

The cost of obtaining specialized cells from a commercial supplier may be significant, ranging from $500 to more than $5000 per vial of frozen cells. It is important to compare this cost to the time and resources required to set up the necessary procedures yourself, as well as any associated procedures such as validation testing (see Section 7.4), before deciding whether

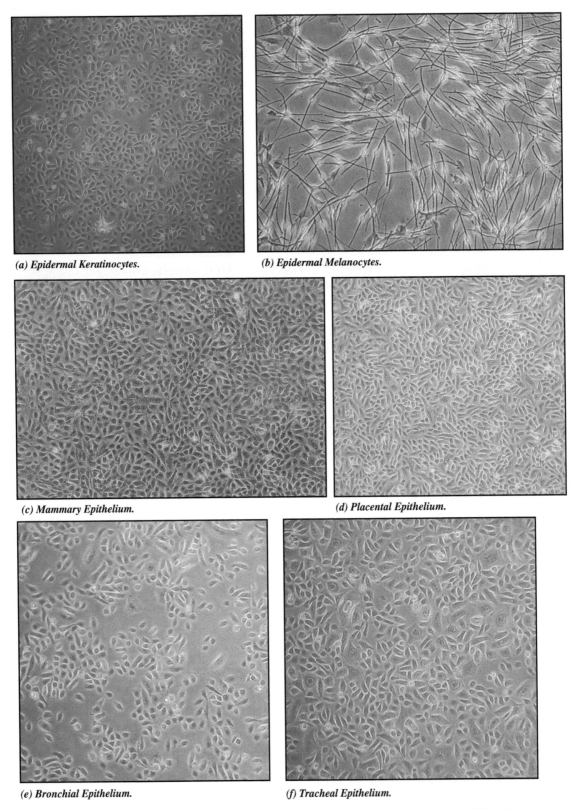

(a) Epidermal Keratinocytes.

(b) Epidermal Melanocytes.

(c) Mammary Epithelium.

(d) Placental Epithelium.

(e) Bronchial Epithelium.

(f) Tracheal Epithelium.

Cells from first or second subcultures. Phase contrast; 10x objective. Source for all images: Courtesy of Cell Applications, Inc.

Fig. 24.1. Human specialized cells in primary culture. Cells are from first or second subcultures. (a) Epidermal keratinocytes. (b) Epidermal melanocytes. (c) Mammary epithelium. (d) Placental epithelium. (e) Bronchial epithelium. (f) Tracheal epithelium. (g) Chondrocytes. (h) Osteoblasts. (i) Internal thoracic smooth muscle myoblasts. (j) Skeletal muscle myoblasts. (k) Pulmonary artery endothelium. (l) Astrocytes. Phase contrast, 10X objective. Source for all images: Courtesy of Cell Applications, Inc.

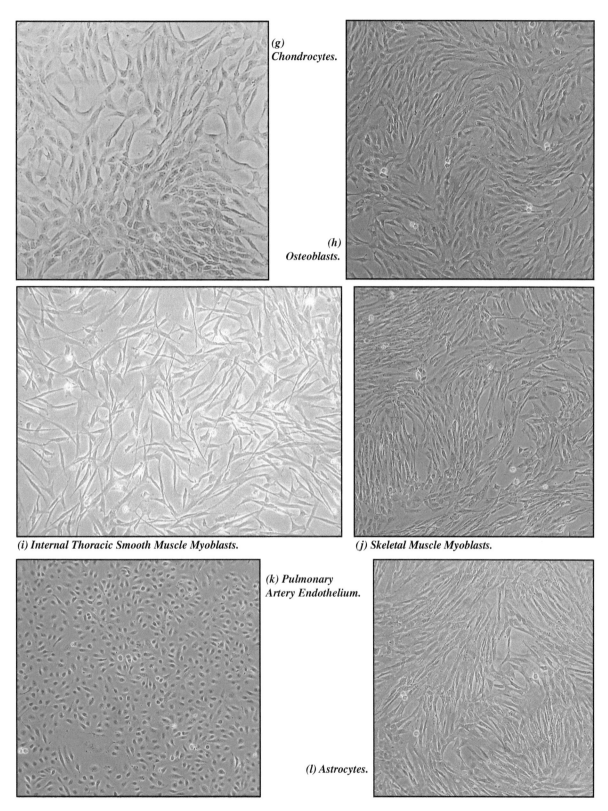

(g)
Chondrocytes.

(h)
Osteoblasts.

(i) Internal Thoracic Smooth Muscle Myoblasts.

(j) Skeletal Muscle Myoblasts.

(k) Pulmonary
Artery Endothelium.

(l) Astrocytes.

Cells from first or second subcultures. Phase contrast; 10x objective. Source for all images: Courtesy of Cell Applications, Inc.

Fig. 24.1. (Continued)

to proceed with the purchase. There may be situations where tissue from a particular species, genetic background, or disease is required. If you have access to the desired material, it may be preferable to undertake the isolation and culture yourself. Laboratories that are already doing specialized cell culture and using selective media are often able to expand into another cell type without a major outlay. You may wish to review the relevant protocol to assess its complexity and the required resources, and then make the choice of in-house preparation or purchase based on that assessment.

Specialized cells may require 3D culture conditions to express differentiated characteristics. For example, "skin equivalents" may be purchased that combine keratinocytes with dermal fibroblasts and collagen in a filter well insert, e.g. Epiderm™ (MatTek Corporation), EpiSkin™ (EPISKIN), or NativeSkin® (Genoskin; see Section 27.7.1). This culture system results in the development of a stratified epidermis and is commonly used for toxicity testing and inflammation research (see Figures 27.15, 29.12). In other cases, genetic modifications may be required to preserve the differentiated phenotype. The *H-2K^b-tsA58* transgenic mouse ("Immortomouse") carries a temperature-sensitive variant of simian virus 40 T-antigen (SV40 TAg) under the control of the *H-2K^b* promoter element (see Section 22.3.2) (Jat et al. 1991). Cells cultured from these animals are already immortalized but still retain some differentiated functions. As the cells carry a temperature-sensitive promoter for the immortalizing gene, it is possible to recover the normal phenotype when cells are grown under non-permissive conditions (39.5 °C without interferon-γ [IFN-γ]). This type of model opens up a wide range of possibilities; for example, cells may be grown from the intestinal tissue of transgenic mice carrying various mutations (Whitehead and Robinson 2009).

Culture of specialized cells can be challenging. There are many reviews and books that are focused on culture techniques for specific cell types (Freshney et al. 1994; Freshney and Freshney 1996, 2002). Such reviews and books are a valuable resource for further information, even if the techniques may sometimes seem a little outdated to the modern reader. Many of today's procedures for handling and primary culture of specific tissue types are based on techniques that were developed many years ago by cell culture practitioners who spent decades developing procedures for specific cell types.

24.2 EPITHELIAL CELLS

Epithelial cells are responsible for the recognized functions of many organs, including polarized transport in the kidney and gut, digestive enzyme secretion in the pancreas, metabolism in the liver, gas exchange in the lung, and barrier protection in skin. They are also of interest as models of differentiation and stem cell kinetics (Clevers and Watt 2018) and are among the principal tissues in which common cancers arise.

The major problem in the culture of pure epithelium is the overgrowth of the culture by stromal cells, such as connective tissue fibroblasts and vascular endothelium. Most variations in technique are aimed at preventing such overgrowth by using selective culture conditions, as described in a previous chapter (see Section 20.7). A summary of methods is provided here that can be used for epithelial cells and some other cell types discussed in this chapter (see Table 24.2). Typically, these methods aim to promote the growth of the undifferentiated epithelium and, preferably, the stem cells. Subsequent modifications may then be employed to enhance epithelial differentiation, although usually at the expense of proliferation (see Section 26.2.2; see also Figure 2.7).

The isolation of epithelial cells from donor tissue is often best performed with collagenase, Dispase, or trypsin at 4 °C (see Section 13.5). These methods disperse the stromal cells but leave the epithelial cells in small clusters, allowing their separation by settling and favoring their subsequent survival. Culture of epithelial cells is best performed using serum-free media. Factors contained in serum — many of them derived from platelets — have a strong mitogenic effect on fibroblasts and tend to inhibit epithelial proliferation by inducing terminal differentiation. Consequently, one of the most significant events in the isolation and propagation of specialized cell cultures has been the development of selective, serum-free media (see Tables 10.1, 10.3).

24.2.1 Epidermis

The epidermis represents one of the most interesting models of epithelial differentiation, partly because of its demonstration of developmental hierarchy by histology and partly because it was one of the first models to lend itself to 3T3 feeder layer-supported growth (see Figure 18.2c). Because of progress in basic cell culture technology and in our understanding of the culture requirements of the various epithelial tissues, keratinocytes of most stratified epithelia can be grown and studied in cell culture. Mostly, the squamous epithelia of the skin and their isolated epithelial cells, the keratinocytes, have been used to study their physiology and pathology *in vitro*. Skin appendages such as the hair follicle have also been isolated and cultured successfully (Fusenig 1994). The growth of knowledge in this area has led to the use of cultured epidermal keratinocytes in grafts for burns or injury repair (Green 2008; Ter Horst et al. 2018).

Cells from different tissues and organs growing in primary or early subculture are commercially available through numerous suppliers (see Table 24.1). Keratinocytes in early subculture that are commercially available are usually derived from neonatal foreskin; these sources often do not discriminate between the ortho-keratinized cells of the outer leaflet and the mucous-type inner leaflet, thus limiting their comparability to keratinocytes derived from adult skin. Isolation and culture of keratinocytes from adult skin is still required for specific research purposes, particularly when dealing with

TABLE 24.2. Inhibition of fibroblastic overgrowth.

Method	Agent	Tissue	References
Selective adhesion and detachment (see Section 20.7.3)	Trypsin	Fetal intestine; cardiac muscle; epidermis	Owens et al. (1974); Milo et al. (1980)
	Collagenase	Breast carcinoma	Freshney (1972); Lasfargues (1973)
Confluent feeder layers (see Section 8.4; see also Protocol P20.7)	Mouse 3T3 feeder cells	Epidermis	Rheinwald and Green (1975)
	Fetal human intestine feeder cells	Normal and malignant breast epithelium	Stampfer et al. (1980)
Selective substrates (see Section 20.7.2)	Matrigel	Hepatocytes	Bissell et al. (1987)
	Collagen, fibronectin, gelatin	Umbilical vein endothelial cells	Relou et al. (1998)
Selective media (see Section 20.7.1; see also Table 10.3)	MCDB 153	Epidermis	See Appendix C
	MCDB 170	Breast	See Appendix C
	Low Ca^{2+}	Epidermal melanocytes	Naeyaert et al. (1991)
Selective inhibitors (see Section 20.7.1)	D-valine	Kidney epithelium	Gilbert and Migeon (1975, 1977)
	Cis-OH-proline	Cell lines	Kao and Prockop (1977)
	Cytosine arabinoside (AraC)	Peripheral nerves	Brockes et al. (1979)
	Ethylmercurithiosalicylate	Neonatal pancreas	Braaten et al. (1974)
	Phenobarbitone	Liver	Fry et al. (1979)
	Antimesodermal antibody	Squamous carcinoma; colonic adenoma	Edwards et al. (1980); Paraskeva and Williams (1992)
	Geneticin	Melanocytes; melanoma	Levin et al. (1995)

diseased skin such as psoriatic or malignant keratinocytes. Procedures for epidermal keratinocytes are available in the Supplementary Material online (see Supp. S24.1).

24.2.2 Cornea

Cultured corneal epithelium is particularly useful for drug toxicity studies. *In vivo* testing (e.g. the Draize test) is problematic due to the number of animals used in each test, their pain and discomfort, and the differences in physiology between the rabbit and human eye (see Section 29.1.1) (Ronkko et al. 2016). Human corneal constructs are available commercially, including the SkinEthic™ HCE (EPISKIN) and EpiOcular™ (MatTek Corporation) models (see Figure 29.12). *In vitro* 3D models can be developed for various uses; for example, immortalized corneal epithelial cells can be grown on curved cellulose filters to study the interactions between contact lenses and disinfecting solutions (Postnikoff et al. 2014). There is also a great deal of interest in developing engineered corneas for transplantation (Wu et al. 2014).

A procedure is provided in the Supplementary Material online for primary culture of expired eye bank tissue (see Supp. S24.2). Although these primary cultures are useful to explore basic cell biological mechanisms and toxicological phenomena, they have a limited lifespan (about 9–10 population doublings). Corneal epithelial cells can be immortalized using SV40 TAg or other methods to provide a reliable source of material that can be shared among laboratories (Ronkko et al. 2016). It is also possible to culture limbal, stromal, and endothelial stem cells from cornea (Du and Funderburgh 2007). Limbal stem cells will differentiate into stratified layers at the air–liquid interface.

24.2.3 Oral Epithelium

Oral keratinocytes are used to study oral epithelium and its disorders, and as transplants to treat limbal stem cell deficiency in the eye (Utheim et al. 2016). Oral keratinocytes are commercially available (see Table 24.1) or may be cultured from oral mucosal biopsy samples. A procedure is provided in the Supplementary Material online for primary culture of oral keratinocytes (see Supp. S24.3). This procedure uses a similar approach to bronchial and tracheal epithelium (see Section 24.2.4); dishes are coated with fibronectin, collagen, and bovine serum albumin (BSA) to promote attachment. Selective media can be used, including MCDB 153 (which is optimized for epidermal keratinocytes) or LHC-9, to inhibit fibroblastic overgrowth (see Section 10.1.2). Transplantable sheets of oral epithelia are usually produced using fetal bovine serum (FBS) and 3T3 feeder layers, as pioneered for oral keratinocytes (Rheinwald and Green 1975; Utheim et al. 2016). Other substrates may be used such as amniotic membrane, fibrin gel, or collagen (Utheim et al. 2016).

24.2.4 Bronchial and Tracheal Epithelium

Normal human bronchial epithelial (NHBE) cells are used to study respiratory biology, disorders such as carcinogenesis, and the consequences of exposure to irritants and carcinogens, e.g. in cigarette smoke condensate (Hellermann et al. 2002). NHBE cells are usually derived from bronchoscopy brushings or biopsies taken from the distal part of the trachea, where it bifurcates to form the primary bronchi (Looi et al. 2011; Davis et al. 2015). As for other epithelial cell types, there are numerous commercial sources for early passage cultures (see Table 24.1). However, if tissue samples are available, laboratories can perform explant culture using coated plastic dishes. Medium should be serum-free; NHBE cells cease to divide when exposed to serum and will terminally differentiate. A protocol for primary culture of airway tissue samples using coated dishes and LHC-9 medium is provided in the Supplementary Material online (see Supp. S24.4). NHBE cells may be cultured on filter well inserts to form an intact epithelium; this is often done using culture at the air–liquid interface (Galietta et al. 1998). Immortalized cells may also be used, due to the limited lifespan of NHBE cultures (up to 30 population doublings). However, 3D filter well culture can be optimized for primary NHBE cells using separate media for expansion and for culture at the air–liquid interface, e.g. PneumaCult™-Ex Plus and PneumaCult-ALI (STEMCELL Technologies) (Rayner et al. 2019).

24.2.5 Gastrointestinal Tract

Although there are numerous reports of continuous cell lines being established from human colon carcinoma, very few reports have described successful culture of normal epithelium from the gut lining (Owens et al. 1976; Quaroni et al. 1979). The greatest success has been reported for the isolation and culture of intact colonic crypts. A procedure to isolate and culture colonic crypts is provided in the Supplementary Material online (see Supp. S24.5). These crypts were originally grown in 3D culture using collagen gels and a feeder layer of mouse fibroblasts or bovine aortic endothelial (BAE) cells (Whitehead et al. 1987). Cells remained viable for prolonged periods but did not actively proliferate (Whitehead and van Eeden 1991). More recently, it was discovered that *Lgr5*-positive cells within these crypt preparations act as adult stem cells, giving rise to the other cell types within the crypt structure (Sato and Clevers 2013). If grown under the correct conditions in organoid culture, *Lgr5*-positive cells continue to proliferate for extended periods in culture and give rise to 3D crypt-like structures (see Section 27.6; see also Figure 1.3a, b) (Sato et al. 2009).

24.2.6 Liver

Although cultures from adult liver do not express all the properties of liver parenchyma, there is no doubt that selective isolation of the different cell types is possible and that the correct lineage of cells may be cultured. Most efforts have been focused on the isolation and culture of hepatocytes. A procedure for isolation of rat hepatocytes is provided in the Supplementary Material online (see Supp. S24.6). This two-step procedure washes the liver with a calcium-free buffer, followed by perfusion with collagenase (Seglen 1976; Guguen-Guillouzo and Guillouzo 1986). The technique can be used to generate a high yield of hepatocytes from various rodent species and has been adapted for the human liver (Guguen-Guillouzo et al. 1982). Primary human hepatocytes are also available from a number of commercial suppliers (see Table 24.1).

Hepatocytes rapidly lose their specialized characteristics (typically 48 hours after plating) and do not proliferate under two-dimensional (2D) culture conditions (Shulman and Nahmias 2013; March et al. 2015). Cells can be grown in various 3D formats, such as a collagen "sandwich" where hepatocytes are positioned between two layers of rat tail collagen (Dunn et al. 1989). Co-culture systems are also used; for example, a mold can be made from polydimethylsiloxane (PDMS) (see Section 8.2.2) and used to generate collagen-coated "islands" for hepatocyte attachment (March et al. 2015). Mouse embryonic fibroblasts (MEFs) (e.g. the Swiss-3T3J2 subclone) are then seeded on the remaining spaces, resulting in a stable and reproducible co-culture system that can last four to six weeks. However, as with intestinal cells, the best method for long-term culture is likely to be organoid culture. Organoids can be established and maintained long-term (more than three months) from fresh or frozen liver biopsies or fine-needle aspirates collected from various species, including human, mouse, dog, and cat (see Figure 24.2) (Huch et al. 2015; Kruitwagen et al. 2017).

24.2.7 Pancreas

Islet and acinar cells can both be grown from pancreatic tissue (Vonen et al. 1992; Bosco et al. 1994). A procedure is provided in the Supplementary Material online for culture of pancreatic epithelium (see Supp. S24.7). However, most cells coming from this method are acinar cells rather than islet cells. There is particular interest in the isolation of pancreatic islets for transplantation into patients with diabetes mellitus (Lakey et al. 2003). Preparation of pure pancreatic islets is challenging, particularly from the human pancreas, where preparations typically consist of about 50% non-islet cells (Pisania et al. 2010). Collagenase and thermolysin (e.g. Liberase™, Sigma-Aldrich) are commonly used to generate islet preparations; collagenase does not affect cell–cell contacts, which improves cell survival and islet yield (Yesil et al. 2009; Vidi et al. 2013). Enzymatic methods are often combined with mechanical dissociation. The Ricordi chamber (or modifications thereof) allows pancreatic islets to be promptly removed from the dissociation chamber, minimizing overdigestion and associated damage (Piemonti and

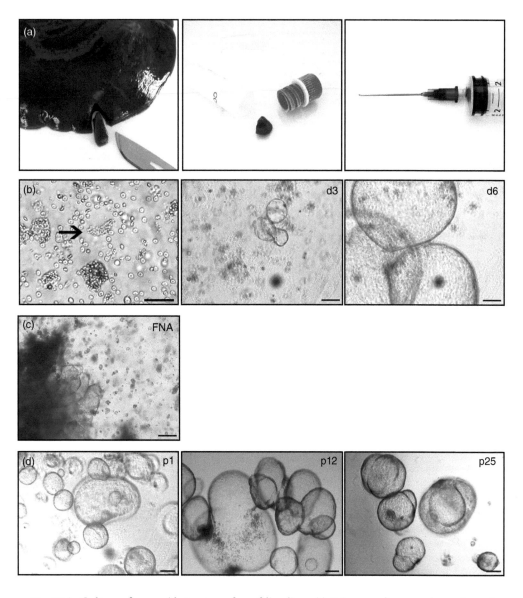

Fig. 24.2. Culture of organoid structures from feline liver. (a) Primary culture can be performed using fresh liver (left), snap-frozen liver (middle), or fine-needle aspirate (right) samples. (b) Representative phase contrast images following enzymatic digestion of liver samples. The arrow (left panel) points to a fragment of biliary duct; spherical structures can be observed after three days (d3) that grow to form large organoids after six days (d6). (c) Representative image of an undigested fine-needle aspirate (FNA); organoids emerge from the tissue fragments after five days. (d) Representative images of feline liver organoid at passages 1 (left), 12 (middle), and 25 (right). Scale bars represent 50 μm (b) or 100 μm (c, d). *Source*: Kruitwagen, H. S., Oosterhoff, L. A., Vernooij, Igwh, et al. (2017). Long-Term Adult Feline Liver Organoid Cultures for Disease Modeling of Hepatic Steatosis. Stem Cell Reports 8 (4):822–830. doi: 10.1016/j.stemcr.2017.02.015., reproduced with permission of Elsevier.

Pileggi 2013). As with other epithelial cell types, expression of differentiated characteristics can be improved using 3D culture. Islet-depleted tissue (following collagenase digestion) can be used to generate organoids with budding structures that are similar to those seen during pancreatic development (Loomans et al. 2018).

24.2.8 Breast

Normal human mammary epithelial cells (HMECs) can be cultured from samples of milk or from normal breast tissue collected during reduction mammoplasty or mastectomy (Buehring 1972; Stampfer et al. 1980; LaBarge et al. 2013). Milk samples give purer cultures of epithelial cells, but they

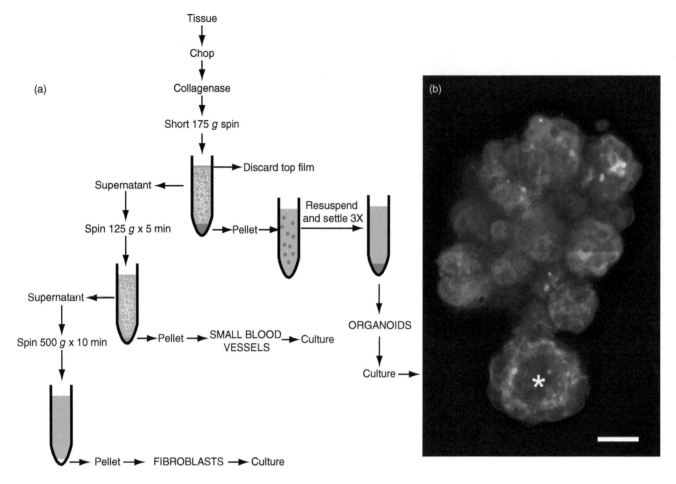

Fig. 24.3. Isolation of organoid structures from breast. (a) Flow chart of isolation of different fractions by selective sedimentation and centrifugation following collagenase digestion. (b) A TDLU-like structure derived from a Muc-/ESA+/CD29hi D920 cell grown in collagen I/laminin-1 3D culture (LaBarge et al. 2007). The structure was stained for keratins-8 and -14 (green and red, respectively, in the electronic version of the chapter) and imaged with a confocal microscope. The presented image is a reconstruction of several optical slices to give the appearance of three dimensions. There is a lumen indicated by a star. Scale bar represents 50 μm. *Source*: Courtesy of Mark LaBarge.

do not survive as long. There are a number of commercial sources for HMECs (see Table 24.1), but primary culture can be readily performed if samples of normal breast tissue are accessible. A procedure to prepare HMECs from reduction mammoplasty specimens is provided in the Supplementary Material online (see Supp. S24.8). Collagenase dissociation is preferred for primary culture and can also be used to solubilize the extracellular matrix (ECM), e.g. if cells are grown in Matrigel (Speirs et al. 1996; Vidi et al. 2013). Confluent feeder layers can be used to repress stromal contamination, including NIH 3T3 cells or fetal human intestinal FHs 74 Int cells (see Protocol P25.2; see also Figures 3.12a, b, 20.1d) (Stampfer et al. 1980; Krasna et al. 2002). An optimized, serum-free medium is also important to discourage fibroblastic overgrowth and to enable serial passaging and cloning (Band and Sager 1989; Pechoux et al. 1999; LaBarge et al.

2013). MCDB 170, which is used as the basis of some commercial media, leads to rapid induction of p16^{INK4a} and the selection of aberrant cells (Hammond et al. 1984; LaBarge et al. 2013).

Depending on the medium formulation and other conditions, HMECs can be grown to at least 50 population doublings (LaBarge et al. 2013). Conditional reprogramming can be used to extend the lifespan of normal HMECs and to preserve epithelial characteristics and cell heterogeneity (see Section 22.3.4) (Jin et al. 2018). However, to maintain the characteristics of the normal breast epithelium (including basoapical polarity and a luminal phenotype), a 3D culture system should be used that allows interaction with the ECM (see Section 26.3.5) (Vidi et al. 2013). When embedded or overlaid with Matrigel, HMECs can generate terminal ductal lobular units (TDLUs) that contain stem cells as well as differentiated acinar epithelium (see Figure 24.3) (LaBarge et al. 2007). Co-culture systems can be established that consist of luminal epithelial cells, myoepithelial cells, and fibroblasts; these systems produce branching structures with an appearance that is strikingly similar to normal breast tissue (Nash et al. 2015).

24.2.9 Cervix

Cervical keratinocytes can be used for a range of purposes, including investigations of viral pathogens, carcinogenesis, and the responses of these cells to mutagens and carcinogens. Although cervical keratinocytes are commercially available (see Table 24.1), their limited lifespan can be challenging (Villa et al. 2018). Primary cultures of cervical keratinocytes can be established using a modification of the method described for epidermal keratinocytes (Rheinwald and Green 1975; Stanley and Parkinson 1979). A procedure for the primary culture of cells from cervical epithelium is provided in the Supplementary Material online (see Supp. S24.9). Cells can be grown in serum-containing medium on Swiss-3T3 feeder layers or using serum-free conditions. The best choice of serum-free medium varies with the anatomical location of the tissue sample and the application, e.g. KGM (Lonza) appears to be the best choice for clonal assays (Deng et al. 2019). Normal cervical cells will grow for approximately 24 population doublings before they undergo senescence (see Figure 22.6) and can be grown in 3D culture using collagen rafts (Villa et al. 2018; Deng et al. 2019).

24.2.10 Prostate

Various cell types can be isolated and cultured from the human prostate, including epithelial cells, fibroblasts, and smooth muscle cells (see Table 24.1). A procedure for the isolation and culture of rat prostatic epithelium is provided in the Supplementary Material online (see Supp. S24.10). Fibroblast growth can be retarded by the choice of medium and supplements, e.g. Keratinocyte-SFM (ThermoFisher Scientific) (Rhim 2013). FGF-1 can be substituted with FGF-7 or FGF-10 to provide an additional form of selection for epithelial cells. Prostate epithelial cells express a specific splice variant, FGF-R2(IIIb), which recognizes FGF-1 and stromal cell-derived FGF-7 and FGF-10, but not FGF-2. Prostate stromal cells express only the *FGFR1* gene, which recognizes FGF-1 and FGF-2, but not FGF-7 or FGF-10. FGF activity can also be potentiated by the addition of 2 μg/ml heparin (Yan et al. 1993).

24.3 MESENCHYMAL CELLS

Mesenchymal cells are derived from the embryonic mesoderm (see Section 2.4.1) and give rise to structural elements (connective tissue, muscle, and bone), vascular cells (blood vessel endothelium and smooth muscle), and lining cells such as mesothelium. The mesoderm also gives rise to the hematopoietic system, which is discussed later in this chapter (see Section 24.5). Whereas epithelial cells are classed as parenchyma – the tissue that is responsible for the specific function of an organ – the mesenchymal cells give rise to much of the stroma (supporting tissue) of the organ.

The term "mesenchymal stromal cell" (MSCs) is commonly used for a type of adult stem cell that can differentiate into various mesodermal lineages, including adipocytes, chondroblasts, and osteoblasts. This term is somewhat misleading, as the MSC probably arises from cell populations that are associated with the blood vessel wall, and its use continues to be debated (see Sections 23.1, 23.6). However, it is important in this context because many of the cell populations described here are difficult to culture due to their scarcity or to loss of differentiated characteristics *in vitro*. MSCs are thus becoming increasingly important as a source of mesenchymal cells for tissue engineering applications (Lennon and Caplan 2006). A protocol to culture MSCs from bone marrow can be found in the Supplementary Material online (see Supp. S23.5).

24.3.1 Connective Tissue

Connective tissue cells (fibroblasts) can be isolated from many different tissues and are assumed to come from fibrocytes, based on similar morphology. However, it is unwise to judge any culture based on morphology alone. Human fibroblasts may have an elongated, spindle-shaped appearance in 2D culture (see Figure 24.4a, b), but they develop a complex network of stellate processes in 3D culture (Grinnell et al. 2003). Human, hamster, and chick fibroblasts assume a spindle-shaped morphology at confluence, producing characteristic parallel arrays of cells, while mouse fibroblasts display a pavement-like appearance (although NIH 3T3 cells may become spindle shaped if allowed to remain at high cell density). Markers that can be used to confirm fibroblast identity include CD90 (Thy-1), fibroblast-activation protein (FAP), fibroblast surface protein (FSP), and S100A4 (fibroblast-specific protein 1). Unfortunately, none of these markers is truly fibroblast-specific; many are expressed by epithelial cell lines, probably due to epithelial–mesenchymal transition (EMT) (see Section 26.2.5) (Kahounova et al. 2018).

The origin and precise identity of "fibroblasts" in culture continue to be debated. MEF cell lines produce types I and III collagen and release it into the medium (e.g. 3T3-Swiss albino and BALB/3T3 clone A31) (Goldberg 1977). Although collagen production is not restricted to fibroblasts, synthesis of type I collagen in relatively large amounts is characteristic of connective tissue. However, some MEFs can be induced to express markers that are typical of adipocytes, chondrocytes, and osteocytes (e.g. NIH 3T3 cells) (Dastagir et al. 2014). Adipose conversion of 3T3-F442A cells has been used as a model for adipose differentiation and is caused by an

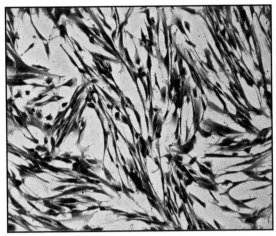

(a) Normal Human Fetal Lung Fibroblasts. Subconfluent culture. Giemsa stained; 10x objective.
Source: R. Ian Freshney.

(b) Confluent Fibroblasts. Dense culture of normal fetal human lung fibroblasts. Phase contrast; 10x objective.
Source: R. Ian Freshney.

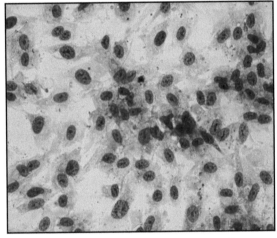

(c) Human Umbilical Cord Endothelium. Cell line from collagenase outwash of human umbilical vein. Giemsa stained; 20x objective. Source: R. Ian Freshney.

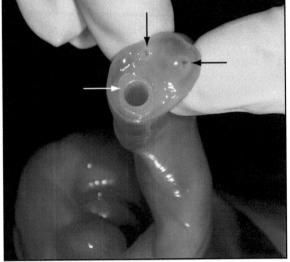

(d) View of Blood Vessels in Human Umbilical Cord.
Preparation for enzymatic isolation of endothelial cells from umbilical vein (see Supp. S24.15). Black arrows: umbilical arteries. White arrow: umbilical vein with Luer adaptor inserted. Source: Courtesy of Charlotte Lawson.

Fig. 24.4. Examples of mesenchymal cultures.

adipogenic factor in serum (Kuri-Harcuch and Green 1978). It seems likely that MEF cell lines (which are often used as feeder layers) consist of primitive mesodermal cells that may be induced to differentiate into more than one lineage. Never assume that fibroblast cultures have identical properties, and always record the origin of the culture, including the species and strain of the donor, their age, and the anatomical location of the sample; all of these properties may affect the culture's morphology and behavior (see Section 3.1).

Fibroblasts are generally regarded as the weeds of the tissue culturist's garden. They survive most mechanical and enzymatic explantation techniques and may be cultured in many simple media, particularly if serum is present. While they are readily available from commercial suppliers (see Table 24.1), primary fibroblast cultures may also be prepared by conventional procedures such as primary explantation, warm and cold trypsinization, and collagenase digestion (see Protocols P13.3–13.6). If samples are very small or precious (e.g. skin fragments from donors with rare diseases), explant culture can be used with serial transfer of the tissue fragments, resulting in an optimal yield of fibroblasts (see Supp. S13.3).

24.3.2 Adipose Tissue

Adipose cells (adipocytes) are differentiated cells whose primary physiological role has classically been described as an energy reservoir for the body. They are a storage depot for triglycerides in times of energy excess and a source of energy in the form of free fatty acids released by lipolysis during times of energy need. Adipocytes are a target of interest in their own right as active regulators of carbohydrate and lipid metabolism; they are also cultured as a source of adipose-derived stem cells (ASCs) for other applications (Badman and Flier 2007; Palumbo et al. 2018). Primary adipocyte cultures are available from commercial suppliers (see Table 24.1) or may be prepared from fatty tissues or liposuction aspirates. A procedure is provided in the Supplementary Material online for the culture of adipose cells from the rat epididymal fat pad (see Supp. S24.11). Procedures for isolation of human adipocytes and ASCs are available elsewhere (Carswell et al. 2012; Palumbo et al. 2015, 2018).

24.3.3 Muscle

Skeletal muscle. It is possible to culture myogenic cells from adult skeletal muscle of several species under conditions in which the cells continue to express at least some of their differentiated traits. The skeletal muscle progenitor cells (satellite cells) are the source of myogenic cells throughout life (Yablonka-Reuveni 2011). When grown *in vitro*, the satellite cells that are associated with the myofiber proliferate and migrate randomly on the substratum, before aligning and undergoing a fusion process to form multinucleated myotubes (Richler and Yaffe 1970; Hartley and Yablonka-Reuveni 1990). Although three to four passages can be performed by means of trypsinization, subculture is no longer possible once differentiation (i.e. fusion) has taken place.

Two protocols are provided in the Supplementary Material online for the primary culture of muscle samples. The first protocol is a traditional method (based on Pronase digestion) that can be used for biopsy samples from adult human muscle (see Supp. S24.12). The second protocol can be used to culture single myofibers and gives a closer simulation of the myogenic process *in vivo* (see Supp. S24.13). For the latter, collagenase is used to digest the connective tissue and single muscle fibers are then removed individually by gentle mechanical trituration. This permits comparisons between the responses of different muscle groups to experimental procedures or between muscles of different ages or genetic origins. Early passage cultures of skeletal myoblasts are also available commercially (see Table 24.1; see also Figure 24.1j).

Smooth muscle. Smooth muscle cells have been grown from vascular tissue and co-cultured with vascular endothelium for the tissue engineering of blood vessels and for many other applications (Subramanian et al. 1991; Klinger and Niklason 2006). Primary cultures of smooth muscle can be established from the aorta and other blood vessels using explant culture or enzymatic digestion with collagenase and elastase. These methods often result in contamination with other cell types; it is important to remove other layers during dissection by scraping to remove the endothelium and by

removing the medial layer from the intimal and adventitial layers (Klinger and Niklason 2006). The blunt back of a pair of ophthalmic curved tweezers can be used to separate the media from the main artery (Chi et al. 2017). Commercially available smooth muscle cells are derived from various blood vessels, including coronary and pulmonary arteries, and from organs such as the bladder, lung, and prostate (see Table 24.1; see also Figure 24.1i).

Cardiac muscle. Cardiac muscle cells (cardiomyocytes) can be established from various animal models; for example, cells can be isolated from avian embryos using cold trypsin disaggregation (see Protocol P13.5) and will display rhythmic contractions if grown on a substrate of suitable elasticity (Engler et al. 2008). Cardiomyocytes from small animals are usually isolated using a Langendorff apparatus, which performs perfusion with an enzymatic solution; procedures for primary culture of cardiomyocytes are available elsewhere (Louch et al. 2011). Cells can also be cultured from human cardiac tissue and are available from several commercial suppliers (see Table 24.1) (Goldman and Wurzel 1992). However, the relative scarcity of human cardiomyocytes is perhaps best addressed by converting human pluripotent stem cells (hPSCs) to cardiac lineages (Parsons 2013).

24.3.4 Cartilage

Chondrocytes are responsible for the synthesis, maintenance, and degradation of the cartilage matrix. A great deal of research in the field of rheumatology has been focused on understanding the mechanisms that induce metabolic changes in articular chondrocytes during osteoarthritis and rheumatoid arthritis (Johnson et al. 2016). Chondrocytes have been cultured from several different sources and early passage cultures are available commercially (see Table 24.1). Articular chondrocytes are a very useful tool, but, cultured in a 2D monolayer, they rapidly divide, become fibroblastic, and lose their biochemical characteristics (Holtzer et al. 1960; von der Mark et al. 1977). As early as the first passage, there is a gradual shift from the synthesis of type II collagen to types I and III collagens, and from aggrecan to low-molecular-weight proteoglycans. Chondrocyte dedifferentiation can be minimized by 3D culture in alginate, which results in a matrix that is similar to that of articular cartilage *in vivo* (Lemare et al. 1998). A procedure for the culture of chondrocytes in alginate beads is provided in the Supplementary Material online (see Supp. S24.14).

24.3.5 Bone

Although bone is mechanically difficult to handle, thin slices can be treated with EDTA and digested in collagenase to generate primary cultures of bone-derived cells. These cells have some functional characteristics of osteocytes, although they rapidly become dedifferentiated when grown as a 2D monolayer (Bard et al. 1972; van der Plas et al. 1994). Human osteoblasts are available commercially or can be isolated from surgical samples, e.g. from the trabecular bone of the humeral head (see Table 24.1) (Taylor et al. 2014). Osteoblasts can also be isolated from the calvaria of chicks and rodents, and from the long bones of neonatal rats and mature mice (Stern et al. 2012; Shah et al. 2016). Detailed procedures for primary osteoblast culture are provided elsewhere (Stern et al. 2012; Taylor et al. 2014; Shah et al. 2016).

24.3.6 Endothelium

Endothelial cell culture can provide models for vascular disease, blood vessel repair, and tumor angiogenesis, as well as providing cells for engineering of blood vessels. Capillary endothelial cells can be induced to form networks of tubes that resemble capillary vascular beds (Folkman and Haudenschild 1980). This discovery suggested that cultured endothelial cells have the ability to reconstruct blood vessels, and was borne out by later work where smooth muscle cells were incorporated into the construct (Klinger and Niklason 2006). Cells were initially grown in conditioned media from tumor cells, leading to the discovery of tumor-derived angiogenesis factors (Zetter 2008). Tube formation was accelerated when cells were grown on laminin-rich matrix, leading to the development of an assay to screen for angiogenic and anti-angiogenic factors (Kubota et al. 1988; Arnaoutova et al. 2009). Tube formation assays typically use human umbilical vein endothelial cells (HUVECs), which are available from a number of commercial suppliers (see Table 24.1). A procedure is provided in the Supplementary Material online for isolation and culture of endothelial cells from human umbilical vein, porcine aorta, and mouse and human heart (see Supp. S24.15; see also Figure 24.4c, d). The tube formation assay is described in detail elsewhere (DeCicco-Skinner et al. 2014).

24.4 NEUROECTODERMAL CELLS

The neuroectoderm in the developing embryo gives rise to the neural tube, which will become the central nervous system (CNS), and the neural crest, which gives rise to the peripheral nervous system and many other specialized cell types. The culture of neuroectodermal lineages – including neurons, glial cells, endocrine cells, and melanocytes – is challenging for many reasons. Cells from these lineages are often difficult to access; they frequently do not proliferate *in vitro* and make up minor populations that are rapidly overgrown by other cell types. Over time, scientists have developed specialized culture vessels, media, and techniques to meet these challenges, starting with Ross Harrison, who studied neuronal cells at the inception of tissue culture in 1909 (Millet and Gillette 2012). Recent techniques, such as the development of cortical organoids (see Figure 1.3c), have added to this classic trend of cell culture innovation.

24.4.1 Neurons

Primary cultures of neuronal cells are available commercially (see Table 24.1) or may be established from brain, spinal cord, dorsal root ganglia, or other neuronal structures. Primary cultures from human brain contain multiple cell types including astrocytes, neurons, microglia, and endothelial cells (Spaethling et al. 2017). Other populations are more homogenous; the granule cells in the cerebellum make up the largest homogenous population in the mammalian brain and are widely used to study neuronal development (Kramer and Minichiello 2010). A procedure for the isolation and culture of rat cerebellar granule cells is provided in the Supplementary Material online (see Supp. S24.16).

Neuronal cells do not survive well on untreated glass or plastic but demonstrate neurite outgrowth on collagen and polylysine. This preference can be utilized to develop strikingly beautiful high-resolution patterns on chemically patterned substrates (Kleinfeld et al. 1988). Outgrowth of neuronal processes can be encouraged by nerve growth factor (NGF) and other factors that are secreted by glial cells, which play an important role in neuronal function (Levi-Montalcini 1982; Meyer and Kaspar 2017). The choice of medium is also important; media formulations such as DMEM have been shown to impair neurophysiological functions, leading to the development of BrainPhys™ neuronal medium (STEMCELL Technologies) (Bardy et al. 2015).

24.4.2 Glial Cells

Astrocytes. Astrocytes are the most abundant cell type in the CNS and the easiest of the glial cells to grow (Meyer and Kaspar 2017). Human primary astrocytes are available commercially (see Table 24.1), or primary cultures can be established from fetal or adult brain samples obtained from various species. A protocol is provided in the Supplementary Material online for the primary culture of human astrocytes (see Supp. S24.17). The field of neuroscience is particularly indebted to Jan Pontén, Bengt Westermark, and colleagues, who established numerous normal and malignant glial cell lines over almost half a century (Ponten and Macintyre 1968; Xie et al. 2015). Many of these are probably astrocytes but do not express the astrocyte marker glial fibrillary acidic protein (GFAP) (Ponten and Macintyre 1968). However, some GFAP-negative cell lines have been shown to express other astrocyte markers, such as glutamine synthetase and β-alanine-sensitive γ-aminobutyric acid (GABA) uptake (Frame et al. 1980). Other markers for the identification of glial cells are A2B5, which marks all glial cells, and galactocerebroside (Gal-C), which is oligodendrocyte-specific (see Table 24.3).

Olfactory ensheathing cells (OECs). The olfactory ensheathing cell (OEC) is a type of glial cell that is located within the olfactory mucosa and olfactory bulb (Higginson and Barnett 2011). Once neuronal death occurs, either during the course of normal cell turnover or after injury, newly

TABLE 24.3. Characterization of glial cells by immunostaining.

Cell type	GFAP	Gal-C	O4	A2B5	NGF
O2-A	−	−	−	+	−
Type 1 astrocyte	+	−	−	−	−
Type 2 astrocyte	+	−	−	+	−
Oligodendrocyte	−	+	+	+[a]	−
OEC	+/−[b]	−	+[c]	−	+

[a] Mature oligodendrocytes lose A2B5 and become Gal-C+ and O4+ only.
[b] GFAP may be fibrous or amorphous. OECs gain GFAP expression in culture.
[c] OECs lose O4 in culture.

generated neurons are recruited from a pool of constantly proliferating putative stem cells in the olfactory epithelium (Schwob 2002). These neurons produce axons that can re-innervate the olfactory bulb and re-establish olfaction. OECs ensheath the olfactory receptor axons and are believed to guide regenerating axons back to the olfactory bulb (Higginson and Barnett 2011). These cells are of considerable interest for use in cellular transplantation following neuronal injury (Raisman et al. 2012). OECs can be isolated from the olfactory bulb of the newborn rat (see Figure 24.5a) and other species. A procedure for the isolation and culture of rat OECs is provided in the Supplementary Material online (see Supp. S24.18). This technique relies on magnetic separation using an EasySep™ kit (STEMCELL Technologies) to select cells that are positive for p75[NTR], resulting in highly purified cultures (see Figure 24.5b–d).

Other glial cell types. Primary oligodendrocytes can be isolated from rodent brain or dorsal root ganglion and enriched using differential adhesion or magnetic separation (O'Meara et al. 2011; Schott et al. 2016). Cells are difficult to subculture but their yield may be improved by the use of papain (Young and Levison 1997). Primary microglia may be cultured from newborn mouse pups; microglia typically grow on the top of the astrocyte layer in mixed culture and may be enriched by tapping the vessel, leading to detachment into the culture medium (Lian et al. 2016). Schwann cells may be isolated from peripheral nerves (e.g. from mouse or rat pups) and enriched by treatment with cytosine arabinoside to remove dividing cell populations (Brockes et al. 1979; Wen et al. 2017).

24.4.3 Endocrine Cells

The problems of culturing endocrine cells (O'Hare et al. 1978) are accentuated for the neuroendocrine cells that arise from the neural crest. The pituitary gland, adrenal gland, and other organs that contain neuroendocrine cells typically give rise to highly heterogeneous primary cultures, with the neuroendocrine cells making up a minor population relative to other parenchymal and stromal cells. Most *in vitro*

neuroendocrine models are cell lines that have been derived from neuroendocrine tumors or (rarely) from normal tissue following immortalization (Grozinsky-Glasberg et al. 2012). Endocrine cell lines often do not represent the full spectrum of cell types within the organ; looking at pituitary cell lines, the majority come from the anterior lobe and there are very few from intermediate or posterior lobes (Ooi et al. 2004). However, enrichment for specific cell types can be performed using magnetic separation or similar techniques, provided candidate antibodies are available for positive or negative selection (see Section 21.3). The monoclonal antibody MOC-1 is one such candidate and has been used to isolate pulmonary neuroendocrine cells from fetal rabbit lung (Speirs et al. 1992). There is a clear need to optimize culture systems for the various endocrine cell types; for example, culture media can be optimized to improve the culture of primary thyroid cells using 2D and 3D approaches (Wang et al. 2016).

24.4.4 Melanocytes

Melanocytes make up approximately 5–10% of human skin, although the density varies with age, sun exposure, and other factors (Thingnes et al. 2012). Primary melanocyte cultures are available commercially from a number of suppliers (see Table 24.1) or can be established from skin samples. Foreskin and scalp hair follicles are often used because these areas have relatively high densities of melanocytes, but facial skin and other regions can be used successfully (Tang et al. 2014). Melanocytes can be recognized in primary culture as small, round cells with dendritic processes; they may have a pigmented appearance due to the presence of melanin granules (see Figure 24.6). Overgrowth by keratinocytes or fibroblasts can be addressed by adding phorbol 12-myristate 13-acetate (PMA), which is toxic to keratinocytes, and by reducing calcium concentration, which reduces fibroblast overgrowth (Eisinger and Marko 1982; Naeyaert et al. 1991). However, the presence of PMA results in reduced melanin synthesis due to downregulation of protein kinase C (PKC) (Park et al. 1993). A hormone-supplemented medium can be used instead to achieve preferential melanocyte attachment and growth, as described in the Supplementary Material online (see Supp. S24.19).

24.5 HEMATOPOIETIC CELLS

24.5.1 Lymphoid Cells

Short-term culture of lymphocytes has been possible for some time and is widely used for DNA extraction, chromosome preparation, or antigenic stimulation. A procedure is given in the Supplementary Material online for preparation of lymphocytes, based on density separation, and for stimulation with phytohemagglutinin (PHA) (see Supp. S24.20). Viral transduction can be performed to extend the lifespan of B-lymphocytes using Epstein-Barr virus (EBV).

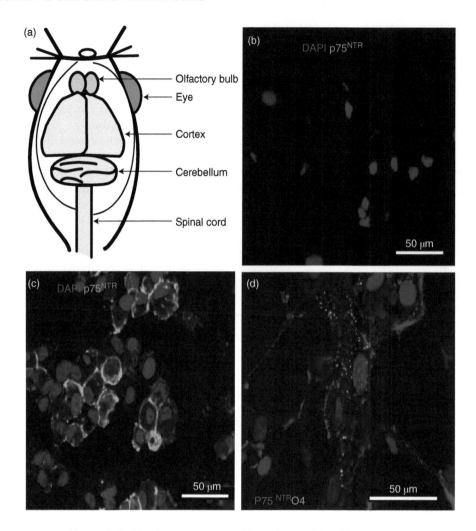

Fig. 24.5. Olfactory bulb. (a) Schematic diagram of the isolation of the olfactory bulbs from the brain of newborn rats. (b–d) Immunostaining of olfactory bulb tissue for p75[NTR] pre- and post-purification using an EasySep kit. (b) After enzymic digestion of the olfactory bulbs, a heterogeneous mix of cells were plated onto poly-L-lysine coated coverslips and allowed to sit for 30 minutes. Immunostaining revealed that very few of these cells expressed p75[NTR]. (c) Immediately after purification using the EasySep kit to select for a p75[NTR]-positive population of cells, approximately 98% of cells expressed this marker, suggesting that this methodology is effective at producing highly purified cultures of OECs. (d) Purified OECs expressing p75[NTR] and the O4 antibody. Scale bars represent 50 μm. *Source*: Courtesy of Susan Barnett.

LCLs are commonly established to increase the amount of DNA available from rare blood samples (e.g. from indigenous populations who live in closed societies), provided the donor's consent allows for further applications (Danjoh et al. 2011). Procedures for LCL generation are widely available and require as little as 0.1 ml blood (Bolton and Spurr 1996; Hui-Yuen et al. 2011; Omi et al. 2017).

Although cell lines that are generated using EBV are often referred to as immortalized (see Section 22.3.1), this is not entirely accurate. The majority of LCLs undergo senescence at an average population doubling level (PDL) of about 90; a minority are truly immortal and continue to grow beyond 160 PDL (Sugimoto et al. 1999). Many LCLs continue to carry the viral genome, which is maintained as a circular episome in the cell nucleus. Thus, LCLs may continue to express EBV and shed infectious virions; these possibilities should be considered during risk assessments for subsequent experimental work (see Sections 6.3.1, 16.5, 22.3.1).

24.5.2 Macrophages and Myeloid Cells

Macrophages may be isolated from many tissues by collecting the cells that attach during enzymatic disaggregation (see Section 20.7.3). The yield is rather low, however, and various techniques have been developed to obtain larger numbers of macrophages. Monocytes from peripheral blood or monocyte/macrophage precursors from bone

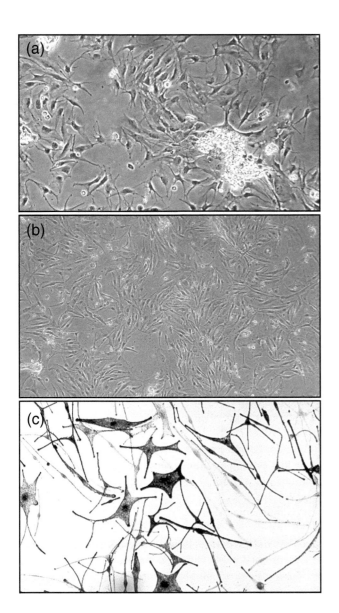

Fig. 24.6. Melanocyte cultures. (a) Culture of Caucasian newborn foreskin-derived melanocytes one week after inoculation. Note the multiple keratinocyte colonies with central stratification and tightly apposed epithelial cells at the periphery. The melanocytes are the relatively small, dark dendritic cells, most of them in contact with the keratinocyte colonies by means of dendritic projections. (b) Ten-day-old primary cultures of Caucasian newborn epidermal melanocytes in medium lacking PMA (also known as TPA). Many cells display branching dendrites and other cells display bipolar to polygonal morphology. (c) Secondary culture of PMA-treated epidermal melanocytes derived from African-American newborn foreskin. Note the dendritic morphology of the cells and their slender spindle shape. In contrast to the melanocytes derived from Caucasian newborn foreskin, these melanocytes display a high level of melanin granules. Seen via phase contrast at 40× (b), 100× (a), and 320× (c) magnification. *Source*: Courtesy of Hee-Young Park.

marrow may be induced to differentiate into macrophages using macrophage colony-stimulating factor (M-CSF) or granulocyte-macrophage colony-stimulating factor (GM-CSF) (Waldo et al. 2008; Jin and Kruth 2016). These two factors give rise to macrophages with different phenotypes (see Figure 24.7) (Waldo et al. 2008).

A number of myeloid cell lines have been developed from murine leukemias and, like some of the human LCLs, have been shown to make globulin chains and (in some cases) α- and γ-globulins (Horibata and Harris 1970; Collins et al. 1977). Some of these lines can be grown in serum-free medium (Iscove and Melchers 1978). The most significant use of myeloid cell lines has been in the generation of hybridomas from myeloma cell lines (see Section 24.5.4).

24.5.3 Erythroid Cells

Erythroid cells have traditionally been difficult to study *in vitro*. An early model system was based on the "Friend virus," which was discovered by Charlotte Friend (Friend 1957). Friend virus was actually a complex of two retroviruses: a spleen focus-forming virus (SSFV) that was replication-defective, and a murine leukemia virus (MLV) that was replication-competent and acted as a helper virus (Ruscetti 1995). When inoculated into susceptible mice, Friend virus led to a multistage disease that included splenic enlargement and subsequent erythroleukemia. Cells taken from the minced spleens of these animals could, in some cases, give rise to continuous cell lines. Friend cells could then be treated with dimethyl sulfoxide (DMSO), sodium butyrate, or other agents to promote erythroid differentiation (Friend 1978). Untreated cells resembled undifferentiated proerythroblasts, whereas treated cells showed nuclear condensation, a reduction in cell size, and an accumulation of hemoglobin to the extent that centrifuged cell pellets were red in color (see Section 26.3.1). Evidence for differentiation could be seen by staining the cells for hemoglobin with benzidine and by performing *in situ* hybridization of globin-specific mRNA (see Figure 26.4a, b).

More recently, progress has been made to establish erythroid progenitor cell lines from embryonic stem cells (ESCs) or hematopoietic stem cells (HSCs). These immortalized cell lines gave rise to enucleated red blood cells (RBCs) (Nakamura et al. 2011; Kurita et al. 2013). Induction of differentiation resulted in the production of hemoglobin, giving an obvious red color when cells were pelleted by centrifugation, as noted previously for Friend cells (see Figure 26.4c, d).

24.5.4 Hybridoma Cells

Monoclonal antibodies are produced by cloned cell lines that are derived from the hybridization of myeloma cells

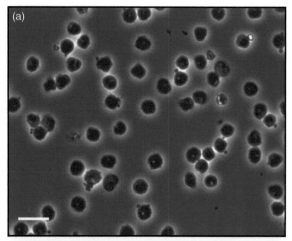

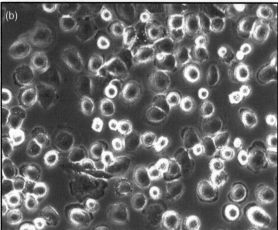

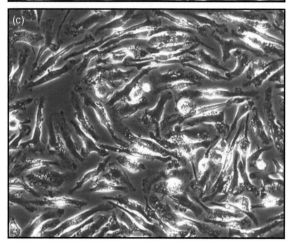

Fig. 24.7. Macrophage cultures. (a) Human monocytes cultured for two hours before induction of differentiation; (b) macrophages following seven days treatment with GM-CSF, resulting in maintenance of a spherical shape; (c) macrophages following seven days treatment with M-CSF, resulting in an elongated appearance with numerous vacuoles. Images are representative of more than 10 monocyte donors. Scale bar represents 10 μm. *Source*: Waldo, S. W., Li, Y., Buono, C., et al. (2008). Heterogeneity of human macrophages in culture and in atherosclerotic plaques. Am J Pathol 172 (4):1112–26. doi: 10.2353/ajpath.2008.070513, reproduced with permission of Elsevier.

with B-lymphocytes (Köhler and Milstein 1975). This results in continually proliferating hybridoma cells that can be induced to express a monoclonal antibody (mAb) targeting a specific antigen. Monoclonal antibodies are commonly used in research laboratories, where they have replaced polyclonal antibodies in many different applications. They are also used for therapeutic purposes; as of 2018, the United States Food and Drug Administration (FDA) had approved 64 monoclonal antibodies for therapeutic use (Tsumoto et al. 2019).

Numerous techniques can be used to generate hybridomas (Tomita and Tsumoto 2011). B-cell targeting (BCT) can be used to increase efficiency; sensitized B-lymphocytes are preselected and induced to form complexes with myeloma cells before fusion is induced by electrical pulses, resulting in a hybridoma that expresses the desired mAb with high efficiency (see Figure 24.8a). A protocol for the BCT technique is provided as Supplementary Material online (see Supp. S24.21). Stereospecific targeting (SST) has been developed more recently; DNA immunization is used to enhance recognition of the antigen and its 3D conformational structure, and antigen-expressing myeloma cells are used to select for B-lymphocytes that express the correct stereospecific monoclonal antibody (ssmAb) (see Figure 24.8b). A protocol for the SST technique is provided as Supplementary Material online (see Supp. S24.22).

Hybridoma cells must be cloned using limiting dilution or soft agar cloning and the clones screened and selected for expansion (see Figure 24.9; see also Protocols P20.1, 20.2) (Wognum and Lee 2013). HAT medium (see Appendix B) is commonly employed for screening, as it selects for cell hybrids that express hypoxanthine guanine phosphoribosyltransferase and thymidine kinase, which are contributed by the two parental cells (Littlefield 1964). Hybridoma media are available from a number of suppliers (see Table 10.3). Different media may be used to encourage cell proliferation, followed by antibody production. Kits are available that contain media for all steps in hybridoma production, e.g. ClonaCell™-HY (STEMCELL Technologies).

24.5.5 Chimeric Antigen Receptor (CAR) T-Cells (CAR T-Cells)

CAR T-cells are T-lymphocytes that have been genetically modified to express a chimeric antigen receptor (CAR). CARs usually consist of (i) an extracellular domain that recognizes a tumor-associated antigen (e.g. CD19, which is expressed by many B-cell malignancies); (ii) a structural region that positions the receptor in the cytoplasmic membrane of the cell; (iii) an intracellular domain that activates the T-cell (e.g. CD3ζ); and (iv) a second intracellular domain that provides co-stimulation (e.g. CD28 or 4-1BB) (Davila and Sadelain 2016; Sadelain 2017). T-lymphocytes are obtained from a patient who has been approved to receive therapy with a specific CAR. Usually, a blood sample is used as the starting point for the procedure and T-lymphocytes are selected

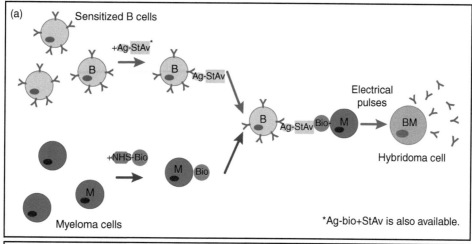

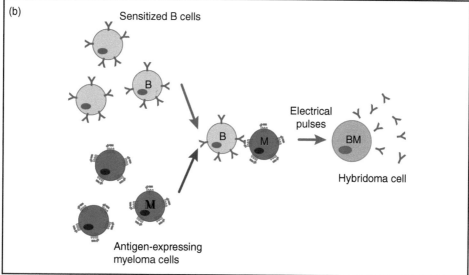

Fig. 24.8. Hybridoma generation techniques. (a) The BCT technique. Sensitized B-lymphocytes after intraperitoneal immunization are selected by biotin–antigen or antigen–streptavidin conjugates. Antigen-selected B-lymphocytes are combined with myeloma cells by biotin–streptavidin interactions. These complexes are fused by electrical pulses to generate hybridoma cells secreting monoclonal antibodies (mAbs). (b) The SST technique. Sensitized B-lymphocytes after DNA and cell immunization are selected by antigen-expressing myeloma cells. The B-cell–myeloma cell complexes are then fused by electrical pulses to generate hybridoma cells secreting stereospecific monoclonal antibodies (ssmAbs).For the associated protocols, see Supp. S24.21 and S24.22. *Source*: Courtesy of M. Tomita.

using magnetic sorting or other separation techniques (Wang and Riviere 2016). The CAR construct is introduced using one of several gene delivery systems, such as lentiviral vectors or electroporation (see Section 22.1); modified T-cells are then expanded and stored for return to the patient. The manufacturing process can take some time (often two to four weeks) and is a specialized procedure; a number of challenges must be overcome, including the need to activate the T-lymphocytes and the choice of platform for expansion (Perica et al. 2018). Despite these challenges, CAR T-cell therapies have had astonishing success to date, with response

rates of more than 80% in some clinical trials (Fesnak et al. 2016). More information on the practical requirements for CAR T-cell manufacture can be found elsewhere (Wang and Riviere 2016; Vormittag et al. 2018).

24.6 CULTURE OF CELLS FROM POIKILOTHERMS

Poikilotherms are organisms whose body temperature changes with their environment. Poikilothermic species include fish, amphibians, reptiles, and many insect species. Cell lines from these species are important models, but their

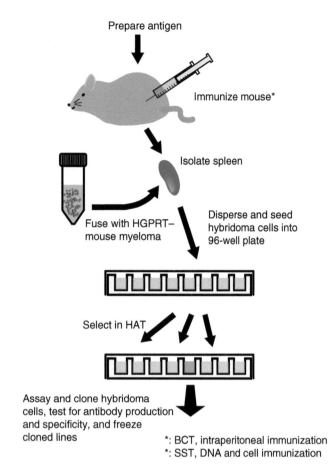

Prepare antigen

Immunize mouse*

Isolate spleen

Fuse with HGPRT–
mouse myeloma

Disperse and seed
hybridoma cells into
96-well plate

Select in HAT

Assay and clone hybridoma
cells, test for antibody production
and specificity, and freeze
cloned lines

*: BCT, intraperitoneal immunization
*: SST, DNA and cell immunization

Fig. 24.9. Fusion and cloning of hybridoma cells. Hybridoma cells secreting mAbs or ssmAbs are screened by enzyme-linked immunosorbent assay (ELISA) or cell-based ELISA (Cell-ELISA). Positive cells are cloned by limiting dilution and tested for affinity and specificity. Cloned hybridoma cells are stored in liquid nitrogen. *Source*: Courtesy of M. Tomita.

culture conditions may come as a surprise to scientists who work with mammalian cells. For example, the RTgill-W1 cell line was established from the gill tissue of normal rainbow trout; explants were left untouched in an office for about a year prior to their successful subculture (Bols et al. 1994; Bols 2017).

The process of primary culture and subculture is similar to that employed for warm-blooded animals, largely because the bulk of present-day experience has been derived from culturing cells from birds and mammals (Wolf 1979). Thus, the dissociation techniques for primary culture use proteolytic enzymes, such as trypsin, with EDTA as a chelating agent. Because of the lower temperature requirement, it may not be feasible to use conventional CO_2 incubators. Leibovitz's L-15 medium is commonly used because it does not require CO_2 in the gas phase (see Section 9.2.3), allowing cells to be maintained at physiological pH in ambient air. FBS appears to substitute well for homologous

serum or hemolymph (and is more readily available) (Wolf 1979).

If a new species is being investigated, optimal conditions may need to be established. Variables for optimization include pH, osmolality, inorganic salts, and amino acid concentrations (Lynn 1996). Try those media and sera that are currently available, assaying for growth, plating efficiency, and specialized functions (see Sections 9.7, 10.6.1). As the development of media for many species is in its infancy, it may prove necessary to develop new formulations if an untried species or tissue type is examined. Temperature should be fixed within the appropriate environmental range; overheating is particularly damaging. Poikilothermic cells have the same cell culture quality concerns as any other species and require validation testing to ensure that they correspond to the correct species and are mycoplasma-free (see Section 7.4).

24.6.1 Fish Cells

Fish cell lines are valuable models that can be used for many applications, including the study of fish pathogens, environmental pollutants, and water quality (Bols et al. 2017). More than 550 fish cell lines have been reported in the scientific literature from a wide range of species including zebrafish (*Danio rerio*), medaka (*Oryzias latipes*), rainbow trout (*Oncorhynchus mykiss*), and fathead minnow (*Pimephales promelas*) (Bairoch 2018). In some cases, cell lines or primary cultures can be used to complement work that is performed in laboratory animals; this is particularly true of the zebrafish, where primary cultures can be established directly from the zebrafish embryo and used for transfection and live cell imaging. Procedures are available elsewhere for the culture of fish cell lines and the establishment of primary cultures from the zebrafish embryo (Lakra et al. 2011; Sassen et al. 2017; Russo et al. 2018).

24.6.2 Insect Cells

The first insect cell lines were established by Thomas D. C. Grace from the Australian emperor gum moth (*Antherea hemolymph*) (Grace 1962). More than 950 insect cell lines have now been reported (Bairoch 2018); these cultures are widely used to study pest control and environmental toxicology and to produce recombinant proteins using baculovirus expression vectors (BEVs) (see Section 22.1.4). Insect cell lines are traditionally cultured in Grace's medium with serum, but serum substitutes and serum-free formulations are also available (see Tables 10.2, 10.3). Cells are usually grown at 26–28 °C in the absence of CO_2 and tend to grow in suspension or are loosely adherent. Procedures can be found elsewhere for the culture of insect cell lines and the isolation of primary cells from insect tissues (Lynn 2016; Lynn and Harrison 2016).

Suppliers.

Supplier[a]	URL
ABM	www.abmgood.com
American Type Culture Collection (ATCC)	www.atcc.org
Biopredic International	http://www.biopredic.com/index.php
Cell Applications Inc.	www.cellapplications.com
Cell Biologics	http://www.cellbiologics.net/index.php?route=common/home
Cellworks	www.cellworks.co.uk/index.php
EPISKIN	www.episkin.com
Genoskin	www.genoskin.com
Lifeline Cell Technology	www.lifelinecelltech.com
LifeNet Health	http://www.lifenethealth.org/institute-regenerative-medicine
Lonza	www.lonza.com
MatTek Corporation	www.mattek.com
Merck Millipore	www.merckmillipore.com
Neuromics	www.neuromics.com
PromoCell	www.promocell.com
ReachBio Research Labs	www.reachbio.com
ScienCell Research Laboratories	www.sciencellonline.com
Sigma-Aldrich (Merck)	http://www.sigmaaldrich.com/life-science/cell-culture.html
STEMCELL Technologies	www.stemcell.com
ThermoFisher Scientific	http://www.thermofisher.com/us/en/home/life-science/cell-culture.html
ZenBio	www.zen-bio.com

[a] The Supplier list for this chapter includes some cell repositories that distribute primary cultures of specialized cells. For a more extensive list of commonly used cell repositories see Table 15.3.

REFERENCES

Arnaoutova, I., George, J., Kleinman, H.K. et al. (2009). The endothelial cell tube formation assay on basement membrane turns 20: state of the science and the art. *Angiogenesis* 12 (3): 267–274. https://doi.org/10.1007/s10456-009-9146-4.

Badman, M.K. and Flier, J.S. (2007). The adipocyte as an active participant in energy balance and metabolism. *Gastroenterology* 132 (6): 2103–2115. https://doi.org/10.1053/j.gastro.2007.03.058.

Bairoch, A. (2018). The Cellosaurus, a cell-line knowledge resource. *J. Biomol. Tech.* 29 (2): 25–38. https://doi.org/10.7171/jbt.18-2902-002.

Band, V. and Sager, R. (1989). Distinctive traits of normal and tumor-derived human mammary epithelial cells expressed in a medium that supports long-term growth of both cell types. *Proc. Natl Acad. Sci. U.S.A.* 86 (4): 1249–1253. https://doi.org/10.1073/pnas.86.4.1249.

Bard, D.R., Dickens, M.J., Smith, A.U. et al. (1972). Isolation of living cells from mature mammalian bone. *Nature* 236 (5345): 314–315. https://doi.org/10.1038/236314a0.

Bardy, C., van den Hurk, M., Eames, T. et al. (2015). Neuronal medium that supports basic synaptic functions and activity of human neurons in vitro. *Proc. Natl Acad. Sci. U.S.A.* 112 (20): E2725–E2734. https://doi.org/10.1073/pnas.1504393112.

Bissell, D.M., Arenson, D.M., Maher, J.J. et al. (1987). Support of cultured hepatocytes by a laminin-rich gel. Evidence for a functionally significant subendothelial matrix in normal rat liver. *J. Clin. Invest.* 79 (3): 801–812. https://doi.org/10.1172/JCI112887.

Bols, N. C. (2017). The invitromatics of RTgill-W1: a star of the rainbow trout invitrome with a bright invitroomic future. https://web.expasy.org/cellosaurus/invitromaticists/bols_nc.pdf (accessed 5 July 2019).

Bols, N.C., Barlian, A., Chirino-Trejo, M. et al. (1994). Development of a cell line from primary cultures of rainbow trout, *Oncorhynchus mykiss* (Walbaum), gills. *J. Fish Dis.* 17 (6): 601–611. https://doi.org/10.1111/j.1365-2761.1994.tb00258.x.

Bols, N.C., Pham, P.H., Dayeh, V.R. et al. (2017). Invitromatics, invitrome, and invitroomics: introduction of three new terms for in vitro biology and illustration of their use with the cell lines from rainbow trout. *In Vitro Cell. Dev. Biol. Anim.* 53 (5): 383–405. https://doi.org/10.1007/s11626-017-0142-5.

Bolton, B.J. and Spurr, N.K. (1996). B-lymphocytes. In: *Culture of Immortalized Cells* (eds. R.I. Freshney and M.G. Freshney), 283–298. New York: Wiley-Liss.

Bosco, D., Soriano, J.V., Chanson, M. et al. (1994). Heterogeneity and contact-dependent regulation of amylase release by individual acinar cells. *J. Cell. Physiol.* 160 (2): 378–388. https://doi.org/10.1002/jcp.1041600219.

Braaten, J.T., Lee, M.J., Schenk, A. et al. (1974). Removal of fibroblastoid cells from primary monolayer cultures of rat neonatal endocrine pancreas by sodium ethylmercurithiosalicylate. *Biochem. Biophys. Res. Commun.* 61 (2): 476–482.

Brockes, J.P., Fields, K.L., and Raff, M.C. (1979). Studies on cultured rat Schwann cells. I. Establishment of purified populations from cultures of peripheral nerve. *Brain Res.* 165 (1): 105–118. https://doi.org/10.1016/0006-8993(79)90048-9.

Buehring, G.C. (1972). Culture of human mammary epithelial cells: keeping abreast with a new method. *J. Natl Cancer Inst.* 49 (5): 1433–1434.

Carswell, K.A., Lee, M.J., and Fried, S.K. (2012). Culture of isolated human adipocytes and isolated adipose tissue. *Methods Mol. Biol.* 806: 203–214. https://doi.org/10.1007/978-1-61779-367-7_14.

Chi, J., Meng, L., Pan, S. et al. (2017). Primary culture of rat aortic vascular smooth muscle cells: a new method. *Med. Sci. Monit.* 23: 4014–4020.

Clevers, H. and Watt, F.M. (2018). Defining adult stem cells by function, not by phenotype. *Annu. Rev. Biochem.* 87: 1015–1027. https://doi.org/10.1146/annurev-biochem-062917-012341.

Collins, S.J., Gallo, R.C., and Gallagher, R.E. (1977). Continuous growth and differentiation of human myeloid leukaemic cells in suspension culture. *Nature* 270 (5635): 347–349. https://doi.org/10.1038/270347a0.

Danjoh, I., Saijo, K., Hiroyama, T. et al. (2011). The Sonoda-Tajima cell collection: a human genetics research resource with emphasis on South American indigenous populations. *Genome Biol. Evol.* 3: 272–283. https://doi.org/10.1093/gbe/evr014.

Dastagir, K., Reimers, K., Lazaridis, A. et al. (2014). Murine embryonic fibroblast cell lines differentiate into three mesenchymal lineages to different extents: new models to investigate differentiation processes. *Cell. Reprogram.* 16 (4): 241–252. https://doi.org/10.1089/cell.2014.0005.

Davila, M.L. and Sadelain, M. (2016). Biology and clinical application of CAR T cells for B cell malignancies. *Int. J. Hematol.* 104 (1): 6–17. https://doi.org/10.1007/s12185-016-2039-6.

Davis, A.S., Chertow, D.S., Moyer, J.E. et al. (2015). Validation of normal human bronchial epithelial cells as a model for influenza a infections in human distal trachea. *J. Histochem. Cytochem.* 63 (5): 312–328. https://doi.org/10.1369/0022155415570968.

DeCicco-Skinner, K.L., Henry, G.H., Cataisson, C. et al. (2014). Endothelial cell tube formation assay for the in vitro study of angiogenesis. *J. Vis. Exp.* 91: e51312. https://doi.org/10.3791/51312.

Deng, H., Mondal, S., Sur, S. et al. (2019). Establishment and optimization of epithelial cell cultures from human ectocervix, transformation zone, and endocervix optimization of epithelial cell cultures. *J. Cell. Physiol.* 234 (6): 7683–7694. https://doi.org/10.1002/jcp.28049.

Du, Y. and Funderburgh, J.L. (2007). Culture of human corneal stem cells. In: *Culture of Human Stem Cells* (eds. R.I. Freshney, G.N. Stacey and J.M. Auerbach), 251–280. Hoboken, NJ: Wiley.

Dunn, J.C., Yarmush, M.L., Koebe, H.G. et al. (1989). Hepatocyte function and extracellular matrix geometry: long-term culture in a sandwich configuration. *FASEB J.* 3 (2): 174–177. https://doi.org/10.1096/fasebj.3.2.2914628.

Edwards, P.A., Easty, D.M., and Foster, C.S. (1980). Selective culture of epithelioid cells from a human squamous carcinoma using a monoclonal antibody to kill fibroblasts. *Cell Biol. Int. Rep.* 4 (10): 917–922.

Eisinger, M. and Marko, O. (1982). Selective proliferation of normal human melanocytes in vitro in the presence of phorbol ester and cholera toxin. *Proc. Natl Acad. Sci. U.S.A.* 79 (6): 2018–2022. https://doi.org/10.1073/pnas.79.6.2018.

Engler, A.J., Carag-Krieger, C., Johnson, C.P. et al. (2008). Embryonic cardiomyocytes beat best on a matrix with heart-like elasticity: scar-like rigidity inhibits beating. *J. Cell Sci.* 121 (Pt 22): 3794–3802. https://doi.org/10.1242/jcs.029678.

Fesnak, A.D., June, C.H., and Levine, B.L. (2016). Engineered T cells: the promise and challenges of cancer immunotherapy. *Nat. Rev. Cancer* 16 (9): 566–581. https://doi.org/10.1038/nrc.2016.97.

Folkman, J. and Haudenschild, C. (1980). Angiogenesis in vitro. *Nature* 288 (5791): 551–556. https://doi.org/10.1038/288551a0.

Frame, M., Freshney, R.I., Shaw, R. et al. (1980). Markers of differentiation in glial cells. *Cell Biol. Int. Rep.* 4 (8): 732. https://doi.org/10.1016/0309-1651(80)90085-5.

Freshney, R.I. (1972). Tumour cells disaggregated in collagenase. *Lancet* 2 (7775): 488–489.

Freshney, R.I. and Freshney, M.G. (1996). *Culture of Immortalized Cells*. New York: Wiley-Liss.

Freshney, R.I. and Freshney, M.G. (2002). *Culture of Epithelial Cells*, 2e. Hoboken, NJ: Wiley-Liss.

Freshney, R.I., Pragnell, I.B., and Freshney, M.G. (1994). *Culture of Hematopoietic Cells*. New York: Wiley-Liss.

Friend, C. (1957). Cell-free transmission in adult Swiss mice of a disease having the character of a leukemia. *J. Exp. Med.* 105 (4): 307–318. https://doi.org/10.1084/jem.105.4.307.

Friend, C. (1978). The phenomenon of differentiation in murine erythroleukemic cells. *Harvey Lect.* 72: 253–281.

Fry, J.R., Wiebkin, P., and Bridges, J.W. (1979). Phenobarbitone induction of microsomal mono-oxygenase activity in primary cultures of adult rat hepatocytes [proceedings]. *Biochem. Soc. Trans.* 7 (1): 119–121.

Fusenig, N.E. (1994). Cell culture models: reliable tools in pharmacotoxicology? In: *Cell Culture in Pharmaceutical Research* (eds. N.E. Fusenig and H. Graf), 1–7. Berlin: Springer.

Galietta, L.J., Lantero, S., Gazzolo, A. et al. (1998). An improved method to obtain highly differentiated monolayers of human bronchial epithelial cells. *In Vitro Cell. Dev. Biol. Anim.* 34 (6): 478–481.

Gilbert, S.F. and Migeon, B.R. (1975). D-valine as a selective agent for normal human and rodent epithelial cells in culture. *Cell* 5 (1): 11–17.

Gilbert, S.F. and Migeon, B.R. (1977). Renal enzymes in kidney cells selected by D-valine medium. *J. Cell. Physiol.* 92 (2): 161–167. https://doi.org/10.1002/jcp.1040920204.

Goldberg, B. (1977). Collagen synthesis as a marker for cell type in mouse 3T3 lines. *Cell* 11 (1): 169–172.

Goldman, B.I. and Wurzel, J. (1992). Effects of subcultivation and culture medium on differentiation of human fetal cardiac myocytes. *In Vitro Cell. Dev. Biol.* 28A (2): 109–119.

Grace, T.D. (1962). Establishment of four strains of cells from insect tissues grown in vitro. *Nature* 195: 788–789.

Green, H. (2008). The birth of therapy with cultured cells. *BioEssays* 30 (9): 897–903. https://doi.org/10.1002/bies.20797.

Grinnell, F., Ho, C.H., Tamariz, E. et al. (2003). Dendritic fibroblasts in three-dimensional collagen matrices. *Mol. Biol. Cell* 14 (2): 384–395. https://doi.org/10.1091/mbc.e02-08-0493.

Grozinsky-Glasberg, S., Shimon, I., and Rubinfeld, H. (2012). The role of cell lines in the study of neuroendocrine tumors. *Neuroendocrinology* 96 (3): 173–187. https://doi.org/10.1159/000338793.

Guguen-Guillouzo, C. and Guillouzo, A. (1986). Methods for preparation of adult and fetal hepatocytes. In: *Isolated and Cultured Hepatocytes* (eds. C. Guguen-Guillouzo and A. Guillouzo), 1–12. Paris: Les Editions INSERM/John Libbey Eurotext.

Guguen-Guillouzo, C., Campion, J.P., Brissot, P. et al. (1982). High yield preparation of isolated human adult hepatocytes by enzymatic perfusion of the liver. *Cell Biol. Int. Rep.* 6 (6): 625–628.

Hammond, S.L., Ham, R.G., and Stampfer, M.R. (1984). Serum-free growth of human mammary epithelial cells: rapid clonal growth in defined medium and extended serial passage with pituitary extract. *Proc. Natl Acad. Sci. U.S.A.* 81 (17): 5435–5439.

Hartley, R.S. and Yablonka-Reuveni, Z. (1990). Long-term maintenance of primary myogenic cultures on a reconstituted basement membrane. *In Vitro Cell. Dev. Biol.* 26 (10): 955–961.

Hellermann, G.R., Nagy, S.B., Kong, X. et al. (2002). Mechanism of cigarette smoke condensate-induced acute inflammatory response in human bronchial epithelial cells. *Respir. Res.* 3: 22.

Higginson, J.R. and Barnett, S.C. (2011). The culture of olfactory ensheathing cells (OECs) – a distinct glial cell type. *Exp. Neurol.* 229 (1): 2–9. https://doi.org/10.1016/j.expneurol.2010.08.020.

Holtzer, H., Abbott, J., Lash, J. et al. (1960). The loss of phenotypic traits by differentiated cells in vitro, I. dedifferentiation of cartilage cells. *Proc. Natl Acad. Sci. U.S.A.* 46 (12): 1533–1542. https://doi.org/10.1073/pnas.46.12.1533.

Horibata, K. and Harris, A.W. (1970). Mouse myelomas and lymphomas in culture. *Exp. Cell Res.* 60 (1): 61–77.

Huch, M., Gehart, H., van Boxtel, R. et al. (2015). Long-term culture of genome-stable bipotent stem cells from adult human liver. *Cell* 160 (1–2): 299–312. https://doi.org/10.1016/j.cell.2014.11.050.

Hui-Yuen, J., McAllister, S., Koganti, S. et al. (2011). Establishment of Epstein-Barr virus growth-transformed lymphoblastoid cell lines. *J. Vis. Exp.* 57: 3321. https://doi.org/10.3791/3321.

Iscove, N.N. and Melchers, F. (1978). Complete replacement of serum by albumin, transferrin, and soybean lipid in cultures of lipopolysaccharide-reactive B lymphocytes. *J. Exp. Med.* 147 (3): 923–933.

Jat, P.S., Noble, M.D., Ataliotis, P. et al. (1991). Direct derivation of conditionally immortal cell lines from an H-2Kb-tsA58 transgenic mouse. *Proc. Natl Acad. Sci. U.S.A.* 88 (12): 5096–5100.

Jin, X. and Kruth, H.S. (2016). Culture of macrophage colony-stimulating factor differentiated human monocyte-derived macrophages. *J. Vis. Exp.* 112: 54244. https://doi.org/10.3791/54244.

Jin, L., Qu, Y., Gomez, L.J. et al. (2018). Characterization of primary human mammary epithelial cells isolated and propagated by conditional reprogrammed cell culture. *Oncotarget* 9 (14): 11503–11514. https://doi.org/10.18632/oncotarget.23817.

Johnson, C.I., Argyle, D.J., and Clements, D.N. (2016). In vitro models for the study of osteoarthritis. *Vet. J.* 209: 40–49. https://doi.org/10.1016/j.tvjl.2015.07.011.

Kahounova, Z., Kurfurstova, D., Bouchal, J. et al. (2018). The fibroblast surface markers FAP, anti-fibroblast, and FSP are expressed by cells of epithelial origin and may be altered during epithelial-to-mesenchymal transition. *Cytometry A* 93 (9): 941–951. https://doi.org/10.1002/cyto.a.23101.

Kao, W.W. and Prockop, D.J. (1977). Proline analogue removes fibroblasts from cultured mixed cell populations. *Nature* 266 (5597): 63–64.

Kleinfeld, D., Kahler, K.H., and Hockberger, P.E. (1988). Controlled outgrowth of dissociated neurons on patterned substrates. *J. Neurosci.* 8 (11): 4098–4120.

Klinger, R.Y. and Niklason, L.E. (2006). Tissue-engineered blood vessels. In: *Culture of Cells for Tissue Engineering* (eds. G. Vunjak-Novakovic and R.I. Freshney), 294–322. Hoboken, NJ: Wiley.

Köhler, G. and Milstein, C. (1975). Continuous cultures of fused cells secreting antibody of predefined specificity. *Nature* 256 (5517): 495–497. https://doi.org/10.1038/256495a0.

Kramer, D. and Minichiello, L. (2010). Cell culture of primary cerebellar granule cells. *Methods Mol. Biol.* 633: 233–239. https://doi.org/10.1007/978-1-59745-019-5_17.

Krasna, L., Dudorkinova, D., Vedralova, J. et al. (2002). Large expansion of morphologically heterogeneous mammary epithelial cells, including the luminal phenotype, from human breast tumours. *Breast Cancer Res. Treat.* 71 (3): 219–235.

Kruitwagen, H.S., Oosterhoff, L.A., Vernooij, I.G.W.H. et al. (2017). Long-term adult feline liver organoid cultures for disease modeling of hepatic steatosis. *Stem Cell Rep.* 8 (4): 822–830. https://doi.org/10.1016/j.stemcr.2017.02.015.

Kubota, Y., Kleinman, H.K., Martin, G.R. et al. (1988). Role of laminin and basement membrane in the morphological differentiation of human endothelial cells into capillary-like structures. *J. Cell Biol.* 107 (4): 1589–1598. https://doi.org/10.1083/jcb.107.4.1589.

Kuri-Harcuch, W. and Green, H. (1978). Adipose conversion of 3T3 cells depends on a serum factor. *Proc. Natl Acad. Sci. U.S.A.* 75 (12): 6107–6109. https://doi.org/10.1073/pnas.75.12.6107.

Kurita, R., Suda, N., Sudo, K. et al. (2013). Establishment of immortalized human erythroid progenitor cell lines able to produce enucleated red blood cells. *PLoS One* 8 (3): e59890. https://doi.org/10.1371/journal.pone.0059890.

LaBarge, M.A., Petersen, O.W., and Bissell, M.J. (2007). Culturing mammary stem cells. In: *Culture of Human Stem Cells* (eds. R.I. Freshney, G.N. Stacey and J.M. Auerbach), 282–302. Hoboken, NJ: Wiley.

LaBarge, M.A., Garbe, J.C., and Stampfer, M.R. (2013). Processing of human reduction mammoplasty and mastectomy tissues for cell culture. *J. Vis. Exp.* 71: 50011. https://doi.org/10.3791/50011.

Lakey, J.R., Burridge, P.W., and Shapiro, A.M. (2003). Technical aspects of islet preparation and transplantation. *Transpl. Int.* 16 (9): 613–632. https://doi.org/10.1007/s00147-003-0651-x.

Lakra, W.S., Swaminathan, T.R., and Joy, K.P. (2011). Development, characterization, conservation and storage of fish cell lines: a review. *Fish Physiol. Biochem.* 37 (1): 1–20. https://doi.org/10.1007/s10695-010-9411-x.

Lasfargues, E.Y. (1973). Human mammary tumors. In: *Tissue Culture: Methods and Applications* (eds. P.F. Kruse Jr. and M.K. Patterson), 45–50. New York: Academic Press.

Lemare, F., Steimberg, N., Le Griel, C. et al. (1998). Dedifferentiated chondrocytes cultured in alginate beads: restoration of the differentiated phenotype and of the metabolic responses to interleukin-1beta. *J. Cell. Physiol.* 176 (2): 303–313. https://doi.org/10.1002/(SICI)1097-4652(199808)176:2<303::AID-JCP8>3.0.CO;2-S.

Lennon, D. and Caplan, A. (2006). Mesenchymal stem cells. In: *Culture of Cells for Tissue Engineering* (eds. G. Vunjak-Novakovic and R.I. Freshney), 23–60. Hoboken, NJ: Wiley-Liss.

Levi-Montalcini, R. (1982). Developmental neurobiology and the natural history of nerve growth factor. *Annu. Rev. Neurosci.* 5: 341–362. https://doi.org/10.1146/annurev.ne.05.030182.002013.

Levin, D.B., Wilson, K., Valadares de Amorim, G. et al. (1995). Detection of p53 mutations in benign and dysplastic nevi. *Cancer Res.* 55 (19): 4278–4282.

Lian, H., Roy, E., and Zheng, H. (2016). Protocol for primary microglial culture preparation. *Bio. Protoc.* 6 (21) https://doi.org/10.21769/BioProtoc.1989.

Littlefield, J.W. (1964). Selection of hybrids from matings of fibroblasts in vitro and their presumed recombinants. *Science* 145 (3633): 709–710.

Looi, K., Sutanto, E.N., Banerjee, B. et al. (2011). Bronchial brushings for investigating airway inflammation and remodelling. *Respirology* 16 (5): 725–737. https://doi.org/10.1111/j.1440-1843.2011.02001.x.

Loomans, C.J.M., Williams Giuliani, N., Balak, J. et al. (2018). Expansion of adult human pancreatic tissue yields organoids harboring progenitor cells with endocrine differentiation potential. *Stem Cell Rep.* 10 (3): 712–724. https://doi.org/10.1016/j.stemcr.2018.02.005.

Louch, W.E., Sheehan, K.A., and Wolska, B.M. (2011). Methods in cardiomyocyte isolation, culture, and gene transfer. *J. Mol. Cell. Cardiol.* 51 (3): 288–298. https://doi.org/10.1016/j.yjmcc.2011.06.012.

Lynn, D.E. (1996). Development and characterization of insect cell lines. *Cytotechnology* 20 (1–3): 3–11. https://doi.org/10.1007/BF00350384.

Lynn, D.E. (2016). Lepidopteran insect cell line isolation from insect tissue. In: *Baculovirus and Insect Cell Expression Protocols* (ed. D.W. Murhammer), 143–159. New York: Springer.

Lynn, D.E. and Harrison, R.L. (2016). Routine maintenance and storage of lepidopteran insect cell lines and baculoviruses. In: *Baculovirus and Insect Cell Expression Protocols* (ed. D.W. Murhammer), 197–221. New York: Springer.

March, S., Ramanan, V., Trehan, K. et al. (2015). Micropatterned coculture of primary human hepatocytes and supportive cells for the study of hepatotropic pathogens. *Nat. Protoc.* 10 (12): 2027–2053. https://doi.org/10.1038/nprot.2015.128.

von der Mark, K., Gauss, V., von der Mark, H. et al. (1977). Relationship between cell shape and type of collagen synthesised as chondrocytes lose their cartilage phenotype in culture. *Nature* 267 (5611): 531–532. https://doi.org/10.1038/267531a0.

Meyer, K. and Kaspar, B.K. (2017). Glia–neuron interactions in neurological diseases: testing non-cell autonomy in a dish. *Brain Res.* 1656: 27–39. https://doi.org/10.1016/j.brainres.2015.12.051.

Millet, L.J. and Gillette, M.U. (2012). Over a century of neuron culture: from the hanging drop to microfluidic devices. *Yale J. Biol. Med.* 85 (4): 501–521.

Milo, G.E., Ackerman, G.A., and Noyes, I. (1980). Growth and ultrastructural characterization of proliferating human keratinocytes in vitro without added extrinsic factors. *In Vitro* 16 (1): 20–30.

Naeyaert, J.M., Eller, M., Gordon, P.R. et al. (1991). Pigment content of cultured human melanocytes does not correlate with tyrosinase message level. *Br. J. Dermatol.* 125 (4): 297–303.

Nakamura, Y., Hiroyama, T., Miharada, K. et al. (2011). Red blood cell production from immortalized progenitor cell line. *Int. J. Hematol.* 93 (1): 5–9. https://doi.org/10.1007/s12185-010-0742-2.

Nash, C.E., Mavria, G., Baxter, E.W. et al. (2015). Development and characterisation of a 3D multi-cellular in vitro model of normal human breast: a tool for cancer initiation studies. *Oncotarget* 6 (15): 13731–13741. https://doi.org/10.18632/oncotarget.3803.

O'Hare, M.J., Ellison, M.L., and Neville, A.M. (1978). Tissue culture in endocrine research: perspectives, pitfalls, and potentials. *Curr. Top. Exp. Endocrinol.* 3: 1–56.

O'Meara, R.W., Ryan, S.D., Colognato, H. et al. (2011). Derivation of enriched oligodendrocyte cultures and oligodendrocyte/neuron myelinating co-cultures from post-natal murine tissues. *J. Vis. Exp.* 54: 3324. https://doi.org/10.3791/3324.

Omi, N., Tokuda, Y., Ikeda, Y. et al. (2017). Efficient and reliable establishment of lymphoblastoid cell lines by Epstein-Barr virus transformation from a limited amount of peripheral blood. *Sci. Rep.* 7: 43833. https://doi.org/10.1038/srep43833.

Ooi, G.T., Tawadros, N., and Escalona, R.M. (2004). Pituitary cell lines and their endocrine applications. *Mol. Cell. Endocrinol.* 228 (1–2): 1–21. https://doi.org/10.1016/j.mce.2004.07.018.

Owens, R.B., Smith, H.S., and Hackett, A.J. (1974). Epithelial cell cultures from normal glandular tissue of mice. *J. Natl Cancer Inst.* 53 (1): 261–269.

Owens, R.B., Smith, H.S., Nelson-Rees, W.A. et al. (1976). Epithelial cell cultures from normal and cancerous human tissues. *J. Natl Cancer Inst.* 56 (4): 843–849. https://doi.org/10.1093/jnci/56.4.843.

Palumbo, P., Miconi, G., Cinque, B. et al. (2015). In vitro evaluation of different methods of handling human liposuction aspirate and their effect on adipocytes and adipose derived stem cells. *J. Cell. Physiol.* 230 (8): 1974–1981. https://doi.org/10.1002/jcp.24965.

Palumbo, P., Lombardi, F., Siragusa, G. et al. (2018). Methods of isolation, characterization and expansion of human adipose-derived stem cells (ASCs): an overview. *Int. J. Mol. Sci.* 19 (7) https://doi.org/10.3390/ijms19071897.

Paraskeva, C. and Williams, A.C. (1992). The colon. In: *Culture of Epithelial Cells* (ed. R.I. Freshney), 82–105. New York: Wiley-Liss.

Park, H.Y., Russakovsky, V., Ohno, S. et al. (1993). The beta isoform of protein kinase C stimulates human melanogenesis by activating tyrosinase in pigment cells. *J. Biol. Chem.* 268 (16): 11742–11749.

Parsons, X.H. (2013). Constraining the pluripotent fate of human embryonic stem cells for tissue engineering and cell therapy – the turning point of cell-based regenerative medicine. *Br. Biotechnol. J.* 3 (4): 424–457. https://doi.org/10.9734/BBJ/2013/4309#sthash.6D8Rulbv.dpuf.

Pechoux, C., Gudjonsson, T., Ronnov-Jessen, L. et al. (1999). Human mammary luminal epithelial cells contain progenitors to myoepithelial cells. *Dev. Biol.* 206 (1): 88–99. https://doi.org/10.1006/dbio.1998.9133.

Perica, K., Curran, K.J., Brentjens, R.J. et al. (2018). Building a CAR garage: preparing for the delivery of commercial CAR T cell products at Memorial Sloan Kettering Cancer Center. *Biol. Blood Marrow Transplant.* 24 (6): 1135–1141. https://doi.org/10.1016/j.bbmt.2018.02.018.

Piemonti, L. and Pileggi, A. (2013). 25 Years of the Ricordi Automated Method for Islet Isolation. CellR4 Repair Replace Regen Reprogram 1 (1).

Pisania, A., Weir, G.C., O'Neil, J.J. et al. (2010). Quantitative analysis of cell composition and purity of human pancreatic islet preparations. *Lab. Investig.* 90 (11): 1661–1675. https://doi.org/10.1038/labinvest.2010.124.

van der Plas, A., Aarden, E.M., Feijen, J.H. et al. (1994). Characteristics and properties of osteocytes in culture. *J. Bone Miner. Res.* 9 (11): 1697–1704. https://doi.org/10.1002/jbmr.5650091105.

Ponten, J. and Macintyre, E.H. (1968). Long term culture of normal and neoplastic human glia. *Acta Pathol. Microbiol. Scand.* 74 (4): 465–486.

Postnikoff, C.K., Pintwala, R., Williams, S. et al. (2014). Development of a curved, stratified, in vitro model to assess ocular biocompatibility. *PLoS One* 9 (5): e96448. https://doi.org/10.1371/journal.pone.0096448.

Quaroni, A., Wands, J., Trelstad, R.L. et al. (1979). Epithelioid cell cultures from rat small intestine. Characterization by morphologic and immunologic criteria. *J. Cell Biol.* 80 (2): 248–265. https://doi.org/10.1083/jcb.80.2.248.

Raisman, G., Barnett, S.C., and Ramon-Cueto, A. (2012). Repair of central nervous system lesions by transplantation of olfactory ensheathing cells. *Handb. Clin. Neurol.* 109: 541–549. https://doi.org/10.1016/B978-0-444-52137-8.00033-4.

Rayner, R.E., Makena, P., Prasad, G.L. et al. (2019). Optimization of normal human bronchial epithelial (NHBE) cell 3D cultures for in vitro lung model studies. *Sci. Rep.* 9 (1): 500. https://doi.org/10.1038/s41598-018-36735-z.

Relou, I.A., Damen, C.A., van der Schaft, D.W. et al. (1998). Effect of culture conditions on endothelial cell growth and responsiveness. *Tissue Cell* 30 (5): 525–530.

Rheinwald, J.G. and Green, H. (1975). Serial cultivation of strains of human epidermal keratinocytes: the formation of keratinizing colonies from single cells. *Cell* 6 (3): 331–343.

Rhim, J.S. (2013). Human prostate epithelial cell cultures. *Methods Mol. Biol.* 946: 383–393. https://doi.org/10.1007/978-1-62703-128-8_24.

Richler, C. and Yaffe, D. (1970). The in vitro cultivation and differentiation capacities of myogenic cell lines. *Dev. Biol.* 23 (1): 1–22.

Ronkko, S., Vellonen, K.S., Jarvinen, K. et al. (2016). Human corneal cell culture models for drug toxicity studies. *Drug Deliv.*

Transl. Res. 6 (6): 660–675. https://doi.org/10.1007/s13346-016-0330-y.

Ruscetti, S.K. (1995). Erythroleukaemia induction by the Friend spleen focus-forming virus. *Baillieres Clin. Haematol.* 8 (1): 225–247.

Russo, G., Lehne, F., Pose Mendez, S.M. et al. (2018). Culture and transfection of zebrafish primary cells. *J. Vis. Exp.* 138: 57872. https://doi.org/10.3791/57872.

Sadelain, M. (2017). CD19 CAR T cells. *Cell* 171 (7): 1471. https://doi.org/10.1016/j.cell.2017.12.002.

Sassen, W.A., Lehne, F., Russo, G. et al. (2017). Embryonic zebrafish primary cell culture for transfection and live cellular and subcellular imaging. *Dev. Biol.* 430 (1): 18–31. https://doi.org/10.1016/j.ydbio.2017.07.014.

Sato, T. and Clevers, H. (2013). Growing self-organizing mini-guts from a single intestinal stem cell: mechanism and applications. *Science* 340 (6137): 1190–1194. https://doi.org/10.1126/science.1234852.

Sato, T., Vries, R.G., Snippert, H.J. et al. (2009). Single Lgr5 stem cells build crypt-villus structures in vitro without a mesenchymal niche. *Nature* 459 (7244): 262–265. https://doi.org/10.1038/nature07935.

Schott, J.T., Kirby, L.A., Calabresi, P.A. et al. (2016). Preparation of rat oligodendrocyte progenitor cultures and quantification of oligodendrogenesis using dual-infrared fluorescence scanning. *J. Vis. Exp.* 108: 53764. https://doi.org/10.3791/53764.

Schwob, J.E. (2002). Neural regeneration and the peripheral olfactory system. *Anat. Rec.* 269 (1): 33–49. https://doi.org/10.1002/ar.10047.

Seglen, P.O. (1976). Preparation of isolated rat liver cells. *Methods Cell Biol.* 13: 29–83.

Shah, K.M., Stern, M.M., Stern, A.R. et al. (2016). Osteocyte isolation and culture methods. *Bonekey Rep.* 5: 838. https://doi.org/10.1038/bonekey.2016.65.

Shulman, M. and Nahmias, Y. (2013). Long-term culture and coculture of primary rat and human hepatocytes. *Methods Mol. Biol.* 945: 287–302. https://doi.org/10.1007/978-1-62703-125-7_17.

Spaethling, J.M., Na, Y.J., Lee, J. et al. (2017). Primary cell culture of live neurosurgically resected aged adult human brain cells and single cell transcriptomics. *Cell Rep.* 18 (3): 791–803. https://doi.org/10.1016/j.celrep.2016.12.066.

Speirs, V., Wang, Y.V., Yeger, H. et al. (1992). Isolation and culture of neuroendocrine cells from fetal rabbit lung using immunomagnetic techniques. *Am. J. Respir. Cell Mol. Biol.* 6 (1): 63–67. https://doi.org/10.1165/ajrcmb/6.1.63.

Speirs, V., White, M.C., and Green, A.R. (1996). Collagenase III: a superior enzyme for complete disaggregation and improved viability of normal and malignant human breast tissue. *In Vitro Cell. Dev. Biol. Anim.* 32 (2): 72–74.

Stampfer, M., Hallowes, R.C., and Hackett, A.J. (1980). Growth of normal human mammary cells in culture. *In Vitro* 16 (5): 415–425.

Stanley, M.A. and Parkinson, E.K. (1979). Growth requirements of human cervical epithelial cells in culture. *Int. J. Cancer* 24 (4): 407–414.

Stern, A.R., Stern, M.M., Van Dyke, M.E. et al. (2012). Isolation and culture of primary osteocytes from the long bones of skeletally mature and aged mice. *BioTechniques* 52 (6): 361–373. https://doi.org/10.2144/0000113876.

Subramanian, M., Madden, J.A., and Harder, D.R. (1991). A method for the isolation of cells from arteries of various sizes. *J. Tissue Cult. Methods* 13 (1): 13–19. https://doi.org/10.1007/bf02388198.

Sugimoto, M., Furuichi, Y., Ide, T. et al. (1999). Incorrect use of "immortalization" for B-lymphoblastoid cell lines transformed by Epstein-Barr virus. *J. Virol.* 73 (11): 9690–9691.

Tang, J., Li, Q., Cheng, B. et al. (2014). Primary culture of human face skin melanocytes for the study of hyperpigmentation. *Cytotechnology* 66 (6): 891–898. https://doi.org/10.1007/s10616-013-9643-6.

Taylor, S.E., Shah, M., and Orriss, I.R. (2014). Generation of rodent and human osteoblasts. *Bonekey Rep.* 3: 585. https://doi.org/10.1038/bonekey.2014.80.

Ter Horst, B., Chouhan, G., Moiemen, N.S. et al. (2018). Advances in keratinocyte delivery in burn wound care. *Adv. Drug Deliv. Rev.* 123: 18–32. https://doi.org/10.1016/j.addr.2017.06.012.

Thingnes, J., Lavelle, T.J., Hovig, E. et al. (2012). Understanding the melanocyte distribution in human epidermis: an agent-based computational model approach. *PLoS One* 7 (7): e40377. https://doi.org/10.1371/journal.pone.0040377.

Tomita, M. and Tsumoto, K. (2011). Hybridoma technologies for antibody production. *Immunotherapy* 3 (3): 371–380. https://doi.org/10.2217/imt.11.4.

Tsumoto, K., Isozaki, Y., Yagami, H. et al. (2019). Future perspectives of therapeutic monoclonal antibodies. *Immunotherapy* 11 (2): 119–127. https://doi.org/10.2217/imt-2018-0130.

Utheim, T.P., Utheim, O.A., Khan, Q.E. et al. (2016). Culture of oral mucosal epithelial cells for the purpose of treating limbal stem cell deficiency. *J. Funct. Biomater.* 7 (1) https://doi.org/10.3390/jfb7010005.

Vidi, P.-A., Bissell, M.J., and Lelievre, S.A. (2013). Three-dimensional culture of human breast epithelial cells: the how and the why. *Methods Mol. Biol.* 945: 193–219. https://doi.org/10.1007/978-1-62703-125-7_13.

Villa, P.L., Jackson, R., Eade, S. et al. (2018). Isolation of biopsy-derived, human cervical keratinocytes propagated as monolayer and organoid cultures. *Sci. Rep.* 8 (1): 17869. https://doi.org/10.1038/s41598-018-36150-4.

Vonen, B., Florholmen, J., Malm, D. et al. (1992). Sorbitol in isolation of rat pancreatic islets. Effects on islet yield, insulin secretion and accumulation of inositol phosphates. *Scand. J. Clin. Lab. Invest.* 52 (4): 237–243.

Vormittag, P., Gunn, R., Ghorashian, S. et al. (2018). A guide to manufacturing CAR T cell therapies. *Curr. Opin. Biotechnol.* 53: 164–181. https://doi.org/10.1016/j.copbio.2018.01.025.

Waldo, S.W., Li, Y., Buono, C. et al. (2008). Heterogeneity of human macrophages in culture and in atherosclerotic plaques. *Am. J. Pathol.* 172 (4): 1112–1126. https://doi.org/10.2353/ajpath.2008.070513.

Wang, X. and Riviere, I. (2016). Clinical manufacturing of CAR T cells: foundation of a promising therapy. *Mol. Ther. Oncolytics.* 3: 16015. https://doi.org/10.1038/mto.2016.15.

Wang, Y., Li, W., Phay, J.E. et al. (2016). Primary cell culture systems for human thyroid studies. *Thyroid* 26 (8): 1131–1140. https://doi.org/10.1089/thy.2015.0518.

Wen, J., Tan, D., Li, L. et al. (2017). Isolation and purification of Schwann cells from spinal nerves of neonatal rat. *Bio-Protocol.* 7 (20): e2588. https://doi.org/10.21769/BioProtoc.2588.

Whitehead, R.H. and Robinson, P.S. (2009). Establishment of conditionally immortalized epithelial cell lines from the intestinal tissue of adult normal and transgenic mice. *Am. J. Physiol. Gastrointest. Liver Physiol.* 296 (3): G455–G460. https://doi.org/10.1152/ajpgi.90381.2008.

Whitehead, R.H. and van Eeden, P.E. (1991). A method for the prolonged culture of colonic epithelial cells. *J. Tissue Cult. Methods* 13 (2): 103–106. https://doi.org/10.1007/bf01666139.

Whitehead, R.H., Brown, A., and Bhathal, P.S. (1987). A method for the isolation and culture of human colonic crypts in collagen gels. *In Vitro Cell. Dev. Biol.* 23 (6): 436–442.

Wognum, B. and Lee, T. (2013). Simultaneous cloning and selection of hybridomas and transfected cell lines in semisolid media. *Methods Mol. Biol.* 946: 133–149. https://doi.org/10.1007/978-1-62703-128-8_9.

Wolf, K. (1979). Cold-blooded vertebrate cell and tissue culture. *Methods Enzymol.* 58: 466–477.

Wu, J., Rnjak-Kovacina, J., Du, Y. et al. (2014). Corneal stromal bioequivalents secreted on patterned silk substrates. *Biomaterials* 35 (12): 3744–3755. https://doi.org/10.1016/j.biomaterials.2013.12.078.

Xie, Y., Bergstrom, T., Jiang, Y. et al. (2015). The human glioblastoma cell culture resource validated cell models representing all molecular subtypes. *EBioMedicine* 2 (10): 1351–1363. https://doi.org/10.1016/j.ebiom.2015.08.026.

Yablonka-Reuveni, Z. (2011). The skeletal muscle satellite cell: still young and fascinating at 50. *J. Histochem. Cytochem.* 59 (12): 1041–1059. https://doi.org/10.1369/0022155411426780.

Yan, G., Fukabori, Y., McBride, G. et al. (1993). Exon switching and activation of stromal and embryonic fibroblast growth factor (FGF)-FGF receptor genes in prostate epithelial cells accompany stromal independence and malignancy. *Mol. Cell. Biol.* 13 (8): 4513–4522. https://doi.org/10.1128/mcb.13.8.4513.

Yesil, P., Michel, M., Chwalek, K. et al. (2009). A new collagenase blend increases the number of islets isolated from mouse pancreas. *Islets* 1 (3): 185–190. https://doi.org/10.4161/isl.1.3.9556.

Young, G.M. and Levison, S.W. (1997). An improved method for propagating oligodendrocyte progenitors in vitro. *J. Neurosci. Methods* 77 (2): 163–168.

Zetter, B.R. (2008). The scientific contributions of M. Judah Folkman to cancer research. *Nat. Rev. Cancer* 8 (8): 647–654. https://doi.org/10.1038/nrc2458.

CHAPTER 25

Culture of Tumor Cells

Malignant tumors have been successfully grown *in vitro* for more than 60 years, starting with the HeLa cell line (Gey et al. 1952). This has led to a steadily increasing number of tumor cell lines for use in cancer research. For example, a large number of glioblastoma cell lines were established from the 1960s onwards by Jan Pontén, Bengt Westermark, and colleagues at the University of Uppsala (Ponten and Macintyre 1968; Xie et al. 2015). Their "U" series of cell lines is widely used and has been incorporated into large panels for the screening of anticancer drugs. Examples include the NCI-60 panel, the Cancer Cell Line Encyclopedia (CCLE), the Genomics of Drug Sensitivity in Cancer (GDSC) panel, and the Catalogue of Somatic Mutations in Cancer (COSMIC) cell line panel (Shoemaker 2006; Barretina et al. 2012; Garnett et al. 2012; Menden et al. 2018). It might be argued that these tumor cell panels are sufficient and new cell lines are not needed. However, the current models do not reflect the full spectrum of tumor subtypes and their behavior do not necessarily represent the original tumors. A recent set of 48 glioblastoma cell lines, established by Westermark and colleagues as the Human Glioblastoma Cell Culture (HGCC) Resource, show the benefits of new tumor models. The HGCC models were established using validated techniques, represent all molecular subtypes of glioma, are available at comparatively early passage, and are accompanied by extensive molecular, phenotypic, and clinical data (Xie et al. 2015). The need for new models is particularly acute for tumor types that have fewer cell lines, such as pediatric brain tumors (Xu et al. 2015). This chapter focuses on the culture of malignant tumor samples in order to generate new models for cancer research.

25.1 CHALLENGES OF TUMOR CELL CULTURE

Many malignant tumors fail to grow *in vitro*, which is surprising considering that they appear to have autonomy from normal regulatory controls *in vivo*. There are many reasons for the failure of tumor cells to survive *in vitro*. Tumor cells must be separated from stromal cells to prevent overgrowth by more rapidly growing populations (see Section 20.7). However, attempts to remove the stroma may deprive the tumor cells of extracellular matrix (ECM), nutrients, or other factors necessary for survival. Dilution of tumor cells to provide a sufficient amount of nutrients per cell may dilute out any growth factors that are produced by the cells themselves (homocrine factors) or by other cell types in their immediate surroundings (paracrine factors; see Section 26.3.4). Any growth factors that arise from nearby cells will require a closely interacting population for the necessary signaling to occur.

Culture conditions for tumor cells are usually based on the requirements of normal cells from the same tissue. However, it may be incorrect to assume that the growth factor dependence of a tumor cell is similar to that of the equivalent normal cell. Tumor cells may produce endogenous autocrine growth factors, such as transforming growth factor-α (TGF-α); exogenous growth factors in the culture medium, such as epidermal growth factor (EGF), may compete for the same receptor. The response of a tumor cell to a growth factor will also depend on the mutational status of the cell. A normal cell, capable of expressing growth suppressor and senescence genes, may respond differently to a cell in which one or more of these genes are inactive or mutated, and in which antagonistic, growth-promoting oncogenes are overexpressed.

Freshney's Culture of Animal Cells: A Manual of Basic Technique and Specialized Applications, Eighth Edition. Amanda Capes-Davis and R. Ian Freshney.
© 2021 John Wiley & Sons Ltd. Published 2021 by John Wiley & Sons Ltd.
Companion website: www.wiley.com/go/freshney/cellculture8

Although the development of selective media for normal cells has advanced considerably (see Sections 10.1.2, 20.7.1), efforts to develop tumor-specific culture conditions have been more limited. This is partly due to variation between samples of tumor tissue, even from the same tumor type. Another reason is the continued use of serum in the culture medium. As many transformed cells are not inhibited by TGF-β, there has been less perceived need to eliminate serum, other than to repress fibroblastic growth. However, serum is known to induce changes in the properties of tumor cultures. For example, glioblastoma cells that are grown in serum-free conditions are more likely to grow as spheroids ("neurospheres") compared to serum controls; spheroid formation is linked to the continued presence of stem cells within the culture (see Section 27.4.1) (Lee et al. 2006). The genotype and phenotype of glioma cells, when grown in serum-free medium, are closer to those of the original tumor (Lee et al. 2006; Xie et al. 2015).

Some progress has been made in developing selective culture media for certain tumor types (see Section 25.4.1). However, there is a need to continue this work of optimization for tumors that remain difficult to grow *in vitro*. Until that optimization has been performed, the most logical approach to tumor cell culture is to adapt the approaches that are utilized for the equivalent normal cells, even if supplemented with minimal amounts of serum or conditioned medium from other cells (Dairkee et al. 1995). This will hopefully provide a sustaining environment, as yet undefined, but nevertheless able to permit the survival of a representative population.

25.2 PRIMARY CULTURE OF TUMOR CELLS

25.2.1 Selection of Representative Cells

Tumor samples are highly heterogeneous (see Section 3.1.1). To isolate the tumor cell population, it is necessary to prevent their overgrowth by other cell types, particularly fibroblasts and vascular cells, which are stimulated to invade and proliferate by many tumors. It may also be necessary to separate tumor cells from adjacent normal cells, which often have similar characteristics. Tumor cells themselves are highly heterogeneous; malignant tumors consist of multiple subclones that display considerable phenotypic diversity (see Section 3.1.2). It is impossible for cultures to be truly representative of these heterogeneous populations. The only way for this to occur would be for the whole tumor to be used and for cell survival to be 100%, which is impossible to achieve in practice. The problems of selection are accentuated when sampling is carried out from secondary metastases, which often grow better, but may not be typical of the primary tumor or of other metastases.

The average tumor culture is a compromise, where the eventual outcome will depend on the cell type or types that you wish to encourage. For example, if the stem cell compartment within the tumor is the main target for therapy, it becomes the key representative population for selection, and the culture conditions should be optimized accordingly to ensure its survival. A compromise must also be reached based on other factors. Many human tumor samples are obtained from surgery or biopsy, where the tissue is used for diagnosis or for therapeutic decision making. In those circumstances, the needs of the patient and the clinical team are paramount. The pathologist is usually the person to decide what is required for diagnostic purposes and what can be used for other applications. It is important to discuss your requirements with the pathologist, who can work with you to identify representative material with the best chance of success for primary culture, e.g. avoiding areas with large amounts of necrosis.

In some cases, the pathologist may be able to give you large amounts of material. Always preserve as much of that material as possible. Cryopreservation of tissue samples can be performed using dimethyl sulfoxide (DMSO) to give viable cells for later culture. Some material should be snap-frozen for storage at −80 °C; this material can be used to extract DNA for authentication or for characterization procedures (see Chapters 17, 18).

PROTOCOL P25.1. FREEZING TUMOR BIOPSIES

Background

Protocols were provided in previous chapters for handling human biopsy samples and for freezing and thawing cryovials (see Protocols P13.2, P15.1, 15.2). Read those protocols for general instructions before proceeding.

This protocol uses liquid nitrogen to snap-freeze some material, e.g. for DNA extraction. Direct immersion into liquid nitrogen is commonly used for snap-freezing, but there is a risk that sealed cryovials will crack or explode during immersion. To help reduce the risk, thermo-conductive equipment is used in this protocol to minimize direct contact. Alternatively, tissue samples may be wrapped in aluminum foil, immersed in liquid nitrogen, and then transferred to cryovials for storage. Whatever method is used, always test that it is effective for subsequent applications and record any major variables, e.g. delays prior to freezing (Grizzle et al. 2010).

Outline

Chop the tumor into small pieces. Expose some pieces to DMSO and freeze at −1 °C/min for later culture. Snap-freeze other pieces for DNA extraction or other applications.

❖ *Safety Note. Pathogens may be present in biopsy material. Human tissue should be handled in a biological safety cabinet (BSC) (see Section*

6.3.4). Equipment and waste must be decontaminated after use or before disposal (see Sections 6.3.5, 6.3.6). All personnel working with liquid nitrogen should be trained to do so safely and wear suitable PPE. Wear a closed lab coat, face shield, and insulated gloves when handling liquid nitrogen or frozen cryovials (see Sections 6.2.3, 6.2.4).

Materials (sterile or aseptically prepared)

- Biopsy sample in collection medium (see Protocol P13.2)
- Dissection Balanced Salt Solution (DBSS; see Appendix B)
- Freezing medium (e.g. growth medium with 10% DMSO; see Section 15.3.2)
- Complete culture medium
- Plastic cryovials, 1.2 ml (see Section 15.2.1)
- Petri dishes, 90 mm (for dissection)
- Instruments (scalpels, forceps, etc., as required for tissue dissection)

Materials (non-sterile)

- BSC and associated consumables (see Protocol P12.1)
- Controlled cooling device (see Section 15.2.2)
- Container rated for liquid nitrogen (e.g. 9-l insulated pan, Corning)
- CoolRack® thermo-conductive cryogenic vial module (e.g. #432049, Corning)
- ThermalTray™ thermo-conductive platform (e.g. #432073, Corning)
- Liquid nitrogen
- Forceps

Procedure

A. Dissection and Washing

1. On receipt of the material in the laboratory, record the key information for each sample and assign a unique sample identifier number (see Protocol P13.2; see also Table 13.3).
2. Prepare snap-freezing equipment:
   (a) Place ThermalTray in the insulated pan.
   (b) Add liquid nitrogen until the level is just below the height of the ThermalTray.
   (c) Place the CoolRack on the ThermalTray.
   (d) Allow approximately 15 minutes to reach approximately −150 °C.
3. Using a BSC, prepare a work space for aseptic technique. Transfer the tissue to fresh, sterile DBSS in a 90-mm Petri dish and rinse. Be careful not to lose the tissue.

4. Transfer the tissue to a second Petri dish. Dissect off unwanted tissue, such as fat or necrotic material, and transfer to a third dish.
5. Chop finely with crossed scalpels (see Figure 13.7a, top) into 2- to 3-mm³ pieces.
6. Transfer the tissue pieces to 1.5-ml cryovials. Place four to five pieces in each cryovial.
7. Optional: wash the tissue pieces by resuspending in DBSS, allowing the pieces to settle and removing the supernatant. Repeat this step two more times.

B. Freezing Samples

8. For the cryovials to be stored for later culture:
   (a) Add 1 mL freezing medium to each cryovial.
   (b) Leave these cryovials for 30 minutes at room temperature.
   (c) Transfer to a controlled cooling device for freezing at −1 °C/min.
   (d) Once cryopreservation has been performed, transfer to the cryofreezer and update the records in the cryofreezer database for future reference (see Table 15.5).
9. For the cryovials to be snap-frozen for DNA extraction or other uses:
   (a) Place the cryovials in the CoolRack, which has been precooled to −150 °C (see step 2).
   (b) Leave the cryovials until the samples are frozen (it is often easiest to leave them in position until ready for storage).
   (c) Remove the cryovials using forceps and transfer to a cryofreezer or −80 °C freezer.

C. Thawing of Samples for Culture

10. Place a frozen cryovial from step 8 to thaw at 37 °C (see Protocol P15.2).
11. Swab the cryovial thoroughly in alcohol and open it within the BSC.
12. Allow the pieces to settle and remove half of the medium.
13. Replace the medium slowly with fresh, DMSO-free medium. Mix by gentle shaking, and allow to stand for 5 minutes.
14. Gradually replace all of the medium with DMSO-free medium and transfer the pieces to a dish or flask.
15. Proceed as for regular primary culture, allowing twice as much material per flask.

Notes

Step 2: Instructions are specific to the CoolRack and ThermalTray (Corning). According to the manufacturer, cryogenic temperature will be reached in approximately 12–14 minutes and will be maintained for as long as liquid nitrogen remains in contact with the fins of the ThermalTray.

Step 7: The extent of washing procedures will depend on the amount of blood and debris that is present; this will vary with the tumor sample.

25.2.2 Disaggregation of Tumor Samples

Some tumors are readily disaggregated by purely mechanical means, such as pipetting and sieving (see Section 13.6). Tumor types that may be disaggregated in this way include human ovarian carcinoma, some gliomas, and many transplantable rodent tumors. Mechanical disaggregation may help to minimize stromal contamination, as stromal cells are often more tightly locked in fibrous connective tissue. However, many of the common human carcinomas are hard or scirrhous, and the tumor cells are contained within large amounts of fibrous stroma, making mechanical disaggregation difficult. Early success was reported by cutting some of these tumors into thin slices using a sharp scalpel blade; this can result in "spillage" of tumor cells, e.g. from breast or lung carcinoma (Lasfargues 1973; Oie et al. 1996). If spillage does not occur, enzymatic disaggregation must be used to release the tumor cells from the surrounding stroma.

Trypsin is commonly used for enzymatic disaggregation (see Section 13.5.1), but its effectiveness against fibrous connective tissue is limited and it can reduce the seeding efficiency of the tumor cells (Lounis et al. 1994). Crude collagenase has been found to be more effective with several different types of tumor (Dairkee et al. 1997). Tumor samples may be incubated in collagenase, using complete growth medium, for up to five days or more for scirrhous tumors (see Protocol P13.6). Repeated collagenase treatment is also helpful to free the epithelial cells in the culture, which are rapidly overtaken by fibroblasts and coated with collagen (Lasfargues 1973).

Extensive necrosis is a common problem when performing primary culture of tumor cells. Usually, the attachment of viable cells allows necrotic material to be removed on subsequent feeding. If the amount of necrotic material is extensive and difficult to remove by washing the tissue samples during dissection, separation using Ficoll/metrizoate may be performed to remove necrotic cells (see Protocol P13.8).

25.3 DEVELOPMENT OF TUMOR CELL LINES

25.3.1 Subculture of Primary Tumor Cultures

Primary cultures of tumor cells do not always take readily to subculture (see Sections 3.2.1, 14.2). Many cells in the primary culture may not be capable of propagation, due to genetic or phenotypic aberrations, terminal differentiation, or nutritional insufficiency. If tumor cells are present after the first subculture, this implies that they have not been overgrown and that they may even have a faster growth rate than contaminating normal cells. Proliferation of the tumor cells in the culture also means that more material is available for further applications. An increase in cell yield is one of the major advantages of subculture. Expanded cultures can be cryopreserved and replicate cultures prepared for characterization and assay of specific parameters such as genomic alterations, changes in gene expression, metabolic pathway analysis, chemosensitivity, and invasiveness. Disadvantages of subculture include evolution away from the phenotype of the tumor, due to the inherent genetic instability of the cells and selective adaptation of the cell line to the culture environment (see Section 3.2).

Cells that survive subculture include tumor cells, adjacent normal cells, stromal cells (e.g. fibroblasts, vascular endothelial cells, and smooth muscle cells), and infiltrating cells (e.g. lymphocytes, granulocytes, and macrophages). The hematopoietic components seldom form cell lines, although hematopoietic cell lines have been derived from small cell lung carcinoma (SCLC); this carcinoma also tends to produce suspension cultures that can express myeloid markers, resulting in considerable confusion (Ruff and Pert 1984). Macrophages and granulocytes are strongly adherent and non-proliferative, meaning that they are generally lost at subculture. Smooth muscle does not propagate readily without the appropriate growth factors, so the major potential contaminants are fibroblasts, endothelial cells, and (for tumor cells) their normal equivalents. Of these cell types, the major problem lies with the fibroblasts, which grow readily in culture and may also respond to tumor-derived mitogenic factors. Selective culture may be required to preserve the tumor cells until the stromal cells become senescent (see Section 25.4).

25.3.2 Continuous Tumor Cell Lines

Continued proliferation is a hallmark of cancer *in vivo*, but this does not necessarily translate to continued growth *in vitro*. An early study found that continuous cell lines were established from only 6% of tumor samples (Giard et al. 1973). Over time, scientists with expertise in specific tumor types improved success rates through a process of optimization, resulting in the development of selective media and techniques (see Sections 25.4.1, 25.5). For some tumors, such as colorectal carcinoma, this has resulted in large cell line panels (> 120 cell lines) that reflect a wide spectrum of tumor subtypes (Mouradov et al. 2014; Wilding and Bodmer 2014). Other tumors, such as breast carcinoma, stubbornly refuse to thrive *in vitro*, leading to a relatively small number of cell lines despite efforts from multiple laboratories (Holliday and Speirs 2011). Their limited numbers make it extremely difficult to generate representative cell line panels for drug screening or other applications (Wilding and Bodmer 2014).

Continuous tumor cell lines can be highly variable in appearance, even when established from the same tumor type (see Figure 25.1). Many continuous cell lines also exhibit signs of transformation, which is a phenotype associated with malignant behavior *in vivo* (see Section 3.6). Transformed cells exhibit genetic instability and aberrant growth control,

and are tumorigenic in nude mice. These characteristics often do not reflect the original tumor. The relationship of a tumor cell line to the primary culture and parent tumor is difficult to assess. The development of a continuous cell line may represent (i) expansion of a specific subset or stem cell population of the tumor; (ii) adaptation to the culture conditions, made possible by their genetic instability, resulting in further evolution; or (iii) both events occurring together. The emergence of a continuous cell line is often from colonies within the monolayer, suggesting that the cell line arises from a minor immortalized subset of the tumor cell population, with cell culture merely providing the appropriate conditions for their expansion. These cells may represent a population of cancer stem cells (CSCs), which have been studied extensively using continuous cell lines as sources of

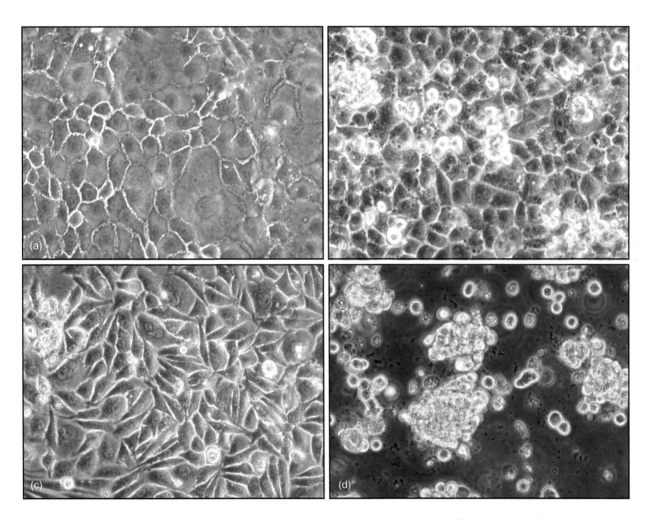

Fig. 25.1. Cell lines from gastric carcinoma. (a–c) Some gastric carcinoma cell lines grow as adherent cultures, showing diffuse spread of tumor cells with fusiform or polygonal contours. Examples include (a) SNU-216, which grows as a well-flattened monolayer; (b) SNU-484, which grows as a pavement-like monolayer; and (c) SNU-668, which is more refractile and has a cobblestone-like appearance. (d) In contrast, the SNU-620 cell line grows as adherent or floating cell aggregates. *Source*: Courtesy of Jae-Gahb Park (Park et al. 2004).

material (see Minireview M25.1). CSCs display considerable plasticity, depending on their environment, which means that their behavior will be heavily dependent on the culture conditions (Batlle and Clevers 2017).

The effectiveness and reliability of continuous tumor cell lines as models continue to be debated, with compelling arguments expressed both for and against their use (Masters 2000; Gillet et al. 2013; Wilding and Bodmer 2014). Although many arguments are focused on the cells themselves, their culture conditions and handling are likely to play a major part in their effectiveness as research models. Variations in handling procedures are often not reported and are difficult to standardize, making this an important source of irreproducibility (see Section 7.1.2) (Niepel et al. 2019).

25.3.3 Validation of Tumor Cell Lines

The need for validation testing of cell lines has been discussed in earlier chapters (see Section 7.4; see also Chapters 16, 17). However, novel tumor cell lines have some additional quality concerns that must be addressed before they are used for experimental work. Quality concerns for novel tumor cell lines include:

(1) *Microbial contamination.* Some tumors arise in association with infectious agents, resulting in cultures that carry these pathogens with them (Uphoff et al. 2010; Shioda et al. 2018). Feeder layers or xenograft hosts may also carry infectious agents that are then transferred to the culture. For example, there is an increased incidence of murine retroviruses in continuous cell lines, particularly in laboratories with access to xenograft facilities (Zhang et al. 2011; Uphoff et al. 2015). Biohazard risk should be assessed to determine the appropriate containment level (see Sections 6.3.2, 6.3.3); additional testing may be advisable for pathogenic viruses (see Section 16.5).

(2) *Cross-contamination.* This may arise due to incomplete inactivation of feeder cells or the presence of mouse stromal cells following xenograft culture (see Sections 25.4.2, 25.4.5). It is essential to confirm the origin of any cell line; additional testing may be needed to detect a mixture of cells from different species (see Section 17.4.1). Contaminating cells from different species can be removed from some cultures, e.g. using the Mouse Cell Depletion Kit (Miltenyi Biotec).

(3) *Incorrect identity.* Cross-contamination will give rise to a misidentified cell line if there is a difference in growth rate between the original, authentic cells and the contaminating cells, resulting in overgrowth by the faster growing population. Xenografts – including patient-derived xenograft (PDX) models – can also become misidentified (Schweppe and Korch 2018). Misidentified cell lines and xenografts can be detected by authentication testing (see Chapter 17).

(4) *Incorrect classification.* A "tumor" cell line may arise from the correct donor but a different cell type to the one expected. This may be due to mistaken diagnosis; for example, the tumor that gave rise to the Hep-G2 cell line was originally reported as a hepatocellular carcinoma, but this was revised to hepatoblastoma after review of the original tumor (Lopez-Terrada et al. 2009). Misclassification can also occur if the tumor cells in the culture are overgrown by a different cell type. It is important to store a sample of the original tumor and copies of the original pathology report and associated images to answer future questions about a cell line's origin (Fogh 1986).

(5) *Spontaneous cell–cell fusion.* Somatic cell hybridization can be performed intentionally using a fusogenic agent (see Section 22.4.1). However, it can also occur spontaneously between tumor cells and stromal cells. Spontaneous cell–cell fusion has been documented in xenograft culture, resulting in more aggressive tumor behavior (Jacobsen et al. 2006). Fusion can also arise *in vitro*; for example, the human NCEB-1 cell line, which was established on a feeder layer of mouse peritoneal macrophages, was unexpectedly found to carry eight mouse chromosomes some years later (Camps et al. 2006; Drexler and MacLeod 2006). Hybrid cell lines can be detected by cytogenetic analysis or other forms of authentication testing (see Section 17.4.3).

25.3.4 Characterization of Tumor Cell Lines

Tumor cells and continuous cell lines are prone to clonal evolution and genomic instability, resulting in phenotypic variation (see Sections 3.2.2, 3.3, 3.4). This tendency is even more marked in tumor cells that demonstrate microsatellite instability (MSI). MSI is commonly associated with Lynch syndrome, which is caused by an inherited mutation in a mismatch repair (MMR) gene. However, it is commonly found in cell lines from hematopoietic malignancies and has been documented in 39 different cancer types (Inoue et al. 2001; Bonneville et al. 2017). Cell lines with MSI are challenging to authenticate, due to increased variation in microsatellite loci (see Section 17.4.2). The presence of MSI is traditionally determined by analysis of variation at five microsatellite markers (BAT25, BAT26, D5S346, D2S123, and D17S250) (Boland et al. 1998; Zeinalian et al. 2018). Some cell line databases provide information on MSI status (see Table 18.1).

Clearly, it is important to characterize tumor cell lines at greater depth if we are to understand their relevance as cancer models. Characterization will vary with the application and may include MSI status, mutation analysis, next-generation sequencing (NGS), and various other forms of molecular characterization (see Chapter 18). It has been suggested that each laboratory's holding of a cell line should be profiled to understand its mutational and pathway status (Hynds et al. 2018). While this would certainly address the need for characterization, many tissue culture laboratories do not have the budget or the capacity to do so. An alternative approach is to invest in a level of characterization that is proportional to its level of use (see Section 18.1). As a cell line becomes more widely used as a cancer model, it should be characterized at increasing depth to ensure that the investment in time and resources is wisely made.

25.4 SELECTIVE CULTURE OF TUMOR CELLS

25.4.1 Selective Media and Techniques

Many selective media depend on the metabolic inhibition of fibroblastic growth and are not specifically optimized for any particular tumor (see Section 20.7.1). Only a few media have been developed as selective agents for tumor cells. HITES medium was developed by Adi Gazdar and colleagues by supplementing RPMI 1640 with hydrocortisone, insulin, transferrin, estradiol, and selenium (see Appendix C) (Carney et al. 1981). Non-small cell lung carcinomas (NSCLCs) do not grow in HITES, leading to the development of ACL-4 by the same group (see Appendix C) (Gazdar and Oie 1986). HITES and ACL-4 media have been used to establish more than 200 lung carcinoma cell lines and can be used for other tumor types, e.g. colon carcinoma (Oie et al. 1996). More recently, Gordon Mills and colleagues developed OCMI medium for ovarian carcinomas; this optimized formulation contains insulin, hydrocortisone, EGF, and cholera toxin (Ince et al. 2015). Other selective media have been used to culture bladder, breast, and prostate carcinoma cells (Messing et al. 1982; Ethier et al. 1993; Uzgare et al. 2004).

While some tumors can grow in serum-free media, many require a small amount of serum, e.g. 2–5% fetal bovine serum (FBS) (Carney et al. 1981; Oie et al. 1996). Serum-free conditions are often selective for cells from normal tissues, but the nutritional and growth factor requirements of tumor cells are more variable and may require the use of serum. However, the presence of serum will allow some normal cells to proliferate; it is often helpful to commence culture using low levels of serum (e.g. 2% FBS) and then increase the serum level as normal cells become less abundant (Ince et al. 2015). Increased serum may also help to overcome crisis in some cultures (Oie et al. 1996).

Optimization should extend to the whole culture system, particularly the substrate on which cells are grown. When using ACL-4 for lung tumors, most cultures initially grow as floating aggregates; collagen can be used to promote attachment and proliferation, but it is not usually required after initial growth has occurred (Gazdar and Oie 1986). When using OCMI for ovarian tumors, success rates can be improved by growing cultures on Primaria™ plasticware (Corning). Primaria plasticware has a lower

level of nitrogen-containing functional groups and mixed positive and negative charge compared to standard tissue culture plasticware, which has a consistently negative charge (see Sections 2.2, 8.3.7) (Ince et al. 2015). Atmospheric conditions are also important. Papillary serous and clear cell ovarian subtypes grow best in 18–21% oxygen, while endometrioid and mucinous subtypes require 5–10% oxygen (Ince et al. 2015).

Tumor cells (particularly clusters of malignant epithelial cells) often take longer to attach compared to other cell populations. This property can be utilized by treating cells with trypsin or collagenase, seeding the resulting cell suspension in a flask, and then performing serial transfers as the various populations adhere (see Section 20.7.3). The adherent cells may be retained and cultured separately, harvested for DNA extraction (e.g. for authentication testing), or used as a feeder layer after irradiation or treatment with mitomycin C (see Sections 20.6.2, 25.4.2). This method of removal by serial transfer is generally only partially successful, but it may be useful as an initial method to enrich certain cell populations, before further separation is performed.

A number of techniques can be used to separate tumor cells from stromal cells. Magnetic separation is perhaps the most widely available technique and can be highly effective for many tumor types. Commercial kits are available to enrich for specific tumor types or deplete other populations, such as the CELLection™ Epithelial Enrich Dynabeads™ kit (ThermoFisher Scientific) or the MACS® Tumor Cell Isolation Kit (Miltenyi Biotec; see Protocol P21.2). The beads in these kits will bind to surface ligands on the tumor cell (positive selection) or on other cell types (negative selection; see Figure 21.6). If these kits are not suitable for selection of a particular tumor type, beads may be purchased for use with your own antibodies (see Section 21.3).

25.4.2 Confluent Feeder Layers

Feeder layers (see Section 8.4) can be used to promote tumor cell adhesion and prevent fibroblastic overgrowth. The latter requires a preformed, confluent monolayer, which can be prepared by seeding feeder cells at a high density or by allowing contact-inhibited cells to grow to confluence (see Figure 25.2). Confluent feeder layers from the human fetal intestinal cell line, FHs 74 Int, have been used to grow breast carcinoma cells (see Figure 20.1c, d) (Lan et al. 1981). This method relied on conditioned medium, with later reports suggesting that selective culture in MCDB 170 was more reproducible (Hammond et al. 1984). Feeder layers have also been prepared from mouse embryonic fibroblasts (MEFs) to grow breast, colon, and basal cell carcinoma cells (Rheinwald and Beckett 1981; Leake et al. 1987). Protocol P25.2 describes how to grow tumor cells from biopsy samples or primary cultures on a confluent feeder layer.

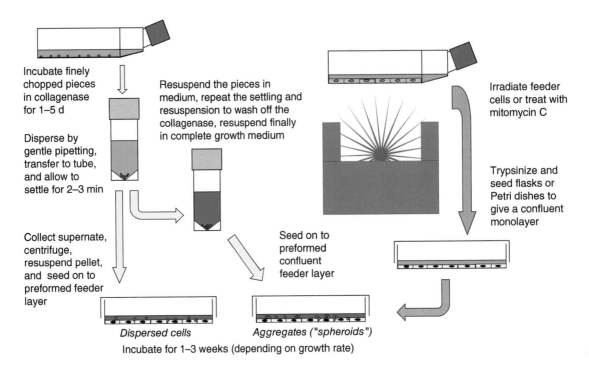

Fig. 25.2. Confluent feeder layers. These can be prepared by seeding cells at a high density or by allowing a normally contact-inhibited cell to grow to confluence. Cells are inactivated using irradiation or mitomycin C and plated at a suitable density to give a confluent monolayer. Tumor fragments are disaggregated using collagenase, followed by gentle pipetting, giving rise to small aggregates ("spheroids"). Dispersed cells can also be generated from the supernatant after disaggregation; although containing more stromal cells, these cells can form colonies on confluent feeder layers with significant restriction of stromal overgrowth. Selection is against stromal components, but not normal epithelium.

PROTOCOL P25.2. GROWTH ON CONFLUENT FEEDER LAYERS

Background

Mitomycin C is employed here to inactivate feeder cells. Alternative methods may be used such as irradiation, as described in a previous protocol (see Protocol P20.7). Collagenase disaggregation of tissue samples has also been described previously (see Protocol P13.6). Refer to both protocols as required.

Outline

Treat feeder cells in the mid-exponential phase with mitomycin C to induce growth arrest. Reseed feeder cells to give a confluent monolayer. Seed tumor cells onto the confluent monolayer. Observe the culture for colonies or infiltration of the feeder layer.

❖ *Safety Note. Mitomycin C is a hazardous chemical. Refer to its Safety Data Sheet for recommended handling and personal protective equipment (PPE).*

Materials (sterile or aseptically prepared)

☐ Feeder cells (e.g. 3T3-Swiss albino, STO, or FHs 74 Int)
☐ Tumor biopsy or primary culture
☐ Growth medium
☐ Dulbecco's phosphate buffered saline without Ca^{2+} and Mg^{2+} (DPBS-A)
☐ Mitomycin C (Sigma-Aldrich), 100 µg/ml stock, in Hanks's balanced salt solution (HBSS) or serum-free medium (store shielded from light)
☐ Collagenase, 2000 U/ml (e.g. Type IA; #C2674 Sigma-Aldrich)
☐ Trypsin, 0.25%, in DPBS-A
☐ Petri dishes and instruments for dissection (if working with a tumor biopsy)
☐ Culture flasks, 75 cm² and 25 cm²

Materials (non-sterile)

☐ BSC and associated consumables (see Protocol P12.1)

Procedure

1. Grow the feeder cells to 80% confluence in six 75-cm^2 flasks.
2. Add mitomycin C to give the appropriate final concentration, usually around 0.25 µg/ml.
3. Incubate the cells overnight ($\sim$ 18 hours) in mitomycin C.
4. If you are using biopsy material, prepare tissue pieces and place in collagenase (see Protocol P13.6) during mitomycin C incubation (step 3).
5. Remove the medium with mitomycin C and wash the monolayer with fresh medium.
6. Incubate the inactivated feeder cells for a further 24–48 hours.
7. Trypsinize the feeder cells (see Protocol P14.2) and reseed in 25-cm^2 flasks at 5×10^5 cells/ml (1×10^5 cells/cm^2). Aim for at least six flasks containing feeder layers.
8. Incubate the feeder layers for 24 hours and check that they are confluent before proceeding.
9. If you are using biopsy material, remove the collagenase from the disaggregated tumor cells, by either repeated settling or centrifugation (see Protocol P13.6).
10. If you are using a primary culture, perform subculture using collagenase or trypsin and count the cells (see Protocols P13.6, P14.2; see also Section 19.1).
11. Resuspend the tumor cells in growth medium, aiming for 1×10^5 cells/ml and a total volume of 12 ml.
12. Seed 5 ml of the cell suspension into two flasks containing confluent feeder layers.
13. Add 4 ml growth medium to two more flasks containing confluent feeder layers. Add 1 ml of the cell suspension to each flask, giving a concentration of 1×10^4 cells/ml.
14. Prepare and set aside control flasks. The third pair of flasks should be kept as controls to guard against feeder cells surviving the mitomycin C treatment. It is also helpful to plate tumor cells onto plastic for comparison to their behavior on feeder layers (see Figure 25.3).
15. Observe flasks for the formation of colonies, which may form in three weeks to three months, or for other changes that may indicate tumor cell growth or continued feeder cell proliferation.

Notes

Step 2: It is advisable to perform a dose-response curve when using feeder cells for the first time, to confirm the dose of mitomycin C that allows the feeder layer to survive for two to three weeks but does not permit further replication in the feeder layer after about two doublings. This helps ensure that no resistant colonies will form and that the confluent feeder layer does not overgrow. The required dose is usually 0.25 µg/ml for overnight exposure or 20 µg/ml for one hour exposure.

Step 7: The suggestion of six flasks is based on two pairs of flasks containing tumor cells at different concentrations. A third pair should be set aside as a control for the mitomycin C treatment. The number of flasks required will vary, depending on the number of tumor cells.

Steps 11–13: It may be impossible to count cells from dissociated tissue samples. If this is the case, aim to seed 20–100 mg/flask, such that each 25-cm^2 flask holds 5 ml of cell suspension.

Step 15: Fibrosarcoma and gliomas do not always form colonies, but may infiltrate the feeder layer and gradually overgrow (see Figure 3.12a, b). The survival of tumor cells will be confirmed only by subculturing the cells without a feeder layer (by which time contaminating stromal cells should have been eliminated).

Confluent feeder layers can be used for selective cloning of certain cell types (see Section 20.6.2). These feeder layers are not selective against normal epithelium, as normal epidermis and normal breast epithelium both form colonies on confluent feeder layers (see Figure 25.3). However, results from glioma suggest that selection against equivalent normal cells may be possible on a homologous feeder layer (MacDonald et al. 1985). Glioma samples grown on normal glial feeder layers should lose any normal glial contaminants. By the same argument, breast carcinoma seeded on confluent cultures of normal breast epithelium – such as from reduction mammoplasty samples (see Supp. S24.8) – could become free of any contaminating normal epithelium.

25.4.3 Suspension Cloning

The transformation of cells *in vitro* leads to a greater ability to grow independently from the culture substrate (see Section 3.6.2). This can be seen when cells are cloned in suspension, using soft agar or Methocel® (see Protocols P20.2, 20.3). Theoretically, it seems reasonable that tumor cells might be cloned or selected using the same approach. It has been shown that tumorigenicity correlates with cloning in Methocel, and it is possible for colonies to form in suspension when grown directly from disaggregated tumors (Freedman and Shin 1974;

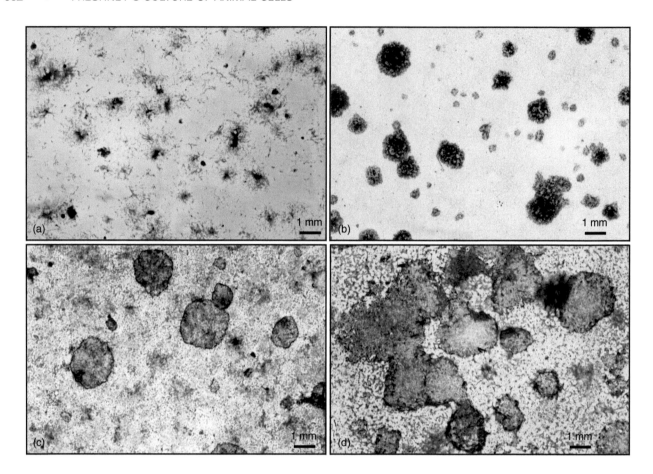

Fig. 25.3. Selective cloning on confluent feeder layers (see also Figure 20.1c, d). (a) Colonies forming on plastic alone after seeding 4000 cells/cm² (2 × 10⁴ cells/ml) from a breast carcinoma culture. Small, dense colonies are epithelial cells; the larger, stellate colonies are fibroblasts. (b) Colonies of cells from the same culture, seeded at 400 cells/cm² (2000 cells/ml) on a confluent feeder layer of FHs 74 Int cells. The epithelial colonies are much larger than those in (a), the plating efficiency is higher, and there are no fibroblastic colonies. (c) Colonies from a different breast carcinoma culture plated onto the same feeder layer. Note the different colony morphology, with a lighter stained center and ring at the point of interaction with the feeder layer. (d) Colonies from normal breast culture seeded onto FHI cells (fetal human intestine; similar to FHs 74 Int). A few small, fibroblastic colonies are present in (c) and (d). *Source*: R. Ian Freshney, experiment based on Smith et al. (1981), and personal communication, A. J. Hackett.

Hamburger and Salmon 1977). However, the colony-forming efficiency when cloning tumor cells is often very low (often <0.1%), and it is not easy to propagate cells isolated from the colonies. Thus, cloning of tumor cells in soft agar or Methocel has its limitations as a method of tumor cell selection, although it has been used for drug screening with tumor biopsies (Pavelic et al. 1980).

Suspension cloning has been proposed in the past as a means of isolating CSCs (Hamburger and Salmon 1977). Why is this method not more effective for tumor cell selection? Considering that many stem cell populations are sensitive to dissociation (see Section 23.4.1), it is possible that CSCs are lost when cultures are dispersed into a single cell suspension for cloning. There has also been some difficulty in propagating cell lines from primary clones, particularly from colonies isolated using the suspension method. Studies of normal keratinocytes have shown that clonal populations vary in their capacity for self-renewal (see Section 20.6). Clonal populations from tumor cell lines exhibit different capacities for self-renewal compared to normal cells, due to variations in the proportion of stem cells within each population (Beaver et al. 2014). A better understanding of CSCs and optimization of their culture conditions would improve the self-renewal capacity of tumor cell cultures, resulting in better *in vitro* targets for drug screening and molecular targeting (see Minireview M25.1).

25.4.4 Spheroid Culture

Tumor cells can be successfully grown in three-dimensional (3D) culture and often form aggregates in suspension (known as "spheroids" in this book; see Section 27.4.1) (Inch et al. 1970; Yuhas et al. 1977; Carlsson et al. 1983). Although spheroid culture is commonly performed using continuous cell lines, spheroids can be established directly from human tumor samples (Darling et al. 1983; Wibe et al. 1984). Unlike suspension cloning, where samples are usually dispersed into a single-cell suspension at low density, spheroid culture brings cells into close association. In some cases, spheroids form spontaneously *in vitro*; breast, ovarian, and some other carcinomas grow as adherent monolayers but may form aggregates that detach into the culture supernatant. In other cases, spheroids do not form spontaneously or may take weeks, or even months, to form. Spheroid formation may be encouraged using "forced floating" techniques that prevent substrate adhesion (see Section 27.4) or by retaining cell–cell contact, e.g. using Liberase (Roche CustomBiotech) for disaggregation (Kondo et al. 2011)

Spheroid culture can be used to separate tumor from stromal cells, which are typically left behind when a spheroid is induced to form spontaneously from an adherent monolayer (although stromal cells may be retained if spheroids are established directly from tumor tissue). It may also be possible to distinguish normal from malignant cells, due to differences in their behavior in 3D culture, particularly if cells are grown in the presence of a scaffold for the development of tissue architecture (see Section 27.5). However, the ability to form spheroids or to develop other 3D structures may be lost with continued culture. A study of normal breast tissue has shown that cells are initially able to form spheroids, but that this ability is lost beyond the fifth passage (Dey et al. 2009). The slow appearance of spheroids in some tumor cell cultures, and the loss of spheroid formation in others, suggests that spheroids are derived from a minority cell population that can be lost from the culture. Considering the close association between sphere formation and stem cell activity (see Section 27.4.1), it seems likely that this minority population consists of CSCs. However, a number of cell populations can form spheroids *in vitro* and not all stem cell populations may be able to do so (Pastrana et al. 2011). It is best to avoid leaping to conclusions regarding the composition of tumor spheroids until CSC populations are better understood (see Minireview M25.1).

Tumor spheroids provide a number of exciting opportunities for cancer research. Spheroids may be embedded in Matrigel, fibrin, collagen, or agarose, resulting in elegant *in vitro* models to study cell invasion or interactions with the tumor stroma (see Protocol P27.1) (Charoen et al. 2014; Vinci et al. 2015; Tevis et al. 2017). Spheroid cultures can be used to study the penetration of cytotoxic drugs or other molecules that are used in targeted therapy, and their subsequent effect on cell cohesion (Sutherland 1988; Virgone-Carlotta et al. 2017). Tumor spheroids have also proved to be useful models to study cell killing by biologically targeted radionuclides

(see Figure 27.10a) (Walker et al. 1988; Boyd et al. 2001; Boyd et al. 2004). Spheroid culture procedures are readily simplified and scaled up for high-throughput screening, enabling cancer drug discovery or personalized therapy (Sant and Johnston 2017; Phan et al. 2019).

25.4.5 Xenografts

When cultures are derived from human tumors, the scarcity of material and the infrequency of re-biopsy make it difficult to repeatedly culture the same tumor. The growth of some tumors in immune-deprived animals provides an alternative approach that makes greater amounts of tumor available (Rygaard and Povlsen 1969; Giovanella et al. 1972). It also favors selection of the tumor cell population, as normally the human stromal cells will not grow and will be replaced by mouse stromal tissue. Establishment of a xenograft, followed by culture of the tumor, can result in greater success rates for cell line establishment (Dangles-Marie et al. 2007). The reasons for greater success are unclear, but probably include the availability of more tissue, progression of the tumor, enrichment of transformed cells, or modification of the tumor cells by the heterologous host (e.g. by cell–cell fusion) (Jacobsen et al. 2006).

The most commonly used animal for xenograft culture is the genetically athymic "nude" mouse, which is T-cell deficient (Giovanella et al. 1972). Success rates for xenograft culture will vary with the animal, with the tumor type, and whether it comes from the primary site or a distant metastasis (Sharkey and Fogh 1984). Implantation with Matrigel or tumor stromal cells has been reported to improve the growth of tumor cells in nude mice (Topley et al. 1993). However, not all tumors grow in nude mice, leading to the use of other animals, such as neonatally thymectomized mice that are subsequently irradiated and treated with cytosine arabinoside (Selby et al. 1980). These animals will regain immune competence and ultimately reject the tumor after a few months.

All research involving animals is subject to ethical regulation; this comes with the obligation to reduce the use of animals in research (see Section 6.4.1). However, some xenograft models are unique and difficult to replicate. For example, the National Cancer Institute (NCI) has developed a Pediatric Preclinical Testing Program (PPTP) panel that includes xenografts and cell lines. The PPTP xenografts include rare pediatric tumors where tissue is difficult to access or cells are difficult to grow *in vitro* (Kang et al. 2011). PDX models transfer tissue samples directly into the mouse without intervening culture; as such, they are outside the scope of this book. Procedures for the establishment of human tumor xenografts are discussed elsewhere (Morton and Houghton 2007).

25.5 SPECIFIC TUMOR TYPES

General protocols for primary culture provide a good starting point for most tumor types; collagenase digestion is particularly helpful (see Protocol P13.6). Selective culture is normally required to encourage tumor cells to grow and to eliminate fibroblasts and other unwanted cell types (see Section 25.4; see also Chapter 21). Some specific examples of tumor culture are discussed below, but it is not possible to explore all tumor types or to provide more than a brief overview in this chapter. For more information, a series of books by John Masters and Bernhard Palsson provides a comprehensive overview of the various tumor types that have been placed in culture and the cell lines established from them (Masters 1991; Masters and Palsson 1999, 2002a,2002b).

25.5.1 Carcinoma

Lung carcinoma. SCLC and NSCLC tumors have been cultured successfully using a combination of mechanical disaggregation, Ficoll-based density gradients, and HITES or ACL-4 selective media (see Sections 13.6, 21.1.1, 25.4.1) (Oie et al. 1996). Adi Gazdar, John Minna, and colleagues established a substantial panel of lung cell lines using this approach; their cell lines usually carry the prefixes NCI- (from the National Cancer Institute) or HCC- (from the Hamon Cancer Center) (Phelps et al. 1996; Gazdar et al. 2010). SCLC cells grow in HITES as aggregates in suspension, while NSCLC cells grow in ACL-4 as adherent monolayers. Cells can be grown on plastic, but ECM has been shown to increase tumor cell growth and used to facilitate culture of lung carcinoma cells from bone marrow micrometastases (Pavelic et al. 1992; Pantel et al. 1995). Although SCLC and NSLC cultures may express differentiated markers, a large number (approximately 30% of SCLC cell lines in one study) have discordant biochemical markers, morphology, or growth compared to "classic" cultures (Gazdar et al. 1985).

Colorectal carcinoma. Colorectal carcinoma has been cultured from both primary tumors and metastases, resulting in the establishment of cell line panels from a wide range of subtypes (Wilding and Bodmer 2014). Primary cultures are best grown at high density using a feeder layer (e.g. 3T3 cells) or collagen, which both promote adhesion (Brattain et al. 2002). Cultures can also be grown as tumor spheroids, provided cell–cell contact is retained (see Section 25.4.4) (Kondo et al. 2011). Density centrifugation on Percoll has been used to purify colonic carcinoma cells for primary culture in conventional medium (RPMI 1640 with 10% FBS) (Csoka et al. 1995). Serum-free conditions have been described for the culture of some human colorectal cancer cell lines, but these conditions are often not suitable for newly isolated carcinoma cultures, which have a greater requirement for serum (Murakami and Masui 1980; Fantini et al. 1987). As with lung carcinoma, some colorectal tumors have neuroendocrine properties; success has been reported for these using HITES medium (Lundqvist et al. 1991). A protocol is provided in the Supplementary Material online for the culture of colorectal tumors (see Supp. S25.1).

Breast carcinoma. Breast carcinoma cells have been cultured from tumor biopsies and pleural effusions, with the latter adapting more readily to culture conditions *in vitro* (Sutherland et al. 2002). Samples can be cultured directly (for effusions) or disaggregated in collagenase or Dispase (for biopsies) (Ethier et al. 1993; Dairkee et al. 1995, 1997). Cell fractionation is useful to separate cell clusters, dispersed epithelial cells, and stromal cells before the desired fractions are propagated on feeder layers or collagen (see Figure 25.4) (Speirs 2004). Many of the selective media used for normal breast may not be optimal for breast carcinoma, leading to the development of tumor-specific media supplements (Ethier et al. 1993). A variety of conditions may need to be tested and the neoplastic origin of the cells confirmed (e.g. using mutations in *TP53* and overexpression of *ERBB2*). Various immortalization techniques have been used to establish breast carcinoma cell lines, including telomerase and conditional reprogramming (see Section 22.3.3, 22.3.4). These techniques appear to be effective for early-stage tumors such as ductal carcinoma *in situ*, where very few cell lines are available (Yong et al. 2014; Brown et al. 2015). A protocol is available in the Supplementary Material online for the culture of mammary tumor cells (see Supp. S25.2).

Ovarian carcinoma. Ovarian carcinoma cells have been cultured from solid tumors, ascites, and pleural effusions, with success rates varying from less than 1% to about 50% (Wilson and Garner 2002; Verschraegen et al. 2003). Samples from ascites and pleural effusions give higher success rates compared to solid tumors; disaggregation with Dispase appears to offer a higher chance of success for the latter (Wilson and Garner 2002). Density centrifugation on Percoll has been used to purify ovarian carcinoma cells for primary culture, as noted previously for colorectal carcinoma (Csoka et al. 1995). Ovarian carcinoma cells can be grown in various media, both serum-containing and serum-free. Culture in selective OCMI medium is reported to give a success rate of more than 95%; cells are initially grown in OCMI with 2% serum, and selective trypsinization is performed to remove stromal cells (see Section 20.7.3) (Ince et al. 2015). Ovarian carcinoma has numerous subtypes, including serous, endometrioid, mucinous, and clear cell carcinoma. Culture in OCMI has been successful for multiple tumor subtypes, but it is important to retain clinical data and histology and perform characterization to be confident that a particular subtype has been retained *in vitro* (Anglesio et al. 2013; Jacob et al. 2014).

Prostate carcinoma. Prostate epithelial cells can be cultured from biopsy, prostatectomy, and tumor metastases (see Section 24.2.10) (Peehl 2003). However, the success rates for tumor cell line establishment are very low. Most prostate carcinoma cell lines come from metastatic deposits and are androgen insensitive, with the exception of LNCaP and a handful of other cell lines (Russell and Kingsley 2003). Primary cultures can be initiated using serum-free media

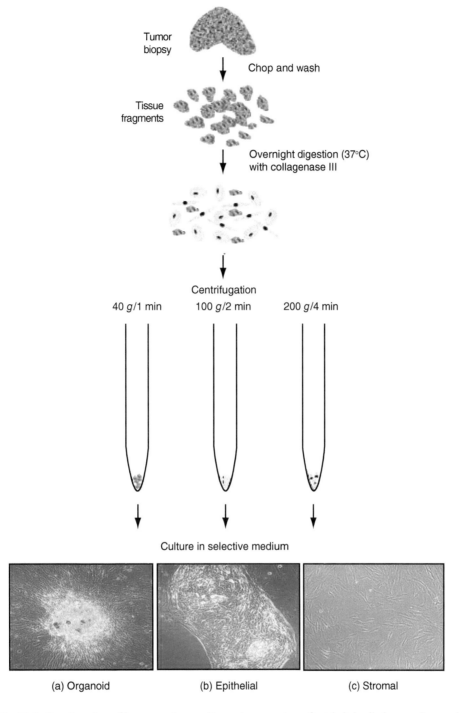

Fig. 25.4. Fractionation of breast carcinoma digest. A suspension of epithelial cell clusters, dispersed epithelial cells, and stromal cells can be segregated into fractions by differential centrifugation, where the epithelial clusters sediment at lower *g*, dispersed epithelial cells at intermediate *g*, and stromal cells at the highest *g* (see Supp. S25.2; see also Figure 24.3). *Source*: Courtesy of Valerie Speirs (Speirs 2004).

such as PFMR-4A, MCDB 105, or Keratinocyte-SFM (ThermoFisher Scientific) (Peehl and Stamey 1986; Peehl 2003; Rhim 2013). The addition of ECM has also proved to successfully encourage tumor cell growth, followed by immortalization (Pantel et al. 1995). Prostate carcinoma cells can be immortalized using various agents including simian virus 40 T antigen (SV40 TAg), human papillomavirus-16 (HPV-16) E6 and E7, or telomerase (Rhim 2013). The latter approach has resulted in a considerable increase in the number of prostate carcinoma cell lines, with the ability to generate cell lines from high-risk populations. Prostate carcinoma cells can also be cultured as organoids using "forced floating" techniques (see Sections 25.4.4, 27.6).

25.5.2 Sarcoma

Sarcomas account for about 1% of adult solid tumors, but they are responsible for up to 15% of malignancies in children and young adults (Gaebler et al. 2017). These tumors are highly heterogeneous and consist of more than 70 different subtypes. For some sarcoma types, multiple cell lines are available and some have been extensively characterized; examples include osteosarcoma and rhabdomyosarcoma (Mohseny et al. 2011; Hinson et al. 2013). Other sarcomas are extremely difficult to establish *in vitro*, such as Kaposi sarcoma, which is represented by less than 10 cell lines (Hattori et al. 2019). The most widely used "Kaposi sarcoma" cell lines, KS Y-1 and SLK, are misidentified and actually come from bladder and renal carcinoma, respectively (Sturzl et al. 2013). Spheroid culture may give improved success rates for sarcomas that are difficult to grow. Because growth stimulation tends to occur when cells are transferred from spheroid to monolayer, it is possible to alternate between the two states until proliferation becomes established (Bruland et al. 1985). Sarcoma samples can now be cultured as organoids using a modified version of this approach (Gaebler et al. 2017).

25.5.3 Melanoma

Melanoma cells have been cultured successfully since the 1940s (Grand and Cameron 1948). It can be argued that melanoma is one of the easiest tumors to grow *in vitro*, although some forms of melanoma are more difficult to establish as cell lines, including uveal melanoma and primary cutaneous tumors (especially thin ones) (Soo et al. 2011; Griewank et al. 2012). Melanoma cells may be cultured in many widely used serum-containing media (e.g. RPMI 1640 with 10% FBS) or in MCDB 153 with additional supplements; the latter has been used successfully for benign and dysplastic nevi (Levin et al. 1995; Soo et al. 2011). Primary uveal melanoma cells appear to proliferate better in media that were developed for the culture of cells from amniotic fluid (Angi et al. 2015). Primary melanomas are often contaminated with fibroblasts, but it may be possible to eliminate the fibroblasts using confluent feeder layers of normal cells (see Protocol P25.2) (Creasey et al. 1979). A feeder layer of mouse keratinocytes can also improve success rates for melanocytes from nevi and primary melanomas (Soo et al. 2011). Cell cultures derived by mechanical spillage can be freed of fibroblasts by treatment with 100 µg/ml G418 (Geneticin) (Halaban 2004).

25.5.4 Lymphoma and Leukemia

The first hematopoietic cell lines were established from Burkitt lymphoma (Pulvertaft 1964; Drexler and MacLeod 2010). More than 500 leukemia-lymphoma (LL) cell lines

have since been established that meet minimal standards of authenticity and characterization (Drexler et al. 2003; Drexler and MacLeod 2010). A great deal of the credit for this belongs to Hans Drexler and colleagues, who have published extensively on LL cell lines, their effectiveness as models, and methods for their establishment and characterization (Drexler 2004, 2010; MacLeod et al. 2008). A panel of 100 LL cell lines has now been extensively characterized by this group that includes representatives from 22 hematopoietic tumor types (Quentmeier et al. 2019). However, there remains an ongoing need for new cell line models – for example, carrying specific disease mutations that can be used to screen anticancer drugs (Drexler and MacLeod 2010). A protocol is provided in the Supplementary Material online for the establishment of LL cell lines (see Supp. S25.3).

25.5.5 Glioma

As noted in the Introduction to this chapter, numerous glioma cell lines have been established but there is an ongoing need for new models (Xie et al. 2015). Cultures of human glioma can be prepared by mechanical disaggregation, enzymatic dissociation using collagenase or Accutase® (Innovative Cell Technologies), or both approaches (Ponten and Macintyre 1968; Freshney 1980; Xie et al. 2015). A previous protocol (see Protocol P13.6) can be used with glioma biopsies and gives a high success rate – about 80% for primary cultures and 60% for early passage cell lines (Freshney 1980). Early passage cultures of human glioma cells will proliferate when plated onto confluent monolayers or normal glial cells, while normal glial cells will not (MacDonald et al. 1985). This opens up a possible selective method to exclude normal glia from glioma cultures. However, most glioma cultures will outgrow the normal glia, especially if grown until they become continuous cell lines (Ponten and Macintyre 1968), which is often achieved without any apparent crisis. Cultures contain multipolar astrocytic cells at early passage, but this appearance is often lost in continuous cell lines, which may contain clonal populations with distinctly different morphologies (see Figure 25.5; see also Figures 14.3c, 20.2a–c). A combination of monolayer culture and culture in neurospheres in serum-free medium has been used to help preserve the *in vivo* genotype in glioblastoma cell lines (Fael Al-Mayhani et al. 2009).

25.6 CANCER STEM CELLS (CSCS)

Minireview M25.1. **Culture of Cancer Stem Cells** Contributed by Tiina Jokela and Mark A. LaBarge, Beckman Research Institute, City of Hope, 1500 E. Duarte Rd, Duarte, CA, 91010 USA.

Cancer stem cells (CSC) are a putative subset of tumor cells that have self-maintenance capacity and the ability to generate the diversity of cells that comprise a given tumor. CSCs have been attributed as the cellular basis of drug resistance,

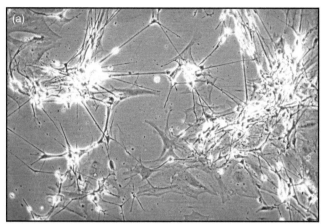

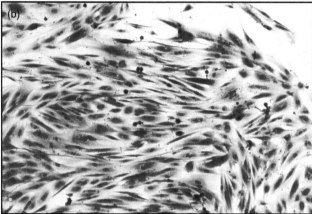

Fig. 25.5. Cultures from human glioma. These two examples are derived from human anaplastic astrocytoma. (a) Primary culture from collagenase digest, showing typical astrocytic cells, which may be differentiated tumor cells or reactive glia. (b) Continuous cell line MOG-G-UVW, showing one of several morphologies found in cell lines from glioma. This example shows a pleomorphic fibroblast-like morphology; there are no typical multipolar astrocytic cells, often lost in continuous cell lines. *Source*: R. Ian Freshney.

metastasis, and relapse. Due to these varied roles, CSC are considered to be important targets for cancer therapy. The possibility of studying CSC in cell culture settings raises a number of challenges.

Sources of CSC

CSC are proposed to arise when oncogenic driver mutations, microenvironments, and altered epigenetic states impose a more plastic phenotype in cancer cells (Chaffer and Weinberg 2015; Poli et al. 2018). According to this definition, CSC could be derived from nearly any heterogenous population of cancer cells, and the specific phenotype of a cancer cell may represent a specific point in an equilibrium that is determined by a combination of genetic, epigenetic, and microenvironment influences.

A majority of CSC studies use cancer cell lines that have been cultured for generations on plastic dishes in a high glucose-

and serum-supplemented media, and the selected cell phenotypes often are not consistent with *in vivo* cell phenotypes. These strongly selective conditions are thought to increase genetic instability in cell lines, even more so than in cancer cells *in vivo*. Cell culture conditions also impose hypomethylation of DNA compared to DNA methylation states in primary malignant tissue (Foksinski et al. 2017). These genetic and epigenetic differences between cell lines and *in vivo* tumor tissue increase the challenge of reproducibility (van Staveren et al. 2009).

Patient-derived tissues are optimal for studying CSC, but availability of material is limited and usually does not allow repetition of experiments at different laboratories. Well-characterized primary culture strains might be a reasonable compromise, where cell material can be expanded in characterized conditions and then tested in PDX models (LaBarge et al. 2013). Cell line culture conditions should be optimized for genetic stability, and a comparison of the cells' epigenomes and transcriptomes *in vivo* versus in culture should be assessed to determine how adaptation to the culture environment has altered the cells (Bentivegna et al. 2013; Foksinski et al. 2017).

Isolation of CSCs

There is yet to be a clearly defined biochemical phenotype of CSC from any tumor type; thus, the only way to identify CSC-enriched subpopulations is to characterize their functional properties. CSC have been enriched by differential expression of marker proteins like CD44, CD38, or CD34, or by intracellular enzyme activity, concentration of reactive oxygen species, mitochondrial membrane potential resistance to cytotoxic compounds or hypoxia, cell adhesion, invasiveness, proliferation, and other physical properties (Lobo et al. 2007; Greve et al. 2012). All these methods are well reviewed in Duan et al. (2013). Enrichment for CSC is ultimately proven with xenografting experiments and results have differed widely among different cancer types and among different markers. For instance, more than 25% of melanoma cells were shown to be tumorigenic with no clear biochemical phenotype (Quintana et al. 2008), whereas multiple reports have claimed that only < 5% of cells in any given tumor are tumorigenic (Lobo et al. 2007). Interpretations of CSC-related experiments depend much upon context, e.g. the length of time one waits to measure tumor formation in mice, the method of CSC enrichment, and the site of tumor engraftment. It has become increasingly clear that the genotype of the host and the length of time CSCs are allowed to form tumors will profoundly affect one's interpretation (LaBarge 2010).

Culture of CSC

Finally, a means of convincingly maintaining CSC in culture is yet to be devised. Normal stem cell activity is maintained

TABLE 25.1. Cancer stem cell (CSC) niches.

Putative properties of CSC niches	Read more …
Physical properties – density	Liu et al. (2015); Brown et al. (2019)
Oxygen tension	Heddleston et al. (2010)
Cell–cell contact	Hale et al. (2012)
Cell–ECM contact	LaBarge et al. (2009); Jokela et al. (2018); Brown et al. (2019)
Paracrine, autocrine, and endocrine signaling molecules – cytokines	Chin and Wang (2014); Ayob and Ramasamy (2018)
Geometric architecture – 3D	Kimlin et al. (2013); Baldan et al. (2019)

Source: Courtesy of Tiina Jokela and Mark LaBarge.

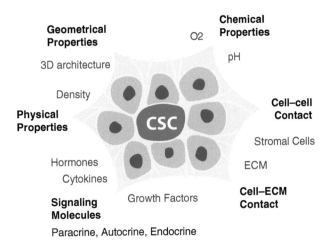

Fig. 25.6. Aspects of CSC culture conditions. The physical, chemical, geometric, cell–cell, and cell–ECM interactions required to maintain the local microenvironment for CSCs. *Source*: Courtesy of Tiina Jokela and Mark LaBarge.

in specific niches (Scadden 2006) and we can assume that CSC are likewise reliant on their niche (LaBarge 2010). Traditional plastic dish monolayer culture conditions probably do not mimic *in vivo* CSC niches. The physical, chemical, geometric, cell–cell, and cell–ECM interactions necessary to retain the CSC phenotype need to be elucidated and recapitulated (see Table 25.1; see also Figure 25.6). If a consensus biochemical phenotype can be agreed upon, it may be beneficial to employ high-throughput screening strategies, which enable probing of common and specific attributes of CSC niches (LaBarge et al. 2009; Ghaemi et al. 2013; Jokela et al. 2018).

There are regulatory signals in the forms of growth factors and oxygen tension that promote acquisition of stem-like states (Heddleston et al. 2010; Ayob and Ramasamy 2018), and even serum-reduced or serum-free culture conditions have been

reported to support stem cell phenotypes, without inducing genomic instability (Loo et al. 1987). 3D culture conditions and specified ECM composition are certainly important elements of the CSC niche (Kimlin et al. 2013; Baldan et al. 2019; Brown et al. 2019; Sukowati et al. 2019). Tumor-specific cell types, like stromal fibroblasts, lymphocytes, macrophages, and adipocytes, also may be important components of the CSC niche (Hale et al. 2012; Ghajar et al. 2013; Ishikawa et al. 2014; Krishnamurthy et al. 2014; Park et al. 2014; Ayob and Ramasamy 2018).

Suppliers

Supplier	URL
Corning	http://www.corning.com/worldwide/en/products/life-sciences/products/cell-culture.html
Innovative Cell Technologies, Inc.	http://www.accutase.com
Miltenyi Biotec	http://www.miltenyibiotec.com/us-en/products/macs-cell-separation.html
Roche CustomBiotech	http://custombiotech.roche.com/home.html
Sigma-Aldrich (Merck)	http://www.sigmaaldrich.com/life-science/cell-culture.html
ThermoFisher Scientific	http://www.thermofisher.com/us/en/home/life-science/cell-culture.html
Worthington Biochemical Corporation	http://www.worthington-biochem.com/default.html

Supp. S25.1 Culture of Colorectal Tumors.

Supp. S25.2 Culture of Mammary Tumor Cells.

Supp. S25.3 Establishment of Continuous Cell Lines from Leukemia-Lymphoma.

REFERENCES

Angi, M., Versluis, M., and Kalirai, H. (2015). Culturing uveal melanoma cells. *Ocul. Oncol. Pathol.* 1 (3): 126–132. https://doi.org/10.1159/000370150.

Anglesio, M.S., Wiegand, K.C., Melnyk, N. et al. (2013). Type-specific cell line models for type-specific ovarian cancer research. *PLoS One* 8 (9): e72162. https://doi.org/10.1371/journal.pone.0072162.

Ayob, A.Z. and Ramasamy, T.S. (2018). Cancer stem cells as key drivers of tumour progression. *J. Biomed. Sci.* 25 (1): 20. https://doi.org/10.1186/s12929-018-0426-4.

Baldan, J., Houbracken, I., Rooman, I. et al. (2019). Adult human pancreatic acinar cells dedifferentiate into an embryonic progenitor-like state in 3D suspension culture. *Sci. Rep.* 9 (1): 4040. https://doi.org/10.1038/s41598-019-40481-1.

Barretina, J., Caponigro, G., Stransky, N. et al. (2012). The Cancer Cell Line Encyclopedia enables predictive modelling of anticancer drug sensitivity. *Nature* 483 (7391): 603–607. https://doi.org/10.1038/nature11003.

Batlle, E. and Clevers, H. (2017). Cancer stem cells revisited. *Nat. Med.* 23 (10): 1124–1134. https://doi.org/10.1038/nm.4409.

Beaver, C.M., Ahmed, A., and Masters, J.R. (2014). Clonogenicity: holoclones and meroclones contain stem cells. *PLoS One* 9 (2): e89834. https://doi.org/10.1371/journal.pone.0089834.

Bentivegna, A., Miloso, M., Riva, G. et al. (2013). DNA methylation changes during in vitro propagation of human mesenchymal stem cells: implications for their genomic stability? *Stem Cells Int.* 2013: 192425. https://doi.org/10.1155/2013/192425.

Boland, C.R., Thibodeau, S.N., Hamilton, S.R. et al. (1998). A National Cancer Institute workshop on microsatellite instability for cancer detection and familial predisposition: development of international criteria for the determination of microsatellite instability in colorectal cancer. *Cancer Res.* 58 (22): 5248–5257.

Bonneville, R., Krook, M.A., Kautto, E.A. et al. (2017). Landscape of microsatellite instability across 39 cancer types. *JCO Precis. Oncol.* 2017. https://doi.org/10.1200/PO.17.00073.

Boyd, M., Mairs, R.J., Cunningham, S.H. et al. (2001). A gene therapy/targeted radiotherapy strategy for radiation cell kill by [^{131}I]meta-iodobenzylguanidine. *J. Gene Med.* 3 (2): 165–172. https://doi.org/10.1002/1521-2254(2000)9999:9999<::AID-JGM158>3.0.CO;2-C.

Boyd, M., Mairs, R.J., Keith, W.N. et al. (2004). An efficient targeted radiotherapy/gene therapy strategy utilising human telomerase promoters and radioastatine and harnessing radiation-mediated bystander effects. *J. Gene Med.* 6 (8): 937–947. https://doi.org/10.1002/jgm.578.

Brattain, M.G., Willson, J.K.V., Koterba, A. et al. (2002). Colorectal cancer. In: *Cancer Cell Lines Part 2* (eds. J.R.W. Masters and B. Palsson), 293–303. New York: Kluwer Academic Publishers.

Brown, D.D., Dabbs, D.J., Lee, A.V. et al. (2015). Developing in vitro models of human ductal carcinoma in situ from primary tissue explants. *Breast Cancer Res. Treat.* 153 (2): 311–321. https://doi.org/10.1007/s10549-015-3551-8.

Brown, Y., Hua, S., and Tanwar, P.S. (2019). Extracellular matrix-mediated regulation of cancer stem cells and chemoresistance. *Int. J. Biochem. Cell Biol.* 109: 90–104. https://doi.org/10.1016/j.biocel.2019.02.002.

Bruland, O., Fodstad, O., and Pihl, A. (1985). The use of multicellular spheroids in establishing human sarcoma cell lines in vitro. *Int. J. Cancer* 35 (6): 793–798. https://doi.org/10.1002/ijc.2910350616.

Camps, J., Salaverria, I., Garcia, M.J. et al. (2006). Genomic imbalances and patterns of karyotypic variability in mantle-cell lymphoma cell lines. *Leuk. Res.* 30 (8): 923–934. https://doi.org/10.1016/j.leukres.2005.11.013.

Carlsson, J., Nilsson, K., Westermark, B. et al. (1983). Formation and growth of multicellular spheroids of human origin. *Int. J. Cancer* 31 (5): 523–533. https://doi.org/10.1002/ijc.2910310502.

Carney, D.N., Bunn, P.A. Jr., Gazdar, A.F. et al. (1981). Selective growth in serum-free hormone-supplemented medium of tumor cells obtained by biopsy from patients with small cell carcinoma of the lung. *Proc. Natl Acad. Sci. U.S.A.* 78 (5): 3185–3189.

Chaffer, C.L. and Weinberg, R.A. (2015). How does multistep tumorigenesis really proceed? *Cancer Discov.* 5 (1): 22–24. https://doi.org/10.1158/2159-8290.CD-14-0788.

Charoen, K.M., Fallica, B., Colson, Y.L. et al. (2014). Embedded multicellular spheroids as a biomimetic 3D cancer model for evaluating drug and drug-device combinations. *Biomaterials* 35 (7): 2264–2271. https://doi.org/10.1016/j.biomaterials.2013.11.038.

Chin, A.R. and Wang, S.E. (2014). Cytokines driving breast cancer stemness. *Mol. Cell. Endocrinol.* 382 (1): 598–602. https://doi.org/10.1016/j.mce.2013.03.024.

Creasey, A.A., Smith, H.S., Hackett, A.J. et al. (1979). Biological properties of human melanoma cells in culture. *In Vitro* 15 (5): 342–350.

Csoka, K., Nygren, P., Graf, W. et al. (1995). Selective sensitivity of solid tumors to suramin in primary cultures of tumor cells from patients. *Int. J. Cancer* 63 (3): 356–360. https://doi.org/10.1002/ijc.2910630309.

Dairkee, S.H., Deng, G., Stampfer, M.R. et al. (1995). Selective cell culture of primary breast carcinoma. *Cancer Res.* 55 (12): 2516–2519.

Dairkee, S.H., Paulo, E.C., Traquina, P. et al. (1997). Partial enzymatic degradation of stroma allows enrichment and expansion of primary breast tumor cells. *Cancer Res.* 57 (8): 1590–1596.

Dangles-Marie, V., Pocard, M., Richon, S. et al. (2007). Establishment of human colon cancer cell lines from fresh tumors versus xenografts: comparison of success rate and cell line features. *Cancer Res.* 67 (1): 398–407. https://doi.org/10.1158/0008-5472.CAN-06-0594.

Darling, J.L., Oktar, N., and Thomas, D.G. (1983). Multicellular tumour spheroids derived from human brain tumours. *Cell Biol. Int. Rep.* 7 (1): 23–30.

Dey, D., Saxena, M., Paranjape, A.N. et al. (2009). Phenotypic and functional characterization of human mammary stem/progenitor cells in long term culture. *PLoS One* 4 (4): e5329. https://doi.org/10.1371/journal.pone.0005329.

Drexler, H.G. (2004). Establishment and culture of human leukemia-lymphoma cell lines. In: *Culture of Human Tumor Cells* (eds. R. Pfragner and R.I. Freshney), 319–348. Hoboken, NJ: Wiley-Liss.

Drexler, H. G. (2010). Guide to leukemia-lymphoma cell lines. https://www.dsmz.de/research/human-and-animal-cell-lines/guide-to-leukemia-lymphoma-cell-lines.html (accessed 24 February 2011).

Drexler, H.G. and MacLeod, R.A. (2006). Mantle cell lymphoma-derived cell lines: unique research tools. *Leuk. Res.* 30 (8): 911–913. https://doi.org/10.1016/j.leukres.2006.02.015.

Drexler, H.G. and MacLeod, R.A. (2010). History of leukemia-lymphoma cell lines. *Hum. Cell* 23 (3): 75–82. https://doi.org/10.1111/j.1749-0774.2010.00087.x.

Drexler, H.G., Dirks, W.G., Matsuo, Y. et al. (2003). False leukemia-lymphoma cell lines: an update on over 500 cell lines. *Leukemia* 17 (2): 416–426. https://doi.org/10.1038/sj.leu.2402799.

Duan, J.J., Qiu, W., Xu, S.L. et al. (2013). Strategies for isolating and enriching cancer stem cells: well begun is half done. *Stem Cells Dev.* 22 (16): 2221–2239. https://doi.org/10.1089/scd.2012.0613.

Ethier, S.P., Mahacek, M.L., Gullick, W.J. et al. (1993). Differential isolation of normal luminal mammary epithelial cells and breast cancer cells from primary and metastatic sites using selective media. *Cancer Res.* 53 (3): 627–635.

Fael Al-Mayhani, T.M., Ball, S.L., Zhao, J.W. et al. (2009). An efficient method for derivation and propagation of glioblastoma cell lines that conserves the molecular profile of their original tumours. *J. Neurosci. Methods* 176 (2): 192–199. https://doi.org/10.1016/j.jneumeth.2008.07.022.

Fantini, J., Galons, J.P., Abadie, B. et al. (1987). Growth in serum-free medium of human colonic adenocarcinoma cell lines on microcarriers: a two-step method allowing optimal cell spreading and growth. *In Vitro Cell. Dev. Biol.* 23 (9): 641–646.

Fogh, J. (1986). Human tumor lines for cancer research. *Cancer Investig.* 4 (2): 157–184.

Foksinski, M., Zarakowska, E., Gackowski, D. et al. (2017). Profiles of a broad spectrum of epigenetic DNA modifications in normal and malignant human cell lines: proliferation rate is not the major factor responsible for the 5-hydroxymethyl-2′-deoxycytidine level in cultured cancerous cell lines. *PLoS One* 12 (11): e0188856. https://doi.org/10.1371/journal.pone.0188856.

Freedman, V.H. and Shin, S.I. (1974). Cellular tumorigenicity in nude mice: correlation with cell growth in semi-solid medium. *Cell* 3 (4): 355–359. https://doi.org/10.1016/0092-8674(74)90050-6.

Freshney, R.I. (1980). Culture of glioma of the brain. In: *Brain Tumours: Scientific Basis, Clinical Investigation and Current Therapy* (eds. D.G.T. Thomas and D.I. Graham), 21–50. London: Butterworths.

Gaebler, M., Silvestri, A., Haybaeck, J. et al. (2017). Three-dimensional patient-derived in vitro sarcoma models: promising tools for improving clinical tumor management. *Front. Oncol.* 7: 203. https://doi.org/10.3389/fonc.2017.00203.

Garnett, M.J., Edelman, E.J., Heidorn, S.J. et al. (2012). Systematic identification of genomic markers of drug sensitivity in cancer cells. *Nature* 483 (7391): 570–575. https://doi.org/10.1038/nature11005.

Gazdar, A.F. and Oie, H.K. (1986). Re: growth of cell lines and clinical specimens of human non-small cell lung cancer in a serum-free defined medium. *Cancer Res.* 46 (11): 6011–6012.

Gazdar, A.F., Carney, D.N., Nau, M.M. et al. (1985). Characterization of variant subclasses of cell lines derived from small cell lung cancer having distinctive biochemical, morphological, and growth properties. *Cancer Res.* 45 (6): 2924–2930.

Gazdar, A.F., Girard, L., Lockwood, W.W. et al. (2010). Lung cancer cell lines as tools for biomedical discovery and research. *J. Natl Cancer Inst.* 102 (17): 1310–1321. https://doi.org/10.1093/jnci/djq279.

Gey, G.O., Coffman, W.D., and Kubicek, M.T. (1952). Tissue culture studies of the proliferative capacity of cervical carcinoma and normal epithelium. *Cancer Res.* 12: 264–265.

Ghaemi, S.R., Harding, F.J., Delalat, B. et al. (2013). Exploring the mesenchymal stem cell niche using high throughput screening. *Biomaterials* 34 (31): 7601–7615. https://doi.org/10.1016/j.biomaterials.2013.06.022.

Ghajar, C.M., Peinado, H., Mori, H. et al. (2013). The perivascular niche regulates breast tumour dormancy. *Nat. Cell Biol.* 15 (7): 807–817. https://doi.org/10.1038/ncb2767.

Giard, D.J., Aaronson, S.A., Todaro, G.J. et al. (1973). In vitro cultivation of human tumors: establishment of cell lines derived from a series of solid tumors. *J. Natl Cancer Inst.* 51 (5): 1417–1423.

Gillet, J.P., Varma, S., and Gottesman, M.M. (2013). The clinical relevance of cancer cell lines. *J. Natl Cancer Inst.* 105 (7): 452–458. https://doi.org/10.1093/jnci/djt007.

Giovanella, B.C., Yim, S.O., Stehlin, J.S. et al. (1972). Development of invasive tumors in the "nude" mouse after injection of cultured human melanoma cells. *J. Natl Cancer Inst.* 48 (5): 1531–1533.

Grand, C.G. and Cameron, G. (1948). Tissue culture studies of pigmented melanomas; fish, mouse, and human. *Ann. N. Y. Acad. Sci.* 4: 171–176.

Greve, B., Kelsch, R., Spaniol, K. et al. (2012). Flow cytometry in cancer stem cell analysis and separation. *Cytometry A* 81 (4): 284–293. https://doi.org/10.1002/cyto.a.22022.

Griewank, K.G., Yu, X., Khalili, J. et al. (2012). Genetic and molecular characterization of uveal melanoma cell lines. *Pigment Cell Melanoma Res.* 25 (2): 182–187. https://doi.org/10.1111/j.1755-148X.2012.00971.x.

Grizzle, W.E., Bell, W.C., and Sexton, K.C. (2010). Issues in collecting, processing and storing human tissues and associated information to support biomedical research. *Cancer Biomark.* 9 (1–6): 531–549. https://doi.org/10.3233/CBM-2011-0183.

Halaban, R. (2004). Culture of melanocytes from normal, benign, and malignant lesions. In: *Culture of Human Tumor Cells* (eds. R. Pfragner and R.I. Freshney), 289–318. Hoboken, NJ: Wiley-Liss.

Hale, J.S., Li, M., and Lathia, J.D. (2012). The malignant social network: cell–cell adhesion and communication in cancer stem cells. *Cell Adhes. Migr.* 6 (4): 346–355. https://doi.org/10.4161/cam.21294.

Hamburger, A.W. and Salmon, S.E. (1977). Primary bioassay of human tumor stem cells. *Science* 197 (4302): 461–463.

Hammond, S.L., Ham, R.G., and Stampfer, M.R. (1984). Serum-free growth of human mammary epithelial cells: rapid clonal growth in defined medium and extended serial passage with pituitary extract. *Proc. Natl Acad. Sci. U.S.A.* 81 (17): 5435–5439.

Hattori, E., Oyama, R., and Kondo, T. (2019). Systematic review of the current status of human sarcoma cell lines. *Cell* 8 (2) https://doi.org/10.3390/cells8020157.

Heddleston, J.M., Li, Z., Lathia, J.D. et al. (2010). Hypoxia inducible factors in cancer stem cells. *Br. J. Cancer* 102 (5): 789–795. https://doi.org/10.1038/sj.bjc.6605551.

Hinson, A.R., Jones, R., Crose, L.E. et al. (2013). Human rhabdomyosarcoma cell lines for rhabdomyosarcoma research: utility and pitfalls. *Front. Oncol.* 3: 183. https://doi.org/10.3389/fonc.2013.00183.

Holliday, D.L. and Speirs, V. (2011). Choosing the right cell line for breast cancer research. *Breast Cancer Res.* 13 (4): 215. https://doi.org/10.1186/bcr2889.

Hynds, R.E., Vladimirou, E., and Janes, S.M. (2018). The secret lives of cancer cell lines. *Dis. Model. Mech.* 11 (11) https://doi.org/10.1242/dmm.037366.

Ince, T.A., Sousa, A.D., Jones, M.A. et al. (2015). Characterization of twenty-five ovarian tumour cell lines that phenocopy primary tumours. *Nat. Commun.* 6: 7419. https://doi.org/10.1038/ncomms8419.

Inch, W.R., McCredie, J.A., and Sutherland, R.M. (1970). Growth of nodular carcinomas in rodents compared with multi-cell spheroids in tissue culture. *Growth* 34 (3): 271–282.

Inoue, K., Kohno, T., Takakura, S. et al. (2001). Corrigendum to: frequent microsatellite instability and BAX mutations in T cell acute lymphoblastic leukemia cell lines. *Leukemia Research* 24 (2000), 255–262. *Leuk. Res.* 25 (3): 275–278.

Ishikawa, M., Inoue, T., Shirai, T. et al. (2014). Simultaneous expression of cancer stem cell-like properties and cancer-associated fibroblast-like properties in a primary culture of breast cancer cells. *Cancers (Basel)* 6 (3): 1570–1578. https://doi.org/10.3390/cancers6031570.

Jacob, F., Nixdorf, S., Hacker, N.F. et al. (2014). Reliable in vitro studies require appropriate ovarian cancer cell lines. *J. Ovarian Res.* 7: 60. https://doi.org/10.1186/1757-2215-7-60.

Jacobsen, B.M., Harrell, J.C., Jedlicka, P. et al. (2006). Spontaneous fusion with, and transformation of mouse stroma by, malignant human breast cancer epithelium. *Cancer Res.* 66 (16): 8274–8279. https://doi.org/10.1158/0008-5472.CAN-06-1456.

Jokela, T.A., Engelsen, A.S.T., Rybicka, A. et al. (2018). Microenvironment-induced non-sporadic expression of the AXL and cKIT receptors are related to epithelial plasticity and drug resistance. *Front. Cell Dev. Biol.* 6: 41. https://doi.org/10.3389/fcell.2018.00041.

Kang, M.H., Smith, M.A., Morton, C.L. et al. (2011). National Cancer Institute pediatric preclinical testing program: model description for in vitro cytotoxicity testing. *Pediatr. Blood Cancer* 56 (2): 239–249. https://doi.org/10.1002/pbc.22801.

Kimlin, L.C., Casagrande, G., and Virador, V.M. (2013). In vitro three-dimensional (3D) models in cancer research: an update. *Mol. Carcinog.* 52 (3): 167–182. https://doi.org/10.1002/mc.21844.

Kondo, J., Endo, H., Okuyama, H. et al. (2011). Retaining cell–cell contact enables preparation and culture of spheroids composed of pure primary cancer cells from colorectal cancer. *Proc. Natl Acad. Sci. U.S.A.* 108 (15): 6235–6240. https://doi.org/10.1073/pnas.1015938108.

Krishnamurthy, S., Warner, K.A., Dong, Z. et al. (2014). Endothelial interleukin-6 defines the tumorigenic potential of primary human cancer stem cells. *Stem Cells* 32 (11): 2845–2857. https://doi.org/10.1002/stem.1793.

LaBarge, M.A. (2010). The difficulty of targeting cancer stem cell niches. *Clin. Cancer Res.* 16 (12): 3121–3129. https://doi.org/10.1158/1078-0432.CCR-09-2933.

LaBarge, M.A., Nelson, C.M., Villadsen, R. et al. (2009). Human mammary progenitor cell fate decisions are products of interactions with combinatorial microenvironments. *Integr. Biol. (Camb).* 1 (1): 70–79. https://doi.org/10.1039/b816472j.

LaBarge, M.A., Garbe, J.C., and Stampfer, M.R. (2013). Processing of human reduction mammoplasty and mastectomy tissues for cell culture. *J. Vis. Exp.* 71: 50011. https://doi.org/10.3791/50011.

Lan, S., Smith, H.S., and Stampfer, M.R. (1981). Clonal growth of normal and malignant human breast epithelia. *J. Surg. Oncol.* 18 (3): 317–322.

Lasfargues, E.Y. (1973). Human mammary tumors. In: *Tissue Culture: Methods and Applications* (eds. P.F. Kruse Jr. and M.K. Patterson), 45–50. New York: Academic Press.

Leake, R.E., Freshney, R.I., and Munir, I. (1987). Steroid responses in vivo and in vitro. In: *Steroid Hormones: A Practical Approach* (eds. B. Green and R.E. Leake), 205–218. Oxford: IRL Press at Oxford University Press.

Lee, J., Kotliarova, S., Kotliarov, Y. et al. (2006). Tumor stem cells derived from glioblastomas cultured in bFGF and EGF more closely mirror the phenotype and genotype of primary tumors than do serum-cultured cell lines. *Cancer Cell* 9 (5): 391–403. https://doi.org/10.1016/j.ccr.2006.03.030.

Levin, D.B., Wilson, K., Valadares de Amorim, G. et al. (1995). Detection of p53 mutations in benign and dysplastic nevi. *Cancer Res.* 55 (19): 4278–4282.

Liu, C., Liu, Y., Xu, X.X. et al. (2015). Potential effect of matrix stiffness on the enrichment of tumor initiating cells under three-dimensional culture conditions. *Exp. Cell Res.* 330 (1): 123–134. https://doi.org/10.1016/j.yexcr.2014.07.036.

Lobo, N.A., Shimono, Y., Qian, D. et al. (2007). The biology of cancer stem cells. *Annu. Rev. Cell Dev. Biol.* 23: 675–699. https://doi.org/10.1146/annurev.cellbio.22.010305.104154.

Loo, D.T., Fuquay, J.I., Rawson, C.L. et al. (1987). Extended culture of mouse embryo cells without senescence: inhibition by serum. *Science* 236 (4798): 200–202.

Lopez-Terrada, D., Cheung, S.W., Finegold, M.J. et al. (2009). Hep G2 is a hepatoblastoma-derived cell line. *Hum. Pathol.* 40 (10): 1512–1515. https://doi.org/10.1016/j.humpath.2009.07.003.

Lounis, H., Provencher, D., Godbout, C. et al. (1994). Primary cultures of normal and tumoral human ovarian epithelium: a powerful tool for basic molecular studies. *Exp. Cell Res.* 215 (2): 303–309. https://doi.org/10.1006/excr.1994.1346.

Lundqvist, M., Mark, J., Funa, K. et al. (1991). Characterisation of a cell line (LCC-18) from a cultured human neuroendocrine-differentiated colonic carcinoma. *Eur. J. Cancer* 27 (12): 1663–1668. https://doi.org/10.1016/0277-5379(91)90441-f.

MacDonald, C.M., Freshney, R.I., Hart, E. et al. (1985). Selective control of human glioma cell proliferation by specific cell interaction. *Exp. Cell Biol.* 53 (3): 130–137.

MacLeod, R.A., Nagel, S., Scherr, M. et al. (2008). Human leukemia and lymphoma cell lines as models and resources. *Curr. Med. Chem.* 15 (4): 339–359.

Masters, J.R.W. (1991). *Human Cancer in Primary Culture, a Handbook*. Dordrecht: Kluwer Academic Publishers.

Masters, J.R. (2000). Human cancer cell lines: fact and fantasy. *Nat. Rev. Mol. Cell Biol.* 1 (3): 233–236. https://doi.org/10.1038/35043102.

Masters, J.R.W. and Palsson, B. (1999). *Cancer Cell Lines Part 1*. New York: Kluwer Academic Publishers.

Masters, J.R.W. and Palsson, B. (2002a). *Cancer Cell Lines Part 2*. New York: Kluwer Academic Publishers.

Masters, J.R.W. and Palsson, B. (2002b). *Cancer Cell Lines Part 3: Leukemias and Lymphomas*. New York: Kluwer Academic Publishers.

Menden, M.P., Casale, F.P., Stephan, J. et al. (2018). The germline genetic component of drug sensitivity in cancer cell lines. *Nat. Commun.* 9 (1): 3385. https://doi.org/10.1038/s41467-018-05811-3.

Messing, E.M., Fahey, J.L., deKernion, J.B. et al. (1982). Serum-free medium for the in vitro growth of normal and malignant urinary bladder epithelial cells. *Cancer Res.* 42 (6): 2392–2397.

Mohseny, A.B., Machado, I., Cai, Y. et al. (2011). Functional characterization of osteosarcoma cell lines provides representative models to study the human disease. *Lab. Investig.* 91 (8): 1195–1205. https://doi.org/10.1038/labinvest.2011.72.

Morton, C.L. and Houghton, P.J. (2007). Establishment of human tumor xenografts in immunodeficient mice. *Nat. Protoc.* 2 (2): 247–250. https://doi.org/10.1038/nprot.2007.25.

Mouradov, D., Sloggett, C., Jorissen, R.N. et al. (2014). Colorectal cancer cell lines are representative models of the main molecular subtypes of primary cancer. *Cancer Res.* 74 (12): 3238–3247. https://doi.org/10.1158/0008-5472.CAN-14-0013.

Murakami, H. and Masui, H. (1980). Hormonal control of human colon carcinoma cell growth in serum-free medium. *Proc. Natl*

Acad. Sci. U.S.A. 77 (6): 3464–3468. https://doi.org/10.1073/pnas.77.6.3464.

Niepel, M., Hafner, M., Mills, C.E. et al. (2019). A multi-center study on the reproducibility of drug-response assays in mammalian cell lines. *Cell Syst.* 9 (1): 35–48. https://doi.org/10.1016/j.cels.2019.06.005.

Oie, H.K., Russell, E.K., Carney, D.N. et al. (1996). Cell culture methods for the establishment of the NCI series of lung cancer cell lines. *J. Cell. Biochem. Suppl.* 24: 24–31.

Pantel, K., Dickmanns, A., Zippelius, A. et al. (1995). Establishment of micrometastatic carcinoma cell lines: a novel source of tumor cell vaccines. *J. Natl Cancer Inst.* 87 (15): 1162–1168. https://doi.org/10.1093/jnci/87.15.1162.

Park, J.G., Ku, J.L., Kim, H.S. et al. (2004). Culture of normal and malignant gastric epithelium. In: *Culture of Human Tumor Cells* (eds. R. Pfragner and R.I. Freshney), 23–66. Hoboken, NJ: Wiley-Liss.

Park, T.S., Donnenberg, V.S., Donnenberg, A.D. et al. (2014). Dynamic interactions between cancer stem cells and their stromal partners. *Curr. Pathobiol. Rep.* 2 (1): 41–52. https://doi.org/10.1007/s40139-013-0036-5.

Pastrana, E., Silva-Vargas, V., and Doetsch, F. (2011). Eyes wide open: a critical review of sphere-formation as an assay for stem cells. *Cell Stem Cell* 8 (5): 486–498. https://doi.org/10.1016/j.stem.2011.04.007.

Pavelic, Z.P., Slocum, H.K., Rustum, Y.M. et al. (1980). Growth of cell colonies in soft agar from biopsies of different human solid tumors. *Cancer Res.* 40 (11): 4151–4158.

Pavelic, K., Antonic, M., Pavelic, L. et al. (1992). Human lung cancers growing on extracellular matrix: expression of oncogenes and growth factors. *Anticancer Res.* 12 (6B): 2191–2196.

Peehl, D.M. (2003). Growth of prostatic epithelial and stromal cells in vitro. In: *Prostate Cancer Methods and Protocols* (eds. P.J. Russell, P. Jackson and E.A. Kingsley), 41–57. Totowa, NJ: Humana Press.

Peehl, D.M. and Stamey, T.A. (1986). Serum-free growth of adult human prostatic epithelial cells. *In Vitro Cell. Dev. Biol.* 22 (2): 82–90.

Phan, N., Hong, J.J., Tofig, B. et al. (2019). A simple high-throughput approach identifies actionable drug sensitivities in patient-derived tumor organoids. *Commun. Biol.* 2: 78. https://doi.org/10.1038/s42003-019-0305-x.

Phelps, R.M., Johnson, B.E., Ihde, D.C. et al. (1996). NCI-navy medical oncology branch cell line data base. *J. Cell. Biochem. Suppl.* 24: 32–91.

Poli, V., Fagnocchi, L., and Zippo, A. (2018). Tumorigenic cell reprogramming and cancer plasticity: interplay between signaling, microenvironment, and epigenetics. *Stem Cells Int.* 2018: 4598195. https://doi.org/10.1155/2018/4598195.

Ponten, J. and Macintyre, E.H. (1968). Long term culture of normal and neoplastic human glia. *Acta Pathol. Microbiol. Scand.* 74 (4): 465–486.

Pulvertaft, R.J. (1964). Phytohaemagglutinin in relation to Burkitt's tumour (African lymphoma). *Lancet* 2 (7359): 552–554. https://doi.org/10.1016/s0140-6736(64)90618-x.

Quentmeier, H., Pommerenke, C., Dirks, W.G. et al. (2019). The LL-100 panel: 100 cell lines for blood cancer studies. *Sci. Rep.* 9 (1): 8218. https://doi.org/10.1038/s41598-019-44491-x.

Quintana, E., Shackleton, M., Sabel, M.S. et al. (2008). Efficient tumour formation by single human melanoma cells. *Nature* 456 (7222): 593–598. https://doi.org/10.1038/nature07567.

Rheinwald, J.G. and Beckett, M.A. (1981). Tumorigenic keratinocyte lines requiring anchorage and fibroblast support cultured from human squamous cell carcinomas. *Cancer Res.* 41 (5): 1657–1663.

Rhim, J.S. (2013). Human prostate epithelial cell cultures. *Methods Mol. Biol.* 946: 383–393. https://doi.org/10.1007/978-1-62703-128-8_24.

Ruff, M.R. and Pert, C.B. (1984). Small cell carcinoma of the lung: macrophage-specific antigens suggest hemopoietic stem cell origin. *Science* 225 (4666): 1034–1036. https://doi.org/10.1126/science.6089338.

Russell, P.J. and Kingsley, E.A. (2003). Human prostate cancer cell lines. *Methods Mol. Med.* 81: 21–39. https://doi.org/10.1385/1-59259-372-0:21.

Rygaard, J. and Povlsen, C.O. (1969). Heterotransplantation of a human malignant tumour to "nude" mice. *Acta Pathol. Microbiol. Scand.* 77 (4): 758–760.

Sant, S. and Johnston, P.A. (2017). The production of 3D tumor spheroids for cancer drug discovery. *Drug Discov. Today Technol.* 23: 27–36. https://doi.org/10.1016/j.ddtec.2017.03.002.

Scadden, D.T. (2006). The stem-cell niche as an entity of action. *Nature* 441 (7097): 1075–1079. https://doi.org/10.1038/nature04957.

Schweppe, R.E. and Korch, C. (2018). Challenges and advances in the development of cell lines and xenografts. *Adv. Mol. Pathol.* 1 (1): 239–251. https://doi.org/10.1016/j.yamp.2018.07.004.

Selby, P.J., Thomas, J.M., Monaghan, P. et al. (1980). Human tumour xenografts established and serially transplanted in mice immunologically deprived by thymectomy, cytosine arabinoside and whole-body irradiation. *Br. J. Cancer* 41 (1): 52–61. https://doi.org/10.1038/bjc.1980.7.

Sharkey, F.E. and Fogh, J. (1984). Considerations in the use of nude mice for cancer research. *Cancer Metastasis Rev.* 3 (4): 341–360.

Shioda, S., Kasai, F., Watanabe, K. et al. (2018). Screening for 15 pathogenic viruses in human cell lines registered at the JCRB cell bank: characterization of in vitro human cells by viral infection. *R. Soc. Open Sci.* 5 (5): 172472. https://doi.org/10.1098/rsos.172472.

Shoemaker, R.H. (2006). The NCI60 human tumour cell line anticancer drug screen. *Nat. Rev. Cancer* 6 (10): 813–823. https://doi.org/10.1038/nrc1951.

Smith, H.S., Lan, S., Ceriani, R. et al. (1981). Clonal proliferation of cultured nonmalignant and malignant human breast epithelia. *Cancer Res.* 41 (11 Pt 1): 4637–4643.

Soo, J.K., Mackenzie Ross, A.D., and Bennett, D.C. (2011). Isolation and culture of melanoma and naevus cells and cell lines. In: *Cancer Cell Culture: Methods and Protocols* (ed. I.A. Cree), 141–150. Totowa, NJ: Humana Press.

Speirs, V. (2004). Primary culture of human mammary tumor cells. In: *Culture of Human Tumor Cells* (eds. R. Pfragner and R.I. Freshney), 205–219. Hoboken, NJ: Wiley-Liss.

van Staveren, W.C., Solis, D.Y., Hebrant, A. et al. (2009). Human cancer cell lines: experimental models for cancer cells in situ? For cancer stem cells? *Biochim. Biophys. Acta* 1795 (2): 92–103. https://doi.org/10.1016/j.bbcan.2008.12.004.

Sturzl, M., Gaus, D., Dirks, W.G. et al. (2013). Kaposi's sarcoma-derived cell line SLK is not of endothelial origin, but is a contaminant from a known renal carcinoma cell line. *Int. J. Cancer* 132 (8): 1954–1958. https://doi.org/10.1002/ijc.27849.

Sukowati, C.H.C., Anfuso, B., Fiore, E. et al. (2019). Hyaluronic acid inhibition by 4-methylumbelliferone reduces the expression of cancer stem cells markers during hepatocarcinogenesis. *Sci. Rep.* 9 (1): 4026. https://doi.org/10.1038/s41598-019-40436-6.

Sutherland, R.M. (1988). Cell and environment interactions in tumor microregions: the multicell spheroid model. *Science* 240 (4849): 177–184.

Sutherland, R.L., Watts, C.K.W., Lee, C.S.L. et al. (2002). Breast cancer. In: *Cancer Cell Lines Part 2* (eds. J.R.W. Masters and B. Palsson), 79–106. New York: Kluwer Academic Publishers.

Tevis, K.M., Colson, Y.L., and Grinstaff, M.W. (2017). Embedded spheroids as models of the cancer microenvironment. *Adv. Biosyst.* 1 (10) https://doi.org/10.1002/adbi.201700083.

Topley, P., Jenkins, D.C., Jessup, E.A. et al. (1993). Effect of reconstituted basement membrane components on the growth of a panel of human tumour cell lines in nude mice. *Br. J. Cancer* 67 (5): 953–958. https://doi.org/10.1038/bjc.1993.176.

Uphoff, C.C., Denkmann, S.A., Steube, K.G. et al. (2010). Detection of EBV, HBV, HCV, HIV-1, HTLV-I and -II, and SMRV in human and other primate cell lines. *J. Biomed. Biotechnol.* 2010: 904767. https://doi.org/10.1155/2010/904767.

Uphoff, C.C., Lange, S., Denkmann, S.A. et al. (2015). Prevalence and characterization of murine leukemia virus contamination in human cell lines. *PLoS One* 10 (4): e0125622. https://doi.org/10.1371/journal.pone.0125622.

Uzgare, A.R., Xu, Y., and Isaacs, J.T. (2004). In vitro culturing and characteristics of transit amplifying epithelial cells from human prostate tissue. *J. Cell. Biochem.* 91 (1): 196–205. https://doi.org/10.1002/jcb.10764.

Verschraegen, C.F., Hu, W., Du, Y. et al. (2003). Establishment and characterization of cancer cell cultures and xenografts derived from primary or metastatic Mullerian cancers. *Clin. Cancer Res.* 9 (2): 845–852.

Vinci, M., Box, C., and Eccles, S.A. (2015). Three-dimensional (3D) tumor spheroid invasion assay. *J. Vis. Exp.* 99: e52686. https://doi.org/10.3791/52686.

Virgone-Carlotta, A., Lemasson, M., Mertani, H.C. et al. (2017). In-depth phenotypic characterization of multicellular tumor spheroids: effects of 5-fluorouracil. *PLoS One* 12 (11): e0188100. https://doi.org/10.1371/journal.pone.0188100.

Walker, K.A., Murray, T., Hilditch, T.E. et al. (1988). A tumour spheroid model for antibody-targeted therapy of micrometastases. *Br. J. Cancer* 58 (1): 13–16. https://doi.org/10.1038/bjc.1988.152.

Wibe, E., Berg, J.P., Tveit, K.M. et al. (1984). Multicellular spheroids grown directly from human tumour material. *Int. J. Cancer* 34 (1): 21–26. https://doi.org/10.1002/ijc.2910340105.

Wilding, J.L. and Bodmer, W.F. (2014). Cancer cell lines for drug discovery and development. *Cancer Res.* 74 (9): 2377–2384. https://doi.org/10.1158/0008-5472.CAN-13-2971.

Wilson, A.P. and Garner, C.M. (2002). Ovarian cancer. In: *Cancer Cell Lines Part 2* (eds. J.R.W. Masters and B. Palsson), 1–53. New York: Kluwer Academic Publishers.

Xie, Y., Bergstrom, T., Jiang, Y. et al. (2015). The human glioblastoma cell culture resource: validated cell models representing all molecular subtypes. *EBioMedicine* 2 (10): 1351–1363. https://doi.org/10.1016/j.ebiom.2015.08.026.

Xu, J., Margol, A., Asgharzadeh, S. et al. (2015). Pediatric brain tumor cell lines. *J. Cell. Biochem.* 116 (2): 218–224. https://doi.org/10.1002/jcb.24976.

Yong, J.W., Choong, M.L., Wang, S. et al. (2014). Characterization of ductal carcinoma in situ cell lines established from breast tumor of a Singapore Chinese patient. *Cancer Cell Int.* 14 (1): 94. https://doi.org/10.1186/s12935-014-0094-8.

Yuhas, J.M., Li, A.P., Martinez, A.O. et al. (1977). A simplified method for production and growth of multicellular tumor spheroids. *Cancer Res.* 37 (10): 3639–3643.

Zeinalian, M., Hashemzadeh-Chaleshtori, M., Salehi, R. et al. (2018). Clinical aspects of microsatellite instability testing in colorectal cancer. *Adv. Biomed. Res.* 7: 28. https://doi.org/10.4103/abr.abr_185_16.

Zhang, Y.A., Maitra, A., Hsieh, J.T. et al. (2011). Frequent detection of infectious xenotropic murine leukemia virus (XMLV) in human cultures established from mouse xenografts. *Cancer Biol. Ther.* 12 (7): 617–628.

CHAPTER 26

Differentiation

We now know that each differentiated cell within an adult organism can trace its lineage back to an embryonic stem cell (ESC) (see Section 2.4.1). Early work on stem cell populations was performed using teratoid tumors, which contain cells from all three germ layers and can be readily maintained as xenografts in nude mice (see Section 2.4.3). In addition to their relevance to stem cell biology, these tumors gave some unexpected and fascinating insights into differentiation. For example, the TERA-2 cell line was derived from a patient with metastatic testicular teratocarcinoma (Fogh and Trempe 1975). Jørgen Fogh, Peter Andrews, and colleagues observed that clonal derivatives of TERA-2 underwent marked differentiation when injected into nude mice (Andrews et al. 1984). One of these clonal derivatives, NTERA-2, has become a classic and widely used *in vitro* model for neuronal differentiation (Andrews 1998; Schwartz et al. 2005). Treatment of NTERA-2 cells with retinoic acid leads to loss of stem cell markers and irreversible differentiation (Andrews 1998). Intriguingly, only a minority (1–5%) of cells display neuronal differentiation; the remaining NTERA-2 cells differentiate into non-neural phenotypes, suggesting that pluripotent cells oscillate between different states that are primed to develop into different lineages (Tonge et al. 2010). Clearly, *in vitro* models can be useful to understand differentiation, but their effectiveness will vary with their inherent biology, the culture conditions, and the presence of inducing agents such as retinoic acid. This chapter focuses on the practical requirements for *in vitro* differentiation and its regulation. Essential concepts and definitions can be found in an earlier chapter, which focuses on the biology of cells in culture (see Chapter 2).

26.1 *IN VITRO* MODELS OF DIFFERENTIATION

Differentiation was initially studied in primary culture, due to the difficulties experienced in establishing cell lines from particular tissues (see Section 13.1). Over time, procedures have been developed for the selective culture of "specialized" cells that continue to express differentiated characteristics (see Chapter 24). In some cases, these procedures have led to the establishment of continuous cell lines that retain many of their differentiated properties (see Table 26.1).

Continuous cell lines can be highly effective models for *in vitro* differentiation, as shown by the HepaRG cell line. HepaRG cells (see Figure 26.1) were originally derived from a hepatocholangiocarcinoma in a female patient with hepatitis C (Gripon et al. 2002). In-depth proteomic analysis has shown that this cell line retains a close similarity to primary human hepatocytes (Tascher et al. 2019). HepaRG cells express liver-specific plasma proteins and liver-specific glycolytic enzymes, and display high expression and inducibility of the major Phase I and Phase II detoxification enzymes (Aninat et al. 2006). The sophistication of this model resides in its ability to rapidly reverse in a few hours to undifferentiated hepatic bipotent progenitors when confluence is broken down. Cells start to divide actively until they reach confluence, at which time they can be shifted to differentiation medium containing dimethyl sulfoxide (DMSO). Cells then re-initiate differentiation programs toward either hepatocyte or biliary cells; a mixed population is established within two weeks when grown in 2% DMSO, approximately half of which are hepatocytes (see Figure 26.1a). The hepatocyte population can be purified to give cultures that contain up

Freshney's Culture of Animal Cells: A Manual of Basic Technique and Specialized Applications, Eighth Edition. Amanda Capes-Davis and R. Ian Freshney.
© 2021 John Wiley & Sons Ltd. Published 2021 by John Wiley & Sons Ltd.
Companion website: www.wiley.com/go/freshney/cellculture8

TABLE 26.1. Continuous cell lines with differentiated properties.

Cell type	Cell line[a]	RRID[b]	Species	Origin	Reported attribute
Endocrine	GH3	CVCL_0273	Rat	Pituitary tumor	Growth hormone production
Endocrine	Y1	CVCL_0585	Mouse	Adrenal cortical carcinoma	Adrenal steroid production
Endothelium	CPAE	CVCL_2877	Cow	Pulmonary artery	ACE activity, vWF expression
Endothelium	SK-HEP-1 [c]	CVCL_0525	Human	Ascites, hepatocellular carcinoma	Tubule formation, vWF expression
Epithelium	Caco-2	CVCL_0025	Human	Colon adenocarcinoma	Transport, dome formation
Epithelium	LNCaP	CVCL_0395	Human	Prostate carcinoma	Hormone-responsive
Epithelium	MCF-7	CVCL_0031	Human	Pleural effusion, breast carcinoma	Hormone-responsive
Epithelium	MDCK	CVCL_0422	Dog	Kidney	Transport, dome formation
Epithelium	LLC-PK1	CVCL_K238	Pig	Kidney	Na$^+$-dependent glucose uptake
Glia	MOG-G-CCM	CVCL_2613	Human	Anaplastic astrocytoma	Glutamyl synthetase production
Glia	C6	CVCL_4060	Rat	Glioma	GFAP, GPDH expression
Hepatocytes	H4-II-E-C3	CVCL_0285	Rat	Hepatocellular carcinoma	Tyrosine aminotransferase production
Hepatocytes	HepaRG	CVCL_9720	Human	Hepatocellular carcinoma	Aldolase B expression
Keratinocytes	HaCaT	CVCL_0038	Human	Epidermis	Cornification
Melanocytes	B16	CVCL_F936	Mouse	Melanoma	Melanin production
Myeloid	Friend cells, e.g. HCD-57 [d]	Multiple including CVCL_5289	Mouse	Spleen	Hemoglobin production
Myeloid	HL-60	CVCL_0002	Human	Acute myeloid leukemia	Morphology (neutrophils), apoptosis
Myeloid	K-562	CVCL_0004	Human	Chronic myelogenous leukemia	Hemoglobin production
Myocytes	C2	CVCL_6812	Mouse	Skeletal muscle	Myotube formation
Myocytes	L6	CVCL_0385	Rat	Skeletal muscle	Myotube formation
Neuroendocrine	PC12	CVCL_0481	Rat	Pheochromo-cytoma	Catecholamine, dopamine, norepinephrine production
Neurons	C-1300	CVCL_4343	Rat	Neuroblastoma	Morphology (neurites)
Type II pneumocyte or Clara cell	A549	CVCL_0023	Human	Lung adenocarcinoma	Surfactant production
Type II pneumocyte	NCI-H441	CVCL_1561	Human	Lung adenocarcinoma	Surfactant production
Various	F9	CVCL_0259	Mouse	Embryonal teratocarcinoma	Laminin, type IV collagen expression
Various (neuronal and non-neural)	NTERA-2	CVCL_0034	Human	Testicular teratocarcinoma	Morphology, βIII tubulin expression

[a]Due to the number of cell lines that are available, it is impossible to provide a comprehensive list. Examples are provided here as a starting point for further investigation. Tables of commonly used cell lines and ESC lines are listed in previous chapters (see Tables 14.2, 23.3).

[b]RRIDs can be used to find additional information in the Cellosaurus knowledge resource, including source publications (https://web.expasy.org/cellosaurus) (Bairoch 2018).

[c]SK-HEP-1 cells were originally thought to be hepatocellular carcinoma, but it is now believed they were misclassified and are of endothelial origin (Heffelfinger et al. 1992).

[d]Friend cells come from mice that have been infected with the Friend virus (see Section 24.5.3; see also Figure 26.4a, b). Some continuous cell lines have been established from these mice, including HCD-57.

ACE, Angiotensin converting enzyme activity; GFAP, glial fibrillary acidic protein; GPDH, glycerol phosphate dehydrogenase; vWF, von Willebrand factor.

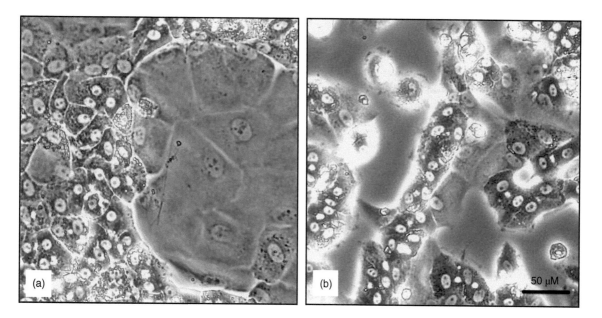

Fig. 26.1. HepaRG cells. (a) Four-week-old culture that has been maintained for two weeks in the presence of 2% DMSO to induce differentiation. Note the mixed population, with highly differentiated hepatocyte colonies bordered by flat biliary cells. (b) Two-day-old culture of selected hepatocytes maintained in the presence of 2% DMSO. Note the purity of the population and the hepatocyte organization in typical hepatic trabeculae. Scale bar as indicated. *Source*: Courtesy of C. Guguen-Guillouzo.

to 80–90% hepatocytes (see Figure 26.1b). A procedure for purification of HepaRG cells is provided as Supplementary Material online (see Supp. S26.1). Because master and working cell banks were prepared for HepaRG cells at early passage (see Section 15.5), they have undergone relatively few passages and continue to maintain their differentiated characteristics (Tascher et al. 2019).

Unfortunately, other hepatic cell lines do not display the same degree of differentiation. Hep-G2 and Huh-7 cell lines are widely used as models for drug screening, but they often fail to predict *in vivo* hepatotoxicity (Nelson et al. 2017). Hep-G2 cells were originally established in 1979 from a patient with hepatoblastoma (Lopez-Terrada et al. 2009). Although early data were promising, Hep-G2 cells gradually lost most of the metabolic properties of normal liver and acquired chromosomal aberrations (Tascher et al. 2019). A subclone (Hep-G2/C3A) has been established that expresses higher drug metabolizing activities, but the C3A subclone appears to demonstrate fewer differentiated properties compared to HepaRG (Kelly and Sussman 2000; Nelson et al. 2017). Huh-7 cells have preserved some characteristics of hepatocytes, which makes this cell line useful for viral hepatic studies, but less so for hepatotoxicity-related studies (Lohmann et al. 1999; Nelson et al. 2017). It should be noted that these studies typically grew cells as adherent monolayers. Three-dimensional (3D) culture systems can promote more physiological behavior when used for the culture of primary liver samples or for hepatic cell lines such as Hep-G2 and Huh-7 (see Figure 24.2) (Godoy et al. 2013).

Clearly, *in vitro* differentiation has a number of challenges in a practical sense. These challenges must be met if we are to use differentiated cells for therapeutic purposes (Parsons 2013). In order to introduce bioengineered cells into the body for regenerative medicine, we must understand what happens to the differentiation status of cells in culture, how to induce differentiation, and how to control differentiation so that specific lineages are generated.

26.2 DIFFERENTIATION STATUS IN CULTURE

26.2.1 Differentiation and Malignancy

The majority of continuous cell lines are established from malignant tumors or show signs of *in vitro* transformation, which is a phenotype that is associated with malignant behavior (see Sections 3.1.2, 3.6). As a tumor progresses, it typically displays poorer differentiation in association with a poorer prognosis. It has been proposed that cancer is principally a failure of cells to differentiate normally (Sell 2010). With these points in mind, it is somewhat surprising to find that many tumor cells successfully differentiate *in vitro* (see Table 26.1). Nevertheless, there appears to be a consistent (inverse) relationship between the expression of differentiated properties and that of malignancy-associated properties. This observation has led to "differentiation therapy," where inducing agents such as retinoic acid, DMSO, steroids, or small molecules are used for cancer treatment (Strickland and Mahdavi 1978; Spremulli and Dexter 1984; Freshney

1985; Kang et al. 2014). While differentiation therapy has had some success (particularly using retinoic acid for acute promyelocytic leukemia), the mechanisms of action of the various differentiation therapies remain somewhat unclear (de The 2018). However, we can conclude that malignant cells are likely to exhibit abnormalities of differentiation, including dedifferentiation and transdifferentiation (see Sections 26.2.3–26.2.5).

26.2.2 Proliferation and Differentiation

All cultures, regardless of their origins, are likely to contain cells at various stages of differentiation in equilibrium with each other (see Section 2.4.4). As differentiation progresses, cell division is reduced and eventually ceases, which means that in many cell systems, proliferation is incompatible with the expression of differentiated properties (see Figure 2.7). Tumor cells can sometimes break this restriction; for example, melanoma cells continue to synthesize melanin while the cells are proliferating (Halaban 2004). However, synthesis of the differentiated product will typically increase when cell division stops. An inverse relationship between proliferation and differentiation is evident in many cell lines, with important implications for cell culture practices, where expansion and propagation are often the main requirements. It is not surprising to find that most cell lines do not express fully differentiated properties and exhibit changes in phenotype when compared to their parental tissues (see Section 3.4).

The relationship between uncontrolled proliferation and poor differentiation was noted many years ago by Honor Fell and other exponents of organ culture (see Section 27.8) (Fell 1972). Fell and colleagues set out to retain high cell density and 3D tissue architecture and to prevent dissociation and selective overgrowth of undifferentiated cells. Organ culture was of considerable value in elucidating cellular interactions regulating differentiation and malignancy, but the approach was technically challenging and could not be used for further propagation. In addition, the heterogeneity of such samples (assumed to be essential for the maintenance of the tissue phenotype) has itself made the biochemical analysis of pure cell populations extremely difficult. Fortunately, more recent 3D techniques have allowed a fresh approach to this topic. It is now possible to re-induce a differentiated phenotype in pure populations of cells by recreating the correct environment and, by doing so, to define individual influences exerted on the induction and maintenance of differentiation. For this work to succeed, it is necessary to create an interactive, high-density cell population where proliferation and differentiation are intrinsically regulated. This typically leads to restriction of cell division to specific areas, such as the outermost layer of a 3D spheroid (see Figure 27.5).

26.2.3 Dedifferentiation

The term "dedifferentiation" refers to loss of the specific phenotypic properties that are associated with the mature,

specialized cell. As dedifferentiation is not always well defined and comprises complex processes with several contributory factors (including cell death, selective overgrowth, and adaptive responses), the term should be used with caution. Early reports of dedifferentiation were found to be associated with overgrowth by other cell types, e.g. fibroblasts (Sato 2008). More recent reports have come from extensive characterization of cell line panels; one such study generated authentication, genotypic, and phenotypic data for 60 thyroid carcinoma cell lines (Landa et al. 2019). All 60 cell lines displayed dedifferentiation, as measured by a thyroid differentiation score that examined expression of 13 tissue-specific genes. Cell lines were shown to carry genomic changes that faithfully recapitulated those seen in primary tumors, including key driver mutations, confirming their origin as thyroid carcinoma cells (Landa et al. 2019). While this study provided clear data on differentiation status, cell lines were grown using recommended media (typically the media used during their establishment or for expansion prior to banking) and it is unclear to what degree their properties might alter if grown under different conditions. Dedifferentiation may be either an adaptive process, implying that the differentiated phenotype may be regained given the right inducers (see Section 2.4.4), or a selective process, implying that a precursor cell has been selected because of its greater proliferative potential. In either case, the precursor cell may be induced to differentiate to a mature cell or revert to a more pluripotent state, depending on the culture conditions.

26.2.4 Transdifferentiation

"Transdifferentiation" refers to an irreversible change in phenotype from one differentiated state to an alternative differentiated fate (Shen et al. 2004; Le Magnen et al. 2018). This change is possible because of lineage plasticity (see Section 2.4.6). A change from one differentiated state to another may occur due to dedifferentiation, where cells revert to a more pluripotent state and then commit to a different embryonic lineage. However, direct conversion can occur from one cell fate to another without the cell reverting to a pluripotent state (Tanabe et al. 2015). This was initially demonstrated in closely related lineages, using the *MyoD* transcription factor to convert mesodermal lineages into muscle cells (Davis et al. 1987; Hollenberg et al. 1993). Distant lineages can also undergo direct conversion, provided suitable factors are supplied. For example, addition of three factors (*Ascl1*, *Brn2*, and *Myt1l*; together referred to as "BAM") will induce neuronal differentiation in cells from distant lineages, including mouse embryonic fibroblasts (MEFs), hESCs, primary human fibroblasts, and differentiated hepatocytes (Vierbuchen et al. 2010; Marro et al. 2011; Pang et al. 2011). A cocktail of small molecules can be added in place of the transcription factors, resulting in a simple, streamlined procedure for neuronal differentiation (Hu et al. 2015; Li et al. 2015). Human fibroblasts can be grown in

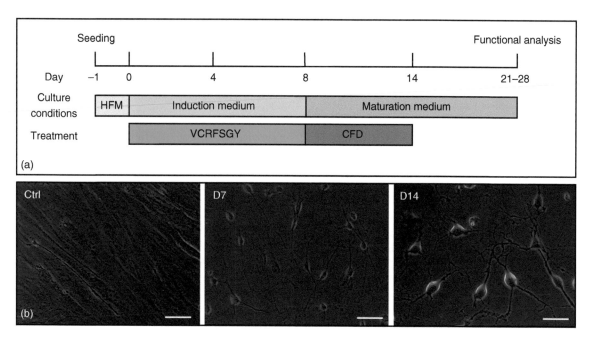

Fig. 26.2. Induction of human neuronal cells by small molecules. (a) Diagram of induction procedure. Human fibroblasts are plated (Day −1) in human fibroblast medium (HFM) and re-fed with induction medium (Day 0) and finally with maturation medium (Day 8). Treatment agents include VPA (V), CHIR99021 (C), RepSox (R), forskolin (F), SP600625 (S), GO6983 (G), Y-27632 (Y), and dorsomorphin (D). Further information on culture conditions can be found in the source reference. (b) Morphology in control cells (untreated fibroblasts, left panel), at Day 7 (middle panel), and at Day 14 (right panel) showing the development of typical neuronal morphology. Scale bars represent 100 μm. *Source*: (a) Hu, W., Qiu, B., Guan, W., et al. (2015). Direct Conversion of Normal and Alzheimer's Disease Human Fibroblasts into Neuronal Cells by Small Molecules. Cell Stem Cell 17 (2):204-12. © 2015 Elsevier. (b) Hu et al. (2015), reproduced with permission of Elsevier.

"induction medium" containing an initial chemical cocktail, followed by "maturation medium" containing a second cocktail to promote neuronal survival and maturation (see Figure 26.2a) (Hu et al. 2015). Induced neuronal (iN) cells develop an obvious neuronal morphology (see Figure 26.2b), express neuronal markers, and develop functional synapses that closely resemble those seen in cortical neuronal cultures (Marro et al. 2011; Pang et al. 2011; Hu et al. 2015).

26.2.5 Epithelial–Mesenchymal Transition (EMT)

The previous examples of transdifferentiation are ones where clear experimental evidence is available to demonstrate their validity. Other, less clearly documented cases exist where cells gained additional characteristics *in vitro* that were not consistent with their reported origins. Some early cases probably occurred due to cross-contamination, but this possibility could be excluded once authentication testing was developed (see Chapter 17). For example, one early study reported that clonal populations of fetal kidney cells changed from an epithelial-like to a fibroblast-like morphology at increasing density; the authors excluded cross-contamination using isoenzyme analysis (Gilbert et al.

1981). In retrospect, this example was probably caused by epithelial–mesenchymal transition (EMT). EMT and its inverse change, mesenchymal–epithelial transition (MET), are transient changes in differentiation status that may occur both *in vivo* and *in vitro*.

EMT and MET are important developmental programs that are employed during embryonic development and cancer (Hay 2005; Zhang and Weinberg 2018). Epithelial cells normally require cell–cell adhesion for optimum survival and develop apical–basal polarity, leading to the formation of an intact sheet of cells with an underlying basement membrane (see Section 2.2). EMT allows polarized epithelial cells to convert to a mesenchymal phenotype, resulting in loss of cell–cell adhesion and the development of front–back polarity. These properties are better suited to cell migration and allow cells to detach from the epithelium and invade the surrounding extracellular matrix (ECM). Once the mesenchymal phenotype is no longer required, cells may revert back to an epithelial phenotype through MET. EMT and MET are required for remodeling and plasticity during embryogenesis, and for invasion and metastasis during cancer development (Le Magnen et al. 2018; Zhang and Weinberg 2018). Research using cancer models suggests that these stages are "plastic," i.e. cells can move freely between epithelial and mesenchymal states and can even exist in a hybrid state expressing both phenotypes (Rhim 2013; Wang and Unternaehrer 2019).

EMT can be induced in human breast epithelial cells *in vitro* by introducing Twist or Snail transcription factors (Mani et al. 2008). It also occurs following *in vitro* transformation (Morel et al. 2008). Transformed cells develop a more fibroblast-like morphology, gain expression of mesenchymal markers, and lose expression of epithelial markers. Unexpectedly, induction of EMT in these studies was associated with development of stem cell-like properties, similar to those seen in cancer stem cells (CSCs) (see Minireview M25.1). The mechanism by which EMT confers a CSC-like phenotype is unclear, but it appears to be a consistent finding for various carcinoma types (Zhang and Weinberg 2018). Studies of the NCI-60 cell line panel have shown that EMT-related genes are upregulated in many tumor cell lines in association with chromosomal instability (Roschke et al. 2008). Transcriptomic analysis has demonstrated a consistent EMT-derived expression signature in these cell lines that can be used to distinguish between epithelial and mesenchymal states in further studies (Klijn et al. 2015).

26.3 INDUCTION OF DIFFERENTIATION

Differentiation is regulated at a number of levels, which allows multiple approaches for its induction (see Figure 26.3). Historically, attempts to induce differentiation have focused on the use of inducing agents such as retinoic acid or DMSO. These exogenous soluble factors continue to be widely used and are discussed first (see Section 26.3.1). However, an array of other methods can be used that will modify the makeup of the cell (endogenous factors), the local microenvironment, or the cell's physical properties (see Sections 26.3.2–26.3.7). Simulating some of these factors *in vitro* can be simple; for example, addition of a hormone or culture to a high density may be sufficient to induce differentiation. Other factors are more complex to model *in vitro*, such as arranging pulsatile flow in a reconstructed blood vessel.

26.3.1 Exogenous Soluble Factors

Physiological inducers. A large number of naturally occurring factors have been used to induce differentiation *in vitro* (see Table 26.2). Some are endocrine factors such as hormones, while others are paracrine factors that would normally be synthesized by adjacent cells, but cannot be supplied by an *in vitro* system that consists of only one cell type (see Sections 10.4.1, 26.3.4).

Physiological inducers of differentiation can be broadly grouped as:

(1) **Steroids.** Steroid hormones such as hydrocortisone are typically endocrine, i.e. they are secreted by distant organs or tissues and reach the target tissue via the vasculature *in vivo*. Other steroids that may induce differentiation include retinoic acid, which is a metabolite of vitamin A$_1$ (retinol) that is essential for embryonic development and

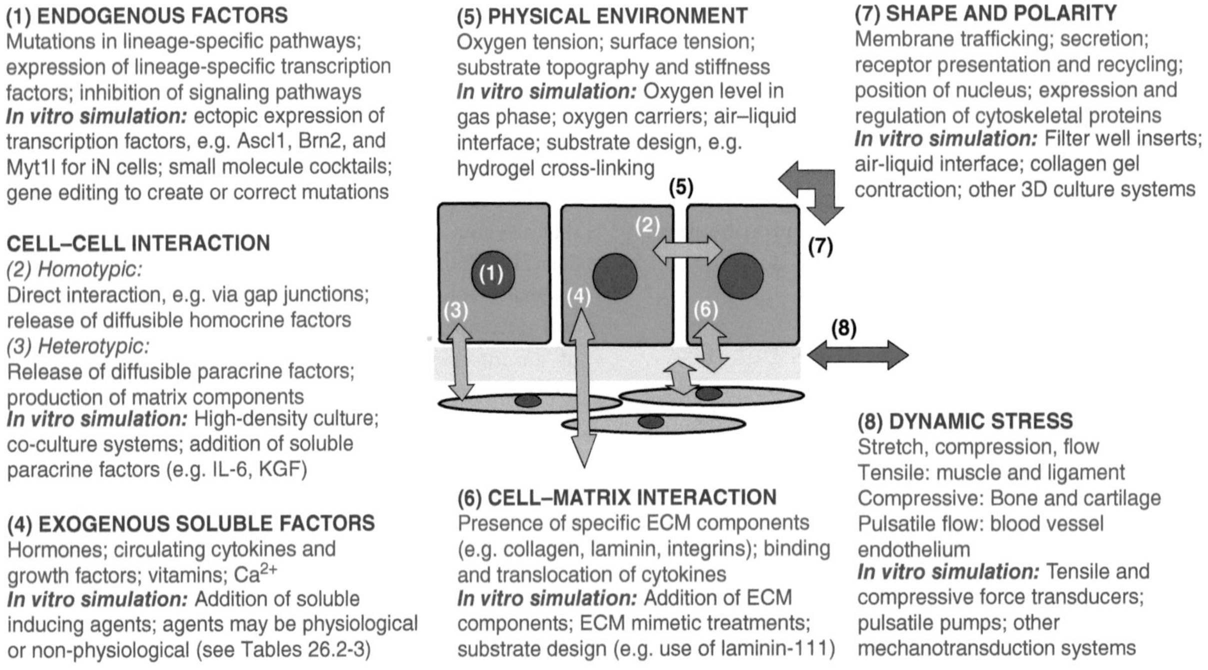

(1) ENDOGENOUS FACTORS
Mutations in lineage-specific pathways; expression of lineage-specific transcription factors; inhibition of signaling pathways
In vitro simulation: ectopic expression of transcription factors, e.g. Ascl1, Brn2, and Myt1l for iN cells; small molecule cocktails; gene editing to create or correct mutations

CELL–CELL INTERACTION
(2) Homotypic:
Direct interaction, e.g. via gap junctions; release of diffusible homocrine factors
(3) Heterotypic:
Release of diffusible paracrine factors; production of matrix components
In vitro simulation: High-density culture; co-culture systems; addition of soluble paracrine factors (e.g. IL-6, KGF)

(4) EXOGENOUS SOLUBLE FACTORS
Hormones; circulating cytokines and growth factors; vitamins; Ca^{2+}
In vitro simulation: Addition of soluble inducing agents; agents may be physiological or non-physiological (see Tables 26.2-3)

(5) PHYSICAL ENVIRONMENT
Oxygen tension; surface tension; substrate topography and stiffness
In vitro simulation: Oxygen level in gas phase; oxygen carriers; air–liquid interface; substrate design, e.g. hydrogel cross-linking

(6) CELL–MATRIX INTERACTION
Presence of specific ECM components (e.g. collagen, laminin, integrins); binding and translocation of cytokines
In vitro simulation: Addition of ECM components; ECM mimetic treatments; substrate design (e.g. use of laminin-111)

(7) SHAPE AND POLARITY
Membrane trafficking; secretion; receptor presentation and recycling; position of nucleus; expression and regulation of cytoskeletal proteins
In vitro simulation: Filter well inserts; air-liquid interface; collagen gel contraction; other 3D culture systems

(8) DYNAMIC STRESS
Stretch, compression, flow
Tensile: muscle and ligament
Compressive: Bone and cartilage
Pulsatile flow: blood vessel endothelium
In vitro simulation: Tensile and compressive force transducers; pulsatile pumps; other mechanotransduction systems

Fig. 26.3. Regulation of differentiation. Parameters controlling the expression of differentiation, with some of the ways to reproduce these conditions *in vitro*.

TABLE 26.2. Soluble inducers of differentiation: physiological.

Inducer	Cell type	References
Steroid and related		
Glucocorticoids (including hydrocortisone and dexamethasone)	Breast epithelial cells	Marte et al. (1994)
	Glial cells, glioma	McLean et al. (1986)
	Hepatocytes	Granner et al. (1968)
	Lung alveolar type II cells	Rooney et al. (1994); McCormick et al. (1995)
	Myeloid leukemia	Sachs (1978)
Retinoids (including retinoic acid and synthetic analogs)	Stem cells (ESCs, MSCs)	Torres et al. (2012); Zhang et al. (2015)
	Colon carcinoma (Caco-2)	McCormack et al. (1996)
	Embryonal carcinoma (NTERA-2)	Andrews (1998)
	Melanoma	Lotan and Lotan (1980); Meyskens and Fuller (1980)
	Myeloid leukemia (HL-60)	Breitman (1990); Degos (1997)
	Neuroblastoma (SK-N-BE)	Ghigo et al. (1998)
Peptide hormones		
EPO	Erythroblasts	Goldwasser (1975)
Insulin	Breast epithelial cells	Rudland (1992); Marte et al. (1994)
	Lung carcinoma (A549)	McCormick et al. (1995)
MSH	Melanocytes	Goding and Fisher (1997)
Prolactin	Breast epithelial cells	Takahashi et al. (1991); Rudland (1992); Marte et al. (1994)
TSH	Thyroid epithelial cells	Chambard et al. (1987)
Cytokines		
BMPs	Stem cells (ESCs, MSCs)	Torres et al. (2012); Beederman et al. (2013)
CNTF	Neurons, type 2 astrocytes	Raff (1989, Talbott et al. (2007)
Endothelin	Melanocytes	Reid et al. (1996)
Epimorphin	Epidermal cells	Okugawa and Hirai (2013)
GMF	Neurons, glial cells (see Figure 26.4e, f)	Lim et al. (1989); Fan et al. (2018)
HGF	Hepatocytes	Montesano et al. (1991)
	Kidney epithelial cells	Balkovetz and Lipschutz (1999)
IFN-α, IFN-γ	Myeloid leukemia (HL-60)	Buessow and Gillespie (1984)
	Lung carcinoma (A549)	McCormick et al. (1995)
	Neuroblastoma (LA-N-5)	Wuarin et al. (1991)
IL-6	Lung carcinoma (A549)	McCormick et al. (1995); McCormick and Freshney (2000)
KGF	Keratinocytes	Marchese et al. (1990)
	Prostate epithelial cells	Thomson et al. (1997); Heer et al. (2006)
NGF	Neurons	Levi-Montalcini (1982)
TGF-β	ESCs, MSCs	Watabe and Miyazono (2009); Krstic et al. (2018)
Vitamins		
Vitamin A	See retinoids above	
Vitamin D_3	Breast epithelial cells	Welsh (2011)
	Keratinocytes	Bikle (2004)
	Colon carcinoma	Palmer et al. (2001)
	Myeloid leukemia	Abe et al. (1981)
	Osteosarcoma	Nozaki et al. (1999)
Vitamin E	Glioma	Prasad et al. (1980)
	Neuroblastoma	Prasad et al. (1980)
Minerals		
Ca^{2+}	Keratinocytes	Boyce and Ham (1983); Bikle (2004)

Abbreviations not included in the main text: CNTF, ciliary neurotrophic factor; GMF, glia maturation factor; NGF, nerve growth factor.

promotes stem cell neural lineage specification (Zhang et al. 2015).

(2) **_Peptide hormones._** This category includes melanocyte-stimulating hormone (MSH), thyroid-stimulating hormone (TSH), erythropoietin (EPO), prolactin, and insulin. Although peptide hormones are typically considered to be endocrine factors, they may also be synthesized locally (Sanders and Harvey 2008).

(3) **_Cytokines._** The term "cytokine" is often used to refer to factors associated with the hematopoietic system, such as interleukins (ILs) and interferons (IFNs). Although these factors play important roles in the development of hematopoietic lineages (Lotem and Sachs 2002), cytokines can induce differentiation in many other tissues (see Table 26.2). Cytokines usually participate in complex signaling networks and, depending on their roles in these networks, can act as mitogenic factors as well as inducing differentiation (see Table 10.1).

(4) **_Vitamins._** Vitamins that have been reported to induce differentiation include vitamin A (which is modified by metabolism to give retinoic acid), vitamin D, and vitamin E (Prasad et al. 2003; Samuel and Sitrin 2008; Gocek and Studzinski 2009). The term "vitamin" usually refers to multiple compounds with similar chemical structures and activities; for example, "vitamin E" may refer to the naturally occurring form (_RRR_-α-tocopherol) or a synthetic version (_all-rac_-α-tocopherol), which is a mixture of approximately eight stereoisomers (Kline et al. 2004). Because these compounds may vary in their activity, it is important to ensure that previous findings are reproducible before using these substances as inducing agents for differentiation.

(5) **_Inorganic ions._** High Ca^{2+} levels are known to be important for keratinocyte differentiation, along with 1,25-dihydroxyvitamin D_3, which is the active metabolite of vitamin D (Bikle 2004). As a result, Ca^{2+} levels can be adjusted _in vitro_ depending on whether proliferation or differentiation are required (see Supp. S24.1). The importance of Ca^{2+} levels probably relates to the role of calcium in cell interaction (since cadherins are calcium-dependent), intracellular signaling, and the membrane flux of calcium in so-called calcium waves that propagate signals from one responding cell to adjacent cells of the same lineage.

Non-physiological inducers. Much of the early work on non-physiological inducers was performed using Friend cells (see Section 24.5.3). This work was initiated by Charlotte Friend, who discovered the importance of DMSO when using it as a vehicle for other compounds. Friend observed that the DMSO alone made the cells turn pink, due to hemoglobin induction (Friend et al. 1971; Diamond 1994). Erythroid differentiation was confirmed by partial purification of hemoglobin and by an increase in globin gene expression (see Figure 26.4a, b) (Scher et al. 1971;

Conkie et al. 1974). DMSO was subsequently used to induce differentiation in other cell types, and attempts to understand its mechanism of action led to the discovery of other inducers (see Table 26.3). A number of these compounds were polar compounds with structural similarities, including hexamethylene-_bis_-acetamide (HMBA) and sodium butyrate, which are both highly active (Reuben et al. 1978). Sodium butyrate is listed here with the non-physiological inducers of differentiation, but butyrate is known to occur naturally (produced by microbial fermentation in the gut) and plays an important part in normal enterocyte differentiation (Haner et al. 2010; Kaiko et al. 2016). Other non-physiological inducers of differentiation include cytotoxic drugs and phorbol esters such as phorbol myristate acetate (PMA) (see Table 26.3).

The action of many non-physiological inducers is unclear, and more than one mechanism of action is likely to be responsible in many cases. Inducing agents may affect (i) membrane fluidity (particularly with DMSO and other polar compounds) (Gurtovenko and Anwar 2007); (ii) signal transduction enzymes such as adenylate cyclase, protein kinase C (PKC), or phospholipase D (PLD); and (iii) the epigenome, due to alterations in DNA methylation or histone acetylation, e.g. due to inhibition of histone deacetylases (HDACs) (Thiagarajan et al. 2014). Many of these agents induce changes in gene expression and cell behavior that are unrelated to differentiation (McLean et al. 1986). This is likely to make them unsuitable for clinical use, although there are some exceptions, e.g. the HDAC inhibitor vorinostat has been approved for use in patients with cutaneous T-cell leukemia (Mann et al. 2007). Another agent, HMBA, was previously found to be effective for approximately a quarter of patients with acute myelogenous leukemia (Marks and Rifkind 1988; Nilsson et al. 2016). Increasing the dosage of HMBA to make it more clinically effective resulted in a number of side-effects, including the development of central nervous system toxicity (Marks and Rifkind 1988).

Although it may not be possible to use some inducing agents in the clinic, they can still have an important role to play in cellular therapies. For example, hESCs rapidly acquire a neuroectodermal phenotype after exposure to retinoic acid and then differentiate to give rise to a cascade of neuronal lineages (Parsons et al. 2011). This means that retinoic acid can be used to convert hESCs directly into neuronal lineages, without the need for embryoid body formation. Direct conversion using retinoic acid generates mature neurons with high efficiency and purity and has considerable potential for tissue engineering (Parsons et al. 2011; Parsons 2016).

26.3.2 Endogenous Factors

Introduction of lineage-specific factors can be used to induce transdifferentiation (see Section 26.2.4). As our understanding of lineage development has grown, this phenomenon has been utilized for direct conversion, e.g. for iN cells

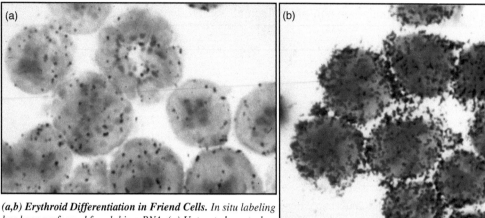

(a,b) Erythroid Differentiation in Friend Cells. *In situ labeling has been performed for globin mRNA. (a) Untreated, control cells; (b) cells exposed to 2% DMSO. Source: courtesy of David Conkie.*

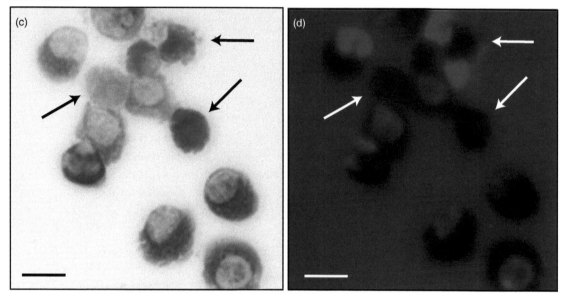

(c,d) Erythroid Differentiation in Immortalized Progenitor Cell Line (HiDEP-1) from Umbilical Cord Blood. *Cells are shown 10 days after induction of differentiation. (c) Benzidine staining for hemoglobin; (d) the same cells, showing DAPI staining. Arrows point to enucleated cells, indicating that mature erythroid cells have been produced. The scale bar represents 10 μm. Source: [Kurita et al. 2013], https://doi.org/10.1371/journal.pone.0059890, licensed under CC BY 4.0.*

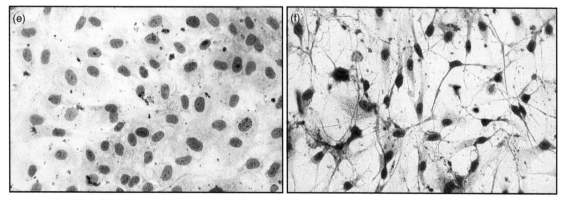

(e,f) Morphological differentiation in human glial cells. *(e) Undifferentiated glial cells from normal human brain; (f) morphological differentiation induced in glial cells from normal human brain by glia maturation factor (GMF). Giemsa stained. Source: R. Ian Freshney.*

Fig. 26.4. Differentiation of erythrocytes and glia.

TABLE 26.3. Soluble inducers of differentiation: non-physiological.

Inducer	Cell type	References
Planar-polar compounds		
DMSO	Breast epithelial cells	Rudland (1992)
	Hepatocyte precursors (HepaRG; see Figure 26.1)	Tascher et al. (2019)
	Erythroleukemia (Friend cells; see Figure 26.4a, b)	Friend et al. (1971)
	Myeloid leukemia (HL-60)	Collins et al. (1979); Breitman (1990)
	Neuroblastoma	Kimhi et al. (1976)
HMBA	Erythroleukemia (Friend cells)	Reuben et al. (1976); Nilsson et al. (2016)
Sodium butyrate	Chronic myelogenous leukemia (K-562)	Andersson et al. (1979)
	Colon carcinoma	Haner et al. (2010)
N-methyl acetamide	Glioma	McLean et al. (1986)
N-methyl formamide, dimethyl formamide	Myeloid leukemia (HL-60)	Breitman (1990)
Butylated hydroxyanisole	Stem cells (ASCs)	Safford and Rice (2007)
Benzodiazepines	Erythroleukemia (Friend cells)	Clarke and Ryan (1980)
Cytotoxic drugs		
Actinomycin D	Myeloid leukemia (HL-60)	Breitman (1990)
Cytosine arabinoside	Stem cells (ESCs)	Jagtap et al. (2011)
	Myeloid leukemia (ML-1)	Takeda et al. (1982)
Genistein	Erythroleukemia (Friend cells)	Watanabe et al. (1991)
Methotrexate	Colorectal carcinoma	Lesuffleur et al. (1990)
Mitomycin C	Melanoma (B16)	Raz (1982)
Histone acetylation		
Valproic acid	Stem cells (ASCs, MSCs, iPSCs)	Safford and Rice (2007); Fu et al. (2014); Kondo et al. (2014)
	Pheochromocytoma (PC12)	Kamata et al. (2007)
SAHA (vorinostat, Zolinza®)	T-cell lymphoma	Marks (2007)
Tetramethylpyrazine	Neuroblastoma (SH-SY5Y)	Yan et al. (2014b)
DNA methylation		
5-Azacytidine	Stem cells (MSCs, endothelial progenitors)	Lopez-Ruiz et al. (2014); Yan et al. (2014a)
	Myeloid leukemia (HL-60)	Christman et al. (1983)
Signal transduction modifiers		
Isobutylmethylxanthine	Stem cells (ASCs)	Safford and Rice (2007)
Forskolin	Stem cells (ASCs)	Safford and Rice (2007)
PMA[a]	Breast epithelial cells	Wada et al. (1994)
	Bronchial epithelial cells	Willey et al. (1984); Masui et al. (1986)
	Colon carcinoma (Caco-2, HT-29)	Pignata et al. (1994); Velcich et al. (1995)
	Chronic myelogenous leukemia (K-562)	Kujoth and Fahl (1997)
	Neuroblastoma	Spinelli et al. (1982)

[a]Phorbol 12-myristate 13-acetate (PMA) is also known as 12-O-tetradecanoylphorbol-13-acetate (TPA).
Abbreviations not included in the main text: ASC, adipose-derived stem cell; SAHA, suberoylanilide hydroxamic acid.

(see Figure 26.2) (Tanabe et al. 2015; Mertens et al. 2016). Induction of a neuronal phenotype was initially achieved by introducing neural-specific genes using lentivirus (see Section 22.1.4), followed by screening for the correct phenotype (Vierbuchen et al. 2010). Combinations of various factors were gradually streamlined until it was possible to induce neuronal differentiation using only three factors. The "BAM" factors that were originally used for iN cells (*Ascl1*, *Brn2*, and *Myt1l*) are transcription factors that are expressed in neuronal cells. Ascl1 appears to act as a "pioneer" factor that opens the chromatin structure; this in turn allows the binding of Brn2, which acts as a secondary factor, while Mytl1 is required for maturation (Mertens et al. 2016). The optimal set of factors varies with the species of cell and the required lineage. The BAM factors are particularly efficient for mouse cells, but Neurogenin 2 appears to be more efficient for human cells, where it acts as a pioneer factor in a similar manner to Ascl1 (Liu et al. 2013). Alternatively, the effects of these factors may be reproduced by using small molecules to modulate specific pathways (Hu et al. 2015; Li et al. 2015). Transcription factors and small molecules can also be used in varying combinations to induce specific neuronal subtypes. The BAM factors can be used in combination with *Lmx1a* and *FoxA2* to generate dopaminergic neurons, while Neurogenin 2, forskolin, and dorsomorphin can be used to generate cholinergic motor neurons (Pfisterer et al. 2011; Liu et al. 2013).

Clearly, it is possible to induce lineage-specific differentiation with a high degree of specificity, provided there is sufficient understanding of which factors are required and screening is performed to optimize their selection. Although specialized knowledge is initially required, non-specialized laboratories will be able to reproduce the technique once it has been published, provided suitable gene delivery reagents and techniques are available (see Section 22.1). As direct conversion techniques become more widespread, scientists should be able to choose between stem cell differentiation and direct conversion, depending on the type of cell that is available and the desired outcome, e.g. whether ongoing proliferation is required (Mertens et al. 2016).

26.3.3 Geometry and Polarity

Differentiated tissues are made up of cells that are in a 3D relationship with one another and with the ECM that surrounds them. Logically, therefore, it is not surprising that 3D culture can induce differentiated characteristics in many cell types. For example, normal breast epithelial cells form ductal structures in 3D culture (but not 2D culture) that are similar to the structures that can be seen in normal breast tissue (Nash et al. 2015). Differentiation may be induced by cell–cell interaction (see Section 26.3.4), but it is also likely to occur due to the cells' ability to develop the correct geometry and polarity when grown in a 3D system.

An early study of adult rat hepatocytes showed that for cells to become fully mature, it was necessary to grow the cells on collagen gel and then release the gel from the bottom of the dish with a spatula or bent Pasteur pipette (Sattler et al. 1978). This process allowed shrinkage of the gel and a change in the shape of the cell from flattened to cuboidal, or even columnar. Accompanying or following the shape change (and possibly due to access to medium through the gel), the cells developed polarity that was visible by electron microscopy. The nucleus became asymmetrically distributed, an active Golgi complex formed, and secretion was observed toward the apical surface. Similar establishment of polarity has been demonstrated in other cell types, including thyroid epithelial cells, hepatocytes, and bronchial epithelium (Chambard et al. 1983; Guguen-Guillouzo and Guillouzo 1986; Saunders et al. 1993).

26.3.4 Cell–Cell Interactions

Homotypic Interactions. Culture of homologous cells at high density is known to induce a more differentiated phenotype in many cell types (e.g. using hollow fiber systems; see Sections 27.5.3, 28.3.4). Homologous cells may interact via gap junctions, allowing the passage of metabolites and second messengers such as cyclic adenosine monophosphate (cAMP), diacylglycerol (DAG), Ca^{2+}, or electrical charge (see Section 2.2.1) (Finbow and Pitts 1981; Herve and Derangeon 2013). Gap junction communication probably harmonizes the expression of differentiation within a population of similar cells, rather than initiating its expression. Homologous cells may also interact through molecules such as cadherins, which are calcium-dependent, and various cellular adhesion molecules (CAMs), which are calcium-independent (see Section 2.2.2). These adhesion molecules promote interaction primarily between like cells via identical, reciprocally acting, extracellular domains. They also appear to have signal transduction potential via phosphorylation of their intracellular domains (Doherty et al. 1991; Gumbiner and Yamada 1995).

Heterotypic Interactions. Co-culture systems that contain multiple cell types can be used to induce differentiation. For example, a 3D co-culture system has been developed to study breast cancer initiation where cells are grown on filter well inserts in a 3D collagen gel (Nash et al. 2015). The breast epithelial HB2 cell line forms tight spheroidal clusters when grown alone in this system or when cultured with the myoepithelial Myo1089 cell line. Addition of a third cell type (fibroblasts from normal breast) results in the appearance of ductal structures with hollow lumens that are strikingly similar to the tissue architecture observed in normal breast tissue. Co-culture systems have also been used to induce differentiation in other cell types, including prostate epithelial cells and mesenchymal stromal cells (MSCs) (Wong et al. 2003; Morita et al. 2015).

Paracrine Factors. Although differentiation may be induced through direct cell–cell contact, it may also arise through exposure to diffusible factors from nearby cells. These factors may come from the same cell ("autocrine" factors), the same lineage ("homocrine" factors), or a different cell lineage to the target cell ("paracrine" factors). The importance of paracrine factors is evident from an *in vitro* model of the human epidermis where keratinocytes and fibroblasts are physically separated from one another (see Figure 27.15) (Stark et al. 2004). Fibroblasts are seeded in a collagen gel and keratinocytes are grown on the gel at the air–liquid interface. The keratinocytes produce IL-1α and IL-1β, which stimulate the fibroblasts within the gel to produce keratinocyte growth factor (KGF) and granulocyte-macrophage colony stimulating factor (GM-CSF) (see Figure 26.5) (Maas-Szabowski et al. 2000, 2002). If fibroblasts are unable to produce KGF and GM-CSF, this results in loss of keratinocyte proliferation and differentiation (Szabowski et al. 2000).

Paracrine factors can act as both mitogens and morphogens (see Table 10.1), resulting in diverse effects on the cells' phenotype and 3D architecture. For example, hepatocyte growth factor (HGF) acts as a mitogen for hepatocytes but can induce tubule formation in canine MDCK kidney cells and rat submandibular gland cultures (Montesano et al. 1991; Orellana et al. 1996; Furue and Saito 1997). HGF is produced primarily by cells of mesenchymal origin and binds to the c-Met receptor, which is found in MDCK cells and a number of other epithelial cell types (Balkovetz 1998). In polarized epithelial cells, c-Met is located primarily in the basolateral plasma membrane, which reinforces the hypothesis that HGF acts as a paracrine factor for epithelial–mesenchymal cell interactions (Crepaldi et al. 1994; Balkovetz 1998). The effect of HGF on tubule formation is phosphorylation-dependent and is at least partly mediated by suppression of IL-8 (Santos et al. 1993;

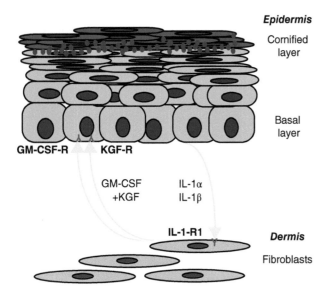

Fig. 26.5. Reciprocal paracrine interactions. Schematic illustration of the reciprocal paracrine pathways of keratinocyte growth regulation in a co-culture model of the epidermis, where keratinocytes and fibroblasts are physically separated from one another. *Source*: R. Ian Freshney, from Maas-Szabowski et al. (2002).

Wells et al. 2013). Interleukins are important paracrine factors for a number of lineages, such as lung adenocarcinoma cells, where members of the IL-6 family are potent inducers of differentiation (McCormick et al. 1995; McCormick and Freshney 2000). Paracrine factors can also have an inhibitory effect on differentiation, as observed when A549 lung adenocarcinoma cells are exposed to transforming growth factor-α (TFG-α), TGF-β, or IL-1α (McCormick et al. 1995).

26.3.5 Cell–Extracellular Matrix (ECM) Interactions

ECM is a complex matrix of glycoproteins and proteoglycans that provides a specific microenvironment for each tissue and even for parts of a tissue (see Section 2.2.4) (Hynes 2009). The importance of cell–matrix interactions can be seen in organs such as the breast, where tissue microenvironment has been extensively studied (Muschler and Streuli 2010; Inman et al. 2015). Breast epithelium consists of branching structures with an inner layer of luminal epithelial cells and an outer layer of basal, myoepithelial cells. The epithelium is separated from the surrounding stroma by a basement membrane, which consists of ECM components that are produced by the myoepithelial and stromal cells. Laminin-111, which is a major basement membrane component, is important for the architecture of the breast epithelium and for the maintenance of breast-specific functions. Laminin-111 binds to integrin receptors on the breast epithelial cells, resulting in upregulation of the milk protein β-casein (Muschler et al. 1999). Tumor-derived myoepithelial cells produce very little laminin-111, and branching structures that are generated from those cells exhibit disordered polarity (Gudjonsson et al. 2002). Clearly, these interactions between the breast epithelium and the ECM are important for normal physiological behavior.

Recreation of the tissue microenvironment is essential to induce differentiation and to understand how stem cells give rise to specific lineages (Gattazzo et al. 2014). But is it possible to mimic the complexity of cell–matrix interactions *in vitro*? Matrix material from the Engelbreth-Holm-Swarm (EHS) sarcoma is commonly used for *in vitro* experiments, of which Matrigel® (Corning) is perhaps the best-known example (see Section 8.3.2). Matrigel is able to promote differentiation in a number of *in vitro* systems, including cell lines (e.g. A549; see Figure 26.6c), primary cultures, and tissue explants (Kleinman and Martin 2005). Although culture using Matrigel is a useful technique, it has the problem of introducing another undefined variable to the culture system. Defined substrates are required that can regulate specific pathways or lineages, based on the biology of the tissue microenvironment. For example, various laminin subtypes have been incorporated into substrates for the culture of human pluripotent stem cells (hPSCs), either as recombinant proteins or as peptide fragments (Lambshead et al. 2013). The use of ECM coatings and mimetic treatments is discussed in more detail in an earlier chapter (see Section 8.3.2–8.3.5).

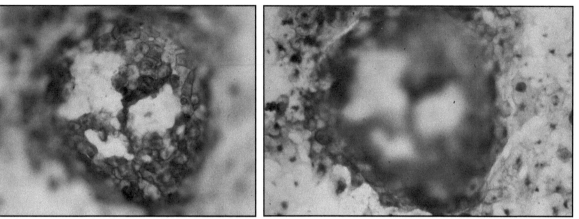

(a, b) Dome Formation in a Monolayer of WIL Lung Adenocarcinoma Cells. Mosaic of CEA-positive and CEA-negative WIL cells. Focused: (a) on top of dome; and (b) on the monolayer at the base of the dome. Staining for CEA with immunoperoxidase; 40x objective. Source: R. Ian Freshney.

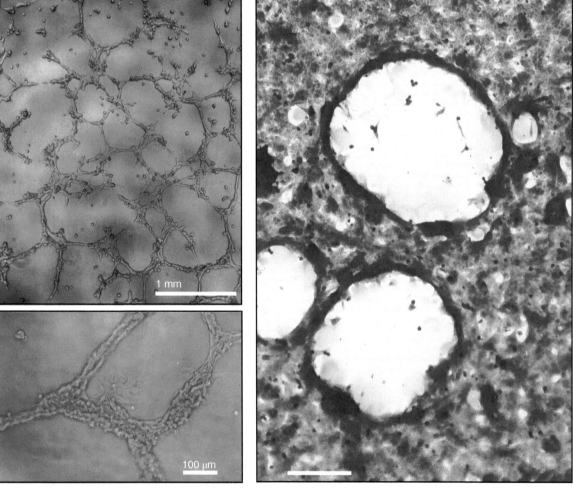

(c) A549 lung adenocarcinoma cells growing on Matrigel. Upper panel, 4x objective; lower panel, 10x objective. Scale bars as indicated. Bright field; unstained. Source: courtesy of Jane Sinclair.

(d) A549 lung adenocarcinoma cells. Cells are grown in a non-tissue culture grade Petri dish, with lung fibroblasts growing on a coverslip in the center of the dish. The scale bar indicates 100 µm. Bright field; 10x objective. Source: R. Ian Freshney.

Fig. 26.6. Differentiation of epithelial cells.

26.3.6 Air–Liquid Interface

Culture at the air–liquid interface can be used to induce differentiation in some cell types. Expression of fully keratinized squamous differentiation in skin requires epidermal cells to be added to an organotypic construct, with the keratinocytes positioned at the air–liquid interface (see Figure 27.15) (Stark et al. 2004). Culture at the air–liquid interface is used to preserve the morphology and barrier function of corneal epithelium (see Supp. S24.2). Alveolar, bronchial, and tracheal epithelial cells can be positioned at the air–liquid interface for optimal differentiation (Speirs et al. 1991; Dobbs et al. 1997). This approach has led to some elegant models of respiratory function; cultures have been used for efficacy testing of inhaled drugs and for toxicity testing of air pollutants (Movia et al. 2018; Upadhyay and Palmberg 2018). Conditional reprogramming has been used to extend the lifespan of bronchial cells derived from patients with asthma and cystic fibrosis, which are then grown at the air–liquid interface to induce differentiation (see Section 22.3.4) (Martinovich et al. 2017; Peters-Hall et al. 2018).

Why is culture at the air–liquid interface successful? Positioning the cells at the interface is believed to enhance gas exchange, facilitating oxygen uptake without raising the partial pressure and risking free radical toxicity. However, the thin film above may mimic the physicochemical conditions *in vivo* (surface tension, lack of nutrients) as well as facilitating oxygenation. As with all cell culture techniques, this approach may require optimization. For example, it may be preferable to use DPBS-A at the apical surface rather than complete medium (Chambard et al. 1983; Chambard et al. 1987). Detailed methods for air–liquid interface culture of human respiratory cells are provided elsewhere (Karp et al. 2002).

26.3.7 Biomechanical Regulation

Differentiation of stem cells into various lineages may be controlled by altering the physical properties of the microenvironment such as mechanical strain, shear stress, or substrate topography and stiffness (see Figure 2.3) (Engler et al. 2006; Kshitiz et al. 2012). The same principles apply to some specialized cell types; it may be necessary to provide tensile stress for skeletal muscle, periodic stress for cardiac muscle, pulsatile flow for endothelium, or compressive stress for bone and cartilage. Provision of these types of physical forces is an important part of tissue engineering and may require quite complex mechanotransduction equipment (see Section 27.7.2) (Vunjak-Novakovic and Freshney 2006; Majkut et al. 2014).

26.4 PRACTICAL ASPECTS

Based on the examples in the previous sections, it is clear that partial, or even complete, differentiation can be achieved *in vitro*, given the correct environmental conditions. How do you select a suitable method if you are a newcomer to this area? As a general approach to promoting differentiation, as opposed to cell proliferation and propagation, the following recommendations may be helpful:

(1) *Understand the requirements of your specific cell type.* What is known about its tissue microenvironment *in vivo*, and how can you mimic that environment *in vitro*? What procedures have been published for that cell type, and can you modify those procedures to induce differentiation for your own application? Procedures are available in the Supplementary Material online that provide a starting point for some specialized cell types (see Supp. 24.1–24.22).

(2) *Choose your starting material with care.* Cell lines are commonly used as *in vitro* models of differentiation (see Table 26.1) but tend to lose differentiated characteristics (see Sections 3.4, 26.2.3). Primary cultures are closer to the original tissue but have a limited period of use, and may display variability between donors, particularly for some cell types (e.g. primary hepatocytes) (Godoy et al. 2013). Stem cells offer an attractive alternative, but their differentiation can be problematic (see Section 26.5.2). If cells from the correct lineage are used as a starting point, induction of differentiation is more likely to be successful.

(3) *Grow the cells to a high cell density (> 1×10^5 cells/cm^2) on the appropriate matrix.* Options for an appropriate matrix include naturally occurring ECM materials (see Table 8.1) and synthetic materials (e.g. poly-D-lysine for neurons; see Section 8.3.5). Many cell types have well-established substrates; for example, normal human bronchial epithelial (NHBE) cells are commonly grown on a mixture of collagen, fibronectin, and bovine serum albumin (BSA) (see Supp. S24.4). Some commercial suppliers offer sample packs, allowing you to test multiple substrates, e.g. BioCoat™ Variety Packs (Corning).

(4) *Change the cells to a differentiation-inducing medium.* Most media formulations were designed to support proliferation rather than differentiation (see Section 9.1). In some cases, simple modifications are sufficient to induce differentiation. For human epidermis, Ca^{2+} concentration should be increased to around 3 mM, and for bronchial mucosa, the addition of serum will initiate differentiation (see Supp. S24.4). In other cases, separate media formulations have been developed for differentiated function, e.g. BrainPhys™ neuronal medium (STEMCELL Technologies; see Section 24.4.1).

(5) *Add differentiation-inducing agents* (see Tables 26.2, 26.3). The tables that are provided in this chapter list examples of cell types where soluble inducers are known to be effective. Discuss with colleagues and search the literature for other publications that relate to your specific cell type.

(6) *Add additional cell types to establish cell–cell interactions* (see Section 26.3.4). The second, interacting cell type should be known to interact with your cell type of interest *in vivo*. It may be added during the growth phase (point 3 in this list), the induction phase (points 4 and 5), or both phases. Selection of the correct cell type is not always obvious, but co-culture has been proven to be effective for some tissues, e.g. epidermis (see Figure 27.15).

(7) *Use a filter well insert* (see Section 27.5.2; see also Figures 27.12, 27.15, 29.12). Elevating the culture in a filter well provides access for the basal surface to nutrients and ligands, the opportunity to establish polarity, and regulation of the nutrient and oxygen concentration at the apical surface. The system also allows a great deal of flexibility, including the ability to establish co-culture on either side of the filter and to adjust the composition and depth of overlying medium.

Not all of these factors may be required, but the sequence in which they are presented is meant to imply some degree of priority. Their timing will also be important. ECM generally turns over slowly, so prolonged exposure to matrix-inducing conditions may be required. The same is true for culture at high density; for example, induction of differentiation in A549 cells requires culture at high density for 21 days (see Figure 3.6). In contrast, soluble factors may require only temporary exposure at a particular stage of differentiation, depending on the presence of the appropriate ECM, cell density, or heterologous cell interaction and on the presence or absence of other soluble factors. This means that the effects of soluble factors may vary with the model system; for example, glucocorticoids promote proliferation at low cell density but induce differentiation at high cell density (Guner et al. 1977; McCormick et al. 1995). For the microenvironment and cell–cell interaction to be optimal, it is likely that a 3D culture system will be required, as discussed in the next chapter (see Chapter 27).

26.5 ONGOING CHALLENGES

26.5.1 Markers of Differentiation

Differentiation may be obvious under the microscope due to changes in cell morphology and architecture. Such changes include outgrowth of neurites by neuronal cultures and the formation of "domes" by some epithelial cultures (see Section 18.3.4; see also Figure 26.6a, b). If the culture does not exhibit changes in morphology, other phenotypic markers must be used to confirm that differentiation has occurred. However, the selection of suitable markers can be a challenging task, due to the variations in differentiation status that occur *in vitro* (see Section 26.2).

Proteins that are expressed early and retained throughout subsequent maturation stages are generally regarded as lineage-specific markers. Examples include intermediate filament proteins, such as the cytokeratins for epithelial cells and glial fibrillary acidic protein (GFAP) for astrocytes (see Section 18.3.2). Markers of the mature, differentiated phenotype are usually specific cell products or enzymes involved in the synthesis of those products. Examples include glycerol phosphate dehydrogenase (GPDH) in oligodendrocytes, transglutaminase in keratinocytes, and hemoglobin in erythrocytes (McCarthy and de Vellis 1980; Candi et al. 2005; Kurita et al. 2013). These properties are often expressed late in the lineage, but they are also more likely to be reversible and under adaptive control by hormones, nutrients, ECM constituents, and cell–cell interaction (see Figure 26.3).

Although changes in phenotype and the expression of these markers may be informative, they can also be misleading. This is perhaps best shown by the MDA-MB-435 cell line, which was originally reported to come from breast carcinoma. Further analysis demonstrated that MDA-MB-435 shares a common donor origin with the M14 melanoma cell line, resulting in debate as to which cell line is misidentified (Korch et al. 2018). MDA-MB-435 cells stain positively for epithelial-specific and breast-specific markers; they can also extend neurites and express melanoma-specific markers, particularly when grown at high density (Zhang et al. 2010; Nerlich and Bachmeier 2013). The debate regarding their origin has now been resolved by authentication testing of early samples, which demonstrated that MDA-MB-435 is misidentified and comes from a male with metastatic melanoma (see Table 17.3) (Korch et al. 2018). However, the reasons for its variable phenotype remain unknown. Melanoma cells can express markers for a number of different lineages *in vivo*, so it is possible that its phenotypic plasticity relates to its tumor origin; culture conditions are also likely to play an important role (Nerlich and Bachmeier 2013; Agaimy et al. 2016).

The story of MDA-MB-435 shows the importance of authentication testing, but it also shows that a limited set of tissue-specific markers is not sufficient to prove that cells belong to a specific lineage when used in isolation. Techniques for analysis of the full transcriptome or proteome are becoming more widely accessible and will help to resolve this problem through more comprehensive characterization (see Section 18.3.3). For example, data from RNA sequencing (RNA-seq) and single nucleotide polymorphism (SNP) analysis can be used to authenticate a cell line and perform hierarchical clustering to determine its differentiation status (Klijn et al. 2015). For this work to be effective, laboratories must pay close attention to the provenance of their cell lines and report on culture conditions as part of publications (see Section 7.3; see also Table 7.3).

26.5.2 Stem Cell Differentiation

Remarkable advances have been made in stem cell culture over the last 30 years, leading to the widespread availability of human MSCs, ESCs, and induced pluripotent stem cells (iPSCs; see Sections 23.2.2, 23.3, 23.6). Most tissue culture laboratories now have access to the protocols and reagents required to maintain stem cells in an undifferentiated state (although the necessary expertise may take longer to develop). Induction of differentiation in these cultures has proven to be more challenging. Differentiation procedures must induce a specific lineage – and not simply a progenitor cell, but a mature specialized cell – in an efficient and reproducible manner. Constraining a pluripotent cell to a specific lineage is more easily said than done, even when work is performed in a laboratory with stem cell expertise. However, the effort is well worth pursuing. Stem cells can be used to generate new *in vitro* models for cell types that are difficult to grow or that rapidly

lose their differentiated characteristics, e.g. hepatocytes (see Section 26.1). Procedures can be used to induce differentiation of iPSCs from specific donors, which is an important advance for personalized therapies (Parsons 2013).

How is stem cell differentiation performed? Usually, an effort is made to mimic the embryonic development of the desired lineage. Many procedures start by generating embryoid bodies, which are then induced to form specific lineages using organoid culture (see Section 27.6). However, embryoid bodies give relatively low efficiency for some cell types and more complicated maneuvers may be required. For hepatocytes, induction of endoderm appears to be the key to generating differentiated cultures with high efficiency (Cheng

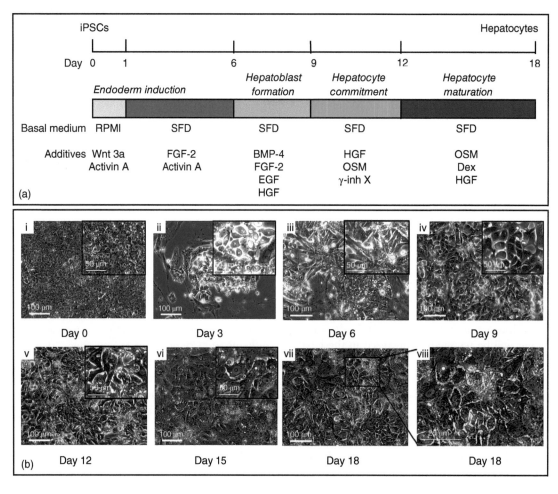

Fig. 26.7. Hepatocyte differentiation from porcine iPSCs. (a) Diagram showing differentiation procedure. Porcine iPSCs (derived from fibroblasts obtained from ear skin) pass through five sets of culture conditions in order to generate endodermal progenitors, hepatic progenitors, and mature hepatocytes. For more information on culture conditions please refer to the source reference. BMP, Bone morphogenetic protein; Dex, dexamethasone; EGF, epidermal growth factor; FGF, fibroblast growth factor; HGF, hepatocyte growth factor; γ-inh X, γ-secretase inhibitor-X; OSM, oncostatin M; SFD, serum-free differentiation medium. (b) Representative phase contrast images, showing changes in cell morphology throughout the procedure. Scale bars as indicated. *Source*: (a) Ao Y, Mich-Basso JD, Lin B, Yang L (2014) High Efficient Differentiation of Functional Hepatocytes from Porcine Induced Pluripotent Stem Cells. PLoS ONE 9(6): e100417. Licensed under CCBY 4.0. (b) Adapted from Ao et al. (2014), https://doi.org/10.1371/journal.pone.0100417, licensed under CC BY 4.0.

et al. 2012; Palakkan et al. 2017). A five-step procedure has been developed that gives rise to endodermal progenitor cells, hepatic progenitor cells, and finally to mature hepatocytes (see Figure 26.7) (Ao et al. 2014). Hepatocytes closely resemble HepaRG cells in their morphology and expression of Phase I and II metabolizing enzymes (see Figure 26.1) (Cheng et al. 2012; Ao et al. 2014).

Although these procedures represent a significant step forward in stem cell differentiation, a number of challenges lie ahead. Many procedures are complex and costly, due to the large number of defined factors that are required. Ideally, procedures can be simplified and streamlined; for example, a shorter, three-step procedure has been developed for hepatocyte differentiation (Chen et al. 2012; Ao et al. 2014). However, hepatocytes from the three-step procedure have subtle differences in morphology and are less mature compared to those from the longer procedure (Ao et al. 2014). The efficiency and reproducibility of differentiation procedures can also vary between laboratories. Differences may relate to variability in earlier reprogramming steps, resulting in reduced efficiency when generating iPSCs and endodermal progenitor cells, and a higher proportion of undifferentiated cells (Palakkan et al. 2017). Differences in reagents or "routine" procedures, such as enzymatic digestion, may also have an impact on differentiation procedures (Osterloh and Mullane 2018). Better characterization and reporting of stem cell culture is an essential part of Good Cell Culture Practice (GCCP) (see Sections 7.2.1, 23.7) and will help to clarify (and hopefully overcome) many of the challenges associated with stem cell differentiation.

Suppliers

Supplier	URL
Advanced BioMatrix	www.advancedbiomatrix.com
Corning	http://www.corning.com/worldwide/en/products/life-sciences/products/cell-culture.html
STEMCELL Technologies	www.stemcell.com

Supp. S26.1. Purification of HepaRG Human Hepatocytes.

REFERENCES

Abe, E., Miyaura, C., Sakagami, H. et al. (1981). Differentiation of mouse myeloid leukemia cells induced by 1 alpha,25-dihydroxyvitamin D3. *Proc. Natl Acad. Sci. U.S.A.* 78 (8): 4990–4994. https://doi.org/10.1073/pnas.78.8.4990.

Agaimy, A., Specht, K., Stoehr, R. et al. (2016). Metastatic malignant melanoma with complete loss of differentiation markers (undifferentiated/dedifferentiated melanoma): analysis of 14 patients emphasizing phenotypic plasticity and the value of molecular testing as surrogate diagnostic marker. *Am. J. Surg. Pathol.* 40 (2): 181–191. https://doi.org/10.1097/PAS.0000000000000527.

Andersson, L.C., Jokinen, M., and Gahmberg, C.G. (1979). Induction of erythroid differentiation in the human leukaemia cell line K562. *Nature* 278 (5702): 364–365. https://doi.org/10.1038/278364a0.

Andrews, P.W. (1998). Teratocarcinomas and human embryology: pluripotent human EC cell lines. Review article. *APMIS* 106 (1): 158–167. discussion 167–8.

Andrews, P.W., Damjanov, I., Simon, D. et al. (1984). Pluripotent embryonal carcinoma clones derived from the human teratocarcinoma cell line Tera-2. Differentiation in vivo and in vitro. *Lab. Investig.* 50 (2): 147–162.

Aninat, C., Piton, A., Glaise, D. et al. (2006). Expression of cytochromes P450, conjugating enzymes and nuclear receptors in human hepatoma HepaRG cells. *Drug Metab. Dispos.* 34 (1): 75–83. https://doi.org/10.1124/dmd.105.006759.

Ao, Y., Mich-Basso, J.D., Lin, B. et al. (2014). High efficient differentiation of functional hepatocytes from porcine induced pluripotent stem cells. *PLoS One* 9 (6): e100417. https://doi.org/10.1371/journal.pone.0100417.

Bairoch, A. (2018). The Cellosaurus, a cell-line knowledge resource. *J. Biomol. Tech.* 29 (2): 25–38. https://doi.org/10.7171/jbt.18-2902-002.

Balkovetz, D.F. (1998). Hepatocyte growth factor and Madin-Darby canine kidney cells: in vitro models of epithelial cell movement and morphogenesis. *Microsc. Res. Tech.* 43 (5): 456–463. https://doi.org/10.1002/(SICI)1097-0029(19981201)43:5<456::AID-JEMT11>3.0.CO;2-2.

Balkovetz, D.F. and Lipschutz, J.H. (1999). Hepatocyte growth factor and the kidney: it is not just for the liver. *Int. Rev. Cytol.* 186: 225–260.

Beederman, M., Lamplot, J.D., Nan, G. et al. (2013). BMP signaling in mesenchymal stem cell differentiation and bone formation. *J. Biomed. Sci. Eng.* 6 (8A): 32–52. https://doi.org/10.4236/jbise.2013.68A1004.

Bikle, D.D. (2004). Vitamin D regulated keratinocyte differentiation. *J. Cell. Biochem.* 92 (3): 436–444. https://doi.org/10.1002/jcb.20095.

Boyce, S.T. and Ham, R.G. (1983). Calcium-regulated differentiation of normal human epidermal keratinocytes in chemically defined clonal culture and serum-free serial culture. *J. Invest. Dermatol.* 81 (1 Suppl): 33s–40s.

Breitman, T.R. (1990). Growth and differentiation of human myeloid leukemia cell line HL60. *Methods Enzymol.* 190: 118–130. https://doi.org/10.1016/0076-6879(90)90016-t.

Buessow, S.C. and Gillespie, G.Y. (1984). Interferon-alpha and -gamma promote myeloid differentiation of HL-60, a human acute promyelocytic leukemia cell line. *J. Biol. Response Mod.* 3 (6): 653–662.

Candi, E., Schmidt, R., and Melino, G. (2005). The cornified envelope: a model of cell death in the skin. *Nat. Rev. Mol. Cell Biol.* 6 (4): 328–340. https://doi.org/10.1038/nrm1619.

Chambard, M., Verrier, B., Gabrion, J. et al. (1983). Polarization of thyroid cells in culture: evidence for the basolateral localization of the iodide "pump" and of the thyroid-stimulating hormone receptor-adenyl cyclase complex. *J. Cell Biol.* 96 (4): 1172–1177. https://doi.org/10.1083/jcb.96.4.1172.

Chambard, M., Mauchamp, J., and Chabaud, O. (1987). Synthesis and apical and basolateral secretion of thyroglobulin by thyroid cell monolayers on permeable substrate: modulation by thyrotropin. *J. Cell. Physiol.* 133 (1): 37–45. https://doi.org/10.1002/jcp.1041330105.

Chen, Y.F., Tseng, C.Y., Wang, H.W. et al. (2012). Rapid generation of mature hepatocyte-like cells from human induced pluripotent stem cells by an efficient three-step protocol. *Hepatology* 55 (4): 1193–1203. https://doi.org/10.1002/hep.24790.

Cheng, X., Ying, L., Lu, L. et al. (2012). Self-renewing endodermal progenitor lines generated from human pluripotent stem cells. *Cell Stem Cell* 10 (4): 371–384. https://doi.org/10.1016/j.stem.2012.02.024.

Christman, J.K., Mendelsohn, N., Herzog, D. et al. (1983). Effect of 5-azacytidine on differentiation and DNA methylation in human promyelocytic leukemia cells (HL-60). *Cancer Res.* 43 (2): 763–769.

Clarke, G.D. and Ryan, P.J. (1980). Tranquillizers can block mitogenesis in 3T3 cells and induce differentiation in Friend cells. *Nature* 287 (5778): 160–161. https://doi.org/10.1038/287160a0.

Collins, S.J., Ruscetti, F.W., Gallagher, R.E. et al. (1979). Normal functional characteristics of cultured human promyelocytic leukemia cells (HL-60) after induction of differentiation by dimethylsulfoxide. *J. Exp. Med.* 149 (4): 969–974.

Conkie, D., Affara, N., Harrison, P.R. et al. (1974). In situ localization of globin messenger RNA formation. II. After treatment of Friend virus-transformed mouse cells with dimethyl sulfoxide. *J. Cell Biol.* 63 (2 Pt 1): 414–419. https://doi.org/10.1083/jcb.63.2.414.

Crepaldi, T., Pollack, A.L., Prat, M. et al. (1994). Targeting of the SF/HGF receptor to the basolateral domain of polarized epithelial cells. *J. Cell Biol.* 125 (2): 313–320. https://doi.org/10.1083/jcb.125.2.313.

Davis, R.L., Weintraub, H., and Lassar, A.B. (1987). Expression of a single transfected cDNA converts fibroblasts to myoblasts. *Cell* 51 (6): 987–1000.

Degos, L. (1997). Differentiation therapy in acute promyelocytic leukemia: European experience. *J. Cell. Physiol.* 173 (2): 285–287. https://doi.org/10.1002/(SICI)1097-4652(199711)173:2<285::AID-JCP37>3.0.CO;2-C.

Diamond, L. (1994). Charlotte Friend: March 11, 1921–January 13, 1987. *Biogr. Mem. Natl Acad. Sci.* 63: 127–148.

Dobbs, L.G., Pian, M.S., Maglio, M. et al. (1997). Maintenance of the differentiated type II cell phenotype by culture with an apical air surface. *Am. J. Phys.* 273 (2 Pt 1): L347–L354. https://doi.org/10.1152/ajplung.1997.273.2.L347.

Doherty, P., Ashton, S.V., Moore, S.E. et al. (1991). Morphoregulatory activities of NCAM and N-cadherin can be accounted for by G protein-dependent activation of L- and N-type neuronal Ca^{2+} channels. *Cell* 67 (1): 21–33. https://doi.org/10.1016/0092-8674(91)90569-k.

Engler, A.J., Sen, S., Sweeney, H.L. et al. (2006). Matrix elasticity directs stem cell lineage specification. *Cell* 126 (4): 677–689. https://doi.org/10.1016/j.cell.2006.06.044.

Fan, J., Fong, T., Chen, X. et al. (2018). Glia maturation factor-beta: a potential therapeutic target in neurodegeneration and neuroinflammation. *Neuropsychiatr. Dis. Treat.* 14: 495–504. https://doi.org/10.2147/NDT.S157099.

Fell, H.B. (1972). Tissue culture and its contribution to biology and medicine. *J. Exp. Biol.* 57 (1): 1–13.

Finbow, M.E. and Pitts, J.D. (1981). Permeability of junctions between animal cells. Intercellular exchange of various metabolites and a vitamin-derived cofactor. *Exp. Cell Res.* 131 (1): 1–13. https://doi.org/10.1016/0014-4827(81)90399-2.

Fogh, J. and Trempe, G. (1975). New human tumor cell lines. In: *Human Tumor Cells In Vitro* (ed. J. Fogh), 115–159. New York: Plenum Press.

Freshney, R.I. (1985). Induction of differentiation in neoplastic cells. *Anticancer Res.* 5 (1): 111–130.

Friend, C., Scher, W., Holland, J.G. et al. (1971). Hemoglobin synthesis in murine virus-induced leukemic cells in vitro: stimulation of erythroid differentiation by dimethyl sulfoxide. *Proc. Natl Acad. Sci. U.S.A.* 68 (2): 378–382. https://doi.org/10.1073/pnas.68.2.378.

Fu, Y., Zhang, P., Ge, J. et al. (2014). Histone deacetylase 8 suppresses osteogenic differentiation of bone marrow stromal cells by inhibiting histone H3K9 acetylation and RUNX2 activity. *Int. J. Biochem. Cell Biol.* 54: 68–77. https://doi.org/10.1016/j.biocel.2014.07.003.

Furue, M. and Saito, S. (1997). Synergistic effect of hepatocyte growth factor and fibroblast growth factor-1 on the branching morphogenesis of rat submandibular gland epithelial cells. *Tissue Cult. Res. Commun.* 16 (4): 189–194.

Gattazzo, F., Urciuolo, A., and Bonaldo, P. (2014). Extracellular matrix: a dynamic microenvironment for stem cell niche. *Biochim. Biophys. Acta* 1840 (8): 2506–2519. https://doi.org/10.1016/j.bbagen.2014.01.010.

Ghigo, D., Priotto, C., Migliorino, D. et al. (1998). Retinoic acid-induced differentiation in a human neuroblastoma cell line is associated with an increase in nitric oxide synthesis. *J. Cell. Physiol.* 174 (1): 99–106. https://doi.org/10.1002/(SICI)1097-4652(199801)174:1<99::AID-JCP11>3.0.CO;2-J.

Gilbert, S.F., Axelman, J., and Migeon, B.R. (1981). Phenotypic heterogeneity within clones of fetal human cells. *Am. J. Hum. Genet.* 33 (6): 950–956.

Gocek, E. and Studzinski, G.P. (2009). Vitamin D and differentiation in cancer. *Crit. Rev. Clin. Lab. Sci.* 46 (4): 190–209. https://doi.org/10.1080/10408360902982128.

Goding, C.R. and Fisher, D.E. (1997). Regulation of melanocyte differentiation and growth. *Cell Growth Differ.* 8 (9): 935–940.

Godoy, P., Hewitt, N.J., Albrecht, U. et al. (2013). Recent advances in 2D and 3D in vitro systems using primary hepatocytes, alternative hepatocyte sources and non-parenchymal liver cells and their use in investigating mechanisms of hepatotoxicity, cell signaling and ADME. *Arch. Toxicol.* 87 (8): 1315–1530. https://doi.org/10.1007/s00204-013-1078-5.

Goldwasser, E. (1975). Erythropoietin and the differentiation of red blood cells. *Fed. Proc.* 34 (13): 2285–2292.

Granner, D.K., Hayashi, S., Thompson, E.B. et al. (1968). Stimulation of tyrosine aminotransferase synthesis by dexamethasone phosphate in cell culture. *J. Mol. Biol.* 35 (2): 291–301.

Gripon, P., Rumin, S., Urban, S. et al. (2002). Infection of a human hepatoma cell line by hepatitis B virus. *Proc. Natl Acad. Sci. U.S.A.* 99 (24): 15655–15660. https://doi.org/10.1073/pnas.232137699.

Gudjonsson, T., Ronnov-Jessen, L., Villadsen, R. et al. (2002). Normal and tumor-derived myoepithelial cells differ in their ability to interact with luminal breast epithelial cells for polarity and basement membrane deposition. *J. Cell Sci.* 115 (Pt 1): 39–50.

Guguen-Guillouzo, C. and Guillouzo, A. (1986). Methods for preparation of adult and fetal hepatocytes. In: *Isolated and Cultured Hepatocytes* (eds. C. Guguen-Guillouzo and A. Guillouzo), 1–12. Paris: Les Editions INSERM, John Libbey Eurotext.

Gumbiner, B.M. and Yamada, K.M. (1995). Cell-to-cell contact and extracellular matrix. *Curr. Opin. Cell Biol.* 7 (5): 615–618.

Guner, M., Freshney, R.I., Morgan, D. et al. (1977). Effects of dexamethasone and betamethasone on in vitro cultures from human astrocytoma. *Br. J. Cancer* 35 (4): 439–447.

Gurtovenko, A.A. and Anwar, J. (2007). Modulating the structure and properties of cell membranes: the molecular mechanism of action of dimethyl sulfoxide. *J. Phys. Chem. B* 111 (35): 10453–10460. https://doi.org/10.1021/jp073113e.

Halaban, R. (2004). Culture of melanocytes from normal, benign, and malignant lesions. In: *Culture of Human Tumor Cells* (eds. R. Pfragner and R.I. Freshney), 289–318. Hoboken, NJ: Wiley-Liss.

Haner, K., Henzi, T., Pfefferli, M. et al. (2010). A bipartite butyrate-responsive element in the human calretinin (CALB2) promoter acts as a repressor in colon carcinoma cells but not in mesothelioma cells. *J. Cell. Biochem.* 109 (3): 519–531. https://doi.org/10.1002/jcb.22429.

Hay, E.D. (2005). The mesenchymal cell, its role in the embryo, and the remarkable signaling mechanisms that create it. *Dev. Dyn.* 233 (3): 706–720. https://doi.org/10.1002/dvdy.20345.

Heer, R., Collins, A.T., Robson, C.N. et al. (2006). KGF suppresses alpha2beta1 integrin function and promotes differentiation of the transient amplifying population in human prostatic epithelium. *J. Cell Sci.* 119 (Pt 7): 1416–1424. https://doi.org/10.1242/jcs.02802.

Heffelfinger, S.C., Hawkins, H.H., Barrish, J. et al. (1992). SK HEP-1: a human cell line of endothelial origin. *In Vitro Cell. Dev. Biol.* 28A (2): 136–142.

Herve, J.C. and Derangeon, M. (2013). Gap-junction-mediated cell-to-cell communication. *Cell Tissue Res.* 352 (1): 21–31. https://doi.org/10.1007/s00441-012-1485-6.

Hollenberg, S.M., Cheng, P.F., and Weintraub, H. (1993). Use of a conditional MyoD transcription factor in studies of MyoD trans-activation and muscle determination. *Proc. Natl Acad. Sci. U.S.A.* 90 (17): 8028–8032. https://doi.org/10.1073/pnas.90.17.8028.

Hu, W., Qiu, B., Guan, W. et al. (2015). Direct conversion of normal and Alzheimer's disease human fibroblasts into neuronal cells by small molecules. *Cell Stem Cell* 17 (2): 204–212. https://doi.org/10.1016/j.stem.2015.07.006.

Hynes, R.O. (2009). The extracellular matrix: not just pretty fibrils. *Science* 326 (5957): 1216–1219. https://doi.org/10.1126/science.1176009.

Inman, J.L., Robertson, C., Mott, J.D. et al. (2015). Mammary gland development: cell fate specification, stem cells and the microenvironment. *Development* 142 (6): 1028–1042. https://doi.org/10.1242/dev.087643.

Jagtap, S., Meganathan, K., Gaspar, J. et al. (2011). Cytosine arabinoside induces ectoderm and inhibits mesoderm expression in human embryonic stem cells during multilineage differentiation. *Br. J. Pharmacol.* 162 (8): 1743–1756. https://doi.org/10.1111/j.1476-5381.2010.01197.x.

Kaiko, G.E., Ryu, S.H., Koues, O.I. et al. (2016). The colonic crypt protects stem cells from microbiota-derived metabolites. *Cell* 167 (4): 1137. https://doi.org/10.1016/j.cell.2016.10.034.

Kamata, Y., Shiraga, H., Tai, A. et al. (2007). Induction of neurite outgrowth in PC12 cells by the medium-chain fatty acid octanoic acid. *Neuroscience* 146 (3): 1073–1081. https://doi.org/10.1016/j.neuroscience.2007.03.001.

Kang, T.W., Choi, S.W., Yang, S.R. et al. (2014). Growth arrest and forced differentiation of human primary glioblastoma multiforme by a novel small molecule. *Sci. Rep.* 4: 5546. https://doi.org/10.1038/srep05546.

Karp, P.H., Moninger, T.O., Pary Weber, S. et al. (2002). An in vitro model of differentiated human airway epithelia. In: *Epithelial Cell Culture Protocols* (ed. C. Wise), 115–137. Totowa, NJ: Humana Press.

Kelly, J.H. and Sussman, N.L. (2000). A fluorescent cell-based assay for cytochrome P-450 isozyme 1A2 induction and inhibition. *J. Biomol. Screen.* 5 (4): 249–254. https://doi.org/10.1177/108705710000500407.

Kimhi, Y., Palfrey, C., Spector, I. et al. (1976). Maturation of neuroblastoma cells in the presence of dimethylsulfoxide. *Proc. Natl Acad. Sci. U.S.A.* 73 (2): 462–466. https://doi.org/10.1073/pnas.73.2.462.

Kleinman, H.K. and Martin, G.R. (2005). Matrigel: basement membrane matrix with biological activity. *Semin. Cancer Biol.* 15 (5): 378–386. https://doi.org/10.1016/j.semcancer.2005.05.004.

Klijn, C., Durinck, S., Stawiski, E.W. et al. (2015). A comprehensive transcriptional portrait of human cancer cell lines. *Nat. Biotechnol.* 33 (3): 306–312. https://doi.org/10.1038/nbt.3080.

Kline, K., Yu, W., and Sanders, B.G. (2004). Vitamin E and breast cancer. *J. Nutr.* 134 (12 Suppl): 3458S–3462S. https://doi.org/10.1093/jn/134.12.3458S.

Kondo, Y., Iwao, T., Yoshihashi, S. et al. (2014). Histone deacetylase inhibitor valproic acid promotes the differentiation of human induced pluripotent stem cells into hepatocyte-like cells. *PLoS One* 9 (8): e104010. https://doi.org/10.1371/journal.pone.0104010.

Korch, C., Hall, E.M., Dirks, W.G. et al. (2018). Authentication of M14 melanoma cell line proves misidentification of MDA-MB-435 breast cancer cell line. *Int. J. Cancer* 142 (3): 561–572. https://doi.org/10.1002/ijc.31067.

Krstic, J., Trivanovic, D., Obradovic, H. et al. (2018). Regulation of mesenchymal stem cell differentiation by transforming growth factor beta superfamily. *Curr. Protein Pept. Sci.* 19 (12): 1138–1154. https://doi.org/10.2174/1389203718666171117103418.

Kshitiz, P.J., Kim, P. et al. (2012). Control of stem cell fate and function by engineering physical microenvironments. *Integr. Biol. (Camb.)* 4 (9): 1008–1018.

Kujoth, G.C. and Fahl, W.E. (1997). C-sis/platelet-derived growth factor-B promoter requirements for induction during the 12-O-tetradecanoylphorbol-13-acetate-mediated megakaryoblastic differentiation of K562 human erythroleukemia cells. *Cell Growth Differ.* 8 (9): 963–977.

Kurita, R., Suda, N., Sudo, K. et al. (2013). Establishment of immortalized human erythroid progenitor cell lines able to produce enucleated red blood cells. *PLoS One* 8 (3): e59890. https://doi.org/10.1371/journal.pone.0059890.

Lambshead, J.W., Meagher, L., O'Brien, C. et al. (2013). Defining synthetic surfaces for human pluripotent stem cell culture. *Cell Regen. (Lond.).* 2 (1): 7. https://doi.org/10.1186/2045-9769-2-7.

Landa, I., Pozdeyev, N., Korch, C., et al. (2019). Comprehensive genetic characterization of human thyroid cancer cell lines:

a validated panel for preclinical studies. *Clin. Cancer Res.* 25: 3141–3151. https://doi.org/10.1158/1078-0432.CCR-18-2953.

Le Magnen, C., Shen, M.M., and Abate-Shen, C. (2018). Lineage plasticity in cancer progression and treatment. *Annu. Rev. Cancer Biol.* 2: 271–289. https://doi.org/10.1146/annurev-cancerbio-030617-050224.

Lesuffleur, T., Barbat, A., Dussaulx, E. et al. (1990). Growth adaptation to methotrexate of HT-29 human colon carcinoma cells is associated with their ability to differentiate into columnar absorptive and mucus-secreting cells. *Cancer Res.* 50 (19): 6334–6343.

Levi-Montalcini, R. (1982). Developmental neurobiology and the natural history of nerve growth factor. *Annu. Rev. Neurosci.* 5: 341–362. https://doi.org/10.1146/annurev.ne.05.030182.002013.

Li, X., Zuo, X., Jing, J. et al. (2015). Small-molecule-driven direct reprogramming of mouse fibroblasts into functional neurons. *Cell Stem Cell* 17 (2): 195–203. https://doi.org/10.1016/j.stem.2015.06.003.

Lim, R., Miller, J.F., and Zaheer, A. (1989). Purification and characterization of glia maturation factor beta: a growth regulator for neurons and glia. *Proc. Natl Acad. Sci. U.S.A.* 86 (10): 3901–3905. https://doi.org/10.1073/pnas.86.10.3901.

Liu, M.L., Zang, T., Zou, Y. et al. (2013). Small molecules enable neurogenin 2 to efficiently convert human fibroblasts into cholinergic neurons. *Nat. Commun.* 4: 2183. https://doi.org/10.1038/ncomms3183.

Lohmann, V., Korner, F., Koch, J. et al. (1999). Replication of subgenomic hepatitis C virus RNAs in a hepatoma cell line. *Science* 285 (5424): 110–113. https://doi.org/10.1126/science.285.5424.110.

Lopez-Ruiz, E., Peran, M., Picon-Ruiz, M. et al. (2014). Cardiomyogenic differentiation potential of human endothelial progenitor cells isolated from patients with myocardial infarction. *Cytotherapy* 16 (9): 1229–1237. https://doi.org/10.1016/j.jcyt.2014.05.012.

Lopez-Terrada, D., Cheung, S.W., Finegold, M.J. et al. (2009). Hep G2 is a hepatoblastoma-derived cell line. *Hum. Pathol.* 40 (10): 1512–1515. https://doi.org/10.1016/j.humpath.2009.07.003.

Lotan, R. and Lotan, D. (1980). Stimulation of melanogenesis in a human melanoma cell line by retinoids. *Cancer Res.* 40 (9): 3345–3350.

Lotem, J. and Sachs, L. (2002). Cytokine control of developmental programs in normal hematopoiesis and leukemia. *Oncogene* 21 (21): 3284–3294. https://doi.org/10.1038/sj.onc.1205319.

Maas-Szabowski, N., Stark, H.J., and Fusenig, N.E. (2000). Keratinocyte growth regulation in defined organotypic cultures through IL-1-induced keratinocyte growth factor expression in resting fibroblasts. *J. Invest. Dermatol.* 114 (6): 1075–1084. https://doi.org/10.1046/j.1523-1747.2000.00987.x.

Maas-Szabowski, N., Stark, H.J., and Fusenig, N.E. (2002). Cell interaction and epithelial differentiation. In: *Culture of Epithelial Cells* (eds. R.I. Freshney and M.G. Freshney), 31–63. Hoboken, NJ: Wiley-Liss.

Majkut, S., Dingal, P.C., and Discher, D.E. (2014). Stress sensitivity and mechanotransduction during heart development. *Curr. Biol.* 24 (10): R495–R501. https://doi.org/10.1016/j.cub.2014.04.027.

Mani, S.A., Guo, W., Liao, M.J. et al. (2008). The epithelial–mesenchymal transition generates cells with properties of stem cells. *Cell* 133 (4): 704–715. https://doi.org/10.1016/j.cell.2008.03.027.

Mann, B.S., Johnson, J.R., Cohen, M.H. et al. (2007). FDA approval summary: vorinostat for treatment of advanced primary cutaneous T-cell lymphoma. *Oncologist* 12 (10): 1247–1252. https://doi.org/10.1634/theoncologist.12-10-1247.

Marchese, C., Rubin, J., Ron, D. et al. (1990). Human keratinocyte growth factor activity on proliferation and differentiation of human keratinocytes: differentiation response distinguishes KGF from EGF family. *J. Cell. Physiol.* 144 (2): 326–332. https://doi.org/10.1002/jcp.1041440219.

Marks, P.A. (2007). Discovery and development of SAHA as an anticancer agent. *Oncogene* 26 (9): 1351–1356. https://doi.org/10.1038/sj.onc.1210204.

Marks, P.A. and Rifkind, R.A. (1988). Hexamethylene bisacetamide-induced differentiation of transformed cells: molecular and cellular effects and therapeutic application. *Int. J. Cell Cloning* 6 (4): 230–240. https://doi.org/10.1002/stem.5530060402.

Marro, S., Pang, Z.P., Yang, N. et al. (2011). Direct lineage conversion of terminally differentiated hepatocytes to functional neurons. *Cell Stem Cell* 9 (4): 374–382. https://doi.org/10.1016/j.stem.2011.09.002.

Marte, B.M., Meyer, T., Stabel, S. et al. (1994). Protein kinase C and mammary cell differentiation: involvement of protein kinase C alpha in the induction of beta-casein expression. *Cell Growth Differ.* 5 (3): 239–247.

Martinovich, K.M., Iosifidis, T., Buckley, A.G. et al. (2017). Conditionally reprogrammed primary airway epithelial cells maintain morphology, lineage and disease specific functional characteristics. *Sci. Rep.* 7 (1): 17971. https://doi.org/10.1038/s41598-017-17952-4.

Masui, T., Lechner, J.F., Yoakum, G.H. et al. (1986). Growth and differentiation of normal and transformed human bronchial epithelial cells. *J. Cell. Physiol. Suppl.* 4: 73–81.

McCarthy, K.D. and de Vellis, J. (1980). Preparation of separate astroglial and oligodendroglial cell cultures from rat cerebral tissue. *J. Cell Biol.* 85 (3): 890–902. https://doi.org/10.1083/jcb.85.3.890.

McCormack, S.A., Viar, M.J., Tague, L. et al. (1996). Altered distribution of the nuclear receptor RAR beta accompanies proliferation and differentiation changes caused by retinoic acid in Caco-2 cells. *In Vitro Cell. Dev. Biol. Anim.* 32 (1): 53–61.

McCormick, C. and Freshney, R.I. (2000). Activity of growth factors in the IL-6 group in the differentiation of human lung adenocarcinoma. *Br. J. Cancer* 82 (4): 881–890. https://doi.org/10.1054/bjoc.1999.1015.

McCormick, C., Freshney, R.I., and Speirs, V. (1995). Activity of interferon alpha, interleukin 6 and insulin in the regulation of differentiation in A549 alveolar carcinoma cells. *Br. J. Cancer* 71 (2): 232–239.

McLean, J.S., Frame, M.C., Freshney, R.I. et al. (1986). Phenotypic modification of human glioma and non-small cell lung carcinoma by glucocorticoids and other agents. *Anticancer Res.* 6 (5): 1101–1106.

Mertens, J., Marchetto, M.C., Bardy, C. et al. (2016). Evaluating cell reprogramming, differentiation and conversion technologies in neuroscience. *Nat. Rev. Neurosci.* 17 (7): 424–437. https://doi.org/10.1038/nrn.2016.46.

Meyskens, F.L. Jr. and Fuller, B.B. (1980). Characterization of the effects of different retinoids on the growth and differentiation of a human melanoma cell line and selected subclones. *Cancer Res.* 40 (7): 2194–2196.

Montesano, R., Matsumoto, K., Nakamura, T. et al. (1991). Identification of a fibroblast-derived epithelial morphogen as hepatocyte growth factor. *Cell* 67 (5): 901–908.

Morel, A.P., Lievre, M., Thomas, C. et al. (2008). Generation of breast cancer stem cells through epithelial–mesenchymal transition. *PLoS One* 3 (8): e2888. https://doi.org/10.1371/journal.pone.0002888.

Morita, Y., Yamamoto, S., and Ju, Y. (2015). Development of a new co-culture system, the "separable-close co-culture system," to enhance stem-cell-to-chondrocyte differentiation. *Biotechnol. Lett.* 37 (9): 1911–1918. https://doi.org/10.1007/s10529-015-1858-5.

Movia, D., Bazou, D., Volkov, Y. et al. (2018). Multilayered cultures of NSCLC cells grown at the air–liquid interface allow the efficacy testing of inhaled anti-cancer drugs. *Sci. Rep.* 8 (1): 12920. https://doi.org/10.1038/s41598-018-31332-6.

Muschler, J. and Streuli, C.H. (2010). Cell–matrix interactions in mammary gland development and breast cancer. *Cold Spring Harb. Perspect. Biol.* 2 (10): a003202. https://doi.org/10.1101/cshperspect.a003202.

Muschler, J., Lochter, A., Roskelley, C.D. et al. (1999). Division of labor among the alpha6beta4 integrin, beta1 integrins, and an E3 laminin receptor to signal morphogenesis and beta-casein expression in mammary epithelial cells. *Mol. Biol. Cell* 10 (9): 2817–2828. https://doi.org/10.1091/mbc.10.9.2817.

Nash, C.E., Mavria, G., Baxter, E.W. et al. (2015). Development and characterisation of a 3D multi-cellular in vitro model of normal human breast: a tool for cancer initiation studies. *Oncotarget* 6 (15): 13731–13741. https://doi.org/10.18632/oncotarget.3803.

Nelson, L.J., Morgan, K., Treskes, P. et al. (2017). Human hepatic HepaRG cells maintain an organotypic phenotype with high intrinsic CYP450 activity/metabolism and significantly outperform standard HepG2/C3A cells for pharmaceutical and therapeutic applications. *Basic Clin. Pharmacol. Toxicol.* 120 (1): 30–37. https://doi.org/10.1111/bcpt.12631.

Nerlich, A.G. and Bachmeier, B.E. (2013). Density-dependent lineage instability of MDA-MB-435 breast cancer cells. *Oncol. Lett.* 5 (4): 1370–1374. https://doi.org/10.3892/ol.2013.1157.

Nilsson, L.M., Green, L.C., Muralidharan, S.V. et al. (2016). Cancer differentiating agent hexamethylene bisacetamide inhibits BET bromodomain proteins. *Cancer Res.* 76 (8): 2376–2383. https://doi.org/10.1158/0008-5472.CAN-15-2721.

Nozaki, K., Kadosawa, T., Nishimura, R. et al. (1999). 1,25-Dihydroxyvitamin D3, recombinant human transforming growth factor-beta 1, and recombinant human bone morphogenetic protein-2 induce in vitro differentiation of canine osteosarcoma cells. *J. Vet. Med. Sci.* 61 (6): 649–656. https://doi.org/10.1292/jvms.61.649.

Okugawa, Y. and Hirai, Y. (2013). Extracellular epimorphin modulates epidermal differentiation signals mediated by epidermal growth factor receptor. *J. Dermatol. Sci.* 69 (3): 236–242. https://doi.org/10.1016/j.jdermsci.2012.11.006.

Orellana, S.A., Neff, C.D., Sweeney, W.E. et al. (1996). Novel Madin Darby canine kidney cell clones exhibit unique phenotypes in response to morphogens. *In Vitro Cell. Dev. Biol. Anim.* 32 (6): 329–339.

Osterloh, J.M. and Mullane, K. (2018). Manipulating cell fate while confronting reproducibility concerns. *Biochem. Pharmacol.* 151: 144–156. https://doi.org/10.1016/j.bcp.2018.01.016.

Palakkan, A.A., Nanda, J., and Ross, J.A. (2017). Pluripotent stem cells to hepatocytes, the journey so far. *Biomed Rep.* 6 (4): 367–373. https://doi.org/10.3892/br.2017.867.

Palmer, H.G., Gonzalez-Sancho, J.M., Espada, J. et al. (2001). Vitamin D(3) promotes the differentiation of colon carcinoma cells by the induction of E-cadherin and the inhibition of beta-catenin signaling. *J. Cell Biol.* 154 (2): 369–387. https://doi.org/10.1083/jcb.200102028.

Pang, Z.P., Yang, N., Vierbuchen, T. et al. (2011). Induction of human neuronal cells by defined transcription factors. *Nature* 476 (7359): 220–223. https://doi.org/10.1038/nature10202.

Parsons, X.H. (2013). Constraining the pluripotent fate of human embryonic stem cells for tissue engineering and cell therapy – the turning point of cell-based regenerative medicine. *Br. Biotechnol. J.* 3 (4): 424–457. https://doi.org/10.9734/BBJ/2013/4309#sthash.6D8Rulbv.dpuf.

Parsons, X.H. (2016). Direct conversion of pluripotent human embryonic stem cells under defined culture conditions into human neuronal or cardiomyocyte cell therapy derivatives. In: *Human Embryonic Stem Cell Protocols* (ed. K. Turksen), 299–318. New York: Springer.

Parsons, X.H., Teng, Y.D., Moore, D.A. et al. (2011). Patents on technologies of human tissue and organ regeneration from pluripotent human embryonic stem cells. *Recent Pat. Regen. Med.* 1 (2): 142–163. https://doi.org/10.2174/2210296511101020142.

Peters-Hall, J.R., Coquelin, M.L., Torres, M.J. et al. (2018). Long-term culture and cloning of primary human bronchial basal cells that maintain multipotent differentiation capacity and CFTR channel function. *Am. J. Physiol. Lung Cell. Mol. Physiol.* 315 (2): L313–L327. https://doi.org/10.1152/ajplung.00355.2017.

Pfisterer, U., Kirkeby, A., Torper, O. et al. (2011). Direct conversion of human fibroblasts to dopaminergic neurons. *Proc. Natl Acad. Sci. U.S.A.* 108 (25): 10343–10348. https://doi.org/10.1073/pnas.1105135108.

Pignata, S., Maggini, L., Zarrilli, R. et al. (1994). The enterocyte-like differentiation of the Caco-2 tumor cell line strongly correlates with responsiveness to cAMP and activation of kinase a pathway. *Cell Growth Differ.* 5 (9): 967–973.

Prasad, K.N., Edwards-Prasad, J., Ramanujam, S. et al. (1980). Vitamin E increases the growth inhibitory and differentiating effects of tumor therapeutic agents on neuroblastoma and glioma cells in culture. *Proc. Soc. Exp. Biol. Med.* 164 (2): 158–163. https://doi.org/10.3181/00379727-164-40840.

Prasad, K.N., Kumar, B., Yan, X.D. et al. (2003). Alpha-tocopheryl succinate, the most effective form of vitamin E for adjuvant cancer treatment: a review. *J. Am. Coll. Nutr.* 22 (2): 108–117.

Raff, M.C. (1989). Glial cell diversification in the rat optic nerve. *Science* 243 (4897): 1450–1455. https://doi.org/10.1126/science.2648568.

Raz, A. (1982). B16 melanoma cell variants: irreversible inhibition of growth and induction of morphologic differentiation by anthracycline antibiotics. *J. Natl Cancer Inst.* 68 (4): 629–638.

Reid, K., Turnley, A.M., Maxwell, G.D. et al. (1996). Multiple roles for endothelin in melanocyte development: regulation of progenitor number and stimulation of differentiation. *Development* 122 (12): 3911–3919.

Reuben, R.C., Wife, R.L., Breslow, R. et al. (1976). A new group of potent inducers of differentiation in murine erythroleukemia cells. *Proc. Natl Acad. Sci. U.S.A.* 73 (3): 862–866. https://doi.org/10.1073/pnas.73.3.862.

Reuben, R.C., Khanna, P.L., Gazitt, Y. et al. (1978). Inducers of erythroleukemic differentiation. Relationship of structure to activity among planar-polar compounds. *J. Biol. Chem.* 253 (12): 4214–4218.

Rhim, A.D. (2013). Epithelial to mesenchymal transition and the generation of stem-like cells in pancreatic cancer. *Pancreatology* 13 (2): 114–117. https://doi.org/10.1016/j.pan.2013.01.004.

Rooney, S.A., Young, S.L., and Mendelson, C.R. (1994). Molecular and cellular processing of lung surfactant. *FASEB J.* 8 (12): 957–967. https://doi.org/10.1096/fasebj.8.12.8088461.

Roschke, A.V., Glebov, O.K., Lababidi, S. et al. (2008). Chromosomal instability is associated with higher expression of genes implicated in epithelial–mesenchymal transition, cancer invasiveness, and metastasis and with lower expression of genes involved in cell cycle checkpoints, DNA repair, and chromatin maintenance. *Neoplasia* 10 (11): 1222–1230.

Rudland, P.S. (1992). Use of peanut lectin and rat mammary stem cell lines to identify a cellular differentiation pathway for the alveolar cell in the rat mammary gland. *J. Cell. Physiol.* 153 (1): 157–168. https://doi.org/10.1002/jcp.1041530120.

Sachs, L. (1978). Control of normal cell differentiation and the phenotypic reversion of malignancy in myeloid leukaemia. *Nature* 274 (5671): 535–539. https://doi.org/10.1038/274535a0.

Safford, K.M. and Rice, H.E. (2007). Tissue culture of adipose-derived stem cells. In: *Culture of Human Stem Cells* (eds. R.I. Freshney, G.N. Stacey and J.M. Auerbach), 303–315. Hoboken, NJ: Wiley.

Samuel, S. and Sitrin, M.D. (2008). Vitamin D's role in cell proliferation and differentiation. *Nutr. Rev.* 66 (10 Suppl 2): S116–S124. https://doi.org/10.1111/j.1753-4887.2008.00094.x.

Sanders, E.J. and Harvey, S. (2008). Peptide hormones as developmental growth and differentiation factors. *Dev. Dyn.* 237 (6): 1537–1552. https://doi.org/10.1002/dvdy.21573.

Santos, O.F., Moura, L.A., Rosen, E.M. et al. (1993). Modulation of HGF-induced tubulogenesis and branching by multiple phosphorylation mechanisms. *Dev. Biol.* 159 (2): 535–548. https://doi.org/10.1006/dbio.1993.1262.

Sato, G. (2008). Tissue culture: the unrealized potential. *Cytotechnology* 57 (2): 111–114. https://doi.org/10.1007/s10616-007-9109-9.

Sattler, C.A., Michalopoulos, G., Sattler, G.L. et al. (1978). Ultrastructure of adult rat hepatocytes cultured on floating collagen membranes. *Cancer Res.* 38 (6): 1539–1549.

Saunders, N.A., Bernacki, S.H., Vollberg, T.M. et al. (1993). Regulation of transglutaminase type I expression in squamous differentiating rabbit tracheal epithelial cells and human epidermal keratinocytes: effects of retinoic acid and phorbol esters. *Mol. Endocrinol.* 7 (3): 387–398. https://doi.org/10.1210/mend.7.3.8097865.

Scher, W., Holland, J.G., and Friend, C. (1971). Hemoglobin synthesis in murine virus-induced leukemic cells in vitro. I. Partial purification and identification of hemoglobins. *Blood* 37 (4): 428–437.

Schwartz, C.M., Spivak, C.E., Baker, S.C. et al. (2005). NTera2: a model system to study dopaminergic differentiation of human embryonic stem cells. *Stem Cells Dev.* 14 (5): 517–534. https://doi.org/10.1089/scd.2005.14.517.

Sell, S. (2010). On the stem cell origin of cancer. *Am. J. Pathol.* 176 (6): 2584–2594. https://doi.org/10.2353/ajpath.2010.091064.

Shen, C.N., Burke, Z.D., and Tosh, D. (2004). Transdifferentiation, metaplasia and tissue regeneration. *Organogenesis* 1 (2): 36–44.

Speirs, V., Ray, K.P., and Freshney, R.I. (1991). Paracrine control of differentiation in the alveolar carcinoma, A549, by human foetal lung fibroblasts. *Br. J. Cancer* 64 (4): 693–699.

Spinelli, W., Sonnenfeld, K.H., and Ishii, D.N. (1982). Effects of phorbol ester tumor promoters and nerve growth factor on neurite outgrowth in cultured human neuroblastoma cells. *Cancer Res.* 42 (12): 5067–5073.

Spremulli, E.N. and Dexter, D.L. (1984). Polar solvents: a novel class of antineoplastic agents. *J. Clin. Oncol.* 2 (3): 227–241. https://doi.org/10.1200/JCO.1984.2.3.227.

Stark, H.J., Szabowski, A., Fusenig, N.E. et al. (2004). Organotypic cocultures as skin equivalents: a complex and sophisticated in vitro system. *Biol. Proced. Online* 6: 55–60. https://doi.org/10.1251/bpo72.

Strickland, S. and Mahdavi, V. (1978). The induction of differentiation in teratocarcinoma stem cells by retinoic acid. *Cell* 15 (2): 393–403. https://doi.org/10.1016/0092-8674(78)90008-9.

Szabowski, A., Maas-Szabowski, N., Andrecht, S. et al. (2000). C-Jun and JunB antagonistically control cytokine-regulated mesenchymal–epidermal interaction in skin. *Cell* 103 (5): 745–755. https://doi.org/10.1016/s0092-8674(00)00178-1.

Takahashi, K., Suzuki, K., Kawahara, S. et al. (1991). Effects of lactogenic hormones on morphological development and growth of human breast epithelial cells cultivated in collagen gels. *Jpn J. Cancer Res.* 82 (5): 553–558. https://doi.org/10.1111/j.1349-7006.1991.tb01886.x.

Takeda, K., Minowada, J., and Bloch, A. (1982). Kinetics of appearance of differentiation-associated characteristics in ML-1, a line of human myeloblastic leukemia cells, after treatment with 12-O-tetradecanoylphorbol-13-acetate, dimethyl sulfoxide or 1-beta-D-arabinofuranosylcytosine. *Cancer Res.* 42 (12): 5152–5158.

Talbott, J.F., Cao, Q., Bertram, J. et al. (2007). CNTF promotes the survival and differentiation of adult spinal cord-derived oligodendrocyte precursor cells in vitro but fails to promote remyelination in vivo. *Exp. Neurol.* 204 (1): 485–489. https://doi.org/10.1016/j.expneurol.2006.12.013.

Tanabe, K., Haag, D., and Wernig, M. (2015). Direct somatic lineage conversion. *Philos. Trans. R. Soc. Lond. Ser. B Biol. Sci.* 370 (1680): 20140368. https://doi.org/10.1098/rstb.2014.0368.

Tascher, G., Burban, A., Camus, S. et al. (2019). In-depth proteome analysis highlights HepaRG cells as a versatile cell system surrogate for primary human hepatocytes. *Cell* 8 (2) https://doi.org/10.3390/cells8020192.

de The, H. (2018). Differentiation therapy revisited. *Nat. Rev. Cancer* 18 (2): 117–127. https://doi.org/10.1038/nrc.2017.103.

Thiagarajan, R.D., Morey, R., and Laurent, L.C. (2014). The epigenome in pluripotency and differentiation. *Epigenomics* 6 (1): 121–137. https://doi.org/10.2217/epi.13.80.

Thomson, A.A., Foster, B.A., and Cunha, G.R. (1997). Analysis of growth factor and receptor mRNA levels during development of the rat seminal vesicle and prostate. *Development* 124 (12): 2431–2439.

Tonge, P.D., Olariu, V., Coca, D. et al. (2010). Prepatterning in the stem cell compartment. *PLoS One* 5 (5): e10901. https://doi.org/10.1371/journal.pone.0010901.

Torres, J., Prieto, J., Durupt, F.C. et al. (2012). Efficient differentiation of embryonic stem cells into mesodermal precursors by BMP, retinoic acid and notch signalling. *PLoS One* 7 (4): e36405. https://doi.org/10.1371/journal.pone.0036405.

Upadhyay, S. and Palmberg, L. (2018). Air–liquid interface: relevant in vitro models for investigating air pollutant-induced pulmonary toxicity. *Toxicol. Sci.* 164 (1): 21–30. https://doi.org/10.1093/toxsci/kfy053.

Velcich, A., Palumbo, L., Jarry, A. et al. (1995). Patterns of expression of lineage-specific markers during the in vitro-induced differentiation of HT29 colon carcinoma cells. *Cell Growth Differ.* 6 (6): 749–757.

Vierbuchen, T., Ostermeier, A., Pang, Z.P. et al. (2010). Direct conversion of fibroblasts to functional neurons by defined factors. *Nature* 463 (7284): 1035–1041. https://doi.org/10.1038/nature08797.

Vunjak-Novakovic, G. and Freshney, R.I. (2006). *Culture of Cells for Tissue Engineering*. Hoboken, NJ: Wiley-Liss.

Wada, T., Darcy, K.M., Guan, X. et al. (1994). Phorbol 12-myristate 13-acetate stimulates proliferation and ductal morphogenesis and inhibits functional differentiation of normal rat mammary epithelial cells in primary culture. *J. Cell. Physiol.* 158 (1): 97–109. https://doi.org/10.1002/jcp.1041580113.

Wang, H. and Unternaehrer, J.J. (2019). Epithelial–mesenchymal transition and cancer stem cells: at the crossroads of differentiation and dedifferentiation. *Dev. Dyn.* 248 (1): 10–20. https://doi.org/10.1002/dvdy.24678.

Watabe, T. and Miyazono, K. (2009). Roles of TGF-beta family signaling in stem cell renewal and differentiation. *Cell Res.* 19 (1): 103–115. https://doi.org/10.1038/cr.2008.323.

Watanabe, T., Kondo, K., and Oishi, M. (1991). Induction of in vitro differentiation of mouse erythroleukemia cells by genistein, an inhibitor of tyrosine protein kinases. *Cancer Res.* 51 (3): 764–768.

Wells, E.K., Yarborough, O. 3rd, Lifton, R.P. et al. (2013). Epithelial morphogenesis of MDCK cells in three-dimensional collagen culture is modulated by interleukin-8. *Am. J. Physiol. Cell Physiol.* 304 (10): C966–C975. https://doi.org/10.1152/ajpcell.00261.2012.

Welsh, J. (2011). Vitamin D metabolism in mammary gland and breast cancer. *Mol. Cell. Endocrinol.* 347 (1–2): 55–60. https://doi.org/10.1016/j.mce.2011.05.020.

Willey, J.C., Moser, C.E. Jr., Lechner, J.F. et al. (1984). Differential effects of 12-O-tetradecanoylphorbol-13-acetate on cultured normal and neoplastic human bronchial epithelial cells. *Cancer Res.* 44 (11): 5124–5126.

Wong, Y.C., Wang, X.H., and Ling, M.T. (2003). Prostate development and carcinogenesis. *Int. Rev. Cytol.* 227: 65–130.

Wuarin, L., Verity, M.A., and Sidell, N. (1991). Effects of interferon-gamma and its interaction with retinoic acid on human neuroblastoma differentiation. *Int. J. Cancer* 48 (1): 136–141. https://doi.org/10.1002/ijc.2910480124.

Yan, X., Ehnert, S., Culmes, M. et al. (2014a). 5-azacytidine improves the osteogenic differentiation potential of aged human adipose-derived mesenchymal stem cells by DNA demethylation. *PLoS One* 9 (6): e90846. https://doi.org/10.1371/journal.pone.0090846.

Yan, Y., Zhao, J., Cao, C. et al. (2014b). Tetramethylpyrazine promotes SH-SY5Y cell differentiation into neurons through epigenetic regulation of topoisomerase IIbeta. *Neuroscience* 278: 179–193. https://doi.org/10.1016/j.neuroscience.2014.08.010.

Zhang, Y. and Weinberg, R.A. (2018). Epithelial-to-mesenchymal transition in cancer: complexity and opportunities. *Front. Med.* 12 (4): 361–373. https://doi.org/10.1007/s11684-018-0656-6.

Zhang, Q., Fan, H., Shen, J. et al. (2010). Human breast cancer cell lines co-express neuronal, epithelial, and melanocytic differentiation markers in vitro and in vivo. *PLoS One* 5 (3): e9712. https://doi.org/10.1371/journal.pone.0009712.

Zhang, J., Gao, Y., Yu, M. et al. (2015). Retinoic acid induces embryonic stem cell differentiation by altering both encoding RNA and microRNA expression. *PLoS One* 10 (7): e0132566. https://doi.org/10.1371/journal.pone.0132566.

PART VIII

Model Environments and Applications

After reading the following chapters in this part of the book, you will be able to:

(27) **Three-Dimensional Culture:**

a) Induce spheroid formation using an ultra-low attachment (ULA) microplate and embed the spheroids in Matrigel at a comparable focal plane for analysis.

b) List examples of other "forced floating" techniques that can be used to induce spheroids and other 3D structures.

c) Perform culture using a filter well insert.

d) List examples of other scaffold-based systems that can be used for 3D culture.

e) Discuss the meanings of "organoid," "organotypic," and "organ" culture and discuss how and why each 3D format is used in tissue culture laboratories.

f) Define "tissue engineering" and describe the basic requirements for tissue engineering to be successful.

(28) **Scale-up and Automation:**

a) Discuss how to perform scale-up of suspension cultures and explain the meanings of related terms such as "spinner culture" and "bioreactor."

b) Provide examples of bioreactor systems that may be used for culture scale-up.

c) Perform scale-up of adherent cultures using a multi-surface propagator (e.g. a Nunc Cell Factory System) or a roller bottle.

d) Explain the purpose of "microcarrier," "fixed bed," and "hollow fiber" systems when performing scale-up of adherent cultures.

e) Discuss how to perform automation of culture handling and list some advantages and disadvantages that must be considered before investing in automation.

f) Provide examples of equipment and consumables that can be used for high-throughput screening and automation of cell-based assays.

(29) **Toxicity Testing:**

a) Provide examples of applications where *in vitro* toxicity testing methods are used and explain why testing must be performed using Good Cell Culture Practice (GCCP) or Good Laboratory Practice (GLP).

b) Provide examples of microtitration assays that can be used to assess cytotoxicity and explain the basis for each assay and the variable(s) assessed.

c) Perform a 3-(4,5-dimethylthiazol-2-yl)-2,5-diphenyltetrazolium bromide (MTT)-based microtitration assay and explain why other assays are often preferred when performing cytotoxicity testing.

d) Perform a clonogenic assay and discuss the additional parameters or variables that can be assessed using this assay compared to a microtitration assay.

e) Discuss why advanced models are required for toxicity testing and how they may be used to replace animal models.

f) Define an "organ-on-chip" system and explain how such a system works.

Freshney's Culture of Animal Cells: A Manual of Basic Technique and Specialized Applications, Eighth Edition. Amanda Capes-Davis and R. Ian Freshney.
© 2021 John Wiley & Sons Ltd. Published 2021 by John Wiley & Sons Ltd.
Companion website: www.wiley.com/go/freshney/cellculture8

CHAPTER 27

Three-Dimensional Culture

The earliest attempts at tissue culture were performed in three dimensions, growing tissue fragments in hanging drops to observe the outgrowth of cells into the surrounding substrate (Harrison 1907; Burrows 1910). The early investigators used these outgrowths to generate large numbers of cells, using flat Petri dishes or flasks. Such two-dimensional (2D) approaches were convenient and became increasingly popular. However, not all early investigators were proponents of the 2D approach. It was observed that outgrowth of cells from a tissue sample resulted in "uncontrolled" growth, whereas culture of the whole sample resulted in "controlled" growth and physiological behavior (Thomson 1914). This early observation led to the development of organ culture, described by Honor Fell as a form of culture where cells are encouraged to form or maintain their normal architecture and to preserve their normal structural and functional relationships (Fell 1972). Organ culture was originally performed using whole tissue samples, but Aron Moscona showed that it was possible to disaggregate tissues and allow cells to reaggregate into clusters, which displayed physiological behavior, e.g. differentiation into the expected cell types for that tissue (Moscona and Moscona 1952). Moscona's work relied on the use of a hollowed slide, where cells sank to the bottom by gravity and thus came into close contact with each other (Fell 1972). Other investigators, such as Joseph Leighton and Wilton Earle, used cellulose sponge or glass rings as three-dimensional (3D) substrates; cells were able to grow into the sponge to form organized 3D structures (Earle et al. 1951; Leighton 1951). Thus, although 3D culture is sometimes described as a recent development, today's techniques are the culmination of more than a century of

work (Simian and Bissell 2017). This chapter aims to describe the various approaches that are available for 3D culture in today's tissue culture laboratories.

27.1 TERMINOLOGY: 3D CULTURE

The long history of 3D culture has resulted in varied, and often inconsistent, terminology. Cell clusters or aggregates that form in suspension culture are referred to as "spheroids" (see Section 27.4.1), but when they arise from specific cell types they may be known by different names, e.g. neural stem cells give rise to "neurospheres" (Sutherland et al. 1971; Azari et al. 2011). Similarly, the term "organoid" has been used in different contexts to describe intracellular organelles, 3D structures grown from primary explants ("tissue organoids"), and 3D structures arising from stem cells ("stem cell organoids") (Leighton 1997; Shamir and Ewald 2014; Simian and Bissell 2017).

Cultures using 3D formats are often described or classified in accordance with their cellular input (Shamir and Ewald 2014). For this book, the following terms are used to describe the cellular input into a 3D culture system (see Figure 27.1):

(1) **Histotypic culture.** Cells from a single lineage are brought into close association (often at high density) to encourage reassembly, differentiation, and the development of 3D structures. Various techniques can be used for histotypic culture, including filter well inserts and hollow fiber systems (see Sections 27.5.2, 27.5.3).

Freshney's Culture of Animal Cells: A Manual of Basic Technique and Specialized Applications, Eighth Edition. Amanda Capes-Davis and R. Ian Freshney.
© 2021 John Wiley & Sons Ltd. Published 2021 by John Wiley & Sons Ltd.
Companion website: www.wiley.com/go/freshney/cellculture8

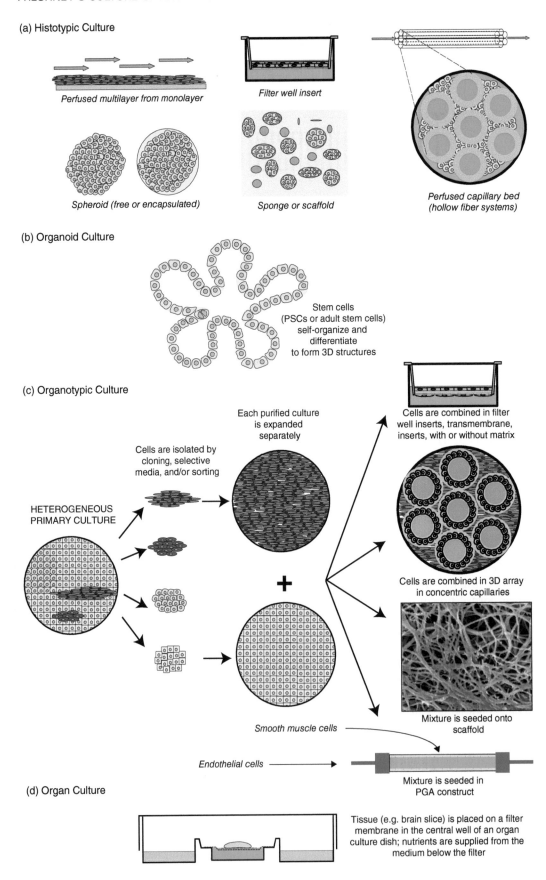

Fig. 27.1. Three-dimensional (3D) culture. Different models for 3D culture are divided by cellular input into: (a) histotypic culture, derived from a single cell type; (b) organoid culture, derived from stem cells; (c) organotypic culture, derived from two or more cell types that are recombined together; (d) organ culture, derived from whole tissue. *Source*: R. Ian Freshney and Amanda Capes-Davis; the PGA construct (bottom right) is simplified from Klinger and Niklason (2006).

(2) ***Organoid culture.*** For this book, organoid culture is defined as arising from a stem cell population. Stem cells are brought into close association, where they self-organize through cell sorting and spatially restricted lineage commitment to form a 3D structure that contains organ-specific cell types (Clevers 2016) (see Section 27.6). The cellular input may consist of pluripotent stem cells (PSCs) (see Section 23.4) or multipotent stem cells (e.g. adult stem cells; see Section 23.6).

(3) ***Organotypic culture.*** Cells from multiple lineages are brought together using culture systems that are designed to recreate the composition of the original tissue. These systems are usually complex, requiring expansion of the various cell types followed by their recombination and co-culture (see Section 27.7).

(4) ***Organ culture.*** The whole organism, organ, or a portion of tissue from an organ is placed into culture and maintained without any effort to dissociate the cells (see Section 27.8).

Although these terms are often useful to describe 3D cultures, the biology of the cell can interfere with the scientist's neat and tidy categories. For example, the colon carcinoma cell line LIM1863 (see Figure 14.11b) spontaneously differentiates into crypt-like structures, mimicking the organoid cultures that arise from intestinal stem cells (Whitehead et al. 1987; Sato et al. 2009). It seems likely that these organoids arise from stem cells within the culture; if so, the parental cell population for LIM1863 appears to be oligopotent (see Section 2.4.2), because its crypt-like structures contain a restricted set of cell lineages and do not fully recapitulate the populations within the normal colonic crypt (Hayward and Whitehead 1992). Cell lines contain heterogeneous populations at various stages of differentiation (see Section 2.4.4). These populations can display lineage plasticity, reverting to a more potent state or undergoing transdifferentiation to a different lineage (see Section 2.4.5). Thus, some cell lines (which would normally be classed as "histotypic") can give rise to organoids under the correct culture conditions.

The traditional terms continue to be used in this book, but an equal emphasis is placed on the culture format (Shamir and Ewald 2014), which will determine the local microenvironment for the cell populations in the culture and thus regulate their behavior. The choice of culture format will depend on the desired outcome and the technologies that are available to each laboratory. Broadly speaking, culture formats can be divided into scaffold-free and scaffold-based systems, as described in the following Minireview.

27.2 TECHNOLOGIES FOR 3D CULTURE

Minireview M27.1. ***Advances in Technologies Enabling 3D Cell Culture and the Formation of Tissue-Like Architecture In Vitro***   Contributed by Professor Stefan Przyborski, Chair

in Cell Technology, School of Biological and Biomedical Sciences, Durham University, Durham, United Kingdom.

Cells adapt to changes in their microenvironment, which can include responses to biological signals and chemical stimuli, or alteration of their physical surroundings. Each has the potential to significantly alter and regulate cell behavior through changes to cell proliferation, viability, differentiation, structure, or function. A major physical difference that exists in cell culture relates to the shape and geometry that cells experience when grown on conventional flat plastic culture plates. Cells adapt to this unnatural 2D substrate and remodel their structure by adopting a flattened cellular morphology. Cells immediately become polarized against the plastic and often lose their natural architecture. Such changes have been shown to alter gene expression, protein synthesis, and thus function (Thomas et al. 2002; Vergani et al. 2004). Conventional 2D cell culture models therefore poorly represent the structure and function of their native counterparts in real tissues. This has a significant impact on cell performance and the outcome of biological assays. For example, 2D culture may result in enhanced proliferation and reduced differentiation (Cukierman et al. 2002) or alter the response to cytotoxic reagents (Sun et al. 2006). It is widely accepted, therefore, that a more appropriately designed cell culture environment could improve the ability of cells to function more normally and would benefit our understanding of cell biology in health and disease.

It has been recognized for many years that culturing cells in 3D can lead to enhanced cell structure and function and improve the physiological relevance of *in vitro* assays. This is achieved by (i) allowing individual cells to establish their normal 3D structure and function with minimal exogenous support and interference; (ii) reducing the stress and artificial responses as a result of cell adaptation to flat, 2D growth surfaces; (iii) encouraging cells to form complex interactions with adjacent cells and to receive and transmit signals; and (iv) providing a physical 3D environment in which cells can spatially organize themselves to form more natural tissue-like structures.

It has only been relatively recently, however, that technologies have become available to enable scientists to practice 3D cell culture routinely. Setting up a 3D model is in general more technically challenging and more expensive than conventional 2D cell culture. While for some this may be considered a barrier against adopting such technology, these issues are outweighed by the benefits of more accurate, predictive, and thus valuable experimental data about cell behavior. Furthermore, 3D cell culture technologies have recently become more widely available and easier to use routinely. Various methods have been developed to meet the growing demand for 3D cell culture, making it easier to practice. It is acknowledged that there is no single solution that satisfies all the needs of 3D cell culture and researchers are required therefore to select the most appropriate model for their cell-based assay.

Technologies for 3D cell culture can broadly be categorized as scaffold-free or scaffold-based culture systems, with scaffolds made from either natural or synthetic materials. The following highlights some of the most popular examples currently used, including 3D aggregate cultures and spheroids, hydrogels, and scaffold-based technologies (see Figure 27.2).

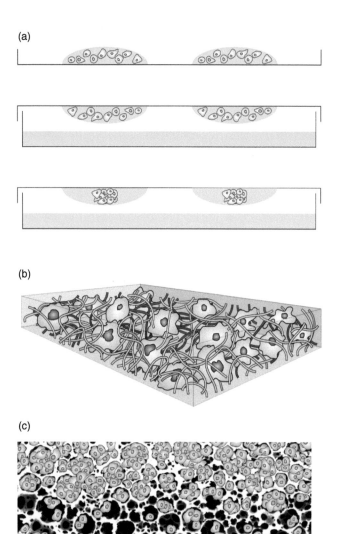

(a)

(b)

(c)

Fig. 27.2. Principal technologies currently used in 3D culture. (a) Formation of spheroids: 3D aggregates of cells develop from a cell suspension maintained in a droplet of medium formed when inverted on the lid of a conventional Petri dish, resulting in cells coming together to form a 3D micro-tissue. (b) Application of hydrogel technology to create a 3D cell construct: cells are shown within a matrix of protein molecules, creating a nano-scale 3D microenvironment mimicking the structure of the ECM. (c) Porous solid scaffold creating a 3D physical microenvironment into which cells can enter and grow to form 3D structures. Cells do not flatten within the scaffold; they retain their more natural 3D shape and form close interactions with adjacent cells, creating 3D tissue-like structures. *Source*: Courtesy of Stefan Przyborski.

Spheroids are often referred to as scaffold-free models that consist of suspended aggregates of multiple cells. Spheroids may consist of a single cell type or multiple different cell types in a co-culture arrangement. Cells produce their own extracellular matrix (ECM) components endogenously and this can result in relatively solid structures of different proportions depending on the cell types used and time in culture. There are different ways in which spheroid cultures can be established, including the classical hanging drop technique and the more recent low adherence surface modified culture plates (see Table 27.1). Inverting the lid of a culture dish produces a hanging drop of culture medium that allows suspended cells to aggregate together in the apex of the drop, forming a 3D spheroid (see Figure 27.2a). This technique has many uses and has often been used for creating embryoid bodies in stem cell research (Keller 1995). More recently, technology has been developed that enables automated and reproducible formation of such droplets for high-throughput applications, particularly in the field of cancer research (Vinci et al. 2012). Low adherence surface modified culture plates such as microfabricated substrates use technology that reduces the adherence of cells to the floor of the culture vessel, resulting in cells growing in suspension (Miyagawa et al. 2011). The suspended cells come together over time, forming multiple 3D aggregates. This is a useful method for producing large amounts of 3D culture material but does not have the same fine control in reproducing aggregate size.

Hydrogels represent a popular option of 3D cell culture through the encapsulation of cells in a hydrogel comprising a loose scaffold framework of cross-linked natural base materials (such as agarose, fibrin, collagen) within an aqueous medium (see Figure 27.2b) (Tibbitt and Anseth 2009). They provide a versatile technology for 3D culture and can be designed to support specific types of cell growth by either trapping cells within an artificial ECM environment (Jongpaiboonkit et al. 2008) or allowing cells to migrate into the interior from the surface (Topman et al. 2013). Components of the ECM may be modified to incorporate biologically active molecules that in turn influence cell behavior directly within the 3D environment. There are essentially two types of technology: (i) natural hydrogels; and (ii) synthetic hydrogels (see Table 27.2). Depending on the nature of the material and method of use, hydrogel cultures enable the growth of cell aggregates, formation of tubules and branching morphogenesis, and construction of layered structures. Mixing alternative cell types and setting up arrangements for co-culture of different cell types is easily achieved.

Solid scaffolds provide a 3D space in which cells can grow, proliferate, and differentiate, allowing them to create more tissue-like 3D structures (see Figure 27.2c). An advantage of inert solid scaffolds is their ability to support 3D culture and produce organized arrangements of cells in a controllable and reproducible fashion using procedures that are more appropriate for routine use (Knight et al. 2011). The open pore structure of a solid scaffold receives cells that then

TABLE 27.1. Alternative aggregate-based technologies.

Technology trade name	Supplier	Description
GravityPLUS™	InSphero	Specially designed culture plate optimized for formation of cellular spheroids
Perfecta 3D® Hanging Drop Plates	3D Biomatrix (now Sigma-Aldrich)	Specially designed culture plate optimized for formation of cellular spheroids
The 3D Petri Dish®	MicroTissues Inc.	Use of micromolds to create agarose wells for the production of cell aggregates
Micro-space Cell Culture Plate	Elplasia (Kuraray Co., Ltd)	Engineered plate surfaces with individual square compartments to encourage spheroid culture
NanoCulture® Plates	Scivax (now MBL International)	Low adherence plates based on nanotechnology that inhibit cell adhesion; thus cells grow as aggregates
Lipidure® Coat	NOF America Corporation	Low cell binding plates and dishes; efficient production of spheroids
BioLevitator™	Hamilton Company	Use of magnetic particles to suspend cell aggregates in media; requires specialized vessels/equipment

Source: Courtesy of Stefan Przyborski.

TABLE 27.2. Alternative hydrogel-based technologies.

Technology trade name	Supplier	Description
Cultrex	Trevigen	Combines spheroid formation ECM and spheroid formation plate resulting in reproducible cell aggregates
Matrigel	Corning	Most popular hydrogel product, derived from mouse sarcoma cells, widely used (see Section 8.3.2)
Extracel line; Hystem line	Glycosan (now BioTime Inc.)	Customized hydrogels for 3D cell culture, defined, growth factor release, physiological, injectable
3D Life Biomimetic Hydrogels	Cellendes	Customized hydrogels for 3D cell culture, defined, control biomimetic modifications, tunable gel stiffness
PureCol	Advanced Biomatrix	Purified collagen gels for tissue engineering and cell culture applications, human and bovine
QGEL 3D Matrix	QGEL	Defined hydrogel technology without RGD motifs (see Section 8.1)

Source: Courtesy of Stefan Przyborski.

enter a physical 3D environment that can be further controlled through surface modification, addition of signaling molecules, mechanical stiffness, and addition of common ECM components as used in conventional cell culture. Commercially available scaffolds enable routine use for 3D culture and overcome the impracticalities and variation associated with user-prepared materials. Synthetic and natural materials can be used to create such scaffolds. Natural or biodegradable materials are useful for tissue engineering applications, while synthetic scaffolds that are inert are more appropriate for *in vitro* assays where experimental variation can be controlled more accurately. Numerous manufacturing methods have

been developed to create such porous materials, each resulting in materials with specific properties and applications to control cell growth in particular ways. For example, thin scaffold membranes are particularly useful for producing tissue-like structures anatomically composed of layers of alternative cell types (see Figure 27.3).

In summary, recent technical advances have provided scientists with new approaches to culturing cells in 3D and creating tissue-like constructs *in vitro*. Such technology in combination with other advancing fields, such as stem cell science, will enable researchers to create more sophisticated *in vitro* models using human cells. This in turn will lead to

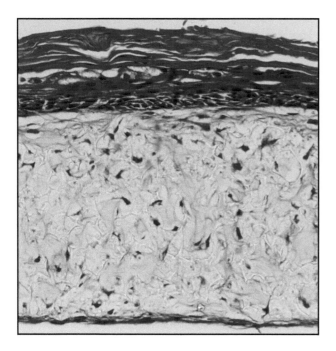

Fig. 27.3. Development of a tissue-like construct from layers of alternative cell types. This example shows an *in vitro* model of human skin grown on a porous scaffold membrane. The construct was produced by first seeding human primary dermal fibroblasts into a 200-μm-thick porous polystyrene scaffold for seven days, followed by seeding human primary keratinocytes onto the surface and maintaining the culture for a further 14 days at the air–liquid interface. As the skin construct matures over time, the keratinocytes form the upper stratified epidermal layer that is supported by the underlying dermal cells. *Source*: Courtesy of N. Robinson, Durham University.

further understanding of molecular and cellular mechanisms in both health and disease, improve the efficiency of the discovery process, and potentially lead to a reduction in animal usage.

27.3 BENEFITS AND LIMITATIONS OF 3D CULTURE

It is clear from Minireview M27.1 that there are significant benefits from 3D culture. While 3D culture has been proposed for many years as a method for enhancing differentiation (see Section 26.3.3), the critical property that these cultures exhibit is the creation of a unique microenvironment or niche. Depending on the matrix and soluble factors applied to the culture, stem cells may be induced to proliferate or to mature into differentiated tissue. Thus, the generation of the correct geometry and cell–cell interaction available in 3D culture may allow that culture to more correctly express a specific phenotype, whether that be a stem, precursor, or differentiated cell.

Advances in technologies have made 3D culture more accessible for use in cell-based assays, e.g. for toxicity screening and drug discovery (Eskes et al. 2017; Langhans 2018). Many of these technologies are also amenable to high-throughput screening. For example, a high-throughput tumor invasion assay can be performed by generating tumor spheroids in microwell plates and measuring invasion using automated image analysis (see Section 27.4.1) (Vinci et al. 2015). Similarly, many 3D technologies can be scaled down for use as microphysiological systems, resulting in organ-on-chip models for the skin, liver, respiratory tract, and other organs (see Section 29.5.2) (Alepee et al. 2014). Although these benefits are increasingly evident, 3D culture systems have limitations to address and (hopefully) to overcome. Limitations of 3D culture systems relate to:

(1) **Diffusion.** Diffusion of nutrient and oxygen into a 3D culture becomes limited as it increases in size, as does outward diffusion of waste products such as lactate and CO_2. This is particularly a concern for organ culture but can affect any form of 3D culture. Reduced diffusion is avoided *in vivo* by the provision of a dense network of capillaries, but limits size *in vitro* to a maximum radius of about 500 μm from the center of the culture. Various solutions exist for this problem, including the fabrication of *in vitro* 3D capillary beds and the use of rotating-wall vessel bioreactors (Chan et al. 2012; Kolesky et al. 2016; DiStefano et al. 2018). In some cases, however, limited diffusion may result in a more physiological model, e.g. for avascular tumors (Verjans et al. 2018).

(2) **Differentiation and maturation.** Organoid or organotypic cultures may fail to fully differentiate or to display mature structures. For example, early studies of retinal organoids established from mouse embryonic stem cells (mESCs) showed that layered retinal structures formed with the correct polarity, but the outer segments failed to form and photoreceptors failed to fully mature to become light-sensitive (Eiraku et al. 2011). Although differentiation was improved in later studies (e.g. using rotating-wall vessel bioreactors to improve diffusion), retinal development did not proceed past the equivalent of postnatal day 6 (DiStefano et al. 2018).

(3) **Characterization.** It can be challenging to monitor growth and viability within a 3D culture or to analyze its response to experimental variables. Confocal microscopy can help to provide a 3D image and will register some changes, e.g. using fluorescence. New techniques are emerging for the characterization of fixed and living 3D constructs that will be discussed in a later section (see Section 27.9).

(4) **Reproducibility.** Data from organoids and other 3D culture formats can be difficult to reproduce (Huch et al. 2017). For example, success rates vary for induction of cerebral organoids even when the same procedure is used by the same laboratory, ranging from 30% to 100% of organoids per batch (Lancaster et al. 2017). Analysis of these cultures showed that they were

virtually indistinguishable at the embryoid body stage but developed variation during neural induction. Variability was believed to be due to the physical conformation of the organoid (low surface area to volume ratio) and was reduced by using floating scaffolds to shape the embryoid body.

(5) *Scalability.* Although many 3D culture technologies can be scaled to a 96-well plate format, increasing throughput to 384-well or 1536-well plates can be challenging (Langhans 2018). Procedures must be adapted for automated liquid handling and image analysis, e.g. positioning spheroids in the same focal plane (Vinci et al. 2015).

(6) *Cost and complexity.* Many 3D culture techniques require a significant investment in time and resources. Laboratory personnel may need to work with novel materials and spend time optimizing procedures for their particular tissue type. For example, if working with organoids, personnel may need to become proficient in a series of experimental procedures to induce pluripotency, generate embryoid bodies, and induce differentiation. Each of these stages may require a different substrate, medium formulation, and method of dissociation.

(7) *Ethics.* Ethical requirements are already in place when working with tissue and cells, including stem cells (see Sections 6.4, 23.2.2). However, as 3D cultures become more complex and develop a closer resemblance to the living organism, new ethical questions will arise (Munsie et al. 2017). For example, cerebral organoids mimic many of the structures of the human cerebral cortex *in vitro*. Despite the common description of these structures as "minibrains," they are incomplete models that do not form mature neural networks and cannot think. However, what will happen if cerebral organoids become more sophisticated and are used in networks of neural structures? It seems likely that this will result in ethical challenges, beginning with the need to communicate such research for informed consent.

27.4 SCAFFOLD-FREE 3D CULTURE SYSTEMS

27.4.1 Spheroid Culture

The term "spheroid" refers to multicellular aggregates that form in suspension culture or in microgravity (see Figure 27.4c) (Mueller-Klieser 1987; Sutherland 1988; Riwaldt et al. 2015). Such aggregates are assumed to be spherical, although closer observation shows that this is not

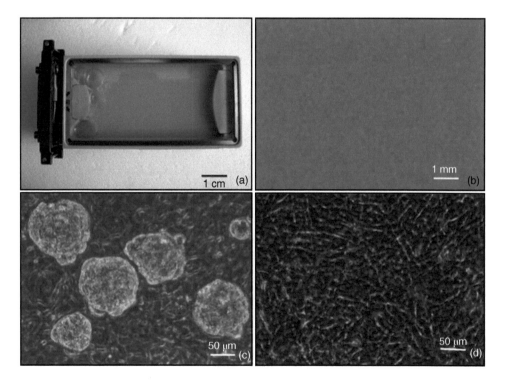

Fig. 27.4. Spheroid formation in microgravity. (a) Cell container suitable for spaceflight, containing human follicular thyroid carcinoma cell line FTC-133. (b) Magnified section of the FTC-133 monolayer. (c) FTC-133 cells formed spheroids when incubated at 1 *g* for two days, followed by five days of simulated microgravity using a random positioning machine. (d) FTC-133 control culture; cells were kept at 1 *g* outside the random positioning machine, resulting in continued growth as a monolayer culture. Scale bars as indicated. *Source*: Riwaldt et al. (2015), reproduced with permission of Wiley.

always the case (Kenny et al. 2007; Breslin and O'Driscoll 2016). Spheroids can be generated from tumor cell lines, xenografts, or (more rarely) primary cultures of normal or tumor cells (Sutherland 1988; Reynolds and Weiss 1992). Some cell lines form spheroids under standard culture conditions; for example, ovarian carcinoma cell lines may grow as adherent monolayers but develop spheroid "buds" that are released into suspension (Pease et al. 2012). However, many cultures that are normally adherent may be induced to form spheroids using conditions that favor cell–cell over cell–matrix interactions, e.g. "forced floating" conditions that prevent substrate adhesion. Some popular methods are discussed here, but there are numerous techniques for spheroid formation that are reviewed elsewhere (Lin and Chang 2008; Cui et al. 2017).

Spheroid culture is commonly used to enhance stem cell behavior and to isolate or enrich stem cell populations. Spheroid culture of mesenchymal stromal cells (MSCs) results in delayed senescence and increased expression of pluripotency-associated genes such as *NANOG*, *SOX2*, and *POU5F1* (see Table 23.2) (Cheng et al. 2013). Adult stem cells within the mammalian central nervous system (CNS) were originally isolated using spheroid culture; the technique continues to be effective for neural stem cells, giving rise to the neurosphere assay (Reynolds and Weiss 1992; Azari et al. 2011). Spheroid formation is particularly popular as a surrogate assay for cancer stem cells (CSCs), although there are some limitations to its effectiveness; for example, the assay culture conditions may not be sufficient for self-renewal of some cell populations (Pastrana et al. 2011; Valent et al. 2012). Tumor spheroids have been widely used as models for tumor cell migration, invasiveness, and responses to cytotoxic treatments (see Section 25.4.4).

Transition to a 3D spheroid alters the morphology and other characteristics of the culture. Cells usually display reduced viability and drug sensitivity, and undergo profound changes in gene expression (Lund-Johansen et al. 1989; Cesarz and Tamama 2016). These changes probably relate to the culture conditions used (Ylostalo et al. 2014), the formation of layers within the spheroid, and the presence of limited diffusion and hypoxia. Spheroids typically consist of (1) an outermost layer, which contains the proliferating cells (see Figure 27.5); (2) a middle layer, which contains cells that are viable but mostly quiescent; and (3) an innermost layer, which receives less oxygen and nutrients. This innermost core becomes necrotic as the spheroid grows in size (>500 μm diameter) (Lin and Chang 2008; Vinci et al. 2012).

Spheroid microplates. The simplest technique for spheroid generation is to plate cells on a non-adhesive substrate in close proximity to one another. A non-adhesive substrate can be achieved by adding a base layer of agar, agarose, or poly(2-hydroxyethyl methacrylate (pHEMA) and adding cells as a "liquid overlay" (Yuhas et al. 1977; Tevis

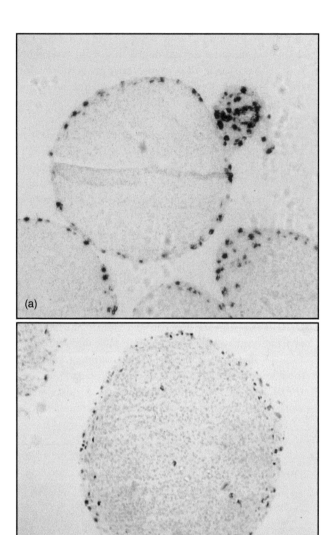

Fig. 27.5. Dividing cells in spheroids. Sections through mature spheroids of approximately 600–800 μm diameter. (a) Autoradiograph labeled with [3H]thymidine, showing restriction of label to periphery; (b) immunoperoxidase staining with an antibody directed against 5′-bromo-2′-deoxyuridine (BrdU), showing similar restriction of label to periphery. *Source*: Courtesy of Ali Neshasterez.

et al. 2017). A protocol to generate spheroids using an agar coating is provided in the Supplementary Material online (see Supp. S27.1). "Spheroid microplates" are now available from commercial suppliers that offer improved standardization; wells usually have a round bottom, allowing cells to settle at close proximity under gravity (see Figure 27.6; see also Table 27.1). These plates can be scaled up to perform rapid, reproducible spheroid culture for cell-based assays. Protocol P27.1 describes how to use a spheroid microplate to generate tumor spheroids and embed the spheroids for further applications.

Fig. 27.6. Spheroid microwell plate. Ultra-low attachment (ULA) plate (Corning), turned over to show round-bottomed wells; these are designed to reduce attachment and ensure that only one spheroid is formed per well. The interior of the well is coated with non-adhesive material to discourage attachment. *Source*: Amanda Capes-Davis; microwell plate provided by Corning.

PROTOCOL P27.1. TUMOR SPHEROID FORMATION AND EMBEDDING

Contributed by Maria Vinci and Suzanne Eccles, The Institute of Cancer Research. Dr Vinci's present address is Bambino Gesù Children's Hospital, Rome, Italy.

Background

The aim of this procedure is to generate tumor spheroids that have a reproducible size and are seeded at a comparable focal plane, with one spheroid per well. The resulting spheroids can be used in assays to assess tumor invasion for drug evaluation or other applications. The full procedure is described in several references, including a video procedure (Vinci et al. 2012, 2013, 2015).

Outline

Subculture the cells and plate at optimal density in an ultra-low attachment (ULA) 96-well plate. Incubate spheroids for four days. Embed spheroids in an ECM coating (e.g. Matrigel®).

Materials (sterile or aseptically prepared)

☐ Tumor cells, e.g. U-87 MG cell line
☐ Complete growth medium, e.g. DMEM +10% fetal bovine serum (FBS)
☐ ECM coating, e.g. Matrigel (Corning)

☐ ULA 96-well round bottom plate (Corning)
☐ Universal containers or centrifuge tubes
☐ Reservoir for pipetting
☐ Microcentrifuge tubes, prechilled (for Matrigel)
☐ Pipettor tips, prechilled (for Matrigel)
☐ Needle

Materials (non-sterile)

☐ Biological safety cabinet (BSC) and associated consumables (see Protocol P12.1)
☐ Cell counting equipment (hemocytometer or automated counter; see Section 19.1)
☐ Pipettors (including P10-P1000 and multichannel pipettors; see Figure 5.3)

Procedure

A. Spheroid Formation

1. Perform subculture of tumor cell culture using standard procedure (e.g. trypsinization; see steps 1–15, Protocol P14.2).
2. Transfer the cell suspension to a centrifuge tube and centrifuge at 500 g for five minutes.
3. Remove the supernatant, tap the tube, and resuspend in 1 ml complete growth medium using a P1000 pipettor. This should yield a single-cell suspension.
4. Count the cells with a hemocytometer or an automated cell counter.
5. Dilute the cell suspension to a suitable cell density (e.g. 0.5×10^4 cells/ml for U-87 MG) using complete growth medium.
6. Transfer the diluted cell suspension into a sterile reservoir.
7. Using a multichannel pipette, dispense 200-µl aliquots of the cell suspension into the ULA 96-well plate.
8. Transfer the plate to an incubator (37 °C, 5% CO_2, 95% humidity) and incubate for four days.
9. Assess spheroid formation after four days before proceeding to further use.

B. Spheroid Embedding

10. One day before embedding commences, place Matrigel at 4 °C to thaw overnight and check that handling consumables (tips, tubes) are prechilled.
11. On the day of the procedure, place the plate containing the four-day-old spheroids on ice.
12. Using a multichannel pipette, gently remove 100 µl/well of growth medium from each well of the ULA 96-well plate. Angle the tip toward the

inside wall of each round-bottom well, avoiding contact with the location of the spheroid.

13. Using prechilled tips, gently add 100 μl Matrigel to each well of the ULA 96-well plate. Angle the tip toward the inside wall of the well, avoiding the location of the spheroid.

14. Use a sterile needle to remove any bubbles that are present.

15. Use a microscope to assess the positions of the spheroids. If positions are not optimal, centrifuge the plate at 300 g for three minutes at 4 °C.

16. Transfer the plate to an incubator (37 °C, 5% CO_2, 95% humidity). Incubate for one hour until the Matrigel has solidified.

17. Using a multichannel pipette, gently add 100 μl/well of complete growth medium.

Notes

Step 5: The optimal cell density is expected to be in the range $0.5–2 \times 10^4$ cells/ml, but must be determined for each cell line (Vinci et al. 2012, 2013). The culture medium and ECM components may also need to be optimized.

Step 9: Four days after seeding, spheroids should have formed and should be approximately 300–500 μm in size.

Step 10: For invasion or chemosensitivity assays, reagents can be added to the Matrigel (2 × final concentration) before it is dispensed into the 96-well plate at step 13. Alternatively, reagents can be added to the complete growth medium (3 × final concentration) at step 17. Use prechilled tips when adding reagents to Matrigel.

Step 13: This step is the most critical for optimal image analysis. Spheroids must remain undisturbed in the center of the well or they will not stay in the same focal plane.

Hanging drop culture. One of the earliest tissue culture techniques, the hanging drop (see Table 1.1), is now staging a comeback for spheroid culture. While the original technique was used to culture explants and observe outgrowth at the glass–liquid interface (held in position by a plasma clot), hanging drops are now used primarily for reaggregation of dissociated cells. Hanging drops can be generated very simply using a Petri dish (Foty 2011). The cell suspension is dropped onto the lid of the dish, which is then inverted and placed onto the base of the dish; phosphate-buffered saline (PBS) is added to the base, which acts as a hydration chamber to prevent the drops on the lid from drying. Alternatively, a hanging drop plate can be used, e.g. the Perfecta3D hanging drop plate (Sigma-Aldrich; see Table 27.1). Spheroid culture can

be scaled up using a 384-well hanging drop plate, allowing the use of automated liquid handling systems; a similar principle is used in some microfluidic devices (Hsiao et al. 2012; Aijian and Garrell 2015). Different plates may result in subtle differences in spheroid formation. For example, a direct comparison of two commercial plates showed that more cells were needed to form a spheroid in the optimal size range (300–500 μm) for the Perfecta3D plate compared to the ULA spheroid microplate (Amaral et al. 2017).

27.4.2 Dynamic Culture Systems

It was recognized some years ago that growing cells in suspension in a rotating chamber can favor aggregation into spheroids (Moscona 1961). Cells which are normally adherent may attach to each other if they are not able to attach to a stationary substrate. Hybridomas will also tend to form aggregates under these conditions. Moscona's discovery was made using Erlenmeyer flasks on a rotary shaker at around 70 rpm. Shaker or spinner culture continues to be used for spheroid generation and can be scaled up to large stirred tank bioreactors; scale-up systems are discussed in a later chapter (see Section 28.2). However, other dynamic culture systems can be used for spheroid formation and more broadly for tissue engineering.

Simulation of microgravity. Cell lines have been exposed to microgravity in orbit since the early years of spaceflight. Approximately 50 cell lines have flown in space, including the HeLa cell line, which accompanied Yuri Gagarin into orbit in Vostok 1 (Dickson 1991; Bairoch 2019). More recent experiments have shown that spheroids and other 3D structures are formed in microgravity (see Figure 27.4) (Riwaldt et al. 2015; Kruger et al. 2019). Unfortunately, few scientists have the opportunity to send an experiment into space and it is necessary to simulate microgravity on the ground. Devices that can be used to simulate microgravity include (Herranz et al. 2013; Grimm et al. 2014, 2018):

(1) **The fast-rotating clinostat.** This device continuously rotates samples to prevent sedimentation, usually in a roller tube or similar vessel (see Section 28.3.1).

(2) **The miniPERM® bioreactor (Sarstedt).** This consists of a rotating two-compartment system (see Figure 27.7). The cell suspension is limited to one compartment, separated from the medium compartment by a permeable membrane. The device can be used as a bioreactor for protein production; the product accumulates in the cellular compartment, while nutrients and waste products diffuse across the semi-permeable membrane to and from the medium compartment. Medium can be sparged or replaced without disturbing the cells or product. This allows the cell concentration to be quite high, as the cells are not diluted by the bulk of the medium. Using a slow rotation speed will

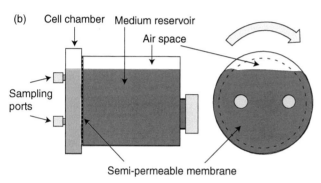

Fig. 27.7. MiniPERM bioreactor (Sarstedt). (a) Photo and (b) sections of the device. The concentrated cell suspension in the left chamber is separated from the medium chamber by a semi-permeable membrane. Products >12.5 kD remain with cells and can be harvested from the sampling ports, while replenishment of the medium is carried out via the right-hand port. Mixing is achieved by rotating the chamber on the Universal Turning Device. *Source*: (a) Courtesy of Sarstedt; (b) R. Ian Freshney.

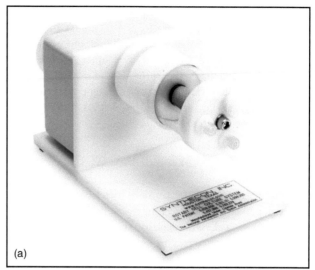

Fig. 27.8. Rotary cell culture system (RCCS) (Synthecon). (a) Reusable cylindrical chamber on RCCS; (b) disposable chambers. Cells are maintained in suspension by adjusting the rotation speed of a cylindrical culture chamber. A gas-permeable silicone membrane core gives the cells ample gas while disallowing shear-causing bubbles. The NASA-designed bioreactor is available from Synthecon Inc., Houston, Texas, and distributors worldwide. *Source*: Courtesy of Synthecon Inc.

tend to favor aggregate formation and this may enhance differentiation.

(3) **The Rotary Cell Culture System (RCCS; Synthecon).** Originally designed by the National Aeronautics and Space Administration (NASA), the RCCS is a rotating vessel that generates very low shear force and turbulence (Begley and Kleis 2000). Gas exchange occurs from the cell-containing cylinder through a concentric silicone membrane core, which also provides rotation (see Figure 27.8). When the rotation stops, the aggregates sediment and the medium can be replaced. The RCCS is available in various formats, including disposable or autoclavable vessels, and can be used for long-term culture of tissue engineering constructs (Grimm et al. 2018).

(4) **The random positioning machine.** This device consists of two frames that are positioned using independent motors, resulting in random changes in the position of the culture (Wuest et al. 2015). Temperature control and CO_2 are typically provided separately, making this a specialized equipment purchase, although a random positioning incubator has been developed (Benavides Damm et al. 2014).

Magnetic levitation. It is possible to position cells above the culture substrate using magnetic levitation. A magnetic nanoparticle assembly is brought together and introduced to the cell, allowing it to become magnetized. Various nanoparticle assemblies can be used; for example, NanoShuttle™ (Greiner Bio-One) consists of iron oxide, gold, and poly-L-lysine, which attaches to the plasma membrane through its surface charge (see Section 8.3.5) (Souza et al. 2010; Haisler et al. 2013). NanoShuttle is supplied in kit form; the spheroid kit comes with a low adhesion plate and an array of magnets that can be positioned below the plate to induce spheroid formation (#655840, Greiner Bio-One). A similar approach can be taken to build more complex 3D structures such as rings, lines, or other patterns (Lin et al. 2008). Magnetic nanoparticles can be used to generate "Janus" spheroids, which contain different structures or zones, and to generate co-culture models (Jaganathan et al. 2014; Mattix et al. 2014).

27.5 SCAFFOLD-BASED 3D CULTURE SYSTEMS

The culture systems that have been described thus far are heavily dependent on cell–cell interactions. However, interactions between cells and their matrix are equally important. Cultured cells are heterogeneous and will evolve depending on their handling and growth conditions (see Sections 3.2–3.4); their behavior is regulated by the microenvironment in a dynamic and reciprocal fashion (Bissell et al. 1982). Scaffold-based 3D culture systems can be used to provide the necessary cell–matrix interactions, usually by mimicking the natural ECM (see Section 2.2.4). Such scaffolds offer structural support and the correct signals to regulate growth and differentiation.

A wide assortment of scaffolds and matrices are used for 3D culture (see Figure 27.9; see also Table 27.2). These are often grouped into "natural" or "synthetic" materials (see Minireview M27.1). Natural materials are a relatively known quantity, although there is a risk that animal-derived materials may carry unknown pathogens. Synthetic materials are often polymers, which provide structural support and the correct physical environment but lack the biological moieties required for cell–matrix interaction (Annabi et al. 2014). More complex materials are now being developed that combine natural and synthetic materials and can form hydrogels that are optimized for specific tissues or cell types. The term "hydrogel" refers to a cross-linked network of polymers that contains a high water content (Tibbitt and Anseth 2009). Rationally designed hydrogels are increasingly used for specialized applications (e.g. tissue engineering) and will no doubt become more accessible to tissue culture laboratories in the next decade.

How do you choose a 3D culture system and which scaffold should you use? Key considerations when choosing a scaffold include (O'Brien 2011; Ruedinger et al. 2015):

(1) *Cell behavior.* The cell type and phenotype of the culture will determine which 3D systems and scaffolds are most effective. For example, normal breast epithelial cells form duct-like structures in scaffold-based systems (see Section 24.2.8), while breast carcinoma cells fail to form organized, polarized structures, and tend to form spheroids in suspension (Breslin and O'Driscoll 2016).

(2) *Biocompatibility.* Cells must be able to adhere, interact, and migrate normally on the material, which should be non-toxic and should not trigger an immune response.

(3) *Biodegradability.* For tissue engineering, the material should be amenable to *in vivo* degradation and replacement without generating toxic byproducts.

(4) *Scaffold architecture.* Various scaffold formats can be used including fibers, meshes, tubes, beads, and gels (see Figure 27.9). At the molecular level, the type of cross-linking, the pore structure, and the pore size of the material are all important.

(5) *Physicochemical properties.* Important characteristics include the elasticity or stiffness of the material, the method used for hydrogel crosslinking, and the presence of cell adhesion ligands such as the RGD motif (see Section 8.1; see also Figure 2.3) (Kshitiz et al. 2012; Li, Sun et al. 2018).

(6) *Tunability.* Tunable materials can vary in their physical, chemical, or biological properties, as determined by the user. For example, tunable hydrogels have been developed that degrade in response to light, allowing patterning, or separation of cell populations (Kloxin et al. 2010; Tamura et al. 2014). Tunable hydrogels can be used to customize the microenvironment for a specific application, such as expansion of intestinal stem cells or formation of organoids (Gjorevski and Lutolf 2017).

27.5.1 Overlay, Embedding, and Encapsulation

Primary cultures and cell lines may be able to form 3D structures if grown within a layer of scaffold material or on top of it. These two approaches have been beautifully demonstrated by Mina Bissell and colleagues, who developed a 3D method to grow normal breast epithelia using a layer of laminin-rich ECM derived from the Engelbreth-Holm-Swarm (EHS) mouse sarcoma, e.g. Matrigel (see Section 8.3.2). Cells are embedded in the ECM or grown on top of a thinner layer of ECM, overlaid with a dilute mixture of ECM and growth medium (described by Bissell as the "3D on-top assay") (Petersen et al. 1992; Lee et al. 2007). Growth on a layer of ECM is sometimes referred to as "2.5D" (i.e. less than the full 3D experience), but this approach can be both convenient and cost-effective. If 3D structures form on top of the ECM layer, they are usually more accessible for molecular analysis and are positioned in a comparable focal plane for imaging (Lee et al. 2007; Shamir and Ewald 2014). Alternatively, structures can be formed in suspension and then embedded for further study (see Protocol P27.1).

Collagen gel (native collagen, as distinct from denatured collagen coating) has been widely used as a matrix for the morphogenesis of primitive epithelial structures. The kidney epithelial cell line MDCK responds to paracrine stimulation from fibroblasts by producing tubular structures, but only in collagen gel (Kenworthy et al. 1992). Neurite outgrowth from sympathetic ganglia neurons growing on collagen gels follows the orientation of the collagen fibers in the gel (Ebendal 1976). Collagen gel is an important constituent of many organotypic models of skin and for demonstration of

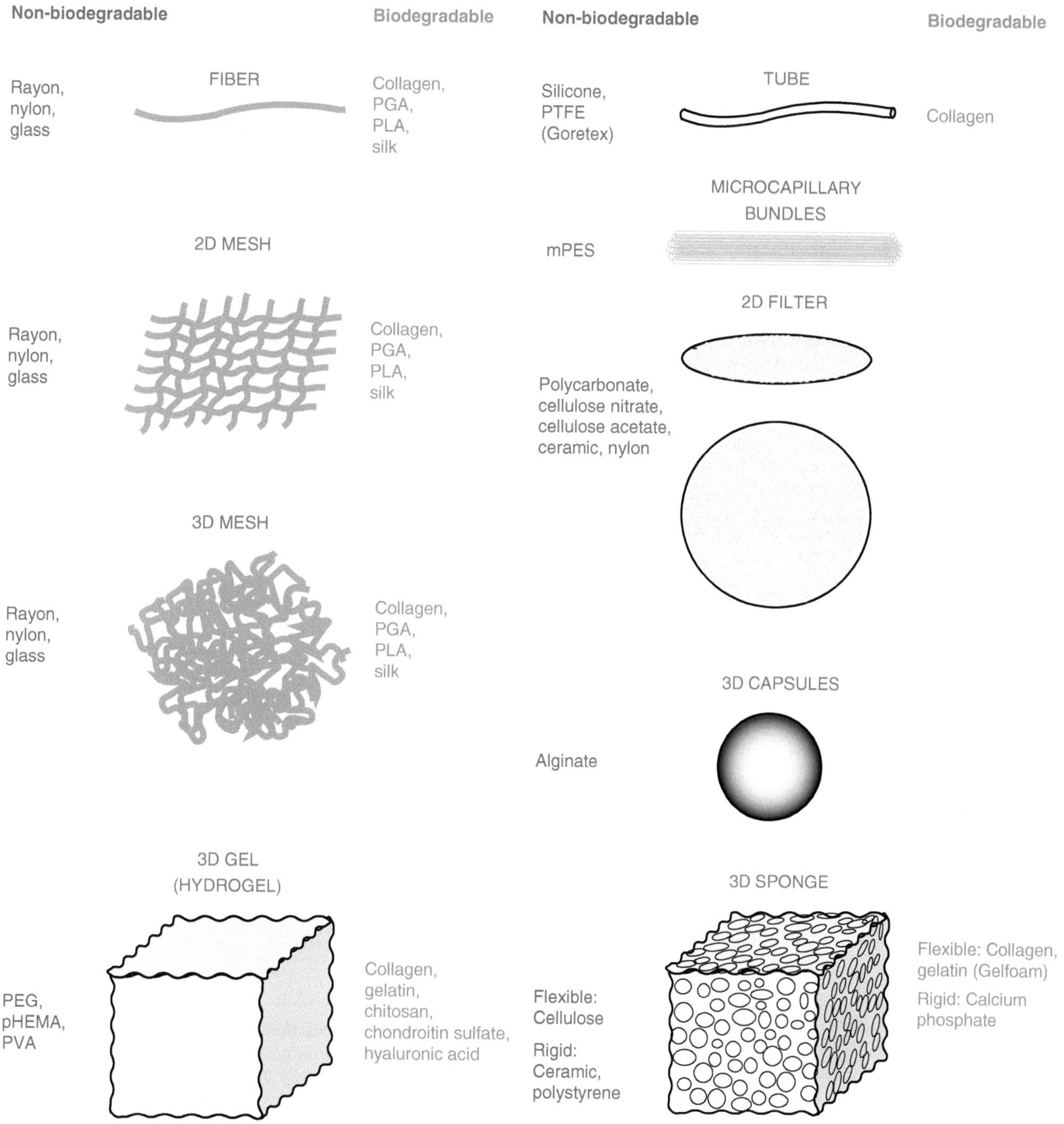

Non-biodegradable   **Biodegradable**   **Non-biodegradable**   **Biodegradable**

Rayon,   FIBER   Collagen,   Silicone,   TUBE   Collagen
nylon,   PGA,   PTFE
glass   PLA,   (Goretex)
                               silk

                                              MICROCAPILLARY
                                              BUNDLES

                                              mPES

2D MESH   2D FILTER

Rayon,   Collagen,   Polycarbonate,
nylon,   PGA,   cellulose nitrate,
glass   PLA,   cellulose acetate,
               silk   ceramic, nylon

3D MESH

Rayon,   Collagen,
nylon,   PGA,
glass   PLA,
               silk

                                              3D CAPSULES

                              Alginate

3D GEL   3D SPONGE
(HYDROGEL)

                                                             Flexible: Collagen,
                                                             gelatin (Gelfoam)

PEG,   Collagen,   Flexible: Rigid: Calcium
pHEMA,   gelatin,   Cellulose   phosphate
PVA   chitosan,
             chondroitin sulfate,   Rigid:
             hyaluronic acid   Ceramic,
                                       polystyrene

Fig. 27.9. Scaffolds and matrices. Many different geometries have been employed, including linear fibers or tubes (top), 2D mesh screens or filters (center), and 3D cubes or spheres (bottom). Although non-biodegradable materials have been used (left-hand labels in each column), the trend is toward biodegradable scaffolds (right-hand labels in each column) and bioengineered matrices (not shown). Abbreviations not in main text: mPES, modified polyethersulfone; PEG, polyethylene glycol; PTFE, polytetrafluoroethylene; PGA, polyglycolic acid; PLA, polylactic acid; PVA, polyvinyl alcohol.

angiogenesis (see Section 27.7.1) (Ment et al. 1997; Stark et al. 2004). The physicochemical properties of the gel can be altered by varying the collagen content (Charoen et al. 2014).

Alginate may be used as a matrix to encapsulate cells for *in vivo* delivery. Encapsulation allows a high level of cell–cell interaction, may facilitate differentiation, and has the potential to provide immune protection and facilitate

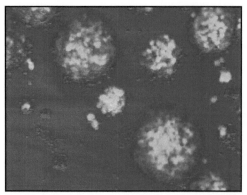

(a) Transfected mosaic spheroids. Cells are derived from the human glioma cell line MOG-G-UVW. The spheroids, ranging in size from 100 to 500 mm diameter, are composed of mixtures of cells transfected with green fluorescent protein (GFP; green) and cells transfected with the noradrenaline transporter gene (NAT; red). Source: Courtesy of Marie Boyd and Rob Mairs.

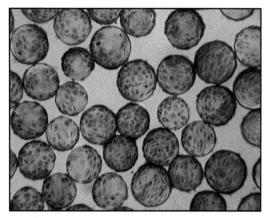

(c) Microcarriers. Vero cells growing on microcarriers. Source: Courtesy of MP Biomedicals.

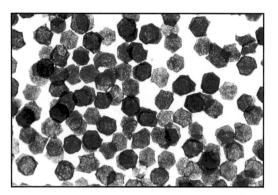

(d) Hexagonal Microcarriers. Nunc Microhex Beads (now discontinued). Source: Courtesy of Nunc.

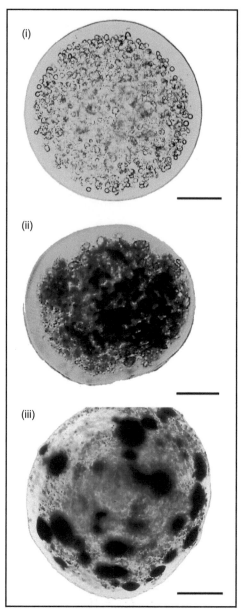

(b) Alginate Encapsulation. Light microscopic images of cells encapsulated in alginate after: (i) 2 hours; (ii) 3 weeks; and (iii) 4 months in vitro. Within the alginate beads, both cell death and cell proliferation will occur, and for many cell lines multicellular spheroids will form inside the beads. The scale bars represent 70 μm. Source: Courtesy of Tracy-Ann Read and Rolf Bjerkvig.

Fig. 27.10. Spheroids, encapsulation, and microcarriers.

scale-up for cell-based therapy. Culture in alginate beads can be used to maintain chondrocyte differentiation, where the matrix is similar to that of native cartilage (see Supp. S24.14; see also Figure 27.10b). Alginate encapsulation has been used to deliver antibody-producing hybridoma cells and insulin-producing pancreatic islet cells in animal models (Dubrot et al. 2010; Bochenek et al. 2018).

27.5.2 Filter Well Inserts

Filter membranes have been used for 3D culture since the 1950s, when 3D structures were induced by growing mouse epithelial cells on top of the membrane, with autologous mesenchymal cells grown below it (Grobstein 1953). The modern filter well insert provides a useful environment to study cell interaction, stratification, polarization, and tissue modeling. Cell populations may be grown on top of the filter (on the membrane itself or on a secondary coating) or beneath the filter (see Figure 27.11). Different cell populations may be grown together or physically separated by the membrane, allowing diffusion of regulatory factors or migration of cells, depending on the porosity of the filter. Filter well inserts allow cells to be maintained at very high, tissue-like densities, with ready access to medium and gas exchange, but in a multi-replicate form. Inserts are easily handled, and cells may be readily viewed under the microscope (see Figure 27.12).

Although initially used for 3D culture, filter well inserts now have a broad range of applications in various fields, including:

(1) **Development of a polarized, functional epithelium.** Some epithelial cells (e.g. MDCK and Caco-2) and endothelial cells (e.g. from umbilical vein) become polarized and form tight junctions several days after reaching confluence on filter membranes (Cereijido et al. 1978; Sambuy et al. 2005). This process can be accelerated by adding a collagen coating. The resulting polarized monolayers can be used for live-cell imaging and many other techniques (Wakabayashi et al. 2007).

(2) **Analysis of permeability.** Once a functional epithelium has formed, transepithelial permeability is restricted to physiologically regulated transport through the cells and pericellular transport falls to near zero. Permeability can be monitored by measuring transepithelial electrical resistance (TEER) or by looking at the transfer of dyes or labeled molecules across the membrane (e.g. Lucifer Yellow or [^{14}C]methylcellulose) (Hidalgo et al. 1989).

(3) **Analysis of polarized transport.** The addition of labeled glucose or amino acid to the upper compartment of the filter well insert will show unidirectional transport to the lower compartment. If the cells possess P-glycoprotein or some other efflux transporter, cytotoxins (e.g. vinblastine) added to the lower compartment will be transported to the upper compartment, but not vice versa.

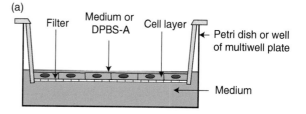

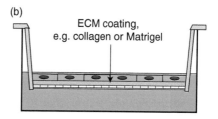

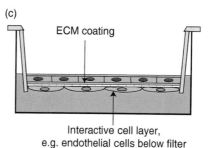

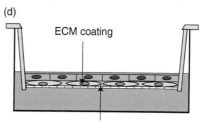

Fig. 27.11. Filter well inserts. Sectional diagram of a hypothetical filter well insert showing: (a) cells grown on top of the filter; (b) cells grown on top of an ECM coating; (c) interactive cell layer added to the underside of the filter; (d) interactive cell layer added to the matrix. Images of cells in these conformations are shown elsewhere (see Figures 27.12, 27.15, 29.12).

(4) **Analysis of cell migration and invasion.** Migration assays can be performed by adding a chemoattractant (e.g. conditioned medium from NIH 3T3 cells); after a period of incubation, cells are stained with crystal violet and counted to determine how many have migrated across the filter toward the chemoattractant (Justus et al. 2014). Invasion assays can be performed by adding a layer of ECM, requiring cells to invade through this barrier before they migrate toward the chemoattractant. Alternatively, a confluent layer of cells may be established as a barrier to invasion, using normal fibroblasts, keratinocytes, endothelium, or other cell types.

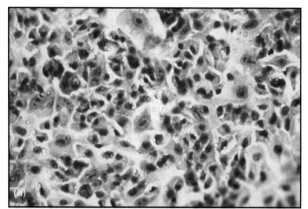

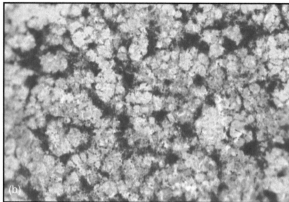

(a) Culture on Filter Well Inserts. A549 human lung adenocarcinoma cells grown on a polycarbonate filter. Holes in the filter (15 μm) are visible, particularly in top right-hand corner space. Giemsa stained; 40x objective.

(b) Human Fetal Lung Epithelial Cells. Cells growing on a filter with human fetal lung fibroblasts on the other side of the filter. Giemsa stained; 4x objective.

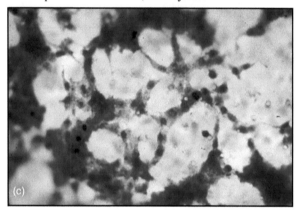

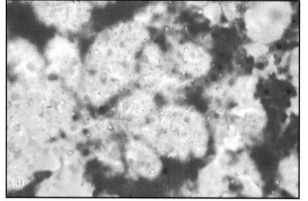

(c,d) Lung Cell Co-culture. Fetal lung epithelial cells and fibroblasts as in (b) but shown at a higher power. Images are focused: (c) above the filter; and (d) below the filter. Giemsa stained; 40x objective.

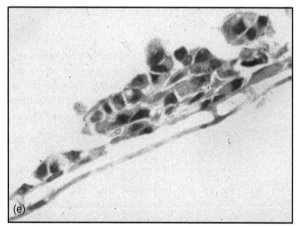

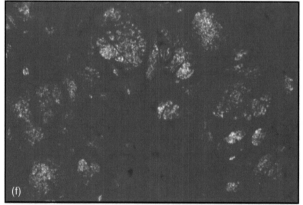

(e) Section of Lung Cell Co-culture. Section through filter with A549 cells and fibroblasts. In this experiment, both cell types were seeded on top, but fibroblasts migrated through the filter first before the A549 cells. Three cells are seen travelling through the pores. H&E stain; 40x objective.

(f) Chick Embryo Lung. High density primary culture of cold trypsin disaggregated 13-day embryonic chick lung. Cells are shown 2 h after seeding at >1 x 106 cells/mL. Giemsa stained; 40x objective. Source for all images: R. Ian Freshney.

Fig. 27.12. Organotypic culture in filter wells.

TABLE 27.3. Types of filter well inserts.

Supplier	Name[a]	Format	Material	Pore size (μm)[b]
Corning	Transwell®	6-, 12-, 24-well plates; 100-mm dish	PC, PET, collagen-coated PTFE	0.4–8.0
	HTS-Transwell	24-, 96-well plates	PC, PET	0.4–8.0
	FluoroBlok	24-, 96-well plates	PET	3.0–8.0
Falcon (Corning)	Falcon Insert System	6-, 12-, 24-, 96-well plates	PET	0.4–8.0
Greiner Bio-One	ThinCert™	6-, 12-, 24-well plates	PET	0.4–8.0
Merck Millipore	Millicell® Hanging Inserts	6-, 12-, 24-well plates	PET	0.4–8.0
	Millicell Standing Inserts	6-, 24-well plates	PC, PTFE, MCE	0.4–12.0
	Millicell Organotypic Insert	6-well plate	PTFE	0.4
Sigma-Aldrich	CellCrown™	6-, 12-, 24-, 48-, 96-well plates	PC, PET, nylon	1.0–8.0

[a]Information here is summarized from supplier websites; always check with the supplier regarding specific products.
[b]Low-porosity filters have a high pore frequency and are likely to be translucent rather than fully transparent.
MCE, Mixed cellulose esters; PC, polycarbonate; PET, polyethylene terephthalate; PTFE, polytetrafluoroethylene.

(5) **Enhanced differentiation.** Some types of cells appear to differentiate better when positioned at the air–liquid interface in filter well inserts (see Section 26.3.6).

(6) **Organotypic culture.** Different cell types can be brought together on a filter well to form more complex structures (see Section 27.7).

Filter well inserts can be purchased in a wide variety of formats, materials, and pore sizes (see Table 27.3). Pore sizes typically range from 0.4 to 8.0 μm (Corning 2016). Smaller pore sizes (e.g. 1.0 μm) allow cell interaction and contact without transit across the filter, while larger pore sizes (e.g. 8.0 μm) allow live cells to cross the filter. Always test the filter material and pore size in your assay before committing to a large purchase (Harouaka et al. 2013). Protocol P27.2 is provided as an example, using uncoated Transwell inserts in a 12-well plate (see Figure 27.13). If cells do not grow well on the filter material, precoated inserts are also available, e.g. BioCoat™ (Corning).

Fig. 27.13. Transwell permeable supports (Corning). Inserts have been added to a 12-well plate; a single insert has been placed alongside the plate. *Source*: R. Ian Freshney; plate and filter well inserts provided by Corning.

PROTOCOL P27.2. CULTURE USING FILTER WELL INSERTS

Outline
Seed cells into filter well inserts and culture the cells in excess medium in multiwell plates.

Materials (sterile or aseptically prepared)

☐ Cell line, e.g. Caco-2
☐ Growth medium (volume will depend on the insert used; refer to supplier instructions)

☐ Filter well inserts, e.g. Transwell, 12 mm diameter, PET with 0.4-μm pore size (Corning)
☐ Multiwell plate, e.g. 12-well plate
☐ Forceps, curved

Materials (non-sterile)

☐ BSC and associated consumables (see Protocol P12.1)

☐ Cell counting equipment (hemocytometer or automated counter; see Section 19.1)
☐ Pipettors

Procedure

1. Subculture the cell line or primary culture to give a cell suspension using standard procedures (e.g. trypsinization; see Protocol P14.2, steps 1–16).
2. Count the cells with a hemocytometer or an automated cell counter.
3. Set up the filter well insert:
   (a) If adding cells to the top of the filter insert: add 1.5 ml medium/well to a 12-well plate and use the sterile forceps to add an insert to each medium-containing well.
   (b) If adding cells to the bottom of the filter insert: use the sterile forceps to add an insert to each well and add the lid. Flip the plate over so that the inserts sit on the lid.
4. Seed the cells onto the filter insert, taking care not to perforate the membrane:
   (a) If adding cells to the top of the filter insert: add 0.5 ml cell suspension to the membrane inside the insert housing.
   (b) If adding cells to the bottom of the filter insert: add the cell suspension to the center of the membrane, allowing the medium to cover the majority of the membrane surface.
5. Place the dish in a humid CO_2 incubator in a protective box (e.g. see Section 12.5.2). It is critical to avoid shaking the box; the cultures should not be moved in the incubator, to avoid spillage and resultant contamination.
6. Check the plate periodically and add fresh medium as required, e.g. three times per week.
7. Cultures may be maintained indefinitely, replacing the medium or transferring the insert to a fresh well every three to five days.

Notes

Step 1: If using Caco-2, subculture cells routinely at 50% confluence to give a more homogenous and polarized monolayer (Natoli et al. 2012).
Step 4: For a long-term experiment using Caco-2 cells and 12-well plates, aim for 0.5 ml per well and a density of 3×10^5 cells/cm^2. The number of cells and the depth of medium will vary with the cell type and application.
Step 5: If adding cells to the bottom of the filter insert, monitor until the cells attach and then invert and set up for continued culture as for the top-loaded cells.

Step 6: Monolayers should become established in 3–5 days, although up to 21 days may be required for differentiation (e.g. polarized transport) to become established.
Step 7: Monolayers can be fixed and stained *in situ*. Inserts can be processed in a multiwell plate or membranes can be removed using a scalpel. Some membranes are fragile and require very careful handling, using a wetted cellulose-based membrane as a support (Corning 2007). Check that the membrane material is compatible with any solvents used.

27.5.3 Hollow Fiber Systems

The hollow fiber technique was developed by Richard Knazek and colleagues to provide an environment in which cells can be grown to tissue-like densities (Knazek et al. 1972). Cells are grown on the outer surface of bundles of plastic microcapillaries, which simulate a bed of capillaries. The potential of these systems for adherent cells lies in the creation of tissue-like cell densities, matrix interactions, and the establishment of cell polarity, all of which may be important for post-translational processing of proteins and for exocytosis. It is claimed that cells in this type of high-density culture behave as they would *in vivo*. Choriocarcinoma cells release more human chorionic gonadotropin (hCG) when grown in hollow fiber systems compared to conventional monolayer culture, and colonic carcinoma cells produce elevated levels of carcinoembryonic antigen (CEA) (Knazek et al. 1974; Rutzky et al. 1979). Thus, hollow fiber systems offer a 3D culture environment that is useful for culture scale-up and to study the synthesis and release of biological products (see Section 28.3.4).

27.5.4 Microcarriers and Macrocarriers

Microcarriers and macrocarriers can be used as substrates for anchorage-dependent cells when propagated in suspension (see Figure 27.10c, d). Microcarriers tend to range from 90 to 300 µm in size; the term macrocarrier is used in this book for carriers that exceed 500 µm in size. Microcarriers and macrocarriers are discussed in more detail in the next chapter (see Section 28.3.3) but are included here because they can act as a substrate for 3D culture. Some porous macrocarriers provide a scaffold of gelatin or of diethylaminoethyl (DEAE) groups bound to cellulose, allowing growth within the interstices of the matrix. Other carriers allow growth on the surface of the bead, which can act as a focus for 3D structures. A 3D *in vitro* angiogenesis assay is provided in the Supplementary Material online that uses Cytodex™ 3 microcarrier beads (GE Healthcare Life Sciences; see Supp. S27.2). This assay seeds endothelial cells onto the surface of the beads, which are then embedded in Matrigel (Corning).

The endothelial cells form a monolayer on the beads, which can then form tubes if treated with appropriate factors (see Figure 3.12d).

27.6 ORGANOID CULTURE

As defined earlier in this chapter (see Section 27.1), organoids arise from a stem cell population, which self-organizes through cell sorting and spatially restricted lineage commitment to form a 3D structure containing organ-specific cell types (Clevers 2016). Organoid culture is currently a "hot topic" with an increasing number of applications (Huch et al. 2017; Artegiani and Clevers 2018). Tissue-specific organoids can be generated from primary cultures, PSCs, or adult stem cells, to be used as models for genetic, neoplastic, or infectious diseases. For example, cerebral organoids can be used as a unique model to study human brain development and its disorders; the human brain displays a complex process of neurogenesis that is distinctly different to that seen in laboratory animals such as the mouse (Lancaster et al. 2013). Human organoids can also be used for disease diagnosis and therapeutic decision-making. Primary cultures from cancer patients can give rise to organoid cultures that recapitulate many of the characteristics of the original tumor, for use in personalized drug screening (van de Wetering et al. 2015; Sachs et al. 2018; Driehuis et al. 2019).

Organoid culture has a long history (Simian and Bissell 2017), but the popular fascination with organoids can be traced back to recent advances in stem cell culture. Yoshiki Sasai and colleagues developed a method to culture ESCs that led to the formation of polarized cortical tissues and optic cup structures *in vitro* (Eiraku et al. 2008, 2011). Sasai used spheroid microplates to induce the formation of embryoid bodies (see Sections 23.2, 27.4.1). When cultured in minimal media, these bodies spontaneously differentiated into neuroectoderm; other lineages could be induced by providing physiologically relevant signaling factors (Eiraku et al. 2008). Using a modified approach, Madeline Lancaster and colleagues were able to generate cerebral organoids from mESCs, hESCs, or induced pluripotent stem cells (iPSCs) (Lancaster et al. 2013). Aggregates were embedded in droplets of Matrigel and cultured using spinner or shaker culture to improve nutrient absorption (see Figure 27.14). The resulting organoids could be maintained for up to one year and developed complex 3D structures (see Figure 1.3c) (Lancaster et al. 2013; Sutcliffe and Lancaster 2017). This overall approach – generation of embryoid bodies from PSCs, followed by differentiation into specific lineages – has been used successfully for the lung, kidney, stomach, and other

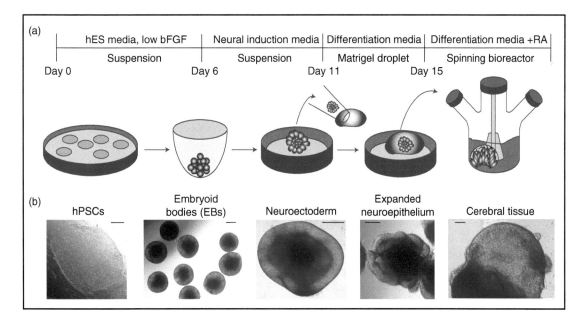

Fig. 27.14. Generation of cerebral organoids. (a) Schematic of the procedure required to generate cerebral organoids. hPSCs are induced to form embryoid bodies in suspension and are transferred to neural induction media, resulting in the development of neuroectodermal tissue. Tissues are then embedded in Matrigel and transferred to a spinning bioreactor, resulting in expanded neuroepithelium and the formation of cerebral tissue. (b) Representative images showing each stage of the procedure (see also Figure 1.3c). *Source*: (a) Lancaster, M. A., Renner, M., Martin, C. A., et al. (2013). Cerebral organoids model human brain development and microcephaly. Nature 501 (7467):373-9. © 2013 Springer Nature. (b) Lancaster et al. (2013), reproduced with permission of Springer Nature.

tissues (Dye et al. 2015; Takasato et al. 2016; Broda et al. 2019).

Further advances in stem cell biology led to the establishment of organoid cultures from adult stem cells. Hans Clevers and colleagues developed a method to establish organoids from intestinal crypts or crypt stem cells; the latter were identified through expression of *LGR5*, which acts as a marker for adult stem cells in many epithelial tissues (Sato et al. 2009; Clevers 2016). To establish epithelial organoids, it is important to (i) activate the Wnt pathway (e.g. using Wnt3a and R-Spondin 1); (ii) activate tyrosine kinase signaling (e.g. using epidermal growth factor [EGF]); (iii) inhibit bone morphogenetic protein (BMP) and transforming growth factor-β (TGF-β) signaling (e.g. using Noggin); and (iv) embed cells in Matrigel (Clevers 2016). If the correct culture conditions are established, intestinal crypt stem cells will give rise to organoids with a characteristic crypt-like appearance (see Figure 1.3a, b) (Sato et al. 2009). This culture system can be used for various tissues and diseases, including prostate carcinoma, breast carcinoma, and metastatic gastrointestinal tumors (Gao et al. 2014; Sachs et al. 2018; Vlachogiannis et al. 2018).

Which method should be used to induce organoid formation? The answer to this question will depend on the samples that you have access to. For example, gastric organoids can be developed from adult stem cells, provided you have access to fresh gastric tissue samples, resulting in explants that can be used as the starting material for organoid formation (Clevers 2016; Broda et al. 2019). If gastric tissue is not available, hPSCs can be used as the starting material and induced to differentiate into various gastric lineages, e.g. epithelial cells from the fundus or antrum of the stomach (Broda et al. 2019). Differentiation into a specific lineage is often a complex process that relies on a good understanding of the developmental pathways that control patterning and morphogenesis for that lineage (see Section 26.5.2) (McCauley and Wells 2017).

27.7 ORGANOTYPIC CULTURE

Organotypic culture systems bring together different cell lineages in a high-density, 3D culture system that is designed to recreate the composition of the original tissue (see Section 27.1). Interactions between different cell types result in signal transduction (via direct cell–cell contact, diffusible paracrine factors, or interaction with the ECM), which in turn gives rise to enhanced differentiation, cell polarity, and the development of a structured microenvironment. Organotypic culture can be created by mixing cells randomly and allowing them to interact; for example, fetal brain cells can be induced to reaggregate if placed at high density above a non-adherent agar coating (Bjerkvig et al. 1986). Alternatively, the construct may be designed to separate different cell

types so that their interactions may be studied, e.g. using filter well inserts (see Section 27.5.2). Organotypic constructs are designed to match the properties of a specific tissue, resulting in (i) an *in vitro* model ("tissue equivalent"); or (ii) an *in vivo* material to replace lost, damaged, or diseased tissue ("tissue engineering").

27.7.1 Tissue Equivalents

The advent of filter well technology, boosted by its commercial availability, has led to the development of some elegant and sophisticated *in vitro* models. This is perhaps best demonstrated by organotypic models of human skin (see Figure 27.15). Skin equivalents can be generated by co-culturing epidermis with dermis or by culturing keratinocytes on a layer of collagen or other suitable substrate; incorporation of dermal fibroblasts into the collagen layer results in enhanced keratinocyte growth and differentiation (Limat et al. 1995; Maas-Szabowski et al. 2002; Stark et al. 2004). Skin and other tissue equivalents can be purchased from commercial suppliers who specialize in tissue models; examples include EPISKIN (EpiSkin™, SkinEthic™), Genoskin (NativeSkin®, HypoSkin®), J-TEC (LabCyte EPI-MODEL), and MatTek Corporation (EpiDerm™, EpiDermFT™). Such models have been developed over many years for use in toxicity testing, e.g. as models for eye and skin irritation (see Section 29.5.1) (Sheasgreen et al. 2009).

A major limitation in creating tissue equivalent constructs is recreating the histology of the original tissue. Where the geometry is simple (e.g. layering epidermal keratinocytes over dermal fibroblasts), this is readily achievable. As the relationship of individual tissue components becomes more complex (e.g. in vascularization or innervation), it becomes more difficult to combine the appropriate cell types with the correct geometry. Recent advances in microfluidics are steadily overcoming this limitation, resulting in the development of devices that can act as microphysiological systems (see Minireview M21.1; see also Section 29.5.2). Bioprinting is likely to provide a further wave of exciting new *in vitro* models (see Section 28.7.3).

27.7.2 Tissue Engineering

Tissue engineering can be defined as a field that applies the principles of engineering and the life sciences to develop biological substitutes that will restore, maintain, or improve tissue function (Vacanti and Langer 1999). Howard Green and colleagues were early pioneers in this area, focusing on skin regeneration (see Section 1.1.2). Green's laboratory established large confluent sheets of epidermis from primary human keratinocytes, which were used to treat patients with burns affecting more than 80% of the total skin surface (Gallico et al. 1984). Approximately 50% of epithelial regeneration in these patients was attributed to cultured

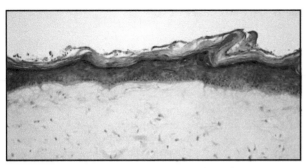

(a) Keratinocyte and Dermal Fibroblast Co-cultures. *Organotypic co-culture after 2 weeks in vitro. H&E stain; 40x objective.*

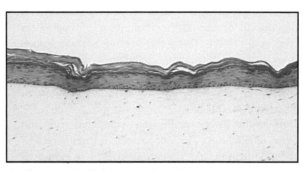

(b) Organotypic Cultures Implanted In Vivo. *Organotypic co-cultures in vivo 3 weeks after transplantation onto the nude mouse. H&E stain; 40x objective.*

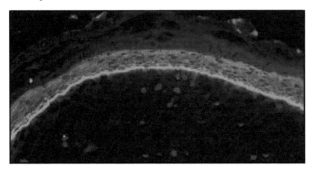

(c) Integrin and Cytokeratin in Co-Cultures In Vitro. *Immunofluorescent staining for α6-integrin (green), keratin 10 (red), and DNA (blue). 40x objective.*

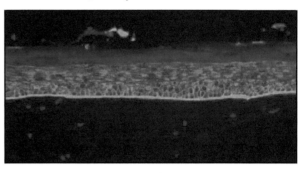

(d) Integrin and Cytokeratin in Organotypic Cultures Implanted In Vivo. *Immunofluorescent staining for α6-integrin (green), keratin 10 (red), and DNA (blue). 40x objective.*

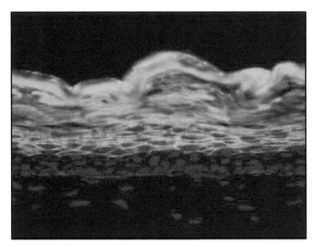

(e) Involucrin and Collagen in Organotypic Co-Cultures at 2 weeks. *Keratinocyte-fibroblast co-cultures with collagen gel stained for involucrin (green) and the basement membrane component collagen VII (red). 40x objective.*

Source for all images: Courtesy of Hans-Jurgen Stark.

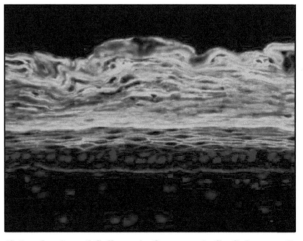

(f) Involucrin and Collagen in Organotypic Co-Cultures at 3 weeks. *Keratinocyte-fibroblast co-cultures with collagen gel stained for involucrin (green) and the basement membrane component collagen VII (red). 40x objective.*

Fig. 27.15. Organotypic culture of skin.

cells (Hynds et al. 2018). Over time, alternative substrates were introduced for keratinocyte culture, e.g. using fibrin (Pellegrini et al. 1999). Tissue engineering is now applied to various tissues and organs, including cartilage, bone, ligament, cardiac and skeletal muscle, blood vessels, liver, and pancreas (Vunjak-Novakovic and Freshney 2006; Lanza et al. 2013; Caddeo et al. 2017).

Constructs for tissue engineering require cell–cell and cell–matrix interactions. This can be seen in the interaction between endothelium and smooth muscle in blood vessel reconstruction (Klinger and Niklason 2006) or between epithelium and smooth muscle in vagina, where it has been possible to reconstruct vagina with normal sexual function (Raya-Rivera et al. 2014). Broadly speaking, the essential requirements for tissue engineering include:

(1) *A source of cells.* Cells must correspond to the correct lineage at a proliferative, precursor stage. Traditionally, cells are obtained from patient tissue samples, but stem cells are an exciting alternative source of material – for example, hPSCs can be induced to differentiate into ventricular cardiomyocytes and maintained in a ventricle-like cardiac organoid chamber (Li, Keung et al. 2018).

(2) *Interactive cells.* Co-culture of different cell types is important to improve the complexity of 3D models, including organoids and organotypic cultures (Holloway et al. 2019). Examples include dermal fibroblasts in skin, smooth muscle cells in blood vessels, and glial cells in neural constructs.

(3) *Scaffold material.* A biodegradable scaffold is necessary to support the structure, using polyglycolic acid (PGA), polylactic acid (PLA), silk, or other materials (see Figure 27.9). Collagen or other ECM components may be required in place of the scaffold for soft tissues or to coat the scaffold to enhance cellular attachment. Matrices must deliver the correct biochemical signals, so it is important to design the matrix to suit a specific tissue type (Gjorevski et al. 2016).

(4) *Mechanical stress.* In addition to biological interactions, some constructs require physical forces. Skeletal muscle needs tensile stress, while bone and cartilage need compressive stress (Mullender et al. 2004; Seidel et al. 2004; Shansky et al. 2006). Vascular endothelium in a blood vessel construct needs pulsatile stimulation (Huang and Niklason 2011).

(5) *Nutrient supply.* Nutrient channels may be provided for the *in vitro* production of large constructs, e.g. cartilage (Nims et al. 2015). However, a lack of vascular tissue represents a major limitation in taking tissue engineering to organ transplantation. Small-diameter vascular grafts can be generated from hPSCs, but some cells within these constructs have differentiated into undesirable lineages, showing that stable lineage commitment is likely to be a concern (Sundaram et al. 2014).

27.8 ORGAN CULTURE

Organ culture is defined here as the culture of the whole organism, organ, or a portion of tissue in its intact form (see Section 27.1). This retains the original structural relationships between cells of the same or different types, and hence their interactive function, in order to study the effect of exogenous stimuli on further development (Lasnitzki 1992). Organ culture was initially used to facilitate histological characterization, but it was discovered that certain elements of phenotypic expression were found only if cells were maintained in close association (Fell 1972). It is particularly useful for complex structures, such as the CNS or the developing embryo. Slices of CNS tissue can be maintained *in vitro* for several months, provided they are (i) prepared at the correct thickness to allow diffusion (usually 100–400 µm); (ii) firmly attached to a suitable substrate; and (iii) provided with a suitable medium and sufficient oxygenation (Gahwiler et al. 1997). For example, a brain slice may be placed at the air–medium interface on a filter well inset, which allows oxygenation from above and supply of nutrients from below the filter membrane. Whole mouse embryos (in a six-day window comprising embryonic days 5.5–10.5) can be cultured in static culture (e.g. a chamber slide) or rotating culture (e.g. a bottle or rube positioned in a rotating drum) (Rivera-Perez et al. 2010). A procedure for chick embryo organ culture is provided in the Supplementary Material online (see Supp. S27.3).

Organ culture can be technically challenging. For example, a study of breast tissue slices found that tumors could only be sectioned into slices after embedding in a supportive matrix such as agarose; slices of normal breast tissue could not be prepared, due to their fatty composition (Holliday et al. 2013). Slices are also difficult to analyze. Biochemical monitoring requires reproducibility between samples, which is less easily achieved in organ culture than in propagated cell lines, because of sampling variation in preparing an organ culture, minor differences in handling and geometry, and variations in the ratios of cell types among cultures. Organ cultures cannot be propagated, and hence each experiment requires recourse to the original donor tissue, making the procedure labor intensive and prone to variation. While organ culture has an important history and continues to have an ongoing role (e.g. for the study of embryogenesis), it is likely to be replaced by organoid and organotypic culture in most applications as these techniques become more widely available.

27.9 CHARACTERIZATION OF 3D CULTURES

While 3D culture offers an exciting opportunity to develop a more physiological culture environment, it is not immune from quality concerns. As organoid cultures become more fully characterized, it has become evident that organoids can be highly heterogeneous and can exhibit strong clonal dynamics (Ben-David et al. 2019). Patient-derived organoids may form from normal cells instead of malignant cells, while PSC-derived organoids may give rise to unintended neural growth instead of the desired cell type (Gao et al. 2014; Broda et al. 2019). Microbial contamination is a risk for 3D cultures, as it is for all forms of cell and tissue culture (see Chapter 16). Cross-contamination may also occur, giving rise to a mixture of cells with different origins (see Chapter 17). It is possible that fewer misidentified cell lines will arise from 3D cultures; misidentification typically occurs when one or more populations in a mixture have a strong growth advantage, which is less likely under the constraints of the 3D environment. However, the close association between different cell types in 3D culture would predispose to other problems, such as spontaneous somatic cell hybridization (see Sections 17.4.3, 22.3.1). Validation testing (see Section 7.4) and broader characterization are essential for 3D cultures.

Imaging and other forms of characterization have traditionally been difficult to perform in 3D culture, where cells form complex structures and may be incorporated into a non-transparent scaffold or spheroid. Sophisticated imaging methods can now be used to visualize organoids and other 3D cultures, including confocal microscopy, two-photon microscopy, and light-sheet microscopy (see Section 18.4.4). High-throughput imaging analysis can be performed using imaging cytometers, such as the ImageXpress® Micro systems (Molecular Devices) and the Celigo Imaging Cytometer (Nexcelom Bioscience). These systems can be used to measure objective characteristics such as spheroid size, invasion into Matrigel, and staining with Hoechst 33342 (for cell nuclei) or Sytox green (for cell death) (Wenzel et al. 2014; Vinci et al. 2015). Such imaging techniques require specialized equipment and software, but if accessible, they offer exciting new opportunities for characterization of 3D models.

Suppliers

Supplier	URL
Corning	http://www.corning.com/worldwide/en/products/life-sciences/products/cell-culture.html
Elplasia (Kuraray Co., Ltd.)	www.elplasia.com
EPISKIN	www.episkin.com
GE Healthcare Life Sciences	http://www.gelifesciences.com/en/it
Genoskin	www.genoskin.com
Greiner Bio-One	http://www.gbo.com/en_int.html
Hamilton Company	www.hamiltoncompany.com
InSphero	www.insphero.com
Japan Tissue Engineering Company (J-TEC)	www.jpte.co.jp/english/index.html
MatTek Corporation	www.mattek.com
MBL International	www.mblintl.com
Merck Millipore	www.merckmillipore.com
MicroTissues Inc.	www.microtissues.com
Molecular Devices	www.moleculardevices.com
Nexcelom Bioscience	www.nexcelom.com
NOF America Corporation	http://www.nofamerica.com/store/index.php
Sarstedt	http://www.sarstedt.com/en/home
Sigma-Aldrich (Merck)	http://www.sigmaaldrich.com/life-science/cell-culture.html
Synthecon	http://www.synthecon.com/pages/default.asp

Supp. S27.1 3D Spheroid Culture Using an Agar Underlay.

Supp. S27.2 *In Vitro* Angiogenesis Assay.

Supp. S27.3 Organ Culture from Chick Embryo.

REFERENCES

Aijian, A.P. and Garrell, R.L. (2015). Digital microfluidics for automated hanging drop cell spheroid culture. *J. Lab. Autom.* 20 (3): 283–295. https://doi.org/10.1177/2211068214562002.

Alepee, N., Bahinski, A., Daneshian, M. et al. (2014). State-of-the-art of 3D cultures (organs-on-a-chip) in safety testing and pathophysiology. *ALTEX* 31 (4): 441–477. https://doi.org/10.14573/altex1406111.

Amaral, R.L.F., Miranda, M., Marcato, P.D. et al. (2017). Comparative analysis of 3D bladder tumor spheroids obtained by forced floating and hanging drop methods for drug screening. *Front. Physiol.* 8: 605. https://doi.org/10.3389/fphys.2017.00605.

Annabi, N., Tamayol, A., Uquillas, J.A. et al. (2014). 25th anniversary article: rational design and applications of hydrogels in regenerative medicine. *Adv. Mater.* 26 (1): 85–123.

Artegiani, B. and Clevers, H. (2018). Use and application of 3D-organoid technology. *Hum. Mol. Genet.* 27 (R2): R99–R107. https://doi.org/10.1093/hmg/ddy187.

Azari, H., Sharififar, S., Rahman, M. et al. (2011). Establishing embryonic mouse neural stem cell culture using the neurosphere assay. *J. Vis. Exp.* 47: 2457. https://doi.org/10.3791/2457.

Bairoch, A. (2019). Cellosaurus micro-review 1: Cellonauts, space-faring cell lines. *OSF Preprints* https://doi.org/10.31219/osf.io/e5fgj.

Begley, C.M. and Kleis, S.J. (2000). The fluid dynamic and shear environment in the NASA/JSC rotating-wall perfused-vessel bioreactor. *Biotechnol. Bioeng.* 70 (1): 32–40.

Benavides Damm, T., Walther, I., Wuest, S.L. et al. (2014). Cell cultivation under different gravitational loads using a novel random positioning incubator. *Biotechnol. Bioeng.* 111 (6): 1180–1190. https://doi.org/10.1002/bit.25179.

Ben-David, U., Beroukhim, R., and Golub, T.R. (2019). Genomic evolution of cancer models: perils and opportunities. *Nat. Rev. Cancer* 19 (2): 97–109. https://doi.org/10.1038/s41568-018-0095-3.

Bissell, M.J., Hall, H.G., and Parry, G. (1982). How does the extracellular matrix direct gene expression? *J. Theor. Biol.* 99 (1): 31–68.

Bjerkvig, R., Steinsvag, S.K., and Laerum, O.D. (1986). Reaggregation of fetal rat brain cells in a stationary culture system. I: methodology and cell identification. *In Vitro Cell. Dev. Biol.* 22 (4): 180–192.

Bochenek, M.A., Veiseh, O., Vegas, A.J. et al. (2018). Alginate encapsulation as long-term immune protection of allogeneic pancreatic islet cells transplanted into the omental bursa of macaques. *Nat. Biomed. Eng.* 2 (11): 810–821. https://doi.org/10.1038/s41551-018-0275-1.

Breslin, S. and O'Driscoll, L. (2016). The relevance of using 3D cell cultures, in addition to 2D monolayer cultures, when evaluating breast cancer drug sensitivity and resistance. *Oncotarget* 7 (29): 45745–45756. https://doi.org/10.18632/oncotarget.9935.

Broda, T.R., McCracken, K.W., and Wells, J.M. (2019). Generation of human antral and fundic gastric organoids from pluripotent stem cells. *Nat. Protoc.* 14 (1): 28–50. https://doi.org/10.1038/s41596-018-0080-z.

Burrows, M.T. (1910). The cultivation of tissues of the chick-embryo outside the body. *JAMA* 55 (24): 2057–2058. https://doi.org/10.1001/jama.1910.04330240035009.

Caddeo, S., Boffito, M., and Sartori, S. (2017). Tissue engineering approaches in the design of healthy and pathological in vitro tissue models. *Front. Bioeng. Biotechnol.* 5: 40. https://doi.org/10.3389/fbioe.2017.00040.

Cereijido, M., Robbins, E.S., Dolan, W.J. et al. (1978). Polarized monolayers formed by epithelial cells on a permeable and translucent support. *J. Cell Biol.* 77 (3): 853–880. https://doi.org/10.1083/jcb.77.3.853.

Cesarz, Z. and Tamama, K. (2016). Spheroid culture of mesenchymal stem cells. *Stem Cells Int.* 2016: 9176357. https://doi.org/10.1155/2016/9176357.

Chan, J.M., Zervantonakis, I.K., Rimchala, T. et al. (2012). Engineering of in vitro 3D capillary beds by self-directed angiogenic sprouting. *PLoS One* 7 (12): e50582. https://doi.org/10.1371/journal.pone.0050582.

Charoen, K.M., Fallica, B., Colson, Y.L. et al. (2014). Embedded multicellular spheroids as a biomimetic 3D cancer model for evaluating drug and drug-device combinations. *Biomaterials* 35 (7): 2264–2271. https://doi.org/10.1016/j.biomaterials.2013.11.038.

Cheng, N.C., Chen, S.Y., Li, J.R. et al. (2013). Short-term spheroid formation enhances the regenerative capacity of adipose-derived stem cells by promoting stemness, angiogenesis, and chemotaxis.

Stem Cells Transl. Med. 2 (8): 584–594. https://doi.org/10.5966/sctm.2013-0007.

Clevers, H. (2016). Modeling development and disease with organoids. *Cell* 165 (7): 1586–1597. https://doi.org/10.1016/j.cell.2016.05.082.

Corning (2007). Transwell permeable supports: instructions for use. http://www.corning.com/catalog/cls/documents/protocols/transwell_instructionmanual.pdf (accessed 9 June 2019).

Corning (2016). Permeable supports selection guide. https://www.corning.com/catalog/cls/documents/selection-guides/selection_guide_cls-cc-027_permeable_supports.pdf (accessed 7 June 2019).

Cui, X., Hartanto, Y., and Zhang, H. (2017). Advances in multicellular spheroids formation. *J. R. Soc. Interface* 14 (127) https://doi.org/10.1098/rsif.2016.0877.

Cukierman, E., Pankov, R., and Yamada, K.M. (2002). Cell interactions with three-dimensional matrices. *Curr. Opin. Cell Biol.* 14 (5): 633–639.

Dickson, K.J. (1991). Summary of biological spaceflight experiments with cells. *ASGSB Bull.* 4 (2): 151–260.

DiStefano, T., Chen, H.Y., Panebianco, C. et al. (2018). Accelerated and improved differentiation of retinal organoids from pluripotent stem cells in rotating-wall vessel bioreactors. *Stem Cell Rep.* 10 (1): 300–313. https://doi.org/10.1016/j.stemcr.2017.11.001.

Driehuis, E., Kolders, S., Spelier, S. et al. (2019). Oral mucosal organoids as a potential platform for personalized cancer therapy. *Cancer Discov.* 9 (7): 852–871. https://doi.org/10.1158/2159-8290.CD-18-1522.

Dubrot, J., Portero, A., Orive, G. et al. (2010). Delivery of immunostimulatory monoclonal antibodies by encapsulated hybridoma cells. *Cancer Immunol. Immunother.* 59 (11): 1621–1631. https://doi.org/10.1007/s00262-010-0888-z.

Dye, B.R., Hill, D.R., Ferguson, M.A. et al. (2015). In vitro generation of human pluripotent stem cell derived lung organoids. *elife* 4: e05098. https://doi.org/10.7554/eLife.05098.

Earle, W.R., Schilling, E.L., and Shannon, J.E. Jr. (1951). Growth of animal tissue cells on three-dimensional substrates. *J. Natl Cancer Inst.* 12 (1): 179–193.

Ebendal, T. (1976). The relative roles of contact inhibition and contact guidance in orientation of axons extending on aligned collagen fibrils in vitro. *Exp. Cell Res.* 98 (1): 159–169.

Eiraku, M., Watanabe, K., Matsuo-Takasaki, M. et al. (2008). Self-organized formation of polarized cortical tissues from ESCs and its active manipulation by extrinsic signals. *Cell Stem Cell* 3 (5): 519–532. https://doi.org/10.1016/j.stem.2008.09.002.

Eiraku, M., Takata, N., Ishibashi, H. et al. (2011). Self-organizing optic-cup morphogenesis in three-dimensional culture. *Nature* 472 (7341): 51–56. https://doi.org/10.1038/nature09941.

Eskes, C., Bostrom, A.C., Bowe, G. et al. (2017). Good cell culture practices & in vitro toxicology. *Toxicol. In Vitro* 45 (Pt 3): 272–277. https://doi.org/10.1016/j.tiv.2017.04.022.

Fell, H.B. (1972). Tissue culture and its contribution to biology and medicine. *J. Exp. Biol.* 57 (1): 1–13.

Foty, R. (2011). A simple hanging drop cell culture protocol for generation of 3D spheroids. *J. Vis. Exp.* 51: 2720. https://doi.org/10.3791/2720.

Gahwiler, B.H., Capogna, M., Debanne, D. et al. (1997). Organotypic slice cultures: a technique has come of age. *Trends Neurosci.* 20 (10): 471–477.

Gallico, G.G. 3rd, O'Connor, N.E., Compton, C.C. et al. (1984). Permanent coverage of large burn wounds with autologous cultured human epithelium. *N. Engl. J. Med.* 311 (7): 448–451. https://doi.org/10.1056/NEJM198408163110706.

Gao, D., Vela, I., Sboner, A. et al. (2014). Organoid cultures derived from patients with advanced prostate cancer. *Cell* 159 (1): 176–187. https://doi.org/10.1016/j.cell.2014.08.016.

Gjorevski, N. and Lutolf, M.P. (2017). Synthesis and characterization of well-defined hydrogel matrices and their application to intestinal stem cell and organoid culture. *Nat. Protoc.* 12 (11): 2263–2274. https://doi.org/10.1038/nprot.2017.095.

Gjorevski, N., Sachs, N., Manfrin, A. et al. (2016). Designer matrices for intestinal stem cell and organoid culture. *Nature* 539 (7630): 560–564. https://doi.org/10.1038/nature20168.

Grimm, D., Wehland, M., Pietsch, J. et al. (2014). Growing tissues in real and simulated microgravity: new methods for tissue engineering. *Tissue Eng. Part B Rev.* 20 (6): 555–566. https://doi.org/10.1089/ten.TEB.2013.0704.

Grimm, D., Egli, M., Kruger, M. et al. (2018). Tissue engineering under microgravity conditions – use of stem cells and specialized cells. *Stem Cells Dev.* 27 (12): 787–804. https://doi.org/10.1089/scd.2017.0242.

Grobstein, C. (1953). Morphogenetic interaction between embryonic mouse tissues separated by a membrane filter. *Nature* 172 (4384): 869–870.

Haisler, W.L., Timm, D.M., Gage, J.A. et al. (2013). Three-dimensional cell culturing by magnetic levitation. *Nat. Protoc.* 8 (10): 1940–1949. https://doi.org/10.1038/nprot.2013.125.

Harouaka, R.A., Nisic, M., and Zheng, S.Y. (2013). Circulating tumor cell enrichment based on physical properties. *J. Lab. Autom.* 18 (6): 455–468. https://doi.org/10.1177/2211068213494391.

Harrison, R.G. (1907). Observations on the living developing nerve fiber. *Proc. Soc. Exp. Biol.* 4 (1): 140–143.

Hayward, I.P. and Whitehead, R.H. (1992). Patterns of growth and differentiation in the colon carcinoma cell line LIM 1863. *Int. J. Cancer* 50 (5): 752–759.

Herranz, R., Anken, R., Boonstra, J. et al. (2013). Ground-based facilities for simulation of microgravity: organism-specific recommendations for their use, and recommended terminology. *Astrobiology* 13 (1): 1–17. https://doi.org/10.1089/ast.2012.0876.

Hidalgo, I.J., Raub, T.J., and Borchardt, R.T. (1989). Characterization of the human colon carcinoma cell line (Caco-2) as a model system for intestinal epithelial permeability. *Gastroenterology* 96 (3): 736–749.

Holliday, D.L., Moss, M.A., Pollock, S. et al. (2013). The practicalities of using tissue slices as preclinical organotypic breast cancer models. *J. Clin. Pathol.* 66 (3): 253–255. https://doi.org/10.1136/jclinpath-2012-201147.

Holloway, E.M., Capeling, M.M., and Spence, J.R. (2019). Biologically inspired approaches to enhance human organoid complexity. *Development* 146 (8) https://doi.org/10.1242/dev.166173.

Hsiao, A.Y., Tung, Y.C., Qu, X. et al. (2012). 384 hanging drop arrays give excellent Z-factors and allow versatile formation of co-culture spheroids. *Biotechnol. Bioeng.* 109 (5): 1293–1304. https://doi.org/10.1002/bit.24399.

Huang, A.H. and Niklason, L.E. (2011). Engineering biological-based vascular grafts using a pulsatile bioreactor. *J. Vis. Exp.* 52: 2646. https://doi.org/10.3791/2646.

Huch, M., Knoblich, J.A., Lutolf, M.P. et al. (2017). The hope and the hype of organoid research. *Development* 144 (6): 938–941. https://doi.org/10.1242/dev.150201.

Hynds, R.E., Bonfanti, P., and Janes, S.M. (2018). Regenerating human epithelia with cultured stem cells: feeder cells, organoids and beyond. *EMBO Mol. Med.* 10 (2): 139–150. https://doi.org/10.15252/emmm.201708213.

Jaganathan, H., Gage, J., Leonard, F. et al. (2014). Three-dimensional in vitro co-culture model of breast tumor using magnetic levitation. *Sci. Rep.* 4: 6468. https://doi.org/10.1038/srep06468.

Jongpaiboonkit, L., King, W.J., Lyons, G.E. et al. (2008). An adaptable hydrogel array format for 3-dimensional cell culture and analysis. *Biomaterials* 29 (23): 3346–3356. https://doi.org/10.1016/j.biomaterials.2008.04.040.

Justus, C.R., Leffler, N., Ruiz-Echevarria, M. et al. (2014). In vitro cell migration and invasion assays. *J. Vis. Exp.* 88: 51046. https://doi.org/10.3791/51046.

Keller, G.M. (1995). In vitro differentiation of embryonic stem cells. *Curr. Opin. Cell Biol.* 7 (6): 862–869.

Kenny, P.A., Lee, G.Y., Myers, C.A. et al. (2007). The morphologies of breast cancer cell lines in three-dimensional assays correlate with their profiles of gene expression. *Mol. Oncol.* 1 (1): 84–96. https://doi.org/10.1016/j.molonc.2007.02.004.

Kenworthy, P., Dowrick, P., Baillie-Johnson, H. et al. (1992). The presence of scatter factor in patients with metastatic spread to the pleura. *Br. J. Cancer* 66 (2): 243–247.

Klinger, R.Y. and Niklason, L.E. (2006). Tissue-engineered blood vessels. In: *Culture of Cells for Tissue Engineering* (eds. G. Vunjak-Novakovic and R.I. Freshney), 294–322. Hoboken, NJ: Wiley.

Kloxin, A.M., Tibbitt, M.W., Kasko, A.M. et al. (2010). Tunable hydrogels for external manipulation of cellular microenvironments through controlled photodegradation. *Adv. Mater.* 22 (1): 61–66. https://doi.org/10.1002/adma.200900917.

Knazek, R.A., Gullino, P.M., Kohler, P.O. et al. (1972). Cell culture on artificial capillaries: an approach to tissue growth in vitro. *Science* 178 (4056): 65–66.

Knazek, R.A., Kohler, P.O., and Gullino, P.M. (1974). Hormone production by cells grown in vitro on artificial capillaries. *Exp. Cell Res.* 84 (1): 251–254.

Knight, E., Murray, B., Carnachan, R. et al. (2011). Alvetex(R): polystyrene scaffold technology for routine three dimensional cell culture. *Methods Mol. Biol.* 695: 323–340. https://doi.org/10.1007/978-1-60761-984-0_20.

Kolesky, D.B., Homan, K.A., Skylar-Scott, M.A. et al. (2016). Three-dimensional bioprinting of thick vascularized tissues. *Proc. Natl Acad. Sci. U.S.A.* 113 (12): 3179–3184. https://doi.org/10.1073/pnas.1521342113.

Kruger, M., Pietsch, J., Bauer, J. et al. (2019). Growth of endothelial cells in space and in simulated microgravity – a comparison on the secretory level. *Cell. Physiol. Biochem.* 52 (5): 1039–1060. https://doi.org/10.33594/000000071.

Kshitiz, Park, J.S., Kim, P. et al. (2012). Control of stem cell fate and function by engineering physical microenvironments. *Integr. Biol. (Camb.)* 4 (9): 1008–1018.

Lancaster, M.A., Renner, M., Martin, C.A. et al. (2013). Cerebral organoids model human brain development and microcephaly. *Nature* 501 (7467): 373–379. https://doi.org/10.1038/nature12517.

Lancaster, M.A., Corsini, N.S., Wolfinger, S. et al. (2017). Guided self-organization and cortical plate formation in human brain organoids. *Nat. Biotechnol.* 35 (7): 659–666. https://doi.org/10.1038/nbt.3906.

Langhans, S.A. (2018). Three-dimensional in vitro cell culture models in drug discovery and drug repositioning. *Front. Pharmacol.* 9: 6. https://doi.org/10.3389/fphar.2018.00006.

Lanza, R., Langer, R., and Vacanti, J. (2013). *Principles of Tissue Engineering*, 4e. London: Academic Press.

Lasnitzki, I. (1992). Organ culture. In: *Animal Cell Culture: A Practical Approach* (ed. R.I. Freshney), 213–261. Oxford: IRL Press at Oxford University Press.

Lee, G.Y., Kenny, P.A., Lee, E.H. et al. (2007). Three-dimensional culture models of normal and malignant breast epithelial cells. *Nat. Methods* 4 (4): 359–365. https://doi.org/10.1038/nmeth1015.

Leighton, J. (1951). A sponge matrix method for tissue culture; formation of organized aggregates of cells in vitro. *J. Natl Cancer Inst.* 12 (3): 545–561.

Leighton, J. (1997). Human mammary cancer cell lines and other epithelial cells cultured as organoid tissue in lenticular pouches of reinforced collagen membranes. *In Vitro Cell Dev. Biol. Anim.* 33 (10): 783–790.

Li, R.A., Keung, W., Cashman, T.J. et al. (2018). Bioengineering an electro-mechanically functional miniature ventricular heart chamber from human pluripotent stem cells. *Biomaterials* 163: 116–127. https://doi.org/10.1016/j.biomaterials.2018.02.024.

Li, X., Sun, Q., Li, Q. et al. (2018). Functional hydrogels with tunable structures and properties for tissue engineering applications. *Front. Chem.* 6: 499. https://doi.org/10.3389/fchem.2018.00499.

Limat, A., Breitkreutz, D., Thiekoetter, G. et al. (1995). Formation of a regular neo-epidermis by cultured human outer root sheath cells grafted on nude mice. *Transplantation* 59 (7): 1032–1038.

Lin, R.Z. and Chang, H.Y. (2008). Recent advances in three-dimensional multicellular spheroid culture for biomedical research. *Biotechnol. J.* 3 (9–10): 1172–1184. https://doi.org/10.1002/biot.200700228.

Lin, R.Z., Chu, W.C., Chiang, C.C. et al. (2008). Magnetic reconstruction of three-dimensional tissues from multicellular spheroids. *Tissue Eng. Part C Methods* 14 (3): 197–205. https://doi.org/10.1089/ten.tec.2008.0061.

Lund-Johansen, M., Bjerkvig, R., and Andersen, K.J. (1989). Multicellular tumor spheroids in serum-free culture. *Anticancer Res.* 9 (2): 413–420.

Maas-Szabowski, N., Stark, H.J., and Fusenig, N.E. (2002). Cell interaction and epithelial differentiation. In: *Culture of Epithelial Cells* (eds. R.I. Freshney and M.G. Freshney), 31–63. Hoboken, NJ: Wiley-Liss.

Mattix, B.M., Olsen, T.R., Casco, M. et al. (2014). Janus magnetic cellular spheroids for vascular tissue engineering. *Biomaterials* 35 (3): 949–960. https://doi.org/10.1016/j.biomaterials.2013.10.036.

McCauley, H.A. and Wells, J.M. (2017). Pluripotent stem cell-derived organoids: using principles of developmental biology to grow human tissues in a dish. *Development* 144 (6): 958–962. https://doi.org/10.1242/dev.140731.

Ment, L.R., Stewart, W.B., Scaramuzzino, D. et al. (1997). An in vitro three-dimensional coculture model of cerebral microvascular angiogenesis and differentiation. *In Vitro Cell Dev Biol Anim* 33 (9): 684–691. https://doi.org/10.1007/s11626-997-0126-y.

Miyagawa, Y., Okita, H., Hiroyama, M. et al. (2011). A microfabricated scaffold induces the spheroid formation of human bone marrow-derived mesenchymal progenitor cells and promotes efficient adipogenic differentiation. *Tissue Eng. Part A* 17 (3–4): 513–521. https://doi.org/10.1089/ten.TEA.2009.0810.

Moscona, A. (1961). Rotation-mediated histogenetic aggregation of dissociated cells. A quantifiable approach to cell interactions in vitro. *Exp. Cell Res.* 22: 455–475.

Moscona, A. and Moscona, H. (1952). The dissociation and aggregation of cells from organ rudiments of the early chick embryo. *J. Anat.* 86 (3): 287–301.

Mueller-Klieser, W. (1987). Multicellular spheroids. A review on cellular aggregates in cancer research. *J. Cancer Res. Clin. Oncol.* 113 (2): 101–122.

Mullender, M., El Haj, A.J., Yang, Y. et al. (2004). Mechanotransduction of bone cells in vitro: mechanobiology of bone tissue. *Med. Biol. Eng. Comput.* 42 (1): 14–21.

Munsie, M., Hyun, I., and Sugarman, J. (2017). Ethical issues in human organoid and gastruloid research. *Development* 144 (6): 942–945. https://doi.org/10.1242/dev.140111.

Natoli, M., Leoni, B.D., D'Agnano, I. et al. (2012). Good Caco-2 cell culture practices. *Toxicol. In Vitro* 26 (8): 1243–1246. https://doi.org/10.1016/j.tiv.2012.03.009.

Nims, R.J., Cigan, A.D., Albro, M.B. et al. (2015). Matrix production in large engineered cartilage constructs is enhanced by nutrient channels and excess media supply. *Tissue Eng. Part C Methods* 21 (7): 747–757. https://doi.org/10.1089/ten.TEC.2014.0451.

O'Brien, F.J. (2011). Biomaterials and scaffolds for tissue engineering. *Mater. Today* 14 (3): 88–95. https://doi.org/10.1016/S1369-7021(11)70058-X.

Pastrana, E., Silva-Vargas, V., and Doetsch, F. (2011). Eyes wide open: a critical review of sphere-formation as an assay for stem cells. *Cell Stem Cell* 8 (5): 486–498. https://doi.org/10.1016/j.stem.2011.04.007.

Pease, J.C., Brewer, M., and Tirnauer, J.S. (2012). Spontaneous spheroid budding from monolayers: a potential contribution to ovarian cancer dissemination. *Biol. Open* 1 (7): 622–628. https://doi.org/10.1242/bio.2012653.

Pellegrini, G., Ranno, R., Stracuzzi, G. et al. (1999). The control of epidermal stem cells (holoclones) in the treatment of massive full-thickness burns with autologous keratinocytes cultured on fibrin. *Transplantation* 68 (6): 868–879.

Petersen, O.W., Ronnov-Jessen, L., Howlett, A.R. et al. (1992). Interaction with basement membrane serves to rapidly distinguish growth and differentiation pattern of normal and malignant human breast epithelial cells. *Proc. Natl Acad. Sci. U.S.A.* 89 (19): 9064–9068. https://doi.org/10.1073/pnas.89.19.9064.

Raya-Rivera, A.M., Esquiliano, D., Fierro-Pastrana, R. et al. (2014). Tissue-engineered autologous vaginal organs in patients: a pilot cohort study. *Lancet* 384 (9940): 329–336. https://doi.org/10.1016/S0140-6736(14)60542-0.

Reynolds, B.A. and Weiss, S. (1992). Generation of neurons and astrocytes from isolated cells of the adult mammalian central nervous system. *Science* 255 (5052): 1707–1710.

Rivera-Perez, J.A., Jones, V., and Tam, P.P. (2010). Culture of whole mouse embryos at early postimplantation to organogenesis stages: developmental staging and methods. *Methods Enzymol.*

476: 185–203. https://doi.org/10.1016/S0076-6879(10)76011-0.

Riwaldt, S., Pietsch, J., Sickmann, A. et al. (2015). Identification of proteins involved in inhibition of spheroid formation under microgravity. *Proteomics* 15 (17): 2945–2952. https://doi.org/10.1002/pmic.201500067.

Ruedinger, F., Lavrentieva, A., Blume, C. et al. (2015). Hydrogels for 3D mammalian cell culture: a starting guide for laboratory practice. *Appl. Microbiol. Biotechnol.* 99 (2): 623–636. https://doi.org/10.1007/s00253-014-6253-y.

Rutzky, L.P., Tomita, J.T., Calenoff, M.A. et al. (1979). Human colon adenocarcinoma cells. III. In vitro organoid expression and carcinoembryonic antigen kinetics in hollow fiber culture. *J. Natl Cancer Inst.* 63 (4): 893–902.

Sachs, N., de Ligt, J., Kopper, O. et al. (2018). A living biobank of breast cancer organoids captures disease heterogeneity. *Cell* 172 (1–2): 373–386. e10. doi: https://doi.org/10.1016/j.cell.2017.11.010.

Sambuy, Y., De Angelis, I., Ranaldi, G. et al. (2005). The Caco-2 cell line as a model of the intestinal barrier: influence of cell and culture-related factors on Caco-2 cell functional characteristics. *Cell Biol. Toxicol.* 21 (1): 1–26. https://doi.org/10.1007/s10565-005-0085-6.

Sato, T., Vries, R.G., Snippert, H.J. et al. (2009). Single Lgr5 stem cells build crypt-villus structures in vitro without a mesenchymal niche. *Nature* 459 (7244): 262–265. https://doi.org/10.1038/nature07935.

Seidel, J.O., Pei, M., Gray, M.L. et al. (2004). Long-term culture of tissue engineered cartilage in a perfused chamber with mechanical stimulation. *Biorheology* 41 (3–4): 445–458.

Shamir, E.R. and Ewald, A.J. (2014). Three-dimensional organotypic culture: experimental models of mammalian biology and disease. *Nat. Rev. Mol. Cell Biol.* 15 (10): 647–664. https://doi.org/10.1038/nrm3873.

Shansky, J., Ferland, P., McGuire, S. et al. (2006). Tissue engineering human skeletal muscle for clinical applications. In: *Culture of Cells for Tissue Engineering* (eds. G. Vunjak-Novakovic and R.I. Freshney), 239–258. Hoboken, NJ: Wiley.

Sheasgreen, J., Klausner, M., Kandarova, H. et al. (2009). The MatTek story – how the three Rs principles led to 3-D tissue success! *Altern. Lab. Anim.* 37 (6): 611–622. https://doi.org/10.1177/026119290903700606.

Simian, M. and Bissell, M.J. (2017). Organoids: a historical perspective of thinking in three dimensions. *J. Cell Biol.* 216 (1): 31–40. https://doi.org/10.1083/jcb.201610056.

Souza, G.R., Molina, J.R., Raphael, R.M. et al. (2010). Three-dimensional tissue culture based on magnetic cell levitation. *Nat. Nanotechnol.* 5 (4): 291–296. https://doi.org/10.1038/nnano.2010.23.

Stark, H.J., Szabowski, A., Fusenig, N.E. et al. (2004). Organotypic cocultures as skin equivalents: a complex and sophisticated in vitro system. *Biol. Proced Online* 6: 55–60. https://doi.org/10.1251/bpo72.

Sun, T., Jackson, S., Haycock, J.W. et al. (2006). Culture of skin cells in 3D rather than 2D improves their ability to survive exposure to cytotoxic agents. *J. Biotechnol.* 122 (3): 372–381. https://doi.org/10.1016/j.jbiotec.2005.12.021.

Sundaram, S., Echter, A., Sivarapatna, A. et al. (2014). Small-diameter vascular graft engineered using human embryonic stem cell-derived mesenchymal cells. *Tissue Eng. Part A* 20 (3–4): 740–750. https://doi.org/10.1089/ten.TEA.2012.0738.

Sutcliffe, M. and Lancaster, M.A. (2017). A simple method of generating 3D brain organoids using standard laboratory equipment. *Methods Mol. Biol.* 1576: 1–12. https://doi.org/10.1007/7651_2017_2.

Sutherland, R.M. (1988). Cell and environment interactions in tumor microregions: the multicell spheroid model. *Science* 240 (4849): 177–184.

Sutherland, R.M., McCredie, J.A., and Inch, W.R. (1971). Growth of multicell spheroids in tissue culture as a model of nodular carcinomas. *J. Natl Cancer Inst.* 46 (1): 113–120.

Takasato, M., Er, P.X., Chiu, H.S. et al. (2016). Generation of kidney organoids from human pluripotent stem cells. *Nat. Protoc.* 11 (9): 1681–1692. https://doi.org/10.1038/nprot.2016.098.

Tamura, M., Yanagawa, F., Sugiura, S. et al. (2014). Optical cell separation from three-dimensional environment in photodegradable hydrogels for pure culture techniques. *Sci. Rep.* 4: 4793. https://doi.org/10.1038/srep04793.

Tevis, K.M., Colson, Y.L., and Grinstaff, M.W. (2017). Embedded spheroids as models of the cancer microenvironment. *Adv. Biosyst.* 1 (10) https://doi.org/10.1002/adbi.201700083.

Thomas, C.H., Collier, J.H., Sfeir, C.S. et al. (2002). Engineering gene expression and protein synthesis by modulation of nuclear shape. *Proc. Natl Acad. Sci. U.S.A.* 99 (4): 1972–1977. https://doi.org/10.1073/pnas.032668799.

Thomson, D. (1914). Controlled growth en masse (somatic growth) of embryonic chick tissue in vitro. *Proc. R. Soc. Med.* 7 (Gen Rep): 71–75.

Tibbitt, M.W. and Anseth, K.S. (2009). Hydrogels as extracellular matrix mimics for 3D cell culture. *Biotechnol. Bioeng.* 103 (4): 655–663. https://doi.org/10.1002/bit.22361.

Topman, G., Shoham, N., Sharabani-Yosef, O. et al. (2013). A new technique for studying directional cell migration in a hydrogel-based three-dimensional matrix for tissue engineering model systems. *Micron* 51: 9–12. https://doi.org/10.1016/j.micron.2013.06.002.

Vacanti, J.P. and Langer, R. (1999). Tissue engineering: the design and fabrication of living replacement devices for surgical reconstruction and transplantation. *Lancet* 354 (Suppl 1): SI32–SI34. https://doi.org/10.1016/s0140-6736(99)90247-7.

Valent, P., Bonnet, D., De Maria, R. et al. (2012). Cancer stem cell definitions and terminology: the devil is in the details. *Nat. Rev. Cancer* 12 (11): 767–775. https://doi.org/10.1038/nrc3368.

Vergani, L., Grattarola, M., and Nicolini, C. (2004). Modifications of chromatin structure and gene expression following induced alterations of cellular shape. *Int. J. Biochem. Cell Biol.* 36 (8): 1447–1461. https://doi.org/10.1016/j.biocel.2003.11.015.

Verjans, E.T., Doijen, J., Luyten, W. et al. (2018). Three-dimensional cell culture models for anticancer drug screening: worth the effort? *J. Cell. Physiol.* 233 (4): 2993–3003. https://doi.org/10.1002/jcp.26052.

Vinci, M., Gowan, S., Boxall, F. et al. (2012). Advances in establishment and analysis of three-dimensional tumor spheroid-based functional assays for target validation and drug evaluation. *BMC Biol.* 10: 29. https://doi.org/10.1186/1741-7007-10-29.

Vinci, M., Box, C., Zimmermann, M. et al. (2013). Tumor spheroid-based migration assays for evaluation of therapeutic agents. *Methods Mol. Biol.* 986: 253–266. https://doi.org/10.1007/978-1-62703-311-4_16.

Vinci, M., Box, C., and Eccles, S.A. (2015). Three-dimensional (3D) tumor spheroid invasion assay. *J. Vis. Exp.* 99: e52686. https://doi.org/10.3791/52686.

Vlachogiannis, G., Hedayat, S., Vatsiou, A. et al. (2018). Patient-derived organoids model treatment response of metastatic gastrointestinal cancers. *Science* 359 (6378): 920–926. https://doi.org/10.1126/science.aao2774.

Vunjak-Novakovic, G. and Freshney, R.I. (2006). *Culture of Cells for Tissue Engineering*. Hoboken, NJ: Wiley-Liss.

Wakabayashi, Y., Chua, J., Larkin, J.M. et al. (2007). Four-dimensional imaging of filter-grown polarized epithelial cells. *Histochem. Cell Biol.* 127 (5): 463–472. https://doi.org/10.1007/s00418-007-0274-x.

Wenzel, C., Riefke, B., Grundemann, S. et al. (2014). 3D high-content screening for the identification of compounds that target cells in dormant tumor spheroid regions. *Exp. Cell Res.* 323 (1): 131–143. https://doi.org/10.1016/j.yexcr.2014.01.017.

van de Wetering, M., Francies, H.E., Francis, J.M. et al. (2015). Prospective derivation of a living organoid biobank of colorectal cancer patients. *Cell* 161 (4): 933–945. https://doi.org/10.1016/j.cell.2015.03.053.

Whitehead, R.H., Jones, J.K., Gabriel, A. et al. (1987). A new colon carcinoma cell line (LIM1863) that grows as organoids with spontaneous differentiation into crypt-like structures in vitro. *Cancer Res.* 47 (10): 2683–2689.

Wuest, S.L., Richard, S., Kopp, S. et al. (2015). Simulated microgravity: critical review on the use of random positioning machines for mammalian cell culture. *Biomed. Res. Int.* 2015: 971474. https://doi.org/10.1155/2015/971474.

Ylostalo, J.H., Bartosh, T.J., Tiblow, A. et al. (2014). Unique characteristics of human mesenchymal stromal/progenitor cells pre-activated in 3-dimensional cultures under different conditions. *Cytotherapy* 16 (11): 1486–1500. https://doi.org/10.1016/j.jcyt.2014.07.010.

Yuhas, J.M., Li, A.P., Martinez, A.O. et al. (1977). A simplified method for production and growth of multicellular tumor spheroids. *Cancer Res.* 37 (10): 3639–3643.

CHAPTER 28

Scale-Up and Automation

Many of the early tissue culture methods were acknowledged to be cumbersome and difficult to perform at a large scale (Fischer 1925; Gey 1933). The tissue culture flask, which was designed to provide a flat surface for microscopy and an angled neck for aseptic handling (see Chapter 8), was not designed to grow large numbers of cells (although modifications have since been made to improve its performance). To address this problem, George Gey developed the roller tube (Gey 1933). Cells were grown on the inner surface of a tube or bottle, which was continuously rotated at a slow speed. By speeding up the rotation in a "tumble tube," Olga Owens was able to successfully grow cells in suspension (Owens et al. 1954). Wilton Earle and colleagues then discovered that larger-scale suspension culture could be achieved using a rotary shaker (Earle et al. 1954). Earle optimized these techniques to perform "massive tissue culture" and to prepare multiple replicates for quantitative studies, using custom-built equipment to control and standardize procedures (Evans et al. 1951, 1954).

Although the techniques developed by these tissue culture pioneers have long since been superseded, they demonstrate the essential requirements for culture scale-up. Such activities typically aim to (i) increase the number of cells produced, e.g. for manufacture of biological products; (ii) increase the number of replicates, e.g. for high-throughput screening; (iii) provide a controlled environment; (iv) optimize cell yield and behavior; and (v) standardize procedures to give a consistent, high-quality outcome. An increase in cell number implies that the volume of the culture must also increase, which introduces complexities of scale that include nutrient supply, gas exchange, and surface area if cells are adherent. An increase in replicates implies that a greater number of samples must

be handled at any one time, which introduces challenges for manual handling. Many of these challenges can be addressed by using automation to replace manual handling, improve reproducibility, and control the culture process.

This chapter initially discusses how to perform scale-up in cell number (see Sections 28.2–28.5) and then discusses how to increase the number of replicates for high-throughput screening and how to incorporate automation into handling procedures (see Sections 28.6, 28.7). Scale-up and automation are specialized areas and it is impossible to explore these topics in depth within a single chapter. This chapter is written primarily from the perspective of a research laboratory where scale-up may be necessary for certain projects. More information on scale-up and automation is available elsewhere (Stacey and Davis 2007; Cabral et al. 2016; Saha et al. 2018).

28.1 TERMINOLOGY: SCALE-UP AND BIOREACTORS

The meaning of cell culture scale-up may vary with the laboratory, which has important implications for the selection of culture vessels and equipment (see Table 28.1). For a basic research laboratory, it may mean the difference between 1×10^6 and 1×10^9 cells per culture, which is readily achievable using conventional tissue culture plasticware, e.g. tissue culture flasks. Handling a greater number of flasks is cumbersome, but the number can be minimized using roller bottles or multisurface flasks (see Sections 28.3.1, 28.3.2). Scale-up between 1×10^9 and 1×10^{10} cells per culture can be performed in suspension, using shaker culture or spinner culture

Freshney's Culture of Animal Cells: A Manual of Basic Technique and Specialized Applications, Eighth Edition. Amanda Capes-Davis and R. Ian Freshney.

TABLE 28.1. Features of most commonly used vessels for scale-up in research laboratories.

Features	T-flasks	e-flasks	Roller bottles	Spinners	Stirred tanks (traditional and single-use)	Rocked system (bag reactors)
Regular volume/unit	10–150 ml	20–3000 ml	100–500 ml	10–5000 ml	0.5–2000 l	0.5–500 l
Recommended maximum production scale	0.1–1 l	1–15 l	1–20 l	1–20 l	Not equipment limited	Up to 100 l
Difficulty of handling	Easy	Easy	Easy	Medium	High	Medium/high
Availability of controls	Temp, (pH)	Temp, (pH)	Temp, (pH)	Temp, DO, (pH)	Full controls (optional)	Full controls (optional)
Estimated set-up cost[a] (USD)	8–10 K (average size incubator)	30–40 K (humidified CO_2 shaker/incubator)	16–20 K (large incubator and roller apparatus)	16–20 K (per 4 units without the cost of the incubator)	50–70 K (for lab scale, i.e. up to 20 l)	35–70 K (for lab scale, i.e. up to 20 l)
Expected yield[b]	Low	Low	Low	Medium	High	Medium/high
Product quality[c]	1–2	2–3	2–3	4	5	4–5
Examples of vendors	Thermo, Corning, Sorfa	Thermo, Corning, Nalgene	Thermo, Corning, Greiner	Thermo, Corning, Eppendorf/NBS, Millipore, Techne	Sartorius, Eppendorf/NBS, Millipore, Applicon	Wave-GE, Sartorius, Applicon

Source: Courtesy of George Lovrecz.
[a] Assuming that services, including gases, to support basic cell culture work are available.
[b] Using similar media, feeding regime, i.e. yield depends on cell density and culture conditions.
[c] Estimated heterogeneity (often depends on controls), ranked 1 to 5 where 5 is best.

(see Sections 8.6.2, 28.2.1). Scale-up between 1×10^{10} and 1×10^{12} cells per culture is likely to require access to a bench-top bioreactor; single-use bioreactors are increasingly used by research laboratories for this purpose (see Section 28.2.2). Work can be performed at laboratory scale to about 20 l capacity (see Table 28.1). Beyond that point, expertise is usually required from a facility that specializes in scale-up and manufacture (see Minireview M28.1). Manufacturing facilities may operate bioreactors of 20 000 l capacity or more using Good Manufacturing Practice (GMP) (see Section 7.2.3) (Nienow 2006).

A number of overlapping terms are used to describe scale-up, including spinner culture, stirrer culture, fermentor, and bioreactor. "Spinner culture" and "stirrer culture" are synonymous terms and tend to be used for suspension culture systems at the low end of the laboratory range of equipment. "Fermentor" and "bioreactor" are not synonymous, but these terms have considerable overlap. Fermentors were originally designed for microbiological culture before being adapted for animal cell culture; the term was commonly used for laboratory or industrial equipment at around 50–100 l capacity. A considerable increase in scale has since occurred (in line with development of the biotechnology industry), coupled with a greater diversity in design to cope with both monolayer and suspension cultures and to provide process control. For this book, the term "bioreactor" is used in preference to "fermentor" and refers to any device that provides a controlled environment for culture – typically for expansion, but potentially for other purposes such as differentiation. In biomanufacturing, the term includes spinner flasks (which may be considered a form of bioreactor if they provide a controlled environment), stainless steel tanks, or single-use disposable systems of various capacities (Gallo-Ramirez et al. 2015). In tissue engineering, the term includes stirred culture of bioengineered constructs using relatively small volumes (Selden and Fuller 2018).

Bioreactors can be operated in various modes including batch, fed-batch, and perfusion (continuous) modes (Butler 2005). In batch mode, cells are maintained within the bioreactor until they reach a desired endpoint, e.g. when production of antibody is deemed to meet requirements. Once that endpoint has been reached, the culture is harvested and the protein product is separated from the cell mass as a batch. While relatively simple, this approach results in a low cell yield and product titer, due to accumulation of toxic metabolites and the depletion of essential nutrients (Xie and Wang 1994). In fed-batch mode, cells are periodically re-fed using media that have been designed to reduce the accumulation of toxic metabolites such as ammonia and lactate, while avoiding critical shortages of glutamine and glucose. Fed-batch mode gives higher cell yields and product titers compared to batch mode and is widely used for antibody and vaccine production (Gallo-Ramirez et al. 2015; Fan et al. 2018). In perfusion (continuous) mode, fresh medium is continuously added and cell-free supernatant is withdrawn using a peristaltic pump or similar device. Perfusion mode is more technically challenging but can give higher productivity compared to fed-batch mode and allows culture for longer periods of time (Butler 2005; Xu et al. 2017).

A dazzling array of bioreactors and other vessels can be used for scale-up, making it challenging for newcomers to decide which system is best suited to their needs. The choice of method will primarily depend on whether cells proliferate in suspension or must be anchored to the substrate. Scale-up in suspension is usually more straightforward compared to monolayer approaches and is dealt with first (see Section 28.2). Monolayer techniques, being dependent on increased provision of growth surface, tend to be more complex and are discussed in a following section (see Section 28.3).

28.2 SCALE-UP IN SUSPENSION

Suspension techniques can be divided into static and dynamic approaches. Static suspension systems include dishes, multi-well plates, and tissue culture flasks. These may be suitable for limited volumes (e.g. as a starting point for scale-up), provided the cells do not adhere to the culture substrate. A dynamic system will become necessary if the cells adhere in a static system or when the depth exceeds 5 mm, to enhance gas exchange and prevent the cells from forming a dense sediment. Above 5–10 cm (depending on the ratio of surface area to volume), mixing and aeration are essential to maintain adequate gas exchange. Typically, this is done by sparging with CO_2 and air, but other approaches may be used, e.g. in airlift bioreactor systems (see Section 28.2.4).

28.2.1 Spinner Culture

In spinner culture, the contents of the culture vessel are agitated using a top-driven suspended paddle (McLimans et al. 1957; Merten 2015). Early designs used a pendulum containing a magnet, whose rotation was driven by a magnetic stirrer (more correctly "stirrer culture," but grouped with the spinner flasks for simplicity here) (see Figure 28.1). Generally, the pendulum design is preferable for minimizing shear, but the paddle becomes preferable as the scale is increased. Stirring is best done slowly, at 25–100 rpm, which is sufficient to prevent cell sedimentation, but not so fast as to create shear forces that would damage the cells. Depending on the agitation speed, a spinner culture can be maintained as a single-cell suspension or as multicellular aggregates (Gupta et al. 2016). The latter approach is particularly useful to grow adherent cells on microcarriers, to perform spheroid culture, or to maintain organoids in suspension (see Sections 27.4.1, 27.6, 28.3.3; see also Figure 27.14) (Merten 2015; Cui et al. 2017).

Spinner flasks are available from a number of suppliers, including Bellco (see Figure 28.1c, d), Cole-Parmer (which now distributes Argos Technologies and Techne products), Corning (e.g. ProCulture® and disposable flasks), Eppendorf

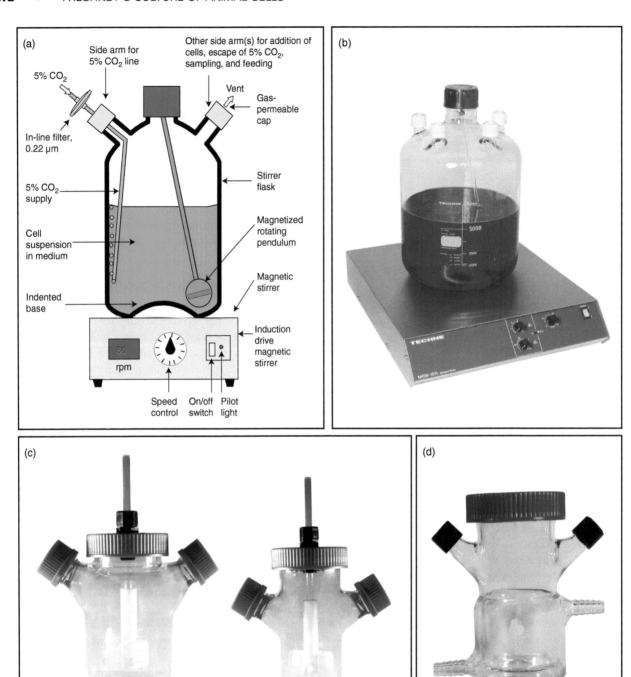

Fig. 28.1. Spinner flasks. (a) Diagram of a flask showing key design features. (b) Flask on a magnetic stirrer (5 l, Techne). This model is no longer available but is included here to show the magnetized rotating pendulum. (c) Spinner flasks used for microcarrier culture (Bellco). Flasks shown here are 100–250 ml capacity; agitation is provided by a top-driven impeller. (d) Flask with water jacket (100 ml, Bellco). *Source*: (a, b) R. Ian Freshney; (c, d) Courtesy of Bellco Glass, with permission.

(e.g. DASGIP® Bioblock Spinner Vessels), STEMCELL Technologies (e.g. StemSpan™ flasks), and Wheaton (e.g. Celstir® flasks). Although spinner flasks have changed in appearance over time, the essential elements remain largely unchanged (see Figure 28.1a). Spinner flasks usually have

two side arms: one for sampling and for addition of cells or media, and the other for perfusion with CO_2 in air. Flasks typically range from about 100 ml to 5 l in capacity and are either reusable (usually borosilicate glass) or disposable (usually polystyrene). Disposable flasks add up to a greater cost over time, but they are typically supplied ready to use and therefore cut down on the preparation time required to clean, sterilize, and assemble a reusable vessel.

28.2.2 Single-Use Bioreactor Systems

Over time, the demand has increased for scale-up at higher volume (e.g. for viral vaccine production) and for greater control over the culture environment (Gallo-Ramirez et al. 2015). Initially, stirred-tank systems were used that were similar to those employed in microbiological culture (Ziegler et al. 1958). These reusable devices were typically made of stainless steel, with a central impeller to agitate the cell suspension. Although stirred-tank systems continue to be widely used for high volumes, they tend to be complex and expensive devices that take considerable time to clean, sterilize, and reassemble before each use. Single-use bioreactors are becoming more widely used, particularly for personalized applications such as stem cell bioengineering (Allison and Richards 2014; Sart et al. 2014). Such applications require flexible systems with minimum down time; they may also require three-dimensional (3D) culture, which can be performed in many single-use bioreactors using spheroid or microcarrier culture (see Sections 27.4.1, 28.3.3).

Single-use bioreactors are available from a large number of suppliers in a broad range of sizes (see Figure 28.2). These devices are often used for GMP applications. As such, they typically comply with various regulatory requirements, e.g. Code of Federal Regulations Title 21 (21 CFR) Part 11 for electronic records (FDA 2019). Always discuss your requirements with suppliers before deciding which bioreactor is best for your particular application.

Single-use stirred-tank systems. Stirred-tank systems have been successfully modified for single use using disposable bags, which are installed prior to use. Examples include the Allegro™ STR bioreactor series (Pall Corporation; see Figure 28.2g), the BIOSTAT STR® bioreactors with Flexsafe STR bags (Sartorius), the HyPerforma™ single-use bioreactor series (ThermoFisher Scientific), the Mobius® bioreactor series (Merck Millipore; see Figure 28.2c), and the Xcellerex™ XDR bioreactor series (GE Healthcare Life Sciences; see Figure 28.2f, h). Other single-use stirred-tank systems have been developed that use disposable plastic vessels. Many of these systems are bench-top devices that are designed to maintain similar conditions to the floor-standing tanks, allowing further scale-up if required. Examples include the UniVessel® SU System (Sartorius) and the BioFlo® system using BioBLU Single Use Vessels (Eppendorf; see Figure 28.2d, e).

Rocking bioreactors. Wave-induced agitation was originally proposed by Vijay Singh as a solution to some of the challenges associated with scale-up of suspension culture, such as cell damage due to mixing and sparging (Singh 1999). The

system developed by Singh, the WAVE Bioreactor™ (GE Healthcare Life Sciences; see Figure 28.3), uses a motorized platform to rock the cells back and forth, generating waves that promote mixing and gas transfer. Cells are grown in a disposable bag that is warmed to the required temperature and partially filled with medium; the remaining space contains gas (e.g. 5% CO_2 in air) that can be introduced through a sterile inlet filter. Samples can be removed for analysis through a separate sampling port. Culture bags vary in their construction and size (up to 200 l capacity, although 25-l bags are commonly used for laboratory bench-top models). Once cells have been harvested, the single-use bag can be discarded and the platform can be rapidly reused for the next application.

Rocking bioreactors that operate according to similar principles are available from a range of commercial suppliers. Examples include the Allegro XRS 25 (Pall Corporation), the BIOSTAT® RM (Sartorius), the CELL-tainer® (Celltainer Biotech), the HyPerforma Rocker Bioreactor (ThermoFisher Scientific), and the Xuri™ Cell Expansion System (GE Healthcare Life Sciences).

Mini- and microscale bioreactors. Although suspension culture is commonly used to scale-up volume, it may also be used to prepare multiple replicates for high-throughput analysis. These typically employ single-use vessels, which act as minireactors (in the milliliter size range) or microreactors (in the microliter or nanoliter size range). At the minireactor end of the scale, volumes can be scaled down to less than a liter using a number of systems, including the ambr® 15 (10–15 ml, Sartorius; see Figure 28.4), the ambr® 250 (100–250 ml, Sartorius; see Figure 28.2b), and the DASbox® using BioBLU 0.3 single-use vessels (60–250 ml, Eppendorf). The ambr® systems are controlled by automated liquid handling platforms, which allow process control and high-throughput screening, e.g. to screen clones for protein production (see Figure 28.4; see also Sections 28.4, 28.6, 28.7).

At the microreactor end of the scale, microfluidic bioreactors have a great deal of potential for the culture and analysis of individual cells or small cell populations. These devices can offer precise control of fluid flow, nutrient supply, and mechanical stress, which is necessary for specialized applications; for example, microfluidic bioreactors can be used for clinical-scale expansion and tissue engineering of mesenchymal stromal cells (MSCs) (Sart et al. 2016). Microfluidic platforms are also readily automated and can be fabricated using 3D printing (see Section 28.7); a design can be assembled from various fluidic parts using customized design software, and then sent to a 3D printer to generate a mold for casting (Shankles et al. 2018). More information on microfluidic culture can be found in an earlier chapter (see Minireview M21.1).

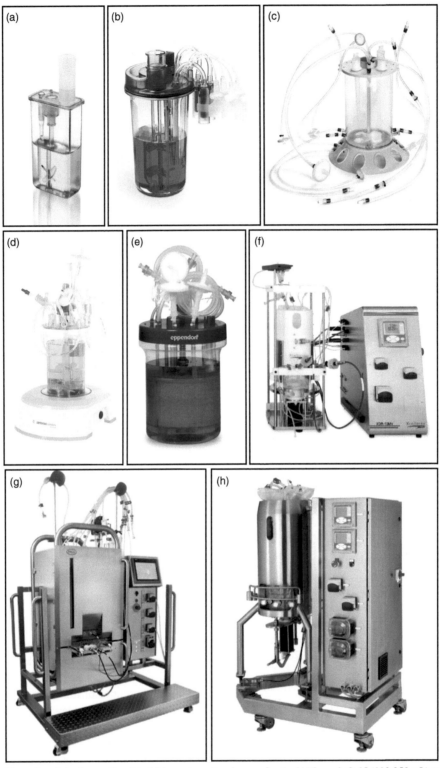

Examples of single-use stirred-tank bioreactors at various scales. *(a, b) The ambr® 15 (100-250 mL) and ambr 250 (100-250 mL), Sartorius; (c) the Mobius® CellReady (3 L), Merck Millipore; (d) the UniVessel® SU (2 L), Sartorius; (e) the BioBLU® packed bed reactor (5 L), Eppendorf; (f) the Xcellerex™ XDR-10 (4.5-10 L), GE Life Sciences; (g) the Allegro™ STR 50 (50 L), Pall Life Sciences; (h) the Xcellerex XDR-50 (50 L), GE Life Sciences. These are representative examples; most suppliers offer a range of bioreactors at various scales. Source: (a, b, d) Courtesy of Sartorius; (c) Courtesy of Merck Millipore; (e) Copyright © 2014 courtesy of Eppendorf AG, Germany; (f, h) Courtesy of GE Healthcare Life Sciences; (g) Courtesy of Pall Life Sciences.*

Fig. 28.2. Single-use stirred-tank bioreactors.

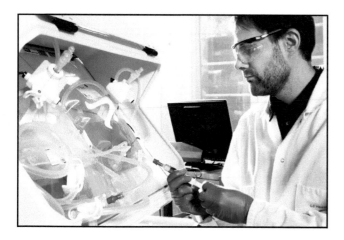

Fig. 28.3. The WAVE 25 bioreactor system (GE Healthcare Life Sciences). Cells are cultured in disposable bags, which are positioned on a motorized stage. Here, the stage is adjusted so that the culture medium in the bag moves to a suitable position for the user to remove a sample for analysis. *Source*: Courtesy of GE Healthcare Life Sciences, with permission.

28.2.3 Scaffold-Free Perfusion Bioreactors

As discussed earlier in this chapter, bioreactors may be operated in various modes (see Section 28.1). While fed-batch mode continues to be widely used, continuous (perfusion) mode has a number of benefits, e.g. for manufacture of unstable products (Croughan et al. 2015). This has resulted in the release of some bioreactors that are designed specifically for perfusion. These systems are sometimes referred to as "cytostats" or "biostats" (Kacmar et al. 2006). The terminology relates to an important outcome of perfusion mode, which is to achieve a defined metabolic state through a steady supply of nutrients and other requirements (Fenge et al. 2018). Maintenance of cells in a steady state is important for some biological products, e.g. proteins that require controlled folding and glycosylation (Curling et al. 1990; Butler 2005).

To operate as a perfusion bioreactor, a cell retention system must be used to ensure that cells are held within the culture vessel when the supernatant is removed without loss of viability (Kwon et al. 2017). Various methods can be used for cell retention, from simple (filtration) to relatively complex (continuous flow centrifugation or acoustic wave separation). Another solution to the problem of cell retention is to use adherent cells, which can be grown on a substrate that is retained in a hollow fiber or fixed bed system. Scaffold-based perfusion systems are discussed later in this chapter (see Section 28.3.4).

The BIOSTAT series (Sartorius; see Figure 28.5) has several models that can be used for perfusion, e.g. the BIOSTAT B. Other bench-top bioreactors that can potentially be used for perfusion include the DASGIP Parallel Bioreactor System (Eppendorf) and the Labfors 5 for cell culture (Infors HT). However, these bioreactors can usually be operated in various

modes, depending on the equipment configuration. Always discuss your requirements with suppliers to understand the bioreactors that may support your application and the various modes of operation that can be achieved.

28.2.4 Other Bioreactor Systems for Suspension Culture

Problems of increased scale in suspension cultures revolve around mixing and gas exchange. Most laboratory-scale bioreactors agitate the cell suspension using a rotating turbine or paddle, but this approach has limitations due to the inherent sensitivity of animal cells to shear damage (Varley and Birch 1999; Nienow 2006). Successful stirred-tank designs usually employ a slowly rotating large-bladed paddle with a relatively high surface area, but it is impossible to completely eliminate shear damage using such a system. Sparging with CO_2 and air also results in shear damage, although this can be minimized through careful design, e.g. by optimization of the bubble size (Nienow 2006). Antifoam agents can be used to minimize the harmful effects of shear within these systems, e.g. Pluronic™ F-68 or carboxymethylcellulose (CMC) (see Sections 9.2.7, 9.2.8). However, some bioreactor systems have been developed for suspension culture that avoid the stirred-tank design in favor of alternative modes of mixing and aeration. These include the rocking bioreactor (see Section 28.2.2.2) and several other designs.

Airlift bioreactor systems. Airlift bioreactors are designed to operate without mechanical mixing. A gas mixture (e.g. 5% CO_2 in air) is introduced into the base of the vessel containing the cell suspension; the movement of the gas bubbles toward the surface leads to mixing of the cell suspension and a high rate of gas transfer (Varley and Birch 1999). Airlift bioreactors have been used successfully for mammalian and insect cell culture for many years, but they are typically installed in industrial facilities, e.g. large airlift bioreactors are installed in Lonza's facilities in the United States and United Kingdom (King et al. 1992; Lonza 2011). Several airlift bioreactors are available for use in research laboratories, e.g. Cellexus offers the CellMaker range of single-use airlift systems. However, these are primarily used for bacteriophage and microbiological culture.

Rotating wall vessel bioreactors. A number of scaffold-free systems have been developed for 3D culture that provide a controlled environment, which means that they are also defined as bioreactors (see Section 28.1). The miniPERM® bioreactor (Sarstedt) is a rotating two-compartment system where the cell suspension is limited to one small compartment, separated from the main medium compartment by a permeable membrane (see Figure 27.7). This allows the cell concentration to be quite high, as the cells are not diluted by the bulk of the medium, and a high rotation speed on the Universal Turning Device prevents aggregation. The system can be used very effectively

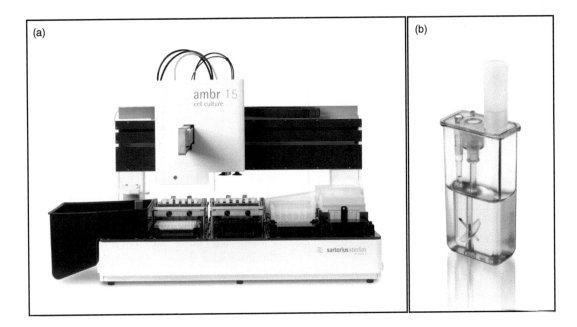

Fig. 28.4. The ambr® 15 cell culture system (Sartorius). (a) The ambr® workstation provides parallel processing, control, and evaluation of up to 48 bioreactors in tandem. (b) The ambr® 15 vessel (also shown in Figure 28.2a) is disposable and includes pH and DO sensors, an impeller for mixing, and a sparge tube for gas delivery. A sample port allows addition of reagents or removal of samples for testing. *Source*: Courtesy of Sartorius, with permission.

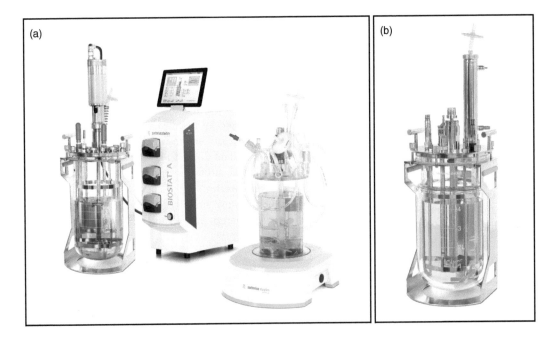

Fig. 28.5. BIOSTAT bioreactors (Sartorius). (a) BIOSTAT A system, showing a control unit with a UniVessel Glass (at left; 5 l capacity) and a disposable UniVessel SU (right; 2 l capacity). The medium reservoir, water jacket, and product receiver are not shown. (b) Closer view of glass vessel in (a); the UniVessel SU is also shown in Figure 28.2d. *Source*: Courtesy of Sartorius, with permission.

for production of biologicals (e.g. antibody generation from a hydridoma). The Rotary Cell Culture System (RCCS; Synthecon) rotates around a central core that includes a gas transfer membrane for aeration (see Figure 27.8). This results in very low turbulence and shear damage, making the RCCS a useful bioreactor for tissue engineering (Crabbe et al. 2015). These bioreactors are discussed in more detail in a previous chapter (see Section 27.4.2).

28.3 SCALE-UP IN MONOLAYER

Most normal, untransformed cells are anchorage-dependent and must therefore grow as an adherent monolayer. For such cultures, it is necessary to increase the surface area of the substrate in proportion to the number of cells and the volume of medium. However, surface area is only one of many challenges when performing scale-up of adherent cells. Stem cells, in particular, have stringent requirements to ensure that their environment provides the necessary cues for proliferation or differentiation. Stem cell culture is increasingly performed in bioreactors, where optimization of cell density, geometry, nutrient supply, gas exchange, and other variables can be performed to mimic the tissue microenvironment (Sart et al. 2014). This section looks at various strategies for scale-up in monolayer, starting with relatively simple approaches that can be achieved in any tissue culture laboratory and then considering more complex bioreactor systems.

28.3.1 Roller Culture

Cells can be seeded into a round tube or bottle that is then rolled around its long axis, dispersing cells evenly over the entire surface area of the bottle (see Figure 28.6a). If the cells are non-adherent, they will be agitated by the rolling action but will remain in suspension in the medium. If the cells are adherent, they will gradually attach to the inner surface of the bottle and grow to form a monolayer. This system has three major advantages over static monolayer culture: (i) the increase in utilizable surface area for a given size of bottle; (ii) the constant, but gentle, agitation of the medium; and (iii) the increased ratio of the medium's surface area to its volume, which allows gas exchange to take place at an increased rate through the thin film of medium over cells not actually submerged in the deep part of the medium.

The roller tube (later modified to give the roller bottle) was the first vessel designed expressly to increase cell yield (see Figure 28.6b) (Gey 1933). Roller bottles were originally glass, but single-use plastic bottles are now available (e.g. from Corning, Greiner Bio-One, or Wheaton). The inner surface of the bottle may be ribbed to increase surface area or treated to improve cell attachment (e.g. Corning roller bottles with CellBIND®; see Section 8.3.7). A roller rack is used to provide the necessary rotation, either in a dedicated roller incubator or in a shared CO_2 incubator with a roller rack (see Figures 5.16, 28.7). Compact systems are available if a laboratory does not have access to a roller rack, e.g. the Argos FlexiRoll cell roller system (Cole-Parmer). Automated systems can be used

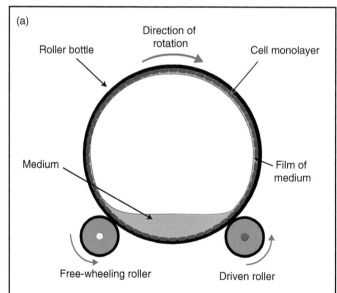

Fig. 28.6. Roller bottle culture. (a) Diagram of a roller bottle showing key design features. The cell monolayer is constantly bathed in liquid but is submerged only for about one-fourth of the cycle, enabling frequent replenishment of the medium and rapid gas exchange. (b) Roller tubes in position on a roller drum. This device (the TC-7 Roller Drum, New Brunswick Scientific) was from an original idea by George Gey. *Source*: (a) R. Ian Freshney; (b) Courtesy of New Brunswick Scientific.

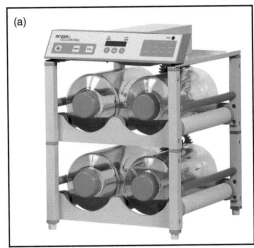 

Fig. 28.7. Roller bottles on racks. (a) CellRoll small bench-top roller rack; (b) large freestanding extendable rack. *Source*: (a) Courtesy of Integra; (b) R. Ian Freshney, courtesy of Beatson Institute.

for large numbers of bottles, such as the RollerCell 40 processing unit (Cellon) or the Cellmate™ (Sartorius; see Section 28.7.1).

Although the increasing availability of other scale-up systems has led to a decline in roller culture, it continues to be used by many laboratories because it is economical, is reliable, and provides an easy and rapid form of culture scale-up. Roller bottles are relatively cheap compared to other scale-up systems and roller racks are usually reported to be reliable, with few mechanical problems. Roller culture is particularly useful for applications where repeated harvesting of cells or supernatant is required, such as antibody production. When hybridomas are grown in roller bottles, they can produce approximately 5–25 μg/ml of antibody, although this is heavily dependent on the cell line, with some hybridomas preferring to grow under static conditions (personal communication, K. Wycherley). A method is provided in the Supplementary Material online for roller bottle culture (see Supp. S28.1).

28.3.2 Multisurface Propagators

Variations on the tissue culture flask have been developed that have an expanded surface area for monolayer culture but do not exceed the footprint of a 175-cm² flask (see Figure 8.7). Multilayered flasks are typically handled in a similar manner to a single-layered flask, with some additional care during pipetting to ensure that liquid is added evenly and removed completely across all layers. Surface areas and estimated cell yields for some commonly used multilayered flasks are provided in a previous chapter (see Table 8.2). If multilayer flasks do not provide sufficient surface area, larger multisurface propagators can be used.

Stacked culture trays. Examples of stacked culture trays include the Nunc Cell Factory™ (ThermoFisher Scientific)

and the CellSTACK® (Corning; see Figure 28.8). These systems are made up of welded stacks of rectangular Petri dish-like units. Up to 40 tiers can be used to give a total surface area of about 25 000 cm² and a yield of 10^9 cells or more (see Table 8.2). The trays are interconnected at two adjacent corners by vertical tubes. Because of the positions of the apertures in the vertical tubes, medium can flow between compartments only when the unit is placed on end. When the unit is rotated and laid flat, the liquid in each compartment is isolated, although the apertures in the interconnecting tubes still allow connection of the gas phase. These systems have the advantage that they are basically similar to a conventional flask or Petri dish, in terms of their geometry and the nature of the substrate; some substrate treatments are also available, e.g. CellBIND (Corning; see Section 8.3.7). The main limitations of these systems are that they are relatively bulky and can take some practice to handle comfortably. They are also difficult to visualize; it is a good idea to seed cells at the same density into a single-layer flask for observation. Protocol P28.1 sets out some of the basic requirements when handling a multisurface propagator.

PROTOCOL P28.1. HANDLING A NUNC CELL FACTORY

Background

Instructions here are written for the Nunc Cell Factory System. The same principles apply to other stacked culture tray systems, but the details may vary depending on the specific product, e.g. the Nunc EasyFill™ Cell Factory has a larger port, allowing it

Fig. 28.8. Multisurface propagators. (a, b) Nunc Cell Factory (ThermoFisher Scientific) showing: (a) a 2-chamber vessel (1264 cm²); a 10-chamber vessel (6320 cm²); and (b) a 40-chamber vessel (25 280 cm²). (c–e) CellSTACK (Corning) showing: (c) a single chamber vessel (636 cm²); (d) a 10-chamber vessel (6360 cm²); and (e) a 40-chamber vessel (254 440 cm²). *Source*: (a, b) Courtesy of Nunc-Thermo Scientific; (c–e) Courtesy of Corning.

to be filled directly by pouring. Always refer to the manufacturer's instructions before proceeding.

Large multisurface propagators such as this were designed primarily for growing large numbers of adherent cells in flask-like conditions. The supernatant can be collected repeatedly for virus or other cell product purification. The collection of cells for analysis depends on the efficiency of trypsinization and can be expensive if the unit is discarded each time cells are collected.

Outline

Prepare a cell suspension in medium and run the suspension into the chambers of the unit, lying on its long edge (see Figure 28.9). Rotate through 90° and then lay the unit flat for incubation.

Materials (sterile or aseptically prepared)

☐ Adherent cell line
☐ Complete growth medium as required for that cell line

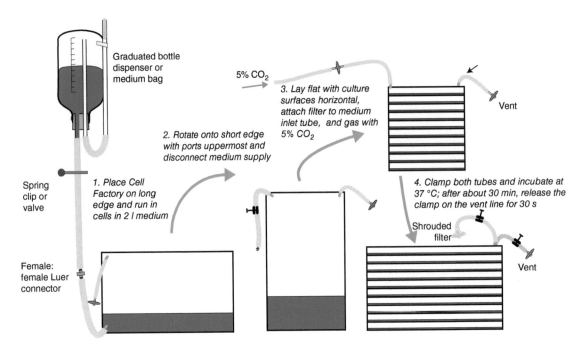

Fig. 28.9. Filling the Nunc cell factory. The device is placed on its side (long edge down) for medium to be added. Once added, the device is placed with its short edge down, allowing the medium supply to be disconnected, and then placed flat for culture.

□ Dulbecco's phosphate buffered saline without Ca^{2+} and Mg^{2+} (DPBS-A)

□ Trypsin, 0.25% (crude; see Section 13.2.1)

□ Nunc Cell Factory System, 10-layer (ThermoFisher Scientific)

□ Bottle or disposable medium bag (for transfer of the cell suspension)

□ Air vent filter assembly (e.g. #179553, ThermoFisher Scientific)

□ Tubing and other accessories to establish a closed, sterile system for transfer of the cell suspension (e.g. #140815, ThermoFisher Scientific)

Materials (non-sterile)

□ Biological safety cabinet (BSC) and associated consumables (see Protocol P12.1)

□ Cell counting equipment (hemocytometer or automated counter; see Section 19.1)

Procedure

A. Seeding the Cell Factory

1. Prepare a work space for aseptic technique using a BSC.

2. Trypsinize the cells to give a cell suspension (see Protocol P14.2, steps 1–15).

3. Count the cells with a hemocytometer or an automated cell counter.

4. Dilute the suspension to give 2×10^4 cells/ml in 2 l medium and transfer to a medium bag or bottle.

5. Open the air vent port on the Cell Factory and attach the air vent filter assembly.

6. Connect the sterile tubing to the other port and to the medium bag or bottle.

7. Place the Cell Factory on a long edge with the air vent at the top and the port for medium transfer at the bottom (see Figure 28.9).

8. Move the bag or bottle so it is above the Cell Factory, allowing it to run into the medium reservoir using gravity. Allow the medium to equilibrate across all layers.

9. Rotate the Cell Factory through 90° in the plane of the monolayer, so that the unit lies on the short edge, with the air vent filter assembly and supply tube at the top.

10. Disconnect the medium delivery tube from the medium reservoir and replace it with a cap. Leave the air filter in position (it can remain during culture).

11. Lay the Cell Factory flat on its base, with the culture surfaces horizontal, ready for culture.
12. Transport the Cell Factory to the incubator, being careful to keep the medium away from the supply and vent ports during movement.

B. Collecting Medium

13. Connect fresh sterile tubing to the port that was used previously for transfer of medium and to a fresh medium bag or bottle.
14. Place the Cell Factory on a long edge with the air vent at the top and the port for medium transfer at the bottom.
15. Move the bag or bottle so it is below the Cell Factory, allowing medium to run out of the Cell Factory and into the bag or bottle using gravity.
16. Tilt the Cell Factory to drain completely.

C. Harvesting Cells

17. Remove the medium as in steps 13–16.
18. Add 500 ml of DPBS-A and then remove it.
19. Add 500 ml of trypsin at 4 °C and remove it after 30 seconds.
20. Incubate the cells with the residual trypsin for 15 minutes.
21. Add 500 ml medium, distribute evenly among the chambers, and rock to resuspend the cells.
22. Run the cells and medium off as in steps 13–16.

Notes

Step 5: This step allows any pressure that builds up during filling to be released.

Step 11: If a CO_2 incubator is not available, purge the unit with 5% CO_2 in air for five minutes (see Section 12.5.3) and then clamp off both the supply and the outlet.

Step 22: The residue may be used to seed the next culture, although this method makes it difficult to control the seeding density. It is better to discard the chamber and start afresh.

Compact multisurface propagators. Although stacked culture trays can be used to expand the surface area, the larger stacks can be difficult to handle and to fit into a CO_2 incubator. Several multisurface propagators have been developed that are sold alongside stacked culture trays as space-saving alternatives. Some of these systems require perfusion, resulting in some overlap with larger perfusion bioreactors (see Section 28.3.4). Examples include:

(1) **The HYPERStack® vessel (Corning).** This vessel uses a gas-permeable film to eliminate the air "headspace"

within each layer, as previously described for the smaller HYPERFlask® (see Section 8.5.4). HYPERStack vessels are available with up to 120 layers, giving a total surface area of up to $60\,000\,cm^2$ and a yield of 10^9 cells or more from a single vessel (see Table 8.2). Several HYPERStack vessels can be set up using a manifold arrangement, allowing operation in series or in parallel. Further information on operation and performance of the HYPERStack vessel is available elsewhere (Titus et al. 2010; Pardo et al. 2011).

(2) **BelloCell aerator culture (Cesco Bioengineering).** This system was designed to offer a compact alternative to roller bottles. The central unit (the BelloStage) has a "bellows" action that forces medium over cells that are anchored in porous matrices and then withdraws it again, resulting in a "breathing" or "tide" effect. Optimum mixing and aeration is claimed with minimum shear, resulting in relatively high densities (10^9 cells from a 500-ml BelloCell bottle) (Ho et al. 2004).

(3) **The CellCube® system (Corning).** This system is essentially a hollow polystyrene cube with multiple inner lamellae of $8500\,cm^2$ growth area, perfused with oxygenated, heated medium. CellCube modules have between 10 and 100 layers, resulting in a total surface area of up to $85\,000\,cm^2$ and an estimated yield of up to 8.5×10^9 cells (Corning 2012). Each module is operated as a closed system using an E-Cube™ Culture System Kit (Corning) and a perfusion pump, which must be supplied by the user. Although this system can be used for intermediate scale-up procedures, some concerns have been expressed regarding non-uniform cell distribution patterns, resulting in unexpected cell behavior (Aunins et al. 2003).

(4) **The Xpansion® multiplate bioreactor system (Pall Corporation).** This system is based on a series of circular plates where the air "headspace" has been eliminated and replaced by an automatically controlled aeration system. A process control system can be used to set the preferred pH and oxygen levels, which are achieved using perfusion (see Section 28.4). The resulting system provides a surface area of up to $122\,400\,cm^2$ in a relatively compact space.

28.3.3 Microcarrier Culture

Microcarriers were first used for the culture of adherent cells by Toon Van Wezel, who successfully grew adherent cells on the surface of chromatography beads made from diethylaminoethyl (DEAE)-Sephadex (van Wezel 1967). Van Wezel's discovery allowed adherent cells to be scaled up for vaccine production (van der Velden-de Groot 1995). Culturing adherent cells on microbeads gives a maximum ratio of the surface area of the culture to volume of the medium (up to $90\,000\,cm^2$/l), depending on the size and density of the beads. Microcarriers are now used for the propagation of many anchorage-dependent cell types, including MSCs and pluripotent stem cells (PSCs) (Sart et al. 2013; Badenes et al. 2016). While microcarriers were originally developed for culture expansion, they are also used for 3D applications, stem cell differentiation, and tissue engineering (Justice et al. 2009; Sart et al. 2013). A protocol for a 3D angiogenesis assay using Cytodex™ 3 microcarriers (GE Healthcare Life Sciences) is provided as Supplementary Material online (see Supp. S27.2).

Microcarriers that are used for cell culture tend to range from 90 to 300 µm in size, or to exceed 500 µm; the larger carriers are often described as macrocarriers (see Table 28.2). Although their sizes may be comparable, microcarriers vary in other physical properties, including their core composition, geometry, and surface charge (see Figure 27.10c, d). Commercially available microcarriers can be broadly grouped into (i) positively charged carriers; (ii) non-charged or negatively charged carriers; (iii) carriers that have been coated with extracellular matrix (ECM) or its constituents; and (iv) porous macrocarriers (Merten 2015). Selection of a suitable carrier will depend on the cell type and the application. For example, culture of two human embryonic stem cell (hESC) lines on 10 different carriers showed that positively charged, cylindrical, or spherical carriers resulted in higher expansion potential compared to negatively charged, small, or porous carriers (Chen et al. 2011). If a carrier is to be used for therapeutic purposes, it is important to employ a rigorous selection process to determine the optimum choice early in the process of development, as the choice must be "locked in" well before clinical trials commence (Rafiq et al. 2016).

Microcarriers can be used in many of the bioreactor systems that are employed for scale-up in suspension (see Section 28.2). For example, induced pluripotent stem cells (iPSCs) can be grown on microcarriers in small (50–100 ml) spinner flasks (see Figure 28.1c); glassware is usually siliconized to prevent adhesion, e.g. using Sigmacote® (Sigma-Aldrich) or Siliclad® (Gelest). Stem cells can maintain their pluripotency under these conditions, although some optimization is required (Bardy et al. 2013; Gupta et al. 2016). Although microcarriers have many benefits, they also have some limitations. Harvesting cells from carrier beads can result in poor recovery, viability, and functionality (Badenes et al. 2016). Several methods have been developed for cell harvesting that are simple and scalable; dissolvable microcarriers can also be used (Nienow et al. 2014; Patel et al. 2017; Rodrigues et al. 2019). Where simple counting is required, a sample of microbeads can be removed and trypsinized prior to counting.

Porous macrocarriers are particularly useful for scale-up of adherent cells that are sensitive to shear damage in suspension. The porous nature of these carriers allows cells to grow in the interstices of the material as well as on its surface, resulting in a fully 3D environment (see Figure 28.10; see also Section 27.5.4). Porous macrocarriers thus offer a greater surface area and greater protection, which is useful for some cell types, and

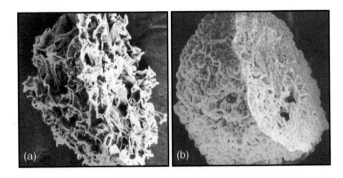

Fig. 28.10. Growth on porous macrocarriers. Carriers viewed by scanning electron microscopy showing: (a) bead alone; (b) bead with adherent CHO cells. *Source*: Courtesy of GE Healthcare.

cells can detach spontaneously and reattach to a fresh carrier bead (Xiao et al. 1999). Such macrobeads can be loaded with cells and stirred in a bioreactor; they can also be perfused in a fixed-bed bioreactor, e.g. for scale-up of hybridomas (personal communication, G. Lovrecz).

Whatever the size of the carrier, there are some key parameters that must be considered for effective microcarrier culture. These include (Badenes et al. 2016):

(1) *Cell seeding procedures.* Cells are typically incubated with the carrier beads under static conditions or at a lower agitation speed. Reducing the volume of the culture is helpful to encourage cell–cell and cell–carrier interaction (Badenes et al. 2016). Stem cells commonly form aggregates surrounding the microcarriers; this phenomenon may be associated with reduced viability. Cells can be seeded in a single-cell suspension to reduce aggregation but Rho kinase (ROCK) inhibitors may be required to reduce apoptosis (see Section 23.4.1) (Ashok et al. 2016).

(2) *Medium composition.* It should be possible to perform microcarrier culture using the optimal medium for the particular cell type. However, conditions must be optimized for each cell line. For example, while serum-free defined media can be used for MSC culture on microbeads, results may not be comparable to serum-containing media or to culture performed under static conditions (Tan et al. 2015).

(3) *Agitation speed.* Stirring or other forms of agitation should be high enough to give a homogenous suspension but low enough to minimize carrier breakage and cell damage. In practice, this means that agitation is often confined to a narrow range; in one study, mouse iPSCs proliferated in a spinner flask operated at 25 rpm but cell number declined when the speed was increased to 28–30 rpm (Gupta et al. 2016). Bead density will also influence the stirring speed, with a higher speed required for higher-density beads.

(4) *Microcarrier coatings.* Coated microcarriers are particularly useful for primary culture and for cell types that are difficult to grow, e.g. stem cells (Merten 2015). Coatings include Matrigel, collagen, laminin, and other ECM constituents

TABLE 28.2. Microcarriers and macrocarriers.

Name[a]	Supplier	Composition	Specific gravity (cm³/g)	Diameter (μm)
Positively charged carriers				
Corning Positively-charged	Corning	Polystyrene	1.09	160–200
Cytodex 1	GE Healthcare Life Sciences	DEAE-dextran	1.03	60–87
SoloHill® Hillex®	Pall Corporation	Polystyrene	1.09–1.15	160–200
SoloHill Plastic Plus	Pall Corporation	Styrene copolymer	1.034–1.046	125–212
SoloHill Star-Plus	Pall Corporation	Polystyrene (modified to improve attachment)	1.02–1.03	125–212
Non-charged or negatively charged carriers				
Corning Enhanced Attachment	Corning	Polystyrene	1.026	125–212
RapidCell™ G	MP Biomedicals	Glass	1.03	150–210
ECM-coated carriers (including constituents and mimetic treatments)				
Corning collagen-coated	Corning	Collagen coating	1.026	125–212
Corning Synthemax II (low and high concentration)	Corning	Synthetic coating for stem cell culture	1.026	125–212
Cytodex 3	GE Healthcare Life Sciences	Gelatin-coated dextran	1.04	175
RapidCell C	MP Biomedicals	Collagen-coated	1.02	90–150
SoloHill collagen-coated	Pall Corporation	Collagen-coated (type I porcine)	1.02–1.03	90–150, 125–212
SoloHill FACT III	Pall Corporation	Collagen-coated (modified to improve attachment)	1.02–1.03	125–212
SphereCol®	Advanced Biomatrix	Collagen-coated	1.03	125–212
Porous carriers and macrocarriers				
Cultispher®-G	Percell Biolytica	Gelatin, macroporous	1.04	130–380
Cultispher-S	Percell Biolytica	Gelatin, microporous (high thermal stability)	1.04	130–380
Cytopore 1	GE Healthcare Life Sciences	Macroporous DEAE-cellulose (low charge density)	1.03	230
Cytopore 2	GE Healthcare Life Sciences	Macroporous DEAE-cellulose (high charge density)	1.03	230
Fibra-Cel	Eppendorf	Treated polyester/polypropylene		6000 diameter disk

[a]Due to the large number of carriers that are now available for cell culture, it is impossible to provide a comprehensive list. This table lists some commonly used examples from commercial suppliers as a starting point for exploration.

(Badenes et al. 2016). However, matrix products may be undefined and suffer from variations between batches (see Sections 8.3.2, 26.3.5). Synthetic surfaces are also available for microcarriers, e.g. Synthemax™ (Corning) (Hervy et al. 2014).

28.3.4 Scaffold-Based Perfusion Bioreactors

Van Wezel's discovery of microcarriers resulted in a significant increase in the available surface area for culture of adherent cells (van der Velden-de Groot 1995). It also enabled the development of new bioreactors where adherent cells could be maintained for prolonged periods of time. This type of system requires perfusion for optimal supply of nutrients and oxygen and for removal of inhibitory metabolites. Culture of adherent cells on a microcarrier or other substrate, combined with perfusion, can lead to substantial increases in capacity, cell yield, and product titer (Merten 2015). These systems can also be used for anchorage-independent cells (e.g. hybridomas), as they may provide a method for cell retention and allow culture at high density for prolonged periods of time.

Fixed bed bioreactors. Fixed bed bioreactors have been used for cell culture since the 1970s, when BHK cells were grown in a bed of glass spheres to produce foot-and-mouth disease virus (Spier and Whiteside 1976). The original vessel for this work contained a deep chamber or bed for the carrier beads; medium and gas were introduced at the base of the bed using an airlift system, and the biological product was collected from the medium reservoir at the top (see Figure 28.11). Subsequent designs were modified to address early problems that arose with scale-up, such as channeling and gradient formation (Griffiths et al. 1987; Warnock et al. 2005). Using a fixed bed bioreactor for hybridoma cells, it became possible to generate 110 g of monoclonal antibody from a single production run using a 15-l bioreactor, which was small enough to be sterilized in an industrial autoclave (Bliem et al. 1990). A fixed bed bioreactor also reduced shear stress, allowing biological production from anchorage-dependent cells that did not grow well in stirred-tank systems (Merten 2015). This has allowed fixed bed bioreactors to be used for tissue engineering, e.g. for culture of hepatocytes in a bioartificial liver (Warnock et al. 2005).

Fixed bed bioreactors are commercially available or can be custom-made by some companies (e.g. BioReactor Sciences). Bioreactors that are commonly used for culture of adherent cells include:

(1) **Bioreactor systems loaded with Fibra-Cel® disks (Eppendorf).** Suitable systems include the CelliGen® 310 (New Brunswick), which has been replaced by the BioFlo 310 (Eppendorf). Disks are retained in a fixed bed using two screens that are positioned on the impeller, which act as a basket (see Figure 28.12). The manufacturer has reported

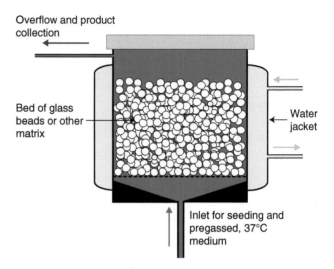

Fig. 28.11. Fixed bed bioreactor. Cells grown on the surface of beads or macrocarriers are perfused with medium. The beads are settled in a dense bed resting on a perforated base at the bottom of the culture vessel, or, if of a lighter material, are restrained within a cage. Once the culture is established, the beads do not move and medium percolates around them.

that anchorage-dependent cells attach to the disks more rapidly than they do to microcarriers; the system can be scaled up to 150 l capacity (Mirro 2011).

(2) **The iCELLis® bioreactor system (Pall Corporation).** This system consists of a single-use bioreactor that contains a proprietary macrocarrier, made from polyester microfibers. The iCELLis Nano is designed for small-scale production (up to 4 m² surface area), while the iCELLis 500+ is designed for industrial manufacture (up to 500 m² surface area).

(3) **The TideCell HD cell culture system (Cesco Bioengineering).** This system is described by the manufacturer as a pilot-scale bioreactor that uses the same patented principle as the BelloCell bottle (see Section 28.3.2), with a fixed bed capacity of about 10–20 l (up to 312 m² surface area) and a yield of up to 10^{11} cells.

Hollow fiber systems. Hollow fiber systems were originally designed to offer a high-density, 3D environment for the culture of anchorage-dependent cells (see Section 27.5.3) (Knazek et al. 1972). These systems operate on the principle of compartmentalization, which means that they can also be used for anchorage-independent cells. Cells are seeded in a hollow fiber cartridge (see Figure 28.13b); typically, the fibers are designed to support cell growth on their outer surfaces and are gas- and nutrient-permeable. High molecular weight products concentrate in the outer space with the cells, while nutrients are supplied and metabolites are removed from the inside of the fiber using a perfusion system. Different plastics and ultrafiltration properties give different molecular

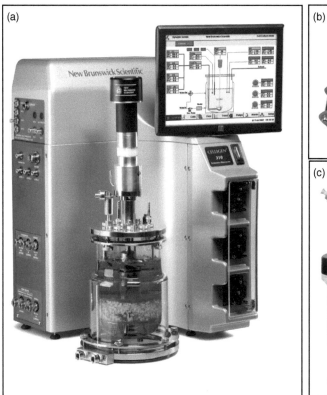
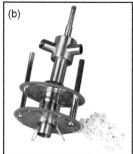

Fig. 28.12. Fixed bed bioreactor using Fibra-Cel disks. (a) Bioreactor chamber with Fibra-Cel disks in medium, surrounded by water jacket connected to pump and control unit. (b) Basket impeller with Fibra-Cel disks lying alongside. (c) BioBLU single-use packed bed bioreactor. *Source*: Courtesy of New Brunswick Scientific.

weight cut-off points, regulating the diffusion of metabolites and products (e.g. monoclonal antibodies) based on their size. The final outcome is a 3D, closed environment where high cell densities can be achieved, similar to those found *in vivo*, and biological products can be generated and harvested for extended periods of time.

Hollow fiber bioreactors are available commercially or can be fabricated by laboratories with the necessary expertise (Storm et al. 2016). These systems can be technically challenging to set up and it may be difficult to sample cells once they are in place, so it is important to discuss with suppliers and compare the performance of their systems to your own requirements. Commercially available hollow fiber systems include:

(4) **FiberCell bioreactor systems.** These systems are designed for operation inside a CO_2 incubator. Hollow fiber cartridges can be purchased separately or as part of a complete system that includes a pump, accessories, and media (see Figure 28.13a). Medium-sized cartridges offer 3000 cm^2 surface area for culture, supporting up to 2×10^9 cells. Cartridges contain three different choices of fiber material (polysulfone, cellulosic, and PS+ fibers)

and have a choice of three molecular weight cut-offs for diffusion (5 kD, 20 kD, and 0.1 μm).

(5) **The Quantum® Cell Expansion System (Terumo BCT).** This system is designed to replace the CO_2 incubator and minimize ongoing handling using a self-contained, automated culture platform. The platform includes a disposable cell expansion set that contains a hollow fiber cartridge, which is loaded as a closed system. Disposable bags are used to supply media and cells and to collect waste and the product harvest.

(6) **HF PRIMER™ (C3).** This system is designed for production of less than 1 g of protein, using a hollow fiber cartridge and accessories that fit inside a CO_2 incubator. Larger models for scale-up are self-contained and include the AutovaxID®, the AcuSyst-Maximizer®, and the AcuSyst-Xcellerator™.

28.4 MONITORING AND PROCESS CONTROL

As cultures are scaled up in size and complexity, it becomes increasingly important to optimize the conditions in which

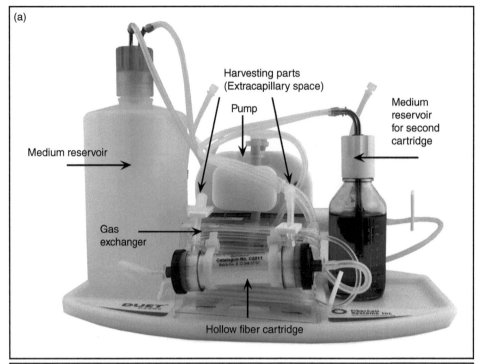

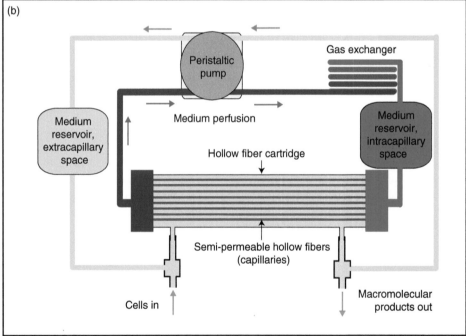

Fig. 28.13. Hollow fiber culture. (a) A hollow fiber culture system set up for operation, with the capacity to operate two cartridges simultaneously (a single cartridge is shown here for clarity); (b) a sectional diagram showing the major components of such a system. Each cartridge is made up of hollow fibers that essentially act as capillaries. Cells are grown within the cartridge in the extracapillary space, while medium is circulated from a reservoir to the intracapillary space via a peristaltic pump. Aeration and CO_2 exchange are via a gas-permeable tubing coil when the assembly is housed in a CO_2 incubator. *Source:* (a) Courtesy of FiberCell Systems, with permission; (b) R. Ian Freshney.

cells are grown, based on monitoring of key variables. For example, if scale-up is performed in suspension using a fed-batch approach (see Section 28.1), it is important to monitor cell density in order to determine the best approach to refeeding. Using cell density measurements to control the process of batch feeding, it is possible to reduce the production of toxic metabolites by 50-fold and increase the production of monoclonal antibody by 10-fold (Xie and Wang 1994). Although this approach is clearly necessary for large industrial bioreactors, it is also important for smaller, bench-top devices. Most bioreactors that are installed in research laboratories come with software from the manufacturer that uses feedback from monitoring of key variables to control the culture environment and associated processes.

This software is usually designed to comply with regulatory requirements in a GMP environment, e.g. 21 CFR Part 11 for electronic records (FDA 2019).

Process control can be defined as the active changing of the process, based on the results of process monitoring (NIST and SEMATECH 2012). Sensors must be installed to monitor key variables which may include temperature, pH, CO_2, dissolved oxygen (DO), osmolarity, and glucose (see Figure 28.14). Single-use bioreactors will typically have non-invasive or disposable sensors, which are supplied with the vessel depending on the user's requirements (Busse et al. 2017). Sensor data will pass to a central computer, where process control software will initiate adjustments to ensure that variables stay within a predetermined range. In a

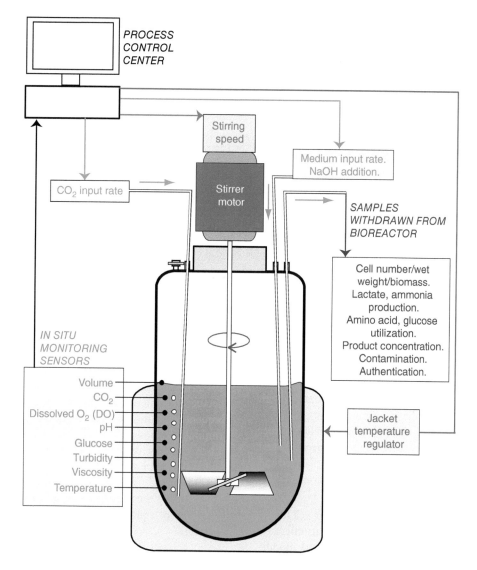

Fig. 28.14. Bioreactor process control. Schematic representation of a paddle-stirred bioreactor with direct-reading probes on the left, feeding to a control unit (top left) that stores the data and also regulates conditions within the bioreactor. A sampling port on the right withdraws the cell suspension from the bioreactor for analysis.

stirred-tank system, adjustments may affect the stirring speed, the input rate of CO_2 or other gases, or the input rate of medium. In a rocking bioreactor, adjustments may affect the rocking speed or angle, which will in turn affect aeration. The user can normally make manual adjustments as required, e.g. after removing samples for cell counting. Depending on the software, the user may also be able to set alarms that will become active if readings pass beyond the set range for a particular variable or to enter instructions that will change the culture conditions in accordance with a standardized protocol.

28.5 SCALE-UP FOR MANUFACTURE

This chapter has concentrated on scale-up at the laboratory level, with some indications of equipment used at pilot-plant or semi-industrial scale, but does not attempt to describe large-scale industrial processes. Minireview M28.1 is presented as an introduction to industrial-scale bioreactors for use in manufacturing of biological products.

Minireview M28.1. ***Culture Scale-Up and Bioreactors*** Contributed by George O. Lovrecz, Adjunct Professor of RMIT and Monash University, Research Group Leader, Protein Production and Fermentation, CSIRO.

Scaling up animal cultures may be difficult as often substantial changes in culture conditions, compared to small-scale work, need to be introduced (see Sections 28.2, 28.3). These changes might include dealing with increased mechanical stress for adequate stirring; higher cell densities to achieve better yields and thus reduce production cost; increased gassing to control pH and DO levels; modified media composition; the addition of various chemicals to protect cells (e.g. Pluronic F68, Sigma-Aldrich), minimize foaming (FoamAway, ThermoFisher Scientific), and/or avoid cell clumping (Anti-Clumping Agent, ThermoFisher Scientific); and new matrixes, e.g. single-use bags (see Section 28.2.2). Anchorage-dependent cells are also subjected to altered geometry when grown in close contact with micro-/macrocarriers and hollow fiber cartridges.

Successful scale-up usually requires a detailed plan to consider certain aspects of cell biology, bioprocess engineering, and protein chemistry. The following steps should be planned and carefully analyzed for large-scale protein production:

(1) Cell and molecular biology (generation of the cell line):
  (a) Nature of the product such as stability, its effect on cell viability, and required post-translational modifications including glycosylation
  (b) Available assays, which should be robust and easy to carry out
  (c) Growth kinetics of host cells and stability of clones (number of generations)

  (d) Expression system including vectors, promoters, and thus formulation of selection media
(2) Test, characterization, and optimization:
  (a) Data collection to establish growth rates, yields, and specific requirements and to identify potential problems (e.g. specific additives, clumping, shear sensitivity)
  (b) Establishment of cell bank to ensure identical starting point for each culture
  (c) Determination of the most suitable operating parameters such as pH, DO, nutrient levels, and length of production
(3) Selection of large-scale processes:
  (a) Methods of production (batch, fed-batch, or continuous)
  (b) Choice of production vessels/bioreactors
(4) Monitoring and control:
  (a) Parameters to be monitored *in situ*, by sampling or by derivation
  (b) Required frequency of sampling
  (c) Level of automation.

The majority of the above issues are discussed earlier in this chapter and elsewhere, so only the selection of large-scale processes will be considered here.

Selection of Large-Scale Processes

The selection of large-scale processes should be based on the results of small-scale development work which is essentially driven by the critical attributes of the final product including biological activity, purity, required sterility, and often cost.

While continuous production methods (maximum yield and minimal product heterogeneity would be achieved under steady-state conditions) are superior compared to batch/fed-batch methods, the most preferred and thus most commonly used production methods are the batch/fed-batch processes. This might be explained by the fact that batch/fed-batch methods are relatively short (10–20 days) and easy to carry out, i.e. often not requiring specific equipment (feed/harvest pumps, storage vessels) or frequent sampling or control measures when compared to continuous processes.

When carefully designed and tested expression systems are available and the method of production (batch or continuous) is selected, the final choice of the production vessel is influenced by the required amount of the final product, the availability of equipment, and the level of experience of the operators.

The Scale-Up Train

Currently, the majority of protein production is carried out in suspension cultures as monolayer (or anchorage-dependent) cultures are difficult to handle and monitor (see Section 28.3).

TABLE 28.3. **Scale-up train for suspension cultures.**

Stage	Initial culture volume	Typical concentrations (×10⁻⁶ cells/ml)		Commonly used equipment
		Starting	Final	
Cell revival (from freeze vial stock)	5–20 ml	0.1–0.5	1.0–3.0	E-flask/T-flask
First passage (no more static cultures)	0.1–0.4 l	0.2–0.5	2.0–4.0	E-flask/roller bottles
Second and subsequent passages	0.5–2.0 l	0.2–0.5	2.0–4.0	E-flask/roller bottles/ Bioreactors
Inoculating production (final) vessel	> 5.0 l	0.2–0.5	2.0–10.0	Spinners, stirred or rocked bag Bioreactors

Source: Courtesy of George Lovrecz.

Suspension culture scale-up is relatively simple: the culture volume has to be increased to achieve the required cell mass, which in turn is capable of producing the amount of protein needed.

The scale-up train – that is, the sequential transfer of cells (subcultures or passages) from small to large volumes – needs to be planned ahead (see Table 28.3). It is desirable to keep the cell concentration of the inoculum low to minimize the number of passages, but it is equally important to avoid slowing down cell growth (e.g. extending the lag phase) due to major changes in the environment. Standard protocols suggest concentrations of 2–5×10^5 viable cells/ml as a minimum inoculum concentration, which normally corresponds to a 5- to 10-fold dilution (i.e. a 5- to 10-fold scale-up), in contrast to microbial cultures where 20- to 100-fold scale-up steps are common. It is also important to understand the minimum volume requirement for the proposed vessel/bioreactor to ensure its proper operation, such as adequate stirring and gassing for pH and DO control. For example, both stirred and rocked (single-use bags) systems normally need a minimum initial volume of 20–40% of the final working volume.

Roller bottles, Erlenmeyer flasks (e-flasks), and Fernbach flasks (see Sections 8.6.2, 28.3.1) are commonly used to generate sufficient cell mass to inoculate spinner cultures, rocked bags (Wave), or smaller stirred-tank bioreactors as an intermediate step. When sufficient cell concentrations (normally 2–5×10^6 viable cells/ml) are achieved in one vessel, the culture may be transferred to the next larger vessel until the final production vessel is inoculated. Table 28.1 lists the most common vessels used in a laboratory environment. Please note that suspension-adapted cells should not be cultivated in static vessels such as tissue culture flasks (T-flasks) or various multistack systems (e.g. Corning CellSTACK or Nunc TripleFlask), as they perform much better in terms of growth and production rates in shaken or stirred vessels.

Bioreactors

There are a great variety of vessels available for the large-scale propagation of animal cells, as discussed above (see Sections 28.2, 28.3) and summarized in Table 28.1. It is highly recommended to engage in scale-up and carry out the production in one vessel as a unit operation, instead of using several similar-sized vessels (scale-out) to obtain the required amount of protein. Using only one vessel or unit increases process consistency and minimizes heterogeneity of the product.

Analysis

Bioreactors may substantially differ regarding their design to achieve adequate mixing and aeration to control pH and DO levels: different vessel geometries, various stirrer blades (such as pitched blades, Rushton turbines, or paddles), and sparger rings are employed and thus generate different flow patterns and shear effects. Often it is difficult to estimate what bioreactor types and final conditions would be optimal for a specific process of a given cell line. For example, the same cell line using the same media cultivated in different vessels might need changes in the experimental conditions and thus require comprehensive studies to settle on the final design (see Table 28.4).

Scale-Up of Stem Cells

Stem cell research and its applications in tissue engineering and cell therapy are growing rapidly. Stem cells have specific features including the potential to differentiate, strong anchorage dependency, elevated sensitivity to shear stress, and unique requirement for growth factors and media conditions compared to common mammalian cells. Therefore stem cell scale-up presents an immense challenge to design a suitable system for scale-up in an undifferentiated state.

Current methods, because of the equipment and media required, are expensive and most likely need substantial improvement to be made widely available. Commonly used

TABLE 28.4. Cell culture parameters for FreeStyle™ 293-F cells using FreeStyle 293 media (ThermoFisher Scientific).

Parameter	Optimum Growth™ flask (Thomson Instrument Company), 2.5–5 l	WAVE Bioreactor System 20/50EH (GE Healthcare Life Sciences), 5–25 l	CelliGen BLU (Eppendorf/New Brunswick), 14 l
Pluronic F-68 (Sigma-Aldrich) concentration	None	0.02%	0.01%
Aeration	Headspace via 0.2-μm vented cap	Headspace with 0.2–0.3 l/min air	Macrosparging with 0.3–0.5 l/min air (plus occasional O_2)

Source: Courtesy of George Lovrecz.

equipment such as plates, carriers, and hollow fiber systems for the scale-up of anchorage-dependent cells (see Section 28.3) need be modified to cater for the particular needs of stem cells. A few systems currently available are listed below.

(1) Static, T-flask-like plates: Xpansion multiplate bioreactor system (Pall Corporation), CellSTACK (Corning)
(2) Carrier-based systems: Mobius CellReady (Merck Millipore)
(3) Hollow fiber-based systems: Quantum Cell Expansion (Terumo BCT), CellMax (Spectrum Labs; now discontinued)
(4) Fibercell Bioreactor System (FiberCellSystems).

Applications

From the late 1980s, pharmaceutical biotechnology enjoyed remarkable growth and reached a worldwide sale of biologics close to USD 200 billion per year. The sale of pharmaceutical drugs, produced by animal cells, is growing 12–17% per year compared to a 4–12% growth for other pharmaceutical agents and over one-third of all biologics are monoclonal antibodies (Evaluate Pharma, McKinsey report). It was expected that eight out of the ten top-selling drugs would be biologics by 2016, further emphasizing the importance of biologics compared to small molecule drugs (BtoBioInnovation).

As the estimated cost of treatment of patients using biologics is almost 20 times higher compared to traditional small molecule drugs (So and Katz 2010), governments are committed to cost-containment policies. This, combined with the fact that drug discovery including clinical trials is an expensive exercise (in the range of USD 1–1.5 billion), has compelled pharmaceutical companies to invest heavily in the development of new methods to ensure maximum yield and consistent production of the target proteins. Modern, genetically engineered cell lines, cultivated in defined media, have high specific productivity (increased from picograms of product to nanograms of product per cell per day), may be grown to higher cell densities ($2–150 \times 10^6$ cell/ml), and have better tolerance and survival compared to cell lines

used a decade ago. As a consequence, routine industrial-scale processes can achieve 3–6 g/l monoclonal antibody yields in fed-batch systems. These remarkable achievements are mainly due to optimized and high-throughput cloning and selection methods, well-researched feeding strategies, and systematic experimental design (see PAT and QbD initiatives discussed in the following section of this Minireview). Higher product yields allow the economical use of smaller, 500–2000 l working volume, bioreactors instead of the traditional > 5000 l bioreactors, i.e. there is only 5–10% difference between small and large bioreactors in terms of cost of goods (COG) when yields are above 4 g/l.

The increased demand for flexibility and faster turn-around time to allow faster route to the market, combined with the economical use of smaller bioreactors, has promoted the development of single-use (or disposable) bioreactors. Several rocked (Wave bag) and stirred (ThermoFisher Scientific, Sartorius, etc.) single-use reactors are available and fully suitable for both research and production work (see Section 28.2.2). Typical research scale, single-use reactor volumes are in the range of 10–100 l both for rocked and stirred bags, but stirred single-use bioreactors are available up to 3000 l for industrial-scale processes. A good summary is available elsewhere (Shukla and Gottschalk 2013).

The obvious advantage of single-use bioreactors is the simplified operation (such as fast set-up, minimal integrity testing, and easy dismantling) and greatly reduced capital investment cost compared to the traditional stainless-steel bioreactors. It is estimated that up to 40% reduction of COG can be achieved when single-use systems are employed for large-scale production of pharmaceuticals.

The advent of single-use systems required changes in sensing and control of certain fermentation process parameters; often pH and DO levels are measured by optical probes and only a heating pad is applied instead of heating and cooling jackets for temperature control.

Process Analytical Technologies (PAT) and Quality by Design (QbD)

From early 2000, it became increasingly obvious that using fixed processes for biological activities, which are inherently variable (scale-dependent, changing cell physiology, and batch-to-batch variations in raw materials), is counterproductive when large-scale production of biopharmaceuticals under GMP conditions must be carried out. Fixed process cannot ensure the required quality of the final product and, as a consequence, may lead to the disposal of the product and thus substantial financial losses. The current view is to carefully research, characterize, and understand the particular fermentation and purification systems and to design a process where the process itself guarantees the quality of the final product. For more details, refer to the case study A-Mab for bioprocess development where big pharma companies such as Pfizer, Genentech, and Amgen formed a working group to showcase this new initiative (CMC Biotech Working Group 2009). Also, refer to the US Food and Drug Administration (FDA) initiative and the European Union's regulatory/harmonization efforts (FDA 2004; ICH 2005).

Future Trends

The pharmaceutical industry has a rather conservative view regarding the use of new cell lines, such as PER.C6® (Crucell), and of novel technologies, as it usually slows down the regulatory approval and thus delays the marketing of product. However, increased pressure to minimize production cost might change this view. It is also expected that the combination of increased yields, the use of single-use bioreactors, and rapid and cost-effective point-of-care diagnostic tools might increase the use of personalized medicine and the application of stem cell therapies (PMC 2014; ISSCR 2016).

28.6 HIGH-THROUGHPUT SCREENING

So far, this chapter has focused on scale-up in the number of cells produced, which is essential when manufacturing biological products. Scale-up in the number of samples is essential for other applications, such as screening transfected cell populations for expression of a transgene (see Section 22.4). This form of scale-up is particularly important when performing cell-based assays for *in vitro* anticancer drug screening or for toxicity testing in a broader sense (see Chapter 29). For example, the 60 tumor cell lines that make up the NCI-60 panel have been used to screen more than 100 000 potential drugs and an even larger number of natural-product extracts (Weinstein 2012). At its peak, testing of the NCI-60 panel generated an estimated 9 million culture units per year; each cell line in the panel was exposed to each compound or extract at five different concentrations in triplicate (Boyd and Paull 1995; Shoemaker 2006).

Testing of the NCI-60 panel was performed using 96-well plates (Boyd and Paull 1995), but a number of other formats can be used for high-throughput screening. Cell lines can be embedded in paraffin for inclusion in "cell microarrays" (Hoos and Cordon-Cardo 2001; Ferrer et al. 2005; Waterworth et al. 2005). Cultures are typically trypsinized to give a cell suspension, fixed in 10% formalin, and suspended in agar to give a block or plug that can be embedded in paraffin. This approach preserves subcellular architecture and staining patterns and is useful for some forms of screening, e.g. selection of iPSC populations (La Spada et al. 2016). However, live cell assays cannot be performed and any 3D architecture is usually lost as part of preparing the culture for fixation.

Spheroid cultures can be adapted for high-throughput screening using spheroid microplates or hanging drop plates; a similar approach can be taken for organoid cultures (see Sections 27.4.1, 27.6). For example, tumor spheroids or organoids can be propagated and used for drug screening in a microwell plate format, usually after embedding in ECM (see Protocol P27.1) (Vinci et al. 2012; Francies et al. 2016). These approaches allow live cell imaging, which enables assays to be extended to incorporate migration, invasion, and various fluorescence- and colorimetric-based measurements of cytotoxicity (Hsiao et al. 2012). Such live cell assays usually require some form of automation, whether this extends to the full culture process or to certain aspects such as image acquisition and analysis (see Section 28.7.2).

Concerns have been raised regarding the validity of high-throughput screening efforts using cell line panels (Hatzis et al. 2014). Data generated from two large panels, the Cancer Cell Line Encyclopedia (CCLE) and the Cancer Genome Project (CGP), gave concordant results with regard to expression profiles but discordant results with regard to drug sensitivity (Haibe-Kains et al. 2013). Discordant results may relate to differences in (i) cell line selection and handling, including the seeding density and incubation period before compounds are added; (ii) compound storage and handling; (iii) culture substrates and geometry; (iv) culture media and serum; and (v) test methods used to count cells or assess cytotoxicity (Weinstein and Lorenzi 2013; Hatzis et al. 2014; Haverty et al. 2016). Clearly, it is necessary to standardize at least some of these variables for future studies. As a starting point, a reference set of cell lines and compounds is needed for comparison of experimental variables and data between laboratories.

28.7 AUTOMATION AND BIOPRINTING

The fundamental purpose of automation is to remove human operator dependency and variation from a process (Thomas and Ratcliffe 2012). Using an automated culture platform results in greater process capability, in terms of both the overall output (e.g. the number of cells or samples) and the proportion of the output that meets specifications (Liu et al. 2010).

Automation is particularly appealing for cell-based therapies, where there is a need to handle cultures using GMP (see Section 7.2.3), to generate large numbers of cells, and to perform complex procedures in a reproducible manner. Various stem cell populations including ESCs, MSCs, and iPSCs have been successfully reprogrammed (in the case of iPSCs), expanded, and induced to differentiate using automated culture platforms (Terstegge et al. 2007; Thomas et al. 2007; Paull et al. 2015).

Why do most tissue culture laboratories continue to rely on manual handling rather than automation? Cost is an obvious reason, but there are other reasons why automated platforms are not more widely used. Automation requires a standardized procedure that can be performed with minimal operator intervention in order to be successful. The protocols that are used in research laboratories are difficult to standardize, due to variable cell line requirements and the need for adaptation as research progresses. Experienced tissue culture personnel develop the ability to rapidly adapt manual handling procedures, based on the behavior of their cultures. This expertise is essential for research laboratories, where cultures often come from different tissues and can display very different growth patterns. Although automation can give a faster and more reproducible result, it can also give an outcome that is "off-center" compared to the manual process, resulting in the need for process development (Liu et al. 2010).

28.7.1 Automation of Culture Handling

The choice of an automated platform will depend on the purpose of the culture process. An increase in cell yield will require process control to automatically adjust the culture conditions in a bioreactor (see Section 28.4). An increase in sample number will require a robotic platform to automatically perform physical manipulations, with less emphasis on process control. If the process yields a biological product that will be used for therapeutic purposes, GMP is also essential (see Section 7.2.3). For research laboratories, the most straightforward approach to GMP is to collaborate with a facility that can perform manufacture at this level. An alternative approach would be to consider purchasing a GMP-compliant device that can be operated as a closed system, e.g. a hollow fiber bioreactor (Haack-Sorensen et al. 2016).

Robotic culture platforms are available as complete systems or can be assembled by combining devices in a modular approach (Paull et al. 2015; Daniszewski et al. 2018). Before searching for a robotic culture platform, it is important to consider whether cells can be grown in suspension and if multiwell plates can be used. If cells can be grown in suspension using multiwell plates, a robotic liquid handling workstation can be used (see Figure 5.9a, b), provided measures are in place to provide laminar flow and sterile handling.

If cells can be grown in suspension but require a dynamic system, an automated bioreactor can be used with stirring and sparging of cultures, e.g. the ambr® 15 system (Sartorius; see Figure 28.4). However, if cells must be grown as adherent monolayers in tissue culture flasks, a robotic platform must be used that is made specifically for this purpose. Tissue culture flasks were originally designed to reduce contamination during manual handling, resulting in angled movements that are not usually achievable with conventional liquid handling workstations.

Commercial robotic platforms that are available for culture of adherent monolayers in tissue culture flasks include (Daniszewski et al. 2018):

(1) **AUTO CULTURE™ (Kawasaki Heavy Industries).** This robotic system can handle tissue culture flasks and dishes. It includes a centrifugal separator, allowing full automation of passaging procedures for adherent cells under GMP (Kami et al. 2013).

(2) **Cellmate (TAP Biosystems, now Sartorius).** This system can handle tissue culture flasks and roller bottles, with subculture by enzymatic or mechanical disaggregation. Some limitations have been noted (e.g. bottlenecks that require manual intervention), which can be addressed by the addition of new workflows (Bernard et al. 2004).

(3) **CompacT SelecT™ (TAP Biosystems, now Sartorius; see Figure 28.15).** This platform can handle multiwell plates, 75-cm² flasks, 175-cm² flasks, and some multilayer flasks (see Section 8.5.4). It is available in various configurations, depending on which plate options are required. The system can be used for subculture by enzymatic or mechanical disaggregation, but centrifugation steps must be substituted or performed manually (Archibald et al. 2016).

28.7.2 Automation of Cell-Based Assays

Research laboratories may not be able to incorporate automation into routine handling procedures, but automation of a standardized assay may be more achievable. Automated liquid handling workstations can result in faster, more reproducible results for any assay, including authentication and other validation testing (see Section 5.1.3; see also Figure 5.9a, b). Automated cell imaging systems can increase speed and data capture for any cell-based assay that requires live-cell imaging, cell counting, or analysis of fluorescent or colorimetric markers. Examples of automated cell imaging systems include the BioSpa Live Cell Analysis System (BioTek), the Celldiscoverer system (Zeiss), the ImageXpress® system (Molecular Devices; see Figure 5.9d), the IncuCyte® live-cell analysis system (Sartorius; see Figure 18.4), and the Operetta (PerkinElmer). Some of these devices may be found in a tissue culture laboratory, while others are high-end devices that are only available through core facilities. It may take time to adapt a manual assay for automation, but the time will be

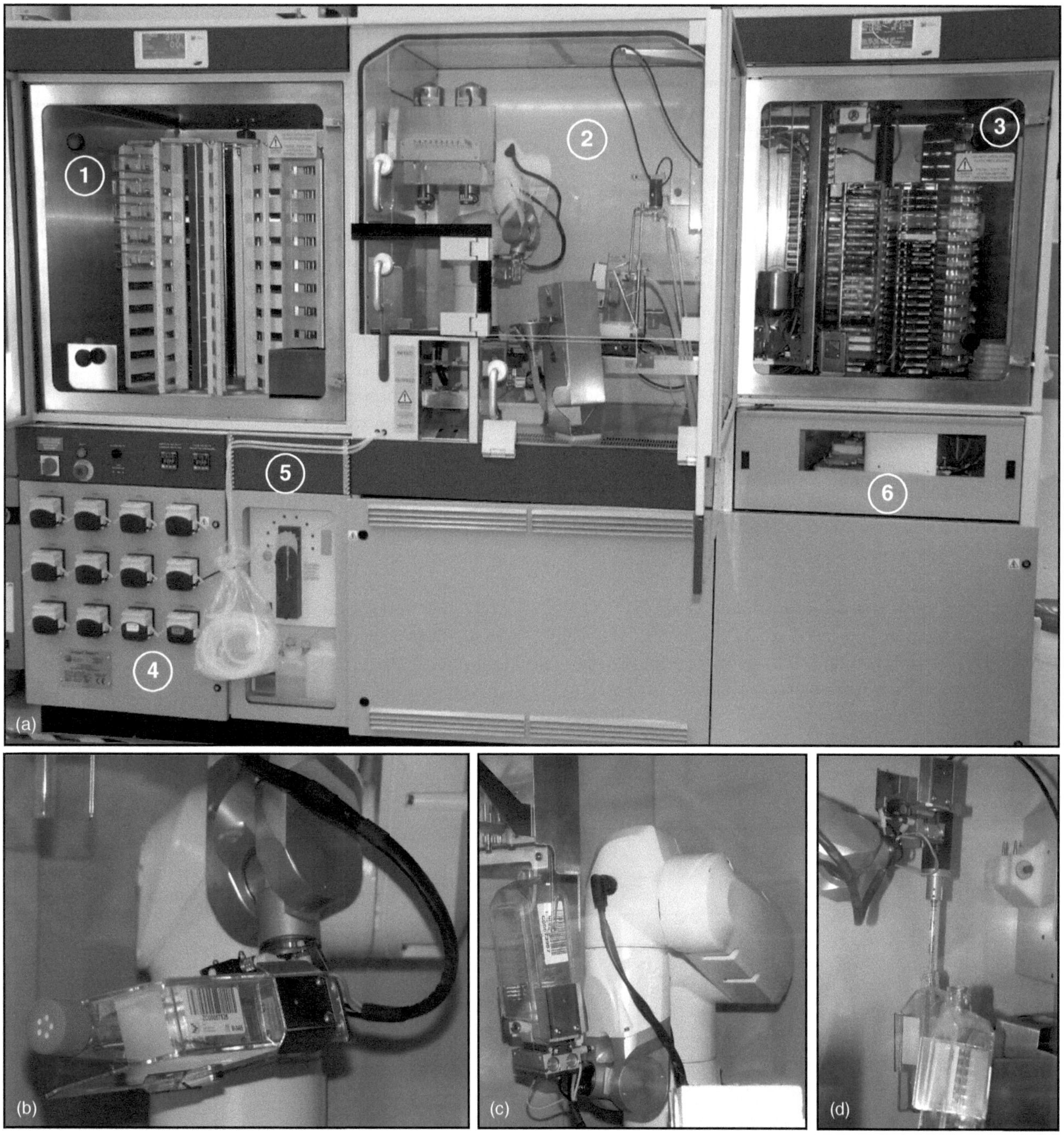

Fig. 28.15. The CompacT SelecT System (Sartorius). (a) Front aspect showing: (1) incubator chamber with flasks on top left of carousel; (2) main handling chamber with robotic arm in center, below left of number, capping/uncapping and dispensing unit to left of number, and pipette container below number; (3) plate incubator chamber; (4) peristaltic pumps for media additions; (5) automated cell counter; and (6) multiwell plate liquid handling unit. (b) Flask in gripper of robotic arm showing flask being rocked; it can also be shaken from side to side. (c) Uncapping; cap remains on device until flask is returned for recapping. (d) Pipetting cell suspension from front flask to rear flask. *Source*: Courtesy of TAP Biosystems, now available from Sartorius.

well spent provided the assay is performed on a regular basis and automation will save time for the operator. These assays are often performed by skilled tissue culture personnel who can use their skills more effectively elsewhere.

28.7.3 3D Bioprinting

Automation of culture handling is constrained by the cells' inherent properties (e.g. anchorage dependence) and by the vessels in which they are grown (see Section 28.7.1). Ideally, such constraints can be overcome by designing the vessel and the culture substrate to fit the application and the cell type. This is now possible using 3D printing. Today's 3D printing technology can be traced back to the pioneering work of Charles Hull, who invented stereolithography – the fabrication of solid objects by successively printing layers of a curable material on top of one another (see Table 28.5) (Hull 1984). The term "3D bioprinting" does not refer to any particular technology and can be defined as a material transfer process that results in the patterning and assembly of biologically relevant materials to accomplish one or more biological functions (Mironov et al. 2006). Bioprinting is performed using a "bioink" composed of cells, a scaffold (although this is not always required), and additional factors, e.g. a cross-linker that results in gelation after printing (Bishop et al. 2017; Gu et al. 2018).

What forms of 3D printing can be used to deliver viable cells? Inkjet printing, extrusion-based printing, and stereolithography are commonly used, with each technology having its own strengths and weaknesses (see Table 28.5) (Placone and Engler 2018). Other key considerations for 3D bioprinting include (i) computer-aided design (CAD) of the structure to be printed, based on suitable images; (ii) access to a bioprinter that can be operated under aseptic conditions; (iii) selection of a bioink that is suitable for printing and can provide a supportive cellular environment; (iv) inclusion of a cross-linker or other process for gelation after printing; and (v) selection of cells to be incorporated into the bioink or seeded in the printed scaffold (Bishop et al. 2017; Gu et al. 2018). Although there is a risk that 3D printing can damage cells, it is possible to achieve good viability and cell function. For example, iPSCs can be printed using extrusion-based printing, resulting in spheroids that maintain viability and pluripotency and that can be removed from the 3D structure for ongoing culture (see Figure 28.1a–c) (Gu et al. 2017).

While 3D bioprinting is primarily associated with tissue engineering, the technology can be used more broadly for 3D culture, automation of cell-based assays, or fabrication of prototypes, e.g. culture vessels for imaging studies (Bishop et al. 2017; Gulyas et al. 2018). Access to a suitable printer is currently one of the major challenges, with many tissue culture laboratories repurposing industrial or hobby 3D printers. This is likely to change in the near future, with an increasing number of purpose-built bioprinters becoming available, e.g. the RASTRUM Bioprinter (Inventia Life Science; see Figure 28.16d, e). If access to a bioprinter and other key requirements can be met, this technology is likely to become essential for many tissue culture laboratories in future.

TABLE 28.5. Commonly used 3D printing technologies.

3D printing technology	Description	Resolution	Fabrication speed	Cell compatible	Support structure required
Extrusion-based printing	A mechanical or pneumatic force is used to dispense bioink through a nozzle	100 µm	Medium	Yes	No
Inkjet printing	A thermal, piezoelectric, or electromagnetic force is used to generate droplets of bioink on demand	~ 30 µm	Fast	Yes	Yes
Stereolithography	A laser is used to cure a liquid polymer, resulting in the formation of solid layers that are built up in succession	20–50 µm	Fast	Yes	Yes
Selective laser sintering	A laser is used to fuse powder granules in a bed of resin or other material, resulting in the formation of solid structures	~ 20–100 µm	Slow	No	No

Source: Placone, J. K. and Engler, A. J. (2018). Recent Advances in Extrusion-Based 3D Printing for Biomedical Applications. Adv Healthc Mater 7 (8):e1701161.© 2018 John Wiley & Sons.

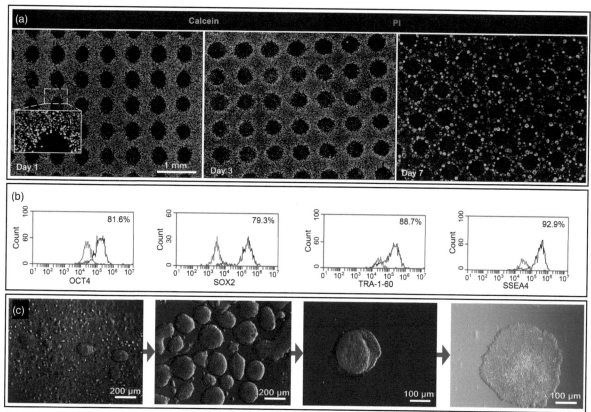

(a-c) Survival and proliferation of iPSCs following 3D bioprinting. *Cells were added to a bioink and printed using an extrusion-based bioprinter as described in the cited reference. (a) Printed constructs were photographed at 1 day (left panel), 3 days (middle panel), and 7 days (right panel) after printing. Cells were stained with calcein AM (showing viable cells in green) and propidium iodide (showing nonviable cells in red; see insert, left panel). Cells are visible at day 1 as evenly distributed single cells; by day 7, large spheroids have formed close to the lumina of the printed scaffold. (b) Cells were extracted from the 3D printed constructs 10 days after printing and flow cytometry was performed for expression of Oct4, Sox2, TRA-1-60, and SSEA4 (black histograms). The red histograms indicate isotype controls. (c) Spheroids were extracted from the 3D printed constructs 11 days after printing and transferred to conventional 2D culture on basement membrane matrix. Typical iPSC colonies were formed that continued to grow in 2D culture. Scale bars are as indicated. Source: [Gu et al. 2017], DOI 10.1002/adhm.201700175, reproduced with permission of John Wiley and Sons.*

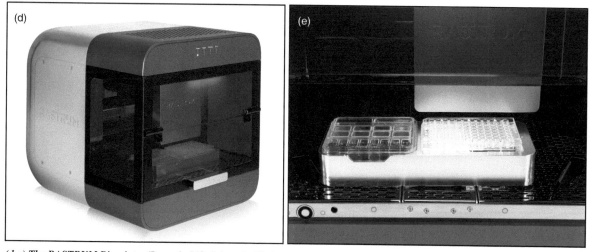

(d-e) The RASTRUM Bioprinter (Inventia Life Science). *(d) The device includes a drop on demand printing technology suitable for 3D bioprinting of viable cells and an integrated biological safety cabinet (BSC). It therefore acts as a compact standalone benchtop unit; external dimensions are 582 mm width x 491 mm height x 560 mm depth (22.9 x 19.3 x 22.0 in). (e) Sample stage showing a ready-to-use bioprinting kit (at left), to which cells are added just prior to use, and a 96-well plate (or other standard multiwell plates; at right), positioned to receive 3D bioprinted constructs from the RASTRUM bioprinter. Source: Courtesy of Inventia Life Science, with permission.*

Fig. 28.16. Three-dimensional (3D) bioprinting.

Suppliers.

Supplier	URL
Advanced BioMatrix	www.advancedbiomatrix.com
Bellco Glass	www.bellcoglass.com
Bioreactor Sciences	www.bioreactorsciences.com
BioTek (Agilent)	www.biotek.com
Cell Culture Company (C3)	cellculturecompany.com
Cellexus	cellexus.com
Cellon	www.cellon.lu/index.html
Celltainer Biotech	celltainer.com
Cesco Bioengineering	www.cescobio.com.tw
Cole-Parmer (includes Argos Technologies, Techne)	www.coleparmer.com
Corning	www.corning.com/worldwide/en/products/life-sciences/products/cell-culture.html
Eppendorf	www.eppendorf.com/oc-en
FiberCellSystems	www.fibercellsystems.com
GE Healthcare Life Sciences	www.gelifesciences.com/en/it
Gelest	www.gelest.com
Greiner Bio-One	www.gbo.com/en_int.html
Infors HT	www.infors-ht.com/en
Inventia Life Science	https://inventia.life/
Kawasaki Heavy Industries	http://global.kawasaki.com/en/
Merck Millipore	www.merckmillipore.com
Molecular Devices	www.moleculardevices.com
MP BioMedicals	www.mpbio.com
Pall Corporation	www.pall.com
Percell Biolytica	www.percell.se
PerkinElmer	www.perkinelmer.com
Sarstedt	www.sarstedt.com/en/home
Sartorius	www.sartorius.com/en
Sigma-Aldrich (Merck)	www.sigmaaldrich.com/life-science/cell-culture.html
STEMCELL Technologies	www.stemcell.com
Synthecon	synthecon.com/pages/default.asp
Terumo BCT	www.terumobct.com
ThermoFisher Scientific	www.thermofisher.com/us/en/home/life-science/cell-culture.html
Thomson Instrument Company	htslabs.com
TPP Techno Plastic Products	www.tpp.ch
Wheaton (DWK Life Sciences)	wheaton.com
Zeiss	www.zeiss.com/microscopy/int/home.html

Supp. S28.1 Roller Bottle Culture.

REFERENCES

Allison, N. and Richards, J. (2014). Current status and future trends for disposable technology in the biopharmaceutical industry. *J. Chem. Technol. Biotechnol.* 89 (9): 1283–1287. https://doi.org/10.1002/jctb.4277.

Archibald, P.R., Chandra, A., Thomas, D. et al. (2016). Comparability of automated human induced pluripotent stem cell culture: a pilot study. *Bioprocess. Biosyst. Eng.* 39 (12): 1847–1858. https://doi.org/10.1007/s00449-016-1659-9.

Ashok, P., Fan, Y., Rostami, M.R. et al. (2016). Aggregate and microcarrier cultures of human pluripotent stem cells in stirred-suspension systems. *Methods Mol. Biol.* 1502: 35–52. https://doi.org/10.1007/7651_2015_312.

Aunins, J.G., Bader, B., Caola, A. et al. (2003). Fluid mechanics, cell distribution, and environment in CellCube bioreactors. *Biotechnol. Progr.* 19 (1): 2–8. https://doi.org/10.1021/bp0256521.

Badenes, S.M., Fernandes, T.G., Rodrigues, C.A.V. et al. (2016). Microcarrier-based platforms for in vitro expansion and differentiation of human pluripotent stem cells in bioreactor culture systems. *J. Biotechnol.* 234: 71–82. https://doi.org/10.1016/j.jbiotec.2016.07.023.

Bardy, J., Chen, A.K., Lim, Y.M. et al. (2013). Microcarrier suspension cultures for high-density expansion and differentiation of human pluripotent stem cells to neural progenitor cells. *Tissue Eng. Part C Methods* 19 (2): 166–180. https://doi.org/10.1089/ten.TEC.2012.0146.

Bernard, C.J., Connors, D., Barber, L. et al. (2004). Adjunct automation to the cellmate™ Cell Culture robot. *J. Assoc. Lab Autom.* 9 (4): 209–217. https://doi.org/10.1016/j.jala.2004.03.004.

Bishop, E.S., Mostafa, S., Pakvasa, M. et al. (2017). 3-D bioprinting technologies in tissue engineering and regenerative medicine: current and future trends. *Genes Dis.* 4 (4): 185–195. https://doi.org/10.1016/j.gendis.2017.10.002.

Bliem, R., Oakley, R., Matsuoka, K. et al. (1990). Antibody production in packed bed reactors using serum-free and protein-free medium. *Cytotechnology* 4 (3): 279–283. https://doi.org/10.1007/bf00563788.

Boyd, M.R. and Paull, K.D. (1995). Some practical considerations and applications of the National Cancer Institute in vitro anticancer drug discovery screen. *Drug Dev. Res.* 34 (2): 91–109. https://doi.org/10.1002/ddr.430340203.

Busse, C., Biechele, P., de Vries, I. et al. (2017). Sensors for disposable bioreactors. *Eng. Life Sci.* 17 (8): 940–952. https://doi.org/10.1002/elsc.201700049.

Butler, M. (2005). Animal cell cultures: recent achievements and perspectives in the production of biopharmaceuticals. *Appl. Microbiol. Biotechnol.* 68 (3): 283–291. https://doi.org/10.1007/s00253-005-1980-8.

Cabral, J.M.S., De Silva, C.L., Chase, L.G. et al. (2016). *Stem Cell Manufacturing*. Amsterdam: Elsevier.

Chen, A.K., Chen, X., Choo, A.B. et al. (2011). Critical microcarrier properties affecting the expansion of undifferentiated human embryonic stem cells. *Stem Cell Res.* 7 (2): 97–111. https://doi.org/10.1016/j.scr.2011.04.007.

CMC Biotech Working Group (2009). A-Mab: a case study in bioprocess development. ISPE. https://ispe.org/publications/guidance-documents/a-mab-case-study-in-bioprocess-development (accessed 2 September 2019).

Corning (2012). Surface areas and recommended medium volumes for Corning cell culture vessels. https://www.corning.com/catalog/cls/documents/application-notes/cls-an-209.pdf (accessed 23 October 2018).

Crabbe, A., Liu, Y., Sarker, S.F. et al. (2015). Recellularization of decellularized lung scaffolds is enhanced by dynamic suspension culture. *PLoS One* 10 (5): e0126846. https://doi.org/10.1371/journal.pone.0126846.

Croughan, M.S., Konstantinov, K.B., and Cooney, C. (2015). The future of industrial bioprocessing: batch or continuous?

Biotechnol. Bioeng. 112 (4): 648–651. https://doi.org/10.1002/bit.25529.

Cui, X., Hartanto, Y., and Zhang, H. (2017). Advances in multicellular spheroids formation. *J. R. Soc. Interface* 14 (127) https://doi.org/10.1098/rsif.2016.0877.

Curling, E.M., Hayter, P.M., Baines, A.J. et al. (1990). Recombinant human interferon-gamma. Differences in glycosylation and proteolytic processing lead to heterogeneity in batch culture. *Biochem. J.* 272 (2): 333–337. https://doi.org/10.1042/bj2720333.

Daniszewski, M., Crombie, D.E., Henderson, R. et al. (2018). Automated cell culture systems and their applications to human pluripotent stem cell studies. *SLAS Technol.* 23 (4): 315–325. https://doi.org/10.1177/2472630317712220.

Earle, W.R., Bryant, J.C., and Schilling, E.L. (1954). Certain factors limiting the size of the tissue culture and the development of massive cultures. *Ann. N. Y. Acad. Sci.* 58 (7): 1000–1011. https://doi.org/10.1111/j.1749-6632.1954.tb45887.x.

Evans, V.J., Earle, W.R., Sanford, K.K. et al. (1951). The preparation and handling of replicate tissue cultures for quantitative studies. *J. Natl Cancer Inst.* 11 (5): 907–927.

Fan, Y., Ley, D., and Andersen, M.R. (2018). Fed-batch CHO cell culture for lab-scale antibody production. *Methods Mol. Biol.* 1674: 147–161. https://doi.org/10.1007/978-1-4939-7312-5_12.

FDA (2004). Pharmaceutical cGMPs for the 21st century: a risk-based approach. https://www.fda.gov/media/77391/download (accessed 2 September 2019).

FDA (2019). Electronic Code of Federal Regulations: Title 21 Food and Drugs. https://www.ecfr.gov/cgi-bin/ecfr?page=browse (accessed 9 April 2018).

Fenge, C., Weyand, J., Greller, G., et al. (2018). Large-scale perfusion and concentrated fed-batch operation of BIOSTAT® STR single-use bioreactor. https://www.sartorius.com/resource/blob/11984/8e3d506edce9939b03efd4e2352d7e6b/appl-large-scale-perfusion-sbt1018-e-data.pdf (accessed 26 August 2019).

Ferrer, B., Bermudo, R., Thomson, T. et al. (2005). Paraffin-embedded cell line microarray (PECLIMA): development and validation of a high-throughput method for antigen profiling of cell lines. *Pathobiology* 72 (5): 225–232. https://doi.org/10.1159/000089416.

Fischer, A. (1925). *Tissue Culture: Studies in Experimental Morphology and General Physiology of Tissue Cells In Vitro*. London: William Heinemann.

Francies, H.E., Barthorpe, A., McLaren-Douglas, A. et al. (2016). Drug sensitivity assays of human cancer organoid cultures. *Methods Mol. Biol.* 1576: 339–351. https://doi.org/10.1007/7651_2016_10.

Gallo-Ramirez, L.E., Nikolay, A., Genzel, Y. et al. (2015). Bioreactor concepts for cell culture-based viral vaccine production. *Expert Rev. Vaccines* 14 (9): 1181–1195. https://doi.org/10.1586/14760584.2015.1067144.

Gey, G.O. (1933). An improved technic for massive tissue culture. *Cancer Res.* 17 (3): 752–756. https://doi.org/10.1158/ajc.1933.752.

Griffiths, J.B., Cameron, D.R., and Looby, D. (1987). A comparison of unit process systems for anchorage dependent cells. *Dev. Biol. Stand.* 66: 331–338.

Gu, Q., Tomaskovic-Crook, E., Wallace, G.G. et al. (2017). 3D bioprinting human induced pluripotent stem cell constructs for in situ cell proliferation and successive multilineage differentiation. *Adv. Healthc. Mater.* 6 (17) https://doi.org/10.1002/adhm.201700175.

Gu, Q., Tomaskovic-Crook, E., Wallace, G.G. et al. (2018). Engineering human neural tissue by 3D bioprinting. In: *Biomaterials for Tissue Engineering: Methods and Protocols* (ed. K. Chawla), 129–138. New York: Springer.

Gulyas, M., Csiszer, M., Mehes, E. et al. (2018). Software tools for cell culture-related 3D printed structures. *PLoS One* 13 (9): e0203203. https://doi.org/10.1371/journal.pone.0203203.

Gupta, P., Ismadi, M.Z., Verma, P.J. et al. (2016). Optimization of agitation speed in spinner flask for microcarrier structural integrity and expansion of induced pluripotent stem cells. *Cytotechnology* 68 (1): 45–59. https://doi.org/10.1007/s10616-014-9750-z.

Haack-Sorensen, M., Follin, B., Juhl, M. et al. (2016). Culture expansion of adipose derived stromal cells. A closed automated quantum cell expansion system compared with manual flask-based culture. *J. Transl. Med.* 14 (1): 319. https://doi.org/10.1186/s12967-016-1080-9.

Haibe-Kains, B., El-Hachem, N., Birkbak, N.J. et al. (2013). Inconsistency in large pharmacogenomic studies. *Nature* 504 (7480): 389–393. https://doi.org/10.1038/nature12831.

Hatzis, C., Bedard, P.L., Birkbak, N.J. et al. (2014). Enhancing reproducibility in cancer drug screening: how do we move forward? *Cancer Res.* 74 (15): 4016–4023. https://doi.org/10.1158/0008-5472.CAN-14-0725.

Haverty, P.M., Lin, E., Tan, J. et al. (2016). Reproducible pharmacogenomic profiling of cancer cell line panels. *Nature* 533 (7603): 333–337. https://doi.org/10.1038/nature17987.

Hervy, M., Weber, J.L., Pecheul, M. et al. (2014). Long term expansion of bone marrow-derived hMSCs on novel synthetic microcarriers in xeno-free, defined conditions. *PLoS One* 9 (3): e92120. https://doi.org/10.1371/journal.pone.0092120.

Ho, L., Greene, C.L., Schmidt, A.W. et al. (2004). Cultivation of HEK 293 cell line and production of a member of the superfamily of G-protein coupled receptors for drug discovery applications using a highly efficient novel bioreactor. *Cytotechnology* 45 (3): 117–123. https://doi.org/10.1007/s10616-004-6402-8.

Hoos, A. and Cordon-Cardo, C. (2001). Tissue microarray profiling of cancer specimens and cell lines: opportunities and limitations. *Lab. Invest.* 81 (10): 1331–1338.

Hsiao, A.Y., Tung, Y.C., Qu, X. et al. (2012). 384 hanging drop arrays give excellent Z-factors and allow versatile formation of co-culture spheroids. *Biotechnol. Bioeng.* 109 (5): 1293–1304. https://doi.org/10.1002/bit.24399.

Hull, C. W. (1984). Patent US4575330A: apparatus for production of three-dimensional objects by stereolithography. https://patents.google.com/patent/us4575330a/en.

ICH (2005). Quality guidelines. https://www.ich.org/products/guidelines/quality/article/quality-guidelines.html.

ISSCR (2016). Guidelines for stem cell research and clinical translation. http://www.isscr.org/membership/policy/2016-guidelines/guidelines-for-stem-cell-research-and-clinical-translation (accessed 30 April 2019).

Justice, B.A., Badr, N.A., and Felder, R.A. (2009). 3D cell culture opens new dimensions in cell-based assays. *Drug Discovery Today* 14 (1–2): 102–107. https://doi.org/10.1016/j.drudis.2008.11.006.

Kacmar, J., Gilbert, A., Cockrell, J. et al. (2006). The cytostat: a new way to study cell physiology in a precisely defined environment. *J. Biotechnol.* 126 (2): 163–172. https://doi.org/10.1016/j.jbiotec.2006.04.015.

Kami, D., Watakabe, K., Yamazaki-Inoue, M. et al. (2013). Large-scale cell production of stem cells for clinical application using the automated cell processing machine. *BMC Biotech.* 13: 102. https://doi.org/10.1186/1472-6750-13-102.

King, G.A., Daugulis, A.J., Faulkner, P. et al. (1992). Recombinant beta-galactosidase production in serum-free medium by insect cells in a 14-L airlift bioreactor. *Biotechnol. Progr.* 8 (6): 567–571. https://doi.org/10.1021/bp00018a015.

Knazek, R.A., Gullino, P.M., Kohler, P.O. et al. (1972). Cell culture on artificial capillaries: an approach to tissue growth in vitro. *Science* 178 (4056): 65–66.

Kwon, T., Prentice, H., Oliveira, J. et al. (2017). Microfluidic cell retention device for perfusion of mammalian suspension culture. *Sci. Rep.* 7 (1): 6703. https://doi.org/10.1038/s41598-017-06949-8.

La Spada, A., Baronchelli, S., Ottoboni, L. et al. (2016). Cell line macroarray: an alternative high-throughput platform to analyze hiPSC lines. *J. Histochem. Cytochem.* 64 (12): 739–751. https://doi.org/10.1369/0022155416673969.

Liu, Y., Hourd, P., Chandra, A. et al. (2010). Human cell culture process capability: a comparison of manual and automated production. *J. Tissue Eng. Regener. Med.* 4 (1): 45–54. https://doi.org/10.1002/term.217.

Lonza (2011). Mammalian cell culture. http://bio.lonza.com/go/literature/4020 (accessed 23 August 2019).

McLimans, W.F., Davis, E.V., Glover, F.L. et al. (1957). The submerged culture of mammalian cells; the spinner culture. *J. Immunol.* 79 (5): 428–433.

Merten, O.W. (2015). Advances in cell culture: anchorage dependence. *Philos. Trans. R. Soc. Lond. B Biol. Sci.* 370 (1661): 20140040. https://doi.org/10.1098/rstb.2014.0040.

Mironov, V., Reis, N., and Derby, B. (2006). Review: bioprinting: a beginning. *Tissue Eng.* 12 (4): 631–634. https://doi.org/10.1089/ten.2006.12.631.

Mirro, R. (2011). An update on the advantages of Fibra-Cel® disks for cell culture. https://handling-solutions.eppendorf.com/fileadmin/community/cell_handling/bioprocess/pdf/application-note_313_fibra-cel_an-update-on-the-adv.pdf (accessed 30 August 2019).

Nienow, A.W. (2006). Reactor engineering in large scale animal cell culture. *Cytotechnology* 50 (1–3): 9–33. https://doi.org/10.1007/s10616-006-9005-8.

Nienow, A.W., Rafiq, Q.A., Coopman, K. et al. (2014). A potentially scalable method for the harvesting of hMSCs from microcarriers. *Biochem. Eng. J.* 85: 79–88. https://doi.org/10.1016/j.bej.2014.02.005.

NIST and SEMATECH (2012). e-Handbook of statistical methods. http://www.itl.nist.gov/div898/handbook (accessed 1 September 2019).

Owens, O. von H., Gey, M.K., and Gey, G.O. (1954). Growth of cells in agitated fluid medium. *Ann. N. Y. Acad. Sci.* 58 (7): 1039–1055. https://doi.org/10.1111/j.1749-6632.1954.tb45891.x.

Pardo, A. M. P., Klimovich, V., Wood, R., et al. (2011). Corning HYPERStack cell culture vessel: performance analysis. https://www.corning.com/catalog/cls/documents/application-

notes/corning_hyperstack_performance_analysis_cls-an-177
.pdf (accessed 28 August 2019).

Patel, G., Liu, J., Splan, D. et al. (2017). Simple and efficient cell harvest methods for microcarrier cultures in bioreactors. *Cytotherapy* 19 (5): S146–S147. https://doi.org/10.1016/j.jcyt.2017.02.215.

Paull, D., Sevilla, A., Zhou, H. et al. (2015). Automated, high-throughput derivation, characterization and differentiation of induced pluripotent stem cells. *Nat. Methods* 12 (9): 885–892. https://doi.org/10.1038/nmeth.3507.

Placone, J.K. and Engler, A.J. (2018). Recent advances in extrusion-based 3D printing for biomedical applications. *Adv. Healthc. Mater.* 7 (8): e1701161. https://doi.org/10.1002/adhm.201701161.

PMC (2014). Personalized medicine 101: the challenges. http://pmc.blueonblue.com/resources/personalized_medicine_101_the_challenges (accessed 2 September 2019).

Rafiq, Q.A., Coopman, K., Nienow, A.W. et al. (2016). Systematic microcarrier screening and agitated culture conditions improves human mesenchymal stem cell yield in bioreactors. *Biotechnol. J.* 11 (4): 473–486. https://doi.org/10.1002/biot.201400862.

Rodrigues, A.L., Rodrigues, C.A.V., Gomes, A.R. et al. (2019). Dissolvable microcarriers allow scalable expansion and harvesting of human induced pluripotent stem cells under xeno-free conditions. *Biotechnol. J.* 14 (4): e1800461. https://doi.org/10.1002/biot.201800461.

Saha, G., Barua, A., and Sinha, S. (2018). *Bioreactors: Animal Cell Culture Control for Bioprocess Engineering*. Boca Raton, FL: CRC Press (Taylor and Francis Group).

Sart, S., Agathos, S.N., and Li, Y. (2013). Engineering stem cell fate with biochemical and biomechanical properties of microcarriers. *Biotechnol. Progr.* 29 (6): 1354–1366. https://doi.org/10.1002/btpr.1825.

Sart, S., Schneider, Y.J., Li, Y. et al. (2014). Stem cell bioprocess engineering towards cGMP production and clinical applications. *Cytotechnology* 66 (5): 709–722. https://doi.org/10.1007/s10616-013-9687-7.

Sart, S., Agathos, S.N., Li, Y. et al. (2016). Regulation of mesenchymal stem cell 3D microenvironment: from macro to microfluidic bioreactors. *Biotechnol. J.* 11 (1): 43–57. https://doi.org/10.1002/biot.201500191.

Selden, C. and Fuller, B. (2018). Role of bioreactor technology in tissue engineering for clinical use and therapeutic target design. *Bioengineering (Basel)* 5 (2) https://doi.org/10.3390/bioengineering5020032.

Shankles, P.G., Millet, L.J., Aufrecht, J.A. et al. (2018). Accessing microfluidics through feature-based design software for 3D printing. *PLoS One* 13 (3): e0192752. https://doi.org/10.1371/journal.pone.0192752.

Shoemaker, R.H. (2006). The NCI60 human tumour cell line anticancer drug screen. *Nat. Rev. Cancer* 6 (10): 813–823. https://doi.org/10.1038/nrc1951.

Shukla, A.A. and Gottschalk, U. (2013). Single-use disposable technologies for biopharmaceutical manufacturing. *Trends Biotechnol.* 31 (3): 147–154. https://doi.org/10.1016/j.tibtech.2012.10.004.

Singh, V. (1999). Disposable bioreactor for cell culture using wave-induced agitation. *Cytotechnology* 30 (1–3): 149–158. https://doi.org/10.1023/A:1008025016272.

So, A. D. and Katz, S. L. 2010. Biologics Boondoggle. New York Times, 8 March 2010. https://www.nytimes.com/2010/03/08/opinion/08so.html.

Spier, R.E. and Whiteside, J.P. (1976). The production of foot-and-mouth disease virus from BHK 21 C 13 cells grown on the surface of glass spheres. *Biotechnol. Bioeng.* 18 (5): 649–657. https://doi.org/10.1002/bit.260180505.

Stacey, G. and Davis, J. (2007). *Medicines from Animal Cell Culture*. Chichester: Wiley.

Storm, M.P., Sorrell, I., Shipley, R. et al. (2016). Hollow fiber bioreactors for in vivo-like mammalian tissue culture. *J. Vis. Exp.* 111: 53431. https://doi.org/10.3791/53431.

Tan, K.Y., Teo, K.L., Lim, J.F. et al. (2015). Serum-free media formulations are cell line-specific and require optimization for microcarrier culture. *Cytotherapy* 17 (8): 1152–1165. https://doi.org/10.1016/j.jcyt.2015.05.001.

Terstegge, S., Laufenberg, I., Pochert, J. et al. (2007). Automated maintenance of embryonic stem cell cultures. *Biotechnol. Bioeng.* 96 (1): 195–201. https://doi.org/10.1002/bit.21061.

Thomas, R. and Ratcliffe, E. (2012). Automated adherent human cell culture (mesenchymal stem cells). *Methods Mol. Biol.* 806: 393–406. https://doi.org/10.1007/978-1-61779-367-7_26.

Thomas, R.J., Chandra, A., Liu, Y. et al. (2007). Manufacture of a human mesenchymal stem cell population using an automated cell culture platform. *Cytotechnology* 55 (1): 31–39. https://doi.org/10.1007/s10616-007-9091-2.

Titus, K., Klimovich, V., Rothenberg, M. et al. (2010). Closed system cell culture protocol using HYPERStack vessels with gas permeable material technology. *J. Vis. Exp.* 45: 2499. https://doi.org/10.3791/2499.

Varley, J. and Birch, J. (1999). Reactor design for large scale suspension animal cell culture. *Cytotechnology* 29 (3): 177–205. https://doi.org/10.1023/A:1008008021481.

van der Velden-de Groot, C.A. (1995). Microcarrier technology, present status and perspective. *Cytotechnology* 18 (1–2): 51–56. https://doi.org/10.1007/BF00744319.

Vinci, M., Gowan, S., Boxall, F. et al. (2012). Advances in establishment and analysis of three-dimensional tumor spheroid-based functional assays for target validation and drug evaluation. *BMC Biol.* 10: 29. https://doi.org/10.1186/1741-7007-10-29.

Warnock, J.N., Bratch, K., and Al-Rubeai, M. (2005). Packed bed bioreactors. In: *Bioreactors for Tissue Engineering: Principles, Design and Operation* (eds. J. Chaudhuri and M. Al-Rubeai), 87–113. Dordrecht: Springer.

Waterworth, A., Hanby, A., and Speirs, V. (2005). A novel cell array technique for high-throughput, cell-based analysis. *In Vitro Cell. Dev. Biol. Anim.* 41 (7): 185–187. https://doi.org/10.1290/0505032.1.

Weinstein, J.N. (2012). Drug discovery: cell lines battle cancer. *Nature* 483 (7391): 544–545. https://doi.org/10.1038/483544a.

Weinstein, J.N. and Lorenzi, P.L. (2013). Cancer: discrepancies in drug sensitivity. *Nature* 504 (7480): 381–383. https://doi.org/10.1038/nature12839.

van Wezel, A.L. (1967). Growth of cell-strains and primary cells on micro-carriers in homogeneous culture. *Nature* 216 (5110): 64–65. https://doi.org/10.1038/216064a0.

Xiao, C., Huang, Z., Li, W. et al. (1999). High density and scale-up cultivation of recombinant CHO cell line and hybridomas with porous microcarrier cytopore. *Cytotechnology* 30 (1–3): 143–147. https://doi.org/10.1023/A:1008038609967.

Xie, L. and Wang, D.I. (1994). Fed-batch cultivation of animal cells using different medium design concepts and feeding strategies. *Biotechnol. Bioeng.* 43 (11): 1175–1189. https://doi.org/10.1002/bit.260431123.

Xu, S., Gavin, J., Jiang, R. et al. (2017). Bioreactor productivity and media cost comparison for different intensified cell culture processes. *Biotechnol. Progr.* 33 (4): 867–878. https://doi.org/10.1002/btpr.2415.

Ziegler, D.W., Davis, E.V., Thomas, W.J. et al. (1958). The propagation of mammalian cells in a 20-liter stainless steel fermentor. *Appl. Microbiol.* 6 (5): 305–310.

CHAPTER 29

Toxicity Testing

As cell culture became accepted as a valid model system for studying biological processes *in vitro*, its role in physiological assays increased significantly. Cell line panels were assembled for "*in vitro* clinical trials" and *in vitro* models were employed for safety assessment following exposure to potentially toxic substances (Ross and Wilson 2014; EURL ECVAM 2019). One of the earliest cell line panels was brought together by the National Cancer Institute (NCI) and comprised 60 human cell lines from nine distinct tumor types (Shoemaker 2006). The NCI-60 panel represented a major step forward in the development of *in vitro* assays. Cell lines were extensively validated using a number of methods, including cytogenetic analysis and short tandem repeat (STR) profiling (Roschke et al. 2003; Lorenzi et al. 2009). NCI scientists developed high-throughput assays and bioinformatics tools that enabled this panel to become an important research platform; for example, the gene for melanocyte inducing transcription factor (*MITF*) was found to act as a melanoma "addiction" oncogene through integrative analysis of data from the NCI-60 panel (Garraway et al. 2005; Ross and Wilson 2014). However, few drug approvals can be traced back to initial screening against the NCI-60 panel (Chabner 2016). This is probably due to its limited size; 60 cell lines translates to about 6–7 cell lines per cancer type, which severely limits its capacity as a screening tool (Wilding and Bodmer 2014). Efforts have since been made to increase the size of cell line panels, to provide new assays for screening, and to develop better *in vitro* model systems. This chapter focuses primarily on assay systems that use cell lines for toxicity testing. Advanced *in vitro* models that provide a more physiological environment are discussed toward the end of the chapter (see Section 29.5).

29.1 *IN VITRO* TOXICITY TESTING

Toxicity testing is essential to detect toxic substances in the environment and to understand the hazardous properties of agricultural or industrial chemicals, pharmaceutical products, food additives, cosmetics, and various personal care products. These substances go through extensive testing before they are released for widespread use (Parasuraman 2011). Toxicity testing has traditionally relied on animal-based screening, but it is important to reduce the use of animals for scientific purposes and to minimize their pain and discomfort (see Section 6.4.1). Animals are in any case imperfect models for toxicity in humans, as shown by the failure of rodents to show major fetal abnormalities when tested with thalidomide (Fratta et al. 1965). It is also important to recognize that the number of new products manufactured each year far exceeds our capacity to assess them using animal models (Nardone 1980). As a result, many countries are moving to replace animals with alternative models for toxicity testing.

The European Union has been particularly proactive in restricting animal-based screening, although other countries are also moving in the same direction (Burden et al. 2015). European Union legislation now prohibits the release of cosmetics where the final product or its ingredients have been the subject of animal testing, if an alternative method has been validated and adopted (EU 2009). Similarly, the Euro-

Freshney's Culture of Animal Cells: A Manual of Basic Technique and Specialized Applications, Eighth Edition. Amanda Capes-Davis and R. Ian Freshney.
© 2021 John Wiley & Sons Ltd. Published 2021 by John Wiley & Sons Ltd.
Companion website: www.wiley.com/go/freshney/cellculture8

pean Union's regulation on the Registration, Evaluation, Authorisation and Restriction of Chemicals (REACH) makes it clear that animal-based testing is a last resort where no other scientifically reliable method can be used (EU 2006). Data collected by the European Chemicals Agency (ECHA) suggest that this stance has resulted in greater use and promotion of non-animal alternatives, although the use of *in vitro* assays as standalone replacements for animal testing remains relatively low (ECHA 2017; Taylor 2018).

29.1.1 Applications

In vitro toxicity assays are particularly common in (or viewed as key requirements for) the following applications:

(1) **Anticancer drug screening.** Drug screening for the identification of new anticancer drugs can be a tedious and often inefficient method for discovering active compounds. Panels of human tumor cell lines can be used to make this process more rapid and efficient using high-throughput screening. Typically, cell lines are grown in multiwell plates and assessed using short-term cytotoxicity assays, although long-term survival assays also give important data (see Sections 29.2.2–29.2.4). A substantial number of cell lines should be used for adequate statistical power and their performance compared to tumor tissue in order to ensure that cell-based assays are clinically relevant (Wilding and Bodmer 2014). Provided sufficient cell lines are used, panels can be stratified into genetically defined subtypes to perform target-orientated drug screening.

(2) **Personalized therapy.** It has been suggested for many years that measurement of the chemosensitivity of cells derived from a patient's tumor might be used in designing a chemotherapeutic regime for the patient (Freshney 1978). The major problem, however, was one of logistics. The number of patients with tumors for which the correct target cells (i) will grow *in vitro* sufficiently to be tested, (ii) can be expected to respond, and (iii) will produce a response that can be followed up is extremely small. Recent advances in cell culture techniques have made personalized therapy more achievable, leading to successful proof-of-concept studies using organoids and other advanced models for drug screening (Artegiani and Clevers 2018).

(3) **Testing for genotoxicity and carcinogenicity.** Substances that cause genetic alterations in somatic or germ cells lead to an increased risk of cancer and congenital abnormalities. *In vitro* assays form an important part of testing for genotoxicity and carcinogenicity, as discussed later in this chapter (see Sections 29.3, 29.4).

(4) **Testing for eye or skin irritation.** Traditionally, the Draize test has been performed to detect eye and skin irritation from pharmaceuticals, cosmetics, and personal care products. This test is painful to the animals that are subjected

to it, and there are a number of inconsistencies between the data generated from animal models and human risk exposure (Barile 2010). It is now possible to use *in vitro* test methods as alternatives to the Draize test, particularly using three-dimensional (3D) "tissue equivalent" constructs (see Sections 27.7.1, 29.5.1).

(5) **Testing for phototoxicity.** In this chapter, "phototoxicity" refers to a toxic response that occurs after skin is exposed to a chemical (either locally or through systemic routes), followed by exposure to light (EURL ECVAM 2019). Acute phototoxicity can be assessed using BALB/3T3 cell lines; cells are exposed to the test chemical and irradiated before cell viability is assessed using neutral red uptake (see Section 19.2.2). This test method is easily automated and has been validated for use in combination with other methods (Ates et al. 2017; EURL ECVAM 2019).

(6) **Testing of aquatic toxicity.** Testing for acute toxicity within ecosystems has primarily used fish as models, with death as the end point of the assay (Tanneberger et al. 2013). Acute fish toxicity can be assessed *in vitro* using the RTgill-W1 cell line, which was established from rainbow trout gill – the primary target and uptake site for water contaminants (Lee et al. 2009). The resulting assay has been validated, showing that it is reproducible across different laboratories and can be used to predict *in vivo* toxicity (Tanneberger et al. 2013; Fischer et al. 2019).

Other *in vitro* applications are currently in development that use cell lines in unexpected and intriguing ways. For example, RTgill-W1 cells can be seeded in a fluidic biochip to give a portable assay for water quality (see Figure 29.1).

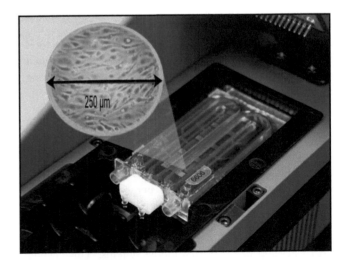

Fig. 29.1. Electric cell-substrate impedance sensing (ECIS) biochip. The fluidic ECIS biochip is positioned in an ECIS reader. The magnified area shows a confluent monolayer of RTgill-W1 cells on a single sensing electrode. *Source*: Brennan et al. (2016), https://doi .org/10.3791/53555, licensed under CC-NC-ND 3.0 with permission of the *Journal of Visualized Experiments*.

Toxicity results in rapid changes in electrical impedance of the cell monolayer, which can be detected using electric cell-substrate impedance sensing (ECIS). ECIS biochips seeded with RTgill-W1 cells can be stored at 6 °C for weeks to months before being used (Brennan et al. 2012, 2016). This is possible because RTgill-W1 cells (which are routinely cultured at 18–22 °C) can survive at 0–28 °C for extended time periods (Bols et al. 2017).

29.1.2 Limitations

Toxicity is a complex event *in vivo*. Exposure of a living organism to a toxic substance may result in (i) direct cellular damage, i.e. "cytotoxicity;" (ii) physiological effects, such as impaired membrane transport in the kidney or neurotoxicity in the brain; (iii) inflammatory effects, ranging from local irritation to pyrexia or vascular dilatation; and (iv) other complex systemic effects, such as the development of congenital abnormalities after *in utero* exposure. It is difficult to monitor complex systemic effects *in vitro*, so most assays determine effects at the cellular level. For example, it would be very difficult to recreate the pharmacokinetics of drug exposure. Between *in vitro* and *in vivo* experiments there usually are significant differences in the exposure time to the drug, its concentration and rate of change, and its metabolism (both activation and detoxification), tissue penetration, clearance, and excretion. It may be possible to simulate these parameters using tumor spheroids for drug penetration or timed perfusion to simulate concentration and time ($C \times T$) effects. However, most studies concentrate on a direct cellular response, thereby gaining simplicity and reproducibility.

Use of cell lines as *in vitro* model systems results in some additional complexity that must be considered. Continuous cell lines can be highly heterogeneous and undergo changes in genotype and phenotype during culture (see Sections 3.3, 3.4). Heterogeneity is particularly obvious when cell lines are obtained from different sources, but can be observed even when a single batch of vials is used for toxicity testing (Kleensang et al. 2016). Clearly, there is a need for characterization that extends beyond the relatively simple validation testing that should be performed by all tissue culture laboratories (see Section 7.4).

Even when toxicity is assessed only at the cellular level, there is a great deal of complexity that must be considered. Cytotoxicity is often assessed by looking for evidence of cell death, which is usually obvious due to changes in morphology or cell number (see Sections 2.5, 29.2). However, these changes come from activation of complex molecular "subroutines," resulting in an array of possible endpoints for assessment (Galluzzi et al. 2018). Cell toxicity may also lead to non-lethal effects such as inhibition of growth, cellular senescence, or terminal differentiation. Determining a suitable endpoint for assessment can be challenging, particularly when there is a need to select a rapid screening method that

can be readily automated. Demonstrating lack of toxicity is also challenging and is likely to require analysis of specific targets based on changes in gene transcription, cell signaling, or cell–cell interaction.

Most cytotoxicity assays oversimplify the events that they measure and are employed because they are cheap, easily quantified, and reproducible. However, it has become increasingly apparent that they are inadequate for modern drug development, which requires greater emphasis on specific molecular targets and precise metabolic regulation. Cytotoxicity assays are still required, but there is a growing need to supplement them with more specific tests of metabolic pathway regulation and signaling. This brings with it the need for advanced analytical software to analyze the ever-increasing output of data and to interpret it in a system-wide context. Perhaps the most obvious need for a system-based approach is to understand the induction of an inflammatory or allergic response, which need not imply cytotoxicity of the allergen and is still one of the hardest results to demonstrate *in vitro*. The need to develop a more systems-based approach to toxicity testing is discussed elsewhere (Hartung et al. 2017).

29.1.3 Requirements

Toxicity testing represents a significant commitment in time and infrastructure. Good Cell Culture Practice (GCCP) must be adopted by all laboratories that are involved in toxicity testing, regardless of application (see Section 7.2.1) (Eskes et al. 2017; OECD 2018). Always screen cell lines for microbial contamination and authenticity prior to commencement; initial screening of the NCI-60 panel demonstrated that a subset of cell lines was misidentified, and similar findings continue to be reported. For example, the Bhas42 cell line (which is used for testing of carcinogenicity) was recently shown to come from a different strain to the one reported (Lorenzi et al. 2009; Uchio-Yamada et al. 2017). Similarly, it is important to screen different assay methods to ensure that a suitable endpoint is assessed (see Section 29.2). If the assay is to be used at large scale or in multiple laboratories, broader validation should be performed to ensure that the method is robust and reproducible (Niepel et al. 2019). More information on validation of high-throughput screening methods is provided elsewhere (Iversen et al. 2012).

If *in vitro* test methods are to be used for human safety assessment, Good Laboratory Practice (GLP) will be required (see Section 7.2.2). A guidance document has been published by the Organization for Economic Co-operation and Development (OECD) that sets out requirements for laboratories that perform such testing (OECD 2018). Test methods require comprehensive validation prior to being adopted. A list of validated test methods is available from the European Union Reference Laboratory for Alternatives to Animal Testing (EURL ECVAM 2019). A test guideline or written standard should be available that specifies how to perform a validated test method. These publications are typically available from OECD, the International Organization for Standardization (ISO), or a similar Standards body (ISO 2019; OECD 2019a).

The extent of validation will partly depend on whether the assay is used as a standalone test to replace animal models. For *in vitro* testing to be accepted as an alternative to animal testing, it must be demonstrated that potential toxins reach the cells *in vitro* in the same form as they would *in vivo*. Some non-toxic substances become toxic after being metabolized by the liver, while other substances that are toxic *in vitro* may be detoxified by liver enzymes. This proof may require additional assays using purified liver microsomal enzyme preparations or co-culture with activated hepatocytes or hepatoma-derived cell lines such as HepaRG (see Section 26.1) (McGregor et al. 1988; Guillouzo and Guguen-Guillouzo 2008; Soldatow et al. 2013).

29.2 CYTOTOXICITY ASSAYS

29.2.1 Selecting a Cytotoxicity Assay

Cytotoxicity assays are typically designed to measure changes in cell proliferation, viability, or survival after cells are incubated with a test substance. Changes can be assessed using cell counting and colony counting at small scale, but it is often difficult to adapt these methods to count small numbers of cells for high-throughput screening (see Section 19.1.3). Because the interpretation of cell counts at a single point in time is often ambiguous, it becomes necessary to count cells at multiple time points to generate a growth curve (see Figures 14.1, 19.7; see also Protocol P19.3). This is cumbersome in a large screen, although some methods of direct counting can make repeated measurements more feasible (e.g. live-cell imaging; see Figure 18.5) (Tahara et al. 2017).

Most cytotoxicity assays that are used for high-throughput screening measure surrogate variables in preference to cell counting (see Table 29.1). Some variables act as effective surrogates for cell number; sulforhodamine B, which was initially used for screening of the NCI-60 panel, binds to cellular protein and thus has a close relationship to the number of cells present (see Table 19.1) (Skehan et al. 1990). Other variables are more effective as surrogates for cell viability or cell death and have a complex relationship to cell number (see Section 19.1). Assays that rely on cell metabolism are essentially cell viability assays, where cytotoxicity results in a decrease in metabolic activity (see Section 29.2.2). Cell death can be examined more directly by looking for loss of membrane integrity or for evidence of apoptosis or other forms of regulated cell death (see Section 29.2.3) (Miret et al. 2006; Riss et al. 2019).

The modern tissue culture laboratory can purchase a large assortment of reagents and kits for use in cytotoxicity assays. Some commonly used assays are described here, but it is impossible to provide a comprehensive list of all approaches. When selecting a cytotoxicity assay, it is important to understand the endpoint that the assay is designed to measure, its sensitivity, and its suitability in other respects (e.g. ease of automation or application to 3D culture). Generally speaking, luminescent assays are more sensitive than fluorescent assays, which are in turn more sensitive than colorimetric assays (Niles et al. 2009). Assay systems will need optimization if they are to be multiplexed or used in 3D culture systems; for example, extended time may be required to lyze cells within spheroids for adenosine triphosphate (ATP)-based assays (Promega 2014).

29.2.2 Assays Based on Cell Metabolism

Tetrazolium reduction. The first rapid, non-radioactive *in vitro* assay for screening of 96-well plates was published in the 1980s and employed the tetrazolium salt 3-(4,5-dimethylthiazol-2-yl)-2,5-diphenyltetrazolium bromide (MTT) (Mosmann 1983; Riss et al. 2011). The tetrazolium ring of this molecule is cleaved by viable cells to form a purple formazan precipitate (see Figure 29.2b). After an additional solubilization step, the amount of formazan product can be assessed by measuring absorbance at 570 nm. Various solubilization methods can be used, including acidification, which has the added advantage that it changes the color of the phenol red in the medium to one that is less likely to interfere with absorbance readings (Riss et al. 2016). Other tetrazolium-based compounds have since been developed that remove the need for solubilization, e.g. 2,3-bis-(2-methoxy-4-nitro-5-sulfophenyl)-2H-tetrazolium -5-carboxanilide (XTT). These reagents are available from a number of commercial suppliers and are also sold in kit form, e.g. the Cell Proliferation Kit I and II (Sigma-Aldrich), CellTiter 96® Non-radioactive Cell Proliferation Assays (Promega), and CyQUANT™ MTT Cell Viability Assay (ThermoFisher Scientific).

MTT requires an incubation step and can be toxic to cells, resulting in obvious changes in the morphology of control cells, e.g. with NIH 3T3 (Riss et al. 2016). The assay is also prone to interference and is less sensitive and more variable compared to the sulforhodamine B, neutral red uptake, and resazurin reduction assays (Niles et al. 2009; van Tonder et al. 2015). However, tetrazolium reduction assays continue to be popular in research laboratories because of their convenience, widespread availability, and low cost.

TABLE 29.1. Cytotoxicity assays for multiwell plates.

Variable assessed	Basis for assay	Assay example[a,b]	Equipment for assay readout
Cell proliferation (see also Table 19.1)			
Sulforhodamine B (SRB) binding	Cells are permeabilized and stained with a reagent that can bind to protein	Skehan et al. (1990); van Tonder et al. (2015)	Spectrophotometer
Cell survival			
Colony formation	Cells are plated at low density and colonies counted as a measure of cell survival	Anderson et al. (2007); Horman et al. (2013)	Imaging analysis or colony counting equipment
Cell metabolism (indicating cell viability)			
Tetrazolium reduction	A tetrazolium compound (e.g. MTS) is reduced to a colored formazan product	CellTiter-96® AQueous One Solution Cell Proliferation Assay	Spectrophotometer
Resazurin (Alamar Blue) reduction	Resazurin is reduced by viable cells to resorufin, which is pink and highly fluorescent	van Tonder et al. (2015); CellTiter-Blue Cell Viability Assay	Fluorometer or spectrophotometer
ATP quantification	Intracellular ATP levels are maintained by viable cells and rapidly decrease following loss of viability	CellTiter-Glo® 2.0	Luminometer
Protease activity	"Live-cell protease" is measured using a substrate that penetrates the cell membrane	CellTiter-Fluor™ Cell Viability Assay	Fluorometer
Cell death (indicating loss of cell viability)			
Uptake of neutral red	Neutral red penetrates viable cells, where it is retained inside lysosomes	van Tonder et al. (2015)	Spectrophotometer
Uptake of a non-permeable, DNA-binding dye	CellTox Green dye is excluded from viable cells but can enter non-viable cells and bind to DNA	CellTox™ Green Cytotoxicity Assay	Fluorometer
LDH activity	LDH is released following loss of membrane integrity	LDH-Glo™ Cytotoxicity Assay	Luminometer
Protease activity	"Dead-cell protease" is released following loss of membrane integrity	CytoTox-Glo™ Cytotoxicity or CytoTox-Fluor™ Cytotoxicity Assays	Luminometer or fluorometer
Apoptosis			
Phosphatidylserine exposure (annexin V binding)	Annexin V binds to phosphatidylserine, which is exposed on the external surface of the cell during apoptosis	RealTime-Glo™ Annexin V Apoptosis and Necrosis Assay	Luminometer for annexin V (fluorometer for necrosis)
Caspase activation	Activity of caspase 3/7 will change during apoptosis	Caspase-Glo® 3/7 or Apo-ONE® Caspase 3/7 Assays	Luminometer or fluorometer
	Activity of other caspases including caspase 2, 6, 8, or 9 may change during apoptosis	Caspase-Glo 2, 6, 8, or 9 Assays	Luminometer

[a]Kits from a single company (Promega) are listed as examples only, to illustrate the range of products available. Various other suppliers provide cytotoxicity assays, including Abcam, Biotium, Lonza, PerkinElmer, R&D Systems, Sigma-Aldrich, and ThermoFisher Scientific.
[b]Only a selection of Promega's kits are included here. Refer to their website for a full list of cell-based assays.

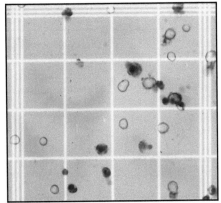

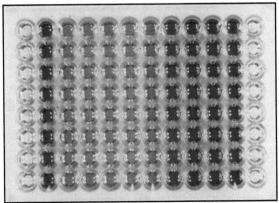

(a) Dye Exclusion, Naphthalene Black.
Hemocytometer slide 200 μm square with viable (unstained) and non-viable (blue stained) cells. 40x objective. Source: R. Ian Freshney.

(b) MTT Assay. *Microtitration plate with cells stained with MTT after 24 h. Extreme left and right hand columns are blanks; 2nd from left and 2nd from right are untreated controls. VP16 (0-10 μM) has been added to columns 3-8. Source: R. Ian Freshney.*

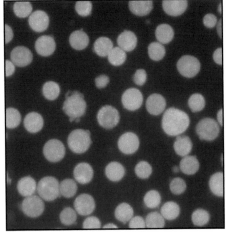

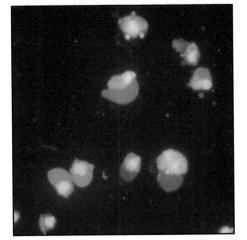

(c,d) Apoptosis in HT29 cells. *Cells have been treated with 0.25 μg/ml TRAIL (TNF-related apoptosis-inducing ligand) for 16 h and stained with acridine orange. (c) Attached cells showing normal morphology. (d) Detached cells from (c) showing chromatin condensation and fragmentation characteristic of apoptosis. 40x objective. Source: Courtesy of Angela Hague.*

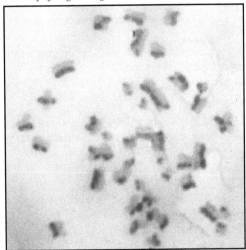

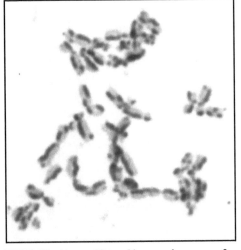

(e,f) Sister Chromatid Exchange (SCE). *(e) Untreated cells: A2780/Cp70 with an additional human chromosome 2 transferred (A2780/cp70 +chr2). (f) Induced SCE. A2780/Cp70 cells treated with 10 μM Cisplatin for 1 h, showing extensive SCE: dark Giemsa staining alternating between strands within individual chromosomes. Source: Courtesy of Robert Brown and Maureen Illand.*

Fig. 29.2. Assessment of viability and cytotoxicity.

PROTOCOL P29.1. MTT-BASED CYTOTOXICITY ASSAY

Contributed by Jane Plumb when at the Cancer Research UK Centre for Oncology and Applied Pharmacology, University of Glasgow, Scotland, United Kingdom.

Background

Cells in the exponential phase of growth are exposed to a cytotoxic drug. The duration of exposure is usually determined based on the time required for maximal damage to occur, but is also influenced by the stability of the drug. After removal of the drug, the cells are allowed to proliferate for 2–3 population doubling times (PDTs) (see Section 19.3.3) in order to distinguish between cells that remain viable and are capable of proliferation and those that remain viable but cannot proliferate. The number of surviving cells is then determined indirectly by MTT dye reduction. The amount of MTT-formazan produced can be determined spectrophotometrically once the MTT-formazan has been dissolved in a suitable solvent.

MTT can be used to determine the number of cells after a variety of treatments other than cytotoxic drug exposure, such as growth factor stimulation. However, in each case, it is essential to ensure that the treatment itself does not affect the ability of the cell to reduce the dye and absorbance remains linear with cell number.

Outline

Incubate monolayer cultures in microwell plates in a range of drug concentrations. Remove the drug and feed the plates daily for 2–3 PDTs; then feed the plates again and add MTT to each well. Incubate the plates in the dark for four hours and then remove the medium and MTT. Dissolve the water-insoluble MTT-formazan crystals in dimethyl sulfoxide (DMSO), add a buffer to adjust the final pH, and record the absorbance at 570 nm using a microplate reader.

❖ *Safety Note. Cytotoxic drugs and MTT are hazardous; read the Safety Data Sheet (SDS) and perform a risk assessment prior to commencing work (see Section 6.1.1). Always wear suitable personal protective equipment (PPE) when handling DMSO and check chemical resistance when selecting gloves.*

Materials (sterile or aseptically prepared)

- ☐ Cell line(s)
- ☐ Cytotoxic drug or other test agent
- ☐ Vehicle in which drug is solubilized, e.g. DMSO or ethanol
- ☐ Growth medium
- ☐ MTT (e.g. Sigma-Aldrich), 50 mg/ml, filter sterilized (see Appendix B)
- ☐ Sorensen's glycine buffer (0.1 M glycine, 0.1 M NaCl adjusted to pH 10.5 with 1 M NaOH)
- ☐ 96-well plates; flat-bottomed for adherent cells, round-bottomed for suspension cells
- ☐ Petri dishes (non-tissue culture-treated), 50 and 90 mm, or pipetting reservoir
- ☐ Universal containers, tubes, or bottles, 30 and 100 ml
- ☐ Pipette tips, preferably in an autoclavable tip box

Materials (non-sterile)

- ☐ Biological safety cabinet (BSC) and associated consumables (see Protocol P12.1)
- ☐ Microplate reader (e.g. SpectraMax spectrophotometer for UV-Vis absorbance detection, Molecular Devices, with SOFTmax PRO software)
- ☐ Plate carrier for centrifuge (for cells growing in suspension)
- ☐ DMSO (e.g. Sigma-Aldrich)
- ☐ Plastic box (clear polystyrene, to hold plates, e.g. a clean lunch box)
- ☐ Multichannel pipettor
- ☐ Microplate dispenser for DMSO (optional; see Figure 5.8)

Procedure

A. Seeding in 96-Well Plates

1. Prepare a work space for aseptic technique using a BSC.
2. Perform subculture using the standard procedure for that cell line (e.g. trypsinization; see Protocol P14.2). Collect the cells in growth medium containing serum.
3. Centrifuge the suspension (5 minutes at 200 g) to pellet the cells. Resuspend the cells in growth medium and count them (see Protocols 19.1, 19.2).
4. Dilute the cells to $2.5–50 \times 10^3$ cells/ml, depending on the growth rate of the cell line, and allowing 20 ml cell suspension per microwell plate.
5. Transfer the cell suspension to a 90-mm Petri dish or a pipetting reservoir. With a multichannel pipettor, add 200 μl of the suspension into each well of the central 10 columns of a flat-bottomed 96-well plate (80 wells per plate), starting with column 2 and ending with column 11, placing $0.5–10 \times 10^3$ cells into each well.

6. Add 200 µl of growth medium to the eight wells in columns 1 and 12. Column 1 will be used to blank the plate reader; column 12 helps to maintain the humidity for column 11 and minimizes the "edge effect."

7. Put the plates in a plastic box and incubate in a humidified atmosphere at 37 °C for one to three days, such that the cells are in the exponential phase of growth at the time that the drug is added.

8. For non-adherent cells, prepare a suspension in fresh growth medium. Dilute the cells to $5-100 \times 10^3$ cells/ml and plate out only 100 µl of the suspension into round-bottomed 96-well plates. Add drug immediately to these plates.

B. Drug Addition

9. Prepare a serial fivefold dilution of the cytotoxic drug in growth medium to give eight concentrations. This set of concentrations should be chosen such that the highest concentration kills most of the cells and the lowest kills none of the cells. Once the toxicity of a drug is known, a smaller range of concentrations can be used. Normally, three plates are used for each drug to give triplicate determinations within one experiment.

10. For adherent cells:
    (a) Remove the medium from the wells in columns 2–11.
    (b) Feed the cells in the eight wells in columns 2 and 11 with 200 µl fresh growth medium containing the solvent for the drug (e.g. DMSO); these cells are the controls.
    (c) Transfer the drug solutions to 50-mm Petri dishes and add 200 µl to each group of four wells with a four-tip pipettor.
    (d) Add the cytotoxic drug to the cells in columns 3–10. Depending on the number of replicated required, it may be possible to use rows A to D for one drug and rows E to H for a second drug.

11. For non-adherent cells, follow steps 10b–d but prepare the drug dilution at twice the desired final concentration and add 100 µl of diluted drug or control medium to the 100 µl of cells already in the wells.

12. Return the plates to the plastic box and incubate them for a defined exposure period.

C. Growth Period

13. At the end of the drug exposure period, remove the medium from all of the wells containing cells and feed the cells with 200 µl fresh medium.

14. For non-adherent cells, centrifuge the plates (5 minutes at 200 g) to pellet the cells. Then remove the medium, using a fine-gauge needle to prevent disturbance of the cell pellet.

15. Feed the plates daily for 2–3 PDTs.

D. MTT Assay

16. Feed the plate with 200 µl fresh medium at the end of the growth period and add 50 µl MTT to all of the wells in columns 1–11. Plates should be wrapped in aluminum foil to protect them from the dark once the MTT (which is sensitive to light) has been added.

17. Incubate the plates for four hours in a humidified atmosphere at 37 °C.

18. Remove the medium and MTT from the wells (centrifuge for non-adherent cells).

19. Add 200 µl DMSO to all of the wells in columns 1–11 to dissolve the remaining MTT-formazan crystals.

20. Add Sorensen's glycine buffer (25 µl/well) to all of the wells containing DMSO.

21. Record absorbance at 570 nm immediately, because the product is unstable. Use the wells in column 1, which contain medium and MTT but no cells, to blank the plate reader.

Notes

Step 6: Some users prefer to leave all outer wells blank and distribute the positive and negative controls randomly throughout the plate (Francies et al. 2016). This is difficult if setting up plates manually but becomes feasible with automated liquid handling (see Sections 5.1.3, 28.6).

Step 12: The duration of exposure will require optimization. Some agents may act more quickly and the exposure period and recovery may be shortened. The cells must remain in exponential growth throughout and the cell concentration at the end should still be within the linear range of the MTT colorimetric assay.

Step 17: Plates can be left for up to eight hours to increase amount of formazan product, provided there is no evidence of toxicity when the cell line is exposed to MTT without the drug.

Step 21: Results can be analyzed by plotting a graph of the absorbance (y axis) against the concentration of drug (x axis) and calculating the IC_{50} (see Section 29.2.5).

Resazurin reduction. Alamar Blue was initially introduced as a non-radioactive alternative to [³H]thymidine for proliferation assays (Fields and Lancaster 1993; Ahmed et al. 1994). The active reagent in this dye, resazurin, is taken up by viable cells, where it is reduced to resorufin, which is pink and highly fluorescent (O'Brien et al. 2000). While product formation can be detected by reading absorbance, measurement of fluorescence results in an easy, sensitive assay; the dye is added to the cells and fluorescence is measured after a period of incubation (usually one to four hours) using a fluorometer equipped with a 560-nm excitation and 590-mm emission filter set (Niles et al. 2009). Alamar Blue and resazurin are commercially available as individual reagents or in kit form, e.g. the alamarBlue™ HS Cell Viability Reagent (ThermoFisher Scientific), CellTiter-Blue® Cell Viability Assay (Promega), Resazurin Cell Viability Assay Kit (Biotium), and PrestoBlue™ HS Cell Viability Reagent (ThermoFisher Scientific). Assays that use Alamar Blue may act differently to those that use resazurin in isolation; always perform optimization if changing from one product to another, e.g. to determine the best incubation time.

Like the tetrazolium reduction assays, resazurin-based assays require a period of incubation, and toxicity may occur if a lengthy incubation time is used (Riss et al. 2016). They are prone to interference by substances that have inherent reductive capacity or that display autofluorescence. However, resazurin-based assays are relatively cheap compared to other cytotoxicity assays and can be multiplexed with other assays that provide a fluorescent or luminescent readout (Miret et al. 2006; Niles et al. 2009). They are also suitable for use in 384-well and 1536-well assay formats (Murray et al. 2016).

Adenosine triphosphate (ATP) quantification. ATP can be used as a surrogate variable for cell viability because it is an essential requirement for living cells (Crouch et al. 1993). Loss of viability results in a rapid decrease in cytoplasmic ATP, which can be measured using luciferase reaction chemistry (Riss et al. 2011). An ATP detection reagent is added to lyze the cells, resulting in the release of intracellular ATP. This reagent also contains luciferin substrate and firefly luciferase enzyme, resulting in the generation of a luminescent signal that is proportional to the amount of ATP that is released from the cells. Modifications of the luciferase enzyme and assay have resulted in improved enzymatic activity and stability; a real-time assay has also been developed that removes the need for cell lysis (Riss et al. 2016). Various ATP-based assays are commercially available, including the ATP-Glo™ Bioluminometric Cell Viability Assay (Biotium), ATPlite™ Luminescence Assay System (PerkinElmer), CellTiter-Glo® 2.0 Cell Viability Assay (Promega), and ViaLight™ Plus BioAssay Kit (Lonza). These assays may give very different results; some kits generate a prolonged "glow-type" luminescent signal, while others generate "flash-type" signals that must be read immediately. Always discuss with suppliers and perform your own optimization when choosing an assay or changing kits.

ATP-based assays avoid many of the disadvantages of reduction assays. An incubation time is not required and there is no interference from reagents with reducing ability. They are rapid and highly sensitive, and can be performed using 384-well and 1536-well plates. However, as with all cytotoxicity assays, there are some limitations that must be considered. Test substances may cause changes in ATP content that are independent of cell viability or number (Niles et al. 2009). For example, etoposide and palbocliclib can alter cell size, increasing the amount of ATP that is present per cell (see Section 19.1) (Niepel et al. 2019). ATP-based assays are also susceptible to temperature variations, which may result in edge effects due to thermal gradients (see Section 8.5.2).

29.2.3 Assays Based on Cell Death

Dye penetration due to loss of membrane integrity. Some cell-based assays rely on the addition of a dye that is excluded from viable cells and can only enter once the cell membrane is damaged. While agents such as trypan blue or naphthalene blue are widely used for cell counting (see Section P19.2.1), they are best avoided in cytotoxicity assays due to their lack of sensitivity and the subjective nature of the staining (see Figure 29.2a) (Riss et al. 2019). Various fluorescent dyes have been developed that will enter non-viable cells and bind to DNA, resulting in greater sensitivity. Some of these dyes are not considered suitable for high-throughput screening because they are taken up by viable cells or cause cytotoxicity after prolonged incubation. However, several non-permeable, DNA-binding dyes have been identified that are suitable for high-throughput assays, including SYTOX Green™ (ThermoFisher Scientific), CellTox™ Green (Promega), GelGreen™, and EvaGreen™ (both from Biotium) (Chiaraviglio and Kirby 2014). A subset of fluorescent dyes can also be used for real-time analysis; for example, the CellTox Green Cytotoxicity Assay (Promega) can be added to cells and fluorescence monitored at any time from 0 to 72 hours (Wlodkowic et al. 2011; Riss et al. 2019).

Marker release due to loss of membrane integrity. Loss of membrane integrity results in the release of intracellular constituents such as lactate dehydrogenase (LDH). This enzyme is widely expressed in many cell types and is relatively stable, making it a popular marker for viability assessment. Although LDH is frequently assessed using a colorimetric assay, it can also be assessed using fluorescent or luminescent assays. The conversion of pyruvate to lactate by LDH results in generation of NADH, which can be detected using a number of secondary, coupled reactions (Kumar et al. 2018; Riss et al. 2019). Numerous kits are available for cytotoxicity testing based on LDH release; examples include the CyQUANT LDH Cytotoxicity Assay (ThermoFisher Scientific), Cytotoxicity Detection Kit^PLUS (LDH; Sigma-Aldrich), Lactate Dehydrogenase (LDH) Assay Kit (Fluorimetric; Abcam), and LDH-Glo™ Cytotoxicity Assay (Promega). Serum may have endogenous LDH activity, while media containing pyruvate may cause a reduction in signal due to inhibition of the LDH enzymatic reaction (see Appendix C) (Riss et al. 2019). Always develop standardized procedures for cytotoxicity assays and include the necessary controls to detect artifacts due to culture conditions or other experimental variables.

Instead of relying on intrinsic markers, reagents can be added that are taken up by viable cells and then released following membrane damage (see Section 19.2.2). Examples include neutral red, calcein-acetoxymethyl (calcein-AM), and diacetyl fluorescein (DAF); the latter two substances are converted to fluorescent products within the cell, and thus also rely on the cell's metabolic activity. Neutral red uptake assays are particularly well established and are used for phototoxicity testing and other applications (Borenfreund and Puerner 1985; Repetto et al. 2008). Protocols for neutral red uptake assays can be found elsewhere (Repetto et al. 2008; Ates et al. 2017).

Apoptosis and regulated cell death. Although loss of membrane integrity indicates that cell death has occurred, it does not provide any information on the mode of cell death or its molecular mechanisms. Classically, cell death has been divided into apoptosis, necrosis, and autophagy based on changes in morphology; apoptosis can be recognized by looking for rounded cells with chromatin condensation and fragmentation (see Section 2.5; see also Figure 29.2c, d) (Crowley et al. 2016). Apoptosis also leads to exposure of phosphatidylserine on the external surface of the cell membrane, which acts as an "eat-me" signal for macrophages and other cells (Yang et al. 2018). Annexin V has high affinity for phosphatidylserine, leading to the use of fluorescently labeled Annexin V as a marker for apoptosis (see Figure 19.6) (Vermes et al. 1995). Caspase activity can also be used to detect apoptosis; at one time, caspase activation and activity was considered to be essential for apoptosis to occur (Cummings and Schnellmann 2004). Caspase assays can be performed in high-throughput formats and can give useful information regarding apoptosis, but results will vary depending on which molecular pathways are activated (Galluzzi et al. 2018). Phosphatidyl serine exposure and caspase activation may also occur in alternative, non-lethal biological processes (Mery et al. 2017).

Although the classic descriptions of cell death continue to be used, more precisely defined modes of cell death have been developed based on molecular characteristics (Galluzzi et al. 2018). High-throughput flow cytometers can be used to study specific molecular pathways as part of cytotoxicity testing, e.g. using the iQue3 flow cytometer (Intellicyt). High-throughput analysis by flow cytometry is a specialized topic that is beyond the scope of this book; more information on this topic is available elsewhere (Black et al. 2011; Murray et al. 2016; Ding et al. 2018).

29.2.4 Assays Based on Cell Survival

The cytotoxicity assays discussed thus far are short-term assays, which are usually quick, easy, and convenient to

perform. Frequently, however, cells that have been subjected to toxic agents (e.g. ionizing radiation or cytotoxic drugs) show delayed effects that require ongoing culture to be detected. Long-term assays typically demonstrate survival rather than toxicity. Survival implies the retention of regenerative capacity and is particularly important when screening anticancer agents or regimens (e.g. using both radiotherapy and chemotherapy), where clonogenic populations must be eliminated in order for treatment to be successful (Zips et al. 2005). These clonogenic assays can provide information that would be difficult to obtain using other methods. For example, let us say that a particular agent results in a decrease in cell number. Using a clonogenic assay, it is possible to distinguish between decreased cell proliferation (resulting in smaller colonies) and decreased cell survival (resulting in fewer colonies). If assays are performed in soft agar, it is also possible to study anchorage-independent growth, which is closely associated with malignant behavior *in vivo* (see Section 3.6.2) (Styles 1977).

Cytotoxicity testing can be performed by exposing cells to the test agent before or after limiting dilution is performed, followed by incubation until colonies form that indicate the presence of clonal populations (see Figure 29.3; see also

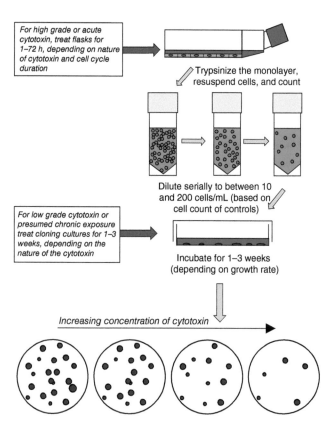

For high grade or acute cytotoxin, treat flasks for 1–72 h, depending on nature of cytotoxin and cell cycle duration

Trypsinize the monolayer, resuspend cells, and count

Dilute serially to between 10 and 200 cells/mL (based on cell count of controls)

For low grade cytotoxin or presumed chronic exposure treat cloning cultures for 1–3 weeks, depending on the nature of the cytotoxin

Incubate for 1–3 weeks (depending on growth rate)

Increasing concentration of cytotoxin

Fig. 29.3. Clonogenic assay for adherent cells. Cells are trypsinized, counted, and diluted as for monolayer dilution cloning (see Protocols P19.4, P20.1). The test substance can be added before trypsinization (see Supp. S29.1) or after seeding for cloning. The colonies are fixed and stained when they reach a reasonable size for counting by eye but before they overlap.

Sections 19.4.1, 20.1). Cells may be plated in flasks, dishes, or multiwell plates; a clonogenic assay for cytotoxicity testing using Petri dishes is provided in the Supplementary Material online (see Supp. S29.1). High-throughput screening can also be performed. For example, a soft agar growth assay has been developed for 384-well plates; a mixture of agar and medium is added to each well to form a base layer, followed by a layer of soft agar that contains a cell suspension (Anderson et al. 2007; Horman et al. 2013). Colony formation can be assessed in high-throughput assays using automated image analysis or an automated colony counter, e.g. the GelCount™ (Oxford Optronix; see Figure 19.12).

Clonogenic assays can be used to assess cell survival by demonstrating long-term retention of self-renewal capacity (five to ten generations or more). However, these assays only apply to the clonogenic component of the culture, which may not be representative of the entire cell population. The question does not arise if controls plate with 100% efficiency, but in practice, colony-forming efficiencies of 20% or less are more likely. Baseline experiments should be performed to assess colony-forming efficiency at various densities; although not ideal, a plating efficiency of over 10% is usually acceptable.

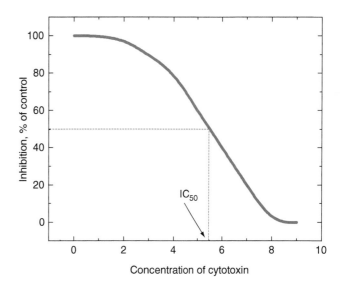

Fig. 29.4. Percentage inhibition curve. Test well values are calculated as a percentage of the controls and plotted against the concentration of cytotoxin. Typically, a sigmoid curve is obtained, and, ideally, the IC_{50} will lie in the center of the inflection of the curve.

29.2.5 Analysis of Cytotoxicity Assays

Microtitration assays. Most short-term cytotoxicity assays that are performed in microwell plates are analyzed by graphing assay data against drug concentration, giving a "microtitration assay." Typically, it is assumed that the readout of a cytotoxicity assay (e.g. an absorbance reading) has a linear relationship to cell number, but this must be explicitly demonstrated during assay selection (see Section 29.2.1). The absolute value of the absorbance should be plotted so that control values may be compared, but the data can then be converted to a percentage of the mean absorbance reading from control wells (cells with vehicle but no active agent), to normalize a series of curves (see Figure 29.4).

The percentage inhibition curve can be used to calculate the IC_{50}, which is the drug concentration at which the cell count is half of the control value. However, there is a great deal of variation in IC_{50} values between different studies, even when the same cell lines and cytotoxic agents are used (Brooks et al. 2019). This is likely to relate to variations in the rate of cell division, e.g. due to different culture conditions. An alternative metric has been developed for cell-based assays that is known as the "normalized growth rate inhibition" (GR) value (Hafner et al. 2016):

$$GR(c) = 2^{k(c)/k(0)} - 1 \qquad (29.1)$$

GR(c) refers to the normalized growth rate inhibition value in the presence of a drug at concentration *c*; *k(0)* refers to the growth rate of control (vehicle-treated) cells, while *k(c)* refers to the growth rate of drug-treated cells at concentration *c* (Hafner et al. 2016; Clark et al. 2017). Thus, the GR

value is the ratio between growth rates under treated and untreated conditions, normalized to a single cell division. An online GR calculator can be used to calculate GR values and to view dose-response datasets from various cell lines (Clark et al. 2017).

Although the adoption of GR metrics will help to remove confounding factors when cells are actively proliferating, these metrics may not be suitable for all culture systems. Primary cultures and some 3D cultures may grow very slowly (or not at all), making it impossible to determine the GR_{50} (Brooks et al. 2019). Different growth rates will also determine experimental choices, such as the duration of the assay (see Figure 29.5). When using a culture system for the first time, parallel plates should be set up for cell counting to generate baseline growth data (see Section 19.1.3). It may also be desirable to sample on each day of drug exposure and recovery. If there is evidence that the cells are growing rapidly and moving into plateau phase, a short form of the assay should be used. With rapidly dividing cells, not only will the cell density increase more rapidly, but also the response to cycle-dependent drugs will be quicker (see Figure 29.6). If growth is slow, a long form of the assay will be required.

The raw data from the assay should be published in addition to any calculated metrics. Computer analysis may make different assumptions or corrections to deal with aberrant data points, which are not apparent unless the raw data are available for scrutiny. Review of the raw data also allows the control wells to be independently examined. For example, values from control wells should be tightly clustered; if they are not, this is taken to indicate uneven plating of cells across the plate.

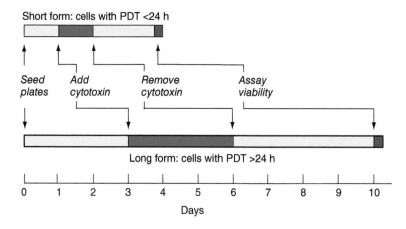

Fig. 29.5. Assay duration. Patterns are shown for short-form and long-form assays. The upper diagram represents an assay that is suitable for cell with a PDT < 24 hours and the bottom diagram represents an assay that is suitable for cells with a PDT > 24 hours; intermediate time scales are also possible.

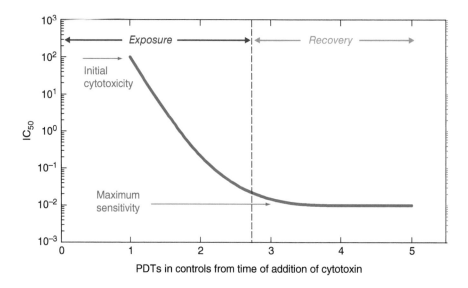

Fig. 29.6. Time course of the fall in IC_{50}. Idealized curve for an agent with a progressive increase in cytotoxicity with time, but eventually reaching a maximum effect after three cell cycles. Not all cytotoxic drugs will conform to this pattern, as shown elsewhere (Freshney et al. 1975).

Clonogenic assays. A survival curve can be generated from a clonogenic assay as follows:

1. Calculate the colony-forming efficiency (the percentage of cells that form colonies) at each drug concentration (see Protocol P19.4).
2. Calculate the relative colony-forming efficiency, i.e. the colony-forming efficiency at each concentration as a fraction of the control; this is the surviving fraction.
3. Plot the surviving fraction on a log scale against the concentration on a linear or log scale, depending on the concentration range used (see Figure 29.7).
4. Determine the IC_{50} or IC_{90}, which is the concentration of compound promoting 50% or 90% inhibition of colony formation, respectively. As this is a semilog plot, the IC_{90} is more appropriate, as it is more likely to fall on

the linear part of the curve, whereas the IC_{50} tends to fall on the knee of the curve, giving a less-stable value.

5. Analyze the curve for differences in sensitivity:

   (a) Slope of the curve and length of the knee. A shallower slope and/or longer knee means reduced sensitivity; a steeper slope and/or shorter knee means increased sensitivity. Both may influence the IC_{50} and the IC_{90}, although a more significant difference can be observed in the IC_{90} (see Figure 29.8).
   (b) Resistant fraction. The fraction of resistant cells is indicated by a flattening of the lower end of the curve.
   (c) Total resistance. This is indicated by the lack of any gradient on the curve.
   (d) Area under the curve. Complex survival curves may be compared by calculating the area under the curve.

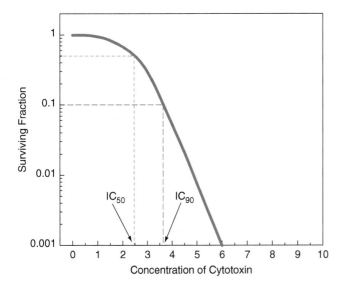

Fig. 29.7. Survival curve. Semilog plot of the surviving fraction of cells (ratio of colonies forming from test cells to colonies forming from control cells) against the concentration of cytotoxin. Typically, the curve has a "knee" and the IC_{90} lies in the linear range of the curve. The IC_{50}, falling on the knee, is a less stable value.

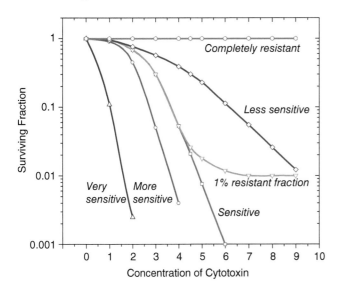

Fig. 29.8. Interpretation of survival curves. Each line represents a separate experiment, with cell survival plotted against the concentration of cytotoxin (semilog plot). The slope increases with increasing sensitivity and decreases with reduced sensitivity until it becomes totally flat for complete resistance. Partial resistance is shown by the curve flattening out at the lower end, resulting in the emergence of a resistant fraction.

Comparison of microtitration and clonogenic agents.
At first glance, microtitration assays using 96-, 384-, or –1536-well plates would seem to be obvious choices for cytotoxicity testing compared to clonogenic assays. Microtitration assays are shorter and more amenable to automated handling, data gathering, and analysis. These assays can also

be multiplexed so that the same cells are examined using multiple assay methods (Murray et al. 2016). However, clonogenic assays give some unique perspectives on the performance of a toxic agent that cannot be examined using a short-term assay. Direct comparison of IC_{50} values derived using microtitration and clonogenic assays has demonstrated a good correlation between the two approaches for the assay of antineoplastic drugs (see Figure 29.9) (Morgan et al. 1983; Plumb et al. 1989). The correlation for IC_{90} was not as tight.

Variables for analysis. Ten basic parameters can be examined in a cytotoxicity assay:

(1) **Concentration of agent.** A wide range of concentrations in log increments (e.g. $1\,\mu M$ to $1\,mM$, and control) should be used for the first attempt and a narrower range (log or linear), based on the results from the first range, for subsequent attempts.

(2) **Number of agents.** The investigation of cytotoxicity often involves the study of the interaction of different drugs. Drug interaction is readily determined by microtitration assays, in which several different ratios of interacting drugs can be examined simultaneously. Analysis of drug interaction can be performed by using an isobologram to interpret the data (Chou 2006, 2010). A rectilinear plot implies an additive response, whereas a curvilinear plot implies synergy if the curve dips below the predicted line and antagonism if it goes above.

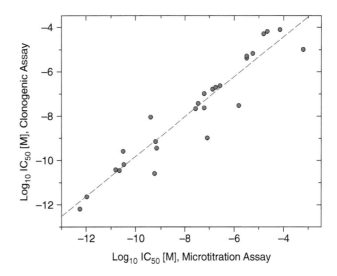

Fig. 29.9. Correlation between microtitration and clonogenic assays. IC_{50} values from each assay are shown, comparing five cell lines from human glioma and six drugs (vincristine, bleomycin, VM-26 epidophyllotoxin, 5-fluorouracil, methyl CCNU, mithramycin). Most of the outlying points were derived from one cell line that later proved to be a mixture of cell types. The broken line is the regression, with the data points from the heterogeneous cell line omitted. *Source:* R. Ian Freshney; microtitration data from Freshney and Morgan (1978).

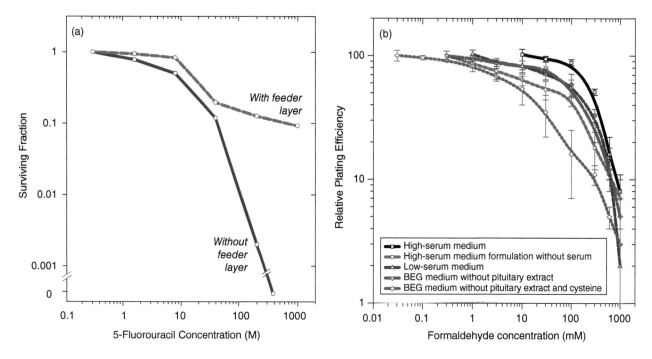

Fig. 29.10. Effect of culture conditions on survival. (a) Human glioma cells were plated in the presence (dashed line) and absence (solid line) of a feeder layer after treatment with 5-fluorouracil. A 10%-resistant fraction is apparent at 1×10^{-4} M drug only in the presence of a feeder layer. In the absence of the feeder layer, the small number of colonies making up the resistant fraction were unable to survive alone. (b) Human oral fibroblasts were grown in high-serum medium (solid line and squares), low-serum medium, and buccal epithelial growth (BEG) medium with supplements. An approximate five-fold increase in the IC_{50} is observed with serum, while removing cysteine from BEG decreases the IC_{50} ten-fold. *Source*: R. Ian Freshney; data in (b) from Nilsson et al. (1998).

(3) *Invariant agent concentrations.* Some conditions that are tested cannot easily be varied, such as the quality of medium, water, or an insoluble plastic. In these cases, the serum concentrations can be varied. As serum may have a masking effect on low-level toxicity, an effect may only be seen in limiting serum (see Figure 29.10b).

(4) *Duration of exposure to agent.* Some agents act rapidly, whereas others act more slowly. For example, exposure to ionizing radiation need last only a matter of minutes to achieve the required dose, whereas testing some cycle-dependent antimetabolic drugs may take several days to achieve a measurable effect. Duration of exposure (T) and drug concentration (C) are related, although $C \times T$ is not always a constant. Prolonging exposure, usually by replacing the drug daily, can increase sensitivity beyond that predicted by $C \times T$, because of cell cycle effects and cumulative damage.

(5) *Time of exposure to agent.* When the agent is soluble and expected to be toxic, it should be added prior to limiting dilution (see Supp. S29.1). However, when the quality of the agent is unknown and stimulation is expected, or only a minor effect is expected (e.g. 20% inhibition rather than several-fold), the agent may be incorporated during clonal growth rather than at preincubation. Confirmation of anticipated toxicity (e.g. for a cytotoxic drug) requires a conservative assay with a minimal drug exposure, as compared to that *in vivo*, applied prior to limiting dilution. Confirmation of the lack of toxicity (e.g. for tap water or a non-toxic pharmaceutical) requires a more stringent assay, with prolonged exposure added at seeding for clonal growth and maintained during clonal growth. The toxin may need to be replaced weekly; daily replacement would, itself, impair cloning efficiency.

(6) *Cell density during exposure.* The density of the cells during exposure to an agent can alter the response of the cells and the agent; for example, HeLa cells are less sensitive to the alkylating agent mustine at high cell densities (Freshney et al. 1975).

(7) *Cell density during cloning.* The number of colonies may fall at high concentrations of a toxic agent, but it is possible to compensate for this effect by seeding more cells so that approximately the same number of colonies form at each concentration. This procedure removes the risk of a low clonal density influencing survival and improves statistical reliability, but means that cells from higher drug concentrations are plated at a higher cell

concentration, a factor that may also influence survival (Pomp et al. 1996). It is preferable to plate cells on a preformed feeder layer, the density of which (5×10^3 cells/cm²) greatly exceeds that of the cloning cells. This step ensures that the cell density is uniform regardless of clonal survival, which contributes little to the total cell density. Note that cloning on a feeder layer can sometimes reveal a resistant fraction of cells that is not apparent without the feeder layer (see Figure 29.10a).

(8) **Colony size.** Some agents are cytostatic (i.e. they inhibit cell proliferation) but not cytotoxic, and during continuous exposure they may reduce the size of colonies in a clonogenic assay without reducing the number. In this case, the size of the colonies should be examined as well as their number. For colony counting, colonies are often grown until they are quite large ($> 1 \times 10^3$ cells), when the growth of larger colonies tends to slow down, and smaller, but still viable, colonies tend to catch up with these larger colonies. For colony sizing, stain the cultures earlier, before the growth rate of larger colonies has slowed down, and score all colonies by manual inspection or using an automated colony counter (see Figure 19.12).

(9) **Effect of medium constituents.** The composition of the culture medium will affect the way that cells respond to a toxin, partly because different media will have different effects on proliferation, but also because individual constituents may affect the stability, binding, and metabolism of the toxin. For example, serum and cysteine reduce the toxicity of formaldehyde on buccal epithelium five- and ten-fold, respectively (see Figure 29.10b). Serum proteins may bind toxins (quite apart from their effect on cell growth) and cysteine and other sulfydryls bind and detoxify reactive chemicals intracellularly and in the medium (Nilsson et al. 1998). As cysteine auto-oxidizes in the medium, the age of the medium will add another variable (personal communication, R. Grafström).

(10) **Vehicle control.** Some agents to be tested have low solubilities in aqueous media, and it may be necessary to use an organic solvent to dissolve them. Ethanol, propylene glycol, and DMSO have been used for this purpose, but may themselves be toxic to cells. Hence, the minimum concentration of solvent should be used to obtain a solution. The agent may be made up at a high concentration in, for example, 100% ethanol, then diluted gradually with balanced salt solution and finally diluted into medium. The final concentration of solvent should be < 0.5% and a vehicle control must be included (i.e. a control with the same final concentration of solvent but without the agent being tested). Always check for compatibility of the solvent with the tubes in which agents are stored. For organic solvents, it is better to use glass with undiluted solvents and to use plastic only when the solvent concentration is < 10%.

29.3 GENOTOXICITY ASSAYS

Genotoxicity implies damage to DNA, which may include changes in nucleotide sequence (mutagenic effects), chromosomal breakage and rearrangements (clastogenic effects), or numerical chromosomal aberrations (aneugenic effects). Genotoxicity testing is an important part of understanding the hazardous properties of all potentially toxic substances, but is of particular importance for cosmetics, hair dyes, and other personal care products, where repeated topical exposure may occur over a period of many years (Nohynek et al. 2010; Chapman et al. 2014). Because different agents can cause different forms of DNA damage, no single method is sufficient, and a tiered approach is required for regulatory purposes (Cimino 2006; Pfuhler et al. 2010). For cosmetic products, the first tier usually comprises at least two tests: an Ames test, which detects mutagenic effects, and an *in vitro* micronucleus assay, which detects clastogenic and aneugenic effects (Pfuhler et al. 2010).

The Ames test uses strains of *Salmonella* that lack the ability to synthesize histidine; novel mutations result in restoration of histidine synthesis, allowing colony formation on low-histidine media (Mortelmans and Zeiger 2000). The *in vitro* micronucleus assay is performed to detect small nuclei that arise in many mammalian cells following DNA damage (see Figure 29.11) (Ye et al. 2019). These micronuclei (also known as Howell-Jolly bodies) come from chromosomal material that is not incorporated into daughter nuclei following cell division, due to chromosomal breakage or problems with attachment to the mitotic spindle during anaphase (Fenech et al. 2011). A number of other genotoxicity assays are available, including the L5178Y mouse lymphoma assay and the sister chromatid exchange assay (Lloyd and Kidd 2012; Stults et al. 2014). These assays have important roles as research-based assays. For example, cells that are deficient in DNA repair display increased levels of chromatid breakage following exposure to ionizing radiation or fluorescent light, which can be detected using the sister chromatid exchange assay (see Figure 29.2e, f) (Parshad and Sanford 2001).

The *in vitro* micronucleus assay is widely used for genotoxicity testing because it is relatively simple to perform, is readily scored and automated, and can detect clastogenic, aneugenic, and some epigenetic effects (Kirsch-Volders et al. 2011). It can be performed using primary cells (e.g. lymphocytes from whole blood), cell lines (e.g. the TK6 lymphoblastoid cell line), or 3D culture systems. If using cell lines, they should have a stable and defined background frequency of micronucleus formation (OECD 2016). The assay uses cytochalasin B to block cytokinesis, which distinguishes between cells that have completed mitosis during exposure to the test substance (resulting in a binucleate appearance) and those that have not. Cells are stained with Giemsa or fluorescent DNA-binding dyes prior to micronucleus scoring. More information on handling of cell lines for genotoxicity testing and protocols for the *in vitro* micronucleus assay are available elsewhere (Lorge et al. 2016; OECD 2016).

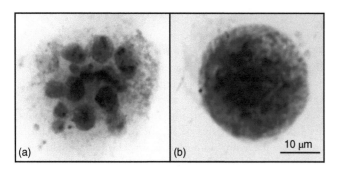

Fig. 29.11. Micronuclei cluster. (a) Mouse ovarian surface epithelial cells derived from a *Brca1*, *Trp53* double knockout animal were stained with DAPI after (a) treatment with 2 µg/ml doxorubicin for two hours; or (b) no treatment. These examples are reverse DAPI. Scale bar is as shown. *Source*: Ye et al. (2019), https://doi.org/10.3390/genes10050366, licensed under CC BY 4.0.

29.4 CARCINOGENICITY ASSAYS

Genotoxicity may cause permanent changes in gene expression and signaling pathways, leading to malignant tumor development (carcinogenic effects). Carcinogenicity testing is important in its own right, but it also acts as a complement to genotoxicity testing. *In vitro* genotoxicity assays have high rates of "false" positive results, which relate to the choice of model, toxicity metrics, and other variables (Fowler et al. 2014). Thus, there is a need to provide supporting data using an assay for carcinogenicity. *In vivo* testing uses large numbers of animals and is prohibited in some countries (see Section 29.1), resulting in the need for *in vitro* assays.

Cell transformation assays use cell lines that are highly sensitive to *in vitro* transformation (see Section 6.3), such as BALB/3T3 or Bhas42; the latter cell line is derived from the Swiss mouse strain and not BALB/c, as originally reported (Uchio-Yamada et al. 2017). Bhas42 is believed to act as a model for initiated cells, according to the classical two-step model of carcinogenesis (Scott et al. 1984). Thus, this assay can be used to assess both genotoxic tumor initiators and non-genotoxic tumor promoters, depending on the design of the experiment (Sakai et al. 2010; Sasaki et al. 2015). Cells are exposed to a test substance for a prolonged period (from days to weeks) and transformed foci are scored at the conclusion of the assay (see Figure 3.9c).

Although the potential for *in vitro* carcinogenicity assays is considerable, this is one area in which testing is far from adequate. Concerns have been expressed regarding the use of cell transformation assays because of their subjective nature and a lack of understanding as to the molecular mechanisms of *in vitro* transformation (Creton et al. 2012). Furthermore, many of the cell lines that are used for these assays are derived from rodent fibroblasts and are already partially transformed; their behavior is unlikely to mimic the various human carcinomas that make up the majority of the common cancers. There is an ongoing need for better models and for "transformics" assays that can provide transcriptome-based data to complement cell transformation assays (Mascolo et al. 2018).

29.5 ADVANCED MODELS FOR TOXICITY TESTING

Although *in vitro* assays are increasingly used for safety testing and screening of potentially therapeutic substances, the existing assays have a number of limitations (see Section 29.1.2). The choice of culture system is particularly important. Currently, most *in vitro* assays are based on human tumor cell lines or rodent fibroblast lines that are prone to transformation, grown on tissue culture plasticware in two dimensions. While these assays are relatively simple, their conditions are far from physiological and their molecular endpoints are often not well understood. There is a need for twenty-first-century toxicology to be matched by twenty-first-century cell culture models (Pamies and Hartung 2017).

Alternative models are now available for *in vitro* toxicity testing, including (i) primary human cells; (ii) cells that have undergone immortalization using reversible techniques, e.g. conditional reprogramming; (iii) stem cell cultures, including cells that have undergone induction of pluripotency using genetic editing; (iv) 3D culture systems, e.g. organoid culture; and (v) microphysiological systems that mimic an organ (or defined parts of an organ) in aspects of structure and function (Francies et al. 2016; Pamies et al. 2018). Many of these culture systems have already been discussed earlier in this book (see Chapters 13, 22, 23, 27). Two systems are highlighted here to illustrate the promise of alternative models: 3D "tissue equivalent" constructs, which have been validated for today's safety testing, and organ-on-chip technologies, which are expected to play a central role in the toxicity testing of the future.

29.5.1 3D Models for Eye and Skin Irritation

There is an increasing need for tissue culture testing to reveal the inflammatory responses that are likely to be induced by pharmaceuticals or cosmetics with topical application. Inflammation is a complex process that involves both local and systemic responses, with the local response involving multiple cell types and structures. For example, topical application of cosmetic ingredients may cause a wide range of biological effects including changes in cell division, differentiation, thickening of the epidermis, increase in permeability of the stratum corneum, and vasodilatation (Nohynek et al. 2010). Eye and skin irritation is traditionally assessed using the Draize test, which subjects a large number of rabbits to a painful and invasive procedure (Lee et al. 2017). A number of alternative models have been developed, including *ex vivo* models (in which isolated eyes or corneas are tested) and cell-based assays; for example, a test substance may be added to an intact monolayer of MDCK cells, followed by the addition of fluorescein to detect damage to the monolayer (OECD 2017). While these assays are all useful, there is a need for better models of the human epidermis and corneal epithelium.

Since the advent of filter well technology, a number of 3D organotypic systems have been developed to act as models for the skin and cornea (see Sections 27.5.2, 27.7.1) (Braa and Triglia 1991; Stark et al. 2004; Ronkko et al. 2016). Examples of commercially available constructs include Epi-Derm™ (MatTek Corporation), LabCyte EPI-MODEL (J-TEC), and SkinEthic™ RHE (EPISKIN) for human epidermis, and EpiOcular™ (MatTek Corporation), Lab-Cyte CORNEA-MODEL (J-TEC), and SkinEthic HCE (EPISKIN) for corneal epithelium. These constructs have undergone validation and can now be used for the assessment of acute eye and skin irritation (OECD 2019b, 2019c). Toxicity is evident by visual inspection and can be measured using viability or other assays (see Figure 29.12) (Kandarova et al. 2009).

29.5.2 Organ-on-Chip Technologies

The term "organ-on-chip" refers to a microfluidic device containing living cells that reconstitutes organ-level structures and functions (Ingber 2018). These devices originated from a longstanding effort to understand how cell shape and function can be regulated through control of the microenvironment (see Sections 2.2.3–2.2.5) (Meyvantsson and Beebe 2008). Typically, cell culture is performed on rigid substrates, but the elasticity of the substrate has an important role to play in cell behavior (see Figure 2.3) and must be incorporated into the model system. Early pioneers in this field developed techniques to study cells on deformable substrates; for example, Albert Harris and colleagues developed a method to generate thin layers of polydimethylsiloxane (PDMS) using a Bunsen burner, which they used to study cell motility and traction (see Section 8.2.2) (Harris et al. 1980). The development of microscale engineering technologies (which were initially used for fabrication of computer chips) allowed such substrates to be manufactured at microscale (Singhvi et al. 1994). These technologies have led to the development of

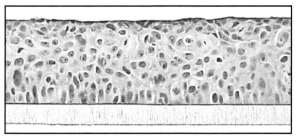

(a) Exposed to Control Solution for 10 min.

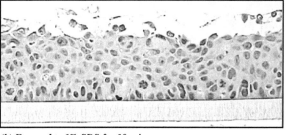

(b) Exposed to 1% SDS for 10 min.

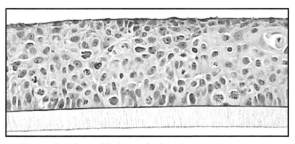

(c) Exposed to Control Solution for 20 min.

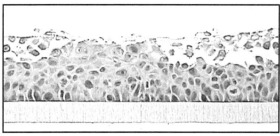

(d) Exposed to 1% SDS for 20 min.

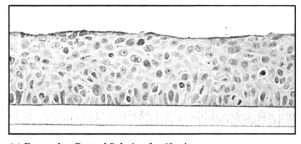

(e) Exposed to Control Solution for 60 min.

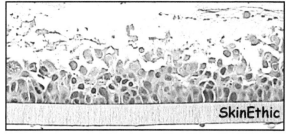

(f) Exposed to 1% SDS for 60 min.

Fig. 29.12. *In vitro* toxicity of human corneal epithelium in filter well inserts. *Source:* Courtesy of SkinEthic.

Fig. 29.13. Small airway chip smoking cigarettes. (a) Photograph of a small airway-on-a-chip device. Scale bar represents 1 cm. (b) Human bronchiolar epithelium is cultured within the device at an air–liquid interface. (c) A programmable smoking machine is loaded with cigarettes to generate fresh smoke. (d) The complete apparatus within a CO_2 incubator, including a microrespirator that is programmed to cyclically "breathe" small volumes of smoke-laden air into the epithelium-lined channel within the chip. *Source*: (a, c, d) Benam et al. (2016), reproduced with permission of Elsevier; (b) Benam, K. H., Novak, R., Nawroth, J., et al. (2016). Matched-Comparative Modeling of Normal and Diseased Human Airway Responses Using a Microengineered Breathing Lung Chip. Cell Syst 3 (5):456–466 e4. © 2016 Elsevier.

microfluidic devices for various applications, particularly in the areas of cell sorting and analysis (see Section 21.5). However, in order to mimic the behavior of the whole organ, there is a need to perform integrated culture using several cell types and to recreate the mechanical microenvironment that is experienced by those cells *in vivo*.

A key advance was made by Donald Ingber and colleagues, who developed a microfluidic device for the integrated culture of human lung alveolar cells (Huh et al. 2010, 2013). The device carried two channels, separated by a thin, flexible membrane made from PDMS. The membrane was coated with extracellular matrix (ECM), allowing alveolar epithelial cells to be grown on one side and pulmonary microvascular endothelial cells on the other. Once the cells became confluent, air was introduced on the epithelial side and cyclic suction was applied, resulting in deformation of the membrane that is similar to the movements that occur during respiration. This "lung-on-a-chip" demonstrated many of the behaviors of the alveolar–capillary interface *in vivo*; for example, introduction of bacteria into the alveolar channel resulted in an inflammatory response on the other side of the membrane (Huh et al. 2010). Many of these behaviors were only seen when the mechanical breathing motions were applied (Ingber 2018).

Ingber's initial proof-of-concept model has led to the development of increasingly sophisticated models that show great promise for toxicity testing (Alepee et al. 2014). For example, a "Small Airway Chip" has been developed by the same group using primary human airway epithelial cells and microvascular endothelial cells (Benam et al. 2016; Benam 2018). Airway cells can be exposed to fresh cigarette smoke (generated from physical or electronic cigarettes) using a programmable machine that mimics human smoking behavior (see Figure 29.13) (Benam et al. 2016). Airway epithelial cells in this microchip form a pseudostratified, ciliated epithelium that show evidence of oxidative damage and ciliary dysfunction after exposure to cigarette smoke; cells can be maintained in the chip for up to 47 days (Benam 2018).

One intriguing aspect of this work is that it does not require cleanroom facilities, allowing organ-on-chip devices to be fabricated in research laboratories (Huh et al. 2013; Novak et al. 2018). The devices discussed here are cast from PDMS using two 3D printed molds, which carry the top and bottom microchannels that make up the two cell compartments. The membrane between these microchannels is cast using a patterned silicon wafer and cured under compression, which results in a flexible membrane of the correct dimensions and porosity (Novak et al. 2018). Although this process is somewhat complex, it is accessible for laboratories that have some degree of fabrication expertise and equipment (or that have engineering colleagues who can supply these requirements). The overall design can be modified for different tissue types (Huh et al. 2013), making it suitable for the study of basic biology and toxicity testing in tissue culture laboratories of the future.

Suppliers.

Supplier	URL
Abcam	www.abcam.com
Biotium	www.biotium.com
EPISKIN	www.episkin.com
Intellicyt (a Sartorius brand)	www.intellicyt.com
Japan Tissue Engineering Company (J-TEC)	www.jpte.co.jp/english/index.html
Lonza	www.lonza.com
MatTek Corporation	www.mattek.com
Molecular Devices	www.moleculardevices.com
PerkinElmer	www.perkinelmer.com
Promega	www.promega.com
R&D Systems	www.rndsystems.com
Sigma-Aldrich (Merck)	www.sigmaaldrich.com/life-science/cell-culture.html
ThermoFisher Scientific	www.thermofisher.com/us/en/home/life-science/cell-culture.html

Supp. S29.1 Clonogenic Assay for Cytotoxicity Testing.

REFERENCES

Ahmed, S.A., Gogal, R.M. Jr., and Walsh, J.E. (1994). A new rapid and simple non-radioactive assay to monitor and determine the proliferation of lymphocytes: an alternative to [3H]thymidine incorporation assay. *J. Immunol. Methods* 170 (2): 211–224. https://doi.org/10.1016/0022-1759(94)90396-4.

Alepee, N., Bahinski, A., Daneshian, M. et al. (2014). State-of-the-art of 3D cultures (organs-on-a-chip) in safety testing and pathophysiology. *ALTEX* 31 (4): 441–477. https://doi.org/10.14573/altex1406111.

Anderson, S.N., Towne, D.L., Burns, D.J. et al. (2007). A high-throughput soft agar assay for identification of anticancer compound. *J. Biomol. Screening* 12 (7): 938–945. https://doi.org/10.1177/1087057107306130.

Artegiani, B. and Clevers, H. (2018). Use and application of 3D-organoid technology. *Hum. Mol. Genet.* 27 (R2): R99–R107. https://doi.org/10.1093/hmg/ddy187.

Ates, G., Vanhaecke, T., Rogiers, V. et al. (2017). Assaying cellular viability using the neutral red uptake assay. *Methods Mol. Biol.* 1601: 19–26. https://doi.org/10.1007/978-1-4939-6960-9_2.

Barile, F.A. (2010). Validating and troubleshooting ocular in vitro toxicology tests. *J. Pharmacol. Toxicol. Methods* 61 (2): 136–145. https://doi.org/10.1016/j.vascn.2010.01.001.

Benam, K.H. (2018). Disrupting experimental strategies for inhalation toxicology: the emergence of microengineered breathing-smoking human lung-on-a-chip. *Appl. In Vitro Toxicol.* 4 (2): 107–114. https://doi.org/10.1089/aivt.2017.0030.

Benam, K.H., Novak, R., Nawroth, J. et al. (2016). Matched-comparative modeling of normal and diseased human airway responses using a microengineered breathing lung chip. *Cell Syst.* 3 (5): 456–466. https://doi.org/10.1016/j.cels.2016.10.003.

Black, C.B., Duensing, T.D., Trinkle, L.S. et al. (2011). Cell-based screening using high-throughput flow cytometry. *Assay Drug Dev. Technol.* 9 (1): 13–20. https://doi.org/10.1089/adt.2010.0308.

Bols, N.C., Pham, P.H., Dayeh, V.R. et al. (2017). Invitromatics, invitrome, and invitroomics: introduction of three new terms for in vitro biology and illustration of their use with the cell lines from rainbow trout. *In Vitro Cell Dev. Biol. Anim.* 53 (5): 383–405. https://doi.org/10.1007/s11626-017-0142-5.

Borenfreund, E. and Puerner, J.A. (1985). A simple quantitative procedure using monolayer cultures for cytotoxicity assays (HTD/NR-90). *J. Tissue Cult. Methods* 9 (1): 7–9. https://doi.org/10.1007/bf01666038.

Braa, S.S. and Triglia, D. (1991). Predicting ocular irritation using three-dimensional human fibroblast cultures. *Cosmet. Toilet.* 106: 55–58.

Brennan, L.M., Widder, M.W., Lee, L.E. et al. (2012). Long-term storage and impedance-based water toxicity testing capabilities of fluidic biochips seeded with RTgill-W1 cells. *Toxicol. In Vitro* 26 (5): 736–745. https://doi.org/10.1016/j.tiv.2012.03.010.

Brennan, L.M., Widder, M.W., McAleer, M.K. et al. (2016). Preparation and testing of impedance-based fluidic biochips with RTgill-W1 cells for rapid evaluation of drinking water samples for toxicity. *J. Vis. Exp.* 109: 53555. https://doi.org/10.3791/53555.

Brooks, E.A., Galarza, S., Gencoglu, M.F. et al. (2019). Applicability of drug response metrics for cancer studies using biomaterials. *Philos. Trans. R. Soc. Lond. B Biol. Sci.* 374 (1779): 20180226. https://doi.org/10.1098/rstb.2018.0226.

Burden, N., Sewell, F., and Chapman, K. (2015). Testing chemical safety: what is needed to ensure the widespread application of non-animal approaches? *PLoS Biol.* 13 (5): e1002156. https://doi.org/10.1371/journal.pbio.1002156.

Chabner, B.A. (2016). NCI-60 cell line screening: a radical departure in its time. *J. Natl Cancer Inst.* 108 (5) https://doi.org/10.1093/jnci/djv388.

Chapman, K.E., Thomas, A.D., Wills, J.W. et al. (2014). Automation and validation of micronucleus detection in the 3D EpiDerm human reconstructed skin assay and correlation with 2D dose responses. *Mutagenesis* 29 (3): 165–175. https://doi.org/10.1093/mutage/geu011.

Chiaraviglio, L. and Kirby, J.E. (2014). Evaluation of impermeant, DNA-binding dye fluorescence as a real-time readout of eukaryotic cell toxicity in a high throughput screening format. *Assay Drug Dev. Technol.* 12 (4): 219–228. https://doi.org/10.1089/adt.2014.577.

Chou, T.C. (2006). Theoretical basis, experimental design, and computerized simulation of synergism and antagonism in drug combination studies. *Pharmacol. Rev.* 58 (3): 621–681. https://doi.org/10.1124/pr.58.3.10.

Chou, T.C. (2010). Drug combination studies and their synergy quantification using the Chou-Talalay method. *Cancer Res.* 70 (2): 440–446. https://doi.org/10.1158/0008-5472.CAN-09-1947.

Cimino, M.C. (2006). Comparative overview of current international strategies and guidelines for genetic toxicology testing for regulatory purposes. *Environ. Mol. Mutagen.* 47 (5): 362–390. https://doi.org/10.1002/em.20216.

Clark, N.A., Hafner, M., Kouril, M. et al. (2017). GRcalculator: an online tool for calculating and mining dose-response data. *BMC Cancer* 17 (1): 698. https://doi.org/10.1186/s12885-017-3689-3.

Creton, S., Aardema, M.J., Carmichael, P.L. et al. (2012). Cell transformation assays for prediction of carcinogenic potential: state of the science and future research needs. *Mutagenesis* 27 (1): 93–101. https://doi.org/10.1093/mutage/ger053.

Crouch, S.P., Kozlowski, R., Slater, K.J. et al. (1993). The use of ATP bioluminescence as a measure of cell proliferation and cytotoxicity. *J. Immunol. Methods* 160 (1): 81–88. https://doi.org/10.1016/0022-1759(93)90011-u.

Crowley, L.C., Marfell, B.J., Scott, A.P. et al. (2016). Dead cert: measuring cell death. *Cold Spring Harb. Protoc.* 2016 (12) https://doi.org/10.1101/pdb.top070318.

Cummings, B.S. and Schnellmann, R.G. (2004). Measurement of cell death in mammalian cells. *Curr. Protoc. Pharmacol.* 25 (1) https://doi.org/10.1002/0471141755.ph1208s25.

Ding, M., Clark, R., Bardelle, C. et al. (2018). Application of high-throughput flow cytometry in early drug discovery: an AstraZeneca perspective. *SLAS Discov.* 23 (7): 719–731. https://doi.org/10.1177/2472555218775074.

ECHA (2017). The use of alternatives to testing on animals for the REACH regulation. https://echa.europa.eu/documents/10162/13639/alternatives_test_animals_2017_en.pdf (accessed 9 October 2019).

Eskes, C., Bostrom, A.C., Bowe, G. et al. (2017). Good cell culture practices & in vitro toxicology. *Toxicol. In Vitro* 45 (Pt 3): 272–277. https://doi.org/10.1016/j.tiv.2017.04.022.

EU (2006). Regulation (EC) No. 1907/2006 of the European Parliament and of the Council of 18 December 2006 concerning the Registration, Evaluation, Authorisation and Restriction of Chemicals (REACH), establishing a European Chemicals Agency, amending Directive 1999/45/EC and repealing Council Regulation (EEC) No. 793/93 and Commission Regulation (EC) No. 1488/94 as well as Council Directive 76/769/EEC and Commission Directives 91/155/EEC, 93/67/EEC, 93/105/EC and 2000/21/EC (Text with EEA relevance)Text with EEA relevance. http://data.europa.eu/eli/reg/2006/1907/2019-07-02 (accessed 10 October 2019).

EU (2009). Regulation (EC) No. 1223/2009 of the European Parliament and of the Council of 20 November 2009 on cosmetic products. http://data.europa.eu/eli/reg/2009/1223/2019-06-17.

EURL ECVAM (2019). Validated test methods. https://ec.europa.eu/jrc/en/eurl/ecvam/alternative-methods-toxicity-testing/validated-test-methods (accessed 10 October 2019).

Fenech, M., Kirsch-Volders, M., Natarajan, A.T. et al. (2011). Molecular mechanisms of micronucleus, nucleoplasmic bridge and nuclear bud formation in mammalian and human cells. *Mutagenesis* 26 (1): 125–132. https://doi.org/10.1093/mutage/geq052.

Fields, R.D. and Lancaster, M.V. (1993). Dual-attribute continuous monitoring of cell proliferation/cytotoxicity. *Am. Biotechnol. Lab.* 11 (4): 48–50.

Fischer, M., Belanger, S.E., Berckmans, P. et al. (2019). Repeatability and reproducibility of the RTgill-W1 cell line assay for predicting fish acute toxicity. *Toxicol. Sci.* 169 (2): 353–364. https://doi.org/10.1093/toxsci/kfz057.

Fowler, P., Smith, R., Smith, K. et al. (2014). Reduction of misleading ("false") positive results in mammalian cell genotoxicity assays. III: sensitivity of human cell types to known genotoxic agents. *Mutat. Res. Genet. Toxicol. Environ. Mutagen.* 767: 28–36. https://doi.org/10.1016/j.mrgentox.2014.03.001.

Francies, H.E., Barthorpe, A., McLaren-Douglas, A. et al. (2016). Drug sensitivity assays of human cancer organoid cultures. *Methods Mol. Biol.* 1576: 339–351. https://doi.org/10.1007/7651_2016_10.

Fratta, I.D., Sigg, E.B., and Maiorana, K. (1965). Teratogenic effects of thalidomide in rabbits, rats, hamsters, and mice. *Toxicol. Appl. Pharmacol.* 7: 268–286. https://doi.org/10.1016/0041-008x(65)90095-5.

Freshney, R.I. (1978). Use of tissue culture in predictive testing of drug sensitivity. *Cancer Top.* 1: 5–7.

Freshney, R.I. and Morgan, D. (1978). Radioisotopic quantitation in microtitration plates by an autofluorographic method. *Cell Biol. Int. Rep.* 2 (4): 375–380. https://doi.org/10.1016/0309-1651(78)90023-1.

Freshney, R.I., Paul, J., and Kane, I.M. (1975). Assay of anti-cancer drugs in tissue culture: conditions affecting their ability to incorporate 3H-leucine after drug treatment. *Br. J. Cancer* 31 (1): 89–99. https://doi.org/10.1038/bjc.1975.11.

Galluzzi, L., Vitale, I., Aaronson, S.A. et al. (2018). Molecular mechanisms of cell death: recommendations of the Nomenclature Committee on Cell Death 2018. *Cell Death Differ.* 25 (3): 486–541. https://doi.org/10.1038/s41418-017-0012-4.

Garraway, L.A., Widlund, H.R., Rubin, M.A. et al. (2005). Integrative genomic analyses identify MITF as a lineage survival oncogene amplified in malignant melanoma. *Nature* 436 (7047): 117–122. https://doi.org/10.1038/nature03664.

Guillouzo, A. and Guguen-Guillouzo, C. (2008). Evolving concepts in liver tissue modeling and implications for in vitro toxicology. *Expert Opin. Drug Metab. Toxicol.* 4 (10): 1279–1294. https://doi.org/10.1517/17425255.4.10.1279.

Hafner, M., Niepel, M., Chung, M. et al. (2016). Growth rate inhibition metrics correct for confounders in measuring sensitivity to cancer drugs. *Nat. Methods* 13 (6): 521–527. https://doi.org/10.1038/nmeth.3853.

Harris, A.K., Wild, P., and Stopak, D. (1980). Silicone rubber substrata: a new wrinkle in the study of cell locomotion. *Science* 208 (4440): 177–179. https://doi.org/10.1126/science.6987736.

Hartung, T., FitzGerald, R.E., Jennings, P. et al. (2017). Systems toxicology: real world applications and opportunities. *Chem. Res. Toxicol.* 30 (4): 870–882. https://doi.org/10.1021/acs.chemrestox.7b00003.

Horman, S.R., To, J., and Orth, A.P. (2013). An HTS-compatible 3D colony formation assay to identify tumor-specific chemotherapeutics. *J. Biomol. Screening* 18 (10): 1298–1308. https://doi.org/10.1177/1087057113499405.

Huh, D., Matthews, B.D., Mammoto, A. et al. (2010). Reconstituting organ-level lung functions on a chip. *Science* 328 (5986): 1662–1668. https://doi.org/10.1126/science.1188302.

Huh, D., Kim, H.J., Fraser, J.P. et al. (2013). Microfabrication of human organs-on-chips. *Nat. Protoc.* 8 (11): 2135–2157. https://doi.org/10.1038/nprot.2013.137.

Ingber, D.E. (2018). From mechanobiology to developmentally inspired engineering. *Philos. Trans. R. Soc. Lond. B Biol. Sci.* 373 (1759) https://doi.org/10.1098/rstb.2017.0323.

ISO (2019). Standards catalogue. https://www.iso.org/standards-catalogue/browse-by-ics.html (accessed 11 October 2019).

Iversen, P.W., Beck, B., Chen, Y.F., et al. (2012). HTS assay validation. https://www.ncbi.nlm.nih.gov/books/nbk83783.

Kandarova, H., Hayden, P., Klausner, M. et al. (2009). An in vitro skin irritation test (SIT) using the EpiDerm reconstructed human epidermal (RHE) model. *J. Vis. Exp.* (29): 1366. https://doi.org/10.3791/1366.

Kirsch-Volders, M., Decordier, I., Elhajouji, A. et al. (2011). In vitro genotoxicity testing using the micronucleus assay in cell lines, human lymphocytes and 3D human skin models. *Mutagenesis* 26 (1): 177–184. https://doi.org/10.1093/mutage/geq068.

Kleensang, A., Vantangoli, M.M., Odwin-DaCosta, S. et al. (2016). Genetic variability in a frozen batch of MCF-7 cells invisible in routine authentication affecting cell function. *Sci. Rep.* 6: 28994. https://doi.org/10.1038/srep28994.

Kumar, P., Nagarajan, A., and Uchil, P.D. (2018). Analysis of cell viability by the lactate dehydrogenase assay. *Cold Spring Harb. Protoc.* 2018 (6) https://doi.org/10.1101/pdb.prot095497.

Lee, L.E., Dayeh, V.R., Schirmer, K. et al. (2009). Applications and potential uses of fish gill cell lines: examples with RTgill-W1. *In Vitro Cell Dev. Biol. Anim.* 45 (3–4): 127–134. https://doi.org/10.1007/s11626-008-9173-2.

Lee, M., Hwang, J.H., and Lim, K.M. (2017). Alternatives to in vivo Draize rabbit eye and skin irritation tests with a focus on 3D reconstructed human cornea-like epithelium and epidermis models. *Toxicol. Res.* 33 (3): 191–203. https://doi.org/10.5487/TR.2017.33.3.191.

Lloyd, M. and Kidd, D. (2012). The mouse lymphoma assay. *Methods Mol. Biol.* 817: 35–54. https://doi.org/10.1007/978-1-61779-421-6_3.

Lorenzi, P.L., Reinhold, W.C., Varma, S. et al. (2009). DNA fingerprinting of the NCI-60 cell line panel. *Mol. Cancer Ther.* 8 (4): 713–724. https://doi.org/10.1158/1535-7163.MCT-08-0921.

Lorge, E., Moore, M.M., Clements, J. et al. (2016). Standardized cell sources and recommendations for good cell culture practices in genotoxicity testing. *Mutat. Res.* 809: 1–15. https://doi.org/10.1016/j.mrgentox.2016.08.001.

Mascolo, M.G., Perdichizzi, S., Vaccari, M. et al. (2018). The transformics assay: first steps for the development of an integrated approach to investigate the malignant cell transformation in vitro. *Carcinogenesis* 39 (7): 955–967. https://doi.org/10.1093/carcin/bgy037.

McGregor, D.B., Edwards, I., Riach, C.G. et al. (1988). Studies of an S9-based metabolic activation system used in the mouse lymphoma L5178Y cell mutation assay. *Mutagenesis* 3 (6): 485–490. https://doi.org/10.1093/mutage/3.6.485.

Mery, B., Guy, J.B., Vallard, A. et al. (2017). In vitro cell death determination for drug discovery: a landscape review of real issues. *J. Cell Death* 10: 1179670717691251. https://doi.org/10.1177/1179670717691251.

Meyvantsson, I. and Beebe, D.J. (2008). Cell culture models in microfluidic systems. *Annu. Rev. Anal. Chem. (Palo Alto Calif.)* 1: 423–449. https://doi.org/10.1146/annurev.anchem.1.031207.113042.

Miret, S., De Groene, E.M., and Klaffke, W. (2006). Comparison of in vitro assays of cellular toxicity in the human hepatic cell line HepG2. *J. Biomol. Screening* 11 (2): 184–193. https://doi.org/10.1177/1087057105283787.

Morgan, D., Freshney, R.I., Darling, J.L. et al. (1983). Assay of anticancer drugs in tissue culture: cell cultures of biopsies from human astrocytoma. *Br. J. Cancer* 47 (2): 205–214. https://doi.org/10.1038/bjc.1983.28.

Mortelmans, K. and Zeiger, E. (2000). The ames *Salmonella*/microsome mutagenicity assay. *Mutat. Res.* 455 (1–2): 29–60. https://doi.org/10.1016/s0027-5107(00)00064-6.

Mosmann, T. (1983). Rapid colorimetric assay for cellular growth and survival: application to proliferation and cytotoxicity assays. *J. Immunol. Methods* 65 (1–2): 55–63. https://doi.org/10.1016/0022-1759(83)90303-4.

Murray, D., McWilliams, L., and Wigglesworth, M. (2016). High-throughput cell toxicity assays. *Methods Mol. Biol.* 1439: 245–262. https://doi.org/10.1007/978-1-4939-3673-1_16.

Nardone, R.M. (1980). The interface of toxicology and tissue culture and reflections on the carnation test. *Toxicology* 17 (2): 105–111. https://doi.org/10.1016/0300-483x(80)90081-5.

Niepel, M., Hafner, M., Mills, C.E. et al. (2019). A multi-center study on the reproducibility of drug-response assays in mammalian cell lines. *Cell Syst.* 9 (1): 35–48. https://doi.org/10.1016/j.cels.2019.06.005.

Niles, A.L., Moravec, R.A., and Riss, T.L. (2009). In vitro viability and cytotoxicity testing and same-well multi-parametric combinations for high throughput screening. *Curr. Chem. Genomics* 3: 33–41. https://doi.org/10.2174/1875397300903010033.

Nilsson, J.A., Zheng, X., Sundqvist, K. et al. (1998). Toxicity of formaldehyde to human oral fibroblasts and epithelial cells: influences of culture conditions and role of thiol status. *J. Dent. Res.* 77 (11): 1896–1903. https://doi.org/10.1177/00220345980770110601.

Nohynek, G.J., Antignac, E., Re, T. et al. (2010). Safety assessment of personal care products/cosmetics and their ingredients. *Toxicol. Appl. Pharmacol.* 243 (2): 239–259. https://doi.org/10.1016/j.taap.2009.12.001.

Novak, R., Didier, M., Calamari, E. et al. (2018). Scalable fabrication of stretchable, dual channel, microfluidic organ chips. *J. Vis. Exp.* (140): 58151. https://doi.org/10.3791/58151.

O'Brien, J., Wilson, I., Orton, T. et al. (2000). Investigation of the Alamar Blue (resazurin) fluorescent dye for the assessment of mammalian cell cytotoxicity. *Eur. J. Biochem.* 267 (17): 5421–5426. https://doi.org/10.1046/j.1432-1327.2000.01606.x.

OECD (2016). Test No. 487: In vitro mammalian cell micronucleus test. https://www.oecd-ilibrary.org/content/publication/9789264264861-en.

OECD (2017). Test No. 460: Fluorescein leakage test method for identifying ocular corrosives and severe irritants. https://www.oecd-ilibrary.org/content/publication/9789264185401-en.

OECD (2018). *Guidance Document on Good in Vitro Method Practices (GIVIMP)*. Paris: OECD Publishing.

OECD (2019a). OECD iLibrary. https://www.oecd-ilibrary.org (accessed 11 October 2019).

OECD (2019b). Test No. 439: In vitro skin irritation: reconstructed human epidermis test method. https://www.oecd-ilibrary.org/content/publication/9789264242845-en.

OECD (2019c). Test No. 492: Reconstructed human cornea-like epithelium (RhCE) test method for identifying chemicals not requiring classification and labelling for eye irritation or serious eye damage. https://www.oecd-ilibrary.org/content/publication/9789264242548-en.

Pamies, D. and Hartung, T. (2017). 21st century cell culture for 21st century toxicology. *Chem. Res. Toxicol.* 30 (1): 43–52. https://doi.org/10.1021/acs.chemrestox.6b00269.

Pamies, D., Bal-Price, A., Chesne, C. et al. (2018). Advanced Good Cell Culture Practice for human primary, stem cell-derived and organoid models as well as microphysiological systems. *ALTEX* 35 (3): 353–378. https://doi.org/10.14573/altex.1710081.

Parasuraman, S. (2011). Toxicological screening. *J. Pharmacol. Pharmacother.* 2 (2): 74–79. https://doi.org/10.4103/0976-500X.81895.

Parshad, R. and Sanford, K.K. (2001). Radiation-induced chromatid breaks and deficient DNA repair in cancer predisposition. *Crit. Rev. Oncol. Hematol.* 37 (2): 87–96.

Pfuhler, S., Kirst, A., Aardema, M. et al. (2010). A tiered approach to the use of alternatives to animal testing for the safety assessment of cosmetics: genotoxicity. A COLIPA analysis. *Regul. Toxicol. Pharm.* 57 (2–3): 315–324. https://doi.org/10.1016/j.yrtph.2010.03.012.

Plumb, J.A., Milroy, R., and Kaye, S.B. (1989). Effects of the pH dependence of 3-(4,5-dimethylthiazol-2-yl)-2,5-diphenyl-tetrazolium bromide-formazan absorption on chemosensitivity determined by a novel tetrazolium-based assay. *Cancer Res.* 49 (16): 4435–4440.

Pomp, J., Wike, J.L., Ouwerkerk, I.J. et al. (1996). Cell density dependent plating efficiency affects outcome and interpretation of colony forming assays. *Radiother. Oncol.* 40 (2): 121–125.

Promega (2014). Webinar: Overview of 3D cell culture model systems and factors to consider when choosing and validating cell-based assays for use with 3D cultures. www.promega.com.au/resources/webinars/worldwide/archive/overview-of-3d-cell-culture-model-systems (accessed 16 October 2019).

Repetto, G., del Peso, A., and Zurita, J.L. (2008). Neutral red uptake assay for the estimation of cell viability/cytotoxicity. *Nat. Protoc.* 3 (7): 1125–1131. https://doi.org/10.1038/nprot.2008.75.

Riss, T.L., Moravec, R.A., and Niles, A.L. (2011). Cytotoxicity testing: measuring viable cells, dead cells, and detecting mechanism of cell death. *Methods Mol. Biol.* 740: 103–114. https://doi.org/10.1007/978-1-61779-108-6_12.

Riss, T.L., Moravec, R.A., Niles, A.L. et al. (2016). Cell viability assays. In: *Assay Guidance Manual [Internet]* (eds. G.S. Sittampalam, A. Grossman and K. Brimacombe). Bethesda, MD: Eli Lilly and Company and the National Center for Advancing Translational Sciences.

Riss, T., Niles, A., Moravec, R. et al. (2019). Cytotoxicity assays: in vitro methods to measure dead cells. In: *Assay Guidance Manual [Internet]* (eds. G.S. Sittampalam, A. Grossman and K. Brimacombe). Bethesda, MD: Eli Lilly and Company and the National Center for Advancing Translational Sciences.

Ronkko, S., Vellonen, K.S., Jarvinen, K. et al. (2016). Human corneal cell culture models for drug toxicity studies. *Drug Deliv. Transl. Res.* 6 (6): 660–675. https://doi.org/10.1007/s13346-016-0330-y.

Roschke, A.V., Tonon, G., Gehlhaus, K.S. et al. (2003). Karyotypic complexity of the NCI-60 drug-screening panel. *Cancer Res.* 63 (24): 8634–8647.

Ross, N.T. and Wilson, C.J. (2014). In vitro clinical trials: the future of cell-based profiling. *Front. Pharmacol.* 5: 121. https://doi.org/10.3389/fphar.2014.00121.

Sakai, A., Sasaki, K., Muramatsu, D. et al. (2010). A Bhas 42 cell transformation assay on 98 chemicals: the characteristics and performance for the prediction of chemical carcinogenicity. *Mutat. Res.* 702 (1): 100–122. https://doi.org/10.1016/j.mrgentox.2010.07.007.

Sasaki, K., Umeda, M., Sakai, A. et al. (2015). Transformation assay in Bhas 42 cells: a model using initiated cells to study mechanisms of carcinogenesis and predict carcinogenic potential of chemicals. *J. Environ. Sci. Health C Environ. Carcinog. Ecotoxicol. Rev.* 33 (1): 1–35. https://doi.org/10.1080/10590501.2014.967058.

Scott, R.E., Wille, J.J. Jr., and Wier, M.L. (1984). Mechanisms for the initiation and promotion of carcinogenesis: a review and a new concept. *Mayo Clin. Proc.* 59 (2): 107–117. https://doi.org/10.1016/s0025-6196(12)60244-4.

Shoemaker, R.H. (2006). The NCI60 human tumour cell line anticancer drug screen. *Nat. Rev. Cancer* 6 (10): 813–823. https://doi.org/10.1038/nrc1951.

Singhvi, R., Kumar, A., Lopez, G.P. et al. (1994). Engineering cell shape and function. *Science* 264 (5159): 696–698. https://doi.org/10.1126/science.8171320.

Skehan, P., Storeng, R., Scudiero, D. et al. (1990). New colorimetric cytotoxicity assay for anticancer-drug screening. *J. Natl Cancer Inst.* 82 (13): 1107–1112. https://doi.org/10.1093/jnci/82.13.1107.

Soldatow, V.Y., Lecluyse, E.L., Griffith, L.G. et al. (2013). In vitro models for liver toxicity testing. *Toxicol. Res. (Camb.)* 2 (1): 23–39. https://doi.org/10.1039/C2TX20051A.

Stark, H.J., Szabowski, A., Fusenig, N.E. et al. (2004). Organotypic cocultures as skin equivalents: a complex and sophisticated in vitro system. *Biol. Proced. Online* 6: 55–60. https://doi.org/10.1251/bpo72.

Stults, D.M., Killen, M.W., and Pierce, A.J. (2014). The sister chromatid exchange (SCE) assay. *Methods Mol. Biol.* 1105: 439–455. https://doi.org/10.1007/978-1-62703-739-6_32.

Styles, J.A. (1977). A method for detecting carcinogenic organic chemicals using mammalian cells in culture. *Br. J. Cancer* 36 (5): 558–563. https://doi.org/10.1038/bjc.1977.231.

Tahara, H., Matsuda, S., Yamamoto, Y. et al. (2017). High-content image analysis (HCIA) assay has the highest correlation with direct counting cell suspension compared to the ATP, WST-8 and Alamar Blue assays for measurement of cytotoxicity. *J. Pharmacol. Toxicol. Methods* 88 (Pt 1): 92–99. https://doi.org/10.1016/j.vascn.2017.08.003.

Tanneberger, K., Knobel, M., Busser, F.J. et al. (2013). Predicting fish acute toxicity using a fish gill cell line-based toxicity assay. *Environ. Sci. Technol.* 47 (2): 1110–1119. https://doi.org/10.1021/es303505z.

Taylor, K. (2018). Ten years of REACH – an animal protection perspective. *Altern. Lab. Anim.* 46 (6): 347–373. https://doi.org/10.1177/026119291804600610.

van Tonder, A., Joubert, A.M., and Cromarty, A.D. (2015). Limitations of the 3-(4,5-dimethylthiazol-2-yl)-2,5-diphenyl-2H-tetrazolium bromide (MTT) assay when compared to three commonly used cell enumeration assays. *BMC Res. Notes* 8: 47. https://doi.org/10.1186/s13104-015-1000-8.

Uchio-Yamada, K., Kasai, F., Ozawa, M. et al. (2017). Incorrect strain information for mouse cell lines: sequential influence of misidentification on sublines. *In Vitro Cell Dev. Biol. Anim.* 53 (3): 225–230. https://doi.org/10.1007/s11626-016-0104-3.

Vermes, I., Haanen, C., Steffens-Nakken, H. et al. (1995). A novel assay for apoptosis. Flow cytometric detection of phosphatidyl serine expression on early apoptotic cells using fluorescein labelled Annexin V. *J. Immunol. Methods* 184 (1): 39–51. https://doi.org/10.1016/0022-1759(95)00072-i.

Wilding, J.L. and Bodmer, W.F. (2014). Cancer cell lines for drug discovery and development. *Cancer Res.* 74 (9): 2377–2384. https://doi.org/10.1158/0008-5472.CAN-13-2971.

Wlodkowic, D., Faley, S., Darzynkiewicz, Z. et al. (2011). Real-time cytotoxicity assays. *Methods Mol. Biol.* 731: 285–291. https://doi.org/10.1007/978-1-61779-080-5_23.

Yang, Y., Lee, M., and Fairn, G.D. (2018). Phospholipid subcellular localization and dynamics. *J. Biol. Chem.* 293 (17): 6230–6240. https://doi.org/10.1074/jbc.R117.000582.

Ye, C.J., Sharpe, Z., Alemara, S. et al. (2019). Micronuclei and genome chaos: changing the system inheritance. *Genes (Basel)* 10 (5) https://doi.org/10.3390/genes10050366.

Zips, D., Thames, H.D., and Baumann, M. (2005). New anticancer agents: in vitro and in vivo evaluation. *In Vivo* 19 (1): 1–7.

PART IX

Teaching and Troubleshooting

After reading the following chapters in this part of the book, you will be able to:

(30) *Training:*

    (a) Describe the primary outcomes when training new personnel and the role of a laboratory "trainer."

    (b) Explain the importance of both written Standard Operating Procedures (SOPs) and hands-on training.

    (c) Perform an induction or orientation session (including a brief tour and essential information) for new personnel in a tissue culture laboratory.

    (d) Develop a tissue culture training program, with reference to essential topics and hands-on training exercises.

(31) *Problem Solving:*

    (a) Explain why both immediate and follow-up actions are required when dealing with common tissue culture problems.

    (b) Identify possible causes for each common tissue culture problem.

    (c) For each cause, perform immediate and follow-up actions to help address the problem at source.

    (d) Refer to other chapters or publications for more information on each problem.

(32) *In Conclusion:*

    (a) Summarize crucial requirements for successful and reproducible cell culture.

Freshney's Culture of Animal Cells: A Manual of Basic Technique and Specialized Applications, Eighth Edition. Amanda Capes-Davis and R. Ian Freshney.
© 2021 John Wiley & Sons Ltd. Published 2021 by John Wiley & Sons Ltd.
Companion website: www.wiley.com/go/freshney/cellculture8

CHAPTER 30

Training

Helen Morton, who played an important role in developing standardized culture media, also participated in the very first tissue culture training course, which was held in Canada in 1948. Morton commented that "virtually everyone using tissue culture in North America gave lectures"; a photograph taken at the course showed that this number included about 30 individuals (Morton 1978). Training extended to preparing media constituents and homemade glassware. Morton described her colleague Joseph Morgan cutting up a truckload of sugar beets to produce his own glutamine, which they used in Medium 199 (see Section 9.4). The experiences of those early tissue culture pioneers are strikingly different from today's practitioners. Commercially prepared media and disposable plasticware have eliminated many of the preparative steps that were once a major requirement for good tissue culture. However, like our predecessors, we continue to struggle to prevent contamination and to provide our cell lines (which may themselves date back to Morton's era) with a physiological environment. Training is a central, and often underrated, part of Good Cell Culture Practice (GCCP) (see Section 7.2.1) (Coecke et al. 2005; Pamies et al. 2018). The need for training is actually far greater than it was in 1948, because of the rapid growth in tissue culture applications and the diverse backgrounds of the newcomers who are now entering the field. This chapter focuses primarily on the principles that underpin tissue culture training (see Section 30.1). Every laboratory should be able to use these principles to develop their own training material to suit their individual requirements. Specific training topics and exercises are also discussed and provided in the Supplementary Material online (see Section 30.2).

30.1 TRAINING PRINCIPLES

New personnel should not be allowed to work in the tissue culture facility until they are deemed to be competent (Geraghty et al. 2014). The primary outcomes when training new personnel are to teach (i) safe practices; (ii) good aseptic technique; and (iii) good awareness of contamination and how to manage it effectively (see Sections 6.1.3, 12.4, 16.2). Although these priorities remain consistent, their implementation will vary with the laboratory and the individual. Some laboratories purchase all media sterile and ready-made; others will prepare their own media and may require new personnel to operate the autoclave for their own sterilization runs. The newcomer may be experienced in tissue culture, but they may not have operated an autoclave before, so it is important to tailor requirements to the individual.

30.1.1 Roles and Responsibilities

Receiving hands-on training from an experienced tissue culture practitioner is the best way to become competent in aseptic technique and other essential procedures. An experienced person should be nominated as the laboratory "trainer," in the same way that a "curator" is appointed to look after an item of equipment (see Section 7.5.4). This does not necessarily mean that the nominated person trains every new arrival; in a busy laboratory, this responsibility may be shared amongst several personnel. The primary reason for nominating a "trainer" is to ensure that training is not forgotten in the midst of other urgent priorities. In many laboratories, the same person will take responsibility for the operation of the tissue culture area

Freshney's Culture of Animal Cells: A Manual of Basic Technique and Specialized Applications, Eighth Edition. Amanda Capes-Davis and R. Ian Freshney.
© 2021 John Wiley & Sons Ltd. Published 2021 by John Wiley & Sons Ltd.
Companion website: www.wiley.com/go/freshney/cellculture8

and will draw up rosters that apportion responsibilities for routine tasks such as cleaning water baths or incubators. New personnel will need to be trained in routine tasks as well as in handling procedures.

It may seem odd asking a highly qualified scientist to clean incubators, but this is the best way to gain a practical understanding of contamination. One episode of fungal contamination in an incubator will teach the newcomer more than any theoretical lecture could achieve. Tissue culture has a long history of involving senior staff in mundane activities. Dr Joseph Morton (who developed Medium 199, the first commercially available medium) would wash the laboratory glassware while planning experiments, while Dr Richard Ham (who developed Ham's F12 medium) would "tune up" his laboratory's old phase microscope and help his students with their metaphase spreads while writing a myriad of journal articles and the textbook *Mechanisms of Development* (Morgan et al. 1950; Morton 1978; Ham and Veomett 1980; McKeehan 2012).

30.1.2 Induction

The newcomer should be taken on a tour of the tissue culture facility and provided with a brief induction or orientation. This allows the trainee to meet other staff, determine their roles and responsibilities, and see the level of preparation that is required. It also allows the trainer to communicate essential information, including:

(1) *Hygiene and personal protective equipment (PPE).* Hand washing, any additional hygiene rules (e.g. the need to tie back hair), and essential PPE should be explained (see Section 12.2.4). The trainee should be shown where to find clean lab coats or gowns and disposables such as gloves.

(2) *Waste disposal and washing up.* The different waste streams should be explained (see Section 6.3.6), with an emphasis on the need to separately dispose of general waste (e.g. used paper towels after washing hands) and biological waste (e.g. used gloves). The trainee should be shown where to put used glassware (and associated requirements, such as adding disinfectant) and where to discard used gowns for laundry or disposal.

(3) *Laboratory layout.* The locations of major equipment items should be noted including biological safety cabinets (BSCs), incubators, and microscopes. Any restrictions on their usage should also be highlighted. For example, some BSCs or incubators may be set aside for the use of specific research groups or for specific purposes, including the quarantine of incoming cell lines (see Sections 4.2.3, 16.2).

(4) *Storage areas.* The locations of storage shelves, refrigerators, and freezers should be noted. The principles of storage should also be explained and attention drawn to the distinctions in location and packaging between sterile and non-sterile stocks, tissue culture grade and non-tissue culture grade plasticware, and fluids stored at room temperature versus those stored at 4 or $-20\,°C$. The trainee should be informed where reagents and consumables can be obtained, who to inform if stocks are close to running out, and how to rotate stocks so that the oldest is used first.

(5) *Cryostorage.* The locations of cryofreezers should be noted and the associated hazards explained when working in cryostorage areas (see Sections 6.2.3, 6.2.4). Trainees should be restricted from working with frozen cryovials until they are considered competent to do so safely. However, safety risks should still be highlighted during induction if trainees are required to work in these areas. For example, if oxygen monitoring is installed, the trainee should be made aware of the reason for the monitoring and the need to evacuate the room if an alarm sounds.

(6) *Safety and security requirements.* The level of biological containment that applies to that particular facility should be noted and it should be explained what that means in practical terms (see Section 6.3.3). For example, personnel may be required to keep facility doors closed and to use a key card or make an entry in a logbook when entering or exiting the facility. Some areas may be off-limits to trainees due to regulatory requirements, e.g. areas where radioisotopes are handled or biosecurity risks have been identified (see Supp. S6.2, S6.3).

(7) *Emergency and first-aid arrangements.* The locations of first-aid kits, eyewash stations, safety showers, and emergency equipment such as fire extinguishers should be noted. Training should be performed separately to cover emergency situations and safety incidents, but it is useful to highlight actions that can be performed within the facility, e.g. the location of "spill kits" to deal with large spills of biological material (see Table 6.1). The location of contact information should be noted for security, emergency services, and key laboratory personnel.

(8) *Facility operation.* Any day-to-day requirements for users of the tissue culture facility should be highlighted, including:
   a) Shared practices, e.g. use of antibiotic-free media (see Section 9.6.2).
   b) Record keeping, e.g. when taking a fresh bottle of medium out of storage.
   c) Inventory systems, e.g. the use of barcodes to trace reagents back to sources.
   d) Booking systems, e.g. when sharing a BSC between multiple users.
   e) Cleaning and maintenance rosters.
   f) Arrangements or restrictions that apply to out-of-hours work.

30.1.3 Training Documents

Trainees will only remember some of the information that is imparted during induction; written material is essential to reinforce key concepts, provide further information, and act as a reference when hands-on training commences. Many laboratories have a printed list of "room rules" that is posted at the entry to the tissue culture area, as a reminder of key requirements. Further information may come from institutional policies or manuals, textbooks such as this one, or guidance documents for GCCP (see Section 7.2.1) (Coecke et al. 2005; Geraghty et al. 2014; OECD 2018). Online information from suppliers or tissue culture practitioners can also be helpful, but some online videos do not show best practice; always watch such videos before you recommend them! It is important to carefully select the information that is given to the trainee. Some topics are particularly crucial, such as contamination, its causes, and its management (see Chapters 16, 17). For this edition of *Freshney's Culture of Animal Cells*, key learning outcomes have been developed for each chapter to help guide selection of material for trainees. Key learning outcomes are listed at the beginning of each part of the book.

Practical training is usually performed with reference to the laboratory's own Standard Operating Procedures (SOPs) (see Section 7.5.1). New personnel can read the relevant SOPs before receiving hands-on instruction; the SOP then acts as a reminder of good practice. However, compliance with written procedures tends to decrease over time, particularly when procedures are complex or require meticulous attention to detail (Geraghty et al. 2014). It is important to assess whether an individual is competent at a particular technique and to provide ongoing training for all personnel. Laboratories should maintain training records and consider how to assess competency and provide ongoing training. For laboratories that operate using Good Laboratory Practice (GLP) or Good Manufacturing Practice (GMP), such ongoing training and competency records make up an integral part of quality assurance (QA) systems (see Sections 7.2.2, 7.2.3, 7.5) (OECD 2018).

30.1.4 Hands-on Training

Tissue culture requires competency in a number of practical techniques and procedures. For the newcomer, the most important technique to master is good aseptic technique, which is essential to manage the ongoing risk of contamination (see Sections 12.1, 16.2). Aseptic technique is best taught by observing an experienced operator and then practicing capping, handling, and pipetting in a BSC (see Protocols P12.1, P12.2). In some laboratories, a spare flask of cells is made available for the trainee to test their aseptic technique by feeding and passaging the culture for several weeks (see Protocols P14.1–P14.3). The best test of competency in aseptic technique is to grow the cells in antibiotic-free medium; routine inspection should demonstrate whether the trainee can maintain a sterile environment. Once the trainee

is proficient in subculture, they can learn how to perform cryopreservation; again, the best test of competency is to thaw a cryovial and assess its viability (see Protocols P15.1, P15.2, P19.1, P19.2). Supervision should be continued until the trainee is deemed competent to work independently. As the trainee continues to work in the tissue culture facility, new opportunities will arise to observe practical techniques and procedures and to develop competency in new skills.

Although this form of hands-on training can be effective, there is a risk that trainees will not develop competency in all of the necessary procedures or will not gain an understanding of why they are performed in a certain way – which becomes important when procedures are modified for a new application. Many laboratories prefer a more structured training program that is conducted by experienced personnel in-house or by an external training provider. Typically, such programs include theoretical knowledge and hands-on training exercises that allow the student to develop competency in each technique and allow the trainer to objectively assess their competency.

30.2 TRAINING PROGRAMS

30.2.1 Topics

A number of learning outcomes (both theoretical and practical) can be conveyed in a tissue culture training program. Broadly speaking, the program can be divided into broad topics that reflect the parts laid out in this book:

(1) **Understanding cell culture.** In addition to basic cell biology, trainees need to understand that a culture will display a process of evolution as it continues to be passaged, resulting in changes in its genotype and phenotype. The concepts of senescence, immortalization, and transformation should be explained.

(2) **Laboratory and regulatory requirements.** For most trainees (who are not required to design a laboratory or purchase large equipment items), requirements will relate to (i) safety; (ii) bioethics; (iii) rigor and reproducibility; and (iv) good practice, particularly GCCP. Cell line provenance and validation testing should be explained.

(3) **Medium and substrate requirements.** A newcomer to tissue culture is likely to begin with a cell line that is grown on plastic in two dimensions, using conventional tissue culture medium. However, the need for serum-free media and physiological environments should be highlighted, with practical advice on their adoption.

(4) **Handling cultures.** This is likely to be the major topic of any practical training course. Trainees should be taught good aseptic technique, refeeding, subculture, cryopreservation, and thawing techniques. Primary culture and other advanced techniques might also be attempted, once trainees are familiar with the basic techniques. Cell

TABLE 30.1. Training exercises.

Number	Topic	Training objectives	Source[a]
S30.1	Washing and sterilizing glassware	Appreciation of preparative practices and quality control (QC) measures performed to ensure that materials are sterile; familiarity with washup and sterilization areas, equipment, and processes	P11.1, P11.2
S30.2	Preparation and sterilization of water	Appreciation of preparative practices carried on outside aseptic area; knowledge of grades of water purity; process of preparing sterile water; process of preparing reagents for an autoclave run; review of sterility indicators after an autoclave run; record keeping when preparing reagents	P11.3
S30.3	Preparation and sterilization of DPBS-A	Constitution of simple salt solution, with buffering and pH control; measurement of pH and other variables (e.g. osmolality); sterilization of heat-stable solutions by autoclaving	P11.4
S30.4	Preparation of pH standards	Familiarization with use of phenol red as a pH indicator; sterilization using autoclaving and filtration; comparison of effect on pH following autoclaving and filtration; handling of tissue culture flasks and syringe-tip filters	S9.1
S30.5	Preparation of basal medium from powder and sterilization by filtration	Familiarization with basal media formulations and formats; process of preparing basal medium from powder; process of assembling a filter flask and connecting to a vacuum pump; operation and use of a BSC or laminar flow hood	P11.5, P11.9
S30.6	Pipetting and transfer of fluids in a BSC	Operation of a BSC; use of liquid handling equipment, particularly pipettes and pipette controllers; familiarization with sterile handling techniques including swabbing, flaming, capping, and pipetting; rapid and accurate transfer of liquid between containers under sterile conditions	P12.1, P12.2
S30.7	Preparation of complete culture medium	Familiarization with basal media formulations and serum-free approaches; process of preparing complete medium (ready for use) from basal medium (1× stock) with additional supplements; reinforcement of good aseptic technique using a BSC	P11.6,P11.7
S30.8	Observation of cultured cells	Operation of microscope and camera; familiarization with the appearance of cultures of different types and at different densities; awareness of microbial contamination, cross-contamination, and misidentification; assessment of growth phase of culture and need for refeeding or subculture	P18.1, P18.2
S30.9	Feeding adherent cultures	Familiarization with changes in morphology after subculture and indicators that medium requires replacement; awareness of microbial contamination, cross-contamination, and misidentification; process of refeeding; further experience of performing sterile handling procedures using a BSC	P14.1
S30.10	Counting cells by hemocytometer and automated cell counter	Familiarization with cell counting methods and assessment of viability; process of trypsinization to produce a cell suspension; cell counting using a hemocytometer or automated cell counter; addition of trypan blue; evaluation of relative merits of hemocytometer and automated counting	P19.1, P19.2
S30.11	Subculture of adherent cultures	Familiarization with changes in morphology and other indicators that subculture is required; process of trypsinization to produce a cell suspension; process of cell counting; reseeding into fresh culture vessels at different split ratios; calculation of cell yield and amount seeded	P14.2
S30.12	Subculture of suspension cultures	Familiarization with suspension mode of growth; process of cell counting using trypan blue; passaging of suspension cultures to allow cells to remain in exponential growth	P14.3
S30.13	Cryopreservation of cultured cells	Familiarization with the principles of cryopreservation and cell banking; process of preparing cell suspensions, cell counting, and volume adjustment; options for freezing medium, controlled cooling devices, and cryofreezers; record keeping for frozen stocks	P15.1
S30.14	Thawing of frozen cryovials	Familiarization with the principles of cryopreservation and cell banking; retrieval of cryovials from a controlled cooling device or cryofreezer; safe handling practices when working with frozen cryovials; record keeping for frozen stocks; process of thawing cryovials; cell counting and viability assessment; optimization of cryopreservation based on viability data	P15.2
S30.15	Primary culture	Familiarization with primary culture and options for explant culture or tissue disaggregation; awareness of origin and diversity of cultured cells; process of explant culture using chick embryos and tissue disaggregation using warm or cold trypsin	S13.1, S13.2, P13.4

[a]The prefix "P" refers to a protocol within the main text, while "S" refers to Supplementary Material that is available through the book companion website.

BSC, Biological safety cabinet; DPBS-A, Dulbecco's phosphate buffered saline without Ca^{2+} and Mg^{2+}.

banking procedures and good record keeping should be explained.

(5) ***Validation and characterization.*** The impact and sources of contamination should be described and management of contamination should be discussed. Trainees should be taught how to count cells, assess viability, detect microbial contamination, and perform mycoplasma and authentication testing.

(6) ***Physical and genetic manipulation.*** Typically, this topic is taught at a more advanced level once the basic techniques have been mastered. Methods for cell cloning, separation and sorting, gene delivery, gene editing, and immortalization should be discussed. The risks of off-target effects and artifacts should be highlighted.

(7) ***Stem cells and differentiated cells.*** Advanced trainees can be introduced to more specialized procedures for handling stem cells, specialized cells, and tumor cells from various sources. Methods to induce or maintain differentiation should be discussed.

(8) ***Model environments and applications.*** Advanced trainees can be introduced to three-dimensional (3D) culture and how it may be achieved using scaffold-free or scaffold-based approaches. Systems that can be used for scale-up (particularly bioreactors) and toxicity testing should be introduced and explained.

(9) ***Teaching and troubleshooting.*** Problem solving should be included as a consistent theme across all topics, with an emphasis on managing common problems.

30.2.2 Exercises

A set of training exercises is available for use with this book. Training exercises are listed here but are provided in full in the Supplementary Material online (see Table 30.1; see also Supp. S30.1–S30.15). These exercises are primarily designed for the newcomer to tissue culture, who must acquire basic

skills before working independently in a tissue culture laboratory. It is assumed that trainees have some experience in general laboratory-based techniques. Competency can be assessed by observing trainees during the procedure, by examining the records generated in the course of the exercise, or by reviewing the analysis and discussion conducted after the procedure has been completed. The exercises are presented in a sequence, starting from the most basic and progressing toward the more complex, in terms of technical manipulation.

Supp. S30.1–S30.15 Training Exercises.

REFERENCES

Coecke, S., Balls, M., Bowe, G. et al. (2005). Guidance on good cell culture practice. A report of the second ECVAM task force on good cell culture practice. *Altern. Lab. Anim.* 33 (3): 261–287.

Geraghty, R.J., Capes-Davis, A., Davis, J.M. et al. (2014). Guidelines for the use of cell lines in biomedical research. *Br. J. Cancer* 111 (6): 1021–1046. https://doi.org/10.1038/bjc.2014.166.

Ham, R.G. and Veomett, M.J. (1980). *Mechanisms of development*. St Louis, MO: Mosby.

McKeehan, W.L. (2012). A tribute to Richard G. Ham (1932–2011). *In Vitro Cell Dev. Biol. Anim.* 48 (5): 259–270. https://doi.org/10.1007/s11626-012-9509-9.

Morgan, J.F., Morton, H.J., and Parker, R.C. (1950). Nutrition of animal cells in tissue culture; initial studies on a synthetic medium. *Proc. Soc. Exp. Biol. Med.* 73 (1): 1–8.

Morton, H.J. (1978). Joseph F. Morgan: the man and his contributions. *In Vitro* 14 (1): 3–10. https://doi.org/10.1007/bf02618169.

OECD (2018). *Guidance document on Good In Vitro Method Practices (GIVIMP)*. Paris: OECD Publishing.

Pamies, D., Bal-Price, A., Chesne, C. et al. (2018). Advanced Good Cell Culture Practice for human primary, stem cell-derived and organoid models as well as microphysiological systems. *ALTEX* 35 (3): 353–378. https://doi.org/10.14573/altex.1710081.

CHAPTER 31

Problem Solving

Problem solving is a necessary, and often frustrating, part of tissue culture. No matter how well a tissue culture laboratory is run, problems arise when new personnel, new techniques, or any other new development destabilizes its normal routine. Some problems, such as incomplete sterilization due to an autoclave fault, can be quickly detected through good equipment design (e.g. generation of error messages) and the inclusion of sterility indicators in Standard Operating Procedures (SOPs) (see Sections 7.5.1, 11.2.7). Such problems only become widespread if the equipment fails to generate error messages or the user fails to follow the SOP. Other problems, such as widespread fungal contamination due to high spore counts, are hard to eliminate and even harder to understand. Problem solving can be divided into immediate and follow-up actions. Immediate actions, such as cleaning water baths and other equipment where fungal contamination is likely to be present, may be sufficient to address the problem. However, follow-up actions will help to understand the problem better and may provide more effective solutions. For example, it is often helpful to collect environmental samples when dealing with widespread fungal contamination. If you can identify the species of the contaminant and where it is found within the laboratory, you can develop a more targeted approach for its early elimination.

The advice given in the preceding chapters has concentrated on practical, "how to do it" instructions, sometimes with indications of what might go wrong. This chapter attempts to summarize these potential problems under topic headings and adds a few more potential difficulties, queries, and (hopefully) solutions.

31.1 MICROBIAL CONTAMINATION

Microbial contamination may be clearly visible or may require specific testing, such as mycoplasma. Problem solving will depend on the type of microbial contamination (see Section 31.1.1), whether it is limited to one person or is more widespread, and whether it is sporadic or repeated (see Sections 31.1.2, 31.1.3). Problems that relate to laminar flow or air quality are particularly challenging and are dealt with separately (see Section 31.1.4).

Freshney's Culture of Animal Cells: A Manual of Basic Technique and Specialized Applications, Eighth Edition. Amanda Capes-Davis and R. Ian Freshney.
Companion website: www.wiley.com/go/freshney/cellculture8

31.1.1 Type of Microbial Contamination

Problem	Cause	Immediate action	Follow-up action	Book section or reference for more information
Bacteria, fungi	Personnel (see Section 31.1.2) Environment, e.g. air quality (see Section 31.1.4) Equipment, e.g. for laminar airflow (see Section 31.1.4) Materials (see Section 31.6) Other cell lines	Discard affected cultures and associated reagents; only attempt elimination if culture is irreplaceable; check other cultures for contamination; wipe down surfaces with disinfectant; record the nature of the contamination for future reference	Grow cells in antibiotic-free media to detect problems early; if repeated or widespread, take samples for sterility testing; plate out onto agar and incubate; if colonies grow, perform species testing	Sections 11.8, 16.1–16.3, 16.6
Mycoplasma	All of above, but other cell lines are most likely	Discard affected cultures; only attempt elimination if culture is irreplaceable	Test all cultures for mycoplasma; return to cryostorage for clean stocks; quarantine incoming cultures until tested; implement regular screening program	Section 16.4
Viruses	All of above, but other cell lines, reagents, and xenograft culture are most likely	Discard affected cultures; do not attempt elimination	If cell line is known to shed virus, either do not use or quarantine to ensure other cell lines are not affected	Section 16.5

31.1.2 Contamination Is Limited to One Person

Problem	Cause	Immediate action	Follow-up action	Book section or reference for more information
Sporadic				
Single species	Occasional lapses in aseptic technique	Check operator technique	Perform refresher training	Sections 12.2–12.2.4, 30.1.4
	Lapses in personal hygiene	Check hand washing and use of personal protective equipment (PPE); tie back hair	Reinforce "room rules," e.g. the need to wash hands on entering the tissue culture facility	Section 12.2.4, 30.1.3
Multiple species	Repeated lapses in aseptic technique	Check that biological safety cabinet (BSC) is uncluttered, materials are swabbed before placing in BSC, and spills are wiped up	Check with other users of the same BSC for problems; BSC may be faulty and need to be serviced	Sections 6.3.4, 12.2.2, 12.2.3

Problem	Cause	Immediate action	Follow-up action	Book section or reference for more information
Repeated				
Single species	Specific reagent	Check for a unique medium, additive, or other reagent that no-one else uses	Check method for preparing reagent; perform sterility testing of reagent; record batch number; discuss with supplier if problem recurs	Sections 11.8, 16.1.4
	Specific cell line	Check if that cell line is used by others; check cryostorage for earlier passage stocks	Test early stocks for contamination; if contaminated, obtain from cell repository	Section 16.1.5
Multiple species	Habitual failure in aseptic technique or personal hygiene, e.g. use of lab coat from microbiological or animal work	Check operator technique for habitual problems; clarify correct procedures	Check with other users for problems; check that sufficient clean gowns or coats are available; review training procedures	Section 12.4, 30.2
	Problems with laminar airflow, e.g. due to overcrowding of BSC	Demonstrate correct layout of BSC; remove unnecessary items	Check equipment for faults or problems	Sections 6.3.4, 12.2.2, 12.2.3
	Non-sterile reagents	Exclude non-sterile reagents from the BSC	Check procedures for preparing sterile reagents	Chapter 11
Continuous				
Single species	Chronic, low-level contamination of one or more cell lines or reagents	Discard all cultures and associated reagents; return to cryostorage for earlier passage stocks	Perform sterility testing to screen all cultures and reagents; grow cells antibiotic-free to pick up early problems	Section 11.8
Multiple species	Total breakdown in aseptic technique	Supervise operator until technique improves	Restrict unsupervised work until operator has completed training	Sections 12.4, 30.2
	Failure of laminar airflow due to incorrect use of BSC	Supervise operation of the BSC	Check equipment for faults or problems	Sections 6.3.4, 12.2.2
	Failure of sterilization due to incorrect use of equipment or procedures	Supervise preparation of media and reagents; demonstrate correct operation of equipment	Check equipment for faults or problems	Chapter 11

31.1.3 Contamination Is Widespread

Problem	Cause	Immediate action	Follow-up action	Book section or reference for more information
Sporadic				
Single species	Contaminated incubator	Clean incubator	Review incubator cleaning regime	Sections 5.3.1, 12.5.1, 16.1.3
	Infrequent use of contaminated medium or other reagent	Record all cases of contamination and look for reagents in common	Perform sterility testing of reagents prior to use	Sections 11.8, 16.1.4
	Low-level, cryptic contamination of cells or reagents	Collect samples from cells and all reagents for testing	Perform sterility testing; if unsuccessful, perform longer assay, e.g. clonogenic assay	Section 11.8
Multiple species	Use of antibiotics in culture media	Use antibiotic-free media; discard affected cultures as they arise	Return to cryostorage for clean stocks; if contaminated, obtain from cell repository	Section 9.6.2
	Lapses in aseptic technique affecting multiple operators	Use antibiotic-free media, allowing problems to be detected early	Clarify correct procedures; perform refresher training	Sections 12.4, 30.2
	Poor air quality, e.g. due to high spore count	See Section 31.1.4		
Repeated				
Single species	Contamination of frequently used reagent	Check frequency of use of reagents among users to narrow the problem down to common reagents	Perform sterility testing of reagents prior to use	Sections 11.8, 16.1.4
	Contaminated incubator after cleaning	Check that SOP is performed correctly; check difficult-to-see areas for contamination; perform second round of heat decontamination	Review SOP to see if cleaning can be improved, e.g. by autoclaving racks and shelves (check with supplier)	Sections 5.3.1, 12.5.1, 16.1.3
	Contaminated BSC due to undetected spills	Clean BSC, including underneath the work surface	If problem persists, arrange service and fumigation	Sections 6.3.4, 6.3.5, 12.2.2
Multiple species	Failure of sterilization due to equipment problem, e.g. autoclave or sterilizing oven	Check equipment for error messages or alarms; check autoclave data to confirm cycle gives correct conditions for sterilization; check integrity of oven door seals and apertures	Arrange service of equipment; include sterility indicators or temperature probes within the load to provide additional data; review and update SOPs	Section 11.2
	Failure of sterilization due to personnel problem	Check with operator that the SOP is correctly followed and the autoclave cycle is correctly selected; check packing for possible overcrowding	Clarify correct procedures; perform random testing of sterilization; perform refresher training	Section 11.2
	Contaminated cold storage equipment	Clean coldroom, fridges, and freezers	Store reagents and samples in movable racks for easier cleaning; review cleaning regime and turnover of sterile stocks; store sterile items in sealed packages	Section 4.2.6
	Overcrowding of the laboratory, e.g. too much traffic	Restrict entry to tissue culture personnel only	Implement booking procedures for use of tissue culture equipment	Section 4.1.1
	Failure of laminar airflow, e.g. due to BSC malfunction	See Section 31.1.4		

31.1.4 Problems with Laminar Flow or Air Quality

Problem	Cause	Immediate action	Follow-up action	Book section or reference for more information
Compromised BSC	Faulty or dirty BSC	(1) Swab work surface and sides of work area with disinfectant; (2) Wash below work surface and swab with disinfectant; (3) Check crevices and ductwork for leaks or contamination; (4) Check whether filters are dirty or blocked, e.g. pressure drop across filter; (5) Check integrity of filter (performed by service engineer)	Arrange for service of BSC; review and update cleaning and maintenance schedules	Sections 6.3.4, 12.2.2, 16.1.3
	Incorrect location	Close door; restrict passing traffic	Move BSC to better location; reorganize layout of laboratory	Sections 4.2.1, 5.1.1
Poor air quality within the room	Focus of contamination inside the room, e.g. fungal contamination in water bath	Swab down all non-sterile items entering the room; clean water bath and other water reservoirs; clean the room	Collect samples for testing; review and update cleaning schedules; if problem persists, consider fumigation	Sections 6.3.4, 6.3.5, 16.1.3, 16.3.1
	Problem with air handling	Discuss with facility staff; schedule check of air filters and air handling equipment; clean the room	Collect samples for testing, including settle plates; review and update maintenance schedules	Section 4.1.4, 16.1.2, 16.3.1; IEST (2018)
Poor air quality beyond the room	Increased dust or high spore count, e.g. due to construction, fires, or seasonal changes	Discuss with facility staff; improve isolation of facility (e.g. by shutting doors); increase frequency of cleaning	Collect samples for testing, including settle plates; reduce contributory factors or schedule away from culture time	Sections 16.1.2, 16.3.1; IEST (2018)

31.2 CROSS-CONTAMINATION AND MISIDENTIFICATION

Cross-contamination or misidentification (see Section 17.1 for terminology) should be excluded on arrival of a new cell line and suspected following any change in phenotype. A more comprehensive list of causes and their prevention can be found in Table 17.2.

Problem	Cause	Immediate action	Follow-up action	Book section or reference for more information
Incorrect source of cell lines	Secondary source, e.g. colleague who is not aware that a cell line is misidentified	Check if cell line is known to be misidentified; discard affected cultures	Do not use known misidentified cell lines; test incoming cell lines for authenticity; obtain from primary sources, i.e. the originator or a cell repository	Sections 15.6, 17.2, 17.3; ICLAC (2019)

Problem	Cause	Immediate action	Follow-up action	Book section or reference for more information
Handling errors	Sharing pipettes or reagents between cell lines; handling more than one cell line at a time; lapses in aseptic technique	Maintain good aseptic technique; discard affected cultures; replace with earlier passage material	Perform cell banking to store frozen stocks at early passage; test stocks for authenticity; return to frozen stocks on a regular basis	Sections 15.5, 17.2, 17.3
Procedural errors	Incorrect inactivation of a feeder layer, resulting in continued proliferation	Follow the correct procedure for feeder layer inactivation	Test stocks for authenticity using a method for species detection	Sections 17.3.3, 20.6.2
Labeling, record keeping, and inventory errors	Mislabeling of flasks or cryovials; freezer inventory errors; selecting the wrong cryovial	Label flasks, vials, etc. before use; double-check labels for accuracy; secure vials correctly in racks; keep good records	Develop a screening program for cell line testing that will pick up occasional errors; conduct random checks of cryostorage	Section 17.2.3

31.3 CHEMICAL CONTAMINATION

Chemical contamination is less common in today's laboratories than it was in previous years, thanks to the quality control (QC) testing performed by reputable suppliers. It is assumed here that supplier QC testing has been performed but there is still reason to suspect a chemical contaminant, e.g. due to slow growth under certain conditions.

Problem	Cause	Immediate action	Follow-up action	Book section or reference for more information
Leachable chemicals	Use of single-use, disposable plastics, e.g. culture bags or upstream storage vessels	Look for common vessels or reagents associated with poor cell behavior	Record batch numbers; discuss with supplier	Sections 7.5.2, 7.5.3
Residue after cleaning	Ineffective washing or rinsing leading to traces of chemicals, detergents, or disinfectants	Perform visual examination to look for obvious residue; check for recent changes in detergent or equipment; check with the operator that the SOP is correctly followed	Keep tissue culture glassware separate from chemical glassware; perform culture testing to assess any changes; review and update SOP for washing glassware	Sections 11.3, 11.8.2
Particulate contamination	Dust accumulation during storage	Foil cap all open bottles; store in dust-free area or container	Plan storage carefully and rotate stocks frequently	Sections 11.3, 12.3.3
Reagent contamination	Dissemination of powders or aerosols during handling	Avoid drafts when weighing powders or dispensing liquids; use a fume hood or cytotoxic drug safety cabinet when handling toxic chemicals	Position equipment carefully to avoid excessive traffic	Sections 4.2.4, 12.2.2

31.4 SLOW CELL GROWTH

The most likely cause of slow cell growth is mycoplasma contamination (see Section 31.1.1). It is assumed here that problem solving for contamination, including mycoplasma testing, has already been performed and this possibility has been excluded.

31.4.1 Problem Is Limited to One Person

Problem	Cause	Immediate action	Follow-up action	Book section or reference for more information
Incorrect culture conditions	Medium, serum, or other reagents	Check culture conditions are correct; check for changes in products or procedure; record batch number for checking with supplier	Test different batch; try different supplier; try other media and sera	Sections 7.5.2, 9.7
	Flask, dish, multiwell plate	Check that dishes are tissue culture treated; record batch number for checking with supplier	Test different batch; try different supplier	Sections 7.5.3, 8.5
Incorrect handling	Trauma during dissociation	Check subculture conditions are correct; pipette more gently; check for toxicity from dissociation agent	Try different temperatures for trypsinization; replace EDTA with EGTA; try different dissociation agents	Section 14.6
Incorrect timing of the cell cycle	See Section 31.8.2			
Finite cell line	See Section 31.8.2			

31.4.2 Problem Is Widespread

Problem	Cause	Immediate action	Follow-up action	Book section or reference for more information
Incubator problems				
Incorrect temperature	Equipment problem, e.g. faulty fan or thermostat	Discuss with users to find affected incubators; check for error messages or alarms	Arrange service of incubator; perform continuous monitoring using independent probe	Sections 4.3.2, 5.3.1
	Usage problem, e.g. incubator accessed too frequently, door left open	Ask users to record usage; remind users of the need to keep doors closed	Perform continuous monitoring to document changes; restrict number of users or cultures	Sections 4.3.2, 5.3.1
Incorrect humidity	Water pan left empty	Fill water pan	Remind users to refill the water pan	Sections 5.3.1, 12.5.1
	Leakage around doors	Check door seals	Arrange service of incubator; check evaporation rate using pre-weighed Petri dish with buffered salt solution	Section 5.3.1
Incorrect CO_2 or other gas mixtures	Loss of gas supply	Check for error messages or alarms; connect backup cylinder or supply	Prepare or revise a contingency plan for loss of gas supply	Sections 4.1.4, 4.3
	Calibration problem, e.g. thermal conductivity (TC) CO_2 sensors need a humid atmosphere	Check water pan; check calibration using an independent sensor	Arrange service of incubator; perform continuous monitoring using independent probe	Sections 4.3.2, 5.3.1

Problem	Cause	Immediate action	Follow-up action	Book section or reference for more information
Recent changes				
Personnel	Change in personnel or training; overcrowding	Review recent changes	Perform refresher training; implement booking procedures	Sections 4.1.1, 30.2
Procedures	Outdated SOP; deviation from current SOP	Discuss current procedures with users	Review and update SOP; perform refresher training	Sections 7.5.1, 30.2
Media, reagents, culture vessels	Change in supplier, product, or storage	Review recent changes; start a log of recurring problems; record associated batch numbers	Review storage arrangements; perform culture testing; discuss problems with supplier	Sections 7.5.2, 7.5.3, 11.8.2
Equipment and apparatus	Undetected faults, e.g. after being moved to a new location	Review recent changes; start a log of recurring problems	Perform regular service and maintenance; review log for trends	Section 7.5.4

31.5 ABNORMAL CELL APPEARANCE

The most likely cause of a change in cell morphology is contamination, due to either mycoplasma or cross-contamination (see Sections 31.1.1, 31.2). It is assumed here that problem solving for contamination, including mycoplasma and authentication testing, has already been performed and these possibilities have been excluded.

Problem	Cause	Immediate action	Follow-up action	Book section or reference for more information
Intracellular granularity or vacuolation	Cell deterioration or death, e.g. autophagy	Record observations and photograph; check culture conditions are correct	If problem continues, discard affected cultures and renew stock from cryostorage	Sections 2.5, 14.4.1
	Senescence (particularly with slow cell growth)	Renew stock from cryostorage	Graph population doubling level (PDL) over time; perform cell banking early	Sections 14.2.2, 15.5; Figure 3.1
Extracellular granularity or debris	Cell deterioration or death	Record observations and photograph; check culture conditions are correct	If problem continues, discard affected cultures and renew stock from cryostorage	Sections 2.5, 14.4.1
	Precipitate from medium or serum	Look for variable particle size; check medium for precipitation (may mean a constituent has been lost); check serum for precipitation (not usually harmful)	Replace medium if precipitate is observed; review storage conditions	Section 9.8
Loss of birefringence (edge of cells becoming indistinct)	Loss of viability, perhaps by drying out	Trypsinize and check viability by dye exclusion and reattachment; discard culture if no cells attach	Check steps of procedure to determine when drying out could occur, e.g. while feeding large numbers of plates	Sections 19.1, 19.2

31.6 PROBLEMS WITH MATERIALS

Culture testing of new materials – including culture vessels, media, and other reagents – has been described previously (see Sections 7.5.3, 11.8). In this section, it is assumed that QC testing has been completed but that problems are reported once materials are in use.

31.6.1 Culture Vessels

Problem	Cause	Immediate action	Follow-up action	Book section or reference for more information
Plasticware, e.g. lack of adhesion	Plastic is not tissue culture grade	Check labels and product information	Review storage arrangements	Section 8.2.1
	Bad batch	Record batch number; check product information	Discuss with supplier; arrange for replacement; perform further testing; if problem continues, change supplier	Section 7.5.3
	Cells do not adhere or display normal morphology	Check culture conditions are correct	Try coatings (e.g. polylysine) or a feeder layer	Sections 8.3–8.5, 31.5
Glassware	Inadequate washing	See Section 31.3		
	Cells do not adhere, e.g. on microscope slides	Perform cleaning step using acid wash	Try different vessels designed for imaging	Sections 8.5, 8.6.1; MacLeod et al. (2017)
Bottle caps	Traces of detergent under liner	Remove liners to wash	Use wadless caps	Section 11.3.3

31.6.2 Medium Formulation and Preparation

Problem	Cause	Immediate action	Follow-up action	Book section or reference for more information
pH fluctuations	Incorrect buffering	Prepare fresh medium and equilibrate overnight	Check the medium composition is correct; if using $NaHCO_3$ solution, prepare fresh stock	Sections 9.2.1–9.2.3; Appendix C
	Equipment problem, e.g. open door, faulty incubator or CO_2 supply	Keep incubator door closed; check CO_2 level	Discuss with other users for problems; recalibrate if necessary; arrange for incubator service	Section 9.2.3
pH too high	Handling problem, e.g. infrequent feeding, mycoplasma contamination	Feed more often	Test for mycoplasma	Sections 14.5, 16.4
	CO_2 concentration too low, e.g. DMEM requires 10% CO_2	Check recommended level for medium formulation	Check CO_2 level and recalibrate if necessary; purchase DMEM without $NaHCO_3$ and add for use	Section 9.2.3; Appendix C
	HCO_3^- concentration too low	Prepare fresh medium and equilibrate overnight	Check the medium composition is correct, particularly $NaHCO_3$	Sections 9.2.1–9.2.3; Appendix C

Problem	Cause	Immediate action	Follow-up action	Book section or reference for more information
Osmolality	Incorrect preparation, e.g. incorrect dilution of concentrate	Prepare fresh medium	Check preparation procedure; check osmolality	Section 9.2.6
	Inappropriate formulation	Check osmolality	Optimize for that cell type	Section 9.2.6
Incorrect medium	Incorrect preparation	Check with other users for problems; clarify correct procedures	Perform refresher training; if made in-house, compare to readymade 1× stock	Sections 11.6, 11.7
	Inappropriate formulation	Revert to previous medium	Match medium to previous type and supplier; perform adaptation if moving to serum-free conditions; perform medium testing before making changes	Sections 9.7, 10.6, 11.8
Deficient medium	Insufficient or defective component, e.g. faulty batch	Prepare fresh medium; record batch numbers; check components for obvious problems, e.g. precipitation	Discuss with supplier; perform medium testing, replacing components one at a time from an alternative source	Section 11.8
	Poor solubility	Ensure all constituents have dissolved before filtration	Clarify correct procedures; review and revise SOP	Sections 11.6, 11.7
	Precipitation on storage	Warm to 37 °C; discard if precipitate does not dissolve	Discuss with supplier; do not refreeze	Section 9.8
Serum	Batch variability	Reserve a batch and perform serum testing	Move toward serum-free media	Sections 9.7.1, 9.7.2, 10.6
	Incorrect concentration	Test range of concentrations	Compare fetal bovine serum (FBS) to other types, e.g. calf serum	Sections 9.5, 9.7.1
Diluent	Ultrapure water (UPW)	Compare to readymade medium	Test UPW; arrange service of water purification system	Sections 11.5.2, 11.5.3
	Balanced salt solution	Compare to readymade stock	Try alternative suppliers or a different formulation	Section 9.3
Antibiotics	Toxicity	Grow cells antibiotic-free	If essential (e.g. for selection), try different antibiotic, dose, or combination	Sections 9.6.2, 16.3.2
	Resistance	Discard contaminated cultures	If essential, identify the species of contaminant and determine its antibiotic sensitivity	Sections 16.2, 16.3.2

31.6.3 Medium Stability and Storage

Problem	Cause	Immediate action	Follow-up action	Book section or reference for more information
Glutamine	Unstable at 37 °C; generates toxic ammonia	Store at −20 °C in aliquots; add to medium prior to use	Consider substituting with a more stable glutamine dipeptide	Sections 9.4.1, 9.8
Serum	Unstable components, e.g. insulin	Store at < −15 °C shielded from light	Perform continuous temperature monitoring	Sections 4.3.2, 9.8
Trypsin	Unstable; undergoes autodigestion	Store at −20 °C in aliquots	Thaw prior to use; do not refreeze	Sections 13.5.1, 14.6.2
Media and reagents	Duration	Check expiry date or preparation date	Revise stock control; define shelf life for in-house reagents, e.g. 6 months at 4 °C	Section 9.8
	Temperature	If stored frozen, prepare single-use aliquots; do not refreeze	Perform continuous monitoring using an independent probe; carry out regular service and maintenance	Sections 4.3.2, 5.5.1
	Light	Store shielded from light	Avoid fluorescent lights; store reagents behind opaque doors	Sections 4.1.4, 9.8

31.6.4 Water Quality

Problem	Cause	Immediate action	Follow-up action	Book section or reference for more information
Inappropriate quality	Incorrect water quality for application, e.g. using Type 2 water to make up media	Use UPW to make up media and other reagents for tissue culture	Include water quality and methods of water purification in training program	Sections 11.5.1, 30.2.2
Poor quality	Prolonged storage	Use UPW immediately	Remind users that UPW should not be stored	Section 11.5.1
	Spent cartridge	Check resistivity (or conductivity) and total organic carbon (TOC); change cartridge	Check for alarms or error messages; implement regular cartridge replacement program	Section 11.5.2, 11.5.3
	Equipment problem, e.g. leakage	Arrange for service of water purification system	Check for alarms or error messages; perform regular maintenance	Section 11.5.3
	Contamination of tubing or other system components	Check by eye for algae; clean and disinfect tubing	Perform regular cleaning and disinfection	Section 11.5.3
	Unidentified cause	Arrange for equipment service; clean and disinfect tubing	Make up medium and test using sensitive method, e.g. cloning efficiency; compare to other UPW	Sections 11.5.3, 11.8.2

31.7 PROBLEMS WITH PRIMARY CULTURE

Primary cultures typically have lower viability (50–90%) compared to passaged cultures. However, this will vary with the sample and tissue type. Protocols for various tissue types are provided as Supplementary Material online (see Supp. S24.1–S24.20, S25.1–S25.3).

31.7.1 Suspected Contamination

Problem	Cause	Immediate action	Follow-up action	Book section or reference for more information
Microbial contamination is clearly visible	Contamination of tissue sample, e.g. intestinal sample	Discard contaminated cultures unless irreplaceable	Use antibiotics in Collection Medium or Dissection Balanced Salt Solution (DBSS); consider washing tissue with 70% alcohol or 0.04% hypochlorite solution	Protocols P13.1, P13.2; Appendix B; Supp. S24.5
Medium is cloudy due to other causes (see also Section 31.7.2)	Debris due to tissue necrosis or non-viable cells	Count cells and check viability; perform enrichment of viable cells	Perform sterility testing on a sample of spent medium to exclude microbial contamination	Protocols P13.8, P19.2; Supp. S11.6
Medium is acidic due to other causes (see also Section 31.6.2)	Higher cell concentration than expected	Count cells and check viability; passage into larger vessel	Perform sterility testing on a sample of spent medium to exclude microbial contamination	Protocols P14.2, P19.2; Supp. S11.6

31.7.2 Poor Take in Primary Culture

Problem	Cause	Immediate action	Follow-up action	Book section or reference for more information
Primary explants do not attach	Explant fragments are too large	Reduce fragment size to 1 mm or less	Practice dissection	Section 13.4; Supp. S30.15
	Explant fragments are poorly adhesive (too fibrous or differentiated)	Scratch the substrate under the explant	Try trapping the explant under a coverslip	Section 13.4; Supp. S24.4
	Conditions are unsuited for adhesion	Start with a small volume of medium; increase serum concentration	Try coatings (e.g. collagen, laminin, fibronectin, polylysine), or a feeder layer	Sections 8.3, 8.4, 25.4.2; Protocol P13.3
Incomplete disaggregation	Inadequate exposure to protease	Increase duration of incubation	Try cold trypsin pretreatment before shorter incubation; try an alternative, or additional, protease	Sections 13.2.1, 13.2.2, 13.5
	Cell lysis produces DNA which promotes reaggregation	Add DNase (2–20 µg/ml) after centrifuging to remove the protease	Improve disaggregation to minimize cell damage	Sections 13.2.1, 13.2.2

Problem	Cause	Immediate action	Follow-up action	Book section or reference for more information
Complete disaggregation but poor attachment	Cells grow in suspension	If cells are viable, propagate in suspension	Check Ca^{2+} in medium (required for adhesion but absent in some media such as RPMI 1640)	Section 9.4.3; Supp. S25.3
	Cell type attaches slowly (may be an advantage for separation from fibroblasts)	If cells are viable, propagate both floating and adherent populations	Try coatings (e.g. collagen, laminin, fibronectin, polylysine) or a feeder layer	Sections 8.3, 8.4, 25.4
Floating cells are mostly non-viable, few cells have attached	Tissue necrosis	Perform enrichment of viable cells; transfer to smaller vessel	Adjust SOP to remove necrotic tissue, increase number of washes	Protocols P13.3, P13.8, P25.1
	Damage from enzymatic or mechanical digestion	Perform enrichment of viable cells; transfer to smaller vessel	Adjust SOP to reduce the protease concentration or exposure time; try cold trypsinization; try an alternative protease or method	Sections 13.2.1, 13.2.2, 13.5, 13.5.6
	Lower cell concentration than expected, e.g. due to incomplete disaggregation	Perform enrichment of viable cells; transfer to smaller vessel	Aim to increase cell concentration up to 1×10^6 cells/ml	Protocol P13.8
	Inappropriate culture conditions (substrate, medium, supplements) for that cell type	Try a range of media and substrates; try conditioned medium	Check the literature for culture conditions used with that cell type	Sections 8.3, 8.4, 9.6.1, 9.7, 10.6.1

31.7.3 Incorrect Phenotype after Primary Culture

Problem	Cause	Immediate action	Follow-up action	Book section or reference for more information
Overgrowth by a different, rapidly growing cell type	Cross-contamination	See Section 31.2		
Overgrowth by stromal cells (fibroblasts) in tissue sample	Stromal cells predominate and grow faster than many other cell types	Use selective culture conditions, e.g. selective medium and a confluent feeder layer; separate desired cell type from fibroblasts using magnetic separation or other techniques	Move toward serum-free media; move away from explant-based primary culture (fibroblasts grow out from explants early and grow well in serum)	Sections 20.7, 21.3, 25.4
Correct cell type but loss of differentiated phenotype	Absence of the physiological environment leading to loss of differentiated characteristics	Use coatings (e.g. collagen, laminin, fibronectin, polylysine) or a feeder layer; grow cells at high density	Use a differentiation-inducing medium if available; try a range of inducing agents; try a three-dimensional (3D) culture system	Sections 2.4.4, 8.3, 8.4, 26.3, 26.4
	Clonal evolution leading to dominance of a phenotype that is related to growth, not differentiation	Thaw fresh stock from cryostorage	Perform cell banking to store frozen stocks at early passage; try cloning early passage material to select that phenotype	Sections 3.2.2, 15.5, 20.2–20.4

31.8 PROBLEMS WITH FEEDING OR SUBCULTURE

Cell lines typically display high viability (90–100%) once their culture conditions and handling procedures have been optimized. In this section, it is assumed that problems have arisen after optimization has been performed.

31.8.1 pH after Feeding

Problem	Cause	Immediate action	Follow-up action	Book section or reference for more information
pH falls rapidly after feeding	Microbial contamination	See Section 31.1.1		
	High cell number leading to exhaustion of medium	Feed more often; count cells at subculture; reduce seeding density	Review or repeat growth curve; check suitability of medium for that cell type	Sections 9.7, 14.4, 14.5, 19.3
	Increased production of lactate by some cell lines	Allow free gas exchange, e.g. using vented cap	Increase buffering capacity with 20 mM HEPES or by increasing CO_2 and HCO_3^- concentrations	Sections 9.2.2, 9.2.3, 9.4.4
pH rises after feeding	Sealed cap in CO_2 incubator, e.g. due to an impermeable cap or a film of liquid blocking a vented cap	For impermeable caps, check cap is loose; for vented caps, look for film of liquid; if found, blot neck with sterile swab or replace cap	Use vented caps; do not pour liquids	Sections 8.5.5, 12.3.5
	Sealed lid in CO_2 incubator, e.g. due to film of liquid near lid	Remove lid, swab rim with 70% alcohol, and replace lid	Avoid medium entering capillary space between base and lid	Section 8.5.5
	Loose lid in ambient air	Tighten cap while outside the incubator	If cells are to be grown in ambient air, check that buffering is correct	Sections 9.2.2, 9.3
pH fluctuations (rise or fall)	See Section 31.6.2			

31.8.2 Poor Take after Subculture

Problem	Cause	Immediate action	Follow-up action	Book section or reference for more information
Incorrect culture conditions	See Section 31.4.1			
Incorrect culture handling	See Section 31.4.1			
Cell counting error leading to incorrect cell concentration	Sampling error	Mix thoroughly before sampling; ensure that cells are singly suspended and are not clumped	Revise disaggregation procedure; if cells must be passaged as aggregates, try a range of seeding densities	Section 14.6.2

Problem	Cause	Immediate action	Follow-up action	Book section or reference for more information
	Problem with manual counting using a hemocytometer	Ensure that coverslip is correctly positioned and counting chamber is not over- or underfilled	Review the number of cells that are counted (should be 100–300 for routine subculture)	Section 19.1.1
	Problem with automated cell counting	Check that instructions are being correctly followed; check with other users for problems	If problem is with user, perform refresher training; if equipment, arrange for service	Section 19.1.2
Incorrect timing of the cell cycle	Seeding density too low, e.g. due to split errors	Count cells at subculture and reseed based on cell count	Try a range of seeding densities	Sections 14.4, 14.6.4
	Too long in plateau phase	Subculture before cells reach confluence	Review or repeat growth curve	Sections 14.6.1, 19.3
	Subculture is too frequent	Time subculture based on cell behavior	Review or repeat growth curve	Sections 14.6.1, 19.3
Finite cell line	Senescence	Renew stock from cryostorage	Graph population doubling level (PDL) over time; perform cell banking early	Sections 14.2.2, 15.5; Figure 3.1
		Perform immortalization at an early passage, e.g. from frozen stocks	Determine consequences of immortalization on phenotype and growth	Sections 3.3–3.5, 22.3, 22.4

31.8.3 Uneven Growth after Subculture

Problem	Cause	Immediate action	Follow-up action	Book section or reference for more information
Obvious growth patterns, e.g. ribbing in flasks, concentric circles in dishes	Vibration from equipment (e.g. fan motor or closing door) or floor	Photograph pattern as a record of the problem; check for vibration	Restrict movements just after subculture; arrange for incubator service	Section 8.5.6; Figure 8.9
More cells at one side of plate, dish, or flask compared to the other	Large number of wells seeded, resulting in drop in concentration	Seed wells using a smaller pipette; mix cells at intervals during seeding	Examine clustering of data from controls for consistency in large assays	Section 29.2.5
	Swelling of sealed flasks due to expansion of air	Look for tilting of flasks, particularly when stacked	Vent large flasks briefly if sealed and stacked	Section 12.5.3; Figure 12.12
	Incubator problem, e.g. uneven level, uneven heating	Check level and stability of incubator and shelves	Check the surface temperature, e.g. by taping a temperature probe to the surface; discuss with supplier	Section 5.3.2

Problem	Cause	Immediate action	Follow-up action	Book section or reference for more information
More cells at edge of dish or well	Meniscus effect; insufficient medium	Increase volume of medium	Adjust cell concentration so cell density remains constant	Sections 8.5, 19.3.2; Table 8.2
More cells at center of dish or well	Swirling of medium during pipetting or mixing or when moving dishes	Add cells and mix randomly without swirling	Avoid tilting, swirling, or shaking dishes just after subculture	Section 12.4.1
	Scouring of cells from edge of well by washing or feeding	Reduce speed of pipetting to give a gentler flow	When pipetting, tilt vessel to add and remove liquid from the side of the well	Section 12.3.6
Wells at edge of microwell plate have fewer cells	"Edge effect," probably due to evaporation from outer wells	Do not use wells at edge of plate	Try plates with improved insulation; replace lid with adhesive film	Section 8.5.2
Random wells throughout plate have fewer cells	Holes in incubator shelving, generating convection currents	Place plates on a tile or tray to reduce local variations	Check for other possible causes, e.g. problem with plates	Sections 5.3.2, 7.5.3

31.9 PROBLEMS WITH CRYOPRESERVATION

Problem solving is particularly challenging for cryopreservation, due to the large number of variables that contribute to loss of viability. In this section, problems with temperature maintenance of frozen stocks are considered first due to their widespread impact on the entire laboratory. Problems with cryopreservation or thawing are considered second.

31.9.1 Loss of Frozen Stocks

Problem	Cause	Immediate action	Follow-up action	Book section or reference for more information
Building-wide disaster	Fire or other disaster leading to destruction of equipment, loss of power, or loss of access	Respond to the emergency (safety is the highest priority); restore power and top up liquid nitrogen once safe to do so	Include cryostorage in contingency planning; arrange backup storage offsite; deposit unique cell lines at a cell repository	Sections 4.3.1, 15.2.3, 15.6
Cryofreezer	Equipment malfunction, e.g. failure of auto-fill system	Arrange for urgent service of cryofreezer; if thawing is likely, transfer contents to other cryofreezers	Perform regular service and maintenance; perform continuous monitoring using an independent probe; store samples across multiple cryofreezers	Sections 4.3.2, 15.2.3
Liquid nitrogen supply	Supplier unable to deliver liquid nitrogen; human errors, e.g. forgetting to top up liquid nitrogen in a cryofreezer	Arrange for urgent fill; keep system closed and minimize evaporation until liquid nitrogen is topped up	Position cryofreezers carefully to allow easy access; implement an equipment alarm system with continuous monitoring and automated alarms	Sections 4.2.6, 4.3
Record keeping and inventory	Last vials removed in error; location of vials unclear	Obtain fresh stock from a cell repository; keep better records	Perform cell banking; reserve seed stocks for this purpose only; restrict user access	Sections 15.5.2, 15.7

Problem	Cause	Immediate action	Follow-up action	Book section or reference for more information
Cryovials	Explosion due to liquid nitrogen drawn into vial	Wear suitable PPE when thawing vials	Move from glass ampoules to plastic cryovials; store vials in vapor phase or isothermal cryofreezer	Sections 15.2.1, 15.2.3
Transport (including shipping)	Unexpected delays leading to loss of dry ice in the shipment	Work with a courier or shipping agent who can top up dry ice in transit	Complete all necessary paperwork before shipping; use correct amount of dry ice for that destination	Supp. S15.1

31.9.2 Poor Viability after Thawing

Problem	Cause	Immediate action	Follow-up action	Book section or reference for more information
Cryoprotectant	DMSO contamination, e.g. dissolved plastic or rubber	Check color (should be colorless)	Use glass or polypropylene tubes, pipettes, and pipette tips	Section 15.1.1
	Glycerol toxicity, e.g. deterioration with prolonged storage	Buy in small amounts that will be used within 3–6 months	Store shielded from light	Section 15.1.1
Freezing medium	Suboptimal composition	Make up fresh; vary cryoprotectant amount; increase serum (if used); try adding ice-cold and then place cells on ice	Try different freezing media, including readymade; for human pluripotent stem cells (hPSCs), add Rho kinase (ROCK) inhibitor	Sections 15.3.2, 23.4.4
Cells	Suboptimal subculture procedure or cell concentration	Increase cell concentration; 1×10^6 to 1×10^7 cells/ml is usually optimum	Optimize subculture prior to cryopreservation	Sections 15.3.3, 23.4.4
Cooling rate	Suboptimal freezing rate	Try different controlled cooling devices	If a programmable rate-controlled freezer is available, use a sample probe to collect temperature data and optimize	Sections 15.1.2, 15.2.2
Storage temperature	Suboptimal temperature, e.g. controlled cooling device left at −80 °C for days or weeks	Transfer vials to cryostorage 24 h after cryopreservation	Maintain storage temperature at < −150 °C once cryopreservation has been completed	Section 15.1.3
Thawing	Cells thawed too slowly	Thaw rapidly at 37 °C using a water bath or bucket	Leave on dry ice until ready to thaw	Section 15.4.3
	Cells diluted too rapidly after thawing	Add medium slowly, starting with drops	Watch someone else do this!	Section 15.4.3
	Cells reseeded at too low a concentration after thawing	Reduce dilution at thawing; pool several cryovials (both may need centrifugation to remove cryoprotectant)	Increase cell concentration at freezing, e.g. try > 5× normal concentration at seeding	

31.9.3 Changed Appearance after Thawing

Problem	Cause	Immediate action	Follow-up action	Book section or reference for more information
Mistaken identity	Poor labeling, mishandling, and inventory control	See Section 31.2		
Microbial contamination (see also Section 31.1.1)	Contamination during cryopreservation, storage, or thawing, e.g. from liquid nitrogen or water bath	Keep cap closed until after thawing (it should be tight but not cause distortion); swab vial before opening in BSC	Add a rack to the water bath to keep vials upright and partially submerged during thawing; store vials in vapor phase	Sections 15.2.3, 15.4.3
Differentiation	DMSO is known to act as an inducing agent	Centrifuge cells after dilution to remove cryoprotectant	Check literature to see if DMSO is known to induce differentiation in that cell type	Sections 15.1.1, 26.3.1
Culture conditions	Changed conditions since cells were last grown, e.g. different medium, plasticware, or serum-free conditions	Revert to previous culture conditions if possible	Test all new culture media and vessels; perform adaptation to serum-free conditions	Sections 10.6.3, 11.8

31.10 PROBLEMS WITH CLONING

Cloning is a challenging technique, whether it is performed as an assay with a defined endpoint or to generate a cloned derivative for further culture. Further information on cloning can be found elsewhere in this book (see Sections 19.4, 20.2–20.4) and in the publications of Richard G. Ham, who successfully optimized conditions for the cloning of numerous cell types (Ham 1974; Ham and McKeehan 1978; Shipley and Ham 1983).

31.10.1 Too Few Colonies per Dish

Problem	Cause	Immediate action	Follow-up action	Book section or reference for more information
Poor colony-forming efficiency	Seeding concentration too low	Increase seeding concentration	Plate at a range of cell concentrations and determine if cloning efficiency is linear	Section 19.4; Figure 19.13
	Procedural problems, e.g. slow preparation of dilutions	Speed up or rationalize procedures	Review and revise SOP	Protocols P19.4, P20.1
	Suboptimal medium	Use a rich medium, e.g. Ham's F12; try supplements, e.g. insulin, transferrin, and selenium (ITS)	Check the literature for media used with that cell type; screen a range of rich media; try conditioned medium	Sections 9.4, 9.6.1, 10.3, 10.4
	Wrong serum	Use FBS (usually better for cloning compared to calf or horse serum)	Move toward serum-free media; if serum essential, screen batches based on the cloning efficiency of the cells that you use	Sections 9.5, 10.6, 11.8.2
	Suboptimal plasticware, e.g. incorrect surface charge	Use a coating, e.g. polylysine, extracellular matrix (ECM)	Screen a range of dishes or plates, including different surface treatments; try a feeder layer	Sections 8.3.2, 8.3.5, 8.3.7, 8.4

Problem	Cause	Immediate action	Follow-up action	Book section or reference for more information
Colonies are too diffuse	Reduction in cell–cell adhesion; increased cell migration	Use a larger dish to allow space for colonies to spread out	Add glucocorticoid at plating, e.g. dexamethasone, 1–10 µM; try a coating	Sections 8.3.2, 8.3.5, 20.6.3
Colonies are poorly attached	Poorly adherent cells	Try cloning in suspension, e.g. using Methocel	Review and revise SOP	Section 20.3; Protocol P20.3
	Suboptimal plasticware	See "Poor colony-forming efficiency"		

31.10.2 Too Many Colonies per Dish

Problem	Cause	Immediate action	Follow-up action	Book section or reference for more information
Colonies are overlapping, making them difficult to count or isolate	Seeding concentration too high	Reduce the seeding concentration or seed the same number into a larger dish	Plate at a range of cell concentrations and determine cloning efficiency	Protocol P19.4
	Incubation period too long	Incubate plates for a shorter time	Plot a time course for optimum colony formation	Protocol P19.4

31.10.3 Non-Random Distribution of Colonies

Problem	Cause	Immediate action	Follow-up action	Book section or reference for more information
Colonies are unevenly spread, e.g. more colonies at the center of the dish	Uneven seeding	Add cells to medium in a bottle, mix cell suspension, and then seed dishes	Review and revise SOP	Protocol P20.1
	Swirling the dish	Do not swirl dish to mix cells or when pipetting	Review and revise SOP	Protocol P20.1
	Medium not covering all of the dish surface	Make sure medium covers all of the bottom of the dish evenly	Ensure dishes are level and incubator is free from vibration	
Vibration (see also Section 31.8.3)	Vibration from equipment (e.g. fan motor or closing door) or floor	Check for vibration; do not move plates unless absolutely necessary; label with "CLONING, DO NOT MOVE"	If possible, restrict access to that incubator; arrange for service of incubator	Section 8.5.6; Figure 8.9

31.10.4 Incubation of Cloning Dishes

Problem	Cause	Immediate action	Follow-up action	Book section or reference for more information
Rapid drying out	Low humidity leading to evaporation	Replace medium; feed more often; check and refill water pan	Incubate dishes in a box containing moistened paper	Section 12.5.2
Microbial contamination	Medium spillage	Discard lids of dishes or plates and replace with fresh lids (particularly if spills are obvious); swab outside of base with 70% alcohol	Feed less often; minimize handling; if using a box, swab before opening	Protocol P20.1
	Contaminated incubator	See Section 31.1.3		
	Low-level, cryptic contamination of cells or reagents (including with mycoplasma)	See Sections 31.1.1, 31.1.3		

REFERENCES

Ham, R.G. (1974). Unique requirements for clonal growth. *J. Natl Cancer Inst.* 53 (5): 1459–1463.

Ham, R.G. and McKeehan, W.L. (1978). Development of improved media and culture conditions for clonal growth of normal diploid cells. *In Vitro* 14 (1): 11–22.

ICLAC (2019). Register of misidentified cell lines. http://iclac.org/databases/cross-contaminations (accessed 11 November 2019).

IEST (2018). IEST-RP-CC018 Cleanroom housekeeping: operating and monitoring procedures. https://www.iest.org/standards-rps/recommended-practices/iest-rp-cc018 (accessed 30 October 2019).

MacLeod, R.A., Kaufmann, M.E., and Drexler, H.G. (2017). Cytogenetic harvesting of cancer cells and cell lines. *Methods Mol. Biol.* 1541: 43–58. https://doi.org/10.1007/978-1-4939-6703-2_5.

Shipley, G.D. and Ham, R.G. (1983). Multiplication of Swiss 3T3 cells in a serum-free medium. *Exp. Cell. Res.* 146 (2): 249–260.

CHAPTER 32

In Conclusion

It has been our intention in the foregoing pages to describe the fundamentals of cell culture in sufficient detail that you can perform practical work with this book as your guide. It is customary, when giving a lecture, to conclude with a summary that highlights the major points raised in the lecture, and that is how we would like to conclude this text. There are certain requirements that are crucial to successful and reproducible cell culture, which we would like to summarize as follows:

(1) Ensure that your work does not compromise your own safety or that of others working around you.

(2) Ensure that your instruction, and the training of anyone who works with you, comes from an experienced source.

(3) Do not mix cell culture with other microbiological work; perform culture handling in a separate laboratory area using separate equipment, glassware, lab coats, etc.

(4) Work in a clean, uncrowded, aseptic environment and clear up when you have finished. Do not use antibiotics for routine culture handling.

(5) To avoid the transfer of contamination, including cross-contamination, do not share media, reagents, cultures, or materials with others.

(6) Do not assume that your work is immune to mycoplasma or other forms of contamination because you have never seen it; test your cells regularly.

(7) Keep adequate records, particularly of changes in procedures, media, or reagents, and keep photographic records of the cell lines that you use.

(8) Work only on cell lines that have been obtained from a properly validated source, such as a reputable cell repository. Distrust any untested cell line (even if it comes from the originator) until you perform your own validation testing to exclude mycoplasma contamination, cross-contamination, and misidentification.

(9) Become familiar with the cell lines that you use – their appearance, growth rate, and special characteristics – so that you can respond immediately to any change.

(10) Preserve cell line stocks in a cryofreezer and replace working stocks regularly.

(11) Protect seed stocks of valuable cell lines and use other stocks for distribution.

(12) Try to work under conditions that are precisely defined, including minimal use of undefined media supplements, and do not change procedures for trivial reasons.

To cover all of the fascinating aspects of cell and tissue culture would take many volumes and defeat the objective of this book. It has been more our intention to provide sufficient information to set up a laboratory and prepare the necessary materials with which to perform basic tissue culture, and to develop some of the more important techniques required for the characterization and understanding of your cultures. In collaboration with experts in their fields, we have also included specialized protocols for many cell culture applications, although space has determined that some of these are available only as Supplementary Material online.

This book may not be sufficient on its own, but with help and advice from colleagues and other laboratories we hope that it will make your introduction to tissue culture easier, more satisfying, and more enjoyable than it otherwise might have been.

Freshney's Culture of Animal Cells: A Manual of Basic Technique and Specialized Applications, Eighth Edition. Amanda Capes-Davis and R. Ian Freshney.
© 2021 John Wiley & Sons Ltd. Published 2021 by John Wiley & Sons Ltd.
Companion website: www.wiley.com/go/freshney/cellculture8

APPENDIX A

Glossary

Tissue culture terminology can be controversial. The overall approach to terminology throughout this book is discussed in the Introduction (see Section 1.1.2). The following list of terms is not comprehensive and is provided to clarify words or phrases that are used in multiple chapters, with an emphasis on practical procedures. Some biological terms are listed here (particularly where different meanings may cause confusion), but it is assumed that readers will already know most biological terms or will have access to another textbook or dictionary where their meanings are given.

Adaptation. A process in which cultured cells adjust to changes in their conditions, e.g. from serum-containing to serum-free media.

Adherent culture. A culture where cells attach to the substrate that makes up the culture vessel (or where they attach to a coating that has been added to the vessel). This usually results in a two-dimensional (2D) monolayer.

Anchorage-dependent. Requiring attachment to a solid substrate for survival or growth. Conversely, anchorage-independent cells do not require such attachment and are able to grow in suspension; anchorage-independence may occur as part of transformation.

Animal product-free medium. Complete medium that does not have any human- or animal-derived components.

Aseptic technique. A set of practices that are used to maintain asepsis, or the absence of microorganisms.

Authentication. A process of testing that is used to confirm or verify the identity of a cell line, demonstrating that it is derived from the correct species and donor.

Autoclaving. Sterilization using pressurized steam.

Balanced salt solution. An isotonic solution of inorganic salts present in approximately the correct physiological concentrations. May contain glucose but is usually free of other organic nutrients.

Basal medium. A defined formulation that is used for standardization of media.

Biohazard. The risk that harm may occur due to exposure to biological material.

Biological product. An item that contains or is derived from cells or tissues, and is produced for use in diagnosis, prevention, or treatment of human disease.

Biological safety cabinet (BSC). Laminar airflow equipment used to provide asepsis (preventing contamination) and containment (preventing exposure to pathogens) when handling living cells or tissues.

Bioreactor. Culture vessel that provides a controlled environment for cultivation of cells. This is commonly used for scale-up but may be used for other purposes, e.g. to induce differentiation. Cells may be cultivated in suspension or anchored to a substrate.

Biosafety. The containment principles, technologies, and practices that are used to prevent unintentional exposure to pathogens and toxins or their accidental release.

Biosecurity. The institutional and personal security measures that are designed to prevent the loss, theft, misuse, diversion, or intentional release of pathogens.

Cancer stem cell (CSC). A putative subset of tumor cells with stem cell capacity, i.e. the ability to perpetuate themselves through self-renewal and the ability to generate the diversity of cells that comprise a given tumor.

Freshney's Culture of Animal Cells: A Manual of Basic Technique and Specialized Applications, Eighth Edition. Amanda Capes-Davis and R. Ian Freshney.
© 2021 John Wiley & Sons Ltd. Published 2021 by John Wiley & Sons Ltd.
Companion website: www.wiley.com/go/freshney/cellculture8

Cell banking. Cultivation of a cell line in order to preserve sufficient numbers of cells for later use, in a manner that preserves early passage material. Typically uses a tiered approach, with preparation of a master cell bank (also known as seed stock) and a working cell bank (also known as working stock or distribution stock).

Cell concentration. Number of cells per milliliter of medium.

Cell counting. Measurement of the number of cells that are present within a culture, using a representative sample. May be performed manually using a hemocytometer or may be automated using cell counting equipment.

Cell culture. Growth of dissociated cells; may have arisen from the parent tissue by spontaneous migration or by enzymatic or mechanical disaggregation.

Cell density. Number of cells per centimeter squared of substrate.

Cell fusion. Formation of a single cell body by the fusion of two other cells, either spontaneously or using a fusogenic agent. Also known as somatic cell hybridization.

Cell line. A culture that has undergone at least one subculture (passage). *See also* Finite cell line *and* Continuous cell line.

Cell repository. A facility or organization that receives cell line deposits, carries out cell banking procedures, and distributes the resulting stocks to other laboratories. May be any size, from small core facilities to large organizations.

Cell strain. A culture where cloning or selection has been performed to isolate individual cells and/or select for specific properties or markers.

Cell viability. The proportion of cells in a culture that are living versus those that are dead (non-viable). Typically measured as part of a cell counting procedure.

Chemically defined medium. Complete medium that is entirely composed of purified recombinant proteins or other components whose chemical compositions or structures are known.

Cleaning. The removal of dirt, organic matter, and stains, e.g. from surfaces or equipment.

Clone. A population of cells that is derived from a single cell by mitosis.

Cloning. The process used to generate a clonal population, usually through limiting dilution or growth in suspension.

Cloning efficiency. The percentage of cells seeded at subculture that form clones at low density.

Clonogenic assay. An assay where cells are plated at low density to measure cloning efficiency; on completion of the assay, visible colonies are counted to determine the colony-forming efficiency. Provided each colony is derived from a single clone, this is equivalent to the cloning efficiency.

Collection medium. A medium used for the collection of tissue samples.

Complete medium. A medium that is sufficient for propagation. Usually comprises a basal medium with additional supplements e.g. serum.

Conditional reprogramming. Extension of cell lifespan by culture under specific conditions; typically uses a feeder layer and Rho kinase (ROCK) inhibitor.

Conditioned medium. Alteration of culture medium by metabolically active cells; traditionally used to improve cell survival or growth when added to a different culture.

Confluent. A complete monolayer of cells; all cells are in contact with other cells around their periphery, and no available substrate is left uncovered.

Contact inhibition. Inhibition of movement and proliferation when cells come into close proximity to one another. Loss of contact inhibition results in cells growing in a disordered manner, often producing visible "transformation foci." *See* Transformation.

Contamination. The presence of adventitious microorganisms or cells within a culture, making it impure.

Continuous cell line. A cell line that has the capacity for unlimited proliferation. Previously known as "established" and often referred to as "immortal."

Controlled cooling device. A device that is used for cryopreservation of cells, allowing a suitable controlled cooling rate to be achieved.

Crisis, or Culture crisis. Decreased proliferation and increased cell death within a culture, resulting from genomic instability.

Cross-contamination. The introduction of cells from another culture, leading to the presence of a mixture of cells from two different sources or a misidentified cell line, where one is overgrown or replaced by the other.

Cryofreezer. Equipment that is used to store material that has undergone cryopreservation; typically uses liquid nitrogen to achieve temperatures less than $-150\,°C$ throughout the storage area.

Cryopreservation. The application of low temperatures to preserve the structural and functional integrity of cells and tissues.

Cytotoxicity. Cellular damage to one or more metabolic pathways, intracellular processes, or structures resulting in impaired function. Often, but not necessarily, linked to loss of viability.

Culture. Used as a verb or noun. The verb refers to the process of propagation of living cells; the noun refers to the sample that is undergoing this process.

Decontamination. A physical or chemical process that renders an area, device, item, or material safe to handle.

Density limitation of growth. Mitotic inhibition correlated with an increase in cell density at confluence.

Derivative cell line. A cell line that has been established from another culture through cloning, selection, or other techniques.

Detergent. A chemical or mixture of chemicals that is used for cleaning.

Differentiation. The presence of specialized properties that are associated with commitment to a particular embryonic lineage, similar to the properties that the cell would have expressed *in vivo*. Related terms include (i) dedifferentiation, in which specialized properties are lost; and (ii) transdifferentiation, in which cells undergo a change from one differentiated state to an alternative differentiated fate.

Disinfectant. A chemical or mixture of chemicals that is used to kill microorganisms, although not necessarily spores.

Disinfection. A physical or chemical means of killing microorganisms, although not necessarily spores.

DNA barcoding. A genotype-based method used for authentication of cell lines. In this context, a short, standardized region of sequence from the mitochondrial cytochrome c oxidase I (CO1) gene is used as a "DNA barcode" for species identity.

Ectoderm. The outer germ layer of the embryo, giving rise to the outer surface epithelia such as the epidermis, buccal epithelium, and outer cervical epithelium.

Electroporation. Delivery of genetic material to cells in a culture or tissue using an electrical impulse.

Embryoid body. A cluster of undifferentiated cells that can differentiate into any of the three germ layers. *See* Ectoderm, Endoderm, *and* Mesoderm.

Embryonic stem cells (ESCs). Pluripotent stem cells (PSCs) isolated from the inner cell mass of an early embryo; can be propagated as cell lines with a wide range of differentiation capabilities. *See also* Potency *and* Stem cell.

Endoderm. The innermost germ layer of the embryo, giving rise to the epithelial component of organs such as the gut, lung, liver, and pancreas.

Epithelial cell. Describes a cell derived from epithelium, but often used more loosely to describe any cell that has a polygonal shape, with regular dimensions and clear, sharp boundaries, that grows in a discrete monolayer alongside other cells. More correctly, cells should be referred to as "epithelial-like" or "epithelioid" to indicate that although they resemble epithelial cells, their origin is unclear.

Epithelial–mesenchymal transition (EMT). A transient change in differentiation status from an epithelial to a mesenchymal phenotype. An inverse change, mesenchymal–epithelial transition (MET), may also occur.

Explant. A fragment of tissue removed from an organism and placed in culture in such a way as to promote its survival and the outgrowth of viable cells.

Explantation. Isolation of tissue for maintenance *in vitro*, strictly as small fragments with accompanying outgrowth (*see* Explant), but also used as a generic term for the isolation of tissue for culture.

Feeder layer. Adherent cells that are used to support the growth of other cell types; traditionally used to improve cell survival or growth. Feeder cells are inactivated to ensure that they do not overgrow and replace the desired cell type.

Feeding. The act of removing medium from a culture and replacing it with fresh medium.

Fibroblast. Describes a proliferating precursor cell of the mature differentiated fibrocyte, which is derived from mesenchyme (stroma). Often used more loosely to describe any migratory cell that has a spindle-shaped (bipolar) or stellate (multipolar) appearance, with a length that is usually more than twice its width. More correctly, cells should be described as "fibroblast-like" or "fibroblastoid" to indicate that although they resemble fibroblasts, their origin is unclear.

Finite cell line. A cell line that has the capacity for only a limited number of population doublings *in vitro* before ceasing to divide.

Freezing medium. A medium in which cells are suspended during cryopreservation.

Genotype. The genetic composition of a cell or culture.

Growth curve. A semi-logarithmic plot of the cell number on a logarithmic scale against time on a linear scale. Usually divided into (i) lag phase, i.e. the phase before growth is initiated; (ii) log phase, i.e. the period of exponential growth; and (iii) plateau phase, i.e. a phase of high cell density when the culture stops growing. Used to determine growth parameters such as the lag time, population doubling time, and saturation density.

Growth medium. A medium that is used to promote both cell survival and proliferation.

Hanging drop. A form of 3D culture in which cells are suspended in a drop of medium (originally a plasma clot).

Hemocytometer. See Cell counting.

Histotypic culture. A form of 3D culture in which cells from a single lineage are brought into close assembly (often at high density) to encourage reassembly, differentiation, and development of 3D structures.

Holding medium. Medium that retains cell survival without proliferation.

Hybrid cell. Mononucleate cell that results from the fusion of two different cells. *See also* Cell fusion.

Hydrogel. A cross-linked network of polymers that contains a high water content; used as a substrate for 3D culture.

Immortalization. The acquisition of an infinite lifespan *in vitro*, either spontaneously or due to genetic modification.

Induced pluripotent stem cell (iPSC). A cell that has been induced to revert (usually from a cell that is fully committed to a specific lineage) to a pluripotent state. Typically requires delivery of a set of reprogramming factors to the cell. *See also* Potency *and* Stem cell.

In vitro. Literally, "in glass," but used conventionally to mean propagated outside of the organism as a culture. May be used to indicate biochemical and molecular reactions carried out in a test tube; those reactions are better referred to as cell-free to avoid confusion.

In vivo. In the living organism.

Karyotype. The number and visual appearance of the chromosomes within a cell or a representative sample of a culture.

Lag phase. *See* Growth curve.

Laminar flow. A flow of air or fluid that closely follows the shape of a streamlined surface without turbulence; used in equipment (e.g. laminar flow hoods or biological safety cabinets [BSCs]) to provide a sterile field for asepsis.

Lineage commitment. Progression from a stem cell to a particular pathway of differentiation, resulting in expression of specialized characteristics.

Lineage plasticity. The ability of a cell to switch to a more potent state in response to physiological demands or insults. *See also* Potency.

Lipofection. Delivery of genetic material to the cell using a "liposome" (a positively charged complex made from cationic lipids), which can enter the cell.

Log phase. *See* Growth curve.

Medium. A mixture of inorganic salts and other nutrients capable of sustaining cell survival *in vitro*. The plural of medium is media.

Mesenchymal stromal cell (MSC). A cell derived from mesenchyme that can differentiate into various mesodermal lineages including adipocytes, chondroblasts, and osteoblasts. Also known by other names, e.g. mesenchymal stem cell.

Mesenchyme. Originally used to describe free cells loosely arranged in a matrix; has come to refer to loosely organized embryonic tissue (mostly derived from the mesoderm) that gives rise to connective tissue and various other tissues.

Mesoderm. A germ layer in the embryo arising between the ectoderm and endoderm and giving rise to connective tissue, vascular tissue, the hematopoietic system, and various other tissues.

Microautoradiography. Localization of radioisotopes in cells and tissue sections, usually by exposure of a photographic emulsion placed in close proximity to the specimen. After development, the specimen may be viewed under a microscope.

Microcarrier. A carrier bead that is used to culture adherent cells in suspension. Microcarriers are typically 90–300 μm in size; larger carriers (>500 μm) are often known as macrocarriers.

Microfluidics. The science and technology of systems that process or manipulate small amounts of fluids, from 10^{-9} to 10^{-18} l, using microchannels (micromolar-sized channels).

Misclassification. The failure of a cell line to correspond to the tissue or disease from which it was reported to be established. May indicate misidentification, but other causes are possible, e.g. incorrect disease diagnosis.

Misidentification. The failure of a cell line to correspond to the donor or species from which it was reported to be established. May arise through various mechanisms, including cross-contamination.

Monoclonal. Derived from a single clone of cells. Monoclonal antibody: an antibody produced by a clone of lymphoid cells either *in vitro* or *in vivo*.

Morphology. The physical appearance of a culture; usually refers to the appearance of cells under the microscope.

Off-target effect. One or more unexpected consequences of genetic modification that occur away from the target region.

Organ culture. A form of 3D culture in which the whole organism, organ, or a portion of tissue from an organ is placed into culture and maintained without any effort to dissociate the cells.

Organoid culture. A form of 3D culture in which stem cells are brought into close association, where they self-organize through cell sorting and spatially restricted lineage commitment to form a 3D structure containing organ-specific cell types.

Organ-on-chip. A microfluidic device containing living cells that reconstitutes organ-level structures and functions.

Organotypic culture. A form of 3D culture in which cells from multiple lineages are brought together using culture systems that are designed to recreate the composition of the original tissue.

Osmolality. The concentration of osmotically active particles in an aqueous solution, expressed in osmoles per kilogram.

Osmolarity. The concentration of osmotically active particles in an aqueous solution, expressed in osmoles per liter.

Parenchyma. That part of a tissue carrying out the major function of the tissue (e.g. the hepatocytes in liver); as distinct from the stroma, such as fibroblastic connective tissue, which is seen as supporting tissue.

Passage number. The number of times a culture has been subcultured.

Passaging. *See* Subculture.

Pathogen. A cell culture, microorganism, or other biological material that is hazardous to human health. A pathogen may cause harm through infection, toxicity, or other mechanisms.

Personal protective equipment (PPE). Equipment that is worn to minimize exposure to hazards that cause workplace injuries and illnesses. For tissue culture, PPE is also used to protect cultures from the operator and the external environment.

Phenotype. The observable characteristics of a cell or culture, including its physical appearance (morphology) and any finer measures of assessment (e.g. of specialized functions).

Plateau phase. *See* Growth curve.

Pluripotent. *See* Potency.

Poikilothermic. An organism that has a body temperature close to that of the environment and not regulated by metabolism.

Population doubling level (PDL). The estimated number of population doublings that the cells have undergone in a culture. Also known as generation number.

Population doubling time (PDT). The interval required for a cell population to double at the middle of the log phase of the cell cycle. *See* Growth curve.

Potency. The range of commitment options that are available to a stem cell. Various terms are used to describe potency, including (i) totipotent, i.e. stem cells can give rise to the entire organism; (ii) pluripotent, i.e. stem cells can form all tissues of an organism (but not the extra-embryonic tissues); (iii) multipotent, i.e. stem cells can give rise to multiple lineages (but not from all three germ layers); (iv) oligopotent, i.e. stem cells can give rise to two or more lineages; and (v) unipotent, i.e. stem cells can give rise to only a single lineage.

Precursor cell. Any ancestral cell in the same lineage as the cell type of interest.

Primary culture. A culture started from cells, tissues, or organs taken directly from an organism, and before the first subculture.

Progenitor cell. Any multipotent cell that is committed to a specific lineage but still retains the ability to divide. *See* Potency.

Provenance. Records that provide information on a cell line's origins, handling, validation, and behavior.

Quality assurance (QA). A program or system that focuses on operational standards for all aspects of activities, to ensure that a procedure or a product consistently meets quality requirements.

Quality control (QC). Testing that is used to assess specific quality outcomes; one component of a QA program. Testing may extend to cell lines (referred to here as validation testing), media and other reagents, culture vessels, equipment, and facilities.

Reagent. A substance (element, compound, or mixture) that participates in a chemical reaction.

Reproducibility. The ability to duplicate the results of a prior study using the same materials that were used by the original investigator. Broader meanings may apply when considering how to improve the quality and validity of life sciences research.

Risk. The possibility that harm (illness, injury, death, or damage) might occur when exposed to a hazard.

Roller culture. A form of adherent culture in which cells are cultivated in a round tube or bottle that is rolled around its long axis, dispersing the cells over its inner surface.

Saturation density. The maximum cell density achieved by a cell line under specific growth conditions; also the cell density at plateau phase. *See* Growth curve.

Seeding density. The cell density chosen when a culture is passaged into a new culture vessel.

Senescence. Loss of proliferative potential that is linked to biological processes, particularly the shortening of the telomeres at the ends of the chromosomes. May also be induced by culture conditions.

Serum-free medium. Complete medium that does not contain serum or unprocessed plasma. It may contain individual proteins or protein fractions.

Shaker culture. A form of suspension culture in which cells are gently shaken, e.g. by a laboratory shaker installed in an incubator. Cells may be cultivated as a single-cell suspension or as aggregates, or grown on a substrate such as a microcarrier.

Short tandem repeat (STR) profiling. A genotype-based method used for authentication of cell lines. STR loci are amplified using a multiplex polymerase chain reaction (PCR) and the fragments are analyzed using capillary electrophoresis; the results are recorded in a tabular format and can be shared between laboratories.

Sparge. Addition of gas to a liquid (e.g. medium containing cells in suspension), resulting in gas exchange and agitation.

Specialized cell. A culture established from a specific cell type that continues to express differentiated characteristics (or can be induced to do so).

Spheroid. A 3D multicellular aggregate. Spheroids may arise spontaneously or may be induced to form using various methods, e.g. by growing cells in microgravity, seeding cells in a spheroid microplate, or performing hanging drop culture.

Spinner culture. A form of suspension culture in which cells are gently agitated using a stirrer or spinning paddle. Cells may be present as a single-cell suspension or as aggregates, or grown on a substrate, e.g. a microcarrier.

Split ratio. The divisor of the dilution ratio of a cell culture at subculture (e.g. one flask divided into four, or 100 ml up to 400 ml, would be a split ratio of 1 : 4).

Stem cell. A cell that can both perpetuate itself through self-renewal and give rise to differentiated cells.

Sterilization. A process that kills or removes all classes of microorganisms and spores.

Stroma. That part of a tissue seen as having a purely supporting role, e.g. fibroblastic connective tissue and its vasculature.

Subconfluent. Less than confluent; not all of the available substrate is covered.

Subculture. The act of transferring cells from one culture vessel to another; usually, but not necessarily, involves the subdivision of a proliferating cell population.

Substrate. The matrix or solid underlay upon which adherent cells grow.

Suspension culture. A culture in which cells are suspended in growth medium, instead of becoming adherent to a substrate.

Synonymous cell line. Any cell line derived from the same individual.

Three-dimensional (3D) culture. Culture of cells as aggregates or in a matrix or scaffold such that the cells are in a 3D array, as distinct from the conventional two-dimensional (2D) array of adherent (monolayer) culture.

Tissue culture. Properly, the maintenance and growth of tissue fragments *in vitro*. The term is used more broadly in this book to refer to the culture of cells derived from a tissue sample using any format. Cells may form part of a tissue fragment, may be dissociated, or may be induced to re-associate to form tissue-like structures.

Tissue engineering. A field that applies the principles of engineering and the life sciences to develop biological substitutes that will restore, maintain, or improve tissue function.

Tissue equivalent. An *in vitro* culture system that approximates the histology of the tissue *in vivo* (though usually lacking vascular, neural, and some other elements).

Transduction. Delivery of genetic material to the cell using viral material. Originally used to refer to whole viruses, but increasingly refers to viral vectors that have been modified to minimize safety concerns and maximize efficacy and specificity. Verb form also used ("transduce").

Transfection. Delivery of genetic material to the cell using nonviral methods, such as calcium phosphate coprecipitation or lipofection. Verb form also used ("transfect").

Transformation. Originally (as used in microbiology), the introduction and stable genomic integration of foreign DNA into the cell, resulting in genetic modification. The meaning has changed over time; used here to refer to an *in vitro* phenotype that is associated with tumorigenicity (*in vivo* tumor formation). The transformed phenotype includes immortalization, genetic instability, and aberrant growth control.

Trypsinization. The act of passaging a culture using trypsin to dissociate the cells from their original substrate.

Validation (of cell lines). Testing for microbial contamination and authenticity, to determine whether a cell line meets minimum quality criteria and is therefore suitable for publication or for other uses.

Vessel. Any container that may be used to cultivate cells; includes flasks, dishes, multiwell plates, bioreactors, etc.

Xeno-free medium. Complete medium that contains human-derived components (e.g. human serum or plasma) but no other animal-derived components.

Xenograft. Transplantation of tissue to a species different from that from which it was derived; often used to describe the implantation of human tumors in athymic (nude), immune-deprived, or immune-suppressed mice.

APPENDIX B

Calculations and Preparation of Reagents

CALCULATIONS

There are a number of simple calculations required while carrying out some of the procedures described in this book. Calculations specific to a particular protocol will be found in that protocol, but some commonly used examples are listed below.

Counting Cells with a Hemocytometer

Chapter 19 includes step-by-step instructions for calculating cell concentration after counting cells with a hemocytometer (see Protocols P19.1, P19.2).

Dilution of a Cell Suspension

Chapter 14 includes step-by-step instructions for generating a cell suspension, e.g. by trypsinization (see Protocol P14.2). To dilute the suspension to a lower concentration, the volume of cell suspension (v) required for dilution can be calculated as follows:

$$\frac{\text{Required concentration}}{\text{Starting concentration}} \times \text{Required volume}$$

e.g. the volume of cell suspension required to dilute 2.36×10^6 to 5×10^4 cells/ml in 50 ml would be:

$$\frac{5 \times 10^4}{2.36 \times 10^6} \times 50 = 1.06\,\text{ml}$$

i.e. dilute 1.06 ml up to 50 ml.

Population Doubling Level (PDL)

Chapter 14 discusses the meaning of population doubling level (PDL), how to calculate it, and how to estimate it using split ratios (see Sections 14.1, 14.2.2, 14.6.4). The following formula can be used to determine PDL at each subculture (Hayflick 1973):

$$n = \frac{\log N_{finish} - \log N_{start}}{\log 2}$$

Logarithms are to the base 10. "N_{finish}" refers to the total number of cells at subculture, "N_{start}" to the number in the starting inoculum, and "n" to the number of cell generations.

Log 2 can be included as a numerical value (0.301), which gives:

$$n = (\log N_{finish} - \log N_{start}) \times 3.32$$

e.g. for 3.2×10^6 cells (N_{finish}) recovered from a seeding of 2.0×10^5 (N_{start}):

$$n = (6.5 - 5.3) \times 3.32 = 4$$

Freshney's Culture of Animal Cells: A Manual of Basic Technique and Specialized Applications, Eighth Edition. Amanda Capes-Davis and R. Ian Freshney.
© 2021 John Wiley & Sons Ltd. Published 2021 by John Wiley & Sons Ltd.
Companion website: www.wiley.com/go/freshney/cellculture8

Alternatively, natural logarithms (base e) may be used with the same result:

$$n = \frac{\ln(N_{finish}/N_{start})}{\ln 2}$$

Molarity

Recipes often quote the concentration of constituents in grams, but this will vary depending on the water of crystallization in a salt, or when quoting the weight of free acid or its salt. Quoting concentrations in molarity (M) avoids this problem:

$$M = \frac{\text{Concentration (g/l)}}{\text{Molecular mass}}$$

e.g. the molarity of 2 g/l of the disodium salt of the dihydrate of ethylene diamine tetra-acetic acid (EDTA) would be:

$$\frac{2}{372.2} = 0.005373\,M \,or\, 5.373\,\text{mM} \cong 5\,\text{mM}$$

Conversely, the concentration in grams per liter is given by multiplying the molarity by the molecular mass:

$$5.373 \times 372.2 = 2.0$$

Percentages and Dilutions

Solutions are often reported as a percent value, e.g. 10% fetal bovine serum (FBS). The percentage value indicates the amount of solid material that is dissolved in the solution (weight/volume; w/v) or the amount of liquid material that is diluted in the solution (volume/volume; v/v).

For w/v solutions, 1% = 1 g in 100 ml
For v/v solutions, 1% = 1 ml in 100 ml

Dilutions quoted as, for example, 1 : 10 or 1 : 100, are v/v and imply that the final volume is 10 or 100 parts, respectively.

Pressure

Pressure is quoted in different units by different disciplines. Fifteen pounds per square inch (15 psi) = 1 atmosphere (1 atm) = 1 bar = 760 mm Hg = 30 in. Hg = 100 kiloPascals (100 kPa) = 100 Newtons/m² (100 N/m²).

Rotor Speed (rpm to *g*)

For convenience, most laboratory personnel will quote the speed of the centrifuge when describing a centrifugation step, but that will only remain consistent if the same centrifuge and rotor are used. Centrifugation is better defined by g, the radial acceleration relative to gravity. The following formula should be used:

$$g = 1.118 \times 10^{-5}$$
$$\times \text{Radius of the rotor to the center of the tube}$$
$$\times \text{rpm}^2$$

e.g. in a centrifuge rotor rotating at 1500 rpm, with a radius of 18 cm from the center of the spindle to the middle depth of the sample in the tube:

$$g = 1.118 \times 10^{-5} \times 18 \times 1500^2 = 452.79 \cong 450$$

PREPARATION OF REAGENTS

The reagents that are listed here are used in several different protocols throughout the book or in protocols where insufficient space is available for reagent preparation. Many are available ready-made from commercial suppliers and details are provided here for historical reference only.

❖ *Safety Note. Some of these reagents are hazardous substances. Always perform a risk assessment and refer to the Safety Data Sheet (SDS) when working with any hazardous substance (see Sections 6.1.1, 6.2.2). If possible, use ready-made solutions to avoid handling hazardous substances in powdered form.*

Acetic Acid: Methanol

Used as a fixative, e.g. mycoplasma detection with Hoechst 33258 (see Protocol P16.4).

(1) Add 1 part glacial acetic acid to 3 parts anhydrous methanol.
(2) Make up fresh each time used and keep on ice.

Agar (2.5%)

Used for multiple applications, including cloning of anchorage-independent cells (see Section 20.3).

Agar	2.5 g
Ultrapure water (UPW)	100 ml

(1) Boil to dissolve agar.
(2) Sterilize by autoclaving or boiling for two minutes.
(3) Store at room temperature.

Alcohol (70%)

Used for disinfection, e.g. swabbing a biological safety cabinet (BSC).

Alcohol (ethanol or isopropanol)	700 ml
Deionized or reverse osmosis (RO) water	300 ml

Bacto™ Peptone (5%)

Traditionally used as a medium supplement (see Section 9.1).

Bacto Peptone (ThermoFisher Scientific)	5 g
Hanks's balanced salt solution (HBSS)	100 ml

(1) Stir to dissolve.
(2) Dispense in aliquots appropriate to a 1 : 50 dilution.
(3) Sterilize by autoclaving.
(4) Store at room temperature.
(5) Dilute 1 : 10 for use.

Balanced Salt Solutions

Used for multiple applications. Formulations for Earle's balanced salt solution (EBSS), HBSS and Dulbecco's phosphate-buffered saline (DPBS) are given in the main text (see Table 9.2).

(1) Dissolve each constituent separately, adding $CaCl_2$ last.
(2) Make up to 1 l.
(3) Adjust to pH 6.5.
(4) Sterilize the solution by autoclaving or filtration. With autoclaving, the pH must be kept below 6.5 to prevent calcium phosphate from precipitating; alternatively, calcium may be omitted and added later. If glucose is included, the solution should be filtered to avoid caramelization of the glucose, or autoclaved separately (see Glucose in this Appendix) at a higher concentration (e.g. 20%) and added later.
(5) Store the solution at room temperature. With autoclaving, mark the level of the liquid beforehand; if evaporation occurs, make up to mark with sterile ultrapure water (UPW) before use.

Carboxymethylcellulose (CMC; 4%)

Used to increase the viscosity of the medium in suspension cultures.

(1) Weigh out 4 g of CMC (Sigma-Aldrich) and place it in a beaker.
(2) Add 90 ml of HBSS and bring the mixture to boil in order to wet the CMC.
(3) Allow the solution to stand overnight at 4 °C to clarify.
(4) Make volume up to 100 ml with HBSS.
(5) Sterilize the solution by autoclaving. It will solidify again but will redissolve at 4 °C.
(6) For use, add 3 ml per 100 ml of growth medium.

Chick Embryo Extract

Traditionally used as a medium supplement. Extracts can be prepared by maceration as follows (Paul 1975):

(1) Remove embryos from eggs (see Supp. S13.1) and place in 90 mm Petri dishes.
(2) Take out the eyes, using two pairs of sterile forceps.
(3) Transfer the embryos to flat or round-bottomed 50-ml containers, two embryos to each container.
(4) Add an equal volume of HBSS to each container.
(5) Using a sterile glass rod that has been previously heated and flattened at one end, mash the embryos in the HBSS until they have broken up.
(6) Let the mixture stand for 30 minutes at room temperature.
(7) Centrifuge the mixture for 15 minutes at 2000 g.
(8) Remove the supernatant and set aside a sample for sterility testing (see Section 11.8.1).
(9) Dispense into aliquots, freeze and thaw (quickly), twice, and store at −20 °C.

Extracts of chick and other tissues may also be prepared by homogenization in a Potter homogenizer or Waring blender (Coon and Cahn 1966):

(1) Homogenize chopped embryos with an equal volume of HBSS.
(2) Transfer the homogenate to centrifuge tubes, and spin at 1000 g for 10 minutes.
(3) Transfer the supernatant to fresh tubes and centrifuge for a further 20 minutes at 10 000 g.
(4) Check the sample for sterility (see Section 11.8.1).
(5) Dispense into aliquots, freeze and thaw (quickly), twice, and store at −20 °C.

Collagenase

Used for tissue disaggregation or cell dissociation (see Protocol P13.6 and Table 14.4). If making up your own collagenase, use a product with known activity (e.g. CLS-1, Worthington Biochemical Corporation) or the equivalent (specific activity 1500–2000 U/mg) and make up at 2000 U/ml in HBSS:

Collagenase	100 000 U
HBSS	50 ml

(1) Stir at 37 °C for two hours or at 4 °C overnight to dissolve.
(2) Sterilize the solution by filtration, as with serum (see Supp. S11.3).
(3) Divide into aliquots, each suitable for one to two weeks of use.
(4) Store at −20°.

Collection Medium

Used for collection of samples when establishing primary cultures (see Chapter 13).

Growth medium, e.g. L-15	500 ml
Penicillin	125 000 units
Streptomycin	125 mg
Kanamycin	50 mg

or

Growth medium	500 ml
Gentamicin	25 mg
Amphotericin	1.25 mg

(1) Select antibiotics based on the most likely contaminants for that sample type.
(2) Store at 4 °C for up to three weeks or at −20 °C for longer periods.

Crystal Violet (0.1%)

Used for staining (see Protocol P18.4), made up in water:

Crystal violet (Sigma-Aldrich)	100 mg
Water	100 ml

(1) Stir to dissolve.
(2) Filter through Whatman No. 1 paper before use.

Alternatively, used for cell lysis prior to counting nuclei, made up in 0.1 M citric acid:

Citric acid	21.0 g
Crystal violet (Sigma-Aldrich)	1.0 g

(3) Make up to 1000 ml with deionized water.
(4) Stir to dissolve.
(5) To clarify, filter the solution through Whatman No. 1 filter paper.

Dexamethasone (1 mg/ml)

Used for multiple applications, including induction of differentiation (see Table 26.1). This reagent comes already sterile from some suppliers, e.g. Merck Millipore, Sigma-Aldrich, STEMCELL Technologies. To make up from a single-use vial:

(1) Add water by syringe to the vial to dissolve contents.
(2) Remove by syringe when dissolved.
(3) Dilute to 1 mg/ml (approximately 2.5 mM).
(4) Divide the solution into aliquots and store at −20 °C.
(5) For use, dilute the solution to give 10–50 nM (physiological concentration range), 0.1–1.0 µM (pharmacological dose range), or 25–100 µM (high dose range).

Dissection Balanced Salt Solution (DBSS)

Used for tissue dissection when establishing primary cultures (see Chapter 13).

HBSS without bicarbonate	100 ml
Penicillin	250 U/ml
Streptomycin	250 µg/ml
Kanamycin	100 µg/ml

or

HBSS without bicarbonate	100 ml
Gentamicin	50 µg/ml
Amphotericin B	2.5 µg/ml

(1) Sterilize the HBSS by autoclaving.
(2) Add the antibiotics under aseptic conditions.
(3) Divide the solution into aliquots and store at −20 °C.

Dulbecco's Phosphate-Buffered Saline Without Ca²⁺ and Mg²⁺ (DPBS-A)

See Table 9.2, Protocol P11.4, and Balanced salt solutions in this Appendix.

EDTA (10 mM in DPBS-A)

Used for multiple applications, including cell dissociation with or without trypsin (see Sections 14.6.2, 23.4.3; see also Tables 13.1, 14.4).

(1) Make up DPBS-A.
(2) Add EDTA disodium salt (0.374 g/l) and stir.
(3) Dispense, autoclave, and store at room temperature or at 4 °C.
(4) Dilute 1 : 10 in DPBS-A to give 1 mM for most applications. Up to 5 mM may be used for high chelating conditions (e.g. trypsinization of Caco-2 cells), but may be toxic. Down to 0.5 mM is recommended for some cell types, e.g. human pluripotent stem cells (hPSCs; used alone without trypsin).

EGTA

As for EDTA, but ethylene glycol tetra-acetic acid (EGTA) may be used at higher concentrations because of its lower toxicity.

Erythrosin B

See Trypan blue.

Gelatin (1%)

Used in a subbing solution for microautoradiography or as a substrate coating (see Supp. S19.3, S24.15).

(1) Prepare a 1% stock solution of gelatin (Sigma-Aldrich) in UHP water.
(2) Autoclave at 121 °C for 20 minutes.
(3) Aliquot and store at 4 °C.
(4) To prepare a 0.1% working solution, warm 1% gelatin to 37 °C until it liquefies and dilute 1 : 10 in sterile DPBS-A.

Giemsa Stain

Used for staining (see Protocol P18.3). Can be applied undiluted and then diluted with buffer or water, or diluted in buffer before use. One of the authors (R. Ian Freshney) found the first method more successful for cultured cells.

$NaH_2HPO_4 \cdot 2H_2O$	1.38 g/l (0.01 M)
$Na_2HPO_4 \cdot 7H_2O$	2.68 g/l (0.01 M)

(1) Prepare buffer as above in correct proportions to give pH 6.5.
(2) Dilute Giemsa concentrate (e.g. #GS500, Sigma-Aldrich) 1 : 10 in 100 ml of buffer.
(3) Filter the solution through Whatman No. 1 filter paper to clarify.
(4) Make up a fresh solution each time, because the concentrate precipitates on storage.

Glucose (20%)

Used for multiple applications, including as a component of balanced salt solutions.

Glucose	20 g
HBSS to	100 ml

(1) Sterilize by autoclaving.
(2) Store at room temperature.

Glutamine, 200 mM

Used as a medium constituent that is added prior to use (see Section 9.4.1).

L-Glutamine	29.2 g
HBSS	1000 ml

(1) Dissolve the glutamine in HBSS and sterilize by filtration.
(2) Dispense the solution into aliquots and store at −20 °C.

Hanks's Balanced Salt Solution (HBSS)

See Table 9.2 and Balanced salt solutions in this Appendix.

HAT Medium

Used in hypoxanthine, aminopterin, and thymidine (HAT) selective medium (see Section 20.7.1).

Drug	Concentration	Dissolve in	Molarity (100× final)
Hypoxanthine (H)	136 mg/ 100 ml	0.05 N HCl	1×10^{-2} M
Aminopterin (A)	1.76 mg/ 100 ml	0.1 N NaOH	4×10^{-5} M
Thymidine (T)	38.7 mg/ 100 ml	HBSS	1.6×10^{-3} M

(1) Mix equal volumes of each and sterilize by filtration.
(2) Add the mixture to medium at 3% v/v.
(3) Store hypoxanthine and thymidine at 4 °C and aminopterin at −20 °C.

HB Medium

Used for culture of human bronchial epithelial cells (see Supp. S24.4) (Lechner and LaVeck 1985). Add the following to CMRL 1066 medium:

Insulin	5 µg/ml
Hydrocortisone	0.36 µg/ml
β-Retinyl acetate	0.1 µg/ml
Glutamine	1.17 mM
Penicillin	50 U/ml
Streptomycin	50 µg/ml
Gentamicin	50 µg/ml
Fungizone	1.0 µg/ml
FBS	1%

HEPES

Used for calcium phosphate co-precipitation (see Protocol P22.1). To make up 2 × 4-(2-hydroxyethyl)-1-piperazine ethanesulfonic acid hydroxyethyl starch (HEPES) solution (final concentrations in parentheses), mix:

HEPES	1 g (42 mM)
NaCl	1.6 g (279 mM)
KCl	0.074 g (10 mM)
Na_2HPO_4	0.02 g (14 mM)
Dextrose	0.2 g (11 mM)
UPW	100 ml total

(1) Adjust the pH to 7.05 ± 0.05 with 1 N NaOH.
(2) Sterilize by filtration.
(3) Store at 4 °C for up to one month or at −20 °C in 1-ml aliquots.

Hoechst 33258

Used for mycoplasma detection (see Protocol P16.4) (Chen 1977).

(1) Make up 1 mg/ml stock in DPBS-A or HBSS without phenol red.
(2) Store the solution at −20 °C.
(3) For use at 50 ng/ml, dilute 1 : 20 000 (1.0 µl in 20 ml) in DPBS-A or HBSS without phenol red at pH 7.0.

❖ *Safety Note. Hoechst 33258 is hazardous. Handle in a fume hood and wear suitable PPE. If possible, substitute with a ready-made solution to reduce handling.*

Media

See Appendix C for commonly used basal media formulations.

2-Mercaptoethanol (β-Mercaptoethanol; 0.1 M)

Used as a medium supplement (see Sections 9.2.4, 10.4.2; see also Supp. S23.1).

2-Mercaptoethanol (e.g. #441433A, BDH; 14.3 M)	200 µl
UPW	28.2 ml

(1) Sterilize by filtration in fume cupboard.
(2) Store aliquots at 4 °C for up to one month.

Methylcellulose (Methocel, 1.6%)

Used for cloning of anchorage-independent cells (see Protocol P20.3).

(1) Weigh out 7.2 g of Methocel (Dow Chemical Company) and add it to a 500-ml bottle containing a large magnetic stirrer bar.
(2) Sterilize by autoclaving.
(3) Add 400 ml of sterile UPW heated to 90 °C to wet the Methocel.
(4) Stir at 4 °C overnight to dissolve; stirring is necessary, as the Methocel will gel if the magnet does not keep stirring.
(5) The resulting solution is now Methocel 2×. For use, it should be diluted with an equal volume of 2× medium of your choice. It is more accurate to use a syringe (without a needle) than a pipette to dispense Methocel.

Mitomycin C (100 µg/ml)

Used for inactivation of feeder layers (see Protocols P20.7, P25.2).

(1) Measure 20 ml of HBSS into a sterile container.
(2) Remove 2 ml of HBSS by syringe and add it to a vial of mitomycin C, e.g. Sigma-Aldrich (2 mg total).
(3) Allow the mixture to dissolve, withdraw the resulting solution, and add it back to the container.
(4) Store for one week only at 4 °C in the dark. For longer periods, store at −20 °C.
(5) Dilute to 0.25 µg/ml for 18 hours exposure or 20 µg/ml for 10 hours exposure.

❖ *Safety Note. Mitomycin C is hazardous. Handle in a fume hood and wear suitable PPE. If possible, substitute with a ready-made solution to reduce handling.*

MTT (50 mg/ml)

Used for 3-(4,5-dimethylthiazol-2-yl)-2,5-diphenyltetrazolium bromide (MTT) assay (see Protocol P29.1). Note that most laboratories now use kits, which typically supply MTT in ready-made form.

(1) Dissolve MTT (Sigma-Aldrich) in DPBS-A to give 50 mg/ml.
(2) Sterilize by filtration.

❖ *Safety Note. MTT is hazardous. Handle in a fume cupboard and wear suitable PPE. If possible, substitute with a ready-made solution to reduce handling.*

N2 Supplement

Used for serum- and feeder-free culture of mouse embryonic stem cells (mESCs) (see Supp. S23.1) (Nichols and Ying 2006).

(1) Make up the following stock solutions:
   a) Apo-transferrin (e.g. #T1147, Sigma-Aldrich); dissolve at 100 mg/ml in filter-sterilized H_2O.
   b) Bovine Albumin Fraction V Solution (e.g. ThermoFisher Scientific, 7.5% solution): as supplied.
   c) Insulin (e.g. #I1882, Sigma-Aldrich); dissolve at 25 mg/ml in 0.01 M filter-sterilized HCl overnight at 4 °C.
   d) Progesterone (e.g. #P8783, Sigma-Aldrich); dissolve at 0.6 mg/ml in ethanol and filter-sterilize.
   e) Putrescine (e.g. #P5780, Sigma-Aldrich); dissolve at 160 mg/ml in H_2O and filter-sterilize.
   f) Sodium selenite (e.g. #S5261, Sigma-Aldrich); dissolve at 3 mM in H_2O and filter-sterilize.
   g) DMEM:F12 (e.g. #42400–010, Invitrogen; now supplied by ThermoFisher Scientific); as supplied.

h) Store DMEM:F12 at 4 °C; aliquot other stock solutions and store at −20 °C.

(2) Combine medium and stock solutions to give N2 supplement (40 ml total):

DMEM:F12	27.5 ml
Bovine Albumin Fraction V	4 ml
Insulin (add 200 µl at a time to prevent precipitation)	4 ml
Apo-transferrin (Sigma-Aldrich)	4 ml
Sodium selenite (Sigma-Aldrich)	40 µl
Putrescine (Sigma-Aldrich)	400 µl
Progesterone (Sigma-Aldrich)	132 µl

a) Store at 4 °C and use within one month.

N2B27 Medium

Used for serum- and feeder-free culture of (mESCs) (see Supp. S23.1) (Nichols and Ying 2006). To make up N2B27 medium:

(1) Prepare DMEM:F12-N2: add 1 ml N2 supplement to 100 ml DMEM:F12. Final concentrations of each constituent are listed in the source reference.

(2) Prepare Neurobasal/B27 medium: add 2 ml B27 supplement (e.g. #17504–044, Invitrogen) and 1 ml 200 mM L-glutamine to 100 ml Neurobasal medium (e.g. #21103–049, Invitrogen; both now supplied by ThermoFisher Scientific).

(3) Mix DMEM:F12-N2 with Neurobasal/B27 in a 1:1 ratio.

(4) Add 2-mercaptoethanol to a final concentration of 0.1 mM from a 0.1-M stock.

(5) Store at 4 °C and use within one month.

Naphthalene Black (Amido Black; 1%)

Used for cell viability assessment as an alternative to trypan blue (see Section 19.2.1).

Naphthalene black	1 g
HBSS	100 ml

(1) Dissolve as much as possible of the stain in the HBSS.

(2) Filter the resulting saturated solution through Whatman No.1 filter paper.

Non-essential Amino Acids (NEAA, 100×)

Used as a medium supplement (see Section 9.4.1; see also Table 9.3). Available as 100× concentrate from media suppliers (e.g. Sigma-Aldrich). If making up from constituents:

L-Alanine	0.89 g
L-Asparagine H_2O	1.50 g
L-Aspartic acid	1.33 g
Glycine	0.75 g
L-Glutamic acid	1.47 g
L-Proline	1.15 g
L-Serine	1.05 g
Water	1000 ml total

(1) Sterilize by filtration.

(2) Store at 4 °C.

(3) Use at a concentration of 1 : 100.

Paraformaldehyde (4%)

Used as a fixative (see Section 18.5.1).

Paraformaldehyde	2 g
DPBS (with Ca^{2+} and Mg^{2+})	50 ml

(1) Heat PBS to 50 °C with stirring. Monitor with a thermometer; paraformaldehyde will degrade at higher temperatures (> 70 °C).

(2) Add paraformaldehyde powder to the prewarmed, stirred PBS solution.

(3) Add drops of 1 N NaOH until the powder dissolves into solution.

(4) Once dissolved, take the solution off the heat and allow to cool.

(5) Check the pH using litmus paper and adjust if needed to pH 7.0–7.5.

(6) Store at 4 °C and use within one week.

❖ *Safety Note. Paraformaldehyde is hazardous. Handle in a fume cupboard (including all preparation steps) and wear suitable PPE.*

Trypan Blue (0.4%)

Used as a dye for viability assessment (see Section 19.2.1). This method can be used for either trypan blue or erythrosin B, which is a less toxic substitute for trypan blue. An aqueous solution of either agent has a pH of about 6.5 and is hypotonic. Stock solutions should be made up as an isotonic salt solution at neutral pH (Phillips 1973).

Trypan blue or erythrosin B	0.4 g
NaCl	0.81 g
KH_2PO_4	0.06 g
Methyl 4-hydroxybenzoate (preservative, optional)	0.05 g
UPW	95 ml (final volume 100 ml)

(1) Mix ingredients and heat to boiling to dissolve.
(2) After cooling, adjust pH using 1 N NaOH to pH 7.2–3.
(3) Adjust volume to 100 ml and filter-sterilize.
(4) Store at room temperature.

❖ *Safety Note. Trypan blue is hazardous. Handle in a fume hood and wear suitable PPE. If possible, substitute with a ready-made solution to reduce handling.*

Trypsin (2.5%)

Used for tissue disaggregation or cell dissociation (see Sections 13.5.1, 14.6.2; see also Tables 13.1, 14.4). Trypsin is usually bought ready-made; suppliers perform quality control (QC) testing to reduce the risk of microbial contamination. Making up your own stock is not recommended for this reason. If you proceed, you must at least perform your own mycoplasma testing. To make up a 2.5% stock:

| Trypsin (crude, 1 : 250; see Note) | 25 g |
| NaCl, 0.85% (0.14 M) | 1 L |

(1) Stir trypsin for 1 hour at room temperature or 10 hours at 4 °C. If the trypsin does not dissolve completely, clarify it by filtration through Whatman No. 1 filter paper.
(2) Sterilize by filtration.
(3) Dispense into 10- to 20-ml aliquots and store at −20 °C.
(4) When ready for use, thaw and dilute 1 : 10 in DPBS-A or in EDTA solution (10 mM in DPBS-A) to give a 0.25% solution.
(5) Store diluted trypsin at 4 °C for a maximum of three weeks.

Note. Trypsin is available as a crude product (usually supplied as 1 : 250 grade), purified preparation, or recombinant protein. Pure and recombinant trypsin have a higher specific activity and should therefore be used at a proportionally lower concentration (e.g. 0.01% or 0.05%). Check the specific activity when you purchase trypsin as it varies between products and suppliers (see Section 13.5.1).

Versene

See EDTA.

Suppliers

Supplier	URL
Abcam	http://www.abcam.com
Dow Chemical Company	http://www.dow.com/en-us/product-search/methocelpremium
Merck Millipore	http://www.merckmillipore.com
Sigma-Aldrich (Merck)	http://www.sigmaaldrich.com/life-science/cell-culture.html
STEMCELL Technologies	http://www.stemcell.com
ThermoFisher Scientific	http://www.thermofisher.com/us/en/home/life-science/cell-culture.html
VWR (Jencons, BDH)	https://uk.vwr.com/store/
Worthington Biochemical Corporation	http://www.worthington-biochem.com/default.html

REFERENCES

Chen, T.R. (1977). In situ detection of mycoplasma contamination in cell cultures by fluorescent Hoechst 33258 stain. *Exp. Cell Res.* 104 (2): 255–262. https://doi.org/10.1016/0014-4827(77)90089-1.

Coon, H.G. and Cahn, R.D. (1966). Differentiation in vitro: effects of Sephadex fractions of chick embryo extract. *Science* 153 (3740): 1116–1119. https://doi.org/10.1126/science.153.3740.1116.

Hayflick, L. (1973). Subculturing human diploid fibroblast cultures. In: *Tissue Culture: Methods and Applications* (eds. P.F. Kruse Jr. and M.K. Patterson), 220–223. New York: Academic Press.

Lechner, J.F. and LaVeck, M.A. (1985). A serum-free method for culturing normal human bronchial epithelial cells at clonal density. *J. Tissue Cult. Methods* 9 (2): 43–48.

Nichols, J. and Ying, Q.L. (2006). Derivation and propagation of embryonic stem cells in serum- and feeder-free culture. *Methods Mol. Biol.* 329: 91–98. https://doi.org/10.1385/1-59745-037-5:91.

Paul, J. (1975). *Cell and Tissue Culture*, 5e. Edinburgh: Churchill Livingstone.

Phillips, H.J. (1973). Dye exclusion tests for cell viability. In: *Tissue Culture: Methods and Applications* (eds. P.F. Kruse Jr. and M.K. Patterson), 406–408. New York: Academic Press.

APPENDIX C

Media Formulations

Commonly used media formulations are listed here, divided into three tables. The "Classic Media" table lists formulations that were developed before 1970 (Morton 1970). The "Serum-Free Complete Media" table lists complete media that were developed for serum-free culture after 1970. The "Serum-Free Supplemented Media" table lists basal media and supplements that can be used together for serum-free culture. Concentrations are molar unless otherwise stated, and computer-style notation is used (e.g. 3.0E-2 = 3.0×10^{-2} M = 30 mM). The units are given in the component column where molarity is not used. Molecular weights may be given for root compounds; although some recipes use salts or hydrated forms, molarities will, of course, remain the same.

Synonyms and abbreviations: AMP, adenosine monophosphate; Arg VP, arginine vasopressin; ATP, adenosine triphosphate; biotin, vitamin H; BPE, bovine pituitary extract; BSA, bovine serum albumin; calciferol, vitamin D2; EGF, epidermal growth factor; FAD, flavine adenine dinucleotide; FBS, fetal bovine serum; FGF, fibroblast growth factor; HSM, hormone-supplemented medium; lipoic acid, thioctic acid; menadione, vitamin K3; MW, molecular weight; myo-inositol, L-inositol; nicotinamide, niacinamide; nicotinic acid, niacin; pyridoxine HCl, vitamin B6; thiamin, vitamin B1; α-tocopherol, vitamin E; retinol, vitamin A1; TPN, triphosphopyridine nucleotide; UTP, uridine triphosphate; vitamin B12, cobalamin.

Freshney's Culture of Animal Cells: A Manual of Basic Technique and Specialized Applications, Eighth Edition. Amanda Capes-Davis and R. Ian Freshney.
© 2021 John Wiley & Sons Ltd. Published 2021 by John Wiley & Sons Ltd.
Companion website: www.wiley.com/go/freshney/cellculture8

CLASSIC MEDIA

Medium	MEM	DMEM	Ham's F12	DMEM/F12	αMEM	CMRL 1066	RPMI 1640	Medium 199	L-15	McCoy's 5A	MB 752/1
References	(Eagle 1959)	(Dulbecco and Freeman 1959)	(Ham 1965)	(Barnes and Sato 1980)	(Stanners et al. 1971)	(Parker et al. 1957)	(Moore et al. 1967)	(Morgan et al. 1950)	(Leibovitz 1963)	(McCoy et al. 1959)	(Waymouth 1959)
Amino acids											
L-Alanine	—	—	1.0E-04	5.0E-05	2.8E-04	2.8E-04	—	2.8E-04	2.5E-03	1.5E-04	—
L-Arginine	6.0E-04	4.0E-04	1.0E-03	7.0E-04	6.0E-04	3.3E-04	1.1E-03	3.3E-04	2.9E-03	2.0E-04	3.6E-04
L-Asparagine	—	—	1.0E-04	5.0E-05	3.3E-04	3.8E-04	1.7E-03	3.0E-04	7.6E-05		—
L-Aspartic acid	—	—	1.0E-04	5.0E-05	2.3E-04	2.3E-04	1.5E-04	2.3E-04	—	1.5E-04	4.5E-04
L-Cysteine	—	—	2.0E-04	1.0E-04	5.7E-04	1.5E-03	—	5.6E-07	9.9E-04	2.0E-04	5.0E-04
L-Cystine	1.0E-04	2.0E-04	—	1.0E-04	1.0E-04	8.3E-05	2.1E-04	9.9E-05	—		6.3E-05
L-Glutamic acid	—	—	1.0E-04	5.0E-05	5.1E-04	5.1E-04	1.4E-04	4.5E-04		1.5E-04	1.0E-03
L-Glutamine	2.0E-03	4.0E-03	1.0E-03	2.5E-03	2.0E-03	6.8E-04	2.1E-03	6.8E-04	2.1E-03	1.5E-03	2.4E-03
Glycine	—	4.0E-04	1.0E-04	2.5E-04	6.7E-04	6.7E-04	1.3E-04	6.7E-04	2.7E-03	1.0E-04	6.7E-04
L-Histidine	2.0E-04	2.0E-04	1.0E-04	1.5E-04	2.0E-04	9.5E-05	9.7E-05	1.0E-04	1.6E-03	1.0E-04	8.3E-04
L-Hydroxyproline	—	—	—	—	—	7.6E-05	1.5E-04	7.6E-05	—	1.5E-04	—
L-Isoleucine	4.0E-04	8.0E-04	3.0E-05	4.2E-04	4.0E-04	1.5E-04	3.8E-04	1.5E-04	9.5E-04	3.0E-04	1.9E-04
L-Leucine	4.0E-04	8.0E-04	1.0E-04	4.5E-04	4.0E-04	4.6E-04	3.8E-04	4.6E-04	9.5E-04	3.0E-04	3.8E-04
L-Lysine HCl	4.0E-04	8.0E-04	2.0E-04	5.0E-04	4.0E-04	3.8E-04	2.2E-04	3.8E-04	5.1E-04	2.0E-04	1.3E-03
L-Methionine	1.0E-04	2.0E-04	3.0E-05	1.2E-04	1.0E-04	1.0E-04	1.0E-04	1.0E-04	5.0E-04	1.0E-04	3.4E-04
L-Phenylalanine	2.0E-04	4.0E-04	3.0E-05	2.2E-04	1.9E-04	1.5E-04	9.1E-05	1.5E-04	7.6E-04	1.0E-04	3.0E-04
L-Proline	—	—	3.0E-04	1.5E-04	3.5E-04	3.5E-04	1.7E-04	3.5E-04	—	1.5E-04	4.3E-04
L-Serine	—	4.0E-04	1.0E-04	2.5E-04	2.4E-04	2.4E-04	2.9E-04	2.4E-04	1.9E-03	2.5E-04	—
L-Threonine	4.0E-04	8.0E-04	1.0E-04	4.5E-04	4.0E-04	2.5E-04	1.7E-04	2.5E-04	2.5E-03	1.5E-04	6.3E-04
L-Tryptophan	4.9E-05	7.8E-05	1.0E-05	4.4E-05	4.9E-05	4.9E-05	2.5E-05	4.9E-05	9.8E-05	1.5E-05	2.0E-04
L-Tyrosine	2.0E-04	4.0E-04	3.0E-05	2.1E-04	2.3E-04	2.2E-04	1.1E-04	2.2E-04	1.7E-03	1.2E-04	2.2E-04
L-Valine	4.0E-04	8.0E-04	1.0E-04	4.5E-04	3.9E-04	2.1E-04	1.7E-04	2.1E-04	8.5E-04	1.5E-04	5.6E-04
Vitamins											
p-Aminobenzoic acid	—	—	—	—	—	3.6E-07	7.3E-06	3.6E-07		7.3E-06	
L-Ascorbic acid	—	—	—	—	2.5E-04	2.8E-04		2.8E-07		3.2E-06	9.9E-05
Biotin	—	—	3.0E-08	1.5E-08	4.1E-07	4.1E-08	8.2E-07	4.1E-08		8.2E-07	8.2E-08
Calciferol	—	—	—	—	—	—	—	2.5E-07			—
Choline chloride	7.1E-06	2.9E-05	1.0E-04	6.4E-05	7.1E-06	3.6E-06	2.1E-05	3.6E-06	7.1E-06	3.6E-05	1.8E-03
Folic acid	2.3E-06	9.1E-06	2.9E-06	6.0E-06	2.3E-06	2.3E-08	2.3E-06	2.3E-08	2.3E-06	2.3E-05	9.1E-07
myo-Inositol	1.1E-05	4.0E-05	1.0E-04	7.0E-05	1.1E-05	2.8E-07	1.9E-04	2.8E-07	1.1E-05	2.0E-04	5.6E-06

Menadione	—	—	—	—	—	—	6.9E-08	—	—	—	—
Nicotinamide	8.2E-06	3.3E-05	3.3E-07	1.7E-05	8.2E-06	8.2E-06	2.0E-07	2.0E-07	8.2E-06	4.1E-06	8.2E-06
Nicotinic acid	—	—	—	—	—	—	2.0E-07	—	—	4.1E-06	—
D-Ca pantothenate	4.2E-06	1.7E-05	2.0E-06	9.4E-06	4.2E-06	4.2E-06	4.2E-08	4.2E-06	4.2E-06	8.4E-07	4.2E-06
Pyridoxal HCl	4.9E-06	2.0E-05	3.0E-07	1.0E-05	4.9E-06	1.1E-06	1.2E-07	1.2E-07	4.2E-06	2.5E-06	—
Pyridoxine HCl	—	—	3.0E-07	1.5E-07	—	4.9E-06	1.2E-07	1.2E-07	4.9E-06	2.4E-06	4.9E-06
Riboflavin	2.7E-07	1.1E-06	1.0E-07	5.8E-07	2.7E-07	5.3E-07	2.7E-08	2.7E-08	1.9E-07	5.3E-07	2.7E-06
Thiamin	3.0E-06	1.2E-05	1.0E-06	6.4E-06	3.0E-06	3.0E-06	3.0E-08	3.0E-08	2.4E-06	5.9E-07	3.0E-05
Thiamin mono PO$_4$	—	—	—	—	—	—	—	—	—	4.8E-06	—
α-Tocopherol	—	—	—	—	—	—	2.3E-08	—	—	—	—
Retinol acetate	—	—	—	—	—	—	3.5E-07	—	—	—	—
Vitamin B$_{12}$	—	1.0E-06	1.0E-06	5.0E-07	1.0E-06	3.7E-09	—	—	—	—	1.5E-07
Antioxidants	—	—	—	—	—	—	—	—	—	—	—
Glutathione	—	—	—	—	—	3.0E-06	1.5E-07	3.0E-05	1.5E-06	1.5E-06	4.5E-05
Inorganic salts											
CaCl$_2$	1.8E-03	1.8E-03	3.0E-04	1.1E-03	1.8E-03	1.8E-03	1.3E-03	1.3E-03	1.3E-03	9.0E-04	8.2E-04
KCl	5.3E-03	5.3E-03	3.0E-03	4.2E-03	5.3E-04	5.3E-03	5.3E-03	5.3E-03	5.3E-03	5.3E-03	2.0E-03
KH$_2$PO$_4$	—	—	—	—	—	—	4.4E-04	4.4E-04	4.4E-04	—	5.9E-04
MgCl$_2$	—	—	—	1.2E-01	—	—	—	—	—	—	1.2E-03
MgSO$_4$	8.1E-04	8.1E-04	6.0E-04	7.0E-04	8.1E-04	4.0E-04	8.1E-04	8.1E-04	8.1E-04	8.1E-04	8.1E-04
NaCl	1.16E-01	1.09E-01	1.28E-01	1.19E-01	1.16E-01	1.03E-01	1.16E-01	1.16E-01	1.37E-01	1.10E-01	1.03E-01
NaHCO$_3$	2.6E-02	4.4E-02	1.4E-02	2.9E-02	2.6E-02	2.3E-02	2.6E-02	2.6E-02	1.3E-02	1.3E-02	2.7E-02
NaH$_2$PO$_4$	1.0E-03	9.1E-04	—	4.5E-04	—	1.0E-03	1.0E-03	—	—	4.2E-03	—
Na$_2$HPO$_4$	—	—	1.0E-03	5.0E-04	—	5.6E-03	4.0E-04	—	1.6E-03	—	2.1E-03
Trace elements											
CuSO$_4$ 5H$_2$O	—	—	1.6E-08	—	7.8E-09	—	1.6E-08	—	—	—	—
Fe(NO$_3$)$_3$ 9H$_2$O	—	2.5E-07	1.2E-07	—	1.2E-07	—	2.0E-07	—	—	—	—
FeSO$_4$ 7H$_2$O	—	—	3.0E-06	—	1.5E-06	—	—	—	—	—	—
ZnSO$_4$ 7H$_2$O	—	—	3.0E-06	—	1.5E-06	—	—	—	—	—	—
Bases, nucleosides, etc.											
Adenine SO$_4$	—	—	—	—	3.7E-05	—	5.4E-05	—	—	—	—
Adenosine	—	—	—	—	—	—	—	—	—	—	—
AMP	—	—	—	—	—	—	5.8E-07	—	—	—	—
ATP	—	—	—	—	—	—	1.8E-05	—	—	—	—
Cytidine	—	—	—	—	4.1E-05	—	—	—	—	—	—
Deoxyadenosine	—	—	4.0E-05	—	4.0E-05	—	—	—	—	—	—
Deoxycytidine	—	—	3.8E-05	—	4.2E-05	—	—	—	—	—	—

(continued)

Medium	MEM	DMEM	Ham's F12	DMEM/F12	αMEM	CMRL 1066	RPMI 1640	Medium 199	L-15	McCoy's 5A	MB 752/1
Deoxyguanosine	—	—	—	—	3.7E-05	3.7E-05	—	—	—	—	—
2-Deoxyribose	—	—	—	—	—	9.5E-06	—	3.7E-06	—	—	—
DPN	—	—	—	—	—	1.2E-06	—	—	—	—	—
FAD	—	—	—	—	—	1.9E-05	—	—	—	—	—
Glucuronate, Na	—	—	—	—	—	—	—	1.6E-06	—	—	—
Guanine	—	—	—	—	—	—	—	—	—	—	—
Guanosine	—	—	—	—	3.5E-05	—	—	—	—	—	—
Hypoxanthine	—	—	3.0E-05	1.5E-05	—	—	—	2.2E-06	—	—	—
5-Me-deoxycytidine	—	—	—	—	—	4.1E-07	—	—	—	—	—
D-Ribose	—	—	—	—	—	—	—	3.3E-06	—	—	—
Thymidine	—	—	3.0E-06	1.5E-06	4.1E-05	4.1E-05	—	—	—	—	—
Thymine	—	—	—	—	—	—	—	2.4E-06	—	—	—
TPN	—	—	—	—	—	1.3E-06	—	—	—	—	—
Uracil	—	—	—	—	—	—	—	2.7E-06	—	—	—
Uridine	—	—	—	—	4.1E-05	—	—	—	—	—	—
UTP	—	—	—	—	—	1.8E-06	—	—	—	—	—
Xanthine	—	—	—	—	—	—	—	2.0E-06	—	—	—
Energy metabolism											
Cocarboxylase	—	—	—	—	—	2.2E-06	—	—	—	—	—
Coenzyme A	—	—	—	—	—	3.3E-06	—	—	—	—	—
D-galactose	—	—	—	—	—	—	—	—	5.0E-02	—	—
D-glucose	5.6E-03	2.5E-02	1.0E-02	1.8E-02	5.6E-03	5.6E-03	1.1E-02	5.6E-03	—	1.7E-02	2.8E-02
Sodium acetate	—	—	—	—	—	6.1E-04	—	4.5E-04	—	—	—
Sodium pyruvate	—	1.0E-03	1.0E-03	1.0E-03	1.0E-03	—	—	—	5.0E-03	—	—
Lipids and precursors											
Cholesterol	—	—	—	—	—	5.2E-07	—	5.2E-07	—	—	—
Ethanol (solvent)	—	—	—	—	—	3.5E-04	—	—	—	—	—
Linoleic acid	—	3.0E-07	—	1.5E-07	—	—	—	—	—	—	—
Lipoic acid	—	—	1.0E-06	5.1E-07	9.7E-07	—	—	—	—	—	8.9E-05
Tween 80	—	—	—	—	—	1.8E-05	—	1.8E-05	—	—	—
Other components											
Peptone, mg/ml	—	—	—	—	—	—	—	—	—	0.6	—
Phenol red	2.7E-05	4.0E-05	3.2E-05	3.6E-05	2.9E-05	5.3E-05	1.3E-05	4.5E-05	2.7E-05	2.9E-05	2.7E-05
Putrescine	—	—	1.0E-06	5.0E-07	—	—	—	—	—	—	—
Gas phase											
CO_2	5%	10%	2%	7%	5%	5%	5%	5%	Air	5%	5%

SERUM-FREE COMPLETE MEDIA

Medium	Molecular weight (MW)	MCDB 110 Human lung fibroblasts (Bettger et al. 1981)	MCDB 131 Human vascular endothelium (Knedler and Ham 1987)	MCDB 153 Human keratinocytes (Peehl and Ham 1980; Boyce and Ham 1983)	MCDB 170 Human mammary epithelium (Hammond et al. 1984)	MCDB 202 Chick embryo fibroblasts (McKeehan and Ham 1976)	MCDB 302 CHO cells (Hamilton and Ham 1977)	MCDB 402 3T3 cells (Shipley and Ham 1983)	WAJC 404 Prostatic epithelium[a] (McKeehan et al. 1984)	IMDM Lymphoid cells (Iscove and Melchers 1978)	LHC-9 Bronchial epithelium (Lechner and LaVeck 1985)
Component											
Amino acids											
L-Alanine	89	1.0E-04	3.0E-05	2.8E-04	1.0E-04	1.0E-04	—	1.0E-04	1.0E-04	1.0E-04	—
L-Arginine	211	1.0E-03	3.0E-04	1.0E-03	3.0E-04	3.0E-04	1.0E-03	3.0E-04	1.0E-03	4.0E-04	2.0E-03
L-Asparagine	132	1.0E-04	1.0E-04	1.0E-04	1.0E-04	1.0E-03	1.1E-04	1.0E-04	1.0E-04	1.9E-04	1.0E-04
L-Aspartic acid	133	1.0E-04	1.0E-04	3.0E-05	1.0E-04	1.0E-04	1.0E-04	1.0E-05	3.0E-05	2.3E-04	3.0E-05
L-Cysteine	176	5.0E-05	2.0E-04	2.4E-04	7.0E-05	2.0E-04	1.0E-04	—	—	—	2.4E-04
L-Cystine	240	—	—	—	2.0E-04	2.0E-04	—	4.0E-04	2.4E-04	2.9E-04	—
L-Glutamic acid	147	1.0E-04	3.0E-05	1.0E-04	1.0E-04	1.0E-04	1.0E-04	1.0E-05	1.0E-04	5.1E-04	1.0E-04
L-Glutamine	146	2.5E-03	1.0E-02	6.0E-03	2.0E-03	1.0E-03	3.0E-03	5.0E-03	6.0E-03	4.0E-03	6.0E-03
Glycine	75	3.0E-04	3.0E-05	1.0E-04	1.0E-04	1.0E-04	1.0E-04	1.0E-04	1.0E-04	4.0E-04	1.0E-04
L-Histidine	210	1.0E-04	2.0E-04	8.0E-05	1.0E-04	1.0E-04	1.0E-04	2.0E-03	8.0E-05	2.0E-04	1.6E-04
L-Isoleucine	131	3.0E-05	5.0E-04	1.5E-05	1.0E-04	1.0E-04	3.0E-05	1.0E-03	1.5E-05	8.0E-04	3.0E-05
L-Leucine	131	1.0E-04	1.0E-03	5.0E-04	3.0E-04	3.0E-04	1.0E-04	2.0E-03	5.0E-04	8.0E-04	1.0E-03
L-Lysine HCl	183	2.0E-04	1.0E-03	1.0E-04	2.0E-04	2.0E-04	2.0E-04	8.0E-04	1.0E-04	8.0E-04	2.0E-04
L-Methionine	149	3.0E-05	1.0E-04	3.0E-05	3.0E-05	3.0E-05	3.0E-05	2.0E-04	3.0E-05	2.0E-04	6.0E-05
L-Phenylalanine	165	3.0E-05	2.0E-04	3.0E-05	3.0E-05	3.0E-05	3.0E-05	3.0E-04	3.0E-05	4.0E-04	6.0E-05
L-Proline	115	3.0E-04	1.0E-04	3.0E-04	5.0E-05	5.0E-05	3.0E-04	—	3.0E-04	3.5E-04	3.0E-04
L-Serine	105	1.0E-04	3.0E-04	6.0E-04	3.0E-04	3.0E-04	1.0E-04	1.0E-04	6.0E-04	4.0E-04	1.2E-03
L-Threonine	119	1.0E-04	1.0E-04	1.0E-04	3.0E-04	3.0E-04	1.0E-04	5.0E-04	1.0E-04	8.0E-04	2.0E-04
L-Tryptophan	204	1.0E-05	2.0E-05	1.5E-05	3.0E-05	3.0E-05	1.0E-05	1.0E-05	1.5E-05	7.8E-05	3.0E-05
L-Tyrosine	181	3.5E-05	1.0E-04	1.5E-05	5.0E-05	5.0E-05	4.4E-05	2.0E-04	1.5E-05	4.6E-04	3.0E-05
L-Valine	117	1.0E-04	1.0E-03	3.0E-04	3.0E-04	3.0E-04	1.0E-04	2.0E-04	3.0E-04	8.0E-04	6.0E-04

(continued)

Medium		MCDB 110	MCDB 131	MCDB 153	MCDB 170	MCDB 202	MCDB 302	MCDB 402	WAJC 404	IMDM	LHC-9
Cell type		Human lung fibroblasts	Human vascular endothelium	Human keratinocytes	Human mammary epithelium	Chick embryo fibroblasts	CHO cells	3T3 cells	Prostatic epithelium[a]	Lymphoid cells	Bronchial epithelium
Vitamins											
Biotin	244	3.0E-08	3.0E-08	6.0E-08	3.0E-08	3.0E-08	3.0E-08	3.0E-08	6.0E-08	5.3E-08	6.0E-08
Choline chloride	140	1.0E-04	—	1.0E-04	1.0E-04	1.0E-04	1.0E-04	1.0E-04	1.0E-04	2.0E-05	2.0E-04
Folic acid	441	—	1.0E-04	1.8E-06	—	—	3.0E-06	—	1.8E-06	9.1E-06	1.8E-06
Folinic acid	512	1.0E-09	1.0E-06	—	1.0E-08	1.0E-08	1.0E-06	1.0E-06	—	—	—
myo-Inositol	180	1.0E-04	4.0E-05	1.0E-04	1.0E-04	1.0E-04	1.0E-04	4.0E-05	1.0E-04	4.0E-05	1.0E-04
Nicotinamide	122	5.0E-05	5.0E-05	3.0E-07	5.0E-05	5.0E-05	3.0E-07	5.0E-05	3.0E-07	3.3E-05	3.0E-07
Pantothenate	238	1.0E-06	5.0E-05	1.0E-06	1.0E-06	1.0E-06	1.0E-06	5.0E-05	1.0E-06	1.7E-05	1.0E-06
Pyridoxal HCl	204	—	—	—	—	—	—	—	—	2.0E-05	—
Pyridoxine HCl	206	3.0E-07	1.0E-05	3.0E-07	3.0E-07	3.0E-07	3.0E-07	1.0E-04	3.0E-07	—	3.0E-07
Riboflavin	376	3.0E-07	1.0E-08	1.0E-07	3.0E-07	3.0E-07	1.0E-07	1.0E-06	1.0E-07	1.1E-06	1.0E-07
Thiamin HCl	337	1.0E-06	1.0E-05	1.0E-06	1.0E-06	1.0E-06	1.0E-06	1.0E-04	1.0E-06	1.2E-05	1.0E-06
α-Tocopherol	430	1.4E-07	—	—	—	—	—	—	—	—	—
Retinoic acid	300	—	—	—	—	—	—	—	—	—	—
Retinol acetate	329	4.2E-07	—	—	—	—	—	—	—	—	3.3E-07
Vitamin B$_{12}$	1355	1.0E-07	1.0E-08	3.0E-07	1.0E-07	1.0E-07	1.0E-07	1.0E-08	3.0E-07	9.6E-09	3.0E-07
Antioxidants											
Dithiothreitol	154	6.5E-06	—	—	—	—	—	—	—	—	—
Glutathione	307	6.5E-07	—	—	—	—	—	—	—	—	—
Inorganic salts											
CaCl$_2$	147	1.0E-03	1.6E-03	3.0E-05	2.0E-03	2.0E-03	6.0E-04	1.6E-03	1.3E-04	1.5E-03	1.1E-04
KCl	75	5.0E-03	4.0E-03	1.5E-03	3.0E-03	3.0E-03	3.0E-03	4.0E-03	1.5E-03	4.4E-03	1.5E-03
KNO$_3$	160	—	—	—	—	—	1.6E-08	—	—	7.5E-07	—
MgCl$_2$	203	—	—	6.0E-04	—	—	6.0E-04	—	6.0E-04	—	2.2E-02
MgSO$_4$	247	1.0E-03	1.0E-02	—	1.5E-03	1.5E-03	6.1E-10	8.0E-04	—	8.1E-04	—
NaCl	58	1.1E-01	1.1E-01	1.2E-01	1.2E-01	1.2E-01	1.3E-01	1.2E-01	1.2E-01	7.7E-02	1.0E-01
NaHCO$_3$	84	—[b]	1.4E-02	1.4E-02	—[b]	—[b]	1.4E-02	1.4E-02	1.4E-02	3.6E-02	1.2E-02
NA$_2$HPO$_4$	120	3.0E-03	5.9E-04	2.0E-03	5.0E-04	5.0E-04	1.2E-03	5.0E-04	2.0E-03	1.0E-03	2.0E-03

Trace elements

Component	MW									
$CuSO_4 \cdot 5H_2O$	160	1.0E-09	7.5E-09	1.1E-08	1.0E-09	1.0E-09	5.0E-09	—	—	1.0E-08
$FeSO_4$	278	5.0E-06	1.0E-06	5.0E-06	5.0E-06	5.0E-06	1.0E-06	3.0E-06	—	5.4E-04
$MnSO_4 \cdot H_2O$	169	1.0E-09	1.2E-09	1.0E-09	5.0E-10	5.0E-10	1.0E-09	—	—	1.0E-09
$(NH_4)_6Mo_7O_{24}$	1236	1.0E-09	3.0E-09	1.0E-09	1.0E-09	1.0E-09	3.0E-09	1.0E-08	—	1.0E-09
$NiCl_2$	238	5.0E-10	4.2E-10	5.0E-10	5.0E-12	5.0E-10	3.0E-10	—	—	5.0E-10
H_2SeO_3	129	3.0E-08	—	3.0E-08	3.0E-08	3.0E-08	1.0E-08	1.3E-08	1.0E-07	3.0E-08
Na_2SiO_3	122	5.0E-07	2.3E-05	5.0E-07	5.0E-07	5.0E-07	1.0E-05	—	—	5.0E-07
$SnCl_2$	190	5.0E-10	—	5.0E-10	5.0E-12	5.0E-12	—	—	—	5.0E-10
NH_4VO_3	117	5.0E-09	5.1E-09	5.0E-09	5.0E-09	5.0E-09	5.0E-09	1.0E-08	—	5.0E-09
$ZnSO_4 \cdot 7H_2O$	288	5.0E-07	1.0E-09	5.0E-07	1.0E-07	1.0E-07	1.0E-06	3.0E-06	—	4.8E-07

Lipids and precursors

Component	MW									
Cholesterol	387	7.6E-06	—	—	—	—	—	—	—	—
Ethanolamine	61	—	—	1.0E-04	1.0E-04	1.0E-04	—	—	—	1.0E-04
Linoleic acid	280	—	—	2.0E-07	2.0E-07	2.0E-07	—	3.0E-07	—	3.0E-07
Lipoic acid	206	1.0E-08	1.0E-06	1.0E-06	1.0E-08	1.0E-08	1.0E-08	9.8E-07	—	1.0E-06
Phosphoethanolamine	141	—	—	—	1.0E-04	1.0E-04	—	1.0E-04	—	—
Soya lecithin, µg/ml	—	6	—	—	—	—	—	—	—	—
Soybean lipid, µg/ml	—	—	—	—	—	—	—	—	50	—
Sphingomyelin, µg/ml	—	1	—	—	—	—	—	—	—	—

Hormones and growth factors

Component	MW									
EGF, ng/ml	—	30	—	25	25	10	—	25	—	5
Epinephrine	183	—	—	—	—	—	—	—	—	2.7E-06
Hydrocortisone^c	362	5.0E-07	—	1.4E-07	1.4E-07	1.4E-07	—	—	5.0E-07	2.0E-07
Insulin, µg/ml	—	1	—	5	5	5	—	5	10	5
Prolactin, µg/ml	—	1	—	5	5	5	—	—	—	—
PGE_1	355	2.5E-08	—	2.5E-08	—	—	—	—	—	—
Triiodo-thyronine	673	—	—	—	—	—	—	—	50	1.0E-08

Nucleosides, etc.

Component	MW									
Adenine SO_4	184	1.0E-05	1.0E-06	1.8E-04	1.0E-06	1.0E-06	1.0E-06	—	1.8E-04	1.8E-04
Hypoxanthine	136	—	—	—	—	—	—	3.0E-05	3.0E-05	—
Thymidine	242	3.0E-07	1.0E-07	3.0E-06	3.0E-07	3.0E-07	1.0E-06	—	3.0E-06	3.0E-06

Energy metabolism

(continued)

Medium		MCDB 110	MCDB 131	MCDB 153	MCDB 170	MCDB 202	MCDB 302	MCDB 402	WAJC 404	IMDM	LHC-9
Cell type		Human lung fibroblasts	Human vascular endothelium	Human keratinocytes	Human mammary epithelium	Chick embryo fibroblasts	CHO cells	3T3 cells	Prostatic epithelium[a]	Lymphoid cells	Bronchial epithelium
D-glucose	180	4.0E-03	5.6E-03	6.0E-03	8.0E-03	8.0E-03	1.0E-02	5.5E-03	6.0E-03	2.5E-02	6.0E-03
Phosphoenol-pyruvate	190	1.0E-05	—	—	—	—	1.0E	—	—	—	—
Sodium acetate $3H_2O$	136	—	—	3.7E-03	—	—	—	—	3.7E-03	—	3.7E-03
Sodium pyruvate	110	1.0E-03	1.0E-03	5.0E-04	1.0E-03	5.0E-04	1.1E+02	—	5.0E-04	1.0E-03	5.0E-04
Other components											
Cholera toxin	~90,000	—	—	—	—	—	—	—	2.0E-10	—	—
HEPES, Na salt	260	3.0E-02	—	2.8E-02	3.0E-02	3.0E-02	—	—	2.8E-02	2.5E-02	2.3E-02
Phenol red	376	3.3E-06	3.3E-05	3.3E-05	3.3E-06	3.3E-06	3.3E-06	3.3E-05	3.3E-05	4.0E-05	3.3E-06
Putrescine 2HCl	161	1.0E-09	1.2E-09	1.0E-06	1.0E-09	1.0E-09	1.0E-06	1.0E-09	1.0E-06	—	1.0E-06
Protein supplements											
BPE, µg/ml[d]		70	—	—	—	—	—	—	25	—	35
BSA, mg/ml		—	—	—	—	—	—	—	—	0.5-10	—
Dialyzed FBS, µg/ml		—	—	1	—	—	—	—	—	—	—
Transferrin, Fe^{3+} saturated, µg/ml		—	—	—	5	—	—	—	—	30-300	10
Gas phase											
CO_2	44	2%	5%	5%	2%	2%	5%	5%	5%	10%	5%

[a]See also complete PFMR-4A (Peehl 2002).
[b]Although no bicarbonate is specified in the formulation, 20 mM NaOH was used to neutralize the medium in a gas phase of 2% CO_2, resulting in 10 mM bicarbonate at pH 7.4.
[c]Soluble analogs of hydrocortisone, such as dexamethasone, can be used.
[d]Ovine prolactin can be substituted for BPE.

SERUM-FREE SUPPLEMENTED MEDIA

Medium	HITES	ACL-4	PEC	K-1	K-2	N3	G3	HSM	E8[a]
Cell type	Human small cell carcinoma	Human nonsmall cell lung carcinoma	Human prostate epithelium	MDCK (dog kidney)	LLC-PK (pig kidney)	LA-N-1 (human neuroblastoma)	Rat glial cells	Human melanocytes	Human pluripotent stem cells (hPSCs)
References	(Carney et al. 1981)	(Gazdar and Oie 1986)	(Roberson and Robertson 1995)	(Taub 1984)	(Taub 1984)	(Bottenstein 1985)	(Bottenstein 1985)	(Naeyaert et al. 1991)	(Chen et al. 2011; Beers et al. 2012)
Basal medium	RPMI 1640	RPMI 1640 or DMEM/F12	αMEM/F12	DMEM/F12	DMEM/F12	DMEM/F12	DMEM	Medium 199	DMEM/F12
Arg VP, μU/ml	—	—	—	—	10	—	—	—	—
BPE, μg/ml	—	—	25	—	—	—	—	—	—
BSA, mg/ml	—	2.0	—	—	—	—	0.3	—	—
Cholera toxin	—	—	—	—	—	—	—	1.0E-09	—
Cholesterol	—	—	—	—	1.0E-08	—	—	—	—
EGF, ng/ml	—	1—	10	—	—	—	—	1	—
Estradiol	1.0E-08	—	—	—	—	—	—	—	—
Ethanolamine	—	1.0E-05	1.0E-03	—	—	—	—	—	—
FGF-2 (basic FGF), ng/ml	—	—	—	—	—	—	—	10	100
Glutamine (additional)	—	2.0E-03	2.0E-03	—	—	—	—	—	—
HEPES (additional)	—	1.0E-02	—	—	—	—	—	—	—
Hydrocortisone[b]	1.0E-08	5.0E-08	1.0E-06	5.0E-08	2.0E-07	—	—	1.4E-06	—
Insulin, μg/ml	5.0	20.0	5.0	5.0	25.0	5.0	0.5	10.0	20
L-ascorbic acid[b]	—	—	—	—	—	—	—	—	2.2E-04
NaCl	—	5.0E-04	—	—	—	—	—	—	1.7E-02
Na pyruvate (additional)	—	—	—	—	—	—	—	—	—
Na₂SeO₃	3.0E-08	2.5E-08	—	—	—	3.0E-08	—	—	8.0E-08

(continued)

SERUM-FREE SUPPLEMENTED MEDIA

Medium	HITES	ACL-4	PEC	K-1	K-2	N3	G3	HSM	E8[a]
Cell type	Human small cell lung carcinoma	Human nonsmall cell lung carcinoma	Human prostate epithelium	MDCK (dog kidney)	LLC-PK (pig kidney)	LA-N-1 (human neuroblastoma)	Rat glial cells	Human melanocytes	Human pluripotent stem cells (hPSCs)
Phosphoethanolamine	—	1.0E-05	1.0E-04	—	—	—	—	—	—
Progesterone	—	—	—	—	—	2.0E-08	2.0E-07	—	—
Prostaglandin E_1	—	—	—	7.0E-08	—	—	—	—	—
Putrescine	—	—	—	—	—	1.0E-04	1.0E-07	—	—
TGF-β1, ng/ml	—	—	—	—	—	—	—	—	1.74[c]
Transferrin, Fe^{3+} saturated, µg/ml	100	10	5	5	10	50	100	10	10
Triiodothyronine	—	1.0E-10	—	5.0E-12	1.0E-09	—	4.9E-07	1.0E-09	—
Thyroxine	—	—	—	—	—	—	4.5E-07	—	—

[a]E8 medium should be adjusted to pH 7.4 using HCl/NaOH; NaCl is used to adjust the osmolarity to 340 mOsm. PSCs have additional specialized requirements (e.g. substrate coatings). Refer to the cited publications for more information (see also Protocols P23.2, P23.3).
[b]Soluble analogs of hydrocortisone (e.g. dexamethasone) and stable forms of L-ascorbic acid (e.g. L-ascorbic acid 2-phosphate magnesium) can be used.
[c]TGF-β1 can be substituted with NODAL (100 ng/ml); if so, this should be specified, e.g. E8 (NODAL).

REFERENCES

Barnes, D. and Sato, G. (1980). Methods for growth of cultured cells in serum-free medium. *Anal. Biochem.* 102 (2): 255–270.

Beers, J., Gulbranson, D.R., George, N. et al. (2012). Passaging and colony expansion of human pluripotent stem cells by enzyme-free dissociation in chemically defined culture conditions. *Nat. Protoc.* 7 (11): 2029–2040. https://doi.org/10.1038/nprot.2012.130.

Bettger, W.J., Boyce, S.T., Walthall, B.J. et al. (1981). Rapid clonal growth and serial passage of human diploid fibroblasts in a lipid-enriched synthetic medium supplemented with epidermal growth factor, insulin, and dexamethasone. *Proc. Natl Acad. Sci. U.S.A.* 78 (9): 5588–5592.

Bottenstein, J.E. (1985). Growth and differentiation of neural cells in defined media. In: *Cell Culture in the Neurosciences* (eds. J.E. Bottenstein and G. Sato), 3–35. New York: Plenum Press.

Boyce, S.T. and Ham, R.G. (1983). Calcium-regulated differentiation of normal human epidermal keratinocytes in chemically defined clonal culture and serum-free serial culture. *J. Invest. Dermatol.* 81 (Suppl. 1): 33s–40s.

Carney, D.N., Bunn, P.A. Jr., Gazdar, A.F. et al. (1981). Selective growth in serum-free hormone-supplemented medium of tumor cells obtained by biopsy from patients with small cell carcinoma of the lung. *Proc. Natl Acad. Sci. U.S.A.* 78 (5): 3185–3189.

Chen, G., Gulbranson, D.R., Hou, Z. et al. (2011). Chemically defined conditions for human iPSC derivation and culture. *Nat. Methods* 8 (5): 424–429. https://doi.org/10.1038/nmeth.1593.

Dulbecco, R. and Freeman, G. (1959). Plaque production by the polyoma virus. *Virology* 8 (3): 396–397.

Eagle, H. (1959). Amino acid metabolism in mammalian cell cultures. *Science* 130 (3373): 432–437.

Gazdar, A.F. and Oie, H.K. (1986). Re: Growth of cell lines and clinical specimens of human non-small cell lung cancer in a serum-free defined medium. *Cancer Res.* 46 (11): 6011–6012.

Ham, R.G. (1965). Clonal growth of mammalian cells in a chemically defined, synthetic medium. *Proc. Natl Acad. Sci. U.S.A.* 53: 288–293.

Hamilton, W.G. and Ham, R.G. (1977). Clonal growth of Chinese hamster cell lines in protein-free media. *In Vitro* 13 (9): 537–547.

Hammond, S.L., Ham, R.G., and Stampfer, M.R. (1984). Serum-free growth of human mammary epithelial cells: rapid clonal growth in defined medium and extended serial passage with pituitary extract. *Proc. Natl Acad. Sci. U.S.A.* 81 (17): 5435–5439.

Iscove, N.N. and Melchers, F. (1978). Complete replacement of serum by albumin, transferrin, and soybean lipid in cultures of lipopolysaccharide-reactive B lymphocytes. *J. Exp. Med.* 147 (3): 923–933.

Knedler, A. and Ham, R.G. (1987). Optimized medium for clonal growth of human microvascular endothelial cells with minimal serum. *In Vitro Cell. Dev. Biol.* 23 (7): 481–491.

Lechner, J.F. and LaVeck, M.A. (1985). A serum-free method for culturing normal human bronchial epithelial cells at clonal density. *J. Tissue Cult. Methods* 9 (2): 43–48.

Leibovitz, A. (1963). The growth and maintenance of tissue-cell cultures in free gas exchange with the atmosphere. *Am. J. Hyg.* 78: 173–180.

McCoy, T.A., Maxwell, M., and Kruse, P.F. Jr. (1959). Amino acid requirements of the Novikoff hepatoma in vitro. *Proc. Soc. Exp. Biol. Med.* 100 (1): 115–118.

McKeehan, W.L., Adams, P.S., and Rosser, M.P. (1984). Direct mitogenic effects of insulin, epidermal growth factor, glucocorticoid, cholera toxin, unknown pituitary factors and possibly prolactin, but not androgen, on normal rat prostate epithelial cells in serum-free, primary cell culture. *Cancer Res.* 44 (5): 1998–2010.

McKeehan, W.L. and Ham, R.G. (1976). Methods for reducing the serum requirement for growth in vitro of nontransformed diploid fibroblasts. *Dev. Biol. Stand.* 37: 97–98.

Moore, G.E., Gerner, R.E., and Franklin, H.A. (1967). Culture of normal human leukocytes. *JAMA* 199 (8): 519–524.

Morgan, J.F., Morton, H.J., and Parker, R.C. (1950). Nutrition of animal cells in tissue culture; initial studies on a synthetic medium. *Proc. Soc. Exp. Biol. Med.* 73 (1): 1–8.

Morton, H.J. (1970). A survey of commercially available tissue culture media. *In Vitro* 6 (2): 89–108.

Naeyaert, J.M., Eller, M., Gordon, P.R. et al. (1991). Pigment content of cultured human melanocytes does not correlate with tyrosinase message level. *Br. J. Dermatol.* 125 (4): 297–303.

Parker, R.C., Castor, L.N., and McCulloch, E.A. (1957). Altered cell strains in continuous culture. Special publications. *N. Y. Acad. Sci.* 5: 303–313.

Peehl, D.M. (2002). Human prostatic epithelial cells. In: *Culture of Epithelial Cells* (eds. R.I. Freshney and M. Freshney), 171–194. Hoboken, NJ: Wiley-Liss.

Peehl, D.M. and Ham, R.G. (1980). Clonal growth of human keratinocytes with small amounts of dialyzed serum. *In Vitro* 16 (6): 526–540.

Roberson, K.M. and Robertson, C.N. (1995). Isolation and growth of human primary prostate epithelial cultures. *Methods Cell Sci.* 17 (3): 177–185.

Shipley, G.D. and Ham, R.G. (1983). Multiplication of Swiss 3T3 cells in a serum-free medium. *Exp. Cell Res.* 146 (2): 249–260.

Stanners, C.P., Eliceiri, G.L., and Green, H. (1971). Two types of ribosome in mouse-hamster hybrid cells. *Nat. New Biol.* 230 (10): 52–54.

Taub, M. (1984). Growth of primary and established kidney cell cultures in serum-free media. In: *Methods for Serum-Free Culture of Epithelial and Fibroblastic Cells* (eds. D.W. Barnes, D.A. Sirbasku and G.H. Sato), 3–24. New York: Alan R. Liss.

Waymouth, C. (1959). Rapid proliferation of sublines of NCTC clone 929 (strain L) mouse cells in a simple chemically defined medium (MB 752/1). *J. Natl Cancer Inst.* 22 (5): 1003–1017.

Index

Indexed terms can be used for keyword searches in the electronic version of the text. Indexed numbers are alphabetized as if they were spelt out (e.g. "3" is listed as if it were spelt "three"). Page numbers refer to the print version. Page numbers in bold suggest where to begin if a topic is covered in more than three locations. Page numbers in italics indicate that the term will be found in a figure or table on that page, not in the main text. Commonly used abbreviations are indexed in preference to full terms; please refer to the Glossary (Appendix A) for definitions of commonly used terms. It is impossible to index all cell lines mentioned within the text, due to space constraints. Please refer to Cellosaurus for information on specific cell lines (https://web.expasy.org/cellosaurus/).

Freshney's Culture of Animal Cells: A Manual of Basic Technique and Specialized Applications, Eighth Edition. Amanda Capes-Davis and R. Ian Freshney.
© 2021 John Wiley & Sons Ltd. Published 2021 by John Wiley & Sons Ltd.
Companion website: www.wiley.com/go/freshney/cellculture8